Table IV Standard Normal Distribution Table (continued)

The entries in this table give the cumulative area under the standard normal curve to the left of *z* with the values of *z* equal to 0 or positive.

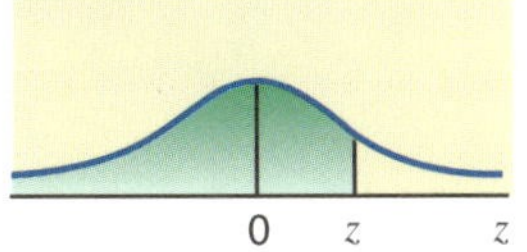

z	.00	.01	.02	.03	.04	.05	.06	.07	.08	.09
0.0	.5000	.5040	.5080	.5120	.5160	.5199	.5239	.5279	.5319	.5359
0.1	.5398	.5438	.5478	.5517	.5557	.5596	.5636	.5675	.5714	.5753
0.2	.5793	.5832	.5871	.5910	.5948	.5987	.6026	.6064	.6103	.6141
0.3	.6179	.6217	.6255	.6293	.6331	.6368	.6406	.6443	.6480	.6517
0.4	.6554	.6591	.6628	.6664	.6700	.6736	.6772	.6808	.6844	.6879
0.5	.6915	.6950	.6985	.7019	.7054	.7088	.7123	.7157	.7190	.7224
0.6	.7257	.7291	.7324	.7357	.7389	.7422	.7454	.7486	.7517	.7549
0.7	.7580	.7611	.7642	.7673	.7704	.7734	.7764	.7794	.7823	.7852
0.8	.7881	.7910	.7939	.7967	.7995	.8023	.8051	.8078	.8106	.8133
0.9	.8159	.8186	.8212	.8238	.8264	.8289	.8315	.8340	.8365	.8389
1.0	.8413	.8438	.8461	.8485	.8508	.8531	.8554	.8577	.8599	.8621
1.1	.8643	.8665	.8686	.8708	.8729	.8749	.8770	.8790	.8810	.8830
1.2	.8849	.8869	.8888	.8907	.8925	.8944	.8962	.8980	.8997	.9015
1.3	.9032	.9049	.9066	.9082	.9099	.9115	.9131	.9147	.9162	.9177
1.4	.9192	.9207	.9222	.9236	.9251	.9265	.9279	.9292	.9306	.9319
1.5	.9332	.9345	.9357	.9370	.9382	.9394	.9406	.9418	.9429	.9441
1.6	.9452	.9463	.9474	.9484	.9495	.9505	.9515	.9525	.9535	.9545
1.7	.9554	.9564	.9573	.9582	.9591	.9599	.9608	.9616	.9625	.9633
1.8	.9641	.9649	.9656	.9664	.9671	.9678	.9686	.9693	.9699	.9706
1.9	.9713	.9719	.9726	.9732	.9738	.9744	.9750	.9756	.9761	.9767
2.0	.9772	.9778	.9783	.9788	.9793	.9798	.9803	.9808	.9812	.9817
2.1	.9821	.9826	.9830	.9834	.9838	.9842	.9846	.9850	.9854	.9857
2.2	.9861	.9864	.9868	.9871	.9875	.9878	.9881	.9884	.9887	.9890
2.3	.9893	.9896	.9898	.9901	.9904	.9906	.9909	.9911	.9913	.9916
2.4	.9918	.9920	.9922	.9925	.9927	.9929	.9931	.9932	.9934	.9936
2.5	.9938	.9940	.9941	.9943	.9945	.9946	.9948	.9949	.9951	.9952
2.6	.9953	.9955	.9956	.9957	.9959	.9960	.9961	.9962	.9963	.9964
2.7	.9965	.9966	.9967	.9968	.9969	.9970	.9971	.9972	.9973	.9974
2.8	.9974	.9975	.9976	.9977	.9977	.9978	.9979	.9979	.9980	.9981
2.9	.9981	.9982	.9982	.9983	.9984	.9984	.9985	.9985	.9986	.9986
3.0	.9987	.9987	.9987	.9988	.9988	.9989	.9989	.9989	.9990	.9990
3.1	.9990	.9991	.9991	.9991	.9992	.9992	.9992	.9992	.9993	.9993
3.2	.9993	.9993	.9994	.9994	.9994	.9994	.9994	.9995	.9995	.9995
3.3	.9995	.9995	.9995	.9996	.9996	.9996	.9996	.9996	.9996	.9997
3.4	.9997	.9997	.9997	.9997	.9997	.9997	.9997	.9997	.9997	.9998

Table IV Standard Normal Distribution Table (continued from previous page)

The entries in this table give the cumulative area under the standard normal curve to the left of *z* with the values of *z* equal to 0 or positive.

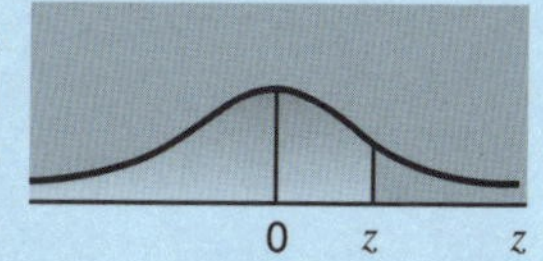

z	.00	.01	.02	.03	.04	.05	.06	.07	.08	.09
0.0	.5000	.5040	.5080	.5120	.5160	.5199	.5239	.5279	.5319	.5359
0.1	.5398	.5438	.5478	.5517	.5557	.5596	.5636	.5675	.5714	.5753
0.2	.5793	.5832	.5871	.5910	.5948	.5987	.6026	.6064	.6103	.6141
0.3	.6179	.6217	.6255	.6293	.6331	.6368	.6406	.6443	.6480	.6517
0.4	.6554	.6591	.6628	.6664	.6700	.6736	.6772	.6808	.6844	.6879
0.5	.6915	.6950	.6985	.7019	.7054	.7088	.7123	.7157	.7190	.7224
0.6	.7257	.7291	.7324	.7357	.7389	.7422	.7454	.7486	.7517	.7549
0.7	.7580	.7611	.7642	.7673	.7704	.7734	.7764	.7794	.7823	.7852
0.8	.7881	.7910	.7939	.7967	.7995	.8023	.8051	.8078	.8106	.8133
0.9	.8159	.8186	.8212	.8238	.8264	.8289	.8315	.8340	.8365	.8389
1.0	.8413	.8438	.8461	.8485	.8508	.8531	.8554	.8577	.8599	.8621
1.1	.8643	.8665	.8686	.8708	.8729	.8749	.8770	.8790	.8810	.8830
1.2	.8849	.8869	.8888	.8907	.8925	.8944	.8962	.8980	.8997	.9015
1.3	.9032	.9049	.9066	.9082	.9099	.9115	.9131	.9147	.9162	.9177
1.4	.9192	.9207	.9222	.9236	.9251	.9265	.9279	.9292	.9306	.9319
1.5	.9332	.9345	.9357	.9370	.9382	.9394	.9406	.9418	.9429	.9441
1.6	.9452	.9463	.9474	.9484	.9495	.9505	.9515	.9525	.9535	.9545
1.7	.9554	.9564	.9573	.9582	.9591	.9599	.9608	.9616	.9625	.9633
1.8	.9641	.9649	.9656	.9664	.9671	.9678	.9686	.9693	.9699	.9706
1.9	.9713	.9719	.9726	.9732	.9738	.9744	.9750	.9756	.9761	.9767
2.0	.9772	.9778	.9783	.9788	.9793	.9798	.9803	.9808	.9812	.9817
2.1	.9821	.9826	.9830	.9834	.9838	.9842	.9846	.9850	.9854	.9857
2.2	.9861	.9864	.9868	.9871	.9875	.9878	.9881	.9884	.9887	.9890
2.3	.9893	.9896	.9898	.9901	.9904	.9906	.9909	.9911	.9913	.9916
2.4	.9918	.9920	.9922	.9925	.9927	.9929	.9931	.9932	.9934	.9936
2.5	.9938	.9940	.9941	.9943	.9945	.9946	.9948	.9949	.9951	.9952
2.6	.9953	.9955	.9956	.9957	.9959	.9960	.9961	.9962	.9963	.9964
2.7	.9965	.9966	.9967	.9968	.9969	.9970	.9971	.9972	.9973	.9974
2.8	.9974	.9975	.9976	.9977	.9977	.9978	.9979	.9979	.9980	.9981
2.9	.9981	.9982	.9982	.9983	.9984	.9984	.9985	.9985	.9986	.9986
3.0	.9987	.9987	.9987	.9988	.9988	.9989	.9989	.9989	.9990	.9990
3.1	.9990	.9991	.9991	.9991	.9992	.9992	.9992	.9992	.9993	.9993
3.2	.9993	.9993	.9994	.9994	.9994	.9994	.9994	.9995	.9995	.9995
3.3	.9995	.9995	.9995	.9996	.9996	.9996	.9996	.9996	.9996	.9997
3.4	.9997	.9997	.9997	.9997	.9997	.9997	.9997	.9997	.9997	.9998

This is Table IV of Appendix C.

Chapter 10 • Estimation and Hypothesis Testing: Two Populations

- Mean of the sampling distribution of $\bar{x}_1 - \bar{x}_2$:

$$\mu_{\bar{x}_1-\bar{x}_2} = \mu_1 - \mu_2$$

- Confidence interval for $\mu_1 - \mu_2$ for two independent samples using the normal distribution when σ_1 and σ_2 are known:

$$(\bar{x}_1 - \bar{x}_2) \pm z\sigma_{\bar{x}_1-\bar{x}_2} \quad \text{where} \quad \sigma_{\bar{x}_1-\bar{x}_2} = \sqrt{\frac{\sigma_1^2}{n_1} + \frac{\sigma_2^2}{n_2}}$$

- Test statistic for a test of hypothesis about $\mu_1 - \mu_2$ for two independent samples using the normal distribution when σ_1 and σ_2 are known:

$$z = \frac{(\bar{x}_1 - \bar{x}_2) - (\mu_1 - \mu_2)}{\sigma_{\bar{x}_1-\bar{x}_2}}$$

- For two independent samples taken from two populations with equal but unknown standard deviations:

  Pooled standard deviation:

$$s_p = \sqrt{\frac{(n_1 - 1)s_1^2 + (n_2 - 1)s_2^2}{n_1 + n_2 - 2}}$$

  Estimate of the standard deviation of $\bar{x}_1 - \bar{x}_2$:

$$s_{\bar{x}_1-\bar{x}_2} = s_p\sqrt{\frac{1}{n_1} + \frac{1}{n_2}}$$

  Confidence interval for $\mu_1 - \mu_2$ using the t distribution:

$$(\bar{x}_1 - \bar{x}_2) \pm ts_{\bar{x}_1-\bar{x}_2}$$

  Test statistic using the t distribution:

$$t = \frac{(\bar{x}_1 - \bar{x}_2) - (\mu_1 - \mu_2)}{s_{\bar{x}_1-\bar{x}_2}}$$

- For two independent samples selected from two populations with unequal and unknown standard deviations:

  Degrees of freedom: $$df = \frac{\left(\dfrac{s_1^2}{n_1} + \dfrac{s_2^2}{n_2}\right)^2}{\dfrac{\left(\dfrac{s_1^2}{n_1}\right)^2}{n_1 - 1} + \dfrac{\left(\dfrac{s_2^2}{n_2}\right)^2}{n_2 - 1}}$$

  Estimate of the standard deviation of $\bar{x}_1 - \bar{x}_2$:

$$s_{\bar{x}_1-\bar{x}_2} = \sqrt{\frac{s_1^2}{n_1} + \frac{s_2^2}{n_2}}$$

  Confidence interval for $\mu_1 - \mu_2$ using the t distribution:

$$(\bar{x}_1 - \bar{x}_2) \pm ts_{\bar{x}_1-\bar{x}_2}$$

  Test statistic using the t distribution:

$$t = \frac{(\bar{x}_1 - \bar{x}_2) - (\mu_1 - \mu_2)}{s_{\bar{x}_1-\bar{x}_2}}$$

- For two paired or matched samples:

  Sample mean for paired differences: $\bar{d} = \Sigma d/n$

  Sample standard deviation for paired differences:

$$s_d = \sqrt{\frac{\Sigma d^2 - \dfrac{(\Sigma d)^2}{n}}{n - 1}}$$

  Mean and standard deviation of the sampling distribution of $\bar{d}$:

$$\mu_{\bar{d}} = \mu_d \quad \text{and} \quad s_{\bar{d}} = s_d/\sqrt{n}$$

  Confidence interval for μ_d using the t distribution:

$$\bar{d} \pm ts_{\bar{d}} \quad \text{where} \quad s_{\bar{d}} = s_d/\sqrt{n}$$

  Test statistic for a test of hypothesis about μ_d using the t distribution:

$$t = \frac{\bar{d} - \mu_d}{s_{\bar{d}}}$$

- For two large and independent samples, confidence interval for $p_1 - p_2$:

$$(\hat{p}_1 - \hat{p}_2) \pm zs_{\hat{p}_1-\hat{p}_2} \quad \text{where} \quad s_{\hat{p}_1-\hat{p}_2} = \sqrt{\frac{\hat{p}_1\hat{q}_1}{n_1} + \frac{\hat{p}_2\hat{q}_2}{n_2}}$$

- For two large and independent samples, for a test of hypothesis about $p_1 - p_2$ with H_0: $p_1 - p_2 = 0$:

  Pooled sample proportion:

$$\bar{p} = \frac{x_1 + x_2}{n_1 + n_2} \quad \text{or} \quad \frac{n_1\hat{p}_1 + n_2\hat{p}_2}{n_1 + n_2}$$

  Estimate of the standard deviation of $\hat{p}_1 - \hat{p}_2$:

$$s_{\hat{p}_1-\hat{p}_2} = \sqrt{\bar{p}\bar{q}\left(\frac{1}{n_1} + \frac{1}{n_2}\right)}$$

  Test statistic: $$z = \frac{(\hat{p}_1 - \hat{p}_2) - (p_1 - p_2)}{s_{\hat{p}_1-\hat{p}_2}}$$

Chapter 11 • Chi-Square Tests

- Expected frequency for a category for a goodness-of-fit test:

$$E = np$$

- Degrees of freedom for a goodness-of-fit test:

  $df = k - 1$ where k is the number of categories

- Expected frequency for a cell for an independence or homogeneity test:

$$E = \frac{\text{(Row total)(Column total)}}{\text{Sample size}}$$

- Degrees of freedom for a test of independence or homogeneity:

$$df = (R - 1)(C - 1)$$

  where R and C are the total number of rows and columns, respectively, in the contingency table

Table IV Standard Normal Distribution Table

The entries in this table give the cumulative area under the standard normal curve to the left of z with the values of z equal to 0 or negative.

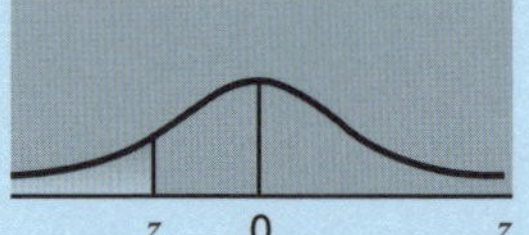

z	.00	.01	.02	.03	.04	.05	.06	.07	.08	.09
−3.4	.0003	.0003	.0003	.0003	.0003	.0003	.0003	.0003	.0003	.0002
−3.3	.0005	.0005	.0005	.0004	.0004	.0004	.0004	.0004	.0004	.0003
−3.2	.0007	.0007	.0006	.0006	.0006	.0006	.0006	.0005	.0005	.0005
−3.1	.0010	.0009	.0009	.0009	.0008	.0008	.0008	.0008	.0007	.0007
−3.0	.0013	.0013	.0013	.0012	.0012	.0011	.0011	.0011	.0010	.0010
−2.9	.0019	.0018	.0018	.0017	.0016	.0016	.0015	.0015	.0014	.0014
−2.8	.0026	.0025	.0024	.0023	.0023	.0022	.0021	.0021	.0020	.0019
−2.7	.0035	.0034	.0033	.0032	.0031	.0030	.0029	.0028	.0027	.0026
−2.6	.0047	.0045	.0044	.0043	.0041	.0040	.0039	.0038	.0037	.0036
−2.5	.0062	.0060	.0059	.0057	.0055	.0054	.0052	.0051	.0049	.0048
−2.4	.0082	.0080	.0078	.0075	.0073	.0071	.0069	.0068	.0066	.0064
−2.3	.0107	.0104	.0102	.0099	.0096	.0094	.0091	.0089	.0087	.0084
−2.2	.0139	.0136	.0132	.0129	.0125	.0122	.0119	.0116	.0113	.0110
−2.1	.0179	.0174	.0170	.0166	.0162	.0158	.0154	.0150	.0146	.0143
−2.0	.0228	.0222	.0217	.0212	.0207	.0202	.0197	.0192	.0188	.0183
−1.9	.0287	.0281	.0274	.0268	.0262	.0256	.0250	.0244	.0239	.0233
−1.8	.0359	.0351	.0344	.0336	.0329	.0322	.0314	.0307	.0301	.0294
−1.7	.0446	.0436	.0427	.0418	.0409	.0401	.0392	.0384	.0375	.0367
−1.6	.0548	.0537	.0526	.0516	.0505	.0495	.0485	.0475	.0465	.0455
−1.5	.0668	.0655	.0643	.0630	.0618	.0606	.0594	.0582	.0571	.0559
−1.4	.0808	.0793	.0778	.0764	.0749	.0735	.0721	.0708	.0694	.0681
−1.3	.0968	.0951	.0934	.0918	.0901	.0885	.0869	.0853	.0838	.0823
−1.2	.1151	.1131	.1112	.1093	.1075	.1056	.1038	.1020	.1003	.0985
−1.1	.1357	.1335	.1314	.1292	.1271	.1251	.1230	.1210	.1190	.1170
−1.0	.1587	.1562	.1539	.1515	.1492	.1469	.1446	.1423	.1401	.1379
−0.9	.1841	.1814	.1788	.1762	.1736	.1711	.1685	.1660	.1635	.1611
−0.8	.2119	.2090	.2061	.2033	.2005	.1977	.1949	.1922	.1894	.1867
−0.7	.2420	.2389	.2358	.2327	.2296	.2266	.2236	.2206	.2177	.2148
−0.6	.2743	.2709	.2676	.2643	.2611	.2578	.2546	.2514	.2483	.2451
−0.5	.3085	.3050	.3015	.2981	.2946	.2912	.2877	.2843	.2810	.2776
−0.4	.3446	.3409	.3372	.3336	.3300	.3264	.3228	.3192	.3156	.3121
−0.3	.3821	.3783	.3745	.3707	.3669	.3632	.3594	.3557	.3520	.3483
−0.2	.4207	.4168	.4129	.4090	.4052	.4013	.3974	.3936	.3897	.3859
−0.1	.4602	.4562	.4522	.4483	.4443	.4404	.4364	.4325	.4286	.4247
0.0	.5000	.4960	.4920	.4880	.4840	.4801	.4761	.4721	.4681	.4641

(continued on next page)

- Test statistic for a goodness-of-fit test and a test of independence or homogeneity:

$$\chi^2 = \Sigma\frac{(O - E)^2}{E}$$

- Confidence interval for the population variance σ^2:

$$\frac{(n-1)s^2}{\chi^2_{\alpha/2}} \text{ to } \frac{(n-1)s^2}{\chi^2_{1-\alpha/2}}$$

- Test statistic for a test of hypothesis about σ^2:

$$\chi^2 = \frac{(n-1)s^2}{\sigma^2}$$

Chapter 12 • Analysis of Variance

Let:

k = the number of different samples (or treatments)
n_i = the size of sample i
T_i = the sum of the values in sample i
n = the number of values in all samples $= n_1 + n_2 + n_3 + \cdots$
Σx = the sum of the values in all samples $= T_1 + T_2 + T_3 + \cdots$
Σx^2 = the sum of the squares of values in all samples

- For the F distribution:
  Degrees of freedom for the numerator $= k - 1$
  Degrees of freedom for the denominator $= n - k$
- Between-samples sum of squares:

$$\text{SSB} = \left(\frac{T_1^2}{n_1} + \frac{T_2^2}{n_2} + \frac{T_3^2}{n_3} + \cdots\right) - \frac{(\Sigma x)^2}{n}$$

- Within-samples sum of squares:

$$\text{SSW} = \Sigma x^2 - \left(\frac{T_1^2}{n_1} + \frac{T_2^2}{n_2} + \frac{T_3^2}{n_3} + \cdots\right)$$

- Total sum of squares:

$$\text{SST} = \text{SSB} + \text{SSW} = \Sigma x^2 - \frac{(\Sigma x)^2}{n}$$

- Variance between samples: $\text{MSB} = \text{SSB}/(k-1)$
- Variance within samples: $\text{MSW} = \text{SSW}/(n-k)$
- Test statistic for a one-way ANOVA test:
  $F = \text{MSB}/\text{MSW}$

Chapter 13 • Simple Linear Regression

- Simple linear regression model: $y = A + Bx + \epsilon$
- Estimated simple linear regression model: $\hat{y} = a + bx$
- Sum of squares of xy, xx, and yy:

$$\text{SS}_{xy} = \Sigma xy - \frac{(\Sigma x)(\Sigma y)}{n}$$

$$\text{SS}_{xx} = \Sigma x^2 - \frac{(\Sigma x)^2}{n} \quad \text{and} \quad \text{SS}_{yy} = \Sigma y^2 - \frac{(\Sigma y)^2}{n}$$

- Least squares estimates of A and B:

$$b = \text{SS}_{xy}/\text{SS}_{xx} \quad \text{and} \quad a = \bar{y} - b\bar{x}$$

- Standard deviation of the sample errors:

$$s_e = \sqrt{\frac{\text{SS}_{yy} - b\,\text{SS}_{xy}}{n-2}}$$

- Error sum of squares: $\text{SSE} = \Sigma e^2 = \Sigma(y - \hat{y})^2$
- Total sum of squares: $\text{SST} = \Sigma y^2 - \frac{(\Sigma y)^2}{n}$
- Regression sum of squares: $\text{SSR} = \text{SST} - \text{SSE}$
- Coefficient of determination: $r^2 = b\,\text{SS}_{xy}/\text{SS}_{yy}$
- Confidence interval for B:

$$b \pm ts_b \quad \text{where} \quad s_b = s_e/\sqrt{\text{SS}_{xx}}$$

- Test statistic for a test of hypothesis about B: $t = \frac{b - B}{s_b}$
- Linear correlation coefficient: $r = \frac{\text{SS}_{xy}}{\sqrt{\text{SS}_{xx}\,\text{SS}_{yy}}}$
- Test statistic for a test of hypothesis about ρ:

$$t = r\sqrt{\frac{n-2}{1-r^2}}$$

- Confidence interval for $\mu_{y|x}$:

$$\hat{y} \pm ts_{\hat{y}_m} \quad \text{where} \quad s_{\hat{y}_m} = s_e\sqrt{\frac{1}{n} + \frac{(x_0 - \bar{x})^2}{\text{SS}_{xx}}}$$

- Prediction interval for y_p:

$$\hat{y} \pm ts_{\hat{y}_p} \quad \text{where} \quad s_{\hat{y}_p} = s_e\sqrt{1 + \frac{1}{n} + \frac{(x_0 - \bar{x})^2}{\text{SS}_{xx}}}$$

Chapter 14 • Multiple Regression

Formulas for Chapter 14 along with the chapter are on the Web site for the text.

Chapter 15 • Nonparametric Methods

Formulas for Chapter 15 along with the chapter are on the Web site for the text.

Table V The *t* Distribution Table

The entries in this table give the critical values of *t* for the specified number of degrees of freedom and areas in the right tail.

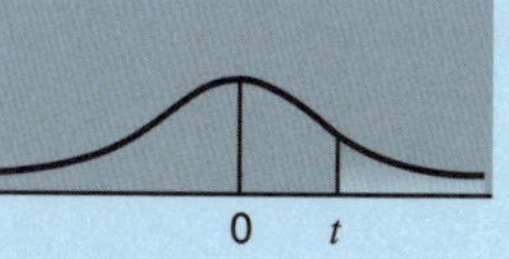

	Area in the Right Tail under the *t* Distribution Curve					
df	.10	.05	.025	.01	.005	.001
1	3.078	6.314	12.706	31.821	63.657	318.309
2	1.886	2.920	4.303	6.965	9.925	22.327
3	1.638	2.353	3.182	4.541	5.841	10.215
4	1.533	2.132	2.776	3.747	4.604	7.173
5	1.476	2.015	2.571	3.365	4.032	5.893
6	1.440	1.943	2.447	3.143	3.707	5.208
7	1.415	1.895	2.365	2.998	3.499	4.785
8	1.397	1.860	2.306	2.896	3.355	4.501
9	1.383	1.833	2.262	2.821	3.250	4.297
10	1.372	1.812	2.228	2.764	3.169	4.144
11	1.363	1.796	2.201	2.718	3.106	4.025
12	1.356	1.782	2.179	2.681	3.055	3.930
13	1.350	1.771	2.160	2.650	3.012	3.852
14	1.345	1.761	2.145	2.624	2.977	3.787
15	1.341	1.753	2.131	2.602	2.947	3.733
16	1.337	1.746	2.120	2.583	2.921	3.686
17	1.333	1.740	2.110	2.567	2.898	3.646
18	1.330	1.734	2.101	2.552	2.878	3.610
19	1.328	1.729	2.093	2.539	2.861	3.579
20	1.325	1.725	2.086	2.528	2.845	3.552
21	1.323	1.721	2.080	2.518	2.831	3.527
22	1.321	1.717	2.074	2.508	2.819	3.505
23	1.319	1.714	2.069	2.500	2.807	3.485
24	1.318	1.711	2.064	2.492	2.797	3.467
25	1.316	1.708	2.060	2.485	2.787	3.450
26	1.315	1.706	2.056	2.479	2.779	3.435
27	1.314	1.703	2.052	2.473	2.771	3.421
28	1.313	1.701	2.048	2.467	2.763	3.408
29	1.311	1.699	2.045	2.462	2.756	3.396
30	1.310	1.697	2.042	2.457	2.750	3.385
31	1.309	1.696	2.040	2.453	2.744	3.375
32	1.309	1.694	2.037	2.449	2.738	3.365
33	1.308	1.692	2.035	2.445	2.733	3.356
34	1.307	1.691	2.032	2.441	2.728	3.348
35	1.306	1.690	2.030	2.438	2.724	3.340

(continued on next page)

Table V The *t* Distribution Table (continued from previous page)

	Area in the Right Tail under the *t* Distribution Curve					
df	.10	.05	.025	.01	.005	.001
36	1.306	1.688	2.028	2.434	2.719	3.333
37	1.305	1.687	2.026	2.431	2.715	3.326
38	1.304	1.686	2.024	2.429	2.712	3.319
39	1.304	1.685	2.023	2.426	2.708	3.313
40	1.303	1.684	2.021	2.423	2.704	3.307
41	1.303	1.683	2.020	2.421	2.701	3.301
42	1.302	1.682	2.018	2.418	2.698	3.296
43	1.302	1.681	2.017	2.416	2.695	3.291
44	1.301	1.680	2.015	2.414	2.692	3.286
45	1.301	1.679	2.014	2.412	2.690	3.281
46	1.300	1.679	2.013	2.410	2.687	3.277
47	1.300	1.678	2.012	2.408	2.685	3.273
48	1.299	1.677	2.011	2.407	2.682	3.269
49	1.299	1.677	2.010	2.405	2.680	3.265
50	1.299	1.676	2.009	2.403	2.678	3.261
51	1.298	1.675	2.008	2.402	2.676	3.258
52	1.298	1.675	2.007	2.400	2.674	3.255
53	1.298	1.674	2.006	2.399	2.672	3.251
54	1.297	1.674	2.005	2.397	2.670	3.248
55	1.297	1.673	2.004	2.396	2.668	3.245
56	1.297	1.673	2.003	2.395	2.667	3.242
57	1.297	1.672	2.002	2.394	2.665	3.239
58	1.296	1.672	2.002	2.392	2.663	3.237
59	1.296	1.671	2.001	2.391	2.662	3.234
60	1.296	1.671	2.000	2.390	2.660	3.232
61	1.296	1.670	2.000	2.389	2.659	3.229
62	1.295	1.670	1.999	2.388	2.657	3.227
63	1.295	1.669	1.998	2.387	2.656	3.225
64	1.295	1.669	1.998	2.386	2.655	3.223
65	1.295	1.669	1.997	2.385	2.654	3.220
66	1.295	1.668	1.997	2.384	2.652	3.218
67	1.294	1.668	1.996	2.383	2.651	3.216
68	1.294	1.668	1.995	2.382	2.650	3.214
69	1.294	1.667	1.995	2.382	2.649	3.213
70	1.294	1.667	1.994	2.381	2.648	3.211
71	1.294	1.667	1.994	2.380	2.647	3.209
72	1.293	1.666	1.993	2.379	2.646	3.207
73	1.293	1.666	1.993	2.379	2.645	3.206
74	1.293	1.666	1.993	2.378	2.644	3.204
75	1.293	1.665	1.992	2.377	2.643	3.202
∞	1.282	1.645	1.960	2.326	2.576	3.090

This is Table V of Appendix C.

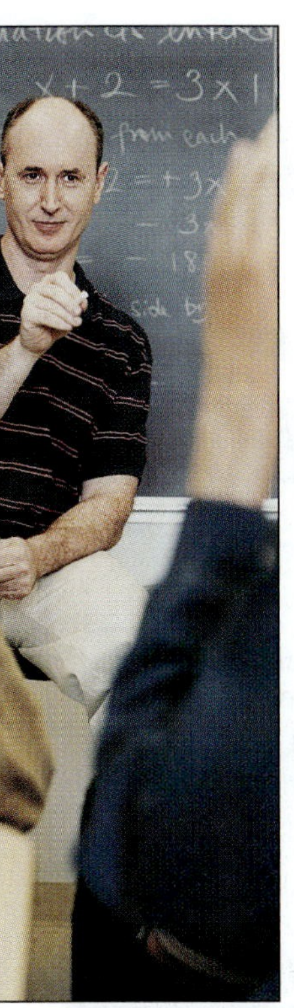

The first person to invent a car that runs on water…

… may be sitting right in your classroom! Every one of your students has the potential to make a difference. And realizing that potential starts right here, in your course.

When students succeed in your course—when they stay on-task and make the breakthrough that turns confusion into confidence—they are empowered to realize the possibilities for greatness that lie within each of them. We know your goal is to create an environment where students reach their full potential and experience the exhilaration of academic success that will last them a lifetime. *WileyPLUS* can help you reach that goal.

Wiley**PLUS** is an online suite of resources—including the complete text—that will help your students:

- come to class better prepared for your lectures
- get immediate feedback and context-sensitive help on assignments and quizzes
- track their progress throughout the course

"I just wanted to say how much this program helped me in studying… I was able to actually see my mistakes and correct them. … I really think that other students should have the chance to use *WileyPLUS*."

Ashlee Krisko, *Oakland University*

---

www.wiley.com/college/wileyplus

80% of students surveyed said it improved their understanding of the material.*

TO THE INSTRUCTOR

WileyPLUS is built around the activities you perform

Prepare & Present

Create outstanding class presentations using a wealth of resources, such as PowerPoint™ slides, image galleries, interactive simulations, and more. Plus you can easily upload any materials you have created into your course, and combine them with the resources Wiley provides you with.

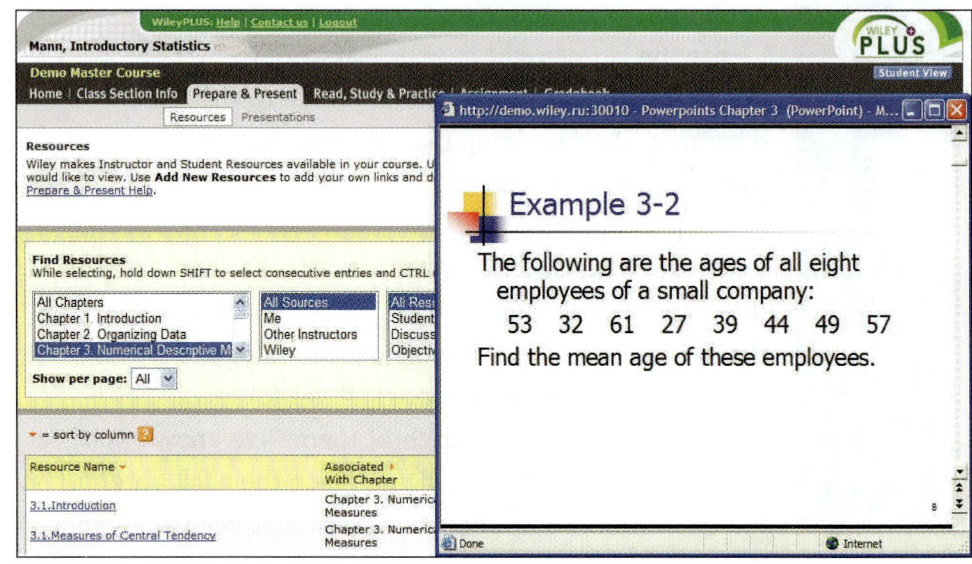

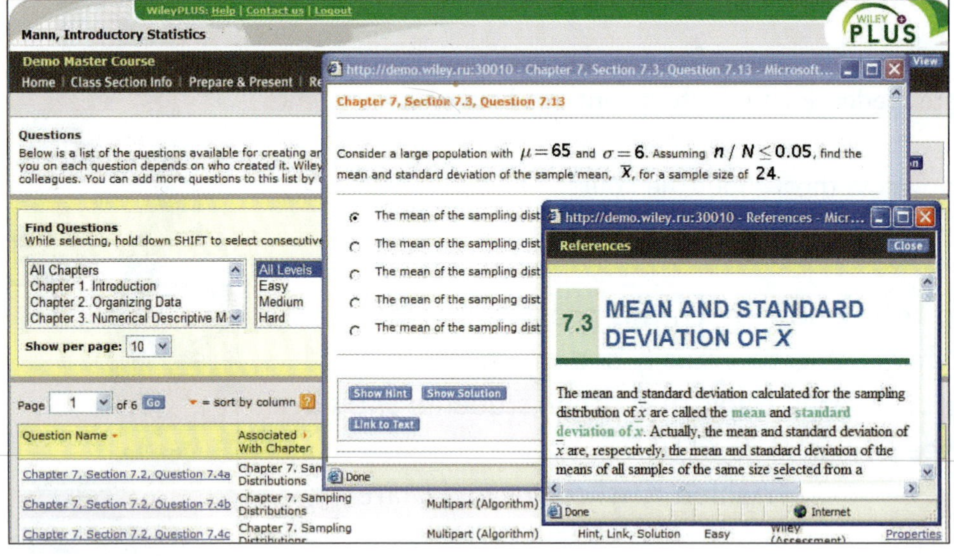

Create Assignments

Automate the assigning and grading of homework or quizzes by using the provided question banks, or by writing your own. Student results will be automatically graded and recorded in your gradebook. *WileyPLUS* also links homework problems to relevant sections of the online text, hints, or solutions—context-sensitive help where students need it most!

* Based on a spring 2005 survey of 972 student users of *WileyPLUS*

in your class each day. With WileyPLUS you can:

Track Student Progress

Keep track of your students' progress via an instructor's gradebook, which allows you to analyze individual and overall class results. This gives you an accurate and realistic assessment of your students' progress and level of understanding.

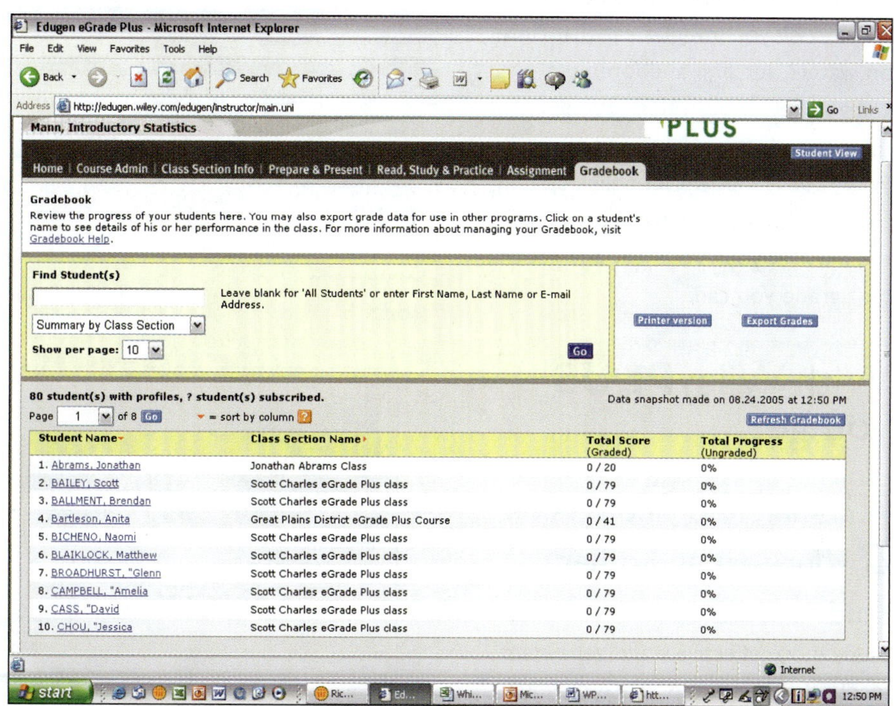

Now Available with WebCT and Blackboard!

Now you can seamlessly integrate all of the rich content and resources available with *WileyPLUS* with the power and convenience of your WebCT or BlackBoard course. You and your students get the best of both worlds with single sign-on, an integrated gradebook, list of assignments and roster, and more. If your campus is using another course management system, contact your local Wiley Representative.

Bb Blackboard

"I studied more for this class than I would have without *WileyPLUS*."

Melissa Lawler, *Western Washington Univ.*

For more information on what *WileyPLUS* can do to help your students reach their potential, please visit

www.wiley.com/college/wileyplus

76% of students surveyed said it made them better prepared for tests. *

TO THE STUDENT

You have the potential to make a difference!

Will you be the first person to land on Mars? Will you invent a car that runs on water? But, first and foremost, will you get through this course?

WileyPLUS is a powerful online system packed with features to help you make the most of your potential, and get the best grade you can!

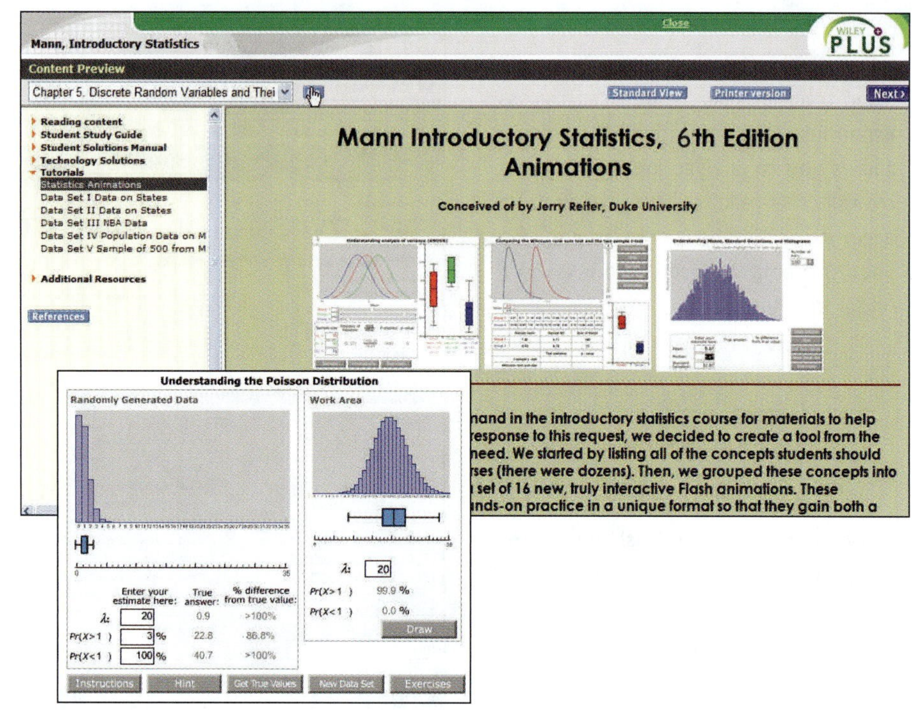

With Wiley**PLUS** you get:

A complete online version of your text and other study resources

Study more effectively and get instant feedback when you practice on your own. Resources like self-assessment quizzes, tutorials, and animations bring the subject matter to life, and help you master the material.

Problem-solving help, instant grading, and feedback on your homework and quizzes

You can keep all of your assigned work in one location, making it easy for you to stay on task. Plus, many homework problems contain direct links to the relevant portion of your text to help you deal with problem-solving obstacles at the moment they come up.

The ability to track your progress and grades throughout the term.

A personal gradebook allows you to monitor your results from past assignments at any time. You'll always know exactly where you stand.

If your instructor uses *WileyPLUS*, you will receive a URL for your class. If not, your instructor can get more information about *WileyPLUS* by visiting www.wiley.com/college/wileyplus

"It has been a great help, and I believe it has helped me to achieve a better grade."
Michael Morris, *Columbia Basin College*

69% of students surveyed said it helped them get a better grade.*

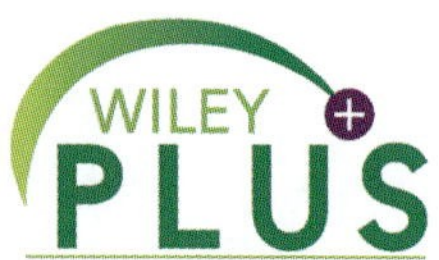

Sixth Edition

INTRODUCTORY STATISTICS

Sixth Edition

INTRODUCTORY STATISTICS

PREM S. MANN

EASTERN CONNECTICUT STATE UNIVERSITY

JOHN WILEY & SONS, INC.

Publisher Laurie Rosatone
Senior Editor Angela Y. Battle
Project Editor Jennifer Battista
Senior Production Editor Valerie A. Vargas
Marketing Manager Amy Sell
Interior Design Karin Kincheloe
Cover Design Madelyn Lesure
Creative Director Harry Nolan
Cover Photo Paolo Curto/Getty Images
Senior Illustration Editor Sandra Rigby
Photo Editor Felicia Ruocco
Production Services mb editorial services
Editorial Assistants Daniel Grace/Danielle Amico
Media Editor Stefanie Liebman

This book was set in 10/12 Times Roman by Techbooks and printed and bound by Courier Companies, Inc. The cover was printed by Courier Companies.

This book is printed on acid free paper. ∞

To order books or for customer service please, call 1-800-CALL WILEY (225-5945).

Library of Congress Cataloging in Publication Data

Mann, Prem S.
Introductory statistics / Prem S. Mann.—6th ed.
p. cm.
Includes index.
ISBN-13: 978-0-471-75530-2 (cloth)
ISBN-10: 0-471-75530-3 (cloth)
1. Statistics. I. Title.

QA276.12.M36 2006
519.5—dc22

2005055154

Printed in the United States of America

10 9 8 7 6 5 4 3 2

To my mother
and
to the memory of my father

PREFACE

After five very successful editions, we are honored to present the sixth edition of *Introductory Statistics* to instructors and students. This new edition features many improvements, changes, and updates.

Introductory Statistics is written for a one- or two-semester first course in applied statistics. The book is intended for students who do not have a strong background in mathematics. The only prerequisite for this text is a knowledge of elementary algebra.

Today, college students from almost all fields of study are required to take at least one course in statistics. Consequently, the study of statistical methods has taken on a prominent role in the education of students majoring in all fields of study. From the very beginning, the goal of this text has been to make the subject of statistics both interesting and accessible to such a wide and varied audience. Three major characteristics of this text support this goal:

1. The realistic content of its examples and exercises that draw on a comprehensive range of applications from all facets of life
2. The clarity and brevity of its presentation
3. The soundness of its pedagogical approach

These characteristics are exhibited through the interplay of a variety of significant text features.

The feedback received from the users of the fifth edition of *Introductory Statistics* has been very supportive and encouraging. This feedback from previous editions serves as evidence that the text is successfully making statistics interesting and accessible—a goal of the author from the very first edition. The author has pursued the same goal through the refinements and updates in this sixth edition so that *Introductory Statistics* will continue to provide a successful experience in statistics to growing numbers of students and professors.

New to the Sixth Edition

The following are some of the many significant changes made in the sixth edition of the text without changing its main features.

- A new section on dotplots has been added to Chapter 2.
- Chapters 8, 9, and 10 have been thoroughly revised to discuss the inference-making topics based on the population standard deviation σ being known and unknown.
- Chapter 9 on hypothesis testing now includes a comprehensive coverage of both the p-value approach and the critical-value approach. This feature is carried over to later chapters.
- A new chapter on multiple regression has been added (Chapter 14), which appears on the text's accompanying Web site.
- The chapter on nonparametric methods is now Chapter 15, and it also appears on the text's accompanying Web site.
- A new feature, *Decide for Yourself,* has been added to almost all chapters in the book. In these sections (which appear just before the Technology Instruction sections) real-world problems are discussed and questions are raised for readers to answer.

- The sections on Uses and Misuses of Statistics have been updated. Many new Uses and Misuses have been added to these sections.
- The Technology sections that appear at the end of chapters have been updated. These sections cover the instructions and screen shots for the TI-84 graphing calculator, MINITAB, and Excel.
- Updated chapter-opening vignettes incorporate real data presented in situations likely to be familiar to students.
- All data sets and a large number of examples and exercises have been revised and updated with current data. Many new data sets have been added. See Appendix B for details.
- Several Case Studies and Mini-Projects have been replaced or revised to reflect current data.
- The text has been rewritten wherever necessary to improve the readability and clarification of the concepts.

Hallmark Features of this Text

Style and Pedagogy

Clear and Concise Exposition The explanation of statistical methods and concepts is clear and concise. Moreover, the style is user-friendly and easy to understand. In chapter introductions and in transitions from section to section, new ideas are related to those discussed earlier.

Thorough Examples

Examples The text contains a wealth of examples, over 200 in 15 chapters and Appendix A. The examples are usually presented in a format showing a problem and its solution. They are well-sequenced and thorough, displaying all facets of concepts. Furthermore, the examples capture students' interest because they cover a wide variety of relevant topics. They are based on situations practicing statisticians encounter every day. Finally, a large number of examples are based on real data taken from sources such as books, government and private data sources and reports, magazines, newspapers, and professional journals.

Step-by-Step Solutions

Solutions A clear, concise solution follows each problem presented in an example. When the solution to an example involves many steps, it is presented in a step-by-step format. For instance, examples related to tests of hypotheses contain five steps that are consistently used to solve such examples in all chapters. Thus, procedures are presented in the concrete settings of applications rather than as isolated abstractions. Frequently, solutions contain highlighted remarks that recall and reinforce ideas critical to the solution of the problem. Such remarks add to the clarity of presentation.

Enlightening Pedagogy

Margin Notes for Examples A margin note appears beside each example that briefly describes what is being done in that example. Students can use these margin notes to assist them as they read through sections and to quickly locate appropriate model problems as they work through exercises.

Frequent Use of Diagrams Concepts can often be made more understandable by describing them visually with the help of diagrams. This text uses diagrams frequently to help students understand concepts and solve problems. For example, tree diagrams are used extensively in Chapters 4 and 5 to assist in explaining probability concepts and in computing probabilities. Similarly, solutions to all examples about tests of hypotheses contain diagrams showing rejection regions, nonrejection regions, and critical values.

Highlighting Definitions of important terms, formulas, and key concepts are enclosed in color boxes so that students can easily locate them.

Cautions Certain items need special attention. These may deal with potential trouble spots that commonly cause errors. Or they may deal with ideas that students often overlook. Special emphasis is placed on such items through the headings: *Remember*, *An Observation*, or *Warning*. An icon is used to identify such items.

Realistic Applications

Case Studies Case studies, which appear in many chapters, provide additional illustrations of the applications of statistics in research and statistical analysis. Most of these case studies are based on articles/snapshots published in journals, magazines, or newspapers. All case studies are based on real data.

Abundant Exercises

Exercises and Supplementary Exercises The text contains an abundance of exercises (excluding Technology Assignments), approximately 1500 in 15 chapters and Appendix A. Moreover, a large number of these exercises contain several parts. Exercise sets appearing at the end of each section (or sometimes at the end of two or three sections) include problems on the topics of that section. These exercises are divided into two parts: **Concepts and Procedures** that emphasize key ideas and techniques, and **Applications** that use these ideas and techniques in concrete settings. Supplementary exercises appear at the end of each chapter and contain exercises on all sections and topics discussed in that chapter. A large number of these exercises are based on real data taken from varied data sources such as books, government and private data sources and reports, magazines, newspapers, and professional journals. Exercises given in the text do not merely provide practice for students, but the real data contained in exercises provide interesting information and insight into economic, political, social, psychological, and other aspects of life. The exercise sets also contain many problems that demand critical thinking skills. The answers to selected odd-numbered exercises appear in the *Answers Section* at the back of the book. **Optional exercises** are indicated by an asterisk (*).

Advanced Exercises All chapters (except Chapters 1 and 14) have a set of exercises that are of greater difficulty. Such exercises appear under the heading *Advanced Exercises* as part of the *Supplementary Exercises*.

Uses and Misuses of Statistics This feature at the end of each chapter (before the Glossary) points out common misconceptions and pitfalls students will encounter in their study of statistics and in everyday life. Subjects highlighted include such diverse topics as the use of the word *average* and grading on a curve.

***New!* Decide for Yourself** This new feature appears at the end of each chapter (except Chapter 1) just before the Technology Instruction section. In this section, a real-world problem is discussed and questions are raised about this problem that readers are required to answer.

Summary and Review

Glossary Each chapter has a glossary that lists the key terms introduced in that chapter, along with a brief explanation of each term. Almost all the terms that appear in boldface type in the text are in the glossary.

Self-Review Tests Each chapter contains a *Self-Review Test*, which appears immediately after the *Supplementary Exercises*. These problems can help students test their grasp of the concepts and skills presented in respective chapters and monitor their understanding of statistical methods. The problems marked by an asterisk (*) in the *Self-Review Tests* are **optional**. The answers to almost all problems of the *Self-Review Tests* appear in the *Answer Section*.

Formula Card A formula card that contains key formulas from all chapters and the normal distribution and t distribution tables is included at the beginning of the book.

Technology Usage

Technology Usage At the end of each chapter is a section covering uses of three major technologies of statistics and probability: the TI-84, MINITAB, and Excel. For each technology, students are guided through performing statistical analyses in a step-by-step fashion, showing them how to enter, revise, format, and save data in a spreadsheet, workbook, or named and unnamed lists, depending on the technology used. Illustrations and screen shots demonstrate the use of these technologies. These sections are followed by *Technology Assignments* for further practice in the use of these technologies. Additional detailed technology instruction is provided in the technology manuals that are online at www.wiley.com/college/mann.

Data Sets Eight data sets appear on the Web site for the text located at www.wiley.com/college/mann. These data sets, collected from different sources, contain information on many variables. They can be used to perform statistical analyses with statistical computer software such as the TI-84 graphing calculator, MINITAB,[1] or Excel. **These data sets are available on the Web site of the text in MINITAB format, Excel format, and in Text format.**

Web Site

http://www.wiley.com/college/mann

The Web site for this text provides additional resources for instructors and students. The following resources/items are available on this Web site:

- Formula card
- Applications index
- Statistical animations
- Computerized Test Bank
- Sample tests
- Discussion questions
- Instructor's solutions manual
- Instructor's resource guide
- Eight large data sets (see Appendix B for a complete list of these data sets)
- Technology Resource Manuals (for all versions of the TI-83 and TI-84, MINITAB, and Excel)
  - **Graphing Calculator Manual** This manual is a basic guide for beginners on the TI-83, TI-83 Plus, and TI-84 graphing calculators. The authors guide students through the important facets of using a graphing calculator in statistics by presenting clear examples, calculator screen captures, and programs specific to each calculator. Note that the instructions included in this manual apply to **all versions of the TI-83 and TI-84** graphing calculators.
  - **MINITAB Manual** This manual provides step-by-step instructions, screen captures, and applications to guide students through using MINITAB for introductory statistics classes.
  - **Excel Manual** This manual contains worked examples using current real data and Excel for data analysis. The manual gives step-by-step instructions, screen captures to guide the students, as well as numerous applications of how to use Excel in statistics.
- Chapter 14: Multiple Regression
- Chapter 15: Nonparametric Methods

The users of this book can download Chapters 14 and 15 along with most of the above items from this Web site. If you need help with downloading or using any of these items please contact the publisher or the author.

[1]MINITAB is a registered trademark of Minitab, Inc., 3081 Enterprise Drive, State College, PA 16801. Phone: 814-238-3280; fax: 814-238-4383; telex: 881612.

Flexible Teaching Options

Because each instructor has different preferences, the text does not indicate optional sections. This decision has been left to the instructor. Instructors may cover the sections or chapters that they think are important. Instructors are most welcome to consult the author in this regard.

Supplements

Complete Learning System

The following supplements are available to accompany this text.

- **Instructor's Solutions Manual (ISBN 0-471-78164-9)** This manual contains complete solutions to all exercises and the *Self-Review Test* problems.
- **Instructor's Resource Guide (on the Web site only)** Sample syllabi, discussion questions, sample exams, suggested homework assignments, and transparency masters are included in this resource. Written for both the experienced and inexperienced instructor, the manual offers a clear overview of the course while outlining major topics and key points of each chapter.
- **Test Bank (ISBN 0-471-78165-7)** The **printed copy** of the test bank contains a large number of multiple-choice questions, essay questions, and quantitative problems for each chapter.
- **Computerized Test Bank** All questions that are in the printed *Test Bank* are available electronically and can be obtained from the publisher.
- **Student Solutions Manual (ISBN 0-471-75531-1)** This manual contains complete solutions to all of the odd-numbered exercises, a few even-numbered exercises, and all the *Self-Review Test* problems.
- **Student Study Guide (ISBN 0-471-75532-X)** This guide contains review material about studying and learning patterns for a first course in statistics. Special attention is given to the critical material of each chapter. Reviews of mathematical notation, formulas, and table reading are also included.
- **WileyPLUS—Expect more from your classroom technology** This text is supported by *WileyPLUS*—a powerful and highly integrated suite of teaching and learning resources. *WileyPLUS* includes a complete online version of the text, algorithmically generated exercises, all of the text supplements, plus course and homework management tools in one easy-to-use Web site. Organized around the everyday activities you perform in class, *WileyPLUS* helps you in many ways:
    - **Prepare and Present:** *WileyPLUS* lets you create class presentations quickly and easily using a wealth of Wiley-provided resources, including an online version of the textbook. You can adapt this content to meet the needs of your course.
    - **Create Assignments:** *WileyPLUS* enables you to automate the process of assigning and grading homework or quizzes. You can use algorithmically generated problems from the text's accompanying test bank, or write your own.
    - **Track Student Progress:** An instructor's gradebook allows you to analyze individual and overall class results to determine students' progress and level of understanding.
    - **Promote Strong Problem-Solving Skills:** *WileyPLUS* can link homework problems to the relevant section of the online text, providing students with context-sensitive help.
    - **Provide numerous practice opportunities:** Algorithmically generated problems provide unlimited self-practice opportunities for students, as well as problems for homework and testing.
    - **Administer Your Course:** You can easily integrate *WileyPLUS* with another course management system, gradebook, or other resources you are using in your class, enabling you to build your course your way.

Acknowledgments

I thank the following reviewers of this, and/or previous editions of this book, whose comments and suggestions were invaluable in improving the text.

Alfred A. Akinsete
Marshall University
Scott S. Albert
College of DuPage
Michael R. Allen
Tennessee Technological University
Peter Arvanites
Rockland Community College
K. S. Asal
Broward Community College
Louise Audette
Manchester Community College
Nicole Betsinger
Arapahoe Community College
Joan Bookbinder
Johnson & Wales University
Dean Burbank
Gulf Coast Community College
Peter A. Carlson
Delta College
Jayanta Chandra
University of Notre Dame
C. K. Chauhan
Indiana-Purdue University at Fort Wayne
James Curl
Modesto Community College
Gregory Daubenmire
Las Positas Community College
Joe DeMaio
Kennesaw State University
Fred H. Dorner
Trinity University, San Antonio
William D. Ergle
Roanoke College, Salem, Virginia
Ruby Evans
Santa Fe Community College
Ronald Ferguson
San Antonio College
James C. Ford
Anda Gadidov
Kennesaw State University
Frank Goulard
Portland Community College
Robert Graham
Jacksonville State University, Jacksonville, Alabama
Larry Griffey
Florida Community College, Jacksonville

Arjun K. Gupta
Bowling Green State University
David Gurney
Southeastern Louisiana University
Daesung Ha
Marshall University
A. Eugene Hileman
Northeastern State University, Tahlequah, Oklahoma
John G. Horner
Cabrillo College
Virginia Horner
Diablo Valley College
Ina Parks S. Howell
Florida International University
Shana Irwin
University of North Texas
Gary S. Itzkowitz
Rowan State College
Jean Johnson
Governors State University
Michael Karelius
American River College, Sacramento
Dix J. Kelly
Central Connecticut State University
Jong Sung Kim
Portland State University
Linda Kohl
University of Michigan, Ann Arbor
Martin Kotler
Pace University, Pleasantville, New York
Marlene Kovaly
Florida Community College, Jacksonville
Hillel Kumin
University of Oklahoma
Carlos de la Lama
San Diego City College
Gaurab Mahapatra
The University of Akron
Richard McGowan
University of Scranton
Daniel S. Miller
Central Connecticut State University
Satya N. Mishra
University of South Alabama
Jeffrey Mock
Diablo Valley College
Luis Moreno
Broome Community College, Binghamton

Robert A. Nagy
University of Wisconsin, Green Bay
Sharon Navard
The College of New Jersey
Nhu T. Nguyen
New Mexico State University
Paul T. Nkansah
Florida Agricultural and Mechanical University
Joyce Oster
Johnson and Wales University
Lindsay Packer
College of Charleston
Mary Parker
Austin Community College
Roger Peck
University of Rhode Island, Kingston
Chester Piascik
Bryant College, Smithfield
Joseph Pigeon
Villanova University
Aaron Robertson
Colgate University
Gerald Rogers
New Mexico State University, Las Cruces
Emily Ross
University of Missouri-St. Louis
Phillis Schumacher
Bryant College, Smithfield
Kathryn Schwartz
Scottsdale Community College
Ronald Schwartz
Wilkes University, Wilkes-Barre
David Stark
University of Akron
Larry Stephens
University of Nebraska, Omaha
Bruce Trumbo
California State University, Hayward
Vasant Waikar
Miami University
Jean Weber
University of Arizona, Tucson
Terry Wilson
San Jacinto College, Pasadena
James Wright
Bucknell University
K. Paul Yoon
Fairleigh Dickinson University, Madison

I express my thanks to Dr. Maryanne Clifford (Eastern Connecticut State University), Professor Gerald Geissert, Professor Daniel S. Miller (Central Connecticut State University), and Dr. David Santana-Ortiz (Rand Organization) for their contributions to different editions of this book that made it better in many ways. I extend my special thanks to Dr. Christopher Lacke of Rowan University who has contributed to this edition in many significant ways. Without his help, this book would not have been in this form. In addition, I thank Eastern Connecticut State University for all the support I received. I thank my student–worker Jessica Vincent for her help during this revision. She is one of the best student–workers I ever have worked with.

It is of utmost importance that a textbook is accompanied by complete and accurate supplements. I take pride in mentioning that the supplements prepared for this text possess these qualities and much more. I appreciatively thank the authors of all these supplements.

It is my pleasure to thank all the professionals at John Wiley and Sons, Inc. with whom I enjoyed working during this revision. Among them are Martha Beyerlein (Full Service Manager), Danielle Amico and Daniel Grace (Editorial Assistants), Jeanine Furino (Production Services Manager), Karin Kincheloe (Interior and Cover Designer), Stefanie Liebman (Media Editor), Amy Sell (Marketing Manager), and Valerie Vargas (Senior Production Editor). Lastly but most importantly I extend my most heartfelt thanks to Angela Y. Battle (Acquisitions Editor) and Jennifer Battista (Project Editor) at John Wiley without whose support this book would not have come out in its present form. It was really a pleasure to work with all of them.

Any suggestions from readers for future revisions would be greatly appreciated. Such suggestions can be sent to the author at mann@easternct.edu or premmann@yahoo.com.

Prem S. Mann
Willimantic, CT 06226
August 2005

CONTENTS

Chapter 1 Introduction 1

1.1 What Is Statistics? 2

1.2 Types of Statistics 2

Case Study 1–1 10 Weeks in Front of the TV 3

Case Study 1–2 Do You Feel Pressured to Come to Work Despite Being Sick? 4

1.3 Population versus Sample 5

Case Study 1–3 Electronic World Swallows Up Kids' Time, Study Finds 7

1.4 Basic Terms 8

1.5 Types of Variables 10

1.6 Cross-Section versus Time-Series Data 13

1.7 Sources of Data 14

1.8 Summation Notation 15

Uses and Misuses / Glossary / Supplementary Exercises / Self-Review Test / Mini-Project / Technology Instruction / Technology Assignments

Chapter 2 Organizing and Graphing Data 26

2.1 Raw Data 27

2.2 Organizing and Graphing Qualitative Data 27

Case Study 2–1 Marrying in the USA 30

Case Study 2–2 Americans Say Keep the Penny 31

2.3 Organizing and Graphing Quantitative Data 34

Case Study 2–3 Hand Hygiene 40

Case Study 2–4 USA Is a Caffeinated Country 44

2.4 Shapes of Histograms 45

Case Study 2–5 Using Truncated Axes 46

2.5 Cumulative Frequency Distributions 51

2.6 Stem-and-Leaf Displays 55

2.7 DOTPLOTS 58

Uses and Misuses / Glossary / Supplementary Exercises / Advanced Exercises / Self-Review Test / Mini-Projects / Decide for Yourself / Technology Instruction / Technology Assignments

Chapter 3 Numerical Descriptive Measures 74

3.1 Measures of Central Tendency for Ungrouped Data 75
Case Study 3–1 High-Priced Tickets in Big Markets 78
Case Study 3–2 Median Annual Starting Salary for MBAs 80
3.2 Measures of Dispersion for Ungrouped Data 87
3.3 Mean, Variance, and Standard Deviation for Grouped Data 94
3.4 Use of Standard Deviation 101
Case Study 3–3 Here Comes the SD 104
3.5 Measures of Position 106
3.6 Box-and-Whisker Plot 111
Uses and Misuses / Glossary / Supplementary Exercises / Advanced Exercises / Appendix 3.1 / Self-Review Test / Mini-Projects / Decide for Yourself / Technology Instruction / Technology Assignments

Chapter 4 Probability 132

4.1 Experiment, Outcomes, and Sample Space 133
4.2 Calculating Probability 138
4.3 Counting Rule 145
4.4 Marginal and Conditional Probabilities 146
Case Study 4–1 American Lefties 149
4.5 Mutually Exclusive Events 150
4.6 Independent versus Dependent Events 151
4.7 Complementary Events 153
4.8 Intersection of Events and the Multiplication Rule 157
Case Study 4–2 Baseball Players Have "Slumps" and "Streaks" 164
4.9 Union of Events and the Addition Rule 167
Uses and Misuses / Glossary / Supplementary Exercises / Advanced Exercises / Self-Review Test / Mini-Projects / Decide for Yourself / Technology Instruction / Technology Assignments

Chapter 5 Discrete Random Variables and Their Probability Distributions 188

5.1 Random Variables 189
5.2 Probability Distribution of a Discrete Random Variable 191
5.3 Mean of a Discrete Random Variable 198
5.4 Standard Deviation of a Discrete Random Variable 199
Case Study 5–1 Aces High Instant Lottery Game—18th Edition 200
5.5 Factorials, Combinations, and Permutations 205
Case Study 5–2 Playing Lotto 209
5.6 The Binomial Probability Distribution 211
5.7 The Hypergeometric Probability Distribution 223
5.8 The Poisson Probability Distribution 227
Case Study 5–3 Ask Mr. Statistics 230
Case Study 5–4 Living and Dying in the USA 233
Uses and Misuses / Glossary / Supplementary Exercises / Advanced Exercises / Self-Review Test / Mini-Projects / Decide for Yourself / Technology Instruction / Technology Assignments

Chapter 6 Continuous Random Variables and the Normal Distribution 247

6.1 Continuous Probability Distribution 248
6.2 The Normal Distribution 251
Case Study 6–1 Distribution of Time Taken to Run a Road Race 252
6.3 The Standard Normal Distribution 256
6.4 Standardizing a Normal Distribution 264
6.5 Applications of the Normal Distribution 270
6.6 Determining the z and x Values When an Area Under the Normal Distribution Curve Is Known 275
6.7 The Normal Approximation to the Binomial Distribution 280
Uses and Misuses / Glossary / Supplementary Exercises / Advanced Exercises / Self-Review Test / Mini-Projects / Decide for Yourself / Technology Instruction / Technology Assignments

Chapter 7 Sampling Distributions 296

7.1 Population and Sampling Distributions 297
7.2 Sampling and Nonsampling Errors 299
7.3 Mean and Standard Deviation of $\bar{x}$ 302
7.4 Shape of the Sampling Distribution of $\bar{x}$ 306
7.5 Applications of the Sampling Distribution of $\bar{x}$ 312
7.6 Population and Sample Proportions 317
7.7 Mean, Standard Deviation, and Shape of the Sampling Distribution of $\hat{p}$ 319
Case Study 7–1 Calling the Vote for Congress 322
7.8 Applications of the Sampling Distribution of $\hat{p}$ 325
Uses and Misuses / Glossary / Supplementary Exercises / Advanced Exercises / Self-Review Test / Mini-Projects / Decide for Yourself / Technology Instruction / Technology Assignments

Chapter 8 Estimation of the Mean and Proportion 337

8.1 Estimation: An Introduction 338
8.2 Point and Interval Estimates 339
8.3 Estimation of a Population Mean: σ Known 341
Case Study 8–1 Education and Earnings 346
8.4 Estimation of a Population Mean: σ Not Known 351
Case Study 8–2 Cardiac Demands of Heavy Snow Shoveling 356
8.5 Estimation of a Population Proportion: Large Samples 360
Case Study 8–3 Buying Habits of Consumers 363
Uses and Misuses / Glossary / Supplementary Exercises / Advanced Exercises / Self-Review Test / Mini-Projects / Decide for Yourself / Technology Instruction / Technology Assignments

Chapter 9 Hypothesis Tests About the Mean and Proportion 378

9.1 Hypothesis Tests: An Introduction 379
9.2 Hypothesis Tests About μ: σ Known 387
Case Study 9–1 The Average Cost of a Wedding 396

9.3 Hypothesis Tests About μ: σ Not Known 401

9.4 Hypothesis Tests About a Population Proportion: Large Samples 411

Case Study 9–2 Coffee or Internet? 417

Uses and Misuses / Glossary / Supplementary Exercises / Advanced Exercises / Self-Review Test / Mini-Projects / Decide for Yourself / Technology Instruction / Technology Assignments

Chapter 10 Estimation and Hypothesis Testing: Two Populations 433

10.1 Inferences About the Difference Between Two Population Means for Independent Samples: σ_1 and σ_2 Known 434

10.2 Inferences About the Difference Between Two Population Means for Independent Samples: σ_1 and σ_2 Unknown but Equal 441

Case Study 10–1 Greater Hunger for News 448

10.3 Inferences About the Difference Between Two Population Means for Independent Samples: σ_1 and σ_2 Unknown and Unequal 451

10.4 Inferences About the Difference Between Two Population Means for Paired Samples 457

10.5 Inferences About the Difference Between Two Population Proportions for Large and Independent Samples 467

Case Study 10–2 Workplace Views Differ By Gender 473

Uses and Misuses / Glossary / Supplementary Exercises / Advanced Exercises / Self-Review Test / Mini-Projects / Decide for Yourself / Technology Instruction / Technology Assignments

Chapter 11 Chi-Square Tests 489

11.1 The Chi-Square Distribution 490

11.2 A Goodness-of-Fit Test 493

Case Study 11–1 Up with 'the Color Purple' 498

11.3 Contingency Tables 502

11.4 A Test of Independence or Homogeneity 502

11.5 Inferences About the Population Variance 514

Uses and Misuses / Glossary / Supplementary Exercises / Advanced Exercises / Self-Review Test / Mini-Projects / Decide for Yourself / Technology Instruction / Technology Assignments

Chapter 12 Analysis of Variance 532

12.1 The *F* Distribution 533

12.2 One-Way Analysis of Variance 535

Uses and Misuses / Glossary / Supplementary Exercises / Advanced Exercises / Self-Review Test / Mini-Projects / Decide for Yourself / Technology Instruction / Technology Assignments

Chapter 13 Simple Linear Regression 555

13.1 Simple Linear Regression Model 556

13.2 Simple Linear Regression Analysis 558

Case Study 13–1 Regression of Heights and Weights of NBA Players 565

13.3 Standard Deviation of Random Errors 571

13.4 Coefficient of Determination 573

13.5 Inferences About *B* 578

13.6 Linear Correlation 583

13.7 Regression Analysis: A Complete Example 589

13.8 Using the Regression Model 596

13.9 Cautions in Using Regression 600

Uses and Misuses / Glossary / Supplementary Exercises / Advanced Exercises / Self-Review Test / Mini-Projects / Decide for Yourself / Technology Instruction / Technology Assignments

Chapter 14 Multiple Regression

This chapter is not included in this text but is available for download on the Web site at www.wiley.com/college/mann.

14.1 Multiple Regression Analysis

14.2 Assumptions of the Multiple Regression Model

14.3 Standard Deviation of Errors

14.4 Coefficient of Multiple Determination

14.5 Computer Solution of Multiple Regression

Exercises / Uses and Misuses / Glossary / Self-Review Test / Mini-Projects / Decide for Yourself

Chapter 15 Nonparametric Methods

This chapter is not included in this text but is available for download on the Web site at www.wiley.com/college/mann.

15.1 The Sign Test

15.2 The Wilcoxon Signed-Rank Test for Two Dependent Samples

15.3 The Wilcoxon Rank Sum Test for Two Independent Samples

15.4 The Kruskal-Wallis Test

15.5 The Spearman Rho Rank Correlation Coefficient Test

15.6 The Runs Test for Randomness

Uses and Misuses / Glossary / Supplementary Exercises / Advanced Exercises / Self-Review Test / Mini-Projects / Decide for Yourself / Technology Instruction / Technology Assignments

Appendix A Sample Surveys, Sampling Techniques, and Design of Experiments A1

A.1 Sources of Data A1
- **A.1.1 Internal Sources A1**
- **A.1.2 External Sources A2**
- **A.1.3 Surveys and Experiments A2**

Case Study A–1 Is It a Simple Question? A3

A.2 Sample Surveys and Sampling Techniques A4
- **A.2.1 Why Sample? A4**
- **A.2.2 Random and Nonrandom Samples A4**
- **A.2.3 Sampling and Nonsampling Errors A5**
- **A.2.4 Random Sampling Techniques A8**

A.3 Design of Experiments A9

Case Study A–2 Do Antibacterial Soaps Work? A13

Exercises / Advanced Exercises / Glossary

Appendix B Explanation of Data Sets B1

Data Set I: City Data B1
Data Set II: Data on States B2
Data Set III: NBA Data B2
Data Set IV: Manchester (Connecticut) Road Race Data B3
Data Set V: Sample of 500 Observations Selected from Data Set IV B3
Data Set VI: Data on Movies B3
Data Set VII: Standard & Poor's 100 Index Data B3
Data Set VIII: McDonald's Data B4

Appendix C Statistical Tables C1

I Table of Binomial Probabilities C2
II Values of $e^{-\lambda}$ C11
III Table of Poisson Probabilities C13
IV Standard Normal Distribution Table C19
V The *t* Distribution Table C21
VI Chi-Square Distribution Table C23
VII The *F* Distribution Table C24

Tables VIII through XII along with Chapters 14 and 15 are available on the Web site of this text.

VIII Critical Values of *X* for the Sign Test
IX Critical Values of *T* for the Wilcoxon Signed-Rank Test
X Critical Values of *T* for the Wilcoxon Rank Sum Test
XI Critical Values for the Spearman Rho Rank Correlation Coefficient Test
XII Critical Values for a Two-Tailed Runs Test with $\alpha = .05$

Answers to Selected Odd-Numbered Exercises and Self-Review Tests AN1
Photo Credits PC1
Index I1

Chapter

1

Introduction

Do you feel pressured by your boss or co-workers to come to work when you are sick with flu? Or are they more considerate and supportive and you do not feel this pressure? In a sample survey, 38% of the respondents said they feel pressured to come to work when they have flu. However, a larger majority, 61%, said that they do not feel pressured. The remaining 1% were not sure. (See Case Study 1–2.)

1.1 What is Statistics?

1.2 Types of Statistics

Case Study 1–1 10 Weeks in Front of the TV

Case Study 1–2 Do You Feel Pressured to Come to Work Despite Being Sick?

1.3 Population versus Sample

Case Study 1–3 Electronic World Swallows Up Kids' Time, Study Finds

1.4 Basic Terms

1.5 Types of Variables

1.6 Cross-Section versus Time-Series Data

1.7 Sources of Data

1.8 Summation Notation

The study of statistics has become more popular than ever over the past four decades or so. The increasing availability of computers and statistical software packages has enlarged the role of statistics as a tool for empirical research. As a result, statistics is used for research in almost all professions, from medicine to sports. Today, college students in almost all disciplines are required to take at least one statistics course. Almost all newspapers and magazines these days contain graphs and stories on statistical studies. After you finish reading this book, it should be much easier to understand these graphs and stories.

Every field of study has its own terminology. Statistics is no exception. This introductory chapter explains the basic terms of statistics. These terms will bridge our understanding of the concepts and techniques presented in subsequent chapters.

1.1 What is Statistics?

The word **statistics** has two meanings. In the more common usage, *statistics* refers to numerical facts. The numbers that represent the income of a family, the age of a student, the percentage of passes completed by the quarterback of a football team, and the starting salary of a typical college graduate are examples of statistics in this sense of the word. A 1988 article in *U.S. News & World Report* declared "Statistics are an American obsession."[1] During the 1988 baseball World Series between the Los Angeles Dodgers and the Oakland A's, the then NBC commentator Joe Garagiola reported to the viewers numerical facts about the players' performances. In response, fellow commentator Vin Scully said, "I love it when you talk statistics." In these examples, the word *statistics* refers to numbers.

After winning the World Series in 1918, the Boston Red Sox did not win the World Series again until 2004. The following table compares a few facts for the years 1918 and 2004. The numbers given in this table can be referred to as *statistics.*

Item	1918	2004
Price of a World Series Ticket	\$3.30	\$140
U.S. Population (in Millions)	106	294
Price of a Quart of Milk	14 Cents	\$1.09
Price of a Loaf of Bread	10 Cents	\$2.19

Source: USA TODAY, October 28, 2004.

The second meaning of *statistics* refers to the field or discipline of study. In this sense of the word, *statistics* is defined as follows.

Definition

Statistics *Statistics* is a group of methods used to collect, analyze, present, and interpret data and to make decisions.

Every day we make decisions that may be personal, business related, or of some other kind. Usually these decisions are made under conditions of uncertainty. Many times, the situations or problems we face in the real world have no precise or definite solution. Statistical methods help us make scientific and intelligent decisions in such situations. Decisions made by using statistical methods are called *educated guesses.* Decisions made without using statistical (or scientific) methods are *pure guesses* and, hence, may prove to be unreliable. For example, opening a large store in an area with or without assessing the need for it may affect its success.

Like almost all fields of study, statistics has two aspects: theoretical and applied. *Theoretical* or *mathematical statistics* deals with the development, derivation, and proof of statistical theorems, formulas, rules, and laws. *Applied statistics* involves the applications of those theorems, formulas, rules, and laws to solve real-world problems. This text is concerned with applied statistics and not with theoretical statistics. By the time you finish studying this book, you will learn how to think statistically and how to make educated guesses.

1.2 Types of Statistics

Broadly speaking, applied statistics can be divided into two areas: *descriptive statistics* and *inferential statistics.*

[1]"The Numbers Racket: How Polls and Statistics Lie," *U.S. News & World Report,* July 11, 1988, pp. 44–47.

10 WEEKS IN FRONT OF THE TV

USA TODAY Snapshots
10 weeks in front of the TV
1,669 hours, or about 70 days
Amount of time the average American will spend watching television in 2004.
Source: Statistical Abstract of the United States, 2003
By Karl Gelles, USA TODAY

As the accompanying chart shows, Americans spent an average of 1669 hours, which is equivalent to almost 70 days or 10 weeks, watching television in 2004. Note that this was an estimate for 2004 made in 2003. This chart describes the data on television watching by using one number–the average. We will learn about the average in Chapter 3.

Source: *USA TODAY*, March 30, 2004. Copyright © 2004, *USA TODAY*. Chart reproduced with permission.

1.2.1 Descriptive Statistics

Suppose we have information on the test scores of students enrolled in a statistics class. In statistical terminology, the whole set of numbers that represents the scores of students is called a **data set**, the name of each student is called an **element**, and the score of each student is called an **observation**. (These terms are defined in more detail in Section 1.4.)

A data set in its original form is usually very large. Consequently, such a data set is not very helpful in drawing conclusions or making decisions. It is easier to draw conclusions from summary tables and diagrams than from the original version of a data set. So, we reduce data to a manageable size by constructing tables, drawing graphs, or calculating summary measures such as averages. The portion of statistics that helps us do this type of statistical analysis is called **descriptive statistics**.

Definition

Descriptive Statistics *Descriptive statistics* consists of methods for organizing, displaying, and describing data by using tables, graphs, and summary measures.

Both Chapters 2 and 3 discuss descriptive statistical methods. In Chapter 2, we learn how to construct tables and how to graph data. In Chapter 3, we learn to calculate numerical summary measures, such as averages.

Case Study 1–1 on this page presents an example of descriptive statistics.

1.2.2 Inferential Statistics

In statistics, the collection of all elements of interest is called a **population**. The selection of a few elements from this population is called a **sample**. (Population and sample are discussed in more detail in Section 1.3.)

A major portion of statistics deals with making decisions, inferences, predictions, and forecasts about populations based on results obtained from samples. For example, we may make some

CASE STUDY

1–2

DO YOU FEEL PRESSURED TO COME TO WORK DESPITE BEING SICK?

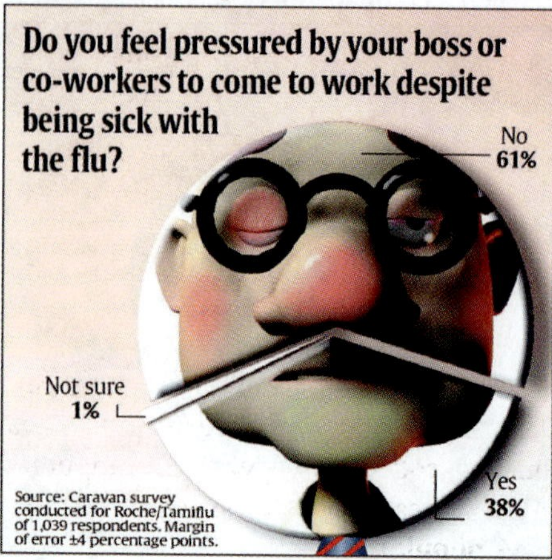
USA TODAY Snapshots®
Do you feel pressured by your boss or co-workers to come to work despite being sick with the flu?
No 61%
Not sure 1%
Yes 38%
Source: Caravan survey conducted for Roche/Tamiflu of 1,039 respondents. Margin of error ±4 percentage points.
By Darryl Haralson and Suzy Parker, USA TODAY

The accompanying chart, reproduced from *USA TODAY*, shows that 61% of workers surveyed said that they are not pressured by their boss or co-workers to come to work when they are sick with flu, 38% said they feel pressured, and 1% said that they are not sure. The chart mentions that there is a ±4% margin of error. We will discuss the concept of margin of error in Chapter 8. But just to give a quick and brief explanation, the margin of error means that the percentages given in the chart can change in the plus or minus direction by 4% when applied to the population.

Source: USA TODAY, February 28, 2005. Copyright © 2005, *USA TODAY*. Chart reproduced with permission.

decisions about the political views of all college and university students based on the political views of 1000 students selected from a few colleges and universities. As another example, we may want to find the starting salary of a typical college graduate. To do so, we may select 2000 recent college graduates, find their starting salaries, and make a decision based on this information. The area of statistics that deals with such decision-making procedures is referred to as **inferential statistics**. This branch of statistics is also called *inductive reasoning* or *inductive statistics.*

Definition

Inferential Statistics *Inferential statistics* consists of methods that use sample results to help make decisions or predictions about a population.

Case Study 1–2 presents an example of inferential statistics. It shows the results of a survey in which people were asked whether or not they are pressured to come to work despite being sick with flu.

Chapters 8 through 13 and parts of Chapter 7 deal with inferential statistics.

Probability, which gives a measurement of the likelihood that a certain outcome will occur, acts as a link between descriptive and inferential statistics. Probability is used to make statements about the occurrence or nonoccurrence of an event under uncertain conditions. Probability and probability distributions are discussed in Chapters 4 through 6 and parts of Chapter 7.

EXERCISES

CONCEPTS AND PROCEDURES

1.1 Briefly describe the two meanings of the word *statistics.*

1.2 Briefly explain the types of statistics.

1.3 Population Versus Sample

We will encounter the terms *population* and *sample* on almost every page of this text.[2] Consequently, understanding the meaning of each of these two terms and the difference between them is crucial.

Suppose a statistician is interested in knowing

1. The percentage of all voters in a city who will vote for a particular candidate in an election
2. The 2005 gross sales of all companies in New York City
3. The prices of all houses in California

In these examples, the statistician is interested in *all* voters, *all* companies, and *all* houses. Each of these groups is called the population for the respective example. In statistics, a population does not necessarily mean a collection of people. It can, in fact, be a collection of people or of any kind of item such as houses, books, television sets, or cars. The population of interest is usually called the **target population**.

Definition

Population or Target Population A *population* consists of all elements—individuals, items, or objects—whose characteristics are being studied. The population that is being studied is also called the *target population.*

Most of the time, decisions are made based on portions of populations. For example, the election polls conducted in the United States to estimate the percentages of voters who favor various candidates in any presidential election are based on only a few hundred or a few thousand voters selected from across the country. In this case, the population consists of all registered voters in the United States. The sample is made up of a few hundred or few thousand voters who are included in an opinion poll. Thus, the collection of a few elements selected from a population is called a **sample**.

Definition

Sample A portion of the population selected for study is referred to as a *sample.*

Figure 1.1 illustrates the selection of a sample from a population.

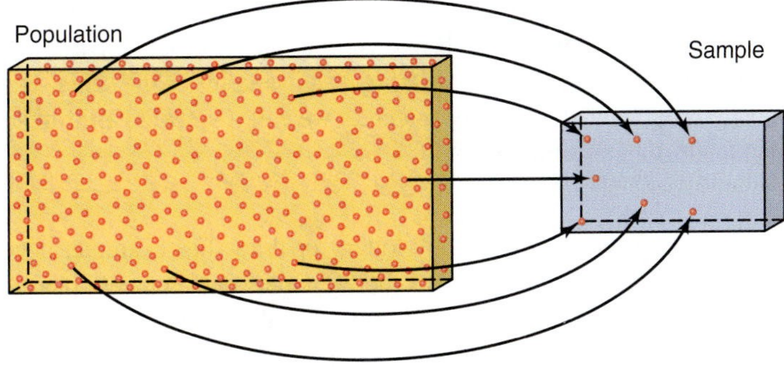

Figure 1.1 Population and sample.

[2]To learn more about sampling and sampling techniques, refer to Appendix A in this text.

The collection of information from the elements of a population or a sample is called a **survey**. A survey that includes every element of the target population is called a **census**. Often the target population is very large. Hence, in practice, a census is rarely taken because it is expensive and time-consuming. In many cases, it is even impossible to identify each element of the target population. Usually, to conduct a survey, we select a sample and collect the required information from the elements included in that sample. We then make decisions based on this sample information. Such a survey conducted on a sample is called a **sample survey**. As an example, if we collect information on the 2005 incomes of all families in Connecticut, it will be referred to as a census. On the other hand, if we collect information on the 2005 incomes of 50 families from Connecticut, it will be called a sample survey.

Definition

Census and Sample Survey A survey that includes every member of the population is called a *census*. The technique of collecting information from a portion of the population is called a *sample survey*.

Case Study 1–3 on page 7 presents an example of a sample survey.

The purpose of conducting a sample survey is to make decisions about the corresponding population. It is important that the results obtained from a sample survey closely match the results that we would obtain by conducting a census. Otherwise, any decision based on a sample survey will not apply to the corresponding population. As an example, to find the average income of families living in New York City by conducting a sample survey, the sample must contain families who belong to different income groups in almost the same proportion as they exist in the population. Such a sample is called a **representative sample**. Inferences derived from a representative sample will be more reliable.

Definition

Representative Sample A sample that represents the characteristics of the population as closely as possible is called a *representative sample*.

A sample may be random or nonrandom. In a **random sample**, each element of the population has a chance of being included in the sample. However, in a nonrandom sample this may not be the case.

Definition

Random Sample A sample drawn in such a way that each element of the population has a chance of being selected is called a *random sample*. If all samples of the same size selected from a population have the same chance of being selected, we call it **simple random sampling**. Such a sample is called a **simple random sample**.

One way to select a random sample is by lottery or draw. For example, if we are to select 5 students from a class of 50, we write each of the 50 names on a separate piece of paper. Then we place all 50 slips in a box and mix them thoroughly. Finally, we randomly draw five slips from the box. The five names drawn give a random sample. On the other hand, if we arrange all 50 names alphabetically and then select the first 5 names on the list, it is a nonrandom sample because the students listed sixth to fiftieth have no chance of being included in the sample.

A sample may be selected with or without replacement. In sampling **with replacement**, each time we select an element from the population, we put it back in the population before we

ELECTRONIC WORLD SWALLOWS UP KIDS' TIME, STUDY FINDS

Children plugged in about 6½ hours a day

The USA's children live in an increasingly heavy stew of media, spending about 6½ hours a day mostly watching TV, using computers and enjoying other electronic activities. And they are spending relatively little time reading or doing homework, a Kaiser Family Foundation survey reported Wednesday.

Kids watch about the same amount of TV—nearly four hours a day—as they did based on a Kaiser survey five years ago, but they're adding newer technology to the mix, such as downloading music and instant-messaging. When multitasking is factored in, children are exposed to 8½ hours of media a day, up about an hour from five years ago.

A record 68% have TVs in their rooms, and an increasing number own DVD and video game players, according to the survey of 2,000 children in grades three through 12.

"We have changed our children's bedrooms into little media arcades," survey co-director Donald Roberts of Stanford University says. "When I was a child, 'Go to your room' was punishment. Now it's 'Go to your room and have a ball.' "

Children with TVs in their rooms watch about 90 minutes more a day and do less reading and homework than those without their own TVs. About half say their families have no TV rules; if there are limits, they're usually not enforced.

"It's alarming, because parents . . . should be setting clear rules and monitoring media use," says Bridget Maher of the Family Research Council, a self-described conservative public policy group.

The survey results come amid concern about the soaring rate of childhood obesity. The more kids watch TV, the more likely they are to be heavy, other studies have shown.

Glorified violence on TV is another concern, Roberts says. And new research suggests that violent video games might be even more likely than TV to spur aggression.

But even more serious could be changes in still-developing brains from the constant multitasking, says psychologist Jane Healy, author of *Endangered Minds.* "When you divide attention like this, it becomes harder to focus deeply on any one thing. They may develop habits of mind that make it hard to do in-depth thinking."

The survey underestimates multi-tasking because kids typically have four screens open on a computer, adds Sherry Turkle, an MIT expert on how technology affects people.

But media psychologist Stuart Fischoff of California State University-Los Angeles says all this concern is "premature hysteria." He says he has seen no changes in students' critical thinking during 38 years as a professor, "and TV and the Internet have been around long enough that it would show up by now."

More watching than reading

Average amount of time 8- to 18-year-olds spend daily:

Watching TV	3 hours, 51 min.
Listening to music	1 hour, 44 min.
Using a computer	1 hour, 2 min.
Playing video games	49 min.
Reading	43 min.
Watching movies	25 min.

Source: Kaiser Family Foundation

By Joni Alexander, USA TODAY

Source: Marilyn Elias, *USA TODAY,* March 10, 2005. Reproduced with Permission.

select the next element. Thus, in sampling with replacement, the population contains the same number of items each time a selection is made. As a result, we may select the same item more than once in such a sample. Consider a box that contains 25 balls of different colors. Suppose we draw a ball, record its color, and put it back in the box before drawing the next ball. Every time we draw a ball from this box, the box contains 25 balls. This is an example of sampling with replacement.

Sampling **without replacement** occurs when the selected element is not replaced in the population. In this case, each time we select an item, the size of the population is reduced by one element. Thus, we cannot select the same item more than once in this type of sampling. Most of the time, samples taken in statistics are without replacement. Consider an opinion poll based on a certain number of voters selected from the population of all eligible voters. In this case, the same voter is not selected more than once. Therefore, this is an example of sampling without replacement.

EXERCISES

CONCEPTS AND PROCEDURES

1.3 Briefly explain the terms *population*, *sample*, *representative sample*, *random sample*, *sampling with replacement*, and *sampling without replacement.*

1.4 Give one example each of sampling with and sampling without replacement.

1.5 Briefly explain the difference between a census and a sample survey. Why is conducting a sample survey preferable to conducting a census?

APPLICATIONS

1.6 Explain whether each of the following constitutes a population or a sample.

- a. Pounds of bass caught by all participants in a bass fishing derby
- b. Credit card debts of 100 families selected from a city
- c. Number of home runs hit by all Major League baseball players in the 2005 season
- d. Number of parole violations by all 2147 parolees in a city
- e. Amount spent on prescription drugs by 200 senior citizens in a large city

1.7 Explain whether each of the following constitutes a population or a sample.

- a. Number of personal fouls committed by all NBA players during the 2005–2006 season
- b. Yield of potatoes per acre for 10 pieces of land
- c. Weekly salaries of all employees of a company
- d. Cattle owned by 100 farmers in Iowa
- e. Number of computers sold during the past week at all computer stores in Los Angeles

1.4 Basic Terms

It is very important to understand the meaning of some basic terms that will be used frequently in this text. This section explains the meaning of an element (or member), a variable, an observation, and a data set. An element and a data set were briefly defined in Section 1.2. This section defines these terms formally and illustrates them with the help of an example.

Table 1.1 gives information on the 2004 profits (in millions of U.S. dollars) of seven U.S. companies. We can call this group of companies a sample of seven companies. Each company listed in this table is called an **element** or a **member** of the sample. Table 1.1 contains information on seven elements. Note that elements are also called *observational units.*

Definition

Element or Member An *element* or *member* of a sample or population is a specific subject or object (for example, a person, firm, item, state, or country) about which the information is collected.

Table 1.1 2004 Profits of Seven U.S. Companies

Company	2004 Profits (millions of dollars) ← Variable
Wal-Mart Stores	10,267
Exxon	25,330
General Electric ← An element or a member	16,593 ← An observation or measurement
Citigroup	17,046
Home Depot	5001
Pfizer	11,361
Target	3198

The *2004 profits* in our example is called a **variable**. The *2004 profits* is a characteristic of companies that we are investigating or studying.

Definition

Variable A *variable* is a characteristic under study that assumes different values for different elements. In contrast to a variable, the value of a *constant* is fixed.

A few other examples of variables are the incomes of households, the number of houses built in a city per month during the past year, the makes of cars owned by people, the gross profits of companies, and the number of insurance policies sold by a salesperson per day during the past month.

In general, a variable assumes different values for different elements, as does the 2004 profits of the seven companies in Table 1.1. For some elements in a data set, however, the value of the variable may be the same. For example, if we collect information on incomes of households, these households are expected to have different incomes, although some of them may have the same income.

A variable is often denoted by x, y, or z. For instance, in Table 1.1, the 2004 profits of companies may be denoted by any one of these letters. Starting with Section 1.8, we will begin to use these letters to denote variables.

Each of the values representing the 2004 profits of the seven companies in Table 1.1 is called an **observation** or **measurement**.

Definition

Observation or Measurement The value of a variable for an element is called an *observation* or *measurement.*

From Table 1.1, the 2004 profits of General Electric were $16,593 million. The value $16,593 million is an observation or measurement. Table 1.1 contains seven observations, one for each of the seven companies.

The information given in Table 1.1 on 2004 profits of companies is called the **data** or a **data set**.

Definition

Data Set A *data set* is a collection of observations on one or more variables.

Other examples of data sets are a list of the prices of 25 recently sold homes, scores of 15 students, opinions of 100 voters, and ages of all employees of a company.

EXERCISES

CONCEPTS AND PROCEDURES

1.8 Explain the meaning of an element, a variable, an observation, and a data set.

APPLICATIONS

1.9 The following table gives the number of dog bites reported to the police last year in six cities.

City	Number of Bites
Center City	47
Elm Grove	32
Franklin	51
Bay City	44
Oakdale	12
Sand Point	3

Briefly explain the meaning of a member, a variable, a measurement, and a data set with reference to this table.

1.10 The following table lists the crude oil reserves (in billions of barrels) for six countries with the largest reserves as of June 2004.

Country	Oil Reserves
Saudi Arabia	261.7
Iraq	112.0
Kuwait	97.7
Iran	94.4
United Arab Emirates	80.3
Venezuela	64.0

Sources: CIA World Fact Book and *USA TODAY* Research, *USA TODAY,* June 7, 2004.

Briefly explain the meaning of a member, a variable, a measurement, and a data set with reference to this table.

1.11 Refer to the data set in Exercise 1.9.
- **a.** What is the variable for this data set?
- **b.** How many observations are in this data set?
- **c.** How many elements does this data set contain?

1.12 Refer to the data set in Exercise 1.10.
- **a.** What is the variable for this data set?
- **b.** How many observations are in this data set?
- **c.** How many elements does this data set contain?

1.5 Types of Variables

In Section 1.4, we learned that a variable is a characteristic under investigation that assumes different values for different elements. The incomes of families, heights of persons, gross sales of companies, prices of college textbooks, makes of cars owned by families, number of accidents, and status (freshman, sophomore, junior, or senior) of students enrolled at a university are a few examples of variables.

A variable may be classified as quantitative or qualitative. These two types of variables are explained next.

1.5.1 Quantitative Variables

Some variables (such as the price of a home) can be measured numerically, whereas others (such as hair color) cannot. The first is an example of a **quantitative variable** and the second that of a qualitative variable.

Definition

Quantitative Variable A variable that can be measured numerically is called a *quantitative variable.* The data collected on a quantitative variable are called *quantitative data.*

Incomes, heights, gross sales, prices of homes, number of cars owned, and number of accidents are examples of quantitative variables because each of them can be expressed numerically. For instance, the income of a family may be $41,520.75 per year, the gross sales for a company may be $567 million for the past year, and so forth. Such quantitative variables may be classified as either *discrete variables* or *continuous variables.*

Discrete Variables

The values that a certain quantitative variable can assume may be countable or noncountable. For example, we can count the number of cars owned by a family, but we cannot count the height of a family member. A variable that assumes countable values is called a **discrete variable**. Note that there are no possible intermediate values between consecutive values of a discrete variable.

Definition

Discrete Variable A variable whose values are countable is called a *discrete variable.* In other words, a discrete variable can assume only certain values with no intermediate values.

For example, the number of cars sold on any day at a car dealership is a discrete variable because the number of cars sold must be 0, 1, 2, 3, . . . and we can count it. The number of cars sold cannot be between 0 and 1, or between 1 and 2. A few other examples of discrete variables are the number of people visiting a bank on any day, the number of cars in a parking lot, the number of cattle owned by a farmer, and the number of students in a class.

Continuous Variables

Some variables cannot be counted, and they can assume any numerical value between two numbers. Such variables are called **continuous variables**.

Definition

Continuous Variable A variable that can assume any numerical value over a certain interval or intervals is called a *continuous variable.*

The time taken to complete an examination is an example of a continuous variable because it can assume any value, let us say, between 30 and 60 minutes. The time taken may be 42.6 minutes, 42.67 minutes, or 42.674 minutes. (Theoretically, we can measure time as precisely as we want.) Similarly, the height of a person can be measured to the tenth of an inch or to the hundredth of an inch. However, neither time nor height can be counted in a discrete fashion. A few other examples of continuous variables are weights of people, amount of soda in a 12-ounce can (note that a can does not contain exactly 12 ounces of soda), and yield of potatoes (in pounds) per acre. Note that any variable that involves money is considered a continuous variable.

1.5.2 Qualitative or Categorical Variables

Variables that cannot be measured numerically but can be divided into different categories are called **qualitative** or **categorical variables**.

> **Definition**
>
> **Qualitative or Categorical Variable** A variable that cannot assume a numerical value but can be classified into two or more nonnumeric categories is called a *qualitative* or *categorical variable.* The data collected on such a variable are called *qualitative data.*

For example, the status of an undergraduate college student is a qualitative variable because a student can fall into any one of four categories: freshman, sophomore, junior, or senior. Other examples of qualitative variables are the gender of a person, hair color, and the make of a car.

Figure 1.2 illustrates the types of variables.

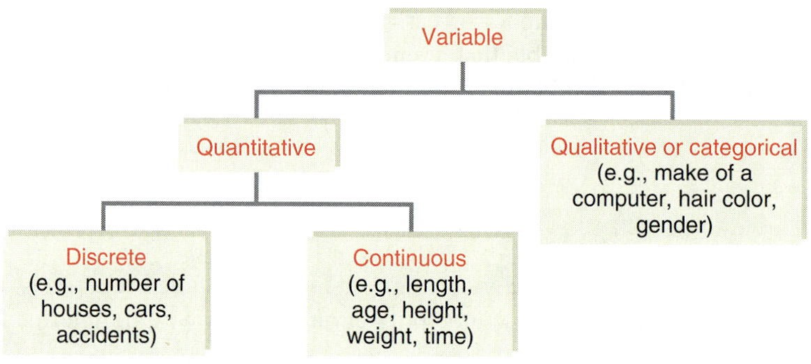

Figure 1.2 Types of variables.

EXERCISES

CONCEPTS AND PROCEDURES

1.13 Explain the meaning of the following terms.

- **a.** Quantitative variable
- **b.** Qualitative variable
- **c.** Discrete variable
- **d.** Continuous variable
- **e.** Quantitative data
- **f.** Qualitative data

APPLICATIONS

1.14 Indicate which of the following variables are quantitative and which are qualitative.

- **a.** Number of persons in a family
- **b.** Color of cars
- **c.** Marital status of people
- **d.** Length of a frog's jump
- **e.** Number of errors in a person's credit report

1.15 Indicate which of the following variables are quantitative and which are qualitative.
  a. Number of typographical errors in newspapers
  b. Monthly TV cable bills
  c. Spring break locations favored by college students
  d. Number of cars owned by families
  e. Lottery revenues of states

1.16 Classify the quantitative variables in Exercise 1.14 as discrete or continuous.

1.17 Classify the quantitative variables in Exercise 1.15 as discrete or continuous.

1.6 Cross-Section Versus Time-Series Data

Based on the time over which they are collected, data can be classified as either cross-section or time-series data.

1.6.1 Cross-Section Data

Cross-section data contain information on different elements of a population or sample for the *same* period of time. The information on incomes of 100 families for 2005 is an example of cross-section data. All examples of data already presented in this chapter have been cross-section data.

Definition

Cross-Section Data Data collected on different elements at the same point in time or for the same period of time are called *cross-section data.*

Table 1.2 shows the leading daytime talk shows based on the average daily number of viewers from September 2003 to July 2004. Because this table presents data on the average daily viewers for seven shows for the same period (September 2003 to July 2004), it is an example of cross-section data.

Table 1.2 Leading Daytime Talk Shows

Talk Show	Average Daily Viewers (millions)
The Oprah Winfrey Show	8.6
Dr. Phil	6.5
Live With Regis and Kelly	4.7
Maury	4.1
Jerry Springer	3.3
Montel Williams	3.2
Ellen	2.2

Source: Neilsen Media Research; Syndicated Network Television Association. Data appeared in *The New York Times,* August 23, 2004.

1.6.2 Time-Series Data

Time-series data contain information on the same element for *different* periods of time. Information on U.S. exports for the years 1983 to 2005 is an example of time-series data.

Definition

Time-Series Data Data collected on the same element for the same variable at different points in time or for different periods of time are called *time-series data.*

The data given in Table 1.3 are an example of time-series data. This table lists the number of collisions between wildlife (mostly birds) and civilian aircraft that were reported for the years 1990, 1995, 2000, and 2002.

Table 1.3 Number of Collisions Between Wildlife and Civilian Aircraft

Year	Number of Collisions
1990	1990
1995	2775
2000	6323
2002	6556

Source: U.S. Air Force, Federal Aviation Administration. *USA Today,* October 15, 2003.

1.7 Sources of Data

The availability of accurate and appropriate data is essential for deriving reliable results.[3] Data may be obtained from internal sources, external sources, or surveys and experiments.

Many times data come from *internal sources*, such as a company's own personnel files or accounting records. For example, a company that wants to forecast the future sales of its product may use the data of past periods from its own records. For most studies, however, all the data that are needed are not usually available from internal sources. In such cases, one may have to depend on outside sources to obtain data. These sources are called *external sources.* For instance, the *Statistical Abstract of the United States* (published annually), which contains various kinds of data on the United States, is an external source of data.

A large number of government and private publications can be used as external sources of data. The following is a list of some of the government publications.

1. *Statistical Abstract of the United States*
2. *Employment and Earnings*
3. *Handbook of Labor Statistics*
4. *Source Book of Criminal Justice Statistics*
5. *Economic Report of the President*
6. *County & City Data Book*
7. *State & Metropolitan Area Data Book*
8. *Digest of Education Statistics*
9. *Health United States*
10. *Agricultural Statistics*

Most of the data contained in these books can be accessed on Internet sites such as www.census.gov (Census Bureau), www.bls.gov (Bureau of Labor Statistics), www.ojp.usdoj.gov/bjs (Office of Justice Program, U.S. Department of Justice, Bureau of Justice Statistics), www.os.dhhs.gov (United States Department of Health and Human Services), and www.usda.gov/nass/pubs/agstats.htm (U.S. Department of Agriculture, Agricultural Statistics).

[3]Sources of data are discussed in more detail in Appendix A.

Besides these government publications, a large number of private publications (e.g., *Standard & Poors' Security Owner's Stock Guide* and *World Almanac and Book of Facts*) and periodicals (e.g., *The Wall Street Journal*, *USA TODAY*, *Fortune*, *Forbes*, and *Business Week*) can be used as external data sources.

Sometimes the needed data may not be available from either internal or external sources. In such cases, the investigator may have to conduct a survey or experiment to obtain the required data. Appendix A discusses surveys and experiments in detail.

EXERCISES

CONCEPTS AND PROCEDURES

1.18 Explain the difference between cross-section and time-series data. Give an example of each of these two types of data.

1.19 Briefly describe internal and external sources of data.

APPLICATIONS

1.20 Classify the following as cross-section or time-series data.

- **a.** Liquor bills of a family for each month of 2005
- **b.** Number of armed robberies each year in Dallas from 1993 to 2005
- **c.** Number of homicides in 40 cities during 2005
- **d.** Gross sales of 200 ice cream parlors in July 2005

1.21 Classify the following as cross-section or time-series data.

- **a.** Average prices of houses in 100 cities
- **b.** Salaries of 50 employees
- **c.** Number of cars sold each year by General Motors from 1980 to 2005
- **d.** Number of employees employed by a company each year from 1985 to 2005

1.8 Summation Notation

Sometimes mathematical notation helps express a mathematical relationship concisely. This section describes the **summation notation** that is used to denote the sum of values.

Suppose a sample consists of five books and the prices of these five books are \$75, \$80, \$35, \$97, and \$88. The variable *price of a book* can be denoted by x. The prices of the five books can be written as follows:

$$\text{Price of the first book} = x_1 = \$75$$

↑ Subscript of x denotes the number of the book

Similarly,

$$\text{Price of the second book} = x_2 = \$80$$
$$\text{Price of the third book} = x_3 = \$35$$
$$\text{Price of the fourth book} = x_4 = \$97$$
$$\text{Price of the fifth book} = x_5 = \$88$$

In this notation, x represents the price, and the subscript denotes a particular book.

Now, suppose we want to add the prices of all five books. We have

$$x_1 + x_2 + x_3 + x_4 + x_5 = 75 + 80 + 35 + 97 + 88 = \$375$$

The uppercase Greek letter Σ (pronounced *sigma*) is used to denote the sum of all values. Using Σ notation, we can write the foregoing sum as follows:

$$\Sigma x = x_1 + x_2 + x_3 + x_4 + x_5 = \$375$$

The notation Σx in this expression represents the sum of all the values of x and is read as "sigma x" or "sum of all values of x."

■ EXAMPLE 1–1

Using summation notation: one variable.

Annual salaries (in thousands of dollars) of four workers are 75, 42, 125, and 61. Find

(a) Σx **(b)** $(\Sigma x)^2$ **(c)** Σx^2

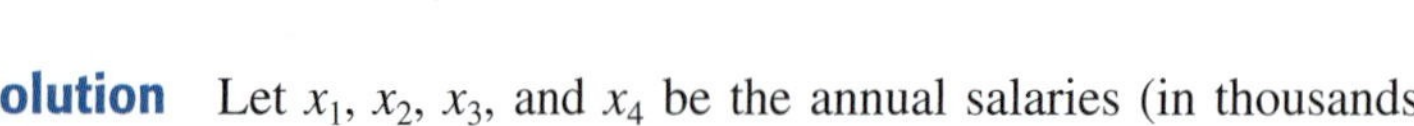

Solution Let x_1, x_2, x_3, and x_4 be the annual salaries (in thousands of dollars) of the first, second, third, and fourth worker, respectively. Then,

$$x_1 = 75, \quad x_2 = 42, \quad x_3 = 125, \quad \text{and} \quad x_4 = 61$$

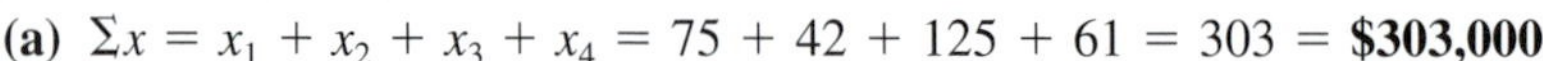

(a) $\Sigma x = x_1 + x_2 + x_3 + x_4 = 75 + 42 + 125 + 61 = 303 = \mathbf{\$303,000}$

(b) Note that $(\Sigma x)^2$ is the square of the sum of all x values. Thus,

$$(\Sigma x)^2 = (303)^2 = \mathbf{91,809}$$

(c) The expression Σx^2 is the sum of the squares of x values. To calculate Σx^2, we first square each of the x values and then sum these squared values. Thus

$$\Sigma x^2 = (75)^2 + (42)^2 + (125)^2 + (61)^2$$
$$= 5625 + 1764 + 15,625 + 3721 = \mathbf{26,735}$$

■

■ EXAMPLE 1–2

Using summation notation: two variables.

The following table lists four pairs of m and f values:

m	12	15	20	30
f	5	9	10	16

Compute the following:

(a) Σm **(b)** Σf^2 **(c)** Σmf **(d)** $\Sigma m^2 f$

Solution We can write

$$m_1 = 12 \quad m_2 = 15 \quad m_3 = 20 \quad m_4 = 30$$
$$f_1 = 5 \quad f_2 = 9 \quad f_3 = 10 \quad f_4 = 16$$

(a) $\Sigma m = 12 + 15 + 20 + 30 = \mathbf{77}$

(b) $\Sigma f^2 = (5)^2 + (9)^2 + (10)^2 + (16)^2 = 25 + 81 + 100 + 256 = \mathbf{462}$

(c) To compute Σmf, we multiply the corresponding values of m and f and then add the products as follows:

$$\Sigma mf = m_1 f_1 + m_2 f_2 + m_3 f_3 + m_4 f_4$$
$$= 12(5) + 15(9) + 20(10) + 30(16) = \mathbf{875}$$

(d) To calculate $\Sigma m^2 f$, we square each m value, then multiply the corresponding m^2 and f values, and add the products. Thus,

$$\Sigma m^2 f = (m_1)^2 f_1 + (m_2)^2 f_2 + (m_3)^2 f_3 + (m_4)^2 f_4$$
$$= (12)^2(5) + (15)^2(9) + (20)^2(10) + (30)^2(16) = \mathbf{21,145}$$

The calculations done in parts (a) through (d) to find the values of Σm, Σf^2, Σmf, and $\Sigma m^2 f$ can be performed in tabular form, as shown in Table 1.4.

Table 1.4

m	f	f^2	mf	m^2f
12	5	$5 \times 5 = 25$	$12 \times 5 = 60$	$12 \times 12 \times 5 = 720$
15	9	$9 \times 9 = 81$	$15 \times 9 = 135$	$15 \times 15 \times 9 = 2025$
20	10	$10 \times 10 = 100$	$20 \times 10 = 200$	$20 \times 20 \times 10 = 4000$
30	16	$16 \times 16 = 256$	$30 \times 16 = 480$	$30 \times 30 \times 16 = 14{,}400$
$\Sigma m = 77$	$\Sigma f = 40$	$\Sigma f^2 = 462$	$\Sigma mf = 875$	$\Sigma m^2 f = 21{,}145$

The columns of Table 1.4 can be explained as follows.

1. The first column lists the values of m. The sum of these values gives $\Sigma m = 77$.
2. The second column lists the values of f. The sum of this column gives $\Sigma f = 40$.
3. The third column lists the squares of the f values. For example, the first value, 25, is the square of 5. The sum of the values in this column gives $\Sigma f^2 = 462$.
4. The fourth column records products of the corresponding m and f values. For example, the first value, 60, in this column is obtained by multiplying 12 by 5. The sum of the values in this column gives $\Sigma mf = 875$.
5. Next, the m values are squared and multiplied by the corresponding f values. The resulting products, denoted by m^2f, are recorded in the fifth column. For example, the first value, 720, is obtained by squaring 12 and multiplying this result by 5. The sum of the values in this column gives $\Sigma m^2 f = 21{,}145$. ■

EXERCISES

CONCEPTS AND PROCEDURES

1.22 The following table lists five pairs of m and f values.

m	5	10	17	20	25
f	12	8	6	16	4

Compute the value of each of the following:

a. Σm b. Σf^2 c. Σmf d. $\Sigma m^2 f$

1.23 The following table lists six pairs of m and f values.

m	3	6	25	12	15	18
f	16	11	16	8	4	14

Calculate the value of each of the following:

a. Σf b. Σm^2 c. Σmf d. $\Sigma m^2 f$

1.24 The following table lists five pairs of x and y values.

x	15	22	11	8	5
y	10	12	14	9	18

Compute

a. Σx b. Σy c. Σxy d. Σx^2 e. Σy^2

1.25 The following table lists six pairs of x and y values.

x	4	18	25	9	12	20
y	12	5	14	7	12	8

Compute

a. Σx b. Σy c. Σxy d. Σx^2 e. Σy^2

APPLICATIONS

1.26 Six adults spent \$20, \$14, \$57, \$23, \$7, and \$102 on lottery tickets last month. Let y denote last month's lottery ticket expenses for an adult. Find

a. Σy b. $(\Sigma y)^2$ c. Σy^2

1.27 The phone bills for January 2006 for four families were \$83, \$205, \$87, and \$154. Let y be the amount of the January 2006 phone bill for a family. Find

a. Σy b. $(\Sigma y)^2$ c. Σy^2

1.28 Prices (in thousands of dollars) of five new cars are 28, 35, 29, 54, and 18. Let x be the price of a new car in this sample. Find

a. Σx b. $(\Sigma x)^2$ c. Σx^2

1.29 The number of students (rounded to the nearest thousand) currently enrolled at seven universities are 7, 39, 21, 16, 3, 43, and 19. Let x be the number of students currently enrolled at a university. Find

a. Σx b. $(\Sigma x)^2$ c. Σx^2

USES AND MISUSES... SPEAKING THE LANGUAGE OF STATISTICS

Have you ever heard the statistic "the average American family has 2.1 children"? What is wrong with this statement, and how do we fix it? How about: "In a representative sample of 10 American families, one can expect there to be 21 children." The statement is wordy but more accurate. Why do we care?

Statisticians pay close attention to definitions because, without them, calculations would be impossible to make and interpretations of the data would be meaningless. Often, when you read statistics reported in the newspaper, the journalist or editor sometimes chooses to describe the results in a way that is easier to understand but that distorts the actual statistical result.

Let's pick apart our example. The word *average* has a very specific meaning in probability (Chapters 4 and 5). The intended meaning of the word here really is *typical.* The adjective *American* helps us define the population. The Census Bureau defines *family* as "a group of two people or more (one of whom is the householder) related by birth, marriage, or adoption and residing together; all such people (including related subfamily members) are considered as members of one family." It defines *children* as "all persons under 18 years, excluding people who maintain households, families, or subfamilies as a reference person or spouse." We understand implicitly that a family cannot have a fractional number of children, so we accept that this discrete variable takes on the properties of a continuous variable when we are talking about the characteristics of a large population. How large does the population need to be before we can derive continuous variables from discrete variables? The answer comes in the chapters that follow.

The moral of the story is that whenever you read a statistical result, be sure that you understand the definitions of the terms used to describe the result and relate those terms to the definitions that you already know. In some cases *year* is a categorical variable, in others it is a discrete variable, in others a continuous variable. Many surveys will report that "respondents feel better, the same, or worse" about a particular subject. Although *better*, *same*, and *worse* have a natural order to them, they do not have numerical values.

Glossary

Census A survey that includes all members of the population.

Continuous variable A (quantitative) variable that can assume any numerical value over a certain interval or intervals.

Cross-section data Data collected on different elements at the same point in time or for the same period of time.

Data or **data set** Collection of observations or measurements on a variable.

Descriptive statistics Collection of methods for organizing, displaying, and describing data using tables, graphs, and summary measures.

Discrete variable A (quantitative) variable whose values are countable.

Element or **member** A specific subject or object included in a sample or population.

Inferential statistics Collection of methods that help make decisions about a population based on sample results.

Observation or **measurement** The value of a variable for an element.

Population or **target population** The collection of all elements whose characteristics are being studied.

Qualitative or **categorical data** Data generated by a qualitative variable.

Qualitative or **categorical variable** A variable that cannot assume numerical values but is classified into two or more categories.

Quantitative data Data generated by a quantitative variable.

Quantitative variable A variable that can be measured numerically.

Random sample A sample drawn in such a way that each element of the population has some chance of being included in the sample.

Representative sample A sample that contains the same characteristics as the corresponding population.

Sample A portion of the population of interest.

Sample survey A survey that includes elements of a sample.

Simple random sampling If all samples of the same size selected from a population have the same chance of being selected, it is called simple random sampling. Such a sample is called a simple random sample.

Statistics Group of methods used to collect, analyze, present, and interpret data and to make decisions.

Survey Collection of data on the elements of a population or sample.

Time-series data Data that give the values of the same variable for the same element at different points in time or for different periods of time.

Variable A characteristic under study or investigation that assumes different values for different elements.

Supplementary Exercises

1.30 The following table gives the average attendance at interleague Major League baseball games for the 1999 to 2004 seasons.

Season	Average Attendance at Interleague Games
1999	33,352
2000	33,212
2001	33,692
2002	31,921
2003	30,894
2004	32,976

Source: USA TODAY, July 8, 2004.

Describe the meaning of a variable, a measurement, and a data set with reference to this table.

1.31 The following table lists the number of Americans who took cruises during 1995 to 2004.

Year	Total Passengers (millions)
1995	4.4
1996	4.7
1997	5.1
1998	5.4
1999	5.9
2000	6.9
2001	6.9
2002	7.6
2003	8.2
2004	9.0

Source: Cruise Lines International Association. *USA TODAY,* February 4, 2005.

Describe the meaning of a variable, a measurement, and a data set with reference to this table.

1.32 Refer to Exercises 1.30 and 1.31. Classify these data sets as either cross-section or time-series.

1.33 Indicate whether each of the following examples refers to a population or to a sample.

a. A group of 25 patients selected to test a new drug
b. Total items produced on a machine for each year from 1995 to 2005
c. Yearly expenditures on clothes for 50 persons
d. Number of houses sold by each of the 10 employees of a real estate agency during 2005

1.34 Indicate whether each of the following examples refers to a population or to a sample.

a. Salaries of CEOs of all companies in New York City
b. Allowances of 1500 sixth-graders selected from Ohio
c. Gross sales for 2005 of four fast-food chains
d. Annual incomes of all 33 employees of a restaurant

1.35 State which of the following is an example of sampling with replacement and which is an example of sampling without replacement.

a. Selecting 10 patients out of 100 to test a new drug
b. Selecting one professor to be a member of the university senate and then selecting one professor from the same group to be a member of the curriculum committee

1.36 State which of the following is an example of sampling with replacement and which is an example of sampling without replacement.

a. Selecting seven cities to market a new deodorant
b. Selecting a high school teacher to drive students to a lecture in March, then selecting a teacher from the same group to chaperone a dance in April

1.37 The number of shoe pairs owned by six women are 8, 14, 3, 7, 10, and 5. Let x denote the number of shoe pairs owned by a woman. Find

a. Σx b. $(\Sigma x)^2$ c. Σx^2

1.38 The number of restaurants in each of five small towns is 4, 12, 8, 10, and 5. Let y denote the number of restaurants in a small town. Find

a. Σy b. $(\Sigma y)^2$ c. Σy^2

1.39 The following table lists five pairs of m and f values.

m	3	16	11	9	20
f	7	32	17	12	34

Compute the value of each of the following:

a. Σm b. Σf^2 c. Σmf d. $\Sigma m^2 f$ e. Σm^2

1.40 The following table lists six pairs of x and y values.

x	7	11	8	4	14	28
y	5	15	7	10	9	19

Compute the value of each of the following:

a. Σy b. Σx^2 c. Σxy d. $\Sigma x^2 y$ e. Σy^2

Self-Review Test

1. A population in statistics means a collection of all
   a. men and women
   b. subjects or objects of interest
   c. people living in a country

2. A sample in statistics means a portion of the
   a. people selected from the population of a country
   b. people selected from the population of an area
   c. population of interest

3. Indicate which of the following is an example of a sample with replacement and which is a sample without replacement.
   a. Five friends go to a livery stable and select five horses to ride (each friend must choose a different horse).
   b. A box contains five balls of different colors. A ball is drawn from this box, its color is recorded, and it is put back into the box before the next ball is drawn. This experiment is repeated 12 times.

4. Indicate which of the following variables are quantitative and which are qualitative. Classify the quantitative variables as discrete or continuous.
   a. Women's favorite TV programs
   b. Salaries of football players
   c. Number of pets owned by families
   d. Favorite breed of dog for each of 20 children

5. The following table lists the total career yards rushed by each of the top ten rushers in the National Football League as of February 4, 2005.

Player	Yards
Emmitt Smith	18,355
Walter Payton	16,726
Barry Sanders	15,269
Curtis Martin	13,336
Jerome Bettis	13,294
Eric Dickerson	13,259
Tony Dorsett	12,739
Jim Brown	12,312
Marcus Allen	12,243
Franco Harris	12,120

Source: USA TODAY, February 4, 2005.

Explain the meaning of a member, a variable, a measurement, and a data set with reference to this table.

6. The number of credit cards possessed by five couples are 2, 5, 3, 12, and 7. Let x be the number of credit cards possessed by a couple. Find:
   a. Σx b. $(\Sigma x)^2$ c. Σx^2

7. The following table lists five pairs of m and f values.

m	3	6	9	12	15
f	15	25	40	20	12

Calculate

a. Σm b. Σf c. Σm^2 d. Σmf e. $\Sigma m^2 f$ f. Σf^2

Mini-Project

■ MINI-PROJECT 1–1

In this mini-project, you are going to obtain a data set of interest to you that you will use for mini-projects in other chapters throughout the course. The data set should contain at least one qualitative variable and one quantitative variable, although having two of each will be necessary in some cases. Ask your instructor how many variables you should have. A good size data set to work with should contain somewhere between 50 and 100 observations.

Here are some examples of the procedures to use to obtain data:

1. Take a random sample of used cars and collect data on them. You may use Web sites like Cars.com, AutoTrader.com, and so forth. Quantitative variables may include the price, mileage, and age of a car. Categorical variables may include the model, drive train (front wheel, rear wheel, and so forth), and type (compact, SUV, minivan, and so forth). You can concentrate on your favorite type of car, or look at a variety of types.

2. Examine the real estate ads in your local newspaper or online and obtain information on rental properties or houses for sale.

3. Use an almanac or go to a government Web site, such as www.census.gov or www.cdc.gov to obtain information for each state. Quantitative variables may include income, birth and death rates, cancer

incidence, and the proportion of people living below the poverty level. Categorical variables may include things like the region of the country where each state is located and which party won the state governorship in the last election. You can also collect this information on a worldwide level and use the continent or world region as a categorical variable.

4. Take a random sample of students and ask them questions such as:

- How much money did you spend on books last semester?
- How many credit hours did you take?
- What is your major?

5. If you are a sports fan, you can use an almanac or sports Web site to obtain statistics on a random sample of athletes. You can look at sport-specific statistics such as home runs, runs batted in, position, left-handed/right-handed, and so forth in baseball, or you could collect information to compare different sports by gathering information on salary, career length, weight, and so forth.

Once you have collected the information, write a brief report that includes answers to the following tasks/questions:

a. Describe the variables that you have collected information on.
b. Describe a reasonable target population for the sample you used.
c. Is your sample a random sample from this target population?
d. Do you feel that your sample is representative of this population?
e. Is this an example of sampling with or without replacement?
f. For each quantitative variable, state whether it is continuous or discrete.
g. Describe the meaning of an element, a variable, and a measurement for this data set.
h. Describe any problems you faced in collecting these data.
i. Were any of the data values unusable? If yes, explain why.

Your instructor will probably want to see a copy of the data you collected. If you are using statistical software in the class, enter the data into that software and submit a copy of the data file. If you are using a handheld technology calculator, such as a graphing calculator, you will probably have to print out a hard copy version of the data set. Save this data set for projects in future chapters.

TECHNOLOGY INSTRUCTION — Entering and Saving Data

Whenever you want to analyze and interpret some data, you need to enter those data in some technology, proofread it, and revise it. If you will be using the data entered into a technology again at a later date, you need to save these data into your technology so that you can retrieve them later.

TI-84

1. On the TI-84, variables are referred to as lists.

2. In order to enter data into the TI-84 calculator, you first need to decide if you will want to save the data for later use or you want to use it only in the immediate future.

3. If you will be using these data only in the immediate future, select **STAT>EDIT>SetUp-Editor**, and then press **Enter**. This will set up the editor to use "scratch" lists **L1, L2, L3, L4, L5,** and **L6** (see **Screen 1.1**). Now select **STAT>EDIT>Edit** and start typing your numeric data into the column or columns and press **Enter** after each entry (see **Screen 1.2**). Note: The TI-84 will not handle non-numeric data.

Screen 1.1

Screen 1.2

4. If you will be using your data at a later date, it is better to give the variables names so that you do not have to reenter the data. Select **STAT>EDIT>SetUpEditor,** and then type in the names of your variables separted by commas (see **Screens 1.3 and 1.4**). Names can be 1 to 5 letters long. These letters can be found in green on the TI-84 keyboard. You can use the green **ALPHA** key with each letter, or press **A-LOCK (2nd>ALPHA)** while you are typing the name. To turn off **A-LOCK**, press **ALPHA**. Note that here we have used the names EX1 and EX2. You can use any names such as Price, Brand, and so forth.

Screen 1.3

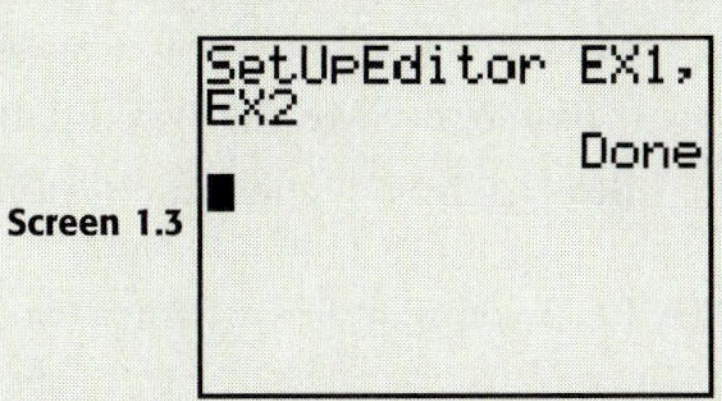

Screen 1.4

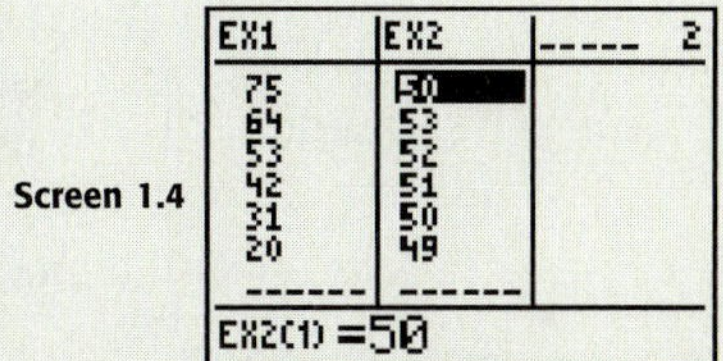

EX1
EX2
75
64
53
42
31
20
50
53
52
51
50
49
EX2(1) =50

5. You can use the arrow keys to move around and go back to a cell to edit its contents. When editing values, you will need to press **Enter** for the changes to take effect.
6. **SetUpEditor** determines what lists are displayed in the editor. Changing what **SetUpEditor** displays does not delete any lists. Your lists remain in storage when the calculator is turned off.

MINITAB

1. Start MINITAB by clicking the MINITAB icon. You will see a session window as well as a worksheet, similar to a spreadsheet, where you will enter your data (see **Screen 1.5**).

Screen 1.5

↓	C1	C2	C3-T
	year	sales	employee
1	2001	35	J. Smith
2	2001	38	A. Jones
3	2002	50	J. Smith
4	2002	48	A. Jones

2. Use the mouse or the arrow keys to select a cell in the MINITAB worksheet where you want to start entering your data. Data can be numeric, text, or date/time, but you can enter only one type of data into a given column since the columns correspond to variables. The rectangles in the Worksheet are called cells, and the cells are organized into columns such as **C1, C2, . . .**, and so forth. Each column has rows numbered **1, 2, . . .**, and so forth. Note that if a column contains text data as column C3 in Screen 1.5, "-T" will be added to the column heading.
3. The blank row between the column labels and row 1 is for variable names.
4. You can change whether you are typing the data across in rows or down in columns by clicking the direction arrow at the top left of the worksheet (see **Screen 1.5**).
5. If you want to type a table in a given array of cells, then select that set of cells and begin typing.
6. If you need to revise an entry, go to that cell with the mouse or the arrow keys and begin typing. Press **Enter** to put the revised entry into the cell.

7. When you are done, select **File>Save Current Worksheet As** to save your work for the first time as a file on your computer. Note that MINITAB will automatically assign the file extension *.mtw* to your work after you choose the filename.

8. Try entering the following data into MINITAB:

January	52	.08
February	48	.06
March	49	.07

Name the columns *Month, Sales, Increase.* Save the typed data as *test.mtw* file.

9. To retrieve the file, select **File>Open** and select the *test.mtw* file.

10. If you are already in MINITAB and you want to start a new worksheet, select **File>New** and choose **Worksheet**.

Excel

1. Start Excel.

2. Use the mouse or the arrow keys to select the cell in the spreadsheet where you want to start entering your data. Data can be numeric or text. The rectangles are called cells, and the cells are collectively known as a spreadsheet.

3. You can format your data by selecting the cells that you want to format, then selecting **Format>Cells**, and then choosing whether you want to format a number, align text, and so forth. For common formatting tasks, you have icons on the toolbar, such as a dollar sign ($) to format currency, a percent sign (%) to format numbers as percents, and icons representing left, center, and right aligned text to change your alignment.

4. If you need to revise an entry, go to that cell with the mouse or the arrow keys. You can retype the entry or you can edit it. To edit it, select the **Formula Bar** at the top of the screen (which contains the contents of the cell) and use the arrow keys and the backspace key to help you revise the entry, then press **Enter** to put the revised entry into the cell.

5. When you are done, select **File>Save As** to save your work for the first time as a file on your computer. Note that Excel will automatically assign the file extension *.xls* to your work after you choose the filename.

6. Try entering the following data into Excel:

January	52	.08
February	48	.06
March	49	.07

Format it to look like the following.

January	$52.00	8%
February	$48.00	6%
March	$49.00	7%

Save the result as *test.xls* file.

7. To retrieve this file, select **File>Open** and select the *test.xls* file.

Screen 1.6 shows the data of Screen 1.5 as entered in Excel.

	A	B	C
1	year	sales	employee
2	2001	35	J. Smith
3	2001	38	A. Jones
4	2002	50	J. Smith
5	2002	48	A. Jones
6			

Screen 1.6

TECHNOLOGY ASSIGNMENTS

TA1.1 The following table gives the names, hours worked, and salary for the past week for five workers.

Name	Hours Worked	Salary
John	42	$725
Shannon	33	1583
Kathy	28	1255
David	47	1090
Steve	40	820

a. Enter these data into the spreadsheet. Save the data file as WORKER. Exit the session or program. Then restart the program or software and retrieve the file WORKER.

b. Print a hard copy of the spreadsheet containing data you entered.

TA1.2 Refer to data on 2004 profits of seven companies given in Table 1.1 on page 9 Enter those data into the spreadsheet and save this file as PROFITS.

Chapter

2

Organizing and Graphing Data

2.1 Raw Data

2.2 Organizing and Graphing Qualitative Data

Case Study 2–1 Marrying in the USA

Case Study 2–2 Americans Say Keep the Penny

2.3 Organizing and Graphing Quantitative Data

Case Study 2–3 Hand Hygiene

Case Study 2–4 USA is a Caffeinated Country

2.4 Shapes of Histograms

Case Study 2–5 Using Truncated Axes

2.5 Cumulative Frequency Distributions

2.6 Stem-and-Leaf Displays

2.7 DOTPLOTS

Are you one of those adults who do not drink caffeinated beverages at all? Or are you one of those who drink four or more cups or cans of caffeinated beverages a day? Or do you fall somewhere in the middle of these two extremes? According to a sample survey of adults, 22% said that they do not drink any caffeinated beverages at all, 25% said they drink four or more cups or cans a day. Of the remaining adults, 16% drink one cup or can a day, 21% drink two cups or cans a day, and 16% drink three cups or cans a day. (See Case Study 2–4.)

In addition to thousands of private organizations and individuals, a large number of U.S. government agencies (such as the Bureau of the Census, the Bureau of Labor Statistics, the National Agricultural Statistics Service, the National Center for Education Statistics, the National Center for Health Statistics, and the Bureau of Justice Statistics) conduct hundreds of surveys every year. The data collected from each of these surveys fill hundreds of thousands of pages. In their original form, these data sets may be so large that they do not make sense to most of us. Descriptive statistics, however, supplies the techniques that help condense large data sets by using tables, graphs, and summary measures. We see such tables, graphs, and summary measures in newspapers and magazines every day. At a glance, these tabular and graphical displays present information on every aspect of life. Consequently, descriptive statistics is of immense importance because it provides efficient and effective methods for summarizing and analyzing information.

This chapter explains how to organize and display data using tables and graphs. We will learn how to prepare frequency distribution tables for qualitative and quantitative data; how to construct bar graphs, pie charts, histograms, and polygons for such data; and how to prepare stem-and-leaf displays.

2.1 Raw Data

When data are collected, the information obtained from each member of a population or sample is recorded in the sequence in which it becomes available. This sequence of data recording is random and unranked. Such data, before they are grouped or ranked, are called **raw data**.

Definition

Raw Data Data recorded in the sequence in which they are collected and before they are processed or ranked are called *raw data*.

Suppose we collect information on the ages (in years) of 50 students selected from a university. The data values, in the order they are collected, are recorded in Table 2.1. For instance, the first student's age is 21, the second student's age is 19 (second number in the first row), and so forth. The data in Table 2.1 are quantitative raw data.

Table 2.1 Ages of 50 Students

21	19	24	25	29	34	26	27	37	33
18	20	19	22	19	19	25	22	25	23
25	19	31	19	23	18	23	19	23	26
22	28	21	20	22	22	21	20	19	21
25	23	18	37	27	23	21	25	21	24

Suppose we ask the same 50 students about their student status. The responses of the students are recorded in Table 2.2. In this table, F, SO, J, and SE are the abbreviations for freshman, sophomore, junior, and senior, respectively. This is an example of qualitative (or categorical) raw data.

Table 2.2 Status of 50 Students

J	F	SO	SE	J	J	SE	J	J	J
F	F	J	F	F	F	SE	SO	SE	J
J	F	SE	SO	SO	F	J	F	SE	SE
SO	SE	J	SO	SO	J	J	SO	F	SO
SE	SE	F	SE	J	SO	F	J	SO	SO

The data presented in Tables 2.1 and 2.2 are also called **ungrouped data**. An ungrouped data set contains information on each member of a sample or population individually.

2.2 Organizing and Graphing Qualitative Data

This section discusses how to organize and display qualitative (or categorical) data. Data sets are organized into tables, and data are displayed using graphs.

2.2.1 Frequency Distributions

A sample of 100 students enrolled at a university were asked what they intended to do after graduation. Forty-four said they wanted to work for private companies/businesses, 16 said they wanted to work for the federal government, 23 wanted to work for state or local governments, and 17 intended to start their own businesses. Table 2.3 lists the types of employment and the number of students who intend to engage in each type of employment. In this table, the variable is the *type of employment*, which is a qualitative variable. The categories (representing the type of employment) listed in the first column are mutually exclusive. In other words, each of the 100 students belongs to one and only one of these categories. The number of students who belong to a certain category is called the *frequency* of that category. A **frequency distribution** exhibits how the frequencies are distributed over various categories. Table 2.3 is called a *frequency distribution table* or simply a *frequency table*.

Table 2.3 Type of Employment Students Intend to Engage In

Type of Employment (← Variable)	Number of Students (← Frequency column)
Private companies/businesses	44
Federal government (← Category)	16 (← Frequency)
State/local government	23
Own business	17
	Sum = 100

Definition

Frequency Distribution for Qualitative Data A *frequency distribution* for qualitative data lists all categories and the number of elements that belong to each of the categories.

Example 2–1 illustrates how a frequency distribution table is constructed for qualitative data.

■ EXAMPLE 2–1

Constructing a frequency distribution table for qualitative data.

A sample of 30 employees from large companies was selected, and these employees were asked how stressful their jobs were. The responses of these employees are recorded below where *very* represents very stressful, *somewhat* means somewhat stressful, and *none* stands for not stressful at all.

somewhat	none	somewhat	very	very	none
very	somewhat	somewhat	very	somewhat	somewhat
very	somewhat	none	very	none	somewhat
somewhat	very	somewhat	somewhat	very	none
somewhat	very	very	somewhat	none	somewhat

Construct a frequency distribution table for these data.

Solution Note that the variable in this example is *how stressful is an employee's job*. This variable is classified into three categories: very stressful, somewhat stressful, and not stressful at all. We record these categories in the first column of Table 2.4. Then we read each employee's response from the given data and mark a *tally*, denoted by the symbol |, in the second column of Table 2.4 next to the corresponding category. For example, the first employee's response is that his or her job is somewhat stressful. We show this in the frequency table by marking a tally in the second column next to the category *somewhat*. Note that the tallies are marked in blocks of five for counting convenience. Finally, we record the total of the tallies for each category in the third column of the table. This column is called the *column of frequencies* and is usually denoted by *f*. The sum of the entries in the frequency column gives the sample size or total frequency. In Table 2.4, this total is 30, which is the sample size.

Table 2.4 Frequency Distribution of Stress on Job

Stress on Job	Tally	Frequency (f)
Very	~~||||~~ ~~||||~~	10
Somewhat	~~||||~~ ~~||||~~ ||||	14
None	~~||||~~		6
		Sum = 30

■

2.2.2 Relative Frequency and Percentage Distributions

The **relative frequency** of a category is obtained by dividing the frequency of that category by the sum of all frequencies. Thus, the relative frequency shows what fractional part or proportion of the total frequency belongs to the corresponding category. A *relative frequency distribution* lists the relative frequencies for all categories.

Calculating Relative Frequency of a Category

$$\text{Relative frequency of a category} = \frac{\text{Frequency of that category}}{\text{Sum of all frequencies}}$$

The **percentage** for a category is obtained by multiplying the relative frequency of that category by 100. A *percentage distribution* lists the percentages for all categories.

Calculating Percentage

$$\text{Percentage} = (\text{Relative frequency}) \cdot 100$$

■ EXAMPLE 2–2

Determine the relative frequency and percentage distributions for the data of Table 2.4.

Constructing relative frequency and percentage distributions.

Solution The relative frequencies and percentages from Table 2.4 are calculated and listed in Table 2.5. Based on this table, we can state that .333 or 33.3% of the employees said that their jobs are very stressful. By adding the percentages for the first two categories, we can state that 80% of the employees said that their jobs are very or somewhat stressful. The other numbers in Table 2.5 can be interpreted the same way.

Notice that the sum of the relative frequencies is always 1.00 (or approximately 1.00 if the relative frequencies are rounded), and the sum of the percentages is always 100 (or approximately 100 if the percentages are rounded).

Table 2.5 Relative Frequency and Percentage Distributions of Stress on Job

Stress on Job	Relative Frequency	Percentage
Very	10/30 = .333	.333(100) = 33.3
Somewhat	14/30 = .467	.467(100) = 46.7
None	6/30 = .200	.200(100) = 20.0
	Sum = 1.000	Sum = 100

■

CASE STUDY

2–1

MARRYING IN THE USA

USA TODAY Snapshots

Marrying in the USA

Marital status of the more than 221 million people age 15 and older:

120 million — Married

41 million — Widowed, divorced, separated

60 million — Never married

Source: Census Bureau   Note: Numbers are rounded.

By Shannon Reilly and Marcy E. Mullins, USA TODAY

The above chart, reproduced from *USA TODAY,* shows a bar graph indicating the marital status of Americans aged 15 and older. According to the Census Bureau, there are around 221 million people in the United States who are 15 years of age or older. Of them, about 120 million are currently married, about 41 million are widowed/divorced/separated, and about 60 million have never married. Using these categories and numbers, we can write a frequency table, and then calculate the relative frequencies and percentages.

Source: USA TODAY, November 10, 2003. Copyright © 2003, *USA TODAY.* Chart reproduced with permission.

2.2.3 Graphical Presentation of Qualitative Data

All of us have heard the adage "a picture is worth a thousand words." A graphic display can reveal at a glance the main characteristics of a data set. The *bar graph* and the *pie chart* are two types of graphs used to display qualitative data.

Bar Graphs

To construct a **bar graph** (also called a *bar chart*), we mark the various categories on the horizontal axis as in Figure 2.1. Note that all categories are represented by intervals of the same width. We mark the frequencies on the vertical axis. Then we draw one bar for each category such that the height of the bar represents the frequency of the corresponding category. We leave a small gap between adjacent bars. Figure 2.1 gives the bar graph for the frequency distribution of Table 2.4.

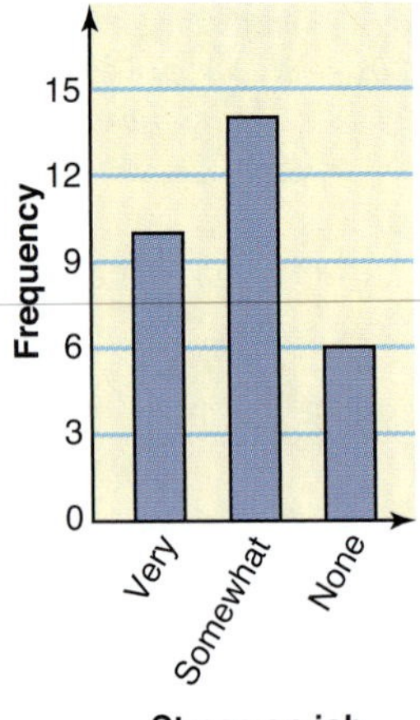

Figure 2.1 Bar graph for the frequency distribution of Table 2.4.

Definition

Bar Graph A graph made of bars whose heights represent the frequencies of respective categories is called a *bar graph.*

The bar graphs for relative frequency and percentage distributions can be drawn simply by marking the relative frequencies or percentages, instead of the class frequencies, on the vertical axis.

Sometimes a bar graph is constructed by marking the categories on the vertical axis and the frequencies on the horizontal axis.

AMERICANS SAY KEEP THE PENNY

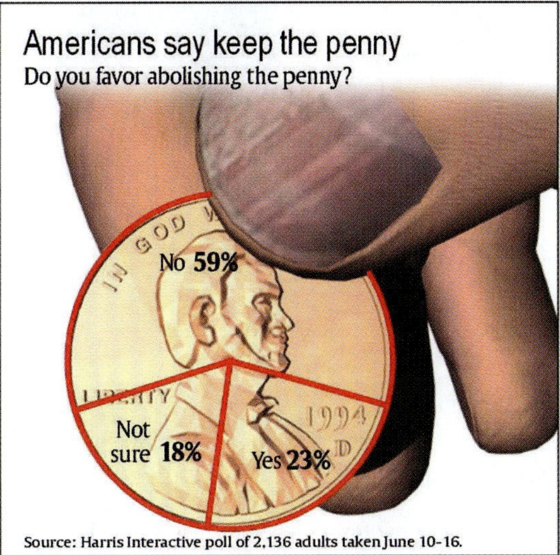
USA TODAY Snapshots
Americans say keep the penny
Do you favor abolishing the penny?
No 59%
Not sure 18%
Yes 23%
Source: Harris Interactive poll of 2,136 adults taken June 10-16.
By Shannon Reilly and Robert W. Ahrens, USA TODAY

The above pie chart shows the opinions of people on the issue of keeping or abolishing the penny. In a Harris Interactive poll of 2136 adults conducted in June 10–16, 2004, 59% of the respondents said that the penny should not be abolished, 23% said it should be abolished, and 18% were not sure. Using these categories and percentages, we can write a percentage distribution table.

Source: USA TODAY, August 5, 2004.

Pie Charts

A **pie chart** is more commonly used to display percentages, although it can be used to display frequencies or relative frequencies. The whole pie (or circle) represents the total sample or population. Then we divide the pie into different portions that represent the different categories.

Definition

Pie Chart A circle divided into portions that represent the relative frequencies or percentages of a population or a sample belonging to different categories is called a *pie chart.*

As we know, a circle contains 360 degrees. To construct a pie chart, we multiply 360 by the relative frequency of each category to obtain the degree measure or size of the angle for the corresponding category. Table 2.6 shows the calculation of angle sizes for the various categories of Table 2.5.

Table 2.6 Calculating Angle Sizes for the Pie Chart

Stress on Job	Relative Frequency	Angle Size
Very	.333	360(.333) = 119.88
Somewhat	.467	360(.467) = 168.12
None	.200	360(.200) = 72.00
	Sum = 1.000	Sum = 360

Figure 2.2 shows the pie chart for the percentage distribution of Table 2.5, which uses the angle sizes calculated in Table 2.6.

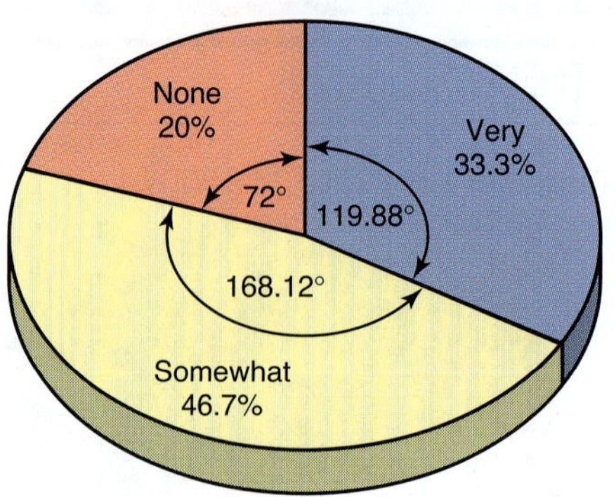

Figure 2.2 Pie chart for the percentage distribution of Table 2.5.

EXERCISES

CONCEPTS AND PROCEDURES

2.1 Why do we need to group data in the form of a frequency table? Explain briefly.

2.2 How are the relative frequencies and percentages of categories obtained from the frequencies of categories? Illustrate with the help of an example.

2.3 The following data give the results of a sample survey. The letters A, B, and C represent the three categories.

A	B	B	A	C	B	C	C	C	A
C	B	C	A	C	C	B	C	C	A
A	B	C	C	B	C	B	A	C	A

a. Prepare a frequency distribution table.
b. Calculate the relative frequencies and percentages for all categories.
c. What percentage of the elements in this sample belong to category B?
d. What percentage of the elements in this sample belong to category A or C?
e. Draw a bar graph for the frequency distribution.

2.4 The following data give the results of a sample survey. The letters Y, N, and D represent the three categories.

D	N	N	Y	Y	Y	N	Y	D	Y
Y	Y	Y	Y	N	Y	Y	N	N	Y
N	Y	Y	N	D	N	Y	Y	Y	Y
Y	Y	N	N	Y	Y	N	N	D	Y

a. Prepare a frequency distribution table.
b. Calculate the relative frequencies and percentages for all categories.
c. What percentage of the elements in this sample belong to category Y?
d. What percentage of the elements in this sample belong to category N or D?
e. Draw a pie chart for the percentage distribution.

APPLICATIONS

2.5 The data on the status of 50 students given in Table 2.2 of Section 2.1 are reproduced here.

J	F	SO	SE	J	J	SE	J	J	J
F	F	J	F	F	F	SE	SO	SE	J
J	F	SE	SO	SO	F	J	F	SE	SE
SO	SE	J	SO	SO	J	J	SO	F	SO
SE	SE	F	SE	J	SO	F	J	SO	SO

a. Prepare a frequency distribution table.
b. Calculate the relative frequencies and percentages for all categories.
c. What percentage of these students are juniors or seniors?
d. Draw a bar graph for the frequency distribution.

2.6 Thirty adults were asked which of the following conveniences they would find most difficult to do without: television (T), refrigerator (R), air conditioning (A), public transportation (P), or microwave (M). Their responses are listed below.

R	A	R	P	P	T	R	M	P	A
A	R	R	T	P	P	T	R	A	A
R	P	A	T	R	P	R	A	P	R

a. Prepare a frequency distribution table.
b. Calculate the relative frequencies and percentages for all categories.
c. What percentage of these adults named refrigerator or air conditioning as the convenience that they would find most difficult to do without?
d. Draw a bar graph for the relative frequency distribution.

2.7 In a 2004 *USA TODAY* survey (*USA TODAY*, July 19, 2004), registered dietitians with the American Dietetic Association were asked, "What is the major reason people want to lose weight?" The responses were classified as *Health* (H), *Cosmetic* (C), and *Other* (O). Suppose a random sample of 20 dietitians is taken and these dietitians are asked the same question. Their responses are as follows.

H	H	C	H	O	C	C	H	C	O
O	H	C	H	H	C	H	H	O	H

a. Prepare a frequency distribution table.
b. Compute the relative frequencies and percentages for all categories.
c. What percentage of these dietitians gave *Health* as the major reason for people to lose weight?
d. Draw a pie chart for the percentage distribution.

2.8 The following data show the method of payment by 16 customers in a supermarket checkout line. Here, C refers to cash, CK to check, CC to credit card, D to debit card, and O stands for other.

C	CK	CK	C	CC	D	O	C
CK	CC	D	CC	C	CK	CK	CC

a. Construct a frequency distribution table.
b. Calculate the relative frequencies and percentages for all categories.
c. Draw a pie chart for the percentage distribution.

2.9 In the *MARS 2004 OTC/DTC* survey, U.S. adults were asked to rate their health. The table below summarizes their responses.

State of Health	Percentage of Responses
Excellent	17.0
Very good	36.2
Good	32.5
Fair	12.0
Poor	2.3

Source: USA TODAY, June 2, 2004.

Draw a pie chart for this percentage distribution.

2.10 In an exit poll taken during the 2004 presidential election, voters were asked to name the issue that most affected their vote for a candidate for presidency. The following table summarizes their responses.

Issue	Percentage of Responses
Moral values	22
Economy/jobs	20
Terrorism	19
Iraq	15
Health care	8
Taxes	5
Education	4

Source: United States General Exit Poll of 13,660 voters. *USA TODAY*, November 5, 2004.

As you will notice, these percentages add up to 93%. Assume that the remaining 7% of these voters named other issues and let us denote these issues as *Other*. Draw a bar graph to display these data.

2.3 Organizing and Graphing Quantitative Data

In the previous section we learned how to group and display qualitative data. This section explains how to group and display quantitative data.

2.3.1 Frequency Distributions

Table 2.7 gives the weekly earnings of 100 employees of a large company. The first column lists the *classes*, which represent the (quantitative) variable *weekly earnings*. For quantitative data, an interval that includes all the values that fall within two numbers, the lower and upper limits, is called a **class**. Note that the classes always represent a variable. As we can observe, the classes are nonoverlapping; that is, each value on earnings belongs to one and only one class. The second column in the table lists the number of employees who have earnings within each class. For example, nine employees of this company earn \$401 to \$600 per week. The numbers listed in the second column are called the **frequencies**, which give the number of values that belong to different classes. The frequencies are denoted by f.

Table 2.7 **Weekly Earnings of 100 Employees of a Company**

	Weekly Earnings (dollars)	Number of Employees f	
Variable ⟶			⟵ Frequency column
	401 to 600	9	
	601 to 800	22	
Third class ⟶	801 to 1000	39	⟵ Frequency of the third class
	1001 to 1200	15	
	1201 to 1400	9	
Lower limit of the sixth class ⟶	1401 to 1600	6	Upper limit of the sixth class

For quantitative data, the frequency of a class represents the number of values in the data set that fall in that class. Table 2.7 contains six classes. Each class has a *lower limit* and an *upper limit*. The values 401, 601, 801, 1001, 1201, and 1401 give the lower limits, and the values 600, 800, 1000, 1200, 1400, and 1600 are the upper limits of the six classes, respectively. The data presented in Table 2.7 are an illustration of a **frequency distribution table** for quantitative data. Whereas the data that list individual values are called ungrouped data, the data presented in a frequency distribution table are called **grouped data**.

Definition

Frequency Distribution for Quantitative Data A *frequency distribution* for quantitative data lists all the classes and the number of values that belong to each class. Data presented in the form of a frequency distribution are called *grouped data*.

To find the midpoint of the upper limit of the first class and the lower limit of the second class in Table 2.7, we divide the sum of these two limits by 2. Thus, this midpoint is

$$\frac{600 + 601}{2} = 600.5$$

The value 600.5 is called the *upper boundary* of the first class and the *lower boundary* of the second class. By using this technique, we can convert the class limits of Table 2.7 to **class boundaries**, which are also called *real class limits*. The second column of Table 2.8 lists the boundaries for Table 2.7.

Definition

Class Boundary The *class boundary* is given by the midpoint of the upper limit of one class and the lower limit of the next class.

The difference between the two boundaries of a class gives the **class width**. The class width is also called the **class size**.

Finding Class Width

$$\text{Class width} = \text{Upper boundary} - \text{Lower boundary}$$

Thus, in Table 2.8,

$$\text{Width of the first class} = 600.5 - 400.5 = 200$$

The class widths for the frequency distribution of Table 2.7 are listed in the third column of Table 2.8. Each class in Table 2.8 (and Table 2.7) has the same width of 200.

The **class midpoint** or **mark** is obtained by dividing the sum of the two limits (or the two boundaries) of a class by 2.

Calculating Class Midpoint or Mark

$$\text{Class midpoint or mark} = \frac{\text{Lower limit} + \text{Upper limit}}{2}$$

Thus, the midpoint of the first class in Table 2.7 or Table 2.8 is calculated as follows:

$$\text{Midpoint of the first class} = \frac{401 + 600}{2} = 500.5$$

The class midpoints for the frequency distribution of Table 2.7 are listed in the fourth column of Table 2.8.

Table 2.8 Class Boundaries, Class Widths, and Class Midpoints for Table 2.7

Class Limits	Class Boundaries	Class Width	Class Midpoint
401 to 600	400.5 to less than 600.5	200	500.5
601 to 800	600.5 to less than 800.5	200	700.5
801 to 1000	800.5 to less than 1000.5	200	900.5
1001 to 1200	1000.5 to less than 1200.5	200	1100.5
1201 to 1400	1200.5 to less than 1400.5	200	1300.5
1401 to 1600	1400.5 to less than 1600.5	200	1500.5

Note that in Table 2.8, when we write classes using class boundaries, we write *to less than* to ensure that each value belongs to one and only one class. As we can see, the upper boundary of the preceding class and the lower boundary of the succeeding class are the same.

2.3.2 Constructing Frequency Distribution Tables

When constructing a frequency distribution table, we need to make the following three major decisions.

Number of Classes

Usually the number of classes for a frequency distribution table varies from 5 to 20, depending mainly on the number of observations in the data set.[1] It is preferable to have more classes as the size of a data set increases. The decision about the number of classes is arbitrarily made by the data organizer.

Class Width

Although it is not uncommon to have classes of different sizes, most of the time it is preferable to have the same width for all classes. To determine the class width when all classes are the same size, first find the difference between the largest and the smallest values in the data. Then, the approximate width of a class is obtained by dividing this difference by the number of desired classes.

Calculation of Class Width

$$\text{Approximate class width} = \frac{\text{Largest value} - \text{Smallest value}}{\text{Number of classes}}$$

Usually this approximate class width is rounded to a convenient number, which is then used as the class width. Note that rounding this number may slightly change the number of classes initially intended.

Lower Limit of the First Class or the Starting Point

Any convenient number that is equal to or less than the smallest value in the data set can be used as the lower limit of the first class.

Example 2–3 illustrates the procedure for constructing a frequency distribution table for quantitative data.

■ EXAMPLE 2–3

Constructing a frequency distribution table for quantitative data.

Table 2.9 (on next page) gives the total home runs hit by all players of each of the 30 Major League Baseball teams during the 2004 season. Construct a frequency distribution table.

Solution In these data, the minimum value is 135 and the maximum value is 242. Suppose we decide to group these data using five classes of equal width. Then,

$$\text{Approximate width of each class} = \frac{242 - 135}{5} = 21.4$$

Now we round this approximate width to a convenient number—say, 22. The lower limit of the first class can be taken as 135 or any number less than 135. Suppose we take 135 as the lower limit of the first class. Then our classes will be

135–156, 157–178, 179–200, 201–222, and 223–244

We record these five classes in the first column of Table 2.10 on page 37.

[1]One rule to help decide on the number of classes is Sturge's formula:

$$c = 1 + 3.3 \log n$$

where c is the number of classes and n is the number of observations in the data set. The value of $\log n$ can be obtained by entering the value of n on the calculator and pressing the *log* key.

Table 2.9 Home Runs Hit by Major League Baseball Teams During the 2004 Season

Team	Home Runs	Team	Home Runs
Arizona	135	Milwaukee	135
Atlanta	178	Minnesota	191
Baltimore	169	Montreal (now Washington)	151
Boston	222	New York Mets	185
Chicago Cubs	235	New York Yankees	242
Chicago White Sox	242	Oakland	189
Cincinnati	194	Philadelphia	215
Cleveland	184	Pittsburgh	142
Colorado	202	St. Louis	214
Detroit	201	San Diego	139
Florida	148	San Francisco	183
Houston	187	Seattle	136
Kansas City	150	Tampa Bay	145
Anaheim Angels[1]	162	Texas	227
Los Angeles Dodgers	203	Toronto	145

[1]In 2005, the Anaheim Angels changed their name to the Los Angeles Angels of Anaheim.

Now we read each value from the given data and mark a tally in the second column of Table 2.10 next to the corresponding class. The first value in our original data is 135, which belongs to the 135–156 class. To record it, we mark a tally in the second column next to the 135–156 class. We continue this process until all the data values have been read and entered in the tally column. Note that tallies are marked in blocks of fives for counting convenience. After the tally column is completed, we count the tally marks for each class and write those numbers in the third column. This gives the column of frequencies. These frequencies represent the number of teams that belong to each of the five different classes representing the total home runs. For example, 10 of the 30 Major League Baseball teams hit a total of 135–156 home runs during the 2004 season.

Table 2.10 Frequency Distribution for the Data of Table 2.9

Total Home Runs	Tally	f
135–156	~~||||~~ ~~||||~~	10
157–178	|||	3
179–200	~~||||~~ ||	7
201–222	~~||||~~ |	6
223–244	||||	4
		$\Sigma f = 30$

In Table 2.10, we can denote the frequencies of the five classes by f_1, f_2, f_3, f_4, and f_5, respectively. Therefore,

$$f_1 = \text{Frequency of the first class} = 10$$

Similarly,

$$f_2 = 3, \quad f_3 = 7, \quad f_4 = 6, \quad \text{and} \quad f_5 = 4$$

Using the Σ notation (see Section 1.8 of Chapter 1), we can denote the sum of the frequencies of all classes by Σf. Hence,

$$\Sigma f = f_1 + f_2 + f_3 + f_4 + f_5 = 10 + 3 + 7 + 6 + 4 = 30$$

The number of observations in a sample is usually denoted by n. Thus, for the sample data, Σf is equal to n. The number of observations in a population is denoted by N. Consequently, Σf is equal to N for population data. Because the data set on the total home runs by Major League Baseball teams in Table 2.10 is for all 30 teams, it represents the population. Therefore, in Table 2.10 we can denote the sum of frequencies by N instead of Σf. ■

Note that when we present the data in the form of a frequency distribution table, as in Table 2.10, we lose the information on individual observations. We cannot know the exact number of home runs hit by any particular Major League Baseball team from Table 2.10. All we know is that the home runs hit by 10 of these teams during the 2004 season are between 135–156, and so forth.

2.3.3 Relative Frequency and Percentage Distributions

Using Table 2.10, we can compute the relative frequency and percentage distributions the same way we did for qualitative data in Section 2.2.2. The relative frequencies and percentages for a quantitative data set are obtained as follows.

Calculating Relative Frequency and Percentage

$$\text{Relative frequency of a class} = \frac{\text{Frequency of that class}}{\text{Sum of all frequencies}} = \frac{f}{\Sigma f}$$

$$\text{Percentage} = (\text{Relative frequency}) \cdot 100$$

Example 2–4 illustrates how to construct relative frequency and percentage distributions.

■ EXAMPLE 2–4

Constructing relative frequency and percentage distributions.

Calculate the relative frequencies and percentages for Table 2.10.

Solution The relative frequencies and percentages for the data in Table 2.10 are calculated and listed in the third and fourth columns, respectively, of Table 2.11 here. Note that the class boundaries are listed in the second column of Table 2.11.

Table 2.11 Relative Frequency and Percentage Distributions for Table 2.10

Total Home Runs	Class Boundaries	Relative Frequency	Percentage
135–156	134.5 to less than 156.5	.333	33.3
157–178	156.5 to less than 178.5	.100	10.0
179–200	178.5 to less than 200.5	.233	23.3
201–222	200.5 to less than 222.5	.200	20.0
223–244	222.5 to less than 244.5	.133	13.3
		Sum = .999	Sum = 99.9%

Using Table 2.11, we can make statements about the percentage of teams with home runs within a certain interval. For example, 33.3% of the Major League Baseball teams in this population hit total home runs between 135–156 during the 2004 season. By adding the percentages for the first two classes, we can state that about 43.3% of these teams hit home runs between 135–178 during the 2004 season. Similarly, by adding the percentages of the last two classes, we can state that about 33.3% of these teams hit home runs between 201–244 during the 2004 season. ■

2.3.4 Graphing Grouped Data

Grouped (quantitative) data can be displayed in a *histogram* or a *polygon*. This section describes how to construct such graphs. We can also draw a pie chart to display the percentage distribution for a quantitative data set. The procedure to construct a pie chart is similar to the one for qualitative data explained in Section 2.2.3; it will not be repeated in this section.

Histograms

A **histogram** can be drawn for a frequency distribution, a relative frequency distribution, or a percentage distribution. To draw a histogram, we first mark classes on the horizontal axis and frequencies (or relative frequencies or percentages) on the vertical axis. Next, we draw a bar for each class so that its height represents the frequency of that class. The bars in a histogram are drawn adjacent to each other with no gap between them. A histogram is called a **frequency histogram**, a **relative frequency histogram**, or a **percentage histogram** depending on whether frequencies, relative frequencies, or percentages are marked on the vertical axis.

Definition

Histogram A *histogram* is a graph in which classes are marked on the horizontal axis and the frequencies, relative frequencies, or percentages are marked on the vertical axis. The frequencies, relative frequencies, or percentages are represented by the heights of the bars. In a histogram, the bars are drawn adjacent to each other.

Figures 2.3 and 2.4 show the frequency and the relative frequency histograms, respectively, for the data of Tables 2.10 and 2.11 of Sections 2.3.2 and 2.3.3. The two histograms look alike because they represent the same data. A percentage histogram can be drawn for the percentage distribution of Table 2.11 by marking the percentages on the vertical axis.

The symbol –//– used in the horizontal axes of Figures 2.3 and 2.4 represents a break, called the **truncation**, in the horizontal axis. It indicates that the entire horizontal axis is not shown in these figures. Notice that the 0 to 134 portion of the horizontal axis has been omitted in each figure.

In Figures 2.3 and 2.4, we have used class limits to mark classes on the horizontal axis. However, we can show the intervals on the horizontal axis by using the class boundaries instead of the class limits.

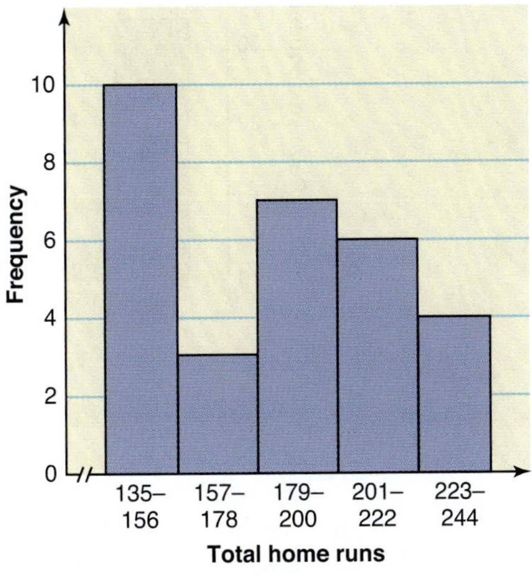

Figure 2.3 Frequency histogram for Table 2.10.

CASE STUDY

2–3

HAND HYGIENE

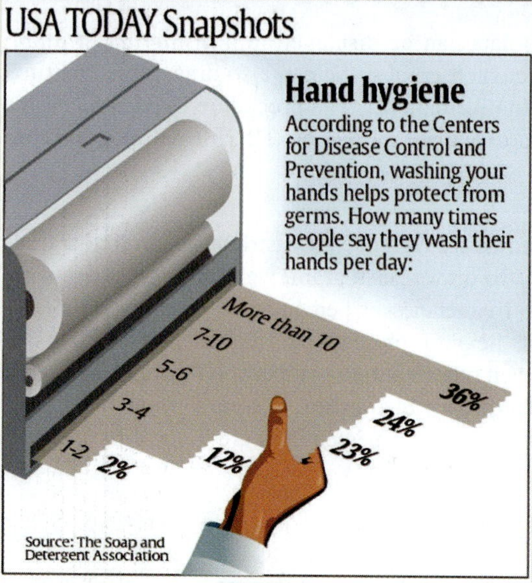

Source: USA TODAY, December 14, 2004. Copyright © 2004, *USA TODAY*. Chart reproduced with permission.

The above chart, reproduced from *USA TODAY*, gives the histogram for the percentage distribution of the number of times people say they wash their hands per day. As we can observe from the chart, 2% of people included in the survey said they wash their hands 1–2 times a day, 12% wash their hands 3–4 times a day, and so on. There are a couple of things you should note about this graph. First, the classes in the chart have different widths. For example, the first three classes (1–2, 3–4, and 5–6) have the same width, which is 2. However, the fourth class (7–10) has a width of 4. The fifth class (more than 10) is called an **open-ended class**. We know it has a lower limit of 11 but it has no upper limit. Second, the percentages for all classes in this chart add up to 97%. The chart does not say anything about the remaining 3% of the people. We can assume that these 3% of the people in the survey either did not wash their hands at all or they did not give an answer.

Figure 2.4 Relative frequency histogram for Table 2.11.

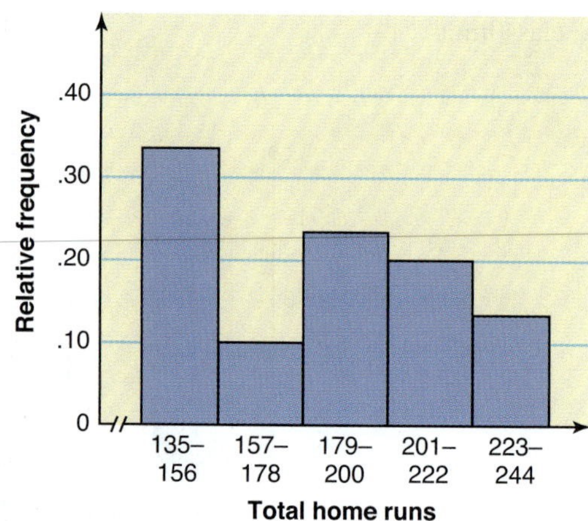

Polygons

A **polygon** is another device that can be used to present quantitative data in graphic form. To draw a **frequency polygon**, we first mark a dot above the midpoint of each class at a height equal to the frequency of that class. This is the same as marking the midpoint at the top of each bar in a histogram. Next we mark two more classes, one at each end, and mark their midpoints. Note that these two classes have zero frequencies. In the last step, we join the adjacent dots with straight lines. The resulting line graph is called a frequency polygon or simply a polygon.

A polygon with relative frequencies marked on the vertical axis is called a *relative frequency polygon*. Similarly, a polygon with percentages marked on the vertical axis is called a *percentage polygon*.

Definition

Polygon A graph formed by joining the midpoints of the tops of successive bars in a histogram with straight lines is called a *polygon*.

Figure 2.5 shows the frequency polygon for the frequency distribution of Table 2.10.

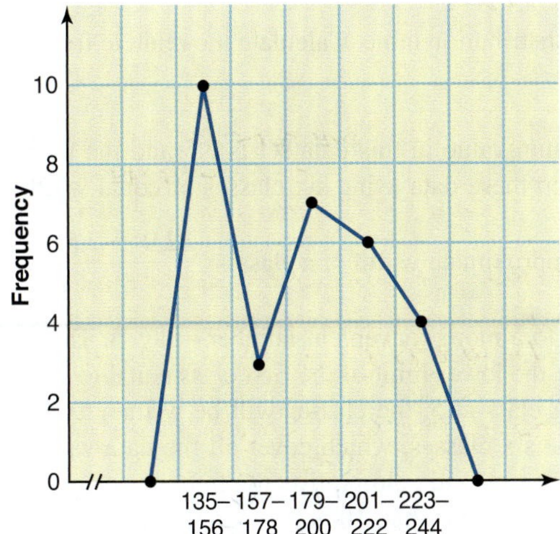

Figure 2.5 Frequency polygon for Table 2.10.

For a very large data set, as the number of classes is increased (and the width of classes is decreased), the frequency polygon eventually becomes a smooth curve. Such a curve is called a *frequency distribution curve* or simply a *frequency curve*. Figure 2.6 shows the frequency curve for a large data set with a large number of classes.

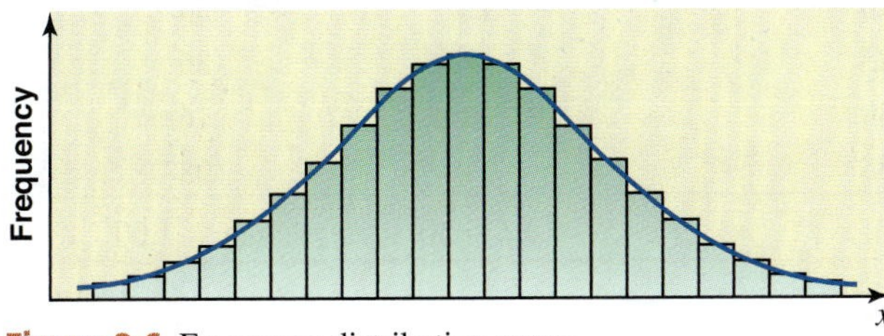

Figure 2.6 Frequency distribution curve.

2.3.5 More on Classes and Frequency Distributions

This section presents two alternative methods for writing classes to construct a frequency distribution for quantitative data.

Less Than Method for Writing Classes

The classes in the frequency distribution given in Table 2.10 for the data on home runs were written as 135–156, 157–178, and so on. Alternatively, we can write the classes in a frequency distribution table using the *less than* method. The technique for writing classes shown in Table 2.10 is more commonly used for data sets that do not contain fractional values. The *less than* method is more appropriate when a data set contains fractional values. Example 2–5 illustrates the *less than* method.

■ EXAMPLE 2–5

Constructing a frequency distribution using the less than method.

According to the American Petroleum Institute, the state taxes (in cents) per gallon of gasoline as of April 1, 2005 for all 50 states[2] are as follows.

18	8	18	21.5	18	22	25	23	14.5	7.5
16	25	19	18	20	24	16	20	25.2	23.5
23.5	19	20	18	17	27.75	25.4	23	18	14.5
17	31.9	26.6	21	26	16	24	31.1	30	16
22	20	20	24.5	20	17.5	28	20.5	32.9	14

Construct a frequency distribution table. Calculate the relative frequencies and percentages for all classes.

Solution The minimum value in this data set is 7.5 and the maximum value is 32.9. Suppose we decide to group these data using six classes of equal width. Then

$$\text{Approximate width of a class} = \frac{32.9 - 7.5}{6} = 4.23$$

We round this number to a more convenient number—say, 5. Then we take 5 as the width of each class. We can take the lower limit of the first class equal to 7.5 or any number lower than 7.5. If we start the first class at 5, the classes will be written as *5 to less than 10, 10 to less than 15,* and so on. The six classes, which cover all the data values, are recorded in the first column of Table 2.12. The second column lists the frequencies of these classes. A value in the data set that is 5 or larger but less than 10 belongs to the first class, a value that is 10 or larger

Table 2.12 Frequency, Relative Frequency, and Percentage Distributions of State Taxes on Gasoline

State Taxes Per Gallon of Gasoline (cents)	f	Relative Frequency	Percentage
5 to less than 10	2	.04	4
10 to less than 15	3	.06	6
15 to less than 20	15	.30	30
20 to less than 25	18	.36	36
25 to less than 30	8	.16	16
30 to less than 35	4	.08	8
	$\Sigma f = 50$	Sum = 1.00	Sum = 100%

[2]The data for the 50 states are entered (by row) in the following order: Alabama, Alaska, Arizona, Arkansas, California, Colorado, Connecticut, Delaware, Florida, Georgia, Hawaii, Idaho, Illinois, Indiana, Iowa, Kansas, Kentucky, Louisiana, Maine, Maryland, Massachusetts, Michigan, Minnesota, Mississippi, Missouri, Montana, Nebraska, Nevada, New Hampshire, New Jersey, New Mexico, New York, North Carolina, North Dakota, Ohio, Oklahoma, Oregon, Pennsylvania, Rhode Island, South Carolina, South Dakota, Tennessee, Texas, Utah, Vermont, Virginia, Washington, West Virginia, Wisconsin, Wyoming.

but less than 15 falls in the second class, and so on. The relative frequencies and percentages for classes are recorded in the third and fourth columns, respectively, of Table 2.12. Note that this table does not contain a column of tallies. ■

A histogram and a polygon for the data of Table 2.12 can be drawn in the same way as for the data of Tables 2.10 and 2.11.

Single-Valued Classes

If the observations in a data set assume only a few distinct (integer) values, it may be appropriate to prepare a frequency distribution table using *single-valued classes*—that is, classes that are made of single values and not of intervals. This technique is especially useful in cases of discrete data with only a few possible values. Example 2–6 exhibits such a situation.

■ EXAMPLE 2–6

Constructing a frequency distribution using single-valued classes.

The administration in a large city wanted to know the distribution of vehicles owned by households in that city. A sample of 40 randomly selected households from this city produced the following data on the number of vehicles owned.

5	1	1	2	0	1	1	2	1	1
1	3	3	0	2	5	1	2	3	4
2	1	2	2	1	2	2	1	1	1
4	2	1	1	2	1	1	4	1	3

Construct a frequency distribution table for these data using single-valued classes.

Solution The observations in this data set assume only six distinct values: 0, 1, 2, 3, 4, and 5. Each of these six values is used as a class in the frequency distribution in Table 2.13, and these six classes are listed in the first column of that table. To obtain the frequencies of these classes, the observations in the data that belong to each class are counted, and the results are recorded in the second column of Table 2.13. Thus, in these data, 2 households own no vehicle, 18 own one vehicle each, 11 own two vehicles each, and so on.

Table 2.13 Frequency Distribution of Vehicles Owned

Vehicles Owned	Number of Households (f)
0	2
1	18
2	11
3	4
4	3
5	2
	$\Sigma f = 40$

■

The data of Table 2.13 can also be displayed in a bar graph, as shown in Figure 2.7 on page 44. To construct a bar graph, we mark the classes, as intervals, on the horizontal axis with a little gap between consecutive intervals. The bars represent the frequencies of respective classes.

CASE STUDY

2–4

USA IS A CAFFEINATED COUNTRY

USA TODAY Snapshots®

USA is a caffeinated country

The number of cups or cans of caffeinated beverages adult Americans say they drink daily:

None	One	Two	Three	Four +
22%	16%	21%	16%	25%

Source: Poll of 1,506 adults for the National Sleep Foundation

By Shannon Reilly and Alejandro Gonzalez, USA TODAY

The above chart, reproduced from *USA TODAY*, shows the percentage distribution of the number of cups or cans of caffeinated beverages that adults in the United States drink per day. This distribution is based on a sample survey of 1506 adults. Here the classes are single-valued except the last one (Four Plus), which is an open-ended multiple-valued class.

Note that the chart does not show the typical bar graph that we learned to make in this chapter. In this chart, classes are not marked on the horizontal axis, percentages are not marked on the vertical axis, and the heights of bars do not show the percentages. Instead, the percentages are shown horizontally and bars show the number of cups or cans that belong to the corresponding percentage. Since 22% of the adults said that they do not drink any caffeinated beverage, there is no cup shown above *None*. For the category *One* cup with 16%, one cup is shown in the chart—the one being held by the person, and so on.

The frequencies of Table 2.13 can be converted to relative frequencies and percentages the same way as in Table 2.11. Then, a bar graph can be constructed to display the relative frequency or percentage distribution by marking the relative frequencies or percentages, respectively, on the vertical axis.

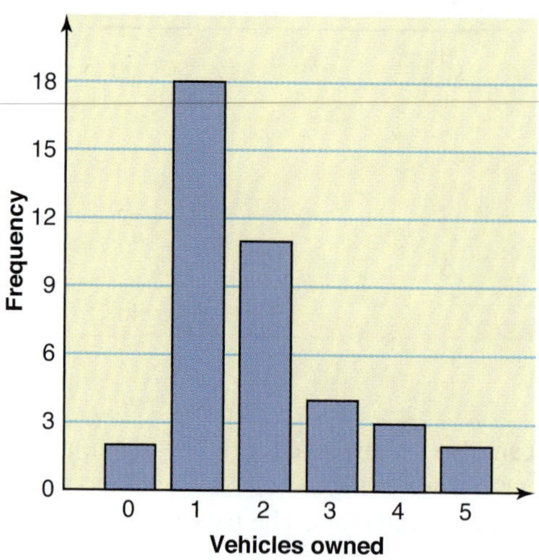

Figure 2.7 Bar graph for Table 2.13.

2.4 Shapes of Histograms

A histogram can assume any one of a large number of shapes. The most common of these shapes are

1. Symmetric
2. Skewed
3. Uniform or rectangular

A **symmetric histogram** is identical on both sides of its central point. The histograms shown in Figure 2.8 are symmetric around the dashed lines that represent their central points.

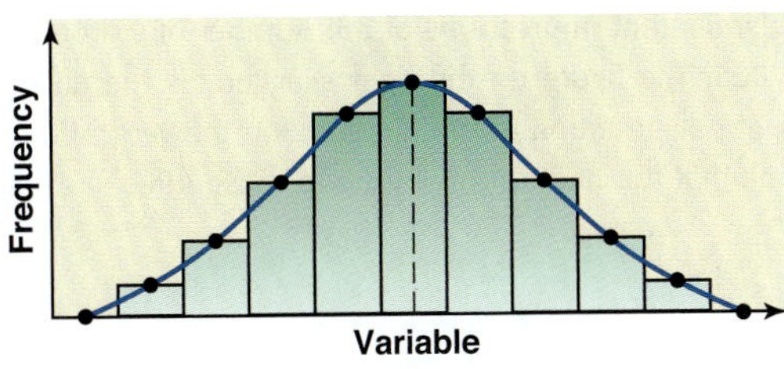

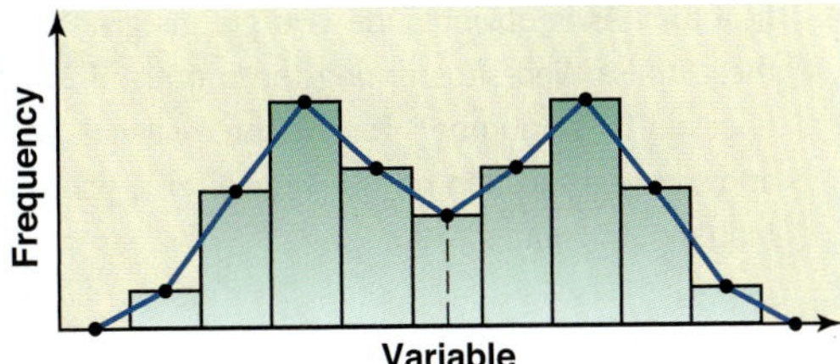

Figure 2.8 Symmetric histograms.

A **skewed histogram** is nonsymmetric. For a skewed histogram, the tail on one side is longer than the tail on the other side. A **skewed-to-the-right histogram** has a longer tail on the right side (see Figure 2.9*a*). A **skewed-to-the-left histogram** has a longer tail on the left side (see Figure 2.9*b*).

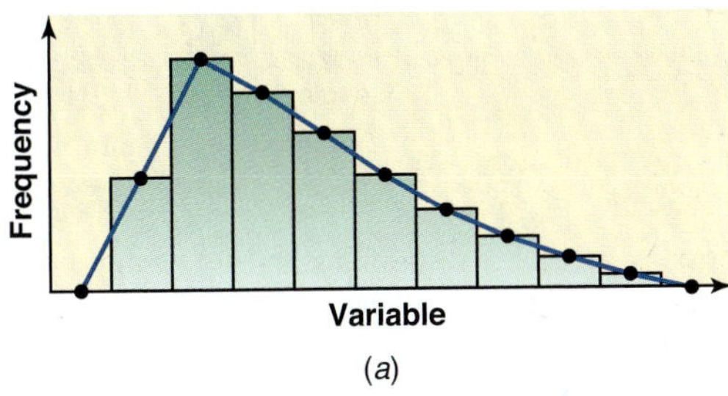

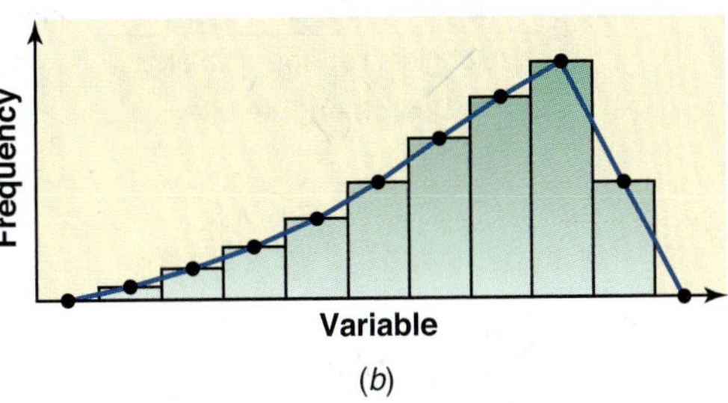

Figure 2.9 (*a*) A histogram skewed to the right. (*b*) A histogram skewed to the left.

A **uniform** or **rectangular histogram** has the same frequency for each class. Figure 2.10 is an illustration of such a case.

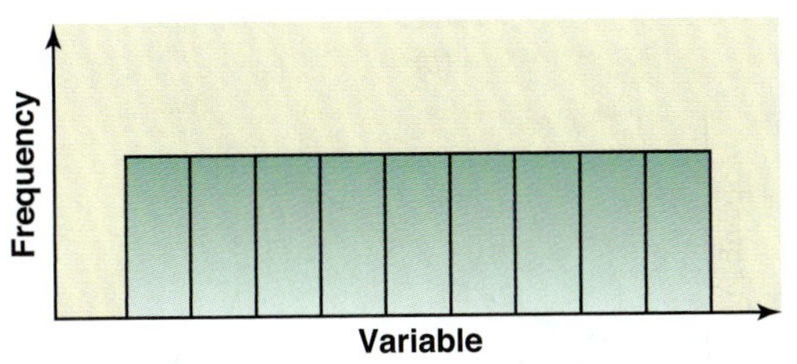

Figure 2.10 A histogram with uniform distribution.

Figures 2.11*a* and 2.11*b* display symmetric frequency curves. Figures 2.11*c* and 2.11*d* show frequency curves skewed to the right and to the left, respectively.

CASE STUDY

2–5

USING TRUNCATED AXES

The following table gives the number of unique visitors (in millions) during January 2005 to the Web sites of three search engines—Google, Yahoo!, and MSN.

Search Engine Site	Unique Visitors During January 2005 (millions)
Google	344
Yahoo!	399
MSN	412

Source: Time, March 21, 2005.

The following two bar graphs are constructed for the data given in the above table. As you would observe, the figure on the left shows the complete vertical axis and the one on the right shows the truncated vertical axis. As is obvious, the two graphs give completely different impressions of the number of visitors to these three Web sites during the month of January 2005. The figure on the right exaggerates the differences in these numbers for the three sites. If you do not pay attention to the numbers on the vertical axis and to the truncation on the vertical axis, you may think that the number of visitors to these three sites in January 2005 varied a lot.

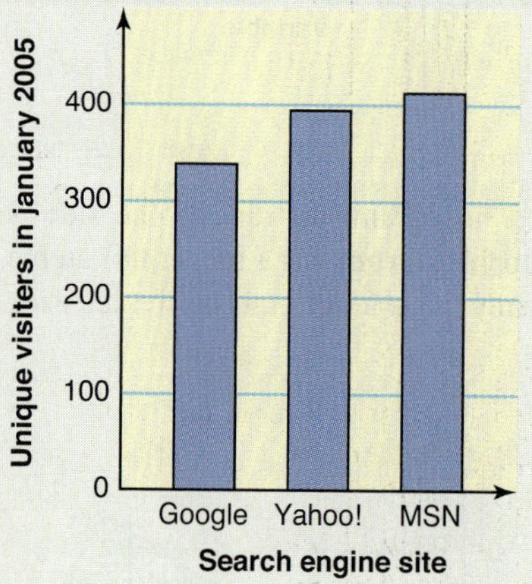

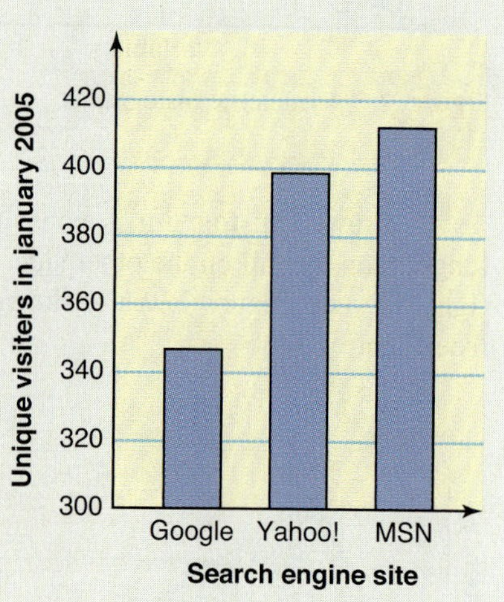

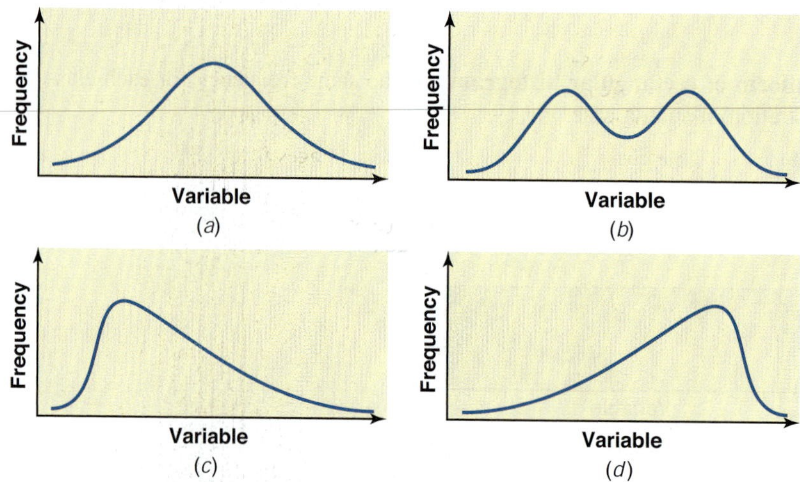

Figure 2.11 (*a*) and (*b*) Symmetric frequency curves. (*c*) Frequency curve skewed to the right. (*d*) Frequency curve skewed to the left.

Describing data using graphs helps give us insights into the main characteristics of the data. But graphs, unfortunately, can also be used, intentionally or unintentionally, to distort the facts and deceive the reader. The following are two ways to manipulate graphs to convey a particular opinion or impression.

◀ *Warning*

1. *Changing the scale* either on one or on both axes—that is, shortening or stretching one or both of the axes.
2. *Truncating the frequency axis*—that is, starting the frequency axis at a number greater than zero.

When interpreting a graph, we should be very cautious. We should observe carefully whether the frequency axis has been truncated or whether any axis has been unnecessarily shortened or stretched. Case Study 2–5 presents an example of this.

EXERCISES

CONCEPTS AND PROCEDURES

2.11 Briefly explain the three decisions that have to be made to group a data set in the form of a frequency distribution table.

2.12 How are the relative frequencies and percentages of classes obtained from the frequencies of classes? Illustrate with the help of an example.

2.13 Three methods—writing classes using limits, using the *less than* method, and grouping data using single-valued classes—were discussed to group quantitative data into classes. Explain these three methods and give one example of each.

APPLICATIONS

2.14 A sample of 80 adults was taken and these adults were asked about the number of credit cards they possess. The following table gives the frequency distribution of their responses.

Number of Credit Cards	Number of Adults
0 to 3	18
4 to 7	26
8 to 11	22
12 to 15	11
16 to 19	3

a. Find the class boundaries and class midpoints.
b. Do all classes have the same width? If so, what is this width?
c. Prepare the relative frequency and percentage distribution columns.
d. What percentage of these adults possess 8 or more credit cards?

2.15 The following table gives the frequency distribution of ages for all 50 employees of a company.

Age	Number of Employees
18 to 30	12
31 to 43	19
44 to 56	14
57 to 69	5

a. Find the class boundaries and class midpoints.
b. Do all classes have the same width? If yes, what is that width?
c. Prepare the relative frequency and percentage distribution columns.
d. What percentage of the employees of this company are age 43 or younger?

2.16 A data set on money spent on lottery tickets during the past year by 200 households has a lowest value of $1 and a highest value of $1167. Suppose we want to group these data into six classes of equal widths.

a. Assuming we take the lower limit of the first class as $1 and the width of each class equal to $200, write the class limits for all six classes.

b. What are the class boundaries and class midpoints?

2.17 A data set on monthly expenditures (rounded to the nearest dollar) incurred on fast food by a sample of 500 households has a minimum value of $3 and a maximum value of $147. Suppose we want to group these data into six classes of equal widths.

a. Assuming we take the lower limit of the first class as $1 and the upper limit of the sixth class as $150, write the class limits for all six classes.

b. Determine the class boundaries and class widths.

c. Find the class midpoints.

2.18 The following table lists the average attendance (rounded to the nearest hundred) per home game for the 2004 season for all 30 Major League Baseball teams. Note that the Montreal Expos are now the Washington Nationals.

Team	Average Attendance	Team	Average Attendance	Team	Average Attendance
A's	27,200	Tigers	24,000	Dodgers	43,100
Angels	41,700	Twins	23,600	Expos	9400
Blue Jays	23,500	White Sox	24,400	Giants	40,200
Devil Rays	16,100	Yankees	47,800	Marlins	22,100
Indians	22,400	Astros	38,100	Mets	29,000
Mariners	36,300	Braves	29,400	Padres	37,200
Orioles	34,300	Brewers	25,500	Phillies	40,600
Rangers	31,800	Cardinals	37,600	Pirates	21,100
Red Sox	35,000	Cubs	39,100	Reds	28,200
Royals	21,000	Diamondbacks	31,100	Rockies	29,600

Source: ESPN Sports Almanac 2005.

a. Construct a frequency distribution table. Take the classes as 9000–16,900, 17,000–24,900, 25,000–32,900, 33,000–40,900, and 41,000–48,900.

b. Calculate the relative frequencies and percentages for all classes.

c. Based on the frequency distribution, can you say whether the data are symmetric or skewed?

d. What percentage of these teams had average attendance of 25,000 or more?

2.19 Nixon Corporation manufactures computer monitors. The following data are the numbers of computer monitors produced at the company for a sample of 30 days.

24	32	27	23	33	33	29	25	23	28
21	26	31	22	27	33	27	23	28	29
31	35	34	22	26	28	23	35	31	27

a. Construct a frequency distribution table using the classes 21–23, 24–26, 27–29, 30–32, and 33–35.

b. Calculate the relative frequencies and percentages for all classes.

c. Construct a histogram and a polygon for the percentage distribution.

d. For what percentage of the days is the number of computer monitors produced in the interval 27–29?

2.20 The following data give the numbers of computer keyboards assembled at the Twentieth Century Electronics Company for a sample of 25 days.

45	52	48	41	56	46	44	42	48	53	51	53	51
48	46	43	52	50	54	47	44	47	50	49	52	

a. Make the frequency distribution table for these data.

b. Calculate the relative frequencies for all classes.

c. Construct a histogram for the relative frequency distribution.

d. Construct a polygon for the relative frequency distribution.

2.21 In its November 29, 2004 issue, *Business Week* magazine presented data on charitable contributions by S&P 500 companies during the 2003 fiscal year. The following table lists the cash contributions (in millions of dollars) of the top 15 companies from this list based on the cash gifts as percentage of their revenues.

Company	Cash Contributions During 2003 (millions of dollars)
Freeport-McMoRan	21.7
Corning	29.0
Avon Products	49.3
Newmont Mining	22.8
Computer Associates	15.3
General Mills	49.3
Fifth Third Bancorp	30.0
M & T Bank	13.7
Eli Lilly & Co.	51.1
Medtronic	31.0
Northern Trust	9.5
Janus Capital Group	3.3
Guidant	12.1
KeyCorp	18.6
Sallie Mae	14.1

Source: Company reports.

a. Construct a frequency distribution table. Take the classes as 2 to less than 12, 12 to less that 22, and so on.
b. Calculate the relative frequencies and percentages for all classes.

Exercises 2.22 through 2.26 are based on the following data.

The following table gives the crime rates per 100,000 people for five types of crimes for 26 states east of the Mississippi River. The rates are based on recent data from FBI *Uniform Crime Reports* that appeared in the *World Almanac and Book of Facts 2005*.

State	Murder Rate	Burglary Rate	Robbery Rate	Aggravated Assault Rate	Motor Vehicle Theft Rate
CT	2.3	493.8	117.3	170.4	334.4
ME	1.1	538.1	20.9	56.8	110.4
MA	2.7	517.2	111.5	342.5	413.6
NH	.9	379.4	32.4	92.9	152.5
RI	3.8	599.7	85.6	158.8	455.8
VT	2.1	565.9	12.5	71.7	124.7
NJ	3.9	511.0	161.9	193.0	416.0
NY	4.7	400.4	191.3	279.7	247.2
PA	5.1	450.8	139.1	227.5	266.0
IL	7.5	643.8	200.6	378.5	356.0
IN	5.9	691.7	107.4	214.1	329.4
MI	6.7	706.1	117.9	362.3	494.7
OH	4.6	868.2	156.5	148.2	374.5
WI	2.8	513.2	86.6	112.7	247.3
DE	3.2	663.3	142.9	408.5	378.6
FL	5.5	1060.5	194.9	529.4	529.6
GA	7.1	863.7	156.9	270.1	444.3

MD	9.4	728.5	245.8	489.5	623.3
NC	6.6	1196.3	146.7	290.5	298.9
SC	7.3	1065.1	140.6	626.5	410.7
VA	5.3	435.4	95.4	165.5	253.3
WV	3.2	537.1	36.5	176.4	216.3
AL	6.8	949.0	132.9	267.5	309.6
KY	4.5	680.6	74.8	173.1	213.8
MS	9.2	1030.5	116.9	178.0	331.6
TN	7.2	1056.5	162.4	507.8	457.8

2.22 **a.** Prepare a frequency distribution table for murder rates using five classes of equal widths.
b. Construct the relative frequency and percentage distribution columns.

2.23 **a.** Prepare a frequency distribution table for burglary rates using five classes of equal widths.
b. Construct the relative frequency and percentage distribution columns.

2.24 **a.** Prepare a frequency distribution table for robbery rates.
b. Construct the relative frequency and percentage distribution columns.
c. Draw a histogram and polygon for the relative frequency distribution.

2.25 **a.** Prepare a frequency distribution table for aggravated assault rates.
b. Calculate the relative frequencies and percentages for all classes.
c. Draw a histogram and a polygon for the percentage distribution.

2.26 **a.** Prepare a frequency distribution table for motor vehicle theft rates. Take 100 as the lower boundary of the first class and 100 as the width of each class.
b. Construct the relative frequency and percentage distribution columns.

2.27 The following table lists the earned run averages (ERAs) for the pitchers of all 16 National League Baseball teams for the 2004 season.

Team	ERA	Team	ERA
Arizona	4.98	Milwaukee	4.24
Atlanta	3.74	Montreal (now Washington)	4.33
Chicago	3.81	New York	4.09
Colorado	5.54	Philadelphia	4.45
Cincinnati	5.19	Pittsburgh	4.29
Florida	4.10	St. Louis	3.75
Houston	4.05	San Diego	4.03
Los Angeles	4.01	San Francisco	4.29

Source: The World Almanac and Book of Facts 2005.

a. Construct a frequency distribution table. Take 3.50 as the lower boundary of the first class and .50 as the width of each class.
b. Prepare the relative frequency and percentage distribution columns for the frequency table of part a.

2.28 The following data give the number of turnovers (fumbles and interceptions) by a college football team for each game in the past two seasons.

3	2	1	4	0	2	2	1	0	3	2	3
0	2	3	1	4	1	3	2	4	0	1	2

a. Prepare a frequency distribution table for these data using single-valued classes.
b. Calculate the relative frequencies and percentages for all classes.
c. In how many games did the team commit two or more turnovers?
d. Draw a bar graph for the frequency distribution of part a.

2.29 According to a survey by the U.S. Public Interest Research Group, about 79% of credit reports contain errors (*USA TODAY,* June 18, 2004). Suppose in a random sample of 25 credit reports, the number of errors found are as listed below.

1	0	2	3	0	1	0	5	4	1	0	2	1
4	1	2	2	0	3	1	0	0	1	2	3	

a. Prepare a frequency distribution table for these data using single-valued classes.
b. Calculate the relative frequencies and percentages for all classes.
c. How many of these reports contained two or more errors?
d. Draw a bar graph for the frequency distribution of part a.

2.30 The following table gives the frequency distribution for the numbers of parking tickets received on the campus of a university during the past week for 200 students.

Number of Tickets	Number of Students
0	59
1	44
2	37
3	32
4	28

Draw two bar graphs for these data, the first without truncating the frequency axis and the second by truncating the frequency axis. In the second case, mark the frequencies on the vertical axis starting with 25. Briefly comment on the two bar graphs.

2.31 Eighty adults were asked to watch a 30-minute infomercial until the presentation ended or until boredom became intolerable. The following table lists the frequency distribution of the times that these adults were able to watch the infomercial.

Time (minutes)	Number of Adults
0 to less than 6	16
6 to less than 12	21
12 to less than 18	18
18 to less than 24	11
24 to less than 30	14

Draw two histograms for these data, the first without truncating the frequency axis. In the second case, mark the frequencies on the vertical axis starting with 10. Briefly comment on the two histograms.

2.5 Cumulative Frequency Distributions

Consider again Example 2–3 of Section 2.3.2 about the home runs hit by Major League Baseball teams. Suppose we want to know how many teams hit a total of 200 or fewer home runs during the 2004 season. Such a question can be answered using a **cumulative frequency distribution**. Each class in a cumulative frequency distribution table gives the total number of values that fall below a certain value. A cumulative frequency distribution is constructed for quantitative data only.

Definition

Cumulative Frequency Distribution A *cumulative frequency distribution* gives the total number of values that fall below the upper boundary of each class.

In a cumulative frequency distribution table, each class has the same lower limit but a different upper limit. Example 2–7 illustrates the procedure to prepare a cumulative frequency distribution.

EXAMPLE 2–7

Constructing a cumulative frequency distribution table.

Using the frequency distribution of Table 2.10, reproduced here, prepare a cumulative frequency distribution for the home runs hit by Major League Baseball teams during the 2004 season.

Total Home Runs	f
135–156	10
157–178	3
179–200	7
201–222	6
223–244	4

Solution Table 2.14 gives the cumulative frequency distribution for the home runs hit by Major League Baseball teams. As we can observe, 135 (which is the lower limit of the first class in Table 2.10) is taken as the lower limit of each class in Table 2.14. The upper limits of all classes in Table 2.14 are the same as those in Table 2.10. To obtain the cumulative frequency of a class, we add the frequency of that class in Table 2.10 to the frequencies of all preceding classes. The cumulative frequencies are recorded in the third column of Table 2.14. The second column of this table lists the class boundaries.

Table 2.14 Cumulative Frequency Distribution of Home Runs by Baseball Teams

Class Limits	Class Boundaries	Cumulative Frequency
135–156	134.5 to less than 156.5	10
135–178	134.5 to less than 178.5	10 + 3 = 13
135–200	134.5 to less than 200.5	10 + 3 + 7 = 20
135–222	134.5 to less than 222.5	10 + 3 + 7 + 6 = 26
135–244	134.5 to less than 244.5	10 + 3 + 7 + 6 + 4 = 30

From Table 2.14, we can determine the number of observations that fall below the upper limit or boundary of each class. For example, 20 Major League Baseball teams hit a total of 200 or fewer home runs. ■

The **cumulative relative frequencies** are obtained by dividing the cumulative frequencies by the total number of observations in the data set. The **cumulative percentages** are obtained by multiplying the cumulative relative frequencies by 100.

Calculating Cumulative Relative Frequency and Cumulative Percentage

$$\text{Cumulative relative frequency} = \frac{\text{Cumulative frequency of a class}}{\text{Total observations in the data set}}$$

$$\text{Cumulative percentage} = (\text{Cumulative relative frequency}) \cdot 100$$

Table 2.15 contains both the cumulative relative frequencies and the cumulative percentages for Table 2.14. We can observe, for example, that 66.7% of the Major League Baseball teams hit 200 or fewer home runs during the 2004 season.

Ogives

When plotted on a diagram, the cumulative frequencies give a curve that is called an **ogive** (pronounced *o-jive*). Figure 2.12 gives an ogive for the cumulative frequency distribution of Table 2.14.

Table 2.15 Cumulative Relative Frequency and Cumulative Percentage Distributions for Home Runs Hit by Baseball Teams

Class Limits	Cumulative Relative Frequency	Cumulative Percentage
135–156	10/30 = .333	33.3
135–178	13/30 = .433	43.3
135–200	20/30 = .667	66.7
135–222	26/30 = .867	86.7
135–244	30/30 = 1.000	100.0

To draw the ogive in Figure 2.12, the variable, which is total home runs, is marked on the horizontal axis and the cumulative frequencies on the vertical axis. Then the dots are marked above the upper boundaries of various classes at the heights equal to the corresponding cumulative frequencies. The ogive is obtained by joining consecutive points with straight lines. Note that the ogive starts at the lower boundary of the first class and ends at the upper boundary of the last class.

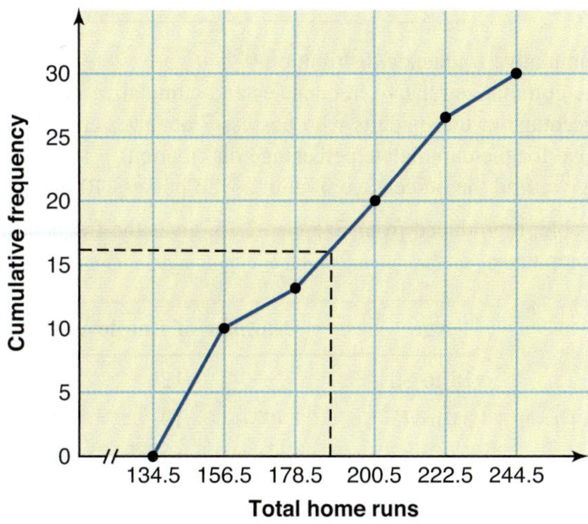

Figure 2.12 Ogive for the cumulative frequency distribution of Table 2.14.

Definition

Ogive An *ogive* is a curve drawn for the cumulative frequency distribution by joining with straight lines the dots marked above the upper boundaries of classes at heights equal to the cumulative frequencies of respective classes.

One advantage of an ogive is that it can be used to approximate the cumulative frequency for any interval. For example, we can use Figure 2.12 to find the number of Major League Baseball teams with 188 or fewer home runs. First, draw a vertical line from 188 on the horizontal axis up to the ogive. Then draw a horizontal line from the point where this line intersects the ogive to the vertical axis. This point gives the cumulative frequency of the class 135–188. In Figure 2.12, this cumulative frequency is (approximately) 16 as shown by the dashed line. Therefore, 16 baseball teams had 188 or fewer home runs during the 2004 season.

We can draw an ogive for cumulative relative frequency and cumulative percentage distributions the same way we did for the cumulative frequency distribution.

EXERCISES

CONCEPTS AND PROCEDURES

2.32 Briefly explain the concept of cumulative frequency distribution. How are the cumulative relative frequencies and cumulative percentages calculated?

2.33 Explain for what kind of frequency distribution an ogive is drawn. Can you think of any use for an ogive? Explain.

APPLICATIONS

2.34 The following table, reproduced from Exercise 2.14, gives the frequency distribution of the number of credit cards possessed by 80 adults.

Number of Credit Cards	Number of Adults
0 to 3	18
4 to 7	26
8 to 11	22
12 to 15	11
16 to 19	3

a. Prepare a cumulative frequency distribution.
b. Calculate the cumulative relative frequencies and cumulative percentages for all classes.
c. Find the percentage of these adults who possess 7 or fewer credit cards.
d. Draw an ogive for the cumulative percentage distribution.
e. Using the ogive, find the percentage of adults who possess 10 or fewer credit cards.

2.35 The following table, reproduced from Exercise 2.15, gives the frequency distribution of ages for all 50 employees of a company.

Age	Number of Employees
18 to 30	12
31 to 43	19
44 to 56	14
57 to 69	5

a. Prepare a cumulative frequency distribution table.
b. Calculate the cumulative relative frequencies and cumulative percentages for all classes.
c. What percentage of the employees of this company are 44 years of age or older?
d. Draw an ogive for the cumulative percentage distribution.
e. Using the ogive, find the percentage of employees who are age 40 or younger.

2.36 Using the frequency distribution table constructed in Exercise 2.18, prepare the cumulative frequency, cumulative relative frequency, and cumulative percentage distributions.

2.37 Using the frequency distribution table constructed in Exercise 2.19, prepare the cumulative frequency, cumulative relative frequency, and cumulative percentage distributions.

2.38 Using the frequency distribution table constructed in Exercise 2.20, prepare the cumulative frequency, cumulative relative frequency, and cumulative percentage distributions.

2.39 Prepare the cumulative frequency, cumulative relative frequency, and cumulative percentage distributions using the frequency distribution constructed in Exercise 2.23.

2.40 Using the frequency distribution table constructed for the data of Exercise 2.25, prepare the cumulative frequency, cumulative relative frequency, and cumulative percentage distributions.

2.41 Refer to the frequency distribution table constructed in Exercise 2.26. Prepare the cumulative frequency, cumulative relative frequency, and cumulative percentage distributions by using that table.

2.42 Using the frequency distribution table constructed for the data of Exercise 2.21, prepare the cumulative frequency, cumulative relative frequency, and cumulative percentage distributions. Draw an ogive for the cumulative frequency distribution. Using the ogive, find the (approximate) number of companies in these 15 who made cash contributions of less than $35 million in 2003.

2.43 Refer to the frequency distribution table constructed in Exercise 2.27. Prepare the cumulative frequency, cumulative relative frequency, and cumulative percentage distributions. Draw an ogive for the cumulative frequency distribution. Using the ogive, find the (approximate) number of teams with an ERA of less than 4.20.

2.6 Stem-and-Leaf Displays

Another technique that is used to present quantitative data in condensed form is the **stem-and-leaf display**. An advantage of a stem-and-leaf display over a frequency distribution is that by preparing a stem-and-leaf display we do not lose information on individual observations. A stem-and-leaf display is constructed only for quantitative data.

Definition

Stem-and-Leaf Display In a *stem-and-leaf display* of quantitative data, each value is divided into two portions—a stem and a leaf. The leaves for each stem are shown separately in a display.

Example 2–8 describes the procedure for constructing a stem-and-leaf display.

■ EXAMPLE 2–8

Constructing a stem-and-leaf display for two-digit numbers.

The following are the scores of 30 college students on a statistics test.

75	52	80	96	65	79	71	87	93	95
69	72	81	61	76	86	79	68	50	92
83	84	77	64	71	87	72	92	57	98

Construct a stem-and-leaf display.

Solution To construct a stem-and-leaf display for these scores, we split each score into two parts. The first part contains the first digit, which is called the *stem*. The second part contains the second digit, which is called the *leaf*. Thus, for the score of the first student, which is 75, 7 is the stem and 5 is the leaf. For the score of the second student, which is 52, the stem is 5 and the leaf is 2. We observe from the data that the stems for all scores are 5, 6, 7, 8, and 9 because all the scores lie in the range 50 to 98. To create a stem-and-leaf display, we draw a vertical line and write the stems on the left side of it, arranged in increasing order, as shown in Figure 2.13.

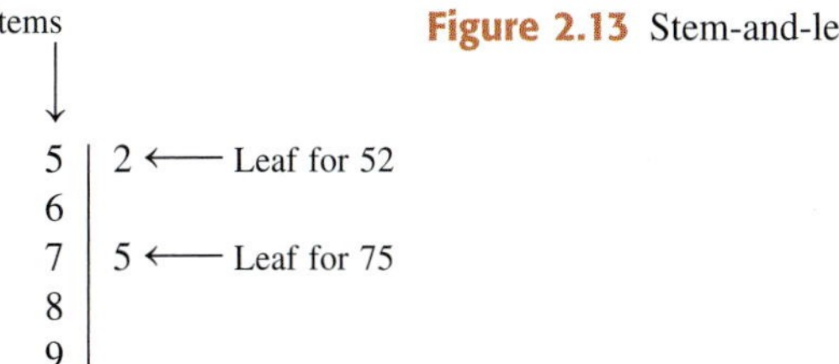

Figure 2.13 Stem-and-leaf display.

After we have listed the stems, we read the leaves for all scores and record them next to the corresponding stems on the right side of the vertical line. For example, for the first score we write the leaf 5 next to the stem 7; for the second score we write the leaf 2 next to the stem 5. The recording of these two scores in a stem-and-leaf display is shown in Figure 2.13.

Now, we read all the scores and write the leaves on the right side of the vertical line in the rows of corresponding stems. The complete stem-and-leaf display for scores is shown in Figure 2.14.

Figure 2.14 Stem-and-leaf display of test scores.

5	2 0 7
6	5 9 1 8 4
7	5 9 1 2 6 9 7 1 2
8	0 7 1 6 3 4 7
9	6 3 5 2 2 8

By looking at the stem-and-leaf display of Figure 2.14, we can observe how the data values are distributed. For example, the stem 7 has the highest frequency, followed by stems 8, 9, 6, and 5.

The leaves for each stem of the stem-and-leaf display of Figure 2.14 are *ranked* (in increasing order) and presented in Figure 2.15.

Figure 2.15 Ranked stem-and-leaf display of test scores.

5	0 2 7
6	1 4 5 8 9
7	1 1 2 2 5 6 7 9 9
8	0 1 3 4 6 7 7
9	2 2 3 5 6 8

■

As already mentioned, one advantage of a stem-and-leaf display is that we do not lose information on individual observations. We can rewrite the individual scores of the 30 college students from the stem-and-leaf display of Figure 2.14 or 2.15. By contrast, the information on individual observations is lost when data are grouped into a frequency table.

■ EXAMPLE 2–9

Constructing a stem-and-leaf display for three-and four-digit numbers.

The following data give the monthly rents paid by a sample of 30 households selected from a small city.

880	1081	721	1075	1023	775	1235	750	965	960
1210	985	1231	932	850	825	1000	915	1191	1035
1151	630	1175	952	1100	1140	750	1140	1370	1280

Construct a stem-and-leaf display for these data.

Solution Each of the values in the data set contains either three or four digits. We will take the first digit for three-digit numbers and the first two digits for four-digit numbers as stems. Then we will use the last two digits of each number as a leaf. Thus for the first value, which is 880, the stem is 8 and the leaf is 80. The stems for the entire data set are 6, 7, 8, 9, 10, 11, 12, and 13. They are recorded on the left side of the vertical line in Figure 2.16. The leaves for the numbers are recorded on the right side.

Figure 2.16 Stem-and-leaf display of rents.

6	30
7	75 50 21 50
8	80 25 50
9	32 52 15 60 85 65
10	23 81 35 75 00
11	91 51 40 75 40 00
12	10 31 35 80
13	70

■

Sometimes a data set may contain too many stems, with each stem containing only a few leaves. In such cases, we may want to condense the stem-and-leaf display by *grouping the stems*. Example 2–10 describes this procedure.

■ EXAMPLE 2–10

Preparing a grouped stem-and-leaf display.

The following stem-and-leaf display is prepared for the number of hours that 25 students spent working on computers during the past month.

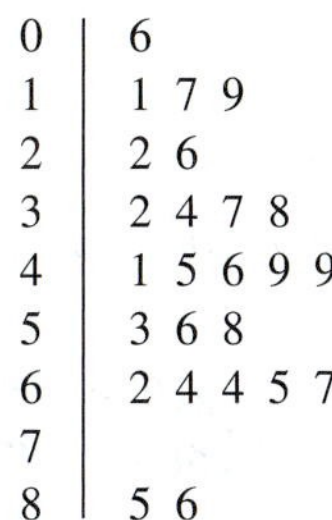

0	6
1	1 7 9
2	2 6
3	2 4 7 8
4	1 5 6 9 9
5	3 6 8
6	2 4 4 5 7
7	
8	5 6

Prepare a new stem-and-leaf display by grouping the stems.

Solution To condense the given stem-and-leaf display, we can combine the first three rows, the middle three rows, and the last three rows, thus getting the stems 0–2, 3–5, and 6–8. The leaves for each stem of a group are separated by an asterisk (*), as shown in Figure 2.17. Thus, the leaf 6 in the first row corresponds to stem 0; the leaves 1, 7, and 9 correspond to stem 1; and leaves 2 and 6 belong to stem 2.

0–2	6 * 1 7 9 * 2 6
3–5	2 4 7 8 * 1 5 6 9 9 * 3 6 8
6–8	2 4 4 5 7 * * 5 6

Figure 2.17 Grouped stem-and-leaf display.

If a stem does not contain a leaf, this is indicated in the grouped stem-and-leaf display by two consecutive asterisks. For example, in the above stem-and-leaf display, there is no leaf for 7; that is, there in no number in the 70s. Hence, in Figure 2.17, we have two asterisks after the leaves for 6 and before the leaves for 8. ■

EXERCISES

■ CONCEPTS AND PROCEDURES

2.44 Briefly explain how to prepare a stem-and-leaf display for a data set. You may use an example to illustrate.

2.45 What advantage does preparing a stem-and-leaf display have over grouping a data set using a frequency distribution? Give one example.

2.46 Consider this stem-and-leaf display.

4	3 6
5	0 1 4 5
6	3 4 6 7 7 7 8 9
7	2 2 3 5 6 6 9
8	0 7 8 9

Write the data set that is represented by the display.

2.47 Consider this stem-and-leaf display.

2–3	18 45 56 * 29 67 83 97
4–5	04 27 33 71 * 23 37 51 63 81 92
6–8	22 36 47 55 78 89 * * 10 41

Write the data set that is represented by the display.

APPLICATIONS

2.48 The following data give the time (in minutes) that each of 20 students waited in line at their bookstore to pay for their textbooks in the beginning of Spring 2006 semester.

15	8	23	21	5	17	31	22	34	6
5	10	14	17	16	25	30	3	31	19

Construct a stem-and-leaf display for these data. Arrange the leaves for each stem in increasing order.

2.49 Following are the total yards gained rushing during the 2005 season by 14 running backs of 14 college football teams.

745	921	1133	1024	848	775	800
1009	1275	857	933	1145	967	995

Prepare a stem-and-leaf display. Arrange the leaves for each stem in increasing order.

2.50 Reconsider the data on the numbers of computer monitors produced at the Nixon Corporation for a sample of 30 days given in Exercise 2.19. Prepare a stem-and-leaf display for those data. Arrange the leaves for each stem in increasing order.

2.51 Reconsider the data on the numbers of computer keyboards assembled at the Twentieth Century Electronics Company given in Exercise 2.20. Prepare a stem-and-leaf display for those data. Arrange the leaves for each stem in increasing order.

2.52 Refer to Exercise 2.18. Rewrite those data by rounding each average attendance to the nearest thousand. For example, an attendance of 27,200 will be rounded to 27 thousand, and 41,700 will be rounded to 42 thousand. Prepare a stem-and-leaf display for these data. Arrange the leaves for each stem in increasing order.

2.53 These data give the times (in minutes) taken to commute from home to work for 20 workers.

10	50	65	33	48	5	11	23	39	26
26	32	17	7	15	19	29	43	21	22

Construct a stem-and-leaf display for these data. Arrange the leaves for each stem in increasing order. (*Note*: To prepare a stem-and-leaf display, each number in this data set can be written as a two-digit number. For example, 5 can be written as 05, for which the stem is 0 and the leaf is 5.)

2.54 The following data give the times served (in months) by 35 prison inmates who were released recently.

37	6	20	5	25	30	24	10	12	20
24	8	26	15	13	22	72	80	96	33
84	86	70	40	92	36	28	90	36	32
72	45	38	18	9					

a. Prepare a stem-and-leaf display for these data.
b. Condense the stem-and-leaf display by grouping the stems as 0–2, 3–5, and 6–9.

2.55 The following data give the money (in dollars) spent on textbooks by 35 students during the 2005–06 academic year.

565	528	270	220	245	368	210	265	350
345	530	705	490	158	320	505	457	487
617	721	635	438	475	702	538	720	460
540	390	560	570	706	430	268	638	

a. Prepare a stem-and-leaf display for these data using the last two digits as leaves.
b. Condense the stem-and-leaf display by grouping the stems as 1–3, 4–5, and 6–7.

2.7 DOTPLOTS

One of the simplest methods for graphing and understanding quantitative data is to create a dotplot. As with most graphs, statistical software should be used to make a dotplot for large data sets. However, Example 2–11 demonstrates how to create a dotplot by hand.

Dotplots can help us detect **outliers** (also called **extreme values**) in a data set. Outliers are the values that are extremely large or extremely small with respect to the rest of the data values.

Definition

Outliers or Extreme Values Values that are very small or very large relative to the majority of the values in a data set are called outliers or extreme values.

■ EXAMPLE 2–11

Creating a dotplot.

Table 2.16 lists the number of runs batted in (RBIs) during the 2004 Major League Baseball playoffs by members of the Boston Red Sox team with at least one at-bat. Create a dotplot for these data.

Table 2.16 Runs Batted In by Boston Red Sox Players in the 2004 Playoffs

Batter	RBIs	Batter	RBIs
D. Mientkiewicz	1	D. Mirabelli	0
D. Ortiz	19	J. Varitek	11
M. Ramirez	11	K. Millar	6
B. Mueller	3	G. Kapler	0
O. Cabrera	11	M. Bellhorn	8
J. Damon	9	P. Reese	0
T. Nixon	8	K. Youkilis	0

Source: http://www.sportsnetwork.com.

Solution Below we show how to make a dotplot for these data on RBIs.

Step 1. The minimum and maximum values in this data set are 0 and 19 RBIs, respectively. First we draw a horizontal line (let us call this the numbers line) with numbers that cover the given data as shown in Figure 2.18. Note that the numbers line in Figure 2.18 shows the values from 0 to 20.

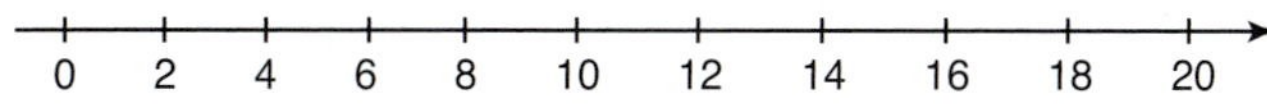

Figure 2.18 Numbers line.

Step 2. Place a dot above the value on the numbers line that represents each RBI listed in the table. For example, Doug Mientkiewicz had 1 RBI in the playoffs. Place a dot above 1 on the numbers line as shown in Figure 2.19. If there are two or more observations with the same value, we stack dots vertically above each other to represent those values. For example, as shown in Table 2.16, three players had 11 RBIs. We stack three dots (one for each player) above 11 on the numbers line as shown in Figure 2.19. Figure 2.19 gives the complete dotplot.

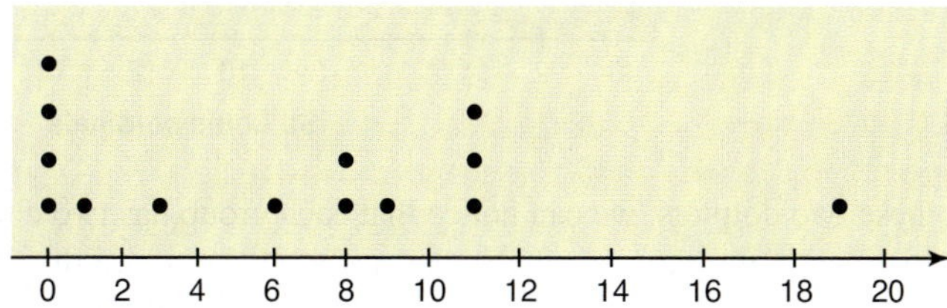

Figure 2.19 Dotplot for RBIs.

As we examine the dotplot of Figure 2.19, we notice that there are two clusters (groups) of data. Approximately half of the players had three or fewer RBIs and approximately half had between 6 and 11 RBIs. In addition, one player (David Ortiz) had 19 RBIs. Graphically, Ortiz's dot is far away from the general pattern of the data. When this occurs, we suspect that the data value could be an outlier. (In Chapter 3 in the box-and-whisker section we will learn a numerical method to determine if a data point should be classified as an outlier.) ■

Dotplots are also very useful for comparing two or more data sets. To do so, we create a dotplot for each data set with numbers lines for all data sets on the same scale. We place these data sets on top of each other, resulting in what are called **stacked dotplots**. Example 2–12 shows this procedure.

■ EXAMPLE 2–12

Comparing two data sets using dotplots.

Refer to Table 2.16 in Example 2–11 that gives the RBIs for players of the Red Sox baseball team during the 2004 playoffs. Table 2.17 below gives the RBIs for the players of the St. Louis Cardinals team during the 2004 playoffs. These two teams played in the 2004 World Series. Make dotplots for both sets of data and compare these two dotplots.

Table 2.17 Runs Batted In by St. Louis Cardinals Players in the 2004 Playoffs

Batter	RBIs	Batter	RBIs
A. Pujols	14	R. Sanders	1
L. Walker	11	M. Anderson	0
E. Renteria	7	S. Taguchi	1
R. Cedeno	1	S. Rolen	7
J. Edmonds	9	Y. Molina	0
T. Womack	2	J. Mabry	1
M. Matheny	7	H. Luna	0

Source: http://www.sportsnetwork.com.

Solution Figure 2.20 shows the dotplots for the RBIs for players of both these teams.

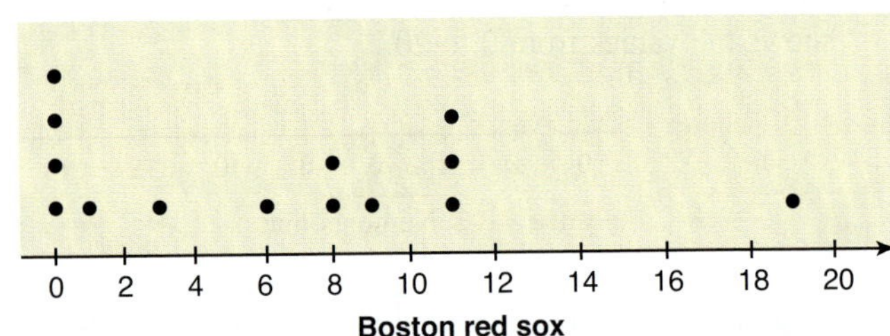

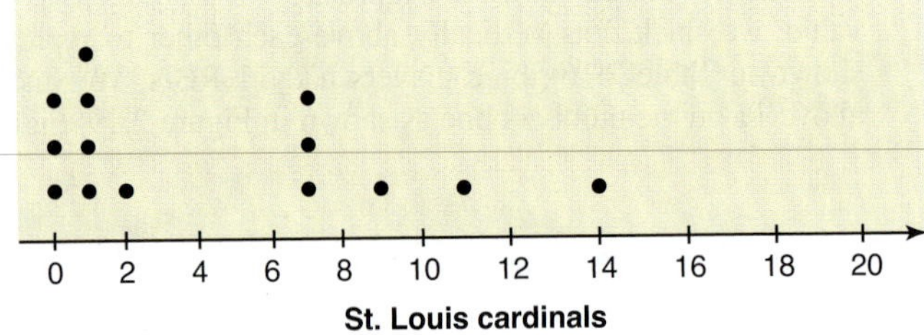

Figure 2.20 Dotplots of RBIs for Boston and St. Louis Baseball Teams.

Looking at these two dotplots, we can notice that each group has two clusters, and the clusters are in approximately the same areas. However, the St. Louis Cardinals had more players who had a lower number of RBIs (the first cluster) than the Boston Red Sox players. We also notice that among the groups of players with higher numbers of RBIs (the second cluster), more Red Sox players are distributed towards the higher end, while more Cardinals are distributed towards the lower end. ■

In practice, dotplots and other statistical graphs will be created using statistical software. The Technology Instruction section at the end of this chapter shows how we can do so.

EXERCISES

CONCEPTS AND PROCEDURES

2.56 Briefly explain how to prepare a dotplot for a data set. You may use an example to illustrate.

2.57 What is a stacked dotplot, and how is it used? Explain.

2.58 Create a dotplot for the following data set.

1	2	0	5	1	1	3	2	0	5
2	1	2	1	2	0	1	3	1	2

APPLICATIONS

2.59 Reconsider the data on the numbers of computer keyboards assembled at the Twentieth Century Electronics Company given in Exercise 2.20. Create a dotplot for those data.

2.60 Create a dotplot for the data on the number of turnovers (fumbles and interceptions) by a college football team for games in the past two seasons given in Exercise 2.28.

2.61 Reconsider the data on the numbers of errors found in 25 randomly selected credit reports given in Exercise 2.29. Create a dotplot for those data.

2.62 The following data give the number of times each of the 30 randomly selected account holders at a bank used that bank's ATM during a 60-day period.

3	2	3	2	2	5	0	4	1	3
2	3	3	5	9	0	3	2	2	15
1	3	2	7	9	3	0	4	2	2

Create a dotplot for these data and point out any clusters or outliers.

2.63 The following data give the number of times each of the 20 randomly selected male students from a state university ate at fast-food restaurants during a seven-day period.

5	8	10	3	5	5	10	7	2	1
10	4	5	0	10	1	2	8	3	5

Create a dotplot for these data and point out any clusters or outliers.

2.64 Reconsider Exercise 2.63. The following data give the number of times each of the 20 randomly selected female students from the same state university ate at fast-food restaurants during the same seven-day period.

0	0	4	2	4	10	2	5	0	5
6	1	1	4	6	2	4	5	6	0

a. Create a dotplot for these data.
b. Use the dotplots for male and female students to compare the two data sets.

2.65 The following table gives the number of stolen bases during the 2004 season by each Boston Red Sox player with 150 or more at-bats.

Player	Stolen Bases	Player	Stolen Bases
J. Damon	19	G. Kapler	5
D. Ortiz	0	P. Reese	6
M. Ramirez	2	O. Cabrera	4
M. Bellhorn	6	K. Youkilis	0
K. Millar	1	D. Mirabelli	0
J. Varitek	10	N. Garciaparra	2
B. Mueller	2	D. McCarty	1
D. Mientkiewicz	2		

Source: Sports Illustrated 2005 Almanac.

Create a dotplot for these data. Mention any clusters and/or outliers you observe.

USES AND MISUSES... BE SENSITIVE

As a budding statistician, the first task that you will perform is organizing your data. As this chapter showed, once organized, it is often convenient to display data in a graphical form. Though you were warned that truncating and changing the scale of axes might distort your data, the simple act of grouping your data can do the very same thing. This is especially important to remember when partitioning your data into classes.

Suppose that you were presented with a set of data on the ages of employees of a company. This phenomenon is known as sensitivity, and your goal is to present results that are not too sensitive to class boundaries and groupings. Let us look at an example to make the notion concrete. Suppose that you are given the following data representing ages of employees at Company X.

37	41	49	23
37	41	51	33
38	43	30	35
38	43	29	52
39	42	23	48
39	42	24	

The histogram in Figure 2.21 was made by partitioning the data by decade, thus getting the classes as 21–30, 31–40, and so on. Simple inspection tells us that most employees of the company are in their 30s and 40s, and that the ages are skewed (a little) to the younger side.

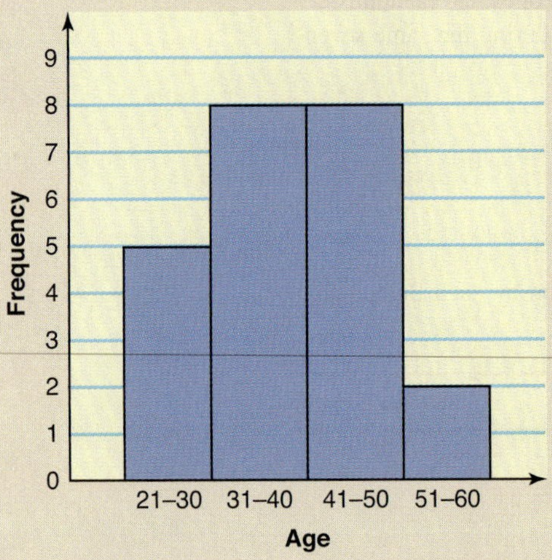

Figure 2.21 Histogram of the Ages of Company X's Workers.

But if we partition the data on the fifth year of the decade, thus getting the classes as 16–25, 26–35, and so on, we get a different picture as shown in Figure 2.22. Twelve of the 23 employees are between the ages of 36 and 45! Which picture is correct? The problem is that they both are.

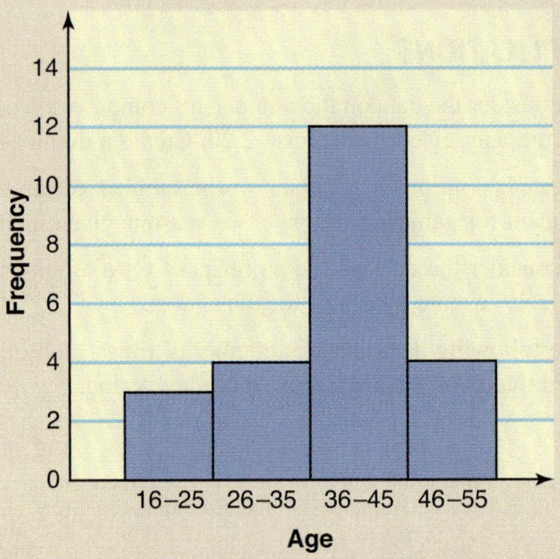

Figure 2.22 Histogram of the Ages of Company X's Workers.

The opposite of sensitivity is called *robustness*, and how do we present results in a robust way? When you are the statistician, be very careful when creating your classes. Follow the basic procedure in the text, rounding to a convenient class width when appropriate, but then look at your data and ask yourself how your data are distributed within the classes you have chosen. If the data seem to be skewed within a class or concentrated at the boundaries, think about how your interpretation would change if you were to shift your class boundaries a little. If your interpretation would change, you need to make a judgment about which set of classes is most appropriate. You might choose to tell your audience that the graph conceals something about the data. When you are reviewing another statistician's graphical results, think about what the choice of classes might be revealing or concealing.

Glossary

Bar graph A graph made of bars whose heights represent the frequencies of respective categories.

Class An interval that includes all the values in a (quantitative) data set that fall within two numbers, the lower and upper limits of the class.

Class boundary The midpoint of the upper limit of one class and the lower limit of the next class.

Class frequency The number of values in a data set that belong to a certain class.

Class midpoint or **mark** The class midpoint or mark is obtained by dividing the sum of the lower and upper limits (or boundaries) of a class by 2.

Class width or **size** The difference between the two boundaries of a class.

Cumulative frequency The frequency of a class that includes all values in a data set that fall below the upper boundary of that class.

Cumulative frequency distribution A table that lists the total number of values that fall below the upper boundary of each class.

Cumulative percentage The cumulative relative frequency multiplied by 100.

Cumulative relative frequency The cumulative frequency of a class divided by the total number of observations.

Frequency distribution A table that lists all the categories or classes and the number of values that belong to each of these categories or classes.

Grouped data A data set presented in the form of a frequency distribution.

Histogram A graph in which classes are marked on the horizontal axis and frequencies, relative frequencies, or percentages are marked on the vertical axis. The frequencies, relative frequencies, or percentages of various classes are represented by bars that are drawn adjacent to each other.

Ogive A curve drawn for a cumulative frequency distribution.

Outliers or **Extreme values** Values that are very small or very large relative to the majority of the values in a data set.

Percentage The percentage for a class or category is obtained by multiplying the relative frequency of that class or category by 100.

Pie chart A circle divided into portions that represent the relative frequencies or percentages of different categories or classes.

Polygon A graph formed by joining the midpoints of the tops of successive bars in a histogram by straight lines.

Raw data Data recorded in the sequence in which they are collected and before they are processed.

Relative frequency The frequency of a class or category divided by the sum of all frequencies.

Skewed-to-the-left histogram A histogram with a longer tail on the left side.

Skewed-to-the-right histogram A histogram with a longer tail on the right side.

Stem-and-leaf display A display of data in which each value is divided into two portions—a stem and a leaf.

Symmetric histogram A histogram that is identical on both sides of its central point.

Ungrouped data Data containing information on each member of a sample or population individually.

Uniform or **rectangular histogram** A histogram with the same frequency for all classes.

Supplementary Exercises

2.66 The following data give the political party of each of the first 30 U.S. presidents. In the data, D stands for Democrat, DR for Democratic Republican, F for Federalist, R for Republican, and W for Whig.

F	F	DR	DR	DR	DR	D	D	W	W
D	W	W	D	D	R	D	R	R	R
R	D	R	D	R	R	R	D	R	R

a. Prepare a frequency distribution table for these data.
b. Calculate the relative frequency and percentage distributions.
c. Draw a bar graph for the relative frequency distribution and a pie chart for the percentage distribution.
d. What percentage of these presidents were Whigs?

2.67 In a survey conducted by Harris Interactive for Tylenol PM, people were asked about how they cope with afternoon drowsiness (*USA TODAY*, October 19, 2004). Of the respondents, 35% said they drink a caffeinated beverage (C). Other responses were: taking a nap (N), going for a walk (W), or eating a sugary snack (S). Suppose that in a recent poll, 30 people were asked which one of the above choices they preferred when dealing with drowsiness. Their responses are given below.

C	C	N	C	W	N	S	C	N	C
S	S	W	C	C	N	S	N	C	C
N	C	C	W	W	C	W	N	C	S

a. Prepare a frequency distribution table for these data.
b. Calculate the relative frequencies and percentages for all classes.
c. Draw a bar graph for the frequency distribution and a pie chart for the percentage distribution.
d. What percentage of these respondents preferred to cope with afternoon drowsiness by taking a nap?

2.68 The following data give the numbers of television sets owned by 40 randomly selected households.

1	1	2	3	2	4	1	3	2	1
3	0	2	1	2	3	2	3	2	2
1	2	1	1	1	3	1	1	1	2
2	4	2	3	1	3	1	2	2	4

a. Prepare a frequency distribution table for these data using single-valued classes.
b. Compute the relative frequency and percentage distributions.
c. Draw a bar graph for the frequency distribution.
d. What percentage of the households own two or more television sets?

2.69 Twenty-four students from universities in Connecticut were asked to name the five current members of the U.S. House of Representatives from Connecticut. The number of correct names supplied by the students are given below.

4	2	3	5	5	4	3	1	5	4	4	3
5	3	2	3	1	3	2	5	2	1	5	0

a. Prepare a frequency distribution for these data using single-valued classes.
b. Compute the relative frequency and percentage distributions.
c. What percentage of the students in this sample named fewer than two of the representatives correctly?
d. Draw a bar graph for the relative frequency distribution.

2.70 The following data give the amounts spent on video rentals (in dollars) during 2005 by 30 households randomly selected from those who rented videos in 2005.

595	24	6	100	100	40	622	405	90
55	155	760	405	90	205	70	180	88
808	100	240	127	83	310	350	160	22
111	70	15						

a. Construct a frequency distribution table. Take \$1 as the lower limit of the first class and \$200 as the width of each class.
b. Calculate the relative frequencies and percentages for all classes.
c. What percentage of the households in this sample spent more than \$400 on video rentals in 2005?

2.71 The following data give the numbers of orders received for a sample of 30 hours at the Time-saver Mail Order Company.

34	44	31	52	41	47	38	35	32	39
28	24	46	41	49	53	57	33	27	37
30	27	45	38	34	46	36	30	47	50

a. Construct a frequency distribution table. Take 23 as the lower limit of the first class and 7 as the width of each class.
b. Calculate the relative frequencies and percentages for all classes.
c. For what percentage of the hours in this sample was the number of orders more than 36?

2.72 The following data give the amounts spent (in dollars) on refreshments by 30 spectators randomly selected from those who patronized the concession stands at a recent Major League Baseball game.

4.95	27.99	8.00	5.80	4.50	2.99	4.85	6.00
9.00	15.75	9.50	3.05	5.65	21.00	16.60	18.00
21.77	12.35	7.75	10.45	3.85	28.45	8.35	17.70
19.50	11.65	11.45	3.00	6.55	16.50		

a. Construct a frequency distribution table using the *less than* method to write classes. Take \$0 as the lower boundary of the first class and \$6 as the width of each class.
b. Calculate the relative frequencies and percentages for all classes.
c. Draw a histogram for the frequency distribution.

2.73 The following data give the repair costs (in dollars) for 30 cars randomly selected from a list of cars that were involved in collisions.

2300	750	2500	410	555	1576
2460	1795	2108	897	989	1866
2105	335	1344	1159	1236	1395
6108	4995	5891	2309	3950	3950
6655	4900	1320	2901	1925	6896

a. Construct a frequency distribution table. Take \$1 as the lower limit of the first class and \$1400 as the width of each class.

b. Compute the relative frequencies and percentages for all classes.

c. Draw a histogram and a polygon for the relative frequency distribution.

d. What are the class boundaries and the width of the fourth class?

2.74 Refer to Exercise 2.70. Prepare the cumulative frequency, cumulative relative frequency, and cumulative percentage distributions by using the frequency distribution table of that exercise.

2.75 Refer to Exercise 2.71. Prepare the cumulative frequency, cumulative relative frequency, and cumulative percentage distributions using the frequency distribution table constructed for the data of that exercise.

2.76 Refer to Exercise 2.72. Prepare the cumulative frequency, cumulative relative frequency, and cumulative percentage distributions using the frequency distribution table constructed for the data of that exercise.

2.77 Construct the cumulative frequency, cumulative relative frequency, and cumulative percentage distributions by using the frequency distribution table constructed for the data of Exercise 2.73.

2.78 Refer to Exercise 2.70. Prepare a stem-and-leaf display for the data of that exercise.

2.79 Construct a stem-and-leaf display for the data given in Exercise 2.71.

2.80 The following table gives the revenues (in millions of dollars) for the seven National Hockey League teams with the largest revenues during the 2003–04 season (*Forbes*, November 29, 2004).

Team	Revenue (millions of dollars)
New York Rangers	118
Toronto Maple Leafs	117
Philadelphia Flyers	106
Dallas Stars	103
Detroit Red Wings	97
Colorado Avalanche	99
Boston Bruins	95

Draw two bar graphs for these data, the first without truncating the axis on which revenues are marked and the second by truncating this axis. In the second graph, mark the revenues on the vertical axis starting with \$90 million. Briefly comment on the two bar graphs.

2.81 The following table lists the average price per gallon for unleaded regular gasoline in the United States from 1997 to 2004. Note that the average price for 2004 is for the months from January to June only.

Year	Average Price Per Gallon (dollars)
1997	1.234
1998	1.059
1999	1.165
2000	1.510
2001	1.461
2002	1.358
2003	1.591
2004	1.819

Source: Energy Information Administration.

Draw two bar graphs for these data, the first without truncating the axis on which the gasoline prices are marked and the second by truncating this axis. In the second graph, mark the prices on the vertical axis, starting with $1.00. Briefly comment on the two bar graphs.

2.82 Reconsider the data on the times (in minutes) taken to commute from home to work for 20 workers given in Exercises 2.53. Create a dotplot for those data.

2.83 Reconsider the data on the numbers of orders received for a sample of 30 hours at the Timesaver Mail Order Company given in Exercise 2.71. Create a dotplot for those data.

2.84 Twenty-four students from a university in Oregon were asked to name the five current members of the U.S. House of Representatives from their state. The following data give the numbers of correct names given by these students.

5	5	1	2	4	5	3	1	5	5	0	1
2	3	5	4	3	1	5	2	5	4	5	3

Create a dotplot for these data.

2.85 The following data give the numbers of visitors during visiting hours on a given evening for each of the 20 randomly selected patients at a hospital.

3	0	1	4	2	0	4	1	1	3
4	2	0	2	2	2	1	1	3	0

Create a dotplot for these data.

Advanced Exercises

2.86 The following frequency distribution table gives the age distribution of drivers who were at fault in auto accidents that occurred during a one-week period in a city.

Age	f
18 to less than 20	7
20 to less than 25	12
25 to less than 30	18
30 to less than 40	14
40 to less than 50	15
50 to less than 60	16
60 and over	35

a. Draw a relative frequency histogram for this table.
b. In what way(s) is this histogram misleading?
c. How can you change the frequency distribution so that the resulting histogram gives a clearer picture?

2.87 Refer to the data presented in Exercise 2.86. Note that there were 50% more accidents in the *25 to less than 30* age group than in the *20 to less than 25* age group. Does this suggest that the older group of drivers in this city is more accident-prone than the younger group? What other explanation might account for the difference in accident rates?

2.88 Suppose a data set contains the ages of 135 autoworkers ranging from 20 to 53.
a. Using Sturge's formula given in footnote 1 on page 36, find an appropriate number of classes for a frequency distribution for this data set.
b. Find an appropriate class width based on the number of classes in part a.

2.89 Statisticians often need to know the shape of a population to make inferences. Suppose that you are asked to specify the shape of the population of weights of all college students.
a. Sketch a graph of what you think the weights of all college students would look like.
b. The following data give the weights (in pounds) of a random sample of 44 college students. (Here, F and M indicate female and male.)

123 F	195 M	138 M	115 F	179 M	119 F	148 F	147 F
180 M	146 F	179 M	189 M	175 M	108 F	193 M	114 F
179 M	147 M	108 F	128 F	164 F	174 M	128 F	159 M
193 M	204 M	125 F	133 F	115 F	168 M	123 F	183 M
116 F	182 M	174 M	102 F	123 F	99 F	161 M	162 M
155 F	202 M	110 F	132 M				

i. Construct a stem-and-leaf display for these data.
ii. Can you explain why these data appear the way they do?

c. Now sketch a new graph of what you think the weights of all college students look like. Is this similar to your sketch in part a?

2.90 Consider the two histograms given in Figure 2.23 that are drawn for the same data set. In this data set, none of the values are integers.

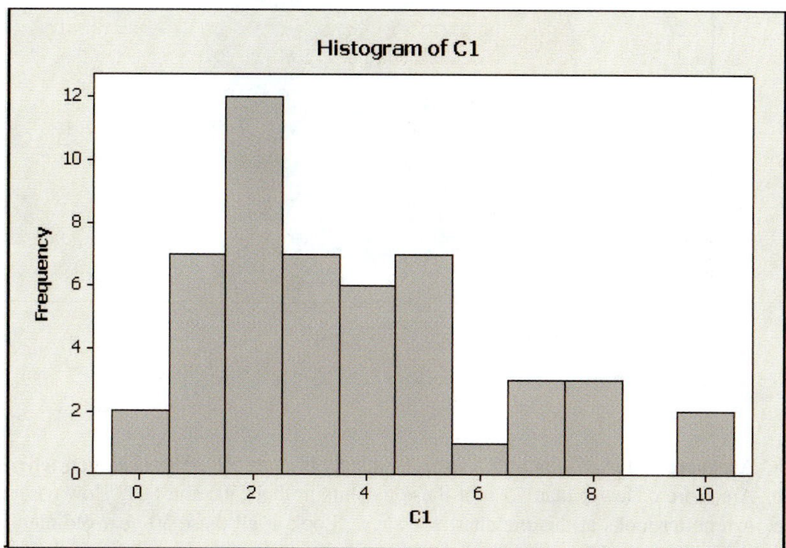

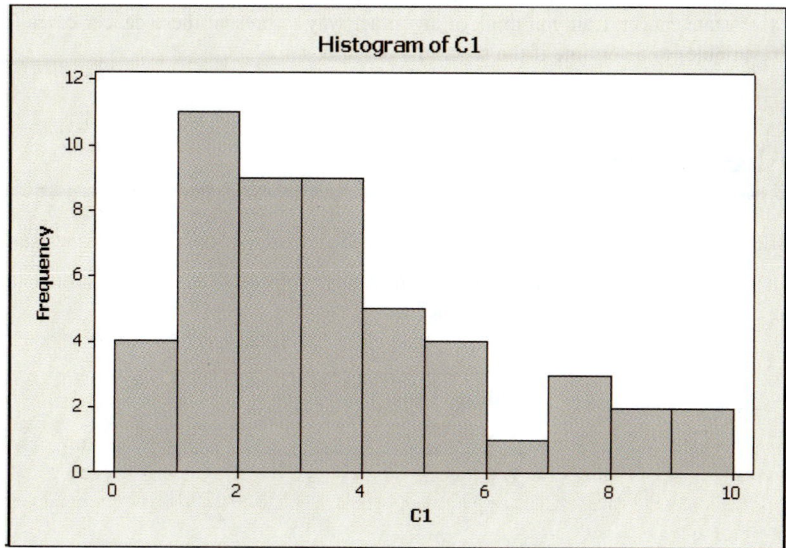

Figure 2.23 Two Histograms for the Same Data.

a. What are the endpoints and widths of classes in each of the two histograms?
b. In the first histogram, of the observations that fall in the interval that is centered at 8, how many are actually between the left endpoint of that interval and 8? Note that you have to consider both histograms to answer this question.
c. Observe the leftmost bars in both histograms. Why is the leftmost bar in the first histogram misleading?

2.91 Refer to the data on weights of 44 college students given in Exercise 2.88. Create a dotplot of all 44 weights. Then create stacked dotplots for the weights of male and female students. Describe the similarities and differences in the distributions of weights of male and female students. Using all three dotplots, explain why you cannot distinguish the lightest males from the heaviest females when you consider only the dotplot of all 44 weights.

2.92 The pie chart in Figure 2.24 shows the percentage distribution of ages (i.e., the percentages of all prostate cancer patients falling in various age groups) for men who were recently diagnosed with prostate cancer.

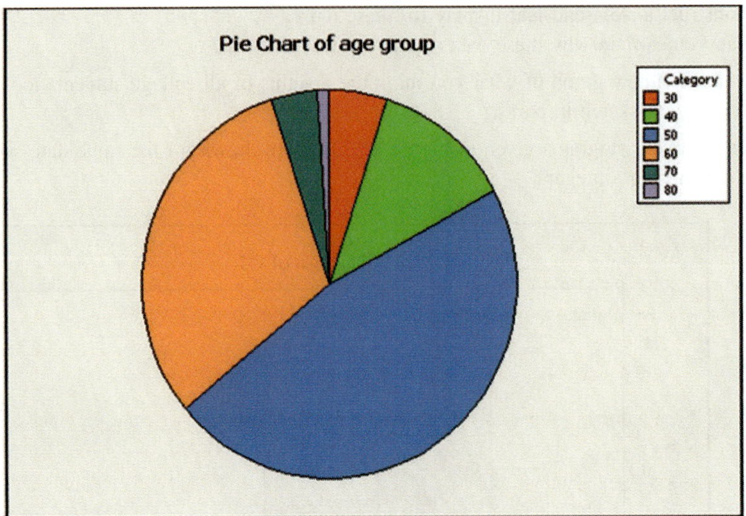

Figure 2.24 Pie Chart of Age Groups.

a. Are more or fewer than 50% of these patients in their 50s? How can you tell?
b. Are more or fewer than 75% of these patients in their 50s and 60s? How can you tell?
c. A reporter looks at this pie chart and says, "Look at all these 50-year-old men who are getting prostate cancer. This is a major concern for a man once he turns 50." Explain why the reporter cannot necessarily conclude from this pie chart that there are a lot of 50-year-old men with prostate cancer. Can you think of any other way to present these cancer cases (both graph and variable) to determine if the reporter's claim is valid?

Self-Review Test

1. Briefly explain the difference between ungrouped and grouped data and give one example of each type.

2. The following table gives the frequency distribution of times (to the nearest hour) that 90 fans spent waiting in line to buy tickets to a rock concert.

Waiting Time (hours)	Frequency
0 to 6	5
7 to 13	27
14 to 20	30
21 to 27	20
28 to 34	8

Circle the correct answer in each of the following statements, which are based on this table.
a. The number of classes in the table is 5, 30, 90
b. The class width is 6, 7, 34
c. The midpoint of the third class is 16.5, 17, 17.5
d. The lower boundary of the second class is 6.5, 7, 7.5
e. The upper limit of the second class is 12.5, 13, 13.5
f. The sample size is 5, 90, 11
g. The relative frequency of the second class is .22, .41, .30

3. Briefly explain and illustrate with the help of graphs a symmetric histogram, a histogram skewed to the right, and a histogram skewed to the left.

4. Twenty elementary school children were asked if they live with both parents (B), father only (F), mother only (M), or someone else (S). The responses of the children follow.

M	B	B	M	F	S	B	M	F	M
B	F	B	M	M	B	B	F	B	M

a. Construct a frequency distribution table.
b. Write the relative frequencies and percentages for all categories.
c. What percentage of the children in this sample live with their mothers only?
d. Draw a bar graph for the frequency distribution and a pie chart for the percentages.

5. A large Midwestern city has been chronically plagued by false fire alarms. The following data set gives the number of false alarms set off each week for a 24-week period in this city.

10	4	8	7	3	7	10	2	6	12	11	8
1	6	5	13	9	7	5	1	14	5	15	3

a. Construct a frequency distribution table. Take 1 as the lower limit of the first class and 3 as the width of each class.
b. Calculate the relative frequencies and percentages for all classes.
c. What percentage of these weeks had 9 or fewer false alarms?
d. Draw the frequency histogram and polygon.

6. Refer to the frequency distribution prepared in Problem 5. Prepare the cumulative percentage distribution using that table. Draw an ogive for the cumulative percentage distribution.

7. Construct a stem-and-leaf display for the following data, which give the times (in minutes) 24 customers spent waiting to speak to a customer service representative when they called about problems with their Internet service provider.

12	15	7	29	32	16	10	14	17	8	19	21
4	14	22	25	18	6	22	16	13	16	12	20

8. Consider this stem-and-leaf display:

3	0 3 7
4	2 4 6 7 9
5	1 3 3 6
6	0 7 7
7	1 9

Write the data set that was used to construct this display.

9. Make a dotplot for the data given in Problem 5.

Mini-Projects

■ MINI-PROJECT 2–1

Using the data you gathered for the mini-project in Chapter 1, prepare a summary of that data set that includes the following.

a. Prepare an appropriate type of frequency distribution table for one of the quantitative variables and then compute relative frequencies and cumulative relative frequencies.
b. Create a histogram, a stem-and-leaf display, and a dotplot of the data. Comment on any symmetry or skewness, and on the presence of clusters and any potential outliers.
c. Make stacked dotplots of the same variable (as in parts a and b) based on the values of one of your categorical variables. For example, if your quantitative variable is GPAs of students, your categorical variable could be gender. Comment on the similarities and differences between the distributions for the different values of your categorical variable.

■ MINI-PROJECT 2–2

Watch four hours of each of two types of television shows (comedy, news, drama, sports, and so forth) and record the duration of each commercial to the nearest second. This can include commercials that are shown between the shows. List these durations of various commercials and the type of the show. Using this information, write a brief report that covers the following.

a. Prepare an appropriate type of frequency distribution table for the quantitative variable and then compute relative frequencies and cumulative relative frequencies.
b. Create a histogram, a stem-and-leaf plot, and a dotplot for these data. Comment on any symmetry or skewness, and on the presence of clusters and any potential outliers.
c. Make stacked dotplots of the same variable for each of the two types of television shows. Comment on the similarities and differences between the distributions for the different types of shows.

DECIDE FOR YOURSELF

Deciding About Statistical Properties

Look around you. Graphs are everywhere. Business reports, newspapers, magazines, and so forth are all loaded with graphs. Unfortunately, some people feel that the primary purpose of graphs is to provide a break from the humdrum text. Executive summaries will often contain graphs so that CEOs and executive vice presidents need only to glance at these graphs to assume that they understand everything without reading more than a paragraph or so of the report. In reality, the usefulness of graphs is somewhere between the fluff of the popular press and the quick answer of the boardroom.

Here you are asked to interpret some graphs, primarily by using them to compare distributions of a variable. As we will discuss in Chapter 3, some of our concerns have to do with the location of the center of a distribution and the variability or spread of a distribution. We can use graphs to compare the centers and variability of two or more distributions.

In practice, the graphs are made using statistical software, so it is important to recognize that computer software is programmed to use the same format for each graph of a specific type, unless you tell the software to do differently. For example, consider the two histograms in Figures 2.25 and 2.26 that are drawn for two different data sets.

1. Examine the two graphs of Figures 2.25 and 2.26.

2. Explain what is meant by the statement "the shapes of the two distributions are the same."

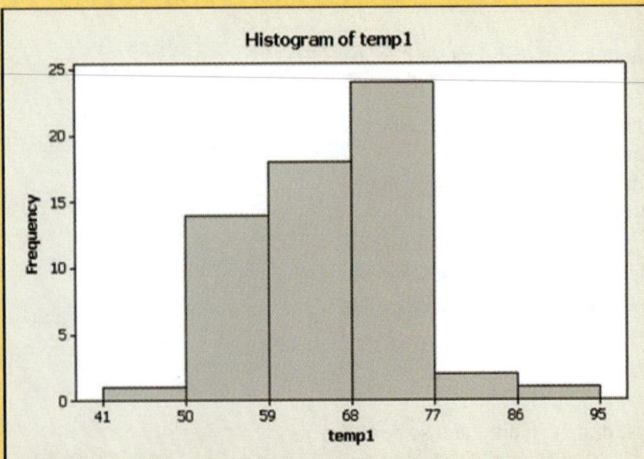

Figure 2.25 Histogram of Data Temp 1.

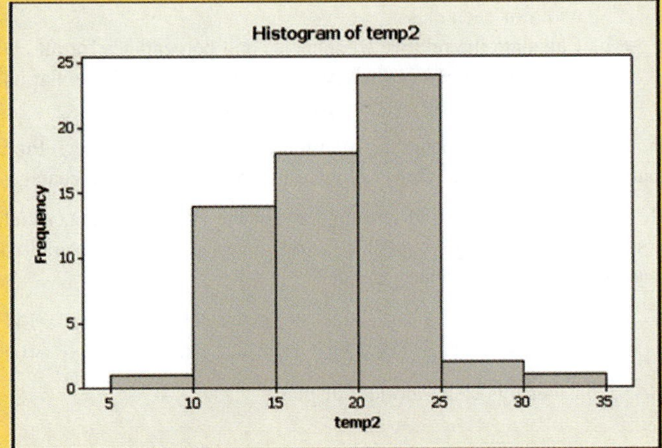

Figure 2.26 Histogram of Data Temp 2.

3. Does the fact that the shapes of the two distributions are the same imply that the centers of the two distributions are the same? Why or why not? Explain.

4. Does the fact that the shapes of the two distributions are the same imply that the spreads of the two distributions are the same? Why or why not? Explain.

5. It turns out that the same variable was represented in the two graphs, but with different units of measurement. Can you figure out the units?

Another situation that is important to compare is when two graphs cover a similar range but have different shapes, such as the histograms in Figures 2.27 and 2.28.

1. Examine the two histograms of Figures 2.27 and 2.28.

2. These two distributions have the same center, but do not have the same spread. Decide which distribution has the larger spread and explain the reasoning behind your decision.

Answer all the above questions again after reading Chapter 3.

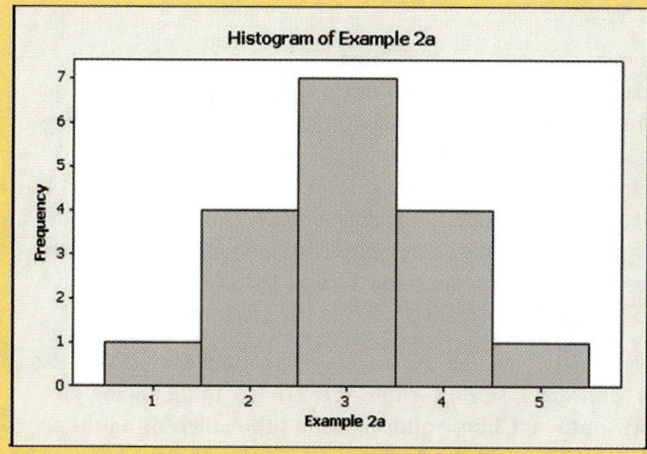

Figure 2.27 Histogram of Example 2a.

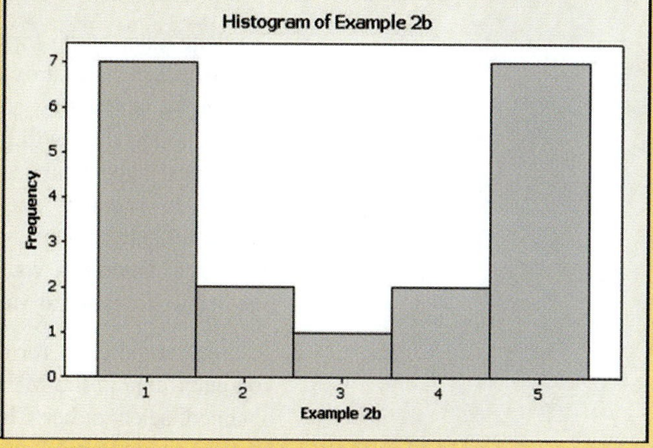

Figure 2.28 Histogram of Example 2b.

TECHNOLOGY INSTRUCTION

Organizing Data

TI-84

Screen 2.1

1. To create a frequency histogram for a list of data, press **STAT PLOT**, which is accessed by pressing **2nd** > **Y=**. The **Y=** key is located at the top left on the calculator.

2. Make sure that only one plot is turned on. If more than one plot is turned on, turn off the unwanted plots by using the following steps. Press the number corresponding to the plot you wish to turn off. A screen similar to **Screen 2.1** will appear. Use the arrow keys to move the cursor to the **Off** button, then press **Enter**. Now use the arrow keys to move to the row with **Plot1**, **Plot2**, and **Plot3**. If there is another plot that you need to turn off, select that plot by moving the cursor to it, press **Enter**, and repeat the previous procedure. If there is no plot that you need to turn off, move the cursor to the plot you want to use and press **Enter**.

3. In the **Type** rows, use the right arrow to move to the third column in the first row, which looks like a histogram, and press **Enter**. Move to **Xlist** to enter the name of the list where the data are located. Press **2nd>Stat**, then use the up and down arrows to move through the list names until you find the list you want to use. Press **Enter**. Leave the **Freq** setting at 1.

4. To see the graph, select **ZOOM** > **9**, where **ZOOM** is the third key in the top row. This sets the window settings to display your graph.

5. If you would like to change the class width, select **WINDOW**, which is the second key in the top row. Change the value of **Xscl** to the desired width, then press **GRAPH**, which is the fifth key in the top row.

6. If you would like to see the interval endpoints and the number of observations in each class (which is given by the height of the corresponding bar), press **TRACE**, then use the left and right arrows to move between bars. When you are done, press **CLEAR**.

MINITAB

1. To create a bar graph for categorical data entered in column C1, select **Graph>Bar Chart**. In the dialog box obtained, select **Counts of unique values** and **Simple** and click **OK**. In the new dialog box you obtain, type **C1** in the box below **Categorical Variables** and click **OK**. This will produce a bar graph for the data.
   If you have categorical data in a frequency table with categories entered in column C1 and frequencies in column C2, select **Graph>Bar Chart**. In the dialog box obtained, select **Values from a table** from choices below **Bars represent**, select **Simple** from graphs, and click **OK**. In the new dialog box you obtain, type **C2** in the box below **Graph variables** and **C1** in the box below **Categorical variable**. Click **OK**. This will produce a bar graph for the data.

2. To create a pie chart for categorical data, you can use raw data or a frequency table. After you enter data in the MINITAB worksheet, select **Graph>Pie Chart**. In the dialog box obtained, select either **Chart raw data** or **Chart values from a table**, then fill in the required column names, and click **OK**. This will produce a pie chart for the data.

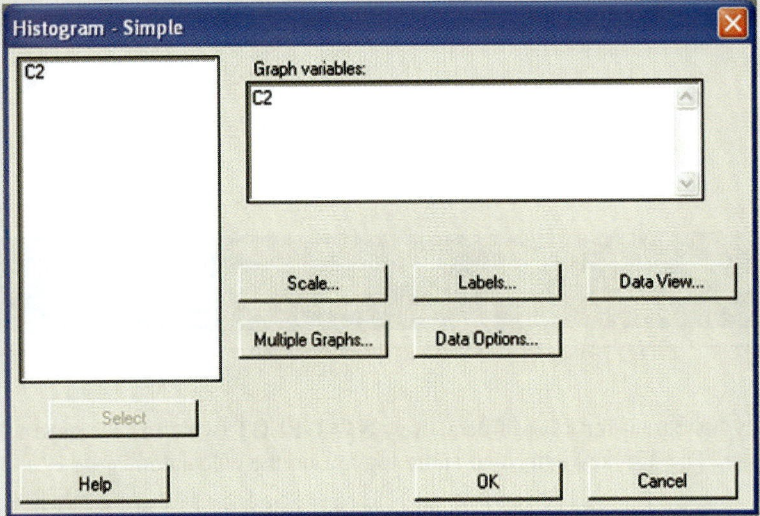

Screen 2.2

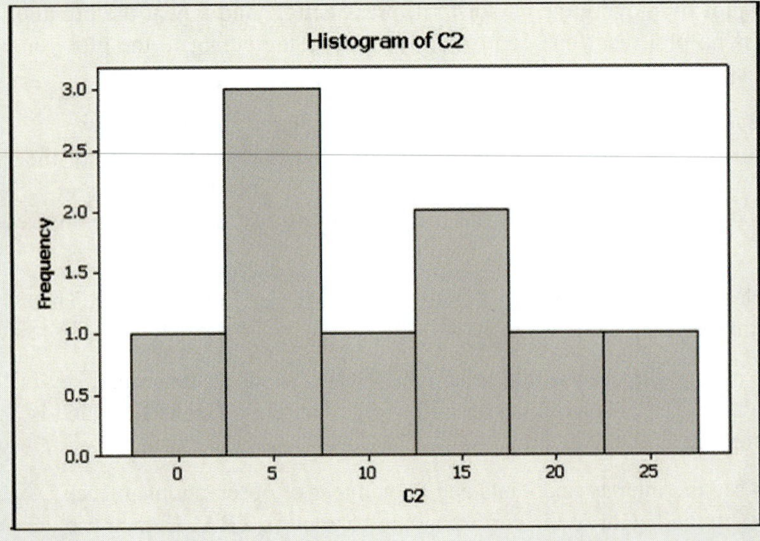

Screen 2.3

3. To create a frequency histogram for a quantitative data set entered in column C2, select **Graph>Histogram**, then select **Simple**, and click **OK**. In the dialog box you obtain, enter the name of your column, e.g., **C2** (see **Screen 2.2**) in the box below **Graph variables** and click **OK** to create the histogram. MINITAB will produce **Screen 2.3** with histogram.

4. To create a stem-and-leaf display for a quantitative data set entered in column C1, select **Graph>Stem-and-Leaf**, enter the name of the column that contains data in the box below **Graph variables** and click **OK** to create the stem-and-leaf display.

5. To create a dotplot for a quantitative data set entered in column C1, select **Graph>Dotplot**, then select the appropriate dotplot graph from the dialog box and click **OK**. In the new dialog box you obtain, enter the name of your column, e.g., **C1** in the box below **Graph variables**, and click **OK** to create the dotplot.

Excel

	A	B	C	D	E
1	Scores	Boundaries	Frequencies		
2					
3	1	2	=(frequency,a3:a12,b3:b6)		
4	2	4			
5	6	6			
6	7	8			
7	6				
8	5				
9	0				
10	2				
11	2				
12	8				

Screen 2.4

1. To create a frequency distribution for a range of numerical data in Excel, decide how many categories you will have. Choose class boundaries between the categories so that you have one fewer boundary than classes. Type the class boundaries into Excel.

2. Select where you want the class frequencies to appear, and select a range of one more cell than the number of boundaries you have.

3. Type **=frequency(**.

4. Select the range of cells of numerical data, and then type a comma.

5. Select the range of class boundaries, and then type a right parenthesis. (See Screen 2.4.)

6. Press **Control-Shift-Enter**. The frequencies should appear.

7. If you would prefer relative frequencies, replace Steps 5 and 6 by the following:
   a. Select the range of class boundaries, then type **)/count(**.
   b. Select the range of cells of numerical data, then type a right parenthesis.
   c. Press **Control-Shift-Enter**. The relative frequencies should appear.

8. To plot frequencies as bar charts, pie charts, and so on for 1, select **Insert>Chart** and follow the instructions in the Chart Wizard.

TECHNOLOGY ASSIGNMENTS

TA2.1 Construct a bar graph and a pie chart for the frequency distribution prepared in Exercise 2.5.

TA2.2 Construct a bar graph and a pie chart for the frequency distribution prepared in Exercise 2.6.

TA2.3 Refer to Data Set V that accompanies this text (see Preface and Appendix B) on the times taken to run the Manchester Road Race for a sample of 500 participants. From that data set, select the 6th value and then select every 10th value after that (i.e., select the 6th, 16th, 26th, 36th . . . values). This subsample will give you 50 measurements. (Such a sample selected from a population is called a *systematic random sample*.) Construct a histogram for these data. Let the software you use decide on classes and class limits.

TA2.4 Refer to Data Set I that accompanies this text on the prices of various products in different cities across the country. Select a subsample of 60 from the column that contains information on telephone charges and then construct a histogram for these data.

TA2.5 Construct a histogram for the data from Exercise 2.20 on the numbers of computer keyboards assembled. Use the classes given in that exercise. Use the midpoints to mark the horizontal axis in the histogram.

TA2.6 Prepare a stem-and-leaf display for the data given in Exercise 2.48.

TA2.7 Prepare a stem-and-leaf display for the data of Exercise 2.53.

TA2.8 Prepare a bar graph for the frequency distribution obtained in Exercise 2.28.

TA2.9 Prepare a bar graph for the frequency distribution obtained in Exercise 2.29.

TA2.10 Make a pie chart for the frequency distribution obtained in Exercise 2.19.

TA2.11 Make a pie chart for the frequency distribution obtained in Exercise 2.29.

TA2.12 Make a dotplot for the data of Exercise 2.64.

TA2.13 Make a dotplot for the data of Exercise 2.65.

Chapter

3

Numerical Descriptive Measures

3.1 Measures of Central Tendency for Ungrouped Data

Case Study 3–1 High-Priced Tickets in Big Markets

Case Study 3–2 Median Annual Starting Salary for MBAs

3.2 Measures of Dispersion for Ungrouped Data

3.3 Mean, Variance, and Standard Deviation for Grouped Data

3.4 Use of Standard Deviation

Case Study 3–3 Here Comes the SD

3.5 Measures of Position

3.6 Box-and-Whisker Plot

Which Major League Baseball team do you think has the highest average ticket price? Do you think it is one of the two New York teams, Mets or Yankees, assuming these teams play in one of the most expensive cities in the world? No, you are not even close. Then, is it one of the teams playing in Los Angeles—Dodgers or Angels? Still wrong. Actually it is the Boston Red Sox that had the highest average ticket price ($40.77) among all Major League Baseball teams in 2004. The next highest was $28.45 for the Chicago Cubs. See Case Study 3–1.

In Chapter 2 we discussed how to summarize data using different methods and to display data using graphs. Graphs are one important component of statistics; however it is also important to numerically describe the main characteristics of a data set. The numerical summary measures, such as the ones that identify the center and spread of a distribution, identify many important features of a distribution. For example, the techniques learned in Chapter 2 can help us graph data on family incomes. However, if we want to know the income of a "typical" family (given by the center of the distribution), the spread of the distribution of incomes, or the relative position of a family with a particular income, the numerical summary measures can provide more detailed information (see Figure 3.1). The measures that we discuss in this chapter include measures of (1) central tendency, (2) dispersion (or spread), and (3) position.

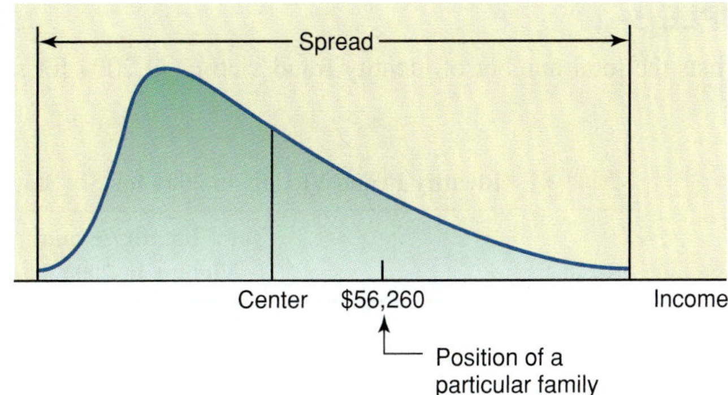

Figure 3.1

3.1 Measures of Central Tendency for Ungrouped Data

We often represent a data set by numerical summary measures, usually called the *typical values*. A **measure of central tendency** gives the center of a histogram or a frequency distribution curve. This section discusses three different measures of central tendency: the mean, the median, and the mode; however, a few other measures of central tendency, such as the trimmed mean, the weighted mean, and the geometric mean, are explained in exercises following this section. We will learn how to calculate each of these measures for ungrouped data. Recall from Chapter 2, the data that give information on each member of the population or sample individually are called *ungrouped data*, whereas *grouped data* are presented in the form of a frequency distribution table.

3.1.1 Mean

The **mean**, also called the *arithmetic mean*, is the most frequently used measure of central tendency. This book will use the words *mean* and *average* synonymously. For ungrouped data, the mean is obtained by dividing the sum of all values by the number of values in the data set.

$$\text{Mean} = \frac{\text{Sum of all values}}{\text{Number of values}}$$

The mean calculated for sample data is denoted by $\bar{x}$ (read as "x bar"), and the mean calculated for population data is denoted by μ (Greek letter *mu*). We know from the discussion in Chapter 2 that the number of values in a data set is denoted by n for a sample and by N for a population. In Chapter 1, we learned that a variable is denoted by x and the sum of all values of x is denoted by Σx. Using these notations, we can write the following formulas for the mean.

Calculating Mean for Ungrouped Data The *mean for ungrouped data* is obtained by dividing the sum of all values by the number of values in the data set. Thus,

$$\text{Mean for population data:} \quad \mu = \frac{\Sigma x}{N}$$

$$\text{Mean for sample data:} \quad \bar{x} = \frac{\Sigma x}{n}$$

where Σx is the sum of all values, N is the population size, n is the sample size, μ is the population mean, and $\bar{x}$ is the sample mean.

■ EXAMPLE 3–1

Calculating the sample mean for ungrouped data.

Table 3.1 lists the total number of identity fraud victims in 2004 for six states.

Table 3.1 Identity Fraud Victims in 2004 for Six States

State	Total Identity Fraud Victims in 2004
California	43,839
Florida	16,062
Illinois	11,138
New York	17,680
Ohio	6956
Texas	26,454

Source: Federal Trade Commission's Identity Theft Data Clearinghouse.

Find the mean number of identity fraud victims in 2004 for these six states.

Solution The variable in this example is the number of identity fraud victims in 2004 for six states. Let us denote it by x. Then, the six values of x are

$$x_1 = 43{,}839, \quad x_2 = 16{,}062, \quad x_3 = 11{,}138, \quad x_4 = 17{,}680, \quad x_5 = 6956, \quad \text{and} \quad x_6 = 26{,}454$$

where $x_1 = 43{,}839$ represents the number of identity fraud victims in 2004 for California, $x_2 = 16{,}062$ represents the number of identity fraud victims in 2004 for Florida, and so on. The sum of the numbers of identity fraud victims for these six states is

$$\begin{aligned}\Sigma x &= x_1 + x_2 + x_3 + x_4 + x_5 + x_6 \\ &= 43{,}839 + 16{,}062 + 11{,}138 + 17{,}680 + 6956 + 26{,}454 = 122{,}129\end{aligned}$$

Note that the given data includes only six states. Hence, it represents a sample. Because the data set contains six values, $n = 6$. Substituting the values of Σx and n in the sample formula, we obtain the mean number of identity fraud victims in 2004 for these six states:

$$\bar{x} = \frac{\Sigma x}{n} = \frac{122{,}129}{6} = \mathbf{20{,}354.83}$$

Thus, the mean number of identity fraud victims in 2004 for these six states is 20,354.83. ■

■ EXAMPLE 3–2

Calculating the population mean for ungrouped data.

The following are the ages of all eight employees of a small company:

53 32 61 27 39 44 49 57

Find the mean age of these employees.

Solution Because the given data set includes *all* eight employees of the company, it represents the population. Hence, $N = 8$.

$$\Sigma x = 53 + 32 + 61 + 27 + 39 + 44 + 49 + 57 = 362$$

The population mean is

$$\mu = \frac{\Sigma x}{N} = \frac{362}{8} = \mathbf{45.25\ years}$$

Thus, the mean age of all eight employees of this company is 45.25 years, or 45 years and 3 months. ■

Reconsider Example 3–2. If we take a sample of three employees from this company and calculate the mean age of those three employees, this mean will be denoted by $\bar{x}$. Suppose the three values included in the sample are 32, 39, and 57. Then, the mean age for this sample is

$$\bar{x} = \frac{32 + 39 + 57}{3} = 42.67 \text{ years}$$

If we take a second sample of three employees of this company, the value of $\bar{x}$ will (most likely) be different. Suppose the second sample includes the values 53, 27, and 44. Then, the mean age for this sample is

$$\bar{x} = \frac{53 + 27 + 44}{3} = 41.33 \text{ years}$$

Consequently, we can state that the value of the population mean μ is constant. However, the value of the sample mean $\bar{x}$ varies from sample to sample. The value of $\bar{x}$ for a particular sample depends on what values of the population are included in that sample.

Sometime a data set may contain a few very small or a few very large values. As mentioned in Chapter 2 on page 58, such values are called *outliers* or *extreme values.*

A major shortcoming of the mean as a measure of central tendency is that it is very sensitive to outliers. Example 3–3 illustrates this point.

■ EXAMPLE 3–3

Table 3.2 lists the total philanthropic givings (in million dollars) by six donors during their lifetimes until 2004.

Illustrating the effect of an outlier on the mean.

Table 3.2 Total Philanthropic Givings in Lifetime

Donors	**Total Philanthropic Giving in Lifetime (millions of dollars)**
Bill and Melinda Gates	27,976
Warren Buffett	2730
George Soros	5171
Michael and Susan Dell	1230
Walton Family	1000
Ted Turner	1200

Source: Business Week, November 29, 2004.

Notice that the lifetime givings of Bill and Melinda Gates are very large compared to the lifetime givings of other donors. Hence, it is an outlier. Show how the inclusion of this outlier affects the value of the mean.

Solution If we do not include the lifetime givings of Bill and Melinda Gates (the outlier), the mean of the lifetime givings of the remaining five donors is

$$\text{Mean} = \frac{2730 + 5171 + 1230 + 1000 + 1200}{5} = \frac{11,331}{5} = \mathbf{\$2266.20 \text{ million}}$$

Now, to see the impact of the outlier on the value of the mean, we include the lifetime givings of Bill and Melinda Gates and find the mean lifetime givings of the six donors. This mean is

$$\text{Mean} = \frac{27,976 + 2730 + 5171 + 1230 + 1000 + 1200}{6} = \frac{39,307}{6} = \mathbf{\$6551.17 \text{ million}}$$

Thus, including the lifetime givings of Bill and Melinda Gates causes almost a threefold increase in the value of the mean, as it changes from $2266.20 million to $6551.17 million. ■

CASE STUDY 3–1

HIGH-PRICED TICKETS IN BIG MARKETS

USA TODAY Snapshots

High-priced tickets in big markets

Major league baseball teams with the highest aver age ticket prices in 2004:

Team	Average ticket price
Boston Red Sox	$40.77
Chicago Cubs	$28.45
Philadelphia Phillies	$26.08
New York Yankees	$24.86
Seattle Mariners	$24.01

Source: AP

By Ellen J. Horrow and Sam Ward, USA TODAY

The above chart, reproduced from *USA TODAY*, shows the average ticket prices of the five Major League Baseball teams that had the highest average ticket prices in 2004. According to the information given in the chart, the highest average price for an MLB team was for the Boston Red Sox, which was \$40.77. The Chicago Cubs had the second highest average ticket price of \$28.45.

Source: USA TODAY, April 26, 2004.

The preceding example should encourage us to be cautious. We should remember that the mean is not always the best measure of central tendency because it is heavily influenced by outliers. Sometimes other measures of central tendency give a more accurate impression of a data set. For example, when a data set has outliers, instead of using the mean, we can use either the trimmed mean (defined in Exercise 3.33) or the median (to be discussed next) as a measure of central tendency.

3.1.2 Median

Another important measure of central tendency is the **median**. It is defined as follows.

Definition

Median The *median* is the value of the middle term in a data set that has been ranked in increasing order.

As is obvious from the definition of the median, it divides a ranked data set into two equal parts. The calculation of the median consists of the following two steps:

1. Rank the data set in increasing order.
2. Find the middle term. The value of this term is the median.[1]

[1]The value of the middle term in a data set ranked in *decreasing* order will also give the value of the median.

Note that if the number of observations in a data set is *odd*, then the median is given by the value of the middle term in the ranked data. However, if the number of observations is *even*, then the median is given by the average of the values of the two middle terms.

■ EXAMPLE 3–4

Calculating the median for ungrouped data: odd number of data values.

The following data give the weight lost (in pounds) by a sample of five members of a health club at the end of two months of membership.

10 5 19 8 3

Find the median.

Solution First, we rank the given data in increasing order as follows:

3 5 8 10 19

Since there are five terms in the data set and the middle term is the third term, the median is given by the value of the third term in the ranked data.

3 5 8 10 19

8 ↑ Median

The median weight loss for this sample of five members of this health club is **8 pounds**. ■

■ EXAMPLE 3–5

Calculating the median for ungrouped data: even number of data values.

Table 3.3 lists the number of car thefts during 2003 in 12 cities.

Table 3.3 Number of Car Thefts in 2003 in 12 Cities

City	Number of Car Thefts
Phoenix-Mesa, Arizona	40,769
Washington, D.C.	33,956
Miami, Florida	21,088
Atlanta, Georgia	29,920
Chicago, Illinois	42,082
Kansas City, Kansas	11,669
Baltimore, Maryland	13,435
Detroit, Michigan	40,197
St. Louis, Missouri	18,215
Las Vegas, Nevada	18,103
Newark, New Jersey	14,413
Dallas, Texas	26,343

Source: National Insurance Crime Bureau.

Find the median for these data.

Solution First we rank the given data on car thefts in increasing order as follows:

11,669 13,435 14,413 18,103 18,215 21,088 26,343 29,920 33,956 40,197 40,769 42,082

There are 12 values in the data set. Because there is an even number of values in the data set, the median will be given by the mean of the two middle values. The two middle values

CASE STUDY 3–2

MEDIAN ANNUAL STARTING SALARY FOR MBAs

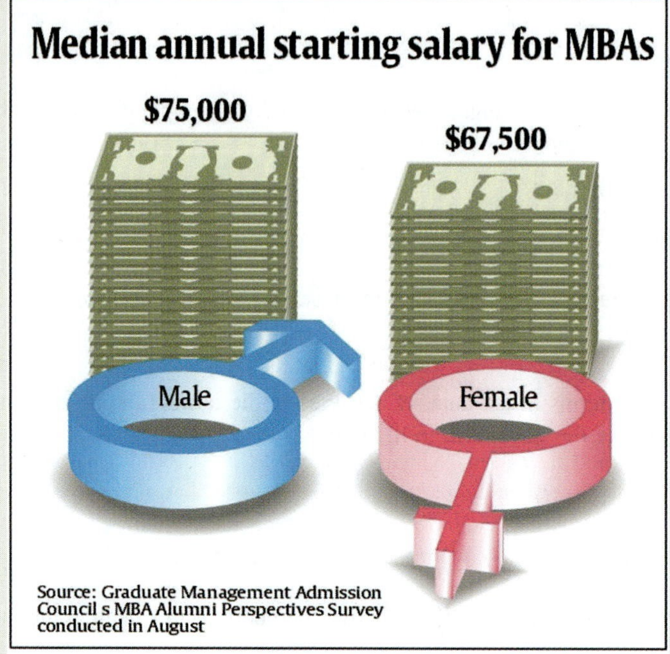

Source: *USA TODAY*, February 4, 2004. Copyright © 2004, *USA TODAY.* Chart reproduced with permission.

The above chart, reproduced from *USA TODAY*, shows the median annual starting salary of MBAs. These salaries are based on a survey conducted in August 2003. According to this survey, the median starting salary for males with an MBA degree was $75,000 and that of females was $67,500.

are the sixth and seventh values in the above arranged data, which are 21,088 and 26,343. The median, which is given by the average of these two values, is calculated below.

11,669 13,435 14,413 18,103 18,215 **21,088 ↑ 26,343** 29,920 33,956 40,197 40,769 42,082

Median

$$\text{Median} = \frac{21{,}088 + 26{,}343}{2} = \frac{47{,}431}{2} = \mathbf{23{,}715.50 \text{ car thefts}}$$

Thus, the median number of car thefts in 2003 for these 12 cities was 23,715.50. ■

The median gives the center of a histogram, with half of the data values to the left of the median and half to the right of the median. The advantage of using the median as a measure of central tendency is that it is not influenced by outliers. Consequently, the median is preferred over the mean as a measure of central tendency for data sets that contain outliers.

3.1.3 Mode

Mode is a French word that means *fashion*—an item that is most popular or common. In statistics, the mode represents the most common value in a data set.

Definition

Mode The *mode* is the value that occurs with the highest frequency in a data set.

■ EXAMPLE 3–6

Calculating the mode for ungrouped data.

The following data give the speeds (in miles per hour) of eight cars that were stopped on I-95 for speeding violations.

77	82	74	81	79	84	74	78

Find the mode.

Solution In this data set, 74 occurs twice and each of the remaining values occurs only once. Because 74 occurs with the highest frequency, it is the mode. Therefore,

Mode = **74 miles per hour** ■

A major shortcoming of the mode is that a data set may have none or may have more than one mode, whereas it will have only one mean and only one median. For instance, a data set with each value occurring only once has no mode. A data set with only one value occurring with the highest frequency has only one mode. The data set in this case is called **unimodal**. A data set with two values that occur with the same (highest) frequency has two modes. The distribution, in this case, is said to be **bimodal**. If more than two values in a data set occur with the same (highest) frequency, then the data set contains more than two modes and it is said to be **multimodal**.

■ EXAMPLE 3–7

Data set with no mode.

Last year's incomes of five randomly selected families were \$46,150, \$95,750, \$64,985, \$87,490, and \$53,740. Find the mode.

Solution Because each value in this data set occurs only once, this data set contains **no mode**. ■

■ EXAMPLE 3–8

Data set with two modes.

The prices of the same brand of television set at eight stores are found to be \$895, \$886, \$903, \$895, \$870, \$905, \$870, and \$899. Find the mode.

Solution In this data set, each of the two values \$895 and \$870 occurs twice and each of the remaining values occurs only once. Therefore, this data set has two modes: **\$895** and **\$870**. ■

■ EXAMPLE 3–9

Data set with three modes.

The ages of 10 randomly selected students from a class are 21, 19, 27, 22, 29, 19, 25, 21, 22, and 30. Find the mode.

Solution This data set has three modes: **19**, **21**, and **22**. Each of these three values occurs with a (highest) frequency of 2. ■

One advantage of the mode is that it can be calculated for both kinds of data, quantitative and qualitative, whereas the mean and median can be calculated for only quantitative data.

■ EXAMPLE 3–10

Finding the mode for qualitative data.

The status of five students who are members of the student senate at a college are senior, sophomore, senior, junior, senior. Find the mode.

Solution Because **senior** occurs more frequently than the other categories, it is the mode for this data set. We cannot calculate the mean and median for this data set. ■

To sum up, we cannot say for sure which of the three measures of central tendency is a better measure overall. Each of them may be better under different situations. Probably the mean is the most used measure of central tendency, followed by the median. The mean has the advantage that its calculation includes each value of the data set. The median is a better measure when a data set includes outliers. The mode is simple to locate, but it is not of much use in practical applications.

3.1.4 Relationships among the Mean, Median, and Mode

As discussed in Chapter 2, two of the many shapes that a histogram or a frequency distribution curve can assume are symmetric and skewed. This section describes the relationships among the mean, median, and mode for three such histograms and frequency distribution curves. Knowing the values of the mean, median, and mode can give us some idea about the shape of a frequency distribution curve.

1. For a symmetric histogram and frequency distribution curve with one peak (see Figure 3.2), the values of the mean, median, and mode are identical, and they lie at the center of the distribution.

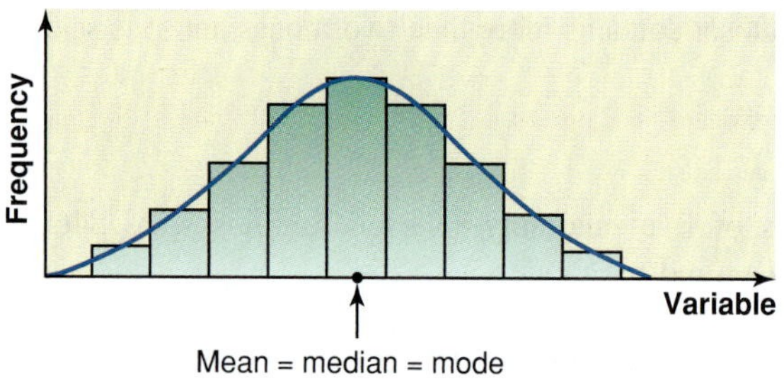

Figure 3.2 Mean, median, and mode for a symmetric histogram and frequency distribution curve.

2. For a histogram and a frequency distribution curve skewed to the right (see Figure 3.3), the value of the mean is the largest, that of the mode is the smallest, and the value of the median lies between these two. (Notice that the mode always occurs at the peak point.) The value of the mean is the largest in this case because it is sensitive to outliers that occur in the right tail. These outliers pull the mean to the right.

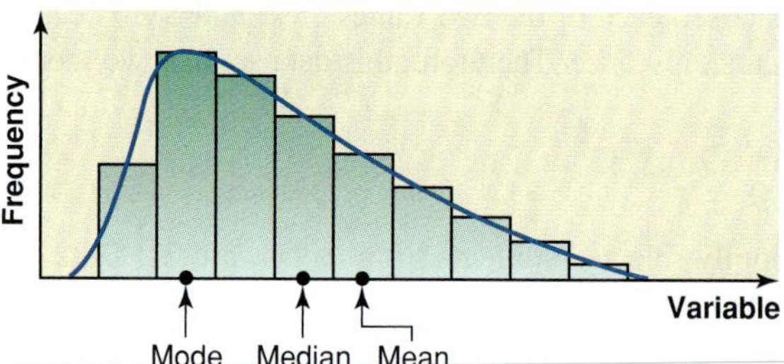

Figure 3.3 Mean, median, and mode for a histogram and frequency distribution curve skewed to the right.

3. If a histogram and a frequency distribution curve are skewed to the left (see Figure 3.4), the value of the mean is the smallest and that of the mode is the largest, with the value of the median lying between these two. In this case, the outliers in the left tail pull the mean to the left.

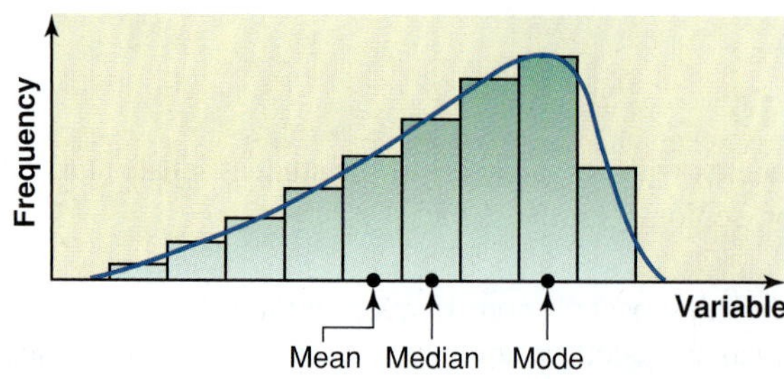

Figure 3.4 Mean, median, and mode for a histogram and frequency distribution curve skewed to the left.

EXERCISES

CONCEPTS AND PROCEDURES

3.1 Explain how the value of the median is determined for a data set that contains an odd number of observations and for a data set that contains an even number of observations.

3.2 Briefly explain the meaning of an outlier. Is the mean or the median a better measure of central tendency for a data set that contains outliers? Illustrate with the help of an example.

3.3 Using an example, show how outliers can affect the value of the mean.

3.4 Which of the three measures of central tendency (the mean, the median, and the mode) can be calculated for quantitative data only, and which can be calculated for both quantitative and qualitative data? Illustrate with examples.

3.5 Which of the three measures of central tendency (the mean, the median, and the mode) can assume more than one value for a data set? Give an example of a data set for which this summary measure assumes more than one value.

3.6 Is it possible for a (quantitative) data set to have no mean, no median, or no mode? Give an example of a data set for which this summary measure does not exist.

3.7 Explain the relationships among the mean, median, and mode for symmetric and skewed histograms. Illustrate these relationships with graphs.

3.8 Prices of cars have a distribution that is skewed to the right with outliers in the right tail. Which of the measures of central tendency is the best to summarize this data set? Explain.

3.9 The following data set belongs to a population:

5 −7 2 0 −9 16 10 7

Calculate the mean, median, and mode.

3.10 The following data set belongs to a sample:

14 18 −10 8 8 −16

Calculate the mean, median, and mode.

APPLICATIONS

Exercises 3.11 and 3.12 are based on the following data.

The following table gives the sticker prices and dealer's prices for base models of 10 two-door small cars as of January 2004.

Make/Model	Sticker Price	Dealer's Cost
Acura RSX	$20,025	$18,261
Chevrolet Cavalier LS	15,820	14,792
Ford Focus ZX3 Comfort	14,495	13,566
Honda Civic EX	16,860	15,410
Hyundai Accent GL	10,899	10,201
Mini Cooper	16,449	14,887
Oldsmobile Alero GL2	21,900	20,039
Pontiac Sunfire	14,930	13,810
Toyota Celica GT	17,390	15,735
Volkswagen Golf GL	15,580	14,593

Sources: MONEY, March 2004.

3.11 Calculate the mean and median for the data on sticker prices for these cars.

3.12 Find the mean and median for the data on dealers' costs for these cars.

3.13 The following data give the number of workers (in thousands) employed by small companies in all 50 states (*USA Today*, June 20, 2005). The data are entered in alphabetic order for states.

786	128	930	476	6800	981	759	171	2900	1500
253	259	2600	1300	642	588	734	853	292	1100
1500	2000	1200	452	1200	210	383	402	302	1800

319	3800	1600	161	2300	645	736	2500	238	739
189	1000	3800	430	162	1400	1200	303	1300	123

a. Calculate the mean and median for these data. Are these values of the mean and median the sample statistics or population parameters?

b. Do these data have a mode? Explain.

3.14 The following data give the 2004 profits (in millions of dollars) of the nine computer and office equipment companies included in the Fortune 1000 (*FORTUNE*, April 18, 2005). The data, entered in that order, are for International Business Machines, Hewlett-Packard, Dell, Xerox, Sun Microsystems, Apple Computer, NCR, Pitney Bowes, and Gateway.

8430	3497	3043	859	−388	276	290	481	−568

Find the mean and median for these data. Do these data have a mode?

3.15 The following data give the annual salaries (in dollars) of governors of 13 western states for 2004 (*Source:* Council of State Governments, *The Book of the States, 2004; The New York Times Almanac,* 2005). The salaries, listed in that order are for AK, HI, CA, OR, WA, ID, MT, WY, CO, UT, NV, AZ, and NM.

85,776	94,780	175,000	93,600	139,087
98,500	93,089	130,000	90,000	100,600
117,000	95,000	110,000		

Find the mean and median for these data.

3.16 The following data give the numbers of car thefts that occurred in a city during the past 12 days.

6	3	7	11	4	3	8	7	2	6	9	15

Find the mean, median, and mode.

3.17 The following data give the revenues (in millions of dollars) for the last available fiscal year for a sample of six charitable organizations that are related to serious diseases (*Forbes*, December 13, 2004). The values listed in that order are for Alzheimer's Association, American Cancer Society, American Diabetes Association, American Heart Association, American Lung Association, and Cystic Fibrosis Foundation.

136	816	192	513	158	152

Compute the mean and median. Do these data have a mode? Why or why not?

3.18 The following table gives the numbers of *takeaways* (recoveries of opponents' fumbles and interceptions of opponents' passes) during the 2004 season for all 16 teams in the National Conference of the National Football League.

Team	Takeaways
Carolina	38
Seattle	35
New Orleans	33
Philadelphia	28
Detroit	24
N.Y. Giants	28
Atlanta	32
Arizona	30
Minnesota	22
Washington	26
Chicago	29
Tampa Bay	27
Green Bay	15
Dallas	22
San Francisco	21
St. Louis	15

Source: USA TODAY, January 5, 2005.

Compute the mean and median for the data on *takeaways*. Do these data have a mode? Why or why not?

3.19 Due to antiquated equipment and frequent windstorms, the town of Oak City often suffers power outages. The following data give the numbers of power outages for the past 12 months.

4 5 7 3 2 0 2 3 2 1 2 4

Compute the mean, median, and mode for these data.

3.20 A brochure from the department of public safety in a northern state recommends that motorists should carry 12 items (flashlights, blankets, and so forth) in their vehicles for emergency use while driving in winter. The following data give the number of items out of these 12 that were carried in their vehicles by 15 randomly selected motorists.

5 3 7 8 0 1 0 5 12 10 7 6 7 11 9

Find the mean, median, and mode for these data. Are the values of these summary measures population parameters or sample statistics? Explain.

3.21 Nixon Corporation manufactures computer monitors. The following data are the numbers of computer monitors produced at the company for a sample of 10 days.

24 32 27 23 35 33 29 40 23 28

Calculate the mean, median, and mode for these data.

3.22 The Tri-City School District has instituted a zero-tolerance policy for students carrying any objects that could be used as weapons. The following data give the number of students suspended during each of the past 12 weeks for violating this school policy.

15 9 12 11 7 6 9 10 14 3 6 5

Calculate the mean, median, and mode for these data.

3.23 The following data give the numbers of casinos in 11 states as of December 21, 2003 (*USA TODAY*, July 16, 2004). The data entered in that order are for CO, IL, IN, IA, LA, MI, MS, MO, NV, NJ, and SD.

44 9 10 13 18 3 29 11 256 12 38

a. Calculate the mean and median for these data.
b. Do these data contain an outlier? If so, drop the outlier and recalculate the mean and median. Which of these two summary measures changes by a larger amount when you drop the outlier?
c. Which is the better summary measure for these data, the mean or the median? Explain.

3.24 The following data, based on the AAA Foundation for Traffic Safety estimates, give the number of fatal crashes caused by road debris from 1999 to 2001 in 10 states with the most such accidents (*USA TODAY*, June 16, 2004). The data entered in that order are for TX, FL, MO, VA, OK, MD, AZ, LA, WI, and IN.

33 17 13 6 6 5 5 4 3 3

Compute the mean and median for these data. Do these data have modes? Why or why not?

***3.25** One property of the mean is that if we know the means and sample sizes of two (or more) data sets, we can calculate the **combined mean** of both (or all) data sets. The combined mean for two data sets is calculated by using the formula

$$\text{Combined mean} = \bar{x} = \frac{n_1\bar{x}_1 + n_2\bar{x}_2}{n_1 + n_2}$$

where n_1 and n_2 are the sample sizes of the two data sets and $\bar{x}_1$ and $\bar{x}_2$ are the means of the two data sets, respectively. Suppose a sample of 10 statistics books gave a mean price of \$95 and a sample of 8 mathematics books gave a mean price of \$104. Find the combined mean. (*Hint:* For this example: $n_1 = 10, n_2 = 8, \bar{x}_1 = \$95, \bar{x}_2 = \$104$.)

***3.26** Twenty business majors and 18 economics majors go bowling. Each student bowls one game. The scorekeeper announces that the mean score for the 18 economics majors is 144 and the mean score for the entire group of 38 students is 150. Find the mean score for the 20 business majors.

***3.27** For any data, the sum of all values is equal to the product of the sample size and mean; that is, $\Sigma x = n\bar{x}$. Suppose the average amount of money spent on shopping by 10 persons during a given week is \$105.50. Find the total amount of money spent on shopping by these 10 persons.

***3.28** The mean 2005 income for five families was \$79,520. What was the total 2005 income of these five families?

***3.29** The mean age of six persons is 46 years. The ages of five of these six persons are 57, 39, 44, 51, and 37 years. Find the age of the sixth person.

*3.30 Seven airline passengers in economy class on the same flight paid an average of \$361 per ticket. Because the tickets were purchased at different times and from different sources, the prices varied. The first five passengers paid \$420, \$210, \$333, \$695, and \$485. The sixth and seventh tickets were purchased by a couple who paid identical fares. What price did each of them pay?

*3.31 Consider the following two data sets.

Data Set I:	12	25	37	8	41
Data Set II:	19	32	44	15	48

Notice that each value of the second data set is obtained by adding 7 to the corresponding value of the first data set. Calculate the mean for each of these two data sets. Comment on the relationship between the two means.

*3.32 Consider the following two data sets.

Data Set I:	4	8	15	9	11
Data Set II:	8	16	30	18	22

Notice that each value of the second data set is obtained by multiplying the corresponding value of the first data set by 2. Calculate the mean for each of these two data sets. Comment on the relationship between the two means.

*3.33 The **trimmed mean** is calculated by dropping a certain percentage of values from each end of a ranked data set. The trimmed mean is especially useful as a measure of central tendency when a data set contains a few outliers at each end. Suppose the following data give the ages of 10 employees of a company:

47	53	38	26	39	49	19	67	31	23

To calculate the 10% trimmed mean, first rank these data values in increasing order; then drop 10% of the smallest values and 10% of the largest values. The mean of the remaining 80% of the values will give the 10% trimmed mean. Note that this data set contains 10 values, and 10% of 10 is 1. Thus, if we drop the smallest value and the largest value from this data set, the mean of the remaining 8 values will be called the 10% trimmed mean. Calculate the 10% trimmed mean for this data set.

*3.34 The following data give the prices (in thousands of dollars) of 20 houses sold recently in a city.

184	297	365	309	245	387	369	438	195	390
323	578	410	679	307	271	457	795	259	590

Find the 20% trimmed mean for this data set.

*3.35 In some applications, certain values in a data set may be considered more important than others. For example, to determine students' grades in a course, an instructor may assign a weight to the final exam twice as much as to each of the other exams. In such cases, it is more appropriate to use the **weighted mean**. In general, for a sequence of n data values $x_1, x_2, \ldots x_n$ that are assigned weights $w_1, w_2, \ldots w_n$, respectively, the ***weighted mean*** is found by the formula

$$\text{Weighted mean} = \frac{\Sigma xw}{\Sigma w}$$

where Σxw is obtained by multiplying each data value by its weight and then adding the products. Suppose an instructor gives two exams and a final, assigning the final exam a weight twice that of each of the other exams. Find the weighted mean for a student who scores 73 and 67 on the first two exams, and 85 on the final. (*Hint:* Here, $x_1 = 73$, $x_2 = 67$, $x_3 = 85$, and $w_1 = w_2 = 1$, and $w_3 = 2$.)

*3.36 When studying phenomena such as inflation or population changes, which involve periodic increases or decreases, the **geometric mean** is used to find the average change over the entire period under study. To calculate the geometric mean of a sequence of n values $x_1, x_2, \ldots x_n$, we multiply them together and then find the nth root of this product. Thus

$$\text{Geometric mean} = \sqrt[n]{x_1 \cdot x_2 \cdot x_3 \cdot \ldots \cdot x_n}$$

Suppose that the inflation rates for the last five years are 4%, 3%, 5%, 6%, and 8%, respectively. Thus at the end of the first year, the price index will be 1.04 times the price index at the beginning of the year, and so on. Find the mean rate of inflation over the five-year period by finding the geometric mean of the data set 1.04, 1.03, 1.05, 1.06, and 1.08. (*Hint:* Here, $n = 5$, $x_1 = 1.04$, $x_2 = 1.03$, etc. Use the $x^{1/n}$ key on your calculator to find the fifth root. Note that the mean inflation rate will be obtained by subtracting 1 from the geometric mean.)

3.2 Measures of Dispersion for Ungrouped Data

The measures of central tendency, such as the mean, median, and mode, do not reveal the whole picture of the distribution of a data set. Two data sets with the same mean may have completely different spreads. The variation among the values of observations for one data set may be much larger or smaller than for the other data set. (Note that the words *dispersion*, *spread*, and *variation* have the same meaning.) Consider the following two data sets on the ages of all workers in each of two small companies.

Company 1:	47	38	35	40	36	45	39
Company 2:		70	33	18	52	27	

The mean age of workers in both these companies is the same, 40 years. If we do not know the ages of individual workers in these two companies and are told only that the mean age of the workers in both companies is the same, we may deduce that the workers in these two companies have a similar age distribution. But as we can observe, the variation in the workers' ages for each of these two companies is very different. As illustrated in the diagram, the ages of the workers in the second company have a much larger variation than the ages of the workers in the first company.

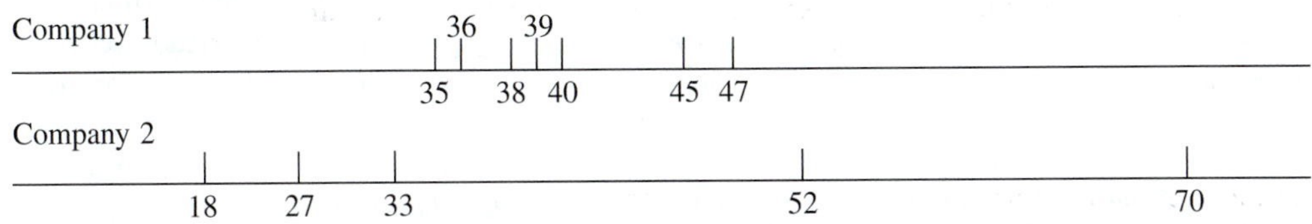

Thus, the mean, median, or mode by itself is usually not a sufficient measure to reveal the shape of the distribution of a data set. We also need a measure that can provide some information about the variation among data values. The measures that help us learn about the spread of a data set are called the **measures of dispersion**. The measures of central tendency and dispersion taken together give a better picture of a data set than the measures of central tendency alone. This section discusses three measures of dispersion: range, variance, and standard deviation.

3.2.1 Range

The **range** is the simplest measure of dispersion to calculate. It is obtained by taking the difference between the largest and the smallest values in a data set.

Finding Range for Ungrouped Data

$$\text{Range} = \text{Largest value} - \text{Smallest value}$$

■ EXAMPLE 3–11

Calculating the range for ungrouped data.

Table 3.4 gives the total areas in square miles of the four western South-Central states of the United States.

Table 3.4

State	Total Area (square miles)
Arkansas	53,182
Louisiana	49,651
Oklahoma	69,903
Texas	267,277

Find the range for this data set.

Solution The maximum total area for a state in this data set is 267,277 square miles, and the smallest area is 49,651 square miles. Therefore,

$$\text{Range} = \text{Largest value} - \text{Smallest value}$$
$$= 267{,}277 - 49{,}651 = \mathbf{217{,}626\ square\ miles}$$

Thus, the total areas of these four states are spread over a range of 217,626 square miles. ■

The range, like the mean, has the disadvantage of being influenced by outliers. In Example 3–11, if the state of Texas with a total area of 267,277 square miles is dropped, the range decreases from 217,626 square miles to 20,252 square miles. Consequently, the range is not a good measure of dispersion to use for a data set that contains outliers.

Another disadvantage of using the range as a measure of dispersion is that its calculation is based on two values only: the largest and the smallest. All other values in a data set are ignored when calculating the range. Thus, the range is not a very satisfactory measure of dispersion.

3.2.2 Variance and Standard Deviation

The **standard deviation** is the most used measure of dispersion. The value of the standard deviation tells how closely the values of a data set are clustered around the mean. In general, a lower value of the standard deviation for a data set indicates that the values of that data set are spread over a relatively smaller range around the mean. In contrast, a larger value of the standard deviation for a data set indicates that the values of that data set are spread over a relatively larger range around the mean.

The *standard deviation is obtained by taking the positive square root of the* **variance**. The variance calculated for population data is denoted by $\boldsymbol{\sigma^2}$ (read as *sigma squared*),[2] and the variance calculated for sample data is denoted by $\boldsymbol{s^2}$. Consequently, the standard deviation calculated for population data is denoted by $\boldsymbol{\sigma}$, and the standard deviation calculated for sample data is denoted by s. Following are what we will call the *basic formulas* that are used to calculate the variance:[3]

$$\sigma^2 = \frac{\Sigma(x - \mu)^2}{N} \quad \text{and} \quad s^2 = \frac{\Sigma(x - \bar{x})^2}{n - 1}$$

where σ^2 is the population variance and s^2 is the sample variance.

The quantity $x - \mu$ or $x - \bar{x}$ in the above formulas is called the *deviation* of the x value from the mean. The sum of the deviations of the x values from the mean is always zero; that is, $\Sigma(x - \mu) = 0$ and $\Sigma(x - \bar{x}) = 0$.

For example, suppose the midterm scores of a sample of four students are 82, 95, 67, and 92. Then, the mean score for these four students is

$$\bar{x} = \frac{82 + 95 + 67 + 92}{4} = 84$$

The deviations of the four scores from the mean are calculated in Table 3.5. As we can observe from the table, the sum of the deviations of the x values from the mean is zero; that is, $\Sigma(x - \bar{x}) = 0$. For this reason we square the deviations to calculate the variance and standard deviation.

Table 3.5

x	$x - \bar{x}$
82	$82 - 84 = -2$
95	$95 - 84 = +11$
67	$67 - 84 = -17$
92	$92 - 84 = +8$
	$\Sigma(x - \bar{x}) = 0$

[2]Note that Σ is uppercase sigma and σ is lowercase sigma of the Greek alphabet.

[3]From the formula for σ^2, it can be stated that the population variance is the mean of the squared deviations of x values from the mean. However, this is not true for the variance calculated for a sample data set.

From the computational point of view, it is easier and more efficient to use *short-cut formulas* to calculate the variance and standard deviation. By using the short-cut formulas, we reduce the computation time and round-off errors. Use of the basic formulas for ungrouped data is illustrated in Section A3.1.1 of Appendix 3.1 of this chapter. The short-cut formulas for calculating the variance and standard deviation are given next.

Short-Cut Formulas for the Variance and Standard Deviation for Ungrouped Data

$$\sigma^2 = \frac{\Sigma x^2 - \frac{(\Sigma x)^2}{N}}{N} \quad \text{and} \quad s^2 = \frac{\Sigma x^2 - \frac{(\Sigma x)^2}{n}}{n-1}$$

where σ^2 is the population variance and s^2 is the sample variance.

The standard deviation is obtained by taking the positive square root of the variance.

Population standard deviation: $\sigma = \sqrt{\sigma^2}$

Sample standard deviation: $s = \sqrt{s^2}$

Note that the denominator in the formula for the population variance is N, but that in the formula for the sample variance it is $n - 1$.[4]

■ EXAMPLE 3–12

Calculating the sample variance and standard deviation for ungrouped data.

The following table, based on *Forbes* magazine's list of the wealthiest people in the world, gives the total wealth (in billions of dollars) of five persons (*USA TODAY*, March 11, 2005).

Billinaire	Total Wealth (billions of dollars)
Bill Gates	46.5
Helen Walton	18.0
Michael Dell	16.0
Keith Rupert Murdoch	7.8
George Soros	7.2

Find the variance and standard deviation for these data.

Solution Let x denote the total wealth (in billions of dollars) of a person. The values of Σx and Σx^2 are calculated in Table 3.6.

Table 3.6

x	x^2
46.5	2162.25
18.0	324.00
16.0	256.00
7.8	60.84
7.2	51.84
$\Sigma x = 95.5$	$\Sigma x^2 = 2854.93$

The calculation of the variance involves the following steps.

[4]The reason that the denominator in the sample formula is $n - 1$ and not n follows: The sample variance underestimates the population variance when the denominator in the sample formula for variance is n. However, the sample variance does not underestimate the population variance if the denominator in the sample formula for variance is $n - 1$. In Chapter 8 we will learn that $n - 1$ is called the degrees of freedom.

Step 1. *Calculate* Σx.

The sum of the values in the first column of Table 3.6 gives the value of Σx, which is 95.5.

Step 2. *Find* Σx^2.

The value of Σx^2 is obtained by squaring each value of x and then adding the squared values. The results of this step are shown in the second column of Table 3.6. Notice that $\Sigma x^2 = 2854.93$.

Step 3. *Determine the variance.*

Substitute all the values in the variance formula and simplify. Because the given data are on the wealth of a sample of five persons, we use the formula for the sample variance.

$$s^2 = \frac{\Sigma x^2 - \frac{(\Sigma x)^2}{n}}{n - 1} = \frac{2854.93 - \frac{(95.5)^2}{5}}{5 - 1} = \frac{2854.93 - 1824.05}{4} = \mathbf{257.72}$$

Step 4. *Obtain the standard deviation.*

The standard deviation is obtained by taking the (positive) square root of the variance.

$$s = \sqrt{257.72} = \mathbf{16.05366} = \mathbf{\$16.05\ billion}$$

Thus, the standard deviation of the wealth of these five individuals is $16.05 billion. ■

Two Observations ▶

1. **The values of the variance and the standard deviation are never negative**. That is, the numerator in the formula for the variance should never produce a negative value. Usually the values of the variance and standard deviation are positive, but if a data set has no variation, then the variance and standard deviation are both zero. For example, if four persons in a group are the same age—say, 35 years—then the four values in the data set are

   35 35 35 35

   If we calculate the variance and standard deviation for these data, their values are zero. This is because there is no variation in the values of this data set.

2. **The measurement units of variance are always the square of the measurement units of the original data**. This is so because the original values are squared to calculate the variance. In Example 3–12, the measurement units of the original data are billions of dollars. However, the measurement units of the variance are squared billions of dollars, which, of course, does not make any sense. Thus, the variance of the wealth of these five persons in Example 3–12 is 257.72 squared billion dollars. But the measurement units of the standard deviation are the same as the measurement units of the original data because the standard deviation is obtained by taking the square root of the variance.

■ EXAMPLE 3–13

Calculating the population variance and standard deviation for ungrouped data.

Following are the 2005 earnings (in thousands of dollars) before taxes for all six employees of a small company.

48.50 38.40 65.50 22.60 79.80 54.60

Calculate the variance and standard deviation for these data.

Solution Let x denote the 2005 earnings before taxes of an employee of this company. The values of Σx and Σx^2 are calculated in Table 3.7.

Table 3.7

x	x^2
48.50	2352.25
38.40	1474.56
65.50	4290.25
22.60	510.76
79.80	6368.04
54.60	2981.16
$\Sigma x = 309.40$	$\Sigma x^2 = 17{,}977.02$

Because the data are on earnings of *all* employees of this company, we use the population formula to compute the variance. Thus, the variance is

$$\sigma^2 = \frac{\Sigma x^2 - \frac{(\Sigma x)^2}{N}}{N} = \frac{17{,}977.02 - \frac{(309.40)^2}{6}}{6} = \mathbf{337.0489}$$

The standard deviation is obtained by taking the (positive) square root of the variance:

$$\sigma = \sqrt{337.0489} = \mathbf{\$18.359\ thousand = \$18{,}359}$$

Thus, the standard deviation of the 2005 earnings of all six employees of this company is $18,359. ■

Note that Σx^2 is not the same as $(\Sigma x)^2$. The value of Σx^2 is obtained by squaring the x values and then adding them. The value of $(\Sigma x)^2$ is obtained by squaring the value of Σx. ◀ *Warning*

The uses of the standard deviation are discussed in Section 3.4. Later chapters explain how the mean and the standard deviation taken together can help in making inferences about the population.

3.2.3 Population Parameters and Sample Statistics

A numerical measure such as the mean, median, mode, range, variance, or standard deviation calculated for a population data set is called a *population parameter*, or simply a **parameter**. A summary measure calculated for a sample data set is called a *sample statistic*, or simply a **statistic**. Thus, μ and σ are population parameters, and $\bar{x}$ and s are sample statistics. As an illustration, $\bar{x} = 20{,}354.83$ in Example 3–1 is a sample statistic, and $\mu = 45.25$ years in Example 3–2 is a population parameter. Similarly, $s = \$16.05$ billion in Example 3–12 is a sample statistic, whereas $\sigma = \$18{,}359$ in Example 3–13 is a population parameter.

EXERCISES

CONCEPTS AND PROCEDURES

3.37 The range, as a measure of spread, has the disadvantage of being influenced by outliers. Illustrate this with an example.

3.38 Can the standard deviation have a negative value? Explain.

3.39 When is the value of the standard deviation for a data set zero? Give one example. Calculate the standard deviation for the example and show that its value is zero.

3.40 Briefly explain the difference between a population parameter and a sample statistic. Give one example of each.

3.41 The following data set belongs to a population:

5 −7 2 0 −9 16 10 7

Calculate the range, variance, and standard deviation.

3.42 The following data set belongs to a sample:

14 18 −10 8 8 −16

Calculate the range, variance, and standard deviation.

APPLICATIONS

3.43 The following data give the number of shoplifters apprehended during each of the past eight weeks at a large department store.

7 10 8 3 15 12 6 11

a. Find the mean for these data. Calculate the deviations of the data values from the mean. Is the sum of these deviations zero?
b. Calculate the range, variance, and standard deviation.

3.44 The following data give the prices of seven textbooks randomly selected from a university bookstore.

$89 $67 $104 $113 $36 $121 $147

a. Find the mean for these data. Calculate the deviations of the data values from the mean. Is the sum of these deviations zero?
b. Calculate the range, variance, and standard deviation.

3.45 The following data give the numbers of car thefts that occurred in a city in the past 12 days.

6 3 7 11 4 3 8 7 2 6 9 15

Calculate the range, variance, and standard deviation.

3.46 During the 2004 presidential election campaign, spending on television commercials was high, particularly in key states where the vote was expected to be close. The following data give the expenditures on television commercials (in millions of dollars) by all candidates in 10 states where such spending was the highest. The data, entered in that order, are for Florida, California, Ohio, Pennsylvania, Missouri, New Jersey, Delaware, Michigan, Wisconsin, and North Carolina (*USA TODAY*, November 26, 2004).

236.7 190.7 166.8 133.9 98.0 88.3 65.3 61.6 54.4 51.6

Find the range, variance, and standard deviation for these data.

3.47 The following data give the numbers of pieces of junk mail received by 10 families during the past month.

41 33 28 21 29 19 14 31 39 36

Find the range, variance, and standard deviation.

3.48 The following data give the number of highway collisions with large wild animals, such as deer or moose, in one of the northeastern states during each week of a nine-week period.

7 10 3 8 2 5 7 4 9

Find the range, variance, and standard deviation.

3.49 Attacks by stinging insects, such as bees or wasps, may become medical emergencies if either the victim is allergic to venom or multiple stings are involved. The following data give the number of patients treated each week for such stings in a large regional hospital during 13 weeks last summer.

1 5 2 3 0 4 1 7 0 1 2 0 1

Compute the range, variance, and standard deviation for these data.

3.50 The following data give the number of hot dogs consumed by 10 participants in a hot-dog-eating contest.

21 17 32 8 20 15 17 23 9 18

Calculate the range, variance, and standard deviation for these data.

3.51 Following are the temperatures (in degrees Fahrenheit) observed during eight wintry days in a midwestern city:

23	14	6	−7	−2	11	16	19

Compute the range, variance, and standard deviation.

3.52 The following data give the numbers of hours spent partying by 10 randomly selected college students during the past week.

7	14	5	0	9	7	10	4	0	8

Compute the range, variance, and standard deviation.

3.53 The following data, based on *Forbes* Magazine's rankings of the wealthiest people in the world, give the net worth (in billions of dollars) of the 10 wealthiest people in the world (*USA TODAY*, March 11, 2005). The data, entered in that order, are for Bill Gates, Warren Buffett, Lakshmi Mittal, Carlos Slim Helu, Prince Alwaleed Bin Talal Alsaud, Ingvar Kamprad, Paul Allen, Karl Albrecht, Lawrence Ellison, and S. Robson Walton.

46.5	44.0	25.0	23.8	23.7	23.0	21.0	18.5	18.4	18.3

Find the range, variance, and standard deviation for these data.

3.54 The following data give the average speeds (rounded to the nearest mile per hour) at the Indianapolis 500 auto race for the years 1995 to 2004 (*The New York Times 2005 Almanac*).

154	148	146	145	153	168	142	166	156	139

Find the range, variance, and standard deviation for these data.

3.55 The following data give the hourly wage rates of eight employees of a company.

12	12	12	12	12	12	12	12

Calculate the standard deviation. Is its value zero? If yes, why?

3.56 The following data are the ages (in years) of six students.

19	19	19	19	19	19

Calculate the standard deviation. Is its value zero? If yes, why?

***3.57** One disadvantage of the standard deviation as a measure of dispersion is that it is a measure of absolute variability and not of relative variability. Sometimes we may need to compare the variability of two different data sets that have different units of measurement. The **coefficient of variation** is one such measure. The coefficient of variation, denoted by CV, expresses standard deviation as a percentage of the mean and is computed as follows:

$$\text{For population data:} \quad CV = \frac{\sigma}{\mu} \times 100\%$$

$$\text{For sample data:} \quad CV = \frac{s}{\bar{x}} \times 100\%$$

The yearly salaries of all employees who work for a company have a mean of $62,350 and a standard deviation of $6820. The years of experience for the same employees have a mean of 15 years and a standard deviation of 2 years. Is the relative variation in the salaries greater or less than that in years of experience for these employees?

***3.58** The SAT scores of 100 students have a mean of 975 and a standard deviation of 105. The GPAs of the same 100 students have a mean of 3.16 and a standard deviation of .22. Is the relative variation in SAT scores greater or less than that in GPAs?

***3.59** Consider the following two data sets.

Data Set I:	12	25	37	8	41
Data Set II:	19	32	44	15	48

Note that each value of the second data set is obtained by adding 7 to the corresponding value of the first data set. Calculate the standard deviation for each of these two data sets using the formula for sample data. Comment on the relationship between the two standard deviations.

***3.60** Consider the following two data sets.

Data Set I:	4	8	15	9	11
Data Set II:	8	16	30	18	22

Note that each value of the second data set is obtained by multiplying the corresponding value of the first data set by 2. Calculate the standard deviation for each of these two data sets using the formula for population data. Comment on the relationship between the two standard deviations.

3.3 Mean, Variance, and Standard Deviation for Grouped Data

In Sections 3.1.1 and 3.2.2, we learned how to calculate the mean, variance, and standard deviation for ungrouped data. In this section, we will learn how to calculate the mean, variance, and standard deviation for grouped data.

3.3.1 Mean for Grouped Data

We learned in Section 3.1.1 that the mean is obtained by dividing the sum of all values by the number of values in a data set. However, if the data are given in the form of a frequency table, we no longer know the values of individual observations. Consequently, in such cases, we cannot obtain the sum of individual values. We find an approximation for the sum of these values using the procedure explained in the next paragraph and example. The formulas used to calculate the mean for grouped data follow.

Calculating Mean for Grouped Data

Mean for population data: $$\mu = \frac{\Sigma mf}{N}$$

Mean for sample data: $$\bar{x} = \frac{\Sigma mf}{n}$$

where m is the midpoint and f is the frequency of a class.

To calculate the mean for grouped data, first find the midpoint of each class and then multiply the midpoints by the frequencies of the corresponding classes. The sum of these products, denoted by Σmf, gives an approximation for the sum of all values. To find the value of the mean, divide this sum by the total number of observations in the data.

■ EXAMPLE 3–14

Calculating the population mean for grouped data.

Table 3.8 gives the frequency distribution of the daily commuting times (in minutes) from home to work for *all* 25 employees of a company.

Table 3.8

Daily Commuting Time (minutes)	Number of Employees
0 to less than 10	4
10 to less than 20	9
20 to less than 30	6
30 to less than 40	4
40 to less than 50	2

Calculate the mean of the daily commuting times.

Solution Note that because the data set includes *all* 25 employees of the company, it represents the population. Table 3.9 shows the calculation of Σmf. Note that in Table 3.9, m denotes the midpoints of the classes.

Table 3.9

Daily Commuting Time (minutes)	f	m	mf
0 to less than 10	4	5	20
10 to less than 20	9	15	135
20 to less than 30	6	25	150
30 to less than 40	4	35	140
40 to less than 50	2	45	90
	$N = 25$		$\Sigma mf = 535$

To calculate the mean, we first find the midpoint of each class. The class midpoints are recorded in the third column of Table 3.9. The products of the midpoints and the corresponding frequencies are listed in the fourth column. The sum of the fourth column values, denoted by Σmf, gives the approximate total daily commuting time (in minutes) for all 25 employees. The mean is obtained by dividing this sum by the total frequency. Therefore,

$$\mu = \frac{\Sigma mf}{N} = \frac{535}{25} = \mathbf{21.40 \text{ minutes}}$$

Thus, the employees of this company spend an average of 21.40 minutes a day commuting from home to work. ■

What do the numbers 20, 135, 150, 140, and 90 in the column labeled mf in Table 3.9 represent? We know from this table that 4 employees spend 0 to less than 10 minutes commuting per day. If we assume that the time spent commuting by these 4 employees is evenly spread in the interval 0 to less than 10, then the midpoint of this class (which is 5) gives the mean time spent commuting by these 4 employees. Hence, $4 \times 5 = 20$ is the approximate total time (in minutes) spent commuting per day by these 4 employees. Similarly, 9 employees spend 10 to less than 20 minutes commuting per day, and the total time spent commuting by these 9 employees is approximately 135 minutes a day. The other numbers in this column can be interpreted the same way. Note that these numbers give the approximate commuting times for these employees based on the assumption of an even spread within classes. The total commuting time for all 25 employees is approximately 535 minutes. Consequently, 21.40 minutes is an approximate and not the exact value of the mean. We can find the exact value of the mean only if we know the exact commuting time for each of the 25 employees of the company.

■ EXAMPLE 3–15

Calculating the sample mean for grouped data.

Table 3.10 gives the frequency distribution of the number of orders received each day during the past 50 days at the office of a mail-order company.

Table 3.10

Number of Orders	Number of Days
10–12	4
13–15	12
16–18	20
19–21	14

Calculate the mean.

Solution Because the data set includes only 50 days, it represents a sample. The value of Σmf is calculated in Table 3.11.

Table 3.11

Number of Orders	f	m	mf
10–12	4	11	44
13–15	12	14	168
16–18	20	17	340
19–21	14	20	280
	$n = 50$		$\Sigma mf = 832$

The value of the sample mean is

$$\bar{x} = \frac{\Sigma mf}{n} = \frac{832}{50} = \textbf{16.64 orders}$$

Thus, this mail-order company received an average of 16.64 orders per day during these 50 days. ■

3.3.2 Variance and Standard Deviation for Grouped Data

Following are what we will call the *basic formulas* used to calculate the population and sample variances for grouped data:

$$\sigma^2 = \frac{\Sigma f(m - \mu)^2}{N} \quad \text{and} \quad s^2 = \frac{\Sigma f(m - \bar{x})^2}{n - 1}$$

where σ^2 is the population variance, s^2 is the sample variance, and m is the midpoint of a class.

In either case, the standard deviation is obtained by taking the positive square root of the variance.

Again, the *short-cut formulas* are more efficient for calculating the variance and standard deviation. Section A3.1.2 of Appendix 3.1 at the end of this chapter shows how to use the basic formulas to calculate the variance and standard deviation for grouped data.

Short-Cut Formulas for the Variance and Standard Deviation for Grouped Data

$$\sigma^2 = \frac{\Sigma m^2 f - \dfrac{(\Sigma mf)^2}{N}}{N} \quad \text{and} \quad s^2 = \frac{\Sigma m^2 f - \dfrac{(\Sigma mf)^2}{n}}{n - 1}$$

where σ^2 is the population variance, s^2 is the sample variance, and m is the midpoint of a class.

The standard deviation is obtained by taking the positive square root of the variance.

Population standard deviation: $\sigma = \sqrt{\sigma^2}$

Sample standard deviation: $s = \sqrt{s^2}$

Examples 3–16 and 3–17 illustrate the use of these formulas to calculate the variance and standard deviation.

Calculating the population variance and standard deviation for grouped data.

■ EXAMPLE 3–16

The following data, reproduced from Table 3.8 of Example 3–14, give the frequency distribution of the daily commuting times (in minutes) from home to work for all 25 employees of a company.

Daily Commuting Time (minutes)	Number of Employees
0 to less than 10	4
10 to less than 20	9
20 to less than 30	6
30 to less than 40	4
40 to less than 50	2

Calculate the variance and standard deviation.

Solution All four steps needed to calculate the variance and standard deviation for grouped data are shown after Table 3.12.

Table 3.12

Daily Commuting Time (minutes)	f	m	mf	m^2f
0 to less than 10	4	5	20	100
10 to less than 20	9	15	135	2025
20 to less than 30	6	25	150	3750
30 to less than 40	4	35	140	4900
40 to less than 50	2	45	90	4050
	$N = 25$		$\Sigma mf = 535$	$\Sigma m^2f = 14{,}825$

Step 1. *Calculate the value of* Σmf.

To calculate the value of Σmf, first find the midpoint m of each class (see the third column in Table 3.12) and then multiply the corresponding class midpoints and class frequencies (see the fourth column). The value of Σmf is obtained by adding these products. Thus,

$$\Sigma mf = 535$$

Step 2. *Find the value of* Σm^2f.

To find the value of Σm^2f, square each m value and multiply this squared value of m by the corresponding frequency (see the fifth column in Table 3.12). The sum of these products (that is, the sum of the fifth column) gives Σm^2f. Hence,

$$\Sigma m^2f = 14{,}825$$

Step 3. *Calculate the variance.*

Because the data set includes all 25 employees of the company, it represents the population. Therefore, we use the formula for the population variance:

$$\sigma^2 = \frac{\Sigma m^2f - \frac{(\Sigma mf)^2}{N}}{N} = \frac{14{,}825 - \frac{(535)^2}{25}}{25} = \frac{3376}{25} = \mathbf{135.04}$$

Step 4. *Calculate the standard deviation.*

To obtain the standard deviation, take the (positive) square root of the variance.

$$\sigma = \sqrt{\sigma^2} = \sqrt{135.04} = \mathbf{11.62\ minutes}$$

Thus, the standard deviation of the daily commuting times for these employees is 11.62 minutes.

Note that the values of the variance and standard deviation calculated in Example 3–16 for grouped data are approximations. The exact values of the variance and standard deviation can be obtained only by using the ungrouped data on the daily commuting times of the 25 employees.

■ EXAMPLE 3–17

Calculating the sample variance and standard deviation for grouped data.

The following data, reproduced from Table 3.10 of Examle 3–15, give the frequency distribution of the number of orders received each day during the past 50 days at the office of a mail-order company.

Number of Orders	f
10–12	4
13–15	12
16–18	20
19–21	14

Calculate the variance and standard deviation.

Solution All the information required for the calculation of the variance and standard deviation appears in Table 3.13.

Table 3.13

Number of Orders	f	m	mf	m^2f
10–12	4	11	44	484
13–15	12	14	168	2352
16–18	20	17	340	5780
19–21	14	20	280	5600
	$n = 50$		$\Sigma mf = 832$	$\Sigma m^2f = 14{,}216$

Because the data set includes only 50 days, it represents a sample. Hence, we use the sample formulas to calculate the variance and standard deviation. By substituting the values into the formula for the sample variance, we obtain

$$s^2 = \frac{\Sigma m^2 f - \frac{(\Sigma mf)^2}{n}}{n - 1} = \frac{14{,}216 - \frac{(832)^2}{50}}{50 - 1} = \mathbf{7.5820}$$

Hence, the standard deviation is

$$s = \sqrt{s^2} = \sqrt{7.5820} = \mathbf{2.75\ orders}$$

Thus, the standard deviation of the number of orders received at the office of this mail-order company during the past 50 days is 2.75. ■

EXERCISES

■ CONCEPTS AND PROCEDURES

3.61 Are the values of the mean and standard deviation that are calculated using grouped data exact or approximate values of the mean and standard deviation, respectively? Explain.

3.62 Using the population formulas, calculate the mean, variance, and standard deviation for the following grouped data.

x	2–4	5–7	8–10	11–13	14–16
f	5	9	14	7	5

3.63 Using the sample formulas, find the mean, variance, and standard deviation for the grouped data displayed in the following table.

x	f
0 to less than 4	17
4 to less than 8	23
8 to less than 12	15
12 to less than 16	11
16 to less than 20	8
20 to less than 24	6

APPLICATIONS

3.64 The following table gives the frequency distribution of the amounts of telephone bills for October 2005 for a sample of 50 families.

Amount of Telephone Bill (dollars)	Number of Families
40 to less than 70	9
70 to less than 100	11
100 to less than 130	16
130 to less than 160	10
160 to less than 190	4

Calculate the mean, variance, and standard deviation.

3.65 The following table gives the frequency distribution of the number of hours spent per week playing video games by all 60 students of the eighth grade at a school.

Hours Per Week	Number of Students
0 to less than 5	7
5 to less than 10	12
10 to less than 15	15
15 to less than 20	13
20 to less than 25	8
25 to less than 30	5

Find the mean, variance, and standard deviation.

3.66 The following table gives the grouped data on the weights of all 100 babies born at a hospital in 2005.

Weight (pounds)	Number of Babies
3 to less than 5	5
5 to less than 7	30
7 to less than 9	40
9 to less than 11	20
11 to less than 13	5

Find the mean, variance, and standard deviation.

3.67 The following table gives the frequency distribution of the total miles driven during 2005 by 300 car owners.

Miles Driven in 2002 (in thousands)	Number of Car Owners
0 to less than 5	7
5 to less than 10	26
10 to less than 15	59
15 to less than 20	71
20 to less than 25	62
25 to less than 30	39
30 to less than 35	22
35 to less than 40	14

Find the mean, variance, and standard deviation. Give a brief interpretation of the values in the column labeled mf in your table of calculations. What does Σmf represent?

3.68 The following table gives information on the amounts (in dollars) of electric bills for August 2005 for a sample of 50 families.

Amount of Electric Bill (dollars)	Number of Families
0 to less than 20	5
20 to less than 40	16
40 to less than 60	11
60 to less than 80	10
80 to less than 100	8

Find the mean, variance, and standard deviation. Give a brief interpretation of the values in the column labeled mf in your table of calculations. What does Σmf represent?

3.69 For 50 airplanes that arrived late at an airport during a week, the time by which they were late was observed. In the following table, x denotes the time (in minutes) by which an airplane was late and f denotes the number of airplanes.

x	f
0 to less than 20	14
20 to less than 40	18
40 to less than 60	9
60 to less than 80	5
80 to less than 100	4

Find the mean, variance, and standard deviation.

3.70 The following table gives the frequency distribution of the number of errors committed by a college baseball team in all of the 45 games that it played during the 2005–2006 season.

Number of Errors	Number of Games
0	11
1	14
2	9
3	7
4	3
5	1

Find the mean, variance, and standard deviation. (*Hint:* The classes in this example are single-valued. These values of classes will be used as values of m in the formulas for the mean, variance, and standard deviation.)

3.71 During fall 2004, oil prices fluctuated a lot due to wars, political unrests, and storm damages in some oil-producing nations. The following data give the spot prices (in dollars) per barrel of crude oil for 15 business days from October 20 to November 9, 2004.

54.92	54.47	55.17	55.18	55.95	52.47	50.93	
51.74	50.14	49.63	50.89	48.83	49.62	49.09	47.38

a. Find the mean for these data.
b. Construct a frequency distribution table for these data using a class width of 2.00 and the lower boundary of the first class equal to 47.00.
c. Using the method of Section 3.3.1, find the mean of the grouped data of part b.
d. Compare your means from parts a and c. If the two means are not equal, then explain why they differ.

3.4 Use of Standard Deviation

By using the mean and standard deviation, we can find the proportion or percentage of the total observations that fall within a given interval about the mean. This section briefly discusses Chebyshev's theorem and the empirical rule, both of which demonstrate this use of the standard deviation.

3.4.1 Chebyshev's Theorem

Chebyshev's theorem gives a lower bound for the area under a curve between two points that are on opposite sides of the mean and at the same distance from the mean.

Definition

Chebyshev's Theorem For any number k greater than 1, at least $(1 - 1/k^2)$ of the data values lie within k standard deviations of the mean.

Figure 3.5 illustrates Chebyshev's theorem.

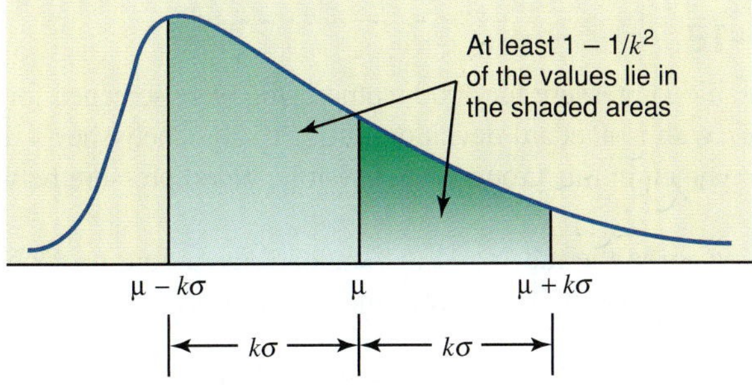

Figure 3.5 Chebyshev's theorem.

Thus, for example, if $k = 2$, then

$$1 - \frac{1}{k^2} = 1 - \frac{1}{(2)^2} = 1 - \frac{1}{4} = 1 - .25 = .75 \text{ or } 75\%$$

Therefore, according to Chebyshev's theorem, at least .75 or 75% of the values of a data set lie within two standard deviations of the mean. This is shown in Figure 3.6 on the next page.

If $k = 3$, then,

$$1 - \frac{1}{k^2} = 1 - \frac{1}{(3)^2} = 1 - \frac{1}{9} = 1 - .11 = .89 \text{ or } 89\% \text{ approximately}$$

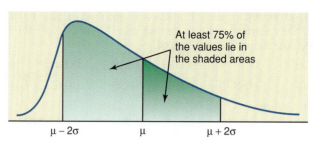

Figure 3.6 Percentage of values within two standard deviations of the mean for Chebyshev's theorem.

According to Chebyshev's theorem, at least .89 or 89% of the values fall within three standard deviations of the mean. This is shown in Figure 3.7.

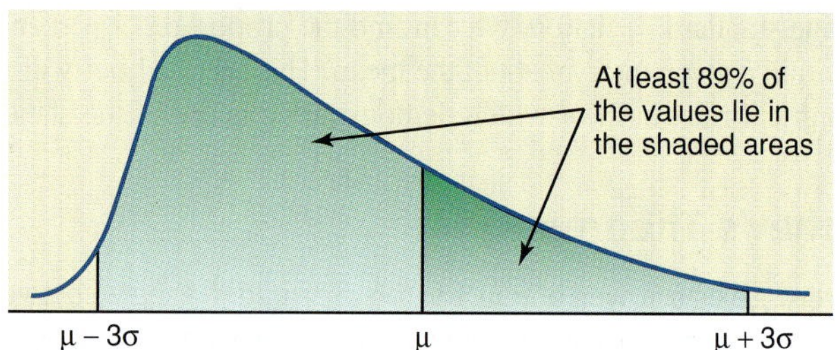

Figure 3.7 Percentage of values within three standard deviations of the mean for Chebyshev's theorem.

Although in Figures 3.5 through 3.7 we have used the population notation for the mean and standard deviation, the theorem applies to both sample and population data. Note that Chebyshev's theorem is applicable to a distribution of any shape. However, Chebyshev's theorem can be used only for $k > 1$. This is so because when $k = 1$, the value of $1 - 1/k^2$ is zero, and when $k < 1$, the value of $1 - 1/k^2$ is negative.

■ EXAMPLE 3–18

Applying Chebyshev's theorem.

The average systolic blood pressure for 4000 women who were screened for high blood pressure was found to be 187 with a standard deviation of 22. Using Chebyshev's theorem, find at least what percentage of women in this group have a systolic blood pressure between 143 and 231.

Solution Let μ and σ be the mean and the standard deviation, respectively, of the systolic blood pressures of these women. Then, from the given information,

$$\mu = 187 \quad \text{and} \quad \sigma = 22$$

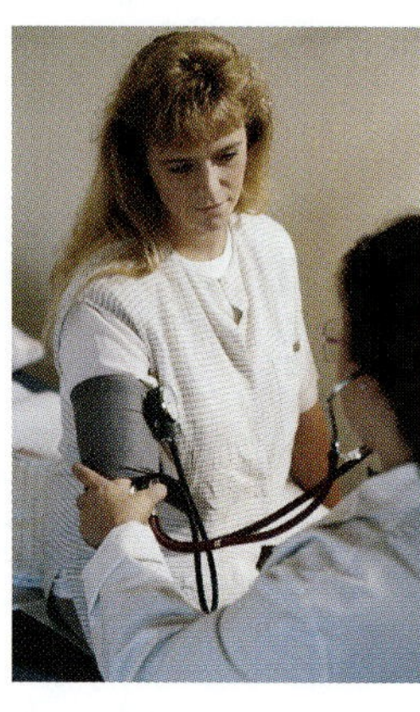

To find the percentage of women whose systolic blood pressures are between 143 and 231, the first step is to determine k. As shown below, each of the two points, 143 and 231, is 44 units away from the mean.

$$|\leftarrow 143 - 187 = -44 \rightarrow|\leftarrow 231 - 187 = 44 \rightarrow|$$

143 $\qquad$ $\mu = 187$ $\qquad$ 231

The value of k is obtained by dividing the distance between the mean and each point by the standard deviation. Thus,

$$k = 44/22 = 2$$

$$1 - \frac{1}{k^2} = 1 - \frac{1}{(2)^2} = 1 - \frac{1}{4} = 1 - .25 = .75 \text{ or } \mathbf{75\%}$$

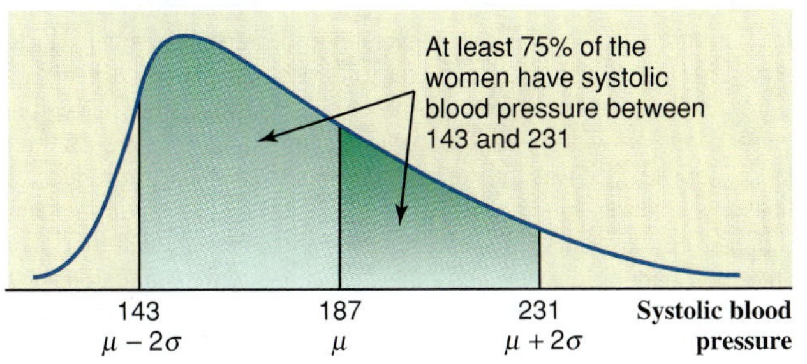

Figure 3.8 Percentage of women with systolic blood pressure between 143 and 231.

Hence, according to Chebyshev's theorem, at least 75% of the women have systolic blood pressure between 143 and 231. This percentage is shown in Figure 3.8. ■

3.4.2 Empirical Rule

Whereas Chebyshev's theorem is applicable to any kind of distribution, the **empirical rule** applies only to a specific type of distribution called a *bell-shaped distribution*, as shown in Figure 3.9. More will be said about such a distribution in Chapter 6, where it is called a *normal curve.* In this section, only the following three rules for the curve are given.

Empirical Rule For a bell-shaped distribution, approximately

1. 68% of the observations lie within one standard deviation of the mean
2. 95% of the observations lie within two standard deviations of the mean
3. 99.7% of the observations lie within three standard deviations of the mean

Figure 3.9 illustrates the empirical rule. Again, the empirical rule applies to population data as well as to sample data.

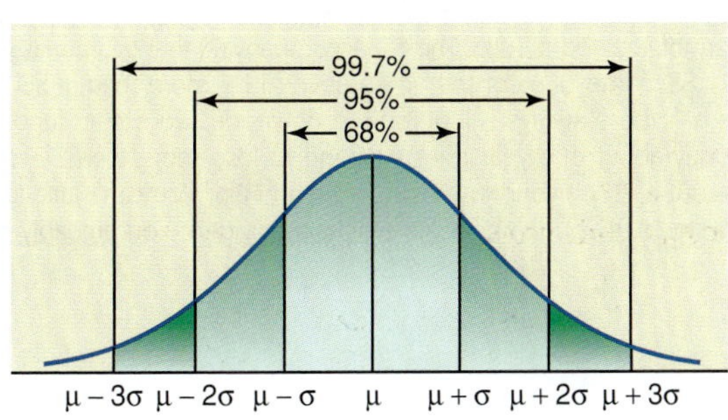

Figure 3.9 Illustration of the empirical rule.

■ EXAMPLE 3–19

Applying the empirical rule.

The age distribution of a sample of 5000 persons is bell-shaped with a mean of 40 years and a standard deviation of 12 years. Determine the approximate percentage of people who are 16 to 64 years old.

Solution We use the empirical rule to find the required percentage because the distribution of ages follows a bell-shaped curve. From the given information, for this distribution,

$$\bar{x} = 40 \text{ years} \quad \text{and} \quad s = 12 \text{ years}$$

Each of the two points, 16 and 64, is 24 units away from the mean. Dividing 24 by 12, we convert the distance between each of the two points and the mean in terms of standard deviations. Thus, the distance between 16 and 40 and between 40 and 64 is each equal to $2s$.

CASE STUDY 3–3

HERE COMES THE SD

When your servant first became a *Fortune* writer several decades ago, it was hard doctrine that "several" meant three to eight, also that writers must not refer to "gross national product" without pausing to define this arcane term. GNP was in fact a relatively new concept at the time, having been introduced to the country only several years previously—in Roosevelt's 1944 budget message—so the presumption that readers had to be told repeatedly it was the "value of all goods and services produced by the economy" seemed entirely reasonable to this young writer, who personally had to look up the definition every time.

Numeracy lurches on. Nowadays the big question for editors is whether an average college-educated bloke needs a handhold when confronted with the term "standard deviation." The SD is suddenly onstage because the Securities and Exchange Commission is wondering aloud whether investment companies should be required to tell investors the standard deviation of their mutual funds' total returns over various past periods. Barry Barbash, SEC director of investment management, favors the requirement but confessed to the Washington *Post* that he worries about investors who will think a standard deviation is the dividing line on a highway or something.

The view around here is that the SEC is performing a noble service, but only partly because the requirement would enhance folks' insights into mutual funds. The commission's underlying idea is to give investors a better and more objective measure than is now available of the risk associated with different kinds of portfolios. The SD is a measure of variability, and funds with unusually variable returns—sometimes very high, sometimes very low—are presumed to be more risky.

What one really likes about the proposal, however, is the prospect that it will incentivize millions of greedy Americans to learn a little elementary statistics. One already has a list of issues that could be discussed much more thrillingly if only your average liberal arts graduate had a glimmer about the SD and the normal curve. The bell-shaped normal curve, or rather, the area underneath the curve, shows you how Providence arranged for things to be distributed in our world—with people's heights, or incomes, or IQs, or investment returns bunched around middling outcomes, and fewer and fewer cases as you move down and out toward the extremes. A line down the center of the curve represents the mean outcome, and deviations from the mean are measured by the SD.

An amazing property of the SD is that exactly 68.26% of all normally distributed data are within one SD of the mean. We once asked a professor of statistics a question that seemed to us quite profound, to wit, why that particular figure? The Prof answered dismissively that God had decided on 68.26% for exactly the same reason He had landed on 3.14 as the ratio between circumferences and diameters—because He just felt like it. The Almighty has also proclaimed that 95.44% of all data are within two SDs of the mean, and 99.73% within three SDs. When you know the mean and SD of some outcome, you can instantly establish the percentage probability of its occurrence. White men's heights in the U.S. average 69.2 inches, with an SD of 2.8 inches (according to the National Center for Health Statistics), which means that a 6-foot-5 chap is in the 99th percentile. In 1994, scores on the verbal portion of the Scholastic Assessment Test had a mean of 423 and an SD of 113, so if you scored 649—two SDs above the mean—you were in the 95th percentile.

As the SEC is heavily hinting, average outcomes are interesting but for many purposes inadequate; one also yearns to know the variability around that average. From 1926 through 1994, the S&P 500 had an average annual return of just about 10%. The SD accompanying that figure was just about 20%. Since returns will be within 1 SD some 68% of the time, they will be more than 1 SD from the mean 32% of the time. And since half these swings will be on the downside, we expect fund owners to lose more than 10% of their money about one year out of six and to lose more than 30% (two SDs below the mean) about one year out of 20. If your time horizon is short and you can't take losses like that, you arguably don't belong in stocks. If you think SDs are highway dividers, you arguably don't belong in cars.

Source: Daniel Seligman, "Here comes the SD," *Fortune*, May 15, 1995.

Figure 3.10 Percentage of people who are 16 to 64 years old.

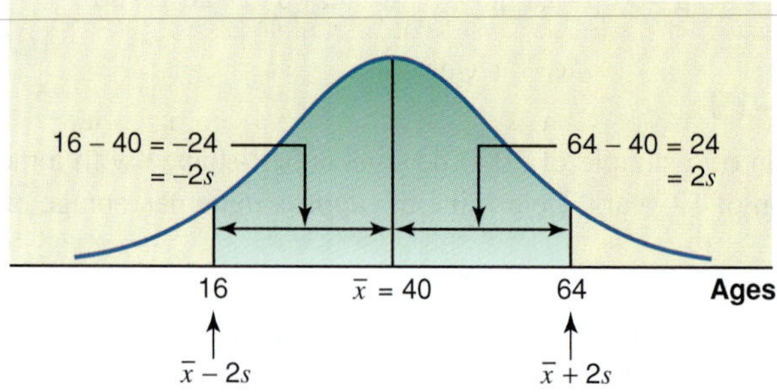

Consequently, as shown in Figure 3.10, the area from 16 to 64 is the area from $\bar{x} - 2s$ to $\bar{x} + 2s$.

Because the area within two standard deviations of the mean is approximately 95% for a bell-shaped curve, approximately **95%** of the people in the sample are 16 to 64 years old. ■

EXERCISES

CONCEPTS AND PROCEDURES

3.72 Briefly explain Chebyshev's theorem and its applications.

3.73 Briefly explain the empirical rule. To what kind of distribution is it applied?

3.74 A sample of 2000 observations has a mean of 74 and a standard deviation of 12. Using Chebyshev's theorem, find at least what percentage of the observations fall in the intervals $\bar{x} \pm 2s$, $\bar{x} \pm 2.5s$, and $\bar{x} \pm 3s$. Note that here $\bar{x} \pm 2s$ represents the interval $\bar{x} - 2s$ to $\bar{x} + 2s$, and so on.

3.75 A large population has a mean of 230 and a standard deviation of 41. Using Chebyshev's theorem, find at least what percentage of the observations fall in the intervals $\mu \pm 2\sigma$, $\mu \pm 2.5\sigma$, and $\mu \pm 3\sigma$.

3.76 A large population has a mean of 310 and a standard deviation of 37. Using the empirical rule, find what percentage of the observations fall in the intervals $\mu \pm 1\sigma$, $\mu \pm 2\sigma$, and $\mu \pm 3\sigma$.

3.77 A sample of 3000 observations has a mean of 82 and a standard deviation of 16. Using the empirical rule, find what percentage of the observations fall in the intervals $\bar{x} \pm 1s$, $\bar{x} \pm 2s$, and $\bar{x} \pm 3s$.

APPLICATIONS

3.78 The mean time taken by all participants to run a road race was found to be 220 minutes with a standard deviation of 20 minutes. Using Chebyshev's theorem, find the percentage of runners who ran this road race in

a. 180 to 260 minutes **b.** 160 to 280 minutes **c.** 170 to 270 minutes

3.79 The 2005 gross sales of all firms in a large city have a mean of \$2.3 million and a standard deviation of \$.6 million. Using Chebyshev's theorem, find at least what percentage of firms in this city had 2005 gross sales of

a. \$1.1 to \$3.5 million **b.** \$.8 to \$3.8 million **c.** \$.5 to \$4.1 million

3.80 Suppose the average credit card debt for households currently is \$9500 with a standard deviation of \$2600.

a. Using Chebyshev's theorem, find at least what percentage of current credit card debts for all households are between

i. \$4300 and \$14,700 **ii.** \$3000 and \$16,000

***b.** Using Chebyshev's theorem, find the interval that contains credit card debts of at least 89% of all households.

3.81 The mean monthly mortgage paid by all home owners in a city is \$2365 with a standard deviation of \$340.

a. Using Chebyshev's theorem, find at least what percentage of all home owners in the city pay a monthly mortgage of

i. \$1685 to \$3045 **ii.** \$1345 to \$3385

***b.** Using Chebyshev's theorem, find the interval that contains the monthly mortgage payments of at least 84% of all home owners.

3.82 The mean life of a certain brand of auto batteries is 44 months with a standard deviation of 3 months. Assume that the lives of all auto batteries of this brand have a bell-shaped distribution. Using the empirical rule, find the percentage of auto batteries of this brand that have a life of

a. 41 to 47 months **b.** 38 to 50 months **c.** 35 to 53 months

3.83 According to Hewitt and Associates (a consulting firm in Lincolnshire, Illinois), the employee share of health insurance premiums at large U.S. companies was expected to be \$1481, on average, in 2005. Suppose the current payments by all such employees toward health insurance premiums have a bell-shaped distribution with a mean of \$1481 per year and a standard deviation of \$355. Using the empirical rule, find the percentage of employees whose annual payments toward such premiums are between

a. \$771 and \$2191 **b.** \$1126 and \$1836 **c.** \$416 and \$2546

3.84 The prices of all college textbooks follow a bell-shaped distribution with a mean of \$105 and a standard deviation of \$20.

a. Using the empirical rule, find the percentage of all college textbooks with their prices between

i. \$85 and \$125 **ii.** \$65 and \$145

***b.** Using the empirical rule, find the interval that contains the prices of 99.7% of college textbooks.

3.85 Suppose that on a certain section of I-95, with a posted speed limit of 65 miles per hour, the speeds of all vehicles have a bell-shaped distribution with a mean of 72 mph and a standard deviation of 3 mph.

- **a.** Using the empirical rule, find the percentage of vehicles with the following speeds on this section of I-95.
  - **i.** 63 to 81 mph **ii.** 69 to 75 mph
- ***b.** Using the empirical rule, find the interval that contains the speeds of 95% of vehicles traveling on this section of I-95.

3.5 Measures of Position

A **measure of position** determines the position of a single value in relation to other values in a sample or a population data set. There are many measures of position; however, only quartiles, percentiles, and percentile rank are discussed in this section.

3.5.1 Quartiles and Interquartile Range

Quartiles are the summary measures that divide a ranked data set into four equal parts. Three measures will divide any data set into four equal parts. These three measures are the **first quartile** (denoted by Q_1), the **second quartile** (denoted by Q_2), and the **third quartile** (denoted by Q_3). The data should be ranked in increasing order before the quartiles are determined. The quartiles are defined as follows.

Definition

Quartiles *Quartiles* are three summary measures that divide a ranked data set into four equal parts. The second quartile is the same as the median of a data set. The first quartile is the value of the middle term among the observations that are less than the median, and the third quartile is the value of the middle term among the observations that are greater than the median.

Figure 3.11 describes the positions of the three quartiles.

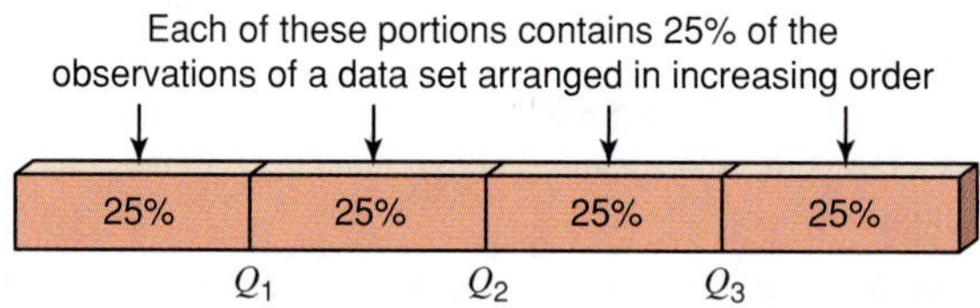

Figure 3.11 Quartiles.

Approximately 25% of the values in a ranked data set are less than Q_1 and about 75% are greater than Q_1. The second quartile, Q_2, divides a ranked data set into two equal parts; hence, the second quartile and the median are the same. Approximately 75% of the data values are less than Q_3 and about 25% are greater than Q_3.

The difference between the third quartile and the first quartile for a data set is called the **interquartile range (IQR)**.

Calculating Interquartile Range The difference between the third and the first quartiles gives the *interquartile range*; that is,

$$\text{IQR} = \text{Interquartile range} = Q_3 - Q_1$$

Examples 3–20 and 3–21 show the calculation of the quartiles and the interquartile range.

■ EXAMPLE 3–20

Finding quartiles and the interquartile range.

Refer to Table 3.3 in Example 3–5 that lists the number of car thefts during 2003 in 12 cities. That table is reproduced below.

City	Number of Car Thefts
Phoenix-Mesa, Arizona	40,769
Washington, D.C.	33,956
Miami, Florida	21,088
Atlanta, Georgia	29,920
Chicago, Illinois	42,082
Kansas City, Kansas	11,669
Baltimore, Maryland	13,435
Detroit, Michigan	40,197
St. Louis, Missouri	18,215
Las Vegas, Nevada	18,103
Newark, New Jersey	14,413
Dallas, Texas	26,343

Source: National Insurance Crime Bureau.

(a) Find the values of the three quartiles. Where does the number of car thefts of 40,197 fall in relation to these quartiles?

(b) Find the interquartile range.

Solution

Finding quartiles for an even number of data values.

(a) First we rank the given data in increasing order. Then we calculate the three quartiles as follows:

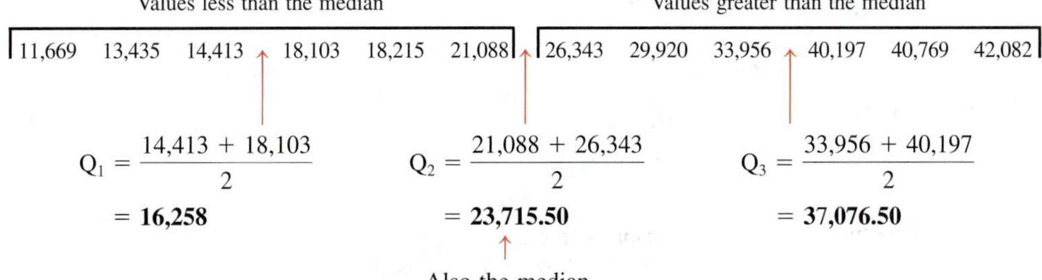

The value of Q_2, which is also the median, is given by the value of the middle term in the ranked data set. For the data of this example, this value is the average of the sixth and seventh terms. Consequently, Q_2 is 23,715.50 car thefts. The value of Q_1 is given by the value of the middle term of the six values that fall below the median (or Q_2). Thus, it is obtained by taking the average of the third and fourth terms. So, Q_1 is 16,258 car thefts. The value of Q_3 is given by the value of the middle term of the six values that fall above the median. For the data of this example, Q_3 is obtained by taking the average of the ninth and tenth terms, and it is 37,076.50 car thefts.

The value of $Q_1 = 16{,}258$ indicates that the number of car thefts in (approximately) 25% of these cities were less than 16,258 in 2003 and those in (approximately) 75% of the cities were greater than this value. Similarly, we can state that the car thefts in about half of these cities were less than 23,715.50 (which is Q_2) in 2003 and those in the other half were greater than this value. The value of $Q_3 = 37{,}076.50$ indicates that the car thefts in (approximately) 75% of the cities in this sample were less than 37,076.50 in 2003 and those in (approximately) 25% of the cities were greater than this value.

By looking at the position of 40,197, we can state that this value lies in the **top 25%** of the car thefts.

Finding the interquartile range.

(b) The interquartile range is given by the difference between the values of the third and the first quartiles. Thus,

$$\text{IQR} = \text{Interquartile range} = Q_3 - Q_1 = 37{,}076.50 - 16{,}258 = \mathbf{20{,}818.50 \text{ car thefts}}$$ ■

■ EXAMPLE 3–21

Finding quartiles and the interquartile range.

The following are the ages of nine employees of an insurance company:

47 28 39 51 33 37 59 24 33

(a) Find the values of the three quartiles. Where does the age of 28 fall in relation to the ages of these employees?

(b) Find the interquartile range.

Solution

Finding quartiles for an odd number of data values.

(a) First we rank the given data in increasing order. Then we calculate the three quartiles as follows:

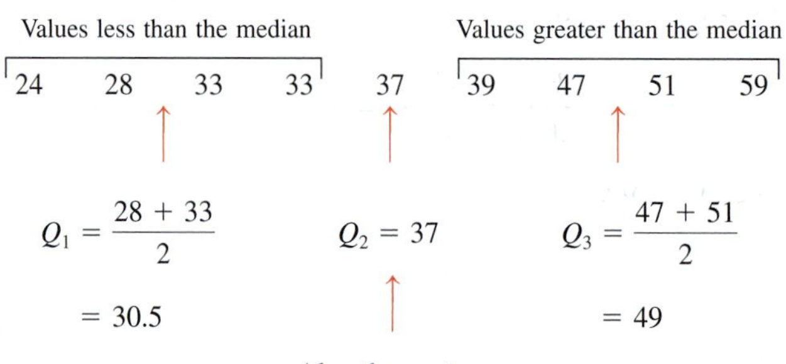

Thus the values of the three quartiles are

$$Q_1 = \mathbf{30.5 \text{ years}}, \quad Q_2 = \mathbf{37 \text{ years}}, \quad \text{and} \quad Q_3 = \mathbf{49 \text{ years}}$$

The age of 28 falls in the **lowest 25%** of the ages.

Finding the interquartile range.

(b) The interquartile range is

$$\text{IQR} = \text{Interquartile range} = Q_3 - Q_1 = 49 - 30.5 = \mathbf{18.5 \text{ years}}$$ ■

3.5.2 Percentiles and Percentile Rank

Percentiles are the summary measures that divide a ranked data set into 100 equal parts. Each (ranked) data set has 99 percentiles that divide it into 100 equal parts. The data should be ranked in increasing order to compute percentiles. The *k*th percentile is denoted by P_k, where k is an integer in the range 1 to 99. For instance, the 25th percentile is denoted by P_{25}. Figure 3.12 shows the positions of the 99 percentiles.

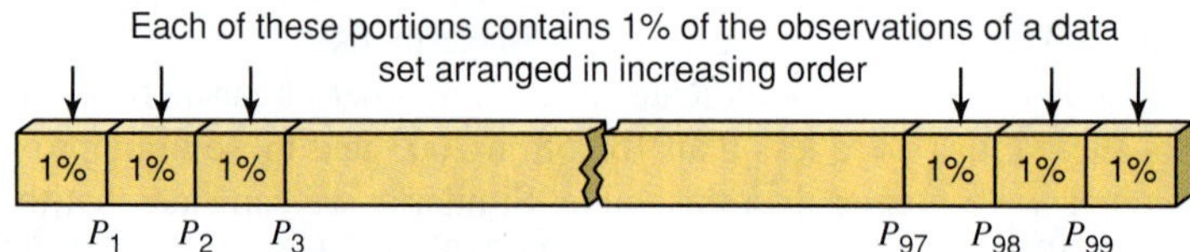

Figure 3.12 Percentiles.

Thus, the *k*th percentile, P_k, can be defined as a value in a data set such that about k% of the measurements are smaller than the value of P_k and about $(100 - k)$% of the measurements are greater than the value of P_k.

The approximate value of the *k*th percentile is determined as explained next.

Calculating Percentiles The (approximate) value of the kth *percentile*, denoted by P_k, is

$$P_k = \text{Value of the} \left(\frac{kn}{100}\right)\text{th term in a ranked data set}$$

where k denotes the number of the percentile and n represents the sample size.

Example 3–22 describes the procedure to calculate the percentiles.

■ EXAMPLE 3–22

Finding the percentile for a data set.

Refer to the data on 2003 car thefts in 12 cities given in Example 3–20. Find the value of the 42nd percentile. Give a brief interpretation of the 42nd percentile.

Solution From Example 3–20, the data arranged in increasing order are as follows:

11,669 13,435 14,413 18,103 18,215 21,088 26,343 29,920 33,956 40,197 40,769 42,082

The position of the 42nd percentile is

$$\frac{kn}{100} = \frac{42(12)}{100} = 5.04\text{th term}$$

The value of the 5.04th term can be approximated by the value of the fifth term in the ranked data. Therefore,

$$P_{42} = \text{42nd percentile} = \textbf{18,215 car thefts}$$

Thus, approximately 42% of these 12 cities had 18,215 or fewer car thefts in 2003 and 58% had higher than 18,215 car thefts. ■

We can also calculate the **percentile rank** for a particular value x_i of a data set by using the formula given below. The percentile rank of x_i gives the percentage of values in the data set that are less than x_i.

Finding Percentile Rank of a Value

$$\text{Percentile rank of } x_i = \frac{\text{Number of values less than } x_i}{\text{Total number of values in the data set}} \times 100$$

Example 3–23 shows how the percentile rank is calculated for a data value.

■ EXAMPLE 3–23

Finding the percentile rank for a data value.

Refer to the data on 2003 car thefts in 12 cities given in Example 3–20. Find the percentile rank for 29,920 car thefts. Give a brief interpretation of this percentile rank.

Solution From Example 3–20, the data arranged in increasing order are as follows:

11,669 13,435 14,413 18,103 18,215 21,088 26,343 29,920 33,956 40,197 40,769 42,082

In this data set, 7 of the 12 values are less than 29,920. Hence,

$$\text{Percentile rank of 29,920} = \frac{7}{12} \times 100 = \textbf{58.33\%}$$

Rounding this answer to the nearest integral value, we can state that about 58% of the cities in these 12 cities had less than 29,920 car thefts in 2003. Hence, about 42% of the 12 cities had 29,920 or higher car thefts in 2003. ■

EXERCISES

CONCEPTS AND PROCEDURES

3.86 Briefly describe how the three quartiles are calculated for a data set. Illustrate by calculating the three quartiles for two examples, the first with an odd number of observations and the second with an even number of observations.

3.87 Explain how the interquartile range is calculated. Give one example.

3.88 Briefly describe how the percentiles are calculated for a data set.

3.89 Explain the concept of the percentile rank for an observation of a data set.

APPLICATIONS

3.90 The following data give the weights (in pounds) lost by 15 members of a health club at the end of two months after joining the club.

5	10	8	7	25	12	5	14
11	10	21	9	8	11	18	

a. Compute the values of the three quartiles and the interquartile range.
b. Calculate the (approximate) value of the 82nd percentile.
c. Find the percentile rank of 10.

3.91 The following data give the speeds of 13 cars, measured by radar, traveling on I-84.

73	75	69	68	78	69	74
76	72	79	68	77	71	

a. Find the values of the three quartiles and the interquartile range.
b. Calculate the (approximate) value of the 35th percentile.
c. Compute the percentile rank of 71.

3.92 The following data give the numbers of computer keyboards assembled at the Twentieth Century Electronics Company for a sample of 25 days.

45	52	48	41	56	46	44	42	48	53
51	53	51	48	46	43	52	50	54	47
44	47	50	49	52					

a. Calculate the values of the three quartiles and the interquartile range.
b. Determine the (approximate) value of the 53rd percentile.
c. Find the percentile rank of 50.

3.93 The following data give the number of runners left on bases by each of the 30 Major League Baseball teams in the games played on August 12, 2004.

6	6	6	7	6	10	6	3	6	8	10	7	18	11	6
9	4	8	9	5	5	4	8	8	8	5	5	5	13	8

a. Calculate the values of the three quartiles and the interquartile range.
b. Find the (approximate) value of the 63rd percentile.
c. Find the percentile rank of 10.

3.94 Refer to Exercise 3.22. The following data give the number of students suspended for bringing weapons to schools in the Tri-City School District for each of the past 12 weeks.

15	9	12	11	7	6	9	10	14	3	6	5

a. Determine the values of the three quartiles and the interquartile range. Where does the value of 10 fall in relation to these quartiles?
b. Calculate the (approximate) value of the 55th percentile.
c. Find the percentile rank of 7.

3.95 Nixon Corporation manufactures computer monitors. The following data give the numbers of computer monitors produced at the company for a sample of 30 days.

24	32	27	23	33	33	29	25	23	36
26	26	31	20	27	33	27	23	28	29
31	35	34	22	37	28	23	35	31	43

a. Calculate the values of the three quartiles and the interquartile range. Where does the value of 31 lie in relation to these quartiles?

b. Find the (approximate) value of the 65th percentile. Give a brief interpretation of this percentile.

c. For what percentage of the days was the number of computer monitors produced 32 or higher? Answer by finding the percentile rank of 32.

3.96 The following data give the numbers of new cars sold at a dealership during a 20-day period.

8	5	12	3	9	10	6	12	8	8
4	16	10	11	7	7	3	5	9	11

a. Calculate the values of the three quartiles and the interquartile range. Where does the value of 4 lie in relation to these quartiles?

b. Find the (approximate) value of the 25th percentile. Give a brief interpretation of this percentile.

c. Find the percentile rank of 10. Give a brief interpretation of this percentile rank.

3.97 According to the National Association of Realtors, the median home price in San Diego for the second quarter of 2003 was $559,700 (*USA TODAY*, August 27, 2004). Suppose the following data give the sale prices (in thousands of dollars) of a random sample of 20 recently sold homes in San Diego.

605	789	550	881	499	675	700	543	910	808
1016	929	544	397	649	752	698	710	495	509

a. Calculate the values of the three quartiles and the interquartile range. Where does the value of 649 fall in relation to these quartiles?

b. Calculate the (approximate) value of the 77th percentile. Give a brief interpretation of this percentile.

c. Find the percentile rank of 700. Give a brief interpretation of this percentile rank.

3.6 Box-and-Whisker Plot

A **box-and-whisker plot** gives a graphic presentation of data using five measures: the median, the first quartile, the third quartile, and the smallest and the largest values in the data set between the lower and the upper inner fences. (The inner fences are explained in Example 3–24 below.) A box-and-whisker plot can help us visualize the center, the spread, and the skewness of a data set. It also helps detect outliers. We can compare different distributions by making box-and-whisker plots for each of them.

Definition

Box-and-Whisker Plot A plot that shows the center, spread, and skewness of a data set. It is constructed by drawing a box and two whiskers that use the median, the first quartile, the third quartile, and the smallest and the largest values in the data set between the lower and the upper inner fences.

Example 3–24 explains all the steps needed to make a box-and-whisker plot.

■ EXAMPLE 3–24

Constructing a box-and-whisker plot.

The following data are the incomes (in thousands of dollars) for a sample of 12 households.

35	29	44	72	34	64	41	50	54	104	39	58

Construct a box-and-whisker plot for these data.

Solution The following five steps are performed to construct a box-and-whisker plot.

Step 1. First, rank the data in increasing order and calculate the values of the median, the first quartile, the third quartile, and the interquartile range. The ranked data are

29 34 35 39 41 44 50 54 58 64 72 104

For these data,

$$\text{Median} = (44 + 50)/2 = 47$$

$$Q_1 = (35 + 39)/2 = 37$$

$$Q_3 = (58 + 64)/2 = 61$$

$$\text{IQR} = Q_3 - Q_1 = 61 - 37 = 24$$

Step 2. Find the points that are $1.5 \times \text{IQR}$ below Q_1 and $1.5 \times \text{IQR}$ above Q_3. These two points are called the **lower** and the **upper inner fences**, respectively.

$$1.5 \times \text{IQR} = 1.5 \times 24 = 36$$

$$\text{Lower inner fence} = Q_1 - 36 = 37 - 36 = 1$$

$$\text{Upper inner fence} = Q_3 + 36 = 61 + 36 = 97$$

Step 3. Determine the smallest and the largest values in the given data set within the two inner fences. These two values for our example are as follows:

Smallest value within the two inner fences = 29

Largest value within the two inner fences = 72

Step 4. Draw a horizontal line and mark the income levels on it such that all the values in the given data set are covered. Above the horizontal line, draw a box with its left side at the position of the first quartile and the right side at the position of the third quartile. Inside the box, draw a vertical line at the position of the median. The result of this step is shown in Figure 3.13.

Figure 3.13

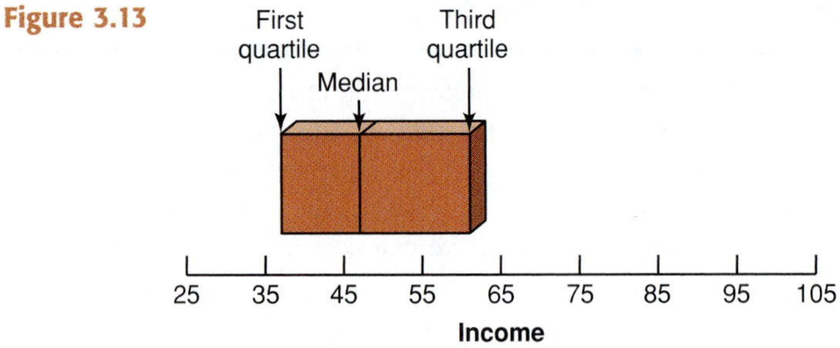

Step 5. By drawing two lines, join the points of the smallest and the largest values within the two inner fences to the box. These values are 29 and 72 in this example as listed in Step 3. The two lines that join the box to these two values are called **whiskers**. A value that falls outside the two inner fences is shown by marking an asterisk and is called an outlier. This completes the box-and-whisker plot, as shown in Figure 3.14.

Figure 3.14

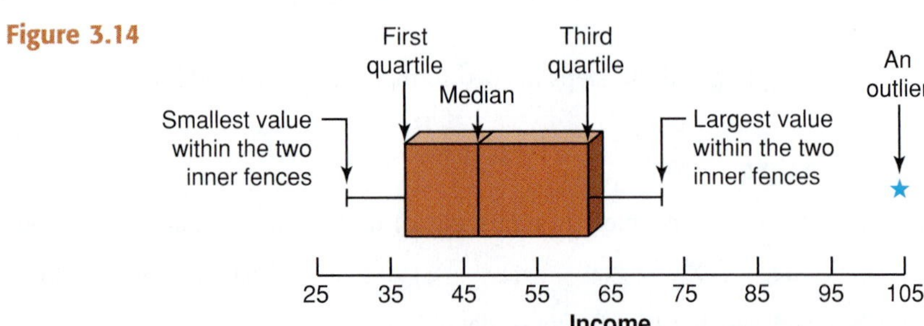

In Figure 3.14, about 50% of the data values fall within the box, about 25% of the values fall on the left side of the box, and about 25% fall on the right side of the box. Also, 50% of the values fall on the left side of the median and 50% lie on the right side of the median. The data of this example are skewed to the right because the lower 50% of the values are spread over a smaller range than the upper 50% of the values. ■

The observations that fall outside the two inner fences are called outliers. These outliers can be classified into two kinds of outliers—mild and extreme outliers. To do so, we define two outer fences—a **lower outer fence** at 3.0 × IQR below the first quartile and an **upper outer fence** at 3.0 × IQR above the third quartile. If an observation is outside either of the two inner fences but within either of the two outer fences, it is called a *mild outlier.* An observation that is outside either of the two outer fences is called an *extreme outlier.* For the previous example, the outer fences are at −35 and 133. Because 104 is outside the upper inner fence but inside the upper outer fence, it is a mild outlier.

For a symmetric data set, the line representing the median will be in the middle of the box and the spread of the values will be over almost the same range on both sides of the box.

EXERCISES

■ CONCEPTS AND PROCEDURES

3.98 Briefly explain what summary measures are used to construct a box-and-whisker plot.

3.99 Prepare a box-and-whisker plot for the following data:

36	43	28	52	41	59	47	61
24	55	63	73	32	25	35	49
31	22	61	42	58	65	98	34

Does this data set contain any outliers?

3.100 Prepare a box-and-whisker plot for the following data:

11	8	26	31	62	19	7	3	14	75
33	30	42	15	18	23	29	13	16	6

Does this data set contain any outliers?

■ APPLICATIONS

3.101 The following data give the time (in minutes) that each of 20 students selected from a university waited in line at their bookstore to pay for their textbooks in the beginning of the Spring 2006 semester.

15	8	23	21	5	17	31	22	34	6
5	10	14	17	16	25	30	3	31	19

Prepare a box-and-whisker plot. Comment on the skewness of these data.

3.102 Refer to Exercise 3.97. The following data give the sale prices (in thousands of dollars) of a random sample of 20 recently sold homes in San Diego.

605	789	550	881	499	675	700	543	910	808
1016	929	544	397	649	752	698	710	495	509

Prepare a box-and-whisker plot. Are the data skewed in any direction?

3.103 The following data give the crude oil reserves (in billions of barrels) of Saudi Arabia, Iraq, Kuwait, Iran, United Arab Emirates, Venezuela, Russia, Libya, Nigeria, China, Mexico, and the United States (*USA TODAY*, June 7, 2004). The reserves for these countries are listed in that order.

261.7	112.0	97.7	94.4	80.3	64.0
51.2	29.8	27.0	26.8	25.0	22.5

Prepare a box-and-whisker plot. Are the data symmetric or skewed?

3.104 The following data give the numbers of computer keyboards assembled at the Twentieth Century Electronics Company for a sample of 25 days.

45	52	48	41	56	46	44	42	48	53
51	53	51	48	46	43	52	50	54	47
44	47	50	49	52					

Prepare a box-and-whisker plot. Comment on the skewness of these data.

3.105 Refer to Exercise 3.93. The following data give the number of runners left on bases by each of the 30 Major League Baseball teams in the games played on August 12, 2004.

6	6	6	7	6	10	6	3	6	8	10	7	18	11	6
9	4	8	9	5	5	4	8	8	8	5	5	5	13	8

Prepare a box-and-whisker plot. Are the data symmetric or skewed?

3.106 Refer to Exercise 3.22. The following data give the number of students suspended for bringing weapons to schools in the Tri-City School District for each of the past 12 weeks.

15	9	12	11	7	6	9	10	14	3	6	5

Make a box-and-whisker plot. Comment on the skewness of these data.

3.107 Nixon Corporation manufactures computer monitors. The following are the numbers of computer monitors produced at the company for a sample of 30 days:

24	32	27	23	33	33	29	25	23	28
21	26	31	20	27	33	27	23	28	29
31	35	34	22	26	28	23	35	31	27

Prepare a box-and-whisker plot. Comment on the skewness of these data.

3.108 The following data give the numbers of new cars sold at a dealership during a 20-day period.

8	5	12	3	9	10	6	12	8	8
4	16	10	11	7	7	3	5	9	11

Make a box-and-whisker plot. Comment on the skewness of these data.

USES AND MISUSES... BEATING THE CURVE

Instructors often grade exams on a curve. The *curve* is a very loose way of saying that a set of exam scores is compared to a bell-shaped curve and grades are assigned by the relationship of any particular score to quantitative statistical measures, such as the mean, standard deviation, quartiles, or quintiles. Knowing a little bit about statistics can help your instructor make an honest assessment.

Quite often, assumptions about the curve itself are flawed. A set of exam scores, when classified and plotted as a histogram, may not resemble a symmetric distribution at all, and in many cases is actually skewed to the left or right. If the distribution of scores is skewed to the left, most of the scores are clustered around the high scores; if the distribution of scores is skewed to the right, most of the scores are clustered around the low scores. Every college or university has several famously difficult classes in which the average exam score is very low: 25 points out of 100, for example. You might hear that a fellow student "beat the curve" on an exam. He or she is the one who scored a 95 when most of the class scored somewhere near 20. This is a way of saying that the student's score is an outlier.

There is nothing wrong with comparing a set of exam scores to a bell curve. A problem can arise when there is an attempt to assign a grade, ultimately a measure of student performance, based on the characteristics of the distribution of scores. If the distribution of scores represents a symmetric curve, the teacher may choose to give the mean (and median, and mode, in this case) a suitably "average" grade of B−. Those students with scores within one standard deviation above the mean get Bs; those students with scores between one and two standard deviations above the mean get As. Similarly, Cs and Ds are given out for scores appropriately below the mean. Students with exceptional scores get A+ while some students fail. Another strategy is to divide the scores into quintiles: the students in the top quintile receive As, those in the second quintile receive Bs, and so on (note that quintiles divide an arranged data set into five equal parts). This technique, called quantitative partitioning of the data, seems to be an objective and reasonable method for assigning grades, but when you as a student are affected by the partitioning technique, issues of fairness arise. For example, is it right that only one-fifth of the class can receive a particular grade?

Glossary

Bimodal distribution A distribution that has two modes.

Box-and-whisker plot A plot that shows the center, spread, and skewness of a data set with a box and two whiskers using the median, the first quartile, the third quartile, and the smallest and the largest values in the data set between the lower and the upper inner fences.

Chebyshev's theorem For any number k greater than 1, at least $(1 - 1/k^2)$ of the values for any distribution lie within k standard deviations of the mean.

Coefficient of variation A measure of relative variability that expresses standard deviation as a percentage of the mean.

Empirical rule For a specific bell-shaped distribution, about 68% of the observations fall in the interval $(\mu - \sigma)$ to $(\mu + \sigma)$, about 95% fall in the interval $(\mu - 2\sigma)$ to $(\mu + 2\sigma)$, and about 99.7% fall in the interval $(\mu - 3\sigma)$ to $(\mu + 3\sigma)$.

First quartile The value in a ranked data set such that about 25% of the measurements are smaller than this value and about 75% are larger. It is the median of the values that are smaller than the median of the whole data set.

Geometric mean Calculated by taking the nth root of the product of all values in a data set.

Interquartile range (IQR) The difference between the third and the first quartiles.

Lower inner fence The value in a data set that is 1.5 × IQR below the first quartile.

Lower outer fence The value in a data set that is 3.0 × IQR below the first quartile.

Mean A measure of central tendency calculated by dividing the sum of all values by the number of values in the data set.

Measures of central tendency Measures that describe the center of a distribution. The mean, median, and mode are three of the measures of central tendency.

Measures of dispersion Measures that give the spread of a distribution. The range, variance, and standard deviation are three such measures.

Measures of position Measures that determine the position of a single value in relation to other values in a data set. Quartiles, percentiles, and percentile rank are examples of measures of position.

Median The value of the middle term in a ranked data set. The median divides a ranked data set into two equal parts.

Mode The value (or values) that occurs with highest frequency in a data set.

Multimodal distribution A distribution that has more than two modes.

Parameter A summary measure calculated for population data.

Percentile rank The percentile rank of a value gives the percentage of values in the data set that are smaller than this value.

Percentiles Ninety-nine values that divide a ranked data set into 100 equal parts.

Quartiles Three summary measures that divide a ranked data set into four equal parts.

Range A measure of spread obtained by taking the difference between the largest and the smallest values in a data set.

Second quartile Middle or second of the three quartiles that divide a ranked data set into four equal parts. About 50% of the values in the data set are smaller and about 50% are larger than the second quartile. The second quartile is the same as the median.

Standard deviation A measure of spread that is given by the positive square root of the variance.

Statistic A summary measure calculated for sample data.

Third quartile Third of the three quartiles that divide a ranked data set into four equal parts. About 75% of the values in a data set are smaller than the value of the third quartile and about 25% are larger. It is the median of the values that are greater than the median of the whole data set.

Trimmed mean The k% trimmed mean is obtained by dropping k% of the smallest values and k% of the largest values from the given data and then calculating the mean of the remaining $(100 - 2k)$% of the values.

Unimodal distribution A distribution that has only one mode.

Upper inner fence The value in a data set that is 1.5 × IQR above the third quartile.

Upper outer fence The value in a data set that is 3.0 × IQR above the third quartile.

Variance A measure of spread.

Weighted mean Mean of a data set whose values are assigned different weights before the mean is calculated.

Supplementary Exercises

3.109 Each year the faculty at Metro Business College chooses 10 members from the current graduating class that they feel are most likely to succeed. The data below give the current annual incomes (in thousands of dollars) of the 10 members of the class of 2005 who were voted most likely to succeed.

59 68 44 68 57 104 56 44 47 40

a. Calculate the mean and median.
b. Does this data set contain any outlier(s)? If yes, drop the outlier(s) and recalculate the mean and median. Which of these measures changes by a greater amount when you drop the outlier(s)?
c. Is the mean or the median a better summary measure for these data? Explain.

3.110 The following data give the weights (in pounds) of the nine running backs selected for *PARADE* magazine's 42nd annual All-America High School Football Team (*PARADE*, January 23, 2005). Note that because this All-America team included only nine running backs, it can be considered the population of running backs for this team.

225	225	210	234	218	188	190	195	185

a. Calculate the mean and the median. Do these data have a mode? Why or why not? Explain.
b. Find the range, variance, and standard deviation.

3.111 The following table gives the total yards gained by each of the top 10 NFL pass receivers in a single game during the 2004 regular National Football League season. Note that the games included in this data set are the ones with the highest total yards for each pass receiver during that season.

Player	Yards Gained
D. Bennett	233
R. Smith	208
J. Walker	200
R. Wayne	184
M. Muhammad	179
T. J. Houshmandzadeh	171
I. Bruce	170
A. Johnson	170
R. Gardner	167
J. Horn	167

a. Calculate the mean and median. Do these data have a mode(s)? Why or why not? Explain.
b. Find the range, variance, and standard deviation.

3.112 The following data give the numbers of driving citations received by 12 drivers.

4	8	0	3	11	7	4	14	8	13	7	9

a. Find the mean, median, and mode for these data.
b. Calculate the range, variance, and standard deviation.
c. Are the values of the summary measures in parts a and b population parameters or sample statistics?

3.113 The following table gives the distribution of the amounts of rainfall (in inches) for July 2005 for 50 cities.

Rainfall	Number of Cities
0 to less than 2	6
2 to less than 4	10
4 to less than 6	20
6 to less than 8	7
8 to less than 10	4
10 to less than 12	3

Find the mean, variance, and standard deviation. Are the values of these summary measures population parameters or sample statistics?

3.114 The following table gives the frequency distribution of the times (in minutes) that 50 commuter students at a large university spent looking for parking spaces on the first day of classes in the Spring semester of 2006.

Time	Number of Students
0 to less than 4	1
4 to less than 8	7
8 to less than 12	15
12 to less than 16	18
16 to less than 20	6
20 to less than 24	3

Find the mean, variance, and standard deviation. Are the values of these summary measures population parameters or sample statistics?

3.115 The mean time taken to learn the basics of a word processor by all students is 200 minutes with a standard deviation of 20 minutes.

a. Using Chebyshev's theorem, find at least what percentage of students will learn the basics of this word processor in

i. 160 to 240 minutes **ii.** 140 to 260 minutes

***b.** Using Chebyshev's theorem, find the interval that contains the time taken by at least 75% of all students to learn this word processor.

3.116 According to the *Statistical Abstract of the United States,* Americans were expected to spend an average of 1669 hours watching television in 2004 (*USA TODAY*, March 30, 2004). Assume that the average time spent watching television by Americans this year will have a distribution that is skewed to the right with a mean of 1750 hours and a standard deviation of 450 hours.

a. Using Chebyshev's theorem, find at least what percentage of Americans will watch television this year for

i. 850 to 2650 hours **ii.** 400 to 3100 hours

***b.** Using Chebyshev's theorem, find the interval that will contain the television viewing times of at least 84% of all Americans.

3.117 Refer to Exercise 3.115. Suppose the times taken to learn the basics of this word processor by all students have a bell-shaped distribution with a mean of 200 minutes and a standard deviation of 20 minutes.

a. Using the empirical rule, find the percentage of students who learn the basics of this word processor in

i. 180 to 220 minutes **ii.** 160 to 240 minutes

***b.** Using the empirical rule, find the interval that contains the time taken by 99.7% of all students to learn this word processor.

3.118 Assume that the annual earnings of all employees with CPA certification and 12 years of experience and working for large firms have a bell-shaped distribution with a mean of $134,000 and a standard deviation of $12,000.

a. Using the empirical rule, find the percentage of all such employees whose annual earnings are between

i. $98,000 and $170,000 **ii.** $110,000 and $158,000

***b.** Using the empirical rule, find the interval that contains the annual earnings of 68% of all such employees.

3.119 Refer to the data of Exercise 3.109 on the current annual incomes (in thousands of dollars) of the 10 members of the class of 2005 of the Metro Business College who were voted most likely to succeed.

59 68 44 68 57 104 56 44 47 40

a. Determine the values of the three quartiles and the interquartile range. Where does the value of 40 fall in relation to these quartiles?

b. Calculate the (approximate) value of the 70th percentile. Give a brief interpretation of this percentile.

c. Find the percentile rank of 47. Give a brief interpretation of this percentile rank.

3.120 Refer to the data given in Exercise 3.111 on the total yards gained by the top 10 NFL pass receivers in single games during the 2004 regular National Football League season.

- **a.** Determine the values of the three quartiles and the interquartile range. Where does the value of 179 lie in relation to these quartiles?
- **b.** Calculate the (approximate) value of the 70th percentile. Give a brief interpretation of this percentile.
- **c.** Find the percentile rank of 171. Give a brief interpretation of this percentile rank.

3.121 A student washes her clothes at a laundromat once a week. The data below give the time (in minutes) she spent in the laundromat for each of 15 randomly selected weeks. Here, time spent in the laundromat includes the time spent waiting for a machine to become available.

75	62	84	73	107	81	93	72
135	77	85	67	90	83	112	

Prepare a box-and-whisker plot. Is the data set skewed in any direction? If yes, is it skewed to the right or to the left? Does this data set contain any outliers?

3.122 The following data give the lengths of time (in weeks) taken to find a full-time job by 18 computer science majors who graduated in 2005 from a small college.

10	3	12	21	15	8	4	2	16
8	9	14	33	7	24	11	42	15

Make a box-and-whisker plot. Comment on the skewness of this data set. Does this data set contain any outliers?

Advanced Exercises

3.123 Melissa's grade in her math class is determined by three 100-point tests and a 200-point final exam. To determine the grade for a student in this class, the instructor will add the four scores together and divide this sum by 5 to obtain a percentage. This percentage must be at least 80 for a grade of B. If Melissa's three test scores are 75, 69, and 87, what is the minimum score she needs on the final exam to obtain a B grade?

3.124 Jeffrey is serving on a six-person jury for a personal-injury lawsuit. All six jurors want to award damages to the plaintiff but cannot agree on the amount of the award. The jurors have decided that each of them will suggest an amount that he or she thinks should be awarded; then they will use the mean of these six numbers as the award to recommend to the plaintiff.

- **a.** Jeffrey thinks the plaintiff should receive $20,000, but he thinks the mean of the other five jurors' recommendations will be about $12,000. He decides to suggest an inflated amount so that the mean for all six jurors is $20,000. What amount would Jeffrey have to suggest?
- **b.** How might this jury revise its procedure to prevent a juror like Jeffrey from having an undue influence on the amount of damages to be awarded to the plaintiff?

3.125 The heights of five starting players on a basketball team have a mean of 76 inches, a median of 78 inches, and a range of 11 inches.

- **a.** If the tallest of these five players is replaced by a substitute who is two inches taller, find the new mean, median, and range.
- **b.** If the tallest player is replaced by a substitute who is four inches shorter, which of the new values (mean, median, range) could you determine, and what would their new values be?

3.126 On a 300-mile auto trip, Lisa averaged 52 miles per hour for the first 100 miles, 65 mph for the second 100 miles, and 58 mph for the last 100 miles.

- **a.** How long did the 300-mile trip take?
- **b.** Could you find Lisa's average speed for the 300-mile trip by calculating $(52 + 65 + 58)/3$? If not, find the correct average speed for the trip.

3.127 A small country bought oil from three different sources in one week, as shown in the following table.

Source	Barrels Purchased	Price Per Barrel
Mexico	1000	$51
Kuwait	200	64
Spot Market	100	70

Find the mean price per barrel for all 1300 barrels of oil purchased in that week.

3.128 During the 2004 winter season, a homeowner received four deliveries of heating oil, as shown in the following table.

Gallons Purchased	Price Per Gallon
198	$1.10
173	1.25
130	1.28
124	1.33

The homeowner claimed that the mean price he paid for oil during the season was $(1.10 + 1.25 + 1.28 + 1.33)/4 = \$1.24$ per gallon. Do you agree with this claim? If not, explain why this method of calculating the mean is not appropriate in this case. Find the correct value of the mean price.

3.129 In the Olympic Games, when events require a subjective judgment of an athlete's performance, the highest and lowest of the judges' scores may be dropped. Consider a gymnast whose performance is judged by seven judges and the highest and the lowest of the seven scores are dropped.

a. Gymnast A's scores in this event are 9.4, 9.7, 9.5, 9.5, 9.4, 9.6, and 9.5. Find this gymnast's mean score after dropping the highest and the lowest scores.

b. The answer to part a is an example of what percentage of trimmed mean?

c. Write another set of scores for a gymnast B so that gymnast A has a higher mean score than gymnast B based on the trimmed mean, but gymnast B would win if all seven scores were counted. Do not use any scores lower than 9.0.

3.130 A survey of young people's shopping habits in a small city during the summer months of 2005 showed the following: Shoppers aged 12–14 took an average of 8 shopping trips per month and spent an average of $14 per trip. Shoppers aged 15–17 took an average of 11 trips per month and spent an average of $18 per trip. Assume that this city has 1100 shoppers aged 12–14 and 900 shoppers aged 15–17.

a. Find the total amount spent per month by all these 2000 shoppers in both age groups.

b. Find the mean number of shopping trips per person per month for these 2000 shoppers.

c. Find the mean amount spent per person per month by shoppers aged 12–17 in this city.

3.131 The following table shows the total population and the number of deaths (in thousands) due to heart attack for two age groups in Countries A and B for 2005.

	Age 30 and Under		Age 31 and Over	
	A	B	A	B
Population	40,000	25,000	20,000	35,000
Deaths due to heart attack	1000	500	2000	3000

a. Calculate the death rate due to heart attack per 1000 population for the 30 and under age group for each of the two countries. Which country has the lower death rate in this age group?

b. Calculate the death rates due to heart attack for the two countries for the 31 and over age group. Which country has the lower death rate in this age group?

c. Calculate the death rate due to heart attack for the entire population of Country A; then do the same for Country B. Which country has the lower overall death rate?

d. How can the country with lower death rate in both age groups have the higher overall death rate? (This phenomenon is known as Simpson's paradox.)

3.132 In a study of distances traveled to a college by commuting students, data from 100 commuters yielded a mean of 8.73 miles. After the mean was calculated, data came in late from three students, with distances of 11.5, 7.6, and 10.0 miles. Calculate the mean distance for all 103 students.

3.133 The test scores for a large statistics class have an unknown distribution with a mean of 70 and a standard deviation of 10.

a. Find k so that at least 50% of the scores are within k standard deviations of the mean.

b. Find k so that at most 10% of the scores are more than k standard deviations above the mean.

3.134 The test scores for a very large statistics class have a bell-shaped distribution with a mean of 70 points.

a. If 16% of all students in the class scored above 85, what is the standard deviation of the scores?

b. If 95% of the scores are between 60 and 80, what is the standard deviation?

3.135 How much does the typical American family spend to go away on vacation each year? Twenty-five randomly selected households reported the following vacation expenditures (rounded to the nearest hundred dollars) during the past year:

2500	500	800	0	100
0	200	2200	0	200
0	1000	900	321,500	400
500	100	0	8200	900
0	1700	1100	600	3400

a. Using both graphical and numerical methods, organize and interpret these data.

b. What measure of central tendency best answers the original question?

3.136 Actuaries at an insurance company must determine a premium for a new type of insurance. A random sample of 40 potential purchasers of this type of insurance were found to have suffered the following values of losses during the past year. These losses would have been covered by the insurance if it were available.

100	32	0	0	470	50	0	14,589	212	93
0	0	1127	421	0	87	135	420	0	250
12	0	309	0	177	295	501	0	143	0
167	398	54	0	141	0	3709	122	0	0

a. Find the mean, median, and mode of these 40 losses.

b. Which of the mean, median, or mode is largest?

c. Draw a box-and-whisker plot for these data, and describe the skewness, if any.

d. Which measure of central tendency should the actuaries use to determine the premium for this insurance?

3.137 A local golf club has men's and women's summer leagues. The following data give the scores for a round of 18 holes of golf for 17 men and 15 women randomly selected from their respective leagues.

Men	87	68	92	79	83	67	71	92	112
	75	77	102	79	78	85	75	72	
Women	101	100	87	95	98	81	117	107	103
	97	90	100	99	94	94			

a. Make a box-and-whisker plot for each of the data sets and use them to discuss the similarities and differences between the scores of the men and women golfers.

b. Compute the various descriptive measures you have learned for each sample. How do they compare?

3.138 Answer the following questions.

a. The total weight of all pieces of luggage loaded onto an airplane is 12,372 pounds, which works out to be an average of 51.55 pounds per piece. How many pieces of luggage are on the plane?

b. A group of seven friends, having just gotten back a chemistry exam, discuss their scores. Six of the students reveal that they received grades of 81, 75, 93, 88, 82, and 85, but the seventh student is reluctant to say what grade she received. After some calculation she announces that the group averaged 81 on the exam. What is her score?

3.139 Suppose that there are 150 freshmen engineering majors at a college and each of them will take the same five courses next semester. Four of these courses will be taught in small sections of 25 students each, whereas the fifth course will be taught in one section containing all 150 freshmen. To accommodate all 150 students, there must be six sections of each of the four courses taught in 25-student sections. Thus, there are 24 classes of 25 students each and one class of 150 students.

a. Find the mean size of these 25 classes.

b. Find the mean class size from a student's point of view, noting that each student has five classes containing 25, 25, 25, 25, and 150 students.

Are the means in parts a and b equal? If not, why not?

3.140 The following data give the weights (in pounds) of a random sample of 44 college students. (Here F and M indicate female and male, respectively.)

123 F	195 M	138 M	115 F	179 M	119 F
148 F	147 F	180 M	146 F	179 M	189 M
175 M	108 F	193 M	114 F	179 M	147 M
108 F	128 F	164 F	174 M	128 F	159 M
193 M	204 M	125 F	133 F	115 F	168 M
123 F	183 M	116 F	182 M	174 M	102 F
123 F	99 F	161 M	162 M	155 F	202 M
110 F	132 M				

Compute the mean, median, and standard deviation for the weights of all students, of men only, and of women only. Of the mean and median, which is the more informative measure of central tendency? Write a brief note comparing the three measures for all students, men only, and women only.

3.141 The distribution of the lengths of fish in a certain lake is not known, but it is definitely not bell-shaped. It is estimated that the mean length is 6 inches with a standard deviation of 2 inches.

a. At least what proportion of fish in the lake are between 3 inches and 9 inches long?

b. What is the smallest interval that will contain the lengths of at least 84% of the fish?

c. Find an interval so that fewer than 36% of the fish have lengths outside this interval.

3.142 The following stem-and-leaf diagram gives the distances (in thousands of miles) driven during the past year by a sample of drivers in a city.

Stem	Leaves
0	3 6 9
1	2 8 5 1 0 5
2	5 1 6
3	8
4	1
5	
6	2

a. Compute the sample mean, median, and mode for the data on distances driven.

b. Compute the range, variance, and standard deviation for these data.

c. Compute the first and third quartiles.

d. Compute the interquartile range. Describe what properties the interquartile range has. When would it be preferable to using the standard deviation when measuring variation?

3.143 Refer to the data in Problem 3.140. Two individuals, one from Canada and one from England, are interested in your analysis of these data but they need your results in different units. The Canadian individual wants the results in grams (1 pound = 435.59 grams). while the English individual wants the results in stone (1 stone = 14 pounds).

a. Convert the data on weights from pounds to grams, and then recalculate the mean, median, and standard deviation of weight for males and females separately. Repeat the procedure, changing the unit from pounds to stones.

b. Convert your answers from Problem 3.140 to grams and stone. What do you notice about these answers and your answers from part a?

c. What happens to the values of the mean, median, and standard deviation when you convert from a larger unit to a smaller unit (e.g., from pounds to grams)? Does the same thing happen if you convert from a smaller unit (e.g., pounds) to a larger unit (e.g., stone)?

d. Figure 3.15 on the next page gives a stacked dotplot of these weights in pounds and stone. Which of these two distributions has more variability? Use your results from parts a to c to explain why this is the case.

e. Now consider the weights in pounds and grams. Make a stacked dotplot for these data and answer part d.

3.144 Although the standard workweek is 40 hours a week, many people work a lot more than 40 hours a week. The data on the next page give the numbers of hours worked last week by 50 people.

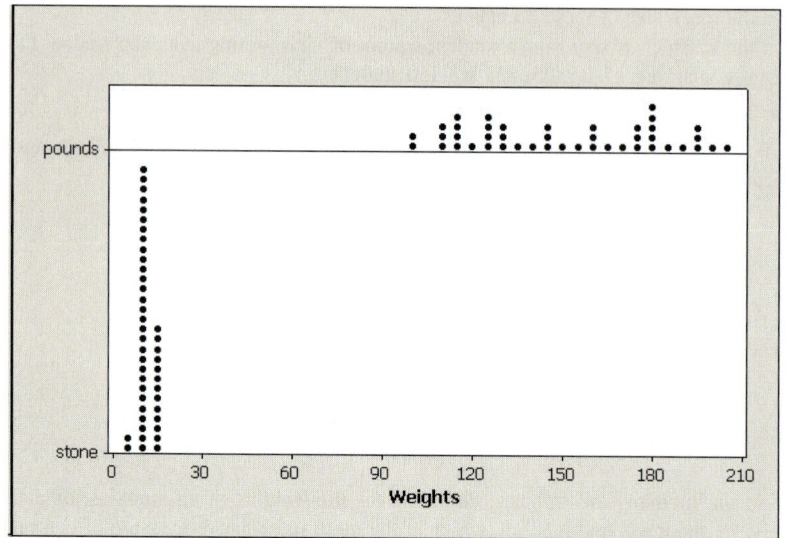

Figure 3.15 Stacked Dotplot of Weights in Stone and Pounds.

40.5	41.3	41.4	41.5	42.0	42.2	42.4	42.4	42.6	43.3
43.7	43.9	45.0	45.0	45.2	45.8	45.9	46.2	47.2	47.5
47.8	48.2	48.3	48.8	49.0	49.2	49.9	50.1	50.6	50.6
50.8	51.5	51.5	52.3	52.3	52.6	52.7	52.7	53.4	53.9
54.4	54.8	55.0	55.4	55.4	55.4	56.2	56.3	57.8	58.7

a. The sample mean and sample standard deviation for this data set are 49.012 and 5.080, respectively. Using the Chebyshev's theorem, calculate the intervals that contain at least 75%, 88.89%, and 93.75% of the data.

b. Determine the actual percentages of the given data values that fall in each of the intervals that you calculated in part a. Also calculate the percentage of the data values that fall within one standard deviation of the mean.

c. Do you think the lower endpoints provided by Chebyshev's Theorem in part a are useful for this problem? Explain your answer.

d. Suppose that the individual with the first number (54.4) in the fifth row of the data is a workaholic who actually worked 84.4 hours last week, and not 54.4 hours. With this change now $\bar{x} = 49.61$ and $s = 7.10$. Recalculate the intervals for part a and the actual percentages for part b. Did your percentages change a lot or a little?

e. How many standard deviations above the mean would you have to go to capture all 50 data values? What is the lower bound for the percentage of the data that should fall in the interval, according to Chebyshev?

3.145 Refer to the women's golf scores in Exercise 3.137. It turns out that 117 was mistakenly entered. Although this person still had the highest score among the 15 women, her score was not a mild or extreme outlier according to the box-and-whisker plot, nor was she tied for the highest score. What are the possible scores that she could have shot?

APPENDIX 3.1

A3.1.1 BASIC FORMULAS FOR THE VARIANCE AND STANDARD DEVIATION FOR UNGROUPED DATA

Example 3–25 illustrates how to use the basic formulas to calculate the variance and standard deviation for ungrouped data. From Section 3.2.2, the basic formulas for variance for ungrouped data are

$$\sigma^2 = \frac{\Sigma(x - \mu)^2}{N} \quad \text{and} \quad s^2 = \frac{\Sigma(x - \bar{x})^2}{n - 1}$$

where σ^2 is the population variance and s^2 is the sample variance.

In either case, the standard deviation is obtained by taking the square root of the variance.

Calculating the variance and standard deviation for ungrouped data using basic formulas.

EXAMPLE 3–25 Refer to Example 3–12, where we used the short-cut formulas to compute the variance and standard deviation for the data on the total wealth (in billions of dollars) of five persons. Calculate the variance and standard deviation for those data using the basic formula.

Solution Let x denote the total wealth (in billions of dollars) of a person. Table 3.14 shows all the required calculations to find the variance and standard deviation.

Table 3.14

x	$(x - \bar{x})$	$(x - \bar{x})^2$
46.5	46.5 − 19.1 = 27.4	750.76
18.0	18.0 − 19.1 = −1.1	1.21
16.0	16.0 − 19.1 = −3.1	9.61
7.8	7.8 − 19.1 = −11.3	127.69
7.2	7.2 − 19.1 = −11.9	141.61
$\Sigma x = 95.5$		$\Sigma(x - \bar{x})^2 = 1030.88$

The following steps are performed to compute the variance and standard deviation.

Step 1. Find the mean as follows:

$$\bar{x} = \frac{\Sigma x}{n} = \frac{95.5}{5} = 19.1$$

Step 2. Calculate $x - \bar{x}$, the deviation of each value of x from the mean. The results are shown in the second column of Table 3.14.

Step 3. Square each of the deviations of x from $\bar{x}$; that is, calculate each of the $(x - \bar{x})^2$ values. These values are called the *squared deviations*, and they are recorded in the third column.

Step 4. Add all the squared deviations to obtain $\Sigma(x - \bar{x})^2$; that is, sum all the values given in the third column of Table 3.14. This gives

$$\Sigma(x - \bar{x})^2 = 1030.88$$

Step 5. Obtain the sample variance by dividing the sum of the squared deviations by $n - 1$. Thus

$$s^2 = \frac{\Sigma(x - \bar{x})^2}{n - 1} = \frac{1030.88}{5 - 1} = 257.72$$

Step 6. Obtain the sample standard deviation by taking the positive square root of the variance. Hence,

$$s = \sqrt{257.72} = \mathbf{16.05366 = \$16.05\ billion}$$

A3.1.2 BASIC FORMULAS FOR THE VARIANCE AND STANDARD DEVIATION FOR GROUPED DATA

Example 3–26 demonstrates how to use the basic formulas to calculate the variance and standard deviation for grouped data. The basic formulas for these calculations are

$$\sigma^2 = \frac{\Sigma f(m - \mu)^2}{N} \quad \text{and} \quad s^2 = \frac{\Sigma f(m - \bar{x})^2}{n - 1}$$

where σ^2 is the population variance, s^2 is the sample variance, m is the midpoint of a class, and f is the frequency of a class.

In either case, the standard deviation is obtained by taking the square root of the variance.

Calculating the variance and standard deviation for grouped data using basic formulas.

EXAMPLE 3–26 In Example 3–17, we used the short-cut formula to compute the variance and standard deviation for the data on the numbers of orders received each day during the past 50 days at the office of a mail-order company. Calculate the variance and standard deviation for those data using the basic formula.

Solution All the required calculations to find the variance and standard deviation appear in Table 3.15.

Table 3.15

Number of Orders	f	m	mf	$m - \bar{x}$	$(m - \bar{x})^2$	$f(m - \bar{x})^2$
10–12	4	11	44	−5.64	31.8096	127.2384
13–15	12	14	168	−2.64	6.9696	83.6352
16–18	20	17	340	.36	.1296	2.5920
19–21	14	20	280	3.36	11.2896	158.0544
	$n = 50$		$\Sigma mf = 832$			$\Sigma f(m - \bar{x})^2 = 371.5200$

The following steps are performed to compute the variance and standard deviation using the basic formula.

Step 1. Find the midpoint of each class. Multiply the corresponding values of m and f. Find Σmf. From Table 3.15, $\Sigma mf = 832$.

Step 2. Find the mean as follows:

$$\bar{x} = \Sigma mf/n = 832/50 = 16.64$$

Step 3. Calculate $m - \bar{x}$, the deviation of each value of m from the mean. These calculations are done in the fifth column of Table 3.15.

Step 4. Square each of the deviations $m - \bar{x}$; that is, calculate each of the $(m - \bar{x})^2$ values. These are called *squared deviations*, and they are recorded in the sixth column.

Step 5. Multiply the squared deviations by the corresponding frequencies (see the seventh column of Table 3.15). Adding the values of the seventh column, we obtain

$$\Sigma f(m - \bar{x})^2 = 371.5200$$

Step 6. Obtain the sample variance by dividing $\Sigma f(m - \bar{x})^2$ by $n - 1$. Thus,

$$s^2 = \frac{\Sigma f(m - \bar{x})^2}{n - 1} = \frac{371.5200}{50 - 1} = \mathbf{7.5820}$$

Step 7. Obtain the standard deviation by taking the positive square root of the variance.

$$s = \sqrt{s^2} = \sqrt{7.5820} = \mathbf{2.75\ orders}$$

Self-Review Test

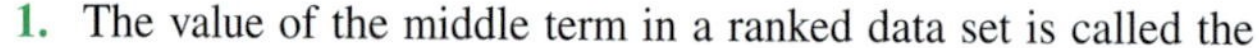

1. The value of the middle term in a ranked data set is called the
   a. mean b. median c. mode
2. Which of the following summary measures is/are influenced by extreme values?
   a. mean b. median c. mode d. range
3. Which of the following summary measures can be calculated for qualitative data?
   a. mean b. median c. mode
4. Which of the following can have more than one value?
   a. mean b. median c. mode
5. Which of the following is obtained by taking the difference between the largest and the smallest values of a data set?
   a. variance b. range c. mean
6. Which of the following is the mean of the squared deviations of x values from the mean?
   a. standard deviation b. population variance c. sample variance
7. The values of the variance and standard deviation are
   a. never negative b. always positive c. never zero

8. A summary measure calculated for the population data is called
   a. a population parameter   b. a sample statistic   c. an outlier

9. A summary measure calculated for the sample data is called a
   a. population parameter   b. sample statistic   c. box-and-whisker plot

10. Chebyshev's theorem can be applied to
   a. any distribution   b. bell-shaped distributions only   c. skewed distributions only

11. The empirical rule can be applied to
   a. any distribution   b. bell-shaped distributions only   c. skewed distributions only

12. The first quartile is a value in a ranked data set such that about
   a. 75% of the values are smaller and about 25% are larger than this value
   b. 50% of the values are smaller and about 50% are larger than this value
   c. 25% of the values are smaller and about 75% are larger than this value

13. The third quartile is a value in a ranked data set such that about
   a. 75% of the values are smaller and about 25% are larger than this value
   b. 50% of the values are smaller and about 50% are larger than this value
   c. 25% of the values are smaller and about 75% are larger than this value

14. The 75th percentile is a value in a ranked data set such that about
   a. 75% of the values are smaller and about 25% are larger than this value
   b. 25% of the values are smaller and about 75% are larger than this value

15. The following data give the numbers of times 10 persons used their credit cards during the past three months.

9 6 28 14 2 18 7 3 16 6

Calculate the mean, median, mode, range, variance, and standard deviation.

16. The mean, as a measure of central tendency, has the disadvantage of being influenced by extreme values. Illustrate this point with an example.

17. The range, as a measure of spread, has the disadvantage of being influenced by extreme values. Illustrate this point with an example.

18. When is the value of the standard deviation for a data set zero? Give one example of such a data set. Calculate the standard deviation for that data set to show that it is zero.

19. The following table gives the frequency distribution of the numbers of computers sold during the past 25 weeks at a computer store.

Computers Sold	Frequency
4 to 9	2
10 to 15	4
16 to 21	10
22 to 27	6
28 to 33	3

   a. What does the frequency column in the table represent?
   b. Calculate the mean, variance, and standard deviation.

20. The cars owned by all people living in a city are, on average, 7.3 years old with a standard deviation of 2.2 years.
   a. Using Chebyshev's theorem, find at least what percentage of the cars in this city are
      i. 1.8 to 12.8 years old   ii. .7 to 13.9 years old
   b. Using Chebyshev's theorem, find the interval that contains the ages of at least 75% of the cars owned by all people in this city.

21. The ages of cars owned by all people living in a city have a bell-shaped distribution with a mean of 7.3 years and a standard deviation of 2.2 years.
   a. Using the empirical rule, find the percentage of cars in this city that are
      i. 5.1 to 9.5 years old   ii. .7 to 13.9 years old
   b. Using the empirical rule, find the interval that contains the ages of 95% of the cars owned by all people in this city.

22. The following data give the number of times the metal detector was set off by passengers at a small airport during 15 consecutive half-hour periods on February 1, 2006.

7	2	12	13	0	8	10	
15	3	5	14	20	1	11	4

a. Calculate the three quartiles and the interquartile range. Where does the value of 4 lie in relation to these quartiles?

b. Find the (approximate) value of the 60th percentile. Give a brief interpretation of this value.

c. Calculate the percentile rank of 12. Give a brief interpretation of this value.

23. Make a box-and-whisker plot for the data on the number of times passengers set off the airport metal detector given in Problem 22. Comment on the skewness of this data set.

*24. The mean weekly wages of a sample of 15 employees of a company are $435. The mean weekly wages of a sample of 20 employees of another company are $490. Find the combined mean for these 35 employees.

*25. The mean GPA of five students is 3.21. The GPAs of four of these five students are 3.85, 2.67, 3.45, and 2.91. Find the GPA of the fifth student.

*26. The following are the prices (in thousands of dollars) of 10 houses sold recently in a city:

179	166	58	207	287	149	193	2534	163	238

Calculate the 10% trimmed mean for this data set. Do you think the 10% trimmed mean is a better summary measure than the (simple) mean (i.e., the mean of all 10 values) for these data? Briefly explain why or why not.

*27. Consider the following two data sets.

Data Set I:	8	16	20	35
Data Set II:	5	13	17	32

Note that each value of the second data set is obtained by subtracting 3 from the corresponding value of the first data set.

a. Calculate the mean for each of these two data sets. Comment on the relationship between the two means.

b. Calculate the standard deviation for each of these two data sets. Comment on the relationship between the two standard deviations.

Mini-Projects

MINI-PROJECT 3–1

Refer to the data you collected for Mini-Project 1–1 of Chapter 1 and analyzed graphically in Mini-Project 2–1 of Chapter 2. Write a report summarizing those data. This report should include answers to at least the following questions.

a. Calculate the summary measures (mean, standard deviation, five-number summary, interquartile range) for the variables you graphed in Mini-Project 2–1. Do this for the entire data set, as well as for the different groups formed by the categorical variable that you used to divide the data set in Mini-Project 2–1.

b. Are the summary measures for the various groups similar to those for the entire data set? If not, which ones differ and how do they differ? Make the same comparisons among the summary measures for various groups. Do the groups have similar levels of variability? Explain how you can determine this from the graphs that you created in Mini-Project 2–1.

c. Draw a box-and-whisker plot for the entire data set. Also draw side-by-side box-and-whisker plots for the various groups. Are there any outliers? If so, are there any values that are outliers in any of the groups but not in the entire data set? Does the plot show any skewness?

d. Discuss which measures for the center and spread would be more appropriate to use to describe your data set. Also, discuss your reasons for using those measures.

MINI-PROJECT 3–2

You are employed as a statistician for a company that makes household products, which are sold by part-time salespersons who work during their spare time. The company has four salespersons employed in a

small town. Let us denote these salespersons by A, B, C, and D. The sales records (in dollars) for the past six weeks for these four salespersons are shown in the following table.

Week	A	B	C	D
1	1774	2205	1330	1402
2	1808	1507	1295	1665
3	1890	2352	1502	1530
4	1932	1939	1104	1826
5	1855	2052	1189	1703
6	1726	1630	1441	1498

Your supervisor has asked you to prepare a brief report comparing the sales volumes and the consistency of sales of these four salespersons. Use the mean sales for each salesperson to compare the sales volumes, and then choose an appropriate statistical measure to compare the consistency of sales. Make the calculations and write a report.

DECIDE FOR YOURSELF

Deciding Where to Live

By the time you get to college, you must have heard it over and over again: "A picture is worth a thousand words." Now we have pictures and numbers discussed in Chapters 2 and 3, respectively. Why both? Well, although each one of them acts as a summary of a data set, it is a combination of the pictures and numbers that tells a big part of the story without having to look at the entire data set. Suppose that you ask a realtor for information on the prices of homes in two different but comparable suburbs. Let us call these Suburbs A and B. The realtor provides you with the following information that is obtained from a random sample of 40 houses in each suburb:

a. The average price of homes in each of the two suburbs
b. The five-number summary of prices of homes in each neighborhood
c. The histogram of the distribution of home prices for each suburb

All the information provided by the realtor is given in the following two tables and two histograms shown in Figures 3.16 and 3.17. Note that the second table gives the minimum and maximum prices of homes (in thousands of dollars) for each suburb along with the values of Q_1, median, and Q_3 (in thousands of dollars).

Suburb	A	B
Average Price (in thousands of dollars)	221.9	220.03

	Minimum	Q_1	Median	Q_3	Maximum
Suburb A	151.0	175.5	188.0	199.5	587.0
Suburb B	187.0	210.0	222.5	228.0	250.0

Before you decide which suburb you should buy the house in, answer the following questions:

1. Examine the summary statistics and graphs given here.
2. Explain how the information given here can help you to make a decision about the suburb where you should look for a house to buy.
3. Explain how and why you might be misled by simply looking at the average prices if you are looking to spend less money to buy a house.
4. Is there any information about the suburbs not given here that you will like to obtain before making a decision about the suburb where you should buy a house?

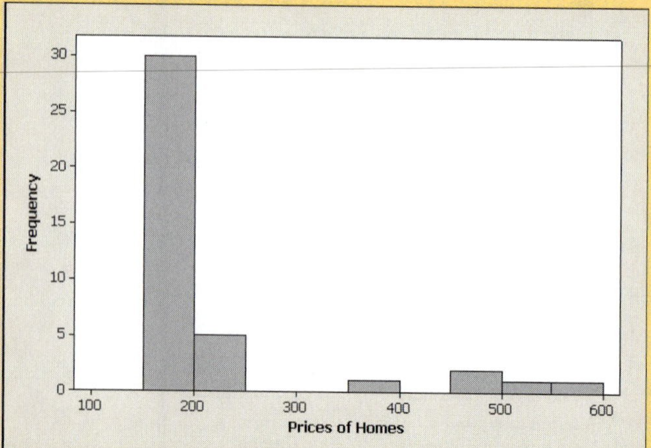

Figure 3.16 Histogram of Prices of Homes in Suburb A.

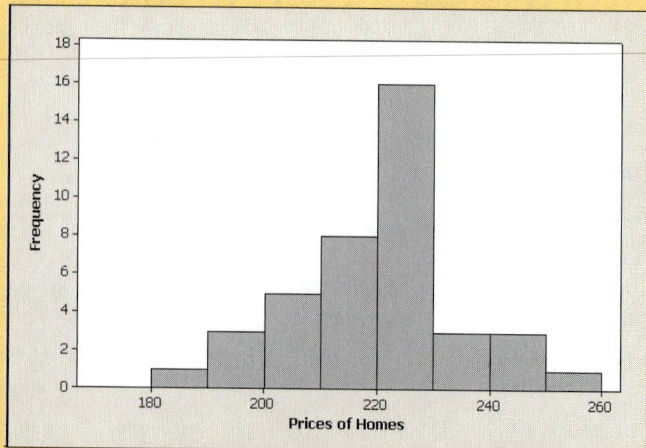

Figure 3.17 Histogram of Prices of Homes in Suburb B.

TECHNOLOGY INSTRUCTION — Numerical Descriptive Measures

TI-84

```
1-Var Stats
 x̄=6.833333333
 Σx=41
 Σx²=377
 Sx=4.400757511
 σx=4.017323598
↓n=6
```

Screen 3.1

```
1-Var Stats
↑n=6
 minX=2
 Q1=3
 Med=6
 Q3=11
 maxX=13
```

Screen 3.2

1. To calculate the **sample statistics** (e.g., mean, standard deviation, and five-number summary), first enter your data into a list such as L1, then select **STAT>CALC>1-Var Stats**, and press **Enter**. Access the name of your list by pressing **2nd>STAT** and scrolling through the list of names until you get to your list name. Press **Enter**. You will obtain the output shown in Screens 3.1 and 3.2.

   Screen 3.1 shows, in this order, the sample mean, the sum of the data values, the sum of the squared data values, the sample standard deviation, the value of the population standard deviation (you will use this only when your data constitute a census instead of a sample), and the number of data values (e.g., the sample or population size). Pressing the downward arrow key will show the five-number summary, which is shown in Screen 3.2.

2. Constructing a box-and-whisker plot is similar to constructing a histogram. First enter your data into a list such as L1, then select **STAT PLOT** and go into one of the three plots. Make sure the plot is turned on. For the type, select the second row, first column (this boxplot will display outliers, if there are any). Enter the name of your list for **XList**. Select **ZOOM>9** to display the plot as shown in Screen 3.3.

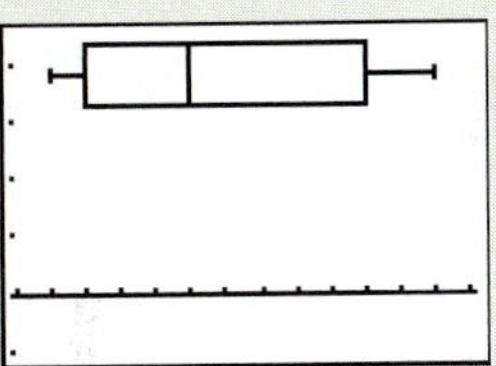

Screen 3.3

MINITAB

1. To find the sample statistics (e.g., the mean, standard deviation, and five-number summary), first enter the given data in a column such as C1, and then select **Stat>Basic Statistics> Display Descriptive Statistics**. In the dialog box you obtain, enter the name of the column where your data are stored in the **Variables** box as shown in Screen 3.4. Click the **Statistics** button in this dialog box and choose the summary measures you want to

Screen 3.4

Display Descriptive Statistics

C1 data

Variables:

data

By variables (optional):

Select

Statistics...

Graphs...

Help

OK

Cancel

calculate in the new dialog box as shown in Screen 3.5. Click **OK** in both dialog boxes. The output will appear in the **Session** window, which is shown in Screen 3.6 here.

Screen 3.5

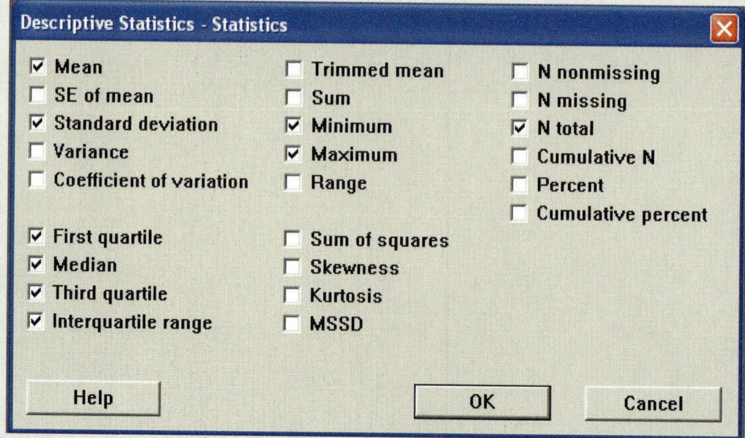

Session

Descriptive Statistics: data

```
         Total
Variable Count  Mean  StDev  Minimum    Q1  Median     Q3  Maximum   IQR
data         6  6.83   4.40     2.00  2.75    6.00  11.50    13.00  8.75
```

Screen 3.6

2. To create a box-and-whisker plot, enter the given data in a column such as C1, select **Graph>Boxplot>Simple**, and click **OK**. In the dialog box you obtain, enter the name of the column with data in the **Graph Variables** box (see Screen 3.7) and click **OK**. The boxplot shown in Screen 3.8 will appear.

Screen 3.7

Boxplot - One Y, Simple

C1 data

Graph variables:

data

Scale... Labels... Data View...

Multiple Graphs... Data Options...

Select

Help OK Cancel

Screen 3.8

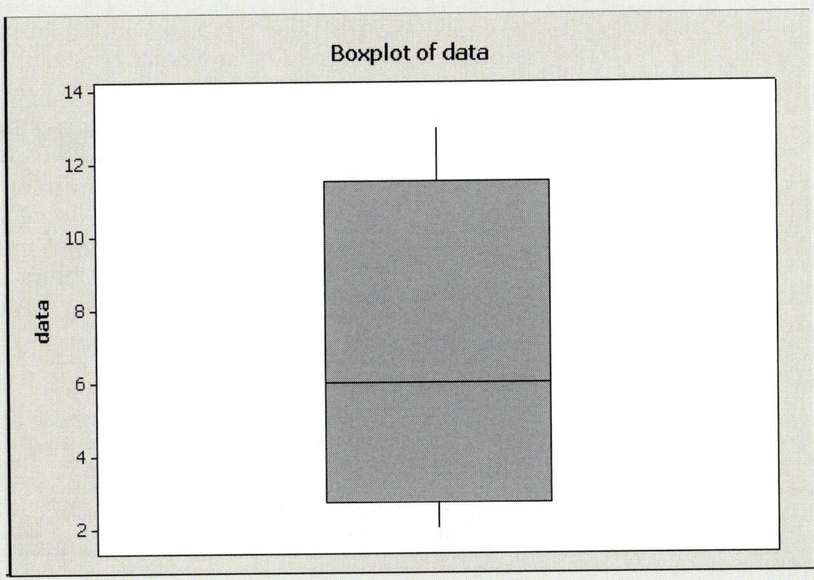

EXCEL

1. For each of the commands in **Excel**,
   a. Type **=command(**
   b. Select the range of data
   c. Type a right parenthesis, and then press **Enter**.
2. To find the mean, use the command **average**. (See Screens 3.9 and 3.10)
3. To find the median, use the command **median**.
4. To find the mode, use the command **mode**.

	A	B	C
1	Data	Average	
2			
3	2	=average(A3:A8)	
4	3		
5	5		
6	7		
7	11		
8	13		

Screen 3.9

	A	B	C
1	Data	Average	
2			
3	2	6.833333	
4	3		
5	5		
6	7		
7	11		
8	13		

Screen 3.10

5. To find the standard deviation, use the command **stdev**.
6. To find the first or third quartiles:
   a. Type **=quartile(**
   b. Select the range of data and then type a comma
   c. Type **1** for the first quartile, **3** for the third quartile
   d. Type a right parenthesis, and then press **Enter**.
7. To find the kth percentile:
   a. Type **=percentile(**
   b. Select the range of data and then type a comma
   c. Type the value of **k** followed by a right parenthesis, and then press **Enter**.

TECHNOLOGY ASSIGNMENTS

TA3.1 Refer to the subsample taken in the Computer Assignment TA2.3 of Chapter 2 from the sample data on the time taken to run the Manchester Road Race. Find the mean, median, range, and standard deviation for those data.

TA3.2 Refer to the data on phone charges given in Data Set I. From that data set select the 4th value and then select every 10th value after that (i.e., select the 4th, 14th, 24th, 34th . . . values). Such a sample taken from a population is called a *systematic random sample.* Find the mean, median, standard deviation, first quartile, and third quartile for the phone charges for this subsample.

TA3.3 Refer to Data Set I on the prices of various products in different cities across the country. Select a subsample of the prices of regular unleaded gas for 40 cities. Find the mean, median, and standard deviation for the data of this subsample.

TA3.4 Refer to Data of TA3.3. Make a box-and-whisker plot for those data.

TA3.5 Refer to Data Set I on the prices of various products in different cities across the country. Make a box-and-whisker plot for the data on the monthly telephone charges.

TA3.6 Refer to the data on the numbers of computer keyboards assembled at the Twentieth Century Electronics Company for a sample of 25 days given in Exercise 3.104. Prepare a box-and-whisker plot for those data.

Chapter

4

Probability

4.1 Experiment, Outcomes, and Sample Space

4.2 Calculating Probability

4.3 Counting Rule

4.4 Marginal and Conditional Probabilities

Case Study 4–1 American Lefties

4.5 Mutually Exclusive Events

4.6 Independent versus Dependent Events

4.7 Complementary Events

4.8 Intersection of Events and the Multiplication Rule

Case Study 4–2 Baseball Players have "Slumps" and "Streaks"

4.9 Union of Events and the Addition Rule

Are you a lefty? Have you ever noticed how many or what percentage of your friends are lefties? What about the past few presidents of the United States? Well, four of the last six presidents (Gerald Ford, Ronald Reagan, George H. Bush, and Bill Clinton) were left-handed. According to a sample survey, (about) 15% of the men and 9% of the women in the United States are lefties. (See Case Study 4–1.)

We often make statements about probability. For example, a weather forecaster may predict that there is an 80% chance of rain tomorrow. A health news reporter may state that a smoker has a much greater chance of getting cancer than a nonsmoker does. A college student may ask an instructor about the chances of passing a course or getting an A if he or she did not do well on the midterm examination.

Probability, which measures the likelihood that an event will occur, is an important part of statistics. It is the basis of inferential statistics, which will be introduced in later chapters. In inferential statistics, we make decisions under conditions of uncertainty. Probability theory is used to evaluate the uncertainty involved in those decisions. For example, estimating next year's sales for a company is based on many assumptions, some of which may happen to be true and others may not. Probability theory will help us make decisions under such conditions of imperfect information and uncertainty. Combining probability and probability distributions (which are discussed in Chapters 5 through 7) with descriptive statistics will help us make decisions about populations based on information obtained from samples. This chapter presents the basic concepts of probability and the rules for computing probability.

4.1 Experiment, Outcomes, and Sample Space

Quality control inspector Jack Cook of Tennis Products Company picks up a tennis ball from the production line to check whether it is good or defective. Cook's act of inspecting a tennis ball is an example of a statistical **experiment**. The result of his inspection will be that the ball is either "good" or "defective." Each of these two observations is called an **outcome** (also called a *basic* or *final outcome*) of the experiment, and these outcomes taken together constitute the **sample space** for this experiment.

Definition

Experiment, Outcomes, and Sample Space An *experiment* is a process that, when performed, results in one and only one of many observations. These observations are called the *outcomes* of the experiment. The collection of all outcomes for an experiment is called a *sample space.*

A sample space is denoted by S. The sample space for the example of inspecting a tennis ball is written as

$$S = \{\text{good, defective}\}$$

The elements of a sample space are called **sample points**.

Table 4.1 lists some examples of experiments, their outcomes, and their sample spaces.

Table 4.1 Examples of Experiments, Outcomes, and Sample Spaces

Experiment	Outcomes	Sample Space
Toss a coin once	Head, Tail	$S = \{\text{Head, Tail}\}$
Roll a die once	1, 2, 3, 4, 5, 6	$S = \{1, 2, 3, 4, 5, 6\}$
Toss a coin twice	*HH, HT, TH, TT*	$S = \{HH, HT, TH, TT\}$
Play lottery	Win, Lose	$S = \{\text{Win, Lose}\}$
Take a test	Pass, Fail	$S = \{\text{Pass, Fail}\}$
Select a student	Male, Female	$S = \{\text{Male, Female}\}$

The sample space for an experiment can also be illustrated by drawing either a Venn diagram or a tree diagram. A **Venn diagram** is a picture (a closed geometric shape such as a rectangle, a square, or a circle) that depicts all the possible outcomes for an experiment. In a **tree diagram**, each outcome is represented by a branch of the tree. Venn and tree diagrams help us understand probability concepts by presenting them visually. Examples 4–1 through 4–3 describe how to draw these diagrams for statistical experiments.

■ EXAMPLE 4–1

Draw the Venn and tree diagrams for the experiment of tossing a coin once.

Drawing Venn and tree diagrams: one toss of a coin.

Solution This experiment has two possible outcomes: head and tail. Consequently, the sample space is given by

$$S = \{H, T\} \qquad \text{where } H = \text{Head} \quad \text{and} \quad T = \text{Tail}$$

To draw a Venn diagram for this example, we draw a rectangle and mark two points inside this rectangle that represent the two outcomes, head and tail. The rectangle is labeled S because it represents the sample space (see Figure 4.1*a*). To draw a tree diagram, we draw two branches starting at the same point, one representing the head and the second representing the tail. The two final outcomes are listed at the ends of the branches (see Figure 4.1*b*).

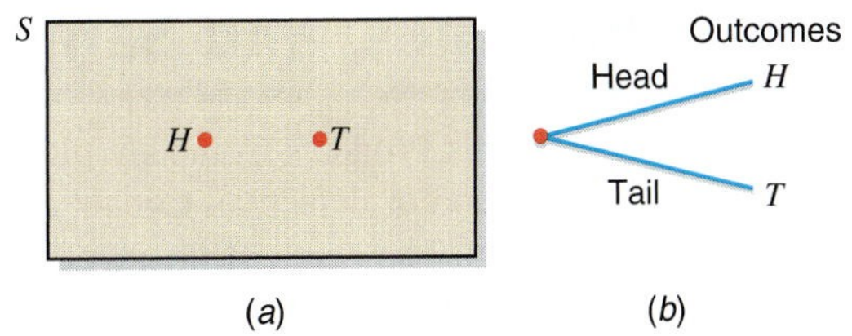

Figure 4.1 (*a*) Venn diagram and (*b*) tree diagram for one toss of a coin.

■ EXAMPLE 4–2

Drawing Venn and tree diagrams: two tosses of a coin.

Draw the Venn and tree diagrams for the experiment of tossing a coin twice.

Solution This experiment can be split into two parts: the first toss and the second toss. Suppose the first time the coin is tossed we obtain a head. Then, on the second toss, we can still obtain a head or a tail. This gives us two outcomes: *HH* (head on both tosses) and *HT* (head on the first toss and tail on the second toss). Now suppose we observe a tail on the first toss. Again, either a head or a tail can occur on the second toss, giving the remaining two outcomes: *TH* (tail on the first toss and head on the second toss) and *TT* (tail on both tosses). Thus, the sample space for two tosses of a coin is

$$S = \{HH, HT, TH, TT\}$$

The Venn and tree diagrams are given in Figure 4.2. Both these diagrams show the sample space for this experiment.

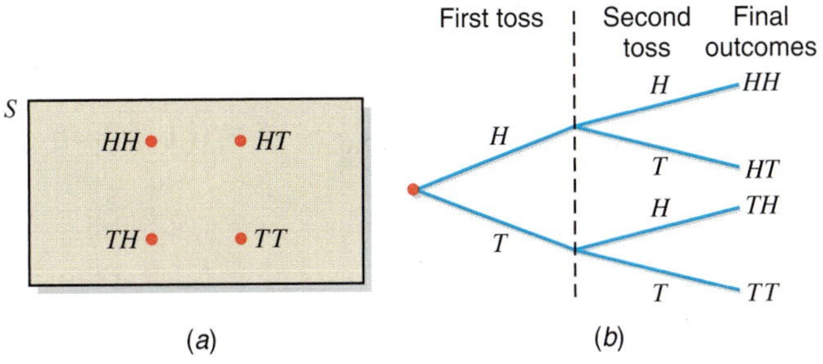

Figure 4.2 (*a*) Venn diagram and (*b*) tree diagram for two tosses of a coin.

■ EXAMPLE 4–3

Drawing Venn and tree diagrams: two selections.

Suppose we randomly select two persons from the members of a club and observe whether the person selected each time is a man or a woman. Write all the outcomes for this experiment. Draw the Venn and tree diagrams for this experiment.

Solution Let us denote the selection of a man by *M* and that of a woman by *W*. We can compare the selection of two persons to two tosses of a coin. Just as each toss of a coin can result in one of two outcomes, head or tail, each selection from the members of this club can result in one of two outcomes, man or woman. As we can see from the Venn and tree diagrams of Figure 4.3, there are four final outcomes: *MM*, *MW*, *WM*, *WW*. Hence, the sample space is written as

$$S = \{MM, MW, WM, WW\}$$

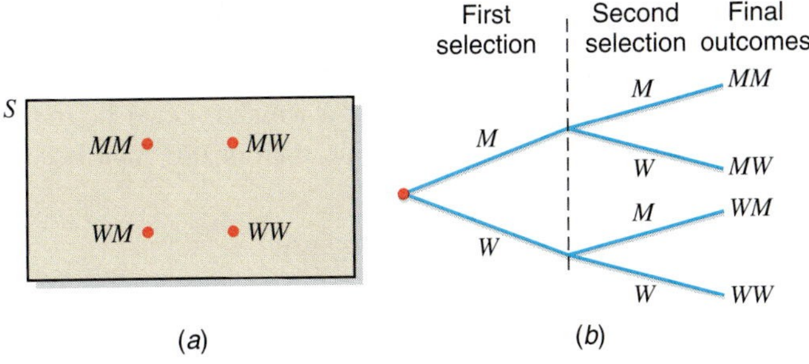

Figure 4.3 (*a*) Venn diagram and (*b*) tree diagram for selecting two persons.

4.1.1 Simple and Compound Events

An **event** consists of one or more of the outcomes of an experiment.

Definition

Event An *event* is a collection of one or more of the outcomes of an experiment.

An event may be a *simple event* or a *compound event.* A simple event is also called an *elementary event*, and a compound event is also called a *composite event.*

Simple Event

Each of the final outcomes for an experiment is called a **simple event**. In other words, a simple event includes one and only one outcome. Usually simple events are denoted by E_1, E_2, E_3, and so forth. However, we can denote them by any of the other letters, too—that is, by A, B, C, and so forth.

Definition

Simple Event An event that includes one and only one of the (final) outcomes for an experiment is called a *simple event* and is usually denoted by E_i.

Example 4–4 describes simple events.

EXAMPLE 4–4

Illustrating simple events.

Reconsider Example 4–3 on selecting two persons from the members of a club and observing whether the person selected each time is a man or a woman. Each of the final four outcomes (MM, MW, WM, and WW) for this experiment is a simple event. These four events can be denoted by E_1, E_2, E_3, and E_4, respectively. Thus,

$$E_1 = (MM),\quad E_2 = (MW),\quad E_3 = (WM),\quad \text{and} \quad E_4 = (WW)$$

Compound Event

A **compound event** consists of more than one outcome.

Definition

Compound Event A *compound event* is a collection of more than one outcome for an experiment.

Compound events are denoted by $A, B, C, D \ldots$ or by $A_1, A_2, A_3 \ldots, B_1, B_2, B_3 \ldots$, and so forth. Examples 4–5 and 4–6 describe compound events.

■ EXAMPLE 4–5

Illustrating a compound event: two selections.

Reconsider Example 4–3 on selecting two persons from the members of a club and observing whether the person selected each time is a man or a woman. Let A be the event that at most one man is selected. Event A will occur if either no man or one man is selected. Hence, the event A is given by

$$A = \{MW, WM, WW\}$$

Because event A contains more than one outcome, it is a compound event. The Venn diagram in Figure 4.4 gives a graphic presentation of compound event A. ■

Figure 4.4 Venn diagram for event A.

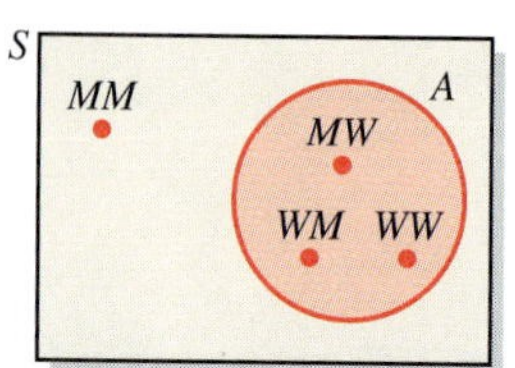

■ EXAMPLE 4–6

Illustrating simple and compound events: two selections.

In a group of people, some are in favor of genetic engineering and others are against it. Two persons are selected at random from this group and asked whether they are in favor of or against genetic engineering. How many distinct outcomes are possible? Draw a Venn diagram and a tree diagram for this experiment. List all the outcomes included in each of the following events and mention whether they are simple or compound events.

(a) Both persons are in favor of genetic engineering.

(b) At most one person is against genetic engineering.

(c) Exactly one person is in favor of genetic engineering.

Solution Let

F = a person is in favor of genetic engineering

A = a person is against genetic engineering

This experiment has the following four outcomes:

FF = both persons are in favor of genetic engineering

FA = the first person is in favor and the second is against

AF = the first person is against and the second is in favor

AA = both persons are against genetic engineering

The Venn and tree diagrams in Figure 4.5 show these four outcomes.

Figure 4.5 Venn and tree diagrams.

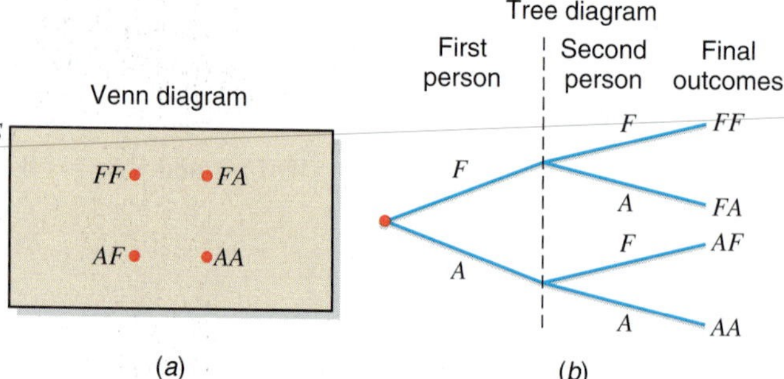

(a) The event "both persons are in favor of genetic engineering" will occur if FF is obtained. Thus,

$$\text{Both persons are in favor of genetic engineering} = \{FF\}$$

Because this event includes only one of the final four outcomes, it is a **simple** event.

(b) The event "at most one person is against genetic engineering" will occur if either none or one of the persons selected is against genetic engineering. Consequently,

$$\text{At most one person is against genetic engineering} = \{FF, FA, AF\}$$

Because this event includes more than one outcome, it is a **compound** event.

(c) The event "exactly one person is in favor of genetic engineering" will occur if one of the two persons selected is in favor and the other is against genetic engineering. Hence, it includes the following two outcomes:

$$\text{Exactly one person is in favor of genetic engineering} = \{FA, AF\}$$

Because this event includes more than one outcome, it is a **compound** event. ■

EXERCISES

CONCEPTS AND PROCEDURES

4.1 Define the following terms: *experiment, outcome, sample space, simple event,* and *compound event.*

4.2 List the simple events for each of the following statistical experiments in a sample space *S*.
a. One roll of a die b. Three tosses of a coin c. One toss of a coin and one roll of a die

4.3 A box contains three items that are labeled A, B, and C. Two items are selected at random (without replacement) from this box. List all the possible outcomes for this experiment. Write the sample space *S*.

APPLICATIONS

4.4 Two students are randomly selected from a statistics class, and it is observed whether or not they suffer from math anxiety. How many total outcomes are possible? Draw a tree diagram for this experiment. Draw a Venn diagram.

4.5 In a group of adults, some are computer literate, and the others are computer illiterate. If two adults are randomly selected from this group, how many total outcomes are possible? Draw a tree diagram for this experiment.

4.6 A test contains two multiple-choice questions. If a student makes a random guess to answer each question, how many outcomes are possible? Depict all these outcomes in a Venn diagram. Also draw a tree diagram for this experiment. (*Hint:* Consider two outcomes for each question—either the answer is correct or it is wrong.)

4.7 A box contains a certain number of computer parts, a few of which are defective. Two parts are selected at random from this box and inspected to determine if they are good or defective. How many total outcomes are possible? Draw a tree diagram for this experiment.

4.8 In a group of people, some are in favor of a tax increase on rich people to reduce the federal deficit and others are against it. (Assume that there is no other outcome such as "no opinion" and "do not know.") Three persons are selected at random from this group and their opinions in favor or against raising such taxes are noted. How many total outcomes are possible? Write these outcomes in a sample space *S*. Draw a tree diagram for this experiment.

4.9 Draw a tree diagram for three tosses of a coin. List all outcomes for this experiment in a sample space *S*.

4.10 Refer to Exercise 4.4. List all the outcomes included in each of the following events. Indicate which are simple and which are compound events.
a. Both students suffer from math anxiety.
b. Exactly one student suffers from math anxiety.
c. The first student does not suffer and the second suffers from math anxiety.
d. None of the students suffers from math anxiety.

4.11 Refer to Exercise 4.5. List all the outcomes included in each of the following events. Indicate which are simple and which are compound events.
a. One person is computer literate and the other is not.
b. At least one person is computer literate.
c. Not more than one person is computer literate.
d. The first person is computer literate and the second is not.

4.12 Refer to Exercise 4.6. List all the outcomes included in each of the following events and mention which are simple and which are compound events.

a. Both answers are correct.
b. At most one answer is wrong.
c. The first answer is correct and the second is wrong.
d. Exactly one answer is wrong.

4.13 Refer to Exercise 4.7. List all the outcomes included in each of the following events. Indicate which are simple and which are compound events.

a. At least one part is good.
b. Exactly one part is defective.
c. The first part is good and the second is defective.
d. At most one part is good.

4.14 Refer to Exercise 4.8. List all the outcomes included in each of the following events and mention which are simple and which are compound events.

a. At most one person is against a tax increase on rich people.
b. Exactly two persons are in favor of a tax increase on rich people.
c. At least one person is against a tax increase on rich people.
d. More than one person is against a tax increase on rich people.

4.2 Calculating Probability

Probability, which gives the likelihood of occurrence of an event, is denoted by *P*. The probability that a simple event E_i will occur is denoted by $P(E_i)$, and the probability that a compound event *A* will occur is denoted by $P(A)$.

Definition

Probability *Probability* is a numerical measure of the likelihood that a specific event will occur.

Two Properties of Probability ▶

1. The probability of an event always lies in the range 0 to 1.

Whether it is a simple or a compound event, the probability of an event is never less than 0 or greater than 1. Using mathematical notation, we can write this property as follows.

First Property of Probability

$$0 \le P(E_i) \le 1$$
$$0 \le P(A) \le 1$$

An event that cannot occur has zero probability; such an event is called an **impossible event**. An event that is certain to occur has a probability equal to 1 and is called a **sure event**. That is,

For an impossible event M: $P(M) = 0$

For a sure event C: $P(C) = 1$

2. The sum of the probabilities of all simple events (or final outcomes) for an experiment, denoted by $\Sigma P(E_i)$, is always 1.

Second Property of Probability For an experiment:

$$\Sigma P(E_i) = P(E_1) + P(E_2) + P(E_3) + \cdots = 1$$

From this property, for the experiment of one toss of a coin,

$$P(H) + P(T) = 1$$

For the experiment of two tosses of a coin,

$$P(HH) + P(HT) + P(TH) + P(TT) = 1$$

For one game of football by a professional team,

$$P(\text{win}) + P(\text{loss}) + P(\text{tie}) = 1$$

4.2.1 Three Conceptual Approaches to Probability

The three conceptual approaches to probability are (1) classical probability, (2) the relative frequency concept of probability, and (3) the subjective probability concept. These three concepts are explained next.

Classical Probability

Many times, various outcomes for an experiment may have the same probability of occurrence. Such outcomes are called **equally likely outcomes**. The classical probability rule is applied to compute the probabilities of events for an experiment for which all outcomes are equally likely.

Definition

Equally Likely Outcomes Two or more outcomes (or events) that have the same probability of occurrence are said to be *equally likely outcomes* (or events).

According to the **classical probability rule**, the probability of a simple event is equal to 1 divided by the total number of outcomes for the experiment. This is obvious because the sum of the probabilities of all final outcomes for an experiment is 1, and all the final outcomes are equally likely. In contrast, the probability of a compound event A is equal to the number of outcomes favorable to event A divided by the total number of outcomes for the experiment.

Classical Probability Rule to Find Probability

$$P(E_i) = \frac{1}{\text{Total number of outcomes for the experiment}}$$

$$P(A) = \frac{\text{Number of outcomes favorable to } A}{\text{Total number of outcomes for the experiment}}$$

Examples 4–7 through 4–9 illustrate how probabilities of events are calculated using the classical probability rule.

■ EXAMPLE 4–7

Calculating the probability of a simple event.

Find the probability of obtaining a head and the probability of obtaining a tail for one toss of a coin.

Solution The two outcomes, head and tail, are equally likely outcomes. Therefore,[1]

$$P(\text{head}) = \frac{1}{\text{Total number of outcomes}} = \frac{1}{2} = \mathbf{.50}$$

Similarly,

$$P(\text{tail}) = \frac{1}{2} = \mathbf{.50}$$

■

■ EXAMPLE 4–8

Calculating the probability of a compound event.

Find the probability of obtaining an even number in one roll of a die.

Solution This experiment has a total of six outcomes: 1, 2, 3, 4, 5, and 6. All these outcomes are equally likely. Let A be an event that an even number is observed on the die. Event A includes three outcomes: 2, 4, and 6; that is,

$$A = \{2, 4, 6\}$$

If any one of these three numbers is obtained, event A is said to occur. Hence,

$$P(A) = \frac{\text{Number of outcomes included in } A}{\text{Total number of outcomes}} = \frac{3}{6} = \mathbf{.50}$$

■

■ EXAMPLE 4–9

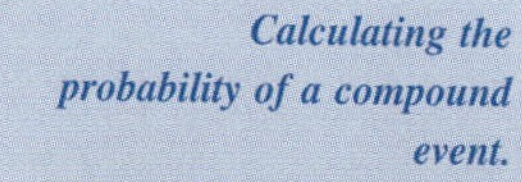

Calculating the probability of a compound event.

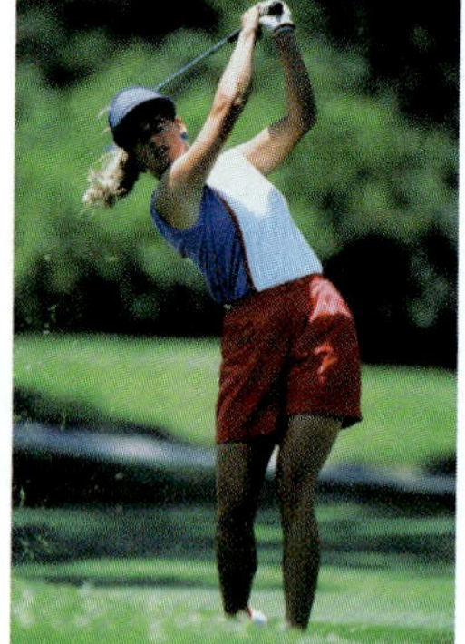

In a group of 500 women, 120 have played golf at least once. Suppose one of these 500 women is randomly selected. What is the probability that she has played golf at least once?

Solution Because the selection is to be made randomly, each of the 500 women has the same probability of being selected. Consequently this experiment has a total of 500 equally likely outcomes. One hundred twenty of these 500 outcomes are included in the event that the selected woman has played golf at least once. Hence,

$$P(\text{selected woman has played golf at least once}) = \frac{120}{500} = \mathbf{.24}$$

■

Relative Frequency Concept of Probability

Suppose we want to calculate the following probabilities:

1. The probability that the next car that comes out of an auto factory is a "lemon"
2. The probability that a randomly selected family owns a home
3. The probability that a randomly selected woman has never smoked
4. The probability that an 80-year-old person will live for at least 1 more year
5. The probability that the tossing of an unbalanced coin will result in a head
6. The probability that a randomly selected person owns a Sport-Utility Vehicle (SUV)

These probabilities cannot be computed using the classical probability rule because the various outcomes for the corresponding experiments are not equally likely. For example, the next car

[1]If the final answer for the probability of an event does not terminate within three decimal places, usually it is rounded to four decimal places.

manufactured at an auto factory may or may not be a lemon. The two outcomes, "it is a lemon" and "it is not a lemon," are not equally likely. If they were, then (approximately) half the cars manufactured by this company would be lemons, and this might prove disastrous to the survival of the firm.

Although the various outcomes for each of these experiments are not equally likely, each of these experiments can be performed again and again to generate data. In such cases, to calculate probabilities, we either use past data or generate new data by performing the experiment a large number of times. The relative frequency of an event is used as an approximation for the probability of that event. This method of assigning a probability to an event is called the **relative frequency concept of probability**. Because relative frequencies are determined by performing an experiment, the probabilities calculated using relative frequencies may change almost each time an experiment is repeated. For example, every time a new sample of 500 cars is selected from the production line of an auto factory, the number of lemons in those 500 cars is expected to be different. However, the variation in the percentage of lemons will be small if the sample size is large. Note that if we are considering the population, the relative frequency will give an exact probability.

Using Relative Frequency as an Approximation of Probability If an experiment is repeated n times and an event A is observed f times, then, according to the relative frequency concept of probability:

$$P(A) = \frac{f}{n}$$

Examples 4–10 and 4–11 illustrate how the probabilities of events are approximated using the relative frequencies.

■ EXAMPLE 4–10

Approximating probability by relative frequency: sample data.

Ten of the 500 randomly selected cars manufactured at a certain auto factory are found to be lemons. Assuming that the lemons are manufactured randomly, what is the probability that the next car manufactured at this auto factory is a lemon?

Solution Let n denote the total number of cars in the sample and f the number of lemons in n. Then,

$$n = 500 \quad \text{and} \quad f = 10$$

Using the relative frequency concept of probability, we obtain

$$P(\text{next car is a lemon}) = \frac{f}{n} = \frac{10}{500} = \mathbf{.02}$$

This probability is actually the relative frequency of lemons in 500 cars. Table 4.2 lists the frequency and relative frequency distributions for this example.

Table 4.2 Frequency and Relative Frequency Distributions for the Sample of Cars

Car	f	Relative Frequency
Good	490	490/500 = .98
Lemon	10	10/500 = .02
	$n = 500$	Sum = 1.00

The column of relative frequencies in Table 4.2 is used as the column of approximate probabilities. Thus, from the relative frequency column,

$$P(\text{next car is a lemon}) = \mathbf{.02}$$

$$P(\text{next car is a good car}) = \mathbf{.98}$$

■

Note that relative frequencies are not probabilities but approximate probabilities. However, if the experiment is repeated again and again, this approximate probability of an outcome obtained from the relative frequency will approach the actual probability of that outcome. This is called the **Law of Large Numbers**.

Definition

Law of Large Numbers If an experiment is repeated again and again, the probability of an event obtained from the relative frequency approaches the actual or theoretical probability.

■ EXAMPLE 4–11

Approximating probability by relative frequency.

Allison wants to determine the probability that a randomly selected family from New York State owns a home. How can she determine this probability?

Solution There are two outcomes for a randomly selected family from New York State: "This family owns a home" and "this family does not own a home." These two events are not equally likely. (Note that these two outcomes will be equally likely if exactly half of the families in New York State own homes and exactly half do not own homes.) Hence, the classical probability rule cannot be applied. However, we can repeat this experiment again and again. In other words, we can select a sample of families from New York State and observe whether or not each of them owns a home. Hence, we will use the relative frequency approach to probability.

Suppose Allison selects a random sample of 1000 families from New York State and observes that 670 of them own homes and 330 do not own homes. Then,

$$n = \text{sample size} = 1000$$

$$f = \text{number of families who own homes} = 670$$

Consequently,

$$P(\text{a randomly selected family owns a home}) = \frac{f}{n} = \frac{670}{1000} = \mathbf{.670}$$

Again, note that .670 is just an approximation of the probability that a randomly selected family from New York State owns a home. Every time Allison repeats this experiment she may obtain a different probability for this event. However, because the sample size (n = 1000) in this example is large, the variation is expected to be very small. ■

Subjective Probability

Many times we face experiments that neither have equally likely outcomes nor can be repeated to generate data. In such cases, we cannot compute the probabilities of events using the classical probability rule or the relative frequency concept. For example, consider the following probabilities of events:

1. The probability that Carol, who is taking statistics, will earn an A in this course
2. The probability that the Dow Jones Industrial Average will be higher at the end of the next trading day

3. The probability that the Miami Dolphins will win the Super Bowl next season
4. The probability that Joe will lose the lawsuit he has filed against his landlord

Neither the classical probability rule nor the relative frequency concept of probability can be applied to calculate probabilities for these examples. All these examples belong to experiments that have neither equally likely outcomes nor the potential of being repeated. For example, Carol, who is taking statistics, will take the test (or tests) only once, and based on that she will either earn an A or not. The two events "she will earn an A" and "she will not earn an A" are not equally likely. The probability assigned to an event in such cases is called **subjective probability**. It is based on the individual's own judgment, experience, information, and belief. Carol may assign a high probability to the event that she will earn an A in statistics, whereas her instructor may assign a low probability to the same event.

Definition

Subjective Probability *Subjective probability* is the probability assigned to an event based on subjective judgment, experience, information, and belief.

Subjective probability is assigned arbitrarily. It is usually influenced by the biases, preferences, and experience of the person assigning the probability.

EXERCISES

CONCEPTS AND PROCEDURES

4.15 Briefly explain the two properties of probability.

4.16 Briefly describe an impossible event and a sure event. What is the probability of the occurrence of each of these two events?

4.17 Briefly explain the three approaches to probability. Give one example of each approach.

4.18 Briefly explain for what kind of experiments we use the classical approach to calculate probabilities of events and for what kind of experiments we use the relative frequency approach.

4.19 Which of the following values cannot be probabilities of events and why?

1/5 .97 −.55 1.56 5/3 0.0 −2/7 1.0

4.20 Which of the following values cannot be probabilities of events and why?

.46 2/3 −.09 1.42 .96 9/4 −1/4 .02

APPLICATIONS

4.21 Suppose a randomly selected passenger is about to go through the metal detector at the JFK New York airport. Consider the following two outcomes: The passenger sets off the metal detector, and the passenger does not set off the metal detector. Are these two outcomes equally likely? Explain why or why not. If you are to find the probability of these two outcomes, would you use the classical approach or the relative frequency approach? Explain why.

4.22 Thirty-two persons have applied for a security guard position with a company. Of them, 7 have previous experience in this area and 25 do not. Suppose one applicant is selected at random. Consider the following two events: This applicant has previous experience, and this applicant does not have previous experience. If you are to find the probabilities of these two events, would you use the classical approach or the relative frequency approach? Explain why.

4.23 The president of a company has a hunch that there is a .80 probability that the company will be successful in marketing a new brand of ice cream. Is this a case of classical, relative frequency, or subjective probability? Explain why.

4.24 The coach of a college football team thinks there is a .75 probability that the team will win the national championship this year. Is this a case of classical, relative frequency, or subjective probability? Explain why.

4.25 A hat contains 40 marbles. Of them, 18 are red and 22 are green. If one marble is randomly selected out of this hat, what is the probability that this marble is

a. red? b. green?

4.26 A die is rolled once. What is the probability that

a. a number less than 5 is obtained?

b. a number 3 to 6 is obtained?

4.27 A random sample of 2000 adults showed that 1120 of them have shopped at least once on the Internet. What is the (approximate) probability that a randomly selected adult has shopped on the Internet?

4.28 In a statistics class of 42 students, 28 have volunteered for community service in the past. Find the probability that a randomly selected student from this class has volunteered for community service in the past.

4.29 In a group of 50 executives, 29 have a type A personality. If one executive is selected at random from this group, what is the probability that this executive has a type A personality?

4.30 Out of the 3000 families who live in an apartment complex in New York City, 600 paid no income tax last year. What is the probability that a randomly selected family from these 3000 families paid income tax last year?

4.31 A multiple-choice question on a test has five answers. If Dianne chooses one answer based on "pure guess," what is the probability that her answer is

a. correct? b. wrong?

Do these two probabilities add up to 1.0? If yes, why?

4.32 There are 1265 eligible voters in a town and 972 of them are registered to vote. If one eligible voter is selected at random, what is the probability that this voter is

a. registered b. not registered?

Do these two probabilities add up to 1.0? If yes, why?

4.33 A company that plans to hire one new employee has prepared a final list of six candidates, all of whom are equally qualified. Four of these six candidates are women. If the company decides to select at random one person out of these six candidates, what is the probability that this person will be a woman? What is the probability that this person will be a man? Do these two probabilities add up to 1.0? If yes, why?

4.34 A sample of 500 large companies showed that 120 of them offer free psychiatric help to their employees who suffer from psychological problems. If one company is selected at random from this sample, what is the probability that this company offers free psychiatric help to its employees who suffer from psychological problems? What is the probability that this company does not offer free psychiatric help to its employees who suffer from psychological problems? Do these two probabilities add up to 1.0? If yes, why?

4.35 A sample of 400 large companies showed that 130 of them offer free health fitness centers to their employees on the company premises. If one company is selected at random from this sample, what is the probability that this company offers a free health fitness center to its employees on the company premises? What is the probability that this company does not offer a free health fitness center to its employees on the company premises? Do these two probabilities add up to 1.0? If yes, why?

4.36 In a large city, 15,000 workers lost their jobs last year. Of them, 7400 lost their jobs because their companies closed down or moved, 4600 lost their jobs due to insufficient work, and the remainder lost their jobs because their positions were abolished. If one of these 15,000 workers is selected at random, find the probability that this worker lost his or her job

a. because the company closed down or moved

b. due to insufficient work

c. because the position was abolished

Do these probabilities add up to 1.0? If so, why?

4.37 A sample of 820 adults showed that 80 of them had no credit cards, 116 had one card each, 94 had two cards each, 77 had three cards each, 43 had four cards each, and 410 had five or more cards each. Write the frequency distribution table for the number of credit cards an adult possesses. Calculate the relative frequencies for all categories. Suppose one adult is randomly selected from these 820 adults. Find the probability that this adult has

a. three credit cards b. five or more credit cards

4.38 In a sample of 500 families, 90 have a yearly income of less than $40,000, 270 have a yearly income of $40,000 to $80,000, and the remaining families have a yearly income of more than $80,000. Write the frequency distribution table for this problem. Calculate the relative frequencies for all classes. Suppose one family is randomly selected from these 500 families. Find the probability that this family has a yearly income of

a. less than $40,000 b. more than $80,000

4.39 Suppose you want to find the (approximate) probability that a randomly selected family from Los Angeles earns more than $125,000 a year. How would you find this probability? What procedure would you use? Explain briefly.

4.40 Suppose you have a loaded die and you want to find the (approximate) probabilities of different outcomes for this die. How would you find these probabilities? What procedure would you use? Explain briefly.

4.3 COUNTING RULE

The experiments dealt with so far in this chapter have had only a few outcomes, which were easy to list. However, for experiments with a large number of outcomes, it may not be easy to list all outcomes. In such cases, we may use the **counting rule** to find the total number of outcomes.

Counting Rule to Find Total Outcomes If an experiment consists of three steps and if the first step can result in m outcomes, the second step in n outcomes, and the third step in k outcomes, then

$$\text{Total outcomes for the experiment} = m \cdot n \cdot k$$

The counting rule can easily be extended to apply to an experiment that has fewer or more than three steps.

■ EXAMPLE 4–12

Applying the counting rule: 3 steps.

Suppose we toss a coin three times. This experiment has three steps: the first toss, the second toss, and the third toss. Each step has two outcomes: a head and a tail. Thus,

$$\text{Total outcomes for three tosses of a coin} = 2 \times 2 \times 2 = \mathbf{8}$$

The eight outcomes for this experiment are *HHH*, *HHT*, *HTH*, *HTT*, *THH*, *THT*, *TTH*, and *TTT*. ■

■ EXAMPLE 4–13

Applying the counting rule: 2 steps.

A prospective car buyer can choose between a fixed and a variable interest rate and can also choose a payment period of 36 months, 48 months, or 60 months. How many total outcomes are possible?

Solution This experiment is made up of two steps: choosing an interest rate and selecting a loan payment period. There are two outcomes (a fixed or a variable interest rate) for the first step and three outcomes (a payment period of 36 months, 48 months, or 60 months) for the second step. Hence,

$$\text{Total outcomes} = 2 \times 3 = \mathbf{6}$$

■

■ EXAMPLE 4–14

Applying the counting rule: 16 steps.

A National Football League team will play 16 games during a regular season. Each game can result in one of three outcomes: a win, a loss, or a tie. The total possible outcomes for 16 games are calculated as follows:

$$\text{Total outcomes} = 3 \cdot 3 \cdot 3 \cdot 3 \cdot 3 \cdot 3 \cdot 3 \cdot 3 \cdot 3 \cdot 3 \cdot 3 \cdot 3 \cdot 3 \cdot 3 \cdot 3 \cdot 3$$
$$= 3^{16} = \mathbf{43{,}046{,}721}$$

One of the 43,046,721 possible outcomes is all 16 wins. ■

4.4 Marginal and Conditional Probabilities

Suppose all 100 employees of a company were asked whether they are in favor of or against paying high salaries to CEOs of U.S. companies. Table 4.3 gives a two-way classification of the responses of these 100 employees.

Table 4.3 Two-Way Classification of Employee Responses

	In Favor	Against
Male	15	45
Female	4	36

Table 4.3 shows the distribution of 100 employees based on two variables or characteristics: gender (male or female) and opinion (in favor or against). Such a table is called a *contingency table.* In Table 4.3, each box that contains a number is called a *cell.* Notice that there are four cells. Each cell gives the frequency for two characteristics. For example, 15 employees in this group possess two characteristics: "male" and "in favor of paying high salaries to CEOs." We can interpret the numbers in other cells the same way.

By adding the row totals and the column totals to Table 4.3, we write Table 4.4.

Table 4.4 Two-Way Classification of Employee Responses with Totals

	In Favor	Against	Total
Male	15	45	60
Female	4	36	40
Total	19	81	100

Suppose one employee is selected at random from these 100 employees. This employee may be classified either on the basis of gender alone or on the basis of opinion. If only one characteristic is considered at a time, the employee selected can be a male, a female, in favor, or against. The probability of each of these four characteristics or events is called **marginal probability** or *simple probability.* These probabilities are called marginal probabilities because they are calculated by dividing the corresponding row margins (totals for the rows) or column margins (totals for the columns) by the grand total.

Definition

Marginal Probability *Marginal probability* is the probability of a single event without consideration of any other event. Marginal probability is also called *simple probability.*

For Table 4.4, the four marginal probabilities are calculated as follows:

$$P(\text{male}) = \frac{\text{Number of males}}{\text{Total number of employees}} = \frac{60}{100} = .60$$

As we can observe, the probability that a male will be selected is obtained by dividing the total of the row labeled "Male" (60) by the grand total (100). Similarly,

$$P(\text{female}) = 40/100 = .40$$

$$P(\text{in favor}) = 19/100 = .19$$

$$P(\text{against}) = 81/100 = .81$$

These four marginal probabilities are shown along the right side and along the bottom of Table 4.5.

Table 4.5 Listing the Marginal Probabilities

	In Favor (*A*)	Against (*B*)	Total	
Male (*M*)	15	45	60	$P(M) = 60/100 = .60$
Female (*F*)	4	36	40	$P(F) = 40/100 = .40$
Total	19	81	100	
	$P(A) = 19/100 = .19$	$P(B) = 81/100 = .81$		

Now suppose that one employee is selected at random from these 100 employees. Furthermore, assume it is known that this (selected) employee is a male. In other words, the event that the employee selected is a male has already occurred. What is the probability that the employee selected is in favor of paying high salaries to CEOs? This probability is written as follows:

$$P(\text{in favor} \mid \text{male})$$

Read as "given"

The event whose probability is to be determined

This event has already occurred

This probability, $P(\text{in favor} \mid \text{male})$, is called the **conditional probability** of "in favor." It is read as "the probability that the employee selected is in favor given that this employee is a male."

Definition

Conditional Probability *Conditional probability* is the probability that an event will occur given that another event has already occurred. If A and B are two events, then the conditional probability of A given B is written as

$$P(A \mid B)$$

and read as "the probability of A given that B has already occurred."

■ EXAMPLE 4–15

Calculating the conditional probability: two-way table.

Compute the conditional probability *P*(in favor | male) for the data on 100 employees given in Table 4.4.

Solution The probability *P*(in favor | male) is the conditional probability that a randomly selected employee is in favor given that this employee is a male. It is known that the event "male" has already occurred. Based on the information that the employee selected is a male, we can infer that the employee selected must be one of the 60 males and, hence, must belong to the first row of Table 4.4. Therefore, we are concerned only with the first row of that table.

	In Favor	Against	Total
Male	15	45	60
	↑ Males who are in favor		↑ Total number of males

The required conditional probability is calculated as follows:

$$P(\text{in favor} \mid \text{male}) = \frac{\text{Number of males who are in favor}}{\text{Total number of males}} = \frac{15}{60} = \mathbf{.25}$$

As we can observe from this computation of conditional probability, the total number of males (the event that has already occurred) is written in the denominator and the number of males who are in favor (the event whose probability we are to find) is written in the numerator. Note that we are considering the row of the event that has already occurred. The tree diagram in Figure 4.6 illustrates this example.

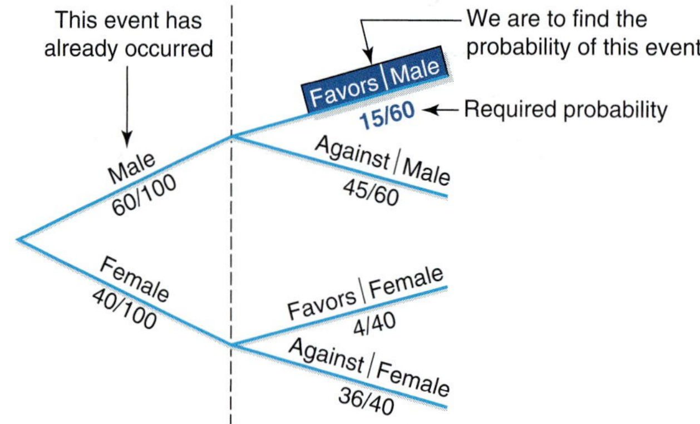

Figure 4.6 Tree diagram.

■

■ EXAMPLE 4–16

Calculating the conditional probability: two-way table.

For the data of Table 4.4, calculate the conditional probability that a randomly selected employee is a female given that this employee is in favor of paying high salaries to CEOs.

Solution We are to compute the probability, *P*(female | in favor). Because it is known that the employee selected is in favor of paying high salaries to CEOs, this employee must belong to the first column (the column labeled "in favor") and must be one of the 19 employees who are in favor.

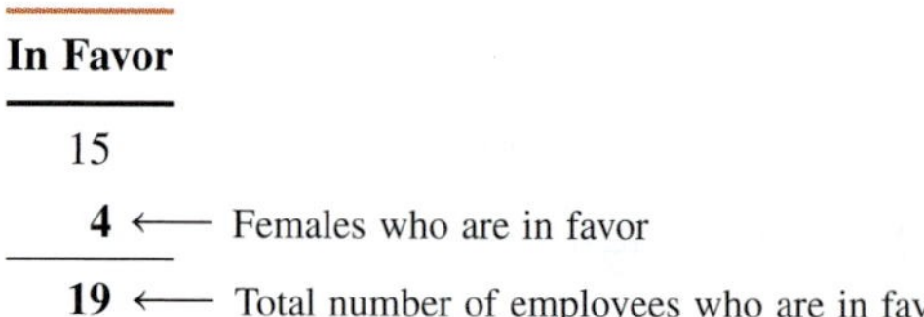

In Favor	
15	
4	⟵ Females who are in favor
19	⟵ Total number of employees who are in favor

CASE STUDY 4–1 AMERICAN LEFTIES

USA TODAY Snapshots®

American lefties

Four of the last six presidents (Ford, Reagan, Bush, Clinton) are left-handed. Percentage of lefties by sex in the USA:

Men 15%

Women 9%

Source: Scripps Survey Research Center Poll

By Allison Gashin and Suzy Parker, USA TODAY

The above chart shows the percentage of men and women who are left-handed. As the percentages in the chart show, more men are left-handed (15%) than women (9%). Thus, if we randomly select one person, there is a higher chance that the selected person is left-handed if this person is a man than if this person is a woman. The percentages given in the chart can be written as conditional probabilities as follows. Suppose one person is selected at random. Then, given that this person is a male, the probability is .15 that he is left-handed. On the other hand, if this selected person is a female, the probability is only .09 that she is left-handed. These probabilities can be written as:

$$P(\text{selected person is left-handed} \mid \text{male}) = .15$$

$$P(\text{selected person is left-handed} \mid \text{female}) = .09$$

Note that these are approximate probabilities because the percentages given in the chart are based on a sample survey.

Hence, the required probability is

$$P(\text{female} \mid \text{in favor}) = \frac{\text{Number of females who are in favor}}{\text{Total number of employees who are in favor}} = \frac{4}{19} = \mathbf{.2105}$$

The tree diagram in Figure 4.7 illustrates this example.

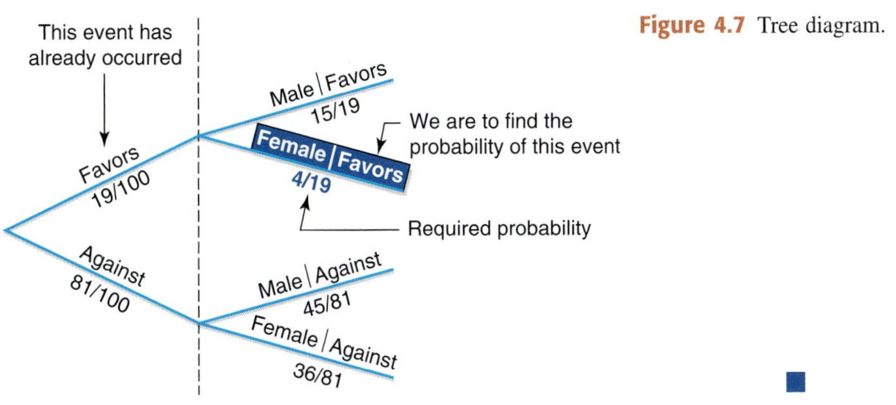

Figure 4.7 Tree diagram.

4.5 Mutually Exclusive Events

Events that cannot occur together are called **mutually exclusive events**. Such events do not have any common outcomes. If two or more events are mutually exclusive, then at most one of them will occur every time we repeat the experiment. Thus the occurrence of one event excludes the occurrence of the other event or events.

Definition

Mutually Exclusive Events Events that cannot occur together are said to be *mutually exclusive events.*

For any experiment, the final outcomes are always mutually exclusive because one and only one of these outcomes is expected to occur in one repetition of the experiment. For example, consider tossing a coin twice. This experiment has four outcomes: *HH*, *HT*, *TH*, and *TT*. These outcomes are mutually exclusive because one and only one of them will occur when we toss this coin twice.

■ EXAMPLE 4–17

Illustrating mutually exclusive and mutually nonexclusive events.

Consider the following events for one roll of a die:

$$A = \text{an even number is observed} = \{2, 4, 6\}$$

$$B = \text{an odd number is observed} = \{1, 3, 5\}$$

$$C = \text{a number less than 5 is observed} = \{1, 2, 3, 4\}$$

Are events A and B mutually exclusive? Are events A and C mutually exclusive?

Solution Figures 4.8 and 4.9 show the diagrams of events A and B and events A and C, respectively.

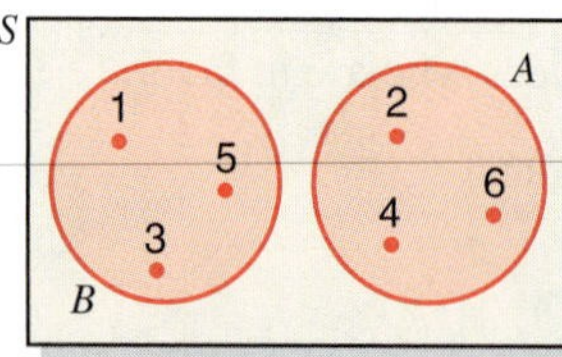

Figure 4.8 Mutually exclusive events A and B.

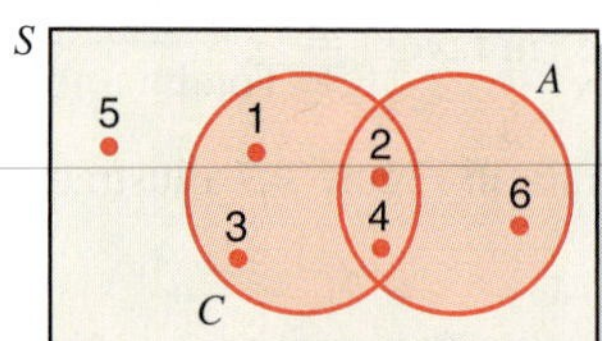

Figure 4.9 Mutually nonexclusive events A and C.

As we can observe from the definitions of events A and B and from Figure 4.8, events A and B have no common element. For one roll of a die, only one of the two events, A and B, can happen. Hence, these are two mutually exclusive events. We can observe from the definitions of events A and C and from Figure 4.9 that events A and C have two common outcomes: 2-spot and 4-spot. Thus, if we roll a die and obtain either a 2-spot or a 4-spot, then A and C happen at the same time. Hence, events A and C are not mutually exclusive. ■

■ EXAMPLE 4–18

Illustrating mutually exclusive events.

Consider the following two events for a randomly selected adult:

Y = this adult has shopped on the Internet at least once

N = this adult has never shopped on the Internet

Are events Y and N mutually exclusive?

Solution Note that event Y consists of all adults who have shopped on the Internet at least once, and event N includes all adults who have never shopped on the Internet. These two events are illustrated in the Venn diagram in Figure 4.10.

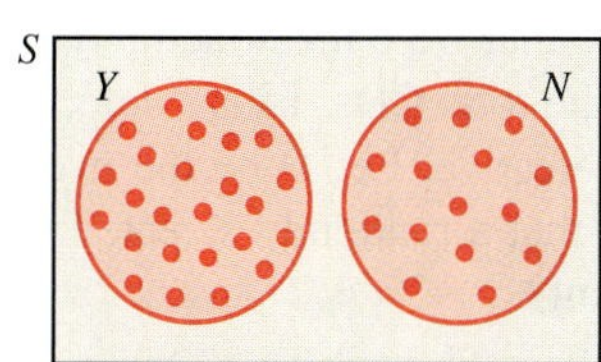

Figure 4.10 Mutually exclusive events Y and N.

As we can observe from the definitions of events Y and N and from Figure 4.10, events Y and N have no common outcome. They represent two distinct sets of adults: the ones who have shopped on the Internet at least once and the ones who have never shopped on the Internet. Hence, these two events are mutually exclusive. ■

4.6 Independent Versus Dependent Events

In the case of two **independent events**, the occurrence of one event does not change the probability of the occurrence of the other event.

Definition

Independent Events Two events are said to be *independent* if the occurrence of one does not affect the probability of the occurrence of the other. In other words, A and B are *independent events* if

$$\text{either} \quad P(A \mid B) = P(A) \quad \text{or} \quad P(B \mid A) = P(B)$$

It can be shown that if one of these two conditions is true, then the second will also be true, and if one is not true, then the second will also not be true.

If the occurrence of one event affects the probability of the occurrence of the other event, then the two events are said to be **dependent events**. In probability notation, the two events are dependent if either $P(A \mid B) \neq P(A)$ or $P(B \mid A) \neq P(B)$.

■ EXAMPLE 4–19

Illustrating two dependent events: two-way table.

Refer to the information on 100 employees given in Table 4.4 in Section 4.4. Are events "female (F)" and "in favor (A)" independent?

Solution Events F and A will be independent if

$$P(F) = P(F \mid A)$$

Otherwise they will be dependent.

Using the information given in Table 4.4, we compute the following two probabilities:

$$P(F) = 40/100 = \mathbf{.40} \quad \text{and} \quad P(F \mid A) = 4/19 = \mathbf{.2105}$$

Because these two probabilities are not equal, the two events are dependent. Here, dependence of events means that the percentages of males who are in favor of and against paying high salaries to CEOs are different from the percentages of females who are in favor and against.

In this example, the dependence of A and F can also be proved by showing that the probabilities $P(A)$ and $P(A \mid F)$ are not equal. ■

■ EXAMPLE 4–20

Illustrating two independent events.

A box contains a total of 100 CDs that were manufactured on two machines. Of them, 60 were manufactured on Machine I. Of the total CDs, 15 are defective. Of the 60 CDs that were manufactured on Machine I, 9 are defective. Let D be the event that a randomly selected CD is defective, and let A be the event that a randomly selected CD was manufactured on Machine I. Are events D and A independent?

Solution From the given information,

$$P(D) = 15/100 = .15 \quad \text{and} \quad P(D \mid A) = 9/60 = .15$$

Hence,

$$P(D) = P(D \mid A)$$

Consequently, the two events, D and A, are independent.

Independence, in this example, means that the probability of any CD being defective is the same, .15, irrespective of the machine on which it is manufactured. In other words, the two machines are producing the same percentage of defective CDs. For example, 9 of the 60 CDs manufactured on Machine I are defective and 6 of the 40 CDs manufactured on Machine II are defective. Thus, for each of the two machines, 15% of the CDs produced are defective.

Actually, using the given information, we can prepare Table 4.6. The numbers in the shaded cells are given to us. The remaining numbers are calculated by doing some arithmetic manipulations.

Table 4.6 Two-Way Classification Table

	Defective (D)	Good (G)	Total
Machine I (A)	9	51	60
Machine II (B)	6	34	40
Total	15	85	100

Using this table, we can find the following probabilities:

$$P(D) = 15/100 = .15$$

$$P(D \mid A) = 9/60 = .15$$

Because these two probabilities are the same, the two events are independent. ■

Two Important Observations ▶ We can make the following two important observations about mutually exclusive, independent, and dependent events.

1. Two events are either mutually exclusive or independent.[2]
   a. Mutually exclusive events are always dependent.
   b. Independent events are never mutually exclusive.
2. Dependent events may or may not be mutually exclusive.

4.7 Complementary Events

Two mutually exclusive events that taken together include all the outcomes for an experiment are called **complementary events**. Note that two complementary events are always mutually exclusive.

Definition

Complementary Events The complement of event A, denoted by $\bar{A}$ and read as "A bar" or "A complement," is the event that includes all the outcomes for an experiment that are not in A.

Events A and $\bar{A}$ are complements of each other. The Venn diagram in Figure 4.11 shows the complementary events A and $\bar{A}$.

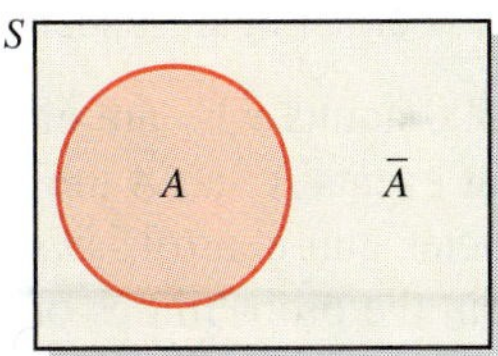

Figure 4.11 Venn diagram of two complementary events.

Because two complementary events, taken together, include all the outcomes for an experiment and because the sum of the probabilities of all outcomes is 1, it is obvious that

$$P(A) + P(\bar{A}) = 1$$

From this equation, we can deduce that

$$P(A) = 1 - P(\bar{A}) \quad \text{and} \quad P(\bar{A}) = 1 - P(A)$$

Thus, if we know the probability of an event, we can find the probability of its complementary event by subtracting the given probability from 1.

■ EXAMPLE 4–21

Calculating probabilities of complementary events.

In a group of 2000 taxpayers, 400 have been audited by the IRS at least once. If one taxpayer is randomly selected from this group, what are the two complementary events for this experiment, and what are their probabilities?

Solution The two complementary events for this experiment are

A = the selected taxpayer has been audited by the IRS at least once

$\bar{A}$ = the selected taxpayer has never been audited by the IRS

Note that here event A includes the 400 taxpayers who have been audited by the IRS at least once, and $\bar{A}$ includes the 1600 taxpayers who have never been audited by the IRS. Hence, the probabilities of events A and $\bar{A}$ are

$$P(A) = 400/2000 = \mathbf{.20} \quad \text{and} \quad P(\bar{A}) = 1600/2000 = \mathbf{.80}$$

[2]The exception to this rule occurs when at least one of the two events has a zero probability.

As we can observe, the sum of these two probabilities is 1. Figure 4.12 shows a Venn diagram for this example.

Figure 4.12 Venn diagram.

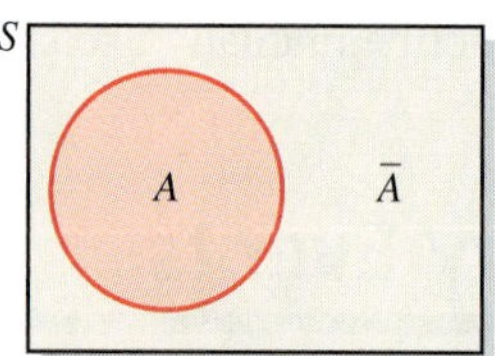

EXAMPLE 4–22

Calculating probabilities of complementary events.

In a group of 5000 adults, 3500 are in favor of stricter gun control laws, 1200 are against such laws, and 300 have no opinion. One adult is randomly selected from this group. Let A be the event that this adult is in favor of stricter gun control laws. What is the complementary event of A? What are the probabilities of the two events?

Solution The two complementary events for this experiment are

A = the selected adult is in favor of stricter gun control laws

$\bar{A}$ = the selected adult is either against such laws or has no opinion

Note that here event $\bar{A}$ includes 1500 adults who are either against stricter gun control laws or have no opinion. Also notice that events A and $\bar{A}$ are complements of each other. Because 3500 adults in the group favor stricter gun control laws and 1500 either are against stricter gun control laws or have no opinion, the probabilities of events A and $\bar{A}$ are

$$P(A) = 3500/5000 = \mathbf{.70} \quad \text{and} \quad P(\bar{A}) = 1500/5000 = \mathbf{.30}$$

As we can observe, the sum of these two probabilities is 1. Also, once we find $P(A)$, we can find the probability of $P(\bar{A})$ as

$$P(\bar{A}) = 1 - P(A) = 1 - .70 = .30$$

Figure 4.13 shows a Venn diagram for this example.

Figure 4.13 Venn diagram.

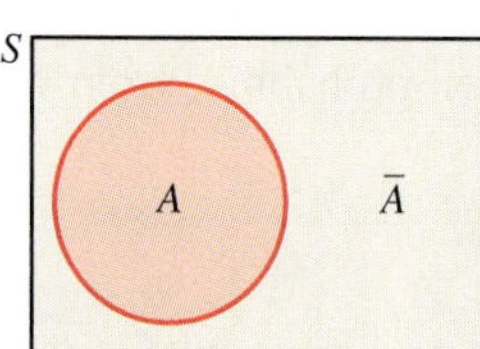

EXERCISES

CONCEPTS AND PROCEDURES

4.41 Briefly explain the difference between the marginal and conditional probabilities of events. Give one example of each.

4.42 What is meant by two mutually exclusive events? Give one example of two mutually exclusive events and another example of two mutually nonexclusive events.

4.43 Briefly explain the meaning of independent and dependent events. Suppose A and B are two events. What formula can you use to prove whether A and B are independent or dependent?

4.44 What is the complement of an event? What is the sum of the probabilities of two complementary events?

4.45 How many different outcomes are possible for four rolls of a die?

4.46 How many different outcomes are possible for 10 tosses of a coin?

4.47 A statistical experiment has eight equally likely outcomes that are denoted by 1, 2, 3, 4, 5, 6, 7, and 8. Let event $A = \{2, 5, 7\}$ and event $B = \{2, 4, 8\}$.

a. Are events A and B mutually exclusive events?
b. Are events A and B independent events?
c. What are the complements of events A and B, respectively, and their probabilities?

4.48 A statistical experiment has 10 equally likely outcomes that are denoted by 1, 2, 3, 4, 5, 6, 7, 8, 9, and 10. Let event $A = \{3, 4, 6, 9\}$ and event $B = \{1, 2, 5\}$.

a. Are events A and B mutually exclusive events?
b. Are events A and B independent events?
c. What are the complements of events A and B, respectively, and their probabilities?

APPLICATIONS

4.49 A small ice cream shop has 10 flavors of ice cream and 5 kinds of toppings for its sundaes. How many different selections of one flavor of ice cream and one kind of topping are possible?

4.50 A man just bought 4 suits, 8 shirts, and 12 ties. All of these suits, shirts, and ties coordinate with each other. If he is to randomly select one suit, one shirt, and one tie to wear on a certain day, how many different outcomes (selections) are possible?

4.51 A restaurant menu has four kinds of soups, eight kinds of main courses, five kinds of desserts, and six kinds of drinks. If a customer randomly selects one item from each of these four categories, how many different outcomes are possible?

4.52 A student is to select three classes for next semester. If this student decides to randomly select one course from each of eight economics classes, six mathematics classes, and five computer classes, how many different outcomes are possible?

4.53 Two thousand randomly selected adults were asked whether or not they have ever shopped on the Internet. The following table gives a two-way classification of the responses.

	Have Shopped	Have Never Shopped
Male	500	700
Female	300	500

a. If one adult is selected at random from these 2000 adults, find the probability that this adult
   i. has never shopped on the Internet
   ii. is a male
   iii. has shopped on the Internet given that this adult is a female
   iv. is a male given that this adult has never shopped on the Internet
b. Are the events "male" and "female" mutually exclusive? What about the events "have shopped" and "male"? Why or why not?
c. Are the events "female" and "have shopped" independent? Why or why not?

4.54 According to the TNS Online Kids Report (*USA TODAY*, May 24, 2004), in April 2004, 660 children aged 6 to 14 years were asked, "Do you worry about having enough money?" If in this survey of 660 children, 330 were boys and 330 were girls, the percentages given in the report would yield the numbers given in the following two-way classification table.

	Yes	No
Boys	201	129
Girls	178	152

a. If one child is selected at random from this group of 660 children, find the probability that this child
   i. worries about having enough money
   ii. is a girl
   iii. does not worry about having enough money given the child is a girl
   iv. is a girl given the child worries about having enough money

b. Are the events *worried about having enough money* and *not worried about having enough money* mutually exclusive? What about the events *worried* and *boys*?
c. Are the events *worried about having enough money* and *girls* independent? Why or why not?

4.55 Two thousand randomly selected adults were asked if they are in favor of or against cloning. The following table gives the responses.

	In Favor	Against	No Opinion
Male	395	405	100
Female	300	680	120

a. If one person is selected at random from these 2000 adults, find the probability that this person is
   i. in favor of cloning
   ii. against cloning
   iii. in favor of cloning given the person is a female
   iv. a male given the person has no opinion
b. Are the events "male" and "in favor" mutually exclusive? What about the events "in favor" and "against"? Why or why not?
c. Are the events "female" and "no opinion" independent? Why or why not?

4.56 Five hundred employees were selected from a city's large private companies, and they were asked whether or not they have any retirement benefits provided by their companies. Based on this information, the following two-way classification table was prepared.

	Have Retirement Benefits	
	Yes	No
Men	225	75
Women	150	50

a. If one employee is selected at random from these 500 employees, find the probability that this employee
   i. is a woman
   ii. has retirement benefits
   iii. has retirement benefits given the employee is a man
   iv. is a woman given that she does not have retirement benefits
b. Are the events "man" and "yes" mutually exclusive? What about the events "yes" and "no"? Why or why not?
c. Are the events "woman" and "yes" independent? Why or why not?

4.57 A consumer agency randomly selected 1700 flights for two major airlines, A and B. The following table gives the two-way classification of these flights based on airline and arrival time. Note that "less than 30 minutes late" includes flights that arrived early or on time.

	Less Than 30 Minutes Late	30 Minutes to 1 Hour Late	More Than 1 Hour Late
Airline A	429	390	92
Airline B	393	316	80

a. If one flight is selected at random from these 1700 flights, find the probability that this flight is
   i. more than 1 hour late
   ii. less than 30 minutes late
   iii. a flight on airline A given that it is 30 minutes to 1 hour late
   iv. more than 1 hour late given that it is a flight on airline B
b. Are the events "airline A" and "more than 1 hour late" mutually exclusive? What about the events "less than 30 minutes late" and "more than 1 hour late"? Why or why not?
c. Are the events "airline B" and "30 minutes to 1 hour late" independent? Why or why not?

4.58 Two thousand randomly selected adults were asked if they think they are financially better off than their parents. The following table gives the two-way classification of the responses based on the education levels of the persons included in the survey and whether they are financially better off, the same, or worse off than their parents.

	Less Than High School	High School	More Than High School
Better off	140	450	420
Same	60	250	110
Worse off	200	300	70

a. If one adult is selected at random from these 2000 adults, find the probability that this adult is
   i. financially better off than his/her parents
   ii. financially better off than his/her parents given he/she has less than high school education
   iii. financially worse off than his/her parents given he/she has high school education
   iv. financially the same as his/her parents given he/she has more than high school education

b. Are the events "better off" and "high school" mutually exclusive? What about the events "less than high school" and "more than high school"? Why or why not?

c. Are the events "worse off" and "more than high school" independent? Why or why not?

4.59 There are a total of 160 practicing physicians in a city. Of them, 75 are female and 25 are pediatricians. Of the 75 females, 20 are pediatricians. Are the events "female" and "pediatrician" independent? Are they mutually exclusive? Explain why or why not.

4.60 Of a total of 100 CDs manufactured on two machines, 20 are defective. Sixty of the total CDs were manufactured on Machine I, and 10 of these 60 are defective. Are the events "machine type" and "defective CDs" independent? (*Note:* Compare this exercise with Example 4–20.)

4.61 A company hired 30 new college graduates last week. Of these, 16 are female and 11 are business majors. Of the 16 females, 9 are business majors. Are the events "female" and "business major" independent? Are they mutually exclusive? Explain why or why not.

4.62 Define the following two events for two tosses of a coin:

$$A = \text{at least one head is obtained}$$

$$B = \text{both tails are obtained}$$

a. Are A and B mutually exclusive events? Are they independent? Explain why or why not.

b. Are A and B complementary events? If yes, first calculate the probability of B and then calculate the probability of A using the complementary event rule.

4.63 Let A be the event that a number less than 3 is obtained if we roll a die once. What is the probability of A? What is the complementary event of A, and what is its probability?

4.64 According to a recent Census American Housing Survey, 72.3 million American households owned their dwelling while 34.0 million rented (*USA TODAY*, October 13, 2004). Assume that all households in the United States are included in these two categories, and that this information is true for the current population. If one household is selected at random, what are the two complementary events and their probabilities?

4.65 The probability that a randomly selected college student attended at least one major league baseball game last year is .12. What is the complementary event? What is the probability of this complementary event?

4.8 Intersection of Events and the Multiplication Rule

This section discusses the intersection of two events and the application of the multiplication rule to compute the probability of the intersection of events.

4.8.1 Intersection of Events

The **intersection of two events** is given by the outcomes that are common to both events.

Definition

Intersection of Events Let A and B be two events defined in a sample space. The *intersection* of A and B represents the collection of all outcomes that are common to both A and B and is denoted by

$$A \text{ and } B$$

The intersection of events A and B is also denoted by either $\boldsymbol{A \cap B}$ or $\boldsymbol{AB}$. Let

A = event that a family owns a DVD player

B = event that a family owns a digital camera

Figure 4.14 illustrates the intersection of events A and B. The shaded area in this figure gives the intersection of events A and B, and it includes all the families who own both a DVD player and a digital camera.

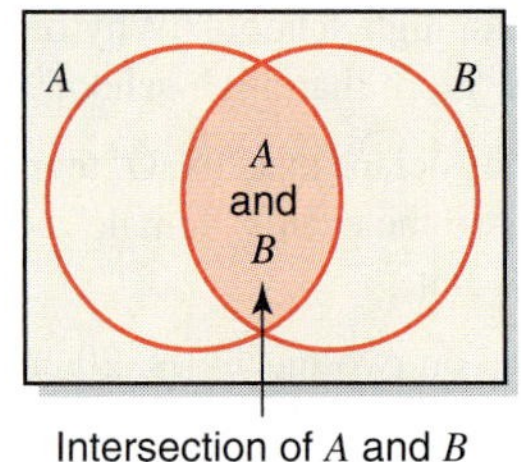

Figure 4.14 Intersection of events A and B.

4.8.2 Multiplication Rule

Sometimes we may need to find the probability of two or more events happening together.

Definition

Joint Probability The probability of the intersection of two events is called their *joint probability.* It is written as

$$P(A \text{ and } B)$$

The probability of the intersection of two events is obtained by multiplying the marginal probability of one event by the conditional probability of the second event. This rule is called the **multiplication rule**.

Multiplication Rule to Find Joint Probability The probability of the intersection of two events A and B is

$$P(A \text{ and } B) = P(A)\, P(B \mid A)$$

The joint probability of events A and B can also be denoted by $P(A \cap B)$ or $P(AB)$.

■ EXAMPLE 4–23

Calculating the joint probability of two events: two-way table.

Table 4.7 gives the classification of all employees of a company by gender and college degree.

Table 4.7 Classification of Employees by Gender and Education

	College Graduate (G)	Not a College Graduate (N)	Total
Male (M)	7	20	27
Female (F)	4	9	13
Total	11	29	40

If one of these employees is selected at random for membership on the employee–management committee, what is the probability that this employee is a female and a college graduate?

Solution We are to calculate the probability of the intersection of the events "female" (denoted by F) and "college graduate" (denoted by G). This probability may be computed using the formula

$$P(F \text{ and } G) = P(F)\,P(G \mid F)$$

The area shaded in red in Figure 4.15 shows the intersection of the events "female" and "college graduate."

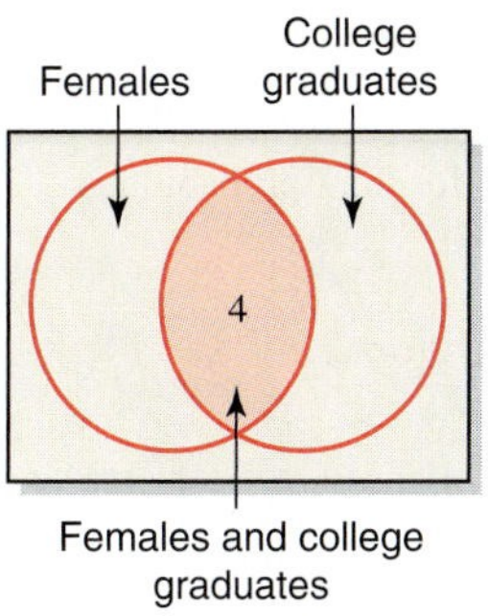

Figure 4.15 Intersection of events F and G.

Notice that there are 13 females among 40 employees. Hence, the probability that a female is selected is

$$P(F) = 13/40$$

To calculate the probability $P(G \mid F)$, we know that F has already occurred. Consequently, the employee selected is one of the 13 females. In the table, there are 4 college graduates among 13 female employees. Hence, the conditional probability of G given F is

$$P(G \mid F) = 4/13$$

The joint probability of F and G is

$$P(F \text{ and } G) = P(F)\,P(G \mid F) = (13/40)(4/13) = \mathbf{.100}$$

Thus, the probability is .100 that a randomly selected employee is a female and a college graduate.

The probability in this example can also be calculated without using the multiplication rule. As we can notice from Figure 4.15 and from the table, 4 employees out of a total of 40 are female and college graduates. Hence, if any of these four employees is selected, the events "female" and "college graduate" both happen. Therefore, the required probability is

$$P(F \text{ and } G) = 4/40 = \mathbf{.100}$$

■

We can compute three other joint probabilities for the table in Example 4–23 as follows:

$$P(M \text{ and } G) = P(M)\,P(G \mid M) = (27/40)(7/27) = .175$$

$$P(M \text{ and } N) = P(M)\,P(N \mid M) = (27/40)(20/27) = .500$$

$$P(F \text{ and } N) = P(F)\,P(N \mid F) = (13/40)(9/13) = .225$$

The tree diagram in Figure 4.16 shows all four joint probabilities for this example. The joint probability of F and G is highlighted.

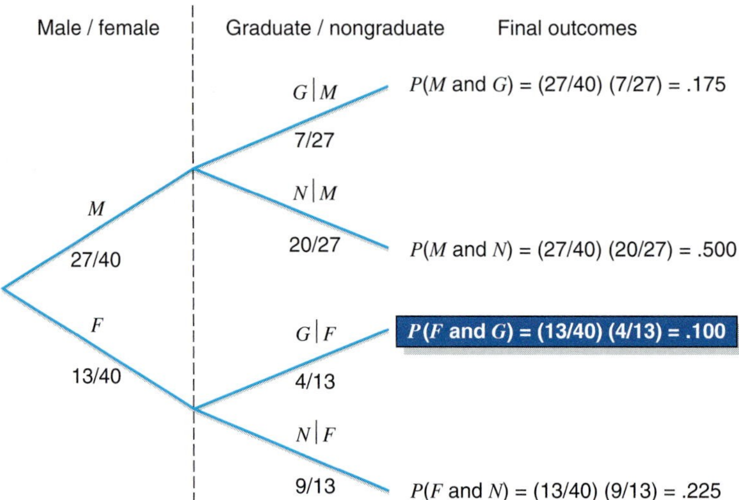

Figure 4.16 Tree diagram for joint probabilities.

■ EXAMPLE 4–24

Calculating the joint probability of two events.

A box contains 20 DVDs, 4 of which are defective. If two DVDs are selected at random (without replacement) from this box, what is the probability that both are defective?

Solution Let us define the following events for this experiment:

G_1 = event that the first DVD selected is good

D_1 = event that the first DVD selected is defective

G_2 = event that the second DVD selected is good

D_2 = event that the second DVD selected is defective

We are to calculate the joint probability of D_1 and D_2, which is given by

$$P(D_1 \text{ and } D_2) = P(D_1)\,P(D_2 \mid D_1)$$

As we know, there are 4 defective DVDs in 20. Consequently, the probability of selecting a defective DVD at the first selection is

$$P(D_1) = 4/20$$

To calculate the probability $P(D_2 \mid D_1)$, we know that the first DVD selected is defective because D_1 has already occurred. Because the selections are made without replacement, there are 19 total DVDs, and 3 of them are defective at the time of the second selection. Therefore,

$$P(D_2 \mid D_1) = 3/19$$

Hence, the required probability is

$$P(D_1 \text{ and } D_2) = P(D_1)\,P(D_2 \mid D_1) = (4/20)(3/19) = \mathbf{.0316}$$

The tree diagram in Figure 4.17 shows the selection procedure and the final four outcomes for this experiment along with their probabilities. The joint probability of D_1 and D_2 is highlighted in the tree diagram.

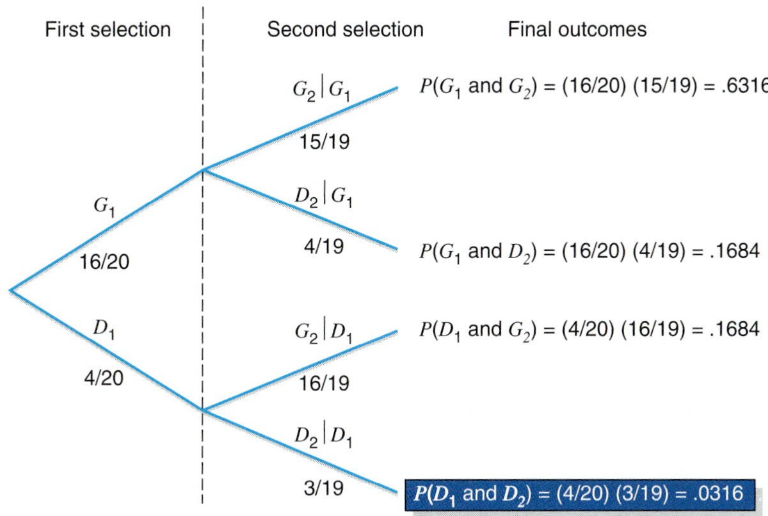

Figure 4.17 Selecting two DVDs.

Conditional probability was discussed in Section 4.4. It is obvious from the formula for joint probability that if we know the probability of an event A and the joint probability of events A and B, then we can calculate the conditional probability of B given A.

Calculating Conditional Probability If A and B are two events, then,

$$P(B \mid A) = \frac{P(A \text{ and } B)}{P(A)} \quad \text{and} \quad P(A \mid B) = \frac{P(A \text{ and } B)}{P(B)}$$

given that $P(A) \neq 0$ and $P(B) \neq 0$.

EXAMPLE 4–25

Calculating the conditional probability of an event.

The probability that a randomly selected student from a college is a senior is .20, and the joint probability that the student is a computer science major and a senior is .03. Find the conditional probability that a student selected at random is a computer science major given that he/she is a senior.

Solution Let us define the following two events:

A = the student selected is a senior

B = the student selected is a computer science major

From the given information,

$$P(A) = .20 \quad \text{and} \quad P(A \text{ and } B) = .03$$

Hence,

$$P(B \mid A) = \frac{P(A \text{ and } B)}{P(A)} = \frac{.03}{.20} = \mathbf{.15}$$

Thus, the (conditional) probability is .15 that a student selected at random is a computer science major given that he or she is a senior.

Multiplication Rule for Independent Events

The foregoing discussion of the multiplication rule was based on the assumption that the two events are dependent. Now suppose that events A and B are independent. Then,

$$P(A) = P(A \mid B) \quad \text{and} \quad P(B) = P(B \mid A)$$

By substituting $P(B)$ for $P(B \mid A)$ into the formula for the joint probability of A and B, we obtain

$$P(A \text{ and } B) = P(A)\, P(B)$$

Multiplication Rule to Calculate the Probability of Independent Events The probability of the intersection of two independent events A and B is

$$P(A \text{ and } B) = P(A)\, P(B)$$

■ EXAMPLE 4–26

Calculating the joint probability of two independent events.

An office building has two fire detectors. The probability is .02 that any fire detector of this type will fail to go off during a fire. Find the probability that both of these fire detectors will fail to go off in case of a fire.

Solution In this example, the two fire detectors are independent because whether or not one fire detector goes off during a fire has no effect on the second fire detector. We define the following two events:

A = the first fire detector fails to go off during a fire

B = the second fire detector fails to go off during a fire

Then, the joint probability of A and B is

$$P(A \text{ and } B) = P(A)\, P(B) = (.02)(.02) = \mathbf{.0004}$$ ■

The multiplication rule can be extended to calculate the joint probability of more than two events. Example 4–27 illustrates such a case for independent events.

■ EXAMPLE 4–27

Calculating the joint probability of three events.

The probability that a patient is allergic to penicillin is .20. Suppose this drug is administered to three patients.

(a) Find the probability that all three of them are allergic to it.

(b) Find the probability that at least one of them is not allergic to it.

Solution

(a) Let A, B, and C denote the events that the first, second, and third patients, respectively, are allergic to penicillin. We are to find the joint probability of A, B, and C. All three events are independent because whether or not one patient is allergic does not depend on whether or not any of the other patients is allergic. Hence,

$$P(A \text{ and } B \text{ and } C) = P(A)\, P(B)\, P(C) = (.20)(.20)(.20) = \mathbf{.008}$$

The tree diagram in Figure 4.18 shows all the outcomes for this experiment. Events $\bar{A}$, $\bar{B}$, and $\bar{C}$ are the complementary events of A, B, and C, respectively. They represent the events that the patients are not allergic to penicillin. Note that the intersection of events A, B, and C is written as ABC in the tree diagram.

(b) Let us define the following events:

$$G = \text{all three patients are allergic}$$

$$H = \text{at least one patient is not allergic}$$

Events G and H are two complementary events. Event G consists of the intersection of events A, B, and C. Hence, from part (a),

$$P(G) = P(A \text{ and } B \text{ and } C) = .008$$

Therefore, using the complementary event rule, we obtain

$$P(H) = 1 - P(G) = 1 - .008 = \mathbf{.992}$$

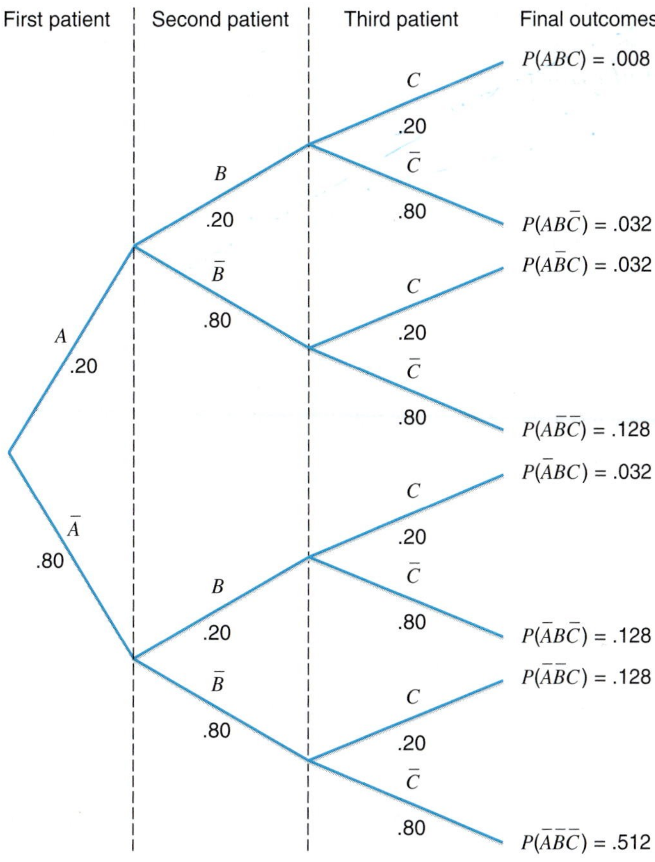

Figure 4.18 Tree diagram for joint probabilities. ■

Case Study 4–2 on page 164 calculates the probability of a hitless streak in baseball by using the multiplication rule.

Joint Probability of Mutually Exclusive Events

We know from an earlier discussion that two mutually exclusive events cannot happen together. Consequently, their joint probability is zero.

Joint Probability of Mutually Exclusive Events The joint probability of two mutually exclusive events is always zero. If A and B are two mutually exclusive events, then,

$$P(A \text{ and } B) = 0$$

CASE STUDY

4–2

BASEBALL PLAYERS HAVE "SLUMPS" AND "STREAKS"

Going "0 for July," as former infielder Bob Aspromonte once put it, is enough to make a baseball player toss out his lucky bat or start seriously searching for flaws in his hitting technique. But the culprit is usually just simple mathematics.

Statistician Harry Roberts of the University of Chicago's Graduate School of Business studied the records of major-league baseball players and found that a batter is no more likely to hit worse when he is in a slump than when he is in a hot streak. The occurrences of hits followed the same pattern as purely random events such as pulling marbles out of a hat. If there were one white marble and three black ones in the hat, for example, then a white marble would come out about one quarter of the time—a .250 average. In the same way, a player who hits .250 will in the long run get a hit every four times at bat.

But that doesn't mean the player will hit the ball exactly every fourth time he comes to the plate—just as it's unlikely that the white marble will come out exactly every fourth time.

Even a batter who goes hitless 10 times in a row might safely be able to pin the blame on statistical fluctuations. The odds of pulling a black marble out of a hat 10 times in a row are about 6 percent—not a frequent occurrence, but not impossible, either. Only in the long run do these statistical fluctuations even out.

As mentioned in the above excerpt from *U.S. News & World Report*, if we assume a player hits .250 in the long run, the probability that this player does not hit during a specific trip to the plate is .75. Hence, we can calculate the probability that he goes hitless 10 times in a row as follows.

$$P(\text{hitless 10 times in a row}) = (.75)(.75)\cdots(.75) \text{ ten times}$$

$$= (.75)^{10} = .0563$$

Note that each trip to the plate is independent, and the probability that a player goes hitless 10 times in a row is given by the intersection of 10 hitless trips. This probability has been rounded off to "about 6%" in this illustration.

Source: U.S. News & World Report, July 11, 1988, p. 46.

■ EXAMPLE 4–28

Illustrating the probability of two mutually exclusive events.

Consider the following two events for an application filed by a person to obtain a car loan:

A = event that the loan application is approved

R = event that the loan application is rejected

What is the joint probability of A and R?

Solution The two events A and R are mutually exclusive. Either the loan application will be approved or it will be rejected. Hence,

$$P(A \text{ and } R) = \mathbf{0}$$ ■

EXERCISES

■ CONCEPTS AND PROCEDURES

4.66 Explain the meaning of the intersection of two events. Give one example.

4.67 What is meant by the joint probability of two or more events? Give one example.

4.68 How is the multiplication rule of probability for two dependent events different from the rule for two independent events?

4.69 What is the joint probability of two mutually exclusive events? Give one example.

4.70 Find the joint probability of A and B for the following.

a. $P(A) = .40$ and $P(B \mid A) = .25$
b. $P(B) = .65$ and $P(A \mid B) = .36$

4.71 Find the joint probability of A and B for the following.
a. $P(B) = .59$ and $P(A \mid B) = .77$
b. $P(A) = .28$ and $P(B \mid A) = .35$

4.72 Given that A and B are two independent events, find their joint probability for the following.
a. $P(A) = .61$ and $P(B) = .27$
b. $P(A) = .39$ and $P(B) = .63$

4.73 Given that A and B are two independent events, find their joint probability for the following.
a. $P(A) = .20$ and $P(B) = .76$
b. $P(A) = .57$ and $P(B) = .32$

4.74 Given that A, B, and C are three independent events, find their joint probability for the following.
a. $P(A) = .20$, $P(B) = .46$, and $P(C) = .25$
b. $P(A) = .44$, $P(B) = .27$, and $P(C) = .43$

4.75 Given that A, B, and C are three independent events, find their joint probability for the following.
a. $P(A) = .49$, $P(B) = .67$, and $P(C) = .75$
b. $P(A) = .71$, $P(B) = .34$, and $P(C) = .45$

4.76 Given that $P(A) = .30$ and $P(A \text{ and } B) = .24$, find $P(B \mid A)$.

4.77 Given that $P(B) = .65$ and $P(A \text{ and } B) = .45$, find $P(A \mid B)$.

4.78 Given that $P(A \mid B) = .40$ and $P(A \text{ and } B) = .36$, find $P(B)$.

4.79 Given that $P(B \mid A) = .80$ and $P(A \text{ and } B) = .58$, find $P(A)$.

APPLICATIONS

4.80 In a sample survey, 1800 senior citizens were asked whether or not they have ever been victimized by a dishonest telemarketer. The following table gives the responses by age group.

		Have Been Victimized	Have Never Been Victimized
Age	60–69 (A)	106	698
	70–79 (B)	145	447
	80 or over (C)	61	343

a. Suppose one person is randomly selected from these senior citizens. Find the following probabilities.
i. *P*(have been victimized *and* C)
ii. *P*(have never been victimized *and* A)
b. Find *P*(B and C). Is this probability zero? Explain why or why not.

4.81 The following table gives a two-way classification of all basketball players at a state university who began their college careers between 1990 and 2000, based on gender and whether or not they graduated.

	Graduated	Did Not Graduate
Male	126	55
Female	133	32

a. If one of these players is selected at random, find the following probabilities.
i. *P*(female *and* graduated)
ii. *P*(male *and* did not graduate)
b. Find *P*(graduated *and* did not graduate). Is this probability zero? If yes, why?

4.82 Five hundred employees were selected from a city's large private companies and asked whether or not they have any retirement benefits provided by their companies. Based on this information, the following two-way classification table was prepared.

	Have Retirement Benefits	
	Yes	No
Men	225	75
Women	150	50

a. Suppose one employee is selected at random from these 500 employees. Find the following probabilities.
   i. Probability of the intersection of events "woman" and "yes"
   ii. Probability of the intersection of events "no" and "man"

b. Mention what other joint probabilities you can calculate for this table and then find them. You may draw a tree diagram to find these probabilities.

4.83 Two thousand randomly selected adults were asked whether or not they have ever shopped on the Internet. The following table gives a two-way classification of the responses obtained.

	Have Shopped	Have Never Shopped
Male	500	700
Female	300	500

a. Suppose one adult is selected at random from these 2000 adults. Find the following probabilities.
   i. *P*(has never shopped on the Internet *and* is a male)
   ii. *P*(has shopped on the Internet *and* is a female)

b. Mention what other joint probabilities you can calculate for this table and then find those. You may draw a tree diagram to find these probabilities.

4.84 A consumer agency randomly selected 1700 flights for two major airlines, A and B. The following table gives the two-way classification of these flights based on airline and arrival time. Note that "less than 30 minutes late" includes flights that arrived early or on time.

	Less Than 30 Minutes Late	30 Minutes to 1 Hour Late	More Than 1 Hour Late
Airline A	429	390	92
Airline B	393	316	80

a. Suppose one flight is selected at random from these 1700 flights. Find the following probabilities.
   i. *P*(more than 1 hour late *and* airline A)
   ii. *P*(airline B *and* less than 30 minutes late)

b. Find the joint probability of events "30 minutes to 1 hour late" and "more than 1 hour late." Is this probability zero? Explain why or why not.

4.85 Two thousand randomly selected adults were asked if they think they are financially better off than their parents. The following table gives the two-way classification of the responses based on the education levels of the persons included in the survey and whether they are financially better off, the same, or worse off than their parents.

	Less Than High School	High School	More Than High School
Better off	140	450	420
Same	60	250	110
Worse off	200	300	70

a. Suppose one adult is selected at random from these 2000 adults. Find the following probabilities.
   i. *P*(better off *and* high school)
   ii. *P*(more than high school *and* worse off)

b. Find the joint probability of the events "worse off" and "better off." Is this probability zero? Explain why or why not.

4.86 In a statistics class of 42 students, 28 have volunteered for community service in the past. If two students are selected at random from this class, what is the probability that both of them have volunteered for community service in the past? Draw a tree diagram for this problem.

4.87 In a political science class of 35 students, 21 favor abolishing the electoral college and thus electing the president of the United States by popular vote. If two students are selected at random from this class, what is the probability that both of them favor abolition of the electoral college? Draw a tree diagram for this problem.

4.88 A company is to hire two new employees. They have prepared a final list of eight candidates, all of whom are equally qualified. Of these eight candidates, five are women. If the company decides to select two persons randomly from these eight candidates, what is the probability that both of them are women? Draw a tree diagram for this problem.

4.89 In a group of 10 persons, 4 have a type A personality and 6 have a type B personality. If two persons are selected at random from this group, what is the probability that the first of them has a type A personality and the second has a type B personality? Draw a tree diagram for this problem.

4.90 The probability is .80 that a senior from a large college in New York State has never gone to Florida for spring break. If two college seniors are selected at random from this college, what is the probability that the first has never gone to Florida for spring break and the second has? Draw a tree diagram for this problem.

4.91 The probability that a student graduating from Suburban State University has student loans to pay off after graduation is .60. If two students are randomly selected from this university, what is the probability that neither of them has student loans to pay off after graduation?

4.92 A contractor has submitted bids for two state construction projects. The probability that he will win any contract is .25, and it is the same for each of the two contracts.

a. What is the probability that he will win both contracts?
b. What is the probability that he will win neither contract?

Draw a tree diagram for this problem.

4.93 Five percent of all items sold by a mail-order company are returned by customers for a refund. Find the probability that of two items sold during a given hour by this company

a. both will be returned for a refund
b. neither will be returned for a refund

Draw a tree diagram for this problem.

4.94 The probability that any given person is allergic to a certain drug is .03. What is the probability that none of three randomly selected persons is allergic to this drug? Assume that all three persons are independent.

4.95 The probability that a farmer is in debt is .80. What is the probability that three randomly selected farmers are all in debt? Assume independence of events.

4.96 The probability that a student graduating from Suburban State University has student loans to pay off after graduation is .60. The probability that a student graduating from this university has student loans to pay off after graduation and is a male is .24. Find the conditional probability that a randomly selected student from this university is a male given that this student has student loans to pay off after graduation?

4.97 The probability that an employee at a company is a female is .36. The probability that an employee is a female and married is .19. Find the conditional probability that a randomly selected employee from this company is married given that she is a female.

4.98 A telephone poll of 1204 adult Americans about their commuting arrangements/habits was conducted by *TNS* for *Time/ABC News/Washington Post* in January 2005 (*Time*, February 21, 2005). Suppose 900 of these 1204 respondents were commuters. Then the percentages given in the magazine article would imply that 756 of these 900 commuters drove alone to work, and 605 of these 900 drove alone and were not interested in carpooling. If one of these 900 commuters was selected at random, what is the probability that the person is not interested in carpooling given that he or she drives alone?

4.99 Suppose that 20% of all adults in a small town live alone, and 8% of the adults live alone and have at least one pet. What is the probability that a randomly selected adult from this town has at least one pet given that this adult lives alone?

4.9 Union of Events and the Addition Rule

This section discusses the union of events and the addition rule that is applied to compute the probability of the union of events.

4.9.1 Union of Events

The **union of two events** A and B includes all outcomes that are either in A or in B or in both A and B.

Definition

Union of Events Let A and B be two events defined in a sample space. The *union of events A and B* is the collection of all outcomes that belong either to A or to B or to both A and B and is denoted by

$$A \text{ or } B$$

The union of events A and B is also denoted by $A \cup B$. Example 4–29 illustrates the union of events A and B.

■ EXAMPLE 4–29

Illustrating the union of two events.

A senior citizens center has 300 members. Of them, 140 are male, 210 take at least one medicine on a permanent basis, and 95 are male *and* take at least one medicine on a permanent basis. Describe the union of the events "male" and "take at least one medicine on a permanent basis."

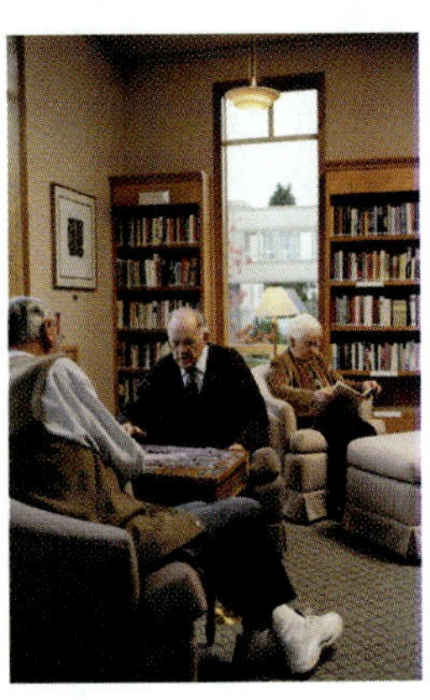

Solution Let us define the following events:

M = a senior citizen is a male

F = a senior citizen is a female

A = a senior citizen takes at least one medicine

B = a senior citizen does not take any medicine

The union of the events "male" and "take at least one medicine" includes those senior citizens who are either male or take at least one medicine or both. The number of such senior citizens is

$$140 + 210 - 95 = 255$$

Why did we subtract 95 from the sum of 140 and 210? The reason is that 95 senior citizens (which represent the intersection of events M and A) are common to both events M and A and, hence, are counted twice. To avoid double counting, we subtracted 95 from the sum of the other two numbers. We can observe this double counting from Table 4.8, which is constructed using the given information. The sum of the numbers in the three shaded cells gives the senior citizens who are either male or take at least one medicine or both. However, if we add the totals of the row labeled M and the column labeled A, we count 95 twice.

Table 4.8

	A	B	Total
M	95	45	140
F	115	45	160
Total	210	90	300

→Counted twice

Figure 4.19 shows the diagram for the union of the events "male" and "take at least one medicine on a permanent basis."

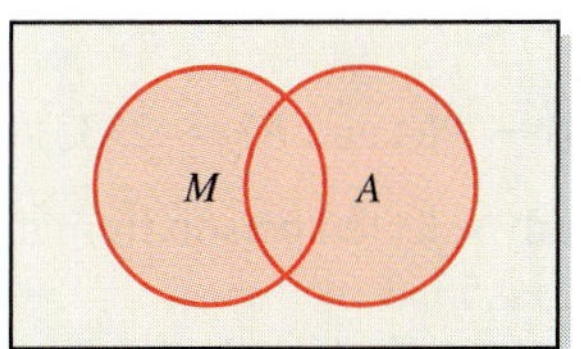

Area shaded in red gives the union of events M and A, and includes 255 senior citizens

Figure 4.19 Union of events M and A.

■

4.9.2 Addition Rule

The method used to calculate the probability of the union of events is called the **addition rule**. It is defined as follows.

Addition Rule to Find the Probability of Union of Events The probability of the union of two events A and B is

$$P(A \text{ or } B) = P(A) + P(B) - P(A \text{ and } B)$$

Thus, to calculate the probability of the union of two events A and B, we add their marginal probabilities and subtract their joint probability from this sum. We must subtract the joint probability of A and B from the sum of their marginal probabilities to avoid double counting because of common outcomes in A and B. This is the case where events A and B are not mutually exclusive.

■ EXAMPLE 4–30

Calculating the probability of the union of two events: two-way table.

A university president has proposed that all students must take a course in ethics as a requirement for graduation. Three hundred faculty members and students from this university were asked about their opinion on this issue. Table 4.9 gives a two-way classification of the responses of these faculty members and students.

Table 4.9 Two-Way Classification of Responses

	Favor	Oppose	Neutral	Total
Faculty	45	15	10	70
Student	90	110	30	230
Total	135	125	40	300

Find the probability that one person selected at random from these 300 persons is a faculty member or is in favor of this proposal.

Solution Let us define the following events:

$$A = \text{the person selected is a faculty member}$$

$$B = \text{the person selected is in favor of the proposal}$$

From the information given in Table 4.9,

$$P(A) = 70/300 = .2333$$

$$P(B) = 135/300 = .4500$$

$$P(A \text{ and } B) = P(A)\,P(B \mid A) = (70/300)(45/70) = .1500$$

Using the addition rule, we have

$$P(A \text{ or } B) = P(A) + P(B) - P(A \text{ and } B) = .2333 + .4500 - .1500 = \mathbf{.5333}$$

Thus, the probability that a randomly selected person from these 300 persons is a faculty member or is in favor of this proposal is .5333.

The probability in this example can also be calculated without using the addition rule. The total number of persons in Table 4.9 who are either faculty members or in favor of this proposal is

$$45 + 15 + 10 + 90 = 160$$

Hence, the required probability is

$$P(A \text{ or } B) = 160/300 = \mathbf{.5333}$$

■

■ EXAMPLE 4–31

Calculating the probability of the union of two events.

In a group of 2500 persons, 1400 are female, 600 are vegetarian, and 400 are female and vegetarian. What is the probability that a randomly selected person from this group is a male or vegetarian?

Solution Let us define the following events:

F = the randomly selected person is a female

M = the randomly selected person is a male

V = the randomly selected person is a vegetarian

N = the randomly selected person is a non-vegetarian

From the given information, we know that there are 1400 female, 600 vegetarian, and 400 female and vegetarian. Hence, there are 1100 male, 1900 nonvegetarian, and 200 male and vegetarian. We are to find the probability $P(M \text{ or } V)$. This probability is obtained as follows:

$$P(M \text{ or } V) = P(M) + P(V) - P(M \text{ and } V)$$

$$= \frac{1100}{2500} + \frac{600}{2500} - \frac{200}{2500}$$

$$= .44 + .24 - .08 = \mathbf{.60}$$

Actually, using the given information, we can prepare Table 4.10 for this example. In the table, the numbers in the shaded cells are given to us. The remaining numbers are calculated by doing some arithmetic manipulations.

Table 4.10 **Two-Way Classification Table**

	Vegetarian (V)	Nonvegetarian (N)	Total
Female (F)	400	1000	1400
Male (M)	200	900	1100
Total	600	1900	2500

Using Table 4.10, the required probability is:

$$P(M \text{ or } V) = P(M) + P(V) - P(M \text{ and } V)$$

$$= \frac{1100}{2500} + \frac{600}{2500} - \frac{200}{2500} = .44 + .24 - .08 = \mathbf{.60}$$

■

Addition Rule for Mutually Exclusive Events

We know from an earlier discussion that the joint probability of two mutually exclusive events is zero. When A and B are mutually exclusive events, the term $P(A \text{ and } B)$ in the addition rule becomes zero and is dropped from the formula. Thus, the probability of the union of two mutually exclusive events is given by the sum of their marginal probabilities.

Addition Rule to Find the Probability of the Union of Mutually Exclusive Events The probability of the union of two mutually exclusive events A and B is

$$P(A \text{ or } B) = P(A) + P(B)$$

■ EXAMPLE 4–32

Calculating the probability of the union of two mutually exclusive events: two-way table.

A university president has proposed that all students must take a course in ethics as a requirement for graduation. Three hundred faculty members and students from this university were asked about their opinion on this issue. The following table, reproduced from Table 4.9 in Example 4–30, gives a two-way classification of the responses of these faculty members and students.

	Favor	Oppose	Neutral	Total
Faculty	45	15	10	70
Student	90	110	30	230
Total	135	125	40	300

What is the probability that a randomly selected person from these 300 faculty members and students is in favor of the proposal or is neutral?

Solution Let us define the following events:

F = the person selected is in favor of the proposal

N = the person selected is neutral

As shown in Figure 4.20, events F and N are mutually exclusive because a person selected can be either in favor or neutral but not both.

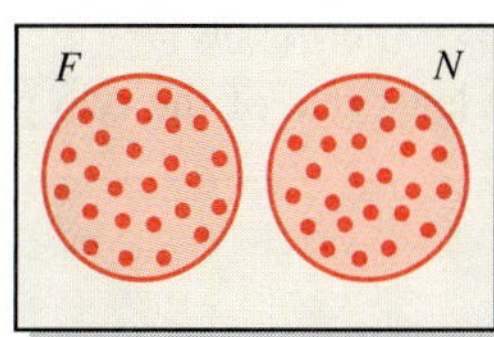

Figure 4.20 Venn diagram of mutually exclusive events.

From the given information,

$$P(F) = 135/300 = .4500$$

$$P(N) = 40/300 = .1333$$

Hence,

$$P(F \text{ or } N) = P(F) + P(N) = .4500 + .1333 = \mathbf{.5833}$$

■

The addition rule formula can easily be extended to apply to more than two events. The following example illustrates this.

■ EXAMPLE 4–33

Calculating the probability of the union of three mutually exclusive events.

Consider the experiment of rolling a die twice. Find the probability that the sum of the numbers obtained on two rolls is 5, 7, or 10.

Solution The experiment of rolling a die twice has a total of 36 outcomes, which are listed in Table 4.11. Assuming that the die is balanced, these 36 outcomes are equally likely.

Table 4.11 Two Rolls of a Die

		Second Roll of the Die					
		1	2	3	4	5	6
First Roll of the Die	1	(1,1)	(1,2)	(1,3)	(1,4)	(1,5)	(1,6)
	2	(2,1)	(2,2)	(2,3)	(2,4)	(2,5)	(2,6)
	3	(3,1)	(3,2)	(3,3)	(3,4)	(3,5)	(3,6)
	4	(4,1)	(4,2)	(4,3)	(4,4)	(4,5)	(4,6)
	5	(5,1)	(5,2)	(5,3)	(5,4)	(5,5)	(5,6)
	6	(6,1)	(6,2)	(6,3)	(6,4)	(6,5)	(6,6)

The events that give the sum of two numbers equal to 5 or 7 or 10 are circled in the table. As we can observe, the three events "the sum is 5," "the sum is 7," and "the sum is 10" are mutually exclusive. Four outcomes give a sum of 5, six give a sum of 7, and three outcomes give a sum of 10. Thus,

$$P(\text{sum is 5 or 7 or 10}) = P(\text{sum is 5}) + P(\text{sum is 7}) + P(\text{sum is 10})$$
$$= 4/36 + 6/36 + 3/36 = 13/36 = \mathbf{.3611}$$ ■

■ EXAMPLE 4–34

Calculating the probability of the union of three mutually exclusive events.

The probability that a person is in favor of genetic engineering is .55 and that a person is against it is .45. Two persons are randomly selected, and it is observed whether they favor or oppose genetic engineering.

(a) Draw a tree diagram for this experiment.

(b) Find the probability that at least one of the two persons favors genetic engineering.

Solution

(a) Let

F = a person is in favor of genetic engineering

A = a person is against genetic engineering

This experiment has four outcomes: both persons are in favor (FF), the first person is in favor and the second is against (FA), the first person is against and the second is in favor (AF), and both persons are against genetic engineering (AA). The tree diagram in Figure 4.21 shows these four outcomes and their probabilities.

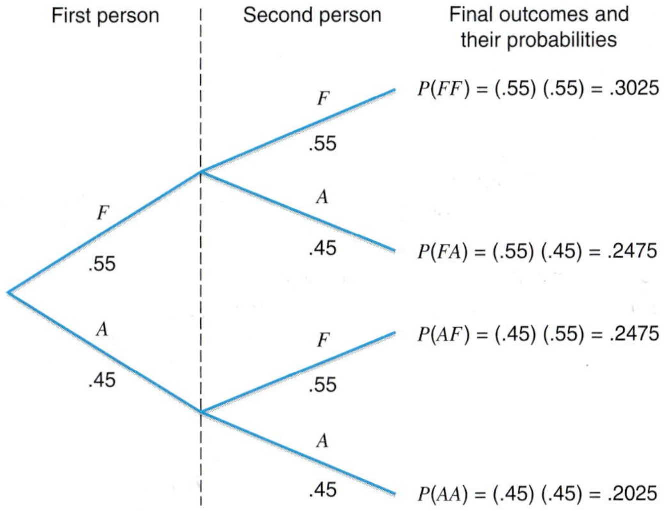

Figure 4.21 Tree diagram.

(b) The probability that at least one person favors genetic engineering is given by the union of events *FF*, *FA*, and *AF*. These three outcomes are mutually exclusive. Hence,

$$\begin{aligned} P(\text{at least one person favors}) &= P(FF \text{ or } FA \text{ or } AF) \\ &= P(FF) + P(FA) + P(AF) \\ &= .3025 + .2475 + .2475 = \mathbf{.7975} \end{aligned}$$

EXERCISES

CONCEPTS AND PROCEDURES

4.100 Explain the meaning of the union of two events. Give one example.

4.101 How is the addition rule of probability for two mutually exclusive events different from the rule for two mutually nonexclusive events?

4.102 Consider the following addition rule to find the probability of the union of two events *A* and *B*:

$$P(A \text{ or } B) = P(A) + P(B) - P(A \text{ and } B)$$

When and why is the term $P(A \text{ and } B)$ subtracted from the sum of $P(A)$ and $P(B)$? Give one example where you might use this formula.

4.103 When is the following addition rule used to find the probability of the union of two events *A* and *B*?

$$P(A \text{ or } B) = P(A) + P(B)$$

Give one example where you might use this formula.

4.104 Find $P(A \text{ or } B)$ for the following.

a. $P(A) = .58$, $P(B) = .66$, and $P(A \text{ and } B) = .57$
b. $P(A) = .72$, $P(B) = .42$, and $P(A \text{ and } B) = .39$

4.105 Find $P(A \text{ or } B)$ for the following.

a. $P(A) = .18$, $P(B) = .49$, and $P(A \text{ and } B) = .11$
b. $P(A) = .73$, $P(B) = .71$, and $P(A \text{ and } B) = .68$

4.106 Given that *A* and *B* are two mutually exclusive events, find $P(A \text{ or } B)$ for the following.

a. $P(A) = .47$ and $P(B) = .32$
b. $P(A) = .16$ and $P(B) = .59$

4.107 Given that A and B are two mutually exclusive events, find $P(A \text{ or } B)$ for the following.

a. $P(A) = .25$ and $P(B) = .27$
b. $P(A) = .58$ and $P(B) = .09$

APPLICATIONS

4.108 In a sample survey, 1800 senior citizens were asked whether or not they have ever been victimized by a dishonest telemarketer. The following table gives the responses by age group.

		Have Been Victimized	Have Never Been Victimized
Age	60–69 (A)	106	698
	70–79 (B)	145	447
	80 or over (C)	61	343

Suppose one person is randomly selected from these senior citizens. Find the following probabilities.

a. P(have been victimized *or* B)
b. P(have never been victimized *or* C)

4.109 The following table gives a two-way classification of all basketball players at a state university who began their college careers between 1990 and 2000, based on gender and whether or not they graduated.

	Graduated	Did Not Graduate
Male	126	55
Female	133	32

If one of these players is selected at random, find the following probabilities.

a. P(female *or* did not graduate)
b. P(graduated *or* male)

4.110 Five hundred employees were selected from a city's large private companies, and they were asked whether or not they have any retirement benefits provided by their companies. Based on this information, the following two-way classification table was prepared.

	Have Retirement Benefits	
	Yes	No
Men	225	75
Women	150	50

Suppose one employee is selected at random from these 500 employees. Find the following probabilities.

a. The probability of the union of events "woman" and "yes"
b. The probability of the union of events "no" and "man"

4.111 Two thousand randomly selected adults were asked whether or not they have ever shopped on the Internet. The following table gives a two-way classification of the responses.

	Have Shopped	Have Never Shopped
Male	500	700
Female	300	500

Suppose that one adult is selected at random from these 2000 adults. Find the following probabilities.

a. P(has never shopped on the Internet *or* is a female)
b. P(is a male *or* has shopped on the Internet)
c. P(has shopped on the Internet *or* has never shopped on the Internet)

4.112 A consumer agency randomly selected 1700 flights for two major airlines, A and B. The following table gives the two-way classification of these flights based on airline and arrival time. Note that "less than 30 minutes late" includes flights that arrived early or on time.

	Less Than 30 Minutes Late	30 Minutes to 1 Hour Late	More Than 1 Hour Late
Airline A	429	390	92
Airline B	393	316	80

If one flight is selected at random from these 1700 flights, find the following probabilities.

a. *P*(more than 1 hour late *or* airline A)
b. *P*(airline B *or* less than 30 minutes late)
c. *P*(airline A *or* airline B)

4.113 Two thousand randomly selected adults were asked if they think they are financially better off than their parents. The following table gives the two-way classification of the responses based on the education levels of the persons included in the survey and whether they are financially better off, the same, or worse off than their parents.

	Less Than High School	High School	More Than High School
Better off	140	450	420
Same	60	250	110
Worse off	200	300	70

Suppose one adult is selected at random from these 2000 adults. Find the following probabilities.

a. *P*(better off *or* high school)
b. *P*(more than high school *or* worse off)
c. *P*(better off *or* worse off)

4.114 There is an area of free (but illegal) parking near an inner-city sports arena. The probability that a car parked in this area will be ticketed by police is .35, that the car will be vandalized is .15, and that it will be ticketed and vandalized is .10. Find the probability that a car parked in this area will be ticketed or vandalized.

4.115 The probability that a family owns a washing machine is .68, that it owns a DVD player is .81, and that it owns both a washing machine and a DVD player is .58. What is the probability that a randomly selected family owns a washing machine or a DVD player?

4.116 Jason and Lisa are planning an outdoor reception following their wedding. They estimate that the probability of bad weather is .25, that of a disruptive incident (a fight breaks out, the limousine is late, etc.) is .15, and that bad weather and a disruptive incident will occur is .08. Assuming these estimates are correct, find the probability that their reception will suffer bad weather or a disruptive incident.

4.117 The probability that a randomly selected elementary or secondary school teacher from a city is a female is .68, holds a second job is .38, and is a female and holds a second job is .29. Find the probability that an elementary or secondary school teacher selected at random from this city is a female or holds a second job.

4.118 According to the U.S. Census Bureau's most recent data on the marital status of the 221 million Americans aged 15 and older, 120 million are currently married and 60 million have never married. If one person is selected at random from these 221 million persons, find the probability that this person is currently married or has never married. Explain why this probability is not equal to 1.0.

4.119 In 2004, the Soap and Detergent Association conducted a survey to find out how often people washed their hands (*USA TODAY*, December 14, 2004). Suppose this survey was based on a sample of 1500 people. Then, the percentages given in the newspaper would show that 540 of these 1500 people washed their hands more than 10 times per day, 360 washed their hands 7 to 10 times a day, and 345 washed their hands 5 or 6 times a day. If one of these 1500 respondents is selected at random, what is the probability that this person washes his/her hands 5 or more times per day? Explain why this probability is not equal to 1.0.

4.120 The probability of a student getting an A grade in an economics class is .24 and that of getting a B grade is .28. What is the probability that a randomly selected student from this class will get an A or B in this class? Explain why this probability is not equal to 1.0.

4.121 Twenty percent of a town's voters favor letting a major discount store move into their neighborhood, 63% are against it, and 17% are indifferent. What is the probability that a randomly selected voter from this town will either be against it or be indifferent? Explain why this probability is not equal to 1.0.

4.122 The probability that a corporation makes charitable contributions is .72. Two corporations are selected at random, and it is noted whether or not they make charitable contributions.

a. Draw a tree diagram for this experiment.

b. Find the probability that at most one corporation makes charitable contributions.

4.123 The probability that an open-heart operation is successful is .84. What is the probability that in two randomly selected open-heart operations at least one will be successful? Draw a tree diagram for this experiment.

USES AND MISUSES...

1. STATISTICS VERSUS PROBABILITY

At this point, you may think that probability and statistics are basically the same things. They both use the term mean, they both report results in terms of percentages, and so on. Do not be fooled: Although they share many of the same mathematical tools, probability and statistics are very different sciences. The first three chapters of the text were very careful to specify whether a particular set of data was a population or a sample. This is because statistics takes a sample of data and, based upon the properties of that sample—mean, median, mode, standard deviation—attempts to say something about a population. Probability does exactly the opposite: In probability, we know the properties of the population based on the sample space and the probability distribution, and we want to make statements about a sample from the population.

Here's an example viewed from a statistical and a probabilistic point of view. A sequence of outcomes from 10 independent coin tosses is {H, T, H, T, H, T, T, H, T, T}. A statistician will ask the question: Based on the observed 4 heads and 6 tails, what combination of heads and tails would he or she expect from 100 or 1000 tosses, and how certain would he or she be of that answer? Someone using probability will ask: If the coin toss was fair (the probability of the event that a single coin toss be a head or tail is .5), what is the probability that the compound event of four heads and six tails will occur? These are significantly different questions.

The distinction between a statistical approach and a probabilistic approach to a problem can be surprising. Imagine that you must determine the average life of an automotive part. One approach would be to take a sample of parts, test each of them until they fail to work, and then perform some calculations regarding the distribution of failures. However, if this particular part has outliers with long life spans (several years), you are going to be spending a lot of time in the laboratory. An approach using probabilistic techniques could develop a hypothetical life span based on the physical properties of the part, the conditions of its use, and the manufacturing characteristics. Then you can use your experimental results over a relatively short period of time—including data on those parts that did not fail—to adjust your prior understanding of what makes the part fail, saving yourself a lot of time.

2. ODDS AND PROBABILITY

One of the first things we learn in probability is that the sum of the probabilities of all outcomes for an experiment must equal one. We also learn about the probabilities that are developed from relative frequencies and about the subjective probabilities. In the latter case, many of the probabilities involve personal opinions, hopefully those of experts in the field. Still, both scenarios (probabilities obtained from relative frequencies and subjective probabilities) require that all probabilities must be non-negative and the sum of the probabilities of all outcomes for an experiment must equal one.

So, while probabilities and probability models are all around us—in weather, medicine, financial markets, and so forth—they are most obvious in the world of gaming and gambling. Sports betting agencies always publish odds of each team winning a specific game or sports title. The table on the next page gives the odds, as of June 14, 2005, of each Major League Baseball team winning the 2005 World Series. These odds are obtained from the Web site http://www.vegas.com/gaming/futures/worldseries.html.

Note that the odds listed in this table are called the odds in favor of winning the World Series. For example, the Los Angeles Angels have 1:8 (which is read as 1 to 8) odds to win the 2005 World Series. If we switch the numbers around, we can state that the odds are 8:1 (or 8 to 1) against the Angels to win the 2005 World Series.

How do we convert these odds into probabilities? Let us consider the Los Angeles Angels. Odds of 1:8 imply that out of 9 chances, there is one chance that the Angels will win the 2005 World Series and 8 chances the Angels will not win the 2005 World Series. Thus, the probability that the Angels will win the 2005 World Series is $\frac{1}{8+1}=\frac{1}{9}=.1111$ and the probability that the Angels will not win the 2005 World Series is $\frac{8}{8+1}=\frac{8}{9}=.8888$. Similarly, for the Boston Red Sox, the probability of winning the 2005 World Series is $\frac{2}{9+2}=\frac{2}{11}=.1818$ and the probability of not winning the 2005 World Series is $\frac{9}{9+2}=\frac{9}{11}=.8181$. We can calculate these probabilities for other teams the same way.

Note that here the outcomes that each team wins the 2005 World Series are mutually exclusive events because it is impossible for two or more teams to win the World Series during the same year. Hence, if we add the probabilities of winning the 2005 World Series for all teams, we are supposed to obtain a value of 1.0. However, if we calculate the probability of winning the 2005 World Series for

Team	Odds	Team	Odds
Arizona Diamondbacks	1:15	Milwaukee Brewers	1:75
Atlanta Braves	1:10	Minnesota Twins	1:8
Baltimore Orioles	1:8	New York Mets	1:20
Boston Red Sox	2:9	New York Yankees	1:5
Chicago Cubs	1:5	Oakland Athletics	1:100
Chicago White Sox	1:5	Philadelphia Phillies	1:30
Cincinnati Reds	1:300	Pittsburgh Pirates	1:150
Cleveland Indians	1:80	San Diego Padres	1:15
Colorado Rockies	1:400	San Francisco Giants	1:70
Detroit Tigers	1:40	Seattle Mariners	1:70
Florida Marlins	1:6	St. Louis Cardinals	1:3
Houston Astros	1:150	Tampa Bay Devil Rays	1:500
Kansas City Royals	1:5000	Texas Rangers	1:20
Los Angeles Angels	1:8	Toronto Blue Jays	1:30
Los Angeles Dodgers	1:15	Washington Nationals	1:75

each team using the odds given in the above table and then add all these probabilities, the sum will be 1.968550707. So, what happened? Did these odds-makers flunk their statistics and probability courses? Probably not.

Casinos and odds-makers, who are in the business of making money, are interested in encouraging people to gamble. These probabilities, which seem to violate one of the basic rules of probability theory, still obey the primary rule for the casinos, which is that, on average, a casino is going to make profits. How does this happen? We will explore this in the Uses and Misuses section of Chapter 5.

Note: When casinos create odds for sports betting, they recognize that many people will bet on one of their favorite teams such as St. Louis or Boston. In order to meet the rule that the sum of all the probabilities is 1.0, the probabilities for the teams more likely to win the World Series would have to be lowered. And lower probabilities correspond to lower odds. For example, if the odds for St. Louis to win the 2005 World Series were lowered from 1:3 to 1:9, the probability for them to win would decrease from .25 to .10. If St. Louis remains the favorite (still has the best odds), many people would bet on them. However, if they win, the casino would have to pay $9 for every one dollar bet instead of paying $3 for every one dollar bet. The casinos do not want to do this and, hence, they ignore the probability rule in order to make more money. However, the casinos cannot do this with their traditional games, which are bound by the standard rules. From a mathematical standpoint, it is not acceptable to ignore this probability rule that the probabilities of all final outcomes for an experiment add up to 1.0.

Glossary

Classical probability rule The method of assigning probabilities to outcomes or events of an experiment with equally likely outcomes.

Complementary events Two events that taken together include all the outcomes for an experiment but do not contain any common outcome.

Compound event An event that contains more than one outcome of an experiment. It is also called a *composite event.*

Conditional probability The probability of an event subject to the condition that another event has already occurred.

Dependent events Two events for which the occurrence of one changes the probability of the occurrence of the other.

Equally likely outcomes Two (or more) outcomes or events that have the same probability of occurrence.

Event A collection of one or more outcomes of an experiment.

Experiment A process with well-defined outcomes that, when performed, results in one and only one of the outcomes per repetition.

Impossible event An event that cannot occur.

Independent events Two events for which the occurrence of one does not change the probability of the occurrence of the other.

Intersection of events The intersection of events is given by the outcomes that are common to two (or more) events.

Joint probability The probability that two (or more) events occur together.

Law of Large Numbers If an experiment is repeated again and again, the probability of an event obtained from the relative frequency approaches the actual or theoretical probability.

Marginal probability The probability of one event or characteristic without consideration of any other event.

Mutually exclusive events Two or more events that do not contain any common outcome and, hence, cannot occur together.

Outcome The result of the performance of an experiment.

Probability A numerical measure of the likelihood that a specific event will occur.

Relative frequency as an approximation of probability Probability assigned to an event based on the results of an experiment or based on historical data.

Sample point An outcome of an experiment.

Sample space The collection of all sample points or outcomes of an experiment.

Simple event An event that contains one and only one outcome of an experiment. It is also called an *elementary event.*

Subjective probability The probability assigned to an event based on the information and judgment of a person.

Sure event An event that is certain to occur.

Tree diagram A diagram in which each outcome of an experiment is represented by a branch of a tree.

Union of two events Given by the outcomes that belong either to one or to both events.

Venn diagram A picture that represents a sample space or specific events.

Supplementary Exercises

4.124 A car rental agency currently has 44 cars available, 18 of which have a GPS navigation system. One of the 44 cars is selected at random. Find the probability that this car

a. has a GPS navigation system

b. does not have a GPS navigation system

4.125 In a class of 35 students, 13 are seniors, 9 are juniors, 8 are sophomores, and 5 are freshmen. If one student is selected at random from this class, what is the probability that this student is

a. a junior?

b. a freshman?

4.126 A random sample of 250 juniors majoring in psychology or communications at a large university is selected. These students are asked whether or not they are happy with their majors. The following table gives the results of the survey. Assume that none of these 250 students is majoring in both areas.

	Happy	Unhappy
Psychology	80	20
Communications	115	35

a. If one student is selected at random from this group, find the probability that this student is

i. happy with the choice of major

ii. a psychology major

iii. a communications major given that the student is happy with the choice of major

iv. unhappy with the choice of major given that the student is a psychology major

v. a psychology major *and* is happy with that major

vi. a communications major *or* is unhappy with his or her major

b. Are the events "psychology major" and "happy with major" independent? Are they mutually exclusive? Explain why or why not.

4.127 A random sample of 250 adults was taken, and they were asked whether they prefer watching sports or opera on television. The following table gives the two-way classification of these adults.

	Prefer Watching Sports	Prefer Watching Opera
Male	96	24
Female	45	85

a. If one adult is selected at random from this group, find the probability that this adult
   i. prefers watching opera
   ii. is a male
   iii. prefers watching sports given that the adult is a female
   iv. is a male given that he prefers watching sports
   v. is a female *and* prefers watching opera
   vi. prefers watching sports *or* is a male

b. Are the events "female" and "prefers watching sports" independent? Are they mutually exclusive? Explain why or why not.

4.128 A random sample of 80 lawyers was taken, and they were asked if they are in favor of or against capital punishment. The following table gives the two-way classification of their responses.

	Favors Capital Punishment	Opposes Capital Punishment
Male	32	24
Female	13	11

a. If one lawyer is randomly selected from this group, find the probability that this lawyer
   i. favors capital punishment
   ii. is a female
   iii. opposes capital punishment given that the lawyer is a female
   iv. is a male given that he favors capital punishment
   v. is a female *and* favors capital punishment
   vi. opposes capital punishment *or* is a male

b. Are the events "female" and "opposes capital punishment" independent? Are they mutually exclusive? Explain why or why not.

4.129 A random sample of 400 college students was asked if college athletes should be paid. The following table gives a two-way classification of the responses.

	Should Be Paid	Should Not Be Paid
Student athlete	90	10
Student nonathlete	210	90

a. If one student is randomly selected from these 400 students, find the probability that this student
   i. is in favor of paying college athletes
   ii. favors paying college athletes given that the student selected is a nonathlete
   iii. is an athlete *and* favors paying student athletes
   iv. is a nonathlete *or* is against paying student athletes

b. Are the events "student athlete" and "should be paid" independent? Are they mutually exclusive? Explain why or why not.

4.130 An appliance repair company that makes service calls to customers' homes has found that 5% of the time there is nothing wrong with the appliance and the problem is due to customer error (appliance unplugged, controls improperly set, etc.). Two service calls are selected at random, and it is observed whether or not the problem is due to customer error. Draw a tree diagram. Find the probability that in this sample of two service calls

a. both problems are due to customer error

b. at least one problem is not due to customer error

4.131 According to data from Watson Wyatt Worldwide, 51% of employees have confidence in their senior management (*Business Week*, January 24, 2005). Assume that the other 49% do not have confidence in their senior management. Further assume that these percentages are true for the current population of all employees. Two employees are selected at random and asked whether or not they have confidence in their senior management. Draw a tree diagram for this problem. Find the probability that in this sample of two employees

a. both have confidence in their senior management

b. at most one has confidence in his or her senior management

4.132 Refer to Exercise 4.124. Two cars are selected at random from these 44 cars. Find the probability that both of these cars have GPS navigation systems.

4.133 Refer to Exercise 4.125. Two students are selected at random from this class of 35 students. Find the probability that the first student selected is a junior and the second is a sophomore.

4.134 A company has installed a generator to back up the power in case there is a power failure. The probability that there will be a power failure during a snowstorm is .30. The probability that the generator will stop working during a snowstorm is .09. What is the probability that during a snowstorm the company will lose both sources of power?

4.135 Terry & Sons makes bearings for autos. The production system involves two independent processing machines so that each bearing passes through these two processes. The probability that the first processing machine is not working properly at any time is .08, and the probability that the second machine is not working properly at any time is .06. Find the probability that both machines will not be working properly at any given time.

Advanced Exercises

4.136 A player plays a roulette game in a casino by betting on a single number each time. Because the wheel has 38 numbers, the probability that the player will win in a single play is $1/38$. Note that each play of the game is independent of all previous plays.

a. Find the probability that the player will win for the first time on the 10th play.
b. Find the probability that it takes the player more than 50 plays to win for the first time.
c. The gambler claims that because he has 1 chance in 38 of winning each time he plays, he is certain to win at least once if he plays 38 times. Does this sound reasonable to you? Find the probability that he will win at least once in 38 plays.

4.137 A certain state's auto license plates have three letters of the alphabet followed by a three-digit number.

a. How many different license plates are possible if all three-letter sequences are permitted and any number from 000 to 999 is allowed?
b. Arnold witnessed a hit-and-run accident. He knows that the first letter on the license plate of the offender's car was a B, that the second letter was an O or a Q, and that the last number was a 5. How many of this state's license plates fit this description?

4.138 The median life of Brand LT5 batteries is 100 hours. What is the probability that in a set of three such batteries, exactly two will last longer than 100 hours?

4.139 Powerball is a game of chance that has generated intense interest because of its large jackpots. To play this game, a player selects five different numbers from 1 through 53, and then picks a powerball number from 1 through 42. The lottery organization randomly draws five different white balls from 53 balls numbered 1 through 53, and then randomly picks a powerball number from 1 through 42. Note that it is possible for the powerball number to be the same as one of the first five numbers.

a. If the player's first five numbers match the numbers on the five white balls drawn by the lottery organization and the player's powerball number matches the powerball number drawn by the lottery organization, the player wins the jackpot. Find the probability that a player who buys one ticket will win the jackpot. (Note that the order in which the five white balls are drawn is unimportant.)
b. If the player's first five numbers match the numbers on the five white balls drawn by the lottery organization but the powerball number does not match the one drawn by the lottery organization, the player wins about $100,000 (or less if several winners must share the prize pool). Find the probability that a player who buys one ticket will win this prize.

4.140 A trimotor plane has three engines—a central engine and an engine on each wing. The plane will crash only if the central engine fails *and* at least one of the two wing engines fails. The probability of failure during any given flight is .005 for the central engine and .008 for each of the wing engines. Assuming that the three engines operate independently, what is the probability that the plane will crash during a flight?

4.141 A box contains 10 red marbles and 10 green marbles.

a. Sampling at random from the box five times with replacement, you have drawn a red marble all five times. What is the probability of drawing a red marble the sixth time?
b. Sampling at random from the box five times without replacement, you have drawn a red marble all five times. Without replacing any of the marbles, what is the probability of drawing a red marble the sixth time?
c. You have tossed a fair coin five times and have obtained heads all five times. A friend argues that according to the law of averages, a tail is due to occur and, hence, the probability of obtaining a head on the sixth toss is less than .50. Is he right? Is coin tossing mathematically equivalent to the procedure mentioned in part a or the procedure mentioned in part b? Explain.

4.142 A gambler has four cards—two diamonds and two clubs. The gambler proposes the following game to you: You will leave the room and the gambler will put the cards face down on a table. When you return to the room, you will pick two cards at random. You will win \$10 if both cards are diamonds, you will win \$10 if both are clubs, and for any other outcome you will lose \$10. Assuming that there is no cheating, should you accept this proposition? Support your answer by calculating your probability of winning \$10.

4.143 A thief has stolen Roger's automatic teller card. The card has a four-digit personal identification number (PIN). The thief knows that the first two digits are 3 and 5, but he does not know the last two digits. Thus, the PIN could be any number from 3500 to 3599. To protect the customer, the automatic teller machine will not allow more than three unsuccessful attempts to enter the PIN. After the third wrong PIN, the machine keeps the card and allows no further attempts.

a. What is the probability that the thief will find the correct PIN within three tries? (Assume that the thief will not try the same wrong PIN twice.)
b. If the thief knew that the first two digits were 3 and 5 and that the third digit was either 1 or 7, what is the probability of guessing the correct PIN in three attempts?

4.144 Consider the following games with two dice.

a. A gambler is going to roll a die four times. If he rolls at least one 6, you must pay him \$5. If he fails to roll a 6 in four tries, he will pay you \$5. Find the probability that you must pay the gambler. Assume that there is no cheating.
b. The same gambler offers to let you roll a pair of dice 24 times. If you roll at least one double 6, he will pay you \$10. If you fail to roll a double 6 in 24 tries, you will pay him \$10. The gambler says that you have a better chance of winning because your probability of success on each of the 24 rolls is $1/36$ and you have 24 chances. Thus, he says, your probability of winning \$10 is $24(1/36) = 2/3$. Do you agree with this analysis? If so, indicate why. If not, point out the fallacy in his argument, and then find the correct probability that you will win.

4.145 A gambler has given you two jars and 20 marbles. Of these 20 marbles, ten are red and ten are green. You must put all 20 marbles in these two jars in such a way that each jar must have at least one marble in it. Then a friend of yours, who is blindfolded, will select one of the two jars at random and then will randomly select a marble from this jar. If the selected marble is red, you and your friend win \$100.

a. If you put five red marbles and five green marbles in each jar, what is the probability that your friend selects a red marble?
b. If you put two red marbles and two green marbles in one jar and the remaining marbles in the other jar, what is the probability that your friend selects a red marble?
c. How should these 20 marbles be distributed among the two jars in order to give your friend the highest possible probability of selecting a red marble?

4.146 A screening test for a certain disease is prone to giving false positives or false negatives. If a patient being tested has the disease, the probability that the test indicates a (false) negative is .13. If the patient does not have the disease, the probability that the test indicates a (false) positive is .10. Assume that 3% of the patients being tested actually have the disease. Suppose that one patient is chosen at random and tested. Find the probability that

a. this patient has the disease and tests positive
b. this patient does not have the disease and tests positive
c. this patient tests positive
d. this patient has the disease given that he/she tests positive

(*Hint:* A tree diagram may be helpful in part c.)

4.147 A pizza parlor has 12 different toppings available for its pizzas, and 2 of these toppings are pepperoni and anchovies. If a customer picks two toppings at random, find the probability that

a. neither topping is anchovies
b. pepperoni is one of the toppings

4.148 An insurance company has information that 93% of its auto policy holders carry collision coverage or uninsured motorist coverage on their policies. Eighty percent of the policy holders carry collision coverage, and 60% have uninsured motorist coverage.

a. What percentage of these policy holders carry both collision and uninsured motorist coverage?
b. What percentage of these policy holders carry neither collision nor uninsured motorist coverage?
c. What percentage of these policy holders carry collision but not uninsured motorist coverage?

4.149 Many states have a lottery game, usually called a Pick-4, in which you pick a four-digit number such as 7359. During the lottery drawing, there are four bins, each containing balls numbered 0 through 9. One ball is drawn from each bin to form the four-digit winning number.

a. You purchase one ticket with one four-digit number. What is the probability that you will win this lottery game?

b. There are many variations of this game. The primary variation allows you to win if the four digits in your number are selected in any order as long as they are the same four digits as obtained by the lottery agency. For example, if you pick four digits making the number 1265, then you will win if 1265, 2615, 5216, 6521, and so forth, are drawn. The variations of the lottery game depend on how many unique digits are in your number. Consider the following four different versions of this game.

   i. All four digits are unique (e.g., 1234)
   ii. Exactly one of the digits appears twice (e.g., 1223 or 9095)
   iii. Two digits each appear twice (e.g., 2121 or 5588)
   iv. One digit appears three times (e.g., 3335 or 2722)

   Find the probability that you will win this lottery in each of these four situations.

4.150 A restaurant chain is planning to purchase 100 ovens from a manufacturer provided that these ovens pass a detailed inspection. Because of high inspection costs, five ovens are selected at random for inspection. These 100 ovens will be purchased if at most one of the five selected ovens fails inspection. Suppose that there are eight defective ovens in this batch of 100 ovens. Find the probability that the batch of ovens is purchased. (Note: In Chapter 5 you will learn another method to solve this problem.)

4.151 A production system has two production lines; each production line performs a two-part process; and each process is completed by a different machine. Thus, there are four machines, which we can identify as two first-level machines and two second-level machines. Each of the first-level machines works properly 98% of the time, and each of the second-level machines works properly 96% of the time. All four machines are independent in regard to working properly or breaking down. Two products enter this production system, one in each production line.

a. Find the probability that both products successfully complete the two-part process (i.e., all four machines are working properly).

b. Find the probability that neither product successfully completes the two-part process (i.e., at least one of the machines in each production line is not working properly).

Self-Review Test

1. The collection of all outcomes for an experiment is called
   a. a sample space b. the intersection of events c. joint probability

2. A final outcome of an experiment is called
   a. a compound event b. a simple event c. a complementary event

3. A compound event includes
   a. all final outcomes b. exactly two outcomes
   c. more than one outcome for an experiment

4. Two equally likely events
   a. have the same probability of occurrence
   b. cannot occur together
   c. have no effect on the occurrence of each other

5. Which of the following probability approaches can be applied only to experiments with equally likely outcomes?
   a. Classical probability b. Empirical probability c. Subjective probability

6. Two mutually exclusive events
   a. have the same probability b. cannot occur together
   c. have no effect on the occurrence of each other

7. Two independent events
   a. have the same probability b. cannot occur together
   c. have no effect on the occurrence of each other

8. The probability of an event is always
   a. less than 0 b. in the range 0 to 1.0 c. greater than 1.0

9. The sum of the probabilities of all final outcomes of an experiment is always
   a. 100 b. 1.0 c. 0

10. The joint probability of two mutually exclusive events is always
   a. 1.0 b. between 0 and 1 c. 0

11. Two independent events are
   a. always mutually exclusive b. never mutually exclusive
   c. always complementary

12. A couple is planning their wedding reception. The bride's parents have given them a choice of four reception facilities, three caterers, five DJs, and two limo services. If the couple randomly selects one reception facility, one caterer, one DJ, and one limo service, how many different outcomes are possible?

13. Lucia graduated this year with an accounting degree from Eastern Connecticut State University. She has received job offers from an accounting firm, an insurance company, and an airline. She cannot decide which of the three job offers she should accept. Suppose she decides to randomly select one of these three job offers. Find the probability that the job offer selected is
   a. from the insurance company
   b. not from the accounting firm

14. There are 200 students in a particular graduate program at a state university. Of them, 110 are female and 125 are out-of-state students. Of the 110 females, 70 are out-of-state students.
   a. Are the events "female" and "out-of-state student" independent? Are they mutually exclusive? Explain why or why not.
   b. If one of these 200 students is selected at random, what is the probability that the student selected is
      i. a male?
      ii. an out-of-state student given that this student is a female?

15. Reconsider Problem 14. If one of these 200 students is selected at random, what is the probability that the selected student is a female *or* an out-of-state student?

16. Reconsider Problem 14. If two of these 200 students are selected at random, what is the probability that both of them are out-of-state students?

17. The probability that an American adult has ever experienced a migraine headache is .35. If two American adults are randomly selected, what is the probability that neither of them has ever experienced a migraine headache?

18. A hat contains five green, eight red, and seven blue marbles. Let A be the event that a red marble is drawn if we randomly select one marble out of this hat. What is the probability of A? What is the complementary event of A, and what is its probability?

19. The probability that a randomly selected student from a college is female is .55 and that a student works for more than 10 hours per week is .62. If these two events are independent, find the probability that a randomly selected student is a
   a. male *and* works for more than 10 hours per week
   b. female *or* works for more than 10 hours per week

20. A sample was selected of 506 workers who currently receive two weeks of paid vacation per year. These workers were asked if they were willing to accept a small pay cut to get an additional week of paid vacation a year. The following table shows the responses of these workers.

	Yes	No	No Response
Man	77	140	32
Woman	104	119	34

   a. If one person is selected at random from these 506 workers, find the following probabilities.
      i. $P(\text{yes})$ ii. $P(\text{yes} \mid \text{woman})$
      iii. $P(\text{woman } and \text{ no})$ iv. $P(\text{no response } or \text{ man})$
   b. Are the events "woman" and "yes" independent? Are they mutually exclusive? Explain why or why not.

Mini-Projects

MINI-PROJECT 4–1

Suppose that a small chest contains three drawers. The first drawer contains two \$1 bills, the second drawer contains two \$100 bills, and the third drawer contains one \$1 bill and one \$100 bill. Suppose that first a drawer is selected at random and then one of the two bills inside that drawer is selected at random. We can define these events:

A = the first drawer is selected	B = the second drawer is selected
C = the third drawer is selected	D = a \$1 bill is selected

a. Suppose when you randomly select one drawer and then one bill from that drawer, the bill you obtain is a \$1 bill. What is the probability that the second bill in this drawer is a \$100 bill? In other words, find the probability $P(C \mid D)$ because for the second bill to be \$100, it has to be the third drawer. Answer this question intuitively without making any calculations.

b. Use the relative frequency concept of probability to estimate $P(C \mid D)$ as follows. First select a drawer by rolling a die once. If either 1 or 2 occurs, the first drawer is selected; if either 3 or 4 occurs, the second drawer is selected; and if either 5 or 6 occurs, the third drawer is selected. Whenever C occurs, then select a bill by tossing a coin once. (Note that if either A or B occurs then you do not need to toss the coin because each of these drawers contains both bills of the same denomination.) If you obtain a head, assume that you select a \$1 bill; if you obtain a tail, assume that you select a \$100 bill. Repeat this process 100 times. How many times in these 100 repetitions did the event D occur? What proportion of the time did C occur when D occurred? Use this proportion to estimate $P(C \mid D)$. Does this estimate support your guess of $P(C \mid D)$ in part a?

c. Calculate $P(C \mid D)$ using the procedures developed in this chapter (a tree diagram may be helpful). Was your estimate in part b close to this value? Explain.

MINI-PROJECT 4–2

Refer to Exercise 4.146. Use the relative frequency concept to estimate the probability that the second child in the family is a girl given that the selected child is a girl. Use the following process to do so. First toss a coin to determine whether the Smith family or the Jones family is chosen. If the Smith family is selected, then record that the second child in this family is a girl given that the selected child is a girl. This is so because both children in this family are girls. If the Jones family is selected, then toss the coin again to select a child and record the gender of the child selected and that of the second child in this family. Repeat this process 50 times, and then use the results to estimate the required probability. How close is your estimate to the probability calculated in Exercise 4.146?

MINI-PROJECT 4–3

The dice game Yahtzee® involves five standard dice. On your turn, you can roll all five or fewer dice up to three times to obtain different sets of numbers on the dice. For example, you will roll all five dice the first time; if you like two of the five numbers obtained, you can roll the other three dice a second time; now if you like three of the five numbers obtained, you can roll the other two dice the third time. Some of the sets of numbers obtained are similar to poker hands (three of a kind, four of a kind, full house, and so forth). However, a few other hands, like five of a kind (called a *yahtzee*), are not poker hands (or at least they are not the hands you would dare to show anyone.)

For the purpose of this project, we will examine the outcomes on the first roll of the five dice. The five scenarios that we will consider are:

i. Three of a kind—three numbers are the same and the remaining two numbers are both different, e.g., 22254

ii. Four of a kind—four numbers are the same and the fifth number is different, e.g., 44442

iii. Full house—three numbers are the same and the other two numbers are the same, e.g., 33366

iv. Large straight—five numbers in a row, e.g., 1 2 3 4 5

v. Yahtzee—all five numbers are the same, e.g., 33333

In the first two cases, the dice that are not part of the three or four of a kind must have different values than those in the three or four of a kind. For example, 22252 cannot be considered three of a kind, but 22254 is three of a kind. (Yahtzee players know that this situation differs from the rules of the actual game, but for the purpose of this project, we will "change" the rules.)

a. Find the probability for each of these five cases for one roll of the five dice.

b. In a regular game, you do not have to roll all five dice on each of the three rolls. You can leave some dice on the table and roll the others in an attempt to improve your score. For example, if you roll 13555 on the first roll of five dice, you can keep the three fives and roll the dice with 1 and 3 outcomes the second time in an attempt to get more fives, or possibly a pair of another number in order to get a full house. After your second roll, you are allowed to pick up any of the dice for your third roll. For example, suppose your first roll is 13644 and you keep the two fours. Then you roll the dice with 1, 3, and 6 outcomes the second time and obtain three fives. Thus, now you have 55544. Although you met your full house requirement before rolling the dice three times, you still need a yahtzee. So, you keep the fives and roll the two dice with fours the third time. Write a paragraph outlining all of the scenarios you will have to consider to calculate the probability of obtaining a yahtzee within your three rolls.

DECIDE FOR YOURSELF

Deciding About Production Processes

Henry Ford was one of the major developers of mass production. Imagine if his factory had only one production line! If any component in that production line would have broken down, all production would have come to a halt. In order for mass production to be successful, the factory must be able to continue production when one or more machines in the production process break down. Automobile factories, like many other forms of production, have multiple production lines running side by side. So if one production line is shut down due to a breakdown, the other production lines can still operate. Probability theory can be used to study the reliability of production systems by determining the likelihood that a system will continue to operate even when some parts of the system fail.

In order to study such systems, we have to consider how they are set up. These systems are comprised of two types of arrangements: series and parallel. In a series system, a process is sequential. One part of the process must be completed before the item can move to the next part of the process. If any part of the system breaks down, none of the tasks that follow can be completed. In the auto example, if something in a series system breaks down while the chassis is being constructed, it will be impossible to install the seats, the windshield, the engine, and so forth.

In a parallel system, various processes work side-by-side. In some cases the processes are like toll collectors at a bridge or on a highway. As long as there is at least one toll collector working, traffic will continue moving, although more toll collectors would certainly speed up the process. In a computer network, different servers are set up in parallel systems. If one server (such as the e-mail server) goes down, people on the network can still access the Web and file servers. However, if the servers were set up in a series system and the e-mail server failed, nobody would be able to do anything.

Let us consider a simplified example. Suppose a production line involves five tasks. Each of the machines that perform these tasks works successfully 97% of the time. In other words, the probability that a specific task can be completed (without interruption) is .97. For the sake of simplicity, let us assume that the machines work and fail independently of each other. Furthermore, suppose that the factory has three of these lines running in a parallel system. Following are some of the questions that arise. Act as if you are in charge of such a production process and try to answer these questions.

1. What is the probability that all five tasks in a single line are completed without interruption?

2. What is the probability that at least one of the three production lines is working properly?

3. Why is the probability that a specific line works properly lower than the probability that at least one of the lines in the factory works properly?

4. What happens to the reliability of the system if an additional task is added to each line?

5. What happens when the number of tasks remains constant, but another line is added?

TECHNOLOGY INSTRUCTION Generating Random Numbers

TI-84

1. To generate a random number (not necessarily an integer) uniformly distributed between m and n, select **MATH>PRB** and type **rand*(n−m)+m**.

2. To generate a random number that is an integer uniformly distributed between m and n, select **MATH>PRB** and type **randInt(m,n)**.

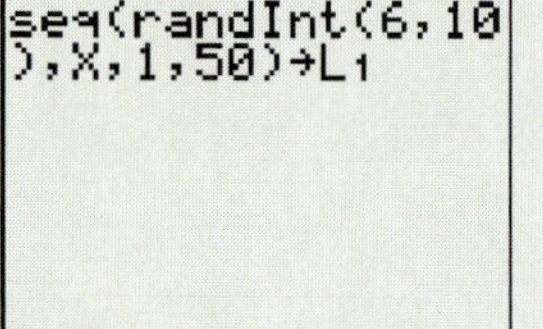

Screen 4.1

3. To create a sequence of random numbers (integer or noninteger) and store them in a list, you will need to use the **seq(** function in conjunction with the appropriate random number function from step 1 or step 2. Specifically, select **2nd>STAT>OPS>SEQ(**, then type the function from step 1 or step 2, then type **,X,1,quantity of random numbers you want)>STO→>L1>ENTER**. These instructions will store the data in list **L1** (see **Screen 4.1**). However, you can replace L1 by any other list you want in the above instructions.

MINITAB

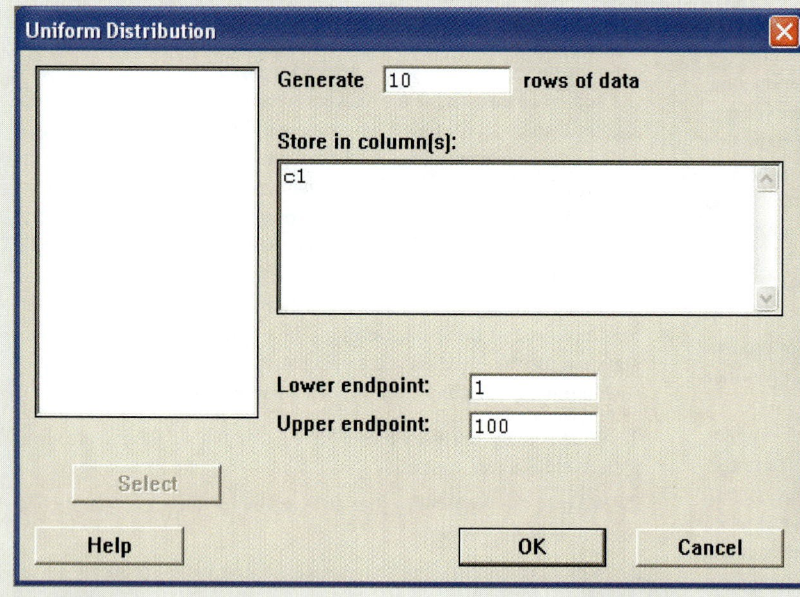

Screen 4.2

1. To generate random numbers (not necessarily integers) uniformly distributed between m and n, select **Calc>Random Data>Uniform**. Enter the number of rows of data, the column where you wish to store the data, and the minimum m and maximum n values for the numbers (see **Screens 4.2** and **4.3**).

2. To generate random integers uniformly distributed between m and n, select **Calc>Random Data >Integer**. Enter the number of rows of data, the column where you wish to store the data, and the minimum m and maximum n values for the integers.

↓	C1
1	82.8024
2	6.5293
3	77.9912
4	2.4420
5	61.5157
6	2.0113
7	52.3235
8	63.2997
9	83.7044
10	26.6449

Screen 4.3

EXCEL

1. To generate a random number (not necessarily an integer) uniformly distributed between m and n, enter the formula **=rand()*(n−m)+m**. If you need more than one random number, copy and paste the formula into as many cells as you need. The numbers will be recalculated every time any cell in the spreadsheet is calculated or recalculated (see **Screen 4.4**).

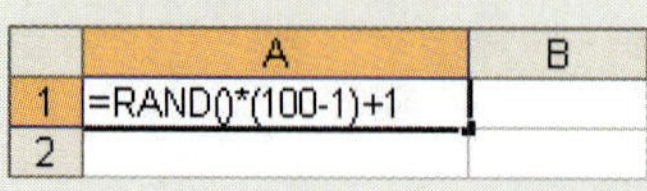

Screen 4.4

2. To generate a random integer uniformly distributed between m and n, enter the formula **=floor(rand()*(n−m+1)+m,1)**. If you need more than one random number, copy and paste the formula into as many cells as you need. The numbers will be recalculated every time any cell in the spreadsheet is calculated or recalculated.

3. To enter a random number (either type) that stays fixed after it is calculated, select the cell containing the formula, select the formula bar, and press **F9**. (Note that this procedure works only one cell at a time.) If you have a bunch of random numbers that you wish to keep fixed after being calculated, highlight all of the numbers, select **Edit>Copy**, go to an empty column, then select **Edit>Paste Special** and check the **Values** box.

TECHNOLOGY ASSIGNMENTS

TA4.1 You want to simulate the tossing of a coin. Assign a value of 0 (zero) to Head and a value of 1 to Tail.

a. Simulate 50 tosses of the coin by generating 50 random (integer) numbers between 0 and 1. Then calculate the mean of these 50 numbers. This mean gives you the proportion of 50 tosses that resulted in tails. Using this proportion, calculate the number of heads and tails you obtained in 50 simulated tosses.

b. Repeat part a by simulating 600 tosses.

c. Repeat part a by simulating 4000 tosses.

Comment on the percentage of Tails obtained as the number of tosses is increased.

TA4.2 You want to simulate the rolling of a die. Assign the values 1 through 6 to the outcomes from 1-spot through 6-spots on the die, respectively.

a. Simulate 200 rolls of the die by generating 200 random (integer) numbers between 1 and 6. Then make a histogram for these 200 numbers.

b. Repeat part a by simulating 1000 rolls of the die.

c. Repeat part a by simulating 6000 rolls of the die.

Comment on the histograms obtained in parts a through c.

Chapter

5

Discrete Random Variables and Their Probability Distributions

5.1 Random Variables

5.2 Probability Distribution of a Discrete Random Variable

5.3 Mean of a Discrete Random Variable

Case Study 5–1 Aces High Instant Lottery Game—18th Edition

5.4 Standard Deviation of a Discrete Random Variable

5.5 Factorials, Combinations, and Permutations

Case Study 5–2 Playing Lotto

5.6 The Binomial Probability Distribution

5.7 The Hypergeometric Probability Distribution

5.8 The Poisson Probability Distribution

Case Study 5–3 Ask Mr. Statistics

Case Study 5–4 Living and Dying in the USA

Now that you know a little about probability, do you feel lucky? If you've got $20 to spend on lunch today, are you willing to spend it all on twenty $1 lotto tickets to increase your chances of winning? What if you know that, depending on what state you are in, it could mean buying as many as 18 million $1 tickets to cover all the possible combinations to have a definite chance to win? (See Case Study 5–2.) That is a lot of lunch! How do we go about determining the outcomes and probabilities in a lotto game?

Chapter 4 discussed the concepts and rules of probability. This chapter extends the concept of probability to explain probability distributions. As we saw in Chapter 4, any given statistical experiment has more than one outcome. It is impossible to predict which of the many possible outcomes will occur if an experiment is performed. Consequently, decisions are made under uncertain conditions. For example, a lottery player does not know in advance whether or not he is going to win that lottery. If he knows that he is not going to win, he will definitely not play. It is the uncertainty about winning (some positive probability of winning) that makes him play. This chapter shows that if the outcomes and their probabilities for a statistical experiment are known, we can find out what will happen, on average, if that experiment is performed many times. For the lottery example, we can find out what a lottery player can expect to win (or lose), on average, if he continues playing this lottery again and again.

In this chapter, random variables and types of random variables are explained. Then, the concept of a probability distribution and its mean and standard deviation for a discrete random variable are discussed. Finally, three special probability distributions for a discrete random variable—the binomial probability distribution, the hypergeometric probability distribution, and the Poisson probability distribution—are developed.

5.1 Random Variables

Suppose Table 5.1 gives the frequency and relative frequency distributions of the number of vehicles owned by all 2000 families living in a small town.

Table 5.1 Frequency and Relative Frequency Distributions of the Number of Vehicles Owned by Families

Number of Vehicles Owned	Frequency	Relative Frequency
0	30	30/2000 = .015
1	470	470/2000 = .235
2	850	850/2000 = .425
3	490	490/2000 = .245
4	160	160/2000 = .080
	$N = 2000$	Sum = 1.000

Suppose one family is randomly selected from this population. The process of randomly selecting a family is called a *random* or *chance experiment*. Let x denote the number of vehicles owned by the selected family. Then x can assume any of the five possible values (0, 1, 2, 3, and 4) listed in the first column of Table 5.1. The value assumed by x depends on which family is selected. Thus, this value depends on the outcome of a random experiment. Consequently, x is called a **random variable** or a **chance variable**. In general, a random variable is denoted by x or y.

Definition

Random Variable A *random variable* is a variable whose value is determined by the outcome of a random experiment.

As will be explained next, a random variable can be discrete or continuous.

5.1.1 Discrete Random Variable

A **discrete random variable** assumes values that can be counted. In other words, the consecutive values of a discrete random variable are separated by a certain gap.

Definition

Discrete Random Variable A *random variable* that assumes countable values is called a *discrete random variable.*

In Table 5.1, *the number of vehicles owned by a family* is an example of a discrete random variable because the values of the random variable x are countable: 0, 1, 2, 3, and 4. Here are some other examples of discrete random variables:

1. The number of cars sold at a dealership during a given month
2. The number of houses in a certain block
3. The number of fish caught on a fishing trip
4. The number of complaints received at the office of an airline on a given day

5. The number of customers who visit a bank during any given hour
6. The number of heads obtained in three tosses of a coin

5.1.2 Continuous Random Variable

A random variable whose values are not countable is called a **continuous random variable**. A continuous random variable can assume any value over an interval or intervals.

Definition

Continuous Random Variable A random variable that can assume any value contained in one or more intervals is called a *continuous random variable*.

Because the number of values contained in any interval is infinite, the possible number of values that a continuous random variable can assume is also infinite. Moreover, we cannot count these values. Consider the life of a battery. We can measure it as precisely as we want. For instance, the life of this battery may be 40 hours, or 40.25 hours, or 40.247 hours. Assume that the maximum life of a battery is 200 hours. Let x denote the life of a randomly selected battery of this kind. Then, x can assume any value in the interval 0 to 200. Consequently, x is a continuous random variable. As shown in the diagram, every point on the line representing the interval 0 to 200 gives a possible value of x.

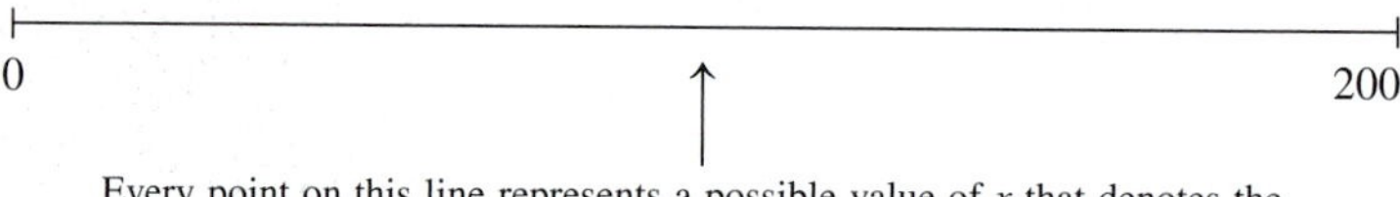

Every point on this line represents a possible value of x that denotes the life of a battery. There are an infinite number of points on this line. The values represented by points on this line are uncountable.

The following are some examples of continuous random variables:

1. The height of a person
2. The time taken to complete an examination
3. The amount of milk in a gallon (Note that we do not expect a gallon to contain exactly one gallon of milk but either slightly more or slightly less than a gallon.)
4. The weight of a fish
5. The price of a house

This chapter is limited to a discussion of discrete random variables and their probability distributions. Continuous random variables will be discussed in Chapter 6.

EXERCISES

CONCEPTS AND PROCEDURES

5.1 Explain the meaning of a random variable, a discrete random variable, and a continuous random variable. Give one example each of a discrete random variable and a continuous random variable.

5.2 Classify each of the following random variables as discrete or continuous.
a. The time left on a parking meter
b. The number of bats broken by a major league baseball team in a season
c. The number of fish in a pond
d. The total pounds of fish caught on a fishing trip
e. The number of gumballs in a vending machine
f. The time spent by a physician examining a patient

5.3 Indicate which of the following random variables are discrete and which are continuous.

a. The number of new accounts opened at a bank during a certain month
b. The time taken to run a marathon
c. The price of a concert ticket
d. The number of rotten eggs in a randomly selected box
e. The points scored in a football game
f. The weight of a randomly selected package

APPLICATIONS

5.4 A household can watch news on any of the three networks—ABC, CBS, or NBC. On a certain day, five households randomly and independently decide which channel to watch. Let x be the number of households among these five that decide to watch news on ABC. Is x a discrete or a continuous random variable? Explain.

5.5 One of the four gas stations located at an intersection of two major roads is a Texaco station. Suppose the next six cars that stop at any of these four gas stations make their selections randomly and independently. Let x be the number of cars in these six that stop at the Texaco station. Is x a discrete or a continuous random variable? Explain.

5.2 Probability Distribution of a Discrete Random Variable

Let x be a discrete random variable. The **probability distribution** of x describes how the probabilities are distributed over all the possible values of x.

Definition

Probability Distribution of a Discrete Random Variable The *probability distribution of a discrete random variable* lists all the possible values that the random variable can assume and their corresponding probabilities.

Example 5–1 illustrates the concept of the probability distribution of a discrete random variable.

EXAMPLE 5–1

Writing the probability distribution of a discrete random variable.

Recall the frequency and relative frequency distributions of the number of vehicles owned by families given in Table 5.1. That table is reproduced below as Table 5.2. Let x be the number of vehicles owned by a randomly selected family. Write the probability distribution of x.

Table 5.2 Frequency and Relative Frequency Distributions of the Number of Vehicles Owned by Families

Number of Vehicles Owned	Frequency	Relative Frequency
0	30	.015
1	470	.235
2	850	.425
3	490	.245
4	160	.080
	$N = 2000$	Sum = 1.000

Solution In Chapter 4, we learned that the relative frequencies obtained from an experiment or a sample can be used as approximate probabilities. However, when the relative frequencies represent the population, as in Table 5.2, they give the actual (theoretical) probabilities of outcomes. Using the relative frequencies of Table 5.2, we can write the *probability distribution* of the discrete random variable x in Table 5.3.

Table 5.3 Probability Distribution of the Number of Vehicles Owned by Families

Number of Vehicles Owned x	Probability $P(x)$
0	.015
1	.235
2	.425
3	.245
4	.080
	$\Sigma P(x) = 1.000$

■

The probability distribution of a discrete random variable possesses the following *two characteristics*.

1. The probability assigned to each value of a random variable x lies in the range 0 to 1; that is, $0 \le P(x) \le 1$ for each x.
2. The sum of the probabilities assigned to all possible values of x is equal to 1.0; that is, $\Sigma P(x) = 1$. (Remember, if the probabilities are rounded, the sum may not be exactly 1.0.)

Two Characteristics of a Probability Distribution The probability distribution of a discrete random variable possesses the following two characteristics.

1. $0 \le P(x) \le 1$ for each value of x
2. $\Sigma P(x) = 1$

These two characteristics are also called the *two conditions* that a probability distribution must satisfy. Notice that in Table 5.3 each probability listed in the column labeled $P(x)$ is between 0 and 1. Also, $\Sigma P(x) = 1.0$. Because both conditions are satisfied, Table 5.3 represents the probability distribution of x.

From Table 5.3, we can read the probability for any value of x. For example, the probability that a randomly selected family from this town owns two vehicles is .425. This probability is written as

$$P(x = 2) = .425 \quad \text{or} \quad P(2) = .425$$

The probability that the selected family owns more than two vehicles is given by the sum of the probabilities of owning three and four vehicles. This probability is $.245 + .080 = .325$, which can be written as

$$P(x > 2) = P(x = 3) + P(x = 4) = P(3) + P(4) = .245 + .080 = .325$$

The probability distribution of a discrete random variable can be presented in the form of a *mathematical formula*, a *table*, or a *graph*. Table 5.3 presented the probability distribution in tabular form. Figure 5.1 shows the graphical presentation of the probability distribution of Table 5.3. In this figure, each value of x is marked on the horizontal axis. The probability for each

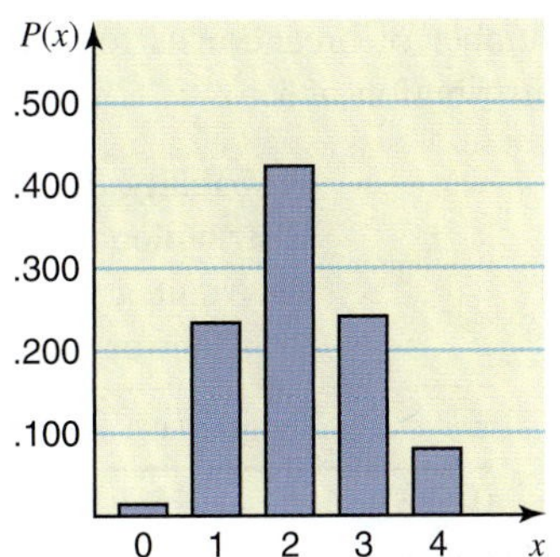

Figure 5.1 Graphical presentation of the probability distribution of Table 5.3.

value of x is exhibited by the height of the corresponding bar. Such a graph is called a **line** or **bar graph**. This section does not discuss the presentation of a probability distribution using a mathematical formula.

■ EXAMPLE 5–2

Verifying the conditions of a probability distribution.

Each of the following tables lists certain values of x and their probabilities. Determine whether or not each table represents a valid probability distribution.

(a)

x	$P(x)$
0	.08
1	.11
2	.39
3	.27

(b)

x	$P(x)$
2	.25
3	.34
4	.28
5	.13

(c)

x	$P(x)$
7	.70
8	.50
9	−.20

Solution

(a) Because each probability listed in this table is in the range 0 to 1, it satisfies the first condition of a probability distribution. However, the sum of all probabilities is not equal to 1.0 because $\Sigma P(x) = .08 + .11 + .39 + .27 = .85$. Therefore, the second condition is not satisfied. Consequently, this table does not represent a valid probability distribution.

(b) Each probability listed in this table is in the range 0 to 1. Also, $\Sigma P(x) = .25 + .34 + .28 + .13 = 1.0$. Consequently, this table represents a valid probability distribution.

(c) Although the sum of all probabilities listed in this table is equal to 1.0, one of the probabilities is negative. This violates the first condition of a probability distribution. Therefore, this table does not represent a valid probability distribution. ■

■ EXAMPLE 5–3

The following table lists the probability distribution of the number of breakdowns per week for a machine based on past data.

Breakdowns per week	0	1	2	3
Probability	.15	.20	.35	.30

(a) Present this probability distribution graphically.

(b) Find the probability that the number of breakdowns for this machine during a given week is

i. exactly 2 **ii.** 0 to 2

iii. more than 1 **iv.** at most 1

Solution Let x denote the number of breakdowns for this machine during a given week. Table 5.4 lists the probability distribution of x.

Table 5.4 Probability Distribution of the Number of Breakdowns

x	$P(x)$
0	.15
1	.20
2	.35
3	.30
	$\Sigma P(x) = 1.00$

Graphing a probability distribution.

(a) Figure 5.2 shows the bar graph of the probability distribution of Table 5.4.

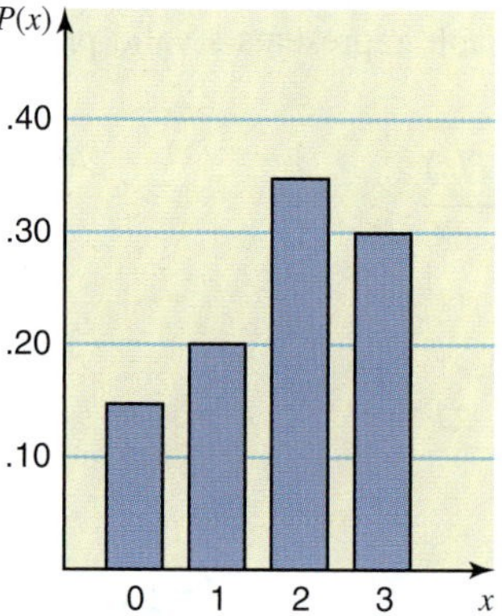

Figure 5.2 Graphical presentation of the probability distribution of Table 5.4.

Finding the probabilities of events for a discrete random variable.

(b) Using Table 5.4, we can calculate the required probabilities as follows.

i. The probability of exactly two breakdowns is

$$P(\text{exactly 2 breakdowns}) = P(x = 2) = \mathbf{.35}$$

ii. The probability of zero to two breakdowns is given by the sum of the probabilities of 0, 1, and 2 breakdowns.

$$\begin{aligned} P(\text{0 to 2 breakdowns}) &= P(0 \leq x \leq 2) \\ &= P(x = 0) + P(x = 1) + P(x = 2) \\ &= .15 + .20 + .35 = \mathbf{.70} \end{aligned}$$

iii. The probability of more than one breakdown is obtained by adding the probabilities of 2 and 3 breakdowns.

$$\begin{aligned} P(\text{more than 1 breakdown}) &= P(x > 1) \\ &= P(x = 2) + P(x = 3) \\ &= .35 + .30 = \mathbf{.65} \end{aligned}$$

iv. The probability of at most one breakdown is given by the sum of the probabilities of 0 and 1 breakdown.

$$\begin{aligned} P(\text{at most 1 breakdown}) &= P(x \leq 1) \\ &= P(x = 0) + P(x = 1) \\ &= .15 + .20 = \mathbf{.35} \end{aligned}$$

■

■ EXAMPLE 5–4

Constructing a probability distribution.

According to a survey, 60% of all students at a large university suffer from math anxiety. Two students are randomly selected from this university. Let x denote the number of students in this sample who suffer from math anxiety. Develop the probability distribution of x.

Solution Let us define the following two events:

N = the student selected does not suffer from math anxiety

M = the student selected suffers from math anxiety

As we can observe from the tree diagram of Figure 5.3, there are four possible outcomes for this experiment: NN (neither of the students suffers from math anxiety), NM (the first student does not suffer from math anxiety and the second does), MN (the first student suffers from math anxiety and the second does not), and MM (both students suffer from math anxiety). The probabilities of these four outcomes are listed in the tree diagram. Because 60% of the students suffer from math anxiety and 40% do not, the probability is .60 that any student selected suffers from math anxiety and .40 that he or she does not.

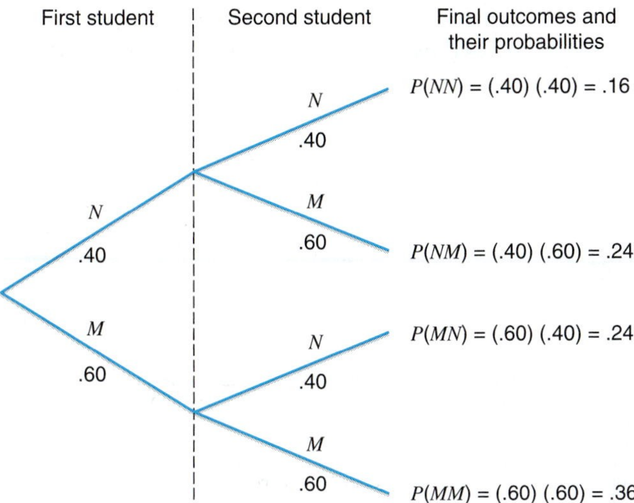

Figure 5.3 Tree diagram.

In a sample of two students, the number who suffer from math anxiety can be 0 (NN), 1 (NM or MN), or 2 (MM). Thus, x can assume any of three possible values: 0, 1, or 2. The probabilities of these three outcomes are calculated as follows:

$$P(x = 0) = P(NN) = .16$$

$$P(x = 1) = P(NM \text{ or } MN) = P(NM) + P(MN) = .24 + .24 = .48$$

$$P(x = 2) = P(MM) = .36$$

Using these probabilities, we can write the probability distribution of x as in Table 5.5.

Table 5.5 Probability Distribution of the Number of Students with Math Anxiety in a Sample of Two Students

x	$P(x)$
0	.16
1	.48
2	.36
	$\Sigma P(x) = 1.00$

■

EXERCISES

CONCEPTS AND PROCEDURES

5.6 Explain the meaning of the probability distribution of a discrete random variable. Give one example of such a probability distribution. What are the three ways to present the probability distribution of a discrete random variable?

5.7 Briefly explain the two characteristics (conditions) of the probability distribution of a discrete random variable.

5.8 Each of the following tables lists certain values of x and their probabilities. Verify whether or not each represents a valid probability distribution.

a.

x	$P(x)$
0	.10
1	.05
2	.45
3	.40

b.

x	$P(x)$
2	.35
3	.28
4	.20
5	.14

c.

x	$P(x)$
7	−.25
8	.85
9	.40

5.9 Each of the following tables lists certain values of x and their probabilities. Determine whether or not each one satisfies the two conditions required for a valid probability distribution.

a.

x	$P(x)$
5	−.36
6	.48
7	.62
8	.26

b.

x	$P(x)$
1	.27
2	.24
3	.49

c.

x	$P(x)$
0	.15
1	.08
2	.20
3	.50

5.10 The following table gives the probability distribution of a discrete random variable x.

x	0	1	2	3	4	5	6
$P(x)$	.11	.19	.28	.15	.12	.09	.06

Find the following probabilities.

a. $P(x = 3)$ b. $P(x \leq 2)$ c. $P(x \geq 4)$ d. $P(1 \leq x \leq 4)$
e. Probability that x assumes a value less than 4
f. Probability that x assumes a value greater than 2
g. Probability that x assumes a value in the interval 2 to 5

5.11 The following table gives the probability distribution of a discrete random variable x.

x	0	1	2	3	4	5
$P(x)$	.03	.17	.22	.31	.15	.12

Find the following probabilities.

a. $P(x = 1)$ b. $P(x \leq 1)$ c. $P(x \geq 3)$ d. $P(0 \leq x \leq 2)$
e. Probability that x assumes a value less than 3
f. Probability that x assumes a value greater than 3
g. Probability that x assumes a value in the interval 2 to 4

APPLICATIONS

5.12 Elmo's Sporting Goods sells exercise machines as well as other sporting goods. On different days, it sells different numbers of these machines. The following table, constructed using past data, lists the probability distribution of the number of exercise machines sold per day at Elmo's.

Machines sold per day	4	5	6	7	8	9	10
Probability	.08	.11	.14	.19	.20	.16	.12

a. Graph the probability distribution.
b. Determine the probability that the number of exercise machines sold at Elmo's on a given day is
i. exactly 6   ii. more than 8   iii. 5 to 8   iv. at most 6

5.13 Stanley Hook, a veteran baseball umpire, is not tolerant of players or managers who dispute his calls. Let x be the number of players and managers ejected from games by Hook during a week. The following table lists the probability distribution of x.

x	0	1	2	3	4	5	6
$P(x)$	.10	.18	.23	.25	.14	.07	.03

a. Draw a bar graph for this probability distribution.
b. Determine the probability that the number of players ejected by Hook during a given week is
i. exactly 3   ii. at least 4   iii. less than 3   iv. 2 to 5

5.14 A consumer agency surveyed all 2500 families living in a small town to collect data on the number of television sets owned by them. The following table lists the frequency distribution of the data collected by this agency.

Number of TV sets owned	0	1	2	3	4
Number of families	120	970	730	410	270

a. Construct a probability distribution table for the numbers of television sets owned by these families. Draw a bar graph of the probability distribution.
b. Are the probabilities listed in the table of part a exact or approximate probabilities of various outcomes? Explain.
c. Let x denote the number of television sets owned by a randomly selected family from this town. Find the following probabilities.
i. $P(x = 1)$ ii. $P(x > 2)$ iii. $P(x \leq 1)$ iv. $P(1 \leq x \leq 3)$

5.15 One of the most profitable items at A1's Auto Security Shop is the remote starting system. Let x be the number of such systems installed on a given day at this shop. The following table lists the frequency distribution of x for the past 80 days.

x	1	2	3	4	5
f	8	20	24	16	12

a. Construct a probability distribution table for the number of remote starting systems installed on a given day. Draw a graph of the probability distribution.
b. Are the probabilities listed in the table of part a exact or approximate probabilities of various outcomes? Explain.
c. Find the following probabilities.
i. $P(x = 3)$ ii. $P(x \geq 3)$ iii. $P(2 \leq x \leq 4)$ iv. $P(x < 4)$

5.16 Five percent of all cars manufactured at a large auto company are lemons. Suppose two cars are selected at random from the production line of this company. Let x denote the number of lemons in this sample. Write the probability distribution of x. Draw a tree diagram for this problem.

5.17 According to data from the Pew Internet & American Life Project, 45% of American adults take prescription drugs regularly (*Time*, October 25, 2004). Assume that this result holds true for the current population of all adults. Suppose that two adults are selected at random. Let x denote the number of adults in this sample who take prescription drugs regularly. Construct the probability distribution table of x. Draw a tree diagram for this problem.

5.18 According to a survey, 30% of adults are against using animals for research. Assume that this result holds true for the current population of all adults. Let x be the number of adults who are against using animals for research in a random sample of two adults. Obtain the probability distribution of x. Draw a tree diagram for this problem.

5.19 In a *USA TODAY*/CNN/Gallop poll of adults, 71% of the respondents said that they would be more likely to see movies in a theater if tickets and concessions cost less (*USA TODAY*, June 23, 2005). Assume that this result holds true for the current population of adults. Suppose that two adults are selected at random. Let x be the number of adults in this sample who hold the above opinion. Write the probability distribution of x. Draw a tree diagram for this problem.

*5.20 In a group of 12 persons, three are left-handed. Suppose that two persons are randomly selected from this group. Let x denote the number of left-handed persons in this sample. Write the probability distribution of x. You may draw a tree diagram and use it to write the probability distribution. (*Hint:* Note that the draws are made without replacement from a small population. Hence, the probabilities of outcomes do not remain constant for each draw.)

*5.21 In a group of 20 athletes, 6 have used performance-enhancing drugs that are illegal. Suppose that two athletes are randomly selected from this group. Let x denote the number of athletes in this sample who have used such illegal drugs. Write the probability distribution of x. You may draw a tree diagram and use that to write the probability distribution. (*Hint:* Note that the draws are made without replacement from a small population. Hence, the probabilities of outcomes do not remain constant for each draw.)

5.3 Mean of a Discrete Random Variable

The **mean of a discrete random variable**, denoted by μ, is actually the mean of its probability distribution. The mean of a discrete random variable x is also called its *expected value* and is denoted by $E(x)$. The mean (or expected value) of a discrete random variable is the value that we expect to observe per repetition, on average, if we perform an experiment a large number of times. For example, we may expect a car salesperson to sell, on average, 2.4 cars per week. This does not mean that every week this salesperson will sell exactly 2.4 cars. (Obviously one cannot sell exactly 2.4 cars.) This simply means that if we observe for many weeks, this salesperson will sell a different number of cars during different weeks; however, the average for all these weeks will be 2.4 cars per week.

To calculate the mean of a discrete random variable x, we multiply each value of x by the corresponding probability and sum the resulting products. This sum gives the mean (or expected value) of the discrete random variable x.

Mean of a Discrete Random Variable The *mean of a discrete random variable x* is the value that is expected to occur per repetition, on average, if an experiment is repeated a large number of times. It is denoted by μ and calculated as

$$\mu = \Sigma x P(x)$$

The mean of a discrete random variable x is also called its expected value and is denoted by $E(x)$; that is,

$$E(x) = \Sigma x P(x)$$

Example 5–5 illustrates the calculation of the mean of a discrete random variable.

■ EXAMPLE 5–5

Calculating and interpreting the mean of a discrete random variable.

Recall Example 5–3 of Section 5.2. The probability distribution Table 5.4 from that example is reproduced below. In this table, x represents the number of breakdowns for a machine during a given week, and $P(x)$ is the probability of the corresponding value of x.

x	$P(x)$
0	.15
1	.20
2	.35
3	.30

Find the mean number of breakdowns per week for this machine.

Solution To find the mean number of breakdowns per week for this machine, we multiply each value of x by its probability and add these products. This sum gives the mean of the probability distribution of x. The products $xP(x)$ are listed in the third column of Table 5.6. The sum of these products gives $\Sigma xP(x)$, which is the mean of x.

Table 5.6 Calculating the Mean for the Probability Distribution of Breakdowns

x	$P(x)$	$xP(x)$
0	.15	0(.15) = .00
1	.20	1(.20) = .20
2	.35	2(.35) = .70
3	.30	3(.30) = .90
		$\Sigma xP(x) = 1.80$

The mean is

$$\mu = \Sigma xP(x) = \mathbf{1.80}$$

Thus, on average, this machine is expected to break down 1.80 times per week over a period of time. In other words, if this machine is used for many weeks, then for certain weeks we will observe no breakdowns; for some other weeks we will observe one breakdown per week; and for still other weeks we will observe two or three breakdowns per week. The mean number of breakdowns is expected to be 1.80 per week for the entire period.

Note that $\mu = 1.80$ is also the expected value of x. It can also be written as

$$E(x) = 1.80$$

■

Case Study 5–1 that appears on the next page illustrates the calculation of the mean amount that an instant lottery player is expected to win.

5.4 Standard Deviation of a Discrete Random Variable

The **standard deviation of a discrete random variable**, denoted by σ, measures the spread of its probability distribution. A higher value for the standard deviation of a discrete random variable indicates that x can assume values over a larger range about the mean. In contrast, a smaller value for the standard deviation indicates that most of the values that x can assume are clustered closely about the mean. The basic formula to compute the standard deviation of a discrete random variable is

$$\sigma = \sqrt{\Sigma[(x - \mu)^2 \cdot P(x)]}$$

However, it is more convenient to use the following shortcut formula to compute the standard deviation of a discrete random variable.

Standard Deviation of a Discrete Random Variable The *standard deviation of a discrete random variable* x measures the spread of its probability distribution and is computed as

$$\sigma = \sqrt{\Sigma x^2 P(x) - \mu^2}$$

CASE STUDY

5–1 ACES HIGH INSTANT LOTTERY GAME–18TH EDITION

Currently (2005) the state of Connecticut has in circulation an instant lottery game called *Aces High*–18th Edition, which is one of the longest running lotteries in the state. The cost of each ticket for this lottery is \$1. A player can instantly win \$1000, \$100, \$40, \$25, \$10, \$4, \$2, or a free ticket (which is equivalent to winning \$1). Each ticket has six erasable spots, one of which contains dealer's card, four spots show player's cards, and one spot indicates the prize won by the player. A player will win the prize shown in the prize spot if any of the player's four cards beats dealer's card.

Based on the information on this lottery, the following table lists the number of tickets with different prizes in a total of 18,000,000 tickets printed. As is obvious from this table, out of a total of 18,000,000 tickets, 14,176,944 are nonwinning tickets (the ones with a prize of \$0 in this table). Of the remaining 3,823,056 tickets with prizes, 1,800,000 have a prize of a free ticket, 1,296,000 have a prize of \$2 each, and so forth.

Note that the variance σ^2 of a discrete random variable is obtained by squaring its standard deviation.

Example 5–6 illustrates how to use the shortcut formula to compute the standard deviation of a discrete random variable.

Calculating the standard deviation of a discrete random variable.

■ EXAMPLE 5–6

Baier's Electronics manufactures computer parts that are supplied to many computer companies. Despite the fact that two quality control inspectors at Baier's Electronics check every

Prize (dollars)	Number of Tickets
0	14,176,944
Free Tickets	1,800,000
2	1,296,000
4	483,480
10	144,000
25	72,000
40	15,552
100	11,880
1000	144
	Total = 18,000,000

The net gain to a player for each of the instant winning tickets is equal to the amount of the prize minus \$1, which is the cost of the ticket. Thus, the net gain for each of the nonwinning tickets is −\$1, which is the cost of the ticket. Let

$$x = \text{the net amount a player wins by playing this lottery}$$

The following table shows the probability distribution of x, and all the calculations required to compute the mean of x for this probability distribution. The probability of an outcome (net winnings) is calculated by dividing the number of tickets with that outcome by the total number of tickets.

x (dollars)	$P(x)$	$xP(x)$
−1	14,176,944/18,000,000 = .787608	−.787608
0	1,800,000/18,000,000 = .100000	.000000
1	1,296,000/18,000,000 = .072000	.072000
3	483,480/18,000,000 = .026860	.080580
9	144,000/18,000,000 = .008000	.072000
24	72,000/18,000,000 = .004000	.096000
39	15,552/18,000,000 = .000864	.033696
99	11,880/18,000,000 = .000660	.065340
999	144/18,000,000 = .000008	.007992
		$\Sigma xP(x) = -.36$

Hence, the mean or expected value of x is

$$\mu = \Sigma xP(x) = -\$.36$$

This mean gives the expected value of the random variable x, that is,

$$E(x) = \Sigma xP(x) = -\$.36$$

Thus, the mean of net winnings for this lottery is −\$.36. In other words, all players taken together will lose an average of \$.36 (or 36 cents) per ticket. This can also be interpreted as follows: Only 100 − 36 = 64% of the total money spent by all players on buying lottery tickets for this lottery will be returned to them in the form of prizes and 36% will not be returned. (The money that will not be returned to players will cover the costs of operating the lottery, the commission paid to agents, and revenue to the state of Connecticut.)

Source: Connecticut Lottery Corporation. Lottery tickets reproduced with permission.

part for defects before it is shipped to another company, a few defective parts do pass through these inspections undetected. Let x denote the number of defective computer parts in a shipment of 400. The following table gives the probability distribution of x.

x	0	1	2	3	4	5
$P(x)$	.02	.20	.30	.30	.10	.08

Compute the standard deviation of x.

Solution Table 5.7 shows all the calculations required for the computation of the standard deviation of x.

Table 5.7 Computations to Find the Standard Deviation

x	$P(x)$	$xP(x)$	x^2	$x^2P(x)$
0	.02	.00	0	.00
1	.20	.20	1	.20
2	.30	.60	4	1.20
3	.30	.90	9	2.70
4	.10	.40	16	1.60
5	.08	.40	25	2.00
		$\Sigma xP(x) = 2.50$		$\Sigma x^2P(x) = 7.70$

We perform the following steps to compute the standard deviation of x.

Step 1. Compute the mean of the discrete random variable.

The sum of the products $xP(x)$, recorded in the third column of Table 5.7, gives the mean of x.

$$\mu = \Sigma xP(x) = 2.50 \text{ defective computer parts in 400}$$

Step 2. Compute the value of $\Sigma x^2P(x)$.

First we square each value of x and record it in the fourth column of Table 5.7. Then we multiply these values of x^2 by the corresponding values of $P(x)$. The resulting values of $x^2P(x)$ are recorded in the fifth column of Table 5.7. The sum of this column is

$$\Sigma x^2P(x) = 7.70$$

Step 3. Substitute the values of μ and $\Sigma x^2P(x)$ in the formula for the standard deviation of x and simplify.

By performing this step, we obtain

$$\sigma = \sqrt{\Sigma x^2P(x) - \mu^2} = \sqrt{7.70 - (2.50)^2} = \sqrt{1.45}$$

$$= \mathbf{1.204} \text{ defective computer parts}$$

Thus, a given shipment of 400 computer parts is expected to contain an average of 2.50 defective parts with a standard deviation of 1.204. ■

Remember ▶ Because the standard deviation of a discrete random variable is obtained by taking the positive square root, its value is never negative.

■ EXAMPLE 5–7

Loraine Corporation is planning to market a new makeup product. According to the analysis made by the financial department of the company, it will earn an annual profit of \$4.5 million if this product has high sales, an annual profit of \$1.2 million if the sales are mediocre, and it will lose \$2.3 million a year if the sales are low. The probabilities of these three scenarios are .32, .51, and .17, respectively.

(a) Let x be the profits (in millions of dollars) earned per annum by the company from this product. Write the probability distribution of x.

(b) Calculate the mean and standard deviation of x.

Solution

Writing the probability distribution of a discrete random variable.

(a) The table below lists the probability distribution of x. Note that because x denotes profits earned by the company, the loss is written as a *negative profit* in the table.

x	$P(x)$
4.5	.32
1.2	.51
−2.3	.17

Calculating the mean and standard deviation of a discrete random variable.

(b) Table 5.8 shows all the calculations needed for the computation of the mean and standard deviation of x.

Table 5.8 Computations to Find the Mean and Standard Deviation

x	$P(x)$	$xP(x)$	x^2	$x^2P(x)$
4.5	.32	1.440	20.25	6.4800
1.2	.51	.612	1.44	.7344
−2.3	.17	−.391	5.29	.8993
		$\Sigma xP(x) = 1.661$		$\Sigma x^2P(x) = 8.1137$

The mean of x is

$$\mu = \Sigma xP(x) = \mathbf{\$1.661\ million}$$

The standard deviation of x is

$$\sigma = \sqrt{\Sigma x^2P(x) - \mu^2} = \sqrt{8.1137 - (1.661)^2} = \mathbf{\$2.314\ million}$$

Thus, it is expected that Loraine Corporation will earn an average of \$1.661 million in profits per year from the new product with a standard deviation of \$2.314 million. ■

▶ Interpretation of the Standard Deviation

The standard deviation of a discrete random variable can be interpreted or used the same way as the standard deviation of a data set in Section 3.4 of Chapter 3. In that section, we learned that according to Chebyshev's theorem, at least $[1 - (1/k^2)] \times 100\%$ of the total area under a curve lies within k standard deviations of the mean, where k is any number greater than 1. Thus, if $k = 2$, then at least 75% of the area under a curve lies between $\mu - 2\sigma$ and $\mu + 2\sigma$. In Example 5–6,

$$\mu = 2.50 \quad \text{and} \quad \sigma = 1.204$$

Hence,

$$\mu - 2\sigma = 2.50 - 2(1.204) = .092$$

$$\mu + 2\sigma = 2.50 + 2(1.204) = 4.908$$

Using Chebyshev's theorem, we can state that at least 75% of the shipments (each containing 400 computer parts) are expected to contain .092 to 4.908 defective computer parts each.

EXERCISES

CONCEPTS AND PROCEDURES

5.22 Briefly explain the concept of the mean and standard deviation of a discrete random variable.

5.23 Find the mean and standard deviation for each of the following probability distributions.

a.

x	$P(x)$
0	.16
1	.27
2	.39
3	.18

b.

x	$P(x)$
6	.40
7	.26
8	.21
9	.13

5.24 Find the mean and standard deviation for each of the following probability distributions.

a.

x	$P(x)$
3	.09
4	.21
5	.34
6	.23
7	.13

b.

x	$P(x)$
0	.43
1	.31
2	.17
3	.09

APPLICATIONS

5.25 Let x be the number of errors that appear on a randomly selected page of a book. The following table lists the probability distribution of x.

x	0	1	2	3	4
$P(x)$	.73	.16	.06	.04	.01

Find the mean and standard deviation of x.

5.26 Let x be the number of magazines a person reads every week. Based on a sample survey of adults, the following probability distribution table was prepared.

x	0	1	2	3	4	5
$P(x)$	.36	.24	.18	.10	.07	.05

Find the mean and standard deviation of x.

5.27 The following table gives the probability distribution of the number of camcorders sold on a given day at an electronics store.

Camcorders sold	0	1	2	3	4	5	6
Probability	.05	.12	.19	.30	.20	.10	.04

Calculate the mean and standard deviation for this probability distribution. Give a brief interpretation of the value of the mean.

5.28 The following table, reproduced from Exercise 5.12, lists the probability distribution of the number of exercise machines sold per day at Elmo's Sporting Goods store.

Machines sold per day	4	5	6	7	8	9	10
Probability	.08	.11	.14	.19	.20	.16	.12

Calculate the mean and standard deviation for this probability distribution. Give a brief interpretation of the value of the mean.

5.29 Let x be the number of heads obtained in two tosses of a coin. The following table lists the probability distribution of x.

x	0	1	2
$P(x)$	.25	.50	.25

Calculate the mean and standard deviation of x. Give a brief interpretation of the value of the mean.

5.30 Let x be the number of potential weapons detected by a metal detector at an airport on a given day. The following table lists the probability distribution of x.

x	0	1	2	3	4	5
$P(x)$	.14	.28	.22	.18	.12	.06

Calculate the mean and standard deviation for this probability distribution and give a brief interpretation of the value of the mean.

5.31 Refer to Exercise 5.14. Find the mean and standard deviation for the probability distribution you developed for the number of television sets owned by all 2500 families in a town. Give a brief interpretation of the values of the mean and standard deviation.

5.32 Refer to Exercise 5.15. Find the mean and standard deviation of the probability distribution you developed for the number of remote starting systems installed per day by Al's Auto Security Shop over the past 80 days. Give a brief interpretation of the values of the mean and standard deviation.

5.33 Refer to the probability distribution you developed in Exercise 5.16 for the number of lemons in two selected cars. Calculate the mean and standard deviation of x for that probability distribution.

5.34 Refer to the probability distribution developed in Exercises 5.17 for the number of adults in a sample of two who take prescription drugs regularly. Compute the mean and standard deviation of x for that probability distribution.

5.35 A contractor has submitted bids on three state jobs: an office building, a theater, and a parking garage. State rules do not allow a contractor to be offered more than one of these jobs. If this contractor is awarded any of these jobs, the profits earned from these contracts are: \$10 million from the office building, \$5 million from the theater, and \$2 million from the parking garage. His profit is zero if he gets no contract. The contractor estimates that the probabilities of getting the office building contract, the theater contract, the parking garage contract, or nothing are .15, .30, .45, and .10, respectively. Let x be the random variable that represents the contractor's profits in millions of dollars. Write the probability distribution of x. Find the mean and standard deviation of x. Give a brief interpretation of the values of the mean and standard deviation.

5.36 An instant lottery ticket costs \$2. Out of a total of 10,000 tickets printed for this lottery, 1000 tickets contain a prize of \$5 each, 100 tickets have a prize of \$10 each, 5 tickets have a prize of \$1000 each, and 1 ticket has a prize of \$5000. Let x be the random variable that denotes the net amount a player wins by playing this lottery. Write the probability distribution of x. Determine the mean and standard deviation of x. How will you interpret the values of the mean and standard deviation of x?

***5.37** Refer to the probability distribution you developed in Exercise 5.20 for the number of left-handed persons in a sample of two persons. Calculate the mean and standard deviation of x for that distribution.

***5.38** Refer to the probability distribution you developed in Exercise 5.21 for the number of athletes in a random sample of two who have used illegal performance-enhancing drugs. Calculate the mean and standard deviation of x for that distribution.

5.5 Factorials, Combinations, and Permutations

This section introduces factorials, combinations, and permutations. Of these, factorials and combinations will be used in the binomial formula discussed in Section 5.6.

5.5.1 Factorials

The symbol ! (read as *factorial*) is used to denote **factorials**. The value of the factorial of a number is obtained by multiplying all the integers from that number to 1. For example, 7! is read as "seven factorial" and is evaluated by multiplying all the integers from 7 to 1.

Definition

Factorials The symbol $n!$, read as "n factorial," represents the product of all the integers from n to 1. In other words,

$$n! = n(n-1)(n-2)(n-3)\cdots 3 \cdot 2 \cdot 1$$

By definition,

$$0! = 1$$

■ EXAMPLE 5–8

Evaluating a factorial.

Evaluate 7!.

Solution To evaluate 7!, we multiply all the integers from 7 to 1.

$$7! = 7 \cdot 6 \cdot 5 \cdot 4 \cdot 3 \cdot 2 \cdot 1 = \mathbf{5040}$$

Thus, the value of 7! is 5040. ■

■ EXAMPLE 5–9

Evaluating a factorial.

Evaluate 10!.

Solution The value of 10! is given by the product of all the integers from 10 to 1. Thus,

$$10! = 10 \cdot 9 \cdot 8 \cdot 7 \cdot 6 \cdot 5 \cdot 4 \cdot 3 \cdot 2 \cdot 1 = \mathbf{3{,}628{,}800}$$ ■

■ EXAMPLE 5–10

Evaluating a factorial of the difference between two numbers.

Evaluate $(12 - 4)!$.

Solution The value of $(12 - 4)!$ is

$$(12 - 4)! = 8! = 8 \cdot 7 \cdot 6 \cdot 5 \cdot 4 \cdot 3 \cdot 2 \cdot 1 = \mathbf{40{,}320}$$ ■

■ EXAMPLE 5–11

Evaluating a factorial of zero.

Evaluate $(5 - 5)!$.

Solution The value of $(5 - 5)!$ is 1.

$$(5 - 5)! = 0! = \mathbf{1}$$

Note that 0! is always equal to 1. ■

Statistical software and most calculators can be used to find the values of factorials. Check if your calculator can evaluate factorials.

5.5.2 Combinations

Quite often we face the problem of selecting a few elements from a large number of distinct elements. For example, a student may be required to attempt any two questions out of four in an examination. As another example, the faculty in a department may need to select 3 professors from 20 to form a committee. Or a lottery player may have to pick 6 numbers from 49. The question arises: In how many ways can we make the selections in each of these examples? For instance, how many possible selections exist for the student who is to choose any two questions out of four? The answer is six. Let the four questions be denoted by the numbers 1, 2, 3, and 4. Then the six selections are

(1 and 2) (1 and 3) (1 and 4) (2 and 3) (2 and 4) (3 and 4)

The student can choose questions 1 and 2, or 1 and 3, or 1 and 4, and so on. Note that in combinations, all selections are made without replacement.

Each of the possible selections in this list is called a **combination**. All six combinations are distinct; that is, each combination contains a different set of questions. It is important to remember that the order in which the selections are made is not important in the case of combinations. Thus, whether we write (1 and 2) or (2 and 1), both these arrangements represent only one combination.

Definition

Combinations Notation *Combinations* give the number of ways x elements can be selected from n elements. The notation used to denote the total number of combinations is

$${}_nC_x$$

which is read as "the number of combinations of n elements selected x at a time."

Suppose there are a total of n elements from which we want to select x elements. Then,

n denotes the total number of elements

${}_nC_x$ = the number of combinations of n elements selected x at a time

x denotes the number of elements selected per selection

Number of Combinations The *number of combinations* for selecting x from n distinct elements is given by the formula

$${}_nC_x = \frac{n!}{x!(n-x)!}$$

where $n!$, $x!$, and $(n - x)!$ are read as "n factorial," "x factorial," and "n minus x factorial," respectively.

In the combinations formula,

$$n! = n(n-1)(n-2)(n-3)\cdots 3\cdot 2\cdot 1$$

$$x! = x(x-1)(x-2)\cdots 3\cdot 2\cdot 1$$

$$(n-x)! = (n-x)(n-x-1)(n-x-2)\cdots 3\cdot 2\cdot 1$$

Note that in combinations, n is always greater than or equal to x. If n is less than x, then we cannot select x distinct elements from n.

■ EXAMPLE 5–12

An ice cream parlor has six flavors of ice cream. Kristen wants to buy two flavors of ice cream. If she randomly selects two flavors out of six, how many possible combinations are there?

Finding the number of combinations using the formula.

Solution For this example,

n = total number of ice cream flavors = 6

x = number of ice cream flavors to be selected = 2

Therefore, the number of ways in which Kristen can select two flavors of ice cream out of six is

$$_6C_2 = \frac{6!}{2!(6-2)!} = \frac{6!}{2!\,4!} = \frac{6 \cdot 5 \cdot 4 \cdot 3 \cdot 2 \cdot 1}{2 \cdot 1 \cdot 4 \cdot 3 \cdot 2 \cdot 1} = \mathbf{15}$$

Thus, there are 15 ways for Kristen to select two ice cream flavors out of six. ■

■ EXAMPLE 5–13

Finding the number of combinations and listing them.

Three members of a jury will be randomly selected from five people. How many different combinations are possible?

Solution There are a total of five persons, and we are to select three of them. Hence,

$$n = 5 \quad \text{and} \quad x = 3$$

Applying the combinations formula, we get

$$_5C_3 = \frac{5!}{3!(5-3)!} = \frac{5!}{3!\,2!} = \frac{120}{6 \cdot 2} = \mathbf{10}$$

If we assume that the five persons are A, B, C, D, and E, then the 10 possible combinations for the selection of three members of the jury are

ABC ABD ABE ACD ACE ADE BCD BCE BDE CDE ■

■ EXAMPLE 5–14

Using the combinations formula.

Marv & Sons advertised to hire a financial analyst. The company has received applications from 10 candidates who seem to be equally qualified. The company manager has decided to call only three of these candidates for an interview. If she randomly selects three candidates from the 10, how many total selections are possible?

Solution The total number of ways to select 3 applicants from 10 is given by $_{10}C_3$. Here, $n = 10$ and $x = 3$. We find the number of combinations as follows.

$$_{10}C_3 = \frac{10!}{3!\,(10-3)!} = \frac{10!}{3!\,7!} = \frac{3{,}628{,}800}{(6)\,(5040)} = \mathbf{120}$$

Thus, the company manager can select 3 applicants from 10 in 120 ways. ■

Statistical software and many calculators can be used to find combinations. Check if your calculator can do so.

Remember ► If the total number of elements and the number of elements to be selected are the same, then there is only one combination. In other words,

$$_nC_n = 1$$

Also, the number of combinations for selecting zero items from n is 1; that is,

$$_nC_0 = 1$$

For example,

$$_5C_5 = \frac{5!}{5!(5-5)!} = \frac{5!}{5!\,0!} = \frac{120}{(120)(1)} = 1$$

$$_8C_0 = \frac{8!}{0!(8-0)!} = \frac{8!}{0!\,8!} = \frac{40{,}320}{(1)(40{,}320)} = 1$$

Case Study 5–2 on the next page describes the number of ways a lottery player can select six numbers in a lotto game.

5–2 PLAYING LOTTO

CASE STUDY

During the past few years, many states have initiated the popular lottery game called lotto. To play lotto, a player picks any 6 numbers from a list of numbers usually starting with 1—for example, from 1 through 49. At the end of the lottery period, the state lottery commission randomly selects six numbers from the same list. If all six numbers picked by a player are the same as the ones randomly selected by the lottery commission, the player wins.

USA TODAY Snapshots®
A look at statistics that shape the nation
Playing to win
Number of $1 tickets someone would have to buy to cover every 6-number combination in selected lotto games. Tickets in millions:
Calif. (51 numbers) 18.0
Fla., Mass. (49) 13.9
Ill., N.Y., Lotto America (54) 1 12.9
Mich., Ohio (47) 10.7
N.J. (46) 9.4
Va., Conn., La. (44) 7.0
1-Players get two plays for $1
Source: USA TODAY research
By Ron Coddington, USA TODAY

The chart shows the number of combinations (in millions) for picking six numbers for lotto games played in a few states in 1992. For example, in California a player has to pick 6 numbers from 1 through 51. As shown in the chart, there are approximately 18 million ways (combinations) to select 6 numbers from 1 through 51. In Florida and Massachusetts, a player has to pick 6 numbers from 1 through 49. For this lotto, there are approximately 13.9 million combinations.

Let us find the probability that a player who picks 6 numbers from 1 through 49 wins this game. The total combinations of selecting 6 numbers from 1 through 49 numbers are obtained as follows:

$$ {}_{49}C_6 = \frac{49!}{6!(49-6)!} = 13{,}983{,}816 $$

Thus, there are a total of 13,983,816 different ways to select 6 numbers from 1 through 49 numbers. Hence, the probability that a player (who plays this lottery once) wins is

$$ P(\text{player wins}) = 1/13{,}983{,}816 = .0000000715 $$

Source: Chart reprinted with permission from *USA TODAY*, February 27, 1992. Copyright © 1992, *USA TODAY*.

5.5.3 Permutations

The concept of permutations is very similar to combinations but with one major difference—here the order of selection is important. Suppose there are three marbles in a jar—red, green, and purple—and we select two marbles from these three. When the order of selection is not important, as we know from the previous section, there are three ways (combinations) to do so. Those three ways are RG, RP, and GP where R represents that a red marble is selected, G means a green marble is selected, and P indicates a purple marble is selected. In these three combinations, the order of selection is not important and, thus, RG and GR represent the same selection. However, if the order of selection is important, then RG and GR are not the same selections but they are two different selections. Similarly, RP and PR are two different selections, and GP and PG are two different selections. Thus, if the order in which the marbles are selected is important, then there are six selections—RG, GR, RP, PR, GP, and PG. These are called six **permutations** or **arrangements**.

Definition

Permutations Notation Permutations give the total selections of x elements from n (different) elements in such a way that the order of selections is important. The notation used to denote the permutations is

$$_nP_x$$

which is read as "the number of permutations of selecting x elements from n elements." Permutations are also called **arrangements**.

Permutations Formula The following formula is used to find the number of permutations or arrangements of selecting x items out of n items. Note that here, the n items should all be different.

$$_nP_x = \frac{n!}{(n-x)!}$$

Example 5–15 shows how to apply this formula.

EXAMPLE 5–15

Finding the number of permutations using the formula.

A club has 20 members. They are to select three office holders—president, secretary, and treasurer—for next year. They always select these office holders by drawing three names randomly from the names of all members. The first person selected becomes the president, the second is the secretary, and the third one takes over as treasurer. Thus, the order in which three names are selected from the 20 names is important. Find the total arrangements of three names from these 20.

Solution For this example,

$$n = \text{total members of the club} = 20$$

$$x = \text{number of names to be selected} = 3$$

Since the order of selections is important, we find the number of permutations or arrangements using the following formula.

$$_nP_x = \frac{n!}{(n-x)!} = \frac{20!}{(20-3)!} = \frac{20!}{17!} = \mathbf{6840}$$

Thus, there are 6840 permutations or arrangements for selecting three names out of 20. ■

Statistical software and many calculators can find permutations. Check if your calculator can do it.

EXERCISES

CONCEPTS AND PROCEDURES

5.39 Determine the value of each of the following using the appropriate formula.

$3!$ $(9-3)!$ $9!$ $(14-12)!$ $_5C_3$ $_7C_4$ $_9C_3$ $_4C_0$ $_3C_3$ $_6P_2$ $_8P_4$

5.40 Find the value of each of the following using the appropriate formula.

$6!$ $11!$ $(7-2)!$ $(15-5)!$ $_8C_2$ $_5C_0$ $_5C_5$ $_6C_4$ $_{11}C_7$ $_9P_6$ $_{12}P_8$

APPLICATIONS

5.41 A ski patrol unit has nine members available for duty and two of them are to be sent to rescue an injured skier. In how many ways can two of these nine members be selected? Suppose the order of selection is important. How many arrangements are possible in this case?

5.42 An ice cream shop offers 25 flavors of ice cream. How many ways are there to select 2 different flavors from these 25 flavors? How many permutations are possible?

5.43 A veterinarian assigned to a racetrack has received a tip that 1 or more of the 12 horses in the third race have been doped. She has time to test only three horses. How many ways are there to randomly select 3 horses from these 12 horses? How many permutations are possible?

5.44 An environmental agency will randomly select 4 houses from a block containing 25 houses for a radon check. How many total selections are possible? How many permutations are possible?

5.45 An investor will randomly select 6 stocks from 20 for an investment. How many total combinations are possible? If the order in which stocks are selected is important, how many permutations will there be?

5.46 A company employs a total of 16 workers. The management has asked these employees to select two workers who will negotiate a new contract with management. The employees have decided to select the two workers randomly. How many total selections are possible? Considering that the order of selection is important, find the number of permutations.

5.47 In how many ways can a sample (without replacement) of 9 items be selected from a population of 20 items?

5.48 In how many ways can a sample (without replacement) of 5 items be selected from a population of 15 items?

5.6 The Binomial Probability Distribution

The **binomial probability distribution** is one of the most widely used discrete probability distributions. It is applied to find the probability that an outcome will occur x times in n performances of an experiment. For example, given that the probability is .05 that a DVD player manufactured at a firm is defective, we may be interested in finding the probability that in a random sample of three DVD players manufactured at this firm, exactly one will be defective. As a second example, we may be interested in finding the probability that a baseball player with a batting average of .250 will have no hits in 10 trips to the plate.

To apply the binomial probability distribution, the random variable x must be a discrete dichotomous random variable. In other words, the variable must be a discrete random variable and each repetition of the experiment must result in one of two possible outcomes. The binomial distribution is applied to experiments that satisfy the four conditions of a *binomial experiment*. (These conditions are described in Section 5.6.1.) Each repetition of a binomial experiment is called a **trial** or a **Bernoulli trial** (after Jacob Bernoulli). For example, if an experiment is defined as one toss of a coin and this experiment is repeated 10 times, then each repetition (toss) is called a trial. Consequently, there are 10 total trials for this experiment.

5.6.1 The Binomial Experiment

An experiment that satisfies the following four conditions is called a **binomial experiment**.

1. There are n identical trials. In other words, the given experiment is repeated n times where n is a positive integer. All these repetitions are performed under identical conditions.
2. Each trial has two and only two outcomes. These outcomes are usually called a *success* and a *failure*.
3. The probability of success is denoted by p and that of failure by q, and $p + q = 1$. The probabilities p and q remain constant for each trial.
4. The trials are independent. In other words, the outcome of one trial does not affect the outcome of another trial.

Conditions of a Binomial Experiment A binomial experiment must satisfy the following four conditions.

1. There are n identical trials.
2. Each trial has only two possible outcomes.
3. The probabilities of the two outcomes remain constant.
4. The trials are independent.

Note that one of the two outcomes of a trial is called a *success* and the other a *failure*. Notice that a success does not mean that the corresponding outcome is considered favorable or desirable. Similarly, a failure does not necessarily refer to an unfavorable or undesirable outcome. Success and failure are simply the names used to denote the two possible outcomes of a trial. The outcome to which the question refers is usually called a success; the outcome to which it does not refer is called a failure.

■ EXAMPLE 5–16

Verifying the conditions of a binomial experiment.

Consider the experiment consisting of 10 tosses of a coin. Determine whether or not it is a binomial experiment.

Solution The experiment consisting of 10 tosses of a coin satisfies all four conditions of a binomial experiment.

1. There are a total of 10 trials (tosses), and they are all identical. All 10 tosses are performed under identical conditions. Here, $n = 10$.
2. Each trial (toss) has only two possible outcomes: a head and a tail. Let a head be called a success and a tail be called a failure.
3. The probability of obtaining a head (a success) is 1/2 and that of a tail (a failure) is 1/2 for any toss. That is,

$$p = P(H) = 1/2 \quad \text{and} \quad q = P(T) = 1/2$$

   The sum of these two probabilities is 1.0. Also, these probabilities remain the same for each toss.
4. The trials (tosses) are independent. The result of any preceding toss has no bearing on the result of any succeeding toss.

Consequently, the experiment consisting of 10 tosses is a binomial experiment. ■

■ EXAMPLE 5–17

Verifying the conditions of a binomial experiment.

Five percent of all DVD players manufactured by a large electronics company are defective. Three DVD players are randomly selected from the production line of this company. The selected DVD players are inspected to determine whether each of them is defective or good. Is this experiment a binomial experiment?

Solution

1. This example consists of three identical trials. A trial represents the selection of a DVD player.
2. Each trial has two outcomes: a DVD player is defective or a DVD player is good. Let a defective DVD player be called a success and a good DVD player be called a failure.
3. Five percent of all DVD players are defective. So, the probability p that a DVD player is defective is .05. As a result, the probability q that a DVD player is good is .95. These two probabilities add up to 1.

4. Each trial (DVD player) is independent. In other words, if one DVD player is defective, it does not affect the outcome of another DVD player being defective or good. This is so because the size of the population is very large compared to the sample size.

Because all four conditions of a binomial experiment are satisfied, this is an example of a binomial experiment. ■

5.6.2 The Binomial Probability Distribution and Binomial Formula

The random variable x that represents the number of successes in n trials for a binomial experiment is called a *binomial random variable*. The probability distribution of x in such experiments is called the **binomial probability distribution** or simply the *binomial distribution*. Thus, the binomial probability distribution is applied to find the probability of x successes in n trials for a binomial experiment. The number of successes x in such an experiment is a discrete random variable. Consider Example 5–17. Let x be the number of defective DVD players in a sample of three. Because we can obtain any number of defective DVD players from zero to three in a sample of three, x can assume any of the values 0, 1, 2, and 3. Since the values of x are countable, it is a discrete random variable.

Binomial Formula For a binomial experiment, the probability of exactly x successes in n trials is given by the binomial formula

$$P(x) = {}_nC_x p^x q^{n-x}$$

where

$$n = \text{total number of trials}$$

$$p = \text{probability of success}$$

$$q = 1 - p = \text{probability of failure}$$

$$x = \text{number of successes in } n \text{ trials}$$

$$n - x = \text{number of failures in } n \text{ trials}$$

In the binomial formula, n is the total number of trials and x is the total number of successes. The difference between the total number of trials and the total number of successes, $n - x$, gives the total number of failures in n trials. The value of ${}_nC_x$ gives the number of ways to obtain x successes in n trials. As mentioned earlier, p and q are the probabilities of success and failure, respectively. Again, although it does not matter which of the two outcomes is called a success and which one a failure, usually the outcome to which the question refers is called a success.

To solve a binomial problem, we determine the values of n, x, $n - x$, p, and q and then substitute these values in the binomial formula. To find the value of ${}_nC_x$, we can use either the combinations formula from Section 5.5.2 or a calculator.

To find the probability of x successes in n trials for a binomial experiment, the only values needed are those of n and p. These are called the *parameters of the binomial probability distribution* or simply the **binomial parameters**. The value of q is obtained by subtracting the value of p from 1.0. Thus, $q = 1 - p$.

Next we solve a binomial problem, first without using the binomial formula and then by using the binomial formula.

■ EXAMPLE 5–18

Calculating the probability using a tree diagram and the binomial formula.

Five percent of all DVD players manufactured by a large electronics company are defective. A quality control inspector randomly selects three DVD players from the production line. What is the probability that exactly one of these three DVD players is defective?

Solution Let

D = a selected DVD player is defective

G = a selected DVD player is good

As the tree diagram in Figure 5.4 shows, there are a total of eight outcomes, and three of them contain exactly one defective DVD player. These three outcomes are

$$DGG, \quad GDG, \quad \text{and} \quad GGD$$

Figure 5.4 Tree diagram for selecting three DVD players.

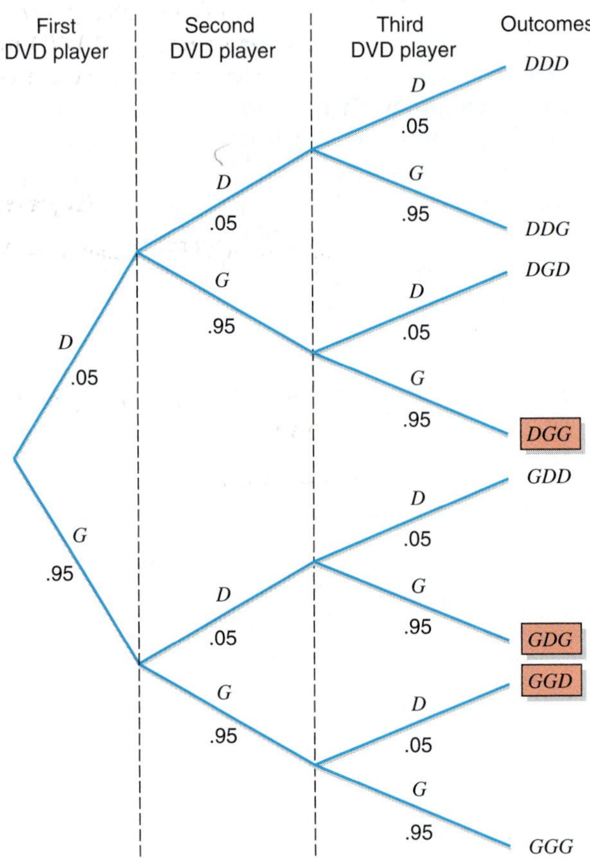

We know that 5% of all DVD players manufactured at this company are defective. As a result, 95% of all DVD players are good. So the probability that a randomly selected DVD player is defective is .05 and the probability that it is good is .95.

$$P(D) = .05 \quad \text{and} \quad P(G) = .95$$

Because the size of the population is large (note that it is a large company), the selections can be considered to be independent. The probability of each of the three outcomes, which give exactly one defective DVD player, is calculated as follows:

$$P(DGG) = P(D) \cdot P(G) \cdot P(G) = (.05)(.95)(.95) = .0451$$

$$P(GDG) = P(G) \cdot P(D) \cdot P(G) = (.95)(.05)(.95) = .0451$$

$$P(GGD) = P(G) \cdot P(G) \cdot P(D) = (.95)(.95)(.05) = .0451$$

Note that DGG is simply the intersection of the three events D, G, and G. In other words, $P(DGG)$ is the joint probability of three events: the first DVD player selected is defective, the second is good, and the third is good. To calculate this probability, we use the multiplication rule for independent events we learned in Chapter 4. The same is true about the probabilities of the other two outcomes: GDG and GGD.

Exactly one defective DVD player will be selected if *DGG* or *GDG* or *GGD* occurs. These are three mutually exclusive outcomes. Therefore, from the addition rule of Chapter 4, the probability of the union of these three outcomes is simply the sum of their individual probabilities.

$$\begin{aligned} P(1 \text{ DVD player in 3 is defective}) &= P(DGG \text{ or } GDG \text{ or } GGD) \\ &= P(DGG) + P(GDG) + P(GGD) \\ &= .0451 + .0451 + .0451 = \mathbf{.1353} \end{aligned}$$

Now let us use the binomial formula to compute this probability. Let us call the selection of a defective DVD player a *success* and the selection of a good DVD player a *failure*. The reason we have called a defective DVD player a *success* is that the question refers to selecting exactly one defective DVD player. Then,

$$\begin{aligned} n &= \text{total number of trials} = 3 \text{ DVD players} \\ x &= \text{number of successes} = \text{number of defective DVD players} = 1 \\ n - x &= \text{number of failures} = \text{number of good DVD players} = 3 - 1 = 2 \\ p &= P(\text{success}) = .05 \\ q &= P(\text{failure}) = 1 - p = .95 \end{aligned}$$

The probability of one success is denoted by $P(x = 1)$ or simply by $P(1)$. By substituting all the values in the binomial formula, we obtain

Number of ways to obtain 1 success in 3 trials → ${}_3C_1$; Number of successes → exponent 1; Number of failures → exponent 2

$$P(x = 1) = {}_3C_1(.05)^1(.95)^2 = (3)(.05)(.9025) = \mathbf{.1354}$$

Probability of success → .05; Probability of failure → .95

Note that the value of ${}_3C_1$ in the formula either can be obtained from a calculator or can be computed as follows:

$${}_3C_1 = \frac{3!}{1!(3-1)!} = \frac{3 \cdot 2 \cdot 1}{1 \cdot 2 \cdot 1} = 3$$

In the above computation, ${}_3C_1$ gives the three ways to select one defective DVD player in three selections. As listed previously, these three ways to select one defective DVD player are *DGG*, *GDG*, and *GGD*. The probability .1354 is slightly different from the earlier calculation (.1353) because of rounding. ■

■ EXAMPLE 5–19

Calculating the probability using the binomial formula.

At the Express House Delivery Service, providing high-quality service to customers is the top priority of the management. The company guarantees a refund of all charges if a package it is delivering does not arrive at its destination by the specified time. It is known from past data that despite all efforts, 2% of the packages mailed through this company do not arrive at their destinations within the specified time. Suppose a corporation mails 10 packages through Express House Delivery Service on a certain day.

(a) Find the probability that exactly one of these 10 packages will not arrive at its destination within the specified time.

(b) Find the probability that at most one of these 10 packages will not arrive at its destination within the specified time.

Solution Let us call it a success if a package does not arrive at its destination within the specified time and a failure if it does arrive within the specified time. Then,

$$n = \text{total number of packages mailed} = 10$$

$$p = P(\text{success}) = .02$$

$$q = P(\text{failure}) = 1 - .02 = .98$$

(a) For this part,

$$x = \text{number of successes} = 1$$

$$n - x = \text{number of failures} = 10 - 1 = 9$$

Substituting all values in the binomial formula, we obtain

$$P(x = 1) = {}_{10}C_1(.02)^1(.98)^9 = \frac{10!}{1!(10-1)!}(.02)^1(.98)^9$$

$$= (10)(.02)(.83374776) = \mathbf{.1667}$$

Thus, there is a .1667 probability that exactly one of the 10 packages mailed will not arrive at its destination within the specified time.

(b) The probability that at most one of the 10 packages will not arrive at its destination within the specified time is given by the sum of the probabilities of $x = 0$ and $x = 1$. Thus,

$$P(x \le 1) = P(x = 0) + P(x = 1)$$

$$= {}_{10}C_0(.02)^0(.98)^{10} + {}_{10}C_1(.02)^1(.98)^9$$

$$= (1)(1)(.81707281) + (10)(.02)(.83374776)$$

$$= .8171 + .1667 = \mathbf{.9838}$$

Thus, the probability that at most one of the 10 packages will not arrive at its destination within the specified time is .9838. ■

■ EXAMPLE 5–20

Constructing a binomial probability distribution and its graph.

According to an Ipsos Global Express survey, 64% of adults in the United States said that "there's never enough time in the day to get things done" (*Business Week,* November 22, 2004). Assume that this result holds true for the current population of adults in the United States. Let x denote the number in a random sample of three adults who hold this opinion. Write the probability distribution of x and draw a bar graph for this probability distribution.

Solution Let x be the number of adults in a sample of three who hold the said opinion. Then, $n - x$ is the number of adults who do not hold this opinion. From the given information,

$$n = \text{total adults in the sample} = 3$$

$$p = P(\text{an adult holds the said opinion}) = .64$$

$$q = P(\text{an adult does not hold the said opinion}) = 1 - .64 = .36$$

The possible values that x can assume are 0, 1, 2, and 3. In other words, the number of adults in a sample of three who hold the said opinion can be 0, 1, 2, or 3. The probability of each of these four outcomes is calculated as follows.

If $x = 0$, then $n - x = 3$. From the binomial formula, the probability of $x = 0$ is

$$P(x = 0) = {}_3C_0(.64)^0(.36)^3 = (1)(1)(.046656) = .0467$$

Note that ${}_3C_0$ is equal to 1 by definition and $(.64)^0$ is equal to 1 because any number raised to the power zero is always 1.

If $x = 1$, then $n - x = 2$. From the binomial formula, the probability of $x = 1$ is

$$P(x = 1) = {}_3C_1(.64)^1(.36)^2 = (3)(.64)(.1296) = .2488$$

Similarly, if $x = 2$, then $n - x = 1$, and if $x = 3$, then $n - x = 0$. The probabilities of $x = 2$ and $x = 3$ are

$$P(x = 2) = {}_3C_2(.64)^2(.36)^1 = (3)(.4096)(.36) = .4424$$

$$P(x = 3) = {}_3C_3(.64)^3(.36)^0 = (1)(.2621)(1) = .2621$$

These probabilities are written in Table 5.9. Figure 5.5 shows the bar graph for the probability distribution of Table 5.9.

Table 5.9 Probability Distribution of x

x	$P(x)$
0	.0467
1	.2488
2	.4424
3	.2621

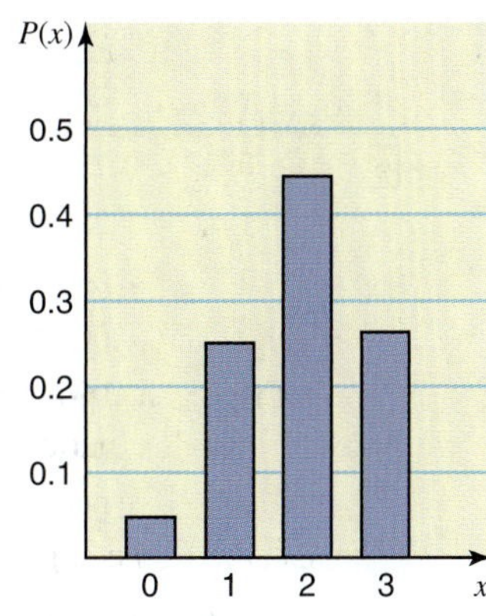

Figure 5.5 Bar graph of the probability distribution of x.

■

5.6.3 Using the Table of Binomial Probabilities

The probabilities for a binomial experiment can also be read from Table I, the table of binomial probabilities, in Appendix C. That table lists the probabilities of x for $n = 1$ to $n = 25$ and for selected values of p. Example 5–21 illustrates how to read Table I.

■ EXAMPLE 5–21

Using the binomial table to find probabilities and to construct the probability distribution and graph.

According to a Lemelson-MIT Index survey, 30% of adults in the United States said that the cell phone is the invention that they hate most but cannot live without (*Business Week*, March 1, 2004). Suppose this result holds true for the current population of adults in the United States. A random sample of six adults is selected. Using Table I of Appendix C, answer the following.

(a) Find the probability that exactly three adults in this sample hold the said opinion.

(b) Find the probability that at most two adults in this sample hold the said opinion.

(c) Find the probability that at least three adults in this sample hold the said opinion.

(d) Find the probability that one to three adults in this sample hold the said opinion.

(e) Let x be the number of adults in this sample who hold the said opinion. Write the probability distribution of x and draw a bar graph for this probability distribution.

Solution

(a) To read the required probability from Table I of Appendix C, we first determine the values of n, x, and p. For this example,

n = number of adults in the sample = 6

x = number of adults in this sample who hold the said opinion = 3

$p = P$(an adult holds the said opinion) = .30

Then we locate $n = 6$ in the column labeled n in Table I. The relevant portion of Table I with $n = 6$ is reproduced on the next page as Table 5.10. Next, we locate 3 in the column for x in the portion of the table for $n = 6$ and locate $p = .30$ in the row for p at the top of the table. The entry at the intersection of the row for $x = 3$ and the column for $p = .30$ gives the probability of three successes in six trials when the probability of success is .30. From Table I or Table 5.10,

$$P(x = 3) = \mathbf{.1852}$$

Table 5.10 Determining $P(x = 3)$ for $n = 6$ and $p = .30$

		p					
n	x	.05	.10	.20	.30	...	.95
6	0	.7351	.5314	.2621	.1176	...	.0000
	1	.2321	.3543	.3932	.3025	...	.0000
	2	.0305	.0984	.2458	.3241	...	.0001
	3	.0021	.0146	.0819	.1852	...	.0021
	4	.0001	.0012	.0154	.0595	...	.0305
	5	.0000	.0001	.0015	.0102	...	.2321
	6	.0000	.0000	.0001	.0007	...	.7351

$n = 6 \longrightarrow$ 6; $x = 3 \longrightarrow$ 3; $p = .30$; $P(x = 3) = .1852$

Using Table I or Table 5.10, we write Table 5.11, which can be used to answer the remaining parts of this example.

Table 5.11 Portion of Table I for $n = 6$ and $p = .30$

		p
n	x	.30
6	0	.1176
	1	.3025
	2	.3241
	3	.1852
	4	.0595
	5	.0102
	6	.0007

(b) The event that at most two adults in this sample hold the said opinion will occur if x is equal to 0, 1, or 2. From Table I of Appendix C or Table 5.11, the required probability is

$$P(\text{at most } 2) = P(0 \text{ or } 1 \text{ or } 2) = P(x = 0) + P(x = 1) + P(x = 2)$$
$$= .1176 + .3025 + .3241 = \mathbf{.7442}$$

(c) The probability that at least three adults in this sample hold the said opinion is given by the sum of the probabilities of 3, 4, 5, or 6. Using Table I of Appendix C or Table 5.11,

$$P(\text{at least } 3) = P(3 \text{ or } 4 \text{ or } 5 \text{ or } 6)$$
$$= P(x = 3) + P(x = 4) + P(x = 5) + P(x = 6)$$
$$= .1852 + .0595 + .0102 + .0007 = \mathbf{.2556}$$

(d) The probability that one to three adults in this sample hold the said opinion is given by the sum of the probabilities of $x = 1, 2,$ or 3. Using Table I of Appendix C or Table 5.11,

$$P(1 \text{ to } 3) = P(x = 1) + P(x = 2) + P(x = 3)$$
$$= .3025 + .3241 + .1852 = \mathbf{.8118}$$

(e) Using Table I of Appendix C or Table 5.11, we list the probability distribution of x for $n = 6$ and $p = .30$ in Table 5.12. Figure 5.6 shows the bar graph of the probability distribution of x.

Table 5.12 Probability Distribution of x for $n = 6$ and $p = .30$

x	$P(x)$
0	.1176
1	.3025
2	.3241
3	.1852
4	.0595
5	.0102
6	.0007

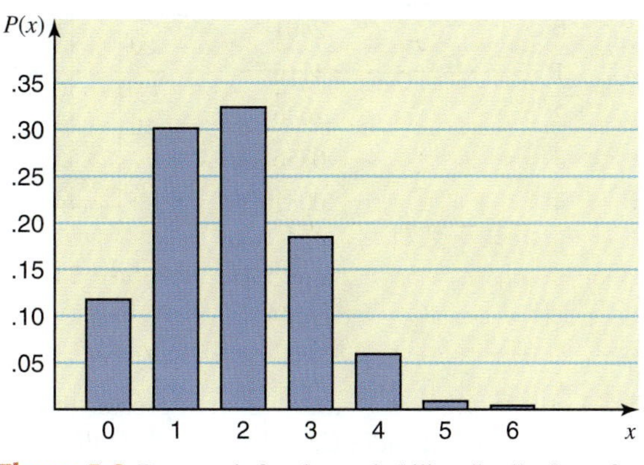

Figure 5.6 Bar graph for the probability distribution of x.

5.6.4 Probability of Success and the Shape of the Binomial Distribution

For any number of trials n:

1. The binomial probability distribution is symmetric if $p = .50$.
2. The binomial probability distribution is skewed to the right if p is less than .50.
3. The binomial probability distribution is skewed to the left if p is greater than .50.

These three cases are illustrated next with examples and graphs.

1. Let $n = 4$ and $p = .50$. Using Table I of Appendix C, we have written the probability distribution of x in Table 5.13 and plotted it in Figure 5.7. As we can observe from Table 5.13 and Figure 5.7, the probability distribution of x is symmetric.

Table 5.13 Probability Distribution of x for $n = 4$ and $p = .50$

x	$P(x)$
0	.0625
1	.2500
2	.3750
3	.2500
4	.0625

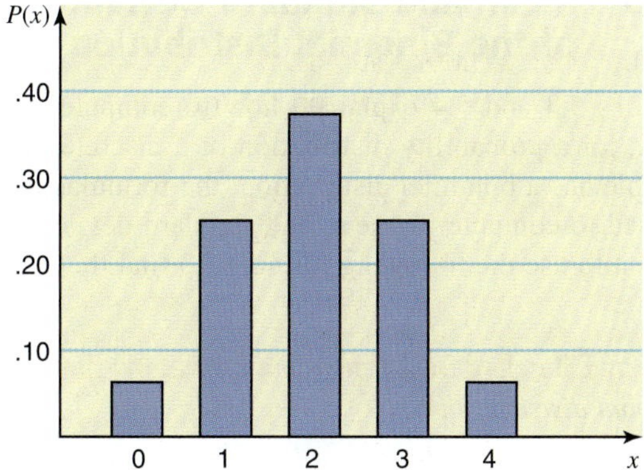

Figure 5.7 Bar graph for the probability distribution of Table 5.13.

2. Let $n = 4$ and $p = .30$ (which is less than .50). Table 5.14, which is written by using Table I of Appendix C, and the graph of the probability distribution in Figure 5.8 show that the probability distribution of x for $n = 4$ and $p = .30$ is skewed to the right.

Table 5.14 Probability Distribution of x for $n = 4$ and $p = .30$

x	$P(x)$
0	.2401
1	.4116
2	.2646
3	.0756
4	.0081

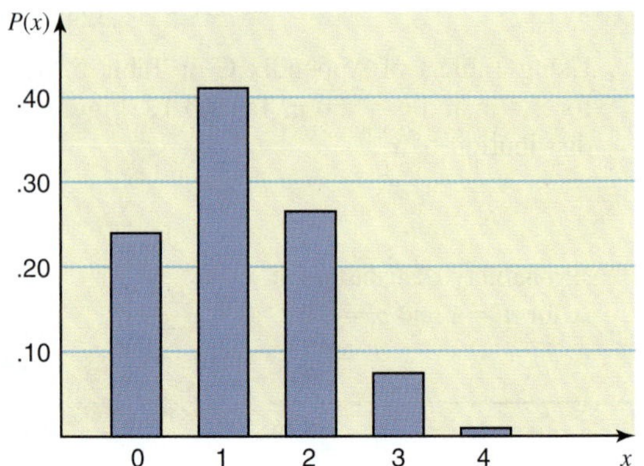

Figure 5.8 Bar graph for the probability distribution of Table 5.14.

3. Let $n = 4$ and $p = .80$ (which is greater than .50). Table 5.15, which is written by using Table I of Appendix C, and the graph of the probability distribution in Figure 5.9 show that the probability distribution of x for $n = 4$ and $p = .80$ is skewed to the left.

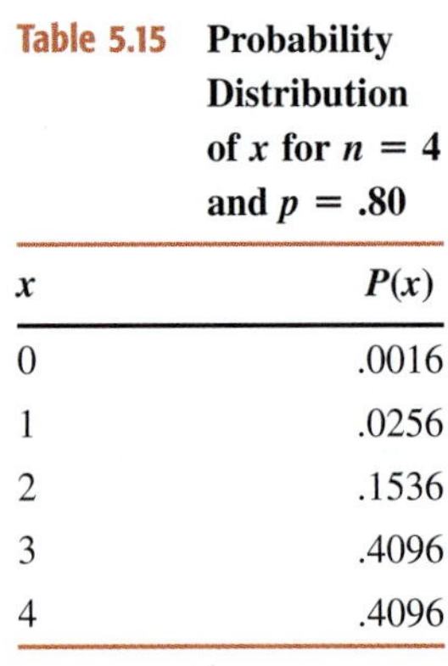

Table 5.15 Probability Distribution of x for $n = 4$ and $p = .80$

x	$P(x)$
0	.0016
1	.0256
2	.1536
3	.4096
4	.4096

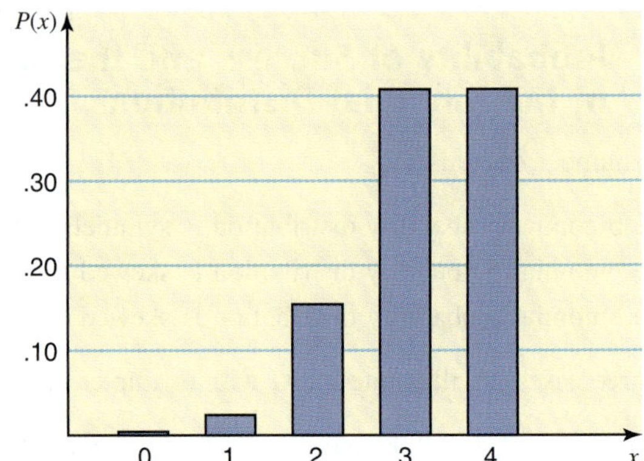

Figure 5.9 Bar graph for the probability distribution of Table 5.15.

5.6.5 Mean and Standard Deviation of the Binomial Distribution

Sections 5.3 and 5.4 explained how to compute the mean and standard deviation, respectively, for a probability distribution of a discrete random variable. When a discrete random variable has a binomial distribution, the formulas learned in Sections 5.3 and 5.4 could still be used to compute its mean and standard deviation. However, it is simpler and more convenient to use the following formulas to find the mean and standard deviation in such cases.

Mean and Standard Deviation of a Binomial Distribution The *mean* and *standard deviation of a binomial distribution* are

$$\mu = np \quad \text{and} \quad \sigma = \sqrt{npq}$$

where n is the total number of trials, p is the probability of success, and q is the probability of failure.

Example 5–22 describes the calculation of the mean and standard deviation of a binomial distribution.

■ EXAMPLE 5–22

Calculating the mean and standard deviation of a binomial random variable.

In a Money/Roper Public Affairs survey of adult Americans with all income levels, 85% said "Money doesn't buy happiness, but it helps" (*Money*, October 2004). Assume that this result is true for the current population of adult Americans. A sample of 60 adult Americans is selected. Let x be the number of adults in this sample who hold the above view. Find the mean and standard deviation of the probability distribution of x.

Solution

This is a binomial experiment with a total of 60 trials. Each trial has two outcomes: (1) the selected adult holds the given opinion, (2) the selected adult does not hold the given opinion. Assume that these are the only two possible outcomes for an adult. The probabilities p and q for these two outcomes are .85 and .15, respectively. Thus,

$$n = 60, \quad p = .85, \quad \text{and} \quad q = .15$$

Using the formulas for the mean and standard deviation of the binomial distribution, we obtain

$$\mu = np = 60\,(.85) = \mathbf{51}$$

$$\sigma = \sqrt{npq} = \sqrt{(60)(.85)(.15)} = \mathbf{2.766}$$

Thus, the mean of the probability distribution of x is 51 and the standard deviation is 2.766. The value of the mean is what we expect to obtain, on average, per repetition of the experiment. In this example, if we select many samples of 60 adult Americans, we expect that all samples will contain an average of 51 adults per sample, with a standard deviation of 2.766, who will possess the given opinion. ■

EXERCISES

■ CONCEPTS AND PROCEDURES

5.49 Briefly explain the following.

a. A binomial experiment **b.** A trial **c.** A binomial random variable

5.50 What are the parameters of the binomial probability distribution and what do they mean?

5.51 Which of the following are binomial experiments? Explain why.

a. Rolling a die many times and observing the number of spots

b. Rolling a die many times and observing whether the number obtained is even or odd

c. Selecting a few voters from a very large population of voters and observing whether or not each of them favors a certain proposition in an election when 54% of all voters are known to be in favor of this proposition.

5.52 Which of the following are binomial experiments? Explain why.

a. Drawing 3 balls with replacement from a box that contains 10 balls, 6 of which are red and 4 are blue, and observing the colors of the drawn balls

b. Drawing 3 balls without replacement from a box that contains 10 balls, 6 of which are red and 4 are blue, and observing the colors of the drawn balls

c. Selecting a few households from New York City and observing whether or not they own stocks when it is known that 28% of all households in New York City own stocks

5.53 Let x be a discrete random variable that possesses a binomial distribution. Using the binomial formula, find the following probabilities.

a. $P(x = 5)$ for $n = 8$ and $p = .70$

b. $P(x = 3)$ for $n = 4$ and $p = .40$

c. $P(x = 2)$ for $n = 6$ and $p = .30$

Verify your answers by using Table I of Appendix C.

5.54 Let x be a discrete random variable that possesses a binomial distribution. Using the binomial formula, find the following probabilities.

a. $P(x = 0)$ for $n = 5$ and $p = .05$
b. $P(x = 4)$ for $n = 7$ and $p = .90$
c. $P(x = 7)$ for $n = 10$ and $p = .60$

Verify your answers by using Table I of Appendix C.

5.55 Let x be a discrete random variable that possesses a binomial distribution.

a. Using Table I of Appendix C, write the probability distribution of x for $n = 7$ and $p = .30$ and graph it.
b. What are the mean and standard deviation of the probability distribution developed in part a?

5.56 Let x be a discrete random variable that possesses a binomial distribution.

a. Using Table I of Appendix C, write the probability distribution of x for $n = 5$ and $p = .80$ and graph it.
b. What are the mean and standard deviation of the probability distribution developed in part a?

5.57 The binomial probability distribution is symmetric for $p = .50$, skewed to the right for $p < .50$, and skewed to the left for $p > .50$. Illustrate each of these three cases by writing a probability distribution table and drawing a graph. Choose any values of n and p and use the table of binomial probabilities (Table I of Appendix C) to write the probability distribution tables.

APPLICATIONS

5.58 According to a Harris Interactive poll, 49% of Americans said that the winter holiday season is the most stressful time of the year (*Time*, April 18, 2005). Suppose that this result is true for the current population of Americans.

a. Let x be a binomial random variable that denotes the number of Americans in a random sample of 12 who hold the above mentioned opinion. What are the possible values that x can assume?
b. Find the probability that exactly eight Americans in a sample of 12 will say that the winter holiday season is the most stressful time of the year

5.59 According to an NPD Group poll, 63% of 18-to-24-year-old women said they shop only at their favorite stores (*Forbes*, November 1, 2004). Assume that this percentage is true for the current population of such women.

a. Let x be a binomial random variable denoting the number of women in a sample of 10 such women. What are the possible values that x can assume?
b. Find the probability that in a random sample of 10 such women, exactly seven will say that they shop only at their favorite stores.

5.60 In a Gallup Poll of adults taken in May 2004, 40% of the respondents indicated that they had *very little* or *no* confidence in health maintenance organizations (*USA TODAY*, June 22, 2004). Assume that this result is true for the current population of adults. Suppose a random sample of 18 adults is selected. Using the binomial probabilities table (Table I of Appendix C), find the probability that the number of adults in this sample who have this opinion is

a. at least 10 b. at most 6 c. 7 to 10

5.61 According to the *Wall Street Journal*, 60% of U.S. companies paid no federal taxes from 1996 to 2000 (*Time*, April 19, 2004). Using the binomial probabilities table (Table I of Appendix C), find the probability that the number of U.S. companies in a random sample of 16 who paid no federal taxes from 1996 to 2000 is

a. at most 7 b. at least 10 c. 8 to 11

5.62 Magnetic resonance imaging (MRI) is a process that produces internal body images using a strong magnetic field. Some patients become claustrophobic and require sedation because they are required to lie within a small, enclosed space during the MRI test. Suppose that 20% of all patients undergoing MRI testing require sedation due to claustrophobia. If five patients are selected at random, find the probability that the number of patients in these five who require sedation is

a. exactly 2 b. none c. exactly 4

5.63 According to a U.S. Government study, 68% of children live in homes with two married parents (*Time*, July 26, 2004). Assuming that this result holds true for all children currently, find the probability that in a random sample of 12 children, the number who live in homes with two married parents is

a. exactly 6 b. none c. exactly 9

5.64 According to Case Study 4–2 in Chapter 4, the probability that a baseball player will have no hits in 10 trips to the plate is .056, given that this player has a batting average of .250. Using the binomial formula, show that this probability is indeed .056.

5.65 A professional basketball player makes 85% of the free throws he tries. Assuming this percentage will hold true for future attempts, find the probability that in the next eight tries, the number of free throws he will make is

a. exactly 8 b. exactly 5

5.66 According to the National Institute of Occupational Safety and Health, at least 20% of workers in the United States work in environments that could endanger their hearing (*Time*, April 5, 2004). Assume that this result is true for the current population of American workers.

a. Using the binomial formula, find the probability that in a sample of 15 American workers, the number who work in such environments is
   i. exactly 5 ii. none
b. Using the binomial probabilities table (Table I of Appendix C), find the probability that in a random sample of 15 American workers, the number who work in such environments is
   i. at least 5 ii. at most 4 iii. 1 to 5

5.67 An office supply company conducted a survey before marketing a new paper shredder designed for home use. In the survey, 80% of the people who used the shredder were satisfied with it. Because of this high acceptance rate, the company decided to market the new shredder. Assume that 80% of all people who will use it will be satisfied. On a certain day, seven customers bought this shredder.

a. Let x denote the number of customers in this sample of seven who will be satisfied with this shredder. Using the binomial probabilities table (Table I, Appendix C), obtain the probability distribution of x and draw a graph of the probability distribution. Find the mean and standard deviation of x.
b. Using the probability distribution of part a, find the probability that exactly four of the seven customers will be satisfied.

5.68 Johnson Electronics makes calculators. Consumer satisfaction is one of the top priorities of the company's management. The company guarantees a refund or a replacement for any calculator that malfunctions within two years from the date of purchase. It is known from past data that despite all efforts, 5% of the calculators manufactured by the company malfunction within a two-year period. The company mailed a package of 10 randomly selected calculators to a store.

a. Let x denote the number of calculators in this package of 10 that will be returned for refund or replacement within a two-year period. Using the binomial probabilities table, obtain the probability distribution of x and draw a graph of the probability distribution. Determine the mean and standard deviation of x.
b. Using the probability distribution of part a, find the probability that exactly 2 of the 10 calculators will be returned for refund or replacement within a 2-year period.

5.69 A fast food chain store conducted a taste survey before marketing a new hamburger. The results of the survey showed that 70% of the people who tried this hamburger liked it. Encouraged by this result, the company decided to market the new hamburger. Assume that 70% of all people like this hamburger. On a certain day, eight customers bought it for the first time.

a. Let x denote the number of customers in this sample of eight who will like this hamburger. Using the binomial probabilities table, obtain the probability distribution of x and draw a graph of the probability distribution. Determine the mean and standard deviation of x.
b. Using the probability distribution of part a, find the probability that exactly three of the eight customers will like this hamburger.

5.7 The Hypergeometric Probability Distribution

In Section 5.6, we learned that one of the conditions required to apply the binomial probability distribution is that the trials are independent so that the probabilities of the two outcomes (success and failure) remain constant. If the trials are not independent, we cannot apply the binomial probability distribution to find the probability of x successes in n trials. In such cases we replace the binomial by the **hypergeometric probability distribution**. Such a case occurs when a sample is drawn without replacement from a finite population.

As an example, suppose 20% of all auto parts manufactured at a company are defective. Four auto parts are selected at random. What is the probability that three of these four parts are good? Note that we are to find the probability that three of the four auto parts are good and one is defective. In this case, the population is very large and the probability of the first, second, third, and fourth auto parts being defective remains the same at .20. Similarly, the probability of any of the parts being good remains unchanged at .80. Consequently, we will apply the binomial probability distribution to find the probability of three good parts in four.

Now suppose this company shipped 25 auto parts to a dealer. Later on, it finds out that five of those parts were defective. By the time the company manager contacts the dealer, four auto parts from that shipment have already been sold. What is the probability that three of those four parts were good parts and one was defective? Here, because the four parts were selected without replacement from a small population, the probability of a part being good changes from the first selection to the second selection, to the third selection, and to the fourth selection. In this case we cannot apply the binomial probability distribution. In such instances, we use the hypergeometric probability distribution to find the required probability.

Hypergeometric Probability Distribution

Let

$$N = \text{total number of elements in the population}$$
$$r = \text{number of successes in the population}$$
$$N - r = \text{number of failures in the population}$$
$$n = \text{number of trials (sample size)}$$
$$x = \text{number of successes in } n \text{ trials}$$
$$n - x = \text{number of failures in } n \text{ trials}$$

The probability of x successes in n trials is given by

$$P(x) = \frac{{}_rC_x \; {}_{N-r}C_{n-x}}{{}_NC_n}$$

Examples 5–23 and 5–24 provide applications of the hypergeometric probability distribution.

■ EXAMPLE 5–23

Calculating probability by using hypergeometric distribution formula.

Brown Manufacturing makes auto parts that are sold to auto dealers. Last week the company shipped 25 auto parts to a dealer. Later on, it found out that five of those parts were defective. By the time the company manager contacted the dealer, four auto parts from that shipment had already been sold. What is the probability that three of those four parts were good parts and one was defective?

Solution Let a good part be called a success and a defective part be called a failure. From the given information,

$$N = \text{total number of elements (auto parts) in the population} = 25$$
$$r = \text{number of successes (good parts) in the population} = 20$$
$$N - r = \text{number of failures (defective parts) in the population} = 5$$
$$n = \text{number of trials (sample size)} = 4$$
$$x = \text{number of successes in four trials} = 3$$
$$n - x = \text{number of failures in four trials} = 1$$

Using the hypergeometric formula, the required probability is calculated as follows.

$$P(x = 3) = \frac{_rC_x \; _{N-r}C_{n-x}}{_NC_n} = \frac{_{20}C_3 \; _5C_1}{_{25}C_4} = \frac{\dfrac{20!}{3!(20-3)!} \cdot \dfrac{5!}{1!(5-1)!}}{\dfrac{25!}{4!(25-4)!}}$$

$$= \frac{(1140)(5)}{12{,}650} = \mathbf{.4506}$$

Thus, the probability that three of the four parts sold are good and one is defective is .4506. In the above calculations, the values of combinations can either be calculated using the formula learned in Section 5.5.2 (as done here) or by using a calculator. ■

■ EXAMPLE 5–24

Calculating probability by using hypergeometric distribution formula.

Dawn Corporation has 12 employees who hold managerial positions. Of them, seven are female and five are male. The company is planning to send three of these 12 managers to a conference. If three managers are randomly selected out of 12,

(a) find the probability that all three of them are female

(b) find the probability that at most one of them is a female

Solution Let the selection of a female be called a success and the selection of a male be called a failure.

(a) From the given information,

N = total number of managers in the population = 12

r = number of successes (females) in the population = 7

$N - r$ = number of failures (males) in the population = 5

n = number of selections (sample size) = 3

x = number of successes (females) in three selections = 3

$n - x$ = number of failures (males) in three selections = 0

Using the hypergeometric formula, the required probability is calculated as follows:

$$P(x = 3) = \frac{_rC_x \; _{N-r}C_{n-x}}{_NC_n} = \frac{_7C_3 \; _5C_0}{_{12}C_3} = \frac{(35)(1)}{220} = \mathbf{.1591}$$

Thus, the probability that all three of the managers selected are female is .1591.

(b) The probability that at most one of them is a female is given by the sum of the probabilities that either none or one of the selected managers is a female.

To find the probability that none of the selected managers is a female,

N = total number of managers in the population = 12

r = number of successes (females) in the population = 7

$N - r$ = number of failures (males) in the population = 5

n = number of selections (sample size) = 3

x = number of successes (females) in three selections = 0

$n - x$ = number of failures (males) in three selections = 3

Using the hypergeometric formula, the required probability is calculated as follows:

$$P(x = 0) = \frac{_rC_x \; _{N-r}C_{n-x}}{_NC_n} = \frac{_7C_0 \; _5C_3}{_{12}C_3} = \frac{(1)(10)}{220} = .0455$$

To find the probability that one of the selected managers is a female,

N = total number of managers in the population = 12

r = number of successes (females) in the population = 7

$N - r$ = number of failures (males) in the population = 5

n = number of selections (sample size) = 3

x = number of successes (females) in three selections = 1

$n - x$ = number of failures (males) in three selections = 2

Using the hypergeometric formula, the required probability is calculated as follows:

$$P(x = 1) = \frac{{}_rC_x \; {}_{N-r}C_{n-x}}{{}_NC_n} = \frac{{}_7C_1 \; {}_5C_2}{{}_{12}C_3} = \frac{(7)(10)}{220} = .3182$$

The probability that at most one of the three managers selected is a female is

$$P(x \leq 1) = P(x = 0) + P(x = 1) = .0455 + .3182 = \mathbf{.3637}$$

EXERCISES

CONCEPTS AND PROCEDURES

5.70 Explain the hypergeometric probability distribution. Under what conditions is this probability distribution applied to find the probability of a discrete random variable x? Give one example of the application of the hypergeometric probability distribution.

5.71 Let $N = 8$, $r = 3$, and $n = 4$. Using the hypergeometric probability distribution formula, find
a. $P(x = 2)$ **b.** $P(x = 0)$ **c.** $P(x \leq 1)$

5.72 Let $N = 14$, $r = 6$, and $n = 5$. Using the hypergeometric probability distribution formula, find
a. $P(x = 4)$ **b.** $P(x = 5)$ **c.** $P(x \leq 1)$

5.73 Let $N = 11$, $r = 4$, and $n = 4$. Using the hypergeometric probability distribution formula, find
a. $P(x = 2)$ **b.** $P(x = 4)$ **c.** $P(x \leq 1)$

5.74 Let $N = 16$, $r = 10$, and $n = 5$. Using the hypergeometric probability distribution formula, find
a. $P(x = 5)$ **b.** $P(x = 0)$ **c.** $P(x \leq 1)$

APPLICATIONS

5.75 An Internal Revenue Service inspector is to select 3 corporations from a list of 15 for tax audit purposes. Of the 15 corporations, 6 earned profits and 9 incurred losses during the year for which the tax returns are to be audited. If the IRS inspector decides to select three corporations randomly, find the probability that the number of corporations in these three that incurred losses during the year for which the tax returns are to be audited is
a. exactly 2 **b.** none **c.** at most 1

5.76 Six jurors are to be selected from a pool of 20 potential candidates to hear a civil case involving a lawsuit between two families. Unknown to the judge or any of the attorneys, 4 of the 20 prospective jurors are potentially prejudiced by being acquainted with one or more of the litigants. They will not disclose this during the jury selection process. If 6 jurors are selected at random from this group of 20, find the probability that the number of potentially prejudiced jurors among the 6 selected jurors is
a. exactly 1 **b.** none **c.** at most 2

5.77 A shop has 11 video games to choose from, and 4 of them contain extreme violence. A customer picks 3 of these 11 games at random. What is the probability that the number of extremely violent games among the three selected games is
a. exactly two **b.** more than one **c.** none

5.78 Bender Electronics buys keyboards for its computers from another company. The keyboards are received in shipments of 100 boxes, each box containing 20 keyboards. The quality control department at Bender Electronics first randomly selects one box from each shipment and then randomly selects five keyboards from that box. The shipment is accepted if not more than one of the five keyboards is defective.

The quality control inspector at Bender Electronics selected a box from a recently received shipment of keyboards. Unknown to the inspector, this box contains six defective keyboards.

a. What is the probability that this shipment will be accepted?
b. What is the probability that this shipment will not be accepted?

5.8 The Poisson Probability Distribution

The **Poisson probability distribution**, named after the French mathematician Simeon D. Poisson, is another important probability distribution of a discrete random variable that has a large number of applications. Suppose a washing machine in a laundromat breaks down an average of three times a month. We may want to find the probability of exactly two breakdowns during the next month. This is an example of a Poisson probability distribution problem. Each breakdown is called an *occurrence* in Poisson probability distribution terminology. The Poisson probability distribution is applied to experiments with random and independent occurrences. The occurrences are random in the sense that they do not follow any pattern, and, hence, they are unpredictable. Independence of occurrences means that one occurrence (or nonoccurrence) of an event does not influence the successive occurrences or nonoccurrences of that event. The occurrences are always considered with respect to an interval. In the example of the washing machine, the interval is one month. The interval may be a time interval, a space interval, or a volume interval. The actual number of occurrences within an interval is random and independent. If the average number of occurrences for a given interval is known, then by using the Poisson probability distribution, we can compute the probability of a certain number of occurrences, x, in that interval. Note that the number of actual occurrences in an interval is denoted by x.

Conditions to Apply the Poisson Probability Distribution The following three conditions must be satisfied to apply the Poisson probability distribution.

1. x is a discrete random variable.
2. The occurrences are random.
3. The occurrences are independent.

The following are three examples of discrete random variables for which the occurrences are random and independent. Hence, these are examples to which the Poisson probability distribution can be applied.

1. Consider the number of telemarketing phone calls received by a household during a given day. In this example, the receiving of a telemarketing phone call by a household is called an occurrence, the interval is one day (an interval of time), and the occurrences are random (that is, there is no specified time for such a phone call to come in). The total number of telemarketing phone calls received by a household during a given day may be 0, 1, 2, 3, 4, and so forth. The independence of occurrences in this example means that the telemarketing phone calls are received individually and none of two (or more) of these phone calls are related.
2. Consider the number of defective items in the next 100 items manufactured on a machine. In this case, the interval is a volume interval (100 items). The occurrences (number of defective items) are random because there may be 0, 1, 2, 3 . . . 100 defective items in 100 items. We can assume the occurrence of defective items to be independent of one another.
3. Consider the number of defects in a five-foot-long iron rod. The interval, in this example, is a space interval (five feet). The occurrences (defects) are random because there may be any number of defects in a five-foot iron rod. We can assume that these defects are independent of one another.

The following examples also qualify for the application of the Poisson probability distribution.

1. The number of accidents that occur on a given highway during a one-week period
2. The number of customers entering a grocery store during a one-hour interval
3. The number of television sets sold at a department store during a given week

In contrast, consider the arrival of patients at a physician's office. These arrivals are nonrandom if the patients have to make appointments to see the doctor. The arrival of commercial airplanes at an airport is nonrandom because all planes are scheduled to arrive at certain times, and airport authorities know the exact number of arrivals for any period (although this number may change slightly because of late or early arrivals and cancellations). The Poisson probability distribution cannot be applied to these examples.

In the Poisson probability distribution terminology, the average number of occurrences in an interval is denoted by $\boldsymbol{\lambda}$ (Greek letter *lambda*). The actual number of occurrences in that interval is denoted by x. Then, using the Poisson probability distribution, we find the probability of x occurrences during an interval given that the mean occurrences during that interval are λ.

Poisson Probability Distribution Formula According to the *Poisson probability distribution*, the probability of x occurrences in an interval is

$$P(x) = \frac{\lambda^x e^{-\lambda}}{x!}$$

where λ (pronounced *lambda*) is the mean number of occurrences in that interval and the value of e is approximately 2.71828.

The mean number of occurrences in an interval, denoted by λ, is called the *parameter of the Poisson probability distribution* or the **Poisson parameter**. As is obvious from the Poisson probability distribution formula, we need to know only the value of λ to compute the probability of any given value of x. We can read the value of $e^{-\lambda}$ for a given λ from Table II of Appendix C. Examples 5–25 through 5–27 illustrate the use of the Poisson probability distribution formula.

■ EXAMPLE 5–25

Using the Poisson formula: x equals a specific value.

On average, a household receives 9.5 telemarketing phone calls per week. Using the Poisson distribution formula, find the probability that a randomly selected household receives exactly six telemarketing phone calls during a given week.

Solution Let λ be the mean number of telemarketing phone calls received by a household per week. Then, $\lambda = 9.5$. Let x be the number of telemarketing phone calls received by a household during a given week. We are to find the probability of $x = 6$. Substituting all the values in the Poisson formula, we obtain

$$P(x = 6) = \frac{\lambda^x e^{-\lambda}}{x!} = \frac{(9.5)^6 e^{-9.5}}{6!} = \frac{(735{,}091.8906)(.00007485)}{720} = \mathbf{.0764}$$

To do these calculations, we can find the value of 6! either by using the factorial key on a calculator or by multiplying all integers from 1 to 6, and we can find the value of $e^{-9.5}$ by using the e^x key on a calculator or from Table II in Appendix C. ■

■ EXAMPLE 5–26

Calculating probabilities using the Poisson formula.

A washing machine in a laundromat breaks down an average of three times per month. Using the Poisson probability distribution formula, find the probability that during the next month this machine will have

(a) exactly two breakdowns **(b)** at most one breakdown

Solution Let λ be the mean number of breakdowns per month, and let x be the actual number of breakdowns observed during the next month for this machine. Then,

$$\lambda = 3$$

(a) The probability that exactly two breakdowns will be observed during the next month is

$$P(x = 2) = \frac{\lambda^x e^{-\lambda}}{x!} = \frac{(3)^2 e^{-3}}{2!} = \frac{(9)(.04978707)}{2} = \mathbf{.2240}$$

(b) The probability that at most one breakdown will be observed during the next month is given by the sum of the probabilities of zero and one breakdown. Thus,

$$P(\text{at most 1 breakdown}) = P(0 \text{ or } 1 \text{ breakdown}) = P(x = 0) + P(x = 1)$$

$$= \frac{(3)^0 e^{-3}}{0!} + \frac{(3)^1 e^{-3}}{1!}$$

$$= \frac{(1)(.04978707)}{1} + \frac{(3)(.04978707)}{1}$$

$$= .0498 + .1494 = \mathbf{.1992}$$

One important point about the Poisson probability distribution is that *the intervals for λ and x must be equal.* If they are not, the mean λ should be redefined to make them equal. Example 5–27 illustrates this point.

◀ *Remember*

■ EXAMPLE 5–27

Calculating a probability using the Poisson formula.

Cynthia's Mail Order Company provides free examination of its products for seven days. If not completely satisfied, a customer can return the product within that period and get a full refund. According to past records of the company, an average of 2 of every 10 products sold by this company are returned for a refund. Using the Poisson probability distribution formula, find the probability that exactly 6 of the 40 products sold by this company on a given day will be returned for a refund.

Solution Let x denote the number of products in 40 that will be returned for a refund. We are to find $P(x = 6)$. The given mean is defined per 10 products, but x is defined for 40 products. As a result, we should first find the mean for 40 products. Because, on average, 2 out of 10 products are returned, the mean number of products returned out of 40 will be 8. Thus, $\lambda = 8$. Substituting $x = 6$ and $\lambda = 8$ in the Poisson probability distribution formula, we obtain

$$P(x = 6) = \frac{\lambda^x e^{-\lambda}}{x!} = \frac{(8)^6 e^{-8}}{6!} = \frac{(262{,}144)(.00033546)}{720} = \mathbf{.1221}$$

Thus, the probability is .1221 that exactly 6 products out of 40 sold on a given day will be returned.

Note that Example 5–27 is actually a binomial problem with $p = 2/10 = .20$, $n = 40$, and $x = 6$. In other words, the probability of success (that is, the probability that a product is returned) is .20 and the number of trials (products sold) is 40. We are to find the probability of six successes (returns). However, we used the Poisson distribution to solve this problem. This is referred to as *using the Poisson distribution as an approximation to the binomial distribution.* We can also use the binomial distribution to find this probability as follows:

$$P(x = 6) = {}_{40}C_6\,(.20)^6\,(.80)^{34} = \frac{40!}{6!(40-6)!}(.20)^6\,(.80)^{34}$$

$$= (3{,}838{,}380)(.000064)(.00050706) = \mathbf{.1246}$$

CASE STUDY

5–3

ASK MR. STATISTICS

Fortune **magazine used to publish a column titled *Ask Mr. Statistics*, which contained questions and answers to statistical problems. The following excerpts are reprinted from one such column.**

Dear Oddgiver: I am in the seafood distribution business and find myself endlessly wrangling with supermarkets about appropriate order sizes, especially with high-end tidbit products like our matjes herring in superspiced wine, which we let them have for $4.25, and still they take only a half-dozen jars, thereby running the risk of getting sold out early in the week and causing the better class of customers to storm out empty-handed. How do I get them to realize that lowballing on inventories is usually bad business, also to at least try a few jars of our pickled crappie balls?

—HEADED FOR A BREAKDOWN

Dear Picklehead: The science of statistics has much to offer people puzzled by seafood inventory problems. Your salvation lies in the Poisson distribution, "poisson" being French for fish and, of arguably greater relevance, the surname of a 19th-century French probabilist.

Simeon Poisson's contribution was to develop a method for calculating the likelihood that a specified number of successes will occur given that (a) the probability of success on any one trial is very low but (b) the number of trials is very high. A real world example often mentioned in the literature concerns the distribution of Prussian cavalry deaths from getting kicked by horses in the period 1875–94.

As you would expect of Teutons, the Prussian military kept meticulous records on horse-kick deaths in each of its army corps, and the data are neatly summarized in a 1963 book called *Lady Luck*, by the late Warren Weaver. There were a total of 196 kicking deaths—these being the, er, "successes." The "trials" were each army corps' observations on the number of kicking deaths sustained in the year. So with 14 army corps and data for 20 years, there were 280 trials. We shall not detain you with the Poisson formula, but it predicts, for example, that there will be 34.1 instances of a corps' having exactly two deaths in a year. In fact, there were 32 such cases. Pretty good, eh?

Back to seafood. The Poisson calculation is appropriate to your case, since the likelihood of any one customer's buying your overspiced herring is extremely small, but the number of trials—i.e., customers in the store during a typical week—is very large. Let us say that one customer in 1,000 deigns to buy the herring, and 6,000 customers visit the store in a week. So six jars are sold in an average week.

But the store manager doesn't care about average weeks. What he's worried about is having too much or not enough. He needs to know the probabilities assigned to different sales levels. Our Poisson distribution shows the following morning line: The chance of fewer than three sales—only 6.2%. Of four to six sales: 45.5%. Chances of losing some sales if the store elects to start the week with six jars because that happens to be the average: 39.4%. If the store wants to be 90% sure of not losing sales, it needs to start with nine jars.

There is no known solution to the problem of pickled crappie balls.

Source: Daniel Seligman, "Ask Mr. Statistics," *Fortune*, March 7, 1994.

Quiz: **Using the Poisson probability distribution, calculate the probabilities mentioned at the end of this case study.**

Thus the probability $P(x = 6)$ is .1246 when we use the binomial distribution.

As we can observe, simplifying the above calculations for the binomial formula is quite complicated when n is large. It is much easier to solve this problem using the Poisson distribution. As a general rule, if it is a binomial problem with $n > 25$ but $\mu \leq 25$, then we can use the Poisson distribution as an approximation to the binomial distribution. However, if $n > 25$ and $\mu > 25$, we prefer to use the normal distribution as an approximation to the binomial. The latter case will be discussed in Chapter 6.

Case Study 5–3 presents applications of the binomial and Poisson probability distributions.

5.8.1 Using the Table of Poisson Probabilities

The probabilities for a Poisson distribution can also be read from Table III, the table of Poisson probabilities, in Appendix C. The following example describes how to read that table.

EXAMPLE 5–28

Using the table of Poisson probabilities.

On average, two new accounts are opened per day at an Imperial Savings Bank branch. Using Table III of Appendix C, find the probability that on a given day the number of new accounts opened at this bank will be

(a) exactly 6 **(b)** at most 3 **(c)** at least 7

Solution Let

λ = mean number of new accounts opened per day at this bank

x = number of new accounts opened at this bank on a given day

(a) The values of λ and x are

$$\lambda = 2 \quad \text{and} \quad x = 6$$

In Table III of Appendix C, we first locate the column that corresponds to $\lambda = 2$. In this column, we then read the value for $x = 6$. The relevant portion of that table is shown here as Table 5.16. The probability that exactly 6 new accounts will be opened on a given day is .0120. Therefore,

$$P(x = 6) = \mathbf{.0120}$$

Table 5.16 **Portion of Table III for λ = 2.0**

		λ			
x	1.1	1.2	. . .	2.0	⟵ $\lambda = 2.0$
0				.1353	
1				.2707	
2				.2707	
3				.1804	
4				.0902	
5				.0361	
$x = 6$ ⟶ 6				.0120	⟵ $P(x = 6)$
7				.0034	
8				.0009	
9				.0002	

Actually, Table 5.16 gives the probability distribution of x for $\lambda = 2.0$. Note that the sum of the 10 probabilities given in Table 5.16 is .9999 and not 1.0. This is so for two reasons. First, these probabilities are rounded to four decimal places. Second, on a given day more than nine new accounts might be opened at this bank. However, the probabilities of 10, 11, 12 . . . new accounts are very small and they are not listed in the table.

(b) The probability that at most three new accounts are opened on a given day is obtained by adding the probabilities of 0, 1, 2, and 3 new accounts. Thus, using Table III of Appendix C or Table 5.16, we obtain

$$P(\text{at most } 3) = P(x = 0) + P(x = 1) + P(x = 2) + P(x = 3)$$
$$= .1353 + .2707 + .2707 + .1804 = \mathbf{.8571}$$

(c) The probability that at least 7 new accounts are opened on a given day is obtained by adding the probabilities of 7, 8, and 9 new accounts. Note that 9 is the last value of x for $\lambda = 2.0$ in Table III of Appendix C or Table 5.16. Hence, 9 is the last value of x whose probability is included in the sum. However, this does not mean that on a given day more than nine new accounts cannot be opened. It simply means that the probability of 10 or more accounts is close to zero. Thus,

$$P(\text{at least } 7) = P(x = 7) + P(x = 8) + P(x = 9)$$
$$= .0034 + .0009 + .0002 = \mathbf{.0045}$$

■ EXAMPLE 5–29

Constructing a Poisson probability distribution and graphing it.

An auto salesperson sells an average of .9 cars per day. Let x be the number of cars sold by this salesperson on any given day. Using the Poisson probability distribution table, write the probability distribution of x. Draw a graph of the probability distribution.

Solution Let λ be the mean number of cars sold per day by this salesperson. Hence, $\lambda = .9$. Using the portion of Table III of Appendix C that corresponds to $\lambda = .9$, we write the probability distribution of x in Table 5.17. Figure 5.10 shows the bar graph for the probability distribution of Table 5.17.

Table 5.17 Probability Distribution of x for $\lambda = .9$

x	$P(x)$
0	.4066
1	.3659
2	.1647
3	.0494
4	.0111
5	.0020
6	.0003

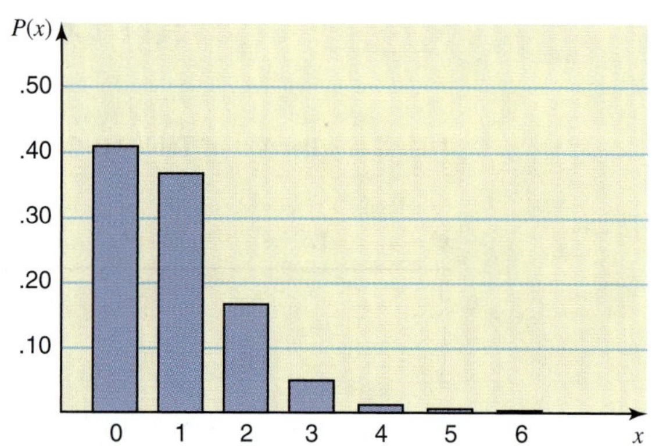

Figure 5.10 Bar graph for the probability distribution of Table 5.17.

Note that 6 is the largest value of x for $\lambda = .9$ listed in Table III for which the probability is greater than zero. However, this does not mean that this salesperson cannot sell more than six cars on a given day. What this means is that the probability of selling seven or more cars is very small. Actually, the probability of $x = 7$ for $\lambda = .9$ calculated by using the Poisson formula is .000039. When rounded to four decimal places, this probability is .0000, as listed in Table III. ■

5.8.2 Mean and Standard Deviation of the Poisson Probability Distribution

For the Poisson probability distribution, the mean and variance both are equal to λ, and the standard deviation is equal to $\sqrt{\lambda}$. That is, for the Poisson probability distribution,

$$\mu = \lambda, \quad \sigma^2 = \lambda, \quad \text{and} \quad \sigma = \sqrt{\lambda}$$

For Example 5–29, $\lambda = .9$. Therefore, for the probability distribution of x in Table 5.17, the mean, variance, and standard deviation are

$$\mu = \lambda = .9 \text{ cars}$$
$$\sigma^2 = \lambda = .9$$
$$\sigma = \sqrt{\lambda} = \sqrt{.9} = .949 \text{ cars}$$

LIVING AND DYING IN THE USA

USA TODAY Snapshots
Living and dying in the USA
294,877,547
The U.S. population as of Nov. 30, 8:07 p.m. ET.
One birth
Every 7 seconds
One death
Every 12 seconds
Source: Census Bureau
By Shannon Reilly and Karl Gelles, USA TODAY

The above chart shows that, on average, one child is born per seven seconds and one person dies per 12 seconds in the United States. The findings are based on Census Bureau data. If we assume that the births and deaths follow the Poisson probability distribution, we can find the probability of any number of births and deaths for a given time interval. Note that there is one birth, on average, per 7 seconds. If x is the actual number of births during any 7-second interval, then x can assume any (integer) value such as 0, 1, 2, 3, ... The same is true for the number of deaths per 12-second interval.

Let x be the number of (actual) births in any given seven-second interval. Then, x is a discrete random variable that can assume any of the values 0, 1, 2, 3, . . ., and so on. We can find the probability of any of these values of x either by using the Poisson formula or by using Table III of Appendix C. For example, from Table III, $P(x = 0)$ is .3679 for $\lambda = 1$, which means that the probability of no birth in a given interval of seven seconds is .3679 when $\lambda = 1$. Using that table, we can construct the probability distribution for x.

Similarly, we can find the probability of any number of deaths in an interval of 12 seconds using Table III when, on average, there is one death per 12 seconds.

Source: *USA TODAY*, December 1, 2004.

EXERCISES

CONCEPTS AND PROCEDURES

www.wiley.com/college/mann

5.79 What are the conditions that must be satisfied to apply the Poisson probability distribution?

5.80 What is the parameter of the Poisson probability distribution, and what does it mean?

5.81 Using the Poisson formula, find the following probabilities.
a. $P(x \leq 1)$ for $\lambda = 5$ **b.** $P(x = 2)$ for $\lambda = 2.5$

Verify these probabilities using Table III of Appendix C.

5.82 Using the Poisson formula, find the following probabilities.
a. $P(x < 2)$ for $\lambda = 3$ **b.** $P(x = 8)$ for $\lambda = 5.5$

Verify these probabilities using Table III of Appendix C.

5.83 Let x be a Poisson random variable. Using the Poisson probabilities table, write the probability distribution of x for each of the following. Find the mean, variance, and standard deviation for each of these probability distributions. Draw a graph for each of these probability distributions.

a. $\lambda = 1.3$ **b.** $\lambda = 2.1$

5.84 Let x be a Poisson random variable. Using the Poisson probabilities table, write the probability distribution of x for each of the following. Find the mean, variance, and standard deviation for each of these probability distributions. Draw a graph for each of these probability distributions.

a. $\lambda = .6$ **b.** $\lambda = 1.8$

APPLICATIONS

5.85 A household receives an average of 1.7 pieces of junk mail per day. Find the probability that this household will receive exactly three pieces of junk mail on a certain day. Use the Poisson probability distribution formula.

5.86 A commuter airline receives an average of 9.7 complaints per day from its passengers. Using the Poisson formula, find the probability that on a certain day this airline will receive exactly six complaints.

5.87 On average, 5.4 shoplifting incidents occur per week at an electronics store. Find the probability that exactly three such incidents will occur during a given week at this store.

5.88 On average, 12.5 rooms stay vacant per day at a large hotel in a city. Find the probability that on a given day exactly three rooms will be vacant. Use the Poisson formula.

5.89 A university police department receives an average of 3.7 reports per week of lost student ID cards.

a. Find the probability that at most one such report will be received during a given week by this police department. Use the Poisson probability distribution formula.

b. Using the Poisson probabilities table, find the probability that during a given week the number of such reports received by this police department is

i. 1 to 4 **ii.** at least 6 **iii.** at most 3

5.90 A large proportion of small businesses in the United States fail during the first few years of operation. On average, 1.6 businesses file for bankruptcy per day in a large city.

a. Using the Poisson formula, find the probability that exactly three businesses will file for bankruptcy on a given day in this city.

b. Using the Poisson probabilities table, find the probability that the number of businesses that will file for bankruptcy on a given day in this city is

i. 2 to 3 **ii.** more than 3 **iii.** less than 3

5.91 Despite all efforts by the quality control department, the fabric made at Benton Corporation always contains a few defects. A certain type of fabric made at this corporation contains an average of .5 defects per 500 yards.

a. Using the Poisson formula, find the probability that a given piece of 500 yards of this fabric will contain exactly one defect.

b. Using the Poisson probabilities table, find the probability that the number of defects in a given 500-yard piece of this fabric will be

i. 2 to 4 **ii.** more than 3 **iii.** less than 2

5.92 A large self-service gas station experiences an average of 1.6 "drive-offs" (a customer drives away without paying) per week.

a. Using the Poisson probability distribution formula, find the probability that exactly two drive-offs will occur during a given week.

b. Using the Poisson probabilities table, find the probability that the number of drive-offs experienced by this gas station during a given week will be

i. less than 3 **ii.** more than 5 **iii.** 2 to 5

5.93 An average of 4.8 customers come to Columbia Savings and Loan every half hour.

a. Find the probability that exactly two customers will come to this savings and loan during a given hour.

b. Find the probability that during a given hour, the number of customers who will come to this savings and loan is

i. 2 or fewer **ii.** 10 or more

5.94 Although Borok's Electronics Company has no openings, it still receives an average of 3.2 unsolicited applications per week from people seeking jobs.

a. Using the Poisson formula, find the probability that this company will receive no applications next week.

b. Let x denote the number of applications this company will receive during a given week. Using the Poisson probabilities table from Appendix C, write the probability distribution table of x.

c. Find the mean, variance, and standard deviation of the probability distribution developed in part b.

5.95 An insurance salesperson sells an average of 1.4 policies per day.

a. Using the Poisson formula, find the probability that this salesperson will sell no insurance policy on a certain day.

b. Let x denote the number of insurance policies that this salesperson will sell on a given day. Using the Poisson probabilities table, write the probability distribution of x.

c. Find the mean, variance, and standard deviation of the probability distribution developed in part b.

5.96 An average of .8 accidents occur per day in a large city.

a. Find the probability that no accident will occur in this city on a given day.

b. Let x denote the number of accidents that will occur in this city on a given day. Write the probability distribution of x.

c. Find the mean, variance, and standard deviation of the probability distribution developed in part b.

***5.97** On average, 20 households in 50 own answering machines.

a. Using the Poisson formula, find the probability that in a random sample of 50 households, exactly 25 will own answering machines.

b. Using the Poisson probabilities table, find the probability that the number of households in 50 who own answering machines is

i. at most 12 ii. 13 to 17 iii. at least 30

***5.98** Twenty percent of the cars passing through a school zone are exceeding the speed limit by more than 10 miles per hour.

a. Using the Poisson formula, find the probability that in a random sample of 100 cars passing through this school zone, exactly 25 will exceed the speed limit by more than 10 miles per hour.

b. Using the Poisson probabilities table, find the probability that the number of cars exceeding the speed limit by more than 10 miles per hour in a random sample of 100 cars passing through this school zone is

i. at most 8 ii. 15 to 20 iii. at least 30

USES AND MISUSES...

1. PUT ON YOUR GAME FACE

Gambling would be nothing without probability. A gambler always has a positive probability of winning. Unfortunately, the house always plays with better odds. A classic discrete probability distribution applies to the hands in straight poker. Using the tools you have learned in this chapter and a bit of creativity, you can derive the probability of being dealt a certain hand. However, this probability distribution is only going to be of limited use when you begin to play poker.

The hands in descending order of rank and increasing order of probability are straight flush, four-of-a-kind, full house, flush, straight, three-of-a-kind, two pair, pair, and high cards. To begin, let us determine how many hands there are. As we know, there are 52 cards in a deck, and any 5 cards can be a valid hand. Using the combinations notation, there are ${}_{52}C_5$ or 2,598,960 hands.

We can count the highest hands based on their composition. The straight flush is any five cards in rank order from the same suit. Because an ace can be high or low, there are 10 straight flushes per suit. Because there are 4 suits, this gives us 40 straight flushes. Once you have chosen your rank for four of a kind, e.g., a jack, there are $52 - 4 = 48$ remaining cards. Hence, there are $13 \times 48 = 624$ possible four-of-a-kind hands.

The rest of the hands require us to use the combinations notation to determine their numbers. A full house is three of a kind and a pair (for example, three kings and a pair of 7s). There are 13 choices for three of a kind (e.g., three aces, three kings, and so on); then there are ${}_4C_3 = 4$ ways to choose each set of three of a kind from 4 cards (e.g., 3 kings out of 4). Once three cards of a kind have been selected, there are 12 possibilities for a pair, and ${}_4C_2 = 6$ ways to choose any 2 cards for a pair out of 4 cards (e.g., two 9s out of 4). Thus, there are $13 \times 4 \times 12 \times 6 = 3744$ full houses. A flush is five cards drawn from the same suit. Hence, we have 4 suits multiplied by ${}_{13}C_5$ ways to choose the members, which gives 5148 flushes. However, 40 of those are straight flushes, so 5108 flushes are not straight flushes. For brevity, we omit the calculation of the remainder of the hands and present the results and the probability of being dealt the hand in the table below.

	Number of Hands	Probability
Straight flush	40	.0000154
Four of a kind	624	.0002401
Full house	3744	.0014406
Flush	5108	.0019654
Straight	10,200	.0039246
Three of a kind	54,912	.0211285
Two pair	123,552	.0475390
Pair	1,098,240	.4225690
High card	1,302,540	.5011774
Total	2,598,960	1.000000

Memorizing this table is only the beginning of poker. Any table entry represents the probability that the five cards you have been dealt constitute one of the nine poker hands. Suppose that you are playing poker with four people and you are dealt a pair of sevens. The probability that you were dealt that hand is .4225690, but that is not the probability in which you are interested. You want to know the probability that the pair of sevens that you hold will beat the hands your opponents were dealt. Despite your intimate knowledge of the probability of your hand, the above table gives information for only one player. Be very careful when working with probability distributions and make sure you understand exactly what the probabilities represent.

2. ODDS AND PROBABILITIES

In Chapter 4 we discussed the fact that casinos and odds-makers do not necessarily follow the basic probability rules when determining odds for sports betting. However, the goal of casinos is to make money, on average, from all forms of betting, including sports gambling. Also, failure to meet the basic probability rules does not affect the profit one expects to make.

When a person places a \$1 bet on a sports team, there are two possibilities: (1) the casino or the betting agency keeps the dollar if the bettor loses; (2) the casino or the betting agency pays the bettor the amount based on the odds. For example, the table (reproduced here from Chapter 4 Uses and Misuses section) shows that the odds for the Milwaukee Brewers to win the 2005 World Series are 1:75. Suppose a bettor places a \$1 bet on the Milwaukee Brewers. The bettor will win a net amount of \$74 (\$75 − \$1) if Milwaukee wins the World Series, and the bettor will lose \$1 if Milwaukee does not win the World Series. Similarly, a person placing a \$1 bet on the Boston Red Sox will either win a net amount of $\$(9/2 - 1) = \$3.50$ or lose the \$1 he/she bet.

As mentioned above, we have reproduced the table of odds from Chapter 4 Uses and Misuses section here. Note that we have added another column to this table, which lists the expected (average) payout for a \$1 bet on each team. This expected payout for each team is calculated by multiplying the probabilities of two outcomes by the amounts won/lost and adding these products. For example, for the Los Angeles Angels, the probability of winning the World Series is 1/9, and the probability of not winning is 8/9 based on the odds of 1:8. On a \$1 bet, the bettor will win a net amount of \$7 (with a net winning of \$7 after deducting \$1) if the Angels win, and the bettor will lose \$1 if the Angels do not win. Thus, the expected net winnings of the bettor on a \$1 bet on the Angels is $\$7(1/9) + \$(-1)(8/9) = -1/9 = -\$.1111$. In other words, the bettor is expected to lose, on average, \$.1111 on a \$1 bet on the Angels. Therefore, the bettor is expected to receive a payout of $\$1 - \$.1111 = \$.8889$ on a \$1 bet. This means that all bettors, on average, are paid back \$.8889 per dollar bet and they lose, on average, \$.1111 per dollar bet, which is the casino's profit.

If we add the expected payouts for all teams, we obtain the expected payout for a \$1 bet on all 30 teams. The sum of all expected

Team	Current Odds	Expected Payout	Team	Current Odds	Expected Payout
Arizona Diamondbacks	1:15	\$.9375	Milwaukee Brewers	1:75	\$.9868
Atlanta Braves	1:10	\$.9091	Minnesota Twins	1:8	\$.8889
Baltimore Orioles	1:8	\$.8889	New York Mets	1:20	\$.9524
Boston Red Sox	2:9	\$.8182	New York Yankees	1:5	\$.8333
Chicago Cubs	1:5	\$.8333	Oakland Athletics	1:100	\$.9901
Chicago White Sox	1:5	\$.8333	Philadelphia Phillies	1:30	\$.9677
Cincinnati Reds	1:300	\$.9967	Pittsburgh Pirates	1:150	\$.9934
Cleveland Indians	1:80	\$.9877	San Diego Padres	1:15	\$.9375
Colorado Rockies	1:400	\$.9975	San Francisco Giants	1:70	\$.9859
Detroit Tigers	1:40	\$.9756	Seattle Mariners	1:70	\$.9859
Florida Marlins	1:6	\$.8571	St. Louis Cardinals	1:3	\$.7500
Houston Astros	1:150	\$.9934	Tampa Bay Devil Rays	1:500	\$.9980
Kansas City Royals	1:5000	\$.9998	Texas Rangers	1:20	\$.9524
Los Angeles Angels	1:8	\$.8889	Toronto Blue Jays	1:30	\$.9677
Los Angeles Dodgers	1:15	\$.9375	Washington Nationals	1:75	\$.9868

payouts listed in the table is \$28.0314. However, the casino received \$30 in \$1 bets on each of the 30 teams. Thus, the casino's expected profit is \$30 − \$28.0314 = \$1.9686. Remember that expected values are based on a large number of repetitions (bets in this case). If the casino received only one \$1 bet on each team, then the casino would be nervous because if any of the 13 teams with odds of 1:31 or worse wins, the casino would lose money. Even if one million people place \$1 bets, the casino can still lose money if 201 or more people bet on the Kansas City Royals and this team wins the 2005 World Series. However, it is highly unlikely that the casino will lose money.

Having said that, the casinos like to offer odds that encourage people to place bets on the teams that are least likely to win, such as the 2005 Kansas City Royals or Tampa Bay Devil Rays. Casinos know that if people bet on these teams, it is very unlikely that they will have to pay these bettors. Many people like to place low bets that have the (extremely small) possibility of a big payoff, which is often the allure of the multistate lotteries such as the Powerball and MegaMillion games. Remember that these odds given in the above table are a snapshot in time. These odds are adjusted as the season goes on. If a team with odds of 1:5000 continues to get worse, the odds will also get worse (e.g., to 1:10,000 from 1:5000). Similarly, teams that are in first place and continue to play very well will see their odds improve.

Glossary

Bernoulli trial One repetition of a binomial experiment. Also called a *trial*.

Binomial experiment An experiment that contains n identical trials such that each of these n trials has only two possible outcomes, the probabilities of these two outcomes remain constant for each trial, and the trials are independent.

Binomial parameters The total trials n and the probability of success p for the binomial probability distribution.

Binomial probability distribution The probability distribution that gives the probability of x successes in n trials when the probability of success is p for each trial of a binomial experiment.

Combinations The number of ways x elements can be selected from n elements. Here order of selection is not important.

Continuous random variable A random variable that can assume any value in one or more intervals.

Discrete random variable A random variable whose values are countable.

Factorial Denoted by the symbol !. The product of all the integers from a given number to 1. For example, $n!$ (read as "n factorial") represents the product of all the integers from n to 1.

Hypergeometric probability distribution The probability distribution that is applied to determine the probability of x successes in n trials when the trials are not independent.

Mean of a discrete random variable The mean of a discrete random variable x is the value that is expected to occur per repetition, on average, if an experiment is performed a large number of times. The mean of a discrete random variable is also called its *expected value*.

Permutations Number of arrangements of x items selected from n items. Here order of selection is important.

Poisson parameter The average occurrences, denoted by λ, during an interval for a Poisson probability distribution.

Poisson probability distribution The probability distribution that gives the probability of x occurrences in an interval when the average occurrences in that interval are λ.

Probability distribution of a discrete random variable A list of all the possible values that a discrete random variable can assume and their corresponding probabilities.

Random variable A variable, denoted by x, whose value is determined by the outcome of a random experiment. Also called a *chance variable*.

Standard deviation of a discrete random variable A measure of spread for the probability distribution of a discrete random variable.

Supplementary Exercises

5.99 Let x be the number of cars that a randomly selected auto mechanic repairs on a given day. The following table lists the probability distribution of x.

x	2	3	4	5	6
$P(x)$	.05	.22	.40	.23	.10

Find the mean and standard deviation of x. Give a brief interpretation of the value of the mean.

5.100 Let x be the number of emergency root canal surgeries performed by Dr. Sharp on a given Monday. The following table lists the probability distribution of x.

x	0	1	2	3	4	5
$P(x)$	.13	.28	.30	.17	.08	.04

Calculate the mean and standard deviation of x. Give a brief interpretation of the value of the mean.

5.101 Based on its analysis of the future demand for its products, the financial department at Tipper Corporation has determined that there is a .17 probability that the company will lose $1.2 million during the next year, a .21 probability that it will lose $.7 million, a .37 probability that it will make a profit of $.9 million, and a .25 probability that it will make a profit of $2.3 million.

a. Let x be a random variable that denotes the profit earned by this corporation during the next year. Write the probability distribution of x.

b. Find the mean and standard deviation of the probability distribution of part a. Give a brief interpretation of the value of the mean.

5.102 GESCO Insurance Company charges a $350 premium per annum for a $100,000 life insurance policy for a 40-year-old female. The probability that a 40-year-old female will die within one year is .002.

a. Let x be a random variable that denotes the gain of the company for next year from a $100,000 life insurance policy sold to a 40-year-old female. Write the probability distribution of x.

b. Find the mean and standard deviation of the probability distribution of part a. Give a brief interpretation of the value of the mean.

5.103 Spoke Weaving Corporation has eight weaving machines of the same kind and of the same age. The probability is .04 that any weaving machine will break down at any time. Find the probability that at any given time

a. all eight weaving machines will be broken down
b. exactly two weaving machines will be broken down
c. none of the weaving machines will be broken down

5.104 At the Bank of California, past data show that 8% of all credit card holders default at some time in their lives. On one recent day, this bank issued 12 credit cards to new customers. Find the probability that of these 12 customers, eventually

a. exactly 3 will default **b.** exactly 1 will default **c.** none will default

5.105 Maine Corporation buys motors for electric fans from another company that guarantees that at most 5% of its motors are defective and that it will replace all defective motors at no cost to Maine Corporation. The motors are received in large shipments. The quality control department at Maine Corporation randomly selects 20 motors from each shipment and inspects them for being good or defective. If this sample contains more than two defective motors, the entire shipment is rejected.

a. Using the appropriate probabilities table from Appendix C, find the probability that a given shipment of motors received by Maine Corporation will be accepted. Assume that 5% of all motors received are defective.

b. Using the appropriate probabilities table from Appendix C, find the probability that a given shipment of motors received by Maine Corporation will be rejected.

5.106 One of the toys made by Dillon Corporation is called Speaking Joe, which is sold only by mail. Consumer satisfaction is one of the top priorities of the company's management. The company guarantees a refund or a replacement for any Speaking Joe toy if the chip that is installed inside becomes defective within one year from the date of purchase. It is known from past data that 10% of these chips become defective within a one-year period. The company sold 15 Speaking Joes on a given day.

a. Let x denote the number of Speaking Joes in these 15 that will be returned for a refund or a replacement within a one-year period. Using the appropriate probabilities table from Appendix C, obtain the probability distribution of x and draw a graph of the probability distribution. Determine the mean and standard deviation of x.

b. Using the probability distribution constructed in part a, find the probability that exactly 5 of the 15 Speaking Joes will be returned for a refund or a replacement within a one-year period.

5.107 In a list of 15 households, 9 own homes and 6 do not own homes. Four households are randomly selected from these 15 households. Find the probability that the number of households in these 4 who own homes is

a. exactly 3 **b.** at most 1 **c.** exactly 4

5.108 Twenty corporations were asked whether or not they provide retirement benefits to their employees. Fourteen of the corporations said they do provide retirement benefits to their employees, and six said they do not. Five corporations are randomly selected from these 20. Find the probability that

a. exactly two of them provide retirement benefits to their employees.
b. none of them provides retirement benefits to their employees.
c. at most one of them provides retirement benefits to employees.

5.109 Uniroyal Electronics Company buys certain parts for its refrigerators from Bob's Corporation. The parts are received in shipments of 400 boxes, each box containing 16 parts. The quality control department at Uniroyal Electronics first randomly selects one box from each shipment and then randomly selects four parts from that box. The shipment is accepted if at most one of the four parts is defective. The quality control inspector at Uniroyal Electronics selected a box from a recently received shipment of such parts. Unknown to the inspector, this box contains three defective parts.

a. What is the probability that this shipment will be accepted?
b. What is the probability that this shipment will not be accepted?

5.110 Alison Bender works for an accounting firm. To make sure her work does not contain errors, her manager randomly checks on her work. Alison recently filled out 12 income tax returns for the company's clients. Unknown to anyone, 2 of these 12 returns have minor errors. Alison's manager randomly selects 3 returns from these 12 returns. Find the probability that

a. exactly one of them contains errors.
b. none of them contains errors.
c. exactly two of them contain errors.

5.111 The student health center at a university treats an average of seven cases of mononucleosis per day during the week of final examinations.

a. Using the appropriate formula, find the probability that on a given day during the finals week exactly four cases of mononucleosis will be treated at this health center.
b. Using the appropriate probabilities table from Appendix C, find the probability that on a given day during the finals week the number of cases of mononucleosis treated at this health center will be
   i. at least 7   ii. at most 3   iii. 2 to 5

5.112 An average of 6.3 robberies occur per day in a large city.

a. Using the Poisson formula, find the probability that on a given day exactly three robberies will occur in this city.
b. Using the appropriate probabilities table from Appendix C, find the probability that on a given day the number of robberies that will occur in this city is
   i. at least 12   ii. at most 3   iii. 2 to 6

5.113 An average of 1.4 private airplanes arrive per hour at an airport.

a. Find the probability that during a given hour no private airplane will arrive at this airport.
b. Let x denote the number of private airplanes that will arrive at this airport during a given hour. Write the probability distribution of x.

5.114 A high school boys' basketball team averages 1.2 technical fouls per game.

a. Using the appropriate formula, find the probability that in a given basketball game this team will commit exactly three technical fouls.
b. Let x denote the number of technical fouls that this team will commit during a given basketball game. Using the appropriate probabilities table from Appendix C, write the probability distribution of x.

Advanced Exercises

5.115 Scott offers you the following game: You will roll two fair dice. If the sum of the two numbers obtained is 2, 3, 4, 9, 10, 11, or 12, Scott will pay you \$20. However, if the sum of the two numbers is 5, 6, 7, or 8, you will pay Scott \$20. Scott points out that you have seven winning numbers and only four losing numbers. Is this game fair to you? Should you accept this offer? Support your conclusion with appropriate calculations.

5.116 Suppose the owner of a salvage company is considering raising a sunken ship. If successful, the venture will yield a net profit of \$10 million. Otherwise, the owner will lose \$4 million. Let p denote the probability of success for this venture. Assume the owner is willing to take the risk to go ahead with this project provided the expected net profit is at least \$500,000.

a. If $p = .40$, find the expected net profit. Will the owner be willing to take the risk with this probability of success?

b. What is the smallest value of p for which the owner will take the risk to undertake this project?

5.117 Two teams, A and B, will play a best-of-seven series, which will end as soon as one of the teams wins four games. Thus, the series may end in four, five, six, or seven games. Assume that each team has an equal chance of winning each game and that all games are independent of one another. Find the following probabilities.

a. Team A wins the series in four games.
b. Team A wins the series in five games.
c. Seven games are required for a team to win the series.

5.118 York Steel Corporation produces a special bearing that must meet rigid specifications. When the production process is running properly, 10% of the bearings fail to meet the required specifications. Sometimes problems develop with the production process that cause the rejection rate to exceed 10%. To guard against this higher rejection rate, samples of 15 bearings are taken periodically and carefully inspected. If more than 2 bearings in a sample of 15 fail to meet the required specifications, production is suspended for necessary adjustments.

a. If the true rate of rejection is 10% (that is, the production process is working properly), what is the probability that the production will be suspended based on a sample of 15 bearings?
b. What assumptions did you make in part a?

5.119 Residents in an inner-city area are concerned about drug dealers entering their neighborhood. Over the past 14 nights, they have taken turns watching the street from a darkened apartment. Drug deals seem to take place randomly at various times and locations on the street and average about three per night. The residents of this street contacted the local police, who informed them that they do not have sufficient resources to set up surveillance. The police suggested videotaping the activity on the street, and if the residents are able to capture five or more drug deals on tape, the police will take action. Unfortunately, none of the residents on this street owns a video camera and, hence, they would have to rent the equipment. Inquiries at the local dealers indicated that the best available rate for renting a video camera is \$75 for the first night and \$40 for each additional night. To obtain this rate, the residents must sign up in advance for a specified number of nights. The residents hold a neighborhood meeting and invite you to help them decide on the length of the rental period. Because it is difficult for them to pay the rental fees, they want to know the probability of taping at least five drug deals on a given number of nights of videotaping.

a. Which of the probability distributions you have studied might be helpful here?
b. What assumption(s) would you have to make?
c. If the residents tape for two nights, what is the probability they will film at least five drug deals?
d. For how many nights must the camera be rented so that there is at least .90 probability that five or more drug deals will be taped?

5.120 A high school history teacher gives a 50-question multiple-choice examination in which each question has four choices. The scoring includes a penalty for guessing. Each correct answer is worth 1 point, and each wrong answer costs 1/2 point. For example, if a student answers 35 questions correctly, 8 questions incorrectly, and does not answer 7 questions, the total score for this student will be $35 - (1/2)(8) = 31$.

a. What is the expected score of a student who answers 38 questions correctly and guesses on the other 12 questions? Assume the student randomly chooses one of the four answers for each of the 12 guessed questions.
b. Does a student increase his expected score by guessing on a question if he has no idea what the correct answer is? Explain.
c. Does a student increase her expected score by guessing on a question for which she can eliminate one of the wrong answers? Explain.

5.121 A baker who makes fresh cheesecakes daily sells an average of five such cakes per day. How many cheesecakes should he make each day so that the probability of running out and losing one or more sales is less than .10? Assume that the number of cheesecakes sold each day follows a Poisson probability distribution. You may use the Poisson probabilities table from Appendix C.

5.122 Suppose that a certain casino has the "money wheel" game. The money wheel is divided into 50 sections, and the wheel has an equal probability of stopping on each of the 50 sections when it is spun. Twenty-two of the sections on this wheel show a \$1 bill, 14 show a \$2 bill, 7 show a \$5 bill, 3 show a \$10 bill, 2 show a \$20 bill, 1 shows a flag, and 1 shows a joker. A gambler may place a bet on any of the seven possible outcomes. If the wheel stops on the outcome that the gambler bet on, he or she wins. The net payoffs for these outcomes for \$1 bets are as follows.

Symbol bet on	$1	$2	$5	$10	$20	Flag	Joker
Payoff (dollars)	1	2	5	10	20	40	40

a. If the gambler bets on the $1 outcome, what is the expected net payoff?
b. Calculate the expected net payoffs for each of the other six outcomes.
c. Which bet(s) is best in terms of expected net payoff? Which is worst?

5.123 A history teacher has given her class a list of seven essay questions to study before the next test. The teacher announced that she will choose four of the seven questions to give on the test, and each student will have to answer three of those four questions.

a. In how many ways can the teacher choose four questions from the set of seven?
b. Suppose that a student has enough time to study only five questions. In how many ways can the teacher choose four questions from the set of seven so that the four selected questions include both questions that the student did not study?
c. What is the probability that the student in part b will have to answer a question that he or she did not study? That is, what is the probability that the four questions on the test will include both questions that the student did not study?

5.124 Consider the following three games. Which one would you be most likely to play? Which one would you be least likely to play? Explain your answer mathematically.

Game I: You toss a fair coin once. If a head appears you receive $3, but if a tail appears you have to pay $1.

Game II: You receive a single ticket for a raffle that has a total of 500 tickets. Two tickets are chosen without replacement from the 500. The holder of the first ticket selected receives $300, and the holder of the second ticket selected receives $150.

Game III: You toss a fair coin once. If a head appears you receive $1,000,002, but if a tail appears you have to pay $1,000,000.

5.125 Brad Henry is a stone products salesman. Let x be the number of contacts he visits on a particular day. The following table gives the probability distribution of x.

x	1	2	3	4
$P(x)$	.12	.25	.56	.07

Let y be the total number of contacts Brad visits on two randomly selected days. Write the probability distribution for y.

5.126 The number of calls that come into a small mail-order company follows a Poisson distribution. Currently, these calls are serviced by a single operator. The manager knows from past experience that an additional operator will be needed if the rate of calls exceeds 20 per hour. The manager observes that 9 calls came into the mail-order company during a randomly selected 15-minute period.

a. If the rate of calls is actually 20 per hour, what is the probability that 9 or more calls will come in during a given 15-minute period?
b. If the rate of calls is really 30 per hour, what is the probability that 9 or more calls will come in during a given 15-minute period?
c. Based on the calculations in parts a and b, do you think that the rate of incoming calls is more likely to be 20 or 30 per hour?
d. Would you advise the manager to hire a second operator? Explain.

5.127 Many of you probably played the game "Rock, Paper, Scissors" as a child. Consider the following variation of that game. Instead of two players, suppose three players play this game, and let us call these players A, B, and C. Each player selects one of these three items—Rock, Paper, or Scissors—independent of each other. Player A will win the game if all three players select the same item, e.g., rock. Player B will win the game if exactly two of the three players select the same item, and the third player selects a different item. Player C will win the game if every player selects a different item. If Player B wins the game, he/she will be paid $1. If Player C wins the game, he/she will be paid $3. Assuming that the expected winnings should be the same for each player to make this a fair game, how much should Player A be paid if he/she wins the game?

5.128 Customers arrive at the checkout counter of a supermarket at an average rate of 10 per hour, and these arrivals follow a Poisson distribution. Using each of the following two methods, find the probability that exactly 4 customers will arrive at this checkout counter during a two-hour period.

a. Use the arrivals in each of the two non-overlapping one-hour periods and then add these. (Note that the numbers of arrivals in two non-overlapping periods are independent of each other.)
b. Use the arrivals in a single two-hour period.

5.129 Consider the Uses and Misuses section (in this chapter) on Poker on page 235 where we learned how to calculate the probabilities of specific poker hands. Find the probability of being dealt

a. three of a kind b. two pairs c. one pair

Self-Review Test

1. Briefly explain the meaning of a random variable, a discrete random variable, and a continuous random variable. Give one example each of a discrete and a continuous random variable.

2. What name is given to a table that lists all the values that a discrete random variable x can assume and their corresponding probabilities?

3. For the probability distribution of a discrete random variable, the probability of any single value of x is always

a. in the range 0 to 1 b. 1.0 c. less than zero

4. For the probability distribution of a discrete random variable, the sum of the probabilities of all possible values of x is always

a. greater than 1 b. 1.0 c. less than 1.0

5. The number of combinations of 10 items selected 7 at a time is

a. 120 b. 200 c. 80

6. State the four conditions of a binomial experiment. Give one example of such an experiment.

7. The parameters of the binomial probability distribution are

a. n, p, and q b. n and p c. n, p, and x

8. The mean and standard deviation of a binomial probability distribution with $n = 25$ and $p = .20$ are

a. 5 and 2 b. 8 and 4 c. 4 and 3

9. The binomial probability distribution is symmetric if

a. $p < .5$ b. $p = .5$ c. $p > .5$

10. The binomial probability distribution is skewed to the right if

a. $p < .5$ b. $p = .5$ c. $p > .5$

11. The binomial probability distribution is skewed to the left if

a. $p < .5$ b. $p = .5$ c. $p > .5$

12. Briefly explain when a hypergeometric probability distribution is used. Give one example of a hypergeometric probability distribution.

13. The parameter/parameters of the Poisson probability distribution is/are

a. λ b. λ and x c. λ and e

14. Describe the three conditions that must be satisfied to apply the Poisson probability distribution.

15. Let x be the number of homes sold per week by all four real estate agents who work at a realty office. The following table lists the probability distribution of x.

x	0	1	2	3	4	5
$P(x)$	.15	.24	.29	.14	.10	.08

Calculate the mean and standard deviation of x. Give a brief interpretation of the value of the mean.

16. According to a survey, 60% of adults believe that all college students should be required to perform a specified number of hours of community service to graduate. Assume that this percentage is true for the current population of all adults.

a. Find the probability that the number of adults in a random sample of 12 who hold this view is
   i. exactly 8 (use the appropriate formula)
   ii. at least 6 (use the appropriate table from Appendix C)
   iii. less than 4 (use the appropriate table from Appendix C)

b. Let x be the number of adults in a random sample of 12 who believe that all college students should be required to perform a specified number of hours of community service to graduate. Using the appropriate table from Appendix C, write the probability distribution of x. Find the mean and standard deviation of x.

17. The Red Cross honors and recognizes its best volunteers from time to time. One of the Red Cross offices has received 12 nominations for the next group of 4 volunteers to be recognized. Eight of these 12 nominated volunteers are female. If the Red Cross office decides to randomly select 4 names out of these 12 nominated volunteers, find the probability that of these 4 volunteers

a. exactly 3 are female.
b. exactly 1 is female.
c. at most 1 is female.

18. The police department in a large city has installed a traffic camera at a busy intersection. Any car that runs a red light will be photographed with its license plate visible, and the driver will receive a citation. Suppose that during the morning rush hour of weekdays, an average of 10 drivers are caught running the red light per day by this system.

a. Find the probability that during the morning rush hour on a given weekday this system will catch
   i. exactly 14 drivers (use the appropriate formula)
   ii. at most 7 drivers (use the appropriate table from Appendix C)
   iii. 13 to 18 drivers (use the appropriate table from Appendix C)
b. Let x be the number of drivers caught by this system during the morning rush hour on a given weekday. Write the probability distribution of x. Use the appropriate table from Appendix C.

19. The binomial probability distribution is symmetric when $p = .50$, it is skewed to the right when $p < .50$, and it is skewed to the left when $p > .50$. Illustrate these three cases by writing three probability distributions and graphing them. Choose any values of n and p and use the table of binomial probabilities (Table I of Appendix C).

Mini-Projects

MINI-PROJECT 5–1

Consider the NBA data given in Data Set III that accompanies this text.

a. What proportion of these players are less than 74 inches tall?
b. Suppose a random sample of 25 of these players is taken and x is the number of players in the sample who are less than 74 inches tall. Find $P(x = 0)$, $P(x = 1)$, $P(x = 2)$, $P(x = 3)$, $P(x = 4)$, and $P(x = 5)$.
c. Note that x in part b has a binomial distribution with $\mu = np$. Use the Poisson probabilities table of Appendix C to approximate $P(x = 0)$, $P(x = 1)$, $P(x = 2)$, $P(x = 3)$, $P(x = 4)$, and $P(x = 5)$.
d. Are the probabilities of parts b and c consistent, or is the Poisson approximation inaccurate? Explain why.

MINI-PROJECT 5–2

Obtain information on the odds and payoffs of one of the instant lottery games in your state or a nearby state. Let the random variable x be the net amount won on one ticket (payoffs minus purchase price). Using the concepts presented in this chapter, find the probability distribution of x. Then calculate the mean and standard deviation of x. What is the player's average net gain (or loss) per ticket purchased?

MINI-PROJECT 5–3

For this project, first collect data by doing the following. Select an intersection in your town that is controlled by traffic light. For a specific time period (e.g., 9–10 A.M. or 5–6 P.M.), count the number of cars that arrive at that intersection from any one direction during each light cycle. Make sure that you do not count a car twice if it has to sit through two red lights before getting through the intersection. Perform the following tasks with your data.

a. Create a graphical display of your data. Describe the shape of the distribution. Also discuss which of the following graphs is more useful for displaying the data you collected: a dotplot, a bar graph, or a histogram.

b. Calculate the mean and variance for your data for light cycles. Note that your sample size is the number of light cycles you observed. Do you notice a relationship between these two summary measures? If so, explain what it is.

c. For each unique number of arrivals in your data, calculate the proportion of light cycles that had that number of arrivals. For example, suppose you collected these data for 100 light cycles, and you observed 8 cars arriving for each of 12 light cycles. Then, $12/100 = .12$ of the light cycles had 8 arrivals. Also calculate the theoretical probabilities for each number of arrivals using the Poisson distribution with λ equal to the sample mean that you obtained in part b. How do the two sets of probabilities compare? Is the Poisson a satisfactory model for your data?

DECIDE FOR YOURSELF

Deciding About Investing

If you are a *traditional* college student, it is quite likely that your financial portfolio includes a checking account and, possibly, a savings account. However, before you know it, you will graduate from college and take a job. On your first day at work, you will meet with your personnel/human resource manager to discuss, among other things, your retirement plans. You may decide to invest a portion of your earnings in a variety of accounts (usually mutual funds) with the hope that you will have enough money to carry you through your *golden years.* But wait, how do you decide what mutual funds to invest in? Moreover, how does this decision relate to the expected value and the variance?

The following table lists the top ten (as on June 2005) mid-cap growth mutual funds based on the three-year average return (*source:* http://biz.yahoo.com/p/tops/mg.html). The table also lists the standard deviations of the annual returns for these funds.

By looking at and analyzing the annual returns and the standard deviations of annual returns for the mutual funds listed in the table, a few questions arise that you should try to answer.

1. If you decide to invest in a mutual fund based solely on these annual returns, which fund would you invest in and why? Is this a wise decision?

2. The Baron iOpportunity fund has the highest average annual return over the three-year period as shown in the table. Does this imply that this fund is still doing better than all the other funds listed in the table? Why or why not? Do you think this fund will continue doing better than other funds in the future?

3. By considering both the average annual return and the standard deviation of the annual returns, why might a person choose to invest in the Bull Moose Growth fund over the Baron iOpportunity fund, even though the average annual return is much lower for the Bull Moose Growth fund?

4. Which of these funds would you invest in and why?

5. People who are in their 20s and 30s can afford to take more risks with their investment portfolios because they have plenty of time to offset short-term losses. However, people who are closer to retirement age are less likely to assume such risks. Which of the mutual funds listed in the table would be better to invest in if you are in your 20s or 30s and why? What if you are close to retirement?

Fund Name	Symbol	Annual Return	Standard Deviation
Baron iOpportunity	BIOPX	20.76%	23.78%
Hodges	HDPMX	19.72%	25.23%
Thornburg Core Growth A	THCGX	15.56%	18.70%
Baron Partners	BPTRX	15.24%	20.48%
Legg Mason Opportunity Institutional	LMNOX	15.06%	25.68%
ING Alliance Mid Cap Growth S	N/A	14.54%	N/A
Thornburg Core Growth C	TCGCX	14.42%	18.58%
Delaware American Services I	DASIX	14.25%	16.34%
Vanguard Capital Opportunity Adm	VHCAX	14.12%	20.56%
Bull Moose Growth	BULLX	14.07%	13.20%

TECHNOLOGY INSTRUCTION — Combinations, Binomial Distribution and Poisson Distribution

TI-84

```
DISTR DRAW
6↑χ²pdf(
7:χ²cdf(
8:Fpdf(
9:Fcdf(
0:binompdf(
A:binomcdf(
B:poissonpdf(
```

Screen 5.1

1. To find the number of ways of choosing x objects out of n, type n, select **MATH>PRB>nCr**, then type x and press **Enter**.

2. To find the binomial probability of x successes out of n trials, with each trial having a probability p of success, select **DISTR>binompdf(*n*, *p*, *x*)**, and press **Enter**. (See **Screen 5.1.**)

3. To find the Poisson probability of x successes when mean is λ, select **DISTR>poissonpdf(λ, *x*)**, and press **Enter**.

MINITAB

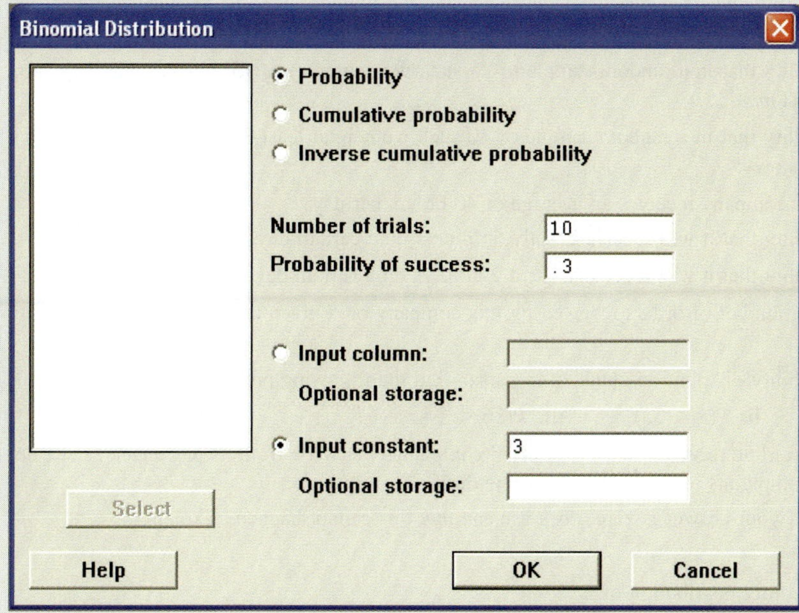

Screen 5.2

1. To find the probability of x successes in n trials for a binomial random variable with probability of success p, select **Calc>Probability Distributions>Binomial**. In the dialog box, make sure that **Probability** is selected, then enter the number of trials n as well as the probability p of success. Select **Input constant** and enter the value of x. (See **Screen 5.2.**)

   If you need to create a table of probabilities for various values of x, first enter them into a column. Again select **Calc>Probability Distributions> Binomial** and enter n and p. Now select **Input column** and enter the name of the column where you entered x. If you wish to store the resulting probabilities, enter the name of a column under **Optional storage**.

2. To find the probability of x for a Poisson random variable, select **Calc>Probability Distributions> Poisson**. In the dialog box, make sure that **Probability** is selected, then enter the value of mean λ. Select **Input constant** and enter x.

   If you need to create a table of probabilities for various values of x, first enter them into a column. Again select **Calc>Probability Distributions>Poisson** and enter λ. Now select **Input column** and enter the name of the column where you entered x. If you wish to store the resulting probabilities, enter the name of a column under **Optional storage**.

Excel

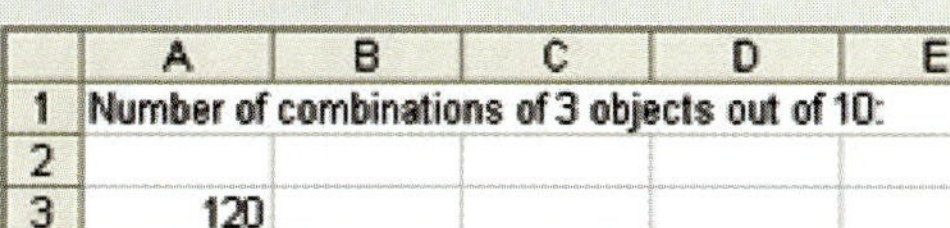

	A	B	C	D	E
1	Number of combinations of 3 objects out of 10:				
2					
3	120				

Screen 5.3

1. To find the number of combinations of x objects chosen out of n, type **=COMBIN(n, x)**. (See **Screens 5.3 and 5.4.**)

2. To find the probability for a binomial random variable to take the value x out of n trials with probability of success p, type **=BINOMDIST(x, n, p, 0)**.

3. To find the probability for a Poisson random variable to take the value x with the average number of occurrences λ, type **=POISSON(x, λ, 0)**.

	A	B	C	D
1	Number of combinations of 3 objects out of 10:			
2				
3	=COMBIN(10,3)			
4	COMBIN(number, number_chosen)			

Screen 5.4

TECHNOLOGY ASSIGNMENTS

TA5.1 Forty-five percent of the adult population in a large city are women. A court is to randomly select a jury of 12 adults from the population of all adults of this city.

a. Find the probability that none of the 12 jurors is a woman.

b. Find the probability that at most 4 of the 12 jurors are women.

c. Let x denote the number of women in 12 adults selected for this jury. Obtain the probability distribution of x.

d. Using the probability distribution obtained in part c, find the following probabilities.

i. $P(x > 6)$ **ii.** $P(x \leq 3)$ **iii.** $P(2 \leq x \leq 7)$

TA5.2 According to an NPD Group poll, 63% of 18-to-24-year-old women said they shop only at their favorite stores (*Forbes*, November 1, 2004). Assume that this percentage is true for the current population of such women.

a. Find the probability that in a random sample of 20 such women, exactly 14 will say that they shop only at their favorite stores.

b. Find the probability that in a random sample of 30 such women, exactly 18 will say that they shop only at their favorite stores.

c. Find the probability that in a random sample of 25 such women, at most 15 will say that they shop only at their favorite stores.

d. Find the probability that in a random sample of 40 such women, at least 30 will say that they shop only at their favorite stores.

TA5.3 A mail-order company receives an average of 40 orders per day.

a. Find the probability that it will receive exactly 55 orders on a certain day.

b. Find the probability that it will receive at most 29 orders on a certain day.

c. Let x denote the number of orders received by this company on a given day. Obtain the probability distribution of x.

d. Using the probability distribution obtained in part c, find the following probabilities.

i. $P(x \geq 45)$ **ii.** $P(x < 33)$ **iii.** $P(36 < x < 52)$

TA5.4 A commuter airline receives an average of 13 complaints per week from its passengers. Let x denote the number of complaints received by this airline during a given week.

a. Find $P(x = 0)$. If your answer is zero, does it mean that this cannot happen? Explain.

b. Find $P(x \leq 10)$.

c. Obtain the probability distribution of x.

d. Using the probability distribution obtained in part c, find the following probabilities.

i. $P(x > 18)$ **ii.** $P(x \leq 9)$ **iii.** $P(10 \leq x \leq 17)$

Chapter

6

Continuous Random Variables and the Normal Distribution

A common excuse for showing up late for an engagement is, "I'm so sorry, I got stuck in traffic!" In a 2000 survey of San Francisco Bay Area residents conducted for the Metropolitan Transportation Commission's Regional Rideshare Program, 73.8% of the participants felt traffic had gotten heavier in the past year, compared to 58.2% in 1998. A particularly bad commute, due to an accident or weather, tends to be a surprise for those afflicted by it because it varies so much from the expected length of time. How early should you leave home so that you are not late for work or your engagement, considering you may face heavy traffic on your way? (See Exercise 6.84.)

6.1 Continuous Probability Distribution

Case Study 6–1 Distribution of Time Taken to Run a Road Race

6.2 The Normal Distribution

6.3 The Standard Normal Distribution

6.4 Standardizing a Normal Distribution

6.5 Applications of the Normal Distribution

6.6 Determining the *z* and *x* Values When an Area Under the Normal Distribution Curve is Known

6.7 The Normal Approximation to the Binomial Distribution

Discrete random variables and their probability distributions were presented in Chapter 5. Section 5.1 defined a continuous random variable as a variable that can assume any value in one or more intervals.

The possible values that a continuous random variable can assume are infinite and uncountable. For example, the variable that represents the time taken by a worker to commute from home to work is a continuous random variable. Suppose 5 minutes is the minimum time and 130 minutes is the maximum time taken by all workers to commute from home to work. Let x be a continuous random variable that denotes the time taken to commute from home to work by a randomly selected worker. Then x can assume any value in the interval 5 to 130 minutes. This interval contains an infinite number of values that are uncountable.

A continuous random variable can possess one of many probability distributions. In this chapter, we discuss the normal probability distribution and the normal distribution as an approximation to the binomial distribution.

6.1 Continuous Probability Distribution

In Chapter 5, we defined a **continuous random variable** as a random variable whose values are not countable. A continuous random variable can assume any value over an interval or intervals. Because the number of values contained in any interval is infinite, the possible number of values that a continuous random variable can assume is also infinite. Moreover, we cannot count these values. In Chapter 5, it was stated that the life of a battery, heights of people, time taken to complete an examination, amount of milk in a gallon, weights of babies, and prices of houses are all examples of continuous random variables. Note that although money can be counted, all variables involving money usually are considered to be continuous random variables. This is so because a variable involving money often has a very large number of outcomes.

Suppose 5000 female students are enrolled at a university, and x is the continuous random variable that represents the heights of these female students. Table 6.1 lists the frequency and relative frequency distributions of x.

Table 6.1 Frequency and Relative Frequency Distributions of Heights of Female Students

Height of a Female Student (inches) x	f	Relative Frequency
60 to less than 61	90	.018
61 to less than 62	170	.034
62 to less than 63	460	.092
63 to less than 64	750	.150
64 to less than 65	970	.194
65 to less than 66	760	.152
66 to less than 67	640	.128
67 to less than 68	440	.088
68 to less than 69	320	.064
69 to less than 70	220	.044
70 to less than 71	180	.036
	$N = 5000$	Sum = 1.0

The relative frequencies given in Table 6.1 can be used as the probabilities of the respective classes. Note that these are exact probabilities because we are considering the population of all female students.

Figure 6.1 displays the histogram and polygon for the relative frequency distribution of Table 6.1. Figure 6.2 shows the smoothed polygon for the data of Table 6.1. The smoothed

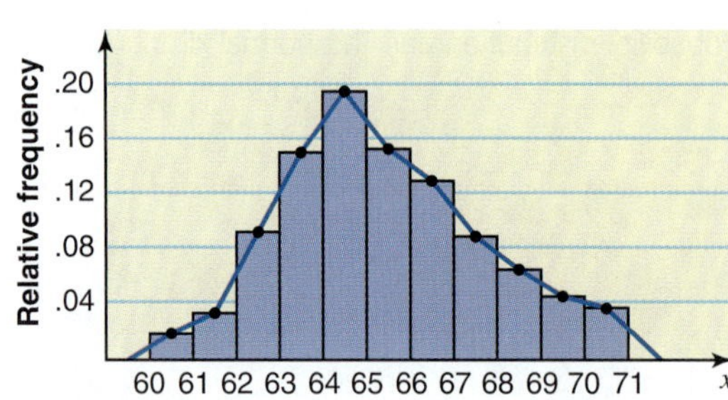

Figure 6.1 Histogram and polygon for Table 6.1.

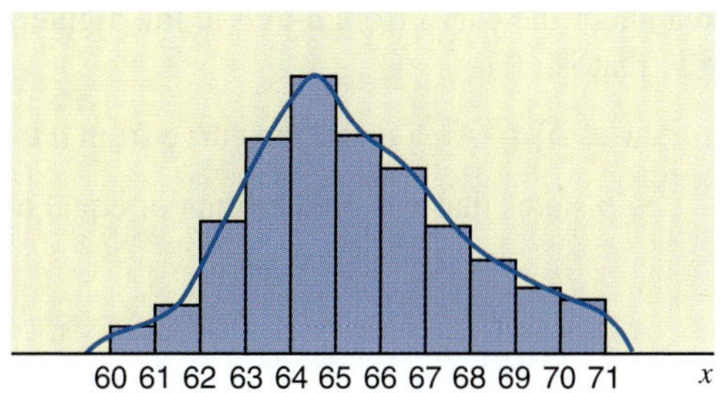

Figure 6.2 Probability distribution curve for heights.

polygon is an approximation of the *probability distribution curve* of the continuous random variable x. Note that each class in Table 6.1 has a width equal to 1 inch. If the width of classes is more than 1 unit, we first obtain the *relative frequency densities* and then graph these relative frequency densities to obtain the distribution curve. The relative frequency density of a class is obtained by dividing the relative frequency of that class by the class width. The relative frequency densities are calculated to make the sum of the areas of all rectangles in the histogram equal to 1.0. Case Study 6–1, which appears later in this section, illustrates this procedure. The probability distribution curve of a continuous random variable is also called its *probability density function.*

The probability distribution of a continuous random variable possesses the following *two characteristics.*

1. The probability that x assumes a value in any interval lies in the range 0 to 1.
2. The total probability of all the (mutually exclusive) intervals within which x can assume a value is 1.0.

The first characteristic states that the area under the probability distribution curve of a continuous random variable between any two points is between 0 and 1, as shown in Figure 6.3. The second characteristic indicates that the total area under the probability distribution curve of a continuous random variable is always 1.0 or 100%, as shown in Figure 6.4.

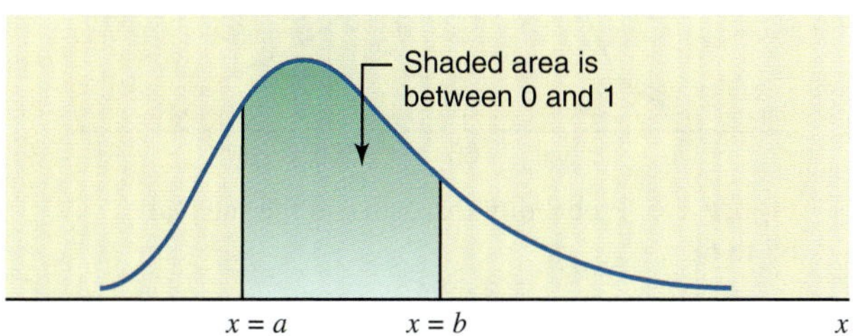

Figure 6.3 Area under a curve between two points.

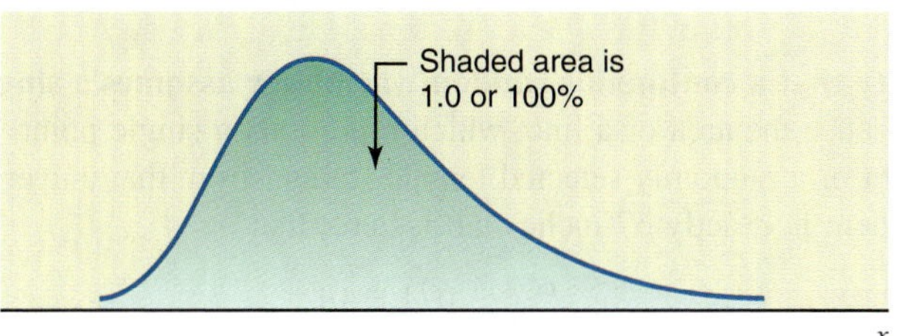

Figure 6.4 Total area under a probability distribution curve.

The probability that a continuous random variable x assumes a value within a certain interval is given by the area under the curve between the two limits of the interval, as shown in

Figure 6.5. The shaded area under the curve from a to b in this figure gives the probability that x falls in the interval a to b; that is,

$$P(a \leq x \leq b) = \text{Area under the curve from } a \text{ to } b$$

Note that the interval $a \leq x \leq b$ states that x is greater than or equal to a but less than or equal to b.

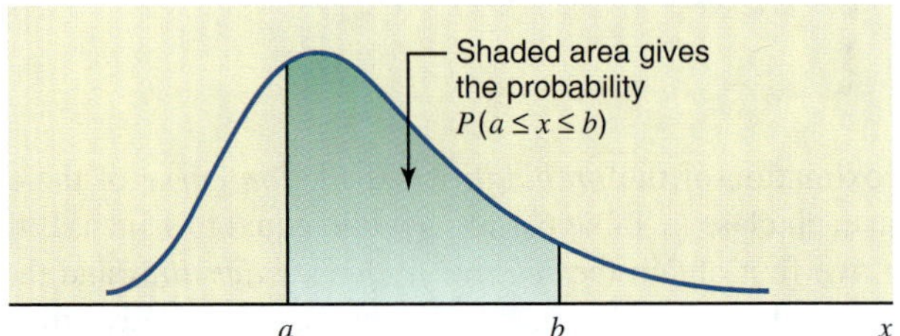

Figure 6.5 Area under the curve as probability.

Reconsider the example on the heights of all female students at a university. The probability that the height of a randomly selected female student from this university lies in the interval 65 to 68 inches is given by the area under the distribution curve of the heights of all female students from $x = 65$ to $x = 68$, as shown in Figure 6.6. This probability is written as

$$P(65 \leq x \leq 68)$$

which states that x is greater than or equal to 65 but less than or equal to 68.

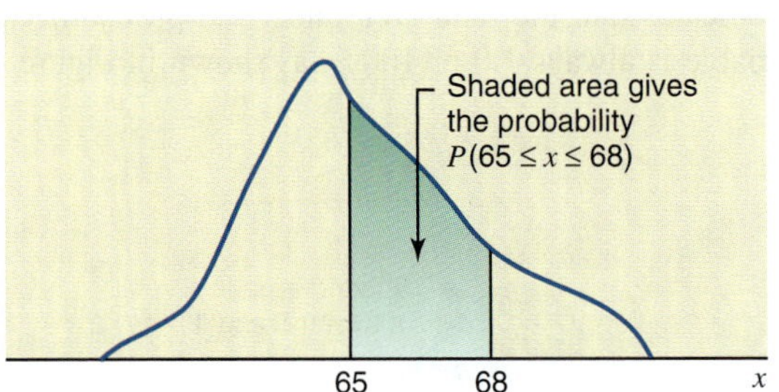

Figure 6.6 Probability that x lies in the interval 65 to 68.

For a continuous probability distribution, the probability is always calculated for an interval. For example, in Figure 6.6, the interval representing the shaded area is from 65 to 68. Consequently, the shaded area in that figure gives the probability for the interval $65 \leq x \leq 68$.

The probability that a continuous random variable x assumes a single value is always zero. This is so because the area of a line, which represents a single point, is zero. For example, if x is the height of a randomly selected female student from that university, then the probability that this student is exactly 67 inches tall is zero; that is,

$$P(x = 67) = 0$$

This probability is shown in Figure 6.7. Similarly, the probability for x to assume any other single value is zero.

In general, if a and b are two of the values that x can assume, then

$$P(a) = 0 \quad \text{and} \quad P(b) = 0$$

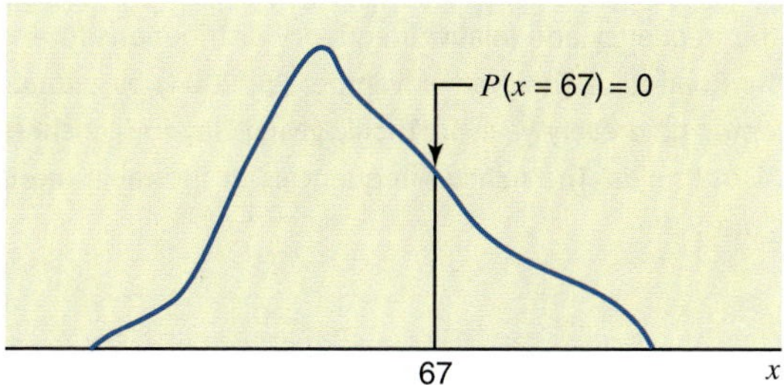

Figure 6.7 The probability of a single value of x is zero.

From this we can deduce that for a continuous random variable,

$$P(a \leq x \leq b) = P(a < x < b)$$

In other words, the probability that x assumes a value in the interval a to b is the same whether or not the values a and b are included in the interval. For the example on the heights of female students, the probability that a randomly selected female student is between 65 and 68 inches tall is the same as the probability that this female is 65 to 68 inches tall. This is shown in Figure 6.8.

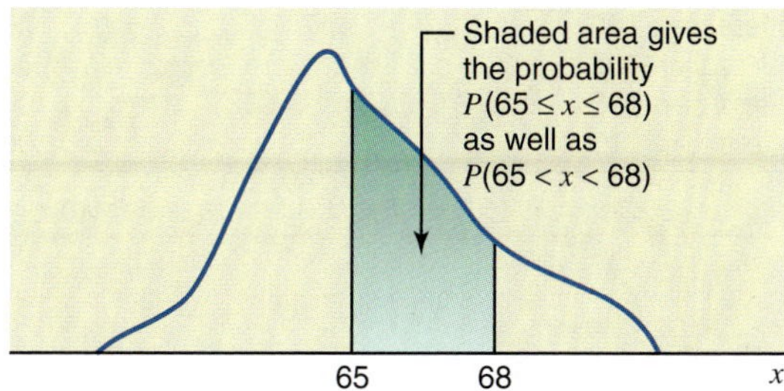

Figure 6.8 Probability "from 65 to 68" and "between 65 and 68."

Note that the interval "between 65 and 68" represents "$65 < x < 68$" and it does not include 65 and 68. On the other hand, the interval "from 65 to 68" represents "$65 \leq x \leq 68$" and it does include 65 and 68. However, as mentioned previously, in the case of a continuous random variable, both of these intervals contain the same probability or area under the curve.

Case Study 6–1 on the next page describes how we obtain the probability distribution curve of a continuous random variable.

6.2 The Normal Distribution

The normal distribution is one of the many probability distributions that a continuous random variable can possess. The normal distribution is the most important and most widely used of all probability distributions. A large number of phenomena in the real world are normally distributed either exactly or approximately. The continuous random variables representing heights and weights of people, scores on an examination, weights of packages (e.g., cereal boxes, boxes of cookies), amount of milk in a gallon, life of an item (such as a light-bulb or a television set), and time taken to complete a certain job have all been observed to have a (approximate) normal distribution.

CASE STUDY

6–1

DISTRIBUTION OF TIME TAKEN TO RUN A ROAD RACE

The following table gives the frequency and relative frequency distributions for the time (in minutes) taken to complete the Manchester Road Race (held on November 25, 2004) by a total of 8911 participants who finished that race. This event is held every year on Thanksgiving Day in Manchester, Connecticut. The total distance of the course is 4.748 miles. The relative frequencies in the table are used to construct the histogram and polygon in Figure 6.9.

Class	Frequency	Relative Frequency
20 to less than 25	26	.0029
25 to less than 30	200	.0224
30 to less than 35	671	.0753
35 to less than 40	1009	.1132
40 to less than 45	1269	.1424
45 to less than 50	1659	.1862
50 to less than 55	1561	.1752
55 to less than 60	950	.1066
60 to less than 65	488	.0548
65 to less than 70	277	.0311
70 to less than 75	204	.0229
75 to less than 80	159	.0178
80 to less than 85	158	.0177
85 to less than 90	120	.0135
90 to less than 95	94	.0105
95 to less than 100	57	.0064
100 to less than 105	6	.0007
105 to less than 110	2	.0002
110 to less than 115	1	.0001
	$\Sigma f = 8911$	Sum = .9999

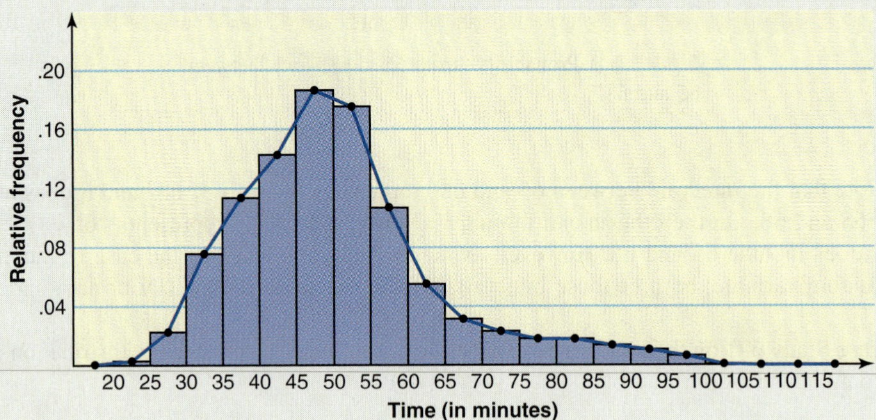

Figure 6.9 Histogram and polygon for the Road Race data.

To derive the probability distribution curve for these data, we calculate the relative frequency densities by dividing the relative frequencies by the class widths. The width of each class in the table is 5. By dividing the relative frequencies by 5, we obtain the relative frequency densities, which are recorded in the next table. Using the relative frequency densities, we draw a histogram and smoothed polygon, as shown in Figure 6.10. The curve in this figure is the probability distribution curve for the Road Race data.

Note that the areas of the rectangles in Figure 6.9 do not give probabilities (which are approximated by relative frequencies). Rather, it is the heights of these rectangles that give the probabilities. This is so because the base of each rectangle is 5 in this histogram. Consequently, the area of any rectangle is given by its height multiplied by 5. Thus, the total area of all the rectangles in Figure 6.9 is 5.0, not 1.0. However, in Figure 6.10, it is the areas, not the heights, of rectangles that give the probabilities of the respective classes. Thus, if we add the areas of all the rectangles in Figure 6.10, we obtain the sum of all probabilities equal to .9999, which is approximately 1.0. Consequently, the total area under the curve is equal to 1.0.

Class	Relative Frequency Density
20 to less than 25	.00058
25 to less than 30	.00448
30 to less than 35	.01506
35 to less than 40	.02264
40 to less than 45	.02848
45 to less than 50	.03724
50 to less than 55	.03504
55 to less than 60	.02132
60 to less than 65	.01096
65 to less than 70	.00622
70 to less than 75	.00458
75 to less than 80	.00356
80 to less than 85	.00354
85 to less than 90	.00270
90 to less than 95	.00210
95 to less than 100	.00128
100 to less than 105	.00014
105 to less than 110	.00004
110 to less than 115	.00002

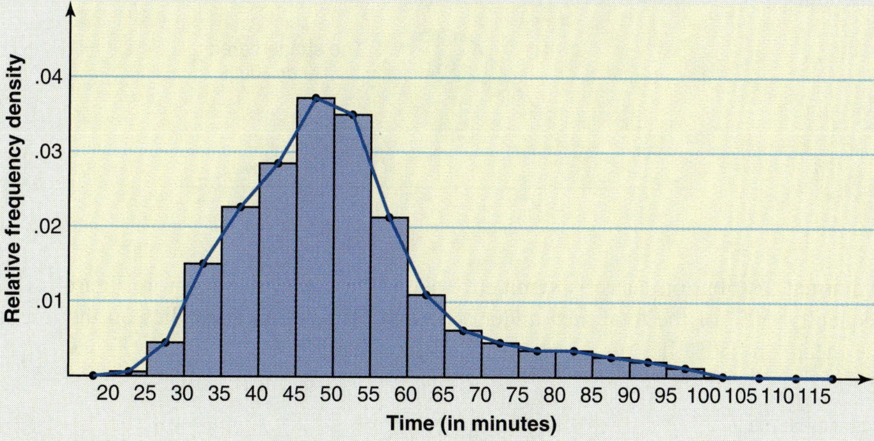

Figure 6.10 Probability distribution curve for the Road Race data.

The probability distribution of a continuous random variable has a mean and a standard deviation, denoted by μ and σ, respectively. The mean and standard deviation of the probability distribution curve of Figure 6.10 are 50.5314 and 13.6353 minutes, respectively. These values of μ and σ are calculated by using the raw data on 8911 participants.

Source: This case study is based on data published on the official Web site of the Manchester Road Race.

The **normal probability distribution** or the *normal curve* is a bell-shaped (symmetric) curve. Such a curve is shown in Figure 6.11. Its mean is denoted by μ and its standard deviation by σ. A continuous random variable x that has a normal distribution is called a *normal random variable.* Note that not all bell-shaped curves represent a normal distribution curve. Only a specific kind of bell-shaped curve represents a normal curve.

Figure 6.11 Normal distribution with mean μ and standard deviation σ.

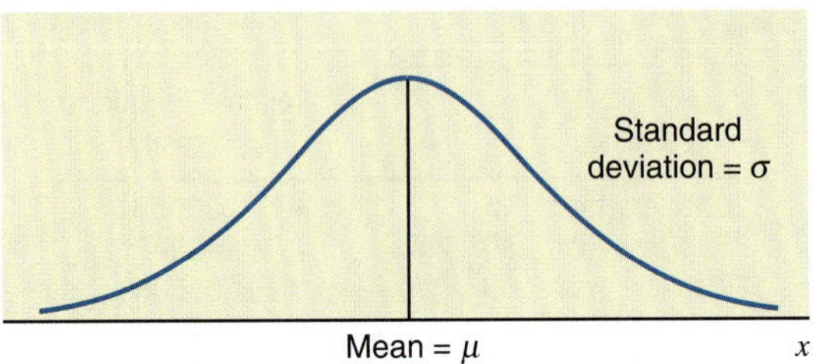

Normal Probability Distribution A *normal probability distribution*, when plotted, gives a bell-shaped curve such that

1. The total area under the curve is 1.0.
2. The curve is symmetric about the mean.
3. The two tails of the curve extend indefinitely.

A normal distribution possesses the following three characteristics.

1. The total area under a normal distribution curve is 1.0 or 100%, as shown in Figure 6.12.

Figure 6.12 Total area under a normal curve.

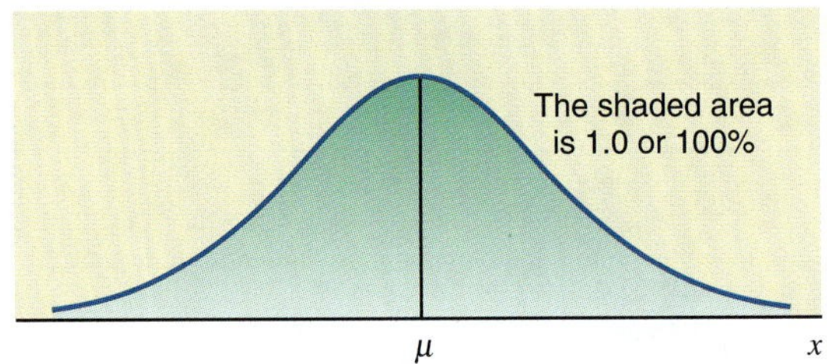

2. A normal distribution curve is symmetric about the mean, as shown in Figure 6.13. Consequently, 50% of the total area under a normal distribution curve lies on the left side of the mean, and 50% lies on the right side of the mean.

Figure 6.13 A normal curve is symmetric about the mean.

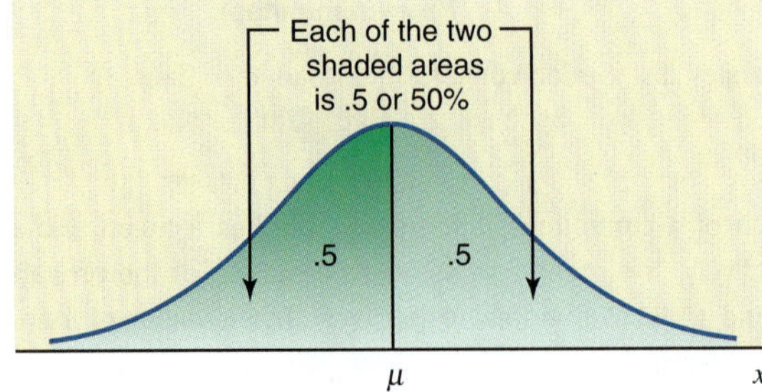

3. The tails of a normal distribution curve extend indefinitely in both directions without touching or crossing the horizontal axis. Although a normal distribution curve never meets the horizontal axis, beyond the points represented by $\mu - 3\sigma$ and $\mu + 3\sigma$ it becomes so close to this axis that the area under the curve beyond these points in both directions can be taken as virtually zero. These areas are shown in Figure 6.14.

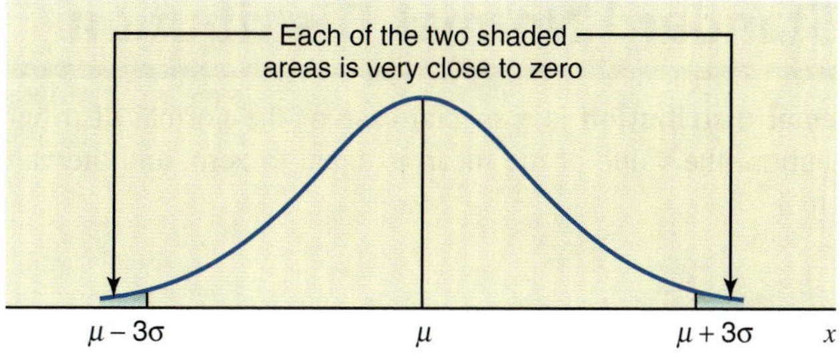

Figure 6.14 Areas of the normal curve beyond $\mu \pm 3\sigma$.

The mean, μ, and the standard deviation, σ, are the *parameters* of the normal distribution. Given the values of these two parameters, we can find the area under a normal distribution curve for any interval. Remember, there is not just one normal distribution curve but a *family* of normal distribution curves. Each different set of values of μ and σ gives a different normal distribution. The value of μ determines the center of a normal distribution curve on the horizontal axis, and the value of σ gives the spread of the normal distribution curve. The three normal distribution curves drawn in Figure 6.15 have the same mean but different standard deviations. By contrast, the three normal distribution curves in Figure 6.16 have different means but the same standard deviation.

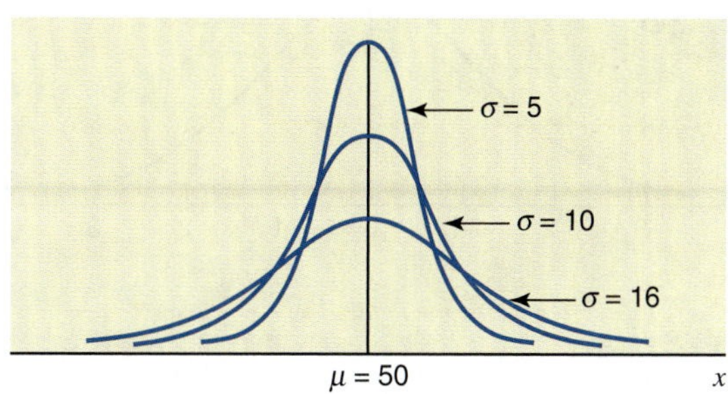

Figure 6.15 Three normal distribution curves with the same mean but different standard deviations.

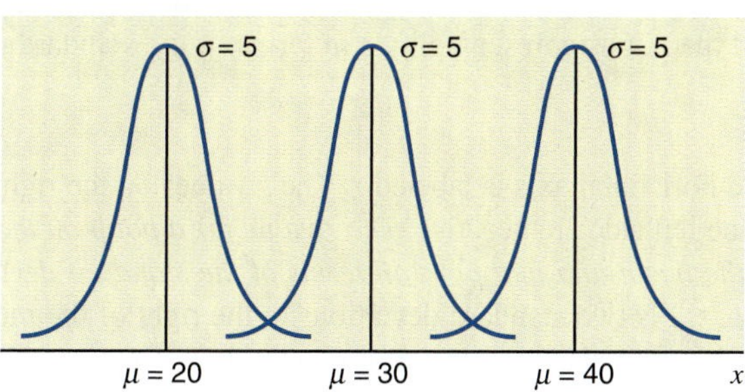

Figure 6.16 Three normal distribution curves with different means but the same standard deviation.

Like the binomial and Poisson probability distributions discussed in Chapter 5, the normal probability distribution can also be expressed by a mathematical equation.[1] However, we will not use this equation to find the area under a normal distribution curve. Instead, we will use Table IV of Appendix C.

[1]The equation of the normal distribution is

$$f(x) = \frac{1}{\sigma\sqrt{2\pi}} e^{-(1/2)[(x-\mu)/\sigma]^2}$$

where $e = 2.71828$ and $\pi = 3.14159$ approximately; $f(x)$, called the probability density function, gives the vertical distance between the horizontal axis and the curve at point x. For the information of those who are familiar with integral calculus, the definite integral of this equation from a to b gives the probability that x assumes a value between a and b.

6.3 The Standard Normal Distribution

The **standard normal distribution** is a special case of the normal distribution. For the standard normal distribution, the value of the mean is equal to zero, and the value of the standard deviation is equal to 1.

Definition

Standard Normal Distribution The normal distribution with $\mu = 0$ and $\sigma = 1$ is called the *standard normal distribution.*

Figure 6.17 displays the standard normal distribution curve. The random variable that possesses the standard normal distribution is denoted by z. In other words, the units for the standard normal distribution curve are denoted by z and are called the **z values** or **z scores**. They are also called *standard units* or *standard scores.*

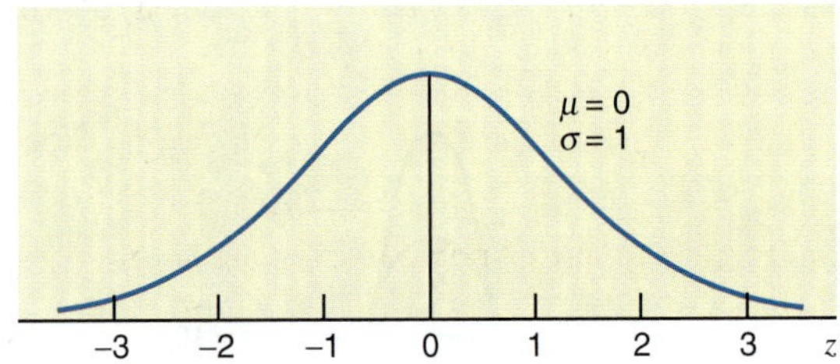

Figure 6.17 The standard normal distribution curve.

Definition

z Values or z Scores The units marked on the horizontal axis of the standard normal curve are denoted by z and are called the *z values* or *z scores*. A specific value of z gives the distance between the mean and the point represented by z in terms of the standard deviation.

In Figure 6.17, the horizontal axis is labeled z. The z values on the right side of the mean are positive and those on the left side are negative. *The z value for a point on the horizontal axis gives the distance between the mean and that point in terms of the standard deviation.* For example, a point with a value of $z = 2$ is two standard deviations to the right of the mean. Similarly, a point with a value of $z = -2$ is two standard deviations to the left of the mean.

The standard normal distribution table, Table IV of Appendix C, lists the areas under the standard normal curve to the left of z-values from -3.49 to 3.49. To read the standard normal distribution table, we look for the given z-value in the table and record the value corresponding to that z-value. As shown in Figure 6.18, Table IV gives what is called the cumulative probability to the left of any z-value.

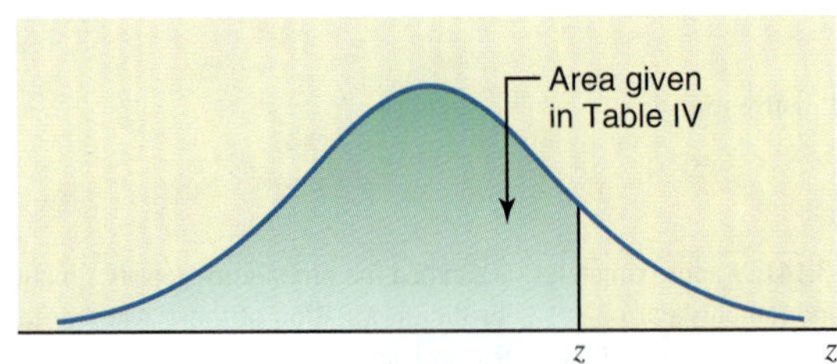

Figure 6.18 Area under the standard normal curve.

Although the values of z on the left side of the mean are negative, the area under the curve is always positive. ◀ *Remember*

The area under the standard normal curve between any two points can be interpreted as the probability that z assumes a value within that interval. Examples 6–1 through 6–4 describe how to read Table IV of Appendix C to find areas under the standard normal curve.

■ EXAMPLE 6–1

Find the area under the standard normal curve to the left of $z = 1.95$.

Finding the area to the left of a positive z.

Solution We divide the given number 1.95 into two portions: 1.9 (the digit before the decimal and one digit after the decimal) and .05 (the second digit after the decimal). (Note that $1.95 = 1.9 + .05$.) To find the required area under the standard normal curve, we locate 1.9 in the column for z on the left side of Table IV and .05 in the row for z at the top of Table IV. The entry where the row for 1.9 and the column for .05 intersect gives the area under the standard normal curve to the left of $z = 1.95$. The relevant portion of Table IV is reproduced as Table 6.2 below. From Table IV or Table 6.2, the entry where the row for 1.9 and the column for .05 cross is .9744. Consequently, the area under the standard normal curve to the left of $z = 1.95$ is .9744. This area is shown in Figure 6.19. (It is always helpful to sketch the curve and mark the area you are determining.)

Table 6.2 **Area Under the Standard Normal Curve to the Left of $z = 1.95$**

z	.00	.01	...	.05	...	.09
−3.4	.0003	.0003	...	.0003	...	.0002
−3.3	.0005	.0005	...	.0004	...	.0003
−3.2	.0007	.0007	...	.0006	...	.0005
.	.	.	...	.	...	.
.	.	.	...	.	...	.
.	.	.	...	.	...	.
1.9	.9713	.9719	...	.9744 ←	...	.9767
.	.	.	...	.	...	.
.	.	.	...	.	...	.
.	.	.	...	.	...	.
3.4	.9997	.9997	...	.9997	...	.9998

Required area

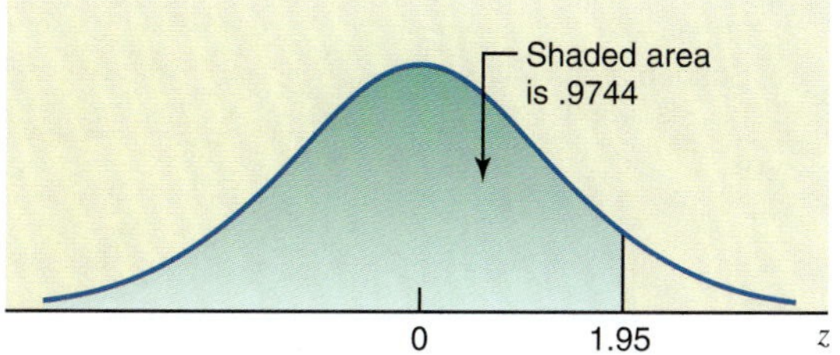

Figure 6.19 Area to the left of $z = 1.95$.

The area to the left of $z = 1.95$ can be interpreted as the probability that z assumes a value less than 1.95; that is,

$$\text{Area to the left of } 1.95 = P(z < 1.95) = \mathbf{.9744}$$

As mentioned in Section 6.1, the probability that a continuous random variable assumes a single value is zero. Therefore,

$$P(z = 1.95) = 0$$

Hence,

$$P(z < 1.95) = P(z \leq 1.95) = \mathbf{.9744}$$

■

■ EXAMPLE 6–2

Finding the area between a negative z and $z = 0$.

Find the area under the standard normal curve from $z = -2.17$ to $z = 0$.

Solution To find the area from $z = -2.17$ to $z = 0$, first we find the areas to the left of $z = 0$ and to the left of $z = -2.17$ in the standard normal distribution table (Table IV). As shown in Table 6.3, these two areas are .5 and .0150, respectively. Next we subtract .0150 from .5 to find the required area.

Table 6.3 Area Under the Standard Normal Curve

z	.00	.01	...	.07	...	.09
−3.4	.0003	.0003	...	.0003	...	.0002
−3.3	.0005	.0005	...	.0004	...	.0003
−3.2	.0007	.0007	...	.0005	...	.0005
.	.	.	...	.	...	.
.	.	.	...	.	...	.
−2.1	.0179	.0174	...	.0150	...	.0143
.	.	.	...	.	...	.
.	.	.	...	.	...	.
0.0	.5000	.5040	...	.5279	...	.5359
.	.	.	...	.	...	.
.	.	.	...	.	...	.
3.4	.9997	.9997	...	.9997	...	.9998

Area to the left of $z = 0$ (.5000); Area to the left of $z = -2.17$ (.0150)

The area from $z = -2.17$ to $z = 0$ gives the probability that z lies in the interval -2.17 to 0; that is,

$$\begin{aligned}\text{Area from } -2.17 \text{ to } 0 &= P(-2.17 \leq z \leq 0)\\ &= P(z \leq 0) - P(z \leq -2.17) = \mathbf{.5000 - .0150 = .4850}\end{aligned}$$

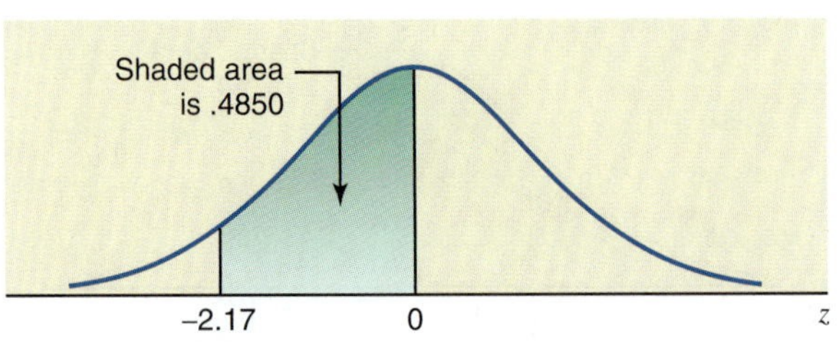

Figure 6.20 Area from $z = -2.17$ to $z = 0$.

■

■ EXAMPLE 6–3

Find the following areas under the standard normal curve.

Finding the areas in the right and left tails.

(a) Area to the right of $z = 2.32$

(b) Area to the left of $z = -1.54$

Solution

(a) As mentioned earlier, the normal distribution table gives the area to the left of a z value. To find the area to the right of $z = 2.32$, first we find the area to the left of $z = 2.32$. Then we subtract this area from 1.0, which is the total area under the curve. From Table IV, the area to the left of $z = 2.32$ is .9898. Consequently, the required area is $1.0 - .9898 = .0102$, as shown in Figure 6.21.

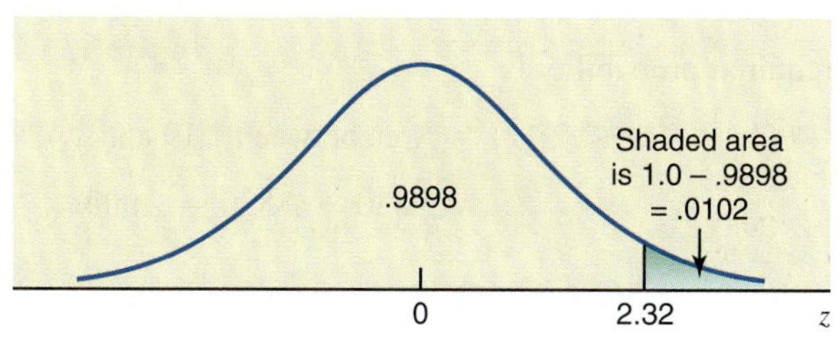

Figure 6.21 Area to the right of $z = 2.32$.

The area to the right of $z = 2.32$ gives the probability that z is greater than 2.32. Thus,

$$\text{Area to the right of } 2.32 = P(z > 2.32) = 1.0 - .9898 = \mathbf{.0102}$$

(b) To find the area under the standard normal curve to the left of $z = -1.54$, we find the area in Table IV that corresponds to -1.5 in the z column and .04 in the top row. This area is .0618. This area is shown in Figure 6.22.

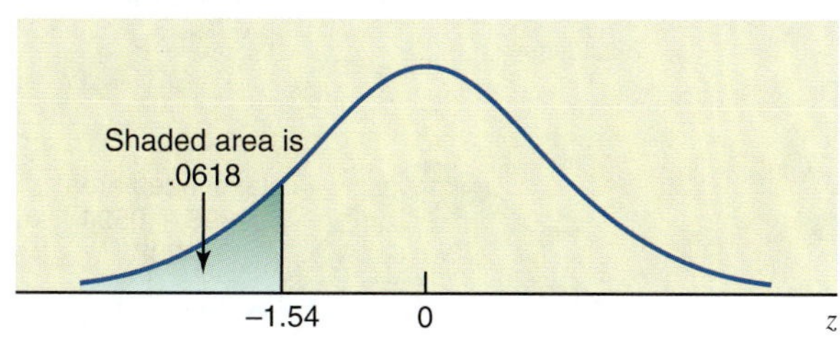

Figure 6.22 Area to the left of $z = -1.54$.

The area to the left of $z = -1.54$ gives the probability that z is less than -1.54. Thus,

$$\text{Area to the left of } -1.54 = P(z < -1.54) = \mathbf{.0618}$$ ■

■ EXAMPLE 6–4

Find the following probabilities for the standard normal curve.

(a) $P(1.19 < z < 2.12)$ **(b)** $P(-1.56 < z < 2.31)$ **(c)** $P(z > -.75)$

Solution

Finding the area between two positive values of z.

(a) The probability $P(1.19 < z < 2.12)$ is given by the area under the standard normal curve between $z = 1.19$ and $z = 2.12$, which is the shaded area in Figure 6.23.

To find the area between $z = 1.19$ and $z = 2.12$, first we find the areas to the left of $z = 1.19$ and $z = 2.12$. Then we subtract the smaller area (to the left of $z = 1.19$) from the larger area (to the left of $z = 2.12$).

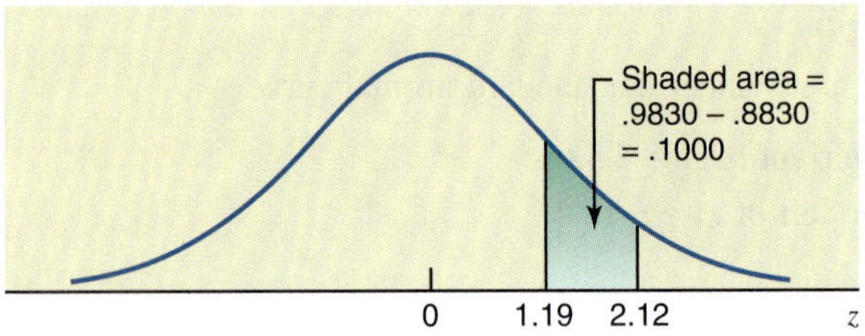

Figure 6.23 Finding $P(1.19 < z < 2.12)$.

From Table IV for the standard normal distribution, we find

$$\text{Area to the left of } 1.19 = .8830$$

$$\text{Area to the left of } 2.12 = .9830$$

Then, the required probability is

$$P(1.19 < z < 2.12) = \text{Area between } 1.19 \text{ and } 2.12$$
$$= .9830 - .8830 = \mathbf{.1000}$$

Finding the area between a positive and a negative value of z.

(b) The probability $P(-1.56 < z < 2.31)$ is given by the area under the standard normal curve between $z = -1.56$ and $z = 2.31$, which is the shaded area in Figure 6.24.

From Table IV for the standard normal distribution, we have

$$\text{Area to the left of } -1.56 = .0594$$

$$\text{Area to the left of } 2.31 = .9896$$

The required probability is

$$P(-1.56 < z < 2.31) = \text{Area between } -1.56 \text{ and } 2.31$$
$$= .9896 - .0594 = \mathbf{.9302}$$

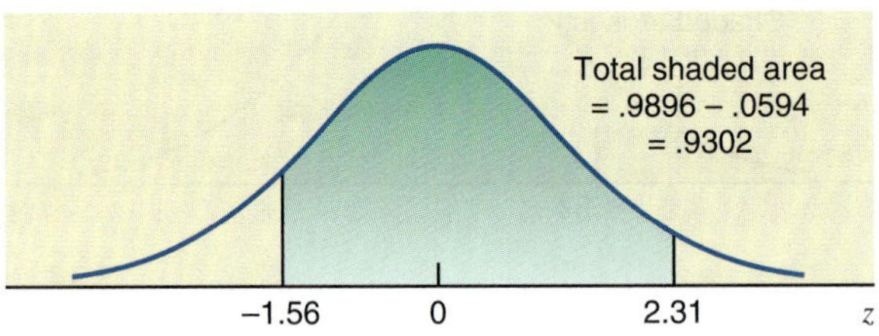

Figure 6.24 Finding $P(-1.56 < z < 2.31)$.

Finding the area to the right of a negative value of z.

(c) The probability $P(z > -.75)$ is given by the area under the standard normal curve to the right of $z = -.75$, which is the shaded area in Figure 6.25.

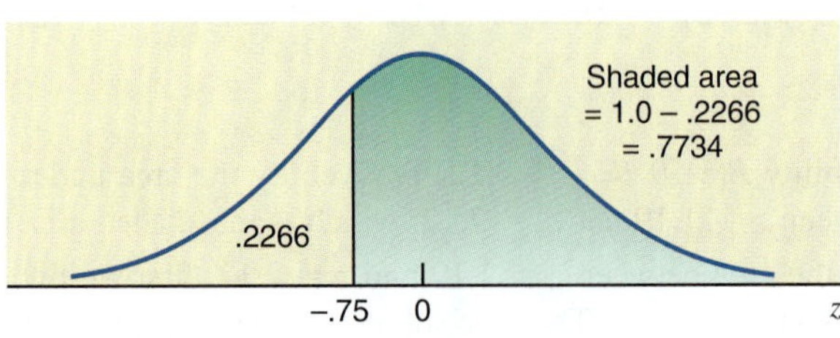

Figure 6.25 Finding $P(z > -.75)$.

From Table IV for the standard normal distribution.

$$\text{Area to the left of } -.75 = .2266$$

The required probability is

$$P(z > -.75) = \text{Area to the right of } -.75 = 1.0 - .2266 = \mathbf{.7734}$$

In the discussion in Section 3.4 of Chapter 3 on the use of the standard deviation, we discussed the empirical rule for a bell-shaped curve. That empirical rule is based on the standard normal distribution. By using the normal distribution table, we can now verify the empirical rule as follows.

1. The total area within one standard deviation of the mean is 68.26%. This area is given by the difference between the area to the left of $z = 1.0$ and the area to the left of $z = -1.0$. As shown in Figure 6.26, this area is .8413 − .1587 = .6826 or 68.26%.

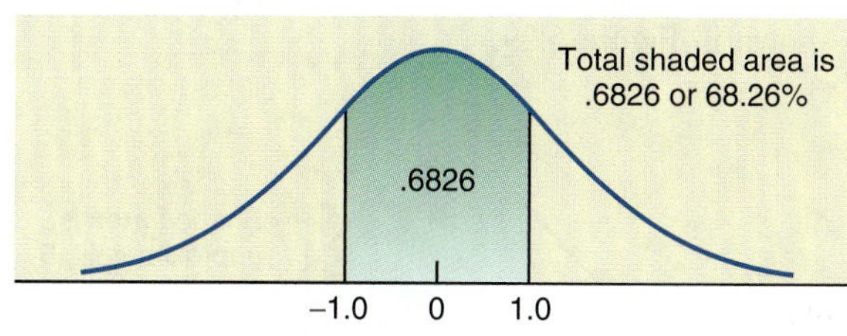

Figure 6.26 Area within one standard deviation of the mean.

2. The total area within two standard deviations of the mean is 95.44%. This area is given by the difference between the area to the left of $z = 2.0$ and the area to the left of $z = -2.0$. As shown in Figure 6.27, this area is .9772 − .0228 = .9544 or 95.44%.

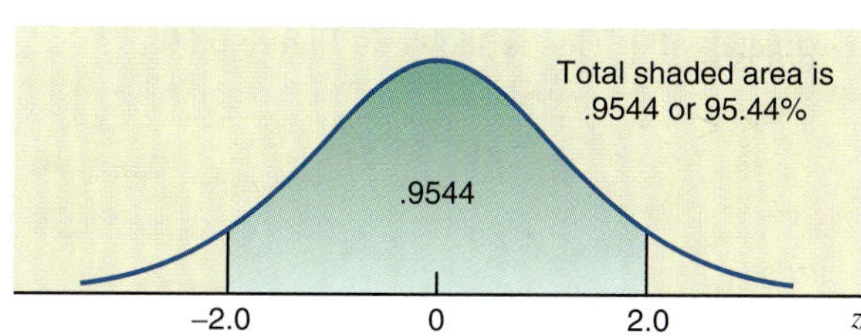

Figure 6.27 Area within two standard deviations of the mean.

3. The total area within three standard deviations of the mean is 99.74%. This area is given by the difference between the area to the left of $z = 3.0$ and the area to the left of $z = -3.0$. As shown in Figure 6.28, this area is .9987 − .0013 = .9974 or 99.74%.

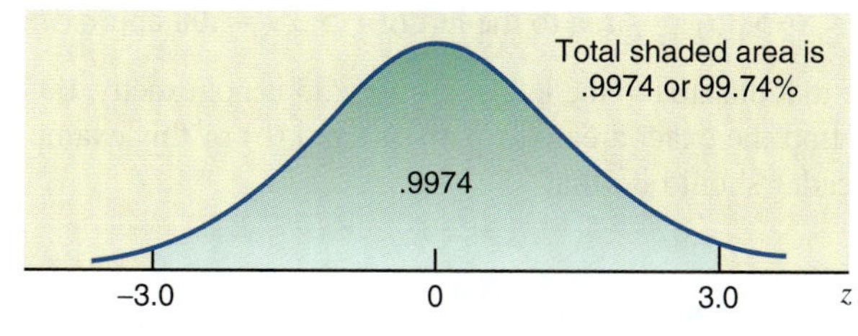

Figure 6.28 Area within three standard deviations of the mean.

Again, note that only a specific bell-shaped curve represents the normal distribution. Now we can state that a bell-shaped curve that contains (about) 68.26% of the total area within one standard deviation of the mean, (about) 95.44% of the total area within two standard deviations of the mean, and (about) 99.74% of the total area within three standard deviations of the mean represents a normal distribution curve.

The standard normal distribution table, Table IV of Appendix C, goes from $z = -3.49$ to $z = 3.49$. Consequently, if we need to find the area to the left of $z = -3.50$ or a smaller value of z, we can assume it to be approximately 0.0. If we need to find the area to the left of $z = 3.50$ or a larger number, we can assume it to be approximately 1.0. Example 6–5 illustrates this procedure.

■ EXAMPLE 6–5

Find the following probabilities for the standard normal curve.

(a) $P(0 < z < 5.67)$ **(b)** $P(z < -5.35)$

Solution

Finding the area between $z = 0$ and a value of z greater than 3.49.

(a) The probability $P(0 < z < 5.67)$ is given by the area under the standard normal curve between $z = 0$ and $z = 5.67$. Because $z = 5.67$ is greater than 3.49 and is not in Table IV, the area under the standard normal curve to the left of $z = 5.67$ can be approximated by 1.0. Also, the area to the left of $z = 0$ is .5. Hence, the required probability is

$$P(0 < z < 5.67) = \text{Area between 0 and 5.67} = 1.0 - .5 = \mathbf{.5} \text{ approximately}$$

Note that the area between $z = 0$ and $z = 5.67$ is not exactly .5 but very close to .5. This area is shown in Figure 6.29.

Figure 6.29 Area between $z = 0$ and $z = 5.67$.

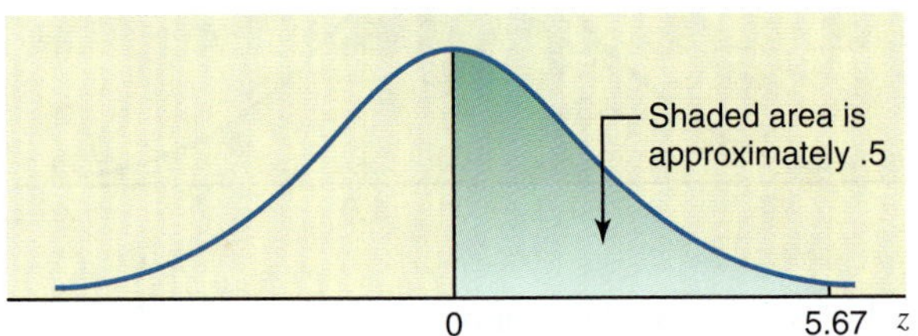

Finding the area to the left of a z that is less than -3.49.

(b) The probability $P(z < -5.35)$ represents the area under the standard normal curve to the left of $z = -5.35$. Since $z = -5.35$ is not in the table, we can assume that this area is approximately .00. This is shown in Figure 6.30.

Figure 6.30 Area to the left of $z = -5.35$.

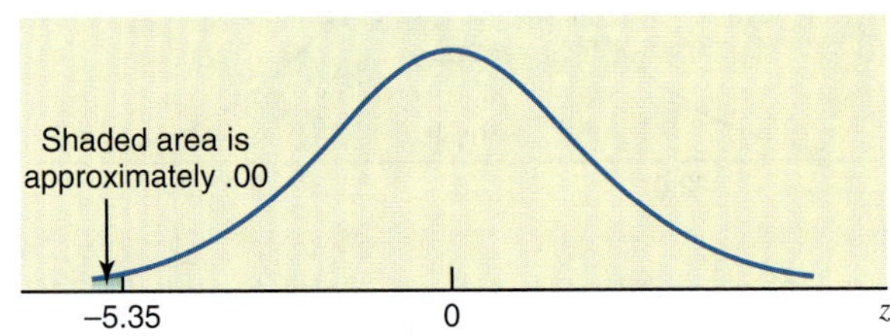

The required probability is

$$P(z < -5.35) = \text{Area to the left of } -5.35 = \mathbf{.00} \text{ approximately}$$

Again, note that the area to the left of $z = -5.35$ is not exactly .00 but very close to .00.

We can find the exact areas for parts (a) and (b) of this example by using technology. The reader should do that. ■

EXERCISES

■ CONCEPTS AND PROCEDURES

6.1 What is the difference between the probability distribution of a discrete random variable and that of a continuous random variable? Explain.

6.2 Let x be a continuous random variable. What is the probability that x assumes a single value, such as a?

6.3 For a continuous probability distribution, why is $P(a < x < b)$ equal to $P(a \leq x \leq b)$?

6.4 Briefly explain the main characteristics of a normal distribution. Illustrate with the help of graphs.

6.5 Briefly describe the standard normal distribution curve.

6.6 What are the parameters of the normal distribution?

6.7 How do the width and height of a normal distribution change when its mean remains the same but its standard deviation decreases?

6.8 Do the width and/or height of a normal distribution change when its standard deviation remains the same but its mean increases?

6.9 For the standard normal distribution, what does z represent?

6.10 For the standard normal distribution, find the area within one standard deviation of the mean—that is, the area between $\mu - \sigma$ and $\mu + \sigma$.

6.11 For the standard normal distribution, find the area within 1.5 standard deviations of the mean—that is, the area between $\mu - 1.5\sigma$ and $\mu + 1.5\sigma$.

6.12 For the standard normal distribution, what is the area within two standard deviations of the mean?

6.13 For the standard normal distribution, what is the area within 2.5 standard deviations of the mean?

6.14 For the standard normal distribution, what is the area within three standard deviations of the mean?

6.15 Find the area under the standard normal curve

a. between $z = 0$ and $z = 1.95$
b. between $z = 0$ and $z = -1.85$
c. between $z = 1.15$ and $z = 2.37$
d. from $z = -1.53$ to $z = -2.88$
e. from $z = -1.67$ to $z = 2.44$

6.16 Find the area under the standard normal curve

a. from $z = 0$ to $z = 2.34$
b. between $z = 0$ and $z = -2.78$
c. from $z = .84$ to $z = 1.95$
d. between $z = -.57$ and $z = -2.39$
e. between $z = -2.15$ and $z = 1.67$

6.17 Find the area under the standard normal curve

a. to the right of $z = 1.56$
b. to the left of $z = -1.97$
c. to the right of $z = -2.05$
d. to the left of $z = 1.86$

6.18 Obtain the area under the standard normal curve

a. to the right of $z = 1.73$
b. to the left of $z = -1.55$
c. to the right of $z = -.65$
d. to the left of $z = .89$

6.19 Find the area under the standard normal curve

a. between $z = 0$ and $z = 4.28$
b. from $z = 0$ to $z = -3.75$
c. to the right of $z = 7.43$
d. to the left of $z = -4.49$

6.20 Find the area under the standard normal curve

a. from $z = 0$ to $z = 3.94$
b. between $z = 0$ and $z = -5.16$
c. to the right of $z = 5.42$
d. to the left of $z = -3.68$

6.21 Determine the following probabilities for the standard normal distribution.

a. $P(-1.83 \le z \le 2.57)$
b. $P(0 \le z \le 2.02)$
c. $P(-1.99 \le z \le 0)$
d. $P(z \ge 1.48)$

6.22 Determine the following probabilities for the standard normal distribution.

a. $P(-2.46 \le z \le 1.68)$
b. $P(0 \le z \le 1.86)$
c. $P(-2.58 \le z \le 0)$
d. $P(z \ge .83)$

6.23 Find the following probabilities for the standard normal distribution.

a. $P(z < -2.14)$
b. $P(.67 \le z \le 2.49)$
c. $P(-2.07 \le z \le -.93)$
d. $P(z < 1.78)$

6.24 Find the following probabilities for the standard normal distribution.

a. $P(z < -1.31)$
b. $P(1.03 \le z \le 2.89)$
c. $P(-2.24 \le z \le -1.09)$
d. $P(z < 2.12)$

6.25 Obtain the following probabilities for the standard normal distribution.

a. $P(z > -.98)$
b. $P(-2.47 \le z \le 1.19)$
c. $P(0 \le z \le 4.25)$
d. $P(-5.36 \le z \le 0)$
e. $P(z > 6.07)$
f. $P(z < -5.27)$

6.26 Obtain the following probabilities for the standard normal distribution.

a. $P(z > -1.06)$
b. $P(-.68 \le z \le 1.84)$
c. $P(0 \le z \le 3.85)$
d. $P(-4.34 \le z \le 0)$
e. $P(z > 4.82)$
f. $P(z < -6.12)$

6.4 Standardizing a Normal Distribution

As was shown in the previous section, Table IV of Appendix C can be used to find areas under the standard normal curve. However, in real-world applications, a (continuous) random variable may have a normal distribution with values of the mean and standard deviation that are different from 0 and 1, respectively. The first step in such a case is to convert the given normal distribution to the standard normal distribution. This procedure is called *standardizing a normal distribution.* The units of a normal distribution (which is not the standard normal distribution) are denoted by x. We know from Section 6.3 that units of the standard normal distribution are denoted by z.

Converting an x Value to a z Value For a normal random variable x, a particular value of x can be converted to its corresponding z value by using the formula

$$z = \frac{x - \mu}{\sigma}$$

where μ and σ are the mean and standard deviation of the normal distribution of x, respectively.

Thus, to find the z value for an x value, we calculate the difference between the given x value and the mean, μ, and divide this difference by the standard deviation, σ. If the value of x is equal to μ, then its z value is equal to zero. Note that we will always round z values to two decimal places.

Remember ▶ The z value for the mean of a normal distribution is always zero.

Examples 6–6 through 6–10 describe how to convert x values to the corresponding z values and how to find areas under a normal distribution curve.

■ EXAMPLE 6–6

Converting x values to the corresponding z values.

Let x be a continuous random variable that has a normal distribution with a mean of 50 and a standard deviation of 10. Convert the following x values to z values and find the probability to the left of these points.

(a) $x = 55$ **(b)** $x = 35$

Solution For the given normal distribution, $\mu = 50$ and $\sigma = 10$.

(a) The z value for $x = 55$ is computed as follows:

$$z = \frac{x - \mu}{\sigma} = \frac{55 - 50}{10} = \mathbf{.50}$$

Thus, the z value for $x = 55$ is .50. The z values for $\mu = 50$ and $x = 55$ are shown in Figure 6.31. Note that the z value for $\mu = 50$ is zero. The value $z = .50$ for $x = 55$ indicates that the distance between $\mu = 50$ and $x = 55$ is 1/2 of the standard deviation, which is 10. Consequently, we can state that the z value represents the distance between μ and x in terms of the standard deviation. Because $x = 55$ is greater than $\mu = 50$, its z value is positive.

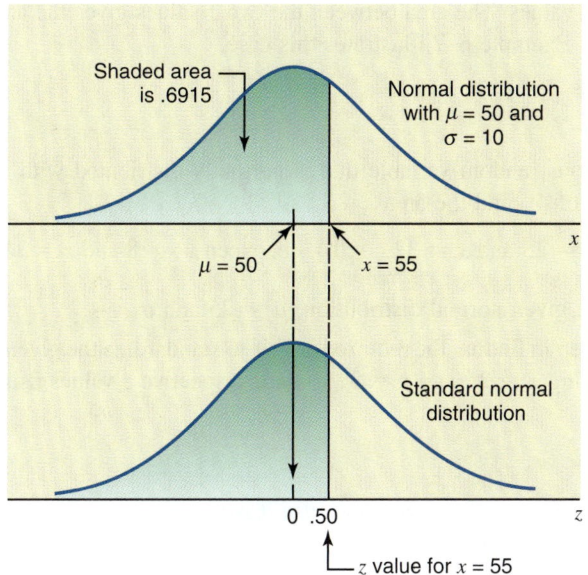

Figure 6.31 z value for $x = 55$.

From this point on, we will usually show only the z axis below the x axis and not the standard normal curve itself.

To find the probability to the left of $x = 55$, we find the probability to the left of $z = .50$ from Table IV. This probability is .6915. Therefore,

$$P(x < 55) = P(z < .50) = \mathbf{.6915}$$

(b) The z value for $x = 35$ is computed as follows and is shown in Figure 6.32:

$$z = \frac{x - \mu}{\sigma} = \frac{35 - 50}{10} = \mathbf{-1.50}$$

Because $x = 35$ is on the left side of the mean (i.e., 35 is less than $\mu = 50$), its z value is negative. As a general rule, whenever an x value is less than the value of μ, its z value is negative.

To find the probability to the left of $x = 35$, we find the area under the normal curve to the left of $z = -1.50$. This area from Table IV is .0668. Hence,

$$P(x < 35) = P(z < -1.50) = \mathbf{.0668}$$

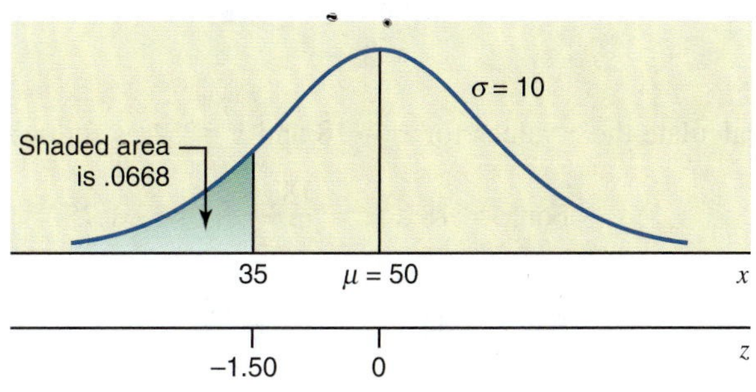

Figure 6.32 z value for $x = 35$.

◀ ***Remember***

The z value for an x value that is greater than μ is positive, the z value for an x value that is equal to μ is zero, and the z value for an x value that is less than μ is negative.

To find the area between two values of x for a normal distribution, we first convert both values of x to their respective z values. Then we find the area under the standard normal curve between those two z values. The area between the two z values gives the area between the corresponding x values. Example 6–7 illustrates this case.

EXAMPLE 6–7

Let x be a continuous random variable that is normally distributed with a mean of 25 and a standard deviation of 4. Find the area

(a) between $x = 25$ and $x = 32$ **(b)** between $x = 18$ and $x = 34$

Solution For the given normal distribution, $\mu = 25$ and $\sigma = 4$.

Finding the area between the mean and a point to its right.

(a) The first step in finding the required area is to standardize the given normal distribution by converting $x = 25$ and $x = 32$ to their respective z values using the formula

$$z = \frac{x - \mu}{\sigma}$$

The z value for $x = 25$ is zero because it is the mean of the normal distribution. The z value for $x = 32$ is

$$z = \frac{32 - 25}{4} = 1.75$$

The area between $x = 25$ and $x = 32$ under the given normal distribution curve is equivalent to the area between $z = 0$ and $z = 1.75$ under the standard normal curve. From Table IV, the area to the left of $z = 1.75$ is .9599, and the area to the left of $z = 0$ is .50. Hence, the required area is $.9599 - .50 = .4599$, which is shown in Figure 6.33.

The area between $x = 25$ and $x = 32$ under the normal curve gives the probability that x assumes a value between 25 and 32. This probability can be written as

$$P(25 < x < 32) = P(0 < z < 1.75) = \mathbf{.4599}$$

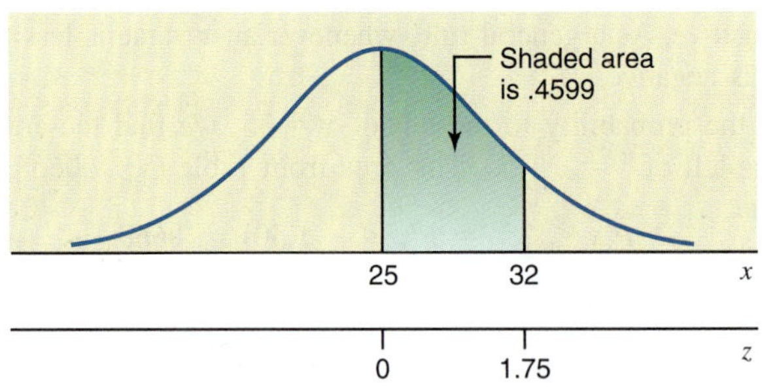

Figure 6.33 Area between $x = 25$ and $x = 32$.

Finding the area between two points on different sides of the mean

(b) First, we calculate the z values for $x = 18$ and $x = 34$ as follows:

$$\text{For } x = 18\text{:} \quad z = \frac{18 - 25}{4} = -1.75$$

$$\text{For } x = 34\text{:} \quad z = \frac{34 - 25}{4} = 2.25$$

The area under the given normal distribution curve between $x = 18$ and $x = 34$ is given by the area under the standard normal curve between $z = -1.75$ and $z = 2.25$. From Table IV, the area to the left of $z = 2.25$ is .9878, and the area to the left of $z = -1.75$ is .0401. Hence, the required area is

$$P(18 < x < 34) = P(-1.75 < z < 2.25) = .9878 - .0401 = \mathbf{.9477}$$

This area is shown in Figure 6.34.

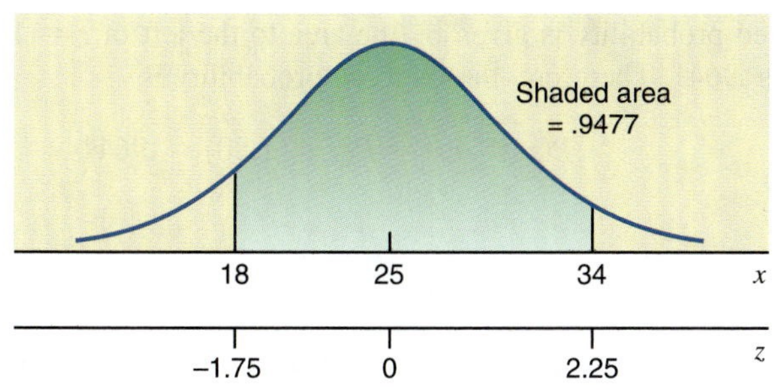

Figure 6.34 Area between $x = 18$ and $x = 34$.

■ EXAMPLE 6–8

Let x be a normal random variable with its mean equal to 40 and standard deviation equal to 5. Find the following probabilities for this normal distribution.

(a) $P(x > 55)$ **(b)** $P(x < 49)$

Solution For the given normal distribution, $\mu = 40$ and $\sigma = 5$.

Calculating the probability of x falling in the right tail.

(a) The probability that x assumes a value greater than 55 is given by the area under the normal distribution curve to the right of $x = 55$, as shown in Figure 6.35. This area is calculated by subtracting the area to the left of $x = 55$ from 1.0, which is the total area under the curve.

$$\text{For } x = 55: \quad z = \frac{55 - 40}{5} = 3.00$$

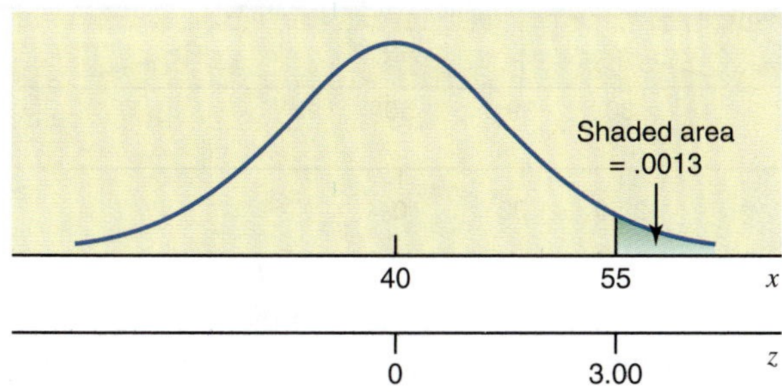

Figure 6.35 Finding $P(x > 55)$.

The required probability is given by the area to the right of $z = 3.00$. To find this area, first we find the area to the left of $z = 3.00$, which is .9987. Then we subtract this area from 1.0. Thus,

$$P(x > 55) = P(z > 3.00) = 1.0 - .9987 = \mathbf{.0013}$$

Calculating the probability that x is less than a value to the right of the mean.

(b) The probability that x will assume a value less than 49 is given by the area under the normal distribution curve to the left of 49, which is the shaded area in Figure 6.36. This area is obtained from Table IV.

$$\text{For } x = 49: \quad z = \frac{49 - 40}{5} = 1.80$$

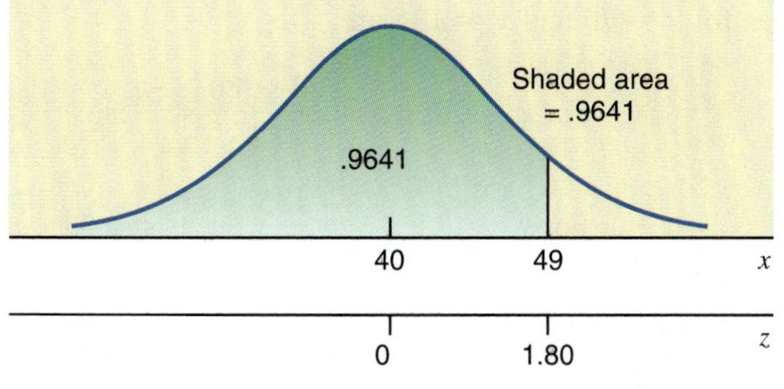

Figure 6.36 Finding $P(x < 49)$.

The required probability is given by the area to the left of $z = 1.80$. This area from Table IV is .9641. Therefore, the required probability is

$$P(x < 49) = P(z < 1.80) = \mathbf{.9641}$$

■ EXAMPLE 6–9

Finding the area between two x values that are less than the mean.

Let x be a continuous random variable that has a normal distribution with $\mu = 50$ and $\sigma = 8$. Find the probability $P(30 \le x \le 39)$.

Solution For this normal distribution, $\mu = 50$ and $\sigma = 8$. The probability $P(30 \le x \le 39)$ is given by the area from $x = 30$ to $x = 39$ under the normal distribution curve. As shown in Figure 6.37, this area is given by the difference between the area to the left of $x = 30$ and the area to the left of $x = 39$.

$$\text{For } x = 30: \quad z = \frac{30 - 50}{8} = -2.50$$

$$\text{For } x = 39: \quad z = \frac{39 - 50}{8} = -1.38$$

Figure 6.37 Finding $P(30 \le x \le 39)$.

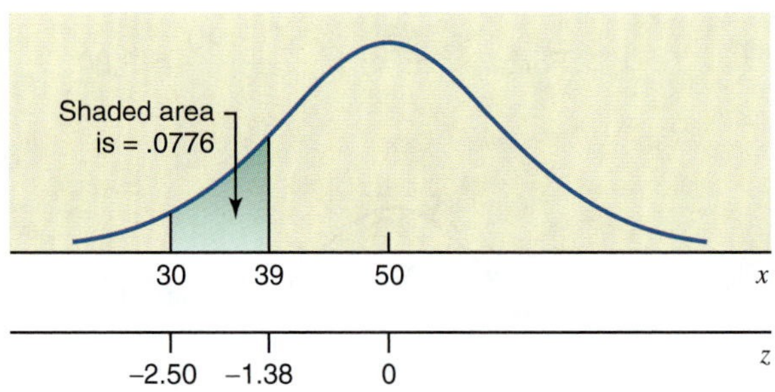

To find the required area, we first find the area to the left of $z = -2.50$, which is .0062. Then, we find the area to the left of $z = -1.38$, which is .0838. The difference between these two areas gives the required probability, which is

$$P(30 \le x \le 39) = P(-2.50 \le z \le -1.38) = .0838 - .0062 = \mathbf{.0776}$$

■ EXAMPLE 6–10

Let x be a continuous random variable that has a normal distribution with a mean of 80 and a standard deviation of 12. Find the area under the normal distribution curve

(a) from $x = 70$ to $x = 135$ **(b)** to the left of 27

Solution For the given normal distribution, $\mu = 80$ and $\sigma = 12$.

Finding the area between two x values that are on different sides of the mean.

(a) The z values for $x = 70$ and $x = 135$ are:

$$\text{For } x = 70: \quad z = \frac{70 - 80}{12} = -.83$$

$$\text{For } x = 135: \quad z = \frac{135 - 80}{12} = 4.58$$

Thus, to find the required area we find the areas to the left of $z = -.83$ and to the left of $z = 4.58$ under the standard normal curve. From Table IV, the area to the left of

Figure 6.38 Area between $x = 70$ and $x = 135$.

$z = -.83$ is .2033 and the area to the left of $z = 4.58$ is approximately 1.0. Note that $z = 4.58$ is not in Table IV.
Hence,

$$P(70 \le x \le 135) = P(-.83 \le z \le 4.58)$$
$$= 1.0 - .2033 = \mathbf{.7967} \text{ approximately}$$

Figure 6.38 shows this area.

(b) First we find the z-value for $x = 27$.

Finding an area in the left tail.

$$\text{For } x = 27: \quad z = \frac{27 - 80}{12} = -4.42$$

As shown in Figure 6.39, the required area is given by the area under the standard normal distribution curve to the left of $z = -4.42$. This area is approximately zero.

$$P(x < 27) = P(z < -4.42) = \mathbf{.00} \text{ approximately}$$

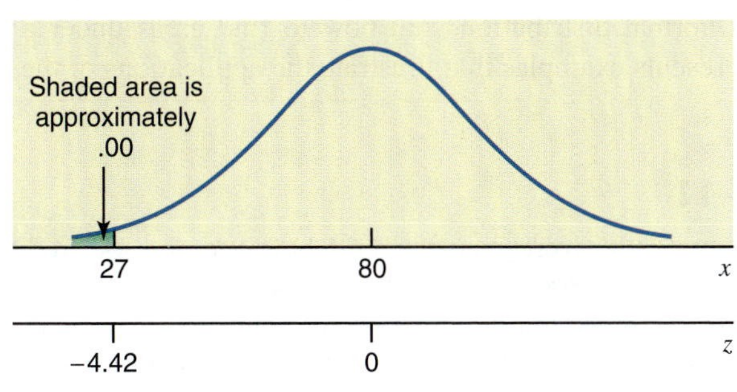

Figure 6.39 Area to the left of $x = 27$.

EXERCISES

CONCEPTS AND PROCEDURES

6.27 Find the z value for each of the following x values for a normal distribution with $\mu = 30$ and $\sigma = 5$.

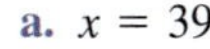

a. $x = 39$ **b.** $x = 17$ **c.** $x = 22$ **d.** $x = 42$

6.28 Determine the z value for each of the following x values for a normal distribution with $\mu = 16$ and $\sigma = 3$.

a. $x = 12$ **b.** $x = 21$ **c.** $x = 19$ **d.** $x = 13$

6.29 Find the following areas under a normal distribution curve with $\mu = 20$ and $\sigma = 4$.

a. Area between $x = 20$ and $x = 27$
b. Area from $x = 23$ to $x = 25$
c. Area between $x = 9.5$ and $x = 17$

6.30 Find the following areas under a normal distribution curve with $\mu = 12$ and $\sigma = 2$.

a. Area between $x = 7.76$ and $x = 12$
b. Area between $x = 14.48$ and $x = 16.54$
c. Area from $x = 8.22$ to $x = 10.06$

6.31 Determine the area under a normal distribution curve with $\mu = 55$ and $\sigma = 7$
a. to the right of $x = 58$
b. to the right of $x = 43$
c. to the left of $x = 67$
d. to the left of $x = 24$

6.32 Find the area under a normal distribution curve with $\mu = 37$ and $\sigma = 3$
a. to the left of $x = 30$
b. to the right of $x = 52$
c. to the left of $x = 44$
d. to the right of $x = 32$

6.33 Let x be a continuous random variable that is normally distributed with a mean of 25 and a standard deviation of 6. Find the probability that x assumes a value
a. between 29 and 36
b. between 22 and 33

6.34 Let x be a continuous random variable that has a normal distribution with a mean of 40 and a standard deviation of 4. Find the probability that x assumes a value
a. between 29 and 35
b. from 34 to 51

6.35 Let x be a continuous random variable that is normally distributed with a mean of 80 and a standard deviation of 12. Find the probability that x assumes a value
a. greater than 69
b. less than 74
c. greater than 101
d. less than 88

6.36 Let x be a continuous random variable that is normally distributed with a mean of 65 and a standard deviation of 15. Find the probability that x assumes a value
a. less than 43
b. greater than 74
c. greater than 56
d. less than 71

6.5 Applications of the Normal Distribution

Sections 6.2 through 6.4 discussed the normal distribution, how to convert a normal distribution to the standard normal distribution, and how to find areas under a normal distribution curve. This section presents examples that illustrate the applications of the normal distribution.

EXAMPLE 6–11

Using the normal distribution: the area between two points on different sides of the mean.

According to an estimate, the average consumer debt (owed on cars, credit cards, and so forth) of U.S. households was \$17,989 in 2004 (*USA TODAY,* October 4, 2004). Suppose such current consumer debts for all U.S. households have a normal distribution with a mean of \$17,989 and a standard deviation of \$3750. Find the probability that such consumer debt of a randomly selected U.S. household is between \$13,000 and \$20,000.

Solution Let x denote the consumer debt of a randomly selected U.S. household. Then, x is normally distributed with

$$\mu = \$17{,}989 \qquad \text{and} \qquad \sigma = \$3750$$

The probability that the consumer debt of a randomly selected U.S. household is between \$13,000 and \$20,000 is given by the area under the normal distribution curve of x that falls between $x = \$13{,}000$ and $x = \$20{,}000$ as shown in Figure 6.40. To find this area, first we find the areas to the left of $x = \$13{,}000$ and $x = \$20{,}000$, respectively, and then take the difference between these two areas.

$$\text{For } x = \$13{,}000: \quad z = \frac{13{,}000 - 17{,}989}{3750} = -1.33$$

$$\text{For } x = \$20{,}000: \quad z = \frac{20{,}000 - 17{,}989}{3750} = .54$$

Thus, the required probability is given by the difference between the areas under the standard normal curve to the left of $z = -1.33$ and to the left of $z = .54$. From Table IV in Appendix C,

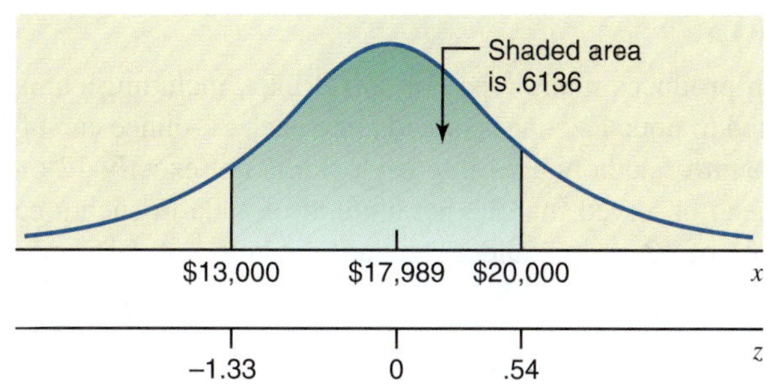

Figure 6.40 Area between x = \$13,000 and x = \$20,000.

the area to the left of $z = -1.33$ is .0918, and the area to the left of $z = .54$ is .7054. Hence, the required probability is

$$P(\$13{,}000 < x < \$20{,}000) = P(-1.33 < z < .54) = .7054 - .0918 = \mathbf{.6136}$$

Thus, the probability is .6136 that the consumer debt of a randomly selected U.S. household is between \$13,000 and \$20,000. Converting this probability into a percentage, we can also state that (about) 61.36% of all U.S. households have consumer debts between \$13,000 and \$20,000. ■

■ EXAMPLE 6–12

Using the normal distribution: probability that x is less than a value that is to the right of the mean.

A racing car is one of the many toys manufactured by Mack Corporation. The assembly times for this toy follow a normal distribution with a mean of 55 minutes and a standard deviation of 4 minutes. The company closes at 5 P.M. every day. If one worker starts to assemble a racing car at 4 P.M., what is the probability that she will finish this job before the company closes for the day?

Solution Let x denote the time this worker takes to assemble a racing car. Then, x is normally distributed with

$$\mu = 55 \text{ minutes} \quad \text{and} \quad \sigma = 4 \text{ minutes}$$

We are to find the probability that this worker can assemble this car in 60 minutes or less (between 4 and 5 P.M.). This probability is given by the area under the normal curve to the left of $x = 60$ as shown in Figure 6.41.

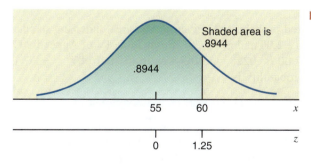

Figure 6.41 Area to the left of $x = 60$.

$$\text{For } x = 60: \quad z = \frac{60 - 55}{4} = 1.25$$

The required probability is given by the area under the standard normal curve to the left of $z = 1.25$, which is .8944 from Table IV of Appendix C. Thus, the required probability is

$$P(x \le 60) = P(z \le 1.25) = \mathbf{.8944}$$

Thus, the probability is .8944 that this worker will finish assembling this racing car before the company closes for the day. ■

■ EXAMPLE 6–13

Using the normal distribution.

Hupper Corporation produces many types of soft drinks, including Orange Cola. The filling machines are adjusted to pour 12 ounces of soda into each 12-ounce can of Orange Cola. However, the actual amount of soda poured into each can is not exactly 12 ounces; it varies from can to can. It has been observed that the net amount of soda in such a can has a normal distribution with a mean of 12 ounces and a standard deviation of .015 ounce.

(a) What is the probability that a randomly selected can of Orange Cola contains 11.97 to 11.99 ounces of soda?

(b) What percentage of the Orange Cola cans contain 12.02 to 12.07 ounces of soda?

Solution Let x be the net amount of soda in a can of Orange Cola. Then, x has a normal distribution with $\mu = 12$ ounces and $\sigma = .015$ ounce.

Calculating the probability between two points that are to the left of the mean.

(a) The probability that a randomly selected can contains 11.97 to 11.99 ounces of soda is given by the area under the normal distribution curve from $x = 11.97$ to $x = 11.99$. This area is shown in Figure 6.42.

$$\text{For } x = 11.97\text{:} \quad z = \frac{11.97 - 12}{.015} = -2.00$$

$$\text{For } x = 11.99\text{:} \quad z = \frac{11.99 - 12}{.015} = -.67$$

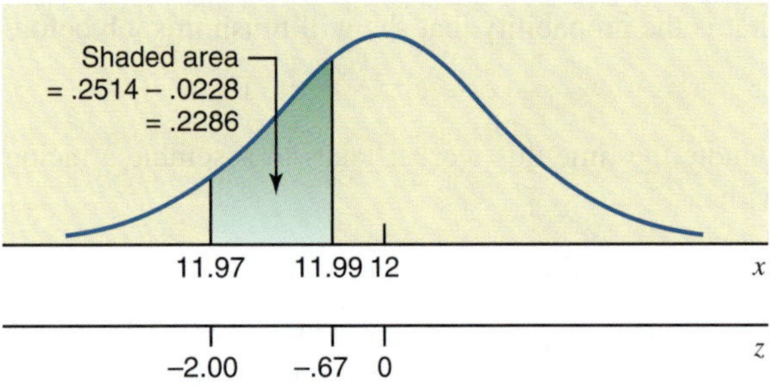

Figure 6.42 Area between $x = 11.97$ and $x = 11.99$.

The required probability is given by the area under the standard normal curve between $z = -2.00$ and $z = -.67$. From Table IV of Appendix C, the area to the left of $z = -2.00$ is .0228, and the area to the left of $z = -.67$ is .2514. Hence, the required probability is

$$P(11.97 \leq x \leq 11.99) = P(-2.00 \leq z \leq -.67) = .2514 - .0228 = \mathbf{.2286}$$

Thus, the probability is .2286 that any randomly selected can of Orange Cola will contain 11.97 to 11.99 ounces of soda. We can also state that about 22.86% of Orange Cola cans contain 11.97 to 11.99 ounces of soda.

Calculating the probability between two points that are to the right of the mean.

(b) The percentage of Orange Cola cans that contain 12.02 to 12.07 ounces of soda is given by the area under the normal distribution curve from $x = 12.02$ to $x = 12.07$, as shown in Figure 6.43.

$$\text{For } x = 12.02\text{:} \quad z = \frac{12.02 - 12}{.015} = 1.33$$

$$\text{For } x = 12.07\text{:} \quad z = \frac{12.07 - 12}{.015} = 4.67$$

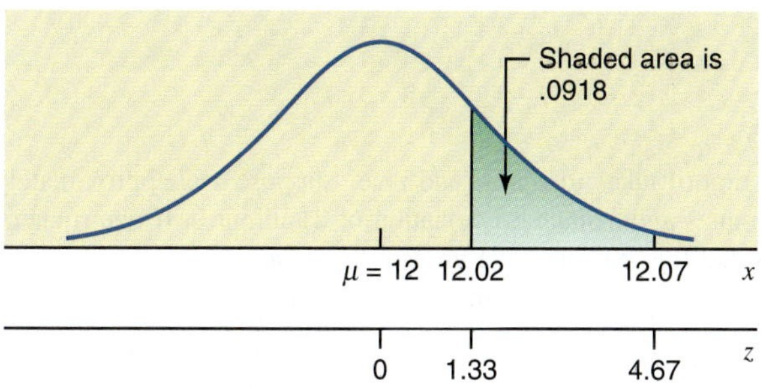

Figure 6.43 Area from x = 12.02 to x = 12.07.

The required probability is given by the area under the standard normal curve between $z = 1.33$ and $z = 4.67$. From Table IV of Appendix C, the area to the left of $z = 1.33$ is .9082, and the area to the left of $z = 4.67$ is approximately 1.0. Hence, the required probability is

$$P(12.02 \leq x \leq 12.07) = P(1.33 \leq z \leq 4.67) = 1.0 - .9082 = \mathbf{.0918}$$

Converting this probability to a percentage, we can state that approximately 9.18% of all Orange Cola cans are expected to contain 12.02 to 12.07 ounces of soda. ■

■ EXAMPLE 6–14

Finding the area to the left of x that is less than the mean.

Suppose the life span of a calculator manufactured by Texas Instruments has a normal distribution with a mean of 54 months and a standard deviation of 8 months. The company guarantees that any calculator that starts malfunctioning within 36 months of the purchase will be replaced by a new one. About what percentage of calculators made by this company are expected to be replaced?

Solution Let x be the life span of such a calculator. Then x has a normal distribution with $\mu = 54$ and $\sigma = 8$ months. The probability that a randomly selected calculator will start to malfunction within 36 months is given by the area under the normal distribution curve to the left of $x = 36$, as shown in Figure 6.44.

$$\text{For } x = 36\text{:} \quad z = \frac{36 - 54}{8} = -2.25$$

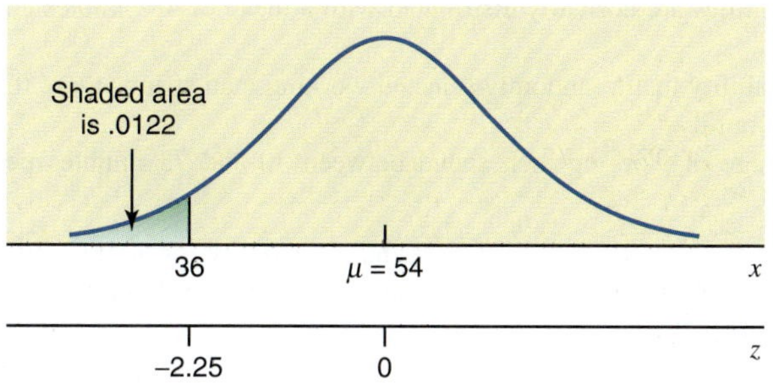

Figure 6.44 Area to the left of x = 36.

The required percentage is given by the area under the standard normal curve to the left of $z = -2.25$. From Table IV of Appendix C, this area is .0122. Hence, the required probability is

$$P(x < 36) = P(z < -2.25) = \mathbf{.0122}$$

The probability that any randomly selected calculator manufactured by Texas Instruments will start to malfunction within 36 months is .0122. Converting this probability to a percentage, we can state that approximately 1.22% of all calculators manufactured by this company are expected to start malfunctioning within 36 months. Hence, 1.22% of the calculators are expected to be replaced. ■

EXERCISES

APPLICATIONS

6.37 Let x denote the time it takes to run a road race. Suppose x is approximately normally distributed with a mean of 190 minutes and a standard deviation of 21 minutes. If one runner is selected at random, what is the probability that this runner will complete this road race

a. in less than 150 minutes? b. in 205 to 245 minutes?

6.38 According to the American Time Use Survey by the Bureau of Labor Statistics, married men aged 25 to 54 who work full time sleep an average of 7.43 hours per night (*The New York Times,* September 15, 2004). Assume that currently the times that such men sleep per night have a normal distribution with a mean of 7.43 hours and a standard deviation of .80 hours. Find the percentage of such men who sleep

a. between 6.5 and 8.5 hours b. less than 6 hours

6.39 According to data from the NPD Group, consumers planned to spend an average of $655 on holiday shopping in 2004 (*USA TODAY,* November 26, 2004). Assume that for the next holiday season, shoppers' planned expenditures will be normally distributed with a mean of $810 and a standard deviation of $155. Find the probability that the planned expenditure for the next holiday season of a randomly selected shopper is

a. more than $1000 b. between $620 and $940

6.40 Tommy Wait, a minor league baseball pitcher, is notorious for taking an excessive amount of time between pitches. In fact, his times between pitches are normally distributed with a mean of 36 seconds and a standard deviation of 2.5 seconds. What percentage of his times between pitches are

a. longer than 40 seconds? b. between 30 and 34 seconds?

6.41 A construction zone on a highway has a posted speed limit of 40 miles per hour. The speeds of vehicles passing through this construction zone are normally distributed with a mean of 46 miles per hour and a standard deviation of 4 miles per hour. Find the percentage of vehicles passing through this construction zone that are

a. exceeding the posted speed limit
b. traveling at speeds between 50 and 55 miles per hour

6.42 The Bank of Connecticut issues Visa and MasterCard credit cards. It is estimated that the balances on all Visa credit cards issued by the Bank of Connecticut have a mean of $845 and a standard deviation of $270. Assume that the balances on all these Visa cards follow a normal distribution.

a. What is the probability that a randomly selected Visa card issued by this bank has a balance between $1000 and $1400?
b. What percentage of the Visa cards issued by this bank have a balance of $750 or more?

6.43 According to the *New York Times,* working men spend an average of 48 minutes per day caring for their families (*Time,* September 27, 2004). Assume that the times that working men currently spend per day caring for their families are normally distributed with a mean of 48 minutes and a standard deviation of 11 minutes.

a. Find the probability that a randomly selected working man spends more than 68 minutes per day caring for his family.
b. What percentage of working men spend between 30 and 73 minutes per day caring for their families?

6.44 The transmission on a model of a specific car has a warranty for 40,000 miles. It is known that the life of such a transmission has a normal distribution with a mean of 72,000 miles and a standard deviation of 12,000 miles.

a. What percentage of the transmissions will fail before the end of the warranty period?
b. What percentage of the transmissions will be good for more than 100,000 miles?

6.45 According to the records of an electric company serving the Boston area, the mean electric consumption for all households during winter is 1650 kilowatt-hours per month. Assume that the monthly electric consumptions during winter by all households in this area have a normal distribution with a mean of 1650 kilowatt-hours and a standard deviation of 320 kilowatt-hours.

a. Find the probability that the monthly electric consumption during winter by a randomly selected household from this area is less than 1850 kilowatt-hours.
b. What percentage of the households in this area have a monthly electric consumption of 900 to 1340 kilowatt-hours?

6.46 The management of a supermarket wants to adopt a new promotional policy of giving a free gift to every customer who spends more than a certain amount per visit at this supermarket. The expectation of the management is that after this promotional policy is advertised, the expenditures for all customers at this supermarket will be normally distributed with a mean of \$95 and a standard deviation of \$21. If the management decides to give free gifts to all those customers who spend more than \$130 at this supermarket during a visit, what percentage of the customers are expected to get free gifts?

6.47 One of the cars sold by Walt's car dealership is a very popular subcompact car called Rhino. The final sale price of the basic model of this car varies from customer to customer depending on the negotiating skills and persistence of the customer. Assume that these sale prices of this car are normally distributed with a mean of \$19,800 and a standard deviation of \$300.

a. Dolores paid \$19,445 for her Rhino. What percentage of Walt's customers paid less than Dolores for a Rhino?

b. Cuthbert paid \$20,300 for a Rhino. What percentage of Walt's customers paid more than Cuthbert for a Rhino?

6.48 A psychologist has devised a stress test for dental patients sitting in the waiting rooms. According to this test, the stress scores (on a scale of 1 to 10) for patients waiting for root canal treatments are found to be approximately normally distributed with a mean of 7.59 and a standard deviation of .73.

a. What percentage of such patients have a stress score lower than 6.0?

b. What is the probability that a randomly selected root canal patient sitting in the waiting room has a stress score between 7.0 and 8.0?

c. The psychologist suggests that any patient with a stress score of 9.0 or higher should be given a sedative prior to treatment. What percentage of patients waiting for root canal treatments would need a sedative if this suggestion is accepted?

6.49 According to the Kaiser Family Foundation, the average amount of time spent watching TV by 8-to-18-year-olds is 231 minutes per day (*USA TODAY*, March 10, 2005). Suppose currently the times spent watching TV by all 8-to-18-year-olds have a normal distribution with a mean of 231 minutes and a standard deviation of 45 minutes. What percentage of the 8-to-18-year-olds watch TV for

a. more than 290 minutes per day?
b. less than 150 minutes per day?
c. 180 to 320 minutes per day?
d. 270 to 350 minutes per day?

6.50 Fast Auto Service guarantees that the maximum waiting time for its customers is 20 minutes for oil and lube service on their cars. It also guarantees that any customer who has to wait longer than 20 minutes for this service will receive a 50% discount on the charges. It is estimated that the mean time taken for oil and lube service at this garage is 15 minutes per car and the standard deviation is 2.4 minutes. Suppose the time taken for oil and lube service on a car follows a normal distribution.

a. What percentage of the customers will receive the 50% discount on their charges?

b. Is it possible that a car may take longer than 25 minutes for oil and lube service? Explain.

6.51 The lengths of 3-inch nails manufactured on a machine are normally distributed with a mean of 3.0 inches and a standard deviation of .009 inches. The nails that are either shorter than 2.98 inches or longer than 3.02 inches are unusable. What percentage of all the nails produced by this machine are unusable?

6.52 The pucks used by the National Hockey League for ice hockey must weigh between 5.5 and 6.0 ounces. Suppose the weights of pucks produced at a factory are normally distributed with a mean of 5.75 ounces and a standard deviation of .11 ounces. What percentage of the pucks produced at this factory cannot be used by the National Hockey League?

6.6 Determining the z and x Values When an Area under the Normal Distribution Curve Is Known

So far in this chapter we have discussed how to find the area under a normal distribution curve for an interval of z or x. Now we reverse this procedure and learn how to find the corresponding value of z or x when an area under a normal distribution curve is known. Examples 6–15 through 6–17 describe this procedure for finding the z value.

■ EXAMPLE 6–15

Find a point z such that the area under the standard normal curve to the left of z is .9251.

Finding z when the area to the left of z is known.

Solution As shown in Figure 6.45, we are to find the z value such that the area to the left of z is .9251. Since this area is greater than .50, z is positive and lies to the right of zero.

Figure 6.45 Finding the z value.

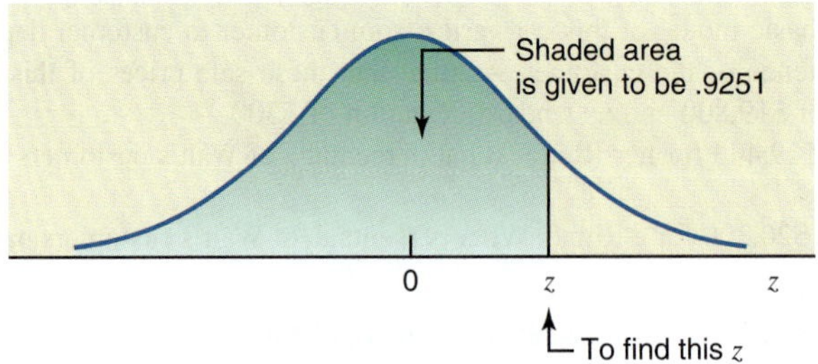

To find the required value of z, we locate .9251 in the body of the normal distribution table, Table IV of Appendix C. The relevant portion of that table is reproduced as Table 6.4 below. Next we read the numbers in the column and row for z that correspond to .9251. As shown in Table 6.4, these numbers are 1.4 and .04, respectively. Combining these two numbers, we obtain the required value of $z = 1.44$.

Table 6.4 **Finding the z Value When Area Is Known**

z	.00	.01	...	.04	...	.09
−3.4	.0003	.0003	...		...	.0002
−3.3	.0005	.0005	...		...	.0003
−3.2	.0007	.0007	...		...	.0005
.	.	.	...		...	.
.	.	.	...		...	.
.	.	.	...		...	.
1.4				.9251	...	...
.	.	.	...	.	...	.
.	.	.	...	.	...	.
.	.	.	...	.	...	.
3.4	.9997	.9997	...	.9997	...	.9998

We locate this value in Table IV of Appendix C

■ EXAMPLE 6–16

Finding z when the area in the right tail is known.

Find the value of z such that the area under the standard normal curve in the right tail is .0050.

Solution To find the required value of z, we first find the area to the left of z. Hence,

$$\text{Area to the left of } z = 1.0 - .0050 = .9950$$

This area is shown in Figure 6.46.

Now we look for .9950 in the body of the normal distribution table. Table IV does not contain .9950. So we find the value closest to .9950, which is either .9949 or .9951. We can use either of these two values. If we choose .9951, the corresponding z value is 2.58. Hence, the required value of z is **2.58**, and the area to the right of $z = 2.58$ is approximately .0050. Note that there is no apparent reason to choose .9951 and not to choose .9949. We can use either of the two values. If we choose .9949, the corresponding z value will be 2.57.

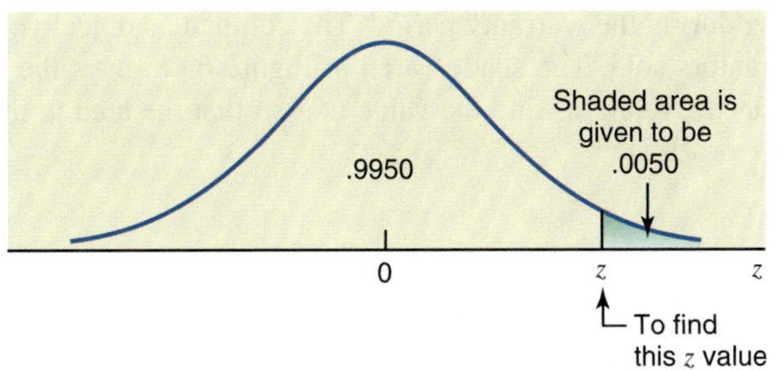

Figure 6.46 Finding the z value. ■

■ EXAMPLE 6–17

Finding z when the area in the left tail is known.

Find the value of z such that the area under the standard normal curve in the left tail is .05.

Solution Because .05 is less than .5 and it is the area in the left tail, the value of z is negative. This area is shown in Figure 6.47.

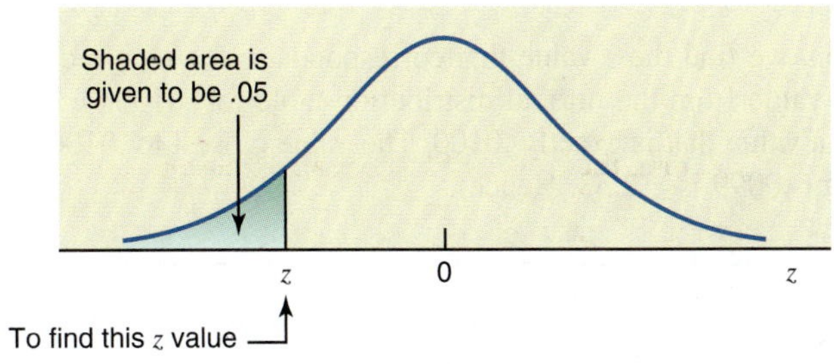

Figure 6.47 Finding the z value.

Next, we look for .0500 in the body of the normal distribution table. The value closest to .0500 in the normal distribution table is either .0505 or .0495. Suppose we use the value .0495. The corresponding z value is -1.65. Thus, the required value of z is $\mathbf{-1.65}$ and the area to the left of $z = -1.65$ is approximately .05. ■

To find an x value when an area under a normal distribution curve is given, first we find the z value corresponding to that x value from the normal distribution table. Then, to find the x value, we substitute the values of μ, σ, and z in the following formula, which is obtained from $z = (x - \mu)/\sigma$ by doing some algebraic manipulations. Also, if we know the values of x, z, and σ, we can find μ using this same formula. Exercises 6.63 and 6.64 present such cases.

Finding an x Value for a Normal Distribution For a normal curve, with known values of μ and σ and for a given area under the curve between the mean and x, the x value is calculated as

$$x = \mu + z\sigma$$

Examples 6–18 and 6–19 illustrate how to find an x value when an area under a normal distribution curve is known.

■ EXAMPLE 6–18

Finding x when the area in the left tail is known.

Recall Example 6–14. It is known that the life of a calculator manufactured by Texas Instruments has a normal distribution with a mean of 54 months and a standard deviation of 8 months. What should the warranty period be to replace a malfunctioning calculator if the company does not want to replace more than 1% of all the calculators sold?

Solution Let x be the life of a calculator. Then, x follows a normal distribution with $\mu = 54$ months and $\sigma = 8$ months. The calculators that would be replaced are the ones that start malfunctioning during the warranty period. The company's objective is to replace at most 1% of all the calculators sold. The shaded area in Figure 6.48 gives the proportion of calculators that are replaced. We are to find the value of x so that the area to the left of x under the normal curve is 1% or .01.

Figure 6.48 Finding an x value.

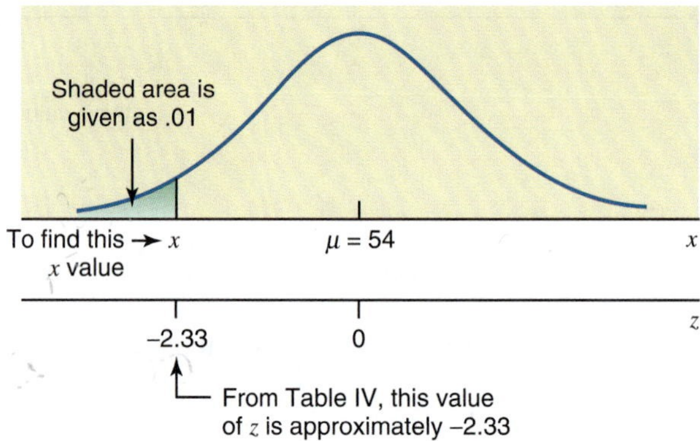

In the first step, we find the z value that corresponds to the required x value.

We find the z value from the normal distribution table for .0100. Table IV of Appendix C does not contain a value that is exactly .0100. The value closest to .0100 in the table is .0099, and the z value for .0099 is -2.33. Hence,

$$z = -2.33$$

Substituting the values of μ, σ, and z in the formula $x = \mu + z\sigma$, we obtain

$$x = \mu + z\sigma = 54 + (-2.33)(8) = 54 - 18.64 = \mathbf{35.36}$$

Thus, the company should replace all the calculators that start to malfunction within 35.36 months (which can be rounded to 35 months) of the date of purchase so that they will not have to replace more than 1% of the calculators. ■

■ EXAMPLE 6–19

Finding x when the area in the right tail is known.

Almost all high school students who intend to go to college take the SAT test. In a recent test, the mean SAT score (in verbal and mathematics) of all students was 1020. Debbie is planning to take this test soon. Suppose the SAT scores of all students who take this test with Debbie will have a normal distribution with a mean of 1020 and a standard deviation of 153. What should her score be on this test so that only 10% of all examinees score higher than she does?

Solution Let x represent the SAT scores of examinees. Then, x follows a normal distribution with $\mu = 1020$ and $\sigma = 153$. We are to find the value of x such that the area under the normal distribution curve to the right of x is 10%, as shown in Figure 6.49.

Figure 6.49 Finding an x value.

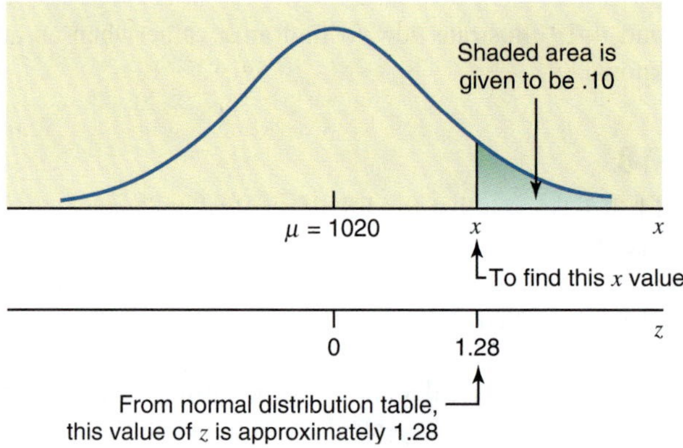

First, we find the area under the normal distribution curve to the left of the x value.

$$\text{Area to the left of the } x \text{ value} = 1.0 - .10 = .9000$$

To find the z value that corresponds to the required x value, we look for .9000 in the body of the normal distribution table. The value closest to .9000 in Table IV is .8997, and the corresponding z value is 1.28. Hence, the value of x is computed as

$$x = \mu + z\sigma = 1020 + 1.28(153) = 1020 + 195.84 = 1215.84 \approx \mathbf{1216}$$

Thus, if Debbie scores 1216 on the SAT, only about 10% of the examinees are expected to score higher than she does. ■

EXERCISES

CONCEPTS AND PROCEDURES

6.53 Find the value of z so that the area under the standard normal curve
- **a.** from 0 to z is .4772 and z is positive
- **b.** between 0 and z is (approximately) .4785 and z is negative
- **c.** in the left tail is (approximately) .3565
- **d.** in the right tail is (approximately) .1530

6.54 Find the value of z so that the area under the standard normal curve
- **a.** from 0 to z is (approximately) .1965 and z is positive
- **b.** between 0 and z is (approximately) .2740 and z is negative
- **c.** in the left tail is (approximately) .2050
- **d.** in the right tail is (approximately) .1053

6.55 Determine the value of z so that the area under the standard normal curve
- **a.** in the right tail is .0500
- **b.** in the left tail is .0250
- **c.** in the left tail is .0100
- **d.** in the right tail is .0050

6.56 Determine the value of z so that the area under the standard normal curve
- **a.** in the right tail is .0250
- **b.** in the left tail is .0500
- **c.** in the left tail is .0010
- **d.** in the right tail is .0100

6.57 Let x be a continuous random variable that follows a normal distribution with a mean of 200 and a standard deviation of 25.
- **a.** Find the value of x so that the area under the normal curve to the left of x is approximately .6330.
- **b.** Find the value of x so that the area under the normal curve to the right of x is approximately .05.
- **c.** Find the value of x so that the area under the normal curve to the right of x is .8051.
- **d.** Find the value of x so that the area under the normal curve to the left of x is .0150.
- **e.** Find the value of x so that the area under the normal curve between μ and x is .4525 and the value of x is less than μ.
- **f.** Find the value of x so that the area under the normal curve between μ and x is approximately .4800 and the value of x is greater than μ.

6.58 Let x be a continuous random variable that follows a normal distribution with a mean of 550 and a standard deviation of 75.
- **a.** Find the value of x so that the area under the normal curve to the left of x is .0250.
- **b.** Find the value of x so that the area under the normal curve to the right of x is .9345.
- **c.** Find the value of x so that the area under the normal curve to the right of x is approximately .0275.
- **d.** Find the value of x so that the area under the normal curve to the left of x is approximately .9600.
- **e.** Find the value of x so that the area under the normal curve between μ and x is approximately .4700 and the value of x is less than μ.
- **f.** Find the value of x so that the area under the normal curve between μ and x is approximately .4100 and the value of x is greater than μ.

APPLICATIONS

6.59 Fast Auto Service provides oil and lube service for cars. It is known that the mean time taken for oil and lube service at this garage is 15 minutes per car and the standard deviation is 2.4 minutes. The management wants to promote the business by guaranteeing a maximum waiting time for its customers. If a customer's car is not serviced within that period, the customer will receive a 50% discount on the charges. The company wants to limit this discount to at most 5% of the customers. What should the maximum guaranteed waiting time be? Assume that the times taken for oil and lube service for all cars have a normal distribution.

6.60 The management of a supermarket wants to adopt a new promotional policy of giving a free gift to every customer who spends more than a certain amount per visit at this supermarket. The expectation of the management is that after this promotional policy is advertised, the expenditures for all customers at this supermarket will be normally distributed with a mean of \$95 and a standard deviation of \$21. If the management wants to give free gifts to at most 10% of the customers, what should the amount be above which a customer would receive a free gift?

6.61 According to the records of an electric company serving the Boston area, the mean electric consumption during winter for all households is 1650 kilowatt-hours per month. Assume that the monthly electric consumptions during winter by all households in this area have a normal distribution with a mean of 1650 kilowatt-hours and a standard deviation of 320 kilowatt-hours. The company sent a notice to Bill Johnson informing him that about 90% of the households use less electricity per month than he does. What is Bill Johnson's monthly electric consumption?

6.62 Rockingham Corporation makes electric shavers. The life (period before which a shaver does not need a major repair) of Model J795 of an electric shaver manufactured by this corporation has a normal distribution with a mean of 70 months and a standard deviation of 8 months. The company is to determine the warranty period for this shaver. Any shaver that needs a major repair during this warranty period will be replaced free by the company.

a. What should the warranty period be if the company does not want to replace more than 1% of the shavers?

b. What should the warranty period be if the company does not want to replace more than 5% of the shavers?

***6.63** A study has shown that 20% of all college textbooks have a price of \$90 or higher. It is known that the standard deviation of the prices of all college textbooks is \$9.50. Suppose the prices of all college textbooks have a normal distribution. What is the mean price of all college textbooks?

***6.64** A machine at Keats Corporation fills 64-ounce detergent jugs. The machine can be adjusted to pour, on average, any amount of detergent into these jugs. However, the machine does not pour exactly the same amount of detergent into each jug; it varies from jug to jug. It is known that the net amount of detergent poured into each jug has a normal distribution with a standard deviation of .35 ounces. The quality control inspector wants to adjust the machine such that at least 95% of the jugs have more than 64 ounces of detergent. What should the mean amount of detergent poured by this machine into these jugs be?

6.7 The Normal Approximation to the Binomial Distribution

Recall from Chapter 5 that

1. The binomial distribution is applied to a discrete random variable.
2. Each repetition, called a trial, of a binomial experiment results in one of two possible outcomes, either a success or a failure.
3. The probabilities of the two (possible) outcomes remain the same for each repetition of the experiment.
4. The trials are independent.

The binomial formula, which gives the probability of x successes in n trials, is

$$P(x) = {}_nC_x\, p^x q^{n-x}$$

The use of the binomial formula becomes very tedious when n is large. In such cases, the normal distribution can be used to approximate the binomial probability. Note that for a binomial problem, the exact probability is obtained by using the binomial formula. If we apply the normal distribution to solve a binomial problem, the probability that we obtain is an approximation to the exact probability. The approximation obtained by using the normal distribution is very close to the exact probability when n is large and p is very close to .50. However, this does not mean that we should not use the normal approximation when p is not close to .50. The reason the approximation is closer to the exact probability when p is close to .50 is that the binomial distribution is symmetric when $p = .50$. The normal distribution is always symmetric. Hence, the two distributions are very close to each other when n is large and p is close to .50. However, this does not mean that whenever $p = .50$, the binomial distribution is the same as the normal distribution because not every symmetric bell-shaped curve is a normal distribution curve.

Normal Distribution as an Approximation to Binomial Distribution Usually, the normal distribution is used as an approximation to the binomial distribution when np and nq are both greater than 5—that is, when

$$np > 5 \quad \text{and} \quad nq > 5$$

Table 6.5 gives the binomial probability distribution of x for $n = 12$ and $p = .50$. This table is constructed using Table I of Appendix C. Figure 6.50 shows the histogram and the smoothed polygon for the probability distribution of Table 6.5. As we can observe, the histogram in Figure 6.50 is symmetric, and the curve obtained by joining the upper midpoints of the rectangles is approximately bell-shaped.

Table 6.5 The Binomial Probability Distribution for $n = 12$ and $p = .50$

x	$P(x)$
0	.0002
1	.0029
2	.0161
3	.0537
4	.1208
5	.1934
6	.2256
7	.1934
8	.1208
9	.0537
10	.0161
11	.0029
12	.0002

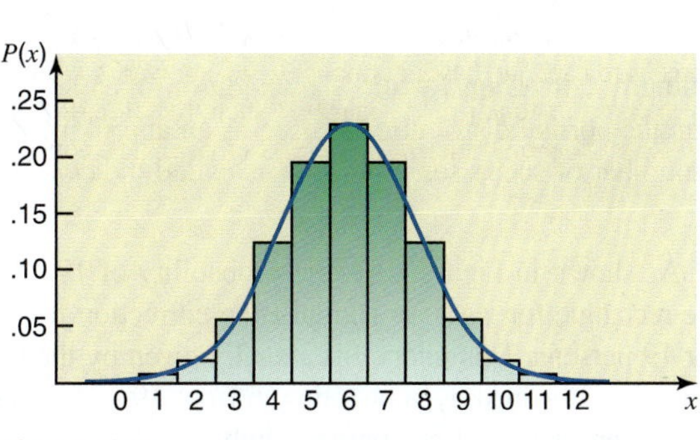

Figure 6.50 Histogram for the probability distribution of Table 6.5.

Examples 6–20 through 6–22 illustrate the application of the normal distribution as an approximation to the binomial distribution.

■ EXAMPLE 6–20

Using the normal approximation to the binomial: x equals a specific value.

According to an estimate, 50% of the people in America have at least one credit card. If a random sample of 30 persons is selected, what is the probability that 19 of them will have at least one credit card?

Solution Let n be the total number of persons in the sample, x be the number of persons in the sample who have at least one credit card, and p be the probability that a person has at least one credit card. Then, this is a binomial problem with

$$n = 30, \quad p = .50, \quad q = 1 - p = .50,$$

$$x = 19, \quad n - x = 30 - 19 = 11$$

From the binomial formula, the exact probability that 19 persons in a sample of 30 have at least one credit card is

$$P(19) = {}_{30}C_{19}(.50)^{19}(.50)^{11} = \mathbf{.0509}$$

Now let us solve this problem using the normal distribution as an approximation to the binomial distribution. For this example,

$$np = 30(.50) = 15 \quad \text{and} \quad nq = 30(.50) = 15$$

Because np and nq are both greater than 5, we can use the normal distribution as an approximation to solve this binomial problem. We perform the following three steps.

Step 1. *Compute μ and σ for the binomial distribution.*

To use the normal distribution, we need to know the mean and standard deviation of the distribution. Hence, the first step in using the normal approximation to the binomial distribution is to compute the mean and standard deviation of the binomial distribution. As we know from Chapter 5, the mean and standard deviation of a binomial distribution are given by np and $\sqrt{npq}$, respectively. Using these formulas, we obtain

$$\mu = np = 30(.50) = 15$$
$$\sigma = \sqrt{npq} = \sqrt{30(.50)(.50)} = 2.73861279$$

Step 2. *Convert the discrete random variable to a continuous random variable.*

The normal distribution applies to a continuous random variable, whereas the binomial distribution applies to a discrete random variable. The second step in applying the normal approximation to the binomial distribution is to convert the discrete random variable to a continuous random variable by making the **correction for continuity**.

Definition

Continuity Correction Factor The addition of .5 and/or subtraction of .5 from the value(s) of x when the normal distribution is used as an approximation to the binomial distribution, where x is the number of successes in n trials, is called the *continuity correction factor.*

As shown in Figure 6.51, the probability of 19 successes in 30 trials is given by the area of the rectangle for $x = 19$. To make the correction for continuity, we use the interval 18.5 to 19.5 for 19 persons. This interval is actually given by the two boundaries of the rectangle for $x = 19$, which are obtained by subtracting .5 from 19 and by adding .5 to 19. Thus, $P(x = 19)$ for the binomial problem will be approximately equal to $P(18.5 \le x \le 19.5)$ for the normal distribution.

Figure 6.51

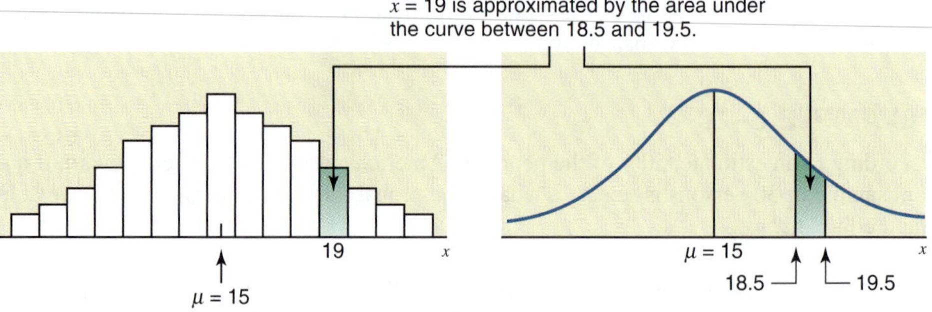
The area contained by the rectangle for x = 19 is approximated by the area under the curve between 18.5 and 19.5.
19
x
μ = 15
μ = 15
x
18.5
19.5

Step 3. *Compute the required probability using the normal distribution.*

As shown in Figure 6.52, the area under the normal distribution curve between $x = 18.5$ and $x = 19.5$ will give us the (approximate) probability that 19 persons have at least one credit card. We calculate this probability as follows:

$$\text{For } x = 18.5: \quad z = \frac{18.5 - 15}{2.73861279} = 1.28$$

$$\text{For } x = 19.5: \quad z = \frac{19.5 - 15}{2.73861279} = 1.64$$

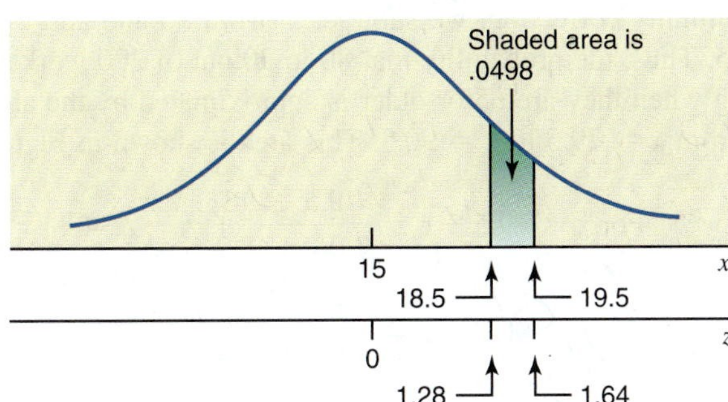

Figure 6.52 Area between $x = 18.5$ and $x = 19.5$.

The required probability is given by the area under the standard normal curve between $z = 1.28$ and $z = 1.64$. This area is obtained by subtracting the area to the left of $z = 1.28$ from the area to the left of $z = 1.64$. From Table IV of Appendix C, the area to the left of $z = 1.28$ is .8997 and the area to the left of $z = 1.64$ is .9495. Hence, the required probability is

$$P(18.5 \leq x \leq 19.5) = P(1.28 \leq z \leq 1.64) = .9495 - .8997 = \mathbf{.0498}$$

Thus, based on the normal approximation, the probability that 19 persons in a sample of 30 will have at least one credit card is approximately .0498. Earlier, using the binomial formula, we obtained the exact probability .0509. The error due to using the normal approximation is $.0509 - .0498 = .0011$. Thus, the exact probability is underestimated by .0011 if the normal approximation is used. ■

◀ *Remember*

When applying the normal distribution as an approximation to the binomial distribution, always make a *correction for continuity.* The continuity correction is made by subtracting .5 from the lower limit of the interval and/or by adding .5 to the upper limit of the interval. For example, the binomial probability $P(7 \leq x \leq 12)$ will be approximated by the probability $P(6.5 \leq x \leq 12.5)$ for the normal distribution; the binomial probability $P(x \geq 9)$ will be approximated by the probability $P(x \geq 8.5)$ for the normal distribution; and the binomial probability $P(x \leq 10)$ will be approximated by the probability $P(x \leq 10.5)$ for the normal distribution. Note that the probability $P(x \geq 9)$ has only the lower limit of 9 and no upper limit, and the probability $P(x \leq 10)$ has only the upper limit of 10 and no lower limit.

■ EXAMPLE 6–21

Using the normal approximation to the binomial: x assumes a value in an interval

According to the Employee Benefit Research Institute, 48% of American workers think that they will still be working when they are 65 or older (*Money*, February 2005). Suppose this result is true for the current population of all American workers. What is the probability that in a random sample of 200 American workers, 80 to 90 will say that they will still be working when they are 65 or older?

Solution Let n be the total number of workers in the sample, x be the number of workers in the sample who say that they will still be working when they are 65 or older, and p be the probability that a worker says that he/she will still be working at age 65 or older. Then, this is a binomial problem with

$$n = 200, \quad p = .48, \quad \text{and} \quad q = 1 - .48 = .52$$

We are to find the probability of 80 to 90 successes in 200 trials. Because n is large, it is easier to apply the normal approximation than to use the binomial formula. We can check that np and nq are both greater than 5. The mean and standard deviation of the binomial distribution are

$$\mu = np = 200(.48) = 96$$
$$\sigma = \sqrt{npq} = \sqrt{200(.48)(.52)} = 7.06540869$$

To make the continuity correction, we subtract .5 from 80 and add .5 to 90 to obtain the interval 79.5 to 90.5. Thus, the probability that 80 to 90 out of 200 workers will say that they will still be working when they are 65 or older is approximated by the area under the normal distribution curve from $x = 79.5$ to $x = 90.5$. This area is shown in Figure 6.53.

$$\text{For } x = 79.5\text{:} \quad z = \frac{79.5 - 96}{7.06540869} = -2.34$$

$$\text{For } x = 90.5\text{:} \quad z = \frac{90.5 - 96}{7.06540869} = -.78$$

Figure 6.53 Area between $x = 79.5$ and $x = 90.5$.

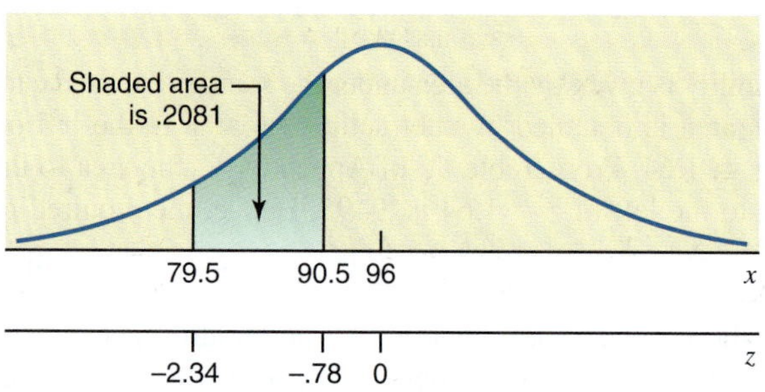

The required probability is given by the area under the standard normal curve between $z = -2.34$ and $z = -.78$. This area is obtained by taking the difference between the areas under the standard normal curve to the left of $z = -2.34$ and to the left of $z = -.78$. From Table IV in Appendix C, the area to the left of $z = -2.34$ is .0096, and the area to the left of $z = -.78$ is .2177. Hence, the required probability is

$$P(79.5 \le x \le 90.5) = P(-2.34 \le z \le -.78) = .2177 - .0096 = \mathbf{.2081}$$

Thus, the probability that 80 to 90 workers in a sample of 200 will say that they will still be working when they are 65 or older is approximately. 2081. ■

■ EXAMPLE 6–22

Using the normal approximation to the binomial: x is greater than or equal to a value.

According to an Accenture survey, 44% of adult Americans said that the service provided by the United States Postal Service is *generally satisfactory* (*Business Week*, November 29, 2004). Assume that this percentage is true for the current population of adult Americans. What is the probability that 140 or more adult Americans in a random sample of 300 will hold this opinion?

Solution Let n be the number of adults in the sample, x be the number of adults in the sample who hold the said opinion, and p be the probability that a randomly selected adult holds the said opinion. Then, this is a binomial problem with

$$n = 300, \quad p = .44, \quad \text{and} \quad q = 1 - .44 = .56$$

We are to find the probability of 140 or more successes in 300 trials. The mean and standard deviation of the binomial distribution are

$$\mu = np = 300(.44) = 132$$
$$\sigma = \sqrt{npq} = \sqrt{300(.44)(.56)} = 8.59767410$$

For the continuity correction, we subtract .5 from 140, which gives 139.5. Thus, the probability that 140 or more of the adults in a random sample of 300 will hold the said opinion is approximated by the area under the normal distribution curve to the right of $x = 139.5$, as shown in Figure 6.54.

$$\text{For } x = 139.5: \quad z = \frac{139.5 - 132}{8.59767410} = .87$$

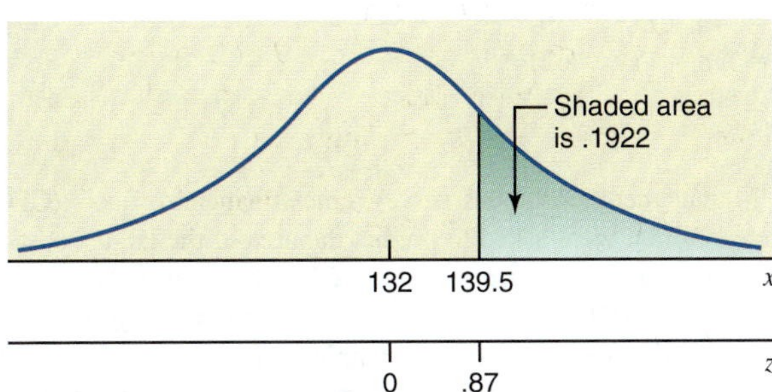

Figure 6.54 Area to the right of $x = 139.5$.

To find the required probability, we find the area to the left of $z = .87$ and subtract this area from 1.0. From Table IV in Appendix C, the area to the left of $z = .87$ is .8078. Hence,

$$P(x \geq 139.5) = P(z \geq .87) = 1.0 - .8078 = \mathbf{.1922}$$

Thus, the probability that 140 or more of adult Americans in a random sample of 300 will say that the service provided by the United States Postal Service is *generally satisfactory* is approximately .1922. ■

EXERCISES

CONCEPTS AND PROCEDURES

6.65 Under what conditions is the normal distribution usually used as an approximation to the binomial distribution?

6.66 For a binomial probability distribution, $n = 20$ and $p = .60$.

- **a.** Find the probability $P(x = 12)$ by using the table of binomial probabilities (Table I of Appendix C).
- **b.** Find the probability $P(x = 12)$ by using the normal distribution as an approximation to the binomial distribution. What is the difference between this approximation and the exact probability calculated in part a?

6.67 For a binomial probability distribution, $n = 25$ and $p = .40$.

- **a.** Find the probability $P(8 \leq x \leq 13)$ by using the table of binomial probabilities (Table I of Appendix C).
- **b.** Find the probability $P(8 \leq x \leq 13)$ by using the normal distribution as an approximation to the binomial distribution. What is the difference between this approximation and the exact probability calculated in part a?

6.68 For a binomial probability distribution, $n = 80$ and $p = .50$. Let x be the number of successes in 80 trials.

- **a.** Find the mean and standard deviation of this binomial distribution.
- **b.** Find $P(x \geq 40)$ using the normal approximation.
- **c.** Find $P(41 \leq x \leq 48)$ using the normal approximation.

6.69 For a binomial probability distribution, $n = 120$ and $p = .60$. Let x be the number of successes in 120 trials.

- **a.** Find the mean and standard deviation of this binomial distribution.
- **b.** Find $P(x \leq 72)$ using the normal approximation.
- **c.** Find $P(67 \leq x \leq 73)$ using the normal approximation.

6.70 Find the following binomial probabilities using the normal approximation.

a. $n = 140, \quad p = .45, \quad P(x = 67)$
b. $n = 100, \quad p = .55, \quad P(52 \le x \le 60)$
c. $n = 90, \quad p = .42, \quad P(x \ge 40)$
d. $n = 104, \quad p = .75, \quad P(x \le 72)$

6.71 Find the following binomial probabilities using the normal approximation.

a. $n = 70, \quad p = .30, \quad P(x = 18)$
b. $n = 200, \quad p = .70, \quad P(133 \le x \le 145)$
c. $n = 85, \quad p = .40, \quad P(x \ge 30)$
d. $n = 150, \quad p = .38, \quad P(x \le 62)$

APPLICATIONS

6.72 In a Robert Half Management Resources survey, chief financial officers (CFOs) at companies with revenues of \$500,000 to \$1 billion were asked to predict the area of the largest cost increase for their companies in the next 12 months. Forty-five percent of these CFOs felt that the largest increase would be from employee health care plans (*USA TODAY*, January 25, 2005). Suppose that this result holds true for all such current CFOs. Find the probability that in a random sample of 400 such CFOs, 187 to 196 believe that the largest cost increase for their companies in the next 12 months will be from employee health care plans.

6.73 According to a study by the nonpartisan group Public Agenda, 34% of teachers in the United States have considered quitting their jobs due to lack of discipline in their schools (*PARADE*, August 1, 2004). Assume that this percentage is true for the current population of teachers in the United States. Find the probability that in a random sample of 300 teachers, 84 to 94 have considered quitting their jobs due to lack of discipline in their schools.

6.74 Many Americans are turning to alternative health care medicines such as prayer, herbs, acupuncture, massage, and yoga to cure diseases. A government survey of 31,000 adults found that 36% of them use various types of *complementary and alternative* therapies such as the ones mentioned above (*The Hartford Courant*, May 28, 2004). The survey was commissioned by the National Center for Complementary and Alternative Medicine of the National Institutes of Health, and was conducted by the National Center for Health Statistics at the Centers for Disease Control and Prevention. Assume that the results of this survey hold true for the current population of adult Americans. Find the probability that in a random sample of 600 adult Americans, the number who use some kind of complementary and alternative therapy is

a. exactly 200   b. 225 to 240   c. at most 230

6.75 In a *CSO Magazine*/U.S. Secret Service Electronic Crimes Task Force/CERT Coordination Center survey of chief security officers (CSOs) conducted in April 2004, 43% of them said that electronic crime had increased at their companies during the previous year (*USA TODAY*, July 8, 2004). Assume that this percentage is true for the current population of CSOs. Find the probability that in a random sample of 400 CSOs, the number who will hold the above mentioned opinion is

a. exactly 175   b. 184 or more   c. 160 to 180

6.76 In a telephone poll conducted by TNS of Horsham, PA, for *Time*/ABC News in May 2004, adult Americans were asked whether or not they supported a law requiring restaurants to list the calorie count and fat content for all items on their menus. Of the respondents, 61% said they support the law *strongly* or *somewhat* (*Time*, June 7, 2004). Assume that this percentage holds true for the current population of adult Americans. Find the probability that in a random sample of 600 adult Americans, the number who favor such a law *strongly* or *somewhat* is

a. exactly 350   b. at most 360   c. 345 to 375

6.77 An office supply company conducted a survey before marketing a new paper shredder designed for home use. In the survey, 80% of the people who tried the shredder were satisfied with it. Because of this high satisfaction rate, the company decided to market the new shredder. Assume that 80% of all people are satisfied with this shredder. During a certain month, 100 customers bought this shredder. Find the probability that of these 100 customers, the number who are satisfied is

a. exactly 75   b. 73 or fewer   c. 74 to 85

6.78 Johnson Electronics makes calculators. Consumer satisfaction is one of the top priorities of the company's management. The company guarantees the refund of money or a replacement for any calculator that malfunctions within two years from the date of purchase. It is known from past data that despite all efforts, 5% of the calculators manufactured by this company malfunction within a two-year period. The company recently mailed 500 such calculators to its customers.

a. Find the probability that exactly 29 of the 500 calculators will be returned for refund or replacement within a two-year period.

b. What is the probability that 27 or more of the 500 calculators will be returned for refund or replacement within a two-year period?

c. What is the probability that 15 to 22 of the 500 calculators will be returned for refund or replacement within a two-year period?

6.79 Hurbert Corporation makes font cartridges for laser printers that it sells to Alpha Electronics Inc. The cartridges are shipped to Alpha Electronics in large volumes. The quality control department at Alpha Electronics randomly selects 100 cartridges from each shipment and inspects them for being good or defective. If this sample contains seven or more defective cartridges, the entire shipment is rejected. Hurbert Corporation promises that of all the cartridges, only 5% are defective.

a. Find the probability that a given shipment of cartridges received by Alpha Electronics will be accepted.

b. Find the probability that a given shipment of cartridges received by Alpha Electronics will not be accepted.

USES AND MISUSES... DON'T LOSE YOUR MEMORY

As discussed in the previous chapter, the Poisson distribution gives the probability of a specified number of events occurring in a time interval. The Poisson distribution provides a model for the number of emails a server might receive during a certain time period or the number of people arriving in line at a bank during lunch hour. These are nice to know for planning purposes, but sometimes we want to know the specific times at which emails or customers arrive. These times are governed by a special continuous probability distribution with certain unusual properties. This distribution is called the *exponential distribution*, and it is derived from the Poisson probability distribution.

Suppose you are a teller at a bank, and a customer has just arrived. You know that the customers arrive according to a Poisson process with a rate of λ customers per hour. Your boss might care how many customers arrive on average during a given time interval to ensure there are enough tellers available to handle the customers efficiently; you are more concerned with the time when the next customer will arrive. Remember that the probability that x customers arrive in an interval of length t is

$$P(x) = \frac{(\lambda t)^x e^{-\lambda t}}{x!}.$$

The probability that a customer arrives within time t is 1 minus the probability that no customer arrives within time t. Hence,

$$P(\text{customer arrives within time } t) = 1 - P(0)$$
$$= 1 - \frac{(\lambda t)^0 e^{-\lambda t}}{0!} = 1 - e^{-\lambda t}$$

If the bank receives an average of 15 customers per hour—an average of one every four minutes—and a customer has just arrived, the probability that a customer arrives within four minutes is $1 - e^{-\lambda t} = 1 - e^{-(15/60)4} = .6321$. The same way, the probability that a customer arrives within eight minutes is .8647.

Let us say that a customer arrived and went to your co-worker's window. No additional customer arrived within the next two minutes—an event with probability .6065—and you dozed off for two more minutes. When you open your eyes, you see that a customer has not arrived yet. What is the probability that a customer arrives within the next four minutes? From the calculation above, you might say that the answer is .8647. After all, you know that a customer arrived eight minutes earlier. But .8647 is not the correct answer.

The exponential distribution, which governs the time between arrivals of a Poisson process, is known as a *memoryless distribution*. For you as a bank teller, this means that if you know a customer has not arrived during the past four minutes, then the clock is reset to zero, as if the previous customer had just arrived. So even after your nap, the probability that a customer arrives within four minutes is .6321. This interesting property reminds us again that we should be careful when we use mathematics to model real-world phenomena.

Glossary

Continuity correction factor Addition of .5 and/or subtraction of .5 from the value(s) of x when the normal distribution is used as an approximation to the binomial distribution, where x is the number of successes in n trials.

Continuous random variable A random variable that can assume any value in one or more intervals.

Normal probability distribution The probability distribution of a continuous random variable that, when plotted, gives a specific bell-shaped curve. The parameters of the normal distribution are the mean μ and the standard deviation σ.

Standard normal distribution The normal distribution with $\mu = 0$ and $\sigma = 1$. The units of the standard normal distribution are denoted by z.

z value or **z score** The units of the standard normal distribution that are denoted by z.

Supplementary Exercises

6.80 The management at Ohio National Bank does not want its customers to wait in line for service for too long. The manager of a branch of this bank estimated that the customers currently have to wait an average of 8 minutes for service. Assume that the waiting times for all customers at this branch have a normal distribution with a mean of 8 minutes and a standard deviation of 2 minutes.

- **a.** Find the probability that a randomly selected customer will have to wait for less than 3 minutes.
- **b.** What percentage of the customers have to wait for 10 to 13 minutes?
- **c.** What percentage of the customers have to wait for 6 to 12 minutes?
- **d.** Is it possible that a customer may have to wait longer than 16 minutes for service? Explain.

6.81 A company that has a large number of supermarket grocery stores claims that customers who pay by personal checks spend an average of $87 on groceries at these stores with a standard deviation of $22. Assume that the expenses incurred on groceries by all such customers at these stores are normally distributed.

- **a.** Find the probability that a randomly selected customer who pays by check spends more than $114 on groceries.
- **b.** What percentage of customers paying by check spend between $40 and $60 on groceries?
- **c.** What percentage of customers paying by check spend between $70 and $105?
- **d.** Is it possible for a customer paying by check to spend more than $185? Explain.

6.82 At Jen and Perry Ice Cream Company, the machine that fills 1-pound cartons of Top Flavor ice cream is set to dispense 16 ounces of ice cream into every carton. However, some cartons contain slightly less than and some contain slightly more than 16 ounces of ice cream. The amounts of ice cream in all such cartons have a normal distribution with a mean of 16 ounces and a standard deviation of .18 ounces.

- **a.** Find the probability that a randomly selected carton contains 16.20 to 16.50 ounces of ice cream.
- **b.** What percentage of such cartons contain less than 15.70 ounces of ice cream?
- **c.** Is it possible for a carton to contain less than 15.20 ounces of ice cream? Explain.

6.83 A machine at Kasem Steel Corporation makes iron rods that are supposed to be 50 inches long. However, the machine does not make all rods of exactly the same length. It is known that the probability distribution of the lengths of rods made on this machine is normal with a mean of 50 inches and a standard deviation of .06 inches. The rods that are either shorter than 49.85 inches or longer than 50.15 inches are discarded. What percentage of the rods made on this machine are discarded?

6.84 Jenn Bard, who lives in San Francisco Bay area, commutes by car from home to work. She has found out that it takes her an average of 28 minutes for this commute in the morning. However, due to the variability in the traffic situation every morning, the standard deviation of these commutes is 5 minutes. Suppose the population of her morning commute times has a normal distribution with a mean of 28 minutes and a standard deviation of 5 minutes. Jenn has to be at work by 8:30 A.M. every morning. By what time must she leave home in the morning so that she is late for work at most 1% of the time?

6.85 The print on the package of 100-watt General Electric soft-white lightbulbs states that these bulbs have an average life of 750 hours. Assume that the lives of all such bulbs have a normal distribution with a mean of 750 hours and a standard deviation of 50 hours.

- **a.** Let x be the life of such a lightbulb. Find x so that only 2.5% of such lightbulbs have lives longer than this value.
- **b.** Let x be the life of such a lightbulb. Find x so that about 80% of such lightbulbs have lives shorter than this value.

6.86 Major League Baseball rules require that the balls used in baseball games must have circumferences between 9 and 9.25 inches. Suppose the balls produced by the factory that supplies balls to Major League Baseball have circumferences normally distributed with a mean of 9.125 inches and a standard deviation of .06 inches. What percentage of these baseballs fail to meet the circumference requirement?

6.87 Mong Corporation makes auto batteries. The company claims that 80% of its LL70 batteries are good for 70 months or longer.

- **a.** What is the probability that in a sample of 100 such batteries, exactly 85 will be good for 70 months or longer?
- **b.** Find the probability that in a sample of 100 such batteries, at most 74 will be good for 70 months or longer.
- **c.** What is the probability that in a sample of 100 such batteries, 75 to 87 will be good for 70 months or longer?
- **d.** Find the probability that in a sample of 100 such batteries, 72 to 77 will be good for 70 months or longer.

6.88 Stress on the job is a major concern of a large number of people who go into managerial positions. It is estimated that 80% of the managers of all companies suffer from job-related stress.

a. What is the probability that in a sample of 200 managers of companies, exactly 150 suffer from job-related stress?
b. Find the probability that in a sample of 200 managers of companies, at least 170 suffer from job-related stress.
c. What is the probability that in a sample of 200 managers of companies, 165 or less suffer from job-related stress?
d. Find the probability that in a sample of 200 managers of companies, 164 to 172 suffer from job-related stress.

Advanced Exercises

6.89 It is known that 15% of all homeowners pay a monthly mortgage of more than \$2500 and that the standard deviation of the monthly mortgage payments of all homeowners is \$350. Suppose that the monthly mortgage payments of all homeowners have a normal distribution. What is the mean monthly mortgage paid by all homeowners?

6.90 At Jen and Perry Ice Cream Company, a machine fills one-pound cartons of Top Flavor ice cream. The machine can be set to dispense, on average, any amount of ice cream into these cartons. However, the machine does not put exactly the same amount of ice cream into each carton; it varies from carton to carton. It is known that the amount of ice cream put into each such carton has a normal distribution with a standard deviation of .18 ounces. The quality control inspector wants to set the machine such that at least 90% of the cartons have more than 16 ounces of ice cream. What should be the mean amount of ice cream put into these cartons by this machine?

6.91 Two companies, A and B, drill wells in a rural area. Company A charges a flat fee of \$3500 to drill a well regardless of its depth. Company B charges \$1000 plus \$12 per foot to drill a well. The depths of wells drilled in this area have a normal distribution with a mean of 250 feet and a standard deviation of 40 feet.

a. What is the probability that Company B would charge more than Company A to drill a well?
b. Find the mean amount charged by Company B to drill a well.

6.92 Otto is trying out for the javelin throw to compete in the Olympics. The lengths of his javelin throws are normally distributed with a mean of 290 feet and a standard deviation of 10 feet. What is the probability that the longest of three of his throws is 320 feet or more?

6.93 Lori just bought a new set of four tires for her car. The life of each tire is normally distributed with a mean of 45,000 miles and a standard deviation of 2000 miles. Find the probability that all four tires will last at least 46,000 miles. Assume that the life of each of these tires is independent of the lives of other tires.

6.94 The Jen and Perry Ice Cream company makes a gourmet ice cream. Although the law allows ice cream to contain up to 50% air, this product is designed to contain only 20% air. Because of variability inherent in the manufacturing process, management is satisfied if each pint contains between 18% and 22% air. Currently two of Jen and Perry's plants are making gourmet ice cream. At Plant A, the mean amount of air per pint is 20% with a standard deviation of 2%. At Plant B, the mean amount of air per pint is 19% with a standard deviation of 1%. Assuming the amount of air is normally distributed at both plants, which plant is producing the greater proportion of pints that contain between 18% and 22% air?

6.95 The highway police in a certain state are using aerial surveillance to control speeding on a highway with a posted speed limit of 55 miles per hour. Police officers watch cars from helicopters above a straight segment of this highway that has large marks painted on the pavement at 1-mile intervals. After the police officers observe how long a car takes to cover the mile, a computer estimates that car's speed. Assume that the errors of these estimates are normally distributed with a mean of 0 and a standard deviation of 2 miles per hour.

a. The state police chief has directed his officers not to issue a speeding citation unless the aerial unit's estimate of speed is at least 65 miles per hour. What is the probability that a car traveling at 60 miles per hour or slower will be cited for speeding?
b. Suppose the chief does not want his officers to cite a car for speeding unless they are 99% sure that it is traveling at 60 miles per hour or faster. What is the minimum estimate of speed at which a car should be cited for speeding?

6.96 Ashley knows that the time it takes her to commute to work is approximately normally distributed with a mean of 45 minutes and a standard deviation of 3 minutes. What time must she leave home in the morning so that she is 95% sure of arriving at work by 9 A.M.?

6.97 A soft-drink vending machine is supposed to pour eight ounces of the drink into a paper cup. However, the actual amount poured into a cup varies. The amount poured into a cup follows a normal distribution with a mean that can be set to any desired amount by adjusting the machine. The standard deviation of the amount poured is always .07 ounces regardless of the mean amount. If the owner of the machine wants to be 99% sure that the amount in each cup is 8 ounces or more, to what level should she set the mean?

6.98 A newspaper article reported that the mean mathematics score on SAT for students from a local high school was 500 and that 20% of the students scored below 430. Assume that the SAT scores for students from this school follow a normal distribution.

a. Find the standard deviation of the mathematics SAT scores for students from this school.
b. Find the percentage of students at this school whose mathematics SAT scores were above 520.

6.99 Alpha Corporation is considering two suppliers to secure the large amounts of steel rods that it uses. Company A produces rods with a mean diameter of 8 mm and a standard deviation of .15 mm and sells 10,000 rods for $400. Company B produces rods with a mean diameter of 8 mm and a standard deviation of .12 mm and sells 10,000 rods for $460. A rod is usable only if its diameter is between 7.8 mm and 8.2 mm. Assume that the diameters of the rods produced by each company have a normal distribution. Which of the two companies should Alpha Corporation use as a supplier? Justify your answer with appropriate calculations.

6.100 A gambler is planning to make a sequence of bets on a roulette wheel. Note that a roulette wheel has 38 numbers of which 18 are red, 18 are black, and 2 are green. Each time the wheel is spun, each of the 38 numbers is equally likely to occur. The gambler will choose one of the following two sequences.

Single-number bet: The gambler will bet $5 on a particular number before each spin. He will win a net amount of $175 if that number comes up and lose $5 otherwise.

Color bet: The gambler will bet $5 on the red color before each spin. He will win a net amount of $5 if a red number comes up and lose $5 otherwise.

a. If the gambler makes a sequence of 25 bets, which of the two betting schemes do you think offers him a better chance of coming out ahead (winning more money than losing) after the 25 bets?
b. Now compute the probability of coming out ahead after 25 single-number bets of $5 each and after 25 color bets of $5 each. Do these results confirm your guess in part a? (Before using an approximation to find either probability, be sure to check whether it is appropriate.)

6.101 A charter bus company is advertising a singles outing on a bus that holds 60 passengers. The company has found that, on average, 10% of ticket holders do not show up for such trips; hence, the company routinely overbooks such trips. Assume that passengers act independently of one another.

a. If the company sells 65 tickets, what is the probability that the bus can hold all the ticket holders who actually show up? In other words, find the probability that 60 or fewer passengers show up.
b. What is the largest number of tickets the company can sell and still be at least 95% sure that the bus can hold all the ticket holders who actually show up?

6.102 The amount of time taken by a bank teller to serve a randomly selected customer has a normal distribution with a mean of 2 minutes and a standard deviation of .5 minutes.

a. What is the probability that both of two randomly selected customers will take less than 1 minute each to be served?
b. What is the probability that at least one of four randomly selected customers will need more than 2.25 minutes to be served?

6.103 Suppose you are conducting a binomial experiment that has 15 trials and the probability of success of .02. According to the sample size requirements, you cannot use the normal distribution to approximate the binomial distribution in this situation. Use the mean and standard deviation of this binomial distribution and the Empirical Rule to explain why there is a problem in this situation. (*Note:* Drawing the graph and marking the values that correspond to the Empirical Rule is a good way to start.)

6.104 A variation of a roulette wheel has slots that are not of equal size. Instead, the width of any slot is proportional to the probability that a standard normal random variable z takes on a value between a and $(a + .1)$, where $a = -3.0, -2.9, -2.8, \ldots, 2.9, 3.0$. In other words, there are slots for the intervals $(-3.0, -2.9)$, $(-2.9, -2.8)$, $(-2.8, -2.7)$ through $(2.9, 3.0)$. There is one more slot that represents the probability that z falls outside the interval $(-3.0, 3.0)$. Find the following probabilities.

a. The ball lands in the slot representing $(.3, .4)$.
b. The ball lands in any of the slots representing $(-.1, .4)$.

c. In at least one out of five games, the ball lands in the slot representing $(-.1, .4)$.
d. In at least 100 out of 500 games, the ball lands in the slot representing $(.4, .5)$.

6.105 Refer to Exercise 6.97. In that exercise, suppose the mean is set to be eight ounces, but the standard deviation is unknown. The cups used in the machine can hold up to 8.2 ounces, but these cups will overflow if more than 8.2 ounces is dispensed by the machine. What is the smallest possible standard deviation that will result in overflows occurring 3% of the time?

Self-Review Test

1. The normal probability distribution is applied to
   a. a continuous random variable b. a discrete random variable c. any random variable
2. For a continuous random variable, the probability of a single value of x is always
   a. zero b. 1.0 c. between 0 and 1
3. Which of the following is not a characteristic of the normal distribution?
   a. The total area under the curve is 1.0.
   b. The curve is symmetric about the mean.
   c. The two tails of the curve extend indefinitely.
   d. The value of the mean is always greater than the value of the standard deviation.
4. The parameters of a normal distribution are
   a. μ, z, and σ b. μ and σ c. μ, x, and σ
5. For the standard normal distribution,
   a. $\mu = 0$ and $\sigma = 1$ b. $\mu = 1$ and $\sigma = 0$ c. $\mu = 100$ and $\sigma = 10$
6. The z value for μ for a normal distribution curve is always
   a. positive b. negative c. 0
7. For a normal distribution curve, the z value for an x value that is less than μ is always
   a. positive b. negative c. 0
8. Usually the normal distribution is used as an approximation to the binomial distribution when
   a. $n \geq 30$ b. $np > 5$ and $nq > 5$ c. $n > 20$ and $p = .50$
9. Find the following probabilities for the standard normal distribution.
   a. $P(.85 \leq z \leq 2.33)$ b. $P(-2.97 \leq z \leq 1.49)$ c. $P(z \leq -1.29)$ d. $P(z > -.74)$
10. Find the value of z for the standard normal curve such that the area
    a. in the left tail is .1000
    b. between 0 and z is .2291 and z is positive
    c. in the right tail is .0500
    d. between 0 and z is .3571 and z is negative
11. According to the Transportation Security Administration, the average waiting time in security lines at the Miami airport was 20.9 minutes on July 5, 2004 (*USA TODAY*, July 13, 2004). Suppose the current waiting times in security lines at the Miami airport are normally distributed with a mean of 20.9 minutes and a standard deviation of 5 minutes.
    a. Find the probability that a randomly selected passenger spends between 15 and 30 minutes waiting in the security line at this airport.
    b. What is the probability that a randomly selected passenger spends less than 13 minutes waiting in the security line at this airport?
    c. Find the probability that a randomly selected passenger spends more than 26 minutes waiting in the security line at this airport.
    d. What is the probability that a randomly selected passenger spends between 30 and 35 minutes waiting in the security line at this airport?
12. Refer to Problem 11.
    a. Suppose that the 10% of passengers who spent the least time waiting in security lines at this airport waited for less than x minutes. Find x.
    b. Suppose that the 5% of passengers who spent the most time waiting in security lines at this airport spent more than x minutes. Find x.
13. According to a Reuters poll, 70% of elderly citizens in the United States had gambled at least once in the past year (*Time*, January 31, 2005). Suppose this percentage is true for the current population of elderly citizens. A random sample of 500 elderly citizens is taken.

a. Find the probability that the number of elderly citizens in this sample who have gambled at least once during the past year is
   i. exactly 345 ii. 338 to 370 iii. At most 342
   iv. at least 335 v. between 322 and 340
b. Find the probability that at most 165 of elderly citizens in this sample have *not* gambled at all during the past year.
c. Find the probability that between 155 and 175 of elderly citizens in this sample have *not* gambled at all during the past year.

Mini-Projects

MINI-PROJECT 6–1

Consider the data on heights of NBA players that accompany this text (see Appendix B).

a. Use statistical software to obtain a histogram. Do these heights appear to be symmetrically distributed? If not, in which direction do they seem to be skewed?
b. Compute μ and σ for heights of all players.
c. What percentage of these heights lie in the interval $\mu - \sigma$ to $\mu + \sigma$? What about in the interval $\mu - 2\sigma$ to $\mu + 2\sigma$? In the interval $\mu - 3\sigma$ to $\mu + 3\sigma$?
d. How do the percentages in part c compare to the corresponding percentages for a normal distribution (68.26%, 95.44%, and 99.74%, respectively)?
e. Use statistical software to select three random samples of 20 players each. Create a histogram and a dotplot of heights for each sample, and calculate the mean and standard deviation of heights for each sample. How well do your graphs and summary statistics match up with the corresponding population graphs and parameter values obtained in earlier parts? Does it seem reasonable that they might not match up very well?

MINI-PROJECT 6–2

Consider the data on weights of NBA players (see Appendix B).

a. Use statistical software to obtain a histogram. Do these weights appear to be symmetrically distributed? If not, in which direction do they seem to be skewed?
b. Compute μ and σ for weights of all players.
c. What percentage of these weights lie in the interval $\mu - \sigma$ to $\mu + \sigma$? What about in the interval $\mu - 2\sigma$ to $\mu + 2\sigma$? In the interval $\mu - 3\sigma$ to $\mu + 3\sigma$?
d. How do the percentages in part c compare to the corresponding percentages for a normal distribution (68.26%, 95.44%, and 99.74%, respectively)?
e. Use statistical software to select three random samples of 20 players each. Create a histogram and a dotplot of weights for each sample, and calculate the mean and standard deviation of weights for each sample. How well do your graphs and summary statistics match up with the corresponding population graphs and parameter values obtained in earlier parts? Does it seem reasonable that they might not match up very well?

MINI-PROJECT 6–3

Using weather records from back issues of a local newspaper or from some other source, record the maximum temperatures in your town for a period of 60 days.

a. Use statistical software to obtain a histogram and a dotplot for your data. Comment on the shape of the distribution as observed from these graphs.
b. Calculate $\bar{x}$ and s.
c. What percentage of the temperatures are in the interval $\bar{x} - s$ to $\bar{x} + s$?
d. What percentage are in the interval $\bar{x} - 2s$ to $\bar{x} + 2s$?
e. How do these percentages compare to the corresponding percentages for a normal distribution (68.26% and 95.44%, respectively)?
f. Now find the minimum temperatures in your town for 60 days by using the same source that you used to find the maximum temperatures or by using a different source. Then repeat parts a through e for this data set.

DECIDE FOR YOURSELF

Deciding About the Shape of a Distribution

Reporting summary measures such as the mean, median, and standard deviation has become very common in modern life. Many companies, government agencies, and so forth will report the mean and standard deviation of a variable, but they will very rarely provide information on the shape of the distribution of that variable. In Chapters 5 and 6, you have learned some basic properties of some distributions that can help you to decide if a specific type of distribution is a good fit for a set of data.

According to the *National Diet and Nutrition Survey: Adults Aged 19 to 64*, British men spend an average of 2.15 hours per day in moderate or high intensity physical activity. The standard deviation of these activity times for this sample was 3.59 hours. (*Source:* http://www.food.gov.uk/multimedia/pdfs/ndnsfour.pdf.) Can we infer that these activity times could follow a normal distribution? The following questions may provide an answer.

1. Sketch a normal curve marking the points representing 1, 2, and 3 standard deviations above and below the mean, and calculate the values at these points using a mean of 2.15 hours and a standard deviation of 3.59 hours.

2. Examine the curve with your calculations. Explain why it is impossible for this distribution to be normal based on your graph and calculations.

3. Considering the variable being measured, is it more likely that the distribution is skewed to the left or that it is skewed to the right? Explain why.

4. Suppose that the standard deviation for this sample was .70 hours instead of 3.59 hours, which makes it numerically possible for the distribution to be normal. Again, considering the variable being measured, explain why the normal distribution is still not a logical choice for this distribution.

TECHNOLOGY INSTRUCTION — Normal and Inverse Normal Probabilities

TI-84

```
normalcdf(-E99,1
25,100,15)
     .9522096696
```

Screen 6.1

1. For a given mean μ and standard deviation σ, to find the probability that a normal random variable x lies below b, select **DISTR>normalcdf(-E99, *b*, μ, σ)**, and press **Enter**. (See **Screen 6.1**.)

2. For a given mean μ and standard deviation σ, to find the probability that a normal random variable x lies above a, select **DISTR>normalcdf(*a*, E99, μ, σ)**, and press **Enter**.

3. For a given mean μ and standard deviation σ, to find the probability that a normal random variable x lies between a and b, select **DISTR>normalcdf(*a*, *b*, μ, σ)**, and press **Enter**.

4. To find a value of a for a normal random variable x with mean μ and standard deviation σ such that the probability of x being less than a is p, select **DISTR>invNorm(*p*, μ, σ)**, and press **Enter**.

Note: To type **E99**, press **2nd>comma key** (which is the key just above the 7 key). The function is labeled EE, but only E is displayed on the screen. Then type 9 twice. For **-E99**, press **(-) key** (which is to the right of the decimal key) before E99.

MINITAB

1. For a given mean μ and standard deviation σ, to find the probability that a normal random variable x lies below a, select **Calc>Probability Distributions>Normal**. Select **Cumulative probability**, enter the mean μ and the standard deviation σ. Select **Input constant** and enter a, then select **OK**. (See **Screens 6.2** and **6.3**.)

2. To find a value of a for a normal random variable x with mean μ and standard deviation σ such that the probability of x being less than a is p, select **Calc>Probability Distributions>Normal**. Select **Inverse cumulative probability**, enter the mean μ and the standard deviation σ. Select **Input constant** and enter a, then select **OK**.

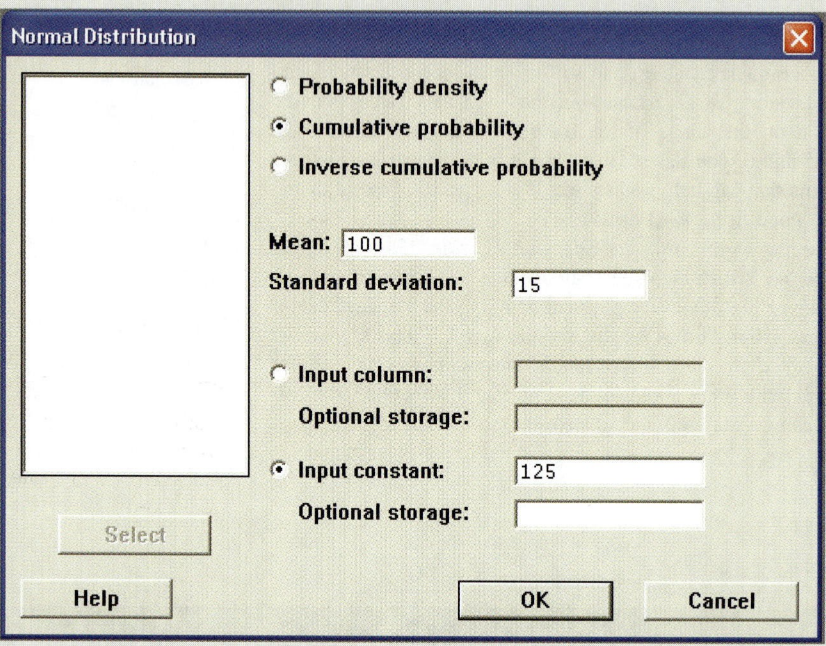

Screen 6.2

```
Cumulative Distribution Function

Normal with mean = 100 and standard deviation = 15

  x   P( X <= x )
125      0.952210
```

Screen 6.3

Excel

1. For a given mean μ and standard deviation σ, to find the probability that a normal random variable x lies below b, type **=NORMDIST(b, μ, σ, 1)**. (See **Screen 6.4.**)
2. For a given mean μ and standard deviation σ, to find the probability that a normal random variable x lies above a, type **=1−NORMDIST(a, μ, σ, 1)**.
3. For a given mean μ and standard deviation σ, to find the probability that a normal random variable x lies between a and b, type **=NORMDIST(b, μ, σ, 1)−NORMDIST(a, μ, σ, 1)**.
4. To find a value of a for a normal random variable x with mean μ and standard deviation σ such that the probability of x being less than a is p, type **=NORMINV(p, λ, σ)**.

	A	B	C	D
1	Mean	100		
2	Std. Dev.	15		
3				
4	P(X<125)	=NORMDIST(125,100,15,1)		

Screen 6.4

TECHNOLOGY ASSIGNMENTS

TA6.1 Find the area under the standard normal curve

a. to the left of $z = -1.94$
b. to the left of $z = .83$
c. to the right of $z = 1.45$
d. to the right of $z = -1.65$
e. between $z = .75$ and $z = 1.90$
f. between $z = -1.20$ and $z = 1.55$

TA6.2 Find the following areas under a normal curve with $\mu = 86$ and $\sigma = 14$.

a. Area to the left of $x = 71$
b. Area to the left of $x = 96$
c. Area to the right of $x = 90$
d. Area to the right of $x = 75$
e. Area between $x = 65$ and $x = 75$
f. Area between $x = 72$ and $x = 95$

TA6.3 According to the *New York Times*, working men spend an average of 48 minutes per day caring for their families (*Time*, September 27, 2004). Assume that the times that working men currently spend per day caring for their families are normally distributed with a mean of 48 minutes and a standard deviation of 11 minutes.

a. Find the probability that a randomly selected working man spends more than 68 minutes per day caring for his family.

b. What percentage of working men spend between 30 and 73 minutes per day caring for their families?

TA6.4 According to the Kaiser Family Foundation, the average amount of time spent watching TV by 8-to-18-year-olds is 231 minutes per day (*USA TODAY*, March 10, 2005). Suppose currently the times spent watching TV by all 8-to-18-year-olds have a normal distribution with a mean of 231 minutes and a standard deviation of 45 minutes. What percentage of the 8-to-18-year-olds watch TV for

a. more than 290 minutes per day?
b. less than 150 minutes per day?
c. 180 to 320 minutes per day?
d. 270 to 350 minutes per day?

TA6.5 The transmission on a particular model of car has a warranty for 40,000 miles. It is known that the life of such a transmission has a normal distribution with a mean of 72,000 miles and a standard deviation of 12,000 miles. Answer the following questions.

a. What percentage of the transmissions will fail before the end of the warranty period?
b. What percentage of the transmissions will be good for more than 100,000 miles?
c. What percentage of the transmissions will be good for 80,000 to 100,000 miles?

Chapter

7

Sampling Distributions

7.1 Population and Sampling Distributions

7.2 Sampling and Nonsampling Errors

7.3 Mean and Standard Deviation of $\bar{x}$

7.4 Shape of the Sampling Distribution of $\bar{x}$

7.5 Applications of the Sampling Distribution of $\bar{x}$

7.6 Population and Sample Proportions

7.7 Mean, Standard Deviation, and Shape of the Sampling Distribution of $\hat{p}$

Case Study 7–1 Calling the Vote for Congress

7.8 Applications of the Sampling Distribution of $\hat{p}$

During the week before the midterm elections in November 2002, most national polls predicted a very close battle for seats in the House of Representatives. The actual election results showed 51.7% of the votes cast for Republican candidates versus 45% for Democrats. The leading national polls showed the following: CBS–*New York Times*, 47% versus 40%; *USA TODAY*–CNN–Gallup, 51% versus 45%; ABC–*Washington Post*, 48% versus 48%; Ipsos-Reid, 44% versus 45%; Zogby, 49% versus 51%; and PSRA-Pew, 42% versus 46%. The data show that different polls predicted different percentages for Republicans and Democrats. Why were the actual votes different from the various polls, and why did the polls differ from each other? (See Case Study 7–1.)

Chapters 5 and 6 discussed probability distributions of discrete and continuous random variables. This chapter extends the concept of probability distribution to that of a sample statistic. As we discussed in Chapter 3, a sample statistic is a numerical summary measure calculated for sample data. The mean, median, mode, and standard deviation calculated for sample data are called *sample statistics*. On the other hand, the same numerical summary measures calculated for population data are called *population parameters*. A population parameter is always a constant, whereas a sample statistic is always a random variable. Because every random variable must possess a probability distribution, each sample statistic possesses a probability distribution. The probability distribution of a sample statistic is more commonly called its *sampling distribution*. This chapter discusses the sampling distributions of the sample mean and the sample proportion. The concepts covered in this chapter are the foundation of the inferential statistics discussed in succeeding chapters.

7.1 Population and Sampling Distributions

This section introduces the concepts of population distribution and sampling distribution. Subsection 7.1.1 explains the population distribution, and Subsection 7.1.2 describes the sampling distribution of $\bar{x}$.

7.1.1 Population Distribution

The **population distribution** is the probability distribution derived from the information on all elements of a population.

Definition

Population Distribution The *population distribution* is the probability distribution of the population data.

Suppose there are only five students in an advanced statistics class and the midterm scores of these five students are

70 78 80 80 95

Let x denote the score of a student. Using single-valued classes (because there are only five data values, there is no need to group them), we can write the frequency distribution of scores as in Table 7.1 along with the relative frequencies of classes, which are obtained by dividing the frequencies of classes by the population size. Table 7.2, which lists the probabilities of various x values, presents the probability distribution of the population. Note that these probabilities are the same as the relative frequencies.

Table 7.1 Population Frequency and Relative Frequency Distributions

x	f	Relative Frequency
70	1	1/5 = .20
78	1	1/5 = .20
80	2	2/5 = .40
95	1	1/5 = .20
	$N = 5$	Sum = 1.00

Table 7.2 Population Probability Distribution

x	$P(x)$
70	.20
78	.20
80	.40
95	.20
	$\Sigma P(x) = 1.00$

The values of the mean and standard deviation calculated for the probability distribution of Table 7.2 give the values of the population parameters μ and σ. These values are $\mu = 80.60$ and $\sigma = 8.09$. The values of μ and σ for the probability distribution of Table 7.2 can be calculated using the formulas given in Sections 5.3 and 5.4 of Chapter 5 (see Exercise 7.6).

7.1.2 Sampling Distribution

As mentioned at the beginning of this chapter, the value of a population parameter is always constant. For example, for any population data set, there is only one value of the population mean, μ. However, we cannot say the same about the sample mean, $\bar{x}$. We would expect different samples of the same size drawn from the same population to yield different values of the sample mean, $\bar{x}$. The value of the sample mean for any one sample will depend on the elements included in that sample. Consequently, *the sample mean, $\bar{x}$, is a random variable*. Therefore, like other random variables, the sample mean possesses a probability distribution, which is more commonly called the **sampling distribution of $\bar{x}$**. Other sample statistics, such as the median, mode, and standard deviation, also possess sampling distributions.

Definition

Sampling Distribution of $\bar{x}$ The probability distribution of $\bar{x}$ is called its sampling distribution. It lists the various values that $\bar{x}$ can assume and the probability of each value of $\bar{x}$.

In general, the probability distribution of a sample statistic is called its *sampling distribution.*

Reconsider the population of midterm scores of five students given in Table 7.1. Consider all possible samples of three scores each that can be selected, without replacement, from that population. The total number of possible samples, given by the combinations formula discussed in Chapter 5, is 10; that is,

$$\text{Total number of samples} = {}_5C_3 = \frac{5!}{3!(5-3)!} = \frac{5 \cdot 4 \cdot 3 \cdot 2 \cdot 1}{3 \cdot 2 \cdot 1 \cdot 2 \cdot 1} = 10$$

Suppose we assign the letters A, B, C, D, and E to the scores of the five students so that

$$\text{A} = 70, \quad \text{B} = 78, \quad \text{C} = 80, \quad \text{D} = 80, \quad \text{E} = 95$$

Then, the 10 possible samples of three scores each are

ABC, ABD, ABE, ACD, ACE, ADE, BCD, BCE, BDE, CDE

These 10 samples and their respective means are listed in Table 7.3. Note that the first two samples have the same three scores. The reason for this is that two of the students (C and D) have the same score and, hence, the samples ABC and ABD contain the same values. The mean of each sample is obtained by dividing the sum of the three scores included in that sample by 3. For instance, the mean of the first sample is $(70 + 78 + 80)/3 = 76$. Note that the values of the means of samples in Table 7.3 are rounded to two decimal places.

By using the values of $\bar{x}$ given in Table 7.3, we record the frequency distribution of $\bar{x}$ in Table 7.4. By dividing the frequencies of the various values of $\bar{x}$ by the sum of all frequencies, we obtain the relative frequencies of classes, which are listed in the third column of Table 7.4. These relative frequencies are used as probabilities and listed in Table 7.5. This table gives the sampling distribution of $\bar{x}$.

If we select just one sample of three scores from the population of five scores, we may draw any of the 10 possible samples. Hence, the sample mean, $\bar{x}$, can assume any of the values listed in Table 7.5 with the corresponding probability. For instance, the probability that the mean of a randomly selected sample of three scores is 81.67 is .20. This probability can be written as

$$P(\bar{x} = 81.67) = .20$$

Table 7.3 All Possible Samples and Their Means When the Sample Size Is 3

Sample	Scores in the Sample	$\bar{x}$
ABC	70, 78, 80	76.00
ABD	70, 78, 80	76.00
ABE	70, 78, 95	81.00
ACD	70, 80, 80	76.67
ACE	70, 80, 95	81.67
ADE	70, 80, 95	81.67
BCD	78, 80, 80	79.33
BCE	78, 80, 95	84.33
BDE	78, 80, 95	84.33
CDE	80, 80, 95	85.00

Table 7.4 Frequency and Relative Frequency Distributions of $\bar{x}$ When the Sample Size Is 3

$\bar{x}$	f	Relative Frequency
76.00	2	2/10 = .20
76.67	1	1/10 = .10
79.33	1	1/10 = .10
81.00	1	1/10 = .10
81.67	2	2/10 = .20
84.33	2	2/10 = .20
85.00	1	1/10 = .10
	$\Sigma f = 10$	Sum = 1.00

Table 7.5 Sampling Distribution of $\bar{x}$ When the Sample Size Is 3

$\bar{x}$	$P(\bar{x})$
76.00	.20
76.67	.10
79.33	.10
81.00	.10
81.67	.20
84.33	.20
85.00	.10
	$\Sigma P(\bar{x}) = 1.00$

7.2 Sampling and Nonsampling Errors

Usually, different samples selected from the same population will give different results because they contain different elements. This is obvious from Table 7.3, which shows that the mean of a sample of three scores depends on which three of the five scores are included in the sample. The result obtained from any one sample will generally be different from the result obtained from the corresponding population. The difference between the value of a sample statistic obtained from a sample and the value of the corresponding population parameter obtained from the population is called the **sampling error**. Note that this difference represents the sampling error only if the sample is random and no nonsampling error has been made. Otherwise, only a part of this difference will be due to the sampling error.

Definition

Sampling Error *Sampling error* is the difference between the value of a sample statistic and the value of the corresponding population parameter. In the case of the mean,

$$\text{Sampling error} = \bar{x} - \mu$$

assuming that the sample is random and no nonsampling error has been made.

It is important to remember that *a sampling error occurs because of chance*. The errors that occur for other reasons, such as errors made during collection, recording, and tabulation of data, are called **nonsampling errors**. These errors occur because of human mistakes, and not chance. Note that there is only one kind of sampling error—the error that occurs due to chance. However, there is not just one nonsampling error but there are many nonsampling errors that may occur for different reasons.

Definition

Nonsampling Errors The errors that occur in the collection, recording, and tabulation of data are called *nonsampling errors*.

The following paragraph, reproduced from the *Current Population Reports* of the U.S. Bureau of the Census, explains how nonsampling errors can occur.

> Nonsampling errors can be attributed to many sources, e.g., inability to obtain information about all cases in the sample, definitional difficulties, differences in the interpretation of questions, inability or unwillingness on the part of the respondents to provide correct information, inability to recall information, errors made in collection such as in recording or coding the data, errors made in processing the data, errors made in estimating values for missing data, biases resulting from the differing recall periods caused by the interviewing pattern used, and failure of all units in the universe to have some probability of being selected for the sample (undercoverage).

The following are the main reasons for the occurrence of nonsampling errors.

1. If a sample is nonrandom (and, hence, nonrepresentative), the sample results may be too different from the census results. The following quote from *U.S. News & World Report* describes how even a randomly selected sample can become nonrandom if some of the members included in the sample cannot be contacted.

   > A test poll conducted in the 1984 presidential election found that if the poll were halted after interviewing only those subjects who could be reached on the first try, Reagan showed a 3-percentage-point lead over Mondale. But when interviewers made a determined effort to reach everyone on their lists of randomly selected subjects—calling some as many as 30 times before finally reaching them—Reagan showed a 13 percent lead, much closer to the actual election result. As it turned out, people who were planning to vote Republican were simply less likely to be at home. ("The Numbers Racket: How Polls and Statistics Lie," *U.S. News & World Report*, July 11, 1988. Copyright © 1988 by U.S. News & World Report, Inc. Reprinted with permission.)

2. The questions may be phrased in such a way that they are not fully understood by the members of the sample or population. As a result, the answers obtained are not accurate.
3. The respondents may intentionally give false information in response to some sensitive questions. For example, people may not tell the truth about their drinking habits, incomes, or opinions about minorities. Sometimes the respondents may give wrong answers because of ignorance. For example, a person may not remember the exact amount he or she spent on clothes during the last year. If asked in a survey, he or she may give an inaccurate answer.
4. The poll taker may make a mistake and enter a wrong number in the records or make an error while entering the data on a computer.

Note that nonsampling errors can occur both in a sample survey and in a census, whereas sampling error occurs only when a sample survey is conducted. Nonsampling errors can be minimized by preparing the survey questionnaire carefully and handling the data cautiously. However, it is impossible to avoid sampling error.

Example 7–1 illustrates the sampling and nonsampling errors using the mean.

■ EXAMPLE 7–1

Illustrating sampling and nonsampling errors.

Reconsider the population of five scores given in Table 7.1. Suppose one sample of three scores is selected from this population, and this sample includes the scores 70, 80, and 95. Find the sampling error.

Solution The scores of the five students are 70, 78, 80, 80, and 95. The population mean is

$$\mu = \frac{70 + 78 + 80 + 80 + 95}{5} = 80.60$$

Now a random sample of three scores from this population is taken and this sample includes the scores 70, 80, and 95. The mean for this sample is

$$\bar{x} = \frac{70 + 80 + 95}{3} = 81.67$$

Consequently,

$$\text{Sampling error} = \bar{x} - \mu = 81.67 - 80.60 = \mathbf{1.07}$$

That is, the mean score estimated from the sample is 1.07 higher than the mean score of the population. Note that this difference occurred due to chance—that is, because we used a sample instead of the population. ■

Now suppose, when we select the sample of three scores, we mistakenly record the second score as 82 instead of 80. As a result, we calculate the sample mean as

$$\bar{x} = \frac{70 + 82 + 95}{3} = 82.33$$

Consequently, the difference between this sample mean and the population mean is

$$\bar{x} - \mu = 82.33 - 80.60 = 1.73$$

However, this difference between the sample mean and the population mean does not represent the sampling error. As we calculated earlier, only 1.07 of this difference is due to the sampling error. The remaining portion, which is equal to 1.73 − 1.07 = .66, represents the nonsampling error because it occurred due to the error we made in recording the second score in the sample. Thus, in this case,

$$\text{Sampling error} = \mathbf{1.07}$$

$$\text{Nonsampling error} = \mathbf{.66}$$

Figure 7.1 shows the sampling and nonsampling errors for these calculations.

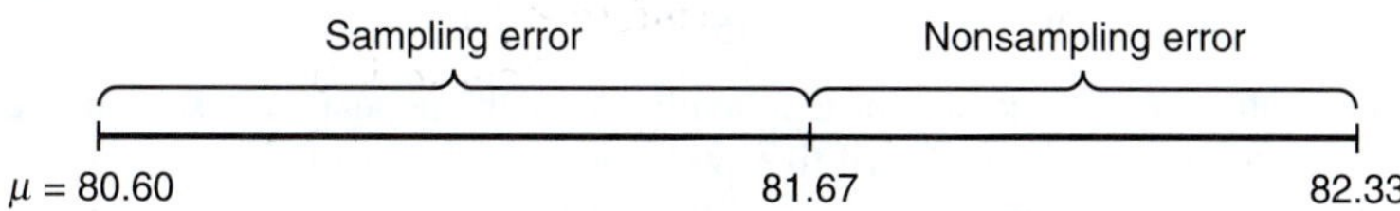

Figure 7.1 Sampling and nonsampling errors.

Thus, the sampling error is the difference between the correct value of $\bar{x}$ and μ, where the correct value of $\bar{x}$ is the value of $\bar{x}$ that does not contain any nonsampling errors. In contrast, the nonsampling error(s) is (are) obtained by subtracting the correct value of $\bar{x}$ from the incorrect value of $\bar{x}$, where the incorrect value of $\bar{x}$ is the value that contains the nonsampling error(s). For our example,

$$\text{Sampling error} = \bar{x} - \mu = 81.67 - 80.60 = 1.07$$

$$\text{Nonsampling error} = \text{Incorrect } \bar{x} - \text{Correct } \bar{x} = 82.33 - 81.67 = .66$$

Note that in the real world we do not know the mean of a population. Hence, we select a sample to use the sample mean as an estimate of the population mean. Consequently, we never know the size of the sampling error.

EXERCISES

CONCEPTS AND PROCEDURES

7.1 Briefly explain the meaning of a population distribution and a sampling distribution. Give an example of each.

7.2 Explain briefly the meaning of sampling error. Give an example. Does such an error occur only in a sample survey or can it occur in both a sample survey and a census?

7.3 Explain briefly the meaning of nonsampling errors. Give an example. Do such errors occur only in a sample survey or can they occur in both a sample survey and a census?

7.4 Consider the following population of six numbers.

15 13 8 17 9 12

a. Find the population mean.
b. Liza selected one sample of four numbers from this population. The sample included the numbers 13, 8, 9, and 12. Calculate the sample mean and sampling error for this sample.
c. Refer to part b. When Liza calculated the sample mean, she mistakenly used the numbers 13, 8, 6, and 12 to calculate the sample mean. Find the sampling and nonsampling errors in this case.
d. List all samples of four numbers (without replacement) that can be selected from this population. Calculate the sample mean and sampling error for each of these samples.

7.5 Consider the following population of 10 numbers.

20 25 13 19 9 15 11 7 17 30

a. Find the population mean.
b. Rich selected one sample of nine numbers from this population. The sample included the numbers 20, 25, 13, 9, 15, 11, 7, 17, and 30. Calculate the sample mean and sampling error for this sample.
c. Refer to part b. When Rich calculated the sample mean, he mistakenly used the numbers 20, 25, 13, 9, 15, 11, 17, 17, and 30 to calculate the sample mean. Find the sampling and nonsampling errors in this case.
d. List all samples of nine numbers (without replacement) that can be selected from this population. Calculate the sample mean and sampling error for each of these samples.

APPLICATIONS

7.6 Using the formulas of Sections 5.3 and 5.4 of Chapter 5 for the mean and standard deviation of a discrete random variable, verify that the mean and standard deviation for the population probability distribution of Table 7.2 are 80.60 and 8.09, respectively.

7.7 The following data give the ages of all six members of a family.

55 53 28 25 21 15

a. Let x denote the age of a member of this family. Write the population distribution of x.
b. List all the possible samples of size five (without replacement) that can be selected from this population. Calculate the mean for each of these samples. Write the sampling distribution of $\bar{x}$.
c. Calculate the mean for the population data. Select one random sample of size five and calculate the sample mean $\bar{x}$. Compute the sampling error.

7.8 The following data give the years of teaching experience for all five faculty members of a department at a university.

7 8 14 7 20

a. Let x denote the years of teaching experience for a faculty member of this department. Write the population distribution of x.
b. List all the possible samples of size four (without replacement) that can be selected from this population. Calculate the mean for each of these samples. Write the sampling distribution of $\bar{x}$.
c. Calculate the mean for the population data. Select one random sample of size four and calculate the sample mean $\bar{x}$. Compute the sampling error.

7.3 Mean and Standard Deviation of $\bar{x}$

The mean and standard deviation calculated for the sampling distribution of $\bar{x}$ are called the **mean** and **standard deviation of $\bar{x}$**. Actually, the mean and standard deviation of $\bar{x}$ are, respectively, the mean and standard deviation of the means of all samples of the same size selected from a population. The standard deviation of $\bar{x}$ is also called the *standard error of $\bar{x}$*.

Definition

Mean and Standard Deviation of $\bar{x}$ The mean and standard deviation of the sampling distribution of $\bar{x}$ are called the *mean and standard deviation of $\bar{x}$* and are denoted by $\mu_{\bar{x}}$ and $\sigma_{\bar{x}}$, respectively.

If we calculate the mean and standard deviation of the 10 values of $\bar{x}$ listed in Table 7.3, we obtain the mean, $\mu_{\bar{x}}$, and the standard deviation, $\sigma_{\bar{x}}$, of $\bar{x}$. Alternatively, we can calculate the mean and standard deviation of the sampling distribution of $\bar{x}$ listed in Table 7.5. These will also be the values of $\mu_{\bar{x}}$ and $\sigma_{\bar{x}}$. From these calculations, we will obtain $\mu_{\bar{x}} = 80.60$ and $\sigma_{\bar{x}} = 3.30$ (see Exercise 7.25 at the end of this section).

The mean of the sampling distribution of $\bar{x}$ is always equal to the mean of the population.

Mean of the Sampling Distribution of $\bar{x}$ The *mean of the sampling distribution of* $\bar{x}$ is always equal to the mean of the population. Thus,

$$\mu_{\bar{x}} = \mu$$

Hence, if we select all possible samples (of the same size) from a population and calculate their means, the mean ($\mu_{\bar{x}}$) of all these sample means will be the same as the mean (μ) of the population. If we calculate the mean for the population probability distribution of Table 7.2 and the mean for the sampling distribution of Table 7.5 by using the formula learned in Section 5.3 of Chapter 5, we get the same value of 80.60 for μ and $\mu_{\bar{x}}$ (see Exercise 7.25).

The sample mean, $\bar{x}$, is called an **estimator** of the population mean, μ. When the expected value (or mean) of a sample statistic is equal to the value of the corresponding population parameter, that sample statistic is said to be an **unbiased estimator**. For the sample mean $\bar{x}$, $\mu_{\bar{x}} = \mu$. Hence, $\bar{x}$ is an unbiased estimator of μ. This is a very important property that an estimator should possess.

However, the standard deviation, $\sigma_{\bar{x}}$, of $\bar{x}$ is not equal to the standard deviation, σ, of the population distribution (unless $n = 1$). The standard deviation of $\bar{x}$ is equal to the standard deviation of the population divided by the square root of the sample size; that is,

$$\sigma_{\bar{x}} = \frac{\sigma}{\sqrt{n}}$$

This formula for the standard deviation of $\bar{x}$ holds true only when the sampling is done either with replacement from a finite population or with or without replacement from an infinite population. These two conditions can be replaced by the condition that the above formula holds true if the sample size is small in comparison to the population size. The sample size is considered to be small compared to the population size if the sample size is equal to or less than 5% of the population size—that is, if

$$\frac{n}{N} \le .05$$

If this condition is not satisfied, we use the following formula to calculate $\sigma_{\bar{x}}$:

$$\sigma_{\bar{x}} = \frac{\sigma}{\sqrt{n}}\sqrt{\frac{N-n}{N-1}}$$

where the factor

$$\sqrt{\frac{N-n}{N-1}}$$

is called the finite population correction factor.

In most practical applications, the sample size is small compared to the population size. Consequently, in most cases, the formula used to calculate $\sigma_{\bar{x}}$ is $\sigma_{\bar{x}} = \sigma/\sqrt{n}$.

Standard Deviation of the Sampling Distribution of $\bar{x}$ The *standard deviation of the sampling distribution of* $\bar{x}$ is

$$\sigma_{\bar{x}} = \frac{\sigma}{\sqrt{n}}$$

where σ is the standard deviation of the population and n is the sample size. This formula is used when $n/N \leq .05$, where N is the population size.

Following are two important observations regarding the sampling distribution of $\bar{x}$.

1. *The spread of the sampling distribution of $\bar{x}$ is smaller than the spread of the corresponding population distribution.* In other words, $\sigma_{\bar{x}} < \sigma$. This is obvious from the formula for $\sigma_{\bar{x}}$. When n is greater than 1, which is usually true, the denominator in $\sigma/\sqrt{n}$ is greater than 1. Hence, $\sigma_{\bar{x}}$ is smaller than σ.
2. *The standard deviation of the sampling distribution of $\bar{x}$ decreases as the sample size increases.* This feature of the sampling distribution of $\bar{x}$ is also obvious from the formula

$$\sigma_{\bar{x}} = \frac{\sigma}{\sqrt{n}}$$

If the standard deviation of a sample statistic decreases as the sample size is increased, that statistic is said to be a **consistent estimator**. This is another important property that an estimator should possess. It is obvious from the above formula for $\sigma_{\bar{x}}$ that as n increases, the value of $\sqrt{n}$ also increases and, consequently, the value of $\sigma/\sqrt{n}$ decreases. Thus, the sample mean $\bar{x}$ is a consistent estimator of the population mean μ. Example 7–2 illustrates this feature.

■ EXAMPLE 7–2

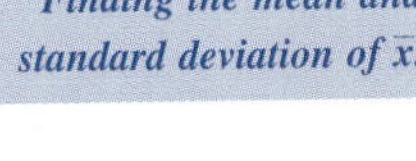

Finding the mean and standard deviation of $\bar{x}$.

The mean wage per hour for all 5000 employees who work at a large company is \$27.50 and the standard deviation is \$3.70. Let $\bar{x}$ be the mean wage per hour for a random sample of certain employees selected from this company. Find the mean and standard deviation of $\bar{x}$ for a sample size of

(a) 30 **(b)** 75 **(c)** 200

Solution From the given information, for the population of all employees,

$$N = 5000, \quad \mu = \$27.50, \quad \text{and} \quad \sigma = \$3.70$$

(a) The mean, $\mu_{\bar{x}}$, of the sampling distribution of $\bar{x}$ is

$$\mu_{\bar{x}} = \mu = \mathbf{\$27.50}$$

In this case, $n = 30$, $N = 5000$, and $n/N = 30/5000 = .006$. Because n/N is less than .05, the standard deviation of $\bar{x}$ is obtained by using the formula $\sigma/\sqrt{n}$. Hence,

$$\sigma_{\bar{x}} = \frac{\sigma}{\sqrt{n}} = \frac{3.70}{\sqrt{30}} = \mathbf{\$.676}$$

Thus, we can state that if we take all possible samples of size 30 from the population of all employees of this company and prepare the sampling distribution of $\bar{x}$, the mean and standard deviation of this sampling distribution of $\bar{x}$ will be \$27.50 and \$.676, respectively.

(b) In this case, $n = 75$ and $n/N = 75/5000 = .015$, which is less than .05. The mean and standard deviation of $\bar{x}$ are

$$\mu_{\bar{x}} = \mu = \mathbf{\$27.50} \quad \text{and} \quad \sigma_{\bar{x}} = \frac{\sigma}{\sqrt{n}} = \frac{3.70}{\sqrt{75}} = \mathbf{\$.427}$$

(c) In this case, $n = 200$ and $n/N = 200/5000 = .04$, which is less than .05. Therefore, the mean and standard deviation of $\bar{x}$ are

$$\mu_{\bar{x}} = \mu = \mathbf{\$27.50} \quad \text{and} \quad \sigma_{\bar{x}} = \frac{\sigma}{\sqrt{n}} = \frac{3.70}{\sqrt{200}} = \mathbf{\$.262}$$

From the preceding calculations we observe that the mean of the sampling distribution of $\bar{x}$ is always equal to the mean of the population whatever the size of the sample. However, the value of the standard deviation of $\bar{x}$ decreases from \$.676 to \$.427 and then to \$.262 as the sample size increases from 30 to 75 and then to 200. ■

EXERCISES

CONCEPTS AND PROCEDURES

7.9 Let $\bar{x}$ be the mean of a sample selected from a population.
  a. What is the mean of the sampling distribution of $\bar{x}$ equal to?
  b. What is the standard deviation of the sampling distribution of $\bar{x}$ equal to? Assume $n/N \leq .05$.

7.10 What is an estimator? When is an estimator unbiased? Is the sample mean, $\bar{x}$, an unbiased estimator of μ? Explain.

7.11 When is an estimator said to be consistent? Is the sample mean, $\bar{x}$, a consistent estimator of μ? Explain.

7.12 How does the value of $\sigma_{\bar{x}}$ change as the sample size increases? Explain.

7.13 Consider a large population with $\mu = 60$ and $\sigma = 10$. Assuming $n/N \leq .05$, find the mean and standard deviation of the sample mean, $\bar{x}$, for a sample size of
  a. 18 b. 90

7.14 Consider a large population with $\mu = 90$ and $\sigma = 18$. Assuming $n/N \leq .05$, find the mean and standard deviation of the sample mean, $\bar{x}$, for a sample size of
  a. 10 b. 35

7.15 A population of $N = 5000$ has $\sigma = 25$. In each of the following cases, which formula will you use to calculate $\sigma_{\bar{x}}$ and why? Using the appropriate formula, calculate $\sigma_{\bar{x}}$ for each of these cases.
  a. $n = 300$ b. $n = 100$

7.16 A population of $N = 100{,}000$ has $\sigma = 40$. In each of the following cases, which formula will you use to calculate $\sigma_{\bar{x}}$ and why? Using the appropriate formula, calculate $\sigma_{\bar{x}}$ for each of these cases.
  a. $n = 2500$ b. $n = 7000$

*7.17 For a population, $\mu = 125$ and $\sigma = 36$.
  a. For a sample selected from this population, $\mu_{\bar{x}} = 125$ and $\sigma_{\bar{x}} = 3.6$. Find the sample size. Assume $n/N \leq .05$.
  b. For a sample selected from this population, $\mu_{\bar{x}} = 125$ and $\sigma_{\bar{x}} = 2.25$. Find the sample size. Assume $n/N \leq .05$.

*7.18 For a population, $\mu = 46$ and $\sigma = 10$.
  a. For a sample selected from this population, $\mu_{\bar{x}} = 46$ and $\sigma_{\bar{x}} = 2.0$. Find the sample size. Assume $n/N \leq .05$.
  b. For a sample selected from this population, $\mu_{\bar{x}} = 46$ and $\sigma_{\bar{x}} = 1.6$. Find the sample size. Assume $n/N \leq .05$.

APPLICATIONS

7.19 According to the Stanford Institute for the Quantitative Study of Society, Internet users in the United States spend an average of three hours per day online (*Time*, January 10, 2005). Suppose currently all Internet users spend an average of three hours per day online with a standard deviation of .80 hours. Let $\bar{x}$ be the mean daily time spent online by 75 randomly selected Internet users. Find the mean and standard deviation of the sampling distribution of $\bar{x}$.

7.20 The living spaces of all homes in a city have a mean of 2300 square feet and a standard deviation of 450 square feet. Let $\bar{x}$ be the mean living space for a random sample of 25 homes selected from this city. Find the mean and standard deviation of the sampling distribution of $\bar{x}$.

7.21 The mean monthly out-of-pocket cost of prescription drugs for all senior citizens in a city is \$320 with a standard deviation of \$72. Let $\bar{x}$ be the mean of such costs for a random sample of 25 senior citizens from this city. Find the mean and standard deviation of the sampling distribution of $\bar{x}$.

7.22 According to the Connecticut Council on Problem Gambling, problem gamblers in Connecticut have an average of \$16,842 in gambling debts (*The Hartford Courant*, August 10, 2004). Assume that currently all problem gamblers in Connecticut have an average of \$16,842 in gambling debts with a standard deviation of \$3600. Let $\bar{x}$ be the mean gambling debt of a random sample of 225 such gamblers from Connecticut. Find the mean and standard deviation of $\bar{x}$.

*7.23 Suppose the standard deviation of recruiting costs per player for all female basketball players recruited by all public universities in the Midwest is \$605. Let $\bar{x}$ be the mean recruiting cost for a sample of a certain number of such players. What sample size will give the standard deviation of $\bar{x}$ equal to \$55?

***7.24** The standard deviation of the 2005 gross sales of all corporations is known to be \$139.50 million. Let $\bar{x}$ be the mean of the 2005 gross sales of a sample of corporations. What sample size will produce the standard deviation of $\bar{x}$ equal to \$15.50 million?

***7.25** Consider the sampling distribution of $\bar{x}$ given in Table 7.5.

a. Calculate the value of $\mu_{\bar{x}}$ using the formula $\mu_{\bar{x}} = \Sigma \bar{x} P(\bar{x})$. Is the value of μ calculated in Exercise 7.6 the same as the value of $\mu_{\bar{x}}$ calculated here?

b. Calculate the value of $\sigma_{\bar{x}}$ by using the formula

$$\sigma_{\bar{x}} = \sqrt{\Sigma \bar{x}^2 P(\bar{x}) - (\mu_{\bar{x}})^2}$$

c. From Exercise 7.6, $\sigma = 8.09$. Also, our sample size is 3 so that $n = 3$. Therefore, $\sigma/\sqrt{n} = 8.09/\sqrt{3} = 4.67$. From part b, you should get $\sigma_{\bar{x}} = 3.30$. Why does $\sigma/\sqrt{n}$ not equal $\sigma_{\bar{x}}$ in this case?

d. In our example (given in the beginning of Section 7.1.1) on scores, $N = 5$ and $n = 3$. Hence, $n/N = 3/5 = .60$. Because n/N is greater than .05, the appropriate formula to find $\sigma_{\bar{x}}$ is

$$\sigma_{\bar{x}} = \frac{\sigma}{\sqrt{n}} \sqrt{\frac{N-n}{N-1}}$$

Show that the value of $\sigma_{\bar{x}}$ calculated by using this formula gives the same value as the one calculated in part b above.

7.4 Shape of the Sampling Distribution of $\bar{x}$

The shape of the sampling distribution of $\bar{x}$ relates to the following two cases.

1. The population from which samples are drawn has a normal distribution.
2. The population from which samples are drawn does not have a normal distribution.

7.4.1 Sampling from a Normally Distributed Population

When the population from which samples are drawn is normally distributed with its mean equal to μ and standard deviation equal to σ, then

1. The mean of $\bar{x}$, $\mu_{\bar{x}}$, is equal to the mean of the population, μ.
2. The standard deviation of $\bar{x}$, $\sigma_{\bar{x}}$, is equal to $\sigma/\sqrt{n}$, assuming $n/N \le .05$.
3. The shape of the sampling distribution of $\bar{x}$ is normal, whatever the value of n.

Sampling Distribution of $\bar{x}$ When the Population has a Normal Distribution If the population from which the samples are drawn is normally distributed with mean μ and standard deviation σ, then the sampling distribution of the sample mean, $\bar{x}$, will also be normally distributed with the following mean and standard deviation, irrespective of the sample size:

$$\mu_{\bar{x}} = \mu \quad \text{and} \quad \sigma_{\bar{x}} = \frac{\sigma}{\sqrt{n}}$$

Remember ► For $\sigma_{\bar{x}} = \sigma/\sqrt{n}$ to be true, n/N must be less than or equal to .05.

Figure 7.2*a* shows the probability distribution curve for a population. The distribution curves in Figure 7.2*b* through Figure 7.2*e* show the sampling distributions of $\bar{x}$ for different sample sizes taken from the population of Figure 7.2*a*. As we can observe, the population has a normal distribution. Because of this, the sampling distribution of $\bar{x}$ is normal for each of

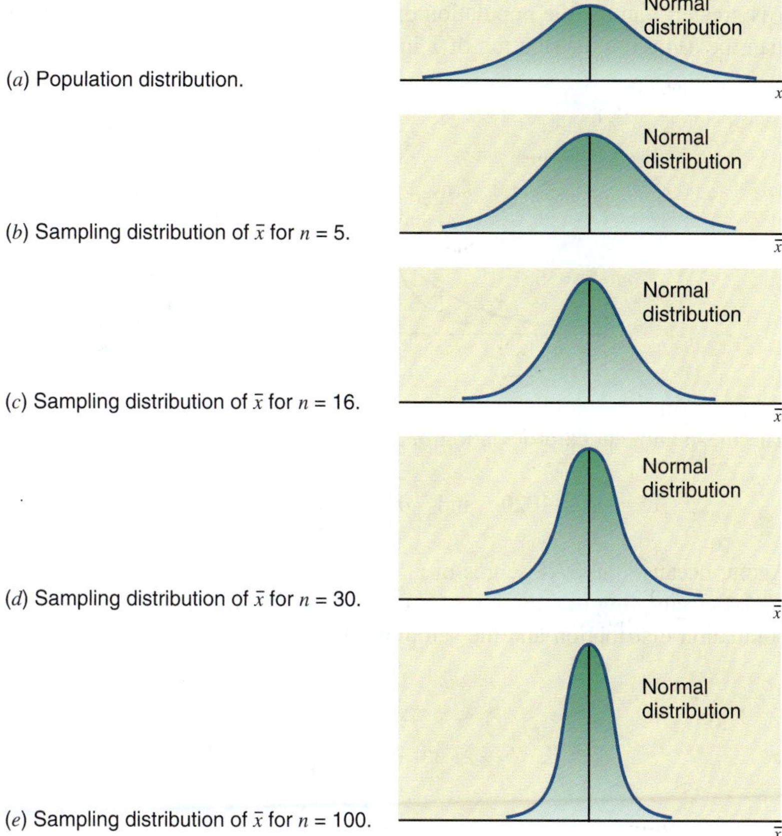

Figure 7.2 Population distribution and sampling distributions of $\bar{x}$.

the four cases illustrated in parts *b* through *e* of Figure 7.2. Also notice from Figure 7.2*b* through Figure 7.2*e* that the spread of the sampling distribution of $\bar{x}$ decreases as the sample size increases.

Example 7–3 illustrates the calculation of the mean and standard deviation of $\bar{x}$ and the description of the shape of its sampling distribution.

■ EXAMPLE 7–3

Finding the mean, standard deviation, and sampling distribution of $\bar{x}$: normal population.

In a recent SAT, the mean score for all examinees was 1020. Assume that the distribution of SAT scores of all examinees is normal with a mean of 1020 and a standard deviation of 153. Let $\bar{x}$ be the mean SAT score of a random sample of certain examinees. Calculate the mean and standard deviation of $\bar{x}$ and describe the shape of its sampling distribution when the sample size is

(a) 16 **(b)** 50 **(c)** 1000

Solution Let μ and σ be the mean and standard deviation of SAT scores of all examinees, and let $\mu_{\bar{x}}$ and $\sigma_{\bar{x}}$ be the mean and standard deviation of the sampling distribution of $\bar{x}$. Then, from the given information,

$$\mu = 1020 \quad \text{and} \quad \sigma = 153$$

(a) The mean and standard deviation of $\bar{x}$ are

$$\mu_{\bar{x}} = \mu = \mathbf{1020} \quad \text{and} \quad \sigma_{\bar{x}} = \frac{\sigma}{\sqrt{n}} = \frac{153}{\sqrt{16}} = \mathbf{38.250}$$

Because the SAT scores of all examinees are assumed to be normally distributed, the sampling distribution of $\bar{x}$ for samples of 16 examinees is also normal. Figure 7.3

shows the population distribution and the sampling distribution of $\bar{x}$. Note that because σ is greater than $\sigma_{\bar{x}}$, the population distribution has a wider spread but smaller height than the sampling distribution of $\bar{x}$ in Figure 7.3.

Figure 7.3

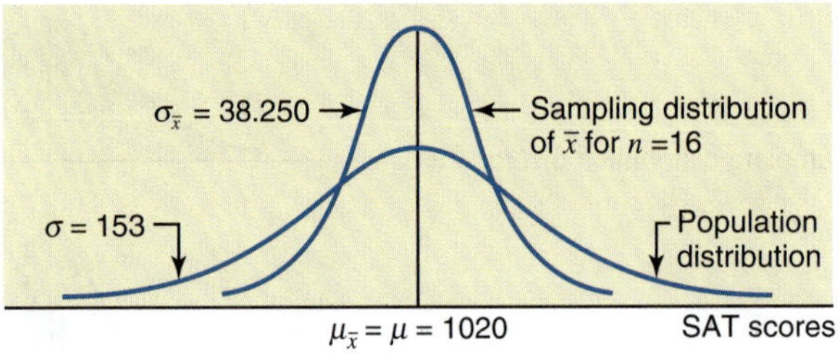

(b) The mean and standard deviation of $\bar{x}$ are

$$\mu_{\bar{x}} = \mu = \mathbf{1020} \quad \text{and} \quad \sigma_{\bar{x}} = \frac{\sigma}{\sqrt{n}} = \frac{153}{\sqrt{50}} = \mathbf{21.637}$$

Again, because the SAT scores of all examinees are assumed to be normally distributed, the sampling distribution of $\bar{x}$ for samples of 50 examinees is also normal. The population distribution and the sampling distribution of $\bar{x}$ are shown in Figure 7.4.

Figure 7.4

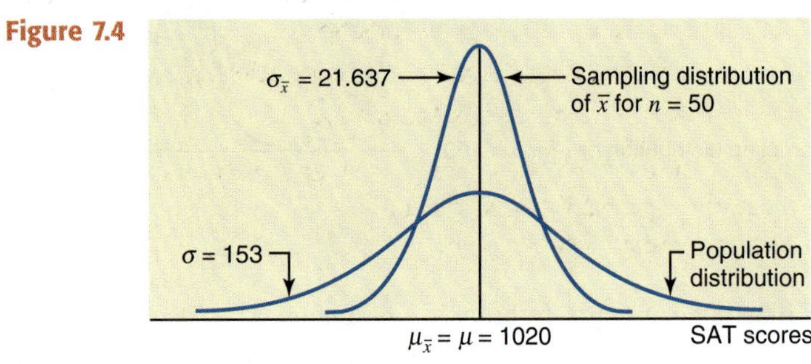

(c) The mean and standard deviation of $\bar{x}$ are

$$\mu_{\bar{x}} = \mu = \mathbf{1020} \quad \text{and} \quad \sigma_{\bar{x}} = \frac{\sigma}{\sqrt{n}} = \frac{153}{\sqrt{1000}} = \mathbf{4.838}$$

Again, because the SAT scores of all examinees are assumed to be normally distributed, the sampling distribution of $\bar{x}$ for samples of 1000 examinees is also normal. The two distributions are shown in Figure 7.5.

Figure 7.5

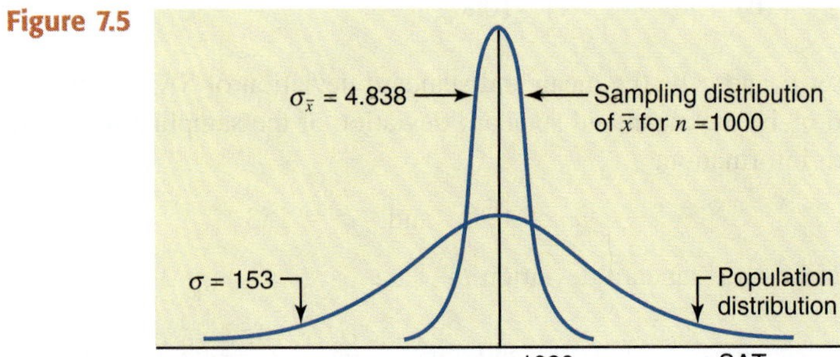

Thus, whatever the sample size, the sampling distribution of $\bar{x}$ is normal when the population from which the samples are drawn is normally distributed. ■

7.4.2 Sampling from a Population That Is Not Normally Distributed

Most of the time the population from which the samples are selected is not normally distributed. In such cases, the shape of the sampling distribution of $\bar{x}$ is inferred from a very important theorem called the **central limit theorem**.

Central Limit Theorem According to the *central limit theorem*, for a large sample size, the sampling distribution of $\bar{x}$ is approximately normal, irrespective of the shape of the population distribution. The mean and standard deviation of the sampling distribution of $\bar{x}$ are

$$\mu_{\bar{x}} = \mu \quad \text{and} \quad \sigma_{\bar{x}} = \frac{\sigma}{\sqrt{n}}$$

The sample size is usually considered to be large if $n \geq 30$.

Note that when the population does not have a normal distribution, the shape of the sampling distribution is not exactly normal but is approximately normal for a large sample size. The approximation becomes more accurate as the sample size increases. Another point to remember is that the central limit theorem applies to *large* samples only. Usually, if the sample size is 30 or more, it is considered sufficiently large to apply the central limit theorem to the sampling distribution of $\bar{x}$. Thus, according to the central limit theorem,

1. When $n \geq 30$, the shape of the sampling distribution of $\bar{x}$ is approximately normal irrespective of the shape of the population distribution.
2. The mean of $\bar{x}$, $\mu_{\bar{x}}$, is equal to the mean of the population, μ.
3. The standard deviation of $\bar{x}$, $\sigma_{\bar{x}}$, is equal to $\sigma/\sqrt{n}$.

Again, remember that for $\sigma_{\bar{x}} = \sigma/\sqrt{n}$ to apply, n/N must be less than or equal to .05.

Figure 7.6*a* shows the probability distribution curve for a population. The distribution curves in Figure 7.6*b* through Figure 7.6*e* show the sampling distributions of $\bar{x}$ for different sample

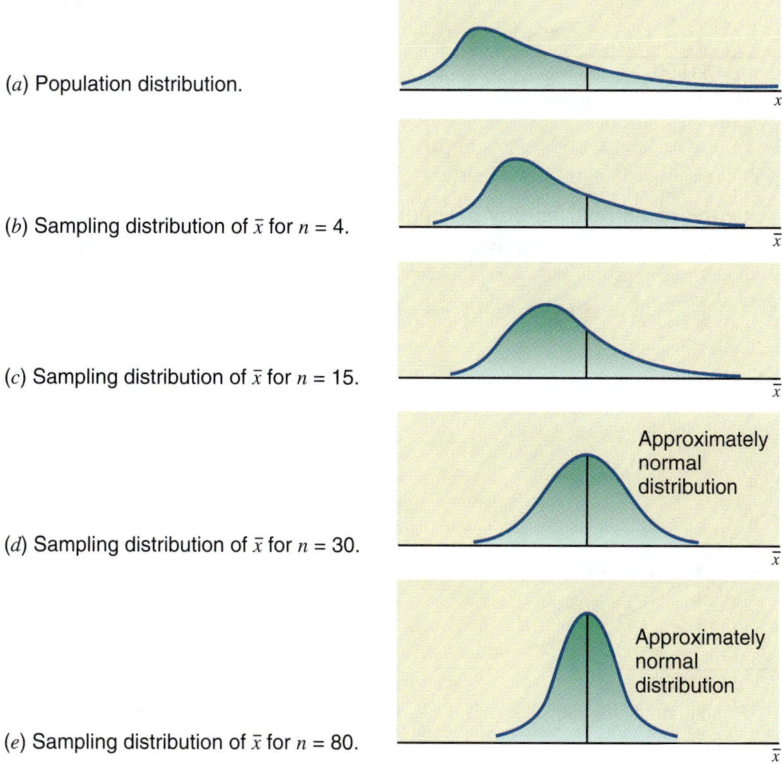

Figure 7.6 Population distribution and sampling distributions of $\bar{x}$.

sizes taken from the population of Figure 7.6*a*. As we can observe, the population is not normally distributed. The sampling distributions of $\bar{x}$ shown in parts *b* and *c*, when $n < 30$, are not normal. However, the sampling distributions of $\bar{x}$ shown in parts *d* and *e*, when $n \geq 30$, are (approximately) normal. Also notice that the spread of the sampling distribution of $\bar{x}$ decreases as the sample size increases.

Example 7–4 illustrates the calculation of the mean and standard deviation of $\bar{x}$ and describes the shape of the sampling distribution of $\bar{x}$ when the sample size is large.

■ EXAMPLE 7–4

Finding the mean, standard deviation, and sampling distribution of $\bar{x}$: nonnormal population.

The mean rent paid by all tenants in a large city is \$1550 with a standard deviation of \$225. However, the population distribution of rents for all tenants in this city is skewed to the right. Calculate the mean and standard deviation of $\bar{x}$ and describe the shape of its sampling distribution when the sample size is

(a) 30 **(b)** 100

Solution Although the population distribution of rents paid by all tenants is not normal, in each case the sample size is large ($n \geq 30$). Hence, the central limit theorem can be applied to infer the shape of the sampling distribution of $\bar{x}$.

(a) Let $\bar{x}$ be the mean rent paid by a sample of 30 tenants. Then, the sampling distribution of $\bar{x}$ is approximately normal with the values of the mean and standard deviation as

$$\mu_{\bar{x}} = \mu = \mathbf{\$1550} \quad \text{and} \quad \sigma_{\bar{x}} = \frac{\sigma}{\sqrt{n}} = \frac{225}{\sqrt{30}} = \mathbf{\$41.079}$$

Figure 7.7 shows the population distribution and the sampling distribution of $\bar{x}$.

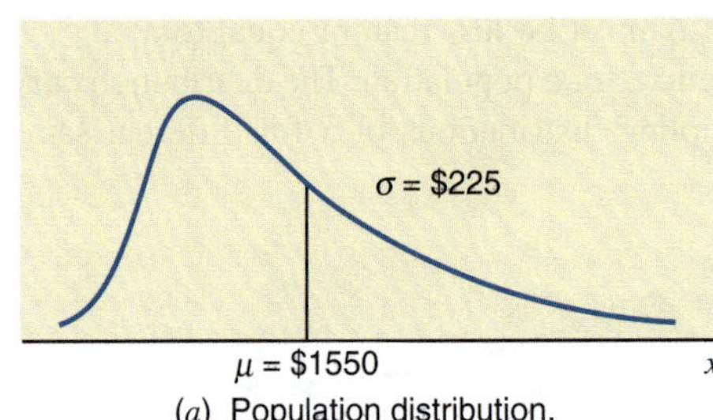

(*a*) Population distribution.

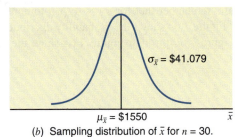

$\sigma_{\bar{x}}$ = \$41.079

$\mu_{\bar{x}}$ = \$1550 $\bar{x}$

(*b*) Sampling distribution of $\bar{x}$ for $n = 30$.

Figure 7.7

(b) Let $\bar{x}$ be the mean rent paid by a sample of 100 tenants. Then, the sampling distribution of $\bar{x}$ is approximately normal with the values of the mean and standard deviation as

$$\mu_{\bar{x}} = \mu = \mathbf{\$1550} \quad \text{and} \quad \sigma_{\bar{x}} = \frac{\sigma}{\sqrt{n}} = \frac{225}{\sqrt{100}} = \mathbf{\$22.500}$$

Figure 7.8 shows the population distribution and the sampling distribution of $\bar{x}$.

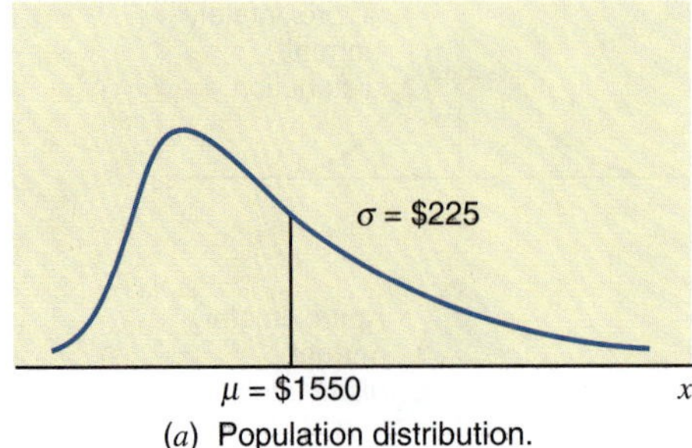

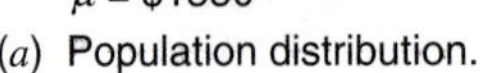
(*a*) Population distribution.

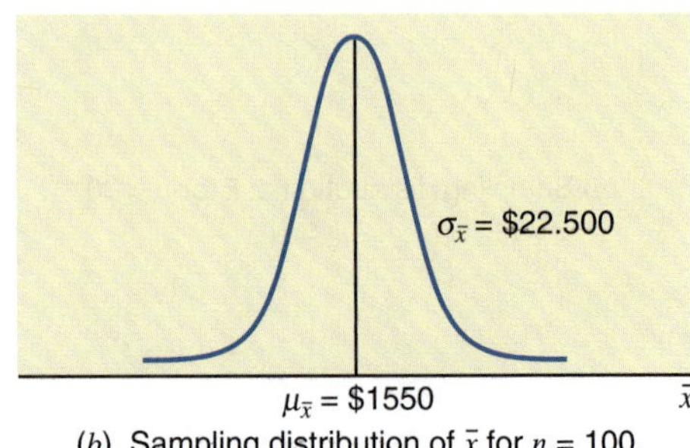

(*b*) Sampling distribution of $\bar{x}$ for $n = 100$.

Figure 7.8 ■

EXERCISES

CONCEPTS AND PROCEDURES

7.26 What condition or conditions must hold true for the sampling distribution of the sample mean to be normal when the sample size is less than 30?

7.27 Explain the central limit theorem.

7.28 A population has a distribution that is skewed to the left. Indicate in which of the following cases the central limit theorem will apply to describe the sampling distribution of the sample mean.

a. $n = 400$ **b.** $n = 25$ **c.** $n = 36$

7.29 A population has a distribution that is skewed to the right. A sample of size n is selected from this population. Describe the shape of the sampling distribution of the sample mean for each of the following cases.

a. $n = 25$ **b.** $n = 80$ **c.** $n = 29$

7.30 A population has a normal distribution. A sample of size n is selected from this population. Describe the shape of the sampling distribution of the sample mean for each of the following cases.

a. $n = 94$ **b.** $n = 11$

7.31 A population has a normal distribution. A sample of size n is selected from this population. Describe the shape of the sampling distribution of the sample mean for each of the following cases.

a. $n = 23$ **b.** $n = 450$

APPLICATIONS

7.32 The delivery times for all food orders at a fast-food restaurant during the lunch hour are normally distributed with a mean of 6.7 minutes and a standard deviation of 2.1 minutes. Let $\bar{x}$ be the mean delivery time for a random sample of 16 orders at this restaurant. Calculate the mean and standard deviation of $\bar{x}$ and describe the shape of its sampling distribution.

7.33 In a highway construction zone with a posted speed limit of 40 miles per hour, the speeds of all vehicles are normally distributed with a mean of 46 mph and a standard deviation of 3 mph. Let $\bar{x}$ be the mean speed of a random sample of 20 vehicles traveling through this construction zone. Calculate the mean and standard deviation of $\bar{x}$ and describe the shape of its sampling distribution.

7.34 The amounts of electric bills for all households in a city have an approximately normal distribution with a mean of \$90 and a standard deviation of \$20. Let $\bar{x}$ be the mean amount of electric bills for a random sample of 25 households selected from this city. Find the mean and standard deviation of $\bar{x}$ and comment on the shape of its sampling distribution.

7.35 The GPAs of all 5540 students enrolled at a university have an approximately normal distribution with a mean of 3.02 and a standard deviation of .29. Let $\bar{x}$ be the mean GPA of a random sample of 48 students selected from this university. Find the mean and standard deviation of $\bar{x}$ and comment on the shape of its sampling distribution.

7.36 The weights of all people living in a town have a distribution that is skewed to the right with a mean of 133 pounds and a standard deviation of 24 pounds. Let $\bar{x}$ be the mean weight of a random sample of 45 persons selected from this town. Find the mean and standard deviation of $\bar{x}$ and comment on the shape of its sampling distribution.

7.37 The amounts of telephone bills for all households in a large city have a distribution that is skewed to the right with a mean of \$96 and a standard deviation of \$27. Let $\bar{x}$ be the mean amount of telephone bills for a random sample of 90 households selected from this city. Calculate the mean and standard deviation of $\bar{x}$ and describe the shape of its sampling distribution.

7.38 Suppose the incomes of all people in America who own hybrid (gas and electric) automobiles are normally distributed with a mean of \$58,000 and a standard deviation of \$8300. Let $\bar{x}$ be the mean income of a random sample of 50 such owners. Calculate the mean and standard deviation of $\bar{x}$ and describe the shape of its sampling distribution.

7.39 According to the Kaiser Family Foundation, children aged 8 to 18 in the United States spend an average of 62 minutes per day using computers, not counting the time they spend using computers to do homework (*Time*, March 21, 2005). Suppose the probability distribution of such times for all children aged 8 to 18 is skewed to the right with a mean of 62 minutes and a standard deviation of 14 minutes. Let $\bar{x}$ be the mean time spent per day using computers by a random sample of 400 children aged 8 to 18. Find the mean and standard deviation of $\bar{x}$ and describe the shape of its sampling distribution.

7.5 Applications of the Sampling Distribution of $\bar{x}$

From the central limit theorem, for large samples, the sampling distribution of $\bar{x}$ is approximately normal with mean μ and standard deviation $\sigma_{\bar{x}} = \sigma/\sqrt{n}$. Based on this result, we can make the following statements about $\bar{x}$ for large samples. The areas under the curve of $\bar{x}$ mentioned in these statements are found from the normal distribution table.

1. *If we take all possible samples of the same (large) size from a population and calculate the mean for each of these samples, then about 68.26% of the sample means will be within one standard deviation of the population mean.* Or we can state that if we take one sample (of $n \geq 30$) from a population and calculate the mean for this sample, the probability that this sample mean will be within one standard deviation of the population mean is .6826. That is,

$$P(\mu - 1\sigma_{\bar{x}} \leq \bar{x} \leq \mu + 1\sigma_{\bar{x}}) = .8413 - .1587 = .6826$$

This probability is shown in Figure 7.9.

Figure 7.9 $P(\mu - 1\sigma_{\bar{x}} \leq \bar{x} \leq \mu + 1\sigma_{\bar{x}})$

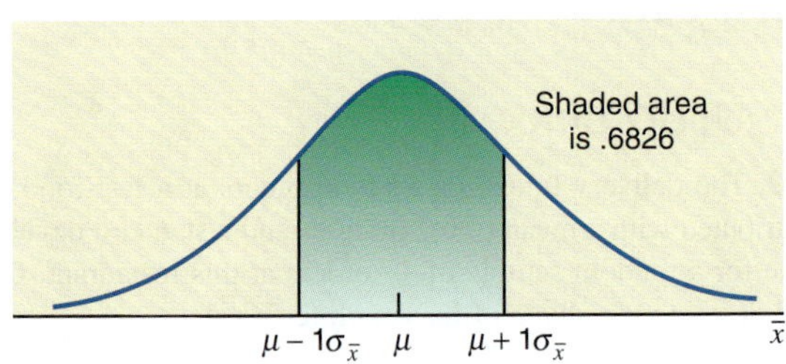

2. *If we take all possible samples of the same (large) size from a population and calculate the mean for each of these samples, then about 95.44% of the sample means will be within two standard deviations of the population mean.* Or we can state that if we take one sample (of $n \geq 30$) from a population and calculate the mean for this sample, the probability that this sample mean will be within two standard deviations of the population mean is .9544. That is,

$$P(\mu - 2\sigma_{\bar{x}} \leq \bar{x} \leq \mu + 2\sigma_{\bar{x}}) = .9772 - .0228 = .9544$$

This probability is shown in Figure 7.10.

Figure 7.10 $P(\mu - 2\sigma_{\bar{x}} \leq \bar{x} \leq \mu + 2\sigma_{\bar{x}})$.

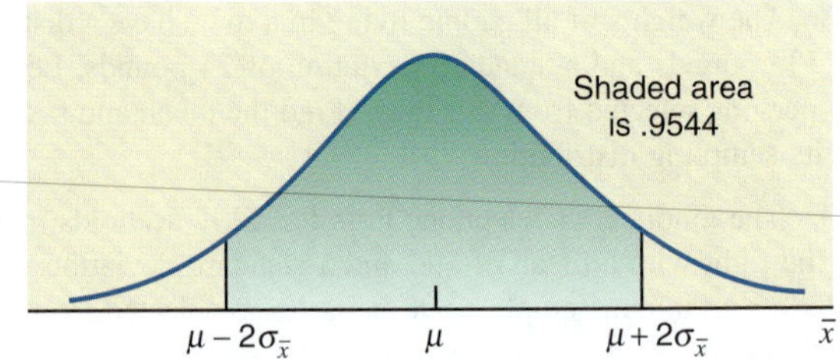

3. *If we take all possible samples of the same (large) size from a population and calculate the mean for each of these samples, then about 99.74% of the sample means will be within three standard deviations of the population mean.* Or we can state that if we take one sample (of $n \geq 30$) from a population and calculate the mean for this sample, the probability that this sample mean will be within three standard deviations of the population mean is .9974. That is,

$$P(\mu - 3\sigma_{\bar{x}} \leq \bar{x} \leq \mu + 3\sigma_{\bar{x}}) = .9987 - .0013 = .9974$$

This probability is shown in Figure 7.11.

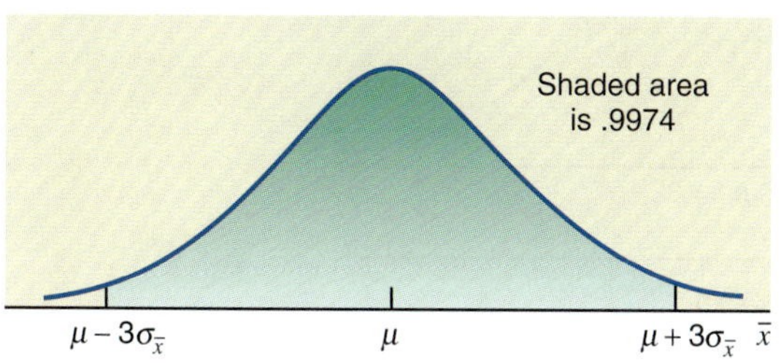

Figure 7.11 $P(\mu - 3\sigma_{\bar{x}} \leq \bar{x} \leq \mu + 3\sigma_{\bar{x}})$

When conducting a survey, we usually select one sample and compute the value of $\bar{x}$ based on that sample. We never select all possible samples of the same size and then prepare the sampling distribution of $\bar{x}$. Rather, we are more interested in finding the probability that the value of $\bar{x}$ computed from one sample falls within a given interval. Examples 7–5 and 7–6 illustrate this procedure.

■ EXAMPLE 7–5

Calculating the probability of $\bar{x}$ in an interval: normal population.

Assume that the weights of all packages of a certain brand of cookies are normally distributed with a mean of 32 ounces and a standard deviation of .3 ounces. Find the probability that the mean weight, $\bar{x}$, of a random sample of 20 packages of this brand of cookies will be between 31.8 and 31.9 ounces.

Solution Although the sample size is small ($n < 30$), the shape of the sampling distribution of $\bar{x}$ is normal because the population is normally distributed. The mean and standard deviation of $\bar{x}$ are

$$\mu_{\bar{x}} = \mu = 32 \text{ ounces} \quad \text{and} \quad \sigma_{\bar{x}} = \frac{\sigma}{\sqrt{n}} = \frac{.3}{\sqrt{20}} = .06708204 \text{ ounces}$$

We are to compute the probability that the value of $\bar{x}$ calculated for one randomly drawn sample of 20 packages is between 31.8 and 31.9 ounces—that is,

$$P(31.8 < \bar{x} < 31.9)$$

This probability is given by the area under the normal distribution curve for $\bar{x}$ between the points $\bar{x} = 31.8$ and $\bar{x} = 31.9$. The first step in finding this area is to convert the two $\bar{x}$ values to their respective z values.

z Value for a Value of $\bar{x}$ The *z value for a value of $\bar{x}$* is calculated as

$$z = \frac{\bar{x} - \mu}{\sigma_{\bar{x}}}$$

The z values for $\bar{x} = 31.8$ and $\bar{x} = 31.9$ are computed next, and they are shown on the z scale below the normal distribution curve for $\bar{x}$ in Figure 7.12.

$$\text{For } \bar{x} = 31.8: \quad z = \frac{31.8 - 32}{.06708204} = -2.98$$

$$\text{For } \bar{x} = 31.9: \quad z = \frac{31.9 - 32}{.06708204} = -1.49$$

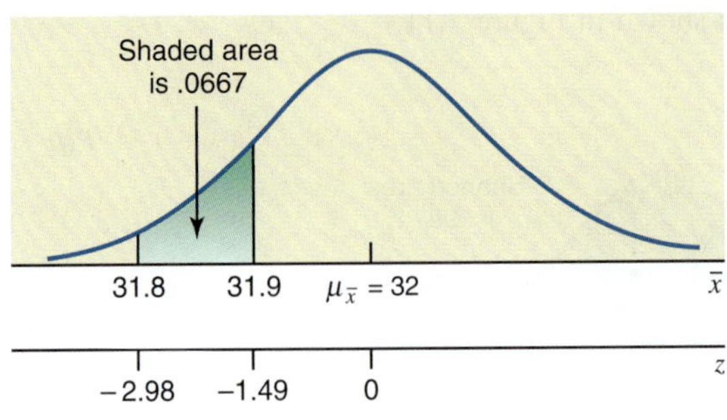

Figure 7.12 $P(31.8 < \bar{x} < 31.9)$

The probability that $\bar{x}$ is between 31.8 and 31.9 is given by the area under the standard normal curve between $z = -2.98$ and $z = -1.49$. Thus, the required probability is

$$\begin{aligned} P(31.8 < \bar{x} < 31.9) &= P(-2.98 < z < -1.49) \\ &= P(z < -1.49) - P(z < -2.98) \\ &= .0681 - .0014 = \mathbf{.0667} \end{aligned}$$

Therefore, the probability is .0667 that the mean weight of a sample of 20 packages will be between 31.8 and 31.9 ounces. ■

■ EXAMPLE 7–6

Calculating the probability of $\bar{x}$ in an interval: $n > 30$.

According to CardWeb, consumers in the United States owed an average of \$7868 on their credit cards in 2004 (*USA TODAY*, July 1, 2005). Suppose the shape of the probability distribution of the current credit card debts of all consumers in the United States is unknown but its mean is \$7868 and the standard deviation is \$2160. Let $\bar{x}$ be the mean credit card debt of a random sample of 81 U.S. consumers.

(a) What is the probability that the mean of the current credit card debts for this sample is within \$440 of the population mean?

(b) What is the probability that the mean of the current credit card debts for this sample is lower than the population mean by \$320 or more?

Solution From the given information, for the current credit card debts of all consumers in the United States,

$$\mu = \$7868 \quad \text{and} \quad \sigma = \$2160$$

Although the shape of the probability distribution of the population (current credit card debts of all consumers in the United States) is unknown, the sampling distribution of $\bar{x}$ is approximately normal because the sample size is large ($n > 30$). Remember that when the sample is large, the central limit theorem applies. The mean and standard deviation of the sampling distribution of $\bar{x}$ are

$$\mu_{\bar{x}} = \mu = \$7868 \quad \text{and} \quad \sigma_{\bar{x}} = \frac{\sigma}{\sqrt{n}} = \frac{2160}{\sqrt{81}} = \$240$$

(a) The probability that the mean of the current credit card debts for this sample is within \$440 of the population mean is written as

$$P(7428 \leq \bar{x} \leq 8308)$$

This probability is given by the area under the normal curve for $\bar{x}$ between $\bar{x} = \$7428$ and $\bar{x} = \$8308$, as shown in Figure 7.13. We find this area as follows.

$$\text{For } \bar{x} = \$7428: \quad z = \frac{\bar{x} - \mu}{\sigma_{\bar{x}}} = \frac{7428 - 7868}{240} = -1.83$$

$$\text{For } \bar{x} = \$8308: \quad z = \frac{\bar{x} - \mu}{\sigma_{\bar{x}}} = \frac{8308 - 7868}{240} = 1.83$$

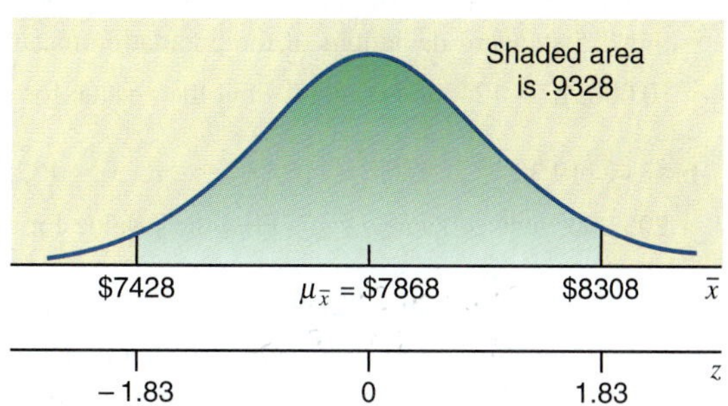

Figure 7.13 $P(\$7428 \leq \bar{x} \leq \$8308)$

Hence, the required probability is

$$\begin{aligned} P(\$7428 \leq \bar{x} \leq \$8308) &= P(-1.83 \leq z \leq 1.83) \\ &= P(z \leq 1.83) - P(z \leq -1.83) \\ &= .9664 - .0336 = \mathbf{.9328} \end{aligned}$$

Therefore, the probability that the mean of the current credit card debts for this sample is within $440 of the population mean is .9328.

(b) The probability that the mean of the current credit card debts for this sample is lower than the population mean by $320 or more is written as

$$P(\bar{x} \leq 7548)$$

This probability is given by the area under the normal curve for $\bar{x}$ to the left of $\bar{x}$ = $7548, as shown in Figure 7.14. We find this area as follows:

$$\text{For } \bar{x} = \$7548: \quad z = \frac{\bar{x} - \mu}{\sigma_{\bar{x}}} = \frac{7548 - 7868}{240} = -1.33$$

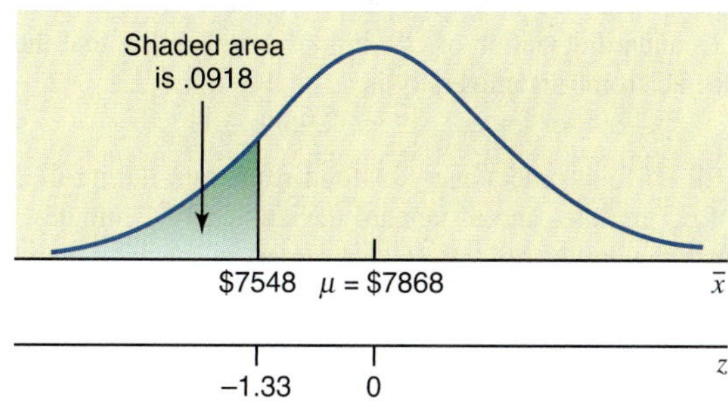

Figure 7.14 $P(\bar{x} \leq 7548)$

Hence, the required probability is

$$P(\bar{x} \leq 7548) = P(z \leq -1.33) = \mathbf{.0918}$$

Therefore, the probability that the mean of the current credit card debts for this sample is lower than the population mean by $320 or more is .0918. ■

EXERCISES

CONCEPTS AND PROCEDURES

7.40 If all possible samples of the same (large) size are selected from a population, what percent of all the sample means will be within 2.5 standard deviations of the population mean?

7.41 If all possible samples of the same (large) size are selected from a population, what percent of all the sample means will be within 1.5 standard deviations of the population mean?

7.42 For a population, $N = 10{,}000$, $\mu = 124$, and $\sigma = 18$. Find the z value for each of the following for $n = 36$.

a. $\bar{x} = 128.60$ **b.** $\bar{x} = 119.30$ **c.** $\bar{x} = 116.88$ **d.** $\bar{x} = 132.05$

7.43 For a population, $N = 205{,}000$, $\mu = 66$, and $\sigma = 7$. Find the z value for each of the following for $n = 49$.

a. $\bar{x} = 68.44$ **b.** $\bar{x} = 58.75$ **c.** $\bar{x} = 62.35$ **d.** $\bar{x} = 71.82$

7.44 Let x be a continuous random variable that has a normal distribution with $\mu = 75$ and $\sigma = 14$. Assuming $n/N \le .05$, find the probability that the sample mean, $\bar{x}$, for a random sample of 20 taken from this population will be

a. between 68.5 and 77.3 **b.** less than 72.4

7.45 Let x be a continuous random variable that has a normal distribution with $\mu = 48$ and $\sigma = 8$. Assuming $n/N \le .05$, find the probability that the sample mean, $\bar{x}$, for a random sample of 16 taken from this population will be

a. between 49.6 and 52.2 **b.** more than 45.7

7.46 Let x be a continuous random variable that has a distribution skewed to the right with $\mu = 60$ and $\sigma = 10$. Assuming $n/N \le .05$, find the probability that the sample mean, $\bar{x}$, for a random sample of 40 taken from this population will be

a. less than 62.20 **b.** between 61.4 and 64.2

7.47 Let x be a continuous random variable that follows a distribution skewed to the left with $\mu = 90$ and $\sigma = 18$. Assuming $n/N \le .05$, find the probability that the sample mean, $\bar{x}$, for a random sample of 64 taken from this population will be

a. less than 82.3 **b.** greater than 86.7

APPLICATIONS

7.48 In a highway construction zone with a posted speed limit of 40 miles per hour, the speeds of all vehicles are normally distributed with a mean of 46 mph and a standard deviation of 3 mph. Find the probability that the mean speed of a random sample of 20 cars traveling through this construction zone is

a. more than 45 mph **b.** less than 45.5 mph **c.** 44.5 to 47 mph

7.49 The GPAs of all students enrolled at a large university have an approximately normal distribution with a mean of 3.02 and a standard deviation of .29. Find the probability that the mean GPA of a random sample of 20 students selected from this university is

a. 3.10 or higher **b.** 2.90 or lower **c.** 2.95 to 3.11

7.50 The delivery times for all food orders at a fast-food restaurant during the lunch hour are normally distributed with a mean of 6.7 minutes and a standard deviation of 2.1 minutes. Find the probability that the mean delivery time for a random sample of 16 such orders at this restaurant is

a. between 7 and 8 minutes
b. within 1 minute of the population mean
c. less than the population mean by 1 minute or more

7.51 According to the Connecticut Council on Problem Gambling, problem gamblers in Connecticut have an average of $16,842 in gambling debts (*The Hartford Courant*, August 10, 2004). Assume that the current gambling debts of all problem gamblers in Connecticut have a normal distribution with a mean of $16,842 and a standard deviation of $3600. Find the probability that the mean gambling debt of a random sample of 225 problem gamblers from Connecticut is

a. between $16,450 and $17,120
b. within $300 of the population mean
c. greater than the population mean by at least $200

7.52 The times that college students spend studying per week have a distribution that is skewed to the right with a mean of 8.4 hours and a standard deviation of 2.7 hours. Find the probability that the mean time spent studying per week for a random sample of 45 students would be

a. between 8 and 9 hours b. less than 8 hours

7.53 The credit card debts of all college students have a distribution that is skewed to the right with a mean of \$1840 and a standard deviation of \$453. Find the probability that the mean credit card debt for a random sample of 36 college students would be

a. between \$1750 and \$1950 b. less than \$1700

7.54 According to the Kaiser Family Foundation, children aged 8 to 18 in the United States spend an average of 62 minutes per day using computers, not counting the time they spend using computers to do homework (*Time*, March 21, 2005). Suppose the probability distribution of such times for all children aged 8 to 18 is skewed to the right with a mean of 62 minutes and a standard deviation of 14 minutes. Find the probability that the mean time spent per day using computers by a random sample of 400 children aged 8 to 18 is

a. between 60.5 and 63 minutes
b. within 2 minutes of the population mean
c. greater than the population mean by 1.5 minutes or more

7.55 The amounts of electric bills for all households in a city have a skewed probability distribution with a mean of \$120 and a standard deviation of \$25. Find the probability that the mean amount of electric bills for a random sample of 75 households selected from this city will be

a. between \$112 and \$117
b. within \$6 of the population mean
c. more than the population mean by at least \$5

7.56 The balances of all savings accounts at a local bank have a distribution that is skewed to the right with its mean equal to \$12,450 and standard deviation equal to \$4300. Find the probability that the mean balance of a sample of 50 savings accounts selected from this bank will be

a. more than \$11,500
b. between \$12,000 and \$13,800
c. within \$1500 of the population mean
d. more than the population mean by at least \$1000

7.57 The heights of all adults in a large city have a distribution that is skewed to the right with a mean of 68 inches and a standard deviation of 4 inches. Find the probability that the mean height of a random sample of 100 adults selected from this city would be

a. less than 67.8 inches
b. between 67.5 inches and 68.7 inches
c. within .6 inches of the population mean
d. less than the population mean by .5 inches or more

7.58 Johnson Electronics Corporation makes electric tubes. It is known that the standard deviation of the lives of these tubes is 150 hours. The company's research department takes a sample of 100 such tubes and finds that the mean life of these tubes is 2250 hours. What is the probability that this sample mean is within 25 hours of the mean life of all tubes produced by this company?

7.59 A machine at Katz Steel Corporation makes 3-inch-long nails. The probability distribution of the lengths of these nails is normal with a mean of 3 inches and a standard deviation of .1 inches. The quality control inspector takes a sample of 25 nails once a week and calculates the mean length of these nails. If the mean of this sample is either less than 2.95 inches or greater than 3.05 inches, the inspector concludes that the machine needs an adjustment. What is the probability that based on a sample of 25 nails the inspector will conclude that the machine needs an adjustment?

7.6 Population and Sample Proportions

The concept of proportion is the same as the concept of relative frequency discussed in Chapter 2 and the concept of probability of success in a binomial experiment. The relative frequency of a category or class gives the proportion of the sample or population that belongs to that category or class. Similarly, the probability of success in a binomial experiment represents the proportion of the sample or population that possesses a given characteristic.

The **population proportion**, denoted by p, is obtained by taking the ratio of the number of elements in a population with a specific characteristic to the total number of elements in the population. The **sample proportion**, denoted by $\hat{p}$ (pronounced *p hat*), gives a similar ratio for a sample.

Population and Sample Proportions The *population* and *sample proportions*, denoted by p and $\hat{p}$, respectively, are calculated as

$$p = \frac{X}{N} \quad \text{and} \quad \hat{p} = \frac{x}{n}$$

where

N = total number of elements in the population
n = total number of elements in the sample
X = number of elements in the population that possess a specific characteristic
x = number of elements in the sample that possess a specific characteristic

Example 7–7 illustrates the calculation of the population and sample proportions.

■ EXAMPLE 7–7

Calculating the population and sample proportions.

Suppose a total of 789,654 families live in a city and 563,282 of them own homes. A sample of 240 families is selected from this city, and 158 of them own homes. Find the proportion of families who own homes in the population and in the sample.

Solution For the population of this city,

$$N = \text{population size} = 789{,}654$$

$$X = \text{families in the population who own homes} = 563{,}282$$

The proportion of all families in this city who own homes is

$$p = \frac{X}{N} = \frac{563{,}282}{789{,}654} = \mathbf{.71}$$

Now, a sample of 240 families is taken from this city and 158 of them are home-owners. Then,

$$n = \text{sample size} = 240$$

$$x = \text{families in the sample who own homes} = 158$$

The sample proportion is

$$\hat{p} = \frac{x}{n} = \frac{158}{240} = \mathbf{.66}$$

■

As in the case of the mean, the difference between the sample proportion and the corresponding population proportion gives the sampling error, assuming that the sample is random and no nonsampling error has been made. That is, in the case of the proportion,

$$\text{Sampling error} = \hat{p} - p$$

For instance, for Example 7–7,

$$\text{Sampling error} = \hat{p} - p = .66 - .71 = -.05$$

7.7 Mean, Standard Deviation, and Shape of the Sampling Distribution of $\hat{p}$

This section discusses the sampling distribution of the sample proportion and the mean, standard deviation, and shape of this sampling distribution.

7.7.1 Sampling Distribution of $\hat{p}$

Just like the sample mean $\bar{x}$, the sample proportion $\hat{p}$ is a random variable. Hence, it possesses a probability distribution, which is called its **sampling distribution**.

Definition

Sampling Distribution of the Sample Proportion, $\hat{p}$ The probability distribution of the sample proportion, $\hat{p}$, is called its *sampling distribution*. It gives the various values that $\hat{p}$ can assume and their probabilities.

The value of $\hat{p}$ calculated for a particular sample depends on what elements of the population are included in that sample. Example 7–8 illustrates the concept of the sampling distribution of $\hat{p}$.

■ EXAMPLE 7–8

Illustrating the sampling distribution of $\hat{p}$.

Boe Consultant Associates has five employees. Table 7.6 gives the names of these five employees and information concerning their knowledge of statistics.

Table 7.6 Information on the Five Employees of Boe Consultant Associates

Name	Knows Statistics
Ally	yes
John	no
Susan	no
Lee	yes
Tom	yes

If we define the population proportion, p, as the proportion of employees who know statistics, then

$$p = 3/5 = .60$$

Now, suppose we draw all possible samples of three employees each and compute the proportion of employees, for each sample, who know statistics. The total number of samples of size three that can be drawn from the population of five employees is

$$\text{Total number of samples} = {}_5C_3 = \frac{5!}{3!(5-3)!} = \frac{5 \cdot 4 \cdot 3 \cdot 2 \cdot 1}{3 \cdot 2 \cdot 1 \cdot 2 \cdot 1} = 10$$

Table 7.7 lists these 10 possible samples and the proportion of employees who know statistics for each of those samples. Note that we have rounded the values of $\hat{p}$ to two decimal places.

Table 7.7 All Possible Samples of Size 3 and the Value of $\hat{p}$ for Each Sample

Sample	Proportion Who Know Statistics $\hat{p}$
Ally, John, Susan	1/3 = .33
Ally, John, Lee	2/3 = .67
Ally, John, Tom	2/3 = .67
Ally, Susan, Lee	2/3 = .67
Ally, Susan, Tom	2/3 = .67
Ally, Lee, Tom	3/3 = 1.00
John, Susan, Lee	1/3 = .33
John, Susan, Tom	1/3 = .33
John, Lee, Tom	2/3 = .67
Susan, Lee, Tom	2/3 = .67

Using Table 7.7, we prepare the frequency distribution of $\hat{p}$ as recorded in Table 7.8, along with the relative frequencies of classes, which are obtained by dividing the frequencies of classes by the population size. The relative frequencies are used as probabilities and listed in Table 7.9. This table gives the sampling distribution of $\hat{p}$.

Table 7.8 Frequency and Relative Frequency Distributions of $\hat{p}$ When the Sample Size Is 3

$\hat{p}$	f	Relative Frequency
.33	3	3/10 = .30
.67	6	6/10 = .60
1.00	1	1/10 = .10
	$\Sigma f = 10$	Sum = 1.00

Table 7.9 Sampling Distribution of $\hat{p}$ When the Sample Size Is 3

$\hat{p}$	$P(\hat{p})$
.33	.30
.67	.60
1.00	.10
	$\Sigma P(\hat{p}) = 1.00$

■

7.7.2 Mean and Standard Deviation of $\hat{p}$

The **mean of $\hat{p}$**, which is the same as the mean of the sampling distribution of $\hat{p}$, is always equal to the population proportion, p, just as the mean of the sampling distribution of $\bar{x}$ is always equal to the population mean, μ.

Mean of the Sample Proportion The *mean of the sample proportion*, $\hat{p}$, is denoted by $\mu_{\hat{p}}$ and is equal to the population proportion, p. Thus,

$$\mu_{\hat{p}} = p$$

The sample proportion, $\hat{p}$, is called an **estimator** of the population proportion, p. As mentioned earlier, when the expected value (or mean) of a sample statistic is equal to the value of the corresponding population parameter, that sample statistic is said to be an **unbiased estimator**. Since for the sample proportion $\mu_{\hat{p}} = p$, $\hat{p}$ is an unbiased estimator of p.

The **standard deviation of $\hat{p}$**, denoted by $\sigma_{\hat{p}}$, is given by the following formula. This formula is true only when the sample size is small compared to the population size. As we know from Section 7.3, the sample size is said to be small compared to the population size if $n/N \leq .05$.

Standard Deviation of the Sample Proportion The *standard deviation of the sample proportion,* $\hat{p}$, is denoted by $\sigma_{\hat{p}}$ and is given by the formula

$$\sigma_{\hat{p}} = \sqrt{\frac{pq}{n}}$$

where p is the population proportion, $q = 1 - p$, and n is the sample size. This formula is used when $n/N \leq .05$, where N is the population size.

However, if n/N is greater than .05, then $\sigma_{\hat{p}}$ is calculated as follows:

$$\sigma_{\hat{p}} = \sqrt{\frac{pq}{n}}\sqrt{\frac{N-n}{N-1}}$$

where the factor

$$\sqrt{\frac{N-n}{N-1}}$$

is called the finite population correction factor.

In almost all cases, the sample size is small compared to the population size and, consequently, the formula used to calculate $\sigma_{\hat{p}}$ is $\sqrt{pq/n}$.

As mentioned earlier, if the standard deviation of a sample statistic decreases as the sample size is increased, that statistic is said to be a **consistent estimator**. It is obvious from the above formula for $\sigma_{\hat{p}}$ that as n increases, the value of $\sqrt{pq/n}$ decreases. Thus, the sample proportion, $\hat{p}$, is a consistent estimator of the population proportion, p.

7.7.3 Shape of the Sampling Distribution of $\hat{p}$

The shape of the sampling distribution of $\hat{p}$ is inferred from the central limit theorem.

Central Limit Theorem for Sample Proportion According to the central limit theorem, the *sampling distribution of $\hat{p}$* is approximately normal for a sufficiently large sample size. In the case of proportion, the sample size is considered to be sufficiently large if np and nq are both greater than 5—that is, if

$$np > 5 \quad \text{and} \quad nq > 5$$

Note that the sampling distribution of $\hat{p}$ will be approximately normal if $np > 5$ and $nq > 5$. This is the same condition that was required for the application of the normal approximation to the binomial probability distribution in Chapter 6.

Example 7–9 shows the calculation of the mean and standard deviation of $\hat{p}$ and describes the shape of its sampling distribution.

■ EXAMPLE 7–9

According to a survey by the Conference Board, 50% of Americans are satisfied with their jobs (*Time,* March 21, 2005). Assume that this result is true for the current population of Americans. Let $\hat{p}$ be the proportion of Americans in a random sample of 1000 who are satisfied with their jobs. Find the mean and standard deviation of $\hat{p}$ and describe the shape of its sampling distribution.

Finding the mean, standard deviation, and shape of the sampling distribution of $\hat{p}$.

CASE STUDY

7–1

CALLING THE VOTE FOR CONGRESS

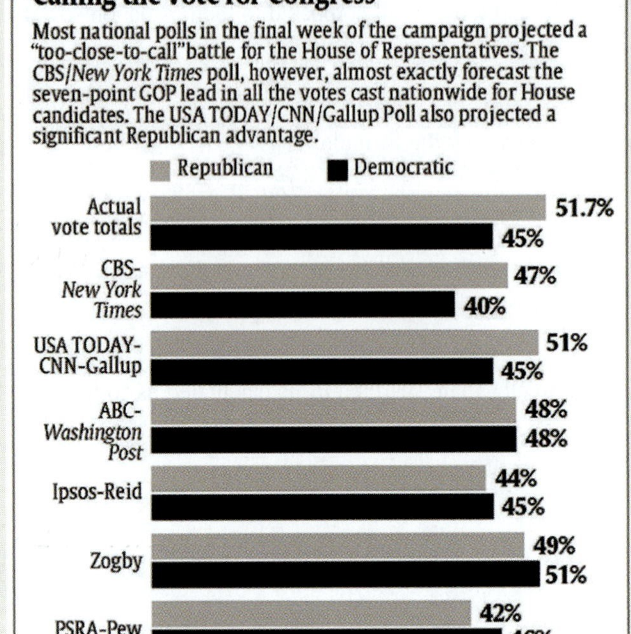
Calling the vote for Congress
Most national polls in the final week of the campaign projected a "too-close-to-call" battle for the House of Representatives. The CBS/New York Times poll, however, almost exactly forecast the seven-point GOP lead in all the votes cast nationwide for House candidates. The USA TODAY/CNN/Gallup Poll also projected a significant Republican advantage.
Republican
Democratic
Actual vote totals
51.7%
45%
CBS-New York Times
47%
40%
USA TODAY-CNN-Gallup
51%
45%
ABC-Washington Post
48%
48%
Ipsos-Reid
44%
45%
Zogby
49%
51%
PSRA-Pew
42%
46%
Source: USA TODAY research
By Quin Tian, USA TODAY

As you may recall, the midterm election for the House of Representatives, part of the Senate, and some state offices were held in November 2002. The above chart, reproduced from *USA TODAY*, shows the actual votes cast for the candidates of the two major parties for the House of Representatives and the predictions by various polls conducted during the week before the election was held. As you can observe from the top two bars in the chart, of all the votes cast, 51.7% went to Republican candidates and 45% were cast for the Democratic candidates. Obviously, the remaining 3.3% were cast for candidates who were neither Republicans nor Democrats. The remaining bars in the chart show the percentages of votes that were predicted by various polling agencies to be cast for the Republican and Democratic candidates for the House. For example, the CBS–*New York Times* poll predicted that 47% of the votes would be cast for the Republican candidates and 40% for the Democratic candidates.

Now just consider the percentage of votes cast for the Republican candidates for the House; the top bar gives the population proportion $p = .517$. In each of the polls, the percentage predicted for the Republican candidates gives $\hat{p}$. For example, the CBS–*New York Times* poll's prediction of 47% of the votes for Republican candidates gives $\hat{p} = .47$. The difference between p and $\hat{p}$ gives the sampling error for this poll, assuming no nonsampling errors were made. Thus, in this case sampling error is $.47 - .517 = -.047$. As we can see, the values of $\hat{p}$ from many polls are close to the population proportion $p = .517$.

Now consider the Democratic candidates for the House; they received 45% of the votes. Here $p = .45$. The percentages predicted for Democratic candidates by various polls give $\hat{p}$. The difference between each of these sample proportions and the population proportion gives the sampling error for the corresponding poll.

Source: Chart reproduced with permission from *USA TODAY*, November 8, 2002.

Solution Let p be the proportion of all Americans who are satisfied with their jobs. Then,

$$p = .50 \quad \text{and} \quad q = 1 - p = 1 - .50 = .50$$

The mean of the sampling distribution of $\hat{p}$ is

$$\mu_{\hat{p}} = p = \mathbf{.50}$$

The standard deviation of $\hat{p}$ is

$$\sigma_{\hat{p}} = \sqrt{\frac{pq}{n}} = \sqrt{\frac{(.50)(.50)}{1000}} = \mathbf{.0158}$$

The values of np and nq are

$$np = 1000(.50) = 500 \qquad \text{and} \qquad nq = 1000(.50) = 500$$

Because np and nq are both greater than 5, we can apply the central limit theorem to make an inference about the shape of the sampling distribution of $\hat{p}$. Therefore, the sampling distribution of $\hat{p}$ is approximately normal with a mean of .50 and a standard deviation of .0158, as shown in Figure 7.15.

Figure 7.15

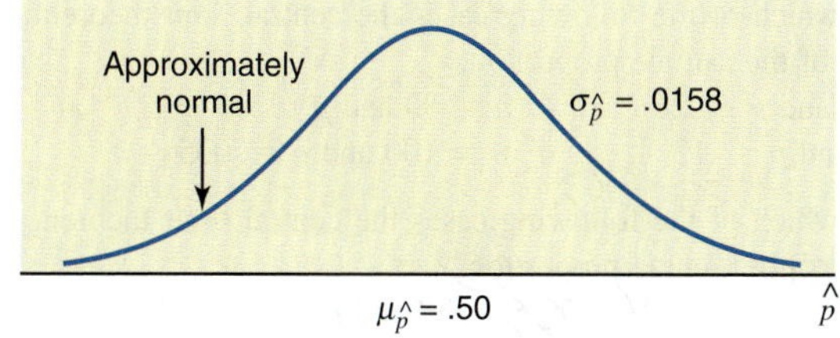

EXERCISES

CONCEPTS AND PROCEDURES

7.60 In a population of 1000 subjects, 640 possess a certain characteristic. A sample of 40 subjects selected from this population has 24 subjects who possess the same characteristic. What are the values of the population and sample proportions?

7.61 In a population of 5000 subjects, 600 possess a certain characteristic. A sample of 120 subjects selected from this population contains 18 subjects who possess the same characteristic. What are the values of the population and sample proportions?

7.62 In a population of 18,700 subjects, 30% possess a certain characteristic. In a sample of 250 subjects selected from this population, 25% possess the same characteristic. How many subjects in the population and sample, respectively, possess this characteristic?

7.63 In a population of 9500 subjects, 75% possess a certain characteristic. In a sample of 400 subjects selected from this population, 78% possess the same characteristic. How many subjects in the population and sample, respectively, possess this characteristic?

7.64 Let $\hat{p}$ be the proportion of elements in a sample that possess a characteristic.

a. What is the mean of $\hat{p}$?
b. What is the standard deviation of $\hat{p}$? Assume $n/N \leq .05$.
c. What condition(s) must hold true for the sampling distribution of $\hat{p}$ to be approximately normal?

7.65 For a population, $N = 12{,}000$ and $p = .71$. A random sample of 900 elements selected from this population gave $\hat{p} = .66$. Find the sampling error.

7.66 For a population, $N = 2800$ and $p = .29$. A random sample of 80 elements selected from this population gave $\hat{p} = .33$. Find the sampling error.

7.67 What is the estimator of the population proportion? Is this estimator an unbiased estimator of p? Explain why or why not.

7.68 Is the sample proportion a consistent estimator of the population proportion? Explain why or why not.

7.69 How does the value of $\sigma_{\hat{p}}$ change as the sample size increases? Explain. Assume $n/N \leq .05$.

7.70 Consider a large population with $p = .63$. Assuming $n/N \leq .05$, find the mean and standard deviation of the sample proportion $\hat{p}$ for a sample size of

a. 100 b. 900

7.71 Consider a large population with $p = .21$. Assuming $n/N \le .05$, find the mean and standard deviation of the sample proportion $\hat{p}$ for a sample size of

a. 400 b. 750

7.72 A population of $N = 4000$ has a population proportion equal to .12. In each of the following cases, which formula will you use to calculate $\sigma_{\hat{p}}$ and why? Using the appropriate formula, calculate $\sigma_{\hat{p}}$ for each of these cases.

a. $n = 800$ b. $n = 30$

7.73 A population of $N = 1400$ has a population proportion equal to .47. In each of the following cases, which formula will you use to calculate $\sigma_{\hat{p}}$ and why? Using the appropriate formula, calculate $\sigma_{\hat{p}}$ for each of these cases.

a. $n = 90$ b. $n = 50$

7.74 According to the central limit theorem, the sampling distribution of $\hat{p}$ is approximately normal when the sample is large. What is considered a large sample in the case of the proportion? Briefly explain.

7.75 Indicate in which of the following cases the central limit theorem will apply to describe the sampling distribution of the sample proportion.

a. $n = 400$ and $p = .28$ b. $n = 80$ and $p = .05$
c. $n = 60$ and $p = .12$ d. $n = 100$ and $p = .035$

7.76 Indicate in which of the following cases the central limit theorem will apply to describe the sampling distribution of the sample proportion.

a. $n = 20$ and $p = .45$ b. $n = 75$ and $p = .22$
c. $n = 350$ and $p = .01$ d. $n = 200$ and $p = .022$

APPLICATIONS

7.77 A company manufactured six television sets on a given day, and these TV sets were inspected for being good or defective. The results of the inspection follow.

Good Good Defective Defective Good Good

a. What proportion of these TV sets are good?
b. How many total samples (without replacement) of size five can be selected from this population?
c. List all the possible samples of size five that can be selected from this population and calculate the sample proportion, $\hat{p}$, of television sets that are good for each sample. Prepare the sampling distribution of $\hat{p}$.
d. For each sample listed in part c, calculate the sampling error.

7.78 Investigation of all five major fires in a western desert during one of the recent summers found the following causes.

Arson Accident Accident Arson Accident

a. What proportion of those fires were due to arson?
b. How many total samples (without replacement) of size three can be selected from this population?
c. List all the possible samples of size three that can be selected from this population and calculate the sample proportion $\hat{p}$ of the fires due to arson for each sample. Prepare the table that gives the sampling distribution of $\hat{p}$.
d. For each sample listed in part c, calculate the sampling error.

7.79 In an Accenture survey of adult Americans, 19% of the respondents said that the services provided by online retailers were *generally satisfactory* (*Business Week*, November 29, 2004). Assume that this result is true for the current population of adult Americans. Let $\hat{p}$ be the proportion of adults in a random sample of 500 who hold this opinion. Find the mean and standard deviation of $\hat{p}$ and describe the shape of its sampling distribution.

7.80 According to data from Ipsos for Findlaw Legal Web site, 56% of the respondents felt that the music industry should not sue people who illegally download music (*USA TODAY*, October 14, 2004). Assume that this percentage is true for the current population of such respondents. Let $\hat{p}$ be the proportion of persons in a random sample of 400 who hold the above opinion. Find the mean and standard deviation of $\hat{p}$ and comment on the shape of its sampling distribution.

7.81 According to a *Newsweek* poll, 17% of Americans believe that the end of the world would occur in their lifetime (*Newsweek*, May 24, 2004). Suppose that this percentage is true for the current population of Americans. Let $\hat{p}$ be the proportion of Americans in a random sample of 60 who hold this opinion. Find the mean and standard deviation of $\hat{p}$ and describe the shape of its sampling distribution.

7.82 In an Impulse Research Corporation survey for Arm & Hammer Baking Soda, mothers were asked how much time, on average, they spend preparing dinner. The most frequent response was *45 minutes or more*, given by 29.2% of the mothers (*USA TODAY*, June 17, 2004). Assume that this percentage is true for the current population of mothers. Let $\hat{p}$ be the proportion of mothers in a random sample of 200 who spend an average of 45 minutes or more preparing dinner. Calculate the mean and standard deviation of $\hat{p}$ and describe the shape of its sampling distribution.

7.8 Applications of the Sampling Distribution of $\hat{p}$

As mentioned in Section 7.5, when we conduct a study, we usually take only one sample and make all decisions or inferences on the basis of the results of that one sample. We use the concepts of the mean, standard deviation, and shape of the sampling distribution of $\hat{p}$ to determine the probability that the value of $\hat{p}$ computed from one sample falls within a given interval. Examples 7–10 and 7–11 illustrate this application.

■ EXAMPLE 7–10

Calculating the probability that $\hat{p}$ is in an interval.

According to an Associated Press poll, circumstances such as income, education, and marital status affect whether or not Americans feel satisfied with their lives. In this poll conducted during August 16–18, 2004, 38% of adult Americans said that they were *very satisfied* with the way things were going in their lives at that time (Yahoo! News, September 24, 2004). Suppose this result is true for the current population of adult Americans. Let $\hat{p}$ be the proportion in a random sample of 1000 adult Americans who will say that they are *very satisfied* with the way things are going in their lives at this time. Find the probability that the value of $\hat{p}$ is between .40 and .42.

Solution From the given information,

$$n = 1000, \qquad p = .38, \qquad \text{and} \qquad q = 1 - p = 1 - .38 = .62$$

where p is the proportion of all adult Americans who will say that they are *very satisfied* with the way things are going in their lives at this time.

The mean of the sample proportion $\hat{p}$ is

$$\mu_{\hat{p}} = p = .38$$

The standard deviation of $\hat{p}$ is

$$\sigma_{\hat{p}} = \sqrt{\frac{pq}{n}} = \sqrt{\frac{(.38)(.62)}{1000}} = .01534927$$

The values of np and nq are

$$np = 1000\,(.38) = 380 \qquad \text{and} \qquad nq = 1000\,(.62) = 620$$

Because np and nq are both greater than 5, we can infer from the central limit theorem that the sampling distribution of $\hat{p}$ is approximately normal. The probability that $\hat{p}$ is between .40 and .42 is given by the area under the normal curve for $\hat{p}$ between $\hat{p} = .40$ and $\hat{p} = .42$, as shown in Figure 7.16.

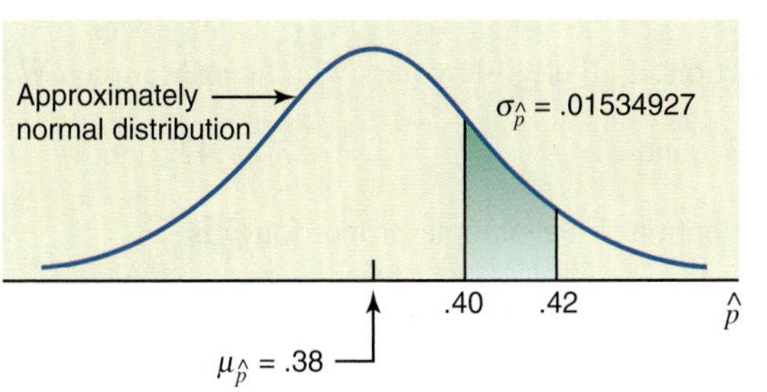

Figure 7.16

The first step in finding the area under the normal curve between $\hat{p} = .40$ and $\hat{p} = .42$ is to convert these two values to their respective z values. The z value for $\hat{p}$ is computed using the following formula.

z Value for a Value of $\hat{p}$ The *z value for a value of $\hat{p}$* is calculated as

$$z = \frac{\hat{p} - p}{\sigma_{\hat{p}}}$$

The two values of $\hat{p}$ are converted to their respective z values and then the area under the normal curve between these two points is found using the normal distribution table.

$$\text{For } \hat{p} = .40: \quad z = \frac{.40 - .38}{.01534927} = 1.30$$

$$\text{For } \hat{p} = .42: \quad z = \frac{.42 - .38}{.01534927} = 2.61$$

Thus, the probability that $\hat{p}$ is between .40 and .42 is given by the area under the standard normal curve between $z = 1.30$ and $z = 2.61$. This area is shown in Figure 7.17. The required probability is

$$P(.40 < \hat{p} < .42) = P(1.30 < z < 2.61) = P(z < 2.61) - P(z < 1.30)$$
$$= .9955 - .9032 = \mathbf{.0923}$$

Figure 7.17 $P(.40 < \hat{p} < .42)$

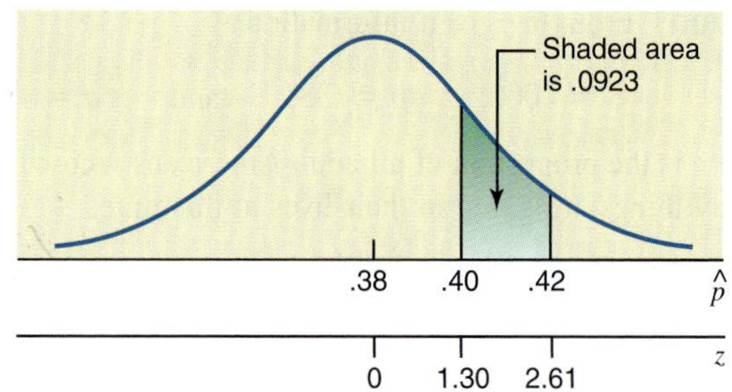

Thus, the probability is .0923 that the proportion of adult Americans in a random sample of 1000 who will say that they are very satisfied with the way things are going in their lives at this time is between .40 and .42. ■

■ EXAMPLE 7–11

Calculating the probability that $\hat{p}$ is less than a certain value.

Maureen Webster, who is running for mayor in a large city, claims that she is favored by 53% of all eligible voters of that city. Assume that this claim is true. What is the probability that in a random sample of 400 registered voters taken from this city, less than 49% will favor Maureen Webster?

Solution Let p be the proportion of all eligible voters who favor Maureen Webster. Then,

$$p = .53 \quad \text{and} \quad q = 1 - p = 1 - .53 = .47$$

The mean of the sampling distribution of the sample proportion $\hat{p}$ is

$$\mu_{\hat{p}} = p = .53$$

The population of all voters is large (because the city is large) and the sample size is small compared to the population. Consequently, we can assume that $n/N \le .05$. Hence, the standard deviation of $\hat{p}$ is calculated as

$$\sigma_{\hat{p}} = \sqrt{\frac{pq}{n}} = \sqrt{\frac{(.53)(.47)}{400}} = .02495496$$

From the central limit theorem, the shape of the sampling distribution of $\hat{p}$ is approximately normal. The probability that $\hat{p}$ is less than .49 is given by the area under the normal distribution curve for $\hat{p}$ to the left of $\hat{p} = .49$, as shown in Figure 7.18. The z value for $\hat{p} = .49$ is

$$z = \frac{\hat{p} - p}{\sigma_{\hat{p}}} = \frac{.49 - .53}{.02495496} = -1.60$$

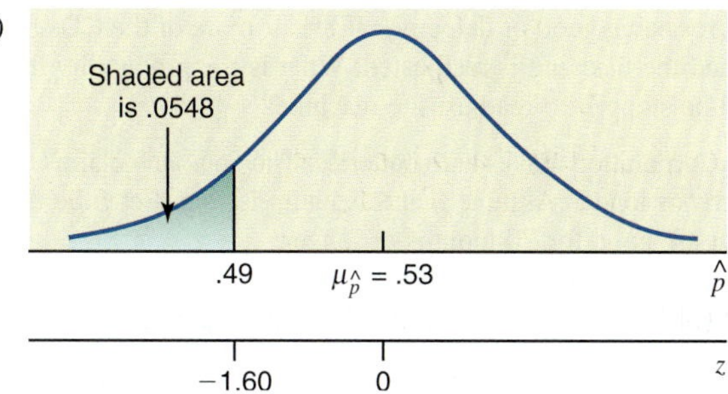

Figure 7.18 $P(\hat{p} < .49)$

Thus, the required probability from table IV is

$$P(\hat{p} < .49) = P(z < -1.60)$$
$$= \mathbf{.0548}$$

Hence, the probability that less than 49% of the voters in a random sample of 400 will favor Maureen Webster is .0548. ■

EXERCISES

■ CONCEPTS AND PROCEDURES

7.83 If all possible samples of the same (large) size are selected from a population, what percentage of all sample proportions will be within 2.0 standard deviations of the population proportion?

7.84 If all possible samples of the same (large) size are selected from a population, what percentage of all sample proportions will be within 3.0 standard deviations of the population proportion?

7.85 For a population, $N = 30{,}000$ and $p = .59$. Find the z value for each of the following for $n = 100$.
a. $\hat{p} = .56$ **b.** $\hat{p} = .68$ **c.** $\hat{p} = .53$ **d.** $\hat{p} = .65$

7.86 For a population, $N = 18{,}000$ and $p = .25$. Find the z value for each of the following for $n = 70$.
a. $\hat{p} = .26$ **b.** $\hat{p} = .32$ **c.** $\hat{p} = .17$ **d.** $\hat{p} = .20$

■ APPLICATIONS

7.87 According to Nielsen Media Research, 25.8% of American men aged 18 to 34 watched television during prime time in July 2004 (*The New York Times*, August 9, 2004). Assume that this percentage is true for the current population of men aged 18 to 34. Let $\hat{p}$ be the proportion of men who watch television during prime time in a random sample of 225 men aged 18 to 34. Find the probability that the value of $\hat{p}$ is
a. between .24 and .30 **b.** greater than .27

7.88 A survey of all medium- and large-sized corporations showed that 64% of them offer retirement plans to their employees. Let $\hat{p}$ be the proportion in a random sample of 50 such corporations that offer retirement plans to their employees. Find the probability that the value of $\hat{p}$ will be

a. between .54 and .61   b. greater than .71

7.89 According to the U.S. Department of Health and Human Services, 44% of Americans take at least one prescription drug regularly (*Newsweek*, December 20, 2004). Assume that this percentage is true for the current population of Americans, and let $\hat{p}$ be the proportion of Americans in a random sample of 900 who take at least one prescription drug regularly. Find the probability that the value of $\hat{p}$ is

a. between .43 and .46   b. greater than .48

7.90 Dartmouth Distribution Warehouse makes deliveries of a large number of products to its customers. It is known that 85% of all the orders it receives from its customers are delivered on time. Let $\hat{p}$ be the proportion of orders in a random sample of 100 that are delivered on time. Find the probability that the value of $\hat{p}$ will be

a. between .81 and .88   b. less than .87

7.91 Brooklyn Corporation manufactures CDs. The machine that is used to make these CDs is known to produce 6% defective CDs. The quality control inspector selects a sample of 100 CDs every week and inspects them for being good or defective. If 8% or more of the CDs in the sample are defective, the process is stopped and the machine is readjusted. What is the probability that based on a sample of 100 CDs the process will be stopped to readjust the machine?

7.92 Mong Corporation makes auto batteries. The company claims that 80% of its LL70 batteries are good for 70 months or longer. Assume that this claim is true. Let $\hat{p}$ be the proportion in a sample of 100 such batteries that are good for 70 months or longer.

a. What is the probability that this sample proportion is within .05 of the population proportion?
b. What is the probability that this sample proportion is less than the population proportion by .06 or more?
c. What is the probability that this sample proportion is greater than the population proportion by .07 or more?

USES AND MISUSES... BEWARE OF BIAS

Mathematics tells us that the sample mean, $\bar{x}$, is an unbiased and consistent estimator for the population mean, μ. This is great news because it allows us to estimate properties of a population based on those of a sample; this is the essence of statistics. But statistics always makes a number of assumptions about the sample from which the mean and standard deviation are calculated. Failure to respect these assumptions can introduce bias in your calculations. In statistics, *bias* means a deviation of the expected value of a statistical estimator from the parameter it estimates.

Let's say you are a quality control manager for a refrigerator parts company. One of the parts that you manufacture has a specification that the length of the part be 2.0 centimeters plus or minus .025 centimeters. The manufacturer expects that the parts it receives have a mean length of 2.0 centimeters and a small variation around that mean. The manufacturing process is to mold the part to something a little bit bigger than necessary—say, 2.1 centimeters—and finish the process by hand. Because the action of cutting material is irreversible, the machinists tend to miss their target by approximately .01 centimeters, so the mean length of the parts is not 2.0 centimeters, but rather 2.01 centimeters. It is your job to catch this.

One of your quality control procedures is to select completed parts randomly and test them against specification. Unfortunately, your measurement device is also subject to variation and might consistently underestimate the length of the parts. If your measurements are consistently .01 centimeters too short, your sample mean will not catch the manufacturing error in the population of parts.

The solution to the manufacturing problem is relatively straightforward: Be certain to calibrate your measurement instrument. Calibration becomes very difficult when working with people. It is known that people tend to overestimate the number of times that they vote and underestimate the time it takes to complete a project. Basing statistical results on this type of data can result in distorted estimates of the properties of your population. It is very important to be careful to weed out bias in your data, because once it gets into your calculations, it is very hard to get it out.

Glossary

Central limit theorem The theorem from which it is inferred that for a large sample size ($n \geq 30$), the shape of the sampling distribution of $\bar{x}$ is approximately normal. Also, by the same theorem, the shape of the sampling distribution of $\hat{p}$ is approximately normal for a sample for which $np > 5$ and $nq > 5$.

Consistent estimator A sample statistic with a standard deviation that decreases as the sample size increases.

Estimator The sample statistic that is used to estimate a population parameter.

Mean of $\hat{p}$ The mean of the sampling distribution of $\hat{p}$, denoted by $\mu_{\hat{p}}$, is equal to the population proportion p.

Mean of $\bar{x}$ The mean of the sampling distribution of $\bar{x}$, denoted by $\mu_{\bar{x}}$, is equal to the population mean μ.

Nonsampling errors The errors that occur during the collection, recording, and tabulation of data.

Population distribution The probability distribution of the population data.

Population proportion p The ratio of the number of elements in a population with a specific characteristic to the total number of elements in the population.

Sample proportion $\hat{p}$ The ratio of the number of elements in a sample with a specific characteristic to the total number of elements in that sample.

Sampling distribution of $\hat{p}$ The probability distribution of all the values of $\hat{p}$ calculated from all possible samples of the same size selected from a population.

Sampling distribution of $\bar{x}$ The probability distribution of all the values of $\bar{x}$ calculated from all possible samples of the same size selected from a population.

Sampling error The difference between the value of a sample statistic calculated from a random sample and the value of the corresponding population parameter. This type of error occurs due to chance.

Standard deviation of $\hat{p}$ The standard deviation of the sampling distribution of $\hat{p}$, denoted by $\sigma_{\hat{p}}$, is equal to $\sqrt{pq/n}$ when $n/N \leq .05$.

Standard deviation of $\bar{x}$ The standard deviation of the sampling distribution of $\bar{x}$, denoted by $\sigma_{\bar{x}}$, is equal to $\sigma/\sqrt{n}$ when $n/N \leq .05$.

Unbiased estimator An estimator with an expected value (or mean) that is equal to the value of the corresponding population parameter.

Supplementary Exercises

7.93 The print on the package of 100-watt General Electric soft-white lightbulbs claims that these bulbs have an average life of 750 hours. Assume that the lives of all such bulbs have a normal distribution with a mean of 750 hours and a standard deviation of 55 hours. Let $\bar{x}$ be the mean life of a random sample of 25 such bulbs. Find the mean and standard deviation of $\bar{x}$ and describe the shape of its sampling distribution.

7.94 The times spent waiting in line by all drivers to get their licenses renewed at the motor vehicle department office in a city have a distribution that is skewed to the right with a mean of 24 minutes and a standard deviation of 7 minutes. Let $\bar{x}$ be the mean waiting time for a random sample of 100 drivers renewing their licenses at this office. Calculate the mean and standard deviation of $\bar{x}$ and comment on the shape of its sampling distribution.

7.95 Refer to Exercise 7.93. The print on the package of 100-watt General Electric soft-white light-bulbs says that these bulbs have an average life of 750 hours. Assume that the lives of all such bulbs have a normal distribution with a mean of 750 hours and a standard deviation of 55 hours. Find the probability that the mean life of a random sample of 25 such bulbs will be

- **a.** greater than 735 hours
- **b.** between 725 and 740 hours
- **c.** within 15 hours of the population mean
- **d.** less than the population mean by 20 hours or more

7.96 Refer to Exercise 7.94. The times spent waiting in line by all drivers to get their licenses renewed at the motor vehicle department office of a large city have a distribution that is skewed to the right with a mean of 24 minutes and a standard deviation of 7 minutes. Find the probability that the mean waiting time for a random sample of 100 drivers to get their licenses renewed is

- **a.** less than 22 minutes
- **b.** between 23 and 26 minutes
- **c.** within 1 minute of the population mean
- **d.** greater than the population mean by 2 minutes or more

7.97 According to the Stanford Institute for the Quantitative Study of Society, Internet users in the United States spend an average of 102 minutes per day watching television (*The New York Times*, December 30, 2004). Suppose currently all Internet users spend an average of 102 minutes watching television per day with a standard deviation of 26 minutes. Find the probability that the mean time spent watching television per day by a random sample of 200 Internet users is

a. greater than 99 minutes
b. between 103 and 106 minutes
c. within 3.5 minutes of the population mean
d. less than the population mean by 3 minutes or more

7.98 A machine at Keats Corporation fills 64-ounce detergent jugs. The probability distribution of the amount of detergent in these jugs is normal with a mean of 64 ounces and a standard deviation of .4 ounces. The quality control inspector takes a sample of 16 jugs once a week and measures the amount of detergent in these jugs. If the mean of this sample is either less than 63.75 ounces or greater than 64.25 ounces, the inspector concludes that the machine needs an adjustment. What is the probability that based on a sample of 16 jugs the inspector will conclude that the machine needs an adjustment when actually it does not?

7.99 Suppose that 88% of the cases of car burglar alarms that go off are false. Let $\hat{p}$ be the proportion of false alarms in a random sample of 80 cases of car burglar alarms that go off. Calculate the mean and standard deviation of $\hat{p}$ and describe the shape of its sampling distribution.

7.100 Seventy percent of adults favor some kind of government control on the prices of medicines. Assume that this percentage is true for the current population of all adults. Let $\hat{p}$ be the proportion of adults in a random sample of 400 who favor government control on the prices of medicines. Calculate the mean and standard deviation of $\hat{p}$ and describe the shape of its sampling distribution.

7.101 Refer to Exercise 7.100. Seventy percent of adults favor some kind of government control on the prices of medicines. Assume that this percentage is true for the current population of all adults.

a. Find the probability that the proportion of adults in a random sample of 400 who favor some kind of government control on the prices of medicines is
   i. less than .65 ii. between .73 and .76
b. What is the probability that the proportion of adults in a random sample of 400 who favor some kind of government control is within .06 of the population proportion?
c. What is the probability that the sample proportion is greater than the population proportion by .05 or more? Assume that sample includes 400 adults.

7.102 According to a recent U.S. government study, 68% of children live in homes with two married parents (*Time*, July 26, 2004). Assume that this percentage is true for the current population of children in the United States. Let $\hat{p}$ be the proportion of children in a random sample of 400 who live in homes with two married parents. Find the probability that the value of $\hat{p}$ is

a. within .04 of the population proportion
b. not within .04 of the population proportion
c. greater than the population proportion by .05 or more
d. less than the population proportion by .035 or more

Advanced Exercises

7.103 Let μ be the mean annual salary of Major League Baseball players for 2005. Assume that the standard deviation of the salaries of these players is \$105,000. What is the probability that the 2005 mean salary of a random sample of 32 baseball players was within \$10,000 of the population mean, μ? Assume that $n/N \leq .05$.

7.104 The test scores for 300 students were entered into a computer, analyzed, and stored in a file. Unfortunately, someone accidentally erased a major portion of this file from the computer. The only information that is available is that 30% of the scores were below 65 and 15% of the scores were above 90. Assuming the scores are normally distributed, find their mean and standard deviation.

7.105 A chemist has a 10-gallon sample of river water taken just downstream from the outflow of a chemical plant. He is concerned about the concentration, c (in parts per million), of a certain toxic substance in the water. He wants to take several measurements, find the mean concentration of the toxic substance for this sample, and have a 95% chance of being within .5 parts per million of the true mean value of c. If the concentration of the toxic substance in all measurements is normally distributed with $\sigma = .8$ parts per million, how many measurements are necessary to achieve this goal?

7.106 A television reporter is covering the election for mayor of a large city and will conduct an exit poll (interviews with voters immediately after they vote) to make an early prediction of the outcome. Assume that the eventual winner of the election will get 60% of the votes.

a. What is the probability that a prediction based on an exit poll of a random sample of 25 voters will be correct? In other words, what is the probability that 13 or more of the 25 voters in the sample will have voted for the eventual winner?

b. How large a sample would the reporter have to take so that the probability of correctly predicting the outcome would be .95 or higher?

7.107 A city is planning to build a hydroelectric power plant. A local newspaper found that 53% of the voters in this city favor the construction of this plant. Assume that this result holds true for the population of all voters in this city.

a. What is the probability that more than 50% of the voters in a random sample of 200 voters selected from this city will favor the construction of this plant?

b. A politician would like to take a random sample of voters in which more than 50% would favor the plant construction. How large a sample should be selected so that the politician is 95% sure of this outcome?

7.108 Refer to Exercise 6.92. Otto is trying out for the javelin throw to compete in the Olympics. The lengths of his javelin throws are normally distributed with a mean of 290 feet and a standard deviation of 10 feet. What is the probability that the total length of three of his throws will exceed 885 feet?

7.109 A certain elevator has a maximum legal carrying capacity of 6000 pounds. Suppose that the population of all people who ride this elevator have a mean weight of 160 pounds with a standard deviation of 25 pounds. If 35 of these people board the elevator, what is the probability that their combined weight will exceed 6000 pounds? Assume that the 35 people constitute a random sample from the population.

7.110 According to a *Newsweek* magazine poll, 74% of Americans surveyed believe that Satan exists (*Newsweek*, May 24, 2004). Suppose that 74% of all Americans currently believe that Satan exists.

a. Suppose that 81.7% in a poll of 60 Americans believe that Satan exists. How likely is it for the sample percentage in a sample of 60 to be 81.7% or more when the population percentage is 74%?

b. Refer to part a. How likely is it for the sample percentage to be 81.7% or more if the sample size is 600?

c. What is the smallest sample size that will produce a sample percentage of 81.7% or higher in only 1% of all sample surveys of that size?

7.111 Refer to the sampling distribution discussed in Section 7.1. Calculate and replace the sample means in Table 7.3 with the sample medians, and then calculate the average of these sample medians. Does this average of the medians equal the population mean? If yes, why does this make sense? If no, how could you change exactly two of the five data values in this example so that the average of the sample medians equals the population mean?

7.112 Suppose you want to calculate $P(a \leq \bar{x} \leq b)$, where a and b are two numbers and x has a distribution with mean μ and standard deviation σ. If $a < \mu < b$ (i.e., μ lies in the interval a to b), what happens to the probability $P(a \leq \bar{x} \leq b)$ as the sample size becomes larger?

Self-Review Test

1. A sampling distribution is the probability distribution of
   a. a population parameter **b.** a sample statistic **c.** any random variable

2. Nonsampling errors are
   a. the errors that occur because the sample size is too large in relation to the population size
   b. the errors made while collecting, recording, and tabulating data
   c. the errors that occur because an untrained person conducts the survey

3. A sampling error is
   a. the difference between the value of a sample statistic based on a random sample and the value of the corresponding population parameter
   b. the error made while collecting, recording, and tabulating data
   c. the error that occurs because the sample is too small

4. The mean of the sampling distribution of $\bar{x}$ is always equal to
a. μ **b.** $\mu - 5$ **c.** $\sigma/\sqrt{n}$

5. The condition for the standard deviation of the sample mean to be $\sigma/\sqrt{n}$ is that
a. $np > 5$ **b.** $n/N \le .05$ **c.** $n > 30$

6. The standard deviation of the sampling distribution of the sample mean decreases when
a. x increases **b.** n increases **c.** n decreases

7. When samples are selected from a normally distributed population, the sampling distribution of the sample mean has a normal distribution
a. if $n \ge 30$ **b.** if $n/N \le .05$ **c.** all the time

8. When samples are selected from a nonnormally distributed population, the sampling distribution of the sample mean has an approximately normal distribution
a. if $n \ge 30$ **b.** if $n/N \le .05$ **c.** always

9. In a sample of 200 customers of a mail-order company, 174 are found to be satisfied with the service they receive from the company. The proportion of customers in this sample who are satisfied with the company's service is
a. .87 **b.** .174 **c.** .148

10. The mean of the sampling distribution of $\hat{p}$ is always equal to
a. p **b.** μ **c.** $\hat{p}$

11. The condition for the standard deviation of the sampling distribution of the sample proportion to be $\sqrt{pq/n}$ is
a. $np > 5$ and $nq > 5$ **b.** $n > 30$ **c.** $n/N \le .05$

12. The sampling distribution of $\hat{p}$ is (approximately) normal if
a. $np > 5$ and $nq > 5$ **b.** $n > 30$ **c.** $n/N \le .05$

13. Briefly state and explain the central limit theorem.

14. The weights of all students at a large university have an approximately normal distribution with a mean of 145 pounds and a standard deviation of 18 pounds. Let $\bar{x}$ be the mean weight of a random sample of certain students selected from this university. Calculate the mean and standard deviation of $\bar{x}$ and describe the shape of its sampling distribution for a sample size of
a. 25 **b.** 100

15. Warning times for tornadoes (from the time the warning is issued until the tornado strikes) have an unknown distribution with a mean of 11 minutes and a standard deviation of 2.7 minutes. Let $\bar{x}$ be the mean warning time for a random sample of a certain number of tornado warnings. Find the mean and standard deviation of $\bar{x}$ and describe the shape of its sampling distribution for a sample size of
a. 25 **b.** 75

16. Tex's Bar has a mechanical bull, which customers may ride. Although most of the riders fall off the bull within 60 seconds, any customer who stays on the bull for 60 seconds is given a free drink. Because of the time limit, the durations of these rides have a distribution that is skewed to the left with a mean of 42 seconds and a standard deviation of 10 seconds. Find the probability that the mean duration of a random sample of 50 such bull rides is
a. between 38 and 41 seconds
b. within 2 seconds of the population mean
c. greater than the population mean by 1 second or more
d. between 39 and 44 seconds
e. less than 43 seconds

17. At Jen and Perry Ice Cream Company, the machine that fills one-pound cartons of Top Flavor ice cream is set to dispense 16 ounces of ice cream into every carton. However, some cartons contain slightly less than and some contain slightly more than 16 ounces of ice cream. The amounts of ice cream in all such cartons have a normal distribution with a mean of 16 ounces and a standard deviation of .18 ounces.
a. Find the probability that the mean amount of ice cream in a random sample of 16 such cartons will be
i. between 15.90 and 15.95 ounces
ii. less than 15.95 ounces
iii. more than 15.97 ounces

b. What is the probability that the mean amount of ice cream in a random sample of 16 such cartons will be within .10 ounces of the population mean?

c. What is the probability that the mean amount of ice cream in a random sample of 16 such cartons will be less than the population mean by .135 ounces or more?

18. According to a Harris Interactive poll, 49% of Americans said that the winter holiday season is the most stressful time of the year (*Time*, April 18, 2005). Let $\hat{p}$ be the proportion of a random sample of Americans who hold this view. Find the mean and standard deviation of $\hat{p}$ and describe the shape of its sampling distribution when the sample size is

a. 40 b. 100 c. 500

19. According to an NPD Group poll, 63% of 18-to-24-year-old women said they shop only at their favorite stores (*Forbes*, November 1, 2004). Assume that this percentage is true for the current population of such women.

a. Find the probability that in a random sample of 300 such women, the proportion who will say they shop only at their favorite stores is

i. greater than .66 ii. between .60 and .67

iii. less than .65 iv. between .57 and .61

b. What is the probability that in a random sample of 300 such women, the proportion who will say they shop only at their favorite stores is within .035 of the population proportion?

c. What is the probability that in a random sample of 300 such women, the proportion who will say they shop only at their favorite stores is less than the population proportion by .025 or more?

d. What is the probability that in a random sample of 300 such women, the proportion who will say they shop only at their favorite stores is greater than the population proportion by .03 or more?

Mini-Projects

MINI-PROJECT 7–1

Consider the data on heights of NBA players.

a. Compute μ and σ for this data set.

b. Take 20 random samples of five players each and find $\bar{x}$ for each sample.

c. Compute the mean and standard deviation of the 20 sample means obtained in part b.

d. Using the formulas given in Section 7.3, find $\mu_{\bar{x}}$ and $\sigma_{\bar{x}}$ for $n = 5$.

e. How do your values of $\mu_{\bar{x}}$ and $\sigma_{\bar{x}}$ in part d compare with those in part c?

f. What percentage of the 20 sample means found in part b lie in the interval $\mu_{\bar{x}} - \sigma_{\bar{x}}$ to $\mu_{\bar{x}} + \sigma_{\bar{x}}$? In the interval $\mu_{\bar{x}} - 2\sigma_{\bar{x}}$ to $\mu_{\bar{x}} + 2\sigma_{\bar{x}}$? In the interval $\mu_{\bar{x}} - 3\sigma_{\bar{x}}$ to $\mu_{\bar{x}} + 3\sigma_{\bar{x}}$?

g. How do the percentages in part f compare to the corresponding percentages for a normal distribution (68%, 95%, and 99.7%, respectively)?

h. Repeat parts b through g using 20 samples of 10 players each.

MINI-PROJECT 7–2

Consider Data Set II, Data on States, that accompanies this text. Let p denote the proportion of the 50 states that have a per capita income of less than $30,000.

a. Find p.

b. Select 20 random samples of five states each and find the sample proportion $\hat{p}$ for each sample.

c. Compute the mean and standard deviation of the 20 sample proportions obtained in part b.

d. Using the formulas given in Section 7.7.2, compute $\mu_{\hat{p}}$ and $\sigma_{\hat{p}}$. Is the finite population correction factor required here?

e. Compare your mean and standard deviation of $\hat{p}$ from part c with the values calculated in part d.

f. Repeat parts b through e using 20 samples of 10 states each.

MINI-PROJECT 7–3

You are to conduct the experiment of sampling 10 times (with replacement) from the digits 0, 1, 2, 3, 4, 5, 6, 7, 8, and 9. You can do this in a variety of ways. One way is to write each digit on a separate piece of paper, place all the slips in a hat, and select 10 times from the hat, returning each selected slip before the next pick. As alternatives, you can use a 10-sided die, statistical software, or a calculator that generates

random numbers. Perform the experiment using any of these methods and compute the sample mean $\bar{x}$ for the 10 numbers obtained. Now repeat the procedure 49 more times. When you are done, you will have 50 sample means.

a. Make a table of the population distribution for the 10 digits and display it using a graph.
b. Make a stem-and-leaf display of your 50 sample means. What shape does it have?
c. What does the central limit theorem say about the shape of the sampling distribution of $\bar{x}$? What mean and standard deviation does the sampling distribution of $\bar{x}$ have in this problem?

MINI-PROJECT 7–4

Reconsider Mini-Project 7–3. Now repeat that project and parts a through c, but this time use a skewed distribution (as explained below), instead of a symmetric distribution, to take samples. This project is more easily performed using a computer or graphing calculator, but it can be done using a hat or a random numbers table. In this project, sample 10 times from a population of digits that contains twenty 0s, fifteen 1s, ten 2s, seven 3s, four 4s, three 5s, two 6s, and one each of the numbers 7, 8, and 9. Select 50 samples of size 10 and repeat parts a through c of Mini–Project 7–3. How do the parts a and b of this project compare to parts a and b of Mini–Project 7–3 in regard to the shapes of the distributions? How does this relate to what the central limit theorem says?

DECIDE FOR YOURSELF

Deciding About Elections

In the first week of November during an election year, you are very likely to hear the following statement on TV news, "We are now able to make a projection. In (*insert the name of the state where you live*), we project that the winner will be (*insert one of the elected officials from your state*)." Many people are aware that news agencies conduct exit polls on election day. A commonly asked question is, "How can an agency call a race based on the results from a sample of only 1200 voters or so, and do this with a high (although not perfect) accuracy level?" Although the actual methods used to make projections based on exit polls are above the level of this book, we will examine a similar but simpler version of the question here. The concepts and logic involved in this process will help you understand the statistical inference concepts discussed in subsequent chapters.

Consider a simple election where there are only two candidates, named A and B. Suppose p and q are the proportions of votes received by candidates A and B, respectively. Suppose we conduct an exit poll based on a simple random sample of 800 voters and determine the mean, standard deviation, and shape of the sampling distribution of $\hat{p}$, where $\hat{p}$ is the proportion of voters in the sample who voted for candidate A.

1. Suppose that 440 of the 800 voters included in the exit poll voted for candidate A, which gives $\hat{p} = .55$. Assuming that each candidate received 50% of the votes (i.e., $p = .50$ and $q = .50$, where p and q are the proportions of votes received by candidates A and B, respectively), what is the probability that at least 440 out of 800 voters in a sample would vote for candidate A?

2. Based on your answer to the above question, the results of the poll make it reasonable to conclude that the proportion of all voters who voted for candidate A is actually higher than .5. Explain why.

3. What implications do the above answers have for the result of the election? Will you make a projection about this election based on the results of this exit poll?

TECHNOLOGY INSTRUCTION Sampling Distribution of Means

TI-84

To create a sampling distribution of a sample mean using the TI-84 requires a good deal of programming, which we will not do here. However, it is quite easy to create a sampling distribution for a sample proportion using the TI-84. Let n and p represent the number of trials and the probability of a success, respectively, for a binomial experiment. On the TI-84, press **2nd>STAT>OPS>seq(>MATH>PRB>randBin(n,p)/n,X,1,100)>STO->L1**

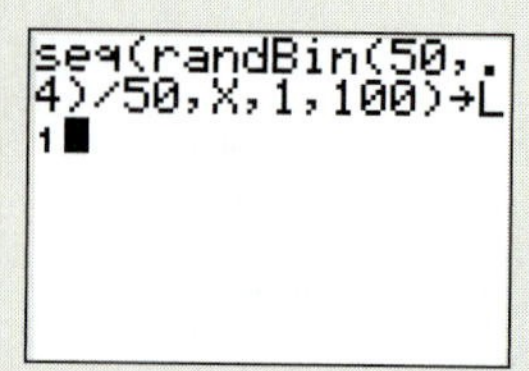

Screen 7.1

(see **Screen 7.1**). This will produce 100 values of $\hat{p}$. If you want more or fewer values of $\hat{p}$, change 100 in the above command to any desired number. Then, you can create a histogram of the data on these $\hat{p}$ values using the technology instructions of Chapter 2.

MINITAB

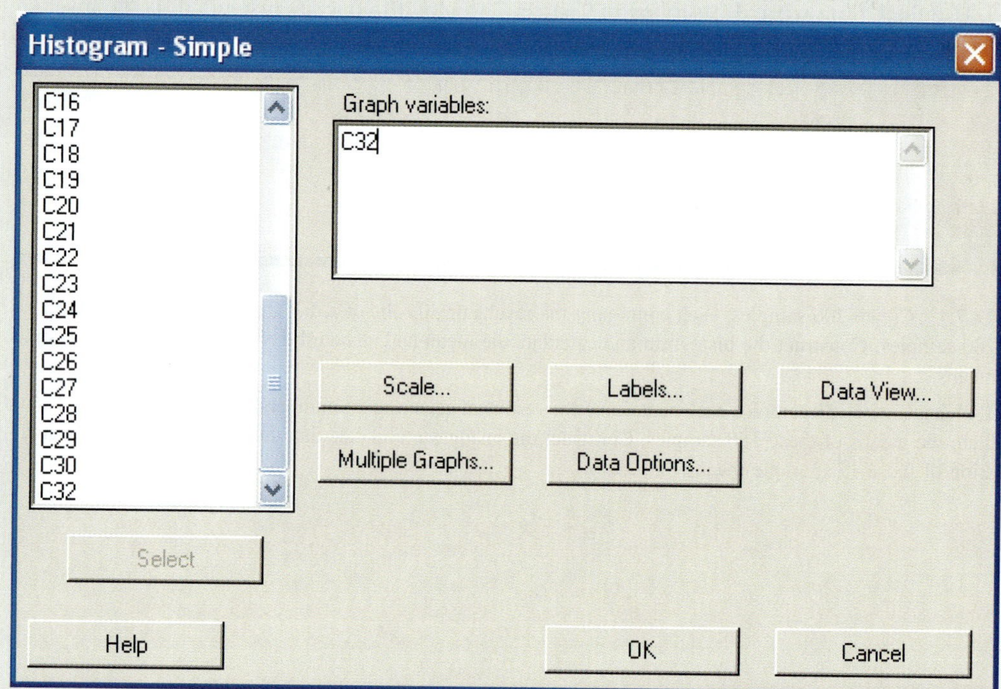

Screen 7.2

1. To see an example of sampling distribution of a sample mean, select **Calc>Random Data>Integer**. We will create 50 samples of size 30, each value a random integer between 0 and 10. Each sample will be a row, so that when we find the mean of each row, the result will go in a column.
2. Enter 50 for **Generate . . . rows of data**.
3. Enter **c1-c30** for **Store in columns**.
4. Enter 0 for **Minimum value** and 10 for **Maximum value**.
5. Select **OK**.
6. Select **Calc>Row Statistics**. Select **Mean** and enter **c1-c30** for **Input variables**. Enter **c32** for **Store results in**.
7. Select **OK**.
8. Select **Graph>Histogram**. Enter **C32** in the **Graph variables:** box (see **Screen 7.2**). Click **OK** to obtain the histogram that will appear in the graph window (see **Screen 7.3**). Is this histogram bell shaped? What do you think is the center of this histogram?

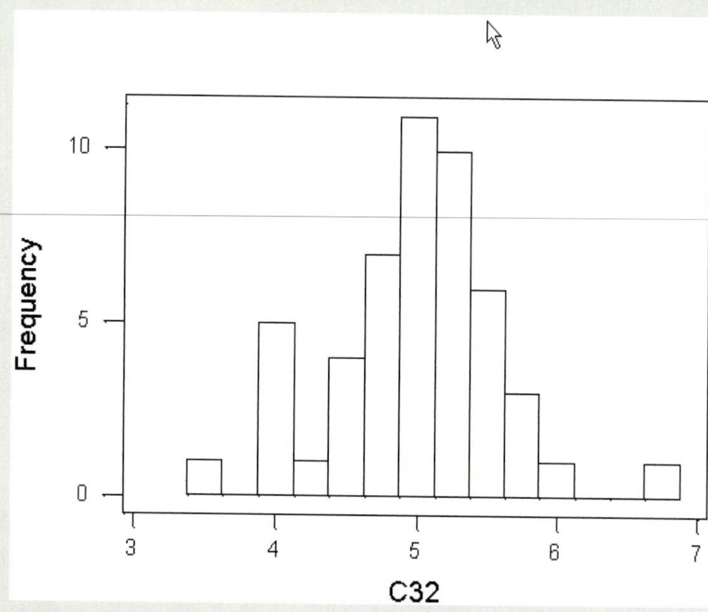

Screen 7.3

Excel

	A	B	C
1	3	=average(a1:a2)	
2	8		
3			

Screen 7.4

1. To see an example of the sampling distribution of means, use the **rand** function described in Chapter 1 to create a sample of two random numbers between 0 and 10 in column A.
2. Use the **average** function described in Chapter 3 to find their mean B. (See **Screen 7.4.**)
3. Cut and paste the pair of random numbers and their mean 30 times.
4. Use the **frequency** function described in Chapter 2 to find the frequency counts between 0 and 1, 1 and 2, 2 and 3, and so forth through 9 and 10.
5. Use the **Chart wizard** described in Chapter 2 to plot a frequency histogram. Is the histogram bell-shaped? Where is it centered?

TECHNOLOGY ASSIGNMENTS

TA7.1 Create 200 samples, each containing the results of 30 rolls of a die. Calculate the means of these 200 samples. Construct the histogram and calculate the mean and standard deviation of these 200 sample means.

TA7.2 Create 150 samples each containing the results of selecting 35 numbers from 1 through 100. Calculate the means of these 150 samples. Construct the histogram and calculate the mean and standard deviation of these 150 sample means.

Chapter

8

Estimation of the Mean and Proportion

Do you know how much more, on average, workers with bachelor's degrees and advanced degrees earn compared to workers without and with high school diplomas? On average, workers with no high school diplomas earn $18,734, and workers with high school diplomas $27,915 per year. Compared to this, college graduates earn an average of $51,206, and workers with advanced degrees earn an average of $74,602. (See Case Study 8–1.) Or what is your shopping personality? Are you a list maker, a spontaneous buyer, a bargain hunter, a window shopper, or some other kind of shopper? See Case Study 8–3 for results of a poll on this issue.

8.1 Estimation: An Introduction

8.2 Point and Interval Estimates

8.3 Estimation of a Population Mean: σ Known

Case Study 8–1 Education and Earnings

8.4 Estimation of a Population Mean: σ Not Known

Case Study 8–2 Cardiac Demands of Heavy Snow Shoveling

8.5 Estimation of a Population Proportion: Large Samples

Case Study 8–3 Buying Habits of Consumers

Now we are entering that part of statistics called *inferential statistics*. In Chapter 1 inferential statistics was defined as the part of statistics that helps us make decisions about some characteristics of a population based on sample information. In other words, inferential statistics uses the sample results to make decisions and draw conclusions about the population from which the sample is drawn. Estimation is the first topic to be considered in our discussion of inferential statistics. Estimation and hypothesis testing (discussed in Chapter 9) taken together are usually referred to as inference making. This chapter explains how to estimate the population mean and population proportion for a single population.

8.1 Estimation: An Introduction

Estimation is a procedure by which a numerical value or values are assigned to a population parameter based on the information collected from a sample.

Definition

Estimation The assignment of value(s) to a population parameter based on a value of the corresponding sample statistic is called *estimation.*

In inferential statistics, μ is called the *true population mean* and p is called the *true population proportion*. There are many other population parameters, such as the median, mode, variance, and standard deviation.

The following are a few examples of estimation: an auto company may want to estimate the mean fuel consumption for a particular model of a car; a manager may want to estimate the average time taken by new employees to learn a job; the U.S. Census Bureau may want to find the mean housing expenditure per month incurred by households; and the AWAH (Association of Wives of Alcoholic Husbands) may want to find the proportion (or percentage) of all husbands who are alcoholic.

The examples about estimating the mean fuel consumption, estimating the average time taken to learn a job by new employees, and estimating the mean housing expenditure per month incurred by households are illustrations of estimating the *true population mean,* μ. The example about estimating the proportion (or percentage) of all husbands who are alcoholic is an illustration of estimating the *true population proportion,* p.

If we can conduct a *census* (a survey that includes the entire population) each time we want to find the value of a population parameter, then the estimation procedures explained in this and subsequent chapters are not needed. For example, if the U.S. Census Bureau can contact every household in the United States to find the mean housing expenditure incurred by households, the result of the survey (which will actually be a census) will give the value of μ, and the procedures learned in this chapter will not be needed. However, it is too expensive, very time consuming, or virtually impossible to contact every member of a population to collect information to find the true value of a population parameter. Therefore, we usually take a sample from the population and calculate the value of the appropriate sample statistic. Then we assign a value or values to the corresponding population parameter based on the value of the sample statistic. This chapter (and subsequent chapters) explains how to assign values to population parameters based on the values of sample statistics.

For example, to estimate the mean time taken to learn a certain job by new employees, the manager will take a sample of new employees and record the time taken by each of these employees to learn the job. Using this information, he or she will calculate the sample mean, $\bar{x}$. Then, based on the value of $\bar{x}$, he or she will assign certain values to μ. As another example, to estimate the mean housing expenditure per month incurred by all households in the United States, the Census Bureau will take a sample of certain households, collect the information on the housing expenditure that each of these households incurs per month, and compute the value of the sample mean, $\bar{x}$. Based on this value of $\bar{x}$, the bureau will then assign values to the population mean, μ. Similarly, the AWAH will take a sample of husbands and determine the value of the sample proportion, $\hat{p}$, which represents the proportion of husbands in the sample who are alcoholic. Using this value of the sample proportion, $\hat{p}$, AWAH will assign values to the population proportion, p.

The value(s) assigned to a population parameter based on the value of a sample statistic is called an **estimate** of the population parameter. For example, suppose the manager takes a sample of 40 new employees and finds that the mean time, $\bar{x}$, taken to learn this job for these employees is 5.5 hours. If he or she assigns this value to the population mean, then 5.5 hours is called an estimate of μ. The sample statistic used to estimate a population parameter is called an **estimator**. Thus, the sample mean, $\bar{x}$, is an estimator of the population mean, μ, and the sample proportion, $\hat{p}$, is an estimator of the population proportion, p.

Definition

Estimate and Estimator The value(s) assigned to a population parameter based on the value of a sample statistic is called an *estimate*. The sample statistic used to estimate a population parameter is called an *estimator*.

The estimation procedure involves the following steps.

1. Select a sample.
2. Collect the required information from the members of the sample.
3. Calculate the value of the sample statistic.
4. Assign value(s) to the corresponding population parameter.

Remember, **the procedures to be learned in this chapter assume that the sample taken is a simple random sample**. If the sample is not a simple random sample (see Appendix A for a few other kinds of samples), then the procedures to be used to estimate a population mean or proportion become more complex. These procedures are outside the scope of this book.

8.2 Point and Interval Estimates

An estimate may be a point estimate or an interval estimate. These two types of estimates are described in this section.

8.2.1 A Point Estimate

If we select a sample and compute the value of the sample statistic for this sample, then this value gives the **point estimate** of the corresponding population parameter.

Definition

Point Estimate The value of a sample statistic that is used to estimate a population parameter is called a *point estimate*.

Thus, the value computed for the sample mean, $\bar{x}$, from a sample is a point estimate of the corresponding population mean, μ. For the example mentioned earlier, suppose the Census Bureau takes a sample of 10,000 households and determines that the mean housing expenditure per month, $\bar{x}$, for this sample is \$1370. Then, using $\bar{x}$ as a point estimate of μ, the bureau can state that the mean housing expenditure per month, μ, for all households is about \$1370. Thus,

Point estimate of a population parameter = Value of the corresponding sample statistic

Each sample selected from a population is expected to yield a different value of the sample statistic. Thus, the value assigned to a population mean, μ, based on a point estimate depends on which of the samples is drawn. Consequently, the point estimate assigns a value to μ that almost always differs from the true value of the population mean.

8.2.2 An Interval Estimate

In the case of **interval estimation**, instead of assigning a single value to a population parameter, an interval is constructed around the point estimate and then a probabilistic statement that this interval contains the corresponding population parameter is made.

Definition

Interval Estimation In *interval estimation*, an interval is constructed around the point estimate, and it is stated that this interval is likely to contain the corresponding population parameter.

For the example about the mean housing expenditure, instead of saying that the mean housing expenditure per month for all households is \$1370, we may obtain an interval by subtracting a number from \$1370 and adding the same number to \$1370. Then we state that this interval contains the population mean, μ. For purposes of illustration, suppose we subtract \$240 from \$1370 and add \$240 to \$1370. Consequently, we obtain the interval (\$1370 – \$240) to (\$1370 + \$240), or \$1130 to \$1610. Then we state that the interval \$1130 to \$1610 is likely to contain the population mean, μ, and that the mean housing expenditure per month for all households in the United States is between \$1130 and \$1610. This procedure is called *interval estimation*. The value \$1130 is called the *lower limit* of the interval and \$1610 is called the *upper limit* of the interval. The number we add to and subtract from the point estimate is called the **margin of error**. Figure 8.1 illustrates the concept of interval estimation.

Figure 8.1 Interval estimation.

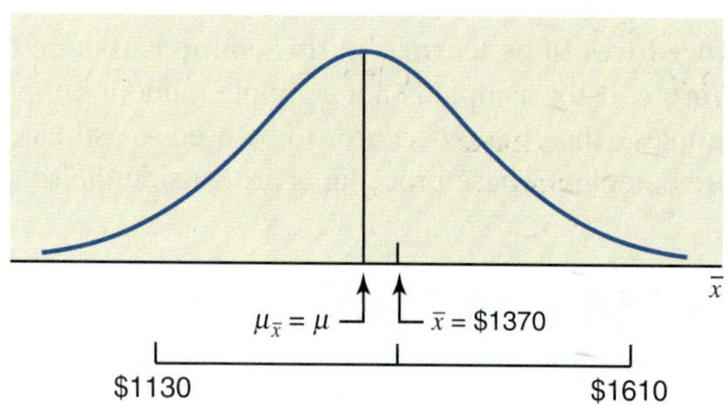

The question arises: What number should we subtract from and add to a point estimate to obtain an interval estimate? The answer to this question depends on two considerations:

1. The standard deviation $\sigma_{\bar{x}}$ of the sample mean, $\bar{x}$
2. The level of confidence to be attached to the interval

First, the larger the standard deviation of $\bar{x}$, the greater is the number subtracted from and added to the point estimate. Thus, it is obvious that if the range over which $\bar{x}$ can assume values is larger, then the interval constructed around $\bar{x}$ must be wider to include μ.

Second, the quantity subtracted and added must be larger if we want to have a higher confidence in our interval. We always attach a probabilistic statement to the interval estimation. This probabilistic statement is given by the **confidence level**. An interval constructed based on this confidence level is called a **confidence interval**.

Definition

Confidence Level and Confidence Interval Each interval is constructed with regard to a given *confidence level* and is called a *confidence interval*. The confidence interval is given as:

$$\text{Point estimate} \pm \text{Margin of Error}$$

The confidence level associated with a confidence interval states how much confidence we have that this interval contains the true population parameter. The confidence level is denoted by $(1 - \alpha)100\%$.

The confidence level is denoted by $\mathbf{(1 - \alpha)100\%}$, where α is the Greek letter *alpha*. When expressed as probability, it is called the *confidence coefficient* and is denoted by $1 - \alpha$. In passing, note that α is called the *significance level*, which will be explained in detail in Chapter 9.

Although any value of the confidence level can be chosen to construct a confidence interval, the more common values are 90%, 95%, and 99%. The corresponding confidence coefficients are .90, .95, and .99. The next section describes how to construct a confidence interval for the population mean for a large sample.

Sections 8.3 and 8.4 discuss the procedures to estimate a population mean μ. In Section 8.3 we assume that the population standard deviation σ is known, and in Section 8.4 we do not assume that the population standard deviation σ is known. In the latter situation, we use the sample standard deviation s instead of σ. In the real world, the population standard deviation σ is almost never known. Consequently, we (almost) always use the sample standard deviation s.

8.3 Estimation of a Population Mean: σ Known

This section explains how to construct a confidence interval for the population mean μ when the population standard deviation σ is known. Here, there are three possible cases that are mentioned below.

Case I. If the following three conditions are fulfilled:

1. The population standard deviation σ is known
2. The sample size is small (i.e., $n < 30$)
3. The population from which the sample is selected is normally distributed,

then we use the normal distribution to make the confidence interval for μ because from Section 7.4.1 of Chapter 7 the sampling distribution of $\bar{x}$ is normal with its mean equal to μ and the standard deviation equal to $\sigma_{\bar{x}} = \sigma/\sqrt{n}$ assuming that $n/N \leq .05$.

Case II. If the following two conditions are fulfilled:

1. The population standard deviation σ is known
2. The sample size is large (i.e., $n \geq 30$),

then, again, we use the normal distribution to make the confidence interval for μ because from Section 7.4.2 of Chapter 7, due to the central limit theorem, the sampling distribution of $\bar{x}$ is (approximately) normal with its mean equal to μ and the standard deviation equal to $\sigma_{\bar{x}} = \sigma/\sqrt{n}$ assuming that $n/N \leq .05$.

Case III. If the following three conditions are fulfilled:

1. The population standard deviation σ is known
2. The sample size is small (i.e., $n < 30$)
3. The population from which the sample is selected is not normally distributed (or its distribution is unknown),

then we use a nonparametric method to make the confidence interval for μ. Such a procedure is not covered in this text.

This section will cover the first two cases. The procedure to make a confidence interval for μ is the same in both these cases. Note that in Case I, the population does not have to be exactly normally distributed. As long as it is close to the normal distribution without any outliers, we can use the normal distribution procedure. In Case II, although 30 is considered a large sample, if the population distribution is very different from the normal distribution, then 30 may not be a large enough sample size for the sampling distribution of $\bar{x}$ to be normal and, hence, to use the normal distribution.

The following chart summarizes the above three cases.

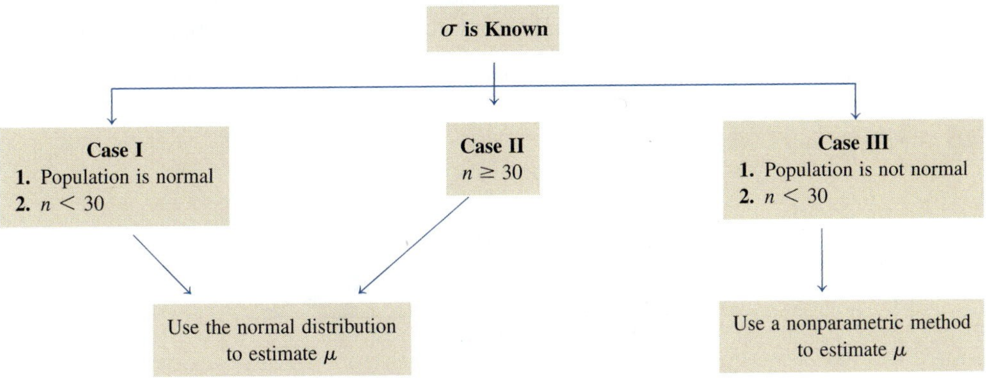

Confidence Interval for μ The $(1 - \alpha)100\%$ *confidence interval for* μ under Cases I and II above is

$$\bar{x} \pm z\sigma_{\bar{x}}$$

where
$$\sigma_{\bar{x}} = \sigma/\sqrt{n}$$

The value of z used here is obtained from the standard normal distribution table (Table IV of Appendix C) for the given confidence level.

The quantity $z\sigma_{\bar{x}}$ in the confidence interval formula is called the **margin of error** and is denoted by E.

Definition

Margin of Error The margin of error for the estimate for μ, denoted by E, is the quantity that is subtracted from and added to the value of $\bar{x}$ to obtain a confidence interval for μ. Thus,

$$E = z\sigma_{\bar{x}}$$

The value of z in the confidence interval formula is obtained from the standard normal distribution table (Table IV of Appendix C) for the given confidence level. To illustrate, suppose we want to construct a 95% confidence interval for μ. A 95% confidence level means that the total area under the normal curve for $\bar{x}$ between two points (at the same distance) on different sides of μ is 95% or .95, as shown in Figure 8.2. Note that we have denoted these two points by z_1 and z_2 in Figure 8.2. To find the value of z for a 95% confidence level, we first find the areas to the left of these two points, z_1 and z_2. Then we find the z values for these two areas from the normal distribution table. Note that these two values of z will be the same but with different signs. To find these values of z, we perform the following two steps:

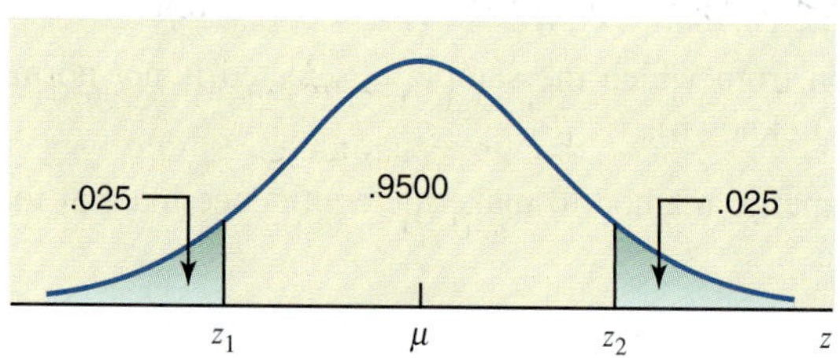

Figure 8.2 Finding z for a 95% confidence level.

1. The first step is to find the areas to the left of z_1 and z_2, respectively. Note that the area between z_1 and z_2 is denoted by $1 - \alpha$. Hence, the total area in the two tails is α, because the total area under the curve is 1.0. Therefore, the area in each tail, as shown in Figure 8.3, is $\alpha/2$. In our example, $1 - \alpha = .95$. Hence, the total area in both tails is $\alpha = 1 - .95 = .05$. Consequently, the area in each tail is $\alpha/2 = .05/2 = .025$. Then, the area to the left of z_1 is .0250 and the area to the left of z_2 is $.0250 + .95 = .9750$.
2. Now find the z values from Table IV of Appendix C such that the areas to the left of z_1 and z_2 are .0250 and .9750, respectively. These z values are -1.96 and 1.96, respectively.

Thus, for a confidence level of 95%, we will use $z = 1.96$ in the confidence interval formula.

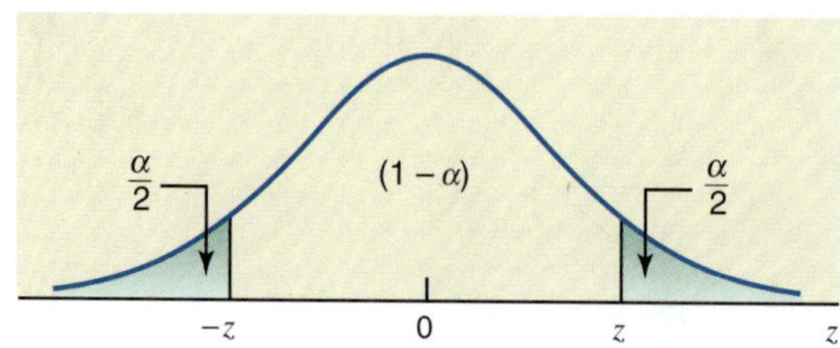

Figure 8.3 Area in the tails.

Table 8.1 lists the z values for some of the most commonly used confidence levels. Note that we always use the positive value of z in the formula.

Table 8.1 z Values for Commonly Used Confidence Levels

Confidence Level	Areas to Look for in Table IV	z Value
90%	.0500 and .9500	1.64 or 1.65
95%	.0250 and .9750	1.96
96%	.0200 and .9800	2.05
97%	.0150 and .9850	2.17
98%	.0100 and .9900	2.33
99%	.0050 and .9950	2.57 or 2.58

Example 8–1 describes the procedure used to construct a confidence interval for μ when σ is known, sample size is small, but the population from which the sample is drawn is normally distributed.

EXAMPLE 8–1

Finding the point estimate and confidence interval for μ: σ known, $n < 30$, and population normal.

A publishing company has just published a new college textbook. Before the company decides the price at which to sell this textbook, it wants to know the average price of all such textbooks in the market. The research department at the company took a sample of 25 comparable textbooks and collected information on their prices. This information produced a mean price of \$90.50 for this sample. It is known that the standard deviation of the prices of all such textbooks is \$7.50 and the population of such prices is normal.

(a) What is the point estimate of the mean price of all such college textbooks?

(b) Construct a 90% confidence interval for the mean price of all such college textbooks.

Solution Here, σ is known and, although $n < 30$, the population is normally distributed. Hence, we can use the normal distribution. From the given information,

$$n = 25, \quad \bar{x} = \$90.50, \quad \text{and} \quad \sigma = \$7.50$$

The standard deviation of $\bar{x}$ is

$$\sigma_{\bar{x}} = \frac{\sigma}{\sqrt{n}} = \frac{7.50}{\sqrt{25}} = \$1.50$$

(a) The point estimate of the mean price of all such college textbooks is \$90.50; that is,

$$\text{Point estimate of } \mu = \bar{x} = \mathbf{\$90.50}$$

(b) The confidence level is 90% or .90. First we find the z value for a 90% confidence level. Here, the area in each tail of the normal distribution curve is $\alpha/2 = (1 - .90)/2 = .05$. Now in Table IV, look for the areas .0500 and .9500 and find the corresponding values of z. These values are $z = -1.65$ and $z = 1.65$.[1]

Next, we substitute all the values in the confidence interval formula for μ. The 90% confidence interval for μ is

$$\bar{x} \pm z\sigma_{\bar{x}} = 90.50 \pm 1.65(1.50) = 90.50 \pm 2.48$$
$$= (90.50 - 2.48) \text{ to } (90.50 + 2.48) = \mathbf{\$88.02 \text{ to } \$92.98}$$

[1]Note that there is no apparent reason for choosing .4505 and not choosing .4495. If we choose .4495, the z value will be 1.64. An alternative is to use the average of 1.64 and 1.65, 1.645, which we will not do in this text.

Thus, we are 90% confident that the mean price of all such college textbooks is between \$88.02 and \$92.98. Note that we cannot say for sure whether the interval \$88.02 to \$92.98 contains the true population mean or not. Because μ is a constant, we cannot say that the probability is .90 that this interval contains μ because either it contains μ or it does not. Consequently, the probability is either 1.0 or 0 that this interval contains μ. All we can say is that we are 90% confident that the mean price of all such college textbooks is between \$88.02 and \$92.98.

In the above estimate, \$2.48 is called the margin of error or give and take figure. ■

How do we interpret a 90% confidence level? In terms of Example 8–1, if we take all possible samples of 25 such college textbooks each and construct a 90% confidence interval for μ around each sample mean, we can expect that 90% of these intervals will include μ and 10% will not. In Figure 8.4 we show means $\bar{x}_1$, $\bar{x}_2$, and $\bar{x}_3$ of three different samples of the same size drawn from the same population. Also shown in this figure are the 90% confidence intervals constructed around these three sample means. As we observe, the 90% confidence intervals constructed around $\bar{x}_1$ and $\bar{x}_2$ include μ, but the one constructed around $\bar{x}_3$ does not. We can state for a 90% confidence level that if we take many samples of the same size from a population and construct 90% confidence intervals around the means of these samples, then 90% of these confidence intervals will be like the ones around $\bar{x}_1$ and $\bar{x}_2$ in Figure 8.4, which include μ, and 10% will be like the one around $\bar{x}_3$, which does not include μ.

Figure 8.4 Confidence intervals.

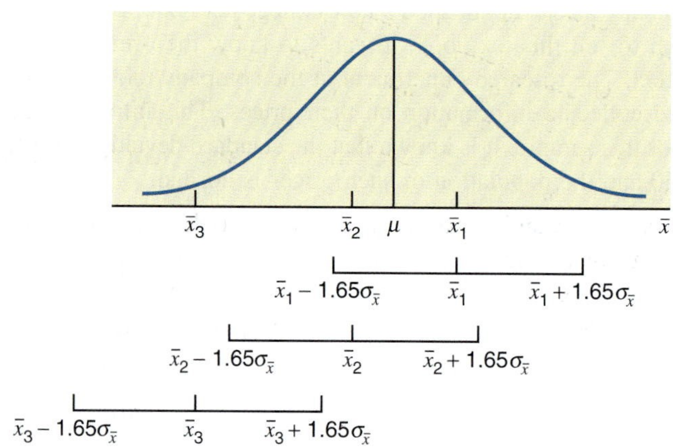

Example 8–2 illustrates how to obtain a confidence interval for μ when σ is known and the sample size is large ($n \geq 30$).

■ EXAMPLE 8–2

Constructing a confidence interval for μ: σ known and $n > 30$.

According to CardWeb.com, the mean bank credit card debt for households was \$7868 in 2004 (*USA TODAY*, July 1, 2005). Assume that this mean was based on a random sample of 900 households and that the standard deviation of such debts for all households in 2004 was \$2070. Make a 99% confidence interval for the 2004 mean bank credit card debt for all households.

Solution From the given information,

$$n = 900, \quad \bar{x} = \$7868, \quad \sigma = \$2070,$$

and

$$\text{Confidence level} = 99\% \text{ or } .99$$

Here, we know the population standard deviation σ. Although the shape of the population distribution is unknown, the sample size is large ($n > 30$). Hence, we can use the normal distribution to make a confidence interval for μ. To make this confidence interval, first we find the standard deviation of $\bar{x}$. The value of $\sigma_{\bar{x}}$ is

$$\sigma_{\bar{x}} = \frac{\sigma}{\sqrt{n}} = \frac{2070}{\sqrt{900}} = \$69$$

To find z for a 99% confidence level, first we find the area in each of the two tails, which is $(1 - .99)/2 = .0050$. Then, we look for .0050 and $.0050 + .99 = .9950$ areas in the normal distribution table to find the two z values. These two z values are (approximately) -2.58 and 2.58. Thus, we will use $z = 2.58$ in the confidence interval formula. Substituting all the values in the formula, the 99% confidence interval for μ is

$$\bar{x} \pm z\sigma_{\bar{x}} = 7868 \pm 2.58(69) = 7868 \pm 178.02 = \mathbf{\$7689.98 \text{ to } \$8046.02}$$

Thus, we can state with 99% confidence that the mean bank credit card debt for all households in 2004 was between \$7689.98 and \$8046.02. In the above calculations, \$178.02 is called the margin of error. ■

The **width of a confidence interval** depends on the size of the margin of error, $z\sigma_{\bar{x}}$, which depends on the values of z, σ, and n because $\sigma_{\bar{x}} = \sigma/\sqrt{n}$. However, the value of σ is not under the control of the investigator. Hence, the width of a confidence interval depends on

1. The value of z, which depends on the confidence level
2. The sample size n

The confidence level determines the value of z, which in turn determines the size of the margin of error. The value of z increases as the confidence level increases, and it decreases as the confidence level decreases. For example, the value of z is approximately 1.65 for a 90% confidence level, 1.96 for a 95% confidence level, and approximately 2.58 for a 99% confidence level. Hence, the higher the confidence level, the larger the width of the confidence interval, other things remaining the same.

For the same value of σ, an increase in the sample size decreases the value of $\sigma_{\bar{x}}$, which in turn decreases the size of the margin of error when the confidence level remains unchanged. Therefore, an increase in the sample size decreases the width of the confidence interval.

Thus, if we want to decrease the width of a confidence interval, we have two choices:

1. Lower the confidence level
2. Increase the sample size

Lowering the confidence level is not a good choice, however, because a lower confidence level may give less reliable results. Therefore, we should always prefer to increase the sample size if we want to decrease the width of a confidence interval. Next we illustrate, using Example 8–2, how either a decrease in the confidence level or an increase in the sample size decreases the width of the confidence interval.

❶ Confidence Level and the Width of the Confidence Interval

Reconsider Example 8–2. Suppose all the information given in that example remains the same. First, let us decrease the confidence level to 95%. From the normal distribution table, $z = 1.96$ for a 95% confidence level. Then, using $z = 1.96$ in the confidence interval for Example 8–2, we obtain

$$\bar{x} \pm z\sigma_{\bar{x}} = 7868 \pm 1.96(69) = 7868 \pm 135.24 = \mathbf{\$7732.76 \text{ to } \$8003.24}$$

Comparing this confidence interval to the one obtained in Example 8–2, we observe that the width of the confidence interval for a 95% confidence level is smaller than the one for a 99% confidence level.

❷ Sample Size and the Width of the Confidence Interval

Consider Example 8–2 again. Now suppose the information given in that example is based on a sample size of 2500. Further assume that all other information given in that example, including

CASE STUDY

8–1

EDUCATION AND EARNINGS

The above chart, based on the Census Bureau data for 2004, gives the average yearly salary of workers aged 18 and older with different levels of education. For example, the average yearly salary of workers aged 18 and older with no high school diploma was $18,734, that of workers with high school diplomas was $27,915, and so on. Of course, these averages are based on sample surveys of workers, but the Census Bureau conducts surveys based on very large samples. If we know the sample sizes and the population standard deviations for these workers with different education levels, we can then find the confidence intervals for the mean salaries of all workers in these groups as shown in the following table.

Category	Mean Salary	Confidence Interval
No high school diploma	$18,734	$\$18{,}734 \pm z\sigma_{\bar{x}}$
High school diploma	$27,915	$\$27{,}915 \pm z\sigma_{\bar{x}}$
Bachelor's degree	$51,206	$\$51{,}206 \pm z\sigma_{\bar{x}}$
Advanced degree	$74,602	$\$74{,}602 \pm z\sigma_{\bar{x}}$

For each confidence interval listed in the table, we can substitute the value of z and the value of $\sigma_{\bar{x}}$, which is calculated as $\frac{\sigma}{\sqrt{n}}$. For example, suppose we want to find the 98% confidence interval for the mean yearly salary of workers aged 18 and older with no high school diploma. Assume that this average yearly salary of $18,734 given in the chart is based on a sample of 4900 such workers and that the population standard deviation for the salaries of such workers is $2100. Then the 98% confidence interval for the corresponding population mean is calculated as follows.

$$\sigma_{\bar{x}} = \frac{\sigma}{\sqrt{n}} = \frac{2100}{\sqrt{4900}} = \$30.00$$

$$\bar{x} \pm z\sigma_{\bar{x}} = 18{,}734 \pm 2.33(30.00) = 18{,}734 \pm 69.90 = \$18{,}664.10 \text{ to } \$18{,}803.90$$

We can find the confidence intervals for the population means of workers with other levels of education mentioned in the chart and table the same way. Note that the sample means given in the table are the point estimates of the corresponding population means.

Source: The chart is reproduced with permission from *USA TODAY*, June 2, 2005.

the confidence level, remains the same. First, we calculate the standard deviation of the sample mean using $n = 2500$:

$$\sigma_{\bar{x}} = \sigma/\sqrt{n} = 2070/\sqrt{2500} = \$41.40$$

Then, the 99% confidence interval for μ is

$$\bar{x} \pm z\sigma_{\bar{x}} = 7868 \pm 2.58(41.40) = 7868 \pm 106.81 = \mathbf{\$7761.19 \text{ to } \$7974.81}$$

Comparing this confidence interval to the one obtained in Example 8–2, we observe that the width of the 99% confidence interval for $n = 2500$ is smaller than the 99% confidence interval for $n = 900$.

8.3.1 Determining the Sample Size for the Estimation of Mean

One reason we usually conduct a sample survey and not a census is that almost always we have limited resources at our disposal. In light of this, if a smaller sample can serve our purpose, then we will be wasting our resources by taking a larger sample. For instance, suppose we want to estimate the mean life of a certain auto battery. If a sample of 40 batteries can give us the confidence interval we are looking for, then we will be wasting money and time if we take a sample of a much larger size—say, 500 batteries. In such cases, if we know the confidence level and the width of the confidence interval that we want, then we can find the (approximate) size of the sample that will produce the required result.

From earlier discussion, we learned that $E = z\sigma_{\bar{x}}$ is called the margin of error of estimate for μ. As we know, the standard deviation of the sample mean is equal to $\sigma/\sqrt{n}$. Therefore, we can write the margin of error of estimate for μ as

$$E = z \cdot \frac{\sigma}{\sqrt{n}}$$

Suppose we predetermine the size of the margin of error, E, and want to find the size of the sample that will yield this margin of error. From the above expression, the following formula is obtained that determines the required sample size n.

Determining the Sample Size for the Estimation of μ Given the confidence level and the standard deviation of the population, the sample size that will produce a predetermined margin of error E of the confidence interval *estimate of* μ is

$$n = \frac{z^2\sigma^2}{E^2}$$

If we do not know σ, we can take a preliminary sample (of any arbitrarily determined size) and find the sample standard deviation, s. Then we can use s for σ in the formula. However, note that using s for σ may give a sample size that eventually may produce an error much larger (or smaller) than the predetermined margin of error. This will depend on how close s and σ are.

Example 8–3 illustrates how we determine the sample size that will produce the margin of error of estimate for μ wihin a certain limit.

■ EXAMPLE 8–3

An alumni association wants to estimate the mean debt of this year's college graduates. It is known that the population standard deviation of the debts of this year's college graduates is \$11,800. How large a sample should be selected so that the estimate with a 99% confidence level is within \$800 of the population mean?

Determining the sample size for the estimation of μ.

Solution The alumni association wants the 99% confidence interval for the mean debt of this year's college graduates to be

$$\bar{x} \pm 800$$

Hence, the maximum size of the margin of error of estimate is to be \$800; that is,

$$E = \$800$$

The value of z for a 99% confidence level is 2.58. The value of σ is given to be \$11,800. Therefore, substituting all values in the formula and simplifying, we obtain

$$n = \frac{z^2\sigma^2}{E^2} = \frac{(2.58)^2(11{,}800)^2}{(800)^2} = 1448.18 \approx \mathbf{1449}$$

Thus, the required sample size is 1449. If the alumni association takes a sample of 1449 of this year's college graduates, computes the mean debt for this sample, and then makes a 99% confidence interval around this sample mean, the margin of error of estimate will be approximately \$800. Note that we have rounded the final answer for the sample size to the next higher integer. This is always the case when determining the sample size. ■

EXERCISES

CONCEPTS AND PROCEDURES

8.1 Briefly explain the meaning of an estimator and an estimate.

8.2 Explain the meaning of a point estimate and an interval estimate.

8.3 What is the point estimator of the population mean, μ? How would you calculate the margin of error for an estimate of μ?

8.4 Explain the various alternatives for decreasing the width of a confidence interval. Which is the best alternative?

8.5 Briefly explain how the width of a confidence interval decreases with an increase in the sample size. Give an example.

8.6 Briefly explain how the width of a confidence interval decreases with a decrease in the confidence level. Give an example.

8.7 Briefly explain the difference between a confidence level and a confidence interval.

8.8 What is the margin of error of estimate for μ when σ is known? How is it calculated?

8.9 How will you interpret a 99% confidence interval for μ? Explain.

8.10 Find z for each of the following confidence levels.

a. 90% **b.** 95% **c.** 96% **d.** 97% **e.** 98% **f.** 99%

8.11 For a data set obtained from a sample, $n = 20$ and $\bar{x} = 24.5$. It is known that $\sigma = 3.1$. The population is normally distributed.

a. What is the point estimate of μ?
b. Make a 99% confidence interval for μ.
c. What is the margin of error of estimate for part b?

8.12 For a data set obtained from a sample, $n = 81$ and $\bar{x} = 48.25$. It is known that $\sigma = 4.8$.

a. What is the point estimate of μ?
b. Make a 95% confidence interval for μ.
c. What is the margin of error of estimate for part b?

8.13 The standard deviation for a population is $\sigma = 15.3$. A sample of 36 observations selected from this population gave a mean equal to 74.8.

a. Make a 90% confidence interval for μ.
b. Construct a 95% confidence interval for μ.
c. Determine a 99% confidence interval for μ.
d. Does the width of the confidence intervals constructed in parts a through c increase as the confidence level increases? Explain your answer.

8.14 The standard deviation for a population is $\sigma = 14.8$. A sample of 25 observations selected from this population gave a mean equal to 143.72. The population is known to have a normal distribution.

a. Make a 99% confidence interval for μ.
b. Construct a 95% confidence interval for μ.
c. Determine a 90% confidence interval for μ.
d. Does the width of the confidence intervals constructed in parts a through c decrease as the confidence level decreases? Explain your answer.

8.15 The standard deviation for a population is $\sigma = 6.30$. A random sample selected from this population gave a mean equal to 81.90. The population is known to be normally distributed.

a. Make a 99% confidence interval for μ assuming $n = 16$.
b. Construct a 99% confidence interval for μ assuming $n = 20$.
c. Determine a 99% confidence interval for μ assuming $n = 25$.
d. Does the width of the confidence intervals constructed in parts a through c decrease as the sample size increases? Explain.

8.16 The standard deviation for a population is $\sigma = 7.14$. A random sample selected from this population gave a mean equal to 48.52.

a. Make a 95% confidence interval for μ assuming $n = 196$.
b. Construct a 95% confidence interval for μ assuming $n = 100$.
c. Determine a 95% confidence interval for μ assuming $n = 49$.
d. Does the width of the confidence intervals constructed in parts a through c increase as the sample size decreases? Explain.

8.17 For a population, the value of the standard deviation is 2.65. A sample of 35 observations taken from this population produced the following data.

42	51	42	31	28	36	49
29	46	37	32	27	33	41
47	41	28	46	34	39	48
26	35	37	38	46	48	39
29	31	44	41	37	38	46

a. What is the point estimate of μ?
b. Make a 98% confidence interval for μ.
c. What is the margin of error of estimate for part b?

8.18 For a population, the value of the standard deviation is 4.96. A sample of 32 observations taken from this population produced the following data.

74	85	72	73	86	81	77	60
83	78	79	88	76	73	84	78
81	72	82	81	79	83	88	86
78	83	87	82	80	84	76	74

a. What is the point estimate of μ?
b. Make a 99% confidence interval for μ.
c. What is the margin of error of estimate for part b?

8.19 For a population data set, $\sigma = 12.5$.

a. How large a sample should be selected so that the margin of error of estimate for a 99% confidence interval for μ is 2.50?
b. How large a sample should be selected so that the margin of error of estimate for a 96% confidence interval for μ is 3.20?

8.20 For a population data set, $\sigma = 14.50$.

a. What should the sample size be for a 98% confidence interval for μ to have a margin of error of estimate equal to 5.50?
b. What should the sample size be for a 95% confidence interval for μ to have a margin of error of estimate equal to 4.25?

8.21 Determine the sample size for the estimate of μ for the following.

a. $E = 2.3$, $\sigma = 15.40$, confidence level = 99%
b. $E = 4.1$, $\sigma = 23.45$, confidence level = 95%
c. $E = 25.9$, $\sigma = 122.25$, confidence level = 90%

8.22 Determine the sample size for the estimate of μ for the following.

a. $E = .17$, $\sigma = .90$, confidence level = 99%
b. $E = 1.45$, $\sigma = 5.82$, confidence level = 95%
c. $E = 5.65$, $\sigma = 18.20$, confidence level = 90%

APPLICATIONS

8.23 A sample of 1500 homes sold recently in a state gave the mean price of homes equal to $269,720. The population standard deviation of the prices of homes in this state is $68,650. Construct a 99% confidence interval for the mean price of all homes in this state.

8.24 In an article by Janet Paskin in *The Journal News* of Westchester, New York, it was stated that baseball players were paid up to \$150 per autograph at a winter show in Secaucus, New Jersey (*USA TODAY*, July 8, 2004). Suppose a random sample of 800 such autographs at many different shows held in 2005 throughout the United States yielded a mean of \$135 per autograph, and that the population standard deviation is \$22. Construct a 95% confidence interval for the corresponding population mean.

8.25 Computer Action Company sells computers and computer parts by mail. The company assures its customers that products are mailed as soon as possible after an order is placed with the company. A sample of 25 recent orders showed that the mean time taken to mail products for these orders was 70 hours. Suppose the population standard deviation is 16 hours and the population distribution is normal.

a. Construct a 95% confidence interval for the mean time taken to mail products for all orders received at the office of this company.
b. Explain why we need to make the confidence interval. Why can we not say that the mean time taken to mail products for all orders received at the office of this company is 70 hours?

8.26 Lazurus Steel Corporation produces iron rods that are supposed to be 36 inches long. The machine that makes these rods does not produce each rod exactly 36 inches long. The lengths of the rods vary slightly. It is known that when the machine is working properly, the mean length of the rods made on this machine is 36 inches. The standard deviation of the lengths of all rods produced on this machine is always equal to .10 inches. The quality control department takes a sample of 20 such rods every week, calculates the mean length of these rods, and makes a 99% confidence interval for the population mean. If either the upper limit of this confidence interval is greater than 36.05 inches or the lower limit of this confidence interval is less than 35.95 inches, the machine is stopped and adjusted. A recent sample of 20 rods produced a mean length of 36.02 inches. Based on this sample, will you conclude that the machine needs an adjustment? Assume that the lengths of all such rods have a normal distribution.

8.27 At Farmer's Dairy, a machine is set to fill 32-ounce milk cartons. However, this machine does not put exactly 32 ounces of milk into each carton; the amount varies slightly from carton to carton. It is known that when the machine is working properly, the mean net weight of these cartons is 32 ounces. The standard deviation of the amounts of milk in all such cartons is always equal to .15 ounces. The quality control department takes a sample of 25 such cartons every week, calculates the mean net weight of these cartons, and makes a 99% confidence interval for the population mean. If either the upper limit of this confidence interval is greater than 32.15 ounces or the lower limit of this confidence interval is less than 31.85 ounces, the machine is stopped and adjusted. A recent sample of 25 such cartons produced a mean net weight of 31.94 ounces. Based on this sample, will you conclude that the machine needs an adjustment? Assume that the amounts of milk put in all such cartons have a normal distribution.

8.28 A consumer agency that proposes that lawyers' rates are too high wanted to estimate the mean hourly rate for all lawyers in New York City. A sample of 70 lawyers taken from New York City showed that the mean hourly rate charged by them is \$420. The population standard deviation of hourly charges for all lawyers in New York City is \$110.

a. Construct a 99% confidence interval for the mean hourly charges for all lawyers in New York City.
b. Suppose the confidence interval obtained in part a is too wide. How can the width of this interval be reduced? Discuss all possible alternatives. Which alternative is the best?

8.29 A bank manager wants to know the mean amount of mortgage paid per month by homeowners in an area. A random sample of 120 homeowners selected from this area showed that they pay an average of \$1575 per month for their mortgages. The population standard deviation of such mortgages is \$215.

a. Find a 97% confidence interval for the mean amount of mortgage paid per month by all homeowners in this area.
b. Suppose the confidence interval obtained in part a is too wide. How can the width of this interval be reduced? Discuss all possible alternatives. Which alternative is the best?

8.30 A marketing researcher wants to find a 95% confidence interval for the mean amount that visitors to a theme park spend per person per day. She knows that the standard deviation of the amounts spent per person per day by all visitors to this park is \$11. How large a sample should the researcher select so that the estimate will be within \$2 of the population mean?

8.31 A company that produces detergents wants to estimate the mean amount of detergent in 64-ounce jugs at a 99% confidence level. The company knows that the standard deviation of the amounts of detergent in all such jugs is .20 ounces. How large a sample should the company take so that the estimate is within .04 ounces of the population mean?

8.32 A department store manager wants to estimate at a 90% confidence level the mean amount spent by all customers at this store. The manager knows that the standard deviation of amounts spent by all customers at this store is \$31. What sample size should he choose so that the estimate is within \$3 of the population mean?

8.33 The principal of a large high school is concerned about the amount of time that his students spend on jobs to pay for their cars, to buy clothes, and so on. He would like to estimate the mean number of hours worked per week by these students. He knows that the standard deviation of the times spent per week on such jobs by all students is 2.5 hours. What sample size should he choose so that the estimate is within .75 hours of the population mean? The principal wants to use a 98% confidence level.

***8.34** You are interested in estimating the mean commuting time from home to school for all commuter students at your school. Briefly explain the procedure you will follow to conduct this study. Collect the required data from a sample of 30 or more such students and then estimate the population mean at a 99% confidence level. Assume that the population standard deviation for such times is 5.5 minutes.

***8.35** You are interested in estimating the mean age of cars owned by all people in the United States. Briefly explain the procedure you will follow to conduct this study. Collect the required data on a sample of 30 or more cars and then estimate the population mean at a 95% confidence level. Assume that the population standard deviation is 2.4 years.

8.4 Estimation of a Population Mean: σ Not Known

This section explains how to construct a confidence interval for the population mean μ when the population standard deviation σ is not known. Here, again, there are three possible cases that are mentioned below.

Case I. If the following three conditions are fulfilled:

1. The population standard deviation σ is not known
2. The sample size is small (i.e., $n < 30$)
3. The population from which the sample is selected is normally distributed,

then we use the t distribution (explained in Section 8.4.1) to make the confidence interval for μ.

Case II. If the following two conditions are fulfilled:

1. The population standard deviation σ is not known
2. The sample size is large (i.e., $n \geq 30$),

then again use the t distribution to make the confidence interval for μ.

Case III. If the following three conditions are fulfilled:

1. The population standard deviation σ is not known
2. The sample size is small (i.e., $n < 30$)
3. The population from which the sample is selected is not normally distributed (or its distribution is unknown),

then we use a nonparametric method to make the confidence interval for μ. Such a procedure is not covered in this text.

The following chart summarizes the above three cases.

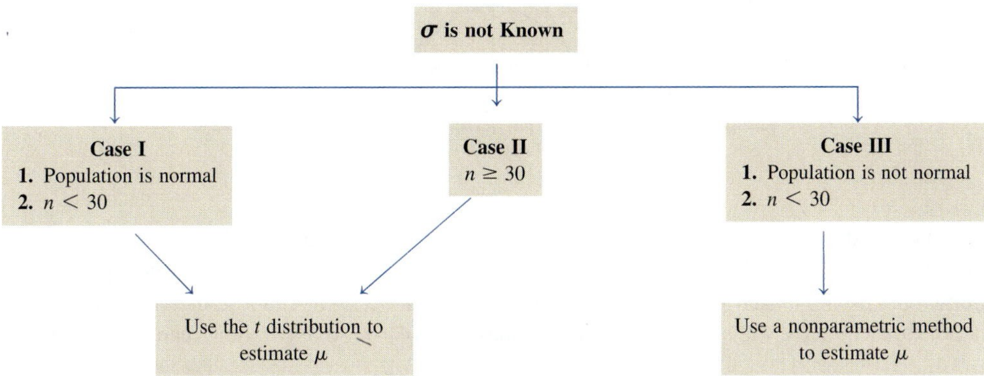

Below in Section 8.4.1 we discuss the t distribution, and then in Section 8.4.2 learn how to use the t distribution to make a confidence interval for μ when σ is not known and conditions of Cases I and II are satisfied.

8.4.1 The *t* Distribution

The ***t* distribution** was developed by W. S. Gosset in 1908 and published under the pseudonym *Student*. As a result, the t distribution is also called *Student's t distribution*. The t distribution is similar to the normal distribution in some respects. Like the normal distribution curve, the t distribution curve is symmetric (bell-shaped) about the mean and never meets the horizontal axis. The total area under a t distribution curve is 1.0 or 100%. However, the t distribution curve is flatter than the standard normal distribution curve. In other words, the t distribution curve has a lower height and a wider spread (or, we can say, larger standard deviation) than the standard normal distribution. However, as the sample size increases, the t distribution approaches the standard normal distribution. The units of a t distribution are denoted by t.

The shape of a particular t distribution curve depends on the number of **degrees of freedom (*df*)**. For the purpose of Chapters 8 and 9, the number of degrees of freedom for a t distribution is equal to the sample size minus one, that is,

$$df = n - 1$$

The number of degrees of freedom is the only parameter of the t distribution. There is a different t distribution for each number of degrees of freedom. Like the standard normal distribution, the mean of the t distribution is 0. But unlike the standard normal distribution, whose standard deviation is 1, the standard deviation of a t distribution is $\sqrt{df/(df - 2)}$, which is always greater than 1. Thus, the standard deviation of a t distribution is larger than the standard deviation of the standard normal distribution.

Definition

The *t* Distribution The *t distribution* is a specific type of bell-shaped distribution with a lower height and a wider spread than the standard normal distribution. As the sample size becomes larger, the t distribution approaches the standard normal distribution. The t distribution has only one parameter, called the degrees of freedom (df). The mean of the t distribution is equal to 0 and its standard deviation is $\sqrt{df/(df - 2)}$.

Figure 8.5 shows the standard normal distribution and the t distribution for 9 degrees of freedom. The standard deviation of the standard normal distribution is 1.0, and the standard deviation of the t distribution is $\sqrt{df/(df - 2)} = \sqrt{9/(9 - 2)} = 1.134$.

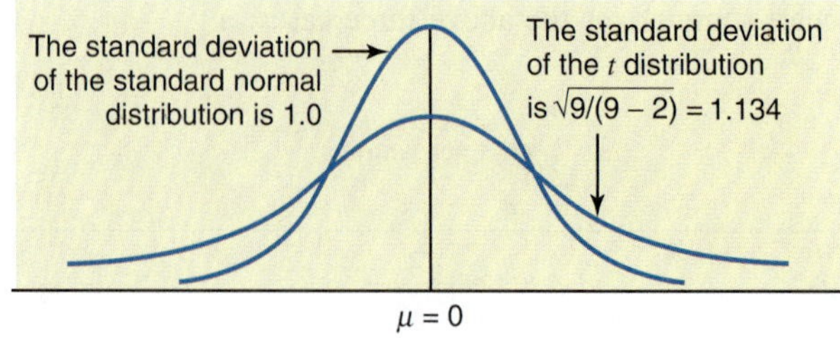
The standard deviation of the standard normal distribution is 1.0
The standard deviation of the t distribution is √9/(9 – 2) = 1.134
μ = 0

Figure 8.5 The t distribution for $df = 9$ and the standard normal distribution.

As stated earlier, the number of degrees of freedom for a t distribution for the purpose of this chapter is $n - 1$. *The number of degrees of freedom is defined as the number of observations that can be chosen freely*. As an example, suppose we know that the mean of four values is 20. Consequently, the sum of these four values is $20(4) = 80$. Now, how many values out of four can we choose freely so that the sum of these four values is 80? The answer is that we can freely choose 4 − 1 = 3 values. Suppose we choose 27, 8, and 19 as the three values. Given these three values and the information that the mean of the four values is 20, the fourth value is 80 − 27 − 8 − 19 = 26. Thus, once we have chosen three values, the fourth value is automatically determined. Consequently, the number of degrees of freedom for this example is

$$df = n - 1 = 4 - 1 = 3$$

We subtract 1 from n because we lose 1 degree of freedom to calculate the mean.

Table V of Appendix C lists the values of t for the given number of degrees of freedom and areas in the right tail of a t distribution. Because the t distribution is symmetric, these are also the values of $-t$ for the same number of degrees of freedom and the same areas in the left tail of the t distribution. Example 8–4 describes how to read Table V of Appendix C.

■ EXAMPLE 8–4

Reading the t distribution table.

Find the value of t for 16 degrees of freedom and .05 area in the right tail of a t distribution curve.

Solution In Table V of Appendix C, we locate 16 in the column of degrees of freedom (labeled df) and .05 in the row of *Area in the right tail under the t distribution curve* at the top of the table. The entry at the intersection of the row of 16 and the column of .05, which is 1.746, gives the required value of t. The relevant portion of Table V of Appendix C is shown here as Table 8.2. The value of t read from the t distribution table is shown in Figure 8.6.

Table 8.2 Determining t for 16 df and .05 Area in the Right Tail

Area in the right tail

	Area in the Right Tail Under the t Distribution Curve				
df	.10	.05	.025	...	.001
1	3.078	6.314	12.706	...	318.309
2	1.886	2.920	4.303	...	22.327
3	1.638	2.353	3.182	...	10.215
.	...	...	...	...	...
.	...	...	...	...	...
.	...	...	...	...	...
df → 16	1.337	1.746	2.120	...	3.686
.	...	...	...	...	...
.	...	...	...	...	...
.	...	...	...	...	...
75	1.293	1.665	1.992	...	3.202
	1.282	1.645	1.960	...	3.090

The required value of t for 16 df and .05 area in the right tail

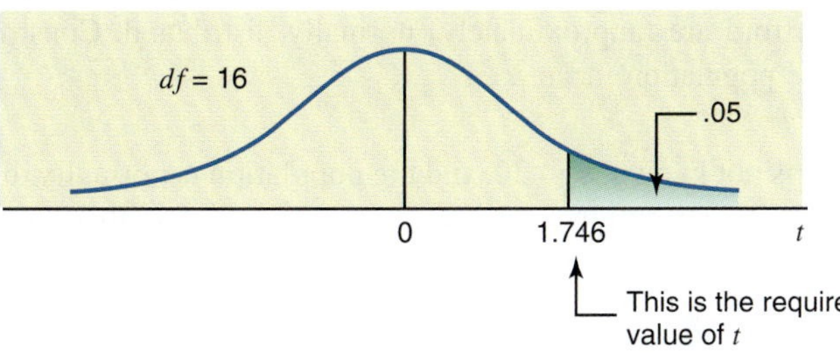

Figure 8.6 The value of t for 16 df and .05 area in the right tail.

■

Because of the symmetric shape of the t distribution curve, the value of t for 16 degrees of freedom and .05 area in the left tail is -1.746. Figure 8.7 illustrates this case.

Figure 8.7 The value of t for 16 df and .05 area in the left tail.

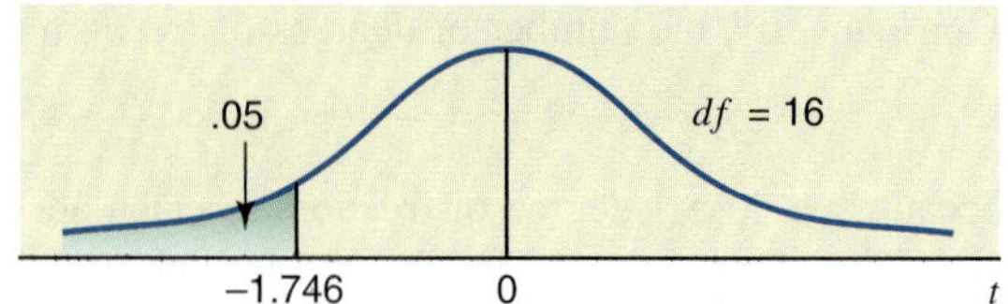

8.4.2 Confidence Interval for μ Using the t Distribution

To reiterate, when the conditions mentioned under Cases I and II in the beginning of Section 8.4 hold true, we use the t distribution to construct a confidence interval for the population mean, μ.

When the population standard deviation σ is not known, then we replace it by the sample standard deviation s, which is its estimator. Consequently, for the standard deviation of $\bar{x}$, we use

$$s_{\bar{x}} = \frac{s}{\sqrt{n}}$$

for $\sigma_{\bar{x}} = \sigma/\sqrt{n}$. Note that the value of $s_{\bar{x}}$ is a point estimate of $\sigma_{\bar{x}}$.

Confidence Interval for μ Using the t Distribution The $(1 - \alpha)100\%$ *confidence interval for* μ is

$$\bar{x} \pm ts_{\bar{x}}$$

where

$$s_{\bar{x}} = \frac{s}{\sqrt{n}}$$

The value of t is obtained from the t distribution table for $n - 1$ degrees of freedom and the given confidence level. Here $ts_{\bar{x}}$ is the margin of error of the estimate, that is:

$$E = ts_{\bar{x}}$$

Examples 8–5 and 8–6 describe the procedure of constructing a confidence interval for μ using the t distribution.

■ EXAMPLE 8–5

Constructing a 95% confidence interval for μ using the t distribution.

Dr. Moore wanted to estimate the mean cholesterol level for all adult men living in Hartford. He took a sample of 25 adult men from Hartford and found that the mean cholesterol level for this sample is 186 with a standard deviation of 12. Assume that the cholesterol levels for all adult men in Hartford are (approximately) normally distributed. Construct a 95% confidence interval for the population mean μ.

Solution Here, σ is not known, $n < 30$, and the population is normally distributed. Therefore, we will use the t distribution to make a confidence interval for μ. From the given information,

$$n = 25, \quad \bar{x} = 186, \quad s = 12,$$

and $$\text{Confidence level} = 95\% \text{ or } .95$$

The value of $s_{\bar{x}}$ is

$$s_{\bar{x}} = \frac{s}{\sqrt{n}} = \frac{12}{\sqrt{25}} = 2.40$$

To find the value of t, we need to know the degrees of freedom and the area under the t distribution curve in each tail.

$$\text{Degrees of freedom} = n - 1 = 25 - 1 = 24$$

To find the area in each tail, we divide the confidence level by 2 and subtract the number obtained from .5. Thus,

$$\text{Area in each tail} = .5 - (.95/2) = .5 - .4750 = .025$$

From the t distribution table, Table V of Appendix C, the value of t for $df = 24$ and .025 area in the right tail is 2.064. The value of t is shown in Figure 8.8.

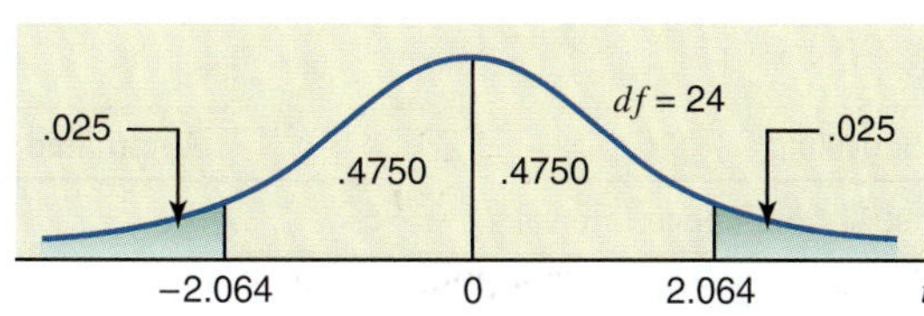

Figure 8.8 The value of t.

When we substitute all values in the formula for the confidence interval for μ, the 95% confidence interval is

$$\bar{x} \pm ts_{\bar{x}} = 186 \pm 2.064(2.40) = 186 \pm 4.95 = \mathbf{181.05 \text{ to } 190.95}$$

Thus, we can state with 95% confidence that the mean cholesterol level for all adult men living in Hartford lies between 181.05 and 190.95.

Note that $\bar{x} = 186$ is a point estimate of μ in this example and 4.95 is the margin of error. ■

■ EXAMPLE 8–6

Constructing a 99% confidence interval for μ using the t distribution.

Sixty-four randomly selected adults who buy books for general reading were asked how much they usually spend on books per year. The sample produced a mean of $1450 and a standard deviation of $300 for such annual expenses. Determine a 99% confidence interval for the corresponding population mean.

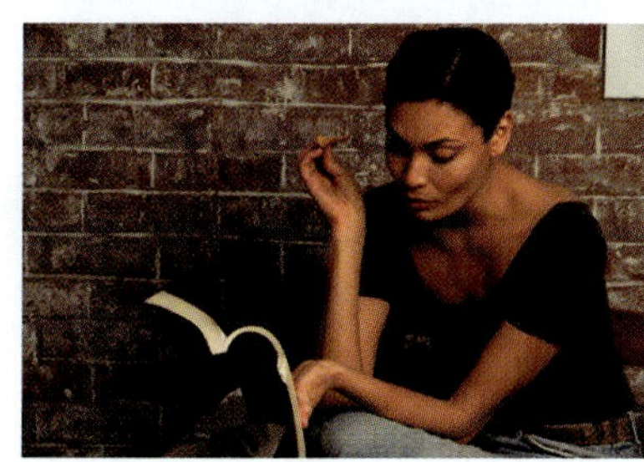

Solution From the given information,

$$n = 64, \quad \bar{x} = \$1450, \quad s = \$300,$$

and $$\text{Confidence level} = 99\% \text{ or } .99$$

Here σ is not known but the sample size is large ($n > 30$). Hence, we will use the t distribution to make a confidence interval for μ. First we calculate the standard deviation of $\bar{x}$, the number of degrees of freedom, and the area in each tail of the t distribution.

$$s_{\bar{x}} = \frac{s}{\sqrt{n}} = \frac{300}{\sqrt{64}} = \$37.50$$

$$df = n - 1 = 64 - 1 = 63$$

$$\text{Area in each tail} = .5 - (.99/2) = .5 - .4950 = .005$$

CASE STUDY

8–2

CARDIAC DEMANDS OF HEAVY SNOW SHOVELING

During the winter months, newspapers often report cases of sudden deaths occurring during snow shoveling. Dr. Barry A. Franklin et al. conducted a controlled experiment to measure the effects of snow shoveling on men. After screening for health problems, the doctors included 10 men, with a mean age of 32.4 years, in the experiment. Each of these men completed exercise tests on a treadmill and on a device known as an arm-crank ergometer. Then, each of these men participated in the snow removal tests. Each man cleared two 15-meter (49.2-feet) long tracts of heavy, wet snow ranging from 3 to 5 inches in depth, using a snow shovel in one phase and an electric snow thrower in the other. The order of these two phases was determined randomly for each man. Each man shoveled snow for 10 minutes at a time, with 10- to 15-minute rest periods. During each 10-minute period of work, heart rates were recorded at 2-minute intervals. Four variables were measured for each man during the experiment: heart rate, systolic blood pressure, oxygen consumption, and the subject's own rating of his level of exertion.

The following table compares the men's reaction to snow removal with shovels and with electric snow throwers. The means and standard deviations are based on the maximum value for each variable for each man during the experiment, and the confidence intervals are calculated using these values of the mean and the standard deviation. It is assumed that the population of each variable has a normal distribution.

Here, heart rate is measured in beats per minute, systolic blood pressure is in millimeters of mercury, oxygen consumption is in metabolic equivalents, and rating of perceived exertion is on a numerical scale ranging from 6 to 20.

	Snow Shoveling			Automated Snow Removal		
Variable	$\bar{x}$	s	90% Confidence Interval	$\bar{x}$	s	90% Confidence Interval
Heart rate	175	15	175 ± 1.833(4.7434)	124	18	124 ± 1.833(5.6921)
Systolic blood pressure	198	17	198 ± 1.833(5.3759)	161	14	161 ± 1.833(4.4272)
Oxygen consumption	5.7	.8	5.7 ± 1.833(.2530)	2.4	.7	2.4 ± 1.833(.2214)
Own rating of exertion	16.7	1.7	16.7 ± 1.833(.5376)	9.9	1.0	9.9 ± 1.833(.3162)

The confidence intervals given in the table are constructed by using the method studied in Section 8.4.2 of this chapter. For example, for the variable on heart rate while shoveling snow manually,

$$\bar{x} = 175, \quad s = 15, \quad \text{and} \quad n = 10$$

Because $n < 30$, we will use the t distribution to make the confidence interval, assuming that the corresponding population has a normal distribution. For the 90% confidence level and $df = 9$, the t value from the t distribution table is 1.833. The standard deviation of $\bar{x}$ is

$$s_{\bar{x}} = \frac{s}{\sqrt{n}} = \frac{15}{\sqrt{10}} = 4.7434$$

Thus, the 90% confidence interval for the corresponding population mean is

$$\bar{x} \pm ts_{\bar{x}} = 175 \pm 1.833(4.7434) = 175 \pm 8.69 = 166.31 \text{ to } 183.69$$

Based on this result, we can state with 90% confidence that the mean heart rate for healthy men in this age group while shoveling snow manually is between 166.31 and 183.69 beats per minute.

Source: Barry A. Franklin et al., "Cardiac Demands of Heavy Snow Shoveling," *The Journal of the American Medical Association* 273[11], March 15, 1995.

From the t distribution table, $t = 2.656$ for 63 degrees of freedom and .005 area in the right tail. The 99% confidence interval for μ is

$$\bar{x} \pm ts_{\bar{x}} = \$1450 \pm 2.656(37.50)$$
$$= \$1450 \pm \$99.60 = \mathbf{\$1350.40 \text{ to } \$1549.60}$$

Thus, we can state with 99% confidence that based on this sample the mean annual expenditure on books by all adults who buy books for general reading is between $1350.40 and $1549.60. ■

Again, we can decrease the width of a confidence interval for μ either by lowering the confidence level or by increasing the sample size, as was done in Section 8.3. However, increasing the sample size is the better alternative.

Note: What if the Sample Size Is Too Large?

In the above section when σ is not known, we used the t distribution to make a confidence interval for μ in Cases I and II. Note that in Case II, the sample size is large. If we have access to technology, it does not matter how large (over 30) the sample size is, we can use the t distribution. However, if we are using the t distribution table (Table V of Appendix C), it may pose a problem. Usually such a table goes only up to a certain number of degrees of freedom. For example, Table V in Appendix C goes only up to 75 degrees of freedom. Thus, if the sample size is larger than 76 in this section, we cannot use Table V to find the t value for the given confidence level to use in the confidence interval in this section. In such a situation when n is too large (for example 500) and is not included in the t distribution table, there are two options:

1. Use the t value from the last row (the row of ∞) in Table V.
2. Use the normal distribution as an approximation to the t distribution.

Note that the t-values you will obtain from the last row of the t-distribution table are the same as obtained from the normal distribution table for the same confidence levels, the only difference being the decimal places. To use the normal distribution as an approximation to the t distribution to make a confidence interval for μ, the procedure is exactly like the one learned in Section 8.3 except that now we will replace σ by s, and $\sigma_{\bar{x}}$ by $s_{\bar{x}}$.

Again, note that here we can use the normal distribution as a convenience and as an approximation but if we can, we should use the t distribution by using technology. Exercises 8.50, 8.51, and 8.62 to 8.65 at the end of this section present such situations.

EXERCISES

CONCEPTS AND PROCEDURES

8.36 Briefly explain the similarities and the differences between the standard normal distribution and the t distribution.

8.37 What are the parameters of a normal distribution and a t distribution? Explain.

8.38 Briefly explain the meaning of the degrees of freedom for a t distribution. Give one example.

8.39 What assumptions must hold true to use the t distribution to make a confidence interval for μ?

8.40 Find the value of t for the t distribution for each of the following.

a. Area in the right tail = .05 and $df = 12$
b. Area in the left tail = .025 and $n = 66$
c. Area in the left tail = .001 and $df = 49$
d. Area in the right tail = .005 and $n = 24$

8.41 a. Find the value of t for a t distribution with a sample size of 21 and area in the left tail equal to .10.

b. Find the value of t for a t distribution with a sample size of 14 and area in the right tail equal to .025.

c. Find the value of t for a t distribution with 45 degrees of freedom and .001 area in the right tail.

d. Find the value of t for a t distribution with 37 degrees of freedom and .005 area in the left tail.

8.42 For each of the following, find the area in the appropriate tail of the t distribution.

a. $t = 2.467$ and $df = 28$
b. $t = -1.672$ and $df = 58$
c. $t = -2.670$ and $n = 55$
d. $t = 2.383$ and $n = 23$

8.43 For each of the following, find the area in the appropriate tail of the t distribution.

a. $t = -1.302$ and $df = 42$
b. $t = 2.797$ and $n = 25$
c. $t = 1.397$ and $n = 9$
d. $t = -2.383$ and $df = 67$

8.44 Find the value of t from the t distribution table for each of the following.

a. Confidence level = 99% and $df = 13$
b. Confidence level = 95% and $n = 36$
c. Confidence level = 90% and $df = 16$

8.45 **a.** Find the value of t from the t distribution table for a sample size of 22 and a confidence level of 95%.
b. Find the value of t from the t distribution table for 60 degrees of freedom and a 90% confidence level.
c. Find the value of t from the t distribution table for a sample size of 24 and a confidence level of 99%.

8.46 A sample of 12 observations taken from a normally distributed population produced the following data.

13	15	9	11	8	19
17	9	10	14	16	12

a. What is the point estimate of μ?
b. Make a 99% confidence interval for μ.
c. What is the margin of error of estimate for μ in part b?

8.47 A sample of 10 observations taken from a normally distributed population produced the following data.

44	52	31	48	46	39	47	36	41	57

a. What is the point estimate of μ?
b. Make a 95% confidence interval for μ.
c. What is the margin of error of estimate for μ in part b?

8.48 Suppose, for a sample selected from a normally distributed population, $\bar{x} = 68.50$ and $s = 8.9$.
a. Construct a 95% confidence interval for μ assuming $n = 16$.
b. Construct a 90% confidence interval for μ assuming $n = 16$. Is the width of the 90% confidence interval smaller than the width of the 95% confidence interval calculated in part a? If yes, explain why.
c. Find a 95% confidence interval for μ assuming $n = 25$. Is the width of the 95% confidence interval for μ with $n = 25$ smaller than the width of the 95% confidence interval for μ with $n = 16$ calculated in part a? If so, why? Explain.

8.49 Suppose, for a sample selected from a population, $\bar{x} = 25.5$ and $s = 4.9$.
a. Construct a 95% confidence interval for μ assuming $n = 47$.
b. Construct a 99% confidence interval for μ assuming $n = 47$. Is the width of the 99% confidence interval larger than the width of the 95% confidence interval calculated in part a? If yes, explain why.
c. Find a 95% confidence interval for μ assuming $n = 32$. Is the width of the 95% confidence interval for μ with $n = 32$ larger than the width of the 95% confidence interval for μ with $n = 47$ calculated in part a? If so, why? Explain.

8.50 **a.** A sample of 100 observations taken from a population produced a sample mean equal to 55.32 and a standard deviation equal to 8.4. Make a 90% confidence interval for μ.
b. Another sample of 100 observations taken from the same population produced a sample mean equal to 57.40 and a standard deviation equal to 7.5. Make a 90% confidence interval for μ.
c. A third sample of 100 observations taken from the same population produced a sample mean equal to 56.25 and a standard deviation equal to 7.9. Make a 90% confidence interval for μ.
d. The true population mean for this population is 55.80. Which of the confidence intervals constructed in parts a through c cover this population mean and which do not?

8.51 **a.** A sample of 400 observations taken from a population produced a sample mean equal to 92.45 and a standard deviation equal to 12.20. Make a 98% confidence interval for μ.
b. Another sample of 400 observations taken from the same population produced a sample mean equal to 91.75 and a standard deviation equal to 14.50. Make a 98% confidence interval for μ.
c. A third sample of 400 observations taken from the same population produced a sample mean equal to 89.63 and a standard deviation equal to 13.40. Make a 98% confidence interval for μ.
d. The true population mean for this population is 90.65. Which of the confidence intervals constructed in parts a through c cover this population mean and which do not?

APPLICATIONS

8.52 A random sample of 16 airline passengers at the Bay City airport showed that the mean time spent waiting in line to check in at the ticket counters was 31 minutes with a standard deviation of 7 minutes. Construct a 99% confidence interval for the mean time spent waiting in line by all passengers at this airport. Assume that such waiting times for all passengers are normally distributed.

8.53 A random sample of 20 acres gave a mean yield of wheat equal to 41.2 bushels per acre with a standard deviation of 3 bushels. Assuming that the yield of wheat per acre is normally distributed, construct a 90% confidence interval for the population mean μ.

8.54 According to Lear Center Local News Archive, the average amount of time that a half-hour local TV news broadcast devotes to U.S. foreign policy, including the war in Iraq, is 38 seconds (*Time*, February 28, 2005). Suppose a random sample of 40 such half-hour news broadcasts shows that an average of 38 seconds are devoted to U.S. foreign policy with a standard deviation of 9 seconds. Find a 95% confidence interval for the mean time that all half-hour local TV news broadcasts devote to U.S. foreign policy.

8.55 A company wants to estimate the mean net weight of its Top Taste cereal boxes. A sample of 36 such boxes produced the mean net weight of 31.98 ounces with a standard deviation of .26 ounces. Make a 95% confidence interval for the mean net weight of all Top Taste cereal boxes.

8.56 The high cost of health care is a matter of major concern for a large number of families. A random sample of 25 families selected from an area showed that they spend an average of \$143 per month on health care with a standard deviation of \$28. Make a 98% confidence interval for the mean health care expenditure per month incurred by all families in this area. Assume that the monthly health care expenditures of all families in this area have a normal distribution.

8.57 Jack's Auto Insurance Company customers sometimes have to wait a long time to speak to a customer service representative when they call regarding disputed claims. A random sample of 25 such calls yielded a mean waiting time of 22 minutes with a standard deviation of 6 minutes. Construct a 99% confidence interval for the population mean of such waiting times. Assume that such waiting times for the population follow a normal distribution.

8.58 A random sample of 36 mid-sized cars tested for fuel consumption gave a mean of 26.4 miles per gallon with a standard deviation of 2.3 miles per gallon.

a. Find a 99% confidence interval for the population mean, μ.

b. Suppose the confidence interval obtained in part a is too wide. How can the width of this interval be reduced? Describe all possible alternatives. Which alternative is the best and why?

8.59 The mean time taken to design a house plan by 40 architects was found to be 23 hours with a standard deviation of 3.75 hours.

a. Construct a 98% confidence interval for the population mean μ.

b. Suppose the confidence interval obtained in part a is too wide. How can the width of this interval be reduced? Describe all possible alternatives. Which alternative is the best and why?

8.60 The following data give the speeds (in miles per hour), as measured by radar, of 10 cars traveling on Interstate I-15.

76	72	80	68	76	74	71	78	82	65

Assuming that the speeds of all cars traveling on this highway have a normal distribution, construct a 90% confidence interval for the mean speed of all cars traveling on this highway.

8.61 A company randomly selected nine office employees and secretly monitored their computers for one month. The times (in hours) spent by these employees using their computers for non–job-related activities (playing games, personal communications, etc.) during this month are given below.

7	12	9	8	11	4	14	1	6

Assuming that such times for all employees are normally distributed, make a 95% confidence interval for the corresponding population mean for all employees of this company.

8.62 According to a 2004 article based on research by *USA TODAY*, the average monthly residential electric bill at that time was \$77 (*USA TODAY*, June 7, 2004). Suppose this mean was based on a random sample of 500 such bills, and the standard deviation for this sample was \$26. Find a 99% confidence interval for the mean of all monthly residential electric bills for 2004.

8.63 According to the Associated Press, the Le Parker Meridien hotel in New York City offers the "Zillion Dollar Frittata," an omelet containing six eggs, lobster meat, and 10 ounces of sevruga caviar, for a price of \$1000 (*Time*, May 31, 2004). Suppose 90 randomly selected people with annual incomes of \$500,000 or higher were asked how much they would be willing to pay for such an omelet if it were offered at an auction to benefit a charity. Suppose the mean of their responses is \$1250 and the standard deviation is \$270. Construct a 95% confidence interval for the corresponding population mean.

8.64 According to Lear Center Local News Archive, a typical half-hour local TV news broadcast devotes 381 seconds (i.e., 6 minutes 21 seconds) to sports and weather (*Time*, February 28, 2005). Suppose a random sample of 80 such half-hour news broadcasts shows that an average of 381 seconds are devoted to sports and weather with a standard deviation of 30 seconds.

a. What is the point estimate of the corresponding population mean?
b. Make a 90% confidence interval for the corresponding population mean. What is the margin of error for this estimate?

8.65 According to Bridal Guide Info Source, the average cost of a wedding in 2003 was \$21,213 (*USA TODAY*, June 22, 2004). Assume that this average is based on a random sample of 1100 weddings and that the sample standard deviation is \$5100.
a. What is the point estimate of the corresponding population mean?
b. Make a 98% confidence interval for the corresponding population mean. What is the margin of error for this estimate?

*8.66 You are working for a supermarket. The manager has asked you to estimate the mean time taken by a cashier to serve customers at this supermarket. Briefly explain how you will conduct this study. Collect data on the time taken by any supermarket cashier to serve 40 customers. Then estimate the population mean. Choose your own confidence level.

*8.67 You are working for a bank. The bank manager wants to know the mean waiting time for all customers who visit this bank. She has asked you to estimate this mean by taking a sample. Briefly explain how you will conduct this study. Collect data on the waiting times for 45 customers who visit a bank. Then estimate the population mean. Choose your own confidence level.

8.5 Estimation of a Population Proportion: Large Samples

Often we want to estimate the population proportion or percentage. (Recall that a percentage is obtained by multiplying the proportion by 100.) For example, the production manager of a company may want to estimate the proportion of defective items produced on a machine. A bank manager may want to find the percentage of customers who are satisfied with the service provided by the bank.

Again, if we can conduct a census each time we want to find the value of a population proportion, there is no need to learn the procedures discussed in this section. However, we usually derive our results from sample surveys. Hence, to take into account the variability in the results obtained from different sample surveys, we need to know the procedures for estimating a population proportion.

Recall from Chapter 7 that the population proportion is denoted by p and the sample proportion is denoted by $\hat{p}$. This section explains how to estimate the population proportion, p, using the sample proportion, $\hat{p}$. The sample proportion, $\hat{p}$, is a sample statistic, and it possesses a sampling distribution. From Chapter 7, we know that for large samples:

1. The sampling distribution of the sample proportion, $\hat{p}$, is (approximately) normal.
2. The mean, $\mu_{\hat{p}}$, of the sampling distribution of $\hat{p}$ is equal to the population proportion, p.
3. The standard deviation, $\sigma_{\hat{p}}$, of the sampling distribution of the sample proportion, $\hat{p}$, is $\sqrt{pq/n}$, where $q = 1 - p$.

Remember ▶ In the case of a proportion, a sample is considered to be large if np and nq are both greater than 5. If p and q are not known, then $n\hat{p}$ and $n\hat{q}$ should each be greater than 5 for the sample to be large.

When estimating the value of a population proportion, we do not know the values of p and q. Consequently, we cannot compute $\sigma_{\hat{p}}$. Therefore, in the estimation of a population proportion, we use the value of $s_{\hat{p}}$ as an estimate of $\sigma_{\hat{p}}$. The value of $s_{\hat{p}}$ is calculated using the following formula.

Estimator of the Standard Deviation of $\hat{p}$ The value of $s_{\hat{p}}$, which gives a point estimate of $\sigma_{\hat{p}}$, is calculated as follows. Here, $s_{\hat{p}}$ is called an estimator of $\sigma_{\hat{p}}$.

$$s_{\hat{p}} = \sqrt{\frac{\hat{p}\hat{q}}{n}}$$

The sample proportion, $\hat{p}$, is the point estimator of the corresponding population proportion, p. Then to find the confidence interval for p, we add to and subtract from $\hat{p}$ a number that is called the **margin of error**, E.

Confidence Interval for the Population Proportion, *p* The $(1 - \alpha)100\%$ *confidence interval for the population proportion, p,* is

$$\hat{p} \pm z\, s_{\hat{p}}$$

The value of z used here is obtained from the standard normal distribution table for the given confidence level, and $s_{\hat{p}} = \sqrt{\hat{p}\hat{q}/n}$. The term $z\, s_{\hat{p}}$ is called the *margin of error, E.*

Examples 8–7 and 8–8 illustrate the procedure for constructing a confidence interval for p.

■ EXAMPLE 8–7

Finding the point estimate and 99% confidence interval for p: large sample.

According to a 2005 survey of 1506 adult Americans conducted by the National Sleep Foundation, 75% of them said that they frequently have symptoms of sleep problems such as frequent waking during the night or snoring (Reuters, March 30, 2005).

(a) What is the point estimate of the population proportion?

(b) Find, with a 99% confidence level, the percentage of all adult Americans who frequently have symptoms of sleep problems such as frequent waking during the night or snoring. What is the margin of error in this estimate?

Solution Let p be the proportion of all adult Americans who frequently have symptoms of sleep problems such as frequent waking during the night or snoring and let $\hat{p}$ be the corresponding sample proportion. From the given information,

$$n = 1506, \quad \hat{p} = .75, \quad \text{and} \quad \hat{q} = 1 - \hat{p} = 1 - .75 = .25$$

First, we calculate the value of the standard deviation of the sample proportion as follows:

$$s_{\hat{p}} = \sqrt{\frac{\hat{p}\hat{q}}{n}} = \sqrt{\frac{(.75)(.25)}{1506}} = .01115805$$

Note that $n\hat{p}$ and $n\hat{q}$ are both greater than 5. (The reader should check this condition.) Consequently, due to the central limit theorem, the sampling distribution of $\hat{p}$ is approximately normal, and we will use the normal distribution to calculate the confidence interval about p.

(a) The point estimate of the proportion of all adult Americans who frequently have symptoms of sleep problems such as frequent waking during the night or snoring is equal to .75; that is,

$$\text{Point estimate of } p = \hat{p} = \mathbf{.75}$$

(b) The confidence level is 99%, or .99. To find z for a 99% confidence level, first we find the area in each of the two tails, which is $(1 - .99)/2 = .0050$. Then, we look for .0050 and $.0050 + .99 = .9950$ areas in the normal distribution table to find the two z values. These two z values are (approximately) -2.58 and 2.58. Thus, we will use $z = 2.58$ in the confidence interval formula. Substituting all the values in the confidence interval formula for p, we obtain,

$$\hat{p} \pm z s_{\hat{p}} = .75 \pm 2.58(.01115805) = .75 \pm .029$$

$$= \mathbf{.721 \text{ to } .779} \text{ or } \mathbf{72.1\% \text{ to } 77.9\%}$$

Thus, we can state with 99% confidence that .721 to .779 or 72.1% to 77.9% of all adult Americans frequently have symptoms of sleep problems such as frequent waking during the night or snoring.

In the above estimate of p, the value .029 or 2.9% is the margin of error. ■

EXAMPLE 8–8

Constructing a 95% confidence interval for p: large sample.

According to an Associated Press poll, circumstances such as income, education, and marital status affect whether or not Americans feel satisfied with their lives. According to this poll conducted by Ipsos-Public Affairs on August 16–18, 2004, 51% of college graduates said that they are *very satisfied* with the way things were going in their lives at that time (Yahoo! News, September 24, 2004). Suppose this poll included 700 college graduates. Construct a 97% confidence interval for the corresponding population proportion.

Solution Let p be the proportion of all college graduates who were *very satisfied* with the way things were going in their lives at the time of the poll, and let $\hat{p}$ be the corresponding sample proportion. From the given information,

$$n = 700, \quad \hat{p} = .51, \quad \hat{q} = 1 - \hat{p} = 1 - .51 = .49$$

and Confidence level = 97% or .97

The standard deviation of the sample proportion is

$$s_{\hat{p}} = \sqrt{\frac{\hat{p}\hat{q}}{n}} = \sqrt{\frac{(.51)(.49)}{700}} = .01889444$$

From the normal distribution table, the value of z for the 97% confidence interval is 2.17. Note that to find this z value, you will look for the areas .0150 and .9850 in Table IV of Appendix C. Substituting all the values in the formula, the 97% confidence interval for p is

$$\hat{p} \pm z s_{\hat{p}} = .51 \pm 2.17(.01889444) = .51 \pm .041$$

$$= \mathbf{.469 \text{ to } .551} \text{ or } \mathbf{46.9\% \text{ to } 55.1\%}$$

Thus, we can state with 97% confidence that the proportion of all college graduates who were *very satisfied* with the way things were going in their lives at the time of the poll is between .469 and .551. The confidence interval can be converted into a percentage interval as 46.9% to 55.1%.

In the above estimate, .51 is the point estimate of p and .041 is the margin of error. ■

Again, we can decrease the width of a confidence interval for p either by lowering the confidence level or by increasing the sample size. However, lowering the confidence level is not a good choice because it simply decreases the likelihood that the confidence interval contains p. Hence, to decrease the width of a confidence interval for p, we should always increase the sample size.

8.5.1 Determining the Sample Size for the Estimation of Proportion

Just as we did with the mean, we can also determine the sample size for estimating the population proportion, p. This sample size will yield an error of estimate that may not be larger than a predetermined margin of error. By knowing the sample size that can give us the required results, we can save our scarce resources by not taking an unnecessarily large sample. From Section 8.5, the margin of error, E, of the interval estimation of the population proportion is

$$E = z s_{\hat{p}} = z \times \sqrt{\frac{\hat{p}\hat{q}}{n}}$$

By manipulating this expression algebraically, we obtain the following formula to find the required sample size given E, $\hat{p}$, $\hat{q}$, and z.

BUYING HABITS OF CONSUMERS

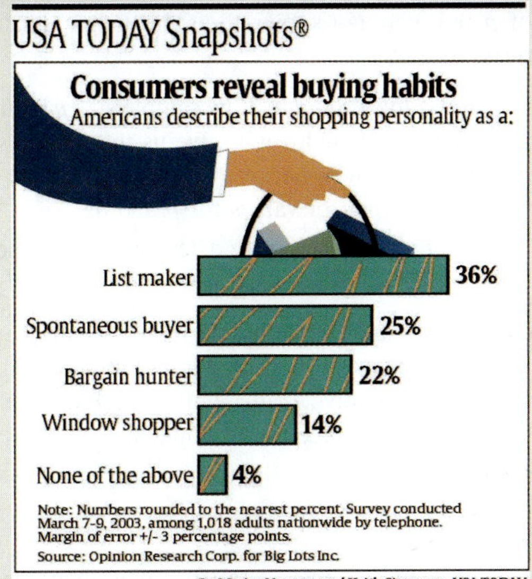
USA TODAY Snapshots®

Consumers reveal buying habits

Americans describe their shopping personality as a:

List maker	36%
Spontaneous buyer	25%
Bargain hunter	22%
Window shopper	14%
None of the above	4%

Note: Numbers rounded to the nearest percent. Survey conducted March 7-9, 2003, among 1,018 adults nationwide by telephone. Margin of error +/- 3 percentage points.

Source: Opinion Research Corp. for Big Lots Inc.

By Marisa Navarro and Keith Simmons, USA TODAY

The above chart shows the buying habits of Americans. Based on a telephone poll of 1018 adults conducted in March 2003 by Opinion Research Corporation, it lists the types of shoppers adult Americans are. For example, according to this poll, 36% of these adults said that they are list makers, i.e., they make a list of what they need to buy before they go shopping. Twenty-five percent of the adults said they are spontaneous buyers, i.e., they see something, they like it or think they need it, and they buy it. Using the procedure learned in this section, we can then find the confidence intervals for the various proportions as shown in the table below. Note that the percentages in the chart add up to 101% because of the rounding.

Shopping Personality	Sample Proportion	Confidence Interval
List maker	.36	$.36 \pm zs_{\hat{p}}$
Spontaneous buyer	.25	$.25 \pm zs_{\hat{p}}$
Bargain hunter	.22	$.22 \pm zs_{\hat{p}}$
Window shopper	.14	$.14 \pm zs_{\hat{p}}$
None of the above	.04	$.04 \pm zs_{\hat{p}}$

For each of the confidence intervals given in the table, we can substitute the value of z and the value of $s_{\hat{p}}$, which is calculated as $\sqrt{\frac{\hat{p}\hat{q}}{n}}$. For example, suppose we want to find a 96% confidence interval for the proportion of all adults who have a shopping personality of a list maker. Then the 96% confidence interval for the corresponding population proportion is determined as follows.

$$s_{\hat{p}} = \sqrt{\frac{\hat{p}\hat{q}}{n}} = \sqrt{\frac{(.36)(.64)}{1018}} = .01504414$$

$$\hat{p} \pm zs_{\hat{p}} = .36 \pm 2.05(.01504414) = .36 \pm .03 = .33 \text{ to } .39$$

Thus, we can expect 33% to 39% of all shoppers to be list makers.

We can find the confidence intervals for the population proportions of other categories the same way.

Source: The chart reproduced with permission from *USA TODAY*, June 12, 2003. Copyright © 2003, *USA TODAY.*

Determining the Sample Size for the Estimation of *p* Given the confidence level and the values of $\hat{p}$ and $\hat{q}$, the sample size that will produce a predetermined margin of error E of the confidence interval *estimate of p* is

$$n = \frac{z^2\hat{p}\hat{q}}{E^2}$$

We can observe from this formula that to find n, we need to know the values of $\hat{p}$ and $\hat{q}$. However, the values of $\hat{p}$ and $\hat{q}$ are not known to us. In such a situation, we can choose one of the following alternatives.

1. We make the *most conservative estimate* of the sample size n by using $\hat{p} = .50$ and $\hat{q} = .50$. For a given E, these values of $\hat{p}$ and $\hat{q}$ will give us the largest sample size in comparison to any other pair of values of $\hat{p}$ and $\hat{q}$ because the product of $\hat{p} = .50$ and $\hat{q} = .50$. is greater than the product of any other pair of values for $\hat{p}$ and $\hat{q}$.
2. We take a *preliminary sample* (of arbitrarily determined size) and calculate $\hat{p}$ and $\hat{q}$ for this sample. Then, we use these values of $\hat{p}$ and $\hat{q}$ to find n.

Examples 8–9 and 8–10 illustrate how to determine the sample size that will produce the error of estimation for the population proportion within a predetermined margin of error value. Example 8–9 gives the most conservative estimate of n, and Example 8–10 uses the results from a preliminary sample to determine the required sample size.

■ EXAMPLE 8–9

Determining the most conservative estimate of n for the estimation of p.

Lombard Electronics Company has just installed a new machine that makes a part that is used in clocks. The company wants to estimate the proportion of these parts produced by this machine that are defective. The company manager wants this estimate to be within .02 of the population proportion for a 95% confidence level. What is the most conservative estimate of the sample size that will limit the margin of error to within .02 of the population proportion?

Solution The company manager wants the 95% confidence interval to be

$$\hat{p} \pm .02$$

Therefore, $E = .02$

The value of z for a 95% confidence level is 1.96. For the most conservative estimate of the sample size, we will use $\hat{p} = .50$ and $\hat{q} = .50$. Hence, the required sample size is

$$n = \frac{z^2 \hat{p} \hat{q}}{E^2} = \frac{(1.96)^2(.50)(.50)}{(.02)^2} = \mathbf{2401}$$

Thus, if the company takes a sample of 2401 parts, there is a 95% chance that the estimate of p will be within .02 of the population proportion. ■

■ EXAMPLE 8–10

Determining n for the estimation of p using preliminary sample results.

Consider Example 8–9 again. Suppose a preliminary sample of 200 parts produced by this machine showed that 7% of them are defective. How large a sample should the company select so that the 95% confidence interval for p is within .02 of the population proportion?

Solution Again, the company wants the 95% confidence interval for p to be

$$\hat{p} \pm .02$$

Hence, $E = .02$

The value of z for a 95% confidence level is 1.96. From the preliminary sample,

$$\hat{p} = .07 \quad \text{and} \quad \hat{q} = 1 - .07 = .93$$

Using these values of $\hat{p}$ and $\hat{q}$, we obtain

$$n = \frac{z^2 \hat{p} \hat{q}}{E^2} = \frac{(1.96)^2(.07)(.93)}{(.02)^2} = \frac{(3.8416)(.07)(.93)}{.0004} = 625.22 \approx \mathbf{626}$$

Thus, if the company takes a sample of 626 items, there is a 95% chance that the estimate of p will be within .02 of the population proportion. However, we should note that this sample size will produce the margin of error within .02 only if $\hat{p}$ is .07 or less for the new sample. If $\hat{p}$ for the new sample happens to be much higher than .07, the margin of error will not be within .02. Therefore, to avoid such a situation, we may be more conservative and take a much larger sample than 626 items. ■

EXERCISES

CONCEPTS AND PROCEDURES

8.68 What assumption(s) must hold true to use the normal distribution to make a confidence interval for the population proportion, p?

8.69 What is the point estimator of the population proportion, p?

8.70 Check if the sample size is large enough to use the normal distribution to make a confidence interval for p for each of the following cases.

a. $n = 50$ and $\hat{p} = .25$ **b.** $n = 160$ and $\hat{p} = .03$
c. $n = 400$ and $\hat{p} = .65$ **d.** $n = 75$ and $\hat{p} = .06$

8.71 Check if the sample size is large enough to use the normal distribution to make a confidence interval for p for each of the following cases.

a. $n = 120$ and $\hat{p} = .04$ **b.** $n = 60$ and $\hat{p} = .08$
c. $n = 40$ and $\hat{p} = .50$ **d.** $n = 900$ and $\hat{p} = .15$

8.72 **a.** A sample of 400 observations taken from a population produced a sample proportion of .63. Make a 95% confidence interval for p.
b. Another sample of 400 observations taken from the same population produced a sample proportion of .59. Make a 95% confidence interval for p.
c. A third sample of 400 observations taken from the same population produced a sample proportion of .67. Make a 95% confidence interval for p.
d. The true population proportion for this population is .65. Which of the confidence intervals constructed in parts a through c cover this population proportion and which do not?

8.73 **a.** A sample of 900 observations taken from a population produced a sample proportion of .32. Make a 90% confidence interval for p.
b. Another sample of 900 observations taken from the same population produced a sample proportion of .36. Make a 90% confidence interval for p.
c. A third sample of 900 observations taken from the same population produced a sample proportion of .30. Make a 90% confidence interval for p.
d. The true population proportion for this population is .34. Which of the confidence intervals constructed in parts a through c cover this population proportion and which do not?

8.74 A sample of 500 observations selected from a population produced a sample proportion equal to .68.
a. Make a 90% confidence interval for p.
b. Construct a 95% confidence interval for p.
c. Make a 99% confidence interval for p.
d. Does the width of the confidence intervals constructed in parts a through c increase as the confidence level increases? If yes, explain why.

8.75 A sample of 200 observations selected from a population gave a sample proportion equal to .27.
a. Make a 99% confidence interval for p.
b. Construct a 97% confidence interval for p.
c. Make a 90% confidence interval for p.
d. Does the width of the confidence intervals constructed in parts a through c decrease as the confidence level decreases? If yes, explain why.

8.76 A sample selected from a population gave a sample proportion equal to .73.
a. Make a 99% confidence interval for p assuming $n = 100$.
b. Construct a 99% confidence interval for p assuming $n = 600$.
c. Make a 99% confidence interval for p assuming $n = 1500$.
d. Does the width of the confidence intervals constructed in parts a through c decrease as the sample size increases? If yes, explain why.

8.77 A sample selected from a population gave a sample proportion equal to .31.

a. Make a 95% confidence interval for p assuming $n = 1200$.

b. Construct a 95% confidence interval for p assuming $n = 500$.

c. Make a 95% confidence interval for p assuming $n = 80$.

d. Does the width of the confidence intervals constructed in parts a through c increase as the sample size decreases? If yes, explain why.

8.78 a. How large a sample should be selected so that the margin of error of estimate for a 99% confidence interval for p is .035 when the value of the sample proportion obtained from a preliminary sample is .29?

b. Find the most conservative sample size that will produce the margin of error for a 99% confidence interval for p equal to .035.

8.79 a. How large a sample should be selected so that the margin of error of estimate for a 98% confidence interval for p is .045 when the value of the sample proportion obtained from a preliminary sample is .53?

b. Find the most conservative sample size that will produce the margin of error for a 98% confidence interval for p equal to .045.

8.80 Determine the most conservative sample size for the estimation of the population proportion for the following.

a. $E = .03$, confidence level = 99%

b. $E = .04$, confidence level = 95%

c. $E = .01$, confidence level = 90%

8.81 Determine the sample size for the estimation of the population proportion for the following, where $\hat{p}$ is the sample proportion based on a preliminary sample.

a. $E = .03$, $\hat{p} = .32$, confidence level = 99%

b. $E = .04$, $\hat{p} = .78$, confidence level = 95%

c. $E = .02$, $\hat{p} = .64$, confidence level = 90%

■ APPLICATIONS

8.82 In a survey of 200 adults, the U.S. Public Interest Group found that the credit reports of (about) 79% of these adults contained errors (*USA TODAY*, June 18, 2004). Assume that these 200 adults make a simple random sample of all adults in the United States.

a. What is the point estimate of the corresponding population proportion?

b. Construct a 99% confidence interval for the proportion of all adults in the United States whose credit reports contain errors. What is the margin of error for this estimate?

8.83 The express check-out lanes at Wally's Supermarket are limited to customers purchasing 12 or fewer items. Cashiers at this supermarket have complained that many customers who use the express lanes have more than 12 items. A recently taken random sample of 200 customers entering express lanes at this supermarket found that 74 of them had more than 12 items.

a. Construct a 98% confidence interval for the percentage of all customers at this supermarket who enter express lanes with more than 12 items.

b. Suppose the confidence interval obtained in part a is too wide. How can the width of this interval be reduced? Discuss all possible alternatives. Which alternative is the best?

8.84 According to data from Ipsos Global Express, 64% of Americans say there is never enough time in the day to get things done (*Business Week*, November 22, 2004). Suppose this percentage is based on a random sample of 900 Americans.

a. What is the point estimate of the corresponding population proportion?

b. Find a 90% confidence interval for the corresponding population proportion. What is the margin of error for this estimate?

8.85 It is said that happy and healthy workers are efficient and productive. A company that manufactures exercising machines wanted to know the percentage of large companies that provide on-site health club facilities. A sample of 240 such companies showed that 96 of them provide such facilities on site.

a. What is the point estimate of the percentage of all such companies that provide such facilities on site?

b. Construct a 97% confidence interval for the percentage of all such companies that provide such facilities on site. What is the margin of error for this estimate?

8.86 A mail-order company promises its customers that the products ordered will be mailed within 72 hours after an order is placed. The quality control department at the company checks from time to time to see if this promise is fulfilled. Recently the quality control department took a sample of 50 orders and found that 35 of them were mailed within 72 hours of the placement of the orders.

a. Construct a 98% confidence interval for the percentage of all orders that are mailed within 72 hours of their placement.

b. Suppose the confidence interval obtained in part a is too wide. How can the width of this interval be reduced? Discuss all possible alternatives. Which alternative is the best?

8.87 In a random sample of 50 homeowners selected from a large suburban area, 19 said that they had serious problems with excessive noise from their neighbors.

a. Make a 99% confidence interval for the percentage of all homeowners in this suburban area who have such problems.

b. Suppose the confidence interval obtained in part a is too wide. How can the width of this interval be reduced? Discuss all possible alternatives. Which option is best?

8.88 According to a *PARADE* and Research! America poll of 1000 Americans, 73% of the respondents claimed that they had changed their behavior in an attempt to lower their risks of heart disease (*PARADE*, February 6, 2005). Assume that these 1000 people make a random sample of Americans.

a. Find a 99% confidence interval for the percentage of all Americans who will say that they have changed their behavior to lower their risks of heart disease.

b. Explain why we need to make this confidence interval. Why can we not say that 73% of all Americans will claim that they have changed their behavior to lower their risks of heart disease?

8.89 In the National Retail Federation 2004 Holiday Consumer Intentions and Actions Survey of 7861 consumers conducted in October 2004 by BIGresearch, 18% of the respondents stated that they had begun their holiday shopping before September of that year (*USA TODAY*, November 10, 2004).

a. Make a 99% confidence interval for the proportion of all consumers who began their 2004 holiday shopping before September.

b. Explain why we need to make the confidence interval. Why can we not simply say that 18% of all consumers began their 2004 holiday shopping before September?

8.90 A researcher wanted to know the percentage of judges who are in favor of the death penalty. He took a random sample of 15 judges and asked them whether or not they favor the death penalty. The responses of these judges are given here.

Yes	No	Yes	Yes	No	No	No	Yes
Yes	No	Yes	Yes	Yes	No	Yes	

a. What is the point estimate of the population proportion?

b. Make a 95% confidence interval for the percentage of all judges who are in favor of the death penalty.

8.91 The management of a health insurance company wants to know the percentage of its policyholders who have tried alternative treatments (such as acupuncture, herbal therapy, etc.). A random sample of 24 of the company's policyholders were asked whether or not they have ever tried such treatments. The following are their responses.

Yes	No	No	Yes	No	Yes	No	No
No	Yes	No	No	Yes	No	Yes	No
No	No	Yes	No	No	No	Yes	No

a. What is the point estimate of the corresponding population proportion?

b. Construct a 99% confidence interval for the percentage of this company's policyholders who have tried alternative treatments.

8.92 Tony's Pizza guarantees all pizza deliveries within 30 minutes of the placement of orders. An agency wants to estimate the proportion of all pizzas delivered within 30 minutes by Tony's. What is the most conservative estimate of the sample size that would limit the margin of error to within .02 of the population proportion for a 99% confidence interval?

8.93 Refer to Exercise 8.92. Assume that a preliminary study has shown that 93% of all Tony's pizzas are delivered within 30 minutes. How large should the sample size be so that the 99% confidence interval for the population proportion has a margin of error of .02?

8.94 A consumer agency wants to estimate the proportion of all drivers who wear seat belts while driving. Assume that a preliminary study has shown that 76% of drivers wear seat belts while driving. How large should the sample size be so that the 99% confidence interval for the population proportion has a margin of error of .03?

8.95 Refer to Exercise 8.94. What is the most conservative estimate of the sample size that would limit the margin of error to within .03 of the population proportion for a 99% confidence interval?

***8.96** You want to estimate the proportion of students at your college who hold off-campus (part-time or full-time) jobs. Briefly explain how you will make such an estimate. Collect data from 40 students at your college on whether or not they hold off-campus jobs. Then calculate the proportion of students in this sample who hold off-campus jobs. Using this information, estimate the population proportion. Select your own confidence level.

***8.97** You want to estimate the percentage of students at your college or university who are satisfied with the campus food services. Briefly explain how you will make such an estimate. Select a sample of 30 students and ask them whether or not they are satisfied with the campus food services. Then calculate the percentage of students in the sample who are satisfied. Using this information, find the confidence interval for the corresponding population percentage. Select your own confidence level.

USES AND MISUSES... NATIONAL VERSUS LOCAL UNEMPLOYMENT RATE

Reading a newspaper article, you learn that the national unemployment rate is 5.75%. The next month you read another article that states that a recent survey in your area, based on a random sample of the labor force, estimates that the local unemployment rate is 5.3% with a margin of error of .5%. Thus, you conclude that the unemployment rate in your area is somewhere between 4.8% and 5.8%.

So, what does this say about the local unemployment picture in your area versus the national unemployment situation? Since a major portion of the interval for the local unemployment rate is below 5.75%, is it reasonable to conclude that the local unemployment rate is below the national unemployment rate? Not really. When looking at the confidence interval, you have some degree of confidence, usually between 90% and 99%. If we use $z = \pm 1.96$ to calculate the margin of error, which is the z value for a 95% confidence level, we can state that there is a 95% chance that the local unemployment rate falls in the interval we obtain by using the margin of error. However, since 5.75% is in the interval for the local unemployment rate, the one thing that you can say is that it appears reasonable to conclude that the local and national unemployment rates are not different. However, if the national rate was 5.95%, then a conclusion that the two rates differ is reasonable because we are confident that the local unemployment rate falls between 4.8% and 5.8%.

When making conclusions based on the types of confidence intervals you have learned and will learn in this class, you will only be able to conclude that either there is a difference or there is not a difference. However, the methods you will learn in Chapter 9 will also allow you to determine the validity of a conclusion that states that the local rate is lower (or higher) than the national rate.

Glossary

Confidence interval An interval constructed around the value of a sample statistic to estimate the corresponding population parameter.

Confidence level Confidence level, denoted by $(1 - \alpha)100\%$, states how much confidence we have that a confidence interval contains the true population parameter.

Degrees of freedom (df) The number of observations that can be chosen freely. For the estimation of μ using the t distribution, the degrees of freedom are $n - 1$.

Estimate The value of a sample statistic that is used to find the corresponding population parameter.

Estimation A procedure by which a numerical value or values are assigned to a population parameter based on the information collected from a sample.

Estimator The sample statistic that is used to estimate a population parameter.

Interval estimate An interval constructed around the point estimate that is likely to contain the corresponding population parameter. Each interval estimate has a confidence level.

Margin of error The quantity that is subtracted from and added to the value of a sample statistic to obtain a confidence interval for the corresponding population parameter.

Point estimate The value of a sample statistic assigned to the corresponding population parameter.

t distribution A continuous distribution with a specific type of bell-shaped curve with its mean equal to 0 and standard deviation equal to $\sqrt{df/(df - 2)}$.

Supplementary Exercises

8.98 Because of inadequate public school budgets and not enough money available to teachers for classroom materials, many teachers often use their own money to buy materials used in the classrooms. A random sample of 100 public school teachers selected from an eastern state showed that they spent an average of $273 on such materials during the 2005 school year. The population standard deviation was $60.

a. What is the point estimate of the mean of such expenses incurred during the 2005 school year by all public school teachers in this state?

b. Make a 95% confidence interval for the corresponding population mean.

8.99 A bank manager wants to know the mean amount owed on credit card accounts that become delinquent. A random sample of 100 delinquent credit card accounts taken by the manager produced a mean amount owed on these accounts equal to $2640. The population standard deviation was $578.

a. What is the point estimate of the mean amount owed on all delinquent credit card accounts at this bank?

b. Construct a 97% confidence interval for the mean amount owed on all delinquent credit card accounts for this bank.

8.100 York Steel Corporation produces iron rings that are supplied to other companies. These rings are supposed to have diameters of 24 inches. The machine that makes these rings does not produce each ring with a diameter of exactly 24 inches. The diameter of each of the rings varies slightly. It is known that when the machine is working properly, the rings made on this machine have a mean diameter of 24 inches. The standard deviation of the diameters of all rings produced on this machine is always equal to .06 inches. The quality control department takes a sample of 25 such rings every week, calculates the mean of the diameters for these rings, and makes a 99% confidence interval for the population mean. If either the lower limit of this confidence interval is less than 23.975 inches or the upper limit of this confidence interval is greater than 24.025 inches, the machine is stopped and adjusted. A recent such sample of 25 rings produced a mean diameter of 24.015 inches. Based on this sample, can you conclude that the machine needs an adjustment? Explain. Assume that the population distribution is normal.

8.101 Yunan Corporation produces bolts that are supplied to other companies. These bolts are supposed to be 4 inches long. The machine that makes these bolts does not produce each bolt exactly 4 inches long. It is known that when the machine is working properly, the mean length of the bolts made on this machine is 4 inches. The standard deviation of the lengths of all bolts produced on this machine is always equal to .04 inches. The quality control department takes a sample of 20 such bolts every week, calculates the mean length of these bolts, and makes a 98% confidence interval for the population mean. If either the upper limit of this confidence interval is greater than 4.02 inches or the lower limit of this confidence interval is less than 3.98 inches, the machine is stopped and adjusted. A recent such sample of 20 bolts produced a mean length of 3.99 inches. Based on this sample, will you conclude that the machine needs an adjustment? Assume that the population distribution is normal.

8.102 A hospital administration wants to estimate the mean time spent by patients waiting for treatment at the emergency room. The waiting times (in minutes) recorded for a random sample of 32 such patients are given below.

110	42	88	19	35	76	10	151
2	44	27	77	53	102	66	39
20	108	92	55	14	52	3	62
78	15	60	121	40	35	11	72

Construct a 98% confidence interval for the corresponding population mean. Use the t distribution.

8.103 A travel magazine wanted to estimate the mean amount of leisure time per week enjoyed by adults. The research department at the magazine took a sample of 36 adults and obtained the following data on the weekly leisure time (in hours).

15	12	18	23	11	21	16	13	9	19	26	14
7	18	11	15	23	26	10	8	17	21	12	7
19	21	11	13	21	16	14	9	15	12	10	14

Construct a 99% confidence interval for the mean leisure time per week enjoyed by all adults. Use the t distribution.

8.104 A random sample of 25 life insurance policyholders showed that the average premium they pay on their life insurance policies is $685 per year with a standard deviation of $74. Assuming that the life insurance policy premiums for all life insurance policyholders have a normal distribution, make a 99% confidence interval for the population mean, μ.

8.105 A drug that provides relief from headaches was tried on 18 randomly selected patients. The experiment showed that the mean time to get relief from headaches for these patients after taking this drug was 24 minutes with a standard deviation of 4.5 minutes. Assuming that the time taken to get relief from a headache after taking this drug is (approximately) normally distributed, determine a 95% confidence interval for the mean relief time for this drug for all patients.

8.106 A survey of 500 randomly selected adult men showed that the mean time they spend per week watching sports on television is 9.75 hours with a standard deviation of 2.2 hours. Construct a 90% confidence interval for the population mean, μ.

8.107 A random sample of 300 female members of health clubs in Los Angeles showed that they spend, on average, 4.5 hours per week doing physical exercise with a standard deviation of .75 hours. Find a 98% confidence interval for the population mean.

8.108 A computer company that recently developed a new software product wanted to estimate the mean time taken to learn how to use this software by people who are somewhat familiar with computers. A random sample of 12 such persons was selected. The following data give the times taken (in hours) by these persons to learn how to use this software.

1.75	2.25	2.40	1.90	1.50	2.75
2.15	2.25	1.80	2.20	3.25	2.60

Construct a 95% confidence interval for the population mean. Assume that the times taken by all persons who are somewhat familiar with computers to learn how to use this software are approximately normally distributed.

8.109 A company that produces eight-ounce low-fat yogurt cups wanted to estimate the mean number of calories for such cups. A random sample of 10 such cups produced the following numbers of calories.

147	159	153	146	144	148	163	153	143	158

Construct a 99% confidence interval for the population mean. Assume that the numbers of calories for such cups of yogurt produced by this company have an approximately normal distribution.

8.110 An insurance company selected a sample of 50 auto claims filed with it and investigated those claims carefully. The company found that 12% of those claims were fraudulent.

a. What is the point estimate of the percentage of all auto claims filed with this company that are fraudulent?

b. Make a 99% confidence interval for the percentage of all auto claims filed with this company that are fraudulent.

8.111 An auto company wanted to know the percentage of people who prefer to own safer cars (that is, cars that possess more safety features) even if they have to pay a few thousand dollars more. A random sample of 500 persons showed that 44% of them will not mind paying a few thousand dollars more to have safer cars.

a. What is the point estimate of the percentage of all people who will not mind paying a few thousand dollars more to have safer cars?

b. Construct a 90% confidence interval for the percentage of all people who will not mind paying a few thousand dollars more to have safer cars.

8.112 A sample of 20 managers was taken, and they were asked whether or not they usually take work home. The responses of these managers are given below, where *yes* indicates they usually take work home and *no* means they do not.

Yes	Yes	No	No	No	Yes	No	No	No	No
Yes	Yes	No	Yes	Yes	No	No	No	No	Yes

Make a 99% confidence interval for the percentage of all managers who take work home.

8.113 Salaried workers at a large corporation receive two weeks' paid vacation per year. Sixteen randomly selected workers from this corporation were asked whether or not they would be willing to take a 3% reduction in their annual salaries in return for two additional weeks of paid vacation. The following are the responses of these workers.

No	Yes	No	No	Yes	No	No	Yes
Yes	No	No	No	Yes	No	No	No

Construct a 97% confidence interval for the percentage of all salaried workers at this corporation who would accept a 3% pay cut in return for two additional weeks of paid vacation.

8.114 A researcher wants to determine a 99% confidence interval for the mean number of hours that adults spend per week doing community service. How large a sample should the researcher select so that the estimate is within 1.2 hours of the population mean? Assume that the standard deviation for time spent per week doing community service by all adults is 3 hours.

8.115 An economist wants to find a 90% confidence interval for the mean sale price of houses in a state. How large a sample should she select so that the estimate is within \$3500 of the population mean? Assume that the standard deviation for the sale prices of all houses in this state is \$31,500.

8.116 A large city with chronic economic problems is considering legalizing casino gambling. The city council wants to estimate the proportion of all adults in the city who favor legalized casino gambling. What is the most conservative estimate of the sample size that would limit the margin of error to be within .05 of the population proportion for a 95% confidence interval?

8.117 Refer to Exercise 8.116. Assume that a preliminary sample has shown that 63% of the adults in this city favor legalized casino gambling. How large should the sample size be so that the 95% confidence interval for the population proportion has a margin of error of .05?

Advanced Exercises

8.118 Let μ be the hourly wage (excluding tips) for workers who provide hotel room service in a large city. A random sample of a number (more than 30) of such workers yielded a 95% confidence interval for μ of \$8.46 to \$9.86 using the normal distribution with a known population standard deviation.

a. Find the value of $\bar{x}$ for this sample.
b. Find the 99% confidence interval for μ based on this sample.

8.119 In November 2004, SRBI Public Affairs conducted a telephone poll of 601 adult Americans aged 18 to 29 for *Time* magazine (*Time*, January 24, 2005). One of the questions asked was: *Which of the following do you consider essential for your job: job security, health benefits, interesting work, good salary?* Respondents could choose more than one answer. Of the respondents, 71% said job security, 63% chose health benefits, 60% mentioned interesting work, and 56% favored good salary. Using these results, find a 95% confidence interval for the corresponding population percentage for each answer. Write a one-page report to present these results to a group of college students who have not taken statistics. Your report should answer questions such as: (1) What is a confidence interval? (2) Why is a range of values more informative than a single percentage? (3) What does 95% confidence mean in this context? (4) What assumptions, if any, are you making when you construct each confidence interval?

8.120 A group of veterinarians wants to test a new canine vaccine for Lyme disease. (Lyme disease is transmitted by the bite of an infected tick.) In an area that has a high incidence of Lyme disease, 100 dogs are randomly selected (with their owners' permission) to receive the vaccine. Over a 12-month period, these dogs are periodically examined by veterinarians for symptoms of Lyme disease. At the end of 12 months, 10 of these 100 dogs are diagnosed with the disease. During the same 12-month period, 18% of the unvaccinated dogs in the area have been found to have Lyme disease. Let p be the proportion of all potential vaccinated dogs who would contract Lyme disease in this area.

a. Find a 95% confidence interval for p.
b. Does 18% lie within your confidence interval of part a? Does this suggest the vaccine might or might not be effective to some degree?
c. Write a brief critique of this experiment, pointing out anything that may have distorted the results or conclusions.

8.121 When one is attempting to determine the required sample size for estimating a population mean and the information on the population standard deviation is not available, it may be feasible to take a small preliminary sample and use the sample standard deviation to estimate the required sample size, n. Suppose that we want to estimate μ, the mean commuting distance for students at a community college, to within 1 mile with a confidence level of 95%. A random sample of 20 students yields a standard deviation of 4.1 miles. Use this value of the sample standard deviation, s, to estimate the required sample size, n. Assume that the corresponding population has a normal distribution.

8.122 A gas station attendant would like to estimate p, the proportion of all households that own more than two vehicles. To obtain an estimate, the attendant decides to ask the next 200 gasoline customers how many vehicles their households own. To obtain an estimate of p, the attendant counts the number of customers who say there are more than two vehicles in their households and then divides this number by 200. How would you critique this estimation procedure? Is there anything wrong with this procedure that would result in sampling and/or nonsampling errors? If so, can you suggest a procedure that would reduce this error?

8.123 A couple considering the purchase of a new home would like to estimate the average number of cars that go past the location per day. The couple guesses that the number of cars passing this location per day has a population standard deviation of 170.

a. On how many randomly selected days should the number of cars passing the location be observed so that the couple can be 99% certain the estimate will be within 100 cars of the true average?

b. Suppose the couple finds out that the population standard deviation of the number of cars passing the location per day is not 170 but is actually 272. If they have already taken a sample of the size computed in part a, what confidence does the couple have that their point estimate is within 100 cars of the true average?

c. If the couple has already taken a sample of the size computed in part a and later finds out that the population standard deviation of the number of cars passing the location per day is actually 130, they can be 99% confident their point estimate is within how much of the true average?

8.124 The U.S. Senate just passed a bill by a vote of 55–45 (with all 100 senators voting). A student who took an elementary statistics course last semester says, "We can use these data to make a confidence interval about p. We have $n = 100$ and $\hat{p} = 55/100 = .55$." Hence, according to him, a 95% confidence interval for p is

$$\hat{p} \pm z\sigma_{\hat{p}} = .55 \pm 1.96\sqrt{\frac{(.55)(.45)}{100}} = .55 \pm .098 = .452 \text{ to } .648$$

Does this make sense? If not, what is wrong with the student's reasoning?

8.125 When calculating a confidence interval for the population mean μ with a known population standard deviation σ, describe the effects of the following two changes on the confidence interval: (1) doubling the sample size, (2) quadrupling (multiplying by 4) the sample size. Give two reasons why this relationship does not hold true if you are calculating a confidence interval for the population mean μ with an unknown population standard deviation.

8.126 At the end of Section 8.3, we noted that we always round up when calculating the minimum sample size for a confidence interval for μ with a specified margin of error and confidence level. Using the formula for the margin of error, explain why we must always round up in this situation.

8.127 Calculating a confidence interval for the proportion requires a minimum sample size. Calculate a confidence interval, using any confidence level, for the population proportion for each of the following.

a. $n = 200$ and $\hat{p} = .01$ b. $n = 160$ and $\hat{p} = .9875$

Explain why these confidence intervals reveal a problem when the conditions for using the normal approximation do not hold.

Self-Review Test

1. Complete the following sentences using the terms *population parameter* and *sample statistic*.
   a. Estimation means assigning values to a __________ based on the value of a __________.
   b. An estimator is the __________ used to estimate a __________.
   c. The value of a __________ is called the point estimate of the corresponding __________.

2. A 95% confidence interval for μ can be interpreted to mean that if we take 100 samples of the same size and construct 100 such confidence intervals for μ, then
   a. 95 of them will not include μ b. 95 will include μ c. 95 will include $\bar{x}$

3. The confidence level is denoted by
   a. $(1 - \alpha)100\%$ b. $100\alpha\%$ c. α

4. The margin of error of the estimate for μ is
   a. $z\sigma_{\bar{x}}$ (or $ts_{\bar{x}}$) b. $\sigma/\sqrt{n}$ (or $s/\sqrt{n}$) c. $\sigma_{\bar{x}}$ (or $s_{\bar{x}}$)

5. Which of the following assumptions is not required to use the t distribution to make a confidence interval for μ?
   a. Either the population from which the sample is taken is (approximately) normally distributed or $n \geq 30$.
   b. The population standard deviation, σ, is not known.
   c. The sample size is at least 10.

6. The parameter(s) of the t distribution is (are)
   a. n b. degrees of freedom c. μ and degrees of freedom

7. A sample of 36 vacation homes built during the past two years in a coastal resort region gave a mean construction cost of $159,000 with a population standard deviation of $27,000.

a. What is the point estimate of the corresponding population mean?
b. Make a 99% confidence interval for the mean construction cost for all vacation homes built in this region during the past two years. What is the margin of error here?

8. A sample of 25 malpractice lawsuits filed against doctors showed that the mean compensation awarded to the plaintiffs was $410,425 with a standard deviation of $74,820. Find a 95% confidence interval for the mean compensation awarded to plaintiffs of all such lawsuits. Assume that the compensations awarded to plaintiffs of all such lawsuits are normally distributed.

9. In November 2004, the Family Credit Counseling Service commissioned a Financial Stress Survey, conducted by Impulse Research Corporation, in which 1590 consumers with credit card debts were asked about physical symptoms that they attributed to stress from their debts. The most frequent complaint, mentioned by 50.8% of these consumers, was that debt-related stress sometimes kept them awake at night (*The Hartford Courant*, February 20, 2005). Assume that these 1590 consumers make a random sample of all consumers with credit card debts.

a. What is the point estimate of the corresponding population proportion?
b. Construct a 95% confidence interval for the proportion of all consumers with credit card debts who are sometimes kept awake by stress from their credit card debts. What is the margin of error for this estimate?

10. A statistician is interested in estimating, at a 95% confidence level, the mean number of houses sold per month by all real estate agents in a large city. From an earlier study, it is known that the standard deviation of the numbers of houses sold per month by all real estate agents in this city is 2.2. How large a sample should be taken so that the estimate is within .65 of the population mean?

11. A college registrar has received numerous complaints about the online registration procedure at her college, alleging that the system is slow, confusing, and error-prone. She wants to estimate the proportion of all students at this college who are dissatisfied with the online registration procedure. What is the most conservative estimate of the sample size that would limit the margin of error to be within .05 of the population proportion for a 90% confidence interval?

12. Refer to Problem 11. Assume that a preliminary study has shown that 70% of the students surveyed at this college are dissatisfied with the current online registration system. How large a sample should be taken in this case so that the margin of error is within .05 of the population proportion for a 90% confidence interval?

13. Dr. Garcia estimated the mean stress score before a statistics test for a random sample of 25 students. She found the mean and standard deviation for this sample to be 7.1 (on a scale of 1 to 10) and 1.2, respectively. She used a 97% confidence level. However, she thinks that the confidence interval is too wide. How can she reduce the width of the confidence interval? Describe all possible alternatives. Which alternative do you think is best and why?

*14. You want to estimate the mean number of hours that students at your college work per week. Briefly explain how you will conduct this study using a small sample. Take a sample of 12 students from your college who hold a job. Collect data on the number of hours that these students spent working last week. Then estimate the population mean. Choose your own confidence level. What assumptions will you make to estimate this population mean?

*15. You want to estimate the proportion of people who are happy with their current jobs. Briefly explain how you will conduct this study. Take a sample of 35 persons and collect data on whether or not they are happy with their current jobs. Then estimate the population proportion. Choose your own confidence level.

Mini-Projects

■ MINI-PROJECT 8–1

Deborah A. Stiles of the School of Education at Webster University and her coauthors reviewed several studies on the occupational goals of adolescent boys. Even though fewer than one in 10,000 American boys ever reach this goal, the most popular occupational choice for these boys in the United States is "professional athlete." In one of the studies, 61 U.S. students (26 boys and 35 girls, aged 12 to 18 years) participated to seek insight into why so many boys aspire to professional athletics. The table, adapted from the paper published by the authors, summarizes the reasons these students offered. Note that each of the 61 students typically gave several reasons or comments. Both boys and girls were asked to answer the question, Why might boys wish to become professional athletes?

Reason/Comment	Number of Comments	Percentage of Total
Well-known and admired	94	35
Money	56	21
Masculine role	29	11
Like sports	27	10
Easy life	24	9
Education not necessary	19	7
Unrealistic fantasy	14	5
Other	5	2

Source: Deborah A. Stiles, Judith L. Gibbons, Daniel L. Sebben, and Deane C. Wiley, "Why Adolescent Boys Dream of Becoming Professional Athletes," *Psychological Reports*, 1999, 84, pp. 1075–1085.

These 61 students, who were attending schools near Boston and near St. Louis, were recruited from two after-school groups and three high school classes, so they were not a random sample of all U.S. students in their age group. If, however, these students had constituted a random sample of all U.S. students aged 12 to 18, explain how you could use the data in the table to construct confidence intervals for the true percentages of total comments for each of the reasons given.

■ MINI-PROJECT 8–2

Consider the data set on the heights of NBA players that accompanies this text.

a. Take a random sample of 15 players and find a 95% confidence interval for μ. Assume the heights of these players are normally distributed.
b. Repeat part a for samples of size 31 and 45, respectively.
c. Compare the widths of your three confidence intervals.
d. Now calculate the mean, μ, of the heights of all players. Do all of your confidence intervals contain this μ? If not, which ones do not contain μ?

■ MINI-PROJECT 8–3

Here is a project that can involve a social activity and also show you the importance of making sure that the underlying requirements are met prior to calculating a confidence interval. Invite some of your friends over and buy a big bag of Milk Chocolate M&Ms. Take at least 40 random samples of 10 M&Ms each from the bag. Note that taking many random samples will reduce the risk of obtaining some extremely odd results. Before eating the candy, calculate the proportion of brown candies for each sample. Then, using each sample proportion, compute a 95% confidence interval for the proportion of brown candies in all M&Ms. According to the company, the population proportion is .13, i.e., 13% of all M&Ms are brown (http://us.mms.com/us/about/products/milkchocolate/). Determine what percentage of the confidence intervals contains the population proportion .13. Is this percentage close to 95%? What happens if you increase your sample size to 20, and then to 50? If you want, you can use technology to simulate those random samples, which makes the process much faster. Besides, the candy will probably be eaten by the time you get ready to take larger samples.

DECIDE FOR YOURSELF

Deciding About the Viability of Poll Results

In the Decide for Yourself feature of Chapter 7, we discussed the idea that underlies the procedures that are used to make projections on election day. Here we discuss the process of collecting data in exit polls.

Instead of selecting a simple random sample and choosing people at random from a list of all voters (imagine the almost impossible process of preparing such a list), exit polls use what is called a multistage sampling procedure. In this sampling technique, the first stage involves randomly selecting a few voter precincts. If the U.S. presidential election were based solely on the popular vote, these precincts could be selected from a list of all precincts in the United

States. However, the presidential election is based on the Electoral College system. Hence, the polling agencies need to select precincts from each state in order to make sure that they have a sufficient sample size from each state. Then, interviewers who are stationed at each selected precinct interview every kth voter, where k is dependent on the expected number of voters at that precinct. Dr. Christian Potholm of Bowdoin College in Brunswick, Maine, cited a problem during the 2004 presidential election. The following excerpt is taken from the Web site http://www.bowdoin.edu/news/archives/1academicnews/001613.shtml that contained an interview with Dr. Potholm. According to Dr. Potholm:

> Those exit polls were really a disservice to polling. All across the country these early polls took a bad sample and exaggerated its impact. Whether it was done maliciously or not, it was just bad polling not to balance your sample. . . . What I think happened in the national polls was that the initial polls were 58 percent women.

1. Explain why taking a poll early in the morning could produce misleading results.

2. As mentioned in the above statement, women made up the majority of the voters in those polls. Was this more likely to overrepresent Kerry voters or Bush voters? Why?

3. Discuss some other potential issues with a *time bias*, which happens if a poll is taken at a specific time of the day.

TECHNOLOGY INSTRUCTION — Confidence Intervals for Population Means and Proportions

TI-84

```
ZInterval
 Inpt:Data Stats
 σ:2
 x̄:11
 n:65
 C-Level:.95
 Calculate
```

Screen 8.1

1. To find a confidence interval for a population mean μ given the population standard deviation σ, select **STAT>TESTS>ZInterval**. If you have the data stored in a list, select **Data** and enter the name of the list. If you have the summary statistics, choose **Stats** and enter the sample mean and size. Enter your value for σ and the confidence level as a decimal as **C-Level**. Select **Calculate**. (See **Screen 8.1**.)
2. To find a confidence interval for a population mean μ without knowing the population standard deviation σ, select **STAT>TESTS>TInterval**. If you have the data stored in a list, select **Data** and enter the name of the list. If you have the summary statistics, choose **Stats** and enter the sample mean, standard deviation, and size. Enter your confidence level as a decimal as **C-Level**. Select **Calculate**.
3. To find a confidence interval for a population proportion p, select **STAT>TESTS> 1-PropZInt**. Enter the number of successes as x and the sample size as n. Enter the confidence level as a decimal as **C-Level**. Select **Calculate**.

MINITAB

1. To find a confidence interval for the population mean μ when the population standard deviation σ is known, select **Stat>Basic Statistics>1-Sample Z**. If you have data on a variable entered in a column of a MINITAB spreadsheet, enter the name of that column in the **Samples in columns**: box. If you know the summary statistics, click next to **Summarized data** and enter the values of the **Sample size** and **Mean** in their respective boxes. In both cases, enter the value of the population standard deviation in the **Standard deviation** box. (See **Screen 8.2**.) Click the **Options** button and enter the **Confidence level**. Now click **OK** in both windows. The confidence interval will appear in the **Session** window.
2. To find a confidence interval for the population mean μ when the population standard deviation σ is not known, select **Stat>Basic Statistics>1-Sample t**. If you have data on a

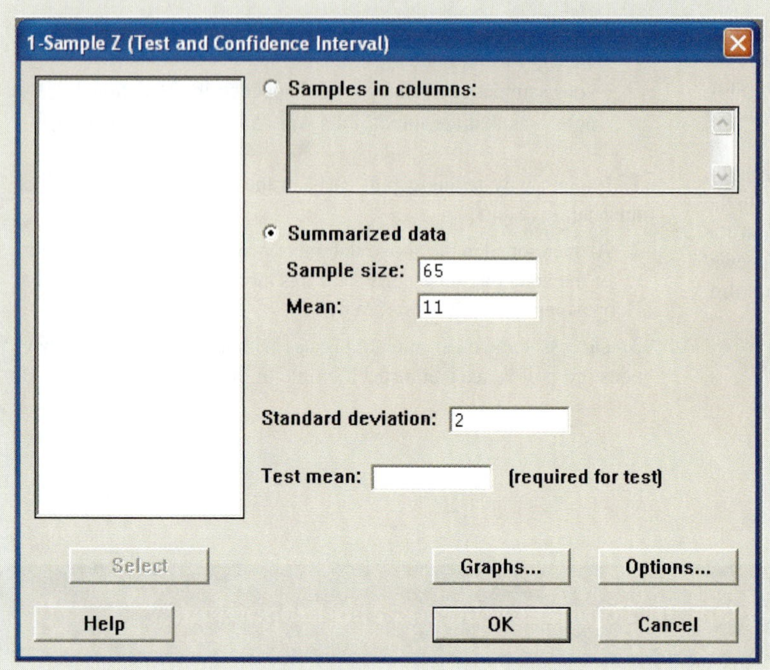

Screen 8.2

variable entered in a column of a MINITAB spreadsheet, enter the name of that column in the **Samples in columns**: box. If you know the summary statistics, click next to **Summarized data** and enter the values of the **Sample size, Mean**, and **Sample standard deviation** in their respective boxes. Click the **Options** button and enter the **Confidence level**. Now click **OK** in both windows. The confidence interval will appear in the **Session** window.

3. To find a confidence interval for a population proportion p, select **Stat>Basic Statistics>1-Proportion**. If you have sample data (consisting of two values for success and failure) entered in a column, select **Samples in columns** and type your column name in the box. If, instead, you have the number of successes and the number of trials, select **Summarized data** and enter them. Click the **Options** button and enter the **Confidence level**. Click **OK** in both boxes. The confidence interval will appear in the **Session** window.

Excel

1. To find the margin of error for a population mean, given the confidence level $1 - \alpha$, standard deviation σ, and sample size n, type **=CONFIDENCE($\boldsymbol{\alpha}$, $\boldsymbol{\sigma}$, $\boldsymbol{n}$)**. Note: For a 95% confidence level, $\alpha = 0.05$. (See **Screens 8.3 and 8.4**.)

	A	B	C	D	E
1	Mean	11			
2	Std. Dev.	2			
3	Size	65			
4	Alpha	0.05			
5					
6	Margin	=CONFIDENCE(0.05,2,65)			
7		CONFIDENCE(alpha, standard_dev, size)			

Screen 8.3

	A	B
1	Mean	11
2	Std. Dev.	2
3	Size	65
4	Alpha	0.05
5		
6	Margin	0.486207

Screen 8.4

TECHNOLOGY ASSIGNMENTS

TA8.1 The following data give the annual incomes (in thousands of dollars) before taxes for a sample of 36 randomly selected families from a city.

21.6	33.0	25.6	37.9	50.0	148.1
50.1	21.5	70.0	72.8	58.2	85.4
91.2	57.0	72.2	45.0	95.0	27.8
92.8	79.4	45.3	76.0	48.6	69.3
40.6	69.0	75.5	57.5	49.7	75.1
96.3	44.5	84.0	43.0	61.7	126.0

Construct a 99% confidence interval for μ assuming that the population standard deviation is $23.75 thousand.

TA8.2 The following data give the checking account balances on a certain day for a randomly selected sample of 30 households.

500	100	650	1917	2200	500	180	3000	1500	1300
319	1500	1102	405	124	1000	134	2000	150	800
200	750	300	2300	40	1200	500	900	20	160

Construct a 97% confidence interval for μ assuming that the population standard deviation is unknown.

TA8.3 Refer to Data Set I (that accompanies this text) on the prices of various products in different cities across the country. Using the data on monthly telephone charges, make a 98% confidence interval for the population mean μ.

TA8.4 Refer to the Manchester Road Race data set (that accompanies this text) for all participants. Take a sample of 100 observations from this data set.

a. Using the sample data, make a 95% confidence interval for the mean time taken to complete this race by all participants.

b. Now calculate the mean time taken to run this race by all participants. Does the confidence interval made in part a include this population mean?

TA8.5 Repeat Technology Assignment TA8.4 for a sample of 25 observations. Assume that the distribution of times taken to run this race by all participants is approximately normal.

TA8.6 The following data give the prices (in thousands of dollars) of 16 recently sold houses in an area.

341	163	327	204	197	203	313	279
456	228	383	289	533	399	271	381

Construct a 99% confidence interval for the mean price of all houses in this area. Assume that the distribution of prices of all houses in the given area is normal.

TA8.7 A researcher wanted to estimate the mean contributions made to charitable causes by major companies. A random sample of 18 companies produced the following data on contributions (in millions of dollars) made by them.

1.8	.6	1.2	.3	2.6	1.9	3.4	2.6	.2
2.4	1.4	2.5	3.1	.9	1.2	2.0	.8	1.1

Make a 98% confidence interval for the mean contributions made to charitable causes by all major companies. Assume that the contributions made to charitable causes by all major companies have a normal distribution.

TA8.8 A mail-order company promises its customers that their orders will be processed and mailed within 72 hours after an order is placed. The quality control department at the company checks from time to time to see if this promise is kept. Recently the quality control department took a sample of 200 orders and found that 176 of them were processed and mailed within 72 hours of the placement of the orders. Make a 98% confidence interval for the corresponding population proportion.

TA8.9 One of the major problems faced by department stores is a high percentage of returns. The manager of a department store wanted to estimate the percentage of all sales that result in returns. A sample of 500 sales showed that 95 of them had products returned within the time allowed for returns. Make a 99% confidence interval for the corresponding population proportion.

TA8.10 One of the major problems faced by auto insurance companies is the filing of fraudulent claims. An insurance company carefully investigated 1000 auto claims filed with it and found 108 of them to be fraudulent. Make a 96% confidence interval for the corresponding population proportion.

Chapter

9

Hypothesis Tests About the Mean and Proportion

9.1 Hypothesis Tests: An Introduction

9.2 Hypothesis Tests About μ: σ Known

Case Study 9–1 The Average Cost of a Wedding

9.3 Hypothesis Tests About μ: σ Not Known

9.4 Hypothesis Tests About a Population Proportion: Large Samples

Case Study 9–2 Coffee or Internet

Suppose the company you work for offers you the following choice: You can either have free morning coffee or you can use the computer at work for personal surfing on the Internet. What will your decision be? Will you give up your morning coffee for being able to use the Internet at work? Or will you give up the ability to use the Internet for personal purposes at work for the free morning coffee? What do you think other workers will decide? According to a survey of workers, 52% of the respondents said they would give up the morning coffee for the ability to use the Internet at work, and 44% said that they will give up the use of the Internet at work for the morning coffee. The remaining 4% were not sure. See Case Study 9–2.

This chapter introduces the second topic in inferential statistics: tests of hypotheses. In a test of hypothesis, we test a certain given theory or belief about a population parameter. We may want to find out, using some sample information, whether or not a given claim (or statement) about a population parameter is true. This chapter discusses how to make such tests of hypotheses about the population mean, μ, and the population proportion, p.

As an example, a soft-drink company may claim that, on average, its cans contain 12 ounces of soda. A government agency may want to test whether or not such cans contain, on average, 12 ounces of soda. As another example, according to the U.S. Bureau of the Census, 15.6% of the population in the United States lacked health insurance in 2003. An economist may want to check if this percentage is still true for this year. In the first of these two examples we are to test a hypothesis about the population mean, μ, and in the second example we are to test a hypothesis about the population proportion, p.

9.1 Hypothesis Tests: An Introduction

Why do we need to perform a test of hypothesis? Reconsider the example about soft-drink cans. Suppose we take a sample of 100 cans of the soft drink under investigation. We then find out that the mean amount of soda in these 100 cans is 11.89 ounces. Based on this result, can we state that, on average, all such cans contain less than 12 ounces of soda and that the company is lying to the public? Not until we perform a test of hypothesis can we make such an accusation. The reason is that the mean, $\bar{x} = 11.89$ ounces, is obtained from a sample. The difference between 12 ounces (the required average amount for the population) and 11.89 ounces (the observed average amount for the sample) may have occurred only because of the sampling error. Another sample of 100 cans may give us a mean of 12.04 ounces. Therefore, we perform a test of hypothesis to find out how large the difference between 12 ounces and 11.89 ounces is and to investigate whether or not this difference has occurred as a result of chance alone. Now, if 11.89 ounces is the mean for all cans and not for just 100 cans, then we do not need to make a test of hypothesis. Instead, we can immediately state that the mean amount of soda in all such cans is less than 12 ounces. We perform a test of hypothesis only when we are making a decision about a population parameter based on the value of a sample statistic.

9.1.1 Two Hypotheses

Consider a nonstatistical example of a person who has been indicted for committing a crime and is being tried in a court. Based on the available evidence, the judge or jury will make one of two possible decisions:

1. The person is not guilty.
2. The person is guilty.

At the outset of the trial, the person is presumed not guilty. The prosecutor's efforts are to prove that the person has committed the crime and, hence, is guilty.

In statistics, *the person is not guilty* is called the **null hypothesis** and *the person is guilty* is called the **alternative hypothesis**. The null hypothesis is denoted by H_0 and the alternative hypothesis is denoted by H_1. In the beginning of the trial it is assumed that the person is not guilty. The null hypothesis is usually the hypothesis that is assumed to be true to begin with. The two hypotheses for the court case are written as follows (notice the colon after H_0 and H_1):

Null hypothesis: H_0: The person is not guilty

Alternative hypothesis: H_1: The person is guilty

In a statistics example, the null hypothesis states that a given claim (or statement) about a population parameter is true. Reconsider the example of the soft-drink company's claim that, on average, its cans contain 12 ounces of soda. In reality, this claim may or may not be true. However, we will initially assume that the company's claim is true (that is, the company is not guilty of cheating and lying). To test the claim of the soft-drink company, the null hypothesis will be that the company's claim is true. Let μ be the mean amount of soda in all cans. The company's claim will be true if $\mu = 12$ ounces. Thus, the null hypothesis will be written as

$$H_0: \mu = 12 \text{ ounces} \quad \text{(The company's claim is true)}$$

In this example, the null hypothesis can also be written as $\mu \geq 12$ ounces because the claim of the company will still be true if the cans contain, on average, more than 12 ounces of soda. The company will be accused of cheating the public only if the cans contain, on average, less than 12 ounces of soda. However, it will not affect the test whether we use an = or a $\geq$ sign in the null hypothesis as long as the alternative hypothesis has a $<$ sign. Remember that in the null hypothesis (and in the alternative hypothesis also) we use the population parameter (such as μ or p), and not the sample statistic (such as $\bar{x}$ or $\hat{p}$).

Definition

Null Hypothesis A *null hypothesis* is a claim (or statement) about a population parameter that is assumed to be true until it is declared false.

The alternative hypothesis in our statistics example will be that the company's claim is false and its soft-drink cans contain, on average, less than 12 ounces of soda—that is, $\mu < 12$ ounces. The alternative hypothesis will be written as

$$H_1\colon \mu < 12 \text{ ounces} \quad \text{(The company's claim is false)}$$

Definition

Alternative Hypothesis An *alternative hypothesis* is a claim about a population parameter that will be true if the null hypothesis is false.

Let us return to the example of the court trial. The trial begins with the assumption that the null hypothesis is true—that is, the person is not guilty. The prosecutor assembles all the possible evidence and presents it in the court to prove that the null hypothesis is false and the alternative hypothesis is true (that is, the person is guilty). In the case of our statistics example, the information obtained from a sample will be used as evidence to decide whether or not the claim of the company is true. In the court case, the decision made by the judge (or jury) depends on the amount of evidence presented by the prosecutor. At the end of the trial, the judge (or jury) will consider whether or not the evidence presented by the prosecutor is sufficient to declare the person guilty. The amount of evidence that will be considered to be sufficient to declare the person guilty depends on the discretion of the judge (or jury).

9.1.2 Rejection and Nonrejection Regions

In Figure 9.1, which represents the court case, the point marked 0 indicates that there is no evidence against the person being tried. The farther we move toward the right on the horizontal axis, the more convincing the evidence is that the person has committed the crime. We have arbitrarily marked a point C on the horizontal axis. Let us assume that a judge (or jury) considers any amount of evidence to the right of point C to be sufficient and any amount of evidence to the left of C to be insufficient to declare the person guilty. Point C is called the **critical value** or **critical point** in statistics. If the amount of evidence presented by the prosecutor falls in the area to the left of point C, the verdict will reflect that there is not enough evidence to declare the person guilty. Consequently, the accused person will be declared *not guilty*. In statistics, this decision is stated as *do not reject* H_0. It is equivalent to saying that there is not enough evidence to declare the null hypothesis false. The area to the left of point C is called the *nonrejection region*; that is, this is the region where the null hypothesis is not

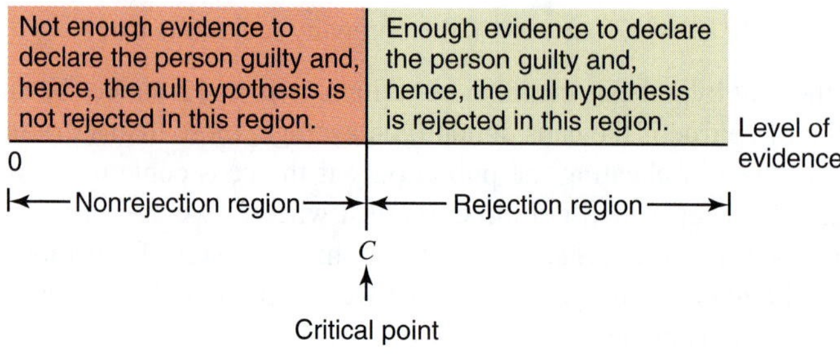

Figure 9.1 Nonrejection and rejection regions for the court case.

rejected. However, if the amount of evidence falls in the area to the right of point C, the verdict will be that there is sufficient evidence to declare the person guilty. In statistics, this decision is stated as *reject* H_0 or *the null hypothesis is false*. Rejecting H_0 is equivalent to saying that *the alternative hypothesis is true*. The area to the right of point C is called the *rejection region*; that is, this is the region where the null hypothesis is rejected.

9.1.3 Two Types of Errors

We all know that a court's verdict is not always correct. If a person is declared guilty at the end of a trial, there are two possibilities.

1. The person has *not* committed the crime but is declared guilty (because of what may be false evidence).
2. The person *has* committed the crime and is rightfully declared guilty.

In the first case, the court has made an error by punishing an innocent person. In statistics, this kind of error is called a **Type I** or an $\boldsymbol{\alpha}$ (*alpha*) **error**. In the second case, because the guilty person has been punished, the court has made the correct decision. The second row in the shaded portion of Table 9.1 shows these two cases. The two columns of Table 9.1, corresponding to *the person is not guilty* and *the person is guilty*, give the two actual situations. Which one of these is true is known only to the person being tried. The two rows in this table, corresponding to *the person is not guilty* and *the person is guilty*, show the two possible court decisions.

Table 9.1

		Actual Situation	
		The Person Is Not Guilty	**The Person Is Guilty**
Court's decision	The person is not guilty	Correct decision	Type II or β error
	The person is guilty	Type I or α error	Correct decision

In our statistics example, a Type I error will occur when H_0 is actually true (that is, the cans do contain, on average, 12 ounces of soda), but it just happens that we draw a sample with a mean that is much less than 12 ounces and we wrongfully reject the null hypothesis, H_0. The value of $\boldsymbol{\alpha}$, called the **significance level** of the test, represents the probability of making a Type I error. In other words, α is the probability of rejecting the null hypothesis, H_0, when in fact it is true.

Definition

Type I Error A *Type I error* occurs when a true null hypothesis is rejected. The value of α represents the probability of committing this type of error; that is,

$$\alpha = P(H_0 \text{ is rejected} \mid H_0 \text{ is true})$$

The value of α represents the *significance level* of the test.

The size of the rejection region in a statistics problem of a test of hypothesis depends on the value assigned to α. In one approach to a test of hypothesis, we assign a value to α before making the test. Although any value can be assigned to α, the commonly used values of α are .01, .025, .05, and .10. Usually the value assigned to α does not exceed .10 (or 10%).

Now, suppose that in the court trial case the person is declared not guilty at the end of the trial. Such a verdict does not indicate that the person has indeed *not* committed the crime. It is possible that the person is guilty but there is not enough evidence to prove the guilt. Consequently, in this situation there are again two possibilities.

1. The person has *not* committed the crime and is declared not guilty.
2. The person *has* committed the crime but, *because of the lack of enough evidence*, is declared not guilty.

In the first case, the court's decision is correct. But in the second case, the court has committed an error by setting a guilty person free. In statistics, this type of error is called a **Type II** or a **β** (the Greek letter *beta*) **error**. These two cases are shown in the first row of the shaded portion of Table 9.1.

In our statistics example, a Type II error will occur when the null hypothesis H_0 is actually false (that is, the soda contained in all cans, on average, is less than 12 ounces), but it happens by chance that we draw a sample with a mean that is close to or greater than 12 ounces and we wrongfully conclude *do not reject* H_0. The value of β represents the probability of making a Type II error. It represents the probability that H_0 is not rejected when actually H_0 is false. The value of $1 - \beta$ is called the **power of the test**. It represents the probability of not making a Type II error.

Definition

Type II Error A *Type II error* occurs when a false null hypothesis is not rejected. The value of β represents the probability of committing a Type II error; that is,

$$\beta = P(H_0 \text{ is not rejected} \mid H_0 \text{ is false})$$

The value of $1 - \beta$ is called the *power of the test*. It represents the probability of not making a Type II error.

The two types of errors that occur in tests of hypotheses depend on each other. We cannot lower the values of α and β simultaneously for a test of hypothesis for a fixed sample size. Lowering the value of α will raise the value of β, and lowering the value of β will raise the value of α. However, we can decrease both α and β simultaneously by increasing the sample size. The explanation of how α and β are related and the computation of β are not within the scope of this text.

Table 9.2, which is similar to Table 9.1, is written for the statistics problem of a test of hypothesis. In Table 9.2 *the person is not guilty* is replaced by *H_0 is true, the person is guilty* by *H_0 is false*, and the *court's decision* by *decision.*

Table 9.2

		Actual Situation	
		H_0 Is True	H_0 Is False
Decision	Do not reject H_0	Correct decision	Type II or β error
	Reject H_0	Type I or α error	Correct decision

9.1.4 Tails of a Test

The statistical hypothesis-testing procedure is similar to the trial of a person in court but with two major differences. The first major difference is that in a statistical test of hypothesis, the partition of the total region into rejection and nonrejection regions is not arbitrary. Instead, it depends on the value assigned to α (Type I error). As mentioned earlier, α is also called the significance level of the test.

The second major difference relates to the rejection region. In the court case, the rejection region is on the right side of the critical point, as shown in Figure 9.1. However, in statistics, the rejection region for a hypothesis-testing problem can be on both sides, with the nonrejection region in the middle, or it can be on the left side or right side of the nonrejection region. These possibilities are explained in the next three parts of this section. A test with two rejection regions is called a **two-tailed test**, and a test with one rejection region is called a **one-tailed test**. The one-tailed test is called a **left-tailed test** if the rejection region is in the left tail of the distribution curve, and it is called a **right-tailed test** if the rejection region is in the right tail of the distribution curve.

Definition

Tails of the Test A *two-tailed test* has rejection regions in both tails, a *left-tailed test* has the rejection region in the left tail, and a *right-tailed test* has the rejection region in the right tail of the distribution curve.

A Two-Tailed Test

According to a Kaiser Family Foundation survey conducted in 2005, 8- to 10-year-old children spent 250 minutes (i.e., 4 hours, 10 minutes) watching television per day (*USA TODAY*, July 12, 2005). A child psychologist wants to check whether or not this mean has changed since 2005. The key word here is *change.* The mean time spent watching television by such children has changed if it has either increased or decreased since 2005. This is an example of a two-tailed test. Let μ be the current mean time spent watching television per day by 8- to 10-year-old children. The two possible decisions are

1. The mean time spent watching television by 8- to 10-year-old children has not changed since 2005, that is, currently $\mu = 250$ minutes.
2. The mean time spent watching television by 8- to 10-year-old children has changed since 2005, that is, currently $\mu \neq 250$ minutes.

We will write the null and alternative hypotheses for this test as follows.

H_0: $\mu = 250$ minutes (The mean time spent watching television has not changed.)

H_1: $\mu \neq 250$ minutes (The mean time spent watching television has changed.)

Whether a test is two-tailed or one-tailed is determined by the sign in the alternative hypothesis. If the alternative hypothesis has a *not equal to* ($\neq$) sign, as in this example, it is a two-tailed test. As shown in Figure 9.2, a two-tailed test has two rejection regions, one in each tail of the distribution curve. Figure 9.2 shows the sampling distribution of $\bar{x}$ assuming it has a normal distribution. Assuming H_0 is true, $\bar{x}$ has a normal distribution with its mean equal to 250 minutes (the value of μ in H_0). In Figure 9.2, the area of each of the two rejection regions

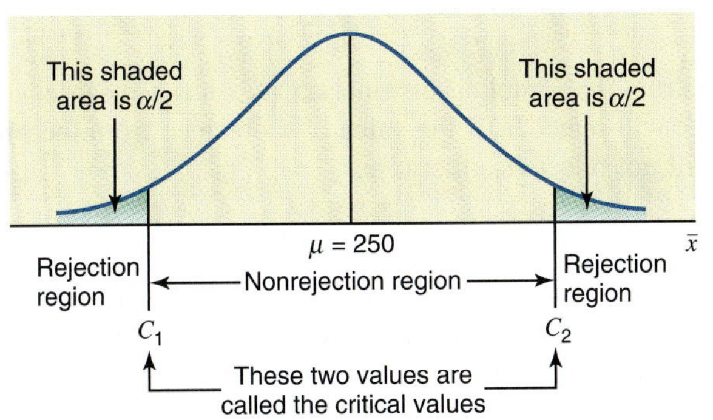

Figure 9.2 A two-tailed test.

is $\alpha/2$ and the total area of both rejection regions is α (the significance level). As shown in this figure, a two-tailed test of hypothesis has two critical values that separate the two rejection regions from the nonrejection region. We will reject H_0 if the value of $\bar{x}$ obtained from the sample falls in either of the two rejection regions. We will not reject H_0 if the value of $\bar{x}$ lies in the nonrejection region. By rejecting H_0, we are saying that the difference between the value of μ stated in H_0 and the value of $\bar{x}$ obtained from the sample is too large to have occurred because of the sampling error alone. Consequently, this difference is real. By not rejecting H_0, we are saying that the difference between the value of μ stated in H_0 and the value of $\bar{x}$ obtained from the sample is small and it may have occurred because of the sampling error alone.

A Left-Tailed Test

Reconsider the example of the mean amount of soda in all soft-drink cans produced by a company. The company claims that these cans, on average, contain 12 ounces of soda. However, if these cans contain less than the claimed amount of soda, then the company can be accused of cheating. Suppose a consumer agency wants to test whether the mean amount of soda per can is less than 12 ounces. Note that the key phrase this time is *less than*, which indicates a left-tailed test. Let μ be the mean amount of soda in all cans. The two possible decisions are

1. The mean amount of soda in all cans is equal to 12 ounces, that is, $\mu = 12$ ounces.
2. The mean amount of soda in all cans is less than 12 ounces, that is, $\mu < 12$ ounces.

The null and alternative hypotheses for this test are written as

$$H_0\colon \mu = 12 \text{ ounces} \quad \text{(The mean is equal to 12 ounces.)}$$

$$H_1\colon \mu < 12 \text{ ounces} \quad \text{(The mean is less than 12 ounces.)}$$

In this case, we can also write the null hypothesis as $H_0\colon \mu \geq 12$. This will not affect the result of the test as long as the sign in H_1 is *less than* (<).

When the alternative hypothesis has a *less than* (<) sign, as in this case, the test is always left-tailed. In a left-tailed test, the rejection region is in the left tail of the distribution curve, as shown in Figure 9.3, and the area of this rejection region is equal to α (the significance level). We can observe from this figure that there is only one critical value in a left-tailed test.

Figure 9.3 A left-tailed test.

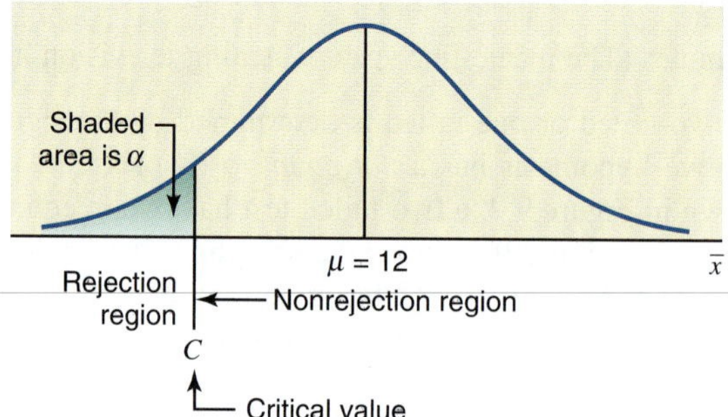

Assuming H_0 is true, the sampling distribution of $\bar{x}$ has a mean equal to 12 ounces (the value of μ in H_0). We will reject H_0 if the value of $\bar{x}$ obtained from the sample falls in the rejection region; we will not reject H_0 otherwise.

A Right-Tailed Test

To illustrate the third case, according to First American RES, the average price of homes in Stamford, Connecticut (with zip code 06903), was \$797,479 in March 2005 (*The Wall Street Journal*, April 15, 2005). Suppose a real estate researcher wants to check if the current mean price of homes in this area is higher than \$797,479. The key phrase in this case is *higher than*, which indicates a right-tailed test. Let μ be the current mean price of homes in this area. The two possible decisions are

1. The current mean price of homes in this area is equal to \$797,479, that is, currently $\mu = \$797{,}479$.
2. The current mean price of homes in this area is higher than \$797,479, that is, currently $\mu > \$797{,}479$.

We will write the null and alternative hypotheses for this test as follows.

H_0: $\mu = \$797{,}479$ (The current mean price of homes in this area is equal to \$797,479.)

H_1: $\mu > \$797{,}479$ (The current mean price of homes in this area is higher than \$797,479.)

Note that here we can also write the null hypothesis as H_0: $\mu \le \$797{,}479$, which states that the current mean price of homes in this area is either equal to or less than \$797,479. Again, the result of the test will not be affected whether we use an *equal to* (=) or a *less than or equal to* ($\le$) sign in H_0 as long as the alternative hypothesis has a *greater than* (>) sign.

When the alternative hypothesis has a *greater than* (>) sign, the test is always right-tailed. As shown in Figure 9.4, in a right-tailed test, the rejection region is in the right tail of the distribution curve. The area of this rejection region is equal to α, the significance level. Like a left-tailed test, a right-tailed test has only one critical value.

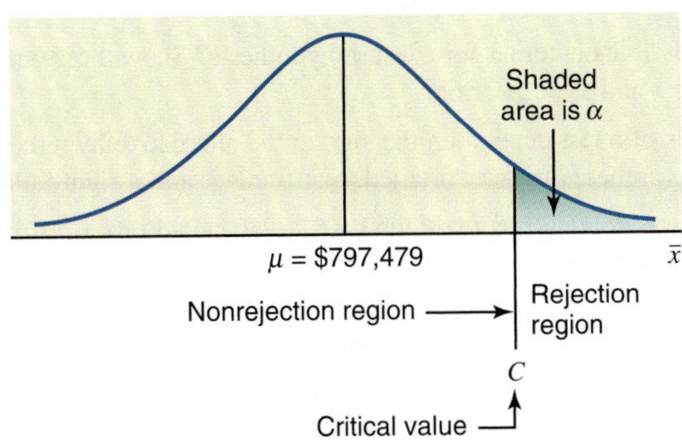

Figure 9.4 A right-tailed test.

Again, assuming H_0 is true, the sampling distribution of $\bar{x}$ has a mean equal to \$797,479 (the value of μ in H_0). We will reject H_0 if the value of $\bar{x}$ obtained from the sample falls in the rejection region. Otherwise, we will not reject H_0.

Table 9.3 summarizes the foregoing discussion about the relationship between the signs in H_0 and H_1 and the tails of a test.

Table 9.3

	Two-Tailed Test	**Left-Tailed Test**	**Right-Tailed Test**
Sign in the null hypothesis H_0	$=$	$=$ or $\ge$	$=$ or $\le$
Sign in the alternative hypothesis H_1	$\ne$	$<$	$>$
Rejection region	In both tails	In the left tail	In the right tail

Note that the null hypothesis always has an *equal to* (=) or a *greater than or equal to* ($\ge$) or a *less than or equal to* ($\le$) sign, and the alternative hypothesis always has a *not equal to* ($\ne$) or a *less than* (<) or a *greater than* (>) sign.

In this text we will use the following two procedures to make tests of hypothesis.

1. **The *p*-value approach** Under this procedure, we calculate what is called the *p*-value for the observed value of the sample statistic. If we have a predetermined significance level, then we compare the *p*-value with this significance level and make a decision. Note that here *p* stands for probability.
2. **The critical-value approach** In this approach, we find the critical value(s) from a table (such as the normal distribution table or the *t* distribution table) and find the value of the test statistic for the observed value of the sample statistic. Then we compare these two values and make a decision.

Remember, **the procedures to be learned in this chapter assume that the sample taken is a simple random sample.**

EXERCISES

CONCEPTS AND PROCEDURES

9.1 Briefly explain the meaning of each of the following terms.

a. Null hypothesis
b. Alternative hypothesis
c. Critical point(s)
d. Significance level
e. Nonrejection region
f. Rejection region
g. Tails of a test
h. Two types of errors

9.2 What are the four possible outcomes for a test of hypothesis? Show these outcomes by writing a table. Briefly describe the Type I and Type II errors.

9.3 Explain how the tails of a test depend on the sign in the alternative hypothesis. Describe the signs in the null and alternative hypotheses for a two-tailed, a left-tailed, and a right-tailed test, respectively.

9.4 Explain which of the following is a two-tailed test, a left-tailed test, or a right-tailed test.

a. H_0: $\mu = 45$, H_1: $\mu > 45$ b. H_0: $\mu = 23$, H_1: $\mu \neq 23$ c. H_0: $\mu \geq 75$, H_1: $\mu < 75$

Show the rejection and nonrejection regions for each of these cases by drawing a sampling distribution curve for the sample mean, assuming that it is normally distributed.

9.5 Explain which of the following is a two-tailed test, a left-tailed test, or a right-tailed test.

a. H_0: $\mu = 12$, H_1: $\mu < 12$ b. H_0: $\mu \leq 85$, H_1: $\mu > 85$ c. H_0: $\mu = 33$, H_1: $\mu \neq 33$

Show the rejection and nonrejection regions for each of these cases by drawing a sampling distribution curve for the sample mean, assuming that it is normally distributed.

9.6 Which of the two hypotheses (null and alternative) is initially assumed to be true in a test of hypothesis?

9.7 Consider H_0: $\mu = 20$ versus H_1: $\mu < 20$.

a. What type of error would you make if the null hypothesis is actually false and you fail to reject it?
b. What type of error would you make if the null hypothesis is actually true and you reject it?

9.8 Consider H_0: $\mu = 55$ versus H_1: $\mu \neq 55$.

a. What type of error would you make if the null hypothesis is actually false and you fail to reject it?
b. What type of error would you make if the null hypothesis is actually true and you reject it?

APPLICATIONS

9.9 Write the null and alternative hypotheses for each of the following examples. Determine if each is a case of a two-tailed, a left-tailed, or a right-tailed test.

a. To test if the mean number of hours spent working per week by college students who hold jobs is different from 20 hours
b. To test whether or not a bank's ATM is out of service for an average of more than 10 hours per month
c. To test if the mean length of experience of airport security guards is different from three years
d. To test if the mean credit card debt of college seniors is less than $1000
e. To test if the mean time a customer has to wait on the phone to speak to a representative of a mail-order company about unsatisfactory service is more than 12 minutes

9.10 Write the null and alternative hypotheses for each of the following examples. Determine if each is a case of a two-tailed, a left-tailed, or a right-tailed test.

a. To test if the mean amount of time spent per week watching sports on television by all adult men is different from 9.5 hours
b. To test if the mean amount of money spent by all customers at a supermarket is less than $105

c. To test whether the mean starting salary of college graduates is higher than $39,000 per year
d. To test if the mean waiting time at the drive-through window at a fast food restaurant during rush hour differs from 10 minutes
e. To test if the mean hours spent per week on house chores by all housewives is less than 30

9.2 Hypothesis Tests About μ: σ Known

This section explains how to perform a test of hypothesis for the population mean μ when the population standard deviation σ is known. As in Section 8.3 of Chapter 8, here also there are three possible cases that are mentioned below.

Case I. If the following three conditions are fulfilled:

1. The population standard deviation σ is known
2. The sample size is small (i.e., $n < 30$)
3. The population from which the sample is selected is normally distributed,

then we use the normal distribution to perform a test of hypothesis about μ because from Section 7.4.1 of Chapter 7 the sampling distribution of $\bar{x}$ is normal with its mean equal to μ and the standard deviation equal to $\sigma_{\bar{x}} = \sigma/\sqrt{n}$ assuming that $n/N \leq .05$.

Case II. If the following two conditions are fulfilled:

1. The population standard deviation σ is known
2. The sample size is large (i.e., $n \geq 30$),

then, again, we use the normal distribution to perform a test of hypothesis about μ because from Section 7.4.2 of Chapter 7, due to the central limit theorem, the sampling distribution of $\bar{x}$ is (approximately) normal with its mean equal to μ and the standard deviation equal to $\sigma_{\bar{x}} = \sigma/\sqrt{n}$ assuming that $n/N \leq .05$.

Case III. If the following three conditions are fulfilled:

1. The population standard deviation σ is known
2. The sample size is small (i.e., $n < 30$)
3. The population from which the sample is selected is not normally distributed (or the shape of its distribution is unknown),

then we use a nonparametric method to perform a test of hypothesis about μ.

This section will cover the first two cases. The procedure to perform a test of hypothesis about μ is the same in both these cases. Note that in Case I, the population does not have to be exactly normally distributed. As long as it is close to the normal distribution without any outliers, we can use the normal distribution procedure. In Case II, although 30 is considered a large sample, if the population distribution is very different from the normal distribution, then 30 may not be a large enough sample size for the sampling distribution of $\bar{x}$ to be normal and, hence, to use the normal distribution.

The following chart summarizes the above three cases.

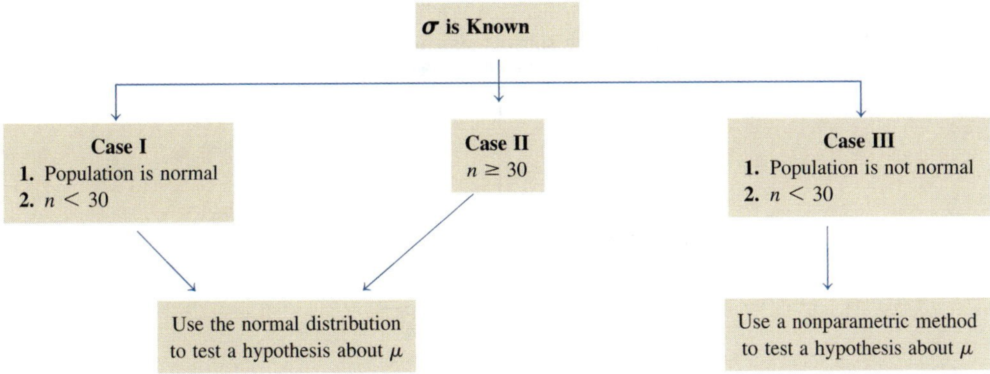

Below we explain two procedures, the p-value approach and the critical-value approach, to test hypotheses about μ under Cases I and II. We will use the normal distribution to perform such tests.

Note that the two approaches—the p-value approach and the critical-value approach—are not mutually exclusive. We do not need to use one or the other. We can use both at the same time.

1. The *p*-Value Approach

In this procedure, we find a probability value such that a given null hypothesis is rejected for any α (significance level) greater than this value and it is not rejected for any α less than this value. The **probability-value approach**, more commonly called the *p-value approach*, gives such a value. In this approach, we calculate the ***p*-value** for the test, which is defined as the smallest level of significance at which the given null hypothesis is rejected. Using this p-value, we state the decision. If we have a predetermined value of α, then we compare the value of p with α and make a decision.

Definition

***p*-Value** Assuming that the null hypothesis is true, the p-value can be defined as the probability that a sample statistic (such as the sample mean) is at least as far away from the hypothesized value in the direction of the alternative hypothesis as the one obtained from the sample data under consideration. Note that the *p-value* is the smallest significance level at which the null hypothesis is rejected.

Using the p-value approach, we reject the null hypothesis if

$$p\text{-value} < \alpha \quad \text{or} \quad \alpha > p\text{-value}$$

and we do not reject the null hypothesis if

$$p\text{-value} \geq \alpha \quad \text{or} \quad \alpha \leq p\text{-value}$$

For a one-tailed test, the p-value is given by the area in the tail of the sampling distribution curve beyond the observed value of the sample statistic. Figure 9.5 shows the p-value for a right-tailed test about μ. For a left-tailed test, the p-value will be the area in the lower tail of the sampling distribution curve to the left of the observed value of $\bar{x}$.

Figure 9.5 The p-value for a right-tailed test.

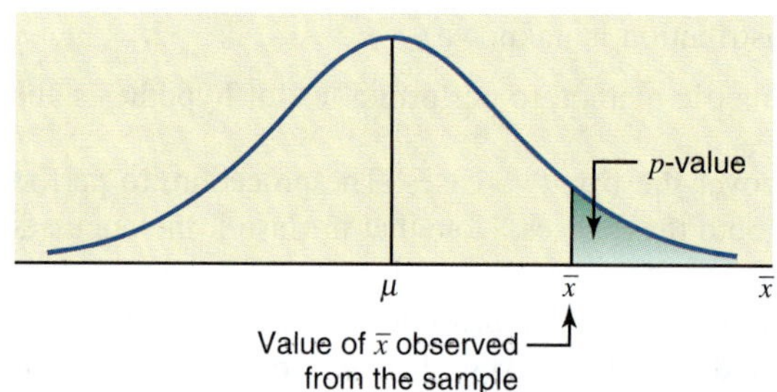

For a two-tailed test, the p-value is twice the area in the tail of the sampling distribution curve beyond the observed value of the sample statistic. Figure 9.6 shows the p-value for a two-tailed test. Each of the areas in the two tails gives one-half the p-value.

Figure 9.6 The p-value for a two-tailed test.

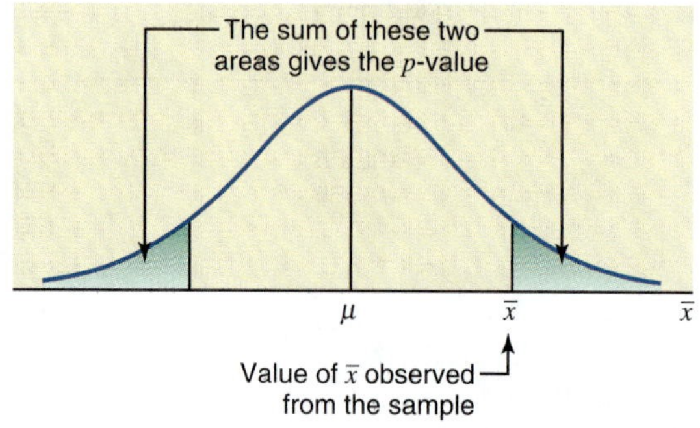

To find the area under the normal distribution curve beyond the sample mean $\bar{x}$, we first find the z value for $\bar{x}$ using the following formula.

Calculating the z Value for $\bar{x}$ When using the normal distribution, *the value of z* for $\bar{x}$ for a test of hypothesis about μ is computed as follows:

$$z = \frac{\bar{x} - \mu}{\sigma_{\bar{x}}} \quad \text{where} \quad \sigma_{\bar{x}} = \frac{\sigma}{\sqrt{n}}$$

The value of z calculated for $\bar{x}$ using this formula is also called the **observed value of z.**

Then we find the area under the tail of the normal distribution curve beyond this value of z. This area gives the p-value or one-half of the p-value depending on whether it is a one-tailed test or a two-tailed test.

A test of hypothesis procedure that uses the p-value approach involves the following four steps.

Steps to Perform a Test of Hypothesis Using the p-Value Approach

1. State the null and alternative hypothesis.
2. Select the distribution to use.
3. Calculate the p-value.
4. Make a decision.

Examples 9–1 and 9–2 illustrate the calculation and use of the p-value to test a hypothesis using the normal distribution.

EXAMPLE 9–1

Performing a hypothesis test using the p-value approach for a two-tailed test with the normal distribution.

At Canon Food Corporation, it took an average of 90 minutes for new workers to learn a food processing job. Recently the company installed a new food processing machine. The supervisor at the company wants to find if the mean time taken by new workers to learn the food processing procedure on this new machine is different from 90 minutes. A sample of 20 workers showed that it took, on average, 85 minutes for them to learn the food processing procedure on the new machine. It is known that the learning times for all new workers are normally distributed with a population standard deviation of 7 minutes. Find the p-value for the test that the mean learning time for the food processing procedure on the new machine is different from 90 minutes. What will your conclusion be if $\alpha = .01$?

Solution Let μ be the mean time (in minutes) taken to learn the food processing procedure on the new machine by all workers, and let $\bar{x}$ be the corresponding sample mean. From the given information,

$$n = 20, \quad \bar{x} = 85 \text{ minutes}, \quad \sigma = 7 \text{ minutes}, \quad \text{and} \quad \alpha = .01$$

To calculate the p-value and make the test, we apply the following four steps.

Step 1. *State the null and alternative hypotheses.*

$$H_0: \mu = 90 \text{ minutes}$$
$$H_1: \mu \neq 90 \text{ minutes}$$

Note that the null hypothesis states that the mean time for learning the food processing procedure on the new machine is 90 minutes, and the alternative hypothesis states that this time is different from 90 minutes.

Step 2. *Select the distribution to use.*

Here, the population standard deviation σ is known, the sample size is small ($n < 30$), but the population distribution is normal. Hence, the sampling distribution of $\bar{x}$ is normal with its mean equal to μ and the standard deviation equal to $\sigma_{\bar{x}} = \sigma/\sqrt{n}$. Consequently, we will use the normal distribution to find the p-value and make the test.

Step 3. *Calculate the p-value.*

The $\neq$ sign in the alternative hypothesis indicates that the test is two-tailed. The p-value is equal to twice the area in the tail of the sampling distribution curve of $\bar{x}$ to the left of $\bar{x} = 85$, as shown in Figure 9.7. To find this area, we first find the z value for $\bar{x} = 85$ as follows:

$$\sigma_{\bar{x}} = \frac{\sigma}{\sqrt{n}} = \frac{7}{\sqrt{20}} = 1.56524758 \text{ minutes}$$

$$z = \frac{\bar{x} - \mu}{\sigma_{\bar{x}}} = \frac{85 - 90}{1.56524758} = -3.19$$

Figure 9.7 The p-value for a two-tailed test.

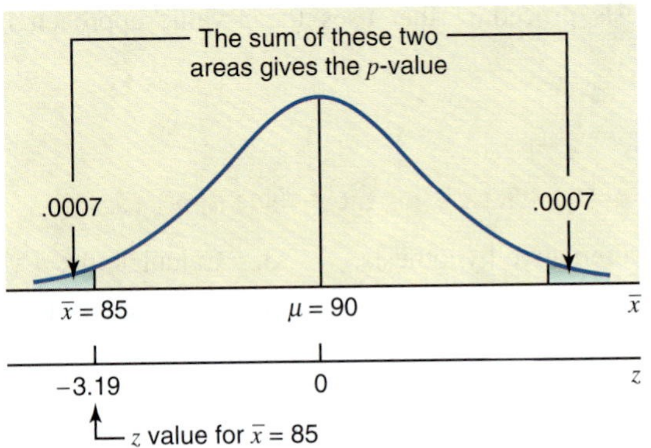

The area to the left of $\bar{x} = 85$ is equal to the area under the standard normal curve to the left of $z = -3.19$. From the normal distribution table, the area to the left of $z = -3.19$ is .0007. Consequently, the p-value is

$$p\text{-value} = 2(.0007) = \mathbf{.0014}$$

Step 4. *Make a decision.*

Thus, based on the p-value of .0014 we can state that for any α (significance level) greater than .0014 we will reject the null hypothesis stated in Step 1 and for any α less than or equal to .0014 we will not reject the null hypothesis.

Because $\alpha = .01$ is greater than the p-value of .0014, we reject the null hypothesis at this significance level. Therefore, we conclude that the mean time for learning the food processing procedure on the new machine is different from 90 minutes. ■

■ EXAMPLE 9–2

Performing a hypothesis test using the p-value approach for a one-tailed test with the normal distribution.

The management of Priority Health Club claims that its members lose an average of 10 pounds or more within the first month after joining the club. A consumer agency that wanted to check this claim took a random sample of 36 members of this health club and found that they lost an average of 9.2 pounds within the first month of membership. The population standard deviation is known to be 2.4 pounds. Find the p-value for this test. What will your decision be if $\alpha = .01$? What if $\alpha = .05$?

Solution Let μ be the mean weight lost during the first month of membership by all members of this health club, and let $\bar{x}$ be the corresponding mean for the sample. From the given information,

$$n = 36, \quad \bar{x} = 9.2 \text{ pounds}, \quad \text{and} \quad \sigma = 2.4 \text{ pounds}$$

The claim of the club is that its members lose, on average, 10 pounds or more within the first month of membership. To perform the test using the *p*-value approach, we apply the following four steps.

Step 1. *State the null and alternative hypotheses.*

$$H_0: \mu \geq 10 \quad \text{(The mean weight lost is 10 pounds or more).}$$

$$H_1: \mu < 10 \quad \text{(The mean weight lost is less than 10 pounds).}$$

Step 2. *Select the distribution to use.*

Here, the population standard deviation σ is known, and the sample size is large $(n > 30)$. Hence, the sampling distribution of $\bar{x}$ is normal with its mean equal to μ and the standard deviation equal to $\sigma_{\bar{x}} = \sigma/\sqrt{n}$. Consequently, we will use the normal distribution to find the *p*-value and perform the test.

Step 3. *Calculate the p-value.*

The $<$ sign in the alternative hypothesis indicates that the test is left-tailed. The *p*-value is given by the area to the left of $\bar{x} = 9.2$ under the sampling distribution curve of $\bar{x}$, as shown in Figure 9.8. To find this area, we first find the z value for $\bar{x} = 9.2$ as follows:

$$\sigma_{\bar{x}} = \frac{\sigma}{\sqrt{n}} = \frac{2.4}{\sqrt{36}} = .40$$

$$z = \frac{\bar{x} - \mu}{\sigma_{\bar{x}}} = \frac{9.2 - 10}{.40} = -2.00$$

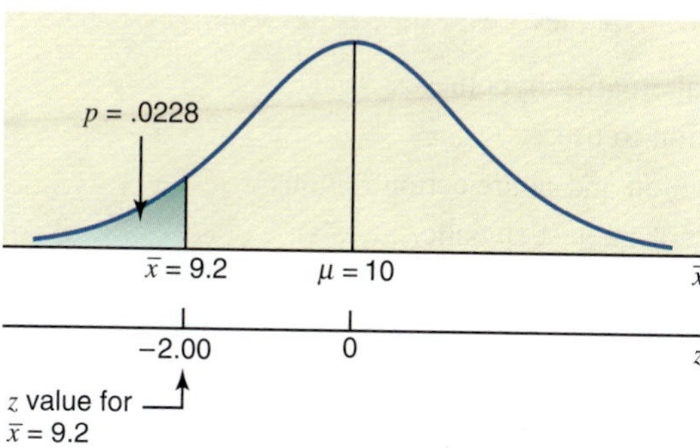

Figure 9.8 The *p*-value for a left-tailed test.

The area to the left of $\bar{x} = 9.2$ under the sampling distribution of $\bar{x}$ is equal to the area under the standard normal curve to the left of $z = -2.00$. From the normal distribution table, the area to the left of $z = -2.00$ is .0228. Consequently,

$$p\text{-value} = \mathbf{.0228}$$

Step 4. *Make a decision.*

Thus, based on the *p*-value of .0228, we can state that for any α (significance level) greater than .0228 we will reject the null hypothesis stated in Step 1, and for any α less than or equal to .0228 we will not reject the null hypothesis.

Since $\alpha = .01$ is less than the *p*-value of .0228, we do not reject the null hypothesis at this significance level. Consequently, we conclude that the mean weight lost within the first month of membership by the members of this club is 10 pounds or more.

Now, because $\alpha = .05$ is greater than the *p*-value of .0228, we reject the null hypothesis at this significance level. Therefore, we conclude that the mean weight lost within the first month of membership by the members of this club is less than 10 pounds. ■

2. The Critical-Value Approach

This is also called the traditional or classical approach. In this procedure, we have a predetermined value of the significance level α. The value of α gives the total area of the rejection region(s). First we find the critical value(s) of z from the normal distribution table for the given significance level. Then we find the value of the test statistic z for the observed value of the sample statistic $\bar{x}$. Finally we compare these two values and make a decision. Remember, if the test is one-tailed, there is only one critical value of z and it is obtained by using the value of α, which gives the area in the left or right tail of the normal distribution curve depending on whether the test is left-tailed or right-tailed, respectively. However, if the test is two-tailed, there are two critical values of z and they are obtained by using $\alpha/2$ area in each tail of the normal distribution curve. The value of the test statistic is obtained as follows.

Test Statistic In tests of hypotheses about μ using the normal distribution, the random variable

$$z = \frac{\bar{x} - \mu}{\sigma_{\bar{x}}} \quad \text{where} \quad \sigma_{\bar{x}} = \frac{\sigma}{\sqrt{n}}$$

is called the *test statistic*. The test statistic can be defined as a rule or criterion that is used to make the decision whether or not to reject the null hypothesis.

A test of hypothesis procedure that uses the critical-value approach involves the following five steps.

Steps to Perform a Test of Hypothesis with the Critical-Value Approach

1. State the null and alternative hypotheses.
2. Select the distribution to use.
3. Determine the rejection and nonrejection regions.
4. Calculate the value of the test statistic.
5. Make a decision.

Examples 9–3 and 9–4 illustrate the use of these five steps to perform tests of hypotheses about the population mean μ. Example 9–3 is concerned with a two-tailed test and Example 9–4 describes a one-tailed test.

■ EXAMPLE 9–3

Conducting a two-tailed test of hypothesis about μ: σ known and $n > 30$.

The TIV Telephone Company provides long-distance telephone service in an area. According to the company's records, the average length of all long-distance calls placed through this company in 2004 was 12.44 minutes. The company's management wanted to check if the mean length of the current long-distance calls is different from 12.44 minutes. A sample of 150 such calls placed through this company produced a mean length of 13.71 minutes. The standard deviation of all such calls is 2.65 minutes. Using the 2% significance level, can you conclude that the mean length of all current long-distance calls is different from 12.44 minutes?

Solution Let μ be the mean length of all current long-distance calls placed through this company and $\bar{x}$ be the corresponding mean for the sample. From the given information,

$$n = 150, \quad \bar{x} = 13.71 \text{ minutes}, \quad \text{and} \quad \sigma = 2.65 \text{ minutes}$$

We are to test whether or not the mean length of all current long-distance calls is different from 12.44 minutes. The significance level α is .02; that is, the probability of rejecting the null hypothesis when it actually is true should not exceed .02. This is the probability of making a Type I error. We perform the test of hypothesis using the five steps as follows.

Step 1. *State the null and alternative hypotheses.*

Notice that we are testing to find whether or not the mean length of all current long-distance calls is different from 12.44 minutes. We write the null and alternative hypotheses as follows.

H_0: $\mu = 12.44$ (The mean length of all current long-distance calls is 12.44 minutes.)

H_1: $\mu \neq 12.44$ (The mean length of all current long-distance calls is different from 12.44 minutes.)

Step 2. *Select the distribution to use.*

Here, the population standard deviation σ is known, and the sample size is large ($n > 30$). Hence, the sampling distribution of $\bar{x}$ is (approximately) normal with its mean equal to μ and the standard deviation equal to $\sigma_{\bar{x}} = \sigma/\sqrt{n}$. Consequently, we will use the normal distribution to perform the test of this example.

Step 3. *Determine the rejection and nonrejection regions.*

The significance level is .02. The $\neq$ sign in the alternative hypothesis indicates that the test is two-tailed with two rejection regions, one in each tail of the normal distribution curve of $\bar{x}$. Because the total area of both rejection regions is .02 (the significance level), the area of the rejection region in each tail is .01; that is,

$$\text{Area in each tail} = \alpha/2 = .02/2 = .01$$

These areas are shown in Figure 9.9. Two critical points in this figure separate the two rejection regions from the nonrejection region. Next, we find the z values for the two critical points using the area of the rejection region. To find the z values for these critical points, we look for .0100 and .9900 areas in the normal distribution table. From Table IV, the z values of the two critical points, as shown in Figure 9.9, are approximately -2.33 and 2.33.

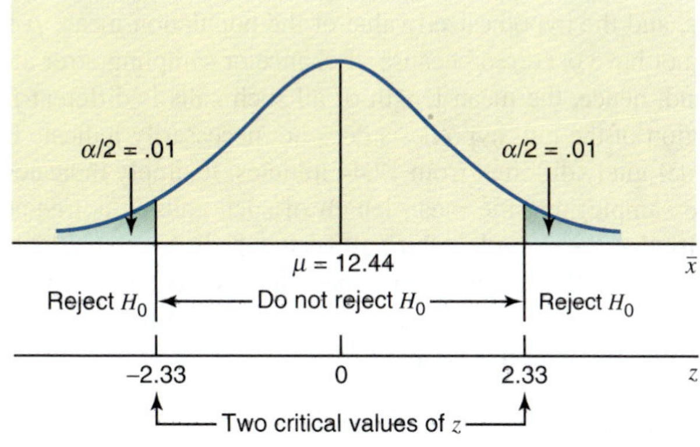

Figure 9.9

Step 4. *Calculate the value of the test statistic.*

The decision to reject or not to reject the null hypothesis will depend on whether the evidence from the sample falls in the rejection or the nonrejection region. If the value of $\bar{x}$ falls in either of the two rejection regions, we reject H_0. Otherwise, we do not reject H_0. The value of $\bar{x}$ obtained from the sample is called the *observed value of* $\bar{x}$. To locate the position of $\bar{x} = 13.71$ on the sampling distribution curve of $\bar{x}$ in Figure 9.9, we first calculate the z value for $\bar{x} = 13.71$. This is called the *value of the test statistic*. Then, we compare the value of the test statistic with the two critical values of z, -2.33 and 2.33, shown in Figure 9.9. If the value of the test statistic is between -2.33 and 2.33, we do not reject H_0. If the value of the test statistic is either greater than 2.33 or less than -2.33, we reject H_0.

Calculating the Value of the Test Statistic When using the normal distribution, *the value of the test statistic z* for $\bar{x}$ for a test of hypothesis about μ is computed as follows:

$$z = \frac{\bar{x} - \mu}{\sigma_{\bar{x}}}$$

where $$\sigma_{\bar{x}} = \frac{\sigma}{\sqrt{n}}$$

This value of z for $\bar{x}$ is also called the **observed value of** z.

The value of $\bar{x}$ from the sample is 13.71. We calculate the z value as follows:

$$\sigma_{\bar{x}} = \frac{\sigma}{\sqrt{n}} = \frac{2.65}{\sqrt{150}} = .21637159$$

From H_0

$$z = \frac{\bar{x} - \mu}{\sigma_{\bar{x}}} = \frac{13.71 - 12.44}{.21637159} = 5.87$$

The value of μ in the calculation of the z value is substituted from the null hypothesis. The value of $z = 5.87$ calculated for $\bar{x}$ is called the *computed value of the test statistic z*. This is the value of z that corresponds to the value of $\bar{x}$ observed from the sample. It is also called the *observed value of z*.

Step 5. *Make a decision.*

In the final step we make a decision based on the location of the value of the test statistic z computed for $\bar{x}$ in Step 4. This value of $z = 5.87$ is greater than the critical value of $z = 2.33$, and it falls in the rejection region in the right tail in Figure 9.9. Hence, we reject H_0 and conclude that based on the sample information, it appears that the mean length of all such calls is not equal to 12.44 minutes.

By rejecting the null hypothesis, we are stating that the difference between the sample mean, $\bar{x} = 13.71$ minutes, and the hypothesized value of the population mean, $\mu = 12.44$ minutes, is too large and may not have occurred because of chance or sampling error alone. This difference seems to be real and, hence, the mean length of all such calls is different from 12.44 minutes. Note that the rejection of the null hypothesis does not necessarily indicate that the mean length of all such calls is definitely different from 12.44 minutes. It simply indicates that there is strong evidence (from the sample) that the mean length of such calls is not equal to 12.44 minutes. There is a possibility that the mean length of all such calls is equal to 12.44 minutes, but by the luck of the draw we selected a sample with a mean that is too far from the hypothesized mean of 12.44 minutes. If so, we have wrongfully rejected the null hypothesis H_0. This is a Type I error and its probability is .02 in this example. ■

We can use the p-value approach to perform the test of hypothesis in Example 9–3. In this example, the test is two-tailed. The p-value is equal to twice the area under the sampling distribution of $\bar{x}$ to the right of $\bar{x} = 13.71$. As calculated in Step 4 above, the z value for $\bar{x} = 13.71$ is 5.87. From the normal distribution table, the area to the right of $z = 5.87$ is (approximately) zero. Hence, the p-value is zero. (If you use technology, you will obtain the p-value of .000.) As we know from earlier discussions, we will reject the null hypothesis for any α (significance level) that is greater than the p-value. Consequently, in this example, we will reject the null hypothesis for any $\alpha > 0$. Since $\alpha = .02$ here, which is greater than zero, we reject the null hypothesis.

■ EXAMPLE 9–4

Conducting a left-tailed test of hypothesis about μ: σ known, $n < 30$, and population normal.

The mayor of a large city claims that the average net worth of families living in this city is at least \$300,000. A random sample of 25 families selected from this city produced a mean net worth of \$288,000. Assume that the net worths of all families in this city have a normal distribution with the population standard deviation of \$80,000. Using the 2.5% significance level, can you conclude that the mayor's claim is false?

Solution Let μ be the mean net worth of families living in this city and $\bar{x}$ be the corresponding mean for the sample. From the given information,

$$n = 25, \quad \bar{x} = \$288,000, \quad \text{and} \quad \sigma = \$80,000$$

The significance level is $\alpha = .025$.

Step 1. *State the null and alternative hypotheses.*

We are to test whether or not the mayor's claim is false. The mayor's claim is that the average net worth of families living in this city is at least \$300,000. Hence, the null and alternative hypotheses are

H_0: $\mu \geq \$300,000$ (The mayor's claim is true. The mean net worth is at least \$300,000.)

H_1: $\mu < \$300,000$ (The mayor's claim is false. The mean net worth is less than \$300,000.)

Step 2. *Select the distribution to use.*

Here, the population standard deviation σ is known, the sample size is small ($n < 30$), but the population distribution is normal. Hence, the sampling distribution of $\bar{x}$ is normal with its mean equal to μ and the standard deviation equal to $\sigma_{\bar{x}} = \sigma/\sqrt{n}$. Consequently, we will use the normal distribution to perform the test.

Step 3. *Determine the rejection and nonrejection regions.*

The significance level is .025. The $<$ sign in the alternative hypothesis indicates that the test is left-tailed with the rejection region in the left tail of the sampling distribution curve of $\bar{x}$. The critical value of z, obtained from the normal table for .0250 area in the left tail, is -1.96, as shown in Figure 9.10.

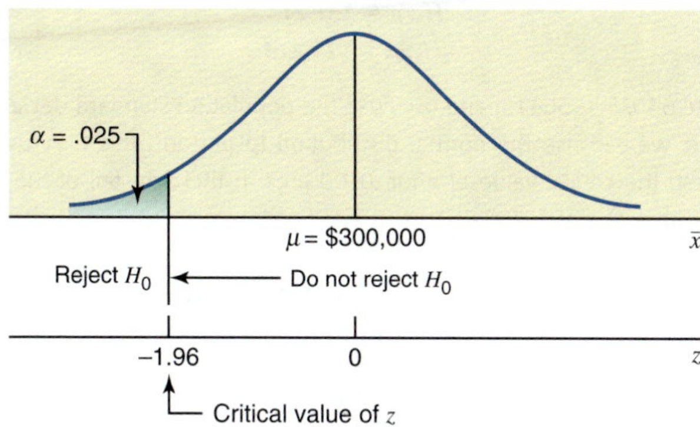

Figure 9.10

Step 4. *Calculate the value of the test statistic.*

The value of the test statistic z for $\bar{x} = \$288,000$ is calculated as follows.

$$\sigma_{\bar{x}} = \frac{\sigma}{\sqrt{n}} = \frac{80,000}{\sqrt{25}} = \$16,000$$

$$z = \frac{\bar{x} - \mu}{\sigma_{\bar{x}}} = \frac{288,000 - 300,000}{16,000} = -.75$$

(300,000: From H_0)

Step 5. *Make a decision.*

The value of the test statistic $z = -.75$ is greater than the critical value of $z = -1.96$, and it falls in the nonrejection region. As a result, we fail to reject H_0. Therefore, we can state that based on the sample information, it appears that the mean net worth of families in this city is not less than \$300,000. Note that we are not concluding that the mean net worth is definitely not less than \$300,000. By not rejecting the null hypothesis, we are saying that the information

CASE STUDY

9–1

THE AVERAGE COST OF A WEDDING

USA TODAY Snapshots

Cost of walk down the aisle rises

Average cost of a wedding:

$20,357 $19,347 $21,213

2001 2002 2003

Source: Bridal Guide InfoSource

By Darryl Haralson and Alejandro Gonzalez, USA TODAY

The above chart shows the average cost of a wedding for the years 2001, 2002, and 2003. According to the information given in this chart, the average cost of a wedding in 2003 was $21,213. Suppose this result is true for all weddings in 2003, and that we want to check if the current mean cost of a wedding is higher than $21,213. Suppose we take a random sample of 700 recent weddings and find out that their average cost was $24,540. Assume that the standard deviation of all current weddings is $6415, and the significance level is 1%. The test is right-tailed. The null and alternative hypotheses are

$$H_0: \mu = \$21{,}213$$
$$H_1: \mu > \$21{,}213$$

Here, $n = 700$, $\bar{x} = \$24{,}540$, $\sigma = \$6415$, and $\alpha = .01$. The population standard deviation is known and the sample is large. Hence, we can use the normal distribution to perform this test. Using the normal distribution to make the test, the critical value of z for .0100 area in the right tail of the normal curve is 2.33. We find the observed value of z as follows.

$$\sigma_{\bar{x}} = \frac{\sigma}{\sqrt{n}} = \frac{6415}{\sqrt{700}} = \$242.4642094$$

$$z = \frac{\bar{x} - \mu}{\sigma_{\bar{x}}} = \frac{24{,}540 - 21{,}213}{242.4642094} = 13.72$$

The value of the test statistic $z = 13.72$ for $\bar{x}$ falls in the rejection region. Consequently, we reject H_0 and conclude that the current average cost of weddings is greater than $21,213.

To use the p-value approach, we find the area under the normal curve to the right of $z = 13.72$ from the normal distribution table. This area is zero. Therefore, the p-value is .0000. Since $\alpha = .01$ is larger than .0000, we reject the null hypothesis.

Source: The chart reproduced with permission from *USA TODAY*, June 22, 2004. Copyright © 2004, *USA TODAY*.

obtained from the sample is not strong enough to reject the null hypothesis and to conclude that the mayor's claim is false. ■

We can use the p-value approach to perform the test of hypothesis in Example 9–4. In this example, the test is left-tailed. The p-value is given by the area under the sampling distribution of $\bar{x}$ to the left of $\bar{x} = \$288{,}000$. As calculated in Step 4 above, the z value for $\bar{x} = \$288{,}000$ is $-.75$. From the normal distribution table, the area to the left of $z = -.75$ is .2266. Hence, the p-value is .2266. We will reject the null hypothesis for any α (significance level) that is greater than the p-value. Consequently, we will reject the null hypothesis in this example for any $\alpha > .2266$. Since, in this example $\alpha = .025$, which is less than .2266, we fail to reject the null hypothesis.

In studies published in various journals, authors usually use the terms *significantly different* and *not significantly different* when deriving conclusions based on hypothesis tests. These terms are short versions of the terms *statistically significantly different* and *statistically not significantly different*. The expression *significantly different* means that the difference between the observed value of the sample mean $\bar{x}$ and the hypothesized value of the population mean μ is so large that it probably did not occur because of the sampling error alone. Consequently, the null hypothesis is rejected. In other words, the difference between $\bar{x}$ and μ is statistically significant. Thus, the statement *significantly different* is equivalent to saying that the *null hypothesis is rejected*. In Example 9–3, we can state as a conclusion that the observed value of $\bar{x} = 13.71$ minutes is significantly different from the hypothesized value of $\mu = 12.44$ minutes. That is, the mean length of all current long-distance calls is different from 12.44 minutes.

On the other hand, the statement *not significantly different* means that the difference between the observed value of the sample mean $\bar{x}$ and the hypothesized value of the population mean μ is so small that it may have occurred just because of chance. Consequently, the null hypothesis is not rejected. Thus, the expression *not significantly different* is equivalent to saying that we *fail to reject the null hypothesis*. In Example 9–4, we can state as a conclusion that the observed value of $\bar{x} = \$288,000$ is not significantly less than the hypothesized value of $\mu = \$300,000$. In other words, the current mean net worth of households in this city is not less than \$300,000.

EXERCISES

CONCEPTS AND PROCEDURES

9.11 What are the five steps of a test of hypothesis using the critical value approach? Explain briefly.

9.12 What does the level of significance represent in a test of hypothesis? Explain.

9.13 By rejecting the null hypothesis in a test of hypothesis example, are you stating that the alternative hypothesis is true?

9.14 What is the difference between the critical value of z and the observed value of z?

9.15 Briefly explain the procedure used to calculate the p-value for a two-tailed and for a one-tailed test, respectively.

9.16 Find the p-value for each of the following hypothesis tests.

a. H_0: $\mu = 23$, H_1: $\mu \neq 23$, $n = 50$, $\bar{x} = 21.25$, $\sigma = 5$
b. H_0: $\mu = 15$, H_1: $\mu < 15$, $n = 80$, $\bar{x} = 13.25$, $\sigma = 5.5$
c. H_0: $\mu = 38$, H_1: $\mu > 38$, $n = 35$, $\bar{x} = 40.25$, $\sigma = 7.2$

9.17 Find the p-value for each of the following hypothesis tests.

a. H_0: $\mu = 46$, H_1: $\mu \neq 46$, $n = 40$, $\bar{x} = 49.60$, $\sigma = 9.7$
b. H_0: $\mu = 26$, H_1: $\mu < 26$, $n = 33$, $\bar{x} = 24.30$, $\sigma = 4.3$
c. H_0: $\mu = 18$, H_1: $\mu > 18$, $n = 55$, $\bar{x} = 20.50$, $\sigma = 7.8$

9.18 Consider H_0: $\mu = 29$ versus H_1: $\mu \neq 29$. A random sample of 25 observations taken from this population produced a sample mean of 25.3. The population is normally distributed with $\sigma = 8$.

a. Calculate the p-value.
b. Considering the p-value of part a, would you reject the null hypothesis if the test were made at the significance level of .05?
c. Considering the p-value of part a, would you reject the null hypothesis if the test were made at the significance level of .01?

9.19 Consider H_0: $\mu = 72$ versus H_1: $\mu > 72$. A random sample of 16 observations taken from this population produced a sample mean of 75.2. The population is normally distributed with $\sigma = 6$.

a. Calculate the p-value.
b. Considering the p-value of part a, would you reject the null hypothesis if the test were made at the significance level of .01?
c. Considering the p-value of part a, would you reject the null hypothesis if the test were made at the significance level of .025?

9.20 For each of the following examples of tests of hypotheses about μ, show the rejection and nonrejection regions on the sampling distribution of the sample mean assuming that it is normal.

a. A two-tailed test with $\alpha = .05$ and $n = 40$
b. A left-tailed test with $\alpha = .01$ and $n = 20$
c. A right-tailed test with $\alpha = .02$ and $n = 55$

9.21 For each of the following examples of tests of hypotheses about μ, show the rejection and nonrejection regions on the sampling distribution of the sample mean assuming it is normal.

a. A two-tailed test with $\alpha = .01$ and $n = 100$
b. A left-tailed test with $\alpha = .005$ and $n = 27$
c. A right-tailed test with $\alpha = .025$ and $n = 36$

9.22 Consider the following null and alternative hypotheses:

$$H_0: \mu = 25 \quad \text{versus} \quad H_1: \mu \neq 25$$

Suppose you perform this test at $\alpha = .05$ and reject the null hypothesis. Would you state that the difference between the hypothesized value of the population mean and the observed value of the sample mean is "statistically significant" or would you state that this difference is "statistically not significant"? Explain.

9.23 Consider the following null and alternative hypotheses:

$$H_0: \mu = 60 \quad \text{versus} \quad H_1: \mu > 60$$

Suppose you perform this test at $\alpha = .01$ and fail to reject the null hypothesis. Would you state that the difference between the hypothesized value of the population mean and the observed value of the sample mean is "statistically significant" or would you state that this difference is "statistically not significant"? Explain.

9.24 For each of the following significance levels, what is the probability of making a Type I error?

a. $\alpha = .025$ b. $\alpha = .05$ c. $\alpha = .01$

9.25 For each of the following significance levels, what is the probability of making a Type I error?

a. $\alpha = .10$ b. $\alpha = .02$ c. $\alpha = .005$

9.26 A random sample of 120 observations produced a sample mean of 32. Find the critical and observed values of z for each of the following tests of hypotheses using $\alpha = .05$. The population standard deviation is known to be 6.

a. $H_0: \mu = 28$ versus $H_1: \mu > 28$
b. $H_0: \mu = 28$ versus $H_1: \mu \neq 28$

9.27 A random sample of 28 observations produced a sample mean of 15. Find the critical and observed values of z for each of the following tests of hypotheses using $\alpha = .01$. It is known that the population has a normal distribution with $\sigma = 4$.

a. $H_0: \mu = 20$ versus $H_1: \mu < 20$
b. $H_0: \mu = 20$ versus $H_1: \mu \neq 20$

9.28 Consider the null hypothesis $H_0: \mu = 50$. Suppose a random sample of 24 observations is taken from a normally distributed population with $\sigma = 7$. Using $\alpha = .05$, show the rejection and nonrejection regions on the sampling distribution curve of the sample mean and find the critical value(s) of z when the alternative hypothesis is

a. $H_1: \mu < 50$ b. $H_1: \mu \neq 50$ c. $H_1: \mu > 50$

9.29 Consider the null hypothesis $H_0: \mu = 35$. Suppose a random sample of 70 observations is taken from a population with $\sigma = 5.5$. Using $\alpha = .01$, show the rejection and nonrejection regions on the sampling distribution curve of the sample mean and find the critical value(s) of z for a

a. left-tailed test b. two-tailed test c. right-tailed test

9.30 Consider $H_0: \mu = 100$ versus $H_1: \mu \neq 100$.

a. A random sample of 64 observations produced a sample mean of 98. Using $\alpha = .01$, would you reject the null hypothesis? The population standard deviation is known to be 12.
b. Another random sample of 64 observations taken from the same population produced a sample mean of 104. Using $\alpha = .01$, would you reject the null hypothesis? The population standard deviation is known to be 12.

Comment on the results of parts a and b.

9.31 Consider H_0: $\mu = 45$ versus H_1: $\mu < 45$.

a. A random sample of 25 observations produced a sample mean of 41.8. Using $\alpha = .025$, would you reject the null hypothesis? The population is known to be normally distributed with $\sigma = 6$.

b. Another random sample of 25 observations taken from the same population produced a sample mean of 43.8. Using $\alpha = .025$, would you reject the null hypothesis? The population is known to be normally distributed with $\sigma = 6$.

Comment on the results of parts a and b.

9.32 Make the following tests of hypotheses.

a. H_0: $\mu = 25$, H_1: $\mu \neq 25$, $n = 81$, $\bar{x} = 28.5$, $\sigma = 3$, $\alpha = .01$

b. H_0: $\mu = 12$, H_1: $\mu < 12$, $n = 45$, $\bar{x} = 11.25$, $\sigma = 4.5$, $\alpha = .05$

c. H_0: $\mu = 40$, H_1: $\mu > 40$, $n = 100$, $\bar{x} = 47$, $\sigma = 7$, $\alpha = .10$

9.33 Make the following tests of hypotheses.

a. H_0: $\mu = 80$, H_1: $\mu \neq 80$, $n = 33$, $\bar{x} = 76.5$, $\sigma = 15$, $\alpha = .10$

b. H_0: $\mu = 32$, H_1: $\mu < 32$, $n = 75$, $\bar{x} = 26.5$, $\sigma = 7.4$, $\alpha = .01$

c. H_0: $\mu = 55$, H_1: $\mu > 55$, $n = 40$, $\bar{x} = 60.5$, $\sigma = 4$, $\alpha = .05$

APPLICATIONS

9.34 A consumer advocacy group suspects that a local supermarket's 10-ounce packages of cheddar cheese actually weigh less than 10 ounces. The group took a random sample of 20 such packages and found that the mean weight for the sample was 9.955 ounces. The population follows a normal distribution with the population standard deviation of .15 ounces.

a. Find the p-value for the test of hypothesis with the alternative hypothesis that the mean weight of all such packages is less than 10 ounces. Will you reject the null hypothesis at $\alpha = .01$?

b. Test the hypothesis of part a using the critical-value approach and $\alpha = .01$.

9.35 The manufacturer of a certain brand of auto batteries claims that the mean life of these batteries is 45 months. A consumer protection agency that wants to check this claim took a random sample of 24 such batteries and found that the mean life for this sample is 43.05 months. The lives of all such batteries have a normal distribution with the population standard deviation of 4.5 months.

a. Find the p-value for the test of hypothesis with the alternative hypothesis that the mean life of these batteries is less than 45 months. Will you reject the null hypothesis at $\alpha = .025$?

b. Test the hypothesis of part a using the critical-value approach and $\alpha = .025$.

9.36 A study claims that all adults spend an average of 14 hours or more on chores during a weekend. A researcher wanted to check if this claim is true. A random sample of 200 adults taken by this researcher showed that these adults spend an average of 14.65 hours on chores during a weekend. The population standard deviation is known to be 3.0 hours.

a. Find the p-value for the hypothesis test with the alternative hypothesis that all adults spend more than 14 hours on chores during a weekend. Will you reject the null hypothesis at $\alpha = .01$?

b. Test the hypothesis of part a using the critical-value approach and $\alpha = .01$.

9.37 According to data from the National Association of Home Builders, the average size of new homes in the United States was 2320 square feet in 2002 (*Money*, June 2004). Suppose a recent random sample of 400 new homes produced a mean size of 2365 square feet. The population standard deviation of the sizes of homes is known to be 312 square feet.

a. Find the p-value for the hypothesis test with the alternative hypothesis that the current mean size of all new homes in the United States exceeds 2320 square feet. Will you reject the null hypothesis at $\alpha = .02$?

b. Test the hypothesis of part a using the critical-value approach and $\alpha = .02$.

9.38 A 2003 study led by Reid Ewing, research professor at the National Center for Smart Growth at the University of Maryland, examined data on over 200,000 Americans living in 448 well-populated counties in the United States. The study showed that people living in the more densely-populated counties tended to weigh less than those living in less densely-populated counties, perhaps because residents of high-density counties were less dependent on vehicles and tended to walk more. In fact, in the 25 most densely-populated counties included in the study, people walked an average of 254 minutes per month, compared to an average of 191 minutes per month in the 25 least densely-populated counties (*Time*, June 7, 2004). Suppose a recent random sample of 400 people from these 25 most densely-populated counties found that they walked an average of 246 minutes per month. The population standard deviation is known to be 64 minutes.

a. Find the p-value for the test with the alternative hypothesis that the current mean time spent walking per month by such people differs from 254 minutes. If $\alpha = .01$, would you reject the null hypothesis based on the p-value you calculated? Explain. What if $\alpha = .02$?
b. Use the critical-value approach to make the hypothesis test of part a. Would you reject the null hypothesis at $\alpha = .01$? What if $\alpha = .02$?

9.39 A telephone company claims that the mean duration of all long-distance phone calls made by its residential customers is 10 minutes. A random sample of 100 long-distance calls made by its residential customers taken from the records of this company showed that the mean duration of calls for this sample is 9.20 minutes. The population standard deviation is known to be 3.80 minutes.

a. Find the p-value for the test that the mean duration of all long-distance calls made by residential customers is different from 10 minutes. If $\alpha = .02$, based on this p-value, would you reject the null hypothesis? Explain. What if $\alpha = .05$?
b. Test the hypothesis of part a using the critical-value approach and $\alpha = .02$. Does your conclusion change if $\alpha = .05$?

9.40 Lazurus Steel Corporation produces iron rods that are supposed to be 36 inches long. The machine that makes these rods does not produce each rod exactly 36 inches long. The lengths of the rods are normally distributed and they vary slightly. It is known that when the machine is working properly, the mean length of the rods is 36 inches. The standard deviation of the lengths of all rods produced on this machine is always equal to .035 inches. The quality control department at the company takes a sample of 20 such rods every week, calculates the mean length of these rods, and tests the null hypothesis $\mu = 36$ inches against the alternative hypothesis $\mu \neq 36$ inches. If the null hypothesis is rejected, the machine is stopped and adjusted. A recent sample of 20 rods produced a mean length of 36.015 inches.

a. Calculate the p-value for this test of hypothesis. Based on this p-value, will the quality control inspector decide to stop the machine and adjust it if he chooses the maximum probability of a Type I error to be .02? What if the maximum probability of a Type I error is .10?
b. Test the hypothesis of part a using the critical-value approach and $\alpha = .02$. Does the machine need to be adjusted? What if $\alpha = .10$?

9.41 At Farmer's Dairy, a machine is set to fill 32-ounce milk cartons. However, this machine does not put exactly 32 ounces of milk into each carton; the amount varies slightly from carton to carton but has a normal distribution. It is known that when the machine is working properly, the mean net weight of these cartons is 32 ounces. The standard deviation of the milk in all such cartons is always equal to .15 ounces. The quality control inspector at this company takes a sample of 25 such cartons every week, calculates the mean net weight of these cartons, and tests the null hypothesis $\mu = 32$ ounces against the alternative hypothesis $\mu \neq 32$ ounces. If the null hypothesis is rejected, the machine is stopped and adjusted. A recent sample of 25 such cartons produced a mean net weight of 31.93 ounces.

a. Calculate the p-value for this test of hypothesis. Based on this p-value, will the quality control inspector decide to stop the machine and readjust it if she chooses the maximum probability of a Type I error to be .01? What if the maximum probability of a Type I error is .05?
b. Test the hypothesis of part a using the critical-value approach and $\alpha = .01$. Does the machine need to be adjusted? What if $\alpha = .05$?

9.42 A study conducted a few years ago claims that adult men spend an average of 11 hours a week watching sports on television. A recent sample of 100 adult men showed that the mean time they spend per week watching sports on television is 9 hours. The population standard deviation is given to be 2.2 hours.

a. Test at the 1% significance level whether currently all adult men spend less than 11 hours per week watching sports on television.
b. What will your decision be in part a if the probability of making a Type I error is zero? Explain.

9.43 A restaurant franchise company has a policy of opening new restaurants only in those areas that have a mean household income of at least $35,000 per year. The company is currently considering an area to open a new restaurant. The company's research department took a sample of 150 households from this area and found that the mean income of these households is $33,400 per year. The population standard deviation of incomes is known to be $5400.

a. Using the 1% significance level, would you conclude that the company should not open a restaurant in this area?
b. What will your decision be in part a if the probability of making a Type I error is zero? Explain.

9.44 A journalist claims that all adults in her city spend an average of 30 hours or more per month on general reading, such as newspapers, magazines, novels, and so forth. A recent sample of 25 adults from this city showed that they spend an average of 27 hours per month on general reading. The population of such times is known to be normally distributed with the population standard deviation of 7 hours.

a. Using the 2.5% significance level, would you conclude that the mean time spent per month on such reading by all adults in this city is less than 30 minutes? Use both procedures—the p-value approach and the critical value approach.

b. Make the test of part a using the 1% significance level. Is your decision different from that of part a? Comment on the results of parts a and b.

9.45 A study claims that all homeowners in a town spend an average of 8 hours or more on house cleaning and gardening during a weekend. A researcher wanted to check if this claim is true. A random sample of 20 homeowners taken by this researcher showed that they spend an average of 7.68 hours on such chores during a weekend. The population of such times for all homeowners in this town is normally distributed with the population standard deviation of 2.1 hours.

a. Using the 1% significance level, can you conclude that the claim that all homeowners spend an average of 8 hours or more on such chores during a weekend is false? Use both approaches.

b. Make the test of part a using a 2.5% significance level. Is your decision different from the one in part a? Comment on the results of parts a and b.

9.46 A company claims that the mean net weight of the contents of its All Taste cereal boxes is at least 18 ounces. Suppose you want to test whether or not the claim of the company is true. Explain briefly how you would conduct this test using a large sample. Assume that $\sigma = .25$ ounces.

9.3 Hypothesis Tests About μ: σ Not Known

This section explains how to perform a test of hypothesis about the population mean μ when the population standard deviation σ is not known. Here, again, there are three possible cases that are mentioned below.

Case I. If the following three conditions are fulfilled:

1. The population standard deviation σ is not known
2. The sample size is small (i.e., $n < 30$)
3. The population from which the sample is selected is normally distributed,

then we use the t distribution to perform a test of hypothesis about μ.

Case II. If the following two conditions are fulfilled:

1. The population standard deviation σ is not known
2. The sample size is large (i.e., $n \geq 30$),

then again we use the t distribution to perform a test of hypothesis about μ.

Case III. If the following three conditions are fulfilled:

1. The population standard deviation σ is not known
2. The sample size is small (i.e., $n < 30$)
3. The population from which the sample is selected is not normally distributed (or the shape of its distribution is unknown),

then we use a nonparametric method to perform a test of hypothesis about μ.

The following chart summarizes the above three cases.

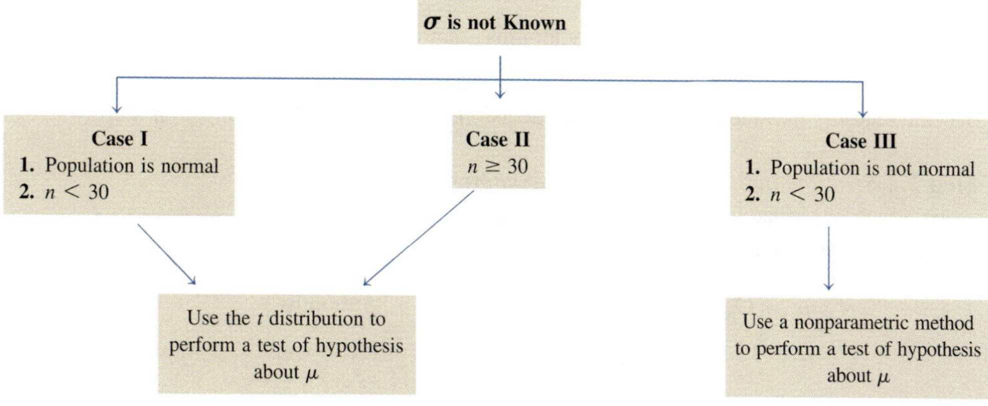

Below we discuss Cases I and II and learn how to use the t distribution to perform a test of hypothesis about μ when σ is not known. When conditions mentioned above for Case I or Case II are satisfied, the random variable

$$t = \frac{\bar{x} - \mu}{s_{\bar{x}}} \quad \text{where} \quad s_{\bar{x}} = \frac{s}{\sqrt{n}}$$

has a t distribution. Here, the t is called the **test statistic** to perform a test of hypothesis about a population mean μ.

Test Statistic The value of the *test statistic* t for the sample mean $\bar{x}$ is computed as

$$t = \frac{\bar{x} - \mu}{s_{\bar{x}}} \quad \text{where} \quad s_{\bar{x}} = \frac{s}{\sqrt{n}}$$

The value of t calculated for $\bar{x}$ by using this formula is also called the **observed value** of t.

In Section 9.2, we discussed two procedures, the p-value approach and the critical-value approach, to test hypotheses about μ when σ is known. In this section also we will use these two procedures to test hypotheses about μ when σ is not known. The steps used in these procedures are the same as in Section 9.2. The only difference is that we will be using the t distribution in place of the normal distribution.

1. The p-Value Approach

To use the p-value approach to perform a test of hypothesis about μ using the t distribution, we will use the same four steps that we used in such a procedure in Section 9.2. Although the p-value can be obtained by using any technology very easily, we can use Table V of Appendix C to find a **range for the p-value** when technology is not available. Note that when using the t distribution and using Table V, we cannot find the exact p-value but only a range within which it falls.

Examples 9–5 and 9–6 illustrate the p-value procedure to test a hypothesis about μ using the t distribution.

■ EXAMPLE 9–5

Finding a p-value and making a decision for a two-tailed test of hypothesis about μ: σ not known, $n < 30$, and population normal.

A psychologist claims that the mean age at which children start walking is 12.5 months. Carol wanted to check if this claim is true. She took a random sample of 18 children and found that the mean age at which these children started walking was 12.9 months with a standard deviation of .80 months. It is known that the ages at which all children start walking are approximately normally distributed. Find the p-value for the test that the mean age at which all children start walking is different from 12.5 months. What will your conclusion be if the significance level is 1%?

Solution Let μ be the mean age at which all children start walking, and $\bar{x}$ be the corresponding mean for the sample. From the given information,

$$n = 18, \quad \bar{x} = 12.9 \text{ months}, \quad \text{and} \quad s = .80 \text{ months}$$

The claim of the psychologist is that the mean age at which children start walking is 12.5 months. To calculate the p-value and to make the decision, we apply the following four steps.

Step 1. *State the null and alternative hypotheses.*

We are to test if the mean age at which all children start walking is different from 12.5 months. Hence, the null and alternative hypotheses are

$$H_0\text{: } \mu = 12.5 \quad \text{(The mean walking age is 12.5 months.)}$$

$$H_1\text{: } \mu \neq 12.5 \quad \text{(The mean walking age is different from 12.5 months.)}$$

Step 2. *Select the distribution to use.*

In this example, we do not know the population standard deviation σ, the sample size is small ($n < 30$), and the population is normally distributed. Hence, it is Case I mentioned in the beginning of this section. Consequently, we will use the t distribution to find the p-value for this test.

Step 3. *Calculate the p-value.*

The $\neq$ sign in the alternative hypothesis indicates that the test is two-tailed. To find the p-value, first we find the degrees of freedom and the t for $\bar{x} = 12.9$ months. Then, the p-value is equal to twice the area in the tail of the t distribution curve to the right of this t for $\bar{x} = 12.9$ months. This p-value is shown in Figure 9.11. We find this p-value as follows.

$$s_{\bar{x}} = \frac{s}{\sqrt{n}} = \frac{.80}{\sqrt{18}} = .18856181$$

$$t = \frac{\bar{x} - \mu}{s_{\bar{x}}} = \frac{12.9 - \overset{\text{From } H_0}{12.5}}{.18856181} = 2.121$$

and

$$df = n - 1 = 18 - 1 = 17$$

Now we can find the range for the p-value. To do so, we go to Table V of Appendix C (the t distribution table) and find the row of $df = 17$. In this row, we find the two values of t that cover $t = 2.121$. From Table V, for $df = 17$, these two values of t are 2.110 and 2.567. The test statistic $t = 2.121$ falls between these two values. Now look in the top row of this table to find the areas in the tail of the t distribution curve that correspond to 2.110 and 2.567. These two areas are .025 and .01, respectively. In other words, the area in the upper tail of the t distribution curve for $df = 17$ and $t = 2.110$ is .025, and the area in the upper tail of the t distribution curve for $df = 17$ and $t = 2.567$ is .01. Because it is a two-tailed test, the p-value for $t = 2.121$ is between $2(.025) = .05$ and $2(.01) = .02$, which can be written as:

$$.02 < p\text{-value} < .05$$

Note that by using Table V of Appendix C, we cannot find the exact p-value but only a range for it. If we have access to technology, we can find the exact p-value by using technology. If we use technology for this example, we will obtain a p-value of .049.

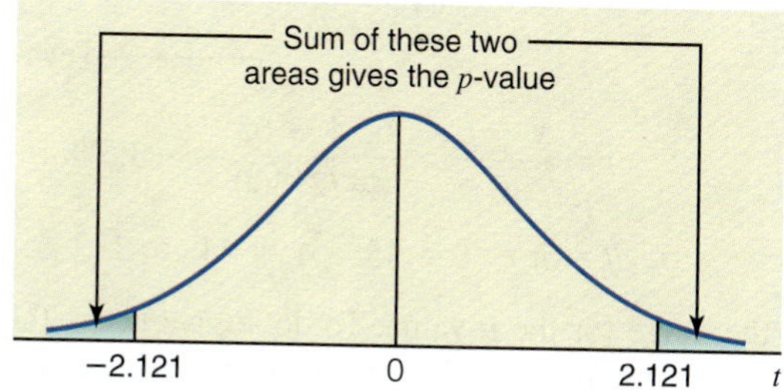

Figure 9.11 The required p-value.

Step 4. *Make a decision.*

Thus, we can state that for any α greater than .05 (the upper limit of the p-value range), we will reject the null hypothesis. For any α less than .02 (the lower limit of the p-value range), we will not reject the null hypothesis. However, if α is between .02 and .05, we cannot make a decision. Note that if we use technology, then the p-value we will obtain for this example is .049 and we can make a decision for any value of α. For our example $\alpha = .01$, which is less than the lower limit of the p-value range of .02. As a result, we fail to reject H_0 and conclude that the mean age at which all children start walking is not different from 12.5 months. As a result, we can state that the difference between the hypothesized population mean and the sample mean is so small that it may have occurred because of sampling error. ■

■ EXAMPLE 9–6

Finding a p-value and making a decision for a left-tailed test of hypothesis about μ: σ not known and $n > 30$.

Grand Auto Corporation produces auto batteries. The company claims that its top-of-the-line Never Die batteries are good, on average, for at least 65 months. A consumer protection agency tested 45 such batteries to check this claim. It found that the mean life of these 45 batteries is 63.4 months and the standard deviation is 3 months. Find the p-value for the test that the mean life of all such batteries is less than 65 months. What will your conclusion be if the significance level is 2.5%?

Solution Let μ be the mean life of all such auto batteries, and $\bar{x}$ be the corresponding mean for the sample. From the given information,

$$n = 45, \quad \bar{x} = 63.4 \text{ months}, \quad \text{and} \quad s = 3 \text{ months}$$

The claim of the company is that the mean life of these batteries is at least 65 months. To calculate the p-value and to make the decision, we apply the following four steps.

Step 1. *State the null and alternative hypotheses.*

We are to test if the mean life of these batteries is at least 65 months. Hence, the null and alternative hypotheses are

H_0: $\mu \geq 65$ (The mean life of batteries is at least 65 months.)

H_1: $\mu < 65$ (The mean life of batteries is less than 65 months.)

Step 2. *Select the distribution to use.*

In this example, we do not know the population standard deviation σ and the sample size is large ($n > 30$). Hence, it is Case II mentioned in the beginning of this section. Consequently, we will use the t distribution to find the p-value for this test.

Step 3. *Calculate the p-value.*

The $<$ sign in the alternative hypothesis indicates that the test is left-tailed. To find the p-value, first we find the degrees of freedom and the t for $\bar{x} = 63.4$ months. Then, the p-value is given by the area in the tail of the t distribution curve to the left of this t for $\bar{x} = 63.4$ months. This p-value is shown in Figure 9.12. We find this p-value as follows.

$$s_{\bar{x}} = \frac{s}{\sqrt{n}} = \frac{3}{\sqrt{45}} = .44721360$$

$$t = \frac{\bar{x} - \mu}{s_{\bar{x}}} = \frac{63.4 - 65}{.44721360} = -3.578 \quad (\mu = 65 \text{ from } H_0)$$

and
$$df = n - 1 = 45 - 1 = 44$$

Now we can find the range for the p-value. To do so, we go to Table V of Appendix C (the t distribution table) and find the row of $df = 44$. In this row, we find the two values of t that cover $t = 3.578$. Note that we use the positive value of the test statistic t although our test statistic has a negative value. From Table V, for $df = 44$, the largest value of t is 3.286 for which the area in the tail of the t distribution is .001. This means that the area to the left of $t = -3.286$ is .001. Because -3.578 is smaller than -3.286, the area to the left of $t = -3.578$

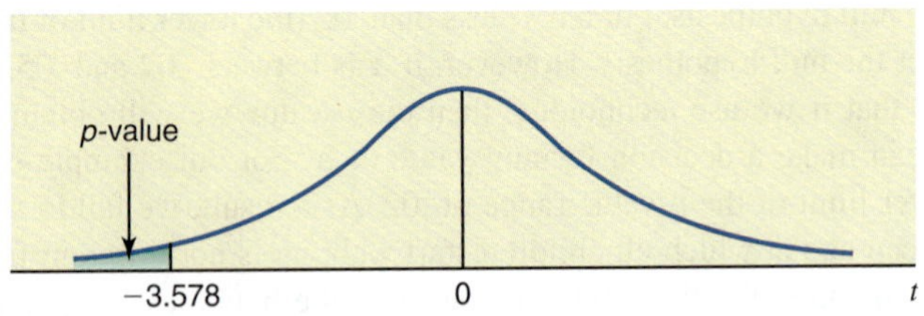

Figure 9.12 The required p-value.

is smaller than .001. Therefore, the p-value for $t = -3.578$ is less than .001, which can be written as:

$$p\text{-value} < .001$$

Thus, here the p-value has only the upper limit of .001. In other words, the p-value for this example is less than .001. If we use technology for this example, we will obtain a p-value of .000.

Step 4. *Make a decision.*

Thus, we can state that for any α greater than .001 (the upper limit of the p-value range), we will reject the null hypothesis. For our example $\alpha = .025$, which is greater than the upper limit of the p-value of .001. As a result, we reject H_0 and conclude that the mean life of such batteries is less than 65 months. Therefore, we can state that the difference between the hypothesized population mean of 65 months and the sample mean of 63.4 is too large to be attributed to sampling error alone. ■

2. The Critical-Value Approach

As mentioned in Section 9.2, this procedure is also called the traditional or classical approach. In this procedure, we have a predetermined value of the significance level α. The value of α gives the total area of the rejection region(s). First we find the critical value(s) of t from the t distribution table in Appendix C for the given degrees of freedom and the significance level. Then we find the value of the test statistic t for the observed value of the sample statistic $\bar{x}$. Finally we compare these two values and make a decision. Remember, if the test is one-tailed, there is only one critical value of t and it is obtained by using the value of α, which gives the area in the left or right tail of the t distribution curve depending on whether the test is left-tailed or right-tailed, respectively. However, if the test is two-tailed, there are two critical values of t and they are obtained by using $\alpha/2$ area in each tail of the t distribution curve. The value of the test statistic t is obtained as mentioned earlier in this section.

Examples 9–7 and 9–8 describe the procedure to test a hypothesis about μ using the critical-value approach and the t distribution.

■ EXAMPLE 9–7

Conducting a two-tailed test of hypothesis about μ: σ unknown, $n < 30$, and population normal.

Refer to Example 9–5. A psychologist claims that the mean age at which children start walking is 12.5 months. Carol wanted to check if this claim is true. She took a random sample of 18 children and found that the mean age at which these children started walking was 12.9 months with a standard deviation of .80 months. Using the 1% significance level, can you conclude that the mean age at which all children start walking is different from 12.5 months? Assume that the ages at which all children start walking have an approximately normal distribution.

Solution Let μ be the mean age at which all children start walking and $\bar{x}$ the corresponding mean for the sample. Then, from the given information,

$$n = 18, \quad \bar{x} = 12.9 \text{ months}, \quad s = .80 \text{ months}, \quad \text{and} \quad \alpha = .01$$

Step 1. *State the null and alternative hypotheses.*

We are to test if the mean age at which all children start walking is different from 12.5 months. The null and alternative hypotheses are

H_0: $\mu = 12.5$ (The mean walking age is 12.5 months.)

H_1: $\mu \neq 12.5$ (The mean walking age is different from 12.5 months.)

Step 2. *Select the distribution to use.*

In this example, the population standard deviation σ is not known, the sample size is small ($n < 30$), and the population is normally distributed. Hence, it is Case I mentioned in the beginning of this section. Consequently, we will use the t distribution to perform the test in this example.

Step 3. *Determine the rejection and nonrejection regions.*

The significance level is .01. The ≠ sign in the alternative hypothesis indicates that the test is two-tailed and the rejection region lies in both tails. The area of the rejection region in each tail of the t distribution curve is

$$\text{Area in each tail} = \alpha/2 = .01/2 = .005$$

$$df = n - 1 = 18 - 1 = 17$$

From the t distribution table, the critical values of t for 17 degrees of freedom and .005 area in each tail of the t distribution curve are −2.898 and 2.898. These values are shown in Figure 9.13.

Figure 9.13 The critical values of t.

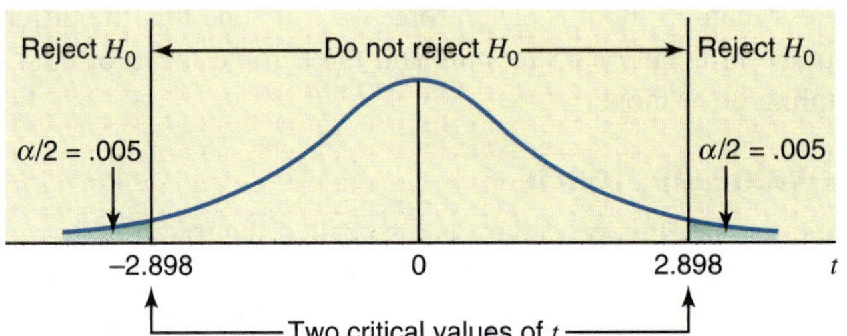

Step 4. *Calculate the value of the test statistic.*

We calculate the value of the test statistic t for $\bar{x} = 12.9$ as follows:

$$s_{\bar{x}} = \frac{s}{\sqrt{n}} = \frac{.80}{\sqrt{18}} = .18856181$$

From H_0

$$t = \frac{\bar{x} - \mu}{s_{\bar{x}}} = \frac{12.9 - 12.5}{.18856181} = 2.121$$

Step 5. *Make a decision.*

The value of the test statistic $t = 2.121$ falls between the two critical points, −2.898 and 2.898, which is the nonrejection region. Consequently, we fail to reject H_0. As a result, we can state that the difference between the hypothesized population mean and the sample mean is so small that it may have occurred because of sampling error. The mean age at which children start walking is not different from 12.5 months. ■

■ EXAMPLE 9–8

Conducting a right-tailed test of hypothesis about μ: σ unknown and $n > 30$.

The management at Massachusetts Savings Bank is always concerned about the quality of service provided to its customers. With the old computer system, a teller at this bank could serve, on average, 22 customers per hour. The management noticed that with this service rate, the waiting time for customers was too long. Recently the management of the bank installed a new computer system in the bank, expecting that it would increase the service rate and consequently make the customers happier by reducing the waiting time. To check if the new computer system is more efficient than the old system, the management of the bank took a random sample of 70 hours and found that during these hours the mean number of customers served by tellers was 27 per hour with a standard deviation of 2.5. Testing at the 1% significance level, would you conclude that the new computer system is more efficient than the old computer system?

Solution Let μ be the mean number of customers served per hour by a teller using the new system, and let $\bar{x}$ be the corresponding mean for the sample. Then, from the given information,

$$n = 70 \text{ hours}, \quad \bar{x} = 27 \text{ customers}, \quad s = 2.5 \text{ customers}, \quad \text{and} \quad \alpha = .01$$

Step 1. *State the null and alternative hypotheses.*

We are to test whether or not the new computer system is more efficient than the old system. The new computer system will be more efficient than the old system if the mean number of customers served per hour by using the new computer system is significantly more than 22; otherwise, it will not be more efficient. The null and alternative hypotheses are

H_0: $\mu = 22$ (The new computer system is not more efficient.)

H_1: $\mu > 22$ (The new computer system is more efficient.)

Step 2. *Select the distribution to use.*

In this example, the population standard deviation σ is not known and the sample size is large ($n > 30$). Hence, it is Case II mentioned in the beginning of this section. Consequently, we will use the t distribution to perform the test for this example.

Step 3. *Determine the rejection and nonrejection regions.*

The significance level is .01. The $>$ sign in the alternative hypothesis indicates that the test is right-tailed and the rejection region lies in the right tail of the t distribution curve.

$$\text{Area in the right tail} = \alpha = .01$$

$$df = n - 1 = 70 - 1 = 69$$

From the t distribution table, the critical value of t for 69 degrees of freedom and .01 area in the right tail is 2.382. This value is shown in Figure 9.14.

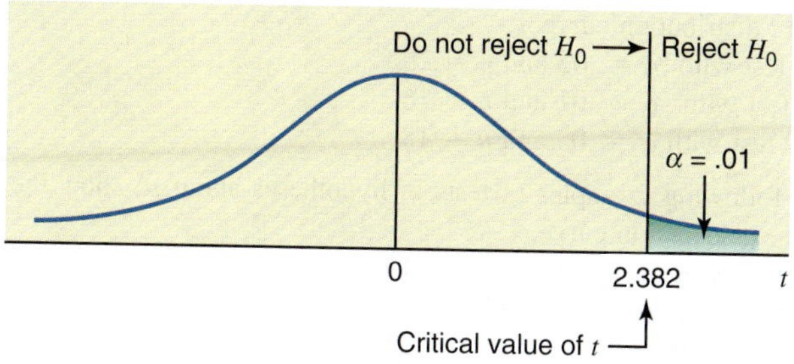

Figure 9.14

Step 4. *Calculate the value of the test statistic.*

The value of the test statistic t for $\bar{x} = 27$ is calculated as follows:

$$s_{\bar{x}} = \frac{s}{\sqrt{n}} = \frac{2.5}{\sqrt{70}} = .29880715$$

$$t = \frac{\bar{x} - \mu}{s_{\bar{x}}} = \frac{27 - \overset{\text{From } H_0}{\overset{\downarrow}{22}}}{.29880715} = 16.733$$

Step 5. *Make a decision.*

The value of the test statistic $t = 16.733$ is greater than the critical value of $t = 2.382$, and it falls in the rejection region. Consequently, we reject H_0. As a result, we conclude that the value of the sample mean is too large compared to the hypothesized value of the population mean, and the difference between the two may not be attributed to chance alone. The mean number of customers served per hour using the new computer system is more than 22. The new computer system is more efficient than the old computer system. ■

Note: What If the Sample Size Is Too Large?

In the above section when σ is not known, we used the t distribution to perform tests of hypothesis about μ in Cases I and II. Note that in Case II, the sample size is large. If we have access to technology, it does not matter how large (over 30) the sample size is, we can use the t distribution. However, if we are using the t distribution table (Table V of Appendix C), it may pose a problem. Usually such a table only goes up to a certain number of degrees of freedom. For example, Table V in Appendix C only goes up to 75 degrees of freedom. Thus, if the sample size is larger than 76 here, we cannot use Table V to find the critical value(s) of t to make a decision in this section. In such a situation when n is too large and is not included in the t distribution table, there are two options:

1. Use the t value from the last row (the row of ∞) in Table V of Appendix C.
2. Use the normal distribution as an approximation to the t distribution.

To use the normal distribution as an approximation to the t distribution to make a test of hypothesis about μ, the procedure is exactly like the one learned in Section 9.2 except that now we will replace σ by s, and $\sigma_{\bar{x}}$ by $s_{\bar{x}}$.

Note that the t-values obtained from the last row of the t distribution table are the same as will be obtained from the normal distribution table for the same areas in the upper tail or lower tail of the distribution. Again, note that here we can use the normal distribution as a convenience and as an approximation but if we can, we should use the t distribution by using technology. Exercises 9.70 and 9.71 at the end of this section present such situations.

EXERCISES

CONCEPTS AND PROCEDURES

9.47 Briefly explain the conditions that must hold true to use the t distribution to make a test of hypothesis about the population mean.

9.48 For each of the following examples of tests of hypotheses about μ, show the rejection and nonrejection regions on the t distribution curve.

a. A two-tailed test with $\alpha = .02$ and $n = 20$
b. A left-tailed test with $\alpha = .01$ and $n = 16$
c. A right-tailed test with $\alpha = .05$ and $n = 18$

9.49 For each of the following examples of tests of hypotheses about μ, show the rejection and nonrejection regions on the t distribution curve.

a. A two-tailed test with $\alpha = .01$ and $n = 15$
b. A left-tailed test with $\alpha = .005$ and $n = 25$
c. A right-tailed test with $\alpha = .025$ and $n = 22$

9.50 A random sample of 25 observations taken from a population that is normally distributed produced a sample mean of 58.5 and a standard deviation of 7.5. Find the ranges for the p-value and the critical and observed values of t for each of the following tests of hypotheses using $\alpha = .01$.

a. H_0: $\mu = 55$ versus H_1: $\mu > 55$
b. H_0: $\mu = 55$ versus H_1: $\mu \neq 55$

9.51 A random sample of 16 observations taken from a population that is normally distributed produced a sample mean of 42.4 and a standard deviation of 8. Find the ranges for the p-value and the critical and observed values of t for each of the following tests of hypotheses using $\alpha = .05$.

a. H_0: $\mu = 46$ versus H_1: $\mu < 46$
b. H_0: $\mu = 46$ versus H_1: $\mu \neq 46$

9.52 Consider the null hypothesis H_0: $\mu = 70$ about the mean of a population. Suppose a random sample of 60 observations is taken from this population to make this test. Using $\alpha = .01$, show the rejection and nonrejection regions and find the critical value(s) of t for a

a. left-tailed test b. two-tailed test c. right-tailed test

9.53 Consider the null hypothesis H_0: $\mu = 35$ about the mean of a population. Suppose a random sample of 52 observations is taken from this population to make this test. Using $\alpha = .05$, show the rejection and nonrejection regions and find the critical value(s) of t for a

a. left-tailed test b. two-tailed test c. right-tailed test

9.54 Consider H_0: $\mu = 80$ versus H_1: $\mu \neq 80$ for a population that is normally distributed.

a. A random sample of 25 observations taken from this population produced a sample mean of 77 and a standard deviation of 8. Using $\alpha = .01$, would you reject the null hypothesis?
b. Another random sample of 25 observations taken from the same population produced a sample mean of 86 and a standard deviation of 6. Using $\alpha = .01$, would you reject the null hypothesis?

Comment on the results of parts a and b.

9.55 Consider H_0: $\mu = 40$ versus H_1: $\mu > 40$.

a. A random sample of 64 observations taken from this population produced a sample mean of 43 and a standard deviation of 5. Using $\alpha = .025$, would you reject the null hypothesis?

b. Another random sample of 64 observations taken from the same population produced a sample mean of 41 and a standard deviation of 7. Using $\alpha = .025$, would you reject the null hypothesis?

Comment on the results of parts a and b.

9.56 Perform the following hypothesis tests.

a. H_0: $\mu = 24$, H_1: $\mu \neq 24$, $n = 35$, $\bar{x} = 28.5$, $s = 4.9$, $\alpha = .01$

b. H_0: $\mu = 30$, H_1: $\mu < 30$, $n = 56$, $\bar{x} = 28.5$, $s = 6.6$, $\alpha = .025$

c. H_0: $\mu = 18$, H_1: $\mu > 18$, $n = 40$, $\bar{x} = 22.5$, $s = 8$, $\alpha = .10$

9.57 Assuming that the respective populations are normally distributed, perform the following hypothesis tests.

a. H_0: $\mu = 60$, H_1: $\mu \neq 60$, $n = 14$, $\bar{x} = 57$, $s = 9$, $\alpha = .05$

b. H_0: $\mu = 35$, H_1: $\mu < 35$, $n = 24$, $\bar{x} = 28$, $s = 5.4$, $\alpha = .005$

c. H_0: $\mu = 47$, H_1: $\mu > 47$, $n = 18$, $\bar{x} = 50$, $s = 6$, $\alpha = .001$

APPLICATIONS

9.58 According to a basketball coach, the mean height of all female college basketball players is 69.5 inches. A random sample of 25 such players produced a mean height of 70.25 inches with a standard deviation of 2.1 inches. Assuming that the heights of all female college basketball players are normally distributed, test at the 2% significance level whether their mean height is different from 69.5 inches. Find the range for the p-value for this test. What will your conclusion be using this p-value range and $\alpha = .01$?

9.59 According to data from Jury Verdict Research, the average jury award in vehicular liability cases was \$220,680 in 2002 (*The Hartford Courant,* February 10, 2005). Suppose in a recent random sample of 16 such cases, the average jury award was \$262,570, with a standard deviation of \$34,256. Assume that jury awards for all such cases have a normal distribution. Using the 1% significance level and the critical value approach, can you conclude that the mean of such awards currently exceeds \$220,680? Find the range for the p-value for this test. What will your conclusion be using this p-value range and $\alpha = .01$?

9.60 The president of a university claims that the mean time spent partying by all students at this university is not more than 7 hours per week. A random sample of 40 students taken from this university showed that they spent an average of 9.50 hours partying the previous week with a standard deviation of 2.3 hours. Test at the 2.5% significance level whether the president's claim is true. Explain your conclusion in words.

9.61 The mean balance of all checking accounts at a bank on December 31, 2005, was \$850. A random sample of 55 checking accounts taken recently from this bank gave a mean balance of \$780 with a standard deviation of \$230. Using the 1% significance level, can you conclude that the mean balance of such accounts has decreased during this period? Explain your conclusion in words. What if $\alpha = .025$?

9.62 A soft-drink manufacturer claims that its 12-ounce cans do not contain, on average, more than 30 calories. A random sample of 64 cans of this soft drink, which were checked for calories, contained a mean of 32 calories with a standard deviation of 3 calories. Does the sample information support the alternative hypothesis that the manufacturer's claim is false? Use a significance level of 5%. Find the range for the p-value for this test. What will your conclusion be using this p-value and $\alpha = .05$?

9.63 According to *USA TODAY* research, the average personal debt (such as loans on cars, credit cards, and so forth) per household in the United States was \$17,989 in 2004 (*USA TODAY,* October 4, 2004). A recent random sample of 75 households from New Hampshire yielded a mean personal debt of \$16,450 with a standard deviation of \$4650. Using the 2% significance level, can you conclude that the current mean personal debt for all households in New Hampshire is different from \$17,989? Use both the p-value approach and the critical-value approach.

9.64 A paint manufacturing company claims that the mean drying time for its paints is not longer than 45 minutes. A random sample of 20 gallons of paints selected from the production line of this company showed that the mean drying time for this sample is 49.50 minutes with a standard deviation of 3 minutes. Assume that the drying times for these paints have a normal distribution.

a. Using the 1% significance level, would you conclude that the company's claim is true?

b. What is the Type I error in this exercise? Explain in words. What is the probability of making such an error?

9.65 The manager of a restaurant in a large city claims that waiters working in all restaurants in his city earn an average of \$150 or more in tips per week. A random sample of 25 waiters selected from restaurants of this city yielded a mean of \$139 in tips per week with a standard deviation of \$28. Assume that the weekly tips for all waiters in this city have a normal distribution.

a. Using the 1% significance level, can you conclude that the manager's claim is true? Use both approaches.
b. What is the Type I error in this exercise? Explain. What is the probability of making such an error?

9.66 A business school claims that students who complete a three-month typing course can type, on average, at least 1200 words an hour. A random sample of 25 students who completed this course typed, on average, 1125 words an hour with a standard deviation of 85 words. Assume that the typing speeds for all students who complete this course have an approximately normal distribution.

a. Suppose the probability of making a Type I error is selected to be zero. Can you conclude that the claim of the business school is true? Answer without performing the five steps of a test of hypothesis.
b. Using the 5% significance level, can you conclude that the claim of the business school is true? Use both approaches.

9.67 According to an estimate, two years ago the average age of all CEOs of medium-sized companies in the United States was 58 years. Jennifer wants to check if this is still true. She took a random sample of 70 such CEOs and found their mean age to be 55 years with a standard deviation of 6 years.

a. Suppose that the probability of making a Type I error is selected to be zero. Can you conclude that the current mean age of all CEOs of medium-sized companies in the Untied States is different from 58 years?
b. Using the 1% significance level, can you conclude that the current mean age of all CEOs of medium-sized companies in the United States is different from 58 years? Use both approaches.

9.68 A past study claims that adults in America spend an average of 18 hours a week on leisure activities. A researcher wanted to test this claim. She took a sample of 10 adults and asked them about the time they spend per week on leisure activities. Their responses (in hours) are as follows.

14	25	22	38	16	26	19	23	41	33

Assume that the times spent on leisure activities by all adults are normally distributed. Using the 5% significance level, can you conclude that the claim of the earlier study is true? (*Hint*: First calculate the sample mean and the sample standard deviation for these data using the formulas learned in Sections 3.1.1 and 3.2.2 of Chapter 3. Then make the test of hypothesis about μ.)

9.69 The past records of a supermarket show that its customers spend an average of \$65 per visit at this store. Recently the management of the store initiated a promotional campaign according to which each customer receives points based on the total money spent at the store, and these points can be used to buy products at the store. The management expects that as a result of this campaign, the customers should be encouraged to spend more money at the store. To check whether this is true, the manager of the store took a sample of 12 customers who visited the store. The following data give the money (in dollars) spent by these customers at this supermarket during their visits.

88	69	141	28	106	45	32	51	78	54	110	83

Assume that the money spent by all customers at this supermarket has a normal distribution. Using the 1% significance level, can you conclude that the mean amount of money spent by all customers at this supermarket after the campaign was started is more than \$65? (*Hint*: First calculate the sample mean and the sample standard deviation for these data using the formulas learned in Sections 3.1.1 and 3.2.2 of Chapter 3. Then make the test of hypothesis about μ.)

9.70 According to *PARADE* magazine, the average salary for nurses in the United States was \$54,574 in 2004 (*PARADE*, March 13, 2005). A recent random sample of 1000 nurses yielded a mean salary of \$56,300 with a standard deviation of \$6500. Does the sample information support the alternative hypothesis that the current mean salary for nurses exceeds \$54,574? Use $\alpha = .025$. Use both the p-value approach and the critical-value approach to make a decision.

9.71 According to a Kaiser Family Foundation survey, children aged 8 to 18 years in the United States spend an average of 231 minutes per day watching television (*USA TODAY*, March 10, 2005). A recent random sample of 800 such children showed that they spend an average of 250 minutes per day watching television with a standard deviation of 55 minutes. At the 2% level of significance, can you conclude that the current mean time spent watching television per day by children in this age group differs from 231 minutes? Use both the p-value approach and the critical-value approach to make a decision.

***9.72** The manager of a service station claims that the mean amount spent on gas by its customers is \$15.90. You want to test if the mean amount spent on gas at this station is different from \$15.90. Briefly explain how you would conduct this test when σ is not known.

***9.73** A tool manufacturing company claims that its top-of-the-line machine that is used to manufacture bolts produces an average of 88 or more bolts per hour. A company that is interested in buying this machine wants to check this claim. Suppose you are asked to conduct this test. Briefly explain how you would do so when σ is not known.

9.4 Hypothesis Tests About a Population Proportion: Large Samples

Often we want to conduct a test of hypothesis about a population proportion. For example, 33% of the students listed in *Who's Who Among American High School Students* said that drugs and alcohol are the most serious problems facing their high schools. A sociologist may want to check if this percentage still holds. As another example, a mail-order company claims that 90% of all orders it receives are shipped within 72 hours. The company's management may want to determine from time to time whether or not this claim is true.

This section presents the procedure to perform tests of hypotheses about the population proportion, p, for large samples. The procedures to make such tests are similar in many respects to the ones for the population mean, μ. Again, the test can be two-tailed or one-tailed. We know from Chapter 7 that when the sample size is large, the sample proportion, $\hat{p}$, is approximately normally distributed with its mean equal to p and standard deviation equal to $\sqrt{pq/n}$. Hence, we use the normal distribution to perform a test of hypothesis about the population proportion, p, for a large sample. As was mentioned in Chapters 7 and 8, in the case of a proportion, the sample size is considered to be large when np and nq are both greater than 5.

Test Statistic The value of the *test statistic* z for the sample proportion, $\hat{p}$, is computed as

$$z = \frac{\hat{p} - p}{\sigma_{\hat{p}}} \quad \text{where} \quad \sigma_{\hat{p}} = \sqrt{\frac{pq}{n}}$$

The value of p used in this formula is the one used in the null hypothesis. The value of q is equal to $1 - p$.

The value of z calculated for $\hat{p}$ using the above formula is also called the **observed value of** z.

In Sections 9.2 and 9.3, we discussed two procedures, the p-value approach and the critical-value approach, to test hypotheses about μ. Here too we will use these two procedures to test hypotheses about p. The steps used in these procedures are the same as in Sections 9.2 and 9.3. The only difference is that we will be making tests of hypotheses about p rather than about μ.

1. The *p*-Value Approach

To use the p-value approach to perform a test of hypothesis about p, we will use the same four steps that we used in such a procedure in Section 9.2. Although the p-value for a test of hypothesis about p can be obtained very easily by using technology, we can use Table IV of Appendix C to find this p-value when technology is not available.

Examples 9–9 and 9–10 illustrate the p-value procedure to test a hypothesis about p for a large sample.

■ EXAMPLE 9–9

Finding a p-value and making a decision for a two-tailed test of hypothesis about p: large sample.

In a *Time* magazine poll of adult Americans conducted by telephone March 15–17, 2005 by SRBI Public Affairs, 66% of the respondents said that there is too much violence on television (*Time*, March 28, 2005). Assume that this result holds true for the 2005 population of all adult Americans. In a recent random sample of 1000 adult Americans, 70% said that there is too much violence on television. Find the p-value to test the hypothesis that the current percentage of adult Americans who think there is too much violence on television is different from that for 2005. If the significance level is 2%, what is your conclusion?

Solution Let p be the current proportion of all adult Americans who think there is too much violence on television, and $\hat{p}$ be the corresponding sample proportion. Then, from the given information,

$$n = 1000, \quad \hat{p} = .70, \quad \text{and} \quad \alpha = .02$$

In 2005, 66% of adult Americans thought there was too much violence on television. Hence,

$$p = .66 \quad \text{and} \quad q = 1 - p = 1 - .66 = .34$$

To calculate the p-value and to make a decision, we apply the following four steps.

Step 1. *State the null and alternative hypotheses.*

The current percentage of all adult Americans who think there is too much violence on television will not be different from that for 2005 if $p = .66$, and the current percentage will be different from that for 2005 if $p \neq .66$. The null and alternative hypotheses are as follows.

H_0: $p = .66$ (The current percentage is not different from 2005.)

H_1: $p \neq .66$ (The current percentage is different from 2005.)

Step 2. *Select the distribution to use.*

To check if the sample is large, we calculate the values of np and nq.

$$np = 1000(.66) = 660 \quad \text{and} \quad nq = 1000(.34) = 340$$

Since np and nq are both greater than 5, we can conclude that the sample size is large. Consequently, we will use the normal distribution to find the p-value for this test.

Step 3. *Calculate the p-value.*

The $\neq$ sign in the alternative hypothesis indicates that the test is two-tailed. The p-value is equal to twice the area in the tail of the normal distribution curve to the right of z for $\hat{p} = .70$. This p-value is shown in Figure 9.15. To find this p-value, first we find the test statistic z for $\hat{p} = .70$ as follows.

$$\sigma_{\hat{p}} = \sqrt{\frac{pq}{n}} = \sqrt{\frac{(.66)(.34)}{1000}} = .01497999$$

$$z = \frac{\hat{p} - p}{\sigma_{\hat{p}}} = \frac{.70 - .66}{.01497999} = 2.67$$

(From H_0: .66)

Now we find the area to the right of $z = 2.67$ from the normal distribution table. This area is $1 - .9962 = .0038$. Consequently, the p-value is

$$p\text{-value} = 2(.0038) = .0076$$

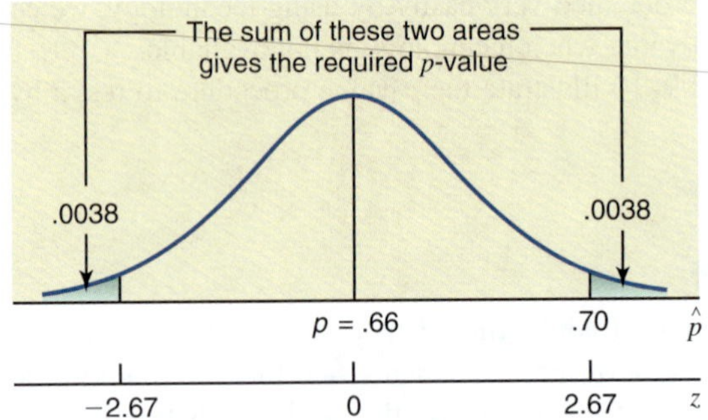

Figure 9.15 The required p-value.

Step 4. *Make a decision.*

Thus, we can state that for any α greater than .0076 we will reject the null hypothesis, and for any α less than or equal to .0076 we will not reject the null hypothesis. For our example

$\alpha = .02$, which is greater than the p-value of .0076. As a result, we reject H_0 and conclude that the current percentage of all adult Americans who think there is too much violence on television is different from .66. Therefore, we can state that the difference between the hypothesized population proportion of .66 and the sample proportion of .70 is too large to be attributed to sampling error alone. ■

■ EXAMPLE 9–10

Finding a p-value and making a decision for a right-tailed test of hypothesis about p: large sample.

When working properly, a machine that is used to make chips for calculators does not produce more than 4% defective chips. Whenever the machine produces more than 4% defective chips, it needs an adjustment. To check if the machine is working properly, the quality control department at the company often takes samples of chips and inspects them to determine if they are good or defective. One such random sample of 200 chips taken recently from the production line contained 12 defective chips. Find the p-value to test the hypothesis whether or not the machine needs an adjustment. What would your conclusion be if the significance level is 2.5%?

Solution Let p be the proportion of defective chips in all chips produced by this machine, and let $\hat{p}$ be the corresponding sample proportion. Then, from the given information,

$$n = 200, \quad \hat{p} = 12/200 = .06, \quad \text{and} \quad \alpha = .025$$

When the machine is working properly, it does not produce more than 4% defective chips. Hence, assuming that the machine is working properly,

$$p = .04 \quad \text{and} \quad q = 1 - p = 1 - .04 = .96$$

To calculate the p-value and to make a decision, we apply the following four steps.

Step 1. *State the null and alternative hypotheses.*

The machine will not need an adjustment if the percentage of defective chips is 4% or less, and it will need an adjustment if this percentage is greater than 4%. Hence, the null and alternative hypotheses are as follows.

$$H_0: p \leq .04 \quad \text{(The machine does not need an adjustment.)}$$

$$H_1: p > .04 \quad \text{(The machine needs an adjustment.)}$$

Step 2. *Select the distribution to use.*

To check if the sample is large, we calculate the values of np and nq.

$$np = 200(.04) = 8 \quad \text{and} \quad nq = 200(.96) = 192$$

Since np and nq are both greater than 5, we can conclude that the sample size is large. Consequently, we will use the normal distribution to find the p-value for this test.

Step 3. *Calculate the p-value.*

The > sign in the alternative hypothesis indicates that the test is right-tailed. The p-value is given by the area in the upper tail of the normal distribution curve to the right of z for $\hat{p} = .06$. This p-value is shown in Figure 9.16. To find this p-value, first we find the test statistic z for $\hat{p} = .06$ as follows.

$$\sigma_{\hat{p}} = \sqrt{\frac{pq}{n}} = \sqrt{\frac{(.04)(.96)}{200}} = .01385641$$

From H_0

$$z = \frac{\hat{p} - p}{\sigma_{\hat{p}}} = \frac{.06 - .04}{.01385641} = 1.44$$

Now we find the area to the right of $z = 1.44$ from the normal distribution table. This area is $1 - .9251 = .0749$. Consequently, the p-value is

$$p\text{-value} = .0749$$

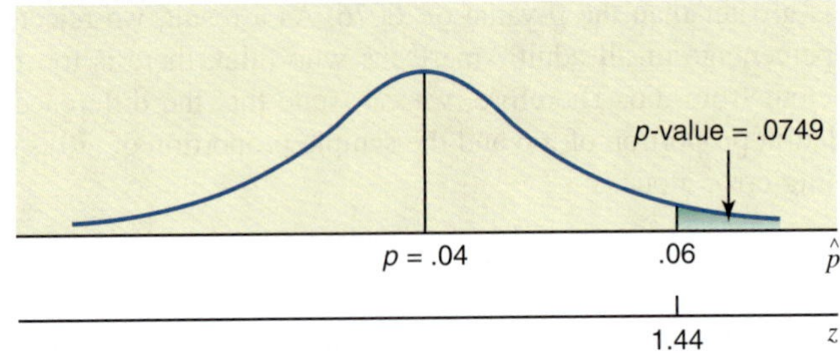

Figure 9.16 The required p-value.

Step 4. *Make a decision.*

Thus, we can state that for any α greater than .0749 we will reject the null hypothesis, and for any α less than or equal to .0749 we will not reject the null hypothesis. For our example $\alpha = .025$, which is less than the p-value of .0749. As a result, we fail to reject H_0 and conclude that the machine does not need an adjustment. ■

2. The Critical-Value Approach

As mentioned in Section 9.2, this procedure is also called the traditional or classical approach. In this procedure, we have a predetermined value of the significance level α. The value of α gives the total area of the rejection region(s). First we find the critical value(s) of z from the normal distribution table for the given significance level. Then we find the value of the test statistic z for the observed value of the sample statistic $\hat{p}$. Finally we compare these two values and make a decision. Remember, if the test is one-tailed, there is only one critical value of z and it is obtained by using the value of α, which gives the area in the left or right tail of the normal distribution curve depending on whether the test is left-tailed or right-tailed, respectively. However, if the test is two-tailed, there are two critical values of z and they are obtained by using $\alpha/2$ area in each tail of the normal distribution curve. The value of the test statistic z is obtained as mentioned earlier in this section.

Examples 9–11 and 9–12 describe the procedure to test a hypothesis about p using the critical-value approach and the normal distribution.

■ EXAMPLE 9–11

Making a two-tailed test of hypothesis about p using the critical-value approach: large sample.

Refer to Example 9–9. In a *Time* magazine poll of adult Americans conducted by telephone March 15–17, 2005 by SRBI Public Affairs, 66% of the respondents said that there is too much violence on television (*Time*, March 28, 2005). Assume that this result holds true for the 2005 population of all adult Americans. In a recent random sample of 1000 adult Americans, 70% said that there is too much violence on television. Using the 2% significance level, can you conclude that the current percentage of adult Americans who think there is too much violence on television is different from that for 2005?

Solution Let p be the current proportion of all adult Americans who think there is too much violence on television, and $\hat{p}$ be the corresponding sample proportion. Then, from the given information,

$$n = 1000, \quad \hat{p} = .70, \quad \text{and} \quad \alpha = .02$$

In 2005, 66% of adult Americans thought there was too much violence on television. Hence,

$$p = .66 \quad \text{and} \quad q = 1 - p = 1 - .66 = .34$$

To calculate the p-value and to make a decision, we apply the following four steps.

Step 1. *State the null and alternative hypotheses.*

The current percentage of all adult Americans who think there is too much violence on television will not be different from that for 2005 if $p = .66$, and the current percentage will be different from that for 2005 if $p \neq .66$. The null and alternative hypotheses are as follows.

$$H_0: p = .66 \quad \text{(The current percentage is not different from 2005.)}$$

$$H_1: p \neq .66 \quad \text{(The current percentage is different from 2005.)}$$

Step 2. *Select the distribution to use.*

To check if the sample is large, we calculate the values of np and nq.

$$np = 1000(.66) = 660 \quad \text{and} \quad nq = 1000(.34) = 340$$

Since np and nq are both greater than 5, we can conclude that the sample size is large. Consequently, we will use the normal distribution to make the test.

Step 3. *Determine the rejection and nonrejection regions.*

The $\neq$ sign in the alternative hypothesis indicates that the test is two-tailed. The significance level is .02. Therefore, the total area of the two rejection regions is .02 and the rejection region in each tail of the sampling distribution of $\hat{p}$ is $\alpha/2 = .02/2 = .01$. The critical values of z, obtained from the standard normal distribution table, are −2.33 and 2.33, as shown in Figure 9.17.

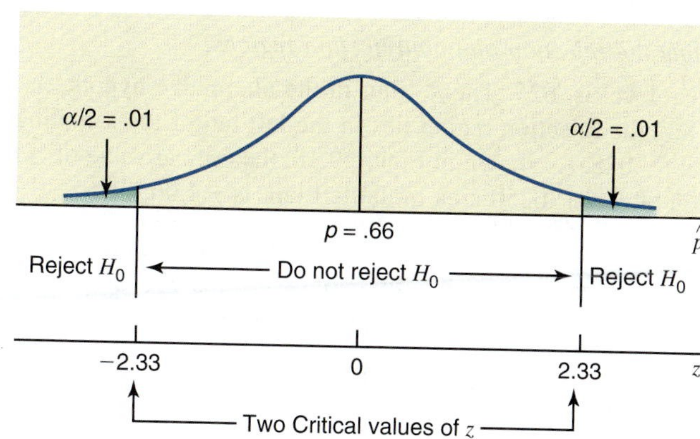

Figure 9.17 The critical values of z.

Step 4. *Calculate the value of the test statistic.*

The value of the test statistic z for $\hat{p} = .70$ is calculated as follows.

$$\sigma_{\hat{p}} = \sqrt{\frac{pq}{n}} = \sqrt{\frac{(.66)(.34)}{1000}} = .01497999$$

From H_0

$$z = \frac{\hat{p} - p}{\sigma_{\hat{p}}} = \frac{.70 - .66}{.01497999} = 2.67$$

Step 5. *Make a decision.*

The value of the test statistic $z = 2.67$ for $\hat{p}$ falls in the rejection region. As a result, we reject H_0 and conclude that the current percentage of all adult Americans who think there is too much violence on television is different from .66. As a result, we can state that the difference between the hypothesized population proportion of .66 and the sample proportion of .70 is too large to be attributed to sampling error alone. ■

■ EXAMPLE 9–12

Direct Mailing Company sells computers and computer parts by mail. The company claims that at least 90% of all orders are mailed within 72 hours after they are received. The quality control department at the company often takes samples to check if this claim is valid. A recently taken sample of 150 orders showed that 129 of them were mailed within 72 hours. Do you think the company's claim is true? Use a 2.5% significance level.

Conducting a left-tailed test of hypothesis about p using the critical-value approach: large sample.

Solution Let p be the proportion of all orders that are mailed by the company within 72 hours and $\hat{p}$ the corresponding sample proportion. Then, from the given information,

$$n = 150, \quad \hat{p} = 129/150 = .86, \quad \text{and} \quad \alpha = .025$$

The company claims that at least 90% of all orders are mailed within 72 hours. Assuming that this claim is true, the values of p and q are

$$p = .90 \quad \text{and} \quad q = 1 - p = 1 - .90 = .10$$

Step 1. *State the null and alternative hypotheses.*

The null and alternative hypotheses are

$$H_0: p \geq .90 \quad \text{(The company's claim is true.)}$$

$$H_1: p < .90 \quad \text{(The company's claim is false.)}$$

Step 2. *Select the distribution to use.*

We first check whether np and nq are both greater than 5.

$$np = 150(.90) = 135 > 5 \quad \text{and} \quad nq = 150(.10) = 15 > 5$$

Consequently, the sample size is large. Therefore, we use the normal distribution to make the hypothesis test about p.

Step 3. *Determine the rejection and nonrejection regions.*

The significance level is .025. The $<$ sign in the alternative hypothesis indicates that the test is left-tailed and the rejection region lies in the left tail of the sampling distribution of $\hat{p}$ with its area equal to .025. As shown in Figure 9.18, the critical value of z, obtained from the normal distribution table for .0250 area in the left tail, is -1.96.

Figure 9.18 Critical value of z.

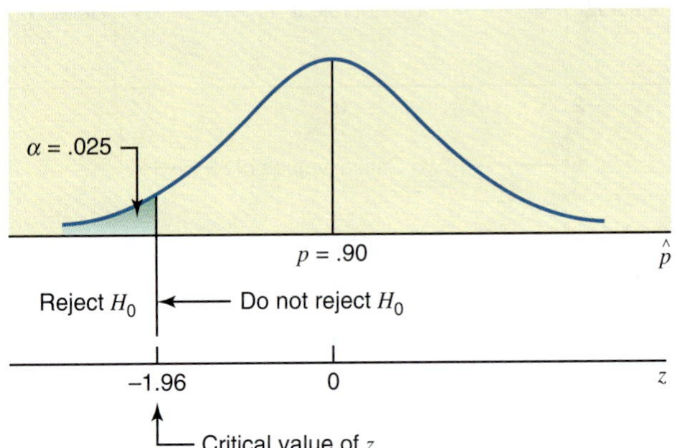

Step 4. *Calculate the value of the test statistic.*

The value of the test statistic z for $\hat{p} = .86$ is calculated as follows:

$$\sigma_{\hat{p}} = \sqrt{\frac{pq}{n}} = \sqrt{\frac{(.90)(.10)}{150}} = .02449490$$

From H_0

$$z = \frac{\hat{p} - p}{\sigma_{\hat{p}}} = \frac{.86 - .90}{.02449490} = -1.63$$

Step 5. *Make a decision.*

The value of the test statistic $z = -1.63$ is greater than the critical value of $z = -1.96$, and it falls in the nonrejection region. Therefore, we fail to reject H_0. We can state that the difference between the sample proportion and the hypothesized value of the population proportion is small and this difference may have occurred owing to chance alone. Therefore, the proportion of all orders that are mailed within 72 hours is at least 90%, and the company's claim is true. ■

9–2 COFFEE OR INTERNET?

USA TODAY Snapshots®

Coffee or Internet?

If you had to, which would you give up, your morning coffee or your ability to use the Internet at work?

Coffee 52%

Internet 44%

Neither or not sure 4%

Source: Websense survey of 500 respondents. Margin of error ±4 percentage points.

By Darryl Haralson and Keith Simmons, USA TODAY

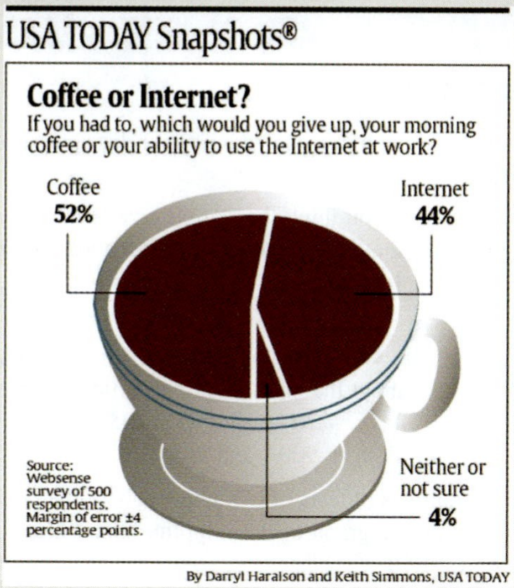

The above chart shows the percentage of people who will choose using the Internet for personal purposes over free morning coffee at work, and vice-versa, if they have to. In this survey, people were asked that if they have to choose between having morning coffee or being able to use the Internet at work, which one will they choose? Fifty-two percent of the respondents said that they will give up morning coffee to be able to use the Internet at work, but 44% said they will be willing to give up Internet use at work for their morning coffee. Note that these percentages are based on a sample survey of 500 workers.

Suppose that these results are true for the July 2005 population of workers, and that we want to check if the percentage of workers who are willing to give up their morning coffee in exchange for using the Internet at work is still 52%. Suppose we take a random sample of 1100 workers and ask them the same question. Of them, 55% say that they are willing to give up their morning coffee in exchange for using the Internet at work. Suppose the significance level is 1%. The test is two-tailed. The null and alternative hypotheses are

$$H_0: p = .52$$
$$H_1: p \neq .52$$

Here, $n = 1100$, $\hat{p} = .55$, and $\alpha = .01$. The sample is large. (The reader should check that np and nq are both greater than 5.) Using the normal distribution to make the test, the critical values of z for .0050 and .9950 areas to the left are -2.58 and 2.58. We find the observed value of z as follows.

$$\sigma_{\hat{p}} = \sqrt{\frac{pq}{n}} = \sqrt{\frac{(.52)(.48)}{1100}} = .01506350$$

$$z = \frac{\hat{p} - p}{\sigma_{\hat{p}}} = \frac{.55 - .52}{.01506350} = 1.99$$

The value of the test statistic $z = 1.99$ for $\hat{p}$ falls in the nonrejection region. Consequently, we fail to reject H_0 and conclude that the percentage of workers who are willing to give up their morning coffee in exchange for using the Internet at work is still 52%.

We can use the p-value approach too. From the normal distribution table, the area under the normal curve to the right of $z = 1.99$ is $1 - .9767 = .0233$. Therefore, the p-value is $2(.0233) = .0466$. Since $\alpha = .01$ is smaller than .0233, we fail to reject the null hypothesis.

Source: The chart reproduced with permission from *USA TODAY*, July 14, 2005. Copyright © 2005, *USA TODAY*.

EXERCISES

CONCEPTS AND PROCEDURES

9.74 Explain when a sample is large enough to use the normal distribution to make a test of hypothesis about the population proportion.

9.75 In each of the following cases, do you think the sample size is large enough to use the normal distribution to make a test of hypothesis about the population proportion? Explain why or why not.

a. $n = 40$ and $p = .11$ **b.** $n = 100$ and $p = .73$
c. $n = 80$ and $p = .05$ **d.** $n = 50$ and $p = .14$

9.76 In each of the following cases, do you think the sample size is large enough to use the normal distribution to make a test of hypothesis about the population proportion? Explain why or why not.

a. $n = 30$ and $p = .65$ **b.** $n = 70$ and $p = .05$
c. $n = 60$ and $p = .06$ **d.** $n = 900$ and $p = .17$

9.77 For each of the following examples of tests of hypotheses about the population proportion, show the rejection and nonrejection regions on the graph of the sampling distribution of the sample proportion.

a. A two-tailed test with $\alpha = .10$
b. A left-tailed test with $\alpha = .01$
c. A right-tailed test with $\alpha = .05$

9.78 For each of the following examples of tests of hypotheses about the population proportion, show the rejection and nonrejection regions on the graph of the sampling distribution of the sample proportion.

a. A two-tailed test with $\alpha = .05$
b. A left-tailed test with $\alpha = .02$
c. A right-tailed test with $\alpha = .025$

9.79 A random sample of 500 observations produced a sample proportion equal to .38. Find the critical and observed values of z for each of the following tests of hypotheses using $\alpha = .05$.

a. H_0: $p = .30$ versus H_1: $p > .30$
b. H_0: $p = .30$ versus H_1: $p \neq .30$

9.80 A random sample of 200 observations produced a sample proportion equal to .60. Find the critical and observed values of z for each of the following tests of hypotheses using $\alpha = .01$.

a. H_0: $p = .63$ versus H_1: $p < .63$
b. H_0: $p = .63$ versus H_1: $p \neq .63$

9.81 Consider the null hypothesis H_0: $p = .65$. Suppose a random sample of 1000 observations is taken to make this test about the population proportion. Using $\alpha = .05$, show the rejection and nonrejection regions and find the critical value(s) of z for a

a. left-tailed test **b.** two-tailed test **c.** right-tailed test

9.82 Consider the null hypothesis H_0: $p = .25$. Suppose a random sample of 400 observations is taken to make this test about the population proportion. Using $\alpha = .01$, show the rejection and nonrejection regions and find the critical value(s) of z for a

a. left-tailed test **b.** two-tailed test **c.** right-tailed test

9.83 Consider H_0: $p = .70$ versus H_1: $p \neq .70$.

a. A random sample of 600 observations produced a sample proportion equal to .68. Using $\alpha = .01$, would you reject the null hypothesis?
b. Another random sample of 600 observations taken from the same population produced a sample proportion equal to .76. Using $\alpha = .01$, would you reject the null hypothesis?

Comment on the results of parts a and b.

9.84 Consider H_0: $p = .45$ versus H_1: $p < .45$.

a. A random sample of 400 observations produced a sample proportion equal to .42. Using $\alpha = .025$, would you reject the null hypothesis?
b. Another random sample of 400 observations taken from the same population produced a sample proportion of .39. Using $\alpha = .025$, would you reject the null hypothesis?

Comment on the results of parts a and b.

9.85 Make the following hypothesis tests about p.

a. H_0: $p = .45$, H_1: $p \neq .45$, $n = 100$, $\hat{p} = .49$, $\alpha = .10$
b. H_0: $p = .72$, H_1: $p < .72$, $n = 700$, $\hat{p} = .64$, $\alpha = .05$
c. H_0: $p = .30$, H_1: $p > .30$, $n = 200$, $\hat{p} = .33$, $\alpha = .01$

9.86 Make the following hypothesis tests about p.

a. H_0: $p = .57$, H_1: $p \neq .57$, $n = 800$, $\hat{p} = .50$, $\alpha = .05$
b. H_0: $p = .26$, H_1: $p < .26$, $n = 400$, $\hat{p} = .23$, $\alpha = .01$
c. H_0: $p = .84$, H_1: $p > .84$, $n = 250$, $\hat{p} = .85$, $\alpha = .025$

APPLICATIONS

9.87 According to a survey by the Centers for Disease Control and Prevention, (about) 22% of high school students in the United States smoked in 2003, down from 29% in 2001 (*USA TODAY*, June 18, 2004). Suppose in a recent random sample of 500 high school students, 95 said they smoke. At the 5% level of significance, can you conclude that the current percentage of U.S. high school students who smoke is different from 22%? Use both the p-value and the critical-value approaches.

9.88 In a November 2003 Lemelson-MIT survey, adult Americans were asked which invention they hated the most but could not live without. The cell phone was the most frequently named such invention, chosen by 30% of the adults (*Business Week*, March 1, 2004). In a recent random sample of 1000 adults, the same question was asked, and 363 of the respondents chose the cell phone. Test at the 5% significance level whether the current percentage of all adults who feel that the cell phone is the invention that they hate the most but cannot live without is different from 30%. Use both the p-value and the critical-value approaches.

9.89 In a *PARADE*/Research! America poll conducted by the Charlton Research Corporation, men were asked why they thought they were less likely than women to seek treatment for depression (*PARADE*, June 20, 2004). Forty-one percent of the men said that it was because of the attitude that a man should "tough it out." Suppose in a recent survey of 900 men, 324 hold this opinion. Using $\alpha = .025$, can you conclude that the current percentage of men who hold this opinion is less than 41%? Use both the p-value and the critical-value approaches.

9.90 In a 2003 Affluent Americans and Their Money (a *Money* magazine/Roper Public Affairs) survey of adults of all income levels, 85% of the respondents agreed with the statement, "Money doesn't buy happiness, but it helps" (*Money*, October 2004). In a recent random sample of 1200 adults, 776 agreed with that statement. At the 1% level of significance, can you conclude that the current percentage of all adults who agree with this statement is less than 85%? Use both the p-value and the critical-value approaches.

9.91 According to *The Wall Street Journal*, 60% of U.S. companies paid no federal taxes from 1996 to 2000 (*Time*, April 19, 2004). In random sample of 300 U.S. companies, 186 paid no federal taxes last year.

a. Test at the 2% level of significance whether the percentage of all U.S. companies who paid no federal taxes last year is higher than 60%.
b. What is the Type I error in part a? What is the probability of making this error?
c. Calculate the p-value for the test of part a. What is your conclusion if $\alpha = .02$?

9.92 According to the U.S. Bureau of Transportation Statistics, 21.9% of airline flights in the United States failed to arrive on time in 2004 (*USA TODAY*, February 4, 2005). This percentage includes the flights that arrived more than 15 minutes after their scheduled arrival time, or were canceled or diverted. Suppose that in a recent random sample of 800 flights, 196 failed to arrive on time.

a. Test at the 2.5% significance level whether the current percentage of U.S. flights that fail to arrive on time exceeds 21.9%.
b. What is the Type I error in part a? What is the probability of making this error?
c. Calculate the p-value for the test of part a. What is your conclusion if $\alpha = .025$?

9.93 A food company is planning to market a new type of frozen yogurt. However, before marketing this yogurt, the company wants to find what percentage of the people like it. The company's management has decided that it will market this yogurt only if at least 35% of the people like it. The company's research department selected a random sample of 400 persons and asked them to taste this yogurt. Of these 400 persons, 112 said they liked it.

a. Testing at the 2.5% significance level, can you conclude that the company should market this yogurt?
b. What will your decision be in part a if the probability of making a Type I error is zero? Explain.
c. Make the test of part a using the p-value approach and $\alpha = .025$.

9.94 A mail-order company claims that at least 60% of all orders are mailed within 48 hours. From time to time the quality control department at the company checks if this promise is fulfilled. Recently the quality control department at this company took a sample of 400 orders and found that 208 of them were mailed within 48 hours of the placement of the orders.

a. Testing at the 1% significance level, can you conclude that the company's claim is true?
b. What will your decision be in part a if the probability of making a Type I error is zero? Explain.
c. Make the test of part a using the p-value approach and $\alpha = .01$.

9.95 Brooklyn Corporation manufactures DVDs. The machine that is used to make these DVDs is known to produce not more than 5% defective DVDs. The quality control inspector selects a sample of 200 DVDs each week and inspects them for being good or defective. Using the sample proportion, the quality control inspector tests the null hypothesis $p \leq .05$ against the alternative hypothesis $p > .05$, where p is the proportion of DVDs that are defective. She always uses a 2.5% significance level. If the null hypothesis is rejected, the production process is stopped to make any necessary adjustments. A recent sample of 200 DVDs contained 17 defective DVDs.

- **a.** Using the 2.5% significance level, would you conclude that the production process should be stopped to make necessary adjustments?
- **b.** Perform the test of part a using a 1% significance level. Is your decision different from the one in part a?

Comment on the results of parts a and b.

9.96 Shulman Steel Corporation makes bearings that are supplied to other companies. One of the machines makes bearings that are supposed to have a diameter of four inches. The bearings that have a diameter of either more or less than four inches are considered defective and are discarded. When working properly, the machine does not produce more than 7% of bearings that are defective. The quality control inspector selects a sample of 200 bearings each week and inspects them for the size of their diameters. Using the sample proportion, the quality control inspector tests the null hypothesis $p \leq .07$ against the alternative hypothesis $p > .07$, where p is the proportion of bearings that are defective. He always uses a 2% significance level. If the null hypothesis is rejected, the machine is stopped to make any necessary adjustments. One sample of 200 bearings taken recently contained 22 defective bearings.

- **a.** Using the 2% significance level, will you conclude that the machine should be stopped to make necessary adjustments?
- **b.** Perform the test of part a using a 1% significance level. Is your decision different from the one in part a?

Comment on the results of parts a and b.

***9.97** Two years ago, 75% of the customers of a bank said that they were satisfied with the services provided by the bank. The manager of the bank wants to know if this percentage of satisfied customers has changed since then. She assigns this responsibility to you. Briefly explain how you would conduct such a test.

***9.98** A study claims that 65% of students at all colleges and universities hold off-campus (part-time or full-time) jobs. You want to check if the percentage of students at your school who hold off-campus jobs is different from 65%. Briefly explain how you would conduct such a test. Collect data from 40 students at your school on whether or not they hold off-campus jobs. Then, calculate the proportion of students in this sample who hold off-campus jobs. Using this information, test the hypothesis. Select your own significance level.

USES AND MISUSES... FOLLOW THE RECIPE

Hypothesis testing is one of the most powerful and dangerous tools of statistics. It allows us to make statements about a population and attach a degree of uncertainty to these statements. Pick up a newspaper and flip through it; rare will be the day when the paper does not contain a story featuring a statistical result, often reported with a significance level. Given that the subjects of these reports—public health, the environment, and so on—are important to our lives, it is critical that we perform the statistical calculations and interpretations properly. The first step, one that you should look for when reading statistical results, is proper formulation/specification.

Formulation or *specification*, simply put, is the list of steps you perform when constructing a hypothesis test. In this chapter, these steps are: stating the null and alternative hypotheses; selecting the appropriate distribution; and determining the rejection and nonrejection regions. Once these steps are performed, all you need to do is to calculate the test statistic to complete the hypothesis test. It is important to beware of traps in the specification.

Though it might seem obvious, stating the hypothesis properly can be difficult. For hypotheses around a population mean, the null and alternative hypotheses are mathematical statements that do not overlap and also provide no holes. Suppose that a confectioner states that the average mass of his chocolate bars is 100 grams. The null hypothesis is that the mass of the bars is 100 grams, and the alternative hypothesis is that the mass of the bars is not 100 grams. When you take a sample of chocolate bars and measure their masses, all possibilities for the sample mean will fall within one of your decision regions. The problem is a little more difficult for hypotheses based on proportions. Make sure that you only have two categories. For example, if you are trying to determine the percentage of the population that has blonde hair, your groups are "blonde" and "not blonde."

You need to decide how to categorize bald people before you conduct this experiment: Do not include bald people in the survey.

Finally, beware of numerical precision. When your sample is large and you assume that it has a normal distribution, the rejection region for a two-tailed test using the normal distribution with a significance level of 5% will be values of the sample mean that are farther than 1.96 standard deviations from the assumed mean. When you perform your calculations, the sample mean may fall on the border of your decision region. Remember that there is measurement error and sample error that you cannot account for. In this case, it is probably best to adjust your significance level so that the sample mean falls squarely in a decision region.

Glossary

α The significance level of a test of hypothesis that denotes the probability of rejecting a null hypothesis when it actually is true. (The probability of committing a Type I error.)

Alternative hypothesis A claim about a population parameter that will be true if the null hypothesis is false.

β The probability of not rejecting a null hypothesis when it actually is false. (The probability of committing a Type II error.)

Critical value or **critical point** One or two values that divide the whole region under the sampling distribution of a sample statistic into rejection and nonrejection regions.

Left-tailed test A test in which the rejection region lies in the left tail of the distribution curve.

Null hypothesis A claim about a population parameter that is assumed to be true until proven otherwise.

Observed value of z or t The value of z or t calculated for a sample statistic such as the sample mean or the sample proportion.

One-tailed test A test in which there is only one rejection region, either in the left tail or in the right tail of the distribution curve.

p-value The smallest significance level at which a null hypothesis can be rejected.

Right-tailed test A test in which the rejection region lies in the right tail of the distribution curve.

Significance level The value of α that gives the probability of committing a Type I error.

Test statistic The value of z or t calculated for a sample statistic such as the sample mean or the sample proportion.

Two-tailed test A test in which there are two rejection regions, one in each tail of the distribution curve.

Type I error An error that occurs when a true null hypothesis is rejected.

Type II error An error that occurs when a false null hypothesis is not rejected.

Supplementary Exercises

9.99 Consider the following null and alternative hypotheses:

$$H_0: \mu = 120 \quad \text{versus} \quad H_1: \mu > 120$$

A random sample of 81 observations taken from this population produced a sample mean of 123.5. The population standard deviation is known to be 15.

a. If this test is made at the 2.5% significance level, would you reject the null hypothesis? Use the critical-value approach.

b. What is the probability of making a Type I error in part a?

c. Calculate the p-value for the test. Based on this p-value, would you reject the null hypothesis if $\alpha = .01$? What if $\alpha = .05$?

9.100 Consider the following null and alternative hypotheses:

$$H_0: \mu = 40 \quad \text{versus} \quad H_1: \mu \neq 40$$

A random sample of 64 observations taken from this population produced a sample mean of 38.4. The population standard deviation is known to be 6.

a. If this test is made at the 2% significance level, would you reject the null hypothesis? Use the critical-value approach.

b. What is the probability of making a Type I error in part a?

c. Calculate the p-value for the test. Based on this p-value, would you reject the null hypothesis if $\alpha = .01$? What if $\alpha = .05$?

9.101 Consider the following null and alternative hypotheses:

$$H_0: p = .82 \quad \text{versus} \quad H_1: p \neq .82$$

A random sample of 600 observations taken from this population produced a sample proportion of .86.

- **a.** If this test is made at the 2% significance level, would you reject the null hypothesis? Use the critical-value approach.
- **b.** What is the probability of making a Type I error in part a?
- **c.** Calculate the p-value for the test. Based on this p-value, would you reject the null hypothesis if $\alpha = .025$? What if $\alpha = .005$?

9.102 Consider the following null and alternative hypotheses:

$$H_0: p = .44 \quad \text{versus} \quad H_1: p < .44$$

A random sample of 450 observations taken from this population produced a sample proportion of .39.

- **a.** If this test is made at the 2% significance level, would you reject the null hypothesis? Use the critical-value approach.
- **b.** What is the probability of making a Type I error in part a?
- **c.** Calculate the p-value for the test. Based on this p-value, would you reject the null hypothesis if $\alpha = .01$? What if $\alpha = .025$?

9.103 According to the information given in the 2004 instruction booklet that accompanied the income tax form 1040, the Internal Revenue Service (IRS) estimated the average time required to prepare this form to be 377 minutes (or 6 hours and 17 minutes). This mean time does not include the time spent on record keeping, learning about the law or the form, copying, assembling, and sending the form to the IRS. Suppose a random sample of 100 tax payers from a small state who filed form 1040 in 2004, showed that it took them an average of 422 minutes to prepare this form. The population standard deviation for such times in this state in known to be 85 minutes.

- **a.** Find the p-value for the test with the alternative hypothesis that the mean time taken by all tax payers in this state to prepare form 1040 in 2004 was different from 377 minutes. What is your conclusion if $\alpha = .05$?
- **b.** Make the test of part a using the critical-value approach. Use $\alpha = .05$.

9.104 The mean consumption of water per household in a city was 1245 cubic feet per month. Due to a water shortage because of a drought, the city council campaigned for water use conservation by households. A few months after the campaign was started, the mean consumption of water for a sample of 100 households was found to be 1175 cubic feet per month. The population standard deviation is given to be 250 cubic feet.

- **a.** Find the p-value for the hypothesis test that the mean consumption of water per household has decreased due to the campaign by the city council. Would you reject the null hypothesis at $\alpha = .025$?
- **b.** Make the test of part a using the critical-value approach and $\alpha = .025$.

9.105 A highway construction zone has a posted speed limit of 40 miles per hour. Workers working at the site claim that the mean speed of vehicles passing through this construction zone is at least 50 miles per hour. A random sample of 36 vehicles passing through this zone produced a mean speed of 48 miles per hour. The population standard deviation is known to be 4 miles per hour.

- **a.** Do you think the sample information is consistent with the workers' claim? Use $\alpha = .025$.
- **b.** What is the Type I error in this case? Explain. What is the probability of making this error?
- **c.** Will your conclusion of part a change if the probability of making a Type I error is zero?
- **d.** Find the p-value for the test of part a. What is your decision if $\alpha = .025$?

9.106 According to an article entitled "Do You Need Long-Term-Care Insurance?" published in the *Consumer Reports* magazine, the average age of admission to nursing homes is 83 years (*Consumer Reports*, November 2003). Suppose that a random sample of 500 recent admissions to nursing homes yielded a mean age of 83.5 years. Assume that the population standard deviation is 3.5 years.

- **a.** Using the critical-value approach, can you conclude that the current mean admission age to nursing homes is different from 83 years? Use $\alpha = .05$.
- **b.** What is the Type I error in part a? Explain. What is the probability of making this error in part a?
- **c.** Will your conclusion of part a change if the probability of making a Type I error is zero?
- **d.** Calculate the p-value for the test of part a. What is your conclusion if $\alpha = .05$?

9.107 A real estate agent claims that the mean living area of all single-family homes in his county is at most 2400 square feet. A random sample of 50 such homes selected from this county produced the mean living area of 2540 square feet and a standard deviation of 472 square feet.

- **a.** Using $\alpha = .05$, can you conclude that the real estate agent's claim is true?
- **b.** What will your conclusion be if $\alpha = .01$?

Comment on the results of parts a and b.

9.108 According to the American Federation of Teachers, during the 2002–2003 school year, the average annual salary of school teachers in California was \$55,693, which was the highest in the nation (*The Hartford*

Courant, July 15, 2004). Suppose that a recent random sample of 65 school teachers from California yielded a mean salary of $58,940 with a standard deviation of $6300.

a. Using $\alpha = .025$, can you conclude that the current mean salary of all California school teachers exceeds $55,693? Use the critical-value approach.

b. Find the range of the p-value for the test of part a. What is your conclusion with $\alpha = .025$?

9.109 Customers often complain about long waiting times at restaurants before the food is served. A restaurant claims that it serves food to its customers, on average, within 15 minutes after the order is placed. A local newspaper journalist wanted to check if the restaurant's claim is true. A sample of 36 customers showed that the mean time taken to serve food to them was 15.75 minutes with a standard deviation of 2.4 minutes. Using the sample mean, the journalist says that the restaurant's claim is false. Do you think the journalist's conclusion is fair to the restaurant? Use the 1% significance level to answer this question.

9.110 The customers at a bank complained about long lines and the time they had to spend waiting for service. It is known that the customers at this bank had to wait 8 minutes, on average, before being served. The management made some changes to reduce the waiting time for its customers. A sample of 60 customers taken after these changes were made produced a mean waiting time of 7.5 minutes with a standard deviation of 2.1 minutes. Using this sample mean, the bank manager displayed a huge banner inside the bank mentioning that the mean waiting time for customers has been reduced by new changes. Do you think the bank manager's claim is justifiable? Use the 2.5% significance level to answer this question. Use both approaches.

9.111 The administrative office of a hospital claims that the mean waiting time for patients to get treatment in its emergency ward is 25 minutes. A random sample of 16 patients who received treatment in the emergency ward of this hospital produced a mean waiting time of 27.5 minutes with a standard deviation of 4.8 minutes. Using the 1% significance level, test whether the mean waiting time at the emergency ward is different from 25 minutes. Assume that the waiting times for all patients at this emergency ward have a normal distribution.

9.112 An earlier study claims that U.S. adults spend an average of 114 minutes with their families per day. A recently taken sample of 25 adults from a city showed that they spend an average of 109 minutes per day with their families. The sample standard deviation is 11 minutes. Assume that the times spent by adults with their families have an approximately normal distribution.

a. Using the 1% significance level, test whether the mean time spent currently by all adults with their families in this city is different from 114 minutes a day.

b. Suppose the probability of making a Type I error is zero. Can you make a decision for the test of part a without going through the five steps of hypothesis testing? If yes, what is your decision? Explain.

9.113 A computer company that recently introduced a new software product claims that the mean time it takes to learn how to use this software is not more than 2 hours for people who are somewhat familiar with computers. A random sample of 12 such persons was selected. The following data give the times taken (in hours) by these persons to learn how to use this software.

1.75	2.25	2.40	1.90	1.50	2.75
2.15	2.25	1.80	2.20	3.25	2.60

Test at the 1% significance level whether the company's claim is true. Assume that the times taken by all persons who are somewhat familiar with computers to learn how to use this software are approximately normally distributed.

9.114 A company claims that its eight-ounce low-fat yogurt cups contain, on average, at most 150 calories per cup. A consumer agency wanted to check whether or not this claim is true. A random sample of 10 such cups produced the following data on calories.

147	159	153	146	144	161	163	153	143	158

Test at the 2.5% significance level whether the company's claim is true. Assume that the numbers of calories for such cups of yogurt produced by this company have an approximately normal distribution.

9.115 According to the Census Bureau, 15.6% of the U.S. population had no health insurance coverage in 2003. Suppose that in a recent random sample of 1200 Americans, 216 had no health insurance coverage.

a. Using the critical-value approach and $\alpha = .02$, test if the current percentage of Americans who have no health insurance coverage exceeds 15.6%.

b. How do you explain the Type I error in part a? What is the probability of making this error in part a?

c. Calculate the p-value for the test of part a. What is your conclusion if $\alpha = .02$?

9.116 In a TD Waterhouse USA survey, investors were asked, "Who should have primary responsibility to protect the rights of individual investors?" Sixty-eight percent of the respondents named industry regulators (such as the Securities and Exchange Commission), while a much smaller percentage of the investors named such entities as Congress or consumer advocacy groups (*USA TODAY*, January 19, 2005). Suppose that in a recent random sample of 900 investors, 648 indicated that industry regulators have primary responsibility to protect investors.

- **a.** Using the critical-value approach and $\alpha = .02$, test whether the current percentage of all investors who will say that industry regulators have primary responsibility to protect investors differs from 68%.
- **b.** How do you explain the Type I error in part a? What is the probability of making this error in part a?
- **c.** Calculate the p-value for the test of part a. What is your conclusion if $\alpha = .02$?

9.117 More and more people are abandoning national brand products and buying store brand products to save money. The president of a company that produces national brand coffee claims that 40% of the people prefer to buy national brand coffee. A random sample of 700 people who buy coffee showed that 259 of them buy national brand coffee. Using $\alpha = .01$, can you conclude that the percentage of people who buy national brand coffee is different from 40%? Use both approaches to make the test.

9.118 In a Caravan survey conducted for Roche/Tamiflu, 38% of the workers indicated that they felt pressured by their bosses or co-workers to come to work when they were ill with the flu (*USA TODAY*, February 28, 2005). Suppose that a recent random sample of 500 workers showed that 35% of them felt this way. At the 2% level of significance, can you conclude that the current percentage of all workers who feel this way is less than 38%? Use both approaches to make the test.

9.119 Mong Corporation makes auto batteries. The company claims that 80% of its LL70 batteries are good for 70 months or longer. A consumer agency wanted to check if this claim is true. The agency took a random sample of 40 such batteries and found that 75% of them were good for 70 months or longer.

- **a.** Using the 1% significance level, can you conclude that the company's claim is false?
- **b.** What will your decision be in part a if the probability of making a Type I error is zero? Explain.

9.120 Dartmouth Distribution Warehouse makes deliveries of a large number of products to its customers. To keep its customers happy and satisfied, the company's policy is to deliver on time at least 90% of all the orders it receives from its customers. The quality control inspector at the company quite often takes samples of orders delivered and checks if this policy is maintained. A recent sample of 90 orders taken by this inspector showed that 75 of them were delivered on time.

- **a.** Using the 2% significance level, can you conclude that the company's policy is maintained?
- **b.** What will your decision be in part a if the probability of making a Type I error is zero? Explain.

Advanced Exercises

9.121 Professor Hansen believes that some people have the ability to predict in advance the outcome of a spin of a roulette wheel. He takes 100 student volunteers to a casino. The roulette wheel has 38 numbers, each of which is equally likely to occur. Of these 38 numbers, 18 are red, 18 are black, and 2 are green. Each student is to place a series of five bets, choosing either a red or a black number before each spin of the wheel. Thus, a student who bets on red has an 18/38 chance of winning that bet. The same is true of betting on black.

- **a.** Assuming random guessing, what is the probability that a particular student will win all five of his or her bets?
- **b.** Suppose for each student we formulate the hypothesis test

  H_0: The student is guessing

  H_1: The student has some predictive ability

  Suppose we reject H_0 only if the student wins all five bets. What is the significance level?
- **c.** Suppose that two of the 100 students win all five of their bets. Professor Hansen says, "For these two students we can reject H_0 and conclude that we have found two students with some ability to predict." What do you make of Professor Hansen's conclusion?

9.122 Acme Bicycle Company makes derailleurs for mountain bikes. Usually no more than 4% of these parts are defective, but occasionally the machines that make them get out of adjustment and the rate of defectives exceeds 4%. To guard against this, the chief quality control inspector takes a random sample of 130 derailleurs each week and checks each one for defects. If too many of these parts are defective, the machines are shut down and adjusted. To decide how many parts must be defective to shut down the machines, the company's statistician has set up the hypothesis test

$$H_0: p \leq .04 \quad \text{versus} \quad H_1: p > .04$$

where p is the proportion of defectives among all derailleurs being made currently. Rejection of H_0 would call for shutting down the machines. For the inspector's convenience, the statistician would like the rejection region to have the form, "Reject H_0 if the number of defective parts is C or more." Find the value of C that will make the significance level (approximately) .05.

9.123 Alpha Airlines claims that only 15% of its flights arrive more than 10 minutes late. Let p be the proportion of all of Alpha's flights that arrive more than 10 minutes late. Consider the hypothesis test

$$H_0: p \leq .15 \quad \text{versus} \quad H_1: p > .15$$

Suppose we take a random sample of 50 flights by Alpha Airlines and agree to reject H_0 if 9 or more of them arrive late. Find the significance level for this test.

9.124 The standard therapy used to treat a disorder cures 60% of all patients in an average of 140 visits. A health care provider considers supporting a new therapy regime for the disorder if it is effective in reducing the number of visits while retaining the cure rate of the standard therapy. A study of 200 patients with the disorder who were treated by the new therapy regime reveals that 108 of them were cured in an average of 132 visits with a standard deviation of 38 visits. What decision should be made using a .01 level of significance?

9.125 The print on the packages of 100-watt General Electric soft-white lightbulbs states that these lightbulbs have an average life of 750 hours. Assume that the standard deviation of the lengths of lives of these lightbulbs is 50 hours. A skeptical consumer does not think these lightbulbs last as long as the manufacturer claims, and she decides to test 64 randomly selected lightbulbs. She has set up the decision rule that if the average life of these 64 lightbulbs is less than 735 hours, then she will conclude that GE has printed too high an average length of life on the packages and will write them a letter to that effect. Approximately what significance level is the consumer using? Approximately what significance level is she using if she decides that GE has printed too high an average length of life on the packages if the average life of the 64 lightbulbs is less than 700 hours? Interpret the values you get.

9.126 Thirty percent of all people who are inoculated with the current vaccine used to prevent a disease contract the disease within a year. The developer of a new vaccine that is intended to prevent this disease wishes to test for significant evidence that the new vaccine is more effective.

a. Determine the appropriate null and alternative hypotheses.

b. The developer decides to study 100 randomly selected people by inoculating them with the new vaccine. If 84 or more of them do not contract the disease within a year, the developer will conclude that the new vaccine is superior to the old one. What significance level is the developer using for the test?

c. Suppose 20 people inoculated with the new vaccine are studied and the new vaccine is concluded to be better than the old one if fewer than 3 people contract the disease within a year. What is the significance level of the test?

9.127 Since 1984, all automobiles have been manufactured with a middle taillight. You have been hired to answer the question, Is the middle taillight effective in reducing the number of rear-end collisions? You have available to you any information you could possibly want about all rear-end collisions involving cars built before 1984. How would you conduct an experiment to answer the question? In your answer, include things like (a) the precise meaning of the unknown parameter you are testing; (b) H_0 and H_1; (c) a detailed explanation of what sample data you would collect to draw a conclusion; and (d) any assumptions you would make, particularly about the characteristics of cars built before 1984 versus those built since 1984.

9.128 Before a championship football game, the referee is given a special commemorative coin to toss to decide which team will kick the ball first. Two minutes before game time, he receives an anonymous tip that the captain of one of the teams may have substituted a biased coin that has a 70% chance of showing heads each time it is tossed. The referee has time to toss the coin 10 times to test it. He decides that if it shows 8 or more heads in 10 tosses, he will reject this coin and replace it with another coin. Let p be the probability that this coin shows heads when it is tossed once.

a. Formulate the relevant null and alternative hypotheses (in terms of p) for the referee's test.

b. Using the referee's decision rule, find α for this test.

9.129 In Las Vegas and Atlantic City, NJ, tests are performed often on the various gaming devices used in casinos. For example, dice are often tested to determine if they are balanced. Suppose you are assigned the task of testing a die, using a two-tailed test to make sure that the probability of a 2-spot is 1/6. Using the 5% significance level, determine how many 2-spots you would have to obtain to reject the null hypothesis when your sample size is

a. 120 **b.** 1200 **c.** 12,000

Calculate the value of $\hat{p}$ for each of these three cases. What can you say about the relationship between (1) the difference between $\hat{p}$ and 1/6 that is necessary to reject the null hypothesis and (2) the sample size as it gets larger?

9.130 A statistician performs the test H_0: $\mu = 15$ versus H_1: $\mu \neq 15$, and finds the p-value to be .4546.

a. The statistician performing the test does not tell you the value of the sample mean and the value of the test statistic. Despite this, you have enough information to determine the pair of p-values associated with the following alternative hypotheses.

i. H_1: $\mu < 15$ ii. H_1: $\mu > 15$

Note that you will need more information to determine which p-value goes with which alternative. Determine the pair of p-values. Here the value of the sample mean is the same in both cases.

b. Suppose the statistician tells you that the value of the test statistic is negative. Match the p-values with the alternative hypotheses.

Note that the result for one of the two alternatives implies that the sample mean is not on the same side of $\mu = 15$ as the rejection region. Although we have not discussed this scenario in the book, it is important to recognize that there are many real-world scenarios in which this type of situation does occur. For example, suppose the EPA is to test whether or not a company is exceeding a specific pollution level. If the average discharge level obtained from the sample falls below the threshold (mentioned in the null hypothesis), then there would be no need to perform the hypothesis test.

9.131 You read an article that states "50 hypothesis tests of H_0: $\mu = 35$ versus H_1: $\mu \neq 35$ were performed using $\alpha = .05$ on 50 different samples taken from the same population with a mean of 35. Of these, 47 tests failed to reject the null hypothesis." Explain why this type of result is not surprising.

Self-Review Test

1. A test of hypothesis is always about
   a. a population parameter b. a sample statistic c. a test statistic
2. A Type I error is committed when
   a. a null hypothesis is not rejected when it is actually false
   b. a null hypothesis is rejected when it is actually true
   c. an alternative hypothesis is rejected when it is actually true
3. A Type II error is committed when
   a. a null hypothesis is not rejected when it is actually false
   b. a null hypothesis is rejected when it is actually true
   c. an alternative hypothesis is rejected when it is actually true
4. A critical value is the value
   a. calculated from sample data
   b. determined from a table (e.g., the normal distribution table or other such tables)
   c. neither a nor b
5. The computed value of a test statistic is the value
   a. calculated for a sample statistic
   b. determined from a table (e.g., the normal distribution table or other such tables)
   c. neither a nor b
6. The observed value of a test statistic is the value
   a. calculated for a sample statistic
   b. determined from a table (e.g., the normal distribution table or other such tables)
   c. neither a nor b
7. The significance level, denoted by α, is
   a. the probability of committing a Type I error
   b. the probability of committing a Type II error
   c. neither a nor b

8. The value of β gives the
   a. probability of committing a Type I error
   b. probability of committing a Type II error
   c. power of the test

9. The value of $1 - \beta$ gives the
   a. probability of committing a Type I error
   b. probability of committing a Type II error
   c. power of the test

10. A two-tailed test is a test with
   a. two rejection regions b. two nonrejection regions c. two test statistics

11. A one-tailed test
   a. has one rejection region b. has one nonrejection region c. both a and b

12. The smallest level of significance at which a null hypothesis is rejected is called
   a. α b. p-value c. β

13. The sign in the alternative hypothesis in a two-tailed test is always
   a. $<$ b. $>$ c. $\neq$

14. The sign in the alternative hypothesis in a left-tailed test is always
   a. $<$ b. $>$ c. $\neq$

15. The sign in the alternative hypothesis in a right-tailed test is always
   a. $<$ b. $>$ c. $\neq$

16. According to the U.S. Department of Transportation, drivers aged 20–24 years spend an average of 52 minutes driving per day (*USA TODAY*, November 7, 2003). Suppose that a recent random sample of 1400 drivers in this age group yielded a mean of 58 minutes of driving per day. The population standard deviation is known to be 12 minutes.
   a. Using the critical-value approach and the 1% significance level, can you conclude that the mean time spent driving per day by this age group currently differs from 52 minutes?
   b. Using the critical-value approach and the 2.5% significance level, can you conclude that the mean time spent driving per day by this age group currently exceeds 52 minutes?
   c. What is the Type I error in parts a and b? What is the probability of making this error in each of parts a and b?
   d. Find the p-value for the test of part a. What is your conclusion using $\alpha = .01$?
   e. Find the p-value for the test of part b. What is your conclusion using $\alpha = .025$?

17. A minor league baseball executive has become concerned about the slow pace of games played in her league, fearing that it will lower attendance. She meets with the league's managers and umpires and discusses guidelines for speeding up the games. Before the meeting, the mean duration of nine-inning games was 3 hours, 5 minutes (i.e., 185 minutes). A random sample of 36 nine-inning games after the meeting showed a mean of 179 minutes with a standard deviation of 12 minutes.
   a. Testing at the 1% significance level, can you conclude that the mean duration of nine-inning games has decreased after the meeting?
   b. What is the Type I error in part a? What is the probability of making this error?
   c. What will your decision be in part a if the probability of making a Type I error is zero? Explain.
   d. Find the range for the p-value for the test of part a. What is your decision based on this p-value?

18. An editor of a New York publishing company claims that the mean time it takes to write a textbook is at least 31 months. A sample of 16 textbook authors found that the mean time taken by them to write a textbook was 25 months with a standard deviation of 7.2 months.
   a. Using the 2.5% significance level, would you conclude that the editor's claim is true? Assume that the time taken to write a textbook is normally distributed for all textbook authors.
   b. What is the Type I error in part a? What is the probability of making this error?
   c. What will your decision be in part a if the probability of making a Type I error is .001?

19. A financial advisor claims that less then 50% of adults in the United States have a will. A random sample of 1000 adults showed that 450 of them have a will.
   a. At the 5% significance level, can you conclude that the percentage of people who have a will is less than 50%?
   b. What is the Type I error in part a? What is the probability of making this error?
   c. What would your decision be in part a if the probability of making a Type I error were zero? Explain.
   d. Find the p-value for the test of hypothesis mentioned in part a. Using this p-value, will you reject the null hypothesis if $\alpha = .05$? What if $\alpha = .01$?

Mini-Projects

MINI-PROJECT 9–1

The mean height of players who were on the rosters of National Basketball Association teams at the beginning of the 2002–2003 season was 79.49 inches. Let μ denote the mean height of NBA players at the beginning of the 2004–2005 season.

a. Take a random sample of 15 players from the NBA data that accompany this text. Test H_0: $\mu = 79.49$ inches against H_1: $\mu \neq 79.49$ inches using $\alpha = .05$. Assume that the population of heights is approximately normal.
b. Repeat part a for samples of 31 and 45 players, respectively.
c. Did any of the three tests in parts a and b lead to the conclusion that the mean height of NBA players in 2004–2005 is different from that in 2002–2003?

MINI-PROJECT 9–2

A thumbtack that is tossed on a desk can land in one of the two ways shown in the illustration.

Brad and Dan cannot agree on the likelihood of obtaining a head or a tail. Brad argues that obtaining a tail is more likely than obtaining a head because of the shape of the tack. If the tack had no point at all, it would resemble a coin that has the same probability of coming up heads or tails when tossed. But the longer the point, the less likely it is that the tack will stand up on its head when tossed. Dan believes that as the tack lands tails, the point causes the tack to jump around and come to rest in the heads position. Brad and Dan need you to settle their dispute. Do you think the tack is equally likely to land heads or tails? To investigate this question, find an ordinary thumbtack and toss it a large number of times (say, 100 times).

a. What is the meaning, in words, of the unknown parameter in this problem?
b. Set up the null and alternative hypotheses and compute the *p*-value based on your results from tossing the tack.
c. How would you answer the original question now? If you decide the tack is not fair, do you side with Brad or Dan?
d. What would you estimate the value of the parameter in part a to be? Find a 90% confidence interval for this parameter.
e. After doing this experiment, do you think 100 tosses are enough to infer the nature of your tack? Using your result as a preliminary estimate, determine how many tosses would be necessary to be 95% certain of having 4% accuracy; that is, the margin of error of estimate is ±4%. Have you observed enough tosses?

MINI-PROJECT 9–3

Collect pennies in the amount of $5. Do not obtain rolls of pennies from a bank because many such rolls will consist solely of new pennies. Treat these 500 pennies as your population. Determine the ages, in years, of all these pennies. Calculate the mean and standard deviation of these ages and denote them by μ and σ.

a. Take a random sample of 10 pennies from these 500. Find the average age of these 10 pennies, which is the value of $\bar{x}$. Perform a test with the null hypothesis that μ is equal to the value obtained for all 500 pennies and the alternative hypothesis that μ is not equal to this value. Use a significance level of .10.
b. Suppose you repeat the procedure of part a 9 more times. How many times would you expect to reject the null hypothesis? Now actually repeat the procedure of part a 9 more times, making sure that you put the ten pennies selected each time back in the population and that you mix all pennies well before taking a sample. How many times did you reject the null hypothesis? Note that

you can enter the ages of these 500 pennies in a technology and then use that technology to take samples and make tests of hypothesis.

c. Repeat parts a and b for a sample size of 25. Did you reject the null hypothesis more often with a sample size of 10 or a sample size of 25?

DECIDE FOR YOURSELF

STATISTICAL AND PRACTICAL SIGNIFICANCE

The hypothesis testing procedure helps us to make a conclusion regarding a claim or statement, and oftentimes this claim or statement is about the value of a parameter or the relationship between two or more parameters. When we reject the null hypothesis, we conclude that the result is statistically significant at the given significance level of α. So, what exactly does the term "statistically significant" mean? Using the single sample analogy, statistically significant implies that the value of a point estimator (such as a sample mean or sample proportion) of a parameter is far enough (in terms of the standard deviation or standard error) from the hypothesized value of the parameter so that it falls in the most extreme $\alpha \times 100\%$ of the area under the sampling distribution curve.

Now the logical follow-up question is: "What does *statistically significant* imply with regard to my specific application?" Unlike the first question, which has a specific answer, the answer to this question is: "It depends." In any hypothesis test, one must consider the practical significance of the result. For example, suppose a new gasoline additive has been invented and the company that produces it claims that it increases average gas mileage. A fleet of cars of a specific model, based on EPA numbers, obtains an average of 448 miles per tank full of gas without this additive. A random sample of 25 such cars is selected. Each car is driven on a tank full of gas with this additive added to the gas. The sample mean for these 25 cars is found to be 453 miles per tank full of gas, with a sample standard deviation of 22 miles. To understand the difference between the statistical significance and practical significance, find answers to the following questions.

1. Perform the appropriate hypothesis test using the t distribution to determine if the average mileage per tank full of gas increases with the additive. Use a 5% significance level. Is this increase statistically significant? Assume that population is normally distributed.

2. Now suppose we use a sample of 100 cars instead of 25 cars but the values of the means and the standard deviation remain the same. Perform the above hypothesis test again and see if your answer changes with this larger sample size.

3. Regardless of the sample size, discuss whether the result (453 miles versus 448 miles) is *practically significant*, that is, whether or not the increase is meaningful to the everyday driver. Suppose it is recommended that the additive should be used every 3000 miles. Assuming that the price of gas is $2.50 per gallon and the gas tank holds 16 gallons of gas, calculate the savings in gas expenditure per mile. Then multiply this number by 3000 to obtain the savings per application of the additive. Assuming that the additive is not free, is it worth using it?

TECHNOLOGY INSTRUCTION — Hypothesis Testing

TI-84

Screen 9.1

1. To test a hypothesis about a population mean μ given the population standard deviation σ, select **STAT>TESTS>ZTest**. If you have the data stored in a list, select **Data** and enter the name of the list. If you have the summary statistics, choose **Stats** and enter the sample mean and size. Enter $\mu 0$, the constant value for the population mean from your null hypothesis. Enter your value for σ and select which alternative hypothesis you are using. Select **Calculate**. (See **Screen 9.1**.)

2. To test a hypothesis about a population mean μ without knowing the population standard deviation σ, select **STAT>TESTS>TTest**. If you have the data stored in a list, select **Data** and enter the name of the list. If you have the summary statistics, choose **Stats** and enter the sample mean, standard deviation, and size. Enter $\mu 0$, the constant value for the population mean from your null hypothesis. Select which alternative hypothesis you are using. Select **Calculate**.

3. To test a hypothesis about a population proportion p, select **STAT>TESTS>1-PropZTest**. Enter the constant value for p from the null hypothesis as $p0$. Enter the number of successes as x and the sample size as n. Select the alternative hypothesis you are using. Select **Calculate**.

MINITAB

1. To perform a hypothesis test for the population mean μ when the population standard deviation σ is given, select **Stat>Basic Statistics>1-Sample Z**. If you have your data entered in a column, enter the name of that column in the **Samples in columns**: box. Instead, if you know the summary statistics, click next to **Summarized data** and enter the values of the **Sample size** and **Mean** in their respective boxes. In both cases, enter the value of the population standard deviation in the **Standard deviation** box. Enter the value of μ from the null hypothesis in the **Test mean**: box (See **Screen 9.2**). Click on the **Options** button and select the appropriate alternative hypothesis from the **Alternative** box. Click **OK** in both windows. The output will appear in the **Session** window, which will give the p-value for the test. Based on this p-value, you can make a decision.

Screen 9.2

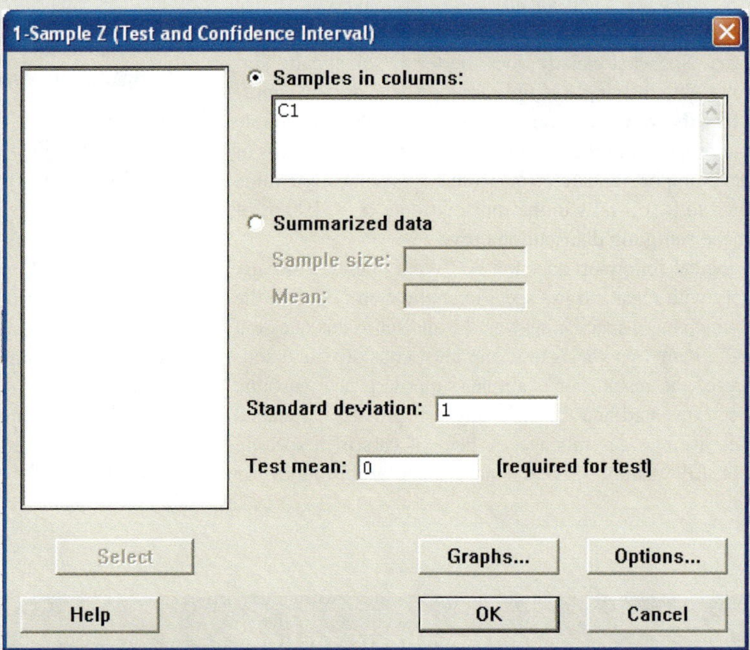

2. To perform a hypothesis test for the population mean μ when the population standard deviation σ is not known, select **Stat>Basic Statistics>1-Sample t**. If you have your data entered in a column, enter the name of that column in the **Samples in columns**: box. Instead, if you know the summary statistics, click next to **Summarized data** and enter the values of the **Sample size**, **Sample standard deviation**, and **Mean** in their respective boxes. Enter the value of μ from the null hypothesis in the **Test mean**: box. Click on the **Options** button and select the appropriate alternative hypothesis from the **Alternative** box. Click **OK** in both windows. The output will appear in the **Session** window, which will give the p-value for the test. Based on this p-value, you can make a decision.

3. To perform a hypothesis test for the population proportion p, select **Stat>Basic Statistics>1 Proportion**. If you have sample data (consisting of values for successes and failures) entered in a column, enter the name of that column in the **Samples in columns**: box. Instead, if you know the number of trials and number of successes, click next to **Summarized data** and enter the required values in the **Number of trials**: and **Number of events**: boxes, respectively. Click on the **Options** button and enter the value of the proportion from the null hypothesis in the **Test proportion**: box. Select the appropriate alternative hypothesis from the **Alternative** box, and check the box next to **Use test and interval based on normal distribution**. Click **OK** in both windows. The output will appear in the **Session** window, which will give the p-value for the test. Based on this p-value, you can make a decision.

EXCEL

	A	B	C	D
1	H0:	mu = 5		
2	H1:	mu != 5		
3	Std. Dev.	3		
4				
5	Data:	a11:a100		
6	p-value:	=ZTEST(a11:a100, 5, 3)		
7				

Screen 9.3

1. To compute a p-value for a two-tailed hypothesis test about a population mean, type **=ZTEST(data, μ0, σ)**, where **data** is the range of sample data, **μ0** is the value for μ in the null hypothesis, and σ is the population standard deviation. (See **Screen 9.3.**)

TECHNOLOGY ASSIGNMENTS

TA9.1 According to an earlier study, the mean amount spent on clothes by American women is \$675 per year. A researcher wanted to check if this result still holds true. A random sample of 39 women taken recently by this researcher produced the following data on the amounts they spent on clothes last year.

671	1284	328	1698	827	921	725	304	382	539
1070	854	669	328	537	849	930	1234	1195	738
341	189	867	923	721	125	298	473	876	932
973	931	460	1430	391	887	958	674	1482	

Test at the 1% significance level whether the mean expenditure on clothes for American women for last year is different from \$675. Assume that the population standard deviation is \$132.

TA9.2 The mean weight of all babies born at a hospital last year was 7.6 pounds. A random sample of 35 babies born at this hospital this year produced the following data.

8.2	9.1	6.9	5.8	6.4	10.3	12.1	9.1	5.9	7.3
11.2	8.3	6.5	7.1	8.0	9.2	5.7	9.5	8.3	6.3
4.9	7.6	10.1	9.2	8.4	7.5	7.2	8.3	7.2	9.7
6.0	8.1	6.1	8.3	6.7					

Test at the 2.5% significance level whether the mean weight of babies born at this hospital this year is more than 7.6 pounds.

TA9.3 The president of a large university claims that the mean time spent partying by all students at the university is not more than 7 hours per week. The following data give the times spent partying during the previous week by a random sample of 16 students taken from this university.

12	9	5	15	11	13	10	6
4	11	6	9	13	6	16	8

Test at the 1% significance level whether the president's claim is true. Assume that the times spent partying by all students at this university have an approximately normal distribution and the population standard deviation is known to be 2 hours.

TA9.4 According to a basketball coach, the mean height of all male college basketball players is 74 inches. A random sample of 25 such players produced the following data on their heights.

68	76	74	83	77	76	69	67	71	74	79	85	69
78	75	78	68	72	83	79	82	76	69	70	81	

Test at the 2% significance level whether the mean height of all male college basketball players is different from 74 inches. Assume that the heights of all male college basketball players are (approximately) normally distributed.

TA9.5 A past study claims that adults in America spend an average of 18 hours a week on leisure activities. A researcher took a sample of 10 adults from a town and asked them about the time they spend per week on leisure activities. Their responses (in hours) follow.

14	25	22	38	16	26	19	23	41	33

Assume that the times spent on leisure activities by all adults are normally distributed and the population standard deviation is 3 hours. Using the 5% significance level, can you conclude that the claim of the earlier study is true?

TA9.6 In a November 2003 Lemelson-MIT survey, adult Americans were asked which invention they hated the most but could not live without. The cell phone was the most frequently named such invention, chosen by 30% of the adults (*Business Week*, March 1, 2004). In a recent random sample of 1000 adults, the same question was asked, and 363 of the respondents chose the cell phone. Test at the 5% significance level whether the current percentage of all adults who feel that the cell phone is the invention that they hate the most but cannot live without is different from 30%.

TA9.7 A mail-order company claims that at least 60% of all orders it receives are mailed within 48 hours. From time to time the quality control department at the company checks if this promise is kept. Recently, the quality control department at this company took a sample of 400 orders and found that 224 of them were mailed within 48 hours of the placement of the orders. Test at the 1% significance level whether or not the company's claim is true.

Chapter

10

Estimation and Hypothesis Testing: Two Populations

Do you watch news on television, read a newspaper, or listen to news on radio? Can you guess how much time, on average, Americans spend on each of these three activities? In 2004, Americans spent an average of 32 minutes per day watching news on television, 17 minutes per day reading a newspaper, and 17 minutes per day listening to news on radio. See Case Study 10–1.

10.1 Inferences About the Difference Between Two Population Means for Independent Samples: σ_1 and σ_2 Known

10.2 Inferences About the Difference Between Two Population Means for Independent Samples: σ_1 and σ_2 Unknown but Equal

Case Study 10–1 Greater Hunger For News

10.3 Inferences About the Difference Between Two Population Means for Independent Samples: σ_1 and σ_2 Unknown and Unequal

10.4 Inferences About the Difference Between Two Population Means for Paired Samples

10.5 Inferences About the Difference Between Two Population Proportions for Large and Independent Samples

Case Study 10–2 Workplace Views Differ by Gender

Chapters 8 and 9 discussed the estimation and hypothesis-testing procedures for μ and p involving a single population. This chapter extends the discussion of estimation and hypothesis-testing procedures to the difference between two population means and the difference between two population proportions. For example, we may want to make a confidence interval for the difference between the mean prices of houses in California and in New York. Or we may want to test the hypothesis that the mean price of houses in California is different from that in New York. As another example, we may want to make a confidence interval for the difference between the proportions of all male and female adults who abstain from drinking. Or we may want to test the hypothesis that the proportion of all adult men who abstain from drinking is different from the proportion of all adult women who abstain from drinking. Constructing confidence intervals and testing hypotheses about population parameters are referred to as *making inferences*.

10.1 Inferences About the Difference Between Two Population Means for Independent Samples: σ_1 and σ_2 Known

Let μ_1 be the mean of the first population and μ_2 be the mean of the second population. Suppose we want to make a confidence interval and test a hypothesis about the difference between these two population means, that is, $\mu_1 - \mu_2$. Let $\bar{x}_1$ be the mean of a sample taken from the first population and $\bar{x}_2$ be the mean of a sample taken from the second population. Then, $\bar{x}_1 - \bar{x}_2$ is the sample statistic that is used to make an interval estimate and to test a hypothesis about $\mu_1 - \mu_2$. This section discusses how to make confidence intervals and test hypotheses about $\mu_1 - \mu_2$ when certain conditions (to be explained later in this section) are satisfied. First the concept of independent and dependent samples is explained below.

10.1.1 Independent versus Dependent Samples

Two samples are **independent** if they are drawn from two different populations and the elements of one sample have no relationship to the elements of the second sample. If the elements of the two samples are somehow related, then the samples are said to be **dependent**. Thus, in two independent samples, the selection of one sample has no effect on the selection of the second sample.

Definition

Independent versus Dependent Samples Two samples drawn from two populations are *independent* if the selection of one sample from one population does not affect the selection of the second sample from the second population. Otherwise, the samples are *dependent*.

Examples 10–1 and 10–2 illustrate independent and dependent samples, respectively.

■ EXAMPLE 10–1

Illustrating two independent samples.

Suppose we want to estimate the difference between the mean salaries of all male and all female executives. To do so, we draw two samples, one from the population of male executives and another from the population of female executives. These two samples are *independent* because they are drawn from two different populations, and the samples have no effect on each other. ■

■ EXAMPLE 10–2

Illustrating two dependent samples.

Suppose we want to estimate the difference between the mean weights of all participants before and after a weight loss program. To accomplish this, suppose we take a sample of 40 participants and measure their weights before and after the completion of this program. Note that these two samples include the same 40 participants. This is an example of two *dependent* samples. Such samples are also called *paired* or *matched samples.* ■

This section and Sections 10.2, 10.3, and 10.5 discuss how to make confidence intervals and test hypotheses about the difference between two population parameters when samples are independent. Section 10.4 discusses how to make confidence intervals and test hypotheses about the difference between two population means when samples are dependent.

10.1.2 Mean, Standard Deviation, and Sampling Distribution of $\bar{x}_1 - \bar{x}_2$

Suppose we select two (independent) samples from two different populations that are referred to as population 1 and population 2. Let

μ_1 = the mean of population 1
μ_2 = the mean of population 2
σ_1 = the standard deviation of population 1
σ_2 = the standard deviation of population 2
n_1 = the size of the sample drawn from population 1
n_2 = the size of the sample drawn from population 2
$\bar{x}_1$ = the mean of the sample drawn from population 1
$\bar{x}_2$ = the mean of the sample drawn from population 2

Then, as we discussed in Chapters 8 and 9, if

1. The standard deviation σ_1 of population 1 is known
2. At least one of the following two conditions is fulfilled:
   i. The sample is large (i.e., $n_1 \geq 30$)
   ii. If the sample size is small, then the population from which the sample is drawn is normally distributed

then the sampling distribution of $\bar{x}_1$ is normal with its mean equal to μ_1 and the standard deviation equal to $\sigma_1/\sqrt{n_1}$ assuming that $n_1/N_1 \leq .05$.

Similarly, if

1. The standard deviation σ_2 of population 2 is known
2. At least one of the following two conditions is fulfilled:
   i. The sample is large (i.e., $n_2 \geq 30$)
   ii. If the sample size is small, then the population from which the sample is drawn is normally distributed

then the sampling distribution of $\bar{x}_2$ is normal with its mean equal to μ_2 and the standard deviation equal to $\sigma_2/\sqrt{n_2}$ assuming that $n_2/N_2 \leq .05$.

Using these results, we can make the following statements about the mean, the standard deviation, and the shape of the sampling distribution of $\bar{x}_1 - \bar{x}_2$.

If the following conditions are satisfied,

1. The two samples are independent
2. The standard deviations σ_1 and σ_2 of the two populations are known
3. At least one of the following two conditions is fulfilled:
   i. Both samples are large (i.e., $n_1 \geq 30$ and $n_2 \geq 30$)
   ii. If either one or both sample sizes are small, then both populations from which the samples are drawn are normally distributed

then the sampling distribution of $\bar{x}_1 - \bar{x}_2$ is (approximately) normally distributed with its mean and standard deviation[1] as:

$$\mu_{\bar{x}_1 - \bar{x}_2} = \mu_1 - \mu_2$$

and

$$\sigma_{\bar{x}_1 - \bar{x}_2} = \sqrt{\frac{\sigma_1^2}{n_1} + \frac{\sigma_2^2}{n_2}}$$

In these cases, we can use the normal distribution to make a confidence interval and test a hypothesis about $\mu_1 - \mu_2$. Figure 10.1 shows the sampling distribution of $\bar{x}_1 - \bar{x}_2$ when the above conditions are fulfilled.

[1]The formula for the standard deviation of $\bar{x}_1 - \bar{x}_2$ can also be written as

$$\sigma_{\bar{x}_1 - \bar{x}_2} = \sqrt{\sigma_{\bar{x}_1}^2 + \sigma_{\bar{x}_2}^2}$$

where $\sigma_{\bar{x}_1} = \sigma_1/\sqrt{n_1}$ and $\sigma_{\bar{x}_2} = \sigma_2/\sqrt{n_2}$.

Figure 10.1

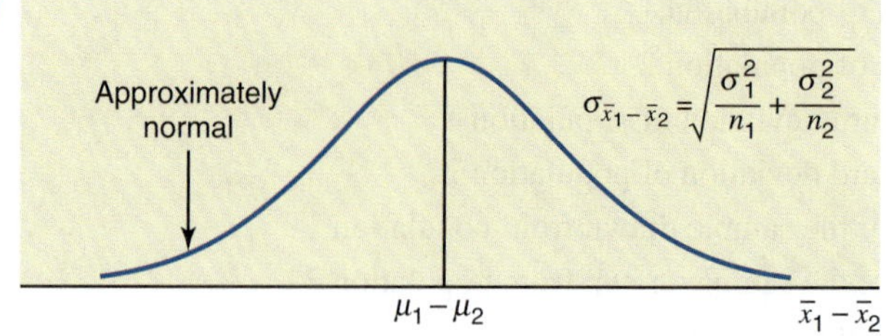

Sampling Distribution, Mean, and Standard Deviation of $\bar{x}_1 - \bar{x}_2$ When the conditions listed on the previous page are satisfied, the *sampling distribution* of $\bar{x}_1 - \bar{x}_2$ is (approximately) normal with its *mean* and *standard deviation* as follows:

$$\mu_{\bar{x}_1 - \bar{x}_2} = \mu_1 - \mu_2 \quad \text{and} \quad \sigma_{\bar{x}_1 - \bar{x}_2} = \sqrt{\frac{\sigma_1^2}{n_1} + \frac{\sigma_2^2}{n_2}}$$

Note that to apply the procedures learned in this chapter, the samples selected must be simple random samples.

10.1.3 Interval Estimation of $\mu_1 - \mu_2$

By constructing a confidence interval for $\mu_1 - \mu_2$, we find the difference between the means of two populations. For example, we may want to find the difference between the mean heights of male and female adults. The difference between the two sample means, $\bar{x}_1 - \bar{x}_2$, is the point estimator of the difference between the two population means, $\mu_1 - \mu_2$. When the conditions mentioned earlier in this section hold true, we use the normal distribution to make a confidence interval for the difference between the two population means. The following formula gives the interval estimation for $\mu_1 - \mu_2$.

Confidence Interval for $\mu_1 - \mu_2$ When using the normal distribution, the $(1 - \alpha)100\%$ *confidence interval for* $\mu_1 - \mu_2$ is

$$(\bar{x}_1 - \bar{x}_2) \pm z\sigma_{\bar{x}_1 - \bar{x}_2}$$

The value of z is obtained from the normal distribution table for the given confidence level. The value of $\sigma_{\bar{x}_1 - \bar{x}_2}$ is calculated as explained earlier. Here, $\bar{x}_1 - \bar{x}_2$ is the point estimator of $\mu_1 - \mu_2$.

Note that in the real world, σ_1 and σ_2 are never known. Consequently we will never use the procedures of this section. But we are discussing these procedures in this book for the information of the readers.

Example 10–3 illustrates the procedure to construct a confidence interval for $\mu_1 - \mu_2$ using the normal distribution.

■ EXAMPLE 10–3

Constructing a confidence interval for $\mu_1 - \mu_2$: σ_1 and σ_2 known, and samples are large.

According to *PARADE* magazine, the average starting salaries for 2004 college graduates with economics and business majors were \$40,906 and \$38,188, respectively (*PARADE*, March 13, 2005). Suppose that these averages were based on random samples of 700 economics majors and 1000 business majors, and that the population standard deviations of the starting salaries of 2004 college graduates with economics and business majors were \$5600 and \$5900, respectively. Let μ_1 and μ_2 be the population means of the starting salaries of 2004 college graduates with economics and business majors, respectively.

(a) What is the point estimate of $\mu_1 - \mu_2$?
(b) Construct a 97% confidence interval for $\mu_1 - \mu_2$.

Solution Let us refer to all 2004 college graduates with economics majors as population 1 and those with business majors as population 2. Then the respective samples are samples 1 and 2. Let $\bar{x}_1$ and $\bar{x}_2$ be the means of the two samples, respectively. From the given information,

For economics majors: $n_1 = 700, \quad \bar{x}_1 = \$40{,}906, \quad \sigma_1 = \$5600$

For business majors: $n_2 = 1000, \quad \bar{x}_2 = \$38{,}188, \quad \sigma_2 = \$5900$

(a) Since $\bar{x}_1 - \bar{x}_2$ is the point estimator of $\mu_1 - \mu_2$, the point estimate of $\mu_1 - \mu_2$ is given by the value of $\bar{x}_1 - \bar{x}_2$. Thus,

$$\text{Point estimate of } \mu_1 - \mu_2 = \$40{,}906 - \$38{,}188 = \mathbf{\$2718}$$

(b) The confidence level is $1 - \alpha = .97$. From the normal distribution table, the values of z for .0150 and .9850 areas to the left are -2.17 and 2.17. Hence, we will use $z = 2.17$ in the confidence interval formula.

The standard deviation of $\bar{x}_1 - \bar{x}_2$ is calculated as follows.

$$\sigma_{\bar{x}_1 - \bar{x}_2} = \sqrt{\frac{\sigma_1^2}{n_1} + \frac{\sigma_2^2}{n_2}} = \sqrt{\frac{(5600)^2}{700} + \frac{(5900)^2}{1000}} = \$282.1524411$$

Finally, substituting all the values in the confidence interval formula, we obtain a 97% confidence interval for $\mu_1 - \mu_2$ as

$$(\bar{x}_1 - \bar{x}_2) \pm z\sigma_{\bar{x}_1 - \bar{x}_2} = (\$40{,}906 - \$38{,}188) \pm 2.17(282.1524411)$$

$$= 2718 \pm 612.27 = \mathbf{\$2105.73 \text{ to } \$3330.27}$$

Thus, with 97% confidence we can state that the difference between the means of starting salaries for the 2004 college graduates with economics and business majors is between \$2105.73 and \$3330.27. The value $z\sigma_{\bar{x}_1 - \bar{x}_2} = \$612.27$ above is called the margin of error for this estimate. ■

Note that in Example 10–3 both sample sizes were large and the population standard deviations were known. If the standard deviations of the two populations are known, at least one of the sample sizes is small, and both populations are normally distributed, we use the normal distribution to make a confidence interval for $\mu_1 - \mu_2$. The procedure in this case is exactly the same as in Example 10–3 above.

10.1.4 Hypothesis Testing About $\mu_1 - \mu_2$

It is often necessary to compare the means of two populations. For example, we may want to know if the mean price of houses in Chicago is the same as that in Los Angeles. Similarly, we may be interested in knowing if, on average, American children spend fewer hours in school than Japanese children do. In both these cases, we will perform a test of hypothesis about $\mu_1 - \mu_2$. The alternative hypothesis in a test of hypothesis may be that the means of the two populations are different, or that the mean of the first population is greater than the mean of the second population, or that the mean of the first population is less than the mean of the second population. These three situations are described next.

1. Testing an alternative hypothesis that the means of two populations are different is equivalent to $\mu_1 \neq \mu_2$, which is the same as $\mu_1 - \mu_2 \neq 0$.
2. Testing an alternative hypothesis that the mean of the first population is greater than the mean of the second population is equivalent to $\mu_1 > \mu_2$, which is the same as $\mu_1 - \mu_2 > 0$.
3. Testing an alternative hypothesis that the mean of the first population is less than the mean of the second population is equivalent to $\mu_1 < \mu_2$, which is the same as $\mu_1 - \mu_2 < 0$.

The procedure followed to perform a test of hypothesis about the difference between two population means is similar to the one used to test hypotheses about single population parameters in Chapter 9. The procedure involves the same five steps for the critical-value approach that were used in Chapter 9 to test hypotheses about μ and p. Here, again, if the following conditions are satisfied, we will use the normal distribution to make a test of hypothesis about $\mu_1 - \mu_2$.

1. The two samples are independent
2. The standard deviations σ_1 and σ_2 of the two populations are known
3. At least one of the following two conditions is fulfilled:
   - **i.** Both samples are large (i.e., $n_1 \geq 30$ and $n_2 \geq 30$)
   - **ii.** If either one or both sample sizes are small, then both populations from which the samples are drawn are normally distributed

Test Statistic z for $\bar{x}_1 - \bar{x}_2$ When using the normal distribution, the value of the *test statistic z for $\bar{x}_1 - \bar{x}_2$* is computed as

$$z = \frac{(\bar{x}_1 - \bar{x}_2) - (\mu_1 - \mu_2)}{\sigma_{\bar{x}_1 - \bar{x}_2}}$$

The value of $\mu_1 - \mu_2$ is substituted from H_0. The value of $\sigma_{\bar{x}_1 - \bar{x}_2}$ is calculated as earlier in this section.

Example 10–4 shows how to make a test of hypothesis about $\mu_1 - \mu_2$.

■ EXAMPLE 10–4

Making a two-tailed test of hypothesis about $\mu_1 - \mu_2$: σ_1 and σ_2 are known, and samples are large.

Refer to Example 10–3 about the average starting salaries for 2004 college graduates with economics and business majors. Test at the 1% significance level if the population means of the starting salaries of 2004 college graduates with economics and business majors are different.

Solution From the information given in Example 10–3,

For economics majors: $n_1 = 700$, $\bar{x}_1 = \$40{,}906$, $\sigma_1 = \$5600$

For business majors: $n_2 = 1000$, $\bar{x}_2 = \$38{,}188$, $\sigma_2 = \$5900$

Let μ_1 and μ_2 be the population means of the starting salaries of 2004 college graduates with economics and business majors, respectively. Let $\bar{x}_1$ and $\bar{x}_2$ be the corresponding sample means.

Step 1. *State the null and alternative hypotheses.*

We are to test if the two population means are different. The two possibilities are

- **i.** The mean starting salaries of 2004 college graduates with economics and business majors are not different. In other words, $\mu_1 = \mu_2$, which can be written as $\mu_1 - \mu_2 = 0$.
- **ii.** The mean starting salaries of 2004 college graduates with economics and business majors are different. That is, $\mu_1 \neq \mu_2$, which can be written as $\mu_1 - \mu_2 \neq 0$.

Considering these two possibilities, the null and alternative hypotheses are

H_0: $\mu_1 - \mu_2 = 0$ (The two population means are not different.)

H_1: $\mu_1 - \mu_2 \neq 0$ (The two population means are different.)

Step 2. *Select the distribution to use.*

Here, the population standard deviations, σ_1 and σ_2, are known and both samples are large ($n_1 > 30$ and $n_2 > 30$). Therefore, the sampling distribution of $\bar{x}_1 - \bar{x}_2$ is approximately normal, and we use the normal distribution to perform the hypothesis test.

Step 3. *Determine the rejection and nonrejection regions.*

The significance level is given to be .01. The $\neq$ sign in the alternative hypothesis indicates that the test is two-tailed. The area in each tail of the normal distribution curve is $\alpha/2 = .01/2 = .005$. The critical values of z for .005 and .9950 areas to the left are (approximately) -2.58 and 2.58 from Table IV of Appendix C. These values are shown in Figure 10.2.

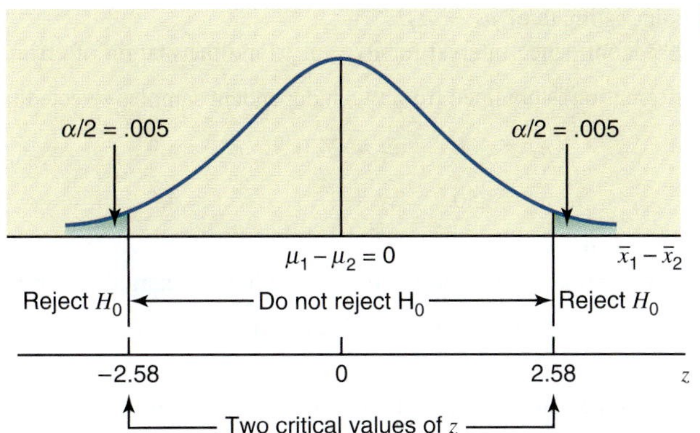

Figure 10.2

Step 4. *Calculate the value of the test statistic.*

The value of the test statistic z for $\bar{x}_1 - \bar{x}_2$ is computed as follows:

$$\sigma_{\bar{x}_1-\bar{x}_2} = \sqrt{\frac{\sigma_1^2}{n_1} + \frac{\sigma_2^2}{n_2}} = \sqrt{\frac{(5600)^2}{700} + \frac{(5900)^2}{1000}} = \$282.1524411$$

$$z = \frac{(\bar{x}_1 - \bar{x}_2) - (\mu_1 - \mu_2)}{\sigma_{\bar{x}_1-\bar{x}_2}} = \frac{(40{,}906 - 38{,}188) - \overset{\text{From } H_0}{0}}{282.1524411} = 9.63$$

Step 5. *Make a decision.*

Because the value of the test statistic $z = 9.63$ falls in the rejection region, we reject the null hypothesis H_0. Therefore, we conclude that the mean starting salaries of 2004 college graduates with economics and business majors are different.

Using the *p*-Value to Make a Decision

We can use the p-value approach to make the above decision. To do so, we keep Steps 1 and 2 above. Then in Step 3, we calculate the value of the test statistic z (as done in Step 4 above) and find the p-value for this z from the normal distribution table. In Step 4 above, the z-value for $\bar{x}_1 - \bar{x}_2$ was calculated to be 9.63. In this example, the test is two-tailed. The p-value is equal to twice the area under the sampling distribution of $\bar{x}_1 - \bar{x}_2$ to the right of $z = 9.63$. Note that $z = 9.63$ is not in the normal distribution table (Table IV). Hence, the area to the right of $z = 9.63$ can be assumed to be (approximately) zero. Therefore, the p-value is zero. As we know from Chapter 9, we will reject the null hypothesis for any α (significance level) that is greater than the p-value. Consequently, in this example, we will reject the null hypothesis for any $\alpha > 0$. Since $\alpha = .01$ here, which is greater than zero, we reject the null hypothesis. ■

EXERCISES

CONCEPTS AND PROCEDURES

10.1 Briefly explain the meaning of independent and dependent samples. Give one example of each.

10.2 Describe the sampling distribution of $\bar{x}_1 - \bar{x}_2$ for two independent samples when σ_1 and σ_2 are known and either both sample sizes are large or both populations are normally distributed. What are the mean and standard deviation of this sampling distribution?

10.3 The following information is obtained from two independent samples selected from two normally distributed populations.

$$n_1 = 24 \quad \bar{x}_1 = 5.56 \quad \sigma_1 = 1.65$$
$$n_2 = 27 \quad \bar{x}_2 = 4.80 \quad \sigma_2 = 1.58$$

a. What is the point estimate of $\mu_1 - \mu_2$?
b. Construct a 99% confidence interval for $\mu_1 - \mu_2$. Find the margin of error for this estimate.

10.4 The following information is obtained from two independent samples selected from two populations.

$$n_1 = 300 \quad \bar{x}_1 = 22.0 \quad \sigma_1 = 4.9$$
$$n_2 = 250 \quad \bar{x}_2 = 27.6 \quad \sigma_2 = 4.5$$

a. What is the point estimate of $\mu_1 - \mu_2$?
b. Construct a 95% confidence interval for $\mu_1 - \mu_2$. Find the margin of error for this estimate.

10.5 Refer to the information given in Exercise 10.3. Test at the 5% significance level if the two population means are different.

10.6 Refer to the information given in Exercise 10.4. Test at the 1% significance level if the two population means are different.

10.7 Refer to the information given in Exercise 10.4. Test at the 5% significance level if μ_1 is less than μ_2.

10.8 Refer to the information given in Exercise 10.3. Test at the 1% significance level if μ_1 is greater than μ_2.

APPLICATIONS

10.9 In parts of the eastern United States, whitetail deer are a major nuisance to farmers and homeowners, frequently damaging crops, gardens, and landscaping. A consumer organization arranges a test of two of the leading deer repellents A and B on the market. Fifty-six unfenced gardens in areas having high concentrations of deer are used for the test. Twenty-nine gardens are chosen at random to receive repellent A, and the other 27 receive repellent B. For each of the 56 gardens, the time elapsed between application of the repellent and the appearance in the garden of the first deer is recorded. For repellent A, the mean time is 101 hours. For repellent B, the mean time is 92 hours. Assume that the two populations of elapsed times have normal distributions with population standard deviations of 15 and 10 hours, respectively.

a. Let μ_1 and μ_2 be the population means of elapsed times for the two repellents, respectively. Find the point estimate of $\mu_1 - \mu_2$.
b. Find a 97% confidence interval for $\mu_1 - \mu_2$.
c. Test at the 2% significance level whether the mean elapsed times for repellents A and B are different. Use both approaches, the critical-value and *p*-value, to perform this test.

10.10 A researcher wanted to investigate if the male and female workers in a city commute the same distance to work. A sample of 25 male workers showed that they commute an average of 21 miles to work. A sample of 22 female workers gave a mean commuting distance of 16 miles. Assume that the two populations of commuting distances have normal distributions with population standard deviations of 5.2 miles and 4.4 miles, respectively.

a. Let μ_1 and μ_2 be the population means of commuting distances for male and female workers, respectively, in this city. What is the point estimate of $\mu_1 - \mu_2$?
b. Construct a 97% confidence interval for $\mu_1 - \mu_2$.
c. Using the 2% significance level, can you conclude that the mean commuting distances are different for male and female workers in this city? Use both approaches to make this test.

10.11 A business consultant wanted to investigate if providing day-care facilities on premises by companies reduces the absentee rate of working mothers with six-year-old or younger children. She took a sample of 45 such mothers from companies that provide day-care facilities on premises. These mothers missed an average of 6.4 days from work last year. Another sample of 50 such mothers taken from companies that do not provide day-care facilities on premises showed that these mothers missed an average of 9.3 days last year. Assume that the standard deviations for the two populations are 1.20 days and 1.85 days, respectively.

a. Construct a 98% confidence interval for the difference between the two population means.
b. Using the 2.5% significance level, can you conclude that the mean number of days missed per year by mothers working for companies that provide day-care facilities on premises is less than the mean number of days missed per year by mothers working for companies that do not provide day-care facilities on premises?
c. What are the Type I error and its probability for the test of hypothesis in part b? Explain.

10.12 Employees of a large corporation are concerned about the declining quality of medical services provided by their group health insurance. A random sample of 100 office visits by employees of this corporation to primary care physicians during 2004 found that the doctors spent an average of 19 minutes with each patient. This year a random sample of 108 such visits showed that doctors spent an average of 15.5 minutes with each patient. Assume that the standard deviations for the two populations are 2.7 and 2.1 minutes, respectively.

a. Construct a 95% confidence interval for the difference between the two population means for these two years.
b. Using the 2.5% level of significance, can you conclude that the mean time spent by doctors with each patient is lower for this year than for 2004?
c. What would your decision be in part b if the probability of making a Type I error were zero? Explain.

10.13 A car magazine is comparing the total repair costs incurred during the first three years on two sports cars, the T-999 and the XPY. Random samples of 45 T-999s and 51 XPYs are taken. All 96 cars are three years old and have similar mileages. The mean of repair costs for the 45 T-999 cars is \$3300 for the first three years. For the 51 XPY cars, this mean is \$3850. Assume that the standard deviations for the two populations are \$800 and \$1000, respectively.

a. Construct a 99% confidence interval for the difference between the two population means.
b. Using the 1% significance level, can you conclude that such mean repair costs are different for these two types of cars?
c. What would your decision be in part b if the probability of making a Type I error were zero? Explain.

10.14 The management at New Century Bank claims that the mean waiting time for all customers at its branches is less than that at the Public Bank, which is its main competitor. A business consulting firm took a sample of 200 customers from the New Century Bank and found that they waited an average of 4.5 minutes before being served. Another sample of 300 customers taken from the Public Bank showed that these customers waited an average of 4.75 minutes before being served. Assume that the standard deviations for the two populations are 1.2 minutes and 1.5 minutes, respectively.

a. Make a 97% confidence interval for the difference between the two population means.
b. Test at the 2.5% significance level whether the claim of the management of the New Century Bank is true.
c. Calculate the *p*-value for the test of part b. Based on this *p*-value, would you reject the null hypothesis if $\alpha = .01$? What if $\alpha = .05$?

10.15 Maine Mountain Dairy claims that its eight-ounce low-fat yogurt cups contain, on average, fewer calories than the eight-ounce low-fat yogurt cups produced by a competitor. A consumer agency wanted to check this claim. A sample of 27 such yogurt cups produced by this company showed that they contained an average of 141 calories per cup. A sample of 25 such yogurt cups produced by its competitor showed that they contained an average of 144 calories per cup. Assume that the two populations are normally distributed with population standard deviations of 5.5 calories and 6.4 calories, repectively.

a. Make a 98% confidence interval for the difference between the mean number of calories in the eight-ounce low-fat yogurt cups produced by the two companies.
b. Test at the 1% significance level whether Maine Mountain Dairy's claim is true.
c. Calculate the *p*-value for the test of part b. Based on this *p*-value, would you reject the null hypothesis if $\alpha = .005$? What if $\alpha = .025$?

10.2 Inferences About the Difference Between Two Population Means for Independent Samples: σ_1 and σ_2 Unknown but Equal

This section discusses making a confidence interval and testing a hypothesis about the difference between the means of two populations, $\mu_1 - \mu_2$, assuming that the standard deviations σ_1 and σ_2 of these populations are not known but they are assumed to be equal. There are some other conditions, explained below, that must be fulfilled to use the procedures discussed in this section.

If the following conditions are satisfied,

1. The two samples are independent
2. The standard deviations σ_1 and σ_2 of the two populations are unknown but they can be assumed to be equal, that is $\sigma_1 = \sigma_2$

3. At least one of the following two conditions is fulfilled:
   i. Both samples are large (i.e., $n_1 \geq 30$ and $n_2 \geq 30$)
   ii. If either one or both sample sizes are small, then both populations from which the samples are drawn are normally distributed

then we use the t distribution to make a confidence interval and test a hypothesis about the difference between the means of two populations, $\mu_1 - \mu_2$.

When the standard deviations of the two populations are equal, we can use σ for both σ_1 and σ_2. Because σ is unknown, we replace it by its point estimator s_p, which is called the **pooled sample standard deviation** (hence, the subscript p). The value of s_p is computed by using the information from the two samples as follows.

Pooled Standard Deviation for Two Samples The *pooled standard deviation for two samples* is computed as

$$s_p = \sqrt{\frac{(n_1 - 1)s_1^2 + (n_2 - 1)s_2^2}{n_1 + n_2 - 2}}$$

where n_1 and n_2 are the sizes of the two samples and s_1^2 and s_2^2 are the variances of the two samples. Here s_p is an estimator of σ.

In this formula, $n_1 - 1$ are the degrees of freedom for sample 1, $n_2 - 1$ are the degrees of freedom for sample 2, and $n_1 + n_2 - 2$ are the *degrees of freedom for the two samples taken together.* Note that s_p is an estimator of the standard deviation, σ, of each of the two populations.

When s_p is used as an estimator of σ, the standard deviation $\sigma_{\bar{x}_1 - \bar{x}_2}$ of $\bar{x}_1 - \bar{x}_2$ is estimated by $s_{\bar{x}_1 - \bar{x}_2}$. The value of $s_{\bar{x}_1 - \bar{x}_2}$ is calculated by using the following formula.

Estimator of the Standard Deviation of $\bar{x}_1 - \bar{x}_2$ The *estimator of the standard deviation of* $\bar{x}_1 - \bar{x}_2$ is

$$s_{\bar{x}_1 - \bar{x}_2} = s_p \sqrt{\frac{1}{n_1} + \frac{1}{n_2}}$$

Now we are ready to discuss the procedures that are used to make confidence intervals and test hypotheses about $\mu_1 - \mu_2$ for small and independent samples selected from two populations with unknown but equal standard deviations.

10.2.1 Interval Estimation of $\mu_1 - \mu_2$

As was mentioned earlier in this chapter, the difference between the two sample means, $\bar{x}_1 - \bar{x}_2$, is the point estimator of the difference between the two population means, $\mu_1 - \mu_2$. The following formula gives the confidence interval for $\mu_1 - \mu_2$ when the t distribution is used and the conditions mentioned earlier in this section are fulfilled.

Confidence Interval for $\mu_1 - \mu_2$ The $(1 - \alpha)100\%$ *confidence interval for* $\mu_1 - \mu_2$ is

$$(\bar{x}_1 - \bar{x}_2) \pm t s_{\bar{x}_1 - \bar{x}_2}$$

where the value of t is obtained from the t distribution table for the given confidence level and $n_1 + n_2 - 2$ degrees of freedom, and $s_{\bar{x}_1 - \bar{x}_2}$ is calculated as explained earlier.

Example 10–5 describes the procedure to make a confidence interval for $\mu_1 - \mu_2$ using the t distribution.

■ EXAMPLE 10–5

Constructing a confidence interval for $\mu_1 - \mu_2$: two independent samples, unknown but equal σ_1 and σ_2.

A consumer agency wanted to estimate the difference in the mean amounts of caffeine in two brands of coffee. The agency took a sample of 15 one-pound jars of Brand I coffee that showed the mean amount of caffeine in these jars to be 80 milligrams per jar with a standard deviation of 5 milligrams. Another sample of 12 one-pound jars of Brand II coffee gave a mean amount of caffeine equal to 77 milligrams per jar with a standard deviation of 6 milligrams. Construct a 95% confidence interval for the difference between the mean amounts of caffeine in one-pound jars of these two brands of coffee. Assume that the two populations are normally distributed and that the standard deviations of the two populations are equal.

Solution Let μ_1 and μ_2 be the mean amounts of caffeine per jar in all one-pound jars of Brands I and II, respectively, and let $\bar{x}_1$ and $\bar{x}_2$ be the means of the two respective samples. From the given information,

$$n_1 = 15 \quad \bar{x}_1 = 80 \text{ milligrams} \quad s_1 = 5 \text{ milligrams}$$

$$n_2 = 12 \quad \bar{x}_2 = 77 \text{ milligrams} \quad s_2 = 6 \text{ milligrams}$$

The confidence level is $1 - \alpha = .95$.

Here, σ_1 and σ_2 are unknown but assumed to be equal, the samples are independent (taken from two different populations), and the sample sizes are small but the two populations are normally distributed. Hence, we will use the t-distribution to make the confidence interval for $\mu_1 - \mu_2$ as all conditions mentioned in the beginning of this section are satisfied.

First we calculate the standard deviation of $\bar{x}_1 - \bar{x}_2$ as follows. Note that since it is assumed that σ_1 and σ_2 are equal, we will use s_p to calculate $s_{\bar{x}_1 - \bar{x}_2}$.

$$s_p = \sqrt{\frac{(n_1 - 1)s_1^2 + (n_2 - 1)s_2^2}{n_1 + n_2 - 2}} = \sqrt{\frac{(15 - 1)(5)^2 + (12 - 1)(6)^2}{15 + 12 - 2}} = 5.46260011$$

$$s_{\bar{x}_1 - \bar{x}_2} = s_p\sqrt{\frac{1}{n_1} + \frac{1}{n_2}} = (5.46260011)\sqrt{\frac{1}{15} + \frac{1}{12}} = 2.11565593$$

Next, to find the t value from the t distribution table, we need to know the area in each tail of the t distribution curve and the degrees of freedom.

$$\text{Area in each tail} = \alpha/2 = (1 - .95)/2 = .025$$

$$\text{Degrees of freedom} = n_1 + n_2 - 2 = 15 + 12 - 2 = 25$$

The t value for $df = 25$ and .025 area in the right tail of the t distribution curve is 2.060. The 95% confidence interval for $\mu_1 - \mu_2$ is

$$(\bar{x}_1 - \bar{x}_2) \pm t s_{\bar{x}_1 - \bar{x}_2} = (80 - 77) \pm 2.060(2.11565593)$$

$$= 3 \pm 4.36 = \mathbf{-1.36 \text{ to } 7.36 \text{ milligrams}}$$

Thus, with 95% confidence we can state that based on these two sample results, the difference in the mean amounts of caffeine in one-pound jars of these two brands of coffee lies between −1.36 and 7.36 milligrams. Because the lower limit of the interval is negative, it is possible that the mean amount of caffeine is greater in the second brand than in the first brand of coffee.

Note that the value of $\bar{x}_1 - \bar{x}_2$, which is $80 - 77 = 3$, gives the point estimate of $\mu_1 - \mu_2$. The value of $t s_{\bar{x}_1 - \bar{x}_2}$, which is 4.36, is the margin of error. ■

10.2.2 Hypothesis Testing About $\mu_1 - \mu_2$

When the conditions mentioned in the beginning of Section 10.2 are satisfied, the t distribution is applied to make a hypothesis test about the difference between two population means. The test statistic in this case is t, which is calculated as follows.

Test Statistic t for $\bar{x}_1 - \bar{x}_2$ The value of the *test statistic t for $\bar{x}_1 - \bar{x}_2$* is computed as

$$t = \frac{(\bar{x}_1 - \bar{x}_2) - (\mu_1 - \mu_2)}{s_{\bar{x}_1 - \bar{x}_2}}$$

The value of $\mu_1 - \mu_2$ in this formula is substituted from the null hypothesis and $s_{\bar{x}_1 - \bar{x}_2}$ is calculated as explained earlier in Section 10.2.1.

Examples 10–6 and 10–7 illustrate how a test of hypothesis about the difference between two population means for small and independent samples that are selected from two populations with equal standard deviations is conducted using the t distribution.

■ EXAMPLE 10–6

Making a two-tailed test of hypothesis about $\mu_1 - \mu_2$: two independent samples, and unknown but equal σ_1 and σ_2.

A sample of 14 cans of Brand I diet soda gave the mean number of calories of 23 per can with a standard deviation of 3 calories. Another sample of 16 cans of Brand II diet soda gave the mean number of calories of 25 per can with a standard deviation of 4 calories. At the 1% significance level, can you conclude that the mean numbers of calories per can are different for these two brands of diet soda? Assume that the calories per can of diet soda are normally distributed for each of the two brands and that the standard deviations for the two populations are equal.

Solution Let μ_1 and μ_2 be the mean numbers of calories per can for diet soda of Brand I and Brand II, respectively, and let $\bar{x}_1$ and $\bar{x}_2$ be the means of the respective samples. From the given information,

$$n_1 = 14 \quad \bar{x}_1 = 23 \quad s_1 = 3$$
$$n_2 = 16 \quad \bar{x}_2 = 25 \quad s_2 = 4$$

The significance level is $\alpha = .01$.

Step 1. *State the null and alternative hypotheses.*

We are to test for the difference in the mean numbers of calories per can for the two brands. The null and alternative hypotheses are

H_0: $\mu_1 - \mu_2 = 0$ (The mean numbers of calories are not different)

H_1: $\mu_1 - \mu_2 \neq 0$ (The mean numbers of calories are different)

Step 2. *Select the distribution to use.*

Here, the two samples are independent, σ_1 and σ_2 are unknown but equal, and the sample sizes are small but both populations are normally distributed. Hence, all conditions mentioned in the beginning of Section 10.2 are fulfilled. Consequently, we will use the t distribution.

Step 3. *Determine the rejection and nonrejection regions.*

The $\neq$ sign in the alternative hypothesis indicates that the test is two-tailed. The significance level is .01. Hence,

$$\text{Area in each tail} = \alpha/2 = .01/2 = .005$$

$$\text{Degrees of freedom} = n_1 + n_2 - 2 = 14 + 16 - 2 = 28$$

The critical values of t for $df = 28$ and .005 area in each tail of the t distribution curve are -2.763 and 2.763, as shown in Figure 10.3.

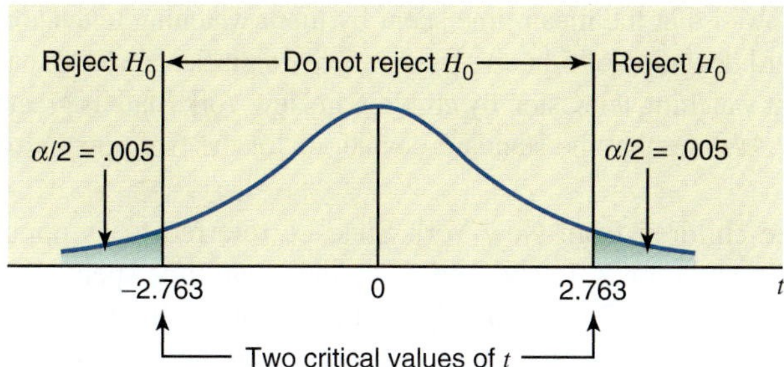

Figure 10.3

Step 4. *Calculate the value of the test statistic.*

The value of the test statistic t for $\bar{x}_1 - \bar{x}_2$ is computed as follows:

$$s_p = \sqrt{\frac{(n_1 - 1)s_1^2 + (n_2 - 1)s_2^2}{n_1 + n_2 - 2}} = \sqrt{\frac{(14 - 1)(3)^2 + (16 - 1)(4)^2}{14 + 16 - 2}} = 3.57071421$$

$$s_{\bar{x}_1 - \bar{x}_2} = s_p\sqrt{\frac{1}{n_1} + \frac{1}{n_2}} = (3.57071421)\sqrt{\frac{1}{14} + \frac{1}{16}} = 1.30674760$$

From H_0

$$t = \frac{(\bar{x}_1 - \bar{x}_2) - (\mu_1 - \mu_2)}{s_{\bar{x}_1 - \bar{x}_2}} = \frac{(23 - 25) - 0}{1.30674760} = -1.531$$

Step 5. *Make a decision.*

Because the value of the test statistic $t = -1.531$ for $\bar{x}_1 - \bar{x}_2$ falls in the nonrejection region, we fail to reject the null hypothesis. Consequently we conclude that there is no difference in the mean numbers of calories per can for the two brands of diet soda. The difference in $\bar{x}_1$ and $\bar{x}_2$ observed for the two samples may have occurred due to sampling error only.

Using the *p*-Value to Make a Decision

We can use the p-value approach to make the above decision. To do so, we keep Steps 1 and 2 of this example. Then in Step 3, we calculate the value of the test statistic t (as done in Step 4 above) and then find the p-value for this t from the t distribution table (Table V of Appendix C) or by using technology. In Step 4 above, the t-value for $\bar{x}_1 - \bar{x}_2$ was calculated to be -1.531. In this example, the test is two-tailed. Therefore, the p-value is equal to twice the area under the t distribution curve to the left of $t = -1.531$. If we have access to technology, we can use it to find the exact p-value, which will be .137. If we use the t distribution table, we can only find the range for the p-value. From Table V of Appendix C, for $df = 28$, the two values that include 1.531 are 1.313 and 1.701. (Note that we use the positive value of t although our t is negative.) Thus, the test statistic $t = -1.531$ falls between -1.313 and -1.701. The areas in the t distribution table that correspond to 1.313 and 1.701 are .10 and .05, respectively. Because it is a two-tailed test, the p-value for $t = -1.531$ is between $2(.10) = .20$ and $2(.05) = .10$, which can be written as:

$$.10 < p\text{-value} < .20$$

As we know from Chapter 9, we will reject the null hypothesis for any α (significance level) that is greater than the p-value. Consequently, in this example, we will reject the null hypothesis for any $\alpha \geq .20$ using the above range and not reject for $\alpha \leq .10$. If we use technology, we will reject the null hypothesis for $\alpha > .137$. Since $\alpha = .01$ in this example, which is smaller than both .10 and .137, we fail to reject the null hypothesis. ■

■ EXAMPLE 10–7

A sample of 40 children from New York State showed that the mean time they spend watching television is 28.50 hours per week with a standard deviation of 4 hours. Another sample of 35 children

Making a right-tailed test of hypothesis about $\mu_1 - \mu_2$: two independent samples, σ_1 and σ_2 unknown but equal, and both samples are large.

from California showed that the mean time spent by them watching television is 23.25 hours per week with a standard deviation of 5 hours. Using a 2.5% significance level, can you conclude that the mean time spent watching television by children in New York State is greater than that for children in California? Assume that the standard deviations for the two populations are equal.

Solution Let the children from New York State be referred to as population 1 and those from California as population 2. Let μ_1 and μ_2 be the mean time spent watching television by children in populations 1 and 2, respectively, and let $\bar{x}_1$ and $\bar{x}_2$ be the mean time spent watching television by children in the respective samples. From the given information,

$$n_1 = 40 \quad \bar{x}_1 = 28.50 \text{ hours} \quad s_1 = 4 \text{ hours}$$

$$n_2 = 35 \quad \bar{x}_2 = 23.25 \text{ hours} \quad s_2 = 5 \text{ hours}$$

The significance level is $\alpha = .025$.

Step 1. *State the null and alternative hypotheses.*

The two possible decisions are:

i. The mean time spent watching television by children in New York State is not greater than that for children in California. This can be written as $\mu_1 = \mu_2$ or $\mu_1 - \mu_2 = 0$.

ii. The mean time spent watching television by children in New York State is greater than that for children in California. This can be written as $\mu_1 > \mu_2$ or $\mu_1 - \mu_2 > 0$.

Hence, the null and alternative hypotheses are

$$H_0: \mu_1 - \mu_2 = 0$$

$$H_1: \mu_1 - \mu_2 > 0$$

Note that the null hypothesis can also be written as $\mu_1 - \mu_2 \leq 0$.

Step 2. *Select the distribution to use.*

Here, the two samples are independent (taken from two different populations), σ_1 and σ_2 are unknown but assumed to be equal, and both samples are large. Hence, all conditions mentioned in the beginning of Section 10.2 are fulfilled. Consequently, we use the t distribution to make the test.

Step 3. *Determine the rejection and nonrejection regions.*

The > sign in the alternative hypothesis indicates that the test is right-tailed. The significance level is .025.

$$\text{Area in the right tail of the } t \text{ distribution} = \alpha = .025$$

$$\text{Degrees of freedom} = n_1 + n_2 - 2 = 40 + 35 - 2 = 73$$

From the t distribution table, the critical value of t for $df = 73$ and .025 area in the right tail of the t distribution is 1.993. This value is shown in Figure 10.4.

Figure 10.4

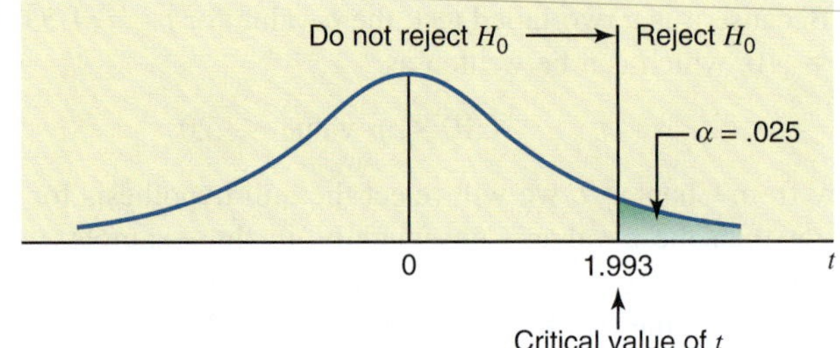

Step 4. *Calculate the value of the test statistic.*

The value of the test statistic t for $\bar{x}_1 - \bar{x}_2$ is computed as follows:

$$s_p = \sqrt{\frac{(n_1 - 1)s_1^2 + (n_2 - 1)s_2^2}{n_1 + n_2 - 2}} = \sqrt{\frac{(40 - 1)(4)^2 + (35 - 1)(5)^2}{40 + 35 - 2}} = 4.49352655$$

$$s_{\bar{x}_1 - \bar{x}_2} = s_p \sqrt{\frac{1}{n_1} + \frac{1}{n_2}} = (4.49352655)\sqrt{\frac{1}{40} + \frac{1}{35}} = 1.04004930$$

From H_0

$$t = \frac{(\bar{x}_1 - \bar{x}_2) - (\mu_1 - \mu_2)}{s_{\bar{x}_1 - \bar{x}_2}} = \frac{(28.50 - 23.25) - 0}{1.04004930} = 5.048$$

Step 5. *Make a decision.*

Because the value of the test statistic $t = 5.048$ for $\bar{x}_1 - \bar{x}_2$ falls in the rejection region, we reject the null hypothesis H_0. Hence, we conclude that children in New York State spend more time, on average, watching TV than children in California.

Using the *p*-Value to Make a Decision

To use the *p*-value approach to make the above decision, we keep Steps 1 and 2 of this example. Then in Step 3, we calculate the value of the test statistic *t* (as done in Step 4 above) and then find the *p*-value for this *t* from the *t* distribution table (Table V of Appendix C) or by using technology. In Step 4 above, the *t*-value for $\bar{x}_1 - \bar{x}_2$ was calculated to be 5.048. In this example, the test is right-tailed. Therefore, the *p*-value is equal to the area under the *t* distribution curve to the right of $t = 5.048$. If we have access to technology, we can use it to find the exact *p*-value, which will be .000. If we use the *t* distribution table, for $df = 73$, the value of the test statistic $t = 5.048$ is larger than 3.206. Therefore, the *p*-value for $t = 5.048$ is less than .001, which can be written as:

$$p\text{-value} < .001$$

Since we will reject the null hypothesis for any α (significance level) greater than the *p*-value, here we reject the null hypothesis because $\alpha = .025$ is greater than both the *p*-values, .001 obtained above from the table and .000 obtained by using technology. Note that obtaining the *p*-value = .000 from technology does not mean that the *p*-value is zero. It means that when it is rounded to three digits after decimal, it is .000. ■

Note: What if the Sample Sizes Are Too Large?

In this section, we used the *t* distribution to make confidence intervals and perform tests of hypothesis about $\mu_1 - \mu_2$. When both sample sizes are large, it does not matter how large (over 30) the sample sizes are if we are using technology. However, if we are using the *t* distribution table (Table V of Appendix C), it may pose a problem if samples are too large. Table V in Appendix C goes up to only 75 degrees of freedom. Thus, if the degrees of freedom are larger than 75, we cannot use Table V to find the critical value(s) of *t*. As mentioned in Chapters 8 and 9, in such a situation, there are two options:

1. Use the *t* value from the last row (the row of ∞) in Table V.
2. Use the normal distribution as an approximation to the *t* distribution.

Exercises 10.31 and 10.32 at the end of this section and Case Study 10–1 present such situations.

EXERCISES

CONCEPTS AND PROCEDURES

10.16 Explain what conditions must hold true to use the *t* distribution to make a confidence interval and to test a hypothesis about $\mu_1 - \mu_2$ for two independent samples selected from two populations with unknown but equal standard deviations.

10.17 The following information was obtained from two independent samples selected from two normally distributed populations with unknown but equal standard deviations.

$$n_1 = 25 \quad \bar{x}_1 = 12.50 \quad s_1 = 3.75$$

$$n_2 = 20 \quad \bar{x}_2 = 14.60 \quad s_2 = 3.10$$

a. What is the point estimate of $\mu_1 - \mu_2$? **b.** Construct a 95% confidence interval for $\mu_1 - \mu_2$.

CASE STUDY

10–1

GREATER HUNGER FOR NEWS

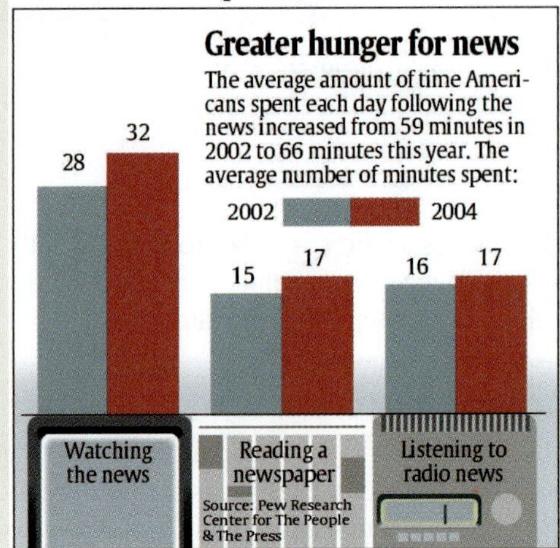
USA TODAY Snapshots
Greater hunger for news
The average amount of time Americans spent each day following the news increased from 59 minutes in 2002 to 66 minutes this year. The average number of minutes spent:
2002
2004
28
32
15
17
16
17
Watching the news
Reading a newspaper
Listening to radio news
Source: Pew Research Center for The People & The Press
By Sam Ward, USA TODAY

The above chart shows the average time spent per day following news by Americans in 2002 and 2004. These results are based on surveys by Pew Research Center for The People & The Press. For example, according to this survey, Americans spent an average of 28 minutes per day watching news on television in 2002, and the corresponding number for 2004 is 32 minutes. If we know the sample sizes and the sample standard deviations for both these years, we can make a confidence interval and perform a hypothesis test for the difference in the mean times spent following news by Americans in 2002 and 2004.

Suppose the sample sizes for 2002 and 2004 are 1200 and 1500, respectively. Let μ_1 and μ_2 be the mean times spent watching news on television by all Americans in 2002 and 2004, respectively. Let $\bar{x}_1$ and $\bar{x}_2$ be the corresponding sample means. Suppose the sample standard deviations for times spent watching news in these two years are 5 minutes and 6 minutes, respectively. Also, assume that although the population standard deviations are not known, they are (approximately) equal. Then, from the given information:

For 2002: $n_1 = 1200$ $\quad \bar{x}_1 = 28$ minutes $\quad s_1 = 5$ minutes

For 2004: $n_2 = 1500$ $\quad \bar{x}_2 = 32$ minutes $\quad s_2 = 6$ minutes

Below we make a confidence interval and perform a hypothesis test about $\mu_1 - \mu_2$ for times spent watching news on television by all Americans in 2002 and 2004.

1. Confidence interval for $\mu_1 - \mu_2$

Suppose we want to make a 98% confidence interval for $\mu_1 - \mu_2$. The area in each tail of the t distribution and the degrees of freedom are

$$\text{Area in each tail} = \alpha/2 = (1 - .98)/2 = .01$$

$$\text{Degrees of freedom} = n_1 + n_2 - 2 = 1200 + 1500 - 2 = 2698$$

Because $df = 2698$ is not in the t distribution table, we will use the last row of Table V of Appendix C to obtain the t value for .01 area in the right tail. This t value is 2.326.

10.18 The following information was obtained from two independent samples selected from two populations with unknown but equal standard deviations.

$$n_1 = 38 \quad \bar{x}_1 = 33.75 \quad s_1 = 5.25$$

$$n_2 = 32 \quad \bar{x}_2 = 28.50 \quad s_2 = 4.55$$

a. What is the point estimate of $\mu_1 - \mu_2$?
b. Construct a 99% confidence interval for $\mu_1 - \mu_2$.

We calculate the standard deviation of $\bar{x}_1 - \bar{x}_2$ as follows.

$$s_p = \sqrt{\frac{(n_1 - 1)s_1^2 + (n_2 - 1)s_2^2}{n_1 + n_2 - 2}} = \sqrt{\frac{(1200 - 1)(5)^2 + (1500 - 1)(6)^2}{1200 + 1500 - 2}} = 5.57777412$$

$$s_{\bar{x}_1 - \bar{x}_2} = s_p\sqrt{\frac{1}{n_1} + \frac{1}{n_2}} = (5.57777412)\sqrt{\frac{1}{1200} + \frac{1}{1500}} = .21602626$$

Hence, the 98% confidence interval for $\mu_1 - \mu_2$ is

$$(\mu_1 - \mu_2) \pm ts_{\bar{x}_1 - \bar{x}_2} = (28 - 32) \pm 2.326(.21602626) = -4 \pm .50$$

$$= \mathbf{-4.50 \text{ to } -3.50 \text{ minutes}}$$

Thus, the 98% confidence interval for $\mu_1 - \mu_2$ is −4.50 to −3.50 minutes.

2. Hypothesis test about $\mu_1 - \mu_2$

Suppose we want to test, at the 1% significance level, whether the mean time spent per day watching news on television by all Americans in 2002 is less than that for 2004. In other words, we are to test if μ_1 is less than μ_2. The null and alternative hypotheses are

$$H_0: \mu_1 = \mu_2 \quad \text{or} \quad \mu_1 - \mu_2 = 0$$

$$H_1: \mu_1 < \mu_2 \quad \text{or} \quad \mu_1 - \mu_2 < 0$$

Note that the test is left-tailed. Because the population standard deviations are not known, we will use the t distribution. The area in the left tail of the t distribution and the degrees of freedom are

Area in the left tail $= \alpha = .01$

Degrees of freedom $= n_1 + n_2 - 2 = 1200 + 1500 - 2 = 2698$

Because $df = 2698$ is not in the t distribution table, we will use the last row of Table V of Appendix C to obtain the t value for .01 area in the left tail. This t value is −2.326.

As calculated above, the standard deviation of $\bar{x}_1 - \bar{x}_2$ is

$$s_{\bar{x}_1 - \bar{x}_2} = .21602626$$

The value of the test statistic t for $\bar{x}_1 - \bar{x}_2$ is computed as follows.

From H_0

$$t = \frac{(\bar{x}_1 - \bar{x}_2) - (\mu_1 - \mu_2)}{s_{\bar{x}_1 - \bar{x}_2}} = \frac{(28 - 32) - 0}{.21602626} = -18.516$$

Because the value of the test statistic $t = -18.516$ is smaller than the critical value of $t = -2.326$, it falls in the rejection region. Consequently, we reject the null hypothesis and conclude that the mean time spent per day watching news on television by Americans in 2002 is less than that for 2004.

We can also use the p-value approach to make this decision. In this example, the test is left-tailed. As calculated above, the t value for $\bar{x}_1 - \bar{x}_2$ is −18.516. From the last row of the t distribution table, −18.516 is less than −3.090. Therefore, the p-value is less than .001. Since $\alpha = .01$ in this example is greater than this p-value of .001, we reject the null hypothesis and conclude that the mean time spent per day watching news on television by all Americans in 2002 is less than that for 2004.

Source: The chart reproduced with permission from *USA TODAY*, July 6, 2004. Copyright © 2004, *USA TODAY.*

10.19 Refer to the information given in Exercise 10.17. Test at the 5% significance level if the two population means are different.

10.20 Refer to the information given in Exercise 10.18. Test at the 1% significance level if the two population means are different.

10.21 Refer to the information given in Exercise 10.17. Test at the 1% significance level if μ_1 is less than μ_2.

10.22 Refer to the information given in Exercise 10.18. Test at the 5% significance level if μ_1 is greater than μ_2.

10.23 The following information was obtained from two independent samples selected from two normally distributed populations with unknown but equal standard deviations.

Sample 1:	27	39	25	33	21	35	30	26	25	31	35	30	28
Sample 2:	24	28	23	25	24	22	29	26	29	28	19	29	

a. Let μ_1 be the mean of population 1 and μ_2 be the mean of population 2. What is the point estimate of $\mu_1 - \mu_2$?
b. Construct a 98% confidence interval for $\mu_1 - \mu_2$.
c. Test at the 1% significance level if μ_1 is greater than μ_2.

10.24 The following information was obtained from two independent samples selected from two normally distributed populations with unknown but equal standard deviations.

Sample 1:	13	14	9	12	8	10	5	10	9	12	16	
Sample 2:	16	18	11	19	14	17	13	16	17	18	22	12

a. Let μ_1 be the mean of population 1 and μ_2 be the mean of population 2. What is the point estimate of $\mu_1 - \mu_2$?
b. Construct a 99% confidence interval for $\mu_1 - \mu_2$.
c. Test at the 2.5% significance level if μ_1 is lower than μ_2.

APPLICATIONS

10.25 The management of a supermarket wanted to investigate whether the male customers spend less money on average than the female customers. A sample of 35 male customers who shopped at this supermarket showed that they spent an average of \$80 with a standard deviation of \$17.50. Another sample of 40 female customers who shopped at the same supermarket showed that they spent an average of \$96 with a standard deviation of \$14.40. Assume that the amounts spent at this supermarket by all male and all female customers have equal but unknown population standard deviations.

a. Construct a 99% confidence interval for the difference between the mean amounts spent by all male and all female customers at this supermarket.
b. Using the 2.5% significance level, can you conclude that the mean amount spent by all male customers at this supermarket is less than that spent by all female customers?

10.26 The department of labor in a state wanted to find the compensation of hotel employees. A random sample of 32 cleaning persons produced the mean hourly earnings (including tips) of \$10.60 with a standard deviation of \$1.02. A random sample of 37 bellhops gave the mean hourly earnings (including tips) of \$11.57 with a standard deviation of \$1.34. Assume that the hourly earnings of both groups have equal but unknown population standard deviations.

a. Construct a 99% confidence interval for the difference between the corresponding population means of the two groups.
b. Using the 1% significance level, can you conclude that the mean hourly earnings of all cleaning persons are lower than those of all bellhops in this state?

10.27 An insurance company wants to know if the average speed at which men drive cars is greater than that of women drivers. The company took a random sample of 27 cars driven by men on a highway and found the mean speed to be 72 miles per hour with a standard deviation of 2.2 miles per hour. Another sample of 18 cars driven by women on the same highway gave a mean speed of 68 miles per hour with a standard deviation of 2.5 miles per hour. Assume that the speeds at which all men and all women drive cars on this highway are both normally distributed with the same population standard deviation.

a. Construct a 98% confidence interval for the difference between the mean speeds of cars driven by all men and all women on this highway.
b. Test at the 1% significance level whether the mean speed of cars driven by all men drivers on this highway is greater than that of cars driven by all women drivers.

10.28 A high school counselor wanted to know if tenth-graders at her high school tend to have more free time than the twelfth-graders. She took random samples of 25 tenth-graders and 23 twelfth-graders. Each student was asked to record the amount of free time he or she had in a typical week. The mean for the tenth-graders was found to be 29 hours of free time per week with a standard deviation of 7.0 hours. For the twelfth-graders, the mean was 22 hours of free time per week with a standard deviation of 6.2 hours. Assume that the two populations are normally distributed with equal but unknown population standard deviations.

a. Make a 90% confidence interval for the difference between the corresponding population means.
b. Test at the 5% significance level whether the two population means are different.

10.29 A company claims that its medicine, Brand A, provides faster relief from pain than another company's medicine, Brand B. A researcher tested both brands of medicine on two groups of randomly selected patients. The results of the test are given in the following table. The mean and standard deviation of relief times are in minutes.

Brand	Sample Size	Mean of Relief Times	Standard Deviation of Relief Times
A	25	44	11
B	22	49	9

a. Construct a 99% confidence interval for the difference between the mean relief times for the two brands of medicine.

b. Test at the 1% significance level whether the mean relief time for Brand A is less than that for Brand B.

Assume that the two populations are normally distributed with unknown but equal standard deviations.

10.30 A consumer organization tested two paper shredders, the Piranha and the Crocodile, designed for home use. Each of 10 randomly selected volunteers shredded 100 sheets of paper with the Piranha, and then another sample of 10 randomly selected volunteers shredded 100 sheets with the Crocodile. The Piranha took an average of 203 seconds to shred 100 sheets with a standard deviation of 6 seconds. The Crocodile took an average of 187 seconds to shred 100 sheets with a standard deviation of 5 seconds. Assume that the shredding times for both machines are normally distributed with equal but unknown standard deviations.

a. Construct a 99% confidence interval for the difference between the two population means.

b. Using the 1% significance level, can you conclude that the mean time taken by the Piranha to shred 100 sheets is greater than that for the Crocodile?

c. What would your decision be in part b if the probability of making a Type I error were zero? Explain.

10.31 Quadro Corporation has two supermarket stores in a city. The company's quality control department wanted to check if the customers are equally satisfied with the service provided at these two stores. A sample of 380 customers selected from Supermarket I produced a mean satisfaction index of 7.6 (on a scale of 1 to 10, 1 being the lowest and 10 being the highest) with a standard deviation of .75. Another sample of 370 customers selected from Supermarket II produced a mean satisfaction index of 8.1 with a standard deviation of .59. Assume that the customer satisfaction index for each supermarket has unknown but same population standard deviation.

a. Construct a 98% confidence interval for the difference between the mean satisfaction indexes for all customers for the two supermarkets.

b. Test at the 1% significance level whether the mean satisfaction indexes for all customers for the two supermarkets are different.

10.32 According to a survey by the Gallup Organization, American men spend an average of 10 hours per week with friends while American women spend an average of 7.5 hours per week with friends (*Psychology Today*, May/June 2005). Suppose that these means are based on random samples of 700 men and 740 women, and that the standard deviations for the times spent per week with friends by men and women included in these samples are 1.9 hours and 1.6 hours, respectively. Assume that the standard deviations of the two populations are unknown but equal.

a. Construct a 95% confidence interval for the difference between the two population means.

b. Test at the 2.5% significance level whether the mean time spent per week with friends by all men is greater than that for all women.

10.3 Inferences About the Difference Between Two Population Means for Independent Samples: σ_1 and σ_2 Unknown and Unequal

Section 10.2 explained how to make inferences about the difference between two population means using the t distribution when the standard deviations of the two populations are unknown but equal and certain other assumptions hold true. Now, what if all other assumptions of Section 10.2 hold true, but the population standard deviations are not only unknown but also unequal? In this case, the procedures used to make confidence intervals and to test hypotheses about $\mu_1 - \mu_2$ remain similar to the ones we learned in Sections 10.2.1 and 10.2.2 except for two differences. When the population standard deviations are unknown and not equal, the degrees of freedom are no longer given by $n_1 + n_2 - 2$ and the standard deviation of $\bar{x}_1 - \bar{x}_2$ is not calculated using the pooled standard deviation s_p.

Degrees of Freedom If

1. The two samples are independent
2. The standard deviations σ_1 and σ_2 of the two populations are unknown and unequal, that is $\sigma_1 \neq \sigma_2$
3. At least one of the following two conditions is fulfilled:
   i. Both samples are large (i.e., $n_1 \geq 30$ and $n_2 \geq 30$)
   ii. If either one or both sample sizes are small, then both populations from which the samples are drawn are normally distributed

then the t distribution is used to make inferences about $\mu_1 - \mu_2$ and the *degrees of freedom* for the t distribution are given by

$$df = \frac{\left(\dfrac{s_1^2}{n_1} + \dfrac{s_2^2}{n_2}\right)^2}{\dfrac{\left(\dfrac{s_1^2}{n_1}\right)^2}{n_1 - 1} + \dfrac{\left(\dfrac{s_2^2}{n_2}\right)^2}{n_2 - 1}}$$

The number given by this formula is always rounded down for df.

Because the standard deviations of the two populations are not known, we use $s_{\bar{x}_1 - \bar{x}_2}$ as a point estimator of $\sigma_{\bar{x}_1 - \bar{x}_2}$. The following formula is used to calculate the standard deviation $s_{\bar{x}_1 - \bar{x}_2}$ of $\bar{x}_1 - \bar{x}_2$.

Estimate of the Standard Deviation of $\bar{x}_1 - \bar{x}_2$ The value of $s_{\bar{x}_1 - \bar{x}_2}$ is calculated as

$$s_{\bar{x}_1 - \bar{x}_2} = \sqrt{\frac{s_1^2}{n_1} + \frac{s_2^2}{n_2}}$$

10.3.1 Interval Estimation of $\mu_1 - \mu_2$

Again, the difference between the two sample means, $\bar{x}_1 - \bar{x}_2$, is the point estimator of the difference between the two population means, $\mu_1 - \mu_2$. The following formula gives the confidence interval for $\mu_1 - \mu_2$ when the t distribution is used and the conditions mentioned earlier in this section are satisfied.

Confidence Interval for $\mu_1 - \mu_2$ The $(1 - \alpha)100\%$ *confidence interval for* $\mu_1 - \mu_2$ is

$$(\bar{x}_1 - \bar{x}_2) \pm t s_{\bar{x}_1 - \bar{x}_2}$$

where the value of t is obtained from the t distribution table for a given confidence level and the degrees of freedom are given by the formula mentioned earlier, and $s_{\bar{x}_1 - \bar{x}_2}$ is also calculated as explained earlier.

Example 10–8 describes how to construct a confidence interval for $\mu_1 - \mu_2$ when the standard deviations of the two populations are unknown and unequal.

■ EXAMPLE 10–8

Constructing a confidence interval for $\mu_1 - \mu_2$: two independent samples, σ_1 and σ_2 unknown and unequal.

According to Example 10–5 of Section 10.2.1, a sample of 15 one-pound jars of coffee of Brand I showed that the mean amount of caffeine in these jars is 80 milligrams per jar with a standard deviation of 5 milligrams. Another sample of 12 one-pound coffee jars of Brand II gave a mean amount of caffeine equal to 77 milligrams per jar with a standard deviation of 6 milligrams. Construct a 95% confidence interval for the difference between the mean amounts of caffeine in one-pound coffee jars of these two brands. Assume that the two populations are normally distributed and that the standard deviations of the two populations are not equal.

Solution Let μ_1 and μ_2 be the mean amounts of caffeine per jar in all one-pound jars of Brands I and II, respectively, and let $\bar{x}_1$ and $\bar{x}_2$ be the means of the two respective samples. From the given information,

$$n_1 = 15 \quad \bar{x}_1 = 80 \text{ milligrams} \quad s_1 = 5 \text{ milligrams}$$

$$n_2 = 12 \quad \bar{x}_2 = 77 \text{ milligrams} \quad s_2 = 6 \text{ milligrams}$$

The confidence level is $1 - \alpha = .95$.

First, we calculate the standard deviation of $\bar{x}_1 - \bar{x}_2$ as follows:

$$s_{\bar{x}_1 - \bar{x}_2} = \sqrt{\frac{s_1^2}{n_1} + \frac{s_2^2}{n_2}} = \sqrt{\frac{(5)^2}{15} + \frac{(6)^2}{12}} = 2.16024690$$

Next, to find the t value from the t distribution table, we need to know the area in each tail of the t distribution curve and the degrees of freedom.

$$\text{Area in each tail} = \alpha/2 = (1 - .95)/2 = .025$$

$$df = \frac{\left(\frac{s_1^2}{n_1} + \frac{s_2^2}{n_2}\right)^2}{\frac{\left(\frac{s_1^2}{n_1}\right)^2}{n_1 - 1} + \frac{\left(\frac{s_2^2}{n_2}\right)^2}{n_2 - 1}} = \frac{\left(\frac{(5)^2}{15} + \frac{(6)^2}{12}\right)^2}{\frac{\left(\frac{(5)^2}{15}\right)^2}{15 - 1} + \frac{\left(\frac{(6)^2}{12}\right)^2}{12 - 1}} = 21.42 \approx 21$$

Note that the degrees of freedom are always rounded down as in this calculation. From the t distribution table, the t value for $df = 21$ and .025 area in the right tail of the t distribution curve is 2.080. The 95% confidence interval for $\mu_1 - \mu_2$ is

$$(\bar{x}_1 - \bar{x}_2) \pm t s_{\bar{x}_1 - \bar{x}_2} = (80 - 77) \pm 2.080(2.16024690)$$

$$= 3 \pm 4.49 = \mathbf{-1.49 \text{ to } 7.49}$$

Thus, with 95% confidence we can state that based on these two sample results, the difference in the mean amounts of caffeine in one-pound jars of these two brands of coffee is between −1.49 and 7.49 milligrams. ■

Comparing this confidence interval with the one obtained in Example 10–5, we observe that the two confidence intervals are very close. From this we can conclude that even if the standard deviations of the two populations are not equal and we use the procedure of Section 10.2.1 to make a confidence interval for $\mu_1 - \mu_2$, the margin of error will be small as long as the difference between the two population standard deviations is not too large.

10.3.2 Hypothesis Testing About $\mu_1 - \mu_2$

When the standard deviations of the two populations are unknown and unequal along with the other conditions of Section 10.3 holding true, we use the t distribution to make a test of hypothesis about $\mu_1 - \mu_2$. This procedure differs from the one in Section 10.2.2 only in the calculation of degrees of freedom for the t distribution and the standard deviation of $\bar{x}_1 - \bar{x}_2$. The df and the standard deviation of $\bar{x}_1 - \bar{x}_2$ in this case are given by the formulas used in Section 10.3.1.

Test Statistic t for $\bar{x}_1 - \bar{x}_2$ The value of the *test statistic t for* $\bar{x}_1 - \bar{x}_2$ is computed as

$$t = \frac{(\bar{x}_1 - \bar{x}_2) - (\mu_1 - \mu_2)}{s_{\bar{x}_1 - \bar{x}_2}}$$

The value of $\mu_1 - \mu_2$ in this formula is substituted from the null hypothesis and $s_{\bar{x}_1 - \bar{x}_2}$ is calculated as explained earlier.

Example 10–9 illustrates the procedure used to conduct a test of hypothesis about $\mu_1 - \mu_2$ when the standard deviations of the two populations are unknown and unequal.

■ EXAMPLE 10–9

Making a two-tailed test of hypothesis about $\mu_1 - \mu_2$: two independent samples, and unknown and unequal σ_1 and σ_2.

According to Example 10–6 of Section 10.2.2, a sample of 14 cans of Brand I diet soda gave the mean number of calories per can of 23 with a standard deviation of 3 calories. Another sample of 16 cans of Brand II diet soda gave the mean number of calories of 25 per can with a standard deviation of 4 calories. Test at the 1% significance level whether the mean numbers of calories per can of diet soda are different for these two brands. Assume that the calories per can of diet soda are normally distributed for each of these two brands and that the standard deviations for the two populations are not equal.

Solution Let μ_1 and μ_2 be the mean numbers of calories for all cans of diet soda of Brand I and Brand II, respectively, and let $\bar{x}_1$ and $\bar{x}_2$ be the means of the respective samples. From the given information,

$$n_1 = 14 \quad \bar{x}_1 = 23 \quad s_1 = 3$$
$$n_2 = 16 \quad \bar{x}_2 = 25 \quad s_2 = 4$$

The significance level is $\alpha = .01$.

Step 1. *State the null and alternative hypotheses.*

We are to test for the difference in the mean numbers of calories per can for the two brands. The null and alternative hypotheses are

$$H_0\text{: } \mu_1 - \mu_2 = 0 \quad \text{(The mean numbers of calories are not different)}$$
$$H_1\text{: } \mu_1 - \mu_2 \neq 0 \quad \text{(The mean numbers of calories are different)}$$

Step 2. *Select the distribution to use.*

Here, the two samples are independent, σ_1 and σ_2 are unknown and unequal, and the sample sizes are small but both populations are normally distributed. Hence, all conditions mentioned in the beginning of Section 10.3 are fulfilled. Consequently, we use the t distribution to make the test.

Step 3. *Determine the rejection and nonrejection regions.*

The $\neq$ sign in the alternative hypothesis indicates that the test is two-tailed. The significance level is .01. Hence,

$$\text{Area in each tail} = \alpha/2 = .01/2 = .005$$

The degrees of freedom are calculated as follows:

$$df = \frac{\left(\frac{s_1^2}{n_1} + \frac{s_2^2}{n_2}\right)^2}{\frac{\left(\frac{s_1^2}{n_1}\right)^2}{n_1 - 1} + \frac{\left(\frac{s_2^2}{n_2}\right)^2}{n_2 - 1}} = \frac{\left(\frac{(3)^2}{14} + \frac{(4)^2}{16}\right)^2}{\frac{\left(\frac{(3)^2}{14}\right)^2}{14 - 1} + \frac{\left(\frac{(4)^2}{16}\right)^2}{16 - 1}} = 27.41 \approx 27$$

From the t distribution table, the critical values of t for $df = 27$ and .005 area in each tail of the t distribution curve are -2.771 and 2.771. These values are shown in Figure 10.5.

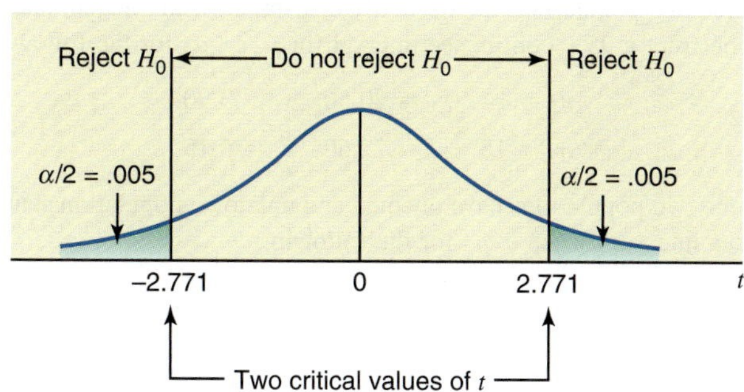

Figure 10.5

Step 4. *Calculate the value of the test statistic.*

The value of the test statistic t for $\bar{x}_1 - \bar{x}_2$ is computed as follows:

$$s_{\bar{x}_1-\bar{x}_2} = \sqrt{\frac{s_1^2}{n_1} + \frac{s_2^2}{n_2}} = \sqrt{\frac{(3)^2}{14} + \frac{(4)^2}{16}} = 1.28173989$$

From H_0

$$t = \frac{(\bar{x}_1 - \bar{x}_2) - (\mu_1 - \mu_2)}{s_{\bar{x}_1-\bar{x}_2}} = \frac{(23 - 25) - 0}{1.28173989} = -1.560$$

Step 5. *Make a decision.*

Because the value of the test statistic $t = -1.560$ for $\bar{x}_1 - \bar{x}_2$ falls in the nonrejection region, we fail to reject the null hypothesis. Hence, there is no difference in the mean numbers of calories per can for the two brands of diet soda. The difference in $\bar{x}_1$ and $\bar{x}_2$ observed for the two samples may have occurred due to sampling error only.

Using the *p*-Value to Make a Decision

We can use the p-value approach to make the above decision. To do so, we keep Steps 1 and 2 of this example. Then in Step 3, we calculate the value of the test statistic t (as done in Step 4 above) and then find the p-value for this t from the t distribution table (Table V of Appendix C) or by using technology. In Step 4 above, the t-value for $\bar{x}_1 - \bar{x}_2$ was calculated to be -1.560. In this example, the test is two-tailed. Therefore, the p-value is equal to twice the area under the t distribution curve to the left of $t = -1.560$. If we have access to technology, we can use it to find the exact p-value, which will be .130. If we use the t distribution table, we can only find the range for the p-value. From Table V of Appendix C, for $df = 27$, the two values that include 1.560 are 1.314 and 1.703. (Note that we use the positive value of t although our t is negative.) Thus, test statistic $t = -1.560$ falls between -1.314 and -1.703. The areas in the t distribution table that correspond to 1.314 and 1.703 are .10 and .05, respectively. Because it is a two-tailed test, the p-value for $t = -1.560$ is between $2(.10) = .20$ and $2(.05) = .10$, which can be written as:

$$.10 < p\text{-value} < .20$$

Since we will reject the null hypothesis for any α (significance level) that is greater than the p-value, we will reject the null hypothesis in this example for any $\alpha \geq .20$ using the above range and not reject for $\alpha \leq .10$. If we use technology, we will reject the null hypothesis for $\alpha > .130$. Since $\alpha = .01$ in this example, which is smaller than both .10 and .130, we fail to reject the null hypothesis. ■

Remember The degrees of freedom for the procedures to make a confidence interval and to test a hypothesis about $\mu_1 - \mu_2$ learned in Sections 10.3.1 and 10.3.2 are always rounded down.

EXERCISES

CONCEPTS AND PROCEDURES

10.33 Assuming that the two populations are normally distributed with unequal and unknown population standard deviations, construct a 95% confidence interval for $\mu_1 - \mu_2$ for the following.

$$n_1 = 24 \quad \bar{x}_1 = 20.50 \quad s_1 = 3.90$$
$$n_2 = 16 \quad \bar{x}_2 = 22.60 \quad s_2 = 5.15$$

10.34 Assuming that the two populations have unequal and unknown population standard deviations, construct a 99% confidence interval for $\mu_1 - \mu_2$ for the following.

$$n_1 = 39 \quad \bar{x}_1 = 52.61 \quad s_1 = 3.55$$
$$n_2 = 36 \quad \bar{x}_2 = 43.75 \quad s_2 = 5.40$$

10.35 Refer to Exercise 10.33. Test at the 5% significance level if the two population means are different.

10.36 Refer to Exercise 10.34. Test at the 1% significance level if the two population means are different.

10.37 Refer to Exercise 10.33. Test at the 1% significance level if μ_1 is less than μ_2.

10.38 Refer to Exercise 10.34. Test at the 2.5% significance level if μ_1 is greater than μ_2.

APPLICATIONS

10.39 According to the information given in Exercise 10.25, a sample of 35 male customers who shopped at a supermarket showed that they spent an average of $80 with a standard deviation of $17.50. Another sample of 40 female customers who shopped at the same supermarket showed that they spent an average of $96 with a standard deviation of $14.40. Assume that the amounts spent at this supermarket by all male and all female customers have unequal and unknown population standard deviations.

- **a.** Construct a 99% confidence interval for the difference between the mean amounts spent by all male and all female customers at this supermarket.
- **b.** Using the 2.5% significance level, can you conclude that the mean amount spent by all male customers at this supermarket is less than that of all female customers?

10.40 As mentioned in Exercise 10.26, the department of labor in a state wanted to find the compensation of hotel employees. A random sample of 32 cleaning persons produced the mean hourly earnings (including tips) of $10.60 with a standard deviation of $1.02. A random sample of 37 bellhops gave the mean hourly earnings (including tips) of $11.57 with a standard deviation of $1.34. Assume that the hourly earnings of both groups have unknown and unequal population standard deviations.

- **a.** Construct a 99% confidence interval for the difference between the corresponding population means of the two groups.
- **b.** Using the 1% significance level, can you conclude that the mean hourly earnings of all cleaning persons are lower than those of all bellhops in this state?

10.41 According to Exercise 10.27, an insurance company wants to know if the average speed at which men drive cars is higher than that of women drivers. The company took a random sample of 27 cars driven by men on a highway and found the mean speed to be 72 miles per hour with a standard deviation of 2.2 miles per hour. Another sample of 18 cars driven by women on the same highway gave a mean speed of 68 miles per hour with a standard deviation of 2.5 miles per hour. Assume that the speeds at which all men and all women drive cars on this highway are both normally distributed with unequal population standard deviations.

- **a.** Construct a 98% confidence interval for the difference between the mean speeds of cars driven by all men and all women on this highway.
- **b.** Test at the 1% significance level whether the mean speed of cars driven by all men drivers on this highway is higher than that of cars driven by all women drivers.

10.42 Refer to Exercise 10.28. Now assume that the two populations are normally distributed with unequal and unknown population standard deviations.

- **a.** Make a 90% confidence interval for the difference between the corresponding population means.
- **b.** Test at the 5% significance level whether the two population means are different.

10.43 As mentioned in Exercise 10.29, a company claims that its medicine, Brand A, provides faster relief from pain than another company's medicine, Brand B. A researcher tested both brands of medicine on two groups of randomly selected patients. The results of the test are given in the following table. The mean and standard deviation of relief times are in minutes.

Brand	Sample Size	Mean of Relief Times	Standard Deviation of Relief Times
A	25	44	11
B	22	49	9

a. Construct a 99% confidence interval for the difference between the mean relief times for the two brands of medicine.

b. Test at the 1% significance level whether the mean relief time for Brand A is less than that for Brand B.

Assume that the two populations are normally distributed with unknown and unequal standard deviations.

10.44 Refer to Exercise 10.30. Now assume that the shredding times for both paper shredders are normally distributed with unequal and unknown standard deviations.

a. Construct a 99% confidence interval for the difference between the two population means.

b. Using the 1% significance level, can you conclude that the mean time taken by the Piranha to shred 100 sheets is greater than that for the Crocodile?

c. What would your decision be in part b if the probability of making a Type I error were zero? Explain.

10.45 As mentioned in Exercise 10.31, Quadro Corporation has two supermarkets in a city. The company's quality control department wanted to check if the customers are equally satisfied with the service provided at these two stores. A sample of 380 customers selected from Supermarket I produced a mean satisfaction index of 7.6 (on a scale of 1 to 10, 1 being the lowest and 10 being the highest) with a standard deviation of .75. Another sample of 370 customers selected from Supermarket II produced a mean satisfaction index of 8.1 with a standard deviation of .59. Assume that the customer satisfaction index for each supermarket has an unknown and different population standard deviation.

a. Construct a 98% confidence interval for the difference between the mean satisfaction indexes for all customers for the two supermarkets.

b. Test at the 1% significance level whether the mean satisfaction indexes for all customers for the two supermarkets are different.

10.46 Refer to Exercise 10.32. According to a survey by the Gallup Organization, American men spend an average of 10 hours per week with friends while American women spend an average of 7.5 hours per week with friends (*Psychology Today*, May/June 2005). Suppose that these means are based on random samples of 700 men and 740 women, and that the sample standard deviations for the times spent per week with friends by men and women are 1.9 hours and 1.6 hours, respectively. Assume that the standard deviations of the two populations are unknown and unequal.

a. Construct a 95% confidence interval for the difference between the two population means.

b. Test at the 2.5% significance level whether the mean time spent per week with friends by all men is greater than that for all women.

10.4 Inferences About the Difference Between Two Population Means for Paired Samples

Sections 10.1, 10.2, and 10.3 were concerned with estimation and hypothesis testing about the difference between two population means when the two samples were drawn independently from two different populations. This section describes estimation and hypothesis-testing procedures for the difference between two population means when the samples are dependent.

In a case of two dependent samples, two data values—one for each sample—are collected from the same source (or element) and, hence, these are also called **paired** or **matched samples**. For example, we may want to make inferences about the mean weight loss for members of a health club after they have gone through an exercise program for a certain period of time. To do so, suppose we select a sample of 15 members of this health club and record their weights before and after the program. In this example, both sets of data are collected from the same 15 persons, once before and once after the program. Thus, although there are two samples, they contain the same 15 persons. This is an example of paired (or dependent or matched) samples. The procedures to make confidence intervals and test hypotheses in the case of paired samples are different from the ones for independent samples discussed in earlier sections of this chapter.

Definition

Paired or Matched Samples Two samples are said to be *paired* or *matched samples* when for each data value collected from one sample there is a corresponding data value collected from the second sample, and both these data values are collected from the same source.

As another example of paired samples, suppose an agronomist wants to measure the effect of a new brand of fertilizer on the yield of potatoes. To do so, he selects 10 pieces of land and divides each piece into two portions. Then he randomly assigns one of the two portions from each piece of land to grow potatoes without using fertilizer (or using some other brand of fertilizer). The second portion from each piece of land is used to grow potatoes with the new brand of fertilizer. Thus, he will have 10 pairs of data values. Then, using the procedure to be discussed in this section, he will make inferences about the difference in the mean yields of potatoes with and without the new fertilizer.

The question arises, why does the agronomist not choose 10 pieces of land on which to grow potatoes without using the new brand of fertilizer and another 10 pieces of land to grow potatoes by using the new brand of fertilizer? If he does so, the effect of the fertilizer might be confused with the effects due to soil differences at different locations. Thus, he will not be able to isolate the effect of the new brand of fertilizer on the yield of potatoes. Consequently, the results will not be reliable. By choosing 10 pieces of land and then dividing each of them into two portions, the researcher decreases the possibility that the difference in the productivities of different pieces of land affects the results.

In paired samples, the difference between the two data values for each element of the two samples is denoted by $\boldsymbol{d}$. This value of d is called the **paired difference**. We then treat all the values of d as one sample and make inferences applying procedures similar to the ones used for one-sample cases in Chapters 8 and 9. Note that because each source (or element) gives a pair of values (one for each of the two data sets), each sample contains the same number of values. That is, both samples are the same size. Therefore, we denote the (common) **sample size** by $\boldsymbol{n}$, which gives the number of paired difference values denoted by d. The **degrees of freedom** for the paired samples are $\boldsymbol{n-1}$. Let

μ_d = the mean of the paired differences for the population

σ_d = the standard deviation of the paired differences for the population, which is usually never known

$\bar{d}$ = the mean of the paired differences for the sample

s_d = the standard deviation of the paired differences for the sample

n = the number of paired difference values

Mean and Standard Deviation of the Paired Differences for Two Samples The values of the mean and standard deviation, $\bar{d}$ and s_d, of paired differences for two samples are calculated as[2]

$$\bar{d} = \frac{\Sigma d}{n}$$

$$s_d = \sqrt{\frac{\Sigma d^2 - \frac{(\Sigma d)^2}{n}}{n-1}}$$

[2]The basic formula to calculate s_d is

$$s_d = \sqrt{\frac{\Sigma(d-\bar{d})^2}{n-1}}$$

However, we will not use this formula to make calculations in this chapter.

In paired samples, instead of using $\bar{x}_1 - \bar{x}_2$ as the sample statistic to make inferences about $\mu_1 - \mu_2$, we use the sample statistic $\bar{d}$ to make inferences about μ_d. Actually the value of $\bar{d}$ is always equal to $\bar{x}_1 - \bar{x}_2$, and the value of μ_d is always equal to $\mu_1 - \mu_2$.

Sampling Distribution, Mean, and Standard Deviation of $\bar{d}$ If σ_d is known and either the sample size is large ($n \geq 30$) or the population is normally distributed, then the *sampling distribution* of $\bar{d}$ is approximately normal with its *mean* and *standard deviation* given as

$$\mu_{\bar{d}} = \mu_d \quad \text{and} \quad \sigma_{\bar{d}} = \frac{\sigma_d}{\sqrt{n}}$$

Thus, if the standard deviation σ_d of the population paired differences is known and either the sample size is large (i.e., $n \geq 30$) or the population of paired differences is normally distributed (with $n < 30$), then the normal distribution can be used to make a confidence interval and test a hypothesis about μ_d. However, usually σ_d is never known. Then, if the standard deviation σ_d of the population paired differences is unknown and either the sample size is large (i.e., $n \geq 30$) or the population of paired differences is normally distributed (with $n < 30$), then the t distribution is used to make a confidence interval and test a hypothesis about μ_d.

Making Inferences About μ_d If

1. The standard deviation σ_d of the population paired differences is unknown
2. At least one of the following two conditions is fulfilled:
   i. The sample size is large (i.e., $n \geq 30$)
   ii. If the sample size is small, then the population of paired differences is normally distributed

then the t distribution is used to make inferences about μ_d. The standard deviation $\sigma_{\bar{d}}$ of $\bar{d}$ is estimated by $s_{\bar{d}}$, which is calculated as

$$s_{\bar{d}} = \frac{s_d}{\sqrt{n}}$$

Sections 10.4.1 and 10.4.2 describe the procedures used to make a confidence interval and test a hypothesis about μ_d under the above conditions. The inferences are made using the t distribution.

10.4.1 Interval Estimation of μ_d

The mean $\bar{d}$ of paired differences for paired samples is the point estimator of μ_d. The following formula is used to construct a confidence interval for μ_d when the t distribution is used.

Confidence Interval for μ_d The $(1 - \alpha)100\%$ *confidence interval for* μ_d is

$$\bar{d} \pm t s_{\bar{d}}$$

where the value of t is obtained from the t distribution table for the given confidence level and $n - 1$ degrees of freedom, and $s_{\bar{d}}$ is calculated as explained above.

Example 10–10 illustrates the procedure to construct a confidence interval for μ_d.

EXAMPLE 10–10

Constructing a confidence interval for μ_d: paired samples, σ_d unknown, and population normal.

A researcher wanted to find the effect of a special diet on systolic blood pressure. She selected a sample of seven adults and put them on this dietary plan for three months. The following table gives the systolic blood pressures of these seven adults before and after the completion of this plan.

Before	210	180	195	220	231	199	224
After	193	186	186	223	220	183	233

Let μ_d be the mean reduction in the systolic blood pressures due to this special dietary plan for the population of all adults. Construct a 95% confidence interval for μ_d. Assume that the population of paired differences is (approximately) normally distributed.

Solution Because the information obtained is from paired samples, we will make the confidence interval for the paired difference mean μ_d of the population using the paired difference mean $\bar{d}$ of the sample. Let d be the difference in the systolic blood pressure of an adult before and after this special dietary plan. Then, d is obtained by subtracting the systolic blood pressure after the plan from the systolic blood pressure before the plan. The third column of Table 10.1 lists the values of d for the seven adults. The fourth column of the table records the values of d^2, which are obtained by squaring each of the d values.

Table 10.1

Before	After	Difference d	d^2
210	193	17	289
180	186	−6	36
195	186	9	81
220	223	−3	9
231	220	11	121
199	183	16	256
224	233	−9	81
		$\Sigma d = 35$	$\Sigma d^2 = 873$

The values of $\bar{d}$ and s_d are calculated as follows:

$$\bar{d} = \frac{\Sigma d}{n} = \frac{35}{7} = 5.00$$

$$s_d = \sqrt{\frac{\Sigma d^2 - \frac{(\Sigma d)^2}{n}}{n-1}} = \sqrt{\frac{873 - \frac{(35)^2}{7}}{7-1}} = 10.78579312$$

Hence, the standard deviation of $\bar{d}$ is

$$s_{\bar{d}} = \frac{s_d}{\sqrt{n}} = \frac{10.78579312}{\sqrt{7}} = 4.07664661$$

Here, σ_d is not known, the sample size is small but the population is normally distributed. Hence, we will use the t distribution to make the confidence interval. For the 95% confidence interval, the area in each tail of the t distribution curve is

$$\text{Area in each tail} = \alpha/2 = (1 - .95)/2 = .025$$

The degrees of freedom are

$$df = n - 1 = 7 - 1 = 6$$

From the t distribution table, the t value for $df = 6$ and .025 area in the right tail of the t distribution curve is 2.447. Therefore, the 95% confidence interval for μ_d is

$$\bar{d} \pm ts_{\bar{d}} = 5.00 \pm 2.447(4.07664661) = 5.00 \pm 9.98 = \mathbf{-4.98 \text{ to } 14.98}$$

Thus, we can state with 95% confidence that the mean difference between systolic blood pressures before and after the given dietary plan for all adult participants is between −4.98 and 14.98.

10.4.2 Hypothesis Testing About μ_d

A hypothesis about μ_d is tested by using the sample statistic $\bar{d}$. This section illustrates the case of the t distribution only. Earlier in this section we learned what conditions should hold true to use the t distribution to test a hypothesis about μ_d. The following formula is used to calculate the value of the test statistic t when testing a hypothesis about μ_d.

Test Statistic t for $\bar{d}$ The value of the *test statistic t for $\bar{d}$* is computed as follows:

$$t = \frac{\bar{d} - \mu_d}{s_{\bar{d}}}$$

The critical value of t is found from the t distribution table for the given significance level and $n - 1$ degrees of freedom.

Examples 10–11 and 10–12 illustrate the hypothesis-testing procedure for μ_d.

■ EXAMPLE 10–11

Conducting a left-tailed test of hypothesis about μ_d for paired samples: σ_d not known, small sample but normally distributed population.

A company wanted to know if attending a course on "how to be a successful salesperson" can increase the average sales of its employees. The company sent six of its salespersons to attend this course. The following table gives the one-week sales of these salespersons before and after they attended this course.

Before	12	18	25	9	14	16
After	18	24	24	14	19	20

Using the 1% significance level, can you conclude that the mean weekly sales for all salespersons increase as a result of attending this course? Assume that the population of paired differences has a normal distribution.

Solution Because the data are for paired samples, we test a hypothesis about the paired differences mean μ_d of the population using the paired differences mean $\bar{d}$ of the sample.

Let

$$d = \text{(Weekly sales before the course)} - \text{(Weekly sales after the course)}$$

In Table 10.2, we calculate d for each of the six salespersons by subtracting the sales after the course from the sales before the course. The fourth column of the table lists the values of d^2.

Table 10.2

Before	After	Difference d	d^2
12	18	−6	36
18	24	−6	36
25	24	1	1
9	14	−5	25
14	19	−5	25
16	20	−4	16
		$\Sigma d = -25$	$\Sigma d^2 = 139$

The values of $\bar{d}$ and s_d are calculated as follows:

$$\bar{d} = \frac{\Sigma d}{n} = \frac{-25}{6} = -4.17$$

$$s_d = \sqrt{\frac{\Sigma d^2 - \frac{(\Sigma d)^2}{n}}{n-1}} = \sqrt{\frac{139 - \frac{(-25)^2}{6}}{6-1}} = 2.63944439$$

The standard deviation of $\bar{d}$ is

$$s_{\bar{d}} = \frac{s_d}{\sqrt{n}} = \frac{2.63944439}{\sqrt{6}} = 1.07754866$$

Step 1. *State the null and alternative hypotheses.*

We are to test if the mean weekly sales for all salespersons increase as a result of taking the course. Let μ_1 be the mean weekly sales for all salespersons before the course and μ_2 the mean weekly sales for all salespersons after the course. Then $\mu_d = \mu_1 - \mu_2$. The mean weekly sales for all salespersons will increase due to attending the course if μ_1 is less than μ_2, which can be written as $\mu_1 - \mu_2 < 0$ or $\mu_d < 0$. Consequently, the null and alternative hypotheses are

H_0: $\mu_d = 0$ ($\mu_1 - \mu_2 = 0$ or the mean weekly sales do not increase)

H_1: $\mu_d < 0$ ($\mu_1 - \mu_2 < 0$ or the mean weekly sales do increase)

Note that we can also write the null hypothesis as $\mu_d \geq 0$.

Step 2. *Select the distribution to use.*

Here σ_d is unknown, and the sample size is small ($n < 30$) but the population of paired differences is normally distributed. Therefore, we use the t distribution to conduct the test.

Step 3. *Determine the rejection and nonrejection regions.*

The $<$ sign in the alternative hypothesis indicates that the test is left-tailed. The significance level is .01. Hence,

$$\text{Area in left tail} = \alpha = .01$$

$$\text{Degrees of freedom} = n - 1 = 6 - 1 = 5$$

The critical value of t for $df = 5$ and .01 area in the left tail of the t distribution curve is -3.365. This value is shown in Figure 10.6.

Figure 10.6

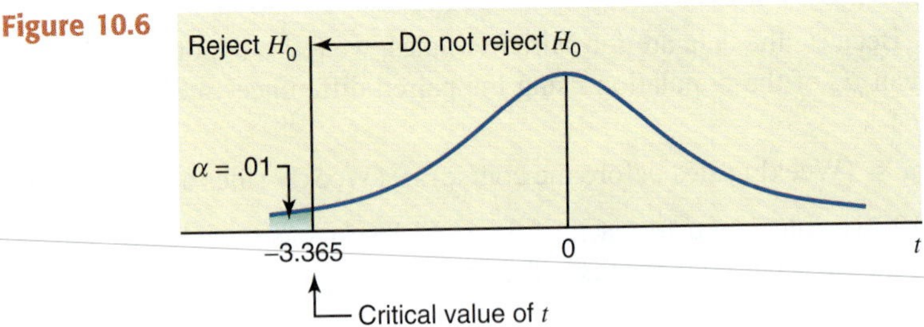

Step 4. *Calculate the value of the test statistic.*

The value of the test statistic t for $\bar{d}$ is computed as follows:

$$t = \frac{\bar{d} - \mu_d}{s_{\bar{d}}} = \frac{-4.17 - 0}{1.07754866} = -3.870$$

(From H_0 — the 0 in the numerator)

Step 5. *Make a decision.*

Because the value of the test statistic $t = -3.870$ for $\bar{d}$ falls in the rejection region, we reject the null hypothesis. Consequently, we conclude that the mean weekly sales for all salespersons increase as a result of this course.

Using the *p*-Value to Make a Decision

We can use the *p*-value approach to make the above decision. To do so, we keep Steps 1 and 2 of this example. Then in Step 3, we calculate the value of the test statistic t for $\bar{d}$ (as done in Step 4 above) and then find the *p*-value for this t from the t distribution table (Table V of Appendix C) or by using technology. If we have access to technology, we can use it to find the exact *p*-value, which will be .006. By using Table V, we can find the range of the *p*-value. From Table V, for $df = 5$, the test statistic $t = -3.870$ falls between -3.365 and -4.032. The areas in the t distribution table that correspond to -3.365 and -4.032 are .01 and .005, respectively. Because it is a left-tailed test, the *p*-value is between .01 and .005, which can be written as:

$$.005 < p\text{-value} < .01$$

Since we will reject the null hypothesis for any α (significance level) that is greater than the *p*-value, we will reject the null hypothesis in this example for any $\alpha > .006$ using the technology and $\alpha \geq .01$ using the above range. Since $\alpha = .01$ in this example, which is larger than .006 obtained from technology, we reject the null hypothesis. Also, because α is equal to .01, using the *p*-value range we reject the null hypothesis. ■

■ EXAMPLE 10–12

Making a two-tailed test of hypothesis about μ_d for paired samples: σ_d not known, small sample but normally distributed population.

Refer to Example 10–10. The table that gives the blood pressures of seven adults before and after the completion of a special dietary plan is reproduced here.

Before	210	180	195	220	231	199	224
After	193	186	186	223	220	183	233

Let μ_d be the mean of the differences between the systolic blood pressures before and after completing this special dietary plan for the population of all adults. Using the 5% significance level, can we conclude that the mean of the paired differences μ_d is different from zero? Assume that the population of paired differences is (approximately) normally distributed.

Solution Table 10.3 gives d and d^2 for each of the seven adults.

Table 10.3

Before	After	Difference d	d^2
210	193	17	289
180	186	−6	36
195	186	9	81
220	223	−3	9
231	220	11	121
199	183	16	256
224	233	−9	81
		$\Sigma d = 35$	$\Sigma d^2 = 873$

The values of $\bar{d}$ and s_d are calculated as follows:

$$\bar{d} = \frac{\Sigma d}{n} = \frac{35}{7} = 5.00$$

$$s_d = \sqrt{\frac{\Sigma d^2 - \frac{(\Sigma d)^2}{n}}{n-1}} = \sqrt{\frac{873 - \frac{(35)^2}{7}}{7-1}} = 10.78579312$$

Hence, the standard deviation of $\bar{d}$ is

$$s_{\bar{d}} = \frac{s_d}{\sqrt{n}} = \frac{10.78579312}{\sqrt{7}} = 4.07664661$$

Step 1. *State the null and alternative hypotheses.*

H_0: $\mu_d = 0$ (The mean of the paired differences is not different from zero)

H_1: $\mu_d \neq 0$ (The mean of the paired differences is different from zero)

Step 2. *Select the distribution to use.*

Here σ_d is unknown, and the sample size is small but the population of paired differences is (approximately) normal. Hence, we use the t distribution to make the test.

Step 3. *Determine the rejection and nonrejection regions.*

The $\neq$ sign in the alternative hypothesis indicates that the test is two-tailed. The significance level is .05.

$$\text{Area in each tail of the curve} = \alpha/2 = .05/2 = .025$$

$$\text{Degrees of freedom} = n - 1 = 7 - 1 = 6$$

The two critical values of t for $df = 6$ and .025 area in each tail of the t distribution curve are -2.447 and 2.447. These values are shown in Figure 10.7.

Figure 10.7

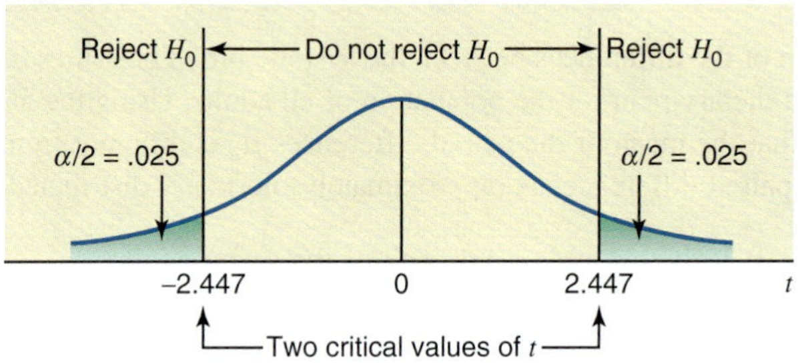

Step 4. *Calculate the value of the test statistic.*

The value of the test statistic t for $\bar{d}$ is computed as follows:

$$t = \frac{\bar{d} - \mu_d}{s_{\bar{d}}} = \frac{5.00 - \overset{\text{From } H_0}{\downarrow} 0}{4.07664661} = 1.226$$

Step 5. *Make a decision.*

Because the value of the test statistic $t = 1.226$ for $\bar{d}$ falls in the nonrejection region, we fail to reject the null hypothesis. Hence, we conclude that the mean of the population paired differences is not different from zero. In other words, we can state that the mean of the differences between the systolic blood pressures before and after completing this special dietary plan for the population of all adults is not different from zero.

Using the p-Value to Make a Decision

We can use the p-value approach to make the above decision. To do so, we keep Steps 1 and 2 of this example. Then in Step 3, we calculate the value of the test statistic t for $\bar{d}$ (as done in Step 4 above) and then find the p-value for this t from the t distribution table (Table V of Appendix C) or by using technology. If we have access to technology, we can use it to find

the exact p-value, which will be .266. By using Table V, we can find the range of the p-value. From Table V, for $df = 6$, the test statistic $t = 1.226$ is less than 1.440. The area in the t distribution table that corresponds to 1.440 is .10. Because it is a two-tailed test, the p-value is greater than $2(.10) = .20$, which can be written as:

$$p\text{-value} > .20$$

Since $\alpha = .05$ in this example, which is smaller than .20 and also .266 (obtained from technology), we fail to reject the null hypothesis. ■

EXERCISES

CONCEPTS AND PROCEDURES

10.47 Explain when you would use the paired samples procedure to make confidence intervals and test hypotheses.

10.48 Find the following confidence intervals for μ_d assuming that the populations of paired differences are normally distributed.

a. $n = 11, \quad \bar{d} = 25.4, \quad s_d = 13.5, \quad$ confidence level = 99%
b. $n = 23, \quad \bar{d} = 13.2, \quad s_d = 4.8, \quad$ confidence level = 95%
c. $n = 18, \quad \bar{d} = 34.6, \quad s_d = 11.7, \quad$ confidence level = 90%

10.49 Find the following confidence intervals for μ_d assuming that the populations of paired differences are normally distributed.

a. $n = 12, \quad \bar{d} = 17.5, \quad s_d = 6.3, \quad$ confidence level = 99%
b. $n = 27, \quad \bar{d} = 55.9, \quad s_d = 14.7, \quad$ confidence level = 95%
c. $n = 16, \quad \bar{d} = 29.3, \quad s_d = 8.3, \quad$ confidence level = 90%

10.50 Perform the following tests of hypotheses assuming that the populations of paired differences are normally distributed.

a. H_0: $\mu_d = 0$, $\quad H_1$: $\mu_d \neq 0$, $\quad n = 9$, $\quad \bar{d} = 6.7$, $\quad s_d = 2.5$, $\quad \alpha = .10$
b. H_0: $\mu_d = 0$, $\quad H_1$: $\mu_d > 0$, $\quad n = 22$, $\quad \bar{d} = 14.8$, $\quad s_d = 6.4$, $\quad \alpha = .05$
c. H_0: $\mu_d = 0$, $\quad H_1$: $\mu_d < 0$, $\quad n = 17$, $\quad \bar{d} = -9.3$, $\quad s_d = 4.8$, $\quad \alpha = .01$

10.51 Conduct the following tests of hypotheses assuming that the populations of paired differences are normally distributed.

a. H_0: $\mu_d = 0$, $\quad H_1$: $\mu_d \neq 0$, $\quad n = 26$, $\quad \bar{d} = 9.6$, $\quad s_d = 3.9$, $\quad \alpha = .05$
b. H_0: $\mu_d = 0$, $\quad H_1$: $\mu_d > 0$, $\quad n = 15$, $\quad \bar{d} = 8.8$, $\quad s_d = 4.7$, $\quad \alpha = .01$
c. H_0: $\mu_d = 0$, $\quad H_1$: $\mu_d < 0$, $\quad n = 20$, $\quad \bar{d} = -7.4$, $\quad s_d = 2.3$, $\quad \alpha = .10$

APPLICATIONS

10.52 A company sent seven of its employees to attend a course in building self-confidence. These employees were evaluated for their self-confidence before and after attending this course. The following table gives the scores (on a scale of 1 to 15, 1 being the lowest and 15 being the highest score) of these employees before and after they attended the course.

Before	8	5	4	9	6	9	5
After	10	8	5	11	6	7	9

a. Construct a 95% confidence interval for the mean μ_d of the population paired differences, where a paired difference is equal to the score of an employee before attending the course minus the score of the same employee after attending the course.
b. Test at the 1% significance level whether attending this course increases the mean score of employees.

Assume that the population of paired differences has a normal distribution.

10.53 Several retired bicycle racers are coaching a large group of young prospects. They randomly select seven of their riders to take part in a test of the effectiveness of a new dietary supplement that is supposed to increase strength and stamina. Each of the seven riders does a time trial on the same course. Then they all take the dietary supplement for four weeks. All other aspects of their training program remain as they were prior to the time trial. At the end of the four weeks, these riders do another time trial on the same

course. The times (in minutes) recorded by each rider for these trials, before and after the four-week period, are shown in the following table.

Before	103	97	111	95	102	96	108
After	100	95	104	101	96	91	101

a. Construct a 99% confidence interval for the mean μ_d of the population paired differences, where a paired difference is equal to the time taken before the dietary supplement minus the time taken after the dietary supplement.

b. Test at the 2.5% significance level whether taking this dietary supplement results in faster times in the time trials.

Assume that the population of paired differences is (approximately) normally distributed.

10.54 A private agency claims that the crash course it offers significantly increases the writing speed of secretaries. The following table gives the scores of eight secretaries before and after they attended this course.

Before	81	75	89	91	65	70	90	64
After	97	72	93	110	78	69	115	72

a. Make a 90% confidence interval for the mean μ_d of the population paired differences, where a paired difference is equal to the score before attending the course minus the score after attending the course.

b. Using the 5% significance level, can you conclude that attending this course increases the writing speed of secretaries?

Assume that the population of paired differences is (approximately) normally distributed.

10.55 A company claims that its 12-week special exercise program significantly reduces weight. A random sample of six persons was selected, and these persons were put on this exercise program for 12 weeks. The following table gives the weights (in pounds) of those six persons before and after the program.

Before	180	195	177	221	208	199
After	183	187	161	204	197	189

a. Make a 95% confidence interval for the mean μ_d of the population paired differences, where a paired difference is equal to the weight before joining this exercise program minus the weight at the end of the 12-week program.

b. Using the 1% significance level, can you conclude that the mean weight loss for all persons due to this special exercise program is greater than zero?

Assume that the population of all paired differences is (approximately) normally distributed.

10.56 The manufacturer of a gasoline additive claims that the use of this additive increases gasoline mileage. A random sample of six cars was selected and these cars were driven for one week without the gasoline additive and then for one week with the gasoline additive. The following table gives the miles per gallon for these cars without and with the gasoline additive.

Without	24.6	28.3	18.9	23.7	15.4	29.5
With	26.3	31.7	18.2	25.3	18.3	30.9

a. Construct a 99% confidence interval for the mean μ_d of the population paired differences, where a paired difference is equal to the miles per gallon without the gasoline additive minus the miles per gallon with the gasoline additive.

b. Using the 2.5% significance level, can you conclude that the use of the gasoline additive increases the gasoline mileage?

Assume that the population of paired differences is (approximately) normally distributed.

10.57 A company is considering installing new machines to assemble its products. The company is considering two types of machines, but it will buy only one type. The company selected eight assembly workers and asked them to use these two types of machines to assemble products. The following table gives the time taken (in minutes) to assemble one unit of the product on each type of machine for each of these eight workers.

Machine I	23	26	19	24	27	22	20	18
Machine II	21	24	23	25	24	28	24	23

a. Construct a 98% confidence interval for the mean μ_d of the population paired differences, where a paired difference is equal to the time taken to assemble a unit of the product on Machine I minus the time taken to assemble a unit of the product on Machine II.

b. Test at the 5% significance level whether the mean times taken to assemble a unit of the product are different for the two types of machines.

Assume that the population of paired differences is (approximately) normally distributed.

10.5 Inferences About the Difference Between Two Population Proportions for Large and Independent Samples

Quite often we need to construct a confidence interval and test a hypothesis about the difference between two population proportions. For instance, we may want to estimate the difference between the proportions of defective items produced on two different machines. If p_1 and p_2 are the proportions of defective items produced on the first and second machine, respectively, then we are to make a confidence interval for $p_1 - p_2$. Or we may want to test the hypothesis that the proportion of defective items produced on Machine I is different from the proportion of defective items produced on Machine II. In this case, we are to test the null hypothesis $p_1 - p_2 = 0$ against the alternative hypothesis $p_1 - p_2 \neq 0$.

This section discusses how to make a confidence interval and test a hypothesis about $p_1 - p_2$ for two large and independent samples. The sample statistic that is used to make inferences about $p_1 - p_2$ is $\hat{p}_1 - \hat{p}_2$, where $\hat{p}_1$ and $\hat{p}_2$ are the proportions for two large and independent samples. As discussed in Chapter 7, we determine a sample proportion by dividing the number of elements in the sample that possess a given attribute by the sample size. Thus,

$$\hat{p}_1 = x_1/n_1 \quad \text{and} \quad \hat{p}_2 = x_2/n_2$$

where x_1 and x_2 are the number of elements that possess a given characteristic in the two samples and n_1 and n_2 are the sizes of the two samples, respectively.

10.5.1 Mean, Standard Deviation, and Sampling Distribution of $\hat{p}_1 - \hat{p}_2$

As discussed in Chapter 7, for a large sample the sample proportion $\hat{p}$ is (approximately) normally distributed with mean p and standard deviation $\sqrt{pq/n}$. Hence, for two large and independent samples of sizes n_1 and n_2, respectively, their sample proportions $\hat{p}_1$ and $\hat{p}_2$ are (approximately) normally distributed with means p_1 and p_2 and standard deviations $\sqrt{p_1q_1/n_1}$ and $\sqrt{p_2q_2/n_2}$, respectively. Using these results, we can make the following statements about the shape of the sampling distribution of $\hat{p}_1 - \hat{p}_2$ and its mean and standard deviation.

Mean, Standard Deviation, and Sampling Distribution of $\hat{p}_1 - \hat{p}_2$ For two large and independent samples, the *sampling distribution* of $\hat{p}_1 - \hat{p}_2$ is (approximately) normal with its *mean* and *standard deviation* given as

$$\mu_{\hat{p}_1 - \hat{p}_2} = p_1 - p_2$$

and

$$\sigma_{\hat{p}_1 - \hat{p}_2} = \sqrt{\frac{p_1q_1}{n_1} + \frac{p_2q_2}{n_2}}$$

respectively, where $q_1 = 1 - p_1$ and $q_2 = 1 - p_2$.

Thus, to construct a confidence interval and test a hypothesis about $p_1 - p_2$ for large and independent samples, we use the normal distribution. As was indicated in Chapter 7, in the case of proportion, the sample is large if np and nq are both greater than 5. In the case of two samples, both sample sizes are large if n_1p_1, n_1q_1, n_2p_2, and n_2q_2 are all greater than 5.

10.5.2 Interval Estimation of $p_1 - p_2$

The difference between two sample proportions $\hat{p}_1 - \hat{p}_2$ is the point estimator for the difference between two population proportions $p_1 - p_2$. Because we do not know p_1 and p_2 when we are making a confidence interval for $p_1 - p_2$, we cannot calculate the value of $\sigma_{\hat{p}_1-\hat{p}_2}$. Therefore, we use $s_{\hat{p}_1-\hat{p}_2}$ as the point estimator of $\sigma_{\hat{p}_1-\hat{p}_2}$ in the interval estimation. We construct the confidence interval for $p_1 - p_2$ using the following formula.

Confidence Interval for $p_1 - p_2$ The $(1 - \alpha)100\%$ *confidence interval for* $p_1 - p_2$ is

$$(\hat{p}_1 - \hat{p}_2) \pm z s_{\hat{p}_1-\hat{p}_2}$$

where the value of z is read from the normal distribution table for the given confidence level, and $s_{\hat{p}_1-\hat{p}_2}$ is calculated as

$$s_{\hat{p}_1-\hat{p}_2} = \sqrt{\frac{\hat{p}_1\hat{q}_1}{n_1} + \frac{\hat{p}_2\hat{q}_2}{n_2}}$$

Example 10–13 describes the procedure used to make a confidence interval for the difference between two population proportions for large samples.

■ EXAMPLE 10–13

Constructing a confidence interval for $p_1 - p_2$: large and independent samples.

A researcher wanted to estimate the difference between the percentages of users of two toothpastes who will never switch to another toothpaste. In a sample of 500 users of Toothpaste A taken by this researcher, 100 said that they will never switch to another toothpaste. In another sample of 400 users of Toothpaste B taken by the same researcher, 68 said that they will never switch to another toothpaste.

(a) Let p_1 and p_2 be the proportions of all users of Toothpastes A and B, respectively, who will never switch to another toothpaste. What is the point estimate of $p_1 - p_2$?

(b) Construct a 97% confidence interval for the difference between the proportions of all users of the two toothpastes who will never switch.

Solution Let p_1 and p_2 be the proportions of all users of Toothpastes A and B, respectively, who will never switch to another toothpaste, and let $\hat{p}_1$ and $\hat{p}_2$ be the respective sample proportions. Let x_1 and x_2 be the number of users of Toothpastes A and B, respectively, in the two samples who said that they will never switch to another toothpaste. From the given information,

$$\text{Toothpaste A:} \quad n_1 = 500 \quad \text{and} \quad x_1 = 100$$
$$\text{Toothpaste B:} \quad n_2 = 400 \quad \text{and} \quad x_2 = 68$$

The two sample proportions are calculated as follows:

$$\hat{p}_1 = x_1/n_1 = 100/500 = .20$$
$$\hat{p}_2 = x_2/n_2 = 68/400 = .17$$

Then,

$$\hat{q}_1 = 1 - .20 = .80 \quad \text{and} \quad \hat{q}_2 = 1 - .17 = .83$$

(a) The point estimate of $p_1 - p_2$ is as follows:

$$\text{Point estimate of } p_1 - p_2 = \hat{p}_1 - \hat{p}_2 = .20 - .17 = \mathbf{.03}$$

(b) The values of $n_1\hat{p}_1$, $n_1\hat{q}_1$, $n_2\hat{p}_2$, and $n_2\hat{q}_2$ are

$$n_1\hat{p}_1 = 500(.20) = 100 \quad n_1\hat{q}_1 = 500(.80) = 400$$

$$n_2\hat{p}_2 = 400(.17) = 68 \quad n_2\hat{q}_2 = 400(.83) = 332$$

Because each of these values is greater than 5, both sample sizes are large. Consequently we use the normal distribution to make a confidence interval for $p_1 - p_2$. The standard deviation of $\hat{p}_1 - \hat{p}_2$ is

$$s_{\hat{p}_1 - \hat{p}_2} = \sqrt{\frac{\hat{p}_1\hat{q}_1}{n_1} + \frac{\hat{p}_2\hat{q}_2}{n_2}} = \sqrt{\frac{(.20)(.80)}{500} + \frac{(.17)(.83)}{400}} = .02593742$$

The z value for a 97% confidence level, obtained from the normal distribution table is 2.17. The 97% confidence interval for $p_1 - p_2$ is

$$(\hat{p}_1 - \hat{p}_2) \pm z s_{\hat{p}_1 - \hat{p}_2} = (.20 - .17) \pm 2.17(.02593742)$$

$$= .03 \pm .056 = \mathbf{-.026 \text{ to } .086}$$

Thus, with 97% confidence we can state that the difference between the two population proportions is between $-.026$ and $.086$.

Note that here $\hat{p}_1 - \hat{p}_2 = .03$ gives the point estimate of $p_1 - p_2$ and $z s_{\hat{p}_1 - \hat{p}_2} = .056$ is the margin of error of the estimate. ■

10.5.3 Hypothesis Testing About $p_1 - p_2$

In this section we learn how to test a hypothesis about $p_1 - p_2$ for two large and independent samples. The procedure involves the same five steps we have used previously. Once again, we calculate the standard deviation of $\hat{p}_1 - \hat{p}_2$ as

$$\sigma_{\hat{p}_1 - \hat{p}_2} = \sqrt{\frac{p_1 q_1}{n_1} + \frac{p_2 q_2}{n_2}}$$

When a test of hypothesis about $p_1 - p_2$ is performed, usually the null hypothesis is $p_1 = p_2$ and the values of p_1 and p_2 are not known. Assuming that the null hypothesis is true and $p_1 = p_2$, a common value of p_1 and p_2, denoted by $\bar{p}$, is calculated by using one of the following two formulas:

$$\bar{p} = \frac{x_1 + x_2}{n_1 + n_2} \quad \text{or} \quad \frac{n_1\hat{p}_1 + n_2\hat{p}_2}{n_1 + n_2}$$

Which of these formulas is used depends on whether the values of x_1 and x_2 or the values of $\hat{p}_1$ and $\hat{p}_2$ are known. Note that x_1 and x_2 are the number of elements in each of the two samples that possess a certain characteristic. This value of $\bar{p}$ is called the **pooled sample proportion**. Using the value of the pooled sample proportion, we compute an estimate of the standard deviation of $\hat{p}_1 - \hat{p}_2$ as follows:

$$s_{\hat{p}_1 - \hat{p}_2} = \sqrt{\bar{p}\bar{q}\left(\frac{1}{n_1} + \frac{1}{n_2}\right)}$$

where $\bar{q} = 1 - \bar{p}$.

Test Statistic z for $\hat{p}_1 - \hat{p}_2$ The value of the *test statistic z for* $\hat{p}_1 - \hat{p}_2$ is calculated as

$$z = \frac{(\hat{p}_1 - \hat{p}_2) - (p_1 - p_2)}{s_{\hat{p}_1 - \hat{p}_2}}$$

The value of $p_1 - p_2$ is substituted from H_0, which usually is zero.

Examples 10–14 and 10–15 illustrate the procedure to test hypotheses about the difference between two population proportions for large samples.

■ EXAMPLE 10–14

Making a right-tailed test of hypothesis about $p_1 - p_2$: large and independent samples.

Reconsider Example 10–13 about the percentages of users of two toothpastes who will never switch to another toothpaste. At the 1% significance level, can we conclude that the proportion of users of Toothpaste A who will never switch to another toothpaste is higher than the proportion of users of Toothpaste B who will never switch to another toothpaste?

Solution Let p_1 and p_2 be the proportions of all users of Toothpastes A and B, respectively, who will never switch to another toothpaste and let $\hat{p}_1$ and $\hat{p}_2$ be the corresponding sample proportions. Let x_1 and x_2 be the number of users of Toothpastes A and B, respectively, in the two samples who said that they will never switch to another toothpaste. From the given information,

$$\text{Toothpaste A:} \quad n_1 = 500 \quad \text{and} \quad x_1 = 100$$

$$\text{Toothpaste B:} \quad n_2 = 400 \quad \text{and} \quad x_2 = 68$$

The significance level is $\alpha = .01$. The two sample proportions are calculated as follows:

$$\hat{p}_1 = x_1/n_1 = 100/500 = .20$$

$$\hat{p}_2 = x_2/n_2 = 68/400 = .17$$

Step 1. *State the null and alternative hypotheses.*

We are to test if the proportion of users of Toothpaste A who will never switch to another toothpaste is higher than the proportion of users of Toothpaste B who will never switch to another toothpaste. In other words, we are to test whether p_1 is greater than p_2. This can be written as $p_1 - p_2 > 0$. Thus, the two hypotheses are

$$H_0: p_1 - p_2 = 0 \quad (p_1 \text{ is not greater than } p_2)$$

$$H_1: p_1 - p_2 > 0 \quad (p_1 \text{ is greater than } p_2)$$

Step 2. *Select the distribution to use.*

As shown in Example 10–13, $n_1\hat{p}_1$, $n_1\hat{q}_1$, $n_2\hat{p}_2$, and $n_2\hat{q}_2$ are all greater than 5. Consequently both samples are large, and we apply the normal distribution to make the test.

Step 3. *Determine the rejection and nonrejection regions.*

The > sign in the alternative hypothesis indicates that the test is right-tailed. From the normal distribution table, for a .01 significance level, the critical value of z is 2.33 for .9900 area to the left. This is shown in Figure 10.8.

Figure 10.8

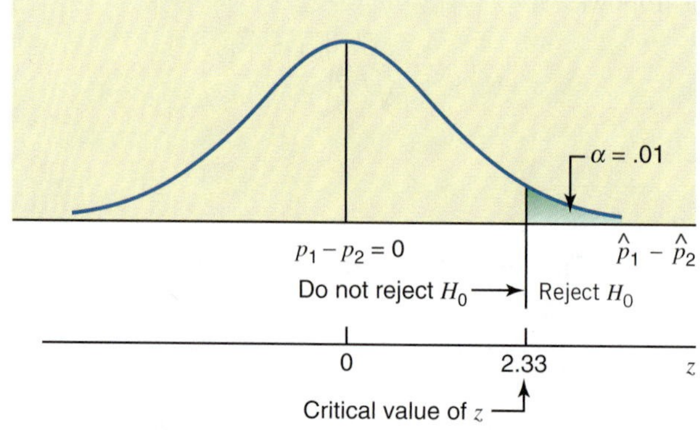

Step 4. *Calculate the value of the test statistic.*

The pooled sample proportion is

$$\bar{p} = \frac{x_1 + x_2}{n_1 + n_2} = \frac{100 + 68}{500 + 400} = .187$$

$$\bar{q} = 1 - \bar{p} = 1 - .187 = .813$$

The estimate of the standard deviation of $\hat{p}_1 - \hat{p}_2$ is

$$s_{\hat{p}_1 - \hat{p}_2} = \sqrt{\bar{p}\bar{q}\left(\frac{1}{n_1} + \frac{1}{n_2}\right)} = \sqrt{(.187)(.813)\left(\frac{1}{500} + \frac{1}{400}\right)} = .02615606$$

The value of the test statistic z for $\hat{p}_1 - \hat{p}_2$ is

$$z = \frac{(\hat{p}_1 - \hat{p}_2) - (p_1 - p_2)}{s_{\hat{p}_1 - \hat{p}_2}} = \frac{(.20 - .17) - 0}{.02615606} = 1.15$$

(From H_0: the 0 in the numerator)

Step 5. *Make a decision.*

Because the value of the test statistic $z = 1.15$ for $\hat{p}_1 - \hat{p}_2$ falls in the nonrejection region, we fail to reject the null hypothesis. Therefore, we conclude that the proportion of users of Toothpaste A who will never switch to another toothpaste is not greater than the proportion of users of Toothpaste B who will never switch to another toothpaste.

Using the *p*-Value to Make a Decision

We can use the p-value approach to make the above decision. To do so, we keep Steps 1 and 2 above. Then in Step 3, we calculate the value of the test statistic z (as done in Step 4 above) and find the p-value for this z from the normal distribution table. In Step 4 above, the z-value for $\hat{p}_1 - \hat{p}_2$ was calculated to be 1.15. In this example, the test is right-tailed. The p-value is given by the area under the normal distribution curve to the right of $z = 1.15$. From the normal distribution table (Table IV of Appendix C), this area is $1 - .8749 = .1251$. Hence, the p-value is .1251. We reject the null hypothesis for any α (significance level) greater than the p-value; in this example, we will reject the null hypothesis for any $\alpha > .1251$ or 12.51%. Because $\alpha = .01$ here, which is less than .1251, we fail to reject the null hypothesis. ■

■ EXAMPLE 10–15

Conducting a two-tailed test of hypothesis about $p_1 - p_2$: large and independent samples.

According to the National Sleep Foundation, 69% of the adults surveyed in 2001 reported symptoms of a sleep problem, and 75% of the adults surveyed in 2005 reported such symptoms (Associated Press, March 29, 2005). The 2005 survey was based on a sample of 1506 adults. Suppose the 2001 survey included 1600 adults. Test whether the percentages of adults with symptoms of a sleep problem are different for these two years. Use the 1% significance level.

Solution Let p_1 and p_2 be the proportions of all adults who had symptoms of a sleep problem in 2001 and 2005, respectively. Let $\hat{p}_1$ and $\hat{p}_2$ be the corresponding sample proportions. From the given information,

For 2001: $n_1 = 1600$ and $\hat{p}_1 = .69$

For 2005: $n_2 = 1506$ and $\hat{p}_2 = .75$

The significance level is $\alpha = .01$.

Step 1. *State the null and alternative hypotheses.*

The null and alternative hypotheses are

H_0: $p_1 - p_2 = 0$ (The two population proportions are not different)

H_1: $p_1 - p_2 \neq 0$ (The two population proportions are different)

CASE STUDY

10–2

WORKPLACE VIEWS DIFFER BY GENDER

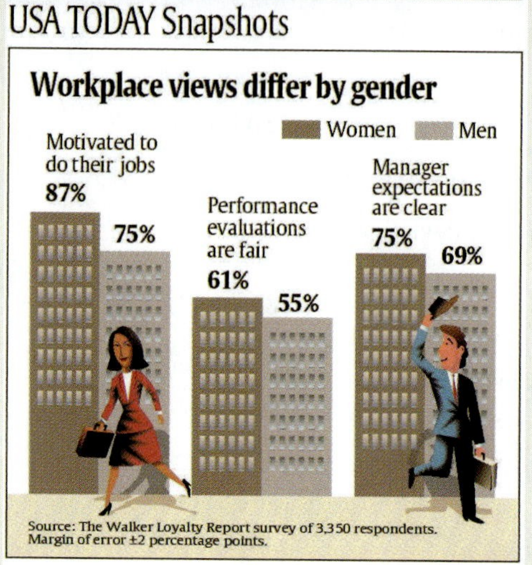

The above chart shows the views of men and women in regard to three work related issues. These results are based on a survey of 3350 respondents. For example, in this survey, 87% of the women and 75% of the men said that they are motivated to do their jobs. If we know the sample sizes for both genders included in this survey, we can make a confidence interval and perform a test of hypothesis for the difference in the percentages for women and men in regard to each of the views listed in the chart.

Suppose, of the 3350 respondents included in the survey, 1700 were women and 1650 were men. Let p_1 and p_2 be the proportions of all women and men, respectively, who would say that they are motivated to do their jobs. Let $\hat{p}_1$ and $\hat{p}_2$ be the corresponding sample proportions. Then, from the given information:

For women: $n_1 = 1700 \quad \hat{p}_1 = .87 \quad \hat{q}_1 = 1 - .87 = .13$

For men: $n_2 = 1650 \quad \hat{p}_2 = .75 \quad \hat{q}_2 = 1 - .75 = .25$

Below we make a confidence interval and test a hypothesis about $p_1 - p_2$ for women and men who hold the above opinion.

1. Confidence interval for $p_1 - p_2$

Suppose we want to make a 97% confidence interval for $p_1 - p_2$. The z value from Table IV of Appendix C for the 97% confidence level is 2.17. The standard deviation of $\hat{p}_1 - \hat{p}_2$ is

$$s_{\hat{p}_1-\hat{p}_2} = \sqrt{\frac{\hat{p}_1\hat{q}_1}{n_1} + \frac{\hat{p}_2\hat{q}_2}{n_2}} = \sqrt{\frac{(.87)(.13)}{1700} + \frac{(.75)(.25)}{1650}} = .01342258$$

Step 2. *Select the distribution to use.*

Because the samples are large and independent, we apply the normal distribution to make the test. (The reader should check that $n_1\hat{p}_1$, $n_1\hat{q}_1$, $n_2\hat{p}_2$, and $n_2\hat{q}_2$ are all greater than 5.)

Step 3. *Determine the rejection and nonrejection regions.*

The $\neq$ sign in the alternative hypothesis indicates that the test is two-tailed. For a 1% significance level, the critical values of z are -2.58 and 2.58. Note that to find these two critical values, we look for .0050 and .9950 areas in Table IV of Appendix C. These values are shown in Figure 10.9.

Hence, the 97% confidence interval for $p_1 - p_2$ is:

$$(\hat{p}_1 - \hat{p}_2) \pm z s_{\hat{p}_1 - \hat{p}_2} = (.87 - .75) \pm 2.17(.01342258) = .12 \pm .029$$

$$= \mathbf{.091 \text{ to } .149 \text{ or } 9.1\% \text{ to } 14.9\%}$$

Thus, we can say with 97% confidence that the difference in the proportions of all women and men who feel motivated to do their jobs is in the interval .091 to .149 or 9.1% to 14.9%.

2. Test of hypothesis about $p_1 - p_2$

Suppose we want to test, at the 1% significance level, whether the proportion of all women who feel motivated to do their jobs is greater than that for all men. In other words, we are to test if p_1 is greater than p_2. The null and alternative hypotheses are

$$H_0: p_1 = p_2 \quad \text{or} \quad p_1 - p_2 = 0$$

$$H_1: p_1 > p_2 \quad \text{or} \quad p_1 - p_2 > 0$$

Note that the test is right-tailed. For $\alpha = .01$, the critical value of z from the normal distribution table for .9900 is 2.33. Thus, we will reject the null hypothesis if the observed value of z is 2.33 or larger. The pooled sample proportion is

$$\bar{p} = \frac{n_1\hat{p}_1 + n_2\hat{p}_2}{n_1 + n_2} = \frac{1700(.87) + 1650(.75)}{1700 + 1650} = .811$$

and $\qquad \bar{q} = 1 - \bar{p} = 1 - .811 = .189$

The estimate of the standard deviation of $\hat{p}_1 - \hat{p}_2$ is

$$s_{\hat{p}_1 - \hat{p}_2} = \sqrt{\bar{p}\,\bar{q}\left(\frac{1}{n_1} + \frac{1}{n_2}\right)} = \sqrt{(.811)(.189)\left(\frac{1}{1700} + \frac{1}{1650}\right)} = .01352998$$

The value of the test statistic z for $\hat{p}_1 - \hat{p}_2$ is

$$z = \frac{(\hat{p}_1 - \hat{p}_2) - (p_1 - p_2)}{s_{\hat{p}_1 - \hat{p}_2}} = \frac{(.87 - .75) - 0}{.01352998} = 8.87$$

Since the observed value of $z = 8.87$ is larger than the critical value of 2.33, we reject the null hypothesis. As a result we conclude that p_1 is greater than p_2, and that the proportion of all women who feel motivated to do their jobs is greater than that for all men.

We can also use the p-value approach to make this decision. In this example, the test is right-tailed. As calculated above, the z value for $\hat{p}_1 - \hat{p}_2$ is 8.87. From the normal distribution table, the area to the right of $z = 8.87$ is (approximately) .0000. Hence, the p-value is .0000. Since, $\alpha = .01$ in this example is greater than .0000, we reject the null hypothesis and conclude that the proportion of all women who feel motivated to do their jobs is greater than that for all men.

Source: The chart reproduced with permission from *USA TODAY*, October 22, 2003. Copyright © 2003, *USA TODAY.*

Figure 10.9

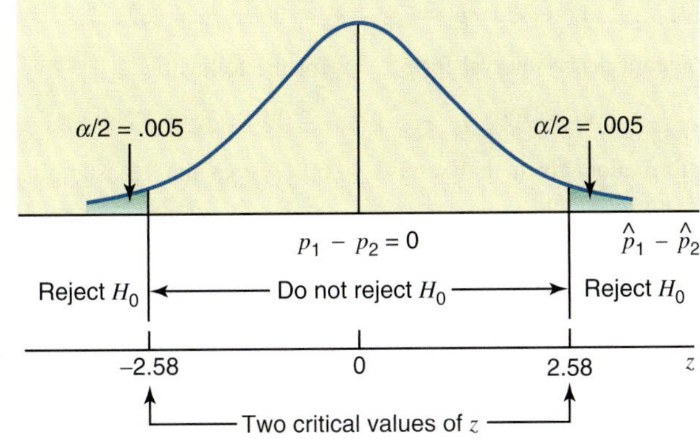

Step 4. *Calculate the value of the test statistic.*

The pooled sample proportion is

$$\bar{p} = \frac{n_1\hat{p}_1 + n_2\hat{p}_2}{n_1 + n_2} = \frac{1600(.69) + 1506(.75)}{1600 + 1506} = .719$$

$$\bar{q} = 1 - \bar{p} = 1 - .719 = .281$$

The estimate of the standard deviation of $\hat{p}_1 - \hat{p}_2$ is

$$s_{\hat{p}_1 - \hat{p}_2} = \sqrt{\bar{p}\bar{q}\left(\frac{1}{n_1} + \frac{1}{n_2}\right)} = \sqrt{(.719)(.281)\left(\frac{1}{1600} + \frac{1}{1506}\right)} = .01613785$$

The value of the test statistic z for $\hat{p}_1 - \hat{p}_2$ is

$$z = \frac{(\hat{p}_1 - \hat{p}_2) - (p_1 - p_2)}{s_{\hat{p}_1 - \hat{p}_2}} = \frac{(.69 - .75) - \overset{\text{From } H_0}{\overset{\downarrow}{0}}}{.01613785} = -3.72$$

Step 5. *Make a decision.*

Because the value of the test statistic $z = -3.72$ falls in the rejection region, we reject the null hypothesis H_0. Therefore, we conclude that the percentages of adults with symptoms of a sleep problem are different for these two years.

Using the *p*-Value to Make a Decision

We can use the p-value approach to make the above decision. To do so, we keep Steps 1 and 2 above. Then in Step 3, we calculate the value of the test statistic z (as done in Step 4 above) and find the p-value for this z from the normal distribution table. In Step 4 above, the z-value for $\hat{p}_1 - \hat{p}_2$ was calculated to be -3.72. In this example, the test is two-tailed. The p-value is given by twice the area under the normal distribution curve to the left of $z = -3.72$. From the normal distribution table (Table IV of Appendix C), the area to the left of $z = -3.72$ is (approximately) zero. Hence, the p-value is $2(.0000) = .0000$. As we know, we will reject the null hypothesis for any α (significance level) greater than the p-value. Since, $\alpha = .01$ in this example, which is greater than .0000, we reject the null hypothesis. ■

EXERCISES

CONCEPTS AND PROCEDURES

10.58 What is the shape of the sampling distribution of $\hat{p}_1 - \hat{p}_2$ for two large samples? What are the mean and standard deviation of this sampling distribution?

10.59 When are the samples considered large enough for the sampling distribution of the difference between two sample proportions to be (approximately) normal?

10.60 Construct a 99% confidence interval for $p_1 - p_2$ for the following.

$$n_1 = 300 \quad \hat{p}_1 = .55 \quad n_2 = 200 \quad \hat{p}_2 = .62$$

10.61 Construct a 95% confidence interval for $p_1 - p_2$ for the following.

$$n_1 = 100 \quad \hat{p}_1 = .81 \quad n_2 = 150 \quad \hat{p}_2 = .77$$

10.62 Refer to the information given in Exercise 10.60. Test at the 1% significance level if the two population proportions are different.

10.63 Refer to the information given in Exercise 10.61. Test at the 5% significance level if $p_1 - p_2$ is different from zero.

10.64 Refer to the information given in Exercise 10.60. Test at the 1% significance level if p_1 is less than p_2.

10.65 Refer to the information given in Exercise 10.61. Test at the 2% significance level if p_1 is greater than p_2.

10.66 A sample of 500 observations taken from the first population gave $x_1 = 305$. Another sample of 600 observations taken from the second population gave $x_2 = 348$.

a. Find the point estimate of $p_1 - p_2$.
b. Make a 97% confidence interval for $p_1 - p_2$.
c. Show the rejection and nonrejection regions on the sampling distribution of $\hat{p}_1 - \hat{p}_2$ for H_0: $p_1 = p_2$ versus H_1: $p_1 > p_2$. Use a significance level of 2.5%.
d. Find the value of the test statistic z for the test of part c.
e. Will you reject the null hypothesis mentioned in part c at a significance level of 2.5%?

10.67 A sample of 1000 observations taken from the first population gave $x_1 = 290$. Another sample of 1200 observations taken from the second population gave $x_2 = 396$.

a. Find the point estimate of $p_1 - p_2$.
b. Make a 98% confidence interval for $p_1 - p_2$.
c. Show the rejection and nonrejection regions on the sampling distribution of $\hat{p}_1 - \hat{p}_2$ for H_0: $p_1 = p_2$ versus H_1: $p_1 < p_2$. Use a significance level of 1%.
d. Find the value of the test statistic z for the test of part c.
e. Will you reject the null hypothesis mentioned in part c at a significance level of 1%?

APPLICATIONS

10.68 According to the Walker Loyalty Report survey of 3350 respondents, 61% of women and 55% of men polled said that performance evaluations at their jobs were fair (*USA TODAY*, October 22, 2003). Assume that these percentages are based on random samples of 1700 men and 1650 women.

a. Determine a 99% confidence interval for the difference between the two population proportions.
b. At the 2.5% significance level, can you conclude that the proportion of all women who think that performance evaluations at their jobs are fair is higher than the proportion of all men who hold that opinion? Use the critical-value approach.
c. Repeat part b using the p-value approach.

10.69 Primary seat belt laws allow police to stop a motorist for not wearing a seat belt. According to the National Highway Traffic Safety Administration, the percentage of drivers wearing seat belts in the state of Tennessee was 68.5% in 2003 and 72% in 2004 (*USA TODAY*, November 23, 2004). (Tennessee passed a primary seat belt law in July 2004.) Assume that these percentages are based on random samples of 1000 drivers in 2003 and 1050 in 2004.

a. Construct a 98% confidence interval for the difference between the two population proportions.
b. Test at the 1% significance level whether the proportion of all drivers in Tennessee who used seat belts in 2003 is lower than that for 2004. Use the critical-value approach.
c. Repeat part b using the p-value approach.

10.70 Bernard Haldane Associates conducted a survey in March 2004 of 1021 workers who held white-collar jobs and had changed jobs in the previous twelve months. Of these workers, 56% of the men and 35% of the women were paid more in their new positions when they changed jobs (*USA TODAY*, April 28, 2004). Suppose that these percentages are based on random samples of 510 men and 511 women white-collar workers.

a. What is the point estimate of the difference between the two population proportions?
b. Construct a 95% confidence interval for the difference between the two population proportions.
c. Using the 2% significance level, can you conclude that the two population proportions are different? Use both the critical-value and the p-value approaches.

10.71 A state that requires periodic emission tests of cars operates two emissions test stations, A and B, in one of its towns. Car owners have complained of lack of uniformity of procedures at the two stations, resulting in different failure rates. A sample of 400 cars at Station A showed that 53 of those failed the test; a sample of 470 cars at Station B found that 51 of those failed the test.

a. What is the point estimate of the difference between the two population proportions?
b. Construct a 95% confidence interval for the difference between the two population proportions.
c. Testing at the 5% significance level, can you conclude that the two population proportions are different? Use both the critical-value and the p-value approaches.

10.72 The management of a supermarket wanted to investigate if the percentages of men and women who prefer to buy national brand products over the store brand products are different. A sample of 600 men shoppers at the company's supermarkets showed that 246 of them prefer to buy national brand products over the store brand products. Another sample of 700 women shoppers at the company's supermarkets showed that 266 of them prefer to buy national brand products over the store brand products.

a. What is the point estimate of the difference between the two population proportions?

b. Construct a 95% confidence interval for the difference between the proportions of all men and all women shoppers at these supermarkets who prefer to buy national brand products over the store brand products.

c. Testing at the 5% significance level, can you conclude that the proportions of all men and all women shoppers at these supermarkets who prefer to buy national brand products over the store brand products are different?

10.73 The lottery commissioner's office in a state wanted to find if the percentages of men and women who play the lottery often are different. A sample of 500 men taken by the commissioner's office showed that 160 of them play the lottery often. Another sample of 300 women showed that 66 of them play the lottery often.

a. What is the point estimate of the difference between the two population proportions?

b. Construct a 99% confidence interval for the difference between the proportions of all men and all women who play the lottery often.

c. Testing at the 1% significance level, can you conclude that the proportions of all men and all women who play the lottery often are different?

10.74 A mail-order company has two warehouses, one on the West Coast and the second on the East Coast. The company's policy is to mail all orders placed with it within 72 hours. The company's quality control department checks quite often whether or not this policy is maintained at the two warehouses. A recently taken sample of 400 orders placed with the warehouse on the West Coast showed that 364 of them were mailed within 72 hours. Another sample of 300 orders placed with the warehouse on the East Coast showed that 279 of them were mailed within 72 hours.

a. Construct a 97% confidence interval for the difference between the proportions of all orders placed at the two warehouses that are mailed within 72 hours.

b. Using the 2.5% significance level, can you conclude that the proportion of all orders placed at the warehouse on the West Coast that are mailed within 72 hours is lower than the corresponding proportion for the warehouse on the East Coast?

10.75 A company that has many department stores in the Southern states wanted to find at two such stores the percentage of sales for which at least one of the items was returned. A sample of 800 sales randomly selected from Store A showed that for 280 of them at least one item was returned. Another sample of 900 sales randomly selected from Store B showed that for 279 of them at least one item was returned.

a. Construct a 98% confidence interval for the difference between the proportions of all sales at the two stores for which at least one item is returned.

b. Using the 1% significance level, can you conclude that the proportions of all sales for which at least one item is returned is higher for Store A than for Store B?

USES AND MISUSES... GRAPES TO GRAPES

Turn on the TV one Sunday morning to one of the news programs in which journalists and writers spar with politicians. It is very common on these programs for someone to claim that events of the previous week are reminiscent of events of the previous decade and that particular actions are thus warranted. It is just as common for a participant in the debate to state that the interpretation is incorrect because the comparison is not "apples to apples, oranges to oranges." Statistical analysis of the differences between two populations requires that the populations be essentially similar, because the methods described in the chapter are for making apples-to-apples and oranges-to-oranges comparisons. The only possible difference between the two populations should be the characteristic that you are studying; remember that the null hypothesis is often that there is no difference between the population means or proportions.

To illustrate, consider an extreme example—wine. The text describes an experiment in which a statistician instructs a farmer to plant two varieties of a crop on 20 distributed plots to ensure that comparisons of the productivity are not affected by peculiarities of the land on which the crops are planted. Winemakers rarely do this because the peculiarities of the land are what they want to emphasize. Suppose that Winemaker A and Winemaker B live in the same region and purchase their vines from the same nursery. Because they plant the same varietals, live near one another, experience the same weather, use similar irrigation and growing systems, harvest the grapes at the same stage of ripeness, and so on, their wines should be very similar. Very similar, that is, but also very different according to the winemakers, because every tiny difference in the land and methods is magnified in the final product. Sometimes the claimed differences are so small as to be silly: In Bordeaux, a wine-producing region in France, a dirt road can separate two vineyards, yet the wine from one may cost several times that of the other.

When performing a statistical analysis of two populations, think about crossing that dirt road and how that tiny difference might cloud your results.

Glossary

d The difference between two matched values in two samples collected from the same source. It is called the paired difference.

$\bar{d}$ The mean of the paired differences for a sample.

Independent samples Two samples drawn from two populations such that the selection of one does not affect the selection of the other.

Paired or **matched samples** Two samples drawn in such a way that they include the same elements and two data values are obtained from each element, one for each sample. Also called **dependent samples**.

μ_d The mean of the paired differences for the population.

s_d The standard deviation of the paired differences for a sample.

σ_d The standard deviation of the paired differences for the population.

Supplementary Exercises

10.76 A consulting agency was asked by a large insurance company to investigate if business majors were better salespersons than those with other majors. A sample of 20 salespersons with a business degree showed that they sold an average of 11 insurance policies per week. Another sample of 25 salespersons with a degree other than business showed that they sold an average of 9 insurance policies per week. Assume that the two populations are normally distributed with population standard deviations of 1.80 and 1.35 policies per week, respectively.

a. Construct a 99% confidence interval for the difference between the two population means.

b. Using the 1% significance level, can you conclude that persons with a business degree are better salespersons than those who have a degree in another area?

10.77 According to an estimate, the average earnings of female workers who are not union members are \$388 per week and those of female workers who are union members are \$505 per week. Suppose that these average earnings are calculated based on random samples of 1500 female workers who are not union members and 2000 female workers who are union members. Further assume that the standard deviations for the two corresponding populations are \$30 and \$35, respectively.

a. Construct a 95% confidence interval for the difference between the two population means.

b. Test at the 2.5% significance level whether the mean weekly earnings of female workers who are not union members are less than those of female workers who are union members.

10.78 A researcher wants to test if the mean GPAs (grade point averages) of all male and all female college students who actively participate in sports are different. She took a random sample of 28 male students and 24 female students who are actively involved in sports. She found the mean GPAs of the two groups to be 2.62 and 2.74, respectively, with the corresponding standard deviations equal to .43 and .38.

a. Test at the 5% significance level whether the mean GPAs of the two populations are different.

b. Construct a 90% confidence interval for the difference between the two population means.

Assume that the GPAs of all male and all female student athletes both are normally distributed with equal but unknown population standard deviations.

10.79 An agency wanted to estimate the difference between the auto insurance premiums paid by drivers insured with two different insurance companies. A random sample of 25 drivers insured with insurance company A showed that they paid an average monthly insurance premium of \$97 with a standard deviation of \$14. Another random sample of 20 drivers insured with insurance company B showed that these drivers paid an average monthly insurance premium of \$89 with a standard deviation of \$12. Assume that the insurance premiums paid by all drivers insured with companies A and B are both normally distributed with equal but unknown population standard deviations.

a. Construct a 99% confidence interval for the difference between the two population means.

b. Test at the 1% significance level whether the mean monthly insurance premium paid by drivers insured with company A is higher than that of drivers insured with company B.

10.80 The manager of a factory has devised a detailed plan for evacuating the building as quickly as possible in the event of a fire or other emergency. An industrial psychologist believes that workers actually leave the factory faster at closing time without following any system. The company holds fire drills periodically in which a bell sounds and workers leave the building according to the system. The evacuation time for each drill is recorded. For comparison, the psychologist also records the evacuation time when the bell sounds for closing time each day. A random sample of 36 fire drills showed a mean evacuation time of 5.1 minutes with a standard deviation of 1.1 minutes. A random sample of 37 days at closing time showed a mean evacuation time of 4.2 minutes with a standard deviation of 1.0 minute.

a. Construct a 99% confidence interval for the difference between the two population means.

b. Test at the 5% significance level whether the mean evacuation time is smaller at closing time than during fire drills.

Assume that the evacuation times at closing time and during fire drills have equal but unknown population standard deviations.

10.81 Based on a nationwide survey of adults by the Travel Industry Association, Americans expected to spend an average of $1019 on their longest vacation trips in 2005, compared to $1101 in 2004 (*USA TODAY*, May 20, 2005). Assume that the average for 2005 is based on a random sample of 1000 adults and the one for 2004 is based on a random sample of 900 adults, and that the sample standard deviations for such expenses are $320 for 2005 and $305 for 2004. Let μ_1 and μ_2 be the population means of such expenses for the years 2005 and 2004, respectively.

a. Construct a 99% confidence interval for $\mu_1 - \mu_2$.

b. Testing at the 1% level of significance, can you conclude that the mean for 2005 is lower than that for 2004?

Assume that such expenses for the years 2005 and 2004 have unknown but equal population standard deviations.

10.82 Repeat Exercise 10.78, but now assume that the GPAs of all male and all female student athletes are both normally distributed with unknown and unequal standard deviations.

10.83 Repeat Exercise 10.79, but now assume that the two populations have normal distributions with unknown and unequal standard deviations.

10.84 Repeat Exercise 10.80, but now assume that the evacuation times under both conditions have unequal and unknown population standard deviations.

10.85 Repeat Exercise 10.81, but now assume that the standard deviations of the two populations are unknown and unequal.

10.86 The owner of a mosquito-infested fishing camp in Alaska wants to test the effectiveness of two rival brands of mosquito repellents, X and Y. During the first month of the season, eight people are chosen at random from those guests who agree to take part in the experiment. For each of these guests, Brand X is randomly applied to one arm and Brand Y is applied to the other arm. These guests fish for four hours, then the owner counts the number of bites on each arm. The table below shows the number of bites on the arm with Brand X and those on the arm with Brand Y for each guest.

Guest	A	B	C	D	E	F	G	H
Brand X	12	23	18	36	8	27	22	32
Brand Y	9	20	21	27	6	18	15	25

a. Construct a 95% confidence interval for the mean μ_d of population paired differences, where a paired difference is defined as the number of bites on the arm with Brand X minus the number of bites on the arm with Brand Y.

b. Test at the 5% significance level whether the mean number of bites on the arm with Brand X and the mean number of bites on the arm with Brand Y are different for all such guests.

Assume that the population of paired differences has a normal distribution.

10.87 A random sample of nine students was selected to test for the effectiveness of a special course designed to improve memory. The following table gives the scores in a memory test given to these students before and after this course.

Before	43	57	48	65	81	49	38	69	58
After	49	56	55	77	89	57	36	64	69

a. Construct a 95% confidence interval for the mean μ_d of the population paired differences, where a paired difference is defined as the difference between the memory test scores of a student before and after attending this course.

b. Test at the 1% significance level whether this course makes any statistically significant improvement in the memory of all students.

Assume that the population of the paired differences has a normal distribution.

10.88 In a random sample of 800 men aged 25 to 35, 24% said they live with one or both parents. In another sample of 850 women of the same age group, 18% said that they live with one or both parents.

a. Construct a 95% confidence interval for the difference between the proportions of all men and all women aged 25 to 35 who live with one or both parents.

b. Test at the 2% significance level whether the two population proportions are different.

c. Repeat the test of part b using the *p*-value approach.

10.89 According to a Scripps Survey Research Center poll, 15% of men and 9% of women in the United States are left-handed (*USA TODAY*, January 13, 2003). Suppose that these percentages are based on random samples of 1200 men and 1080 women.

a. Construct a 99% confidence interval for the difference between the two population proportions.

b. At the 1% significance level, can you conclude that the percentage of left-handed men in the United States is higher than the percentage of left-handed women?

c. Repeat the test of part b using the *p*-value approach.

10.90 In the Pew Internet and American Life Project, 809 artists and 2755 musicians were questioned on some issues involving the Internet. (Here, *artists* included people in performing and visual arts and creative writing.) In response to the question whether or not it should be illegal to burn a copy of a CD or DVD for a friend, 48% of the artists and 41% of the musicians indicated that it should be illegal (*The New York Times*, December 6, 2004). Assume that the 809 artists and 2755 musicians surveyed make random samples from their respective populations.

a. Find a 98% confidence interval for the difference between the two population proportions.

b. At the 2% significance level, can you conclude that the two population proportions are different?

10.91 In a *Newsweek* telephone poll conducted by Princeton Survey Research Associates in February 2005, adults were asked about their attitudes toward social security and possible changes in the system. Forty-two percent of 45- to 54-year-olds expected social security would be able to pay all benefits to which they were entitled under the law at that time, while only 32% of 18- to 34-year-olds held that opinion (*Newsweek*, February 14, 2005). Suppose that these percentages were based on random samples of 800 adults from each of these two age groups.

a. Construct a 95% confidence interval for the difference between the two population proportions.

b. Using the 2.5% significance level, can you conclude that the proportion of 18- to 34-year-olds who expected social security to pay all such benefits is lower than the percentage of 45- to 54-year-olds who held this opinion?

Advanced Exercises

10.92 Manufacturers of two competing automobile models, Gofer and Diplomat, each claim to have the lowest mean fuel consumption. Let μ_1 be the mean fuel consumption in miles per gallon (mpg) for the Gofer and μ_2 the mean fuel consumption in mpg for the Diplomat. The two manufacturers have agreed to a test in which several cars of each model will be driven on a 100-mile test run. Then the fuel consumption, in mpg, will be calculated for each test run. The average of the mpg for all 100-mile test runs for each model gives the corresponding mean. Assume that for each model the gas mileages for the test runs are normally distributed with $\sigma = 2$ mpg. Note that each car is driven for one and only one 100-mile test run.

a. How many cars (i.e., sample size) for each model are required to estimate $\mu_1 - \mu_2$ with a 90% confidence level and with a margin of error of estimate of 1.5 mpg? Use the same number of cars (i.e., sample size) for each model.

b. If μ_1 is actually 33 mpg and μ_2 is actually 30 mpg, what is the probability that five cars for each model would yield $\bar{x}_1 \geq \bar{x}_2$?

10.93 Maria and Ellen both specialize in throwing the javelin. Maria throws the javelin a mean distance of 200 feet with a standard deviation of 10 feet, whereas Ellen throws the javelin a mean distance of 210 feet with a standard deviation of 12 feet. Assume that the distances each of these athletes throws the javelin are normally distributed with these population means and standard deviations. If Maria and Ellen each throw the javelin once, what is the probability that Maria's throw is longer than Ellen's?

10.94 A new type of sleeping pill is tested against an older standard pill. Two thousand insomniacs are randomly divided into two equal groups. The first group is given the old pill, and the second group receives the new pill. The time required to fall asleep after the pill is administered is recorded for each person. The results of the experiment are given in the following table, where $\bar{x}$ and s represent the mean and standard deviation, respectively, for the times required to fall asleep for people in each group after the pill is taken.

	Group 1 (Old Pill)	Group 2 (New Pill)
n	1000	1000
$\bar{x}$	15.4 minutes	15.0 minutes
s	3.5 minutes	3.0 minutes

Consider the test of hypothesis H_0: $\mu_1 - \mu_2 = 0$ versus H_1: $\mu_1 - \mu_2 > 0$, where μ_1 and μ_2 are the mean times required for all potential users to fall asleep using the old pill and the new pill, respectively.

a. Find the p-value for this test.

b. Does your answer to part a indicate that the result is statistically significant? Use $\alpha = .025$.

c. Find the 95% confidence interval for $\mu_1 - \mu_2$.

d. Does your answer to part c imply that this result is of great *practical* significance?

10.95 Gamma Corporation is considering the installation of governors on cars driven by its sales staff. These devices would limit the car speeds to a preset level, which is expected to improve fuel economy. The company is planning to test several cars for fuel consumption without governors for one week. Then governors would be installed in the same cars, and fuel consumption will be monitored for another week. Gamma Corporation wants to estimate the mean difference in fuel consumption with a margin of error of estimate of 2 mpg with a 90% confidence level. Assume that the differences in fuel consumption are normally distributed and that previous studies suggest that an estimate of $s_d = 3$ mpg is reasonable. How many cars should be tested? (Note that the critical value of t will depend on n, so it will be necessary to use trial and error.)

10.96 Refer to Exercise 10.95. Suppose Gamma Corporation decides to test governors on seven cars. However, the management is afraid that the speed limit imposed by the governors will reduce the number of contacts the salespersons can make each day. Thus, both the fuel consumption and the number of contacts made are recorded for each car/salesperson for each week of the testing period, both before and after the installation of governors.

	Number of Contacts		Fuel Consumption (mpg)	
Salesperson	**Before**	**After**	**Before**	**After**
A	50	49	25	26
B	63	60	21	24
C	42	47	27	26
D	55	51	23	25
E	44	50	19	24
F	65	60	18	22
G	66	58	20	23

Suppose that as a statistical analyst with the company, you are directed to prepare a brief report that includes statistical analysis and interpretation of the data. Management will use your report to help decide whether or not to install governors on all salespersons' cars. Use 90% confidence intervals and .05 significance levels for any hypothesis tests to make suggestions. Assume that the differences in fuel consumption and the differences in the number of contacts are both normally distributed.

10.97 Two competing airlines, Alpha and Beta, fly a route between Des Moines, Iowa, and Wichita, Kansas. Each airline claims to have a lower percentage of flights that arrive late. Let p_1 be the proportion of Alpha's flights that arrive late and p_2 the proportion of Beta's flights that arrive late.

a. You are asked to observe a random sample of arrivals for each airline to estimate $p_1 - p_2$ with a 90% confidence level and a margin of error of estimate of .05. How many arrivals for each airline would you have to observe? (Assume that you will observe the same number of arrivals, n, for each airline. To be sure of taking a large enough sample, use $p_1 = p_2 = .50$ in your calculations for n.)

b. Suppose that p_1 is actually .30 and p_2 is actually .23. What is the probability that a sample of 100 flights for each airline (200 in all) would yield $\hat{p}_1 \geq \hat{p}_2$?

10.98 Refer to Exercise 10.56, in which a random sample of six cars was selected to test a gasoline additive. The six cars were driven for one week without the gasoline additive and then for one week with the additive. The data reproduced here from that exercise show miles per gallon without and with the additive.

Without	24.6	28.3	18.9	23.7	15.4	29.5
With	26.3	31.7	18.2	25.3	18.3	30.9

Suppose that instead of the study with 6 cars, a random sample of 12 cars is selected and these cars are divided randomly into two groups of 6 cars each. The cars in the first group are driven for one week without the additive, and the cars in the second group are driven for one week with the additive. Suppose that the top row of the table lists the gas mileages for the six cars without the additive, and the bottom row gives the gas mileages for the cars with the additive. Assume that the distributions of the gas mileages with or without the additive are (approximately) normal with equal but unknown standard deviations.

a. Would a paired sample test as described in Section 10.4 be appropriate in this case? Why or why not? Explain.

b. If the paired sample test is inappropriate here, carry out a suitable test of whether the mean gas mileage is lower without the additive. Use $\alpha = .025$.

c. Compare your conclusion in part b with the result of the hypothesis test in Exercise 10.56.

10.99 Does the use of cellular telephones increase the risk of brain tumors? Suppose that a manufacturer of cell phones hires you to answer this question because of concern about public liability suits. How would you conduct an experiment to address this question? Be specific. Explain how you would observe, how many observations you would take, and how you would analyze the data once you collect them. What are your null and alternative hypotheses? Would you want to use a higher or a lower significance level for the test? Explain.

10.100 We wish to estimate the difference between the mean scores on a standardized test of students taught by Instructors A and B. The scores of all students taught by Instructor A have a normal distribution with a standard deviation of 15, and the scores of all students taught by Instructor B have a normal distribution with a standard deviation of 10. To estimate the difference between the two means, you decide that the same number of students from each instructor's class should be observed.

a. Assuming that the sample size is the same for each instructor's class, how large a sample should be taken from each class to estimate the difference between the mean scores of the two populations to within 5 points with 90% confidence?

b. Suppose that samples of the size computed in part a will be selected in order to test for the difference between the two population mean scores using a .05 level of significance. How large does the difference between the two sample means have to be for us to conclude that the two population means are different?

c. Explain why a paired samples design would be inappropriate for comparing the scores of Instructor A versus Instructor B.

10.101 The weekly weight losses of all dieters on Diet I have a normal distribution with a mean of 1.3 pounds and a standard deviation of .4 pounds. The weekly weight losses of all dieters on Diet II have a normal distribution with a mean of 1.5 pounds and a standard deviation of .7 pounds. A random sample of 25 dieters on Diet I and another sample of 36 dieters on Diet II are observed.

a. What is the probability that the difference between the two sample means, $\bar{x}_1 - \bar{x}_2$, will be within $-.15$ to .15, i.e., $-.15 < \bar{x}_1 - \bar{x}_2 < .15$?

b. What is the probability that the average weight loss $\bar{x}_1$ for dieters on Diet I will be greater than the average weight loss $\bar{x}_2$ for dieters on Diet II?

c. If the average weight loss of the 25 dieters using Diet I is computed to be 2.0 pounds, what is the probability that the difference between the two sample means, $\bar{x}_1 - \bar{x}_2$, will be within $-.15$ to .15, i.e., $-.15 < \bar{x}_1 - \bar{x}_2 < .15$?

d. Suppose you conclude that the assumption $-.15 < \mu_1 - \mu_2 < .15$ is reasonable. What does this mean to a person who chooses one of these diets?

10.102 Sixty-five percent of all male voters and 40% of all female voters favor a particular candidate. A sample of 100 male voters and another sample of 100 female voters will be polled. What is the probability that at least 10 more male voters than female voters will favor this candidate?

Self-Review Test

1. To test the hypothesis that the mean blood pressure of university professors is lower than that of company executives, which of the following would you use?

   a. A left-tailed test **b.** A two-tailed test **c.** A right-tailed test

2. Briefly explain the meaning of independent and dependent samples. Give one example of each of these cases.

3. A company psychologist wanted to test if company executives have job-related stress scores higher than those of university professors. He took a sample of 40 executives and 50 professors and tested them for job-related stress. The sample of 40 executives gave a mean stress score of 7.6. The sample of 50 professors produced a mean stress score of 5.4. Assume that the standard deviations of the two populations are .8 and 1.3, respectively.
   a. Construct a 99% confidence interval for the difference between the mean stress scores of all executives and all professors.
   b. Test at the 2.5% significance level whether the mean stress score of all executives is higher than that of all professors.

4. A sample of 20 alcoholic fathers showed that they spend an average of 2.3 hours per week playing with their children with a standard deviation of .54 hours. A sample of 25 nonalcoholic fathers gave a mean of 4.6 hours per week with a standard deviation of .8 hours.
   a. Construct a 95% confidence interval for the difference between the mean times spent per week playing with their children by all alcoholic and all nonalcoholic fathers.
   b. Test at the 1% significance level whether the mean time spent per week playing with their children by all alcoholic fathers is less than that of nonalcoholic fathers.

   Assume that the times spent per week playing with their children by all alcoholic and all nonalcoholic fathers both are normally distributed with equal but unknown standard deviations.

5. Repeat Problem 4 assuming that the times spent per week playing with their children by all alcoholic and all nonalcoholic fathers both are normally distributed with unequal and unknown standard deviations.

6. Lake City has two shops, Zeke's and Elmer's, that handle the majority of the town's auto body repairs. Seven cars that were damaged in collisions were taken to both shops for written estimates of the repair costs. These estimates (in dollars) are shown in the following table.

Zeke's	$1058	544	1349	1296	676	998	1698
Elmer's	$995	540	1175	1350	605	970	1520

   a. Construct a 99% confidence interval for the mean μ_d of the population paired differences, where a paired difference is equal to Zeke's estimate minus Elmer's estimate.
   b. Test at the 5% significance level whether the mean μ_d of the population paired differences is different from zero.

   Assume that the population of paired differences is (approximately) normally distributed.

7. A sample of 500 male registered voters showed that 57% of them voted in the last presidential election. Another sample of 400 female registered voters showed that 55% of them voted in the same election.
   a. Construct a 97% confidence interval for the difference between the proportions of all male and all female registered voters who voted in the last presidential election.
   b. Test at the 1% significance level whether the proportion of all male voters who voted in the last presidential election is different from that of all female voters.

Mini-Projects

■ MINI-PROJECT 10–1

Suppose that a new cold-prevention drug was tested in a randomized, placebo-controlled, double-blind experiment during the month of January. One thousand healthy adults were randomly divided into two groups of 500 each, a treatment group and a control group. The treatment group was given the new drug, and the control group received a placebo. During the month, 40 people in the treatment group and 120 people in the control group caught a cold. Explain how to construct a 95% confidence interval for the difference between the relevant population proportions. Also describe an appropriate hypothesis test, using the given data, to evaluate the effectiveness of this new drug for cold prevention.

Find a similar article in a journal of medicine, psychology, or other field that lends itself to confidence intervals and hypothesis tests for differences in two means or proportions. First explain how to make the confidence intervals and hypothesis tests; then do so using the data given in the article.

MINI-PROJECT 10–2

A researcher conjectures that cities in the more populous states of the United States tend to have higher costs for hospital rooms. Using "CITY DATA" that accompany this text, select a random sample of 10 cities from the 6 most populous states (California, Texas, New York, Florida, Pennsylvania, and Illinois). Then take a random sample of 10 cities from the remaining states in the data set. For each of the 20 cities, record the average daily cost of a private hospital room. Assume that such costs are approximately normally distributed for all cities in each of the two groups of states. Further assume that the cities you selected make random samples of all cities for the two groups of states. Assume that the standard deviations for the two groups are unequal and unknown.

a. Construct a 95% confidence interval for the difference in the means of such hospital costs for all cities in the two groups of states.

b. At the 5% level of significance, can you conclude that the average daily cost of a private hospital room for all cities in the six most populous states is higher than that of such a room for all cities in the remaining states?

MINI-PROJECT 10–3

Many different kinds of analyses have been performed on the salaries of professional athletes. Perform a hypothesis test of whether or not the average salaries of players in two sports are different by taking independent random samples of 35 players each from any two sports of your choice from Major League Baseball, the National Football League, the National Basketball Association, and the National Hockey League. (Note: A good Internet reference for such data is http://www.usatoday.com/sports/salaries/index.htm.) After you take samples, do the following.

a. For each player, calculate the weekly salary. For your information, the approximate length (in weeks) of a season is 32.5 for MLB, 22.5 for the NFL, 28 for the NBA, and 29.5 for the NHL. This length of a season does not include the playoffs but it does include training camp and the preseason games because each player is expected to participate in these events. Players may receive bonuses for making the playoffs, but these are not included in their base salaries. You may ignore such bonuses.

b. Perform a hypothesis test to determine if the average weekly salaries are the same for the two sports that you selected. Use a significance level of 5%. Make certain to indicate whether you decide to use the pooled variance assumption or not and justify your selection.

c. Perform a hypothesis test on the same data to determine if the average annual salaries are the same for the two sports that you selected. Explain why you could get a different answer (in regard to rejecting or failing to reject the null hypothesis) when using the weekly salaries versus the annual salaries.

DECIDE FOR YOURSELF

Deciding About How to Design a Study

By now, you might feel that you have learned almost everything there is to know about statistics. In some ways, you have learned a great deal. When using the p-value approach, the rule to reject a null hypothesis whenever the p-value is less than the significance level never changes. If you know this rule, you do not have to worry about changing it. You have also learned the basic concept of a confidence interval, which will also never change. However, one of the most important lessons to learn in statistics is to conduct a valid study. Design of experiments and sampling design are two areas of statistics that are dedicated to determining the proper way to plan a study before any data are collected. Without a proper plan, the time and money spent on the study could be a complete waste if the results are not valid.

Consider the example of gasoline additive mentioned in the Decide for Yourself section of Chapter 9. In that section, we discussed performing a single-sample procedure. However, the same problem could be addressed by using some of the procedures learned in this chapter.

1. Describe how that analysis could be performed by selecting two independent samples of cars. Be specific about how the treatments are applied/assigned to the cars, whether there are any special considerations as to how the cars are selected, and the specific measurements that would be compared.

2. Answer question 1 assuming that we use a paired-sample procedure instead of a two independent samples procedure.

3. Discuss the strengths and weaknesses of the three procedures (including the single sample procedure discussed in Chapter 9). Which method would you prefer and why? Explain.

TECHNOLOGY INSTRUCTION — Confidence Intervals and Hypothesis Tests for Two Populations

TI-84

```
2-SampTTest
 Inpt:Data Stats
 List1:L1
 List2:L2
 Freq1:1
 Freq2:1
 μ1:≠μ2 <μ2 >μ2
↓Pooled:No Yes
```

Screen 10.1

```
2-SampTTest
 μ1≠μ2
 t=-1.204796141
 p=.2522070299
 df=11.63973844
 x̄1=22.375
↓x̄2=24
```

Screen 10.2

1. To perform a hypothesis test about the difference between the means of two populations with independent samples, select **STAT>TESTS>2-SampTtest**. If the data are stored in lists, select **Data** and enter the names of the lists. If, instead, you have summary statistics for the two samples, select **Stats** and enter the mean, standard deviation, and sample size for each sample. Choose the form of the alternative hypothesis. If you are assuming that the standard deviations are equal for the two populations, select **Yes** for **Pooled**; otherwise select **No**. Select **Calculate** to find the p-value. (See **Screens 10.1** and **10.2**.)

2. To perform a hypothesis test about the proportions of two populations using independent samples, select **STAT>TESTS>2-PropZTest**. Enter the successes and trials (as x and n respectively) for each of the two samples. Select the alternative hypothesis and then **Calculate** to find the p-value of the test. Be careful to distinguish between the p-value and the sample proportions, which have hats above them.

3. To find a confidence interval for the difference of the means of two populations using independent samples, select **STAT>TESTS>2-SampTInt**. If the data is stored in lists, select **Data** and enter the names of the lists. If, instead, you have summary statistics for the two samples, select **Stats** and enter the mean, standard deviation, and sample size for each sample. Enter the confidence level as the **C-Level**. If you are assuming that the standard deviations are equal for the two populations, select **Yes** for **Pooled**; otherwise select **No**. Select **Calculate** to find the confidence interval.

4. To find a confidence interval for the difference between two population proportions, select **STATS>TESTS>2-PropZInt**. Enter the successes and trials (as x and n respectively) for each of the two samples. Enter the confidence level and then select **Calculate** to find the confidence interval.

MINITAB

1. To find a confidence interval for $\mu_1 - \mu_2$ for two independent populations with unknown but equal standard deviations as discussed in Section 10.2, select **Stat>Basic Statistics> 2-Sample t**. In the dialog box you obtain, select **Summarized data**, and enter the values of the **Sample sizes**, (sample) **Means** and **Standard deviations** for the two samples. Check the box next to **Assume equal variances**. Click the **Options** button and enter the value of the **Confidence level** in the new dialog box. Click **OK** in both boxes. The output containing the confidence interval will appear in the session window.

   If instead of summary measures, you have data on the two samples, enter these data in columns **C1** and **C2** of the MINITAB spreadsheet. In the dialog box, click next to **Samples in different columns** and enter the column names for two samples. (See **Screens 10.3** and **10.4**.) The rest of the procedure is the same as above.

2. To perform a hypothesis test about $\mu_1 - \mu_2$ for two independent populations with unknown but equal standard deviations as discussed in Section 10.2, select **Stat>Basic Statistics> 2-Sample t**. In the dialog box you obtain, select **Summarized data** and enter the values of the **Sample sizes**, (sample) **Means**, and **Standard deviations** for the two samples. Check the box next to **Assume equal variances**. Click the **Options** button. In the new dialog box you obtain, enter **0** for the **Test difference** and select the appropriate **Alternative** hypothesis.

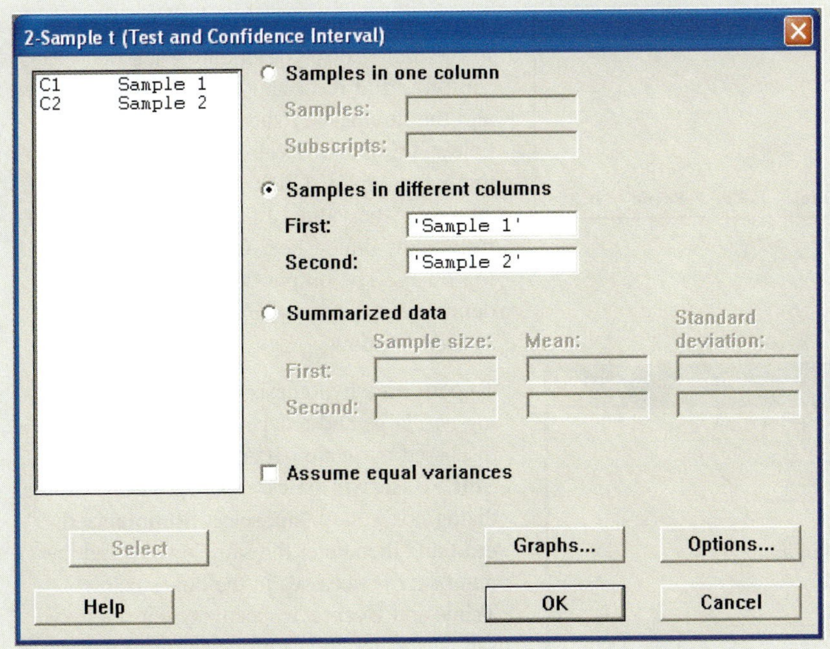

Screen 10.3

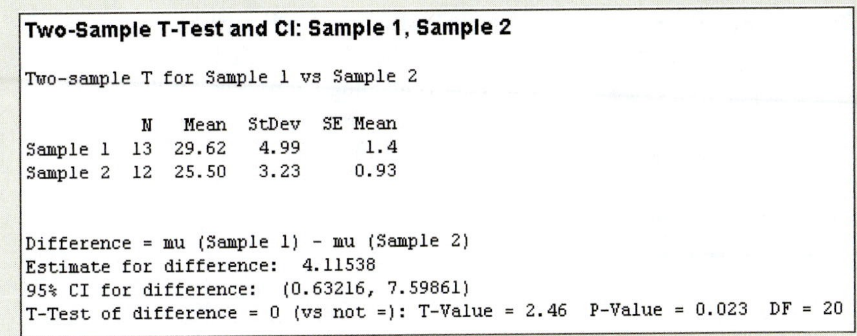

Two-Sample T-Test and CI: Sample 1, Sample 2

```
Two-sample T for Sample 1 vs Sample 2

           N   Mean  StDev  SE Mean
Sample 1  13  29.62   4.99      1.4
Sample 2  12  25.50   3.23     0.93


Difference = mu (Sample 1) - mu (Sample 2)
Estimate for difference:  4.11538
95% CI for difference:  (0.63216, 7.59861)
T-Test of difference = 0 (vs not =): T-Value = 2.46  P-Value = 0.023  DF = 20
```

Screen 10.4

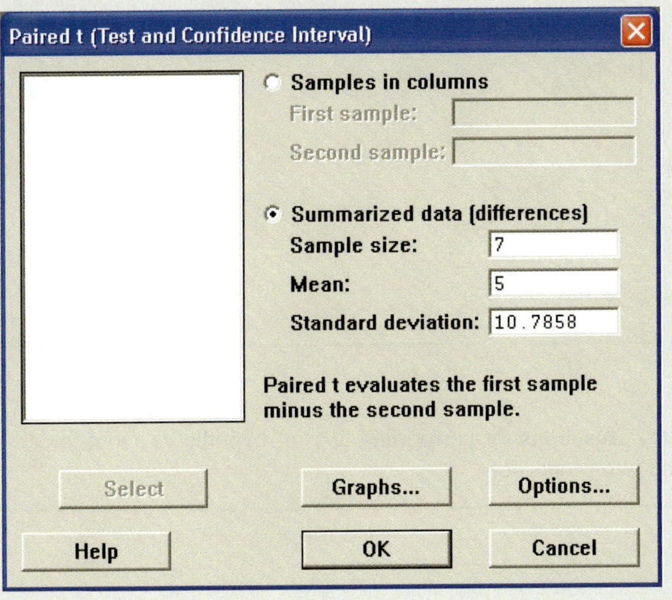

Screen 10.5

Click **OK** in both boxes. The output containing the p-value will appear in the Session window.

If instead of summary measures, you have data on the two samples, enter these data in columns **C1** and **C2** of the MINITAB spreadsheet. In the dialog box, click next to **Samples in different columns** and enter the column names for two samples. The rest of the procedure is the same as above.

3. To find a confidence interval for $\mu_1 - \mu_2$ or to perform a hypothesis test about $\mu_1 - \mu_2$ for two independent populations with unknown and unequal standard deviations discussed in Section 10.3, the procedures are the same as in numbers 1 and 2 above, respectively, except that you do not check next to **Assume equal variances**.

4. To find a confidence interval for μ_d for paired data discussed in Section 10.4, enter the *Before* and *After* data into columns **C1** and **C2**, respectively. Select **Stat>Basic Statistics>Paired t**. In the dialog box you obtain, select **Samples in columns** and enter the column names C1 and C2 in the boxes next to **First sample** and **Second sample**. Click the **Options** button and enter the value of the **Confidence level** in the new dialog box. Click **OK** in both boxes. The output containing the confidence interval will appear in the session window. Note that the confidence interval here is for the mean of the differences given by C1 − C2, which represents Before − After.

5. To perform a hypothesis test about μ_d for paired data discussed in Section 10.4, enter the *Before* and *After* data into columns **C1** and **C2,** respectively. Select **Stat>Basic Statistics>Paired t**. In the dialog box you obtain, select **Samples in columns** and enter the column names C1 and C2 in the boxes next to **First sample** and **Second sample**. Click the **Options** button. In the new dialog box you obtain, enter **0** for the **Test mean** and select the appropriate **Alternative** hypothesis. Click **OK** in both boxes. The output containing the p-value will appear in the session window. Note that the hypothesis test here is for the mean of the differences given by C1 − C2, which represents Before − After. You need to keep this in mind when determining your alternative hypothesis. (See **Screens 10.5** and **10.6**.)

6. To find a confidence interval for $p_1 - p_2$ using two large and independent samples as discussed in Section 10.5,

```
Paired T-Test and CI

              N      Mean     StDev  SE Mean
Difference    7   5.00000  10.78580  4.07665

95% CI for mean difference: (-4.97520, 14.97520)
T-Test of mean difference = 0 (vs not = 0): T-Value = 1.23  P-Value = 0.266
```

Screen 10.6

select **Stat>Basic Statistics>2 Proportions**. In the dialog box you obtain, click on **Summarized data** and enter the sample sizes and the numbers of successes in the boxes below **Trials** and **Events**, respectively, for the two samples. Click the **Options** button and enter the value of the **Confidence Level** in the new dialog box. Click **OK** in both dialog boxes. The output containing the confidence interval for $p_1 - p_2$ will appear in the session window.

7. To perform a hypothesis test about $p_1 - p_2$ using two large and independent samples as discussed in Section 10.5, select **Stat>Basic Statistics>2 Proportions**. In the dialog box you obtain, select **Summarized data** and then enter the sample sizes and the numbers of successes in the boxes below **Trials** and **Events,** respectively, for the two samples. Click the **Options** button. Set **Test difference** to **0,** select the appropriate **Alternative** hypothesis, and check next to **Use pooled estimate of p for test** in the new dialog box. Click **OK** in both dialog boxes. The output containing the *p*-value for the test will appear in the session window. (See **Screens 10.7** and **10.8**.)

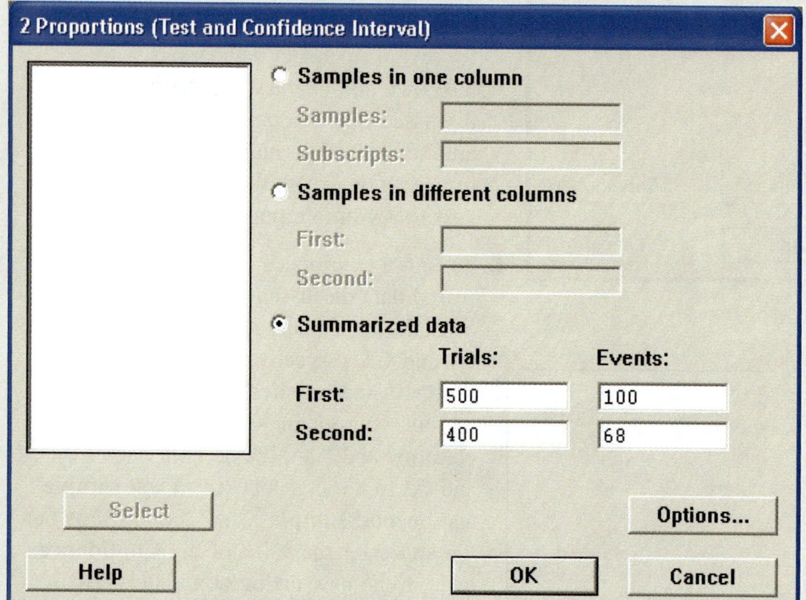

Screen 10.7

```
Test and CI for Two Proportions

Sample    X    N  Sample p
1       100  500  0.200000
2        68  400  0.170000

Difference = p (1) - p (2)
Estimate for difference:  0.03
95% CI for difference:  (-0.0208364, 0.0808364)
Test for difference = 0 (vs not = 0):  Z = 1.16  P-Value = 0.247
```

Screen 10.8

Excel

Excel does not contain any built-in functions for performing tests of hypotheses about the difference between two means and two proportions.

TECHNOLOGY ASSIGNMENTS

TA10.1 A random sample of 13 male college students who hold jobs gave the following data on their GPAs.

3.12	2.84	2.43	2.15	3.92	2.45	2.73
3.06	2.36	1.93	2.81	3.27	1.83	

Another random sample of 16 female college students who also hold jobs gave the following data on their GPAs.

2.76	3.84	2.24	2.81	1.79	3.89	2.96	3.77
2.36	2.81	3.29	2.08	3.11	1.69	2.84	3.02

a. Construct a 99% confidence interval for the difference between the mean GPAs of all male and all female college students who hold jobs.

b. Test at the 5% significance level whether the mean GPAs of all male and all female college students who hold jobs are different.

Assume that the GPAs of all such male and female college students are normally distributed with equal but unknown population standard deviations.

TA10.2 A company recently opened two supermarkets in two different areas. The management wants to know if the mean sales per day for these two supermarkets are different. A sample of 10 days for Supermarket A produced the following data on daily sales (in thousand dollars).

47.56	57.66	51.23	58.29	43.71
49.33	52.35	50.13	47.45	53.86

A sample of 12 days for Supermarket B produced the following data on daily sales (in thousand dollars).

56.34	63.55	61.64	63.75	54.78	58.19
55.40	59.44	62.33	67.82	56.65	67.90

Assume that the daily sales of the two supermarkets are both normally distributed with equal but unknown standard deviations.

a. Construct a 99% confidence interval for the difference between the mean daily sales for these two supermarkets.

b. Test at the 1% significance level whether the mean daily sales for these two supermarkets are different.

TA10.3 Refer to Technology Assignment TA10.1. Now do that assignment assuming the GPAs of all such male and female college students are normally distributed with unequal and unknown population standard deviations.

TA10.4 Refer to Technology Assignment TA10.2. Now do that assignment assuming the daily sales of the two supermarkets are both normally distributed with unequal and unknown standard deviations.

TA10.5 The manufacturer of a gasoline additive claims that the use of this additive increases gasoline mileage. A random sample of six cars was selected. These cars were driven for one week without the gasoline additive and then for one week with the gasoline additive. The table gives the miles per gallon for these cars without and with the gasoline additive.

Without	24.6	28.3	18.9	23.7	15.4	29.5
With	26.3	31.7	18.2	25.3	18.3	30.9

a. Construct a 99% confidence interval for the mean μ_d of the population paired differences.

b. Test at the 1% significance level whether the use of the gasoline additive increases the gasoline mileage.

Assume that the population of paired differences is (approximately) normally distributed.

TA10.6 A company is considering installing new machines to assemble its products. The company is considering two types of machines, but it will buy only one type. The company selected eight assembly workers and asked them to use these two types of machines to assemble products. The following table gives the time taken (in minutes) to assemble one unit of the product on each type of machine for each of these eight workers.

Machine I	23	26	19	24	27	22	20	18
Machine II	21	24	23	25	24	28	24	23

a. Construct a 98% confidence interval for the mean μ_d of the population paired differences, where a paired difference is equal to the time taken to assemble a unit of the product on Machine I minus the time taken to assemble a unit of the product on Machine II by the same worker.

b. Test at the 5% significance level whether the mean time taken to assemble a unit of the product is different for the two types of machines.

Assume that the population of paired differences is (approximately) normally distributed.

TA10.7 A company has two restaurants in two different areas of New York City. The company wants to estimate the percentages of patrons who think that the food and service at each of these restaurants are excellent. A sample of 200 patrons taken from the restaurant in Area A showed that 118 of them think that the food and service are excellent at this restaurant. Another sample of 250 patrons selected from the restaurant in Area B showed that 160 of them think that the food and service are excellent at this restaurant.

a. Construct a 97% confidence interval for the difference between the two population proportions.

b. Testing at the 2.5% significance level, can you conclude that the proportion of patrons at the restaurant in Area A who think that the food and service are excellent is lower than the corresponding proportion at the restaurant in Area B?

TA10.8 The management of a supermarket wanted to investigate whether the percentages of all men and women who prefer to buy national-brand products over the store-brand products are different. A sample of 600 men shoppers at the company's supermarkets showed that 246 of them prefer to buy national-brand products over the store-brand products. Another sample of 700 women shoppers at the company's supermarkets showed that 266 of them prefer to buy national-brand products over the store-brand products.

a. Construct a 99% confidence interval for the difference between the proportions of all men and all women shoppers at these supermarkets who prefer to buy national brand products over the store brand products.

b. Testing at the 2% significance level, can you conclude that the proportions of all men and all women shoppers at these supermarkets who prefer to buy national brand products over the store brand products are different?

Chapter

11

Chi-Square Tests

When you went to school or to work today, did you drive by yourself or take public transportation? The 2000 census reported that nearly 76% of Americans drove alone to their jobs, up from 73% in 1990. Slightly less than 5% of Americans commuted to work via public transportation in 2000, compared to slightly more than 5% in 1990. Only 3% of Americans worked at home in 2000, which is 800,000 more than in 1990. Do you think these trends of transportation modes will continue in the same way over the next decade? How do you test hypotheses about such changes?

11.1 The Chi-Square Distribution

11.2 A Goodness-of-Fit Test

Case Study 11–1 Up with 'the Color Purple'

11.3 Contingency Tables

11.4 A Test of Independence or Homogeneity

11.5 Inferences About the Population Variance

The tests of hypotheses about the mean, the difference between two means, the proportion, and the difference between two proportions were discussed in Chapters 9 and 10. The tests about proportions dealt with countable or categorical data. In the case of a proportion and the difference between two proportions, the tests concerned experiments with only two categories. Recall from Chapter 5 that such experiments are called binomial experiments.

This chapter describes three types of tests:

1. Tests of hypotheses for experiments with more than two categories, called goodness-of-fit tests
2. Tests of hypotheses about contingency tables, called independence and homogeneity tests
3. Tests of hypotheses about the variance and standard deviation of a single population

All of these tests are performed by using the **chi-square distribution**, which is sometimes written as χ^2 *distribution* and is read as "chi-square distribution." The symbol χ is the Greek letter **chi**, pronounced "kī." The values of a chi-square distribution are denoted by the symbol χ^2 (read as "chi-square"), just as the values of the standard normal distribution and the t distribution are denoted by z and t, respectively. Section 11.1 describes the chi-square distribution.

11.1 The Chi-Square Distribution

Like the t distribution, the chi-square distribution has only one parameter called the degrees of freedom (df). The shape of a specific chi-square distribution depends on the number of degrees of freedom.[1] (The degrees of freedom for a chi-square distribution are calculated by using different formulas for different tests. This will be explained when we discuss those tests.) The random variable χ^2 assumes nonnegative values only. Hence, a chi-square distribution curve starts at the origin (zero point) and lies entirely to the right of the vertical axis. Figure 11.1 shows three chi-square distribution curves. They are for 2, 7, and 12 degrees of freedom.

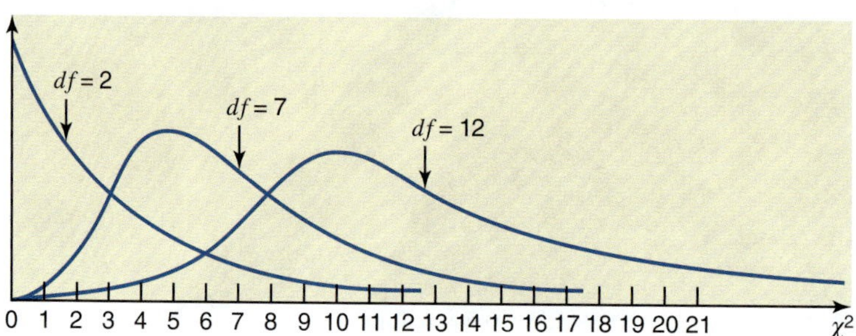

Figure 11.1 Three chi-square distribution curves.

As we can see from Figure 11.1, the shape of a chi-square distribution curve is skewed for very small degrees of freedom, and it changes drastically as the degrees of freedom increase. Eventually, for large degrees of freedom, the chi-square distribution curve looks like a normal distribution curve. The peak (or mode) of a chi-square distribution curve with 1 or 2 degrees of freedom occurs at zero and for a curve with 3 or more degrees of freedom at $df - 2$. For instance, the peak of the chi-square distribution curve with $df = 2$ in Figure 11.1 occurs at zero. The peak for the curve with $df = 7$ occurs at $7 - 2 = 5$. Finally, the peak for the curve with $df = 12$ occurs at $12 - 2 = 10$. Like all other continuous distribution curves, the total area under a chi-square distribution curve is 1.0.

Definition

The Chi-Square Distribution The *chi-square distribution* has only one parameter, called the degrees of freedom. The shape of a chi-square distribution curve is skewed to the right for small df and becomes symmetric for large df. The entire chi-square distribution curve lies to the right of the vertical axis. The chi-square distribution assumes nonnegative values only, and these are denoted by the symbol χ^2 (read as "chi-square").

If we know the degrees of freedom and the area in the right tail of a chi-square distribution curve, we can find the value of χ^2 from Table VI of Appendix C. Examples 11–1 and 11–2 show how to read that table.

■ EXAMPLE 11–1

Reading the chi-square distribution table: area in the right tail known.

Find the value of χ^2 for 7 degrees of freedom and an area of .10 in the right tail of the chi-square distribution curve.

Solution To find the required value of χ^2, we locate 7 in the column for df and .100 in the top row in Table VI of Appendix C. The required χ^2 value is given by the entry at the

[1]The mean of a chi-square distribution is equal to its df, and the standard deviation is equal to $\sqrt{2\,df}$.

intersection of the row for 7 and the column for .100. This value is 12.017. The relevant portion of Table VI is presented below as Table 11.1.

Table 11.1 χ^2 **for** $df = 7$ **and .10 Area in the Right Tail**

	Area in the Right Tail Under the Chi-Square Distribution Curve				
df	**.995**	...	**.100**	...	**.005**
1	0.000	...	2.706	...	7.879
2	0.010	...	4.605	...	10.597
.	...	...	...	...	...
.	...	...	...	...	...
.	...	...	...	...	...
7	0.989	...	12.017 ← Required value of χ^2	...	20.278
.	...	...	...	...	...
.	...	...	...	...	...
.	...	...	...	...	...
100	67.328	...	118.498	...	140.169

As shown in Figure 11.2, the χ^2 value for $df = 7$ and an area of .10 in the right tail of the chi-square distribution curve is **12.017**.

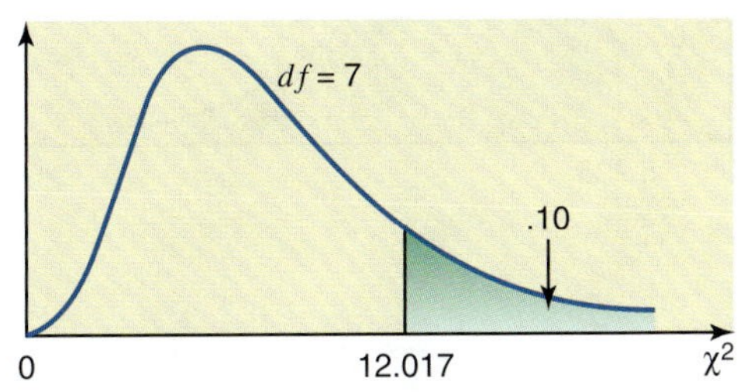

Figure 11.2

■ EXAMPLE 11–2

Find the value of χ^2 for 12 degrees of freedom and an area of .05 in the left tail of the chi-square distribution curve.

Reading the chi-square distribution table: area in the left tail known.

Solution We can read Table VI of Appendix C only when an area in the right tail of the chi-square distribution curve is known. When the given area is in the left tail, as in this example, the first step is to find the area in the right tail of the chi-square distribution curve as follows.

$$\text{Area in the right tail} = 1 - \text{Area in the left tail}$$

Therefore, for our example,

$$\text{Area in the right tail} = 1 - .05 = .95$$

Next, we locate 12 in the column for *df* and .950 in the top row in Table VI of Appendix C. The required value of χ^2, given by the entry at the intersection of the row for 12 and the column for .950, is 5.226. The relevant portion of Table VI is presented as Table 11.2.

Table 11.2 χ^2 for df = 12 and .95 Area in the Right Tail

	Area in the Right Tail Under the Chi-Square Distribution Curve				
df	.995	...	.950	...	.005
1	0.000	...	0.004	...	7.879
2	0.010	...	0.103	...	10.597
.	...	...	...	...	...
.	...	...	...	...	...
.	...	...	...	...	...
12	3.074	...	5.226 ←	...	28.300
.	...	...	...	...	...
.	...	...	...	...	...
.	...	...	...	...	...
100	67.328	...	77.929	...	140.169

Required value of χ^2

As shown in Figure 11.3, the χ^2 value for df = 12 and .05 area in the left tail is **5.226**.

Figure 11.3

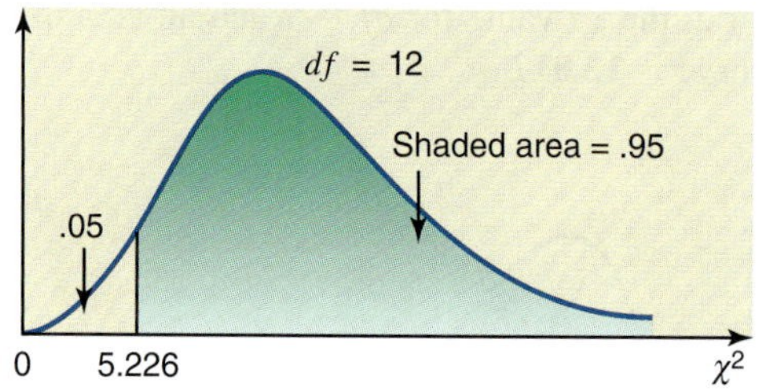

EXERCISES

CONCEPTS AND PROCEDURES

11.1 Describe the chi-square distribution. What is the parameter (parameters) of such a distribution?

11.2 Find the value of χ^2 for 12 degrees of freedom and an area of .025 in the right tail of the chi-square distribution curve.

11.3 Find the value of χ^2 for 28 degrees of freedom and an area of .05 in the right tail of the chi-square distribution curve.

11.4 Determine the value of χ^2 for 14 degrees of freedom and an area of .10 in the left tail of the chi-square distribution curve.

11.5 Determine the value of χ^2 for 23 degrees of freedom and an area of .990 in the left tail of the chi-square distribution curve.

11.6 Find the value of χ^2 for 4 degrees of freedom and

a. .005 area in the right tail of the chi-square distribution curve

b. .05 area in the left tail of the chi-square distribution curve

11.7 Determine the value of χ^2 for 13 degrees of freedom and

a. .025 area in the left tail of the chi-square distribution curve

b. .995 area in the right tail of the chi-square distribution curve

11.2 A Goodness-of-Fit Test

This section explains how to make tests of hypotheses about experiments with more than two possible outcomes (or categories). Such experiments, called **multinomial experiments**, possess four characteristics. Note that a binomial experiment is a special case of a multinomial experiment.

Definition

A Multinomial Experiment An experiment with the following characteristics is called a *multinomial experiment*.

1. It consists of n identical trials (repetitions).
2. Each trial results in one of k possible outcomes (or categories), where $k > 2$.
3. The trials are independent.
4. The probabilities of the various outcomes remain constant for each trial.

An experiment of many rolls of a die is an example of a multinomial experiment. It consists of many identical rolls (trials); each roll (trial) results in one of the six possible outcomes; each roll is independent of the other rolls; and the probabilities of the six outcomes remain constant for each roll.

As a second example of a multinomial experiment, suppose we select a random sample of people and ask them whether or not the quality of American cars is better than that of Japanese cars. The response of a person can be *yes*, *no*, or *does not know*. Each person included in the sample can be considered as one trial (repetition) of the experiment. There will be as many trials for this experiment as the number of persons selected. Each person can belong to any of the three categories—*yes*, *no*, or *does not know*. The response of each selected person is independent of the responses of other persons. Given that the population is large, the probabilities of a person belonging to the three categories remain the same for each trial. Consequently, this is an example of a multinomial experiment.

The frequencies obtained from the actual performance of an experiment are called the **observed frequencies**. In a **goodness-of-fit test**, we test the null hypothesis that the observed frequencies for an experiment follow a certain pattern or theoretical distribution. The test is called a goodness-of-fit test because the hypothesis tested is how *good* the observed frequencies *fit* a given pattern.

For our first example involving the experiment of many rolls of a die, we may test the null hypothesis that the given die is fair. The die will be fair if the observed frequency for each outcome is close to one-sixth of the total number of rolls.

For our second example involving opinions of people on the quality of American cars, suppose such a survey was conducted in 2005 and in that survey 41% of the people said *yes*, 48% said *no*, and 11% said *do not know*. We want to test if these percentages still hold true. Suppose we take a random sample of 1000 adults and observe that 536 of them think that the quality of American cars is better than that of Japanese cars, 362 say it is worse, and 102 have no opinion. The frequencies 536, 362, and 102 are the observed frequencies. These frequencies are obtained by actually performing the survey. Now, assuming that the 2005 percentages are still true (which will be our null hypothesis), in a sample of 1000 adults we will expect 410 to say *yes*, 480 to say *no*, and 110 to say *do not know*. These frequencies are obtained by multiplying the sample size (1000) by the 2005 proportions. These frequencies are called the **expected frequencies**. Then, we will make a decision to reject or not to reject the null hypothesis based on how large the difference between the observed frequencies and the expected frequencies is. To perform this test, we will use the chi-square distribution. Note that in this case we are testing the null hypothesis that all three percentages (or proportions) are unchanged. However, if we want to make a test for only one of the three proportions, we use the procedure learned in Section 9.4 of Chapter 9. For example, if we are testing the hypothesis that the percentage of people who think the quality of American cars is better than that of the Japanese cars is different from 41%, then we will test the null hypothesis $H_0: p = .41$ against the alternative hypothesis $H_1: p \neq .41$. This test will be conducted using the procedure discussed in Section 9.4 of Chapter 9.

As mentioned earlier, the frequencies obtained from the performance of an experiment are called the observed frequencies. They are denoted by $\boldsymbol{O}$. To make a goodness-of-fit test, we calculate the expected frequencies for all categories of the experiment. The expected frequency for a category, denoted by $\boldsymbol{E}$, is given by the product of n and p, where n is the total number of trials and p is the probability for that category.

Definition

Observed and Expected Frequencies The frequencies obtained from the performance of an experiment are called the *observed frequencies* and are denoted by O. The *expected frequencies*, denoted by E, are the frequencies that we expect to obtain if the null hypothesis is true. The expected frequency for a category is obtained as

$$E = np$$

where n is the sample size and p is the probability that an element belongs to that category if the null hypothesis is true.

Degrees of Freedom for a Goodness-of-Fit Test In a goodness-of-fit test, the *degrees of freedom* are

$$df = k - 1$$

where k denotes the number of possible outcomes (or categories) for the experiment.

The procedure to make a goodness-of-fit test involves the same five steps that we used in the preceding chapters. *The chi-square goodness-of-fit test is always a right-tailed test.*

Test Statistic for a Goodness-of-Fit Test The *test statistic for a goodness-of-fit test* is χ^2 and its value is calculated as

$$\chi^2 = \sum \frac{(O - E)^2}{E}$$

where O = observed frequency for a category

E = expected frequency for a category = np

Remember that a chi-square goodness-of-fit test is always a right-tailed test.

Whether or not the null hypothesis is rejected depends on how much the observed and expected frequencies differ from each other. To find how large the difference between the observed frequencies and the expected frequencies is, we do not look at just $\Sigma(O - E)$ because some of the $O - E$ values will be positive and others will be negative. The net result of the sum of these differences will always be zero. Therefore, we square each of the $O - E$ values to obtain $(O - E)^2$, and then we weight them according to the reciprocals of their expected frequencies. The sum of the resulting numbers gives the computed value of the test statistic χ^2.

To make a goodness-of-fit test, the sample size should be large enough so that the expected frequency for each category is at least 5. If there is a category with an expected frequency of less than 5, either increase the sample size or combine two or more categories to make each expected frequency at least 5.

Examples 11–3 and 11–4 describe the procedure for performing goodness-of-fit tests using the chi-square distribution.

■ EXAMPLE 11–3

Conducting a goodness-of-fit test: equal proportions for all categories.

A bank has an ATM installed inside the bank, and it is available to its customers only from 7 AM to 6 PM Monday through Friday. The manager of the bank wanted to investigate if the percentage of transactions made on this ATM is the same for each of the five days (Monday through Friday) of the week. She randomly selected one week and counted the number of transactions made on this ATM on each of the five days during this week. The information she obtained is given in the following table, where the number of users represents the number of transactions on this ATM on these days. For convenience, we will refer to these transactions as "people" or "users."

Day	Monday	Tuesday	Wednesday	Thursday	Friday
Number of users	253	197	204	279	267

At the 1% level of significance, can we reject the null hypothesis that the proportion of people who use this ATM each of the five days of the week is the same? Assume that this week is typical of all weeks in regard to the use of this ATM.

Solution To conduct this test of hypothesis, we proceed as follows.

Step 1. *State the null and alternative hypotheses.*

Because there are five categories (days) as listed in the table, the proportion of ATM users will be the same for each of these five days if one-fifth (or 20%) of all users use the ATM each day. The null and alternative hypotheses are as follows.

H_0: The proportion of people using the ATM is the same for all five days of the week.

H_1: The proportion of people using the ATM is not the same for all five days of the week.

If the proportion of people using this ATM is the same for all five days of the week, then .20 of the users will use this ATM on any of the five days of the week. Let p_1, p_2, p_3, p_4, and p_5 be the proportions of people who use this ATM on Monday, Tuesday, Wednesday, Thursday, or Friday, respectively. Then, the null and alternative hypotheses can also be written as

$$H_0: p_1 = p_2 = p_3 = p_4 = p_5 = .20$$

H_1: At least two of the five proportions are not equal to .20

Step 2. *Select the distribution to use.*

Because there are five categories (i.e., five days on which the ATM is used), this is a multinomial experiment. Consequently, we use the chi-square distribution to make this test.

Step 3. *Determine the rejection and nonrejection regions.*

The significance level is given to be .01, and the goodness-of-fit test is always right-tailed. Therefore, the area in the right tail of the chi-square distribution curve is

$$\text{Area in the right tail} = \alpha = .01$$

The degrees of freedom are calculated as follows.

$$k = \text{number of categories} = 5$$

$$df = k - 1 = 5 - 1 = 4$$

From the chi-square distribution table (Table VI of Appendix C), the critical value of χ^2 for $df = 4$ and .01 area in the right tail of the chi-square distribution curve is 13.277, as shown in Figure 11.4.

Figure 11.4

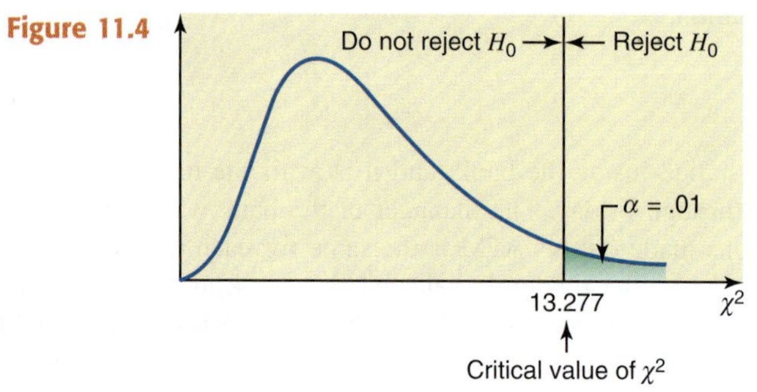

Step 4. *Calculate the value of the test statistic.*

Table 11.3

Category (Day)	Observed Frequency O	p	Expected Frequency $E = np$	$(O - E)$	$(O - E)^2$	$\frac{(O - E)^2}{E}$
Monday	253	.20	1200(.20) = 240	13	169	.704
Tuesday	197	.20	1200(.20) = 240	−43	1849	7.704
Wednesday	204	.20	1200(.20) = 240	−36	1296	5.400
Thursday	279	.20	1200(.20) = 240	39	1521	6.338
Friday	267	.20	1200(.20) = 240	27	729	3.038
	$n = 1200$					Sum = 23.184

All the required calculations to find the value of the test statistic χ^2 are shown in Table 11.3. The calculations made in Table 11.3 are explained next.

1. The first two columns of Table 11.3 list the five categories (days) and the observed frequencies for the sample of 1200 persons who used the ATM during each of the five days of the selected week. The third column contains the probabilities for the five categories assuming that the null hypothesis is true.

2. The fourth column contains the expected frequencies. These frequencies are obtained by multiplying the sample size ($n = 1200$) by the probabilities listed in the third column. If the null hypothesis is true (i.e., the ATM users are equally distributed over all five days), then we will expect 240 out of 1200 persons to use the ATM each day. Consequently, each category in the fourth column has the same expected frequency.

3. The fifth column lists the differences between the observed and expected frequencies, that is, $O - E$. These values are squared and recorded in the sixth column.

4. Finally, we divide the squared differences (that appear in the sixth column) by the corresponding expected frequencies (listed in the fourth column) and write the resulting numbers in the seventh column.

5. The sum of the seventh column gives the value of the test statistic χ^2. Thus,

$$\chi^2 = \sum \frac{(O - E)^2}{E} = 23.184$$

Step 5. *Make a decision.*

The value of the test statistic $\chi^2 = 23.184$ is larger than the critical value of $\chi^2 = 13.277$ and it falls in the rejection region. Hence, we reject the null hypothesis and state that the proportion of persons who use this ATM is not the same for the five days of the week. In other words, we conclude that a higher percentage of users of this ATM use this machine on one or more of these days.

If you make this chi-square test using any of the statistical software packages, you will obtain a p-value for the test. In this case you can compare the p-value obtained in the computer output with the level of significance and make a decision. As you know from Chapter 9, you will reject the null hypothesis if α (significance level) is greater than the p-value and not reject it otherwise. ■

■ EXAMPLE 11–4

Conducting a goodness-of-fit test: testing if results of a survey fit a given distribution.

In a Duffey Communications survey of business and political leaders conducted in 2004, the participants were asked whether or not the schools were preparing students to meet employers' needs. Seventy percent of the respondents said *No*, 20% replied *Yes*, and 10% said *Somewhat* (*USA TODAY*, December 28, 2004). Assume that these percentages hold true for the 2004 population of business and political leaders. Recently 1000 randomly selected business and political leaders were asked the same question. The following table lists the number of leaders in this sample who gave each response.

Response	No	Yes	Somewhat
Frequency	675	215	110

Test at the 2.5% level of significance whether the current distribution of responses is different from that for 2004.

Solution We perform the following five steps for this test of hypothesis.

Step 1. *State the null and alternative hypotheses.*

The null and alternative hypotheses are

H_0: The current percentage distribution of responses is the same as for 2004.

H_1: The current percentage distribution of responses is different from that for 2004.

Step 2. *Select the distribution to use.*

Because this experiment has three categories as listed in the table, it is a multinomial experiment. Consequently we use the chi-square distribution to make this test.

Step 3. *Determine the rejection and nonrejection regions.*

The significance level is given to be .025, and because the goodness-of-fit test is always right-tailed, the area in the right tail of the chi-square distribution curve is

$$\text{Area in the right tail} = \alpha = .025$$

The degrees of freedom are calculated as follows.

$$k = \text{number of categories} = 3$$

$$df = k - 1 = 3 - 1 = 2$$

CASE STUDY 11–1

UP WITH 'THE COLOR PURPLE'

USA TODAY Snapshots®

Up with 'the color purple'

Purple got 41% of the vote for M&M's new color. No longer will brown occupy 30% of the bag of candy. New bag breakdown:

Purple 20%
Red 20%
Yellow 20%
Green 10%
Orange 10%
Brown 10%
Blue 10%

Source: DL Blair for Master Foods Inc.

By Charmere Gatson and Sam Ward, USA TODAY

The pie chart shows the percentage distribution of colors of candies in M&M bags. According to this distribution, all such M&M bags contain, on average, 20% purple candies, 20% red, 20% yellow, and so on. Suppose when the machines that put candies in such M&M bags are working properly these percentages are true for all bags. However, sometimes the machines may start malfunctioning and put different percentages of these colored candies in bags. The quality control inspectors may take samples of bags from time to time and check for the percentages of candies of different colors to make sure that the machines are working properly. Suppose an inspector wants to check one such machine. In this case the null and alternative hypotheses will be:

H_0: The machine is working properly.

H_1: The machine is not working properly.

From the chi-square distribution table (Table VI of Appendix C), the critical value of χ^2 for $df = 2$ and .025 area in the right tail of the chi-square distribution curve is 7.378, as shown in Figure 11.5.

Figure 11.5

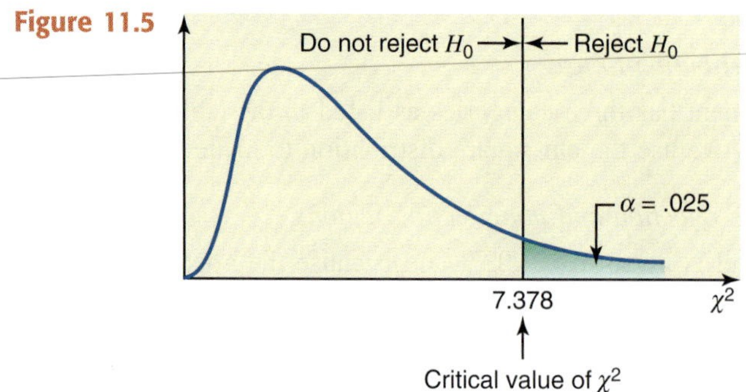

Step 4. *Calculate the value of the test statistic.*

All the required calculations to find the value of the test statistic χ^2 are shown in Table 11.4. Note that the percentages for 2004 have been converted into probabilities and recorded

Here the null hypothesis states that the percentage distribution of colored candies put in the bags by this machine is the same as described in the pie chart, and the alternative hypothesis states that this percentage distribution of colored candies put in the bags by this machine is not the same as described in the pie chart.

Suppose the inspector takes a sample of a few M&M bags filled by this machine, opens those bags, and counts the candies of different colors in these bags. Suppose these bags contain a total of 10,000 candies and the numbers of these candies observed with different colors are given in the second column of the following table.

Category	Observed Frequency O	p	Expected Frequency $E = np$	$(O - E)$	$(O - E)^2$	$\frac{(O - E)^2}{E}$
Purple	2080	.20	10,000(.20) = 2000	80	6400	3.200
Red	1900	.20	10,000(.20) = 2000	−100	10,000	5.000
Yellow	1920	.20	10,000(.20) = 2000	−80	6400	3.200
Green	980	.10	10,000(.10) = 1000	−20	400	.400
Orange	1090	.10	10,000(.10) = 1000	90	8100	8.100
Brown	1100	.10	10,000(.10) = 1000	100	10,000	10.000
Blue	930	.10	10,000(.10) = 1000	−70	4900	4.900
	n = 10,000					Sum = 34.800

Suppose we use the 5% significance level to perform this test. Then, for $df = 7 - 1 = 6$ and .05 area in the right tail, the critical value of χ^2 from Table VI in Appendix C is 12.592. Thus, we will reject the null hypothesis if the observed value of χ^2 is 12.592 or larger. The above table shows all the calculations to find the observed value of χ^2. Because the observed value of $\chi^2 = 34.800$ is larger than the critical value of $\chi^2 = 12.592$, we reject the null hypothesis. Thus, we can conclude that the machine is not working properly, and it may need an adjustment.

Source: The chart is reproduced with permission from *USA TODAY*, July 23, 2002. Copyright © 2002, *USA TODAY.*

in the third column of Table 11.4. The value of the test statistic χ^2 is given by the sum of the last column in this table. Thus,

$$\chi^2 = \sum \frac{(O - E)^2}{E} = 3.018$$

Table 11.4 Calculating the Value of the Test Statistic

Category (Response)	Observed Frequency O	p	Expected Frequency $E = np$	$(O - E)$	$(O - E)^2$	$\frac{(O - E)^2}{E}$
No	675	.70	1000(.70) = 700	−25	625	.893
Yes	215	.20	1000(.20) = 200	15	225	1.125
Somewhat	110	.10	1000(.10) = 100	10	100	1.000
	n = 1000					Sum = 3.018

Step 5. *Make a decision.*

The value of the test statistic $\chi^2 = 3.018$ is smaller than the critical value of $\chi^2 = 7.378$ and it falls in the nonrejection region. Hence, we fail to reject the null hypothesis and state that the current percentage distribution of responses is the same as for 2004.

If you make this chi-square test using any of the statistical software packages, you will obtain a p-value for the test. In this case you can compare the p-value obtained in the computer output with the level of significance and make a decision. As you know from Chapter 9, you will reject the null hypothesis if α (significance level) is greater than the p-value, and not reject it otherwise. ■

EXERCISES

CONCEPTS AND PROCEDURES

11.8 Describe the four characteristics of a multinomial experiment.

11.9 What is a goodness-of-fit test and when is it applied? Explain.

11.10 Explain the difference between the observed and expected frequencies for a goodness-of-fit test.

11.11 How is the expected frequency of a category calculated for a goodness-of-fit test? What are the degrees of freedom for such a test?

11.12 To make a goodness-of-fit test, what should be the minimum expected frequency for each category? What are the alternatives if this condition is not satisfied?

11.13 The following table lists the frequency distribution for 60 rolls of a die.

Outcome	1-spot	2-spot	3-spot	4-spot	5-spot	6-spot
Frequency	7	12	8	15	11	7

Test at the 5% significance level whether the null hypothesis that the given die is fair is true.

APPLICATIONS

11.14 In an August 2004 online survey of households conducted by IPSOS INSIGHT U.S. EXPRESS OMNIBUS, respondents were asked if they had to live on a deserted island and they could choose only one from a list of media related items to take with them, what would they choose. The most popular choice was computer and Internet access, which was favored by 64% of the respondents. Of the remaining respondents, 18% said that they would take a large supply of books with them, 6% chose television, 6% wanted to take radio with them, and 4% chose a cell phone (*The Atlantic Monthly*, January/February 2005). Assume that the remaining 2% chose other media items. Suppose we denote the above responses by A, B, C, D, E, and F, respectively. Recently 500 randomly selected persons were asked the same question. Their responses are summarized in the following table.

Response	A	B	C	D	E	F
Frequency	346	78	25	19	26	6

Test at the 5% significance level whether the current distribution of responses differs from that of August 2004.

11.15 In a 2004 survey conducted by the Educational Testing Service, American adults were asked to assign letter grades to the nation's schools indicating their views about the quality of schools (*USA TODAY*, June 22, 2004). The following table lists the percentage distribution of the grades assigned to the nation's schools by these adults. Assume that these percentages were true for all adults in 2004.

Grade	A	B	C	D	F	None
Percentage	2	20	47	15	4	12

The following table gives the frequency distribution of grades given to the nation's schools in a recent random sample of 700 American adults.

Grade	A	B	C	D	F	None
Frequency	16	132	314	121	35	82

At the 5% significance level, can you conclude that the current distribution of grades differs from that for 2004?

11.16 In May 2004, a Gallup Poll of adults' attitudes toward Health Maintenance Organization (HMOs) found that 40% of adults had little or no confidence in HMOs, 39% had some confidence, 18% had a great deal or quite a lot of confidence, and 3% had no opinion (*USA TODAY*, June 22, 2004). Let us denote these outcomes as L, S, G, and N, respectively. A recent random sample of 500 adults yielded the frequency distribution given in the following table.

Response	L	S	G	N
Frequency	212	198	82	8

Using the 1% significance level, can you conclude that the current distribution of opinions differs from the distribution of May 2004?

11.17 Home Mail Corporation sells products by mail. The company's management wants to find out if the number of orders received at the company's office on each of the five days of the week is the same. The company took a sample of 400 orders received during a four-week period. The following table lists the frequency distribution for these orders by the day of the week.

Day of the week	Mon	Tue	Wed	Thu	Fri
Number of orders received	92	71	65	83	89

Test at the 5% significance level whether the null hypothesis that the orders are evenly distributed over all days of the week is true.

11.18 Over the past three years, Art's Supermarket has observed the following distribution of modes of payment in the express lines: cash (C) 41%, checks (CK) 24%, credit or debit cards (D) 26%, and other (N) 9%. In an effort to make express checkout more efficient, Art's has just begun offering a 1% discount for cash payment in the express checkout line. The following table lists the frequency distribution of the modes of payment for a sample of 500 express-line customers after the discount went into effect.

Mode of payment	C	CK	D	N
Number of customers	240	104	111	45

Test at the 1% significance level whether the distribution of modes of payment in the express checkout line changed after the discount went into effect.

11.19 The following table lists the frequency distribution of cars sold at an auto dealership during the past 12 months.

Month	Jan	Feb	Mar	Apr	May	Jun	Jul	Aug	Sep	Oct	Nov	Dec
Cars sold	23	17	15	10	14	12	13	15	23	26	27	29

Using the 10% significance level, will you reject the null hypothesis that the number of cars sold at this dealership is the same for each month?

11.20 Of all students enrolled at a large undergraduate university, 19% are seniors, 23% are juniors, 27% are sophomores, and 31% are freshmen. A sample of 200 students taken from this university by the student senate to conduct a survey includes 50 seniors, 46 juniors, 55 sophomores, and 49 freshmen. Using the 10% significance level, test the null hypothesis that this sample is a random sample. (*Hint:* This sample will be a random sample if it includes approximately 19% seniors, 23% juniors, 27% sophomores, and 31% freshmen.)

11.21 Chance Corporation produces beauty products. Two years ago the quality control department at the company conducted a survey of users of one of the company's products. The survey revealed that 53% of the users said the product was excellent, 31% said it was satisfactory, 7% said it was unsatisfactory, and 9% had no opinion. Assume that these percentages were true for the population of all users of this product at that time. After this survey was conducted, the company redesigned this product. A recent survey of 800 users of the redesigned product conducted by the quality control department at the company showed that 495 of the users think the product is excellent, 255 think it is satisfactory, 35 think it is unsatisfactory, and

15 have no opinion. Do you think the percentage distribution of the opinions of users of the redesigned product is different from the percentage distribution of users of this product before it was redesigned? Use $\alpha = .025$.

11.22 Henderson Corporation makes metal sheets, among other products. When the process that is used to make metal sheets works properly, 92% of the metal sheets contain no defects, 5% have one defect each, and 3% have two or more defects each. The quality control inspectors at the company take samples of metal sheets quite often and check them for defects. If the distribution of defects for a sample is significantly different from the above mentioned percentage distribution, the process is stopped and adjusted. A recent sample of 300 sheets produced the frequency distribution of defects listed in the following table.

Number of defects	None	One	Two or More
Number of metal sheets	262	24	14

Does the evidence from this sample suggest that the process needs an adjustment? Use $\alpha = .01$.

11.3 Contingency Tables

Often we may have information on more than one variable for each element. Such information can be summarized and presented using a two-way classification table, which is also called a *contingency table* or *cross-tabulation*. Suppose a university has a total of 20,758 students enrolled. By classifying these students based on gender and whether these students are full-time or part-time, we can prepare Table 11.5, which provides an example of a contingency table. Table 11.5 has two rows (one for males and the second for females) and two columns (one for full-time and the second for part-time students). Hence, it is also called a 2×2 (read as "two by two") contingency table.

Table 11.5 Total Enrollment at a University

	Full-time	Part-time
Male	6768	2615 ← Students who are male and enrolled part-time
Female	7658	3717

A contingency table can be of any size. For example, it can be 2×3, 3×2, 3×3, or 4×2. Note that in these notations, the first digit refers to the number of rows in the table, and the second digit refers to the number of columns. For example, a 3×2 table will contain three rows and two columns. In general, an $R \times C$ table contains R rows and C columns.

Each of the four boxes that contain numbers in Table 11.5 is called a *cell*. The number of cells in a contingency table is obtained by multiplying the number of rows by the number of columns. Thus, Table 11.5 contains $2 \times 2 = 4$ cells. The subjects that belong to a cell of a contingency table possess two characteristics. For example, 2615 students listed in the second cell of the first row in Table 11.5 are *male* and *part-time*. The numbers written inside the cells are usually called the *joint frequencies*. For example, 2615 students belong to the joint category of *male* and *part-time*. Hence, it is referred to as the joint frequency of this category.

11.4 A Test of Independence or Homogeneity

This section is concerned with tests of independence and homogeneity, which are performed using contingency tables. Except for a few modifications, the procedure used to make such tests is almost the same as the one applied in Section 11.2 for a goodness-of-fit test.

11.4.1 A Test of Independence

In a **test of independence** for a contingency table, we test the null hypothesis that the two attributes (characteristics) of the elements of a given population are not related (that is, they are independent) against the alternative hypothesis that the two characteristics are related (that is, they are dependent). For example, we may want to test if the affiliation of people with the Democratic and Republican parties is independent of their income levels. We perform such a test by using the chi-square distribution. As another example, we may want to test if there is an association between being a man or a woman and having a preference for watching sports or soap operas on television.

Definition

Degrees of Freedom for a Test of Independence A test of independence involves a test of the null hypothesis that two attributes of a population are not related. The *degrees of freedom for a test of independence* are

$$df = (R - 1)(C - 1)$$

where R and C are the number of rows and the number of columns, respectively, in the given contingency table.

The value of the test statistic χ^2 in a test of independence is obtained using the same formula as in the goodness-of-fit test described in Section 11.2.

Test Statistic for a Test of Independence The value of the *test statistic* χ^2 *for a test of independence* is calculated as

$$\chi^2 = \sum \frac{(O - E)^2}{E}$$

where O and E are the observed and expected frequencies, respectively, for a cell.

The null hypothesis in a test of independence is always that the two attributes are not related. The alternative hypothesis is that the two attributes are related.

The frequencies obtained from the performance of an experiment for a contingency table are called the **observed frequencies**. The procedure to calculate the expected frequencies for a contingency table for a test of independence is different from the one for a goodness-of-fit test. Example 11–5 describes this procedure.

■ EXAMPLE 11–5

Calculating expected frequencies for a test of independence.

Violence and lack of discipline have become major problems in schools in the United States. A random sample of 300 adults was selected, and these adults were asked if they favor giving more freedom to schoolteachers to punish students for violence and lack of discipline. The two-way classification of the responses of these adults is presented in the following table.

	In Favor (F)	Against (A)	No Opinion (N)
Men (M)	93	70	12
Women (W)	87	32	6

Calculate the expected frequencies for this table assuming that the two attributes, gender and opinions on the issue, are independent.

Solution The preceding table is reproduced as Table 11.6 below. Note that Table 11.6 includes the row and column totals.

Table 11.6

	In Favor (F)	Against (A)	No Opinion (N)	Row Totals
Men (M)	93	70	12	175
Women (W)	87	32	6	125
Column Totals	180	102	18	300

The numbers 93, 70, 12, 87, 32, and 6 listed inside the six cells of Table 11.6 are called the *observed frequencies* of the respective cells.

As mentioned earlier, the null hypothesis in a test of independence is that the two attributes (or classifications) are independent. In an independence test of hypothesis, first we assume that the null hypothesis is true and that the two attributes are independent. Assuming that the null hypothesis is true and that gender and opinions are not related in this example, the expected frequency for the cell corresponding to *Men* and *In Favor* is calculated as shown next. From Table 11.6,

$$P(\text{a person is a } Man) = P(M) = 175/300$$

$$P(\text{a person is } In\ Favor) = P(F) = 180/300$$

Because we are assuming that M and F are independent (by assuming that the null hypothesis is true), from the formula learned in Chapter 4, the joint probability of these two events is

$$P(M \text{ and } F) = P(M) \times P(F) = (175/300) \times (180/300)$$

Then, assuming that M and F are independent, the number of persons expected to be *Men* and *In Favor* in a sample of 300 is

$$E \text{ for } Men \text{ and } In\ Favor = 300 \times P(M \text{ and } F)$$

$$= 300 \times \frac{175}{300} \times \frac{180}{300} = \frac{175 \times 180}{300}$$

$$= \frac{(\text{Row total})(\text{Column total})}{\text{Sample size}}$$

Thus, the rule for obtaining the expected frequency for a cell is to divide the product of the corresponding row and column totals by the sample size.

Expected Frequencies for a Test of Independence The expected frequency E for a cell is calculated as

$$E = \frac{(\text{Row total})(\text{Column total})}{\text{Sample size}}$$

Using this rule, we calculate the expected frequencies of the six cells of Table 11.6 as follows.

$$E \text{ for } Men \text{ and } In\ Favor \text{ cell} = (175)(180)/300 = \mathbf{105.00}$$

$$E \text{ for } Men \text{ and } Against \text{ cell} = (175)(102)/300 = \mathbf{59.50}$$

$$E \text{ for } Men \text{ and } No\ Opinion \text{ cell} = (175)(18)/300 = \mathbf{10.50}$$

$$E \text{ for } \textit{Women} \text{ and } \textit{In Favor} \text{ cell} = (125)(180)/300 = \mathbf{75.00}$$

$$E \text{ for } \textit{Women} \text{ and } \textit{Against} \text{ cell} = (125)(102)/300 = \mathbf{42.50}$$

$$E \text{ for } \textit{Women} \text{ and } \textit{No Opinion} \text{ cell} = (125)(18)/300 = \mathbf{7.50}$$

The expected frequencies are usually written in parentheses below the observed frequencies within the corresponding cells, as shown in Table 11.7.

Table 11.7

	In Favor (F)	Against (A)	No Opinion (N)	Row Totals
Men (M)	93 (105.00)	70 (59.50)	12 (10.50)	175
Women (W)	87 (75.00)	32 (42.50)	6 (7.50)	125
Column Totals	180	102	18	300

■

Like a goodness-of-fit test, *a test of independence is always right-tailed.* To apply a chi-square test of independence, *the sample size should be large enough so that the expected frequency for each cell is at least 5.* If the expected frequency for a cell is not at least 5, we either increase the sample size or combine some categories. Examples 11–6 and 11–7 describe the procedure to make tests of independence using the chi-square distribution.

■ EXAMPLE 11–6

Making a test of independence: 2 × 3 table.

Reconsider the two-way classification table given in Example 11–5. In that example, a random sample of 300 adults was selected, and they were asked if they favor giving more freedom to schoolteachers to punish students for violence and lack of discipline. Based on the results of the survey, a two-way classification table was prepared and presented in Example 11–5. Does the sample provide sufficient evidence to conclude that the two attributes, gender and opinions of adults, are dependent? Use a 1% significance level.

Solution The test involves the following five steps.

Step 1. *State the null and alternative hypotheses.*

As mentioned earlier, the null hypothesis must be that the two attributes are independent. Consequently, the alternative hypothesis is that these attributes are dependent.

H_0: Gender and opinions of adults are independent.

H_1: Gender and opinions of adults are dependent.

Step 2. *Select the distribution to use.*

We use the chi-square distribution to make a test of independence for a contingency table.

Step 3. *Determine the rejection and nonrejection regions.*

The significance level is 1%. Because a test of independence is always right-tailed, the area of the rejection region is .01, and it falls in the right tail of the chi-square distribution curve. The contingency table contains two rows (*Men* and *Women*) and three columns (*In Favor, Against*, and *No Opinion*). Note that we do not count the row and column of totals. The degrees of freedom are

$$df = (R - 1)(C - 1) = (2 - 1)(3 - 1) = 2$$

From Table VI of Appendix C, the critical value of χ^2 for $df = 2$ and $\alpha = .01$ is 9.210. This value is shown in Figure 11.6.

Figure 11.6

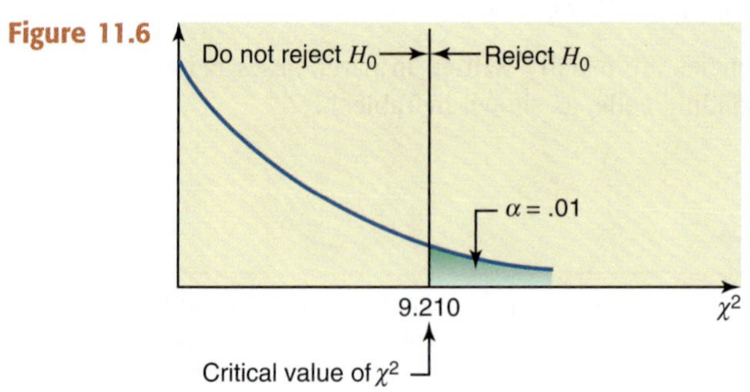

Step 4. *Calculate the value of the test statistic.*

Table 11.7, with the observed and expected frequencies constructed in Example 11–5, is reproduced as Table 11.8.

Table 11.8

	In Favor (*F*)	Against (*A*)	No Opinion (*N*)	Row Totals
Men (*M*)	93 (105.00)	70 (59.50)	12 (10.50)	175
Women (*W*)	87 (75.00)	32 (42.50)	6 (7.50)	125
Column Totals	180	102	18	300

To compute the value of the test statistic χ^2, we take the difference between each pair of observed and expected frequencies listed in Table 11.8, square those differences, and then divide each of the squared differences by the respective expected frequencies. The sum of the resulting numbers gives the value of the test statistic χ^2. All these calculations are made as follows.

$$\chi^2 = \sum \frac{(O - E)^2}{E}$$

$$= \frac{(93 - 105.00)^2}{105.00} + \frac{(70 - 59.50)^2}{59.50} + \frac{(12 - 10.50)^2}{10.50}$$

$$+ \frac{(87 - 75.00)^2}{75.00} + \frac{(32 - 42.50)^2}{42.50} + \frac{(6 - 7.50)^2}{7.50}$$

$$= 1.371 + 1.853 + .214 + 1.920 + 2.594 + .300 = 8.252$$

Step 5. *Make a decision.*

The value of the test statistic $\chi^2 = 8.252$ is less than the critical value of $\chi^2 = 9.210$, and it falls in the nonrejection region. Hence, we fail to reject the null hypothesis and state that there is not enough evidence from the sample to conclude that the two characteristics, *gender* and *opinions of adults*, are dependent for this issue. ■

Making a test of independence: 2 × 2 table.

■ EXAMPLE 11–7

A researcher wanted to study the relationship between gender and owning cell phones. She took a sample of 2000 adults and obtained the information given in the following table.

	Own Cell Phones	Do Not Own Cell Phones
Men	640	450
Women	440	470

At the 5% level of significance, can you conclude that gender and owing a cell phone are related for all adults?

Solution We perform the following five steps to make this test of hypothesis.

Step 1. *State the null and alternative hypotheses.*

The null and alternative hypotheses are

H_0: Gender and owning a cell phone are not related.

H_1: Gender and owning a cell phone are related.

Step 2. *Select the distribution to use.*

Because we are performing a test of independence, we use the chi-square distribution to make the test.

Step 3. *Determine the rejection and nonrejection regions.*

With a significance level of 5%, the area of the rejection region is .05, and it falls into the right tail of the chi-square distribution curve. The contingency table contains two rows (*men* and *women*) and two columns (*own cell phones* and *do not own cell phones*). The degrees of freedom are

$$df = (R - 1)(C - 1) = (2 - 1)(2 - 1) = 1$$

From Table VI of Appendix C, the critical value of χ^2 for $df = 1$ and $\alpha = .05$ is 3.841. This value is shown in Figure 11.7.

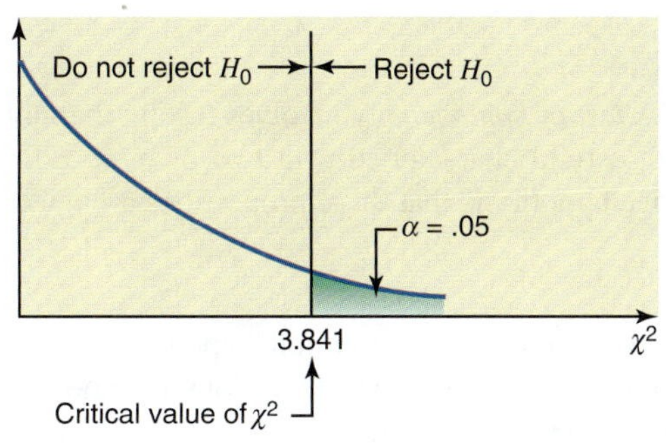

Figure 11.7

Step 4. *Calculate the value of the test statistic.*

The expected frequencies for the various cells are calculated as shown on the next page, and they are listed within parentheses in Table 11.9.

Table 11.9

	Own Cell Phones (Y)	Do Not Own Cell Phones (N)	Row Totals
Men (M)	640 (588.60)	450 (501.40)	1090
Women (W)	440 (491.40)	470 (418.60)	910
Column Totals	1080	920	2000

$$E \text{ for } \textit{men} \text{ and } \textit{own cell phones} \text{ cell} = (1090)(1080)/2000 = 588.60$$
$$E \text{ for } \textit{men} \text{ and } \textit{do not own cell phones} \text{ cell} = (1090)(920)/2000 = 501.40$$
$$E \text{ for } \textit{women} \text{ and } \textit{own cell phones} \text{ cell} = (910)(1080)/2000 = 491.40$$
$$E \text{ for } \textit{women} \text{ and } \textit{do not own cell phones} \text{ cell} = (910)(920)/2000 = 418.60$$

The value of the test statistic χ^2 is calculated as follows.

$$\chi^2 = \sum \frac{(O - E)^2}{E}$$
$$= \frac{(640 - 588.60)^2}{588.60} + \frac{(450 - 501.40)^2}{501.40} + \frac{(440 - 491.40)^2}{491.40} + \frac{(470 - 418.60)^2}{418.60}$$
$$= 4.489 + 5.269 + 5.376 + 6.311 = 21.445$$

Step 5. *Make a decision.*

The value of the test statistic $\chi^2 = 21.445$ is larger than the critical value of $\chi^2 = 3.841$ and it falls into the rejection region. Hence, we reject the null hypothesis and state that there is strong evidence from the sample to conclude that the two characteristics, *gender* and *owning cell phones*, are related for all adults. ■

11.4.2 A Test of Homogeneity

In a **test of homogeneity**, we test if two (or more) populations are homogeneous (similar) with regard to the distribution of a certain characteristic. For example, we might be interested in testing the null hypothesis that the proportions of households that belong to different income groups are the same in California and Wisconsin. Or we may want to test whether or not the preferences of people in Florida, Arizona, and Vermont are similar with regard to Coke, Pepsi, and 7-Up.

Definition

A Test of Homogeneity A *test of homogeneity* involves testing the null hypothesis that the proportions of elements with certain characteristics in two or more different populations are the same against the alternative hypothesis that these proportions are not the same.

Let us consider the example of testing the null hypothesis that the proportions of households in California and Wisconsin who belong to various income groups are the same. (Note that in a test of homogeneity, the null hypothesis is always that the proportions of elements with certain characteristics are the same in two or more populations. The alternative hypothesis is that these proportions are not the same.) Suppose we define three income strata: high income group (with an income of more than \$100,000), medium income group (with an income of \$50,000 to \$100,000), and low income group (with an income of less than \$50,000). Furthermore, assume that we take one sample of 250 households from California and another sample of 150 households from Wisconsin, collect the information on the incomes of these households, and prepare the contingency Table 11.10.

Table 11.10

	California	Wisconsin	Row Totals
High income	70	34	104
Medium income	80	40	120
Low income	100	76	176
Column Totals	250	150	400

Note that in this example the column totals are fixed. That is, we decided in advance to take samples of 250 households from California and 150 from Wisconsin. However, the row totals (of 104, 120, and 176) are determined randomly by the outcomes of the two samples. If we compare this example to the one about violence and lack of discipline in schools in the previous section, we will notice that neither the column nor the row totals were fixed in that example. Instead, the researcher took just one sample of 300 adults, collected the information on gender and opinions, and prepared the contingency table. Thus, in that example, the row and column totals were all determined randomly. Thus, when both the row and column totals are determined randomly, we make a test of independence. However, when either the column totals or the row totals are fixed, we make a test of homogeneity. In the case of income groups in California and Wisconsin, we will make a test of homogeneity to test for the similarity of income groups in the two states.

The procedure to make a test of homogeneity is similar to the procedure used to make a test of independence discussed earlier. Like a test of independence, a test of homogeneity is right-tailed. Example 11–8 illustrates the procedure to make a homogeneity test.

■ EXAMPLE 11–8

Making a test of homogeneity.

Consider the data on income distributions for households in California and Wisconsin given in Table 11.10. Using the 2.5% significance level, test the null hypothesis that the distribution of households with regard to income levels is similar (homogeneous) for the two states.

Solution We perform the following five steps to make this test of hypothesis.

Step 1. *State the null and alternative hypotheses.*

The two hypotheses are[2]

H_0: The proportions of households that belong to different income groups are the same in both states.

H_1: The proportions of households that belong to different income groups are not the same in both states.

Step 2. *Select the distribution to use.*

We use the chi-square distribution to make a homogeneity test.

Step 3. *Determine the rejection and nonrejection regions.*

The significance level is 2.5%. Because the homogeneity test is right-tailed, the area of the rejection region is .025, and it lies in the right tail of the chi-square distribution curve. The contingency table for income groups in California and Wisconsin contains three rows and two columns. Hence, the degrees of freedom are

$$df = (R - 1)(C - 1) = (3 - 1)(2 - 1) = 2$$

From Table VI of Appendix C, the value of χ^2 for $df = 2$ and .025 area in the right tail of the chi-square distribution curve is 7.378. This value is shown in Figure 11.8.

[2]Let p_{HC}, p_{MC}, and p_{LC} be the proportions of households in California who belong to high, middle, and low income groups, respectively. Let p_{HW}, p_{MW}, and p_{LW} be the corresponding proportions for Wisconsin. Then we can also write the null hypothesis as

$$H_0\text{: } p_{HC} = p_{HW}, p_{MC} = p_{MW}, \text{ and } p_{LC} = p_{LW}$$

and the alternative hypothesis as

$$H_1\text{: At least two of the equalities mentioned in } H_0 \text{ are not true.}$$

Figure 11.8

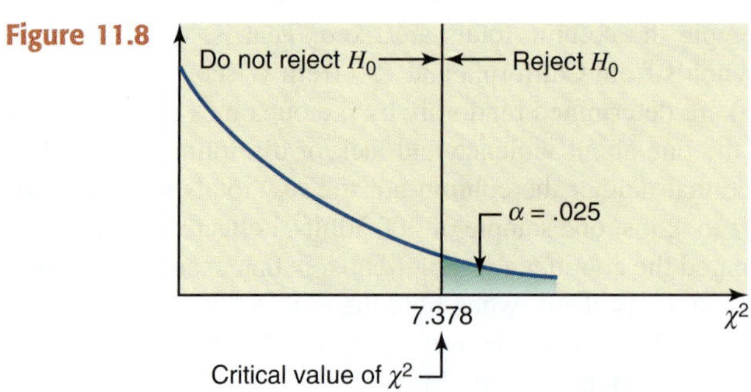

Step 4. *Calculate the value of the test statistic.*

To compute the value of the test statistic χ^2, we need to calculate the expected frequencies first. Table 11.11 lists the observed as well as the expected frequencies. The numbers in parentheses in this table are the expected frequencies, which are calculated using the formula

$$E = \frac{(\text{Row total})(\text{Column total})}{\text{Total of both samples}}$$

Thus, for instance,

$$E \text{ for } \textit{High income} \text{ and } \textit{California} \text{ cell} = \frac{(104)(250)}{400} = 65$$

Table 11.11

	California	Wisconsin	Row Totals
High income	70 (65)	34 (39)	104
Medium income	80 (75)	40 (45)	120
Low income	100 (110)	76 (66)	176
Column Totals	250	150	400

The remaining expected frequencies are calculated in the same way. Note that the expected frequencies in a test of homogeneity are calculated in the same way as in a test of independence. The value of the test statistic χ^2 is computed as follows:

$$\chi^2 = \sum \frac{(O-E)^2}{E}$$

$$= \frac{(70-65)^2}{65} + \frac{(34-39)^2}{39} + \frac{(80-75)^2}{75} + \frac{(40-45)^2}{45}$$

$$+ \frac{(100-110)^2}{110} + \frac{(76-66)^2}{66}$$

$$= .385 + .641 + .333 + .556 + .909 + 1.515 = \mathbf{4.339}$$

Step 5. *Make a decision.*

The value of the test statistic $\chi^2 = 4.339$ is less than the critical value of $\chi^2 = 7.378$, and it falls in the nonrejection region. Hence, we fail to reject the null hypothesis and state that the distribution of households with regard to income appears to be similar (homogeneous) in California and Wisconsin. ■

EXERCISES

CONCEPTS AND PROCEDURES

11.23 Describe in your own words a test of independence and a test of homogeneity. Give one example of each.

11.24 Explain how the expected frequencies for cells of a contingency table are calculated in a test of independence or homogeneity. How do you find the degrees of freedom for such tests?

11.25 To make a test of independence or homogeneity, what should be the minimum expected frequency for each cell? What are the alternatives if this condition is not satisfied?

11.26 Consider the following contingency table that is based on a sample survey.

	Column 1	Column 2	Column 3
Row 1	137	64	105
Row 2	98	71	65
Row 3	115	81	115

a. Write the null and alternative hypotheses for a test of independence for this table.
b. Calculate the expected frequencies for all cells assuming that the null hypothesis is true.
c. For $\alpha = .01$, find the critical value of χ^2. Show the rejection and nonrejection regions on the chi-square distribution curve.
d. Find the value of the test statistic χ^2.
e. Using $\alpha = .01$, would you reject the null hypothesis?

11.27 Consider the following contingency table that records the results obtained for four samples of fixed sizes selected from four populations.

	Sample Selected From			
	Population 1	Population 2	Population 3	Population 4
Row 1	24	81	60	121
Row 2	46	64	91	72
Row 3	20	37	105	93

a. Write the null and alternative hypotheses for a test of homogeneity for this table.
b. Calculate the expected frequencies for all cells assuming that the null hypothesis is true.
c. For $\alpha = .025$, find the critical value of χ^2. Show the rejection and nonrejection regions on the chi-square distribution curve.
d. Find the value of the test statistic χ^2.
e. Using $\alpha = .025$, would you reject the null hypothesis?

APPLICATIONS

11.28 During the recent economic recession, many families faced hard times financially. Some studies observed that more people stopped buying name brand products and started buying less expensive store brand products instead. Data produced by a recent sample of 700 adults on whether they usually buy store brand or name brand products are recorded in the following table.

	More Often Buy	
	Name Brand	Store Brand
Men	170	145
Women	182	203

Using the 1% significance level, can you reject the null hypothesis that the two attributes, gender and buying name or store brand products, are independent?

11.29 One hundred auto drivers who were stopped by police for some violation were also checked to see if they were wearing seat belts. The following table records the results of this survey.

	Wearing Seat Belt	Not Wearing Seat Belt
Men	34	21
Women	32	13

Test at the 2.5% significance level whether being a man or a woman and wearing or not wearing a seat belt are related.

11.30 Many students graduate from college deeply in debt from student loans, credit card debts, and so on. A sociologist took a random sample of 401 single persons, classified them by gender, and asked, "Would you consider marrying someone who was $25,000 or more in debt?" The results of this survey are shown in the following table.

	Yes	No	Uncertain
Women	125	59	21
Men	101	79	16

Test at the 1% significance level whether gender and response are related.

11.31 According to the TNS Online Kids Report based on a survey conducted in April 2004 of 660 children 6- to 14-years old, children were asked whether they worried about having enough money (*USA TODAY*, May 24, 2004). Assuming that the sample consisted of 325 boys and 335 girls, the percentages given in the newspaper would yield the numbers shown in the following table.

		Boys	Girls
Worried About	Yes	198	181
Money?	No	127	154

At the 5% significance level, can you conclude that worries about money and gender are related?

11.32 The following table gives the two-way classification of 400 randomly selected persons based on their status as a smoker or a nonsmoker and on the number of visits they made to their physicians last year.

	Visits to the Physician		
	0–1	2–4	≥ 5
Smoker	25	60	75
Nonsmoker	110	90	40

Test at the 5% significance level whether smoking and visits to the physician are related for all persons.

11.33 A forestry official is comparing the causes of forest fires in two regions, A and B. The following table shows the causes of fire for 76 randomly selected recent fires in these two regions.

	Arson	Accident	Lightning	Unknown
Region A	6	9	6	10
Region B	7	14	15	9

Test at the 5% significance level whether causes of fire and regions of fires are related.

11.34 National Electronics Company buys parts from two subsidiaries. The quality control department at this company wanted to check if the distribution of good and defective parts is the same for the supplies of parts received from both subsidiaries. The quality control inspector selected a sample of 300 parts received from Subsidiary A and a sample of 400 parts received from Subsidiary B. These parts were checked for being good or defective. The following table records the results of this investigation.

	Subsidiary A	Subsidiary B
Good	284	381
Defective	16	19

Using the 5% significance level, test the null hypothesis that the distributions of good and defective parts are the same for both subsidiaries.

11.35 Two drugs were administered to two groups of randomly assigned 60 and 40 patients, respectively, to cure the same disease. The following table gives information about the number of patients who were cured and not cured by each of the two drugs.

	Cured	Not Cured
Drug I	44	16
Drug II	18	22

Test at the 1% significance level whether or not the two drugs are similar in curing and not curing the patients.

11.36 A company introduced a new product in the market a few months ago. The management wants to determine the reaction of customers in different regions to this product. The research department at the company selected four different samples of 400 users of this product from four regions—East, South, Midwest, and West. The users of the product were asked whether or not they like the product. The responses of these people are recorded in the following table.

	East	South	Midwest	West
Like	274	206	291	254
Do not like	126	194	109	146

Based on the evidence from these samples, can you conclude that the distributions of opinions of users of this product are not homogeneous for all four regions with regard to liking and not liking this product? Use $\alpha = .01$.

11.37 In a Bernard Haldane Associates survey conducted in March 2004, white-collar workers who had changed jobs in the past 12 months were asked whether their new positions paid more, less, or the same as their previous jobs (*USA TODAY*, April 28, 2004). Assuming that the survey included randomly selected samples of 240 men and 240 women who had changed their jobs in the previous 12 months, the percentages given in the newspaper would yield the following table.

	Pay at New Job		
	More	**Less**	**Same**
Men	140	50	50
Women	93	104	43

Using the 1% significance level, test the null hypothesis that the changes in pay for workers who change jobs are similar for both men and women.

11.38 The following table gives the distributions of grades for three professors for a few randomly selected classes that each of them taught during the past two years.

		Professor		
		Miller	**Smith**	**Moore**
Grade	A	18	36	20
	B	25	44	15
	C	85	73	82
	D&F	17	12	8

Using the 2.5% significance level, test the null hypothesis that the grade distributions are homogeneous for these three professors.

11.39 Two random samples, one of 95 blue-collar workers and a second of 50 white-collar workers, were taken from a large company. These workers were asked about their views on a certain company issue. The following table gives the results of the survey.

	Opinion		
	Favor	**Oppose**	**Uncertain**
Blue-collar workers	44	39	12
White-collar workers	21	26	3

Using the 2.5% significance level, test the null hypothesis that the distributions of opinions are homogeneous for the two groups of workers.

11.5 Inferences About the Population Variance

Earlier chapters explained how to make inferences (confidence intervals and hypothesis tests) about the population mean and population proportion. However, we may often need to control the variance (or standard deviation). Consequently, there may be a need to estimate and to test a hypothesis about the population variance σ^2. Section 11.5.1 describes how to make a confidence interval for the population variance (or standard deviation). Section 11.5.2 explains how to test a hypothesis about the population variance.

As an example, suppose a machine is set up to fill packages of cookies so that the net weight of cookies per package is 32 ounces. Note that the machine will not put exactly 32 ounces of cookies into each package. Some of the packages will contain less and some will contain more than 32 ounces. However, if the variance (and, hence, the standard deviation) is too large, some of the packages will contain quite a bit less than 32 ounces of cookies and some others will contain quite a bit more than 32 ounces. The manufacturer will not want a large variation in the amounts of cookies put into different packages. To keep this variation within some specified acceptable limit, the machine will be adjusted from time to time. Before the manager decides to adjust the machine at any time, he must estimate the variance or test a hypothesis or do both to find out if the variance exceeds the maximum acceptable value.

Like every sample statistic, the sample variance is a random variable, and it possesses a sampling distribution. If all the possible samples of a given size are taken from a population and their variances are calculated, the probability distribution of these variances is called the *sampling distribution of the sample variance.*

Sampling Distribution of $(n-1)s^2/\sigma^2$ If the population from which the sample is selected is (approximately) normally distributed, then

$$\frac{(n-1)s^2}{\sigma^2}$$

has a chi-square distribution with $n-1$ degrees of freedom.

Thus, the chi-square distribution is used to construct a confidence interval and test a hypothesis about the population variance σ^2.

11.5.1 Estimation of the Population Variance

The value of the sample variance s^2 is a point estimate of the population variance σ^2. The $(1 - \alpha)100\%$ confidence interval for σ^2 is given by the following formula.

Confidence Interval for the Population Variance σ^2 Assuming that the population from which the sample is selected is (approximately) normally distributed, the $(1 - \alpha)100\%$ *confidence interval for the population variance* σ^2 is

$$\frac{(n-1)s^2}{\chi^2_{\alpha/2}} \quad \text{to} \quad \frac{(n-1)s^2}{\chi^2_{1-\alpha/2}}$$

where $\chi^2_{\alpha/2}$ and $\chi^2_{1-\alpha/2}$ are obtained from the chi-square distribution table for $\alpha/2$ and $1 - \alpha/2$ areas in the right tail of the chi-square distribution curve, respectively, and for $n - 1$ degrees of freedom.

The confidence interval for the population standard deviation can be obtained by simply taking the positive square roots of the two limits of the confidence interval for the population variance.

The procedure for making a confidence interval for σ^2 involves the following three steps.

1. Take a sample of size n and compute s^2 using the formula learned in Chapter 3. However, if n and s^2 are given, then perform only steps 2 and 3.
2. Calculate $\alpha/2$ and $1 - \alpha/2$. Find two values of χ^2 from the chi-square distribution table (Table VI of Appendix C): one for $\alpha/2$ area in the right tail of the chi-square distribution curve and $df = n - 1$, and the second for $1 - \alpha/2$ area in the right tail and $df = n - 1$.
3. Substitute all the values in the formula for the confidence interval for σ^2 and simplify.

Example 11–9 illustrates the estimation of the population variance and population standard deviation.

■ EXAMPLE 11–9

Constructing confidence intervals for σ^2 and σ.

One type of cookie manufactured by Haddad Food Company is Cocoa Cookies. The machine that fills packages of these cookies is set up in such a way that the average net weight of these packages is 32 ounces with a variance of .015 square ounces. From time to time the quality control inspector at the company selects a sample of a few such packages, calculates the variance of the net weights of these packages, and constructs a 95% confidence interval for the population variance. If either both or one of the two limits of this confidence interval is not in the interval .008 to .030, the machine is stopped and adjusted. A recently taken random sample of 25 packages from the production line gave a sample variance of .029 square ounces. Based on this sample information, do you think the machine needs an adjustment? Assume that the net weights of cookies in all packages are normally distributed.

Solution The following three steps are performed to estimate the population variance and to make a decision.

Step 1. From the given information, $n = 25$ and $s^2 = .029$

Step 2. The confidence level is $1 - \alpha = .95$. Hence, $\alpha = 1 - .95 = .05$. Therefore,

$$\alpha/2 = .05/2 = .025$$

$$1 - \alpha/2 = 1 - .025 = .975$$

$$df = n - 1 = 25 - 1 = 24$$

From Table VI of Appendix C,

$$\chi^2 \text{ for 24 } df \text{ and .025 area in the right tail} = 39.364$$

$$\chi^2 \text{ for 24 } df \text{ and .975 area in the right tail} = 12.401$$

These values are shown in Figure 11.9.

Figure 11.9

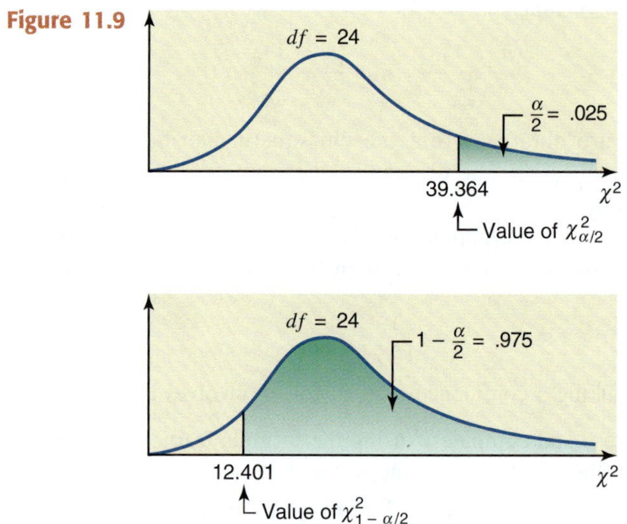

Step 3. The 95% confidence interval for σ^2 is

$$\frac{(n-1)s^2}{\chi^2_{\alpha/2}} \quad \text{to} \quad \frac{(n-1)s^2}{\chi^2_{1-\alpha/2}}$$

$$\frac{(25-1)(.029)}{39.364} \quad \text{to} \quad \frac{(25-1)(.029)}{12.401}$$

$$\mathbf{.0177 \quad to \quad .0561}$$

Thus, with 95% confidence, we can state that the variance for all packages of Cocoa Cookies lies between .0177 and .0561 square ounces. Note that the lower limit (.0177) of this confidence interval is between .008 and .030, but the upper limit (.0561) is larger than .030 and falls outside the interval .008 to .030. Because the upper limit is larger than .030, we can state that the machine needs to be stopped and adjusted.

We can obtain the confidence interval for the population standard deviation σ by taking the positive square roots of the two limits of the above confidence interval for the population variance. Thus, a 95% confidence interval for the population standard deviation is

$$\sqrt{.0177} \text{ to } \sqrt{.0561} \quad \text{or} \quad \mathbf{.133 \text{ to } .237}$$

Hence, the standard deviation of all packages of Cocoa Cookies is between .133 and .237 ounces at a 95% confidence level. ■

11.5.2 Hypothesis Tests About the Population Variance

A test of hypothesis about the population variance can be one-tailed or two-tailed. To make a test of hypothesis about σ^2, we perform the same five steps we have used earlier in hypothesis-testing examples. The procedure to test a hypothesis about σ^2 discussed in this section is applied only when the population from which a sample is selected is (approximately) normally distributed.

Test Statistic for a Test of Hypothesis About σ^2 The value of the *test statistic* χ^2 is calculated as

$$\chi^2 = \frac{(n-1)s^2}{\sigma^2}$$

where s^2 is the sample variance, σ^2 is the hypothesized value of the population variance, and $n - 1$ represents the degrees of freedom. The population from which the sample is selected is assumed to be (approximately) normally distributed.

Examples 11–10 and 11–11 illustrate the procedure to make tests of hypothesis about σ^2.

EXAMPLE 11–10

Making a right-tailed test of hypothesis about σ^2.

One type of cookie manufactured by Haddad Food Company is Cocoa Cookies. The machine that fills packages of these cookies is set up in such a way that the average net weight of these packages is 32 ounces with a variance of .015 square ounces. From time to time the quality control inspector at the company selects a sample of a few such packages, calculates the variance of the net weights of these packages, and makes a test of hypothesis about the population variance. She always uses $\alpha = .01$. The acceptable value of the population variance is .015 square ounces or less. If the conclusion from the test of hypothesis is that the population variance is not within the acceptable limit, the machine is stopped and adjusted. A recently taken random sample of 25 packages from the production line gave a sample variance of .029 square ounces. Based on this sample information, do you think the machine needs an adjustment? Assume that the net weights of cookies in all packages are normally distributed.

Solution From the given information,

$$n = 25, \quad \alpha = .01, \quad \text{and} \quad s^2 = .029$$

The population variance should not exceed .015 square ounce.

Step 1. *State the null and alternative hypotheses.*

We are to test whether or not the population variance is within the acceptable limit. The population variance is within the acceptable limit if it is less than or equal to .015; otherwise, it is not. Thus, the two hypotheses are

H_0: $\sigma^2 \le .015$ (The population variance is within the acceptable limit)

H_1: $\sigma^2 > .015$ (The population variance exceeds the acceptable limit)

Step 2. *Select the distribution to use.*

We use the chi-square distribution to test a hypothesis about σ^2.

Step 3. *Determine the rejection and nonrejection regions.*

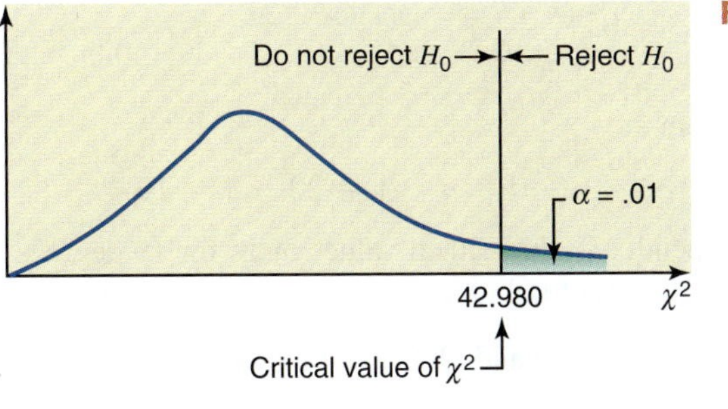

Figure 11.10

The significance level is 1% and, because of the > sign in H_1, the test is right-tailed. The rejection region lies in the right tail of the chi-square distribution curve with its area equal to .01. The degrees of freedom for a chi-square test about σ^2 are $n - 1$; that is,

$$df = n - 1 = 25 - 1 = 24$$

From Table VI of Appendix C, the critical value of χ^2 for 24 degrees of freedom and .01 area in the right tail is 42.980. This value is shown in Figure 11.10.

Step 4. *Calculate the value of the test statistic.*

The value of the test statistic χ^2 for the sample variance is calculated as follows:

$$\chi^2 = \frac{(n-1)s^2}{\sigma^2} = \frac{(25-1)(.029)}{\underset{\substack{\uparrow \\ \text{From } H_0}}{.015}} = 46.400$$

Step 5. *Make a decision.*

The value of the test statistic $\chi^2 = 46.400$ is greater than the critical value of $\chi^2 = 42.980$, and it falls in the rejection region. Consequently, we reject H_0 and conclude that the population variance is not within the acceptable limit. The machine should be stopped and adjusted. ■

■ EXAMPLE 11–11

Conducting a two-tailed test of hypothesis about σ^2.

The variance of scores on a standardized mathematics test for all high school seniors was 150 in 2004. A sample of scores for 20 high school seniors who took this test this year gave a variance of 170. Test at the 5% significance level if the variance of current scores of all high school seniors on this test is different from 150. Assume that the scores of all high school seniors on this test are (approximately) normally distributed.

Solution From the given information,

$$n = 20, \quad \alpha = .05, \quad \text{and} \quad s^2 = 170$$

The population variance was 150 in 2004.

Step 1. *State the null and alternative hypotheses.*

The null and alternative hypotheses are

H_0: $\sigma^2 = 150$ (The population variance is not different from 150)

H_1: $\sigma^2 \neq 150$ (The population variance is different from 150)

Step 2. *Select the distribution to use.*

We use the chi-square distribution to test a hypothesis about σ^2.

Step 3. *Determine the rejection and nonrejection regions.*

The significance level is 5%. The $\neq$ sign in H_1 indicates that the test is two-tailed. The rejection region lies in both tails of the chi-square distribution curve with its total area equal to .05. Consequently, the area in each tail of the distribution curve is .025. The values of $\alpha/2$ and $1 - \alpha/2$ are

$$\frac{\alpha}{2} = \frac{.05}{2} = .025 \quad \text{and} \quad 1 - \frac{\alpha}{2} = 1 - .025 = .975$$

The degrees of freedom are

$$df = n - 1 = 20 - 1 = 19$$

From Table VI of Appendix C, the critical values of χ^2 for 19 degrees of freedom and for $\alpha/2$ and $1 - \alpha/2$ areas in the right tail are

χ^2 for 19 df and .025 area in the right tail = 32.852

χ^2 for 19 df and .975 area in the right tail = 8.907

These two values are shown in Figure 11.11.

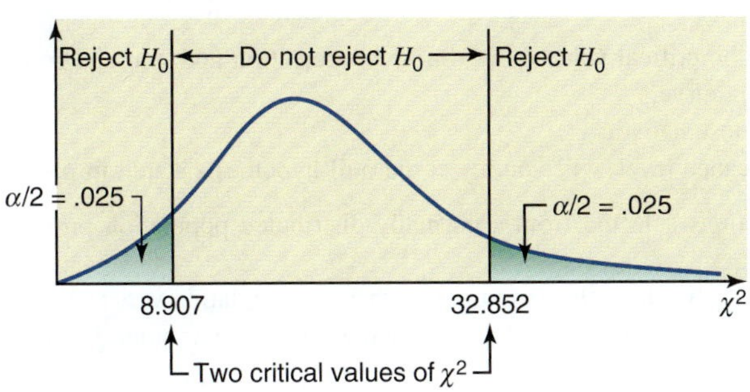

Figure 11.11

Step 4. *Calculate the value of the test statistic.*

The value of the test statistic χ^2 for the sample variance is calculated as follows.

$$\chi^2 = \frac{(n-1)s^2}{\sigma^2} = \frac{(20-1)(170)}{150} = 21.533$$

From H_0 (the 150 in the denominator)

Step 5. *Make a decision.*

The value of the test statistic $\chi^2 = 21.533$ is between the two critical values of χ^2, 8.907 and 32.852, and it falls in the nonrejection region. Consequently, we fail to reject H_0 and conclude that the population variance of the current scores of high school seniors on this standardized mathematics test does not appear to be different from 150. ■

Note that we can make a test of hypothesis about the population standard deviation σ using the same procedure as that for the population variance σ^2. To make a test of hypothesis about σ, the only change will be mentioning the values of σ in H_0 and H_1. The rest of the procedure remains the same as in case of σ^2.

EXERCISES

CONCEPTS AND PROCEDURES

11.40 A sample of certain observations selected from a normally distributed population produced a sample variance of 46. Construct a 95% confidence interval for σ^2 for each of the following cases and comment on what happens to the confidence interval of σ^2 when the sample size increases.

a. $n = 12$ **b.** $n = 16$ **c.** $n = 25$

11.41 A sample of 25 observations selected from a normally distributed population produced a sample variance of 35. Construct a confidence interval for σ^2 for each of the following confidence levels and comment on what happens to the confidence interval of σ^2 when the confidence level decreases.

a. $1 - \alpha = .99$ **b.** $1 - \alpha = .95$ **c.** $1 - \alpha = .90$

11.42 A sample of 22 observations selected from a normally distributed population produced a sample variance of 18.

a. Write the null and alternative hypotheses to test whether the population variance is different from 14.

b. Using $\alpha = .05$, find the critical values of χ^2. Show the rejection and nonrejection regions on a chi-square distribution curve.

c. Find the value of the test statistic χ^2.

d. Using the 5% significance level, will you reject the null hypothesis stated in part a?

11.43 A sample of 16 observations selected from a normally distributed population produced a sample variance of 1.10.

a. Write the null and alternative hypotheses to test whether the population variance is greater than .80.
b. Using $\alpha = .01$, find the critical value of χ^2. Show the rejection and nonrejection regions on a chi-square distribution curve.
c. Find the value of the test statistic χ^2.
d. Using the 1% significance level, will you reject the null hypothesis stated in part a?

11.44 A sample of 25 observations selected from a normally distributed population produced a sample variance of .70.

a. Write the null and alternative hypotheses to test whether the population variance is less than 1.25.
b. Using $\alpha = .025$, find the critical value of χ^2. Show the rejection and nonrejection regions on a chi-square distribution curve.
c. Find the value of the test statistic χ^2.
d. Using the 2.5% significance level, will you reject the null hypothesis stated in part a?

11.45 A sample of 18 observations selected from a normally distributed population produced a sample variance of 4.6.

a. Write the null and alternative hypotheses to test whether the population variance is different from 2.2.
b. Using $\alpha = .05$, find the critical values of χ^2. Show the rejection and nonrejection regions on a chi-square distribution curve.
c. Find the value of the test statistic χ^2.
d. Using the 5% significance level, will you reject the null hypothesis stated in part a?

APPLICATIONS

11.46 The management of a soft-drink company does not want the variance of the amounts of soda in 12-ounce cans to be more than .01 square ounces. (Recall from Chapter 3 that the variance is always in square units.) The company manager takes a sample of certain cans and estimates the population variance quite often. A random sample of twenty 12-ounce cans taken from the production line of this company showed that the variance for this sample was .014 square ounces.

a. Construct the 99% confidence intervals for the population variance and standard deviation. Assume that the amounts of soda in all 12-ounce cans have a normal distribution.
b. Test at the 2.5% significance level whether the variance of the amounts of soda in all 12-ounce cans for this company is greater than .01 square ounces.

11.47 Professor Fox's "50-minute" lectures vary in length. Professor Fox claims that the variance of the lengths of his lectures is within 2 square minutes. A random sample of 23 of these lectures was timed, and the variance of the lengths of these lectures was found to be 2.7 square minutes. Assume that the lengths of all such lectures by Professor Fox are (approximately) normally distributed.

a. Make the 98% confidence intervals for the variance and standard deviation of the lengths of all 50-minute lectures by Professor Fox.
b. Test at the 1% significance level whether the variance of the lengths of all such lectures by Professor Fox exceeds 2 square minutes.

11.48 An auto manufacturing company wants to estimate the variance of miles per gallon for its auto model AST727. A random sample of 22 cars of this model showed that the variance of miles per gallon for these cars is .62.

a. Construct the 95% confidence intervals for the population variance and standard deviation. Assume that the miles per gallon for all such cars are (approximately) normally distributed.
b. Test at the 1% significance level whether the sample result indicates that the population variance is different from .30.

11.49 The manufacturer of a certain brand of lightbulbs claims that the variance of the lives of these bulbs is 4200 square hours. A consumer agency took a random sample of 25 such bulbs and tested them. The variance of the lives of these bulbs was found to be 5200 square hours. Assume that the lives of all such bulbs are (approximately) normally distributed.

a. Make the 99% confidence intervals for the variance and standard deviation of the lives of all such bulbs.
b. Test at the 5% significance level whether the variance of such bulbs is different from 4200 square hours.

USES AND MISUSES... DON'T FEED THE ANIMALS

You are a wildlife enthusiast studying African wildlife: gnus, zebras, and gazelles. You know that a herd of each species will visit one of three watering places in a region every day, but you do not know the distribution of choices that the animals make or whether these choices are dependent. You have observed that the animals sometimes drink together and sometimes do not. A statistician offers to help and says that he will perform a test for independence of watering place choices based on your observations of the animals' behavior over the past several months. The statistician performs some calculations and says that he has answered your question because his chi-square test of the independence of watering place choices, at a 5% significance level, told him to reject the null hypothesis. He has also performed a goodness-of-fit test on the hypothesis that the animals are equally likely to choose any watering place, and he has rejected that hypothesis as well.

The statistician barely helped you. In the first case, you know a single piece of information: the choice of a watering place for the three groups of animals is dependent. Another way of stating the result is that your data indicate that the choice of watering places for at least one of the animals is not independent of the others. Perhaps the zebras get up early, and the gnus and gazelles follow, making the gnus and gazelles dependent on the choice of the zebras. Or perhaps the animals choose the watering place of the day independent of the other animals, but always avoid the watering place at which the lions are drinking. Regarding the goodness-of-fit test, all you know is that the hypothesis that the animals equally favor the three watering places was wrong. But you do not know what the expected distribution should be. In short, the rejection of the null hypothesis raises more questions than it answers.

Glossary

Chi-square distribution A distribution, with degrees of freedom as the only parameter, that is skewed to the right for small df and looks like a normal curve for large df.

Expected frequencies The frequencies for different categories of a multinomial experiment or for different cells of a contingency table that are expected to occur when a given null hypothesis is true.

Goodness-of-fit test A test of the null hypothesis that the observed frequencies for an experiment follow a certain pattern or theoretical distribution.

Multinomial experiment An experiment with n trials for which (1) the trials are identical, (2) there are more than two possible outcomes per trial, (3) the trials are independent, and (4) the probabilities of the various outcomes remain constant for each trial.

Observed frequencies The frequencies actually obtained from the performance of an experiment.

Test of homogeneity A test of the null hypothesis that the proportions of elements that belong to different groups in two (or more) populations are similar.

Test of independence A test of the null hypothesis that two attributes of a population are not related.

Supplementary Exercises

11.50 The U.S. Census American Housing Survey in 2001 collected data on modes of transportation used for commuting by American workers. For those who rented their dwellings, 68.1% drove themselves to work, 13.1% carpooled, 9.4% took mass transit, and 5.3% walked (*USA TODAY*, February 25, 2003). Assume that the remaining 4.1% used other modes of transportation such as bicycles. Further assume that these percentages were true for the population of all workers who rented their dwellings at the time of the survey. A recent random sample of 1100 workers who rented their dwellings yielded the following distribution of modes of transportation.

Mode of Transportation	Frequency
Drive themselves	702
Carpool	176
Mass transit	106
Walk	56
Other	60

Test at the 1% significance level whether the current distribution of modes of transportation to commute to work by renters differs from that for 2001.

11.51 One of the products produced by Branco Food Company is All-Bran Cereal, which competes with three other brands of similar all-bran cereals. The company's research office wants to investigate if the percentage of people who consume all-bran cereal is the same for each of these four brands. Let us denote the four brands of cereal by A, B, C, and D. A sample of 1000 persons who consume all-bran cereal was taken, and they were asked which brand they most often consume. Of the respondents, 212 said they usually consume Brand A, 284 consume Brand B, 254 consume Brand C, and 250 consume Brand D. Does the sample provide enough evidence to reject the null hypothesis that the percentage of people who consume all-bran cereal is the same for all four brands? Use $\alpha = .05$.

11.52 In the MARS 2004 OTC/DTC Survey, American adults were asked to rate their current states of health (*USA TODAY*, June 2, 2004). The following table records the distribution of their responses. Suppose these results were true for the 2004 population of all American adults.

State of Health	Percentage
Excellent	17.0
Very good	36.2
Good	32.5
Fair	12.0
Poor	2.3

A recent random sample of 1000 adults yielded the frequencies for the responses listed in the table as 175, 346, 335, 112, and 32, respectively. Does the sample information provide sufficient evidence to reject the null hypothesis that the current distribution of health ratings by all American adults is the same as the one in 2004? Use the 2.5% level of significance.

11.53 In 2004, a poll commissioned by *PARADE* and Research! America was conducted by the Charlton Research Co. to examine men's attitudes towards several health issues. In one part of the study, men were asked what concerned them the most about their own health. The following table lists the percentage of responses for each concern given by these men (*PARADE*, June 20, 2004). Assume that these results were true for the 2004 population of all men.

Health Concern	Percentage
No Concerns	27
Cancer	20
Heart	16
Diabetes	7
Blood Pressure	6
Weight/Diet	5
Other	19

Recently 800 randomly selected men were asked to name their most important health concerns. The men who mentioned each of the concerns listed in the table are 185, 155, 138, 72, 47, 52, and 151, respectively. At the 5% level of significance, can you conclude that the current distribution of men's health concerns differs from that of 2004?

11.54 During a bear market, 140 investors were asked how they were adjusting their portfolios to protect themselves. Some of these investors were keeping most of their money in stocks, whereas others were shifting large amounts of money to bonds, real estate, or cash (such as money market accounts). The results of the survey are shown in the following table.

Favored Choice	Stocks	Bonds	Real Estate	Cash
Number of Investors	46	41	32	21

Using the 2.5% significance level, test the null hypothesis that the percentages of investors favoring the four choices are equal.

11.55 A randomly selected sample of 100 persons who suffer from allergies were asked during what season they suffer the most. The results of the survey are recorded in the following table.

Season	Fall	Winter	Spring	Summer
Persons allergic	18	13	31	38

Using the 1% significance level, test the null hypothesis that the proportions of all allergic persons are the same over the four seasons.

11.56 All shoplifting cases in the town of Seven Falls are randomly assigned to either Judge Stark or Judge Rivera. A citizens group wants to know whether either of the two judges is more likely to sentence the offenders to jail time. A sample of 180 recent shoplifting cases produced the following two-way table.

	Jail	**Other Sentence**
Judge Stark	27	65
Judge Rivera	31	57

Test at the 5% significance level whether the type of sentence for shoplifting depends on which judge tries the case.

11.57 A 2004 newspaper article, based on U.S. Census Bureau data, gave a regional breakdown of percentages of Americans who were not covered by health insurance (*USA TODAY*, August 27, 2004). Assuming that the data were based on random samples of 1000 people from each of the four regions considered, the percentages given in the newspaper would yield the following table.

	Region			
	Northeast	**Midwest**	**South**	**West**
Number Insured	871	880	820	824
Number Uninsured	129	120	180	176

Using the 2.5% significance level, can you conclude that the percentages of insured and uninsured people are the same for all four regions?

11.58 A random sample of 100 jurors was selected and asked whether or not each of them had ever been a victim of crime. The jurors were also asked whether they are strict, fair, or lenient regarding punishment for crime. The following table gives the results of the survey.

	Strict	**Fair**	**Lenient**
Have been a victim	20	8	3
Have never been a victim	22	33	14

Test at the 5% significance level whether the two attributes for all jurors are dependent.

11.59 Recent recession and bad economic conditions forced many people to hold more than one job to make ends meet. A sample of 500 persons who held more than one job produced the following two-way table.

	Single	**Married**	**Other**
Male	72	209	39
Female	33	102	45

Test at the 10% significance level whether gender and marital status are related for all people who hold more than one job.

11.60 ATVs (all-terrain vehicles) have become a source of controversy. Some people feel that their use should be tightly regulated, while others prefer fewer restrictions. Suppose a survey consisting of a random sample of 200 people aged 18 to 27 and another random sample of 210 people aged 28 to 37 was conducted and these people were asked whether they favored more restrictions on ATVs, fewer restrictions, or no change. The results of this survey are summarized in the following table.

		More Restrictions	Fewer Restrictions	No Change
Age	18 to 27	40	92	68
	28 to 37	55	68	87

Test at the 2.5% significance level whether the distribution of opinions in regard to ATVs are the same for both age groups.

11.61 A random sample of 100 persons was selected from each of four regions in the United States. These people were asked whether or not they support a certain farm subsidy program. The results of the survey are summarized in the following table.

	Favor	Oppose	Uncertain
Northeast	56	33	11
Midwest	73	23	4
South	67	28	5
West	59	35	6

Using the 1% significance level, test the null hypothesis that the percentages of people with different opinions are similar for all four regions.

11.62 Construct the 98% confidence intervals for the population variance and standard deviation for the following data assuming that the respective populations are (approximately) normally distributed.

a. $n = 21, s^2 = 9.2$ **b.** $n = 17, s^2 = 1.7$

11.63 Construct the 95% confidence intervals for the population variance and standard deviation for the following data assuming that the respective populations are (approximately) normally distributed.

a. $n = 10, s^2 = 7.2$ **b.** $n = 18, s^2 = 14.8$

11.64 Refer to Exercise 11.62a. Test at the 5% significance level if the population variance is different from 6.5.

11.65 Refer to Exercise 11.62b. Test at the 2.5% significance level if the population variance is greater than 1.1.

11.66 Refer to Exercise 11.63a. Test at the 1% significance level if the population variance is greater than 4.2.

11.67 Refer to Exercise 11.63b. Test at the 5% significance level if the population variance is different from 10.4.

11.68 Usually people do not like waiting in line a long time for service. A bank manager does not want the variance of the waiting times for her customers to be greater than 4.0 square minutes. A random sample of 25 customers taken from this bank gave the variance of the waiting times equal to 8.3 square minutes.

a. Test at the 1% significance level whether the variance of the waiting times for all customers at this bank is greater than 4.0 square minutes. Assume that the waiting times for all customers are normally distributed.

b. Construct a 99% confidence interval for the population variance.

11.69 The variance of the SAT scores for all students who took that test this year is 5000. The variance of the SAT scores for a random sample of 20 students from one school is equal to 3175.

a. Test at the 2.5% significance level whether the variance of the SAT scores for students from this school is lower than 5000. Assume that the SAT scores for all students at this school are (approximately) normally distributed.

b. Construct the 98% confidence intervals for the variance and the standard deviation of SAT scores for all students at this school.

11.70 A company manufactures ball bearings that are supplied to other companies. The machine that is used to manufacture these ball bearings produces them with a variance of diameters of .025 square millimeters or less. The quality control officer takes a sample of such ball bearings quite often and checks, using confidence intervals and tests of hypotheses, whether or not the variance of these bearings is within .025 square millimeters. If it is not, the machine is stopped and adjusted. A recently taken random sample of 23 ball bearings gave a variance of the diameters equal to .034 square millimeters.

a. Using the 5% significance level, can you conclude that the machine needs an adjustment? Assume that the diameters of all ball bearings have a normal distribution.
b. Construct a 95% confidence interval for the population variance.

11.71 A random sample of 25 students taken from a university gave the variance of their GPAs equal to .19.

a. Construct the 99% confidence intervals for the population variance and standard deviation. Assume that the GPAs of all students are (approximately) normally distributed.
b. The variance of GPAs of all students at this university was .13 two years ago. Test at the 1% significance level whether the variance of GPAs now is different from .13.

11.72 A sample of seven passengers boarding a domestic flight produced the following data on weights (in pounds) of their carry-on bags.

22 17 29 19 12 25 16

a. Using the formula from Chapter 3, find the sample variance s^2 for these data.
b. Make the 98% confidence intervals for the population variance and standard deviation. Assume that the population from which this sample is selected is normally distributed.
c. Test at the 1% significance level whether the population variance is less than 150 square pounds.

11.73 The following are the prices of the same brand of camcorder found at eight stores in Los Angeles.

\$755 815 789 799 732 835 799 769

a. Using the formula from Chapter 3, find the sample variance s^2 for these data.
b. Make the 95% confidence intervals for the population variance and standard deviation. Assume that the prices of this camcorder at all stores in Los Angeles follow a normal distribution.
c. Test at the 5% significance level whether the population variance is different from 500 square dollars.

Advanced Exercises

11.74 As part of the High School Survey of Student Engagement, a survey conducted at Indiana University in Bloomington investigated the amount of time spent preparing for class per week by high school students in four instructional tracks: general, special education, college, and vocational (*USA TODAY*, May 9, 2005). Although this study used very large samples, suppose (for illustrative purposes) the study included a random sample of 200 students for each of the four tracks. Based on the article, the percentages of students in various tracks who spent at least seven hours per week preparing for classes would have yielded the numbers listed in the table below.

Track	Frequency
General	32
Special education	28
College	74
Vocational	22

Using the 1% significance level, test the null hypothesis that the percentages of students in these four instructional tracks who spend at least seven hours per week preparing for classes are the same.

11.75 A chemical manufacturing company wants to locate a hazardous waste disposal site near a city of 50,000 residents and has offered substantial financial inducements to the city. Two hundred adults (110 women and 90 men) who are residents of this city are chosen at random. Sixty percent of these adults oppose the site, 32% are in favor, and 8% are undecided. Of those who oppose the site, 65% are women; of those in favor, 62.5% are men. Using the 5% level of significance, can you conclude that opinions on the disposal site are dependent on gender?

11.76 A student who needs to pass an elementary statistics course wonders whether it will make a difference if she takes the course with instructor A rather than instructor B. Observing the final grades given by each instructor in a recent elementary statistics course, she finds that Instructor A gave

48 passing grades in a class of 52 students and Instructor B gave 44 passing grades in a class of 54 students. Assume that these classes and grades make simple random samples of all classes and grades of these instructors.

a. Compute the value of the standard normal test statistic z of Section 10.5.3 for the data and use it to find the p-value when testing for the difference between the proportions of passing grades given by these instructors.

b. Construct a 2×2 contingency table for these data. Compute the value of the χ^2 test statistic for the test of homogeneity and use it to find the p-value.

c. How do the test statistics in parts a and b compare? How do the p-values for the tests in parts a and b compare? Do you think this is a coincidence, or do you think this will always happen?

11.77 Each of five boxes contains a large (but unknown) number of red and green marbles. You have been asked to find if the proportions of red and green marbles are the same for each of the five boxes. You sample 50 times, with replacement, from each of the five boxes and observe 20, 14, 23, 30, and 18 red marbles, respectively. Can you conclude that the five boxes have the same proportions of red and green marbles? Use a .05 level of significance.

11.78 Suppose that you have a two-way table with the following row and column totals.

		Variable 1			
		A	B	C	Total
Variable 2	X				120
	Y				205
	Z				175
	Total	165	140	195	500

The observed values in the cells must be counts, which are non-negative integers. Calculate the expected counts for the cells under the assumption that the two variables are independent. Based on your calculations, explain why it is impossible for the test statistic to have a value of zero.

11.79 You have collected data on a variable and you want to determine if a normal distribution is a reasonable model for these data. The following table shows how many of the values fall within certain ranges of z-values for these data.

Category	Count
z score below -2	48
z score from -2 to less than -1.5	67
z score from -1.5 to less than -1	146
z score from -1 to less than -0.5	248
z score from -0.5 to less than 0	187
z score from 0 to less than 0.5	125
z score from 0.5 to less than 1	88
z score from 1 to less than 1.5	47
z score from 1.5 to less than 2	25
z score of 2 or above	19
Total	1000

Perform a hypothesis test to determine if a normal distribution is an appropriate model for these data. Use a significance level of 5%.

11.80 Refer to Problem 11.61. Explain why the hypothesis test in that problem is a test of homogeneity as opposed to a test of independence. What feature of the data would change if you were to collect data in order to test for independence?

11.81 You are performing a goodness-of-fit test with four categories, all of which are supposed to be equally likely. You have a total of 100 observations. The observed frequencies are 21, 26, 31, and 22, respectively, for the four categories.

a. Show that you would fail to reject the null hypothesis for these data for any reasonable significance level.

b. The sum of the absolute differences (between the expected and the observed frequencies) for these data is 14 (i.e., $4 + 1 + 6 + 3 = 14$). Is it possible to have different observed frequencies keeping the sum at 14 so that you get a p-value of .10 or less?

Self-Review Test

1. The random variable χ^2 assumes only
   a. positive b. nonnegative c. nonpositive values

2. The parameter(s) of the chi-square distribution is (are)
   a. degrees of freedom b. df and n c. χ^2

3. Which of the following is *not* a characteristic of a multinomial experiment?
   a. It consists of n identical trials.
   b. There are k possible outcomes for each trial and $k > 2$.
   c. The trials are random.
   d. The trials are independent.
   e. The probabilities of outcomes remain constant for each trial.

4. The observed frequencies for a goodness-of-fit test are
   a. the frequencies obtained from the performance of an experiment
   b. the frequencies given by the product of n and p
   c. the frequencies obtained by adding the results of a and b

5. The expected frequencies for a goodness-of-fit test are
   a. the frequencies obtained from the performance of an experiment
   b. the frequencies given by the product of n and p
   c. the frequencies obtained by adding the results of a and b

6. The degrees of freedom for a goodness-of-fit test are
   a. $n - 1$ b. $k - 1$ c. $n + k - 1$

7. The chi-square goodness-of-fit test is always
   a. two-tailed b. left-tailed c. right-tailed

8. To apply a goodness-of-fit test, the expected frequency of each category must be at least
   a. 10 b. 5 c. 8

9. The degrees of freedom for a test of independence are
   a. $(R - 1)(C - 1)$ b. $n - 2$ c. $(n - 1)(k - 1)$

10. In a *USA TODAY* survey in 2004, 126 registered dietitians with the American Dietetic Association were asked to name the biggest mistake that people make when trying to lose weight. Of them, 27% said going on a fad diet, 20% said not limiting portions, 15% said setting inappropriate goals (such as losing weight too quickly), 13% said not exercising, and 25% gave other reasons (*USA TODAY*, July 19, 2004). Suppose these results were true for the 2004 population of all dietitians. In a recent random sample of 150 dietitians who were asked the same question, 48 said fad diets, 41 mentioned not limiting portions, 17 said inappropriate goals, 12 mentioned not exercising, and 32 gave other reasons. At the 5% significance level, can you conclude that the current distribution of all dietitians' responses differs from that for 2004?

11. The following table gives the two-way classification of 1000 persons who have been married at least once. They are classified by educational level and marital status.

	Educational Level			
	Less Than High School	**High School Degree**	**Some College**	**College Degree**
Divorced	173	158	95	53
Never divorced	162	126	110	123

Test at the 1% significance level whether educational level and ever being divorced are dependent.

12. A researcher wanted to investigate if people who belong to different income groups are homogeneous with regard to playing lotteries. She took a sample of 600 people from the low income group, another sample of 500 people from the middle income group, and a third sample of 400 people from the high income group. All these people were asked whether they play the lottery often, sometimes, or never. The results of the survey are summarized in the following table.

	Income Group		
	Low	**Middle**	**High**
Play often	174	163	90
Play sometimes	286	217	120
Never play	140	120	190

Using the 5% significance level, can you reject the null hypothesis that the percentages of people who play the lottery often, sometimes, and never are the same for each income group?

13. The owner of an ice cream parlor is concerned about consistency in the amount of ice cream his servers put in each cone. He would like the variance of all such cones to be no more than .25 square ounces. He decides to weigh each double-dip cone just before it is given to the customer. For a sample of 20 double-dip cones, the weights were found to have a variance of .48 square ounces. Assume that the weights of all such cones are (approximately) normally distributed.

a. Construct the 99% confidence intervals for the population variance and the population standard deviation.

b. Test at the 1% significance level whether the variance of the weights of all such cones exceeds .25 square ounces.

Mini-Projects

MINI-PROJECT 11–1

In recent years drivers have become careless about signaling their turns. To study this problem, go to a busy intersection and observe at least 75 vehicles that make left turns. Divide these vehicles into three or four classes. For example, you might use cars, trucks, and others, where "others" include minivans and sport-utility vehicles, as classes. For each left turn made by a vehicle, record the type of vehicle and whether or not the driver used the left turn signal before making this turn. It would be better to avoid intersections that have designated left-turn lanes or green arrows for left turns because drivers in these situations often assume that their intent to turn left is obvious. Carry out an appropriate test at the 1% level of significance to determine if signaling behavior and vehicle type are dependent.

MINI-PROJECT 11–2

Refer to Case Study 11–1. Buy enough M&M bags with such candies to provide at least 2000 pieces of candy, put all the pieces in a box, then count the number of candies of each color in the box. Carry out an appropriate test at the 1% level of significance to determine whether the current distribution of colors of M&M candies in all such bags is the same as the distribution of colors given in the chart in Case Study 11–1.

MINI-PROJECT 11–3

One day during lunch, visit your school cafeteria, observe at least 100 people, and write down what they are drinking. Categorize the drinks as soft drink (soda, fruit punch, or lemonade), iced tea, milk or juice, hot drink, and water. Also identify the gender of each person. Perform a hypothesis test to determine if the type of drink and gender are independent.

DECIDE FOR YOURSELF

Testing for the Fairness of Gambling Equipment

Casino gambling has grown rapidly in the United States. Native American tribes have opened casinos on reservations, many horse racing tracks have been allowed to add slot machines on site, and riverboat/lakefront casinos have also been opened in recent years. States with casino gambling have state agencies that are responsible for verifying and making sure that the games and equipment are fair, and not fixed. In many states, such an agency is called the Division of Gaming Enforcement. New Jersey and Nevada have two of the largest such agencies, given the presence of Atlantic City and Las Vegas in these states. The chi-square procedures that you have learned in this chapter can be used to test the validity of the *fairness* assumption in regard to the gaming equipment.

A simple example would involve checking to see whether or not a given die is balanced. Under the null hypothesis, we would assume that the probability of a specific side coming up when we roll this die is 1/6. To test this notion, we can roll the given die a specific number of times and observe the frequency for each outcome. Suppose we roll this die 180 times and obtain the frequencies for various outcomes as listed in the following table.

Outcome	1-spot	2-spots	3-spots	4-spots	5-spots	6-spots
Frequency	26	31	29	33	26	35

1. Theoretically, how often would you expect each outcome to occur if we roll this die 180 times assuming it is a fair die?

2. Perform the appropriate hypothesis test to determine the p-value with the null hypothesis that the die is fair. What is your conclusion?

3. How much do you have to change the frequencies for various outcomes in the above table to obtain a conclusion for the hypothesis test of question 2 that is the opposite of the one you obtained above? Does your conclusion switch faster if you make a big change to one frequency and small changes to the others or if you make moderate changes to all of the categories? (Remember that the sum of all frequencies has to remain 180.)

TECHNOLOGY INSTRUCTION — Chi-Square Tests

TI-84

X²-Test
Observed: [A]
Expected: [B]
Calculate Draw

Screen 11.1

1. To perform an independence or homogeneity test on a contingency table, enter the actual data and the expected values as matrices. To do so, select **MATRX>EDIT** and use the arrow key to select the name of your matrix. Press **Enter** and then type in the number of rows, the number of columns, and the entries for each matrix.
2. Select **STAT>TESTS>χ^2-Test**. You will need to enter the names of the **Observed** and the **Expected** data matrices. For each entry, position the cursor and then select **MATRX>NAMES** and use the arrow keys to choose the appropriate name, and then press **Enter**. (See **Screen 11.1.**) After entering the matrix names, press **Enter**. The result includes the value of χ^2, the p-value, and the degrees of freedom. (See **Screen 11.2.**)

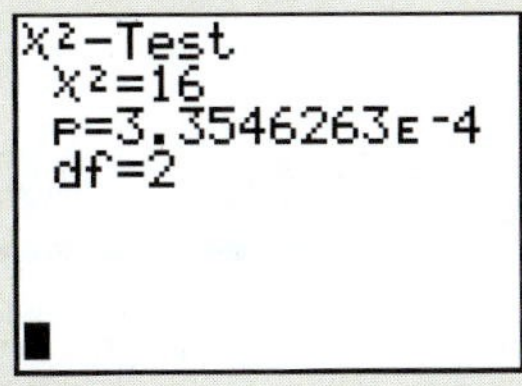

Screen 11.2

MINITAB

1. To perform an independence or homogeneity test on a contingency table, enter the actual data into columns, then select **Stat>Tables>Chi-square Test**.

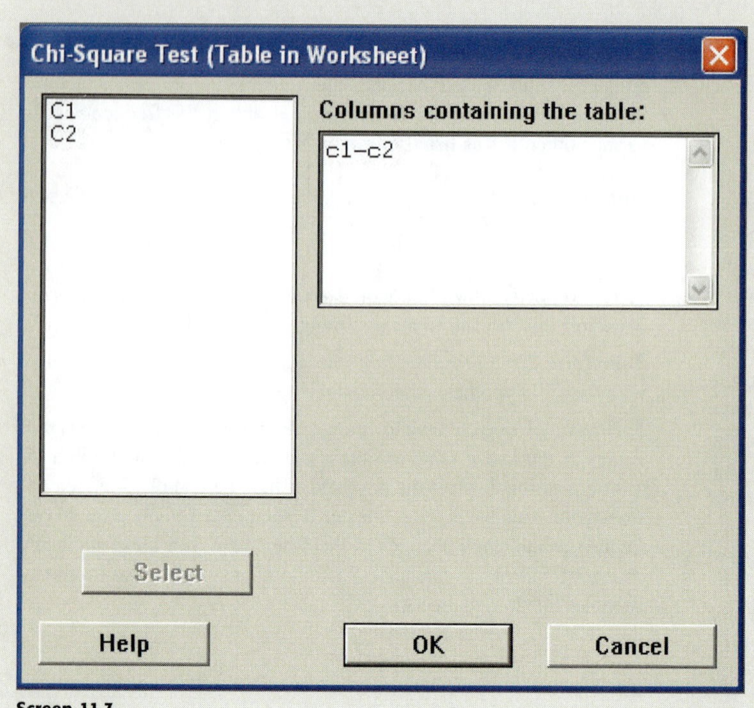

Screen 11.3

2. Enter the names of the columns containing the table and select **OK**. (See **Screen 11.3**.) The result includes the expected values, the degrees of freedom, the value of chi-square, and the *p*-value.

Excel

	A	B	C	D	E
1	Actual			Expected	
2					
3	20	80		25	75
4	40	160		50	150
5	40	60		25	75
6					
7	p-value	=CHITEST(A3:B5,D3:E5)			
8		CHITEST(actual_range, expected_range)			

Screen 11.4

1. To perform a goodness-of-fit or independence test on a contingency table, enter the actual data in a range of cells and the expected data in another range of cells with the same number of rows and columns.

2. Type **=CHITEST(actual range, expected range)** and press **Enter**. The result is the *p*-value of the test. (See **Screen 11.4**.)

TECHNOLOGY ASSIGNMENTS

TA11.1 Each day, a large industrial city announces the Air Quality Index (AQI), which measures the level of air pollution. From 1998 through 2002, the AQI has been "good" on 37% of the days, "moderate" on 30% of the days, "unhealthy" on 20% of the days, and "very unhealthy" on 13% of the days. The following table gives the distribution of the AQI for a sample of 40 days in 2005.

AQI	Good	Moderate	Unhealthy	Very Unhealthy
Number of days	18	13	6	3

Test at the 5% significance level whether the distribution of the AQI in 2005 differed from the distribution of 1998 through 2002.

TA11.2 A sample of 4000 persons aged 18 and older produced the following two-way classification table.

	Men	Women
Single	531	357
Married	1375	1179
Widowed	55	195
Divorced	139	169

Test at the 10% significance level whether gender and marital status are dependent for all persons aged 18 and older.

TA11.3 Two samples, one of 3000 students from urban high schools and another of 2000 students from rural high schools, were taken. These students were asked if they have ever smoked. The following table lists the summary of the results.

	Urban	Rural
Have never smoked	1448	1228
Have smoked	1552	772

Using the 5% significance level, test the null hypothesis that the proportions of urban and rural students who have smoked and who have never smoked are homogeneous.

Chapter

12

Analysis of Variance

12.1 The *F* Distribution

12.2 One-Way Analysis of Variance

Trying something new can be risky, whether it is attempting to snowboard for the first time or trying a new method of teaching in the classroom. In 1989 the National Council of Teachers of Mathematics released its Curriculum and Evaluations Standards for School Mathematics, guidelines for a new way of teaching mathematics. Those teachers who were brave enough to try the new methods were well rewarded: The top five states in fourth-grade mathematics achievement as reported by *Education Week* in its Quality Counts 2000—Connecticut, Minnesota, Maine, Wisconsin, and Texas—were all leaders in adopting the spirit and methods of the standards. But how could teachers know that these guidelines would be successful? Are there procedures to compare the new ways of teaching with the traditional ways?

Chapter 10 described the procedures that are used to test hypotheses about the difference between two population means using the normal and t distributions. Also described in that chapter were the hypothesis-testing procedures for the difference between two population proportions using the normal distribution. Then, Chapter 11 explained the procedures used to test hypotheses about the equality of more than two population proportions using the chi-square distribution.

This chapter explains how to test the null hypothesis that the means of more than two populations are equal. For example, suppose that teachers at a school have devised three different methods to teach arithmetic. They want to find out if these three methods produce different mean scores. Let μ_1, μ_2, and μ_3 be the mean scores of all students who will be taught by Methods I, II, and III, respectively. To test whether or not the three teaching methods produce the same mean, we test the null hypothesis

$$H_0: \mu_1 = \mu_2 = \mu_3 \quad \text{(All three population means are equal)}$$

against the alternative hypothesis

$$H_1: \text{Not all three population means are equal}$$

We use the analysis of variance procedure to perform such a test of hypothesis.

Note that the analysis of variance procedure can be used to compare two population means. However, the procedures learned in Chapter 10 are more efficient for performing tests of hypotheses about the difference between two population means; the analysis of variance procedure, to be discussed in this chapter, is used to compare three or more population means.

An *analysis of variance* test is performed using the *F* distribution. First, the *F* distribution is described in Section 12.1 of this chapter. Then, Section 12.2 discusses the application of the one-way analysis of variance procedure to perform tests of hypotheses.

12.1 The *F* Distribution

Like the *t* and chi-square distributions, the shape of a particular ***F* distribution**[1] curve depends on the number of degrees of freedom. However, the *F* distribution has *two* numbers of degrees of freedom: *degrees of freedom for the numerator* and *degrees of freedom for the denominator*. These two numbers representing two types of degrees of freedom are the *parameters of the F distribution*. Each combination of degrees of freedom for the numerator and for the denominator gives a different *F* distribution curve. The units of an *F* distribution are denoted by *F*, which assumes only nonnegative values. Like the normal, *t*, and chi-square distributions, the *F* distribution is a continuous distribution. The shape of an *F* distribution curve is skewed to the right, but the skewness decreases as the number of degrees of freedom increases.

Definition

The *F* Distribution

1. The *F distribution* is continuous and skewed to the right.
2. The *F* distribution has two numbers of degrees of freedom: *df* for the numerator and *df* for the denominator.
3. The units of an *F* distribution, denoted by *F*, are nonnegative.

For an *F* distribution, degrees of freedom for the numerator and degrees of freedom for the denominator are usually written as follows:

$$df = (8, 14)$$

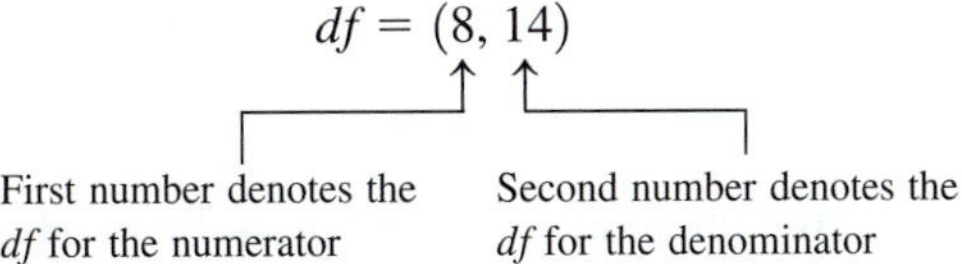
First number denotes the df for the numerator
Second number denotes the df for the denominator

Figure 12.1 (on the next page) gives three *F* distribution curves for three sets of degrees of freedom for the numerator and for the denominator. In the figure, the first number gives the degrees of freedom associated with the numerator and the second number gives the degrees of freedom associated with the denominator. We can observe from this figure that as the degrees of freedom increase, the peak of the curve moves to the right; that is, the skewness decreases.

Table VII in Appendix C lists the values of *F* for the *F* distribution. To read Table VII, we need to know three quantities: the degrees of freedom for the numerator, the degrees of freedom for the denominator, and an area in the right tail of an *F* distribution curve. Note that the *F* distribution table (Table VII) is read only for an area in the right tail of the *F* distribution curve. Also note that Table VII has four parts. These four parts give the *F* values for areas of .01, .025,

[1]The *F* distribution is named after Sir Ronald Fisher.

Figure 12.1 Three F distribution curves.

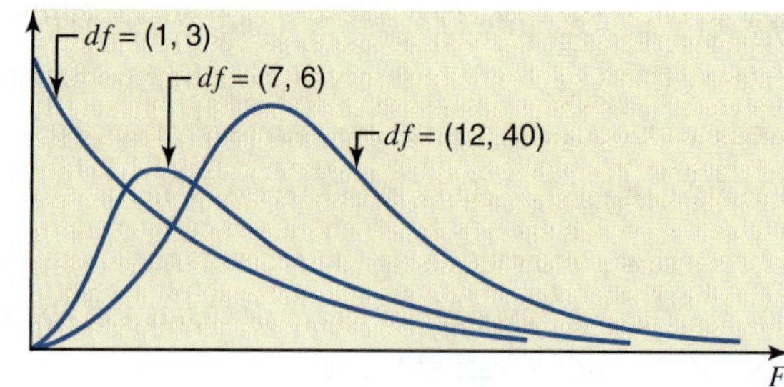

.05, and .10, respectively, in the right tail of the F distribution curve. We can make the F distribution table for other values in the right tail. Example 12–1 illustrates how to read Table VII.

EXAMPLE 12–1

Reading the F distribution table.

Find the F value for 8 degrees of freedom for the numerator, 14 degrees of freedom for the denominator, and .05 area in the right tail of the F distribution curve.

Solution To find the required value of F, we use the portion of Table VII of Appendix C that corresponds to .05 area in the right tail of the F distribution curve. The relevant portion of that table is shown here as Table 12.1. To find the required F value, we locate 8 in the row

Table 12.1

Degrees of Freedom for the Denominator	Degrees of Freedom for the Numerator					
	1	**2**	...	**8**	...	**100**
1	161.5	199.5	...	238.9	...	253.0
2	18.51	19.00	...	19.37	...	19.49
.	...	...	...	...	...	...
.	...	...	...	...	...	...
14	4.60	3.74	...	**2.70**	...	2.19
.	...	...	...	...	...	...
.	...	...	...	...	...	...
100	3.94	3.09	...	2.03	...	1.39

The F value for 8 df for the numerator, 14 df for the denominator, and .05 area in the right tail

for degrees of freedom for the numerator (at the top of Table VII) and 14 in the column for degrees of freedom for the denominator (the first column on the left side in Table VII). The entry where the column for 8 and the row for 14 intersect gives the required F value. This value of F is **2.70**, as shown in Table 12.1 and Figure 12.2. The F value taken from this table for a test of hypothesis is called the critical value of F.

Figure 12.2 The critical value of F for 8 df for the numerator, 14 df for the denominator, and .05 area in the right tail.

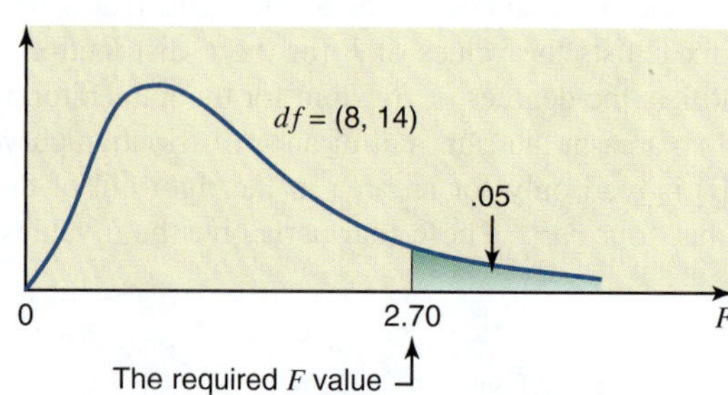

EXERCISES

CONCEPTS AND PROCEDURES

12.1 Describe the main characteristics of an F distribution.

12.2 Find the critical value of F for the following.
a. $df = (6, 12)$ and area in the right tail $= .05$
b. $df = (4, 18)$ and area in the right tail $= .05$
c. $df = (12, 6)$ and area in the right tail $= .05$

12.3 Find the critical value of F for the following.
a. $df = (6, 12)$ and area in the right tail $= .025$
b. $df = (4, 18)$ and area in the right tail $= .025$
c. $df = (12, 6)$ and area in the right tail $= .025$

12.4 Determine the critical value of F for the following.
a. $df = (7, 11)$ and area in the right tail $= .01$
b. $df = (5, 12)$ and area in the right tail $= .01$
c. $df = (15, 6)$ and area in the right tail $= .01$

12.5 Determine the critical value of F for the following.
a. $df = (4, 14)$ and area in the right tail $= .10$
b. $df = (9, 11)$ and area in the right tail $= .10$
c. $df = (11, 5)$ and area in the right tail $= .10$

12.6 Find the critical value of F for an F distribution with $df = (3, 12)$ and
a. area in the right tail $= .05$ b. area in the right tail $= .10$

12.7 Find the critical value of F for an F distribution with $df = (11, 5)$ and
a. area in the right tail $= .01$ b. area in the right tail $= .025$

12.8 Find the critical value of F for an F distribution with .025 area in the right tail and
a. $df = (4, 11)$ b. $df = (15, 3)$

12.9 Find the critical value of F for an F distribution with .01 area in the right tail and
a. $df = (10, 10)$ b. $df = (9, 25)$

12.2 One-Way Analysis of Variance

As mentioned in the beginning of this chapter, the analysis of variance procedure is used to test the null hypothesis that the means of three or more populations are the same against the alternative hypothesis that not all population means are the same. The analysis of variance procedure can be used to compare two population means. However, the procedures learned in Chapter 10 are more efficient for performing tests of hypotheses about the difference between two population means; the analysis of variance procedure is used to compare three or more population means.

Reconsider the example of teachers at a school who have devised three different methods to teach arithmetic. They want to find out if these three methods produce different mean scores. Let μ_1, μ_2, and μ_3 be the mean scores of all students who are taught by Methods I, II, and III, respectively. To test if the three teaching methods produce different means, we test the null hypothesis

$$H_0: \mu_1 = \mu_2 = \mu_3 \quad \text{(All three population means are equal)}$$

against the alternative hypothesis

$$H_1: \text{Not all three population means are equal}$$

One method to test such a hypothesis is to test the three hypotheses H_0: $\mu_1 = \mu_2$, H_0: $\mu_1 = \mu_3$, and H_0: $\mu_2 = \mu_3$ separately using the procedure discussed in Chapter 10. Besides being time consuming, such a procedure has other disadvantages. First, if we reject even one of these three hypotheses, then we must reject the null hypothesis H_0: $\mu_1 = \mu_2 = \mu_3$. Second, combining the Type I error probabilities for the three tests (one for each test) will give a very large Type I error probability for the test H_0: $\mu_1 = \mu_2 = \mu_3$. Hence, we should prefer a procedure that can test the equality of three means in one test. The **ANOVA**, short for **analysis of variance**, provides such a procedure. It is used to compare three or more population means in a single test.

Definition

ANOVA *ANOVA* is a procedure used to test the null hypothesis that the means of three or more populations are equal.

This section discusses the **one-way ANOVA** procedure to make tests comparing the means of several populations. By using a one-way ANOVA test, we analyze only one factor or variable. For instance, in the example of testing for the equality of mean arithmetic scores of students taught by each of the three different methods, we are considering only one factor, which is the effect of different teaching methods on the scores of students. Sometimes we may analyze the effects of two factors. For example, if different teachers teach arithmetic using these three methods, we can analyze the effects of teachers and teaching methods on the scores of students. This is done by using a two-way ANOVA. The procedure under discussion in this chapter is called the analysis of variance because the test is based on the analysis of variation in the data obtained from different samples. The application of one-way ANOVA requires that the following assumptions hold true.

Assumptions of One-Way ANOVA The following assumptions must hold true to use *one-way ANOVA*.

1. The populations from which the samples are drawn are (approximately) normally distributed.
2. The populations from which the samples are drawn have the same variance (or standard deviation).
3. The samples drawn from different populations are random and independent.

For instance, in the example about three methods of teaching arithmetic, we first assume that the scores of all students taught by each method are (approximately) normally distributed. Second, the means of the distributions of scores for the three teaching methods may or may not be the same, but all three distributions have the same variance σ^2. Third, when we take samples to make an ANOVA test, these samples are drawn independently and randomly from three different populations.

The ANOVA test is applied by calculating two estimates of the variance, σ^2, of population distributions: the **variance between samples** and the **variance within samples**. The variance between samples is also called the **mean square between samples** or **MSB**. The variance within samples is also called the **mean square within samples** or **MSW**.

The variance between samples, MSB, gives an estimate of σ^2 based on the variation among the means of samples taken from different populations. For the example of three teaching methods, MSB will be based on the values of the mean scores of three samples of students taught by three different methods. If the means of all populations under consideration are equal, the means of the respective samples will still be different but the variation among them is expected to be small and, consequently, the value of MSB is expected to be small. However, if the means of populations under consideration are not all equal, the variation among the means of respective samples is expected to be large and, consequently, the value of MSB is expected to be large.

The variance within samples, MSW, gives an estimate of σ^2 based on the variation within the data of different samples. For the example of three teaching methods, MSW will be based on the scores of individual students included in the three samples taken from three populations. The concept of MSW is similar to the concept of the pooled standard deviation, s_p, for two samples discussed in Section 10.2 of Chapter 10.

The one-way ANOVA test is always right-tailed with the rejection region in the right tail of the F distribution curve. The hypothesis-testing procedure using ANOVA involves the same five steps that were used in earlier chapters. The next subsection explains how to calculate the value of the test statistic F for an ANOVA test.

12.2.1 Calculating the Value of the Test Statistic

The value of the test statistic F for a test of hypothesis using ANOVA is given by the ratio of two variances, the variance between samples (MSB) and the variance within samples (MSW).

Test Statistic F for a One-Way ANOVA Test The value of the *test statistic F* for an ANOVA test is calculated as

$$F = \frac{\text{Variance between samples}}{\text{Variance within samples}} \quad \text{or} \quad \frac{\text{MSB}}{\text{MSW}}$$

The calculation of MSB and MSW is explained in Example 12–2.

Example 12–2 describes the calculation of MSB, MSW, and the value of the test statistic F. Since the basic formulas are laborious to use, they are not presented here. We have used only the short-cut formulas to make calculations in this chapter.

■ EXAMPLE 12–2

Calculating the value of the test statistic F.

Fifteen fourth-grade students were randomly assigned to three groups to experiment with three different methods of teaching arithmetic. At the end of the semester, the same test was given to all 15 students. The table gives the scores of students in the three groups.

Method I	Method II	Method III
48	55	84
73	85	68
51	70	95
65	69	74
87	90	67

Calculate the value of the test statistic F. Assume that all the required assumptions mentioned in Section 12.2 hold true.

Solution In ANOVA terminology, the three methods used to teach arithmetic are called **treatments**. The table contains data on the scores of fourth-graders included in the three samples. Each sample of students is taught by a different method. Let

x = the score of a student

k = the number of different samples (or treatments)

n_i = the size of sample i

T_i = the sum of the values in sample i

n = the number of values in all samples $= n_1 + n_2 + n_3 + \cdots$

Σx = the sum of the values in all samples $= T_1 + T_2 + T_3 + \cdots$

Σx^2 = the sum of the squares of the values in all samples

To calculate MSB and MSW, we first compute the **between-samples sum of squares** denoted by **SSB** and the **within-samples sum of squares** denoted by **SSW**. The sum of SSB and SSW is called the **total sum of squares** and is denoted by **SST**; that is,

$$\text{SST} = \text{SSB} + \text{SSW}$$

The values of SSB and SSW are calculated using the following formulas.

Between- and Within-Samples Sums of Squares The *between-samples sum of squares*, denoted by SSB, is calculated as

$$\text{SSB} = \left(\frac{T_1^2}{n_1} + \frac{T_2^2}{n_2} + \frac{T_3^2}{n_3} + \cdots\right) - \frac{(\Sigma x)^2}{n}$$

The *within-samples sum of squares*, denoted by SSW, is calculated as

$$\text{SSW} = \Sigma x^2 - \left(\frac{T_1^2}{n_1} + \frac{T_2^2}{n_2} + \frac{T_3^2}{n_3} + \cdots\right)$$

Table 12.2 lists the scores of 15 students who were taught arithmetic by each of the three different methods; the values of T_1, T_2, and T_3; and the values of n_1, n_2, and n_3.

Table 12.2

Method I	Method II	Method III
48	55	84
73	85	68
51	70	95
65	69	74
87	90	67
$T_1 = 324$	$T_2 = 369$	$T_3 = 388$
$n_1 = 5$	$n_2 = 5$	$n_3 = 5$

In Table 12.2, T_1 is obtained by adding the five scores of the first sample. Thus, $T_1 = 48 + 73 + 51 + 65 + 87 = 324$. Similarly, the sums of the values in the second and third samples give $T_2 = 369$ and $T_3 = 388$, respectively. Because there are five observations in each sample, $n_1 = n_2 = n_3 = 5$. The values of Σx and n are

$$\Sigma x = T_1 + T_2 + T_3 = 324 + 369 + 388 = 1081$$

$$n = n_1 + n_2 + n_3 = 5 + 5 + 5 = 15$$

To calculate Σx^2, we square all the scores included in all three samples and then add them. Thus,

$$\Sigma x^2 = (48)^2 + (73)^2 + (51)^2 + (65)^2 + (87)^2 + (55)^2 + (85)^2 + (70)^2 + (69)^2 + (90)^2 + (84)^2 + (68)^2 + (95)^2 + (74)^2 + (67)^2 = 80{,}709$$

Substituting all the values in the formulas for SSB and SSW, we obtain the following values of SSB and SSW:

$$\text{SSB} = \left(\frac{(324)^2}{5} + \frac{(369)^2}{5} + \frac{(388)^2}{5}\right) - \frac{(1081)^2}{15} = 432.1333$$

$$\text{SSW} = 80{,}709 - \left(\frac{(324)^2}{5} + \frac{(369)^2}{5} + \frac{(388)^2}{5}\right) = 2372.8000$$

The value of SST is obtained by adding the values of SSB and SSW. Thus,

$$\text{SST} = 432.1333 + 2372.8000 = 2804.9333$$

The variance between samples (MSB) and the variance within samples (MSW) are calculated using the following formulas.

Calculating the Values of MSB and MSW MSB and MSW are calculated as

$$\text{MSB} = \frac{\text{SSB}}{k-1} \quad \text{and} \quad \text{MSW} = \frac{\text{SSW}}{n-k}$$

where $k - 1$ and $n - k$ are, respectively, the *df* for the numerator and the *df* for the denominator for the F distribution. Remember, k is the number of different samples.

Consequently, the variance between samples is

$$\text{MSB} = \frac{\text{SSB}}{k-1} = \frac{432.1333}{3-1} = 216.0667$$

The variance within samples is

$$\text{MSW} = \frac{\text{SSW}}{n-k} = \frac{2372.8000}{15-3} = 197.7333$$

The value of the test statistic F is given by the ratio of MSB and MSW. Therefore,

$$F = \frac{\text{MSB}}{\text{MSW}} = \frac{216.0667}{197.7333} = \mathbf{1.09}$$

For convenience, all these calculations are often recorded in a table called the *ANOVA table*. Table 12.3 gives the general form of an ANOVA table.

Table 12.3 ANOVA Table

Source of Variation	Degrees of Freedom	Sum of Squares	Mean Square	Value of the Test Statistic
Between	$k - 1$	SSB	MSB	$F = \frac{\text{MSB}}{\text{MSW}}$
Within	$n - k$	SSW	MSW	
Total	$n - 1$	SST		

Substituting the values of the various quantities into Table 12.3, we write the ANOVA table for our example as Table 12.4.

Table 12.4 ANOVA Table for Example 12–2

Source of Variation	Degrees of Freedom	Sum of Squares	Mean Square	Value of the Test Statistic
Between	2	432.1333	216.0667	$F = \frac{216.0667}{197.7333} = 1.09$
Within	12	2372.8000	197.7333	
Total	14	2804.9333		

■

12.2.2 One-Way ANOVA Test

Now suppose we want to test the null hypothesis that the mean scores are equal for all three groups of fourth-graders taught by three different methods of Example 12–2 against the alternative hypothesis that the mean scores of all three groups are not equal. Note that in a one-way ANOVA test, the null hypothesis is that the means for all populations are equal. The alternative hypothesis is that not all population means are equal. In other words, the alternative hypothesis states that at least one of the population means is different from the others. Example 12–3 demonstrates how we use the one-way ANOVA procedure to make such a test.

■ EXAMPLE 12–3

Performing a one-way ANOVA test: all samples the same size.

Reconsider Example 12–2 about the scores of 15 fourth-grade students who were randomly assigned to three groups in order to experiment with three different methods of teaching arithmetic. At the 1% significance level, can we reject the null hypothesis that the mean arithmetic score of all fourth-grade students taught by each of these three methods is the same? Assume that all the assumptions required to apply the one-way ANOVA procedure hold true.

Solution To make a test about the equality of the means of three populations, we follow our standard procedure with five steps.

Step 1. *State the null and alternative hypotheses.*

Let μ_1, μ_2, and μ_3 be the mean arithmetic scores of all fourth-grade students who are taught, respectively, by Methods I, II, and III. The null and alternative hypotheses are

H_0: $\mu_1 = \mu_2 = \mu_3$ (The mean scores of the three groups are equal.)

H_1: Not all three means are equal

Note that the alternative hypothesis states that at least one population mean is different from the other two.

Step 2. *Select the distribution to use.*

Because we are comparing the means for three normally distributed populations, we use the F distribution to make this test.

Step 3. *Determine the rejection and nonrejection regions.*

The significance level is .01. Because a one-way ANOVA test is always right-tailed, the area in the right tail of the F distribution curve is .01, which is the rejection region in Figure 12.3.

Next we need to know the degrees of freedom for the numerator and the denominator. In our example, the students were assigned to three different methods. As mentioned earlier, these methods are called treatments. The number of treatments is denoted by k. The total number of observations in all samples taken together is denoted by n. Then, the number of degrees of freedom for the numerator is equal to $k - 1$ and the number of degrees of freedom for the denominator is equal to $n - k$. In our example, there are 3 treatments (methods of teaching) and 15 total observations (total number of students) in all 3 samples. Thus,

$$\text{Degrees of freedom for the numerator} = k - 1 = 3 - 1 = 2$$

$$\text{Degrees of freedom for the denominator} = n - k = 15 - 3 = 12$$

From Table VII of Appendix C, we find the critical value of F for 2 df for the numerator, 12 df for the denominator, and .01 area in the right tail of the F distribution curve. This value is shown in Figure 12.3. The required value of F is 6.93.

Thus, we will fail to reject H_0 if the calculated value of the test statistic F is less than 6.93 and we will reject H_0 if it is greater than 6.93.

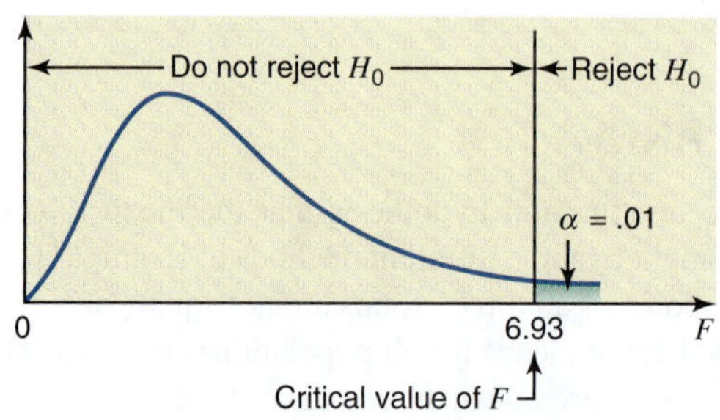

Figure 12.3 Critical value of F for $df = (2, 12)$ and $\alpha = .01$.

Step 4. *Calculate the value of the test statistic.*

We computed the value of the test statistic F for these data in Example 12–2. This value is

$$F = 1.09$$

Step 5. *Make a decision.*

Because the value of the test statistic $F = 1.09$ is less than the critical value of $F = 6.93$, it falls in the nonrejection region. Hence, we fail to reject the null hypothesis and conclude that the means of the three populations are equal. In other words, the three different methods of teaching arithmetic do not seem to affect the mean scores of students. The difference in the three mean scores in the case of our three samples occurred only because of sampling error. ■

In Example 12–3, the sample sizes were the same for all treatments. Example 12–4 describes a case in which the sample sizes are not the same for all treatments.

■ EXAMPLE 12–4

Performing a one-way ANOVA test: all samples not the same size.

From time to time, unknown to its employees, the research department at Post Bank observes various employees for their work productivity. Recently this department wanted to check whether the four tellers at a branch of this bank serve, on average, the same number of customers per hour. The research manager observed each of the four tellers for a certain number of hours. The following table gives the number of customers served by the four tellers during each of the observed hours.

Teller A	Teller B	Teller C	Teller D
19	14	11	24
21	16	14	19
26	14	21	21
24	13	13	26
18	17	16	20
	13	18	

At the 5% significance level, test the null hypothesis that the mean number of customers served per hour by each of these four tellers is the same. Assume that all the assumptions required to apply the one-way ANOVA procedure hold true.

Solution To make a test about the equality of means of four populations, we follow our standard procedure with five steps.

Step 1. *State the null and alternative hypotheses.*

Let μ_1, μ_2, μ_3, and μ_4 be the mean number of customers served per hour by tellers A, B, C, and D, respectively. The null and alternative hypotheses are

H_0: $\mu_1 = \mu_2 = \mu_3 = \mu_4$ (The mean number of customers served per hour by each of the four tellers is the same.)

H_1: Not all four population means are equal

Step 2. *Select the distribution to use.*

Because we are testing for the equality of four means for four normally distributed populations, we use the F distribution to make the test.

Step 3. *Determine the rejection and nonrejection regions.*

The significance level is .05, which means the area in the right tail of the F distribution curve is .05. In this example, there are 4 treatments (tellers) and 22 total observations in all four samples. Thus,

$$\text{Degrees of freedom for the numerator} = k - 1 = 4 - 1 = 3$$

$$\text{Degrees of freedom for the denominator} = n - k = 22 - 4 = 18$$

The critical value of F from Table VII for 3 df for the numerator, 18 df for the denominator, and .05 area in the right tail of the F distribution curve is 3.16. This value is shown in Figure 12.4.

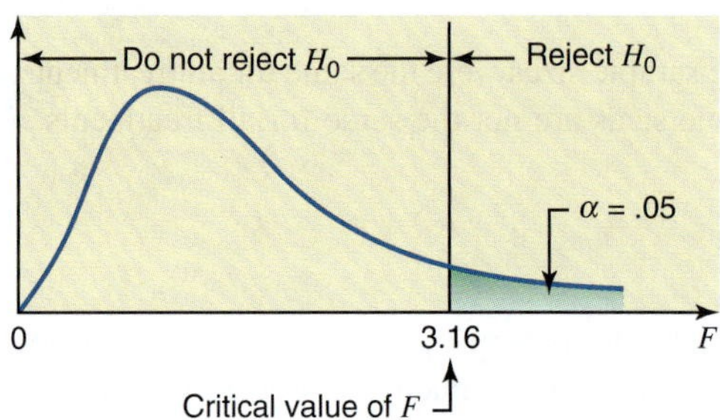

Figure 12.4 Critical value of F for $df = (3, 18)$ and $\alpha = .05$.

Step 4. *Calculate the value of the test statistic.*

First we calculate SSB and SSW. Table 12.5 lists the numbers of customers served by the four tellers during the selected hours; the values of T_1, T_2, T_3, and T_4; and the values of n_1, n_2, n_3, and n_4.

Table 12.5

Teller A	Teller B	Teller C	Teller D
19	14	11	24
21	16	14	19
26	14	21	21
24	13	13	26
18	17	16	20
	13	18	
$T_1 = 108$	$T_2 = 87$	$T_3 = 93$	$T_4 = 110$
$n_1 = 5$	$n_2 = 6$	$n_3 = 6$	$n_4 = 5$

The values of Σx and n are

$$\Sigma x = T_1 + T_2 + T_3 + T_4 = 108 + 87 + 93 + 110 = 398$$

$$n = n_1 + n_2 + n_3 + n_4 = 5 + 6 + 6 + 5 = 22$$

The value of Σx^2 is calculated as follows:

$$\begin{aligned}\Sigma x^2 &= (19)^2 + (21)^2 + (26)^2 + (24)^2 + (18)^2 + (14)^2 + (16)^2 + (14)^2 \\ &\quad + (13)^2 + (17)^2 + (13)^2 + (11)^2 + (14)^2 + (21)^2 + (13)^2 \\ &\quad + (16)^2 + (18)^2 + (24)^2 + (19)^2 + (21)^2 + (26)^2 + (20)^2 \\ &= 7614\end{aligned}$$

Substituting all the values in the formulas for SSB and SSW, we obtain the following values of SSB and SSW:

$$\text{SSB} = \left(\frac{T_1^2}{n_1} + \frac{T_2^2}{n_2} + \frac{T_3^2}{n_3} + \frac{T_4^2}{n_4}\right) - \frac{(\Sigma x)^2}{n}$$

$$= \left(\frac{(108)^2}{5} + \frac{(87)^2}{6} + \frac{(93)^2}{6} + \frac{(110)^2}{5}\right) - \frac{(398)^2}{22} = 255.6182$$

$$\text{SSW} = \Sigma x^2 - \left(\frac{T_1^2}{n_1} + \frac{T_2^2}{n_2} + \frac{T_3^2}{n_3} + \frac{T_4^2}{n_4}\right)$$

$$= 7614 - \left(\frac{(108)^2}{5} + \frac{(87)^2}{6} + \frac{(93)^2}{6} + \frac{(110)^2}{5}\right) = 158.2000$$

Hence, the variance between samples MSB and the variance within samples MSW are

$$\text{MSB} = \frac{\text{SSB}}{k-1} = \frac{255.6182}{4-1} = 85.2061$$

$$\text{MSW} = \frac{\text{SSW}}{n-k} = \frac{158.2000}{22-4} = 8.7889$$

The value of the test statistic F is given by the ratio of MSB and MSW, which is

$$F = \frac{\text{MSB}}{\text{MSW}} = \frac{85.2061}{8.7889} = 9.69$$

Writing the values of the various quantities in the ANOVA table, we obtain Table 12.6.

Table 12.6 ANOVA Table for Example 12–4

Source of Variation	Degrees of Freedom	Sum of Squares	Mean Square	Value of the Test Statistic
Between	3	255.6182	85.2061	$F = \frac{85.2061}{8.7889} = 9.69$
Within	18	158.2000	8.7889	
Total	21	413.8182		

Step 5. *Make a decision.*

Because the value of the test statistic $F = 9.69$ is greater than the critical value of $F = 3.16$, it falls in the rejection region. Consequently, we reject the null hypothesis and conclude that the mean number of customers served per hour by each of the four tellers is not the same. In other words, at least one of the four means is different from the other three. ■

EXERCISES

CONCEPTS AND PROCEDURES

12.10 Briefly explain when a one-way ANOVA procedure is used to make a test of hypothesis.

12.11 Describe the assumptions that must hold true to apply the one-way analysis of variance procedure to test hypotheses.

12.12 Consider the following data obtained for two samples selected at random from two populations that are independent and normally distributed with equal variances.

Sample I	Sample II
32	27
26	35
31	33
20	40
27	38
34	31

a. Calculate the means and standard deviations for these samples using the formulas from Chapter 3.
b. Using the procedure learned in Section 10.2 of Chapter 10, test at the 1% significance level whether the means of the populations from which these samples are drawn are equal.
c. Using the one-way ANOVA procedure, test at the 1% significance level whether the means of the populations from which these samples are drawn are equal.
d. Are the conclusions reached in parts b and c the same?

12.13 Consider the following data obtained for two samples selected at random from two populations that are independent and normally distributed with equal variances.

Sample I	Sample II
14	11
21	8
11	12
9	18
13	15
20	7
17	6

a. Calculate the means and standard deviations for these samples using the formulas from Chapter 3.
b. Using the procedure learned in Section 10.2 of Chapter 10, test at the 5% significance level whether the means of the populations from which these samples are drawn are equal.
c. Using the one-way ANOVA procedure, test at the 5% significance level whether the means of the populations from which these samples are drawn are equal.
d. Are the conclusions reached in parts b and c the same?

12.14 The following ANOVA table, based on information obtained for three samples selected from three independent populations that are normally distributed with equal variances, has a few missing values.

Source of Variation	Degrees of Freedom	Sum of Squares	Mean Square	Value of the Test Statistic
Between	2		19.2813	
Within		89.3677		$F = \text{——} =$
Total	12			

a. Find the missing values and complete the ANOVA table.
b. Using $\alpha = .01$, what is your conclusion for the test with the null hypothesis that the means of the three populations are all equal against the alternative hypothesis that the means of the three populations are not all equal?

12.15 The following ANOVA table, based on information obtained for four samples selected from four independent populations that are normally distributed with equal variances, has a few missing values.

Source of Variation	Degrees of Freedom	Sum of Squares	Mean Square	Value of the Test Statistic
Between				
Within	15		9.2154	$F = \text{———} = 4.07$
Total	18			

a. Find the missing values and complete the ANOVA table.
b. Using $\alpha = .05$, what is your conclusion for the test with the null hypothesis that the means of the four populations are all equal against the alternative hypothesis that the means of the four populations are not all equal?

■ APPLICATIONS

For the following exercises assume that all the assumptions required to apply the one-way ANOVA procedure hold true.

12.16 A dietitian wanted to test three different diets to find out if the mean weight loss for each of these diets is the same. She randomly selected 21 overweight persons, randomly divided them into three groups, and put each group on one of the three diets. The following table records the weights (in pounds) lost by these persons after being on these diets for two months.

Diet I	Diet II	Diet III
15	11	9
8	16	17
17	9	11
7	16	8
26	24	15
12	20	6
8	19	14

a. We are to test if the mean weight lost by all persons on each of the three diets is the same. Write the null and alternative hypotheses.
b. Show the rejection and nonrejection regions on the F distribution curve for $\alpha = .025$.
c. Calculate SSB, SSW, and SST.
d. What are the degrees of freedom for the numerator and the denominator?
e. Calculate the between-samples and within-samples variances.
f. What is the critical value of F for $\alpha = .025$?
g. What is the calculated value of the test statistic F?
h. Write the ANOVA table for this exercise.
i. Will you reject the null hypothesis stated in part a at a significance level of 2.5%?

12.17 The following table gives the numbers of classes missed during one semester by 25 randomly selected college students drawn from three different age groups.

Below 25	25 to 30	31 and Above
19	9	5
13	6	8
25	11	2
10	14	3
19	5	10
4	9	9
15	3	
10	11	
16	18	
9		

a. We are to test if the mean number of classes missed during the semester by all students in each of these three age groups is the same. Write the null and alternative hypotheses.
b. What are the degrees of freedom for the numerator and the denominator?
c. Calculate SSB, SSW, and SST.
d. Show the rejection and nonrejection regions on the F distribution curve for $\alpha = .01$.
e. Calculate the between-samples and within-samples variances.
f. What is the critical value of F for $\alpha = .01$?
g. What is the calculated value of the test statistic F?
h. Write the ANOVA table for this exercise.
i. Will you reject the null hypothesis stated in part a at a significance level of 1%?

12.18 A consumer agency wanted to investigate if four insurance companies differed with regard to the premiums they charge for auto insurance. The agency randomly selected a few auto drivers who were insured by each of these four companies and had similar driving records, autos, and insurance policies. The following table gives the premiums paid per month by these drivers insured with these four insurance companies.

Company A	Company B	Company C	Company D
75	59	65	76
83	75	70	60
68	100	97	52
52		90	58
		73	

Using the 1% significance level, test the null hypothesis that the mean auto insurance premium paid per month by all drivers insured by each of these four companies is the same.

12.19 A university employment office wants to compare the time taken by graduates with three different majors to find their first job after graduation. The following table lists the time (in days) taken to find their first full-time job after graduation for a random sample of eight business majors, seven computer science majors, and six engineering majors who graduated in May 2005.

Business	Computer Science	Engineering
36	56	26
62	13	51
35	24	63
80	28	46
48	44	78
27	47	34
76	20	
44		

At the 5% significance level, can you conclude that the mean time taken to find their first job for all 2005 graduates in these fields is the same?

12.20 A consumer agency wanted to find out if the mean time it takes for each of three brands of medicines to provide relief from a headache is the same. The first drug was administered to six randomly selected patients, the second to four randomly selected patients, and the third to five randomly selected patients. The following table gives the time (in minutes) taken by each patient to get relief from a headache after taking the medicine.

Drug I	Drug II	Drug III
25	15	44
38	21	39
42	19	54
65	25	58
47		73
52		

At the 2.5% significance level, will you conclude that the mean time taken to provide relief from a headache is the same for each of the three drugs?

12.21 A large company buys thousands of lightbulbs every year. The company is currently considering four brands of lightbulbs to choose from. Before the company decides which lightbulbs to buy, it wants to investigate if the mean lifetimes of the four types of lightbulbs are the same. The company's research department randomly selected a few bulbs of each type and tested them. The following table lists the number of hours (in thousands) that each of the bulbs in each brand lasted before being burned out.

Brand I	Brand II	Brand III	Brand IV
23	19	23	26
24	23	27	24
19	18	25	21
26	24	26	29
22	20	23	28
23	22	21	24
25	19	27	28

At the 2.5% significance level, test the null hypothesis that the mean lifetime of bulbs for each of these four brands is the same.

USES AND MISUSES... DON'T BE LATE

Imagine that working at your company requires that staff travel frequently. You want to determine if the on-time performance of any one airline is sufficiently different from the remaining airlines to warrant a preferred status with your company. The local airport Web site publishes the scheduled and actual departure and arrival times for the four airlines that service it. You decide to perform an ANOVA test on the mean delay times for all airline carriers at the airport. The null hypothesis here is that the mean delay times for Airlines A, B, C, and D are all the same. The results of the ANOVA test tell you to accept the null hypothesis: All airline carriers have the same mean departure and arrival delay times, so that adopting a preferred status based on the on-time performance is not warranted.

When your boss tells you to redo your analysis, you should not be surprised. The choice to study flights only at the local airport was a good one, because your company should be concerned about the performance of an airline at the most convenient airport. A regional airport will have a much different on-time performance profile than a large hub airport. But by mixing both arrival and departure data, you violated the assumption that the populations are normally distributed. For arrival data, this assumption could be valid: The influence of high-altitude winds, local weather, and the fact that the arrival time is an estimate in the first place result in a distribution of arrival times around the predicted arrival times. However, departure delays are not normally distributed. Because a flight does not leave before its departure time but can leave after, departure delays are skewed to the right. As the statistical methods become more sophisticated, so do the assumptions regarding the characteristics of the data. Careful attention to these assumptions is required.

Glossary

Analysis of variance (ANOVA) A statistical technique used to test whether the means of three or more populations are equal.

***F* distribution** A continuous distribution that has two parameters: *df* for the numerator and *df* for the denominator.

Mean square between samples or **MSB** A measure of the variation among the means of samples taken from different populations.

Mean square within samples or **MSW** A measure of the variation within the data of all samples taken from different populations.

One-way ANOVA The analysis of variance technique that analyzes one variable only.

SSB The sum of squares between samples. Also called the sum of squares of the factor or treatment.

SST The total sum of squares given by the sum of SSB and SSW.

SSW The sum of squares within samples. Also called the sum of squares of errors.

Supplementary Exercises

For the following exercises, assume that all the assumptions required to apply the one-way ANOVA procedure hold true.

12.22 The following table lists the numbers of violent crimes reported to police on randomly selected days for this year. The data are taken from three large cities of about the same size.

City A	City B	City C
5	2	8
9	4	12
12	1	10
3	13	3
9	7	9
7	6	14
13		

Using the 5% significance level, test the null hypothesis that the mean number of violent crimes reported per day is the same for each of these three cities.

12.23 A consumer agency wants to check if the mean lives of four brands of auto batteries, which sell for nearly the same price, are the same. The agency randomly selected a few batteries of each brand and tested them. The following table gives the lives of these batteries in thousands of hours.

Brand A	Brand B	Brand C	Brand D
74	53	57	56
78	51	71	51
51	47	81	49
56	59	77	43
65		68	

a. At the 5% significance level, will you reject the null hypothesis that the mean lifetime of each of these four brands of batteries is the same?

b. What is the Type I error in this case and what is the probability of committing such an error? Explain.

12.24 A travel magazine took a random sample of students from a midwestern university and asked whether they went to California, Texas, or Florida for 2005 spring break. The students who went somewhere else or did not go anywhere were excluded from the survey. The following table shows the amount of money spent on this trip by a sample of six students for each of these three destinations.

California	Texas	Florida
$ 850	$775	$668
904	808	810
1045	810	844
882	912	787
910	855	705
946	680	908

a. At the 1% level of significance, can you conclude that the mean amount spent by all such students who went to each of these three destinations for 2005 spring break is the same?

b. What is the Type I error in this case and what is the probability of committing such an error? Explain.

12.25 A farmer wants to test three brands of weight-gain diets for chickens to determine if the mean weight gain for each of these brands is the same. He selected 15 chickens and randomly put each of them on one of these three brands of diet. The following table lists the weights (in pounds) gained by these chickens after a period of one month.

Brand A	Brand B	Brand C
.8	.6	1.2
1.3	1.3	.8
1.7	.6	.7
.9	.4	1.5
.6	.7	.9

a. At the 1% significance level, can you conclude that the mean weight gain for all chickens is the same for each of these three diets?

b. If you did not reject the null hypothesis in part a, explain the Type II error that you may have made in this case. Note that you cannot calculate the probability of committing a Type II error without additional information.

12.26 The following table lists the prices of certain randomly selected college textbooks in statistics, psychology, economics, and business.

Statistics	Psychology	Economics	Business
102	88	75	84
81	101	99	115
93	92	80	94
78	86	105	103
112		91	87

a. Using the 5% significance level, test the null hypothesis that the mean prices of college textbooks in statistics, psychology, economics, and business are all equal.

b. If you did not reject the null hypothesis in part a, explain the Type II error that you may have made in this case. Note that you cannot calculate the probability of committing a Type II error without additional information.

12.27 A resort area has three seafood restaurants, which employ students during the summer season. The local chamber of commerce took a random sample of five servers from each restaurant and recorded the tips they received on a recent Friday night. The results of the survey are shown in the table below. Assume that the Friday night for which the data were collected is typical of all Friday nights of the summer season.

Barzini's	Hwang's	Jack's
$ 97	$ 67	$ 93
114	85	102
105	92	98
85	78	80
120	90	91

a. Would a student seeking a server's job at one of these three restaurants conclude that the mean tips on a Friday night are the same for all three restaurants? Use the 5% level of significance.

b. What will your decision be in part a if the probability of making a Type I error is zero? Explain.

12.28 A student who has a 9 A.M. class on Monday, Wednesday, and Friday mornings wants to know if the mean time taken by students to find parking spaces just before 9 A.M. is the same for each of these three days of the week. He randomly selects five weeks and records the time taken to find a parking space on Monday, Wednesday, and Friday of each of these five weeks. These times (in minutes) are given in the following table. Assume that this student is representative of all students who need to find a parking space just before 9 A.M. on these three days.

Monday	Wednesday	Friday
6	9	3
12	12	2
15	5	10
14	14	7
10	13	5

At the 5% significance level, test the null hypothesis that the mean time taken to find a parking space just before 9 A.M. on Monday, Wednesday, and Friday is the same for all students.

Advanced Exercises

12.29 A billiards parlor in a small town is open just four days per week—Thursday through Sunday. Revenues vary considerably from day to day and week to week, so the owner is not sure whether some days of the week are more profitable than others. He takes random samples of five Thursdays, five Fridays, five Saturdays, and five Sundays from last year's records and lists the revenues for these 20 days. His bookkeeper finds the average revenue for each of the four samples, and then calculates Σx^2. The results are shown in the following table. The value of the Σx^2 came out to be 2,890,000.

Day	Mean Revenue	Sample Size
Thursday	$295	5
Friday	380	5
Saturday	405	5
Sunday	345	5

Assume that the revenues for each day of the week are normally distributed and that the standard deviations are equal for all four populations. At the 1% level of significance, can you conclude that the mean revenue is the same for each of the four days of the week?

12.30 Suppose that you are a reporter for a newspaper whose editor has asked you to compare the hourly wages of carpenters, plumbers, electricians, and masons in your city. Since many of these workers are not union members, the wages vary considerably among individuals in the same trade.

a. What data should you gather and how would you collect them? What statistics would you present in your article and how would you calculate them? Assume that your newspaper is not intended for technical readers.

b. Suppose that you must submit your findings to a technical journal that requires statistical analysis of your data. If you want to determine whether or not the mean hourly wages are the same for all four trades, briefly describe how you would analyze the data. Assume that hourly wages in each trade are normally distributed and that the four variances are equal.

12.31 The editor of an automotive magazine has asked you to compare the mean gas mileages of city driving for three makes of compact cars. The editor has made available to you one car of each of the three makes, three drivers, and a budget sufficient to buy gas and pay the drivers for approximately 500 miles of city driving for each car.

a. Explain how you would conduct an experiment and gather the data for a magazine article comparing the gas mileage.

b. Suppose you wish to test the null hypothesis that the mean gas mileages of city driving are the same for all three makes. Outline the procedure for using your data to conduct this test. Assume that the assumptions for applying analysis of variance are satisfied.

12.32 Do rock music CDs and country music CDs give the consumers the same amount of music listening time? A sample of 12 randomly selected single rock music CDs and a sample of 14 randomly selected single country music CDs have the following total lengths (in minutes).

Rock Music	Country Music
43.0	45.3
44.3	40.2
63.8	42.8
32.8	33.0
54.2	33.5
51.3	37.7
64.8	36.8
36.1	34.6
33.9	33.4
51.7	36.5
36.5	43.3
59.7	31.7
	44.0
	42.7

Assume that the two populations are normally distributed with equal standard deviations.

a. Compute the value of the test statistic t for testing the null hypothesis that the mean lengths of the rock and country music single CDs are the same against the alternative hypothesis that these mean lengths are not the same. Use the value of this t statistic to compute the (approximate) p-value.

b. Compute the value of the (one-way ANOVA) test statistic F for performing the test of equality of the mean lengths of the rock and country music single CDs and use it to find the (approximate) p-value.

c. How do the test statistics in parts a and b compare? How do the p-values computed in parts a and b compare? Do you think that this is a coincidence, or will this always happen?

12.33 Suppose you are performing a one-way ANOVA with only the information given in the following table.

Source of Variation	Degrees of Freedom	Sum of Squares
Between	4	200
Within	45	3547

a. Suppose the sample sizes for all groups are equal. How many groups are there? What are the group sample sizes?

b. The p-value for the test of the equality of the means of all populations is calculated to be .6406. Suppose you plan to increase the sample sizes for all groups but keep them all equal. However, when you do this, the sum of squares within samples and the sum of squares between samples (magically) remains the same. What are the smallest sample sizes for groups that would make this result significant at the 5% significance level?

Self-Review Test

1. The F distribution is
   a. continuous b. discrete c. neither
2. The F distribution is always
   a. symmetric b. skewed to the right c. skewed to the left
3. The units of the F distribution, denoted by F, are always
   a. nonpositive b. positive c. nonnegative
4. The one-way ANOVA test analyzes only one
   a. variable b. population c. sample

5. The one-way ANOVA test is always
   a. right-tailed   b. left-tailed   c. two-tailed

6. For a one-way ANOVA with k treatments and n observations in all samples taken together, the degrees of freedom for the numerator are
   a. $k - 1$ b. $n - k$ c. $n - 1$

7. For a one-way ANOVA with k treatments and n observations in all samples taken together, the degrees of freedom for the denominator are
   a. $k - 1$ b. $n - k$ c. $n - 1$

8. The ANOVA test can be applied to compare
   a. three or more population means
   b. more than four population means only
   c. more than three population means only

9. Briefly describe the assumptions that must hold true to apply the one-way ANOVA procedure as mentioned in this chapter.

10. A small college town has four pizza parlors that make deliveries. A student doing a research paper for her business management class decides to compare how promptly the four parlors deliver. On six randomly chosen nights, she orders a large pepperoni pizza from each establishment, then records the elapsed time until the pizza is delivered to her apartment. Assume that her apartment is approximately the same distance from the four pizza parlors. The following table shows the times (in minutes) for these deliveries. Assume that all the assumptions required to apply the one-way ANOVA procedure hold true.

Tony's	Luigi's	Angelo's	Kowalski's
20.0	22.1	22.3	23.9
24.0	27.0	26.0	24.1
18.3	20.2	24.0	25.8
22.0	32.0	30.1	29.0
20.8	26.0	28.0	25.0
19.0	24.8	25.8	24.2

   a. Using the 5% significance level, test the null hypothesis that the mean delivery time is the same for each of the four pizza parlors.
   b. Is it a Type I error or a Type II error that may have been committed in part a? Explain.

Mini-Projects

■ MINI-PROJECT 12–1

Are some days of the week busier than others on the New York Stock Exchange? Record the number of shares traded on the NYSE each day for a period of six weeks (round the number of shares to the nearest million). You will have five samples—first for shares traded on six Mondays, second for shares traded on six Tuesdays, and so forth. Assume that these days make up random samples for the respective populations. Further assume that each of the five populations from which these five samples are taken follows a normal distribution with the same variance. Test if the mean number of shares traded is the same for each of the five populations. Use a 1% significance level.

■ MINI-PROJECT 12–2

Pick at least 30 students at random and divide them randomly into three groups (A, B, and C) of approximately equal size. Take the students one by one, ring a bell, and 17 seconds later ring another bell. Then ask the students to estimate the elapsed time between the first and second rings. For group A, tell each student before the experiment starts that people tend to underestimate the elapsed time. Tell each student in group B that people tend to overestimate the time. Do not make any such statement to the students in group C. Record the estimates for all students, and then conduct an appropriate hypothesis test to see if the mean estimates of elapsed time are all equal for the populations represented by these groups. Use the 5% level of significance and assume that the three populations of elapsed time are normally distributed with equal standard deviations.

MINI-PROJECT 12–3

Obtain a Wiffle™ ball, a plastic golf ball with dimples and no holes, and a plastic golf ball with holes instead of dimples. Throw each ball 20 times and measure the distances. Perform a hypothesis test to determine if the average distance is the same for each type of ball. Use a significance level of 5%.

DECIDE FOR YOURSELF

Deciding About Heights of Basketball Players and Where They Come From

One-way ANOVA has given you a method/procedure to compare three or more means obtained from independent samples to make a decision about the corresponding population means. If you fail to reject the null hypothesis, you conclude that the assumption that the means of all populations under consideration are equal is a reasonable assumption. However, if you reject the null hypothesis, you conclude that at least two of the population means are different. Of course, there is still a glaring piece of information that you need in the latter case. If at least two means are different, which ones are different?

To determine which two means are different requires what is called a *pairwise comparison* procedure. This type of procedure compares each pair of means to determine whether or not they are equal. There are many such procedures available that can be used to make these pairwise comparisons. Some of these procedures are the Tukey HSD, Bonferroni, Scheffe, and Tamhane T2. To select the method that should be used depends on conditions such as whether or not the sample sizes are equal and whether or not using a pooled variance is reasonable.

There are a few informal (or *ad hoc*) methods that can be used to have an idea about what might happen with the *pairwise comparisons.* It is very important to note that the results from these procedures depend on how well the data meet the assumptions of an ANOVA, so these methods are not a substitute for a formal statistical process. These informal methods are simply graphical methods that can help you understand what is going on in a data set.

The accompanying figure gives a side-by-side plot of 95% confidence intervals for the mean heights of NBA players who entered the league from one of three sources—colleges, foreign countries, or directly from high schools. The horizontal lines in these intervals represent the ends of the intervals, while the circles identify the values of the sample means for the three groups. These confidence intervals are based on random samples of 15 players selected from each of the three groups. It is important to note that the condition $\frac{n}{N} \leq .05$ is not met for the high school and foreign players, but we will not address this issue at this time. Answer the following questions.

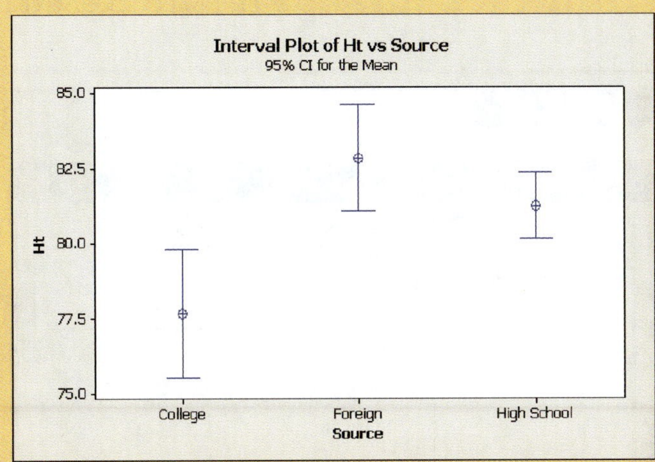

1. From the graph, we observe that players from one source seem to be significantly taller or shorter, on average, than players from the other two sources. Identify the source for which this is the case, the specific difference (taller or shorter), and what characteristic of the graph led you to make this conclusion.

2. The confidence interval for the NBA players who entered the league from high school is much narrower than the confidence intervals for the players who entered the league from the other two sources, yet the standard deviations of heights for all players in each of these three groups are relatively close. What does this tell you about the affect of random sampling on summary statistics?

3. What does your conclusion in question 1 imply about the types of players (centers, forwards, or guards) who come from these three sources?

TECHNOLOGY INSTRUCTION — Analysis of Variance

TI-84

```
ANOVA(L1,L2,L3)
```

Screen 12.1

1. To perform a one-way analysis of variance on a collection of samples, store the sample data in lists.
2. Select **STAT>TESTS>ANOVA(**.
3. Enter the names of the lists, separated by commas, and then type a right parenthesis. Press **Enter**. (See **Screen 12.1**.)
4. The results include the F statistic for performing the test, as well as the p-value. (See **Screen 12.2**.)

```
One-way ANOVA
 F=4.890046614
 p=.0279723575
 Factor
  df=2
  SS=1188.93333
↓ MS=594.466667
```

Screen 12.2

MINITAB

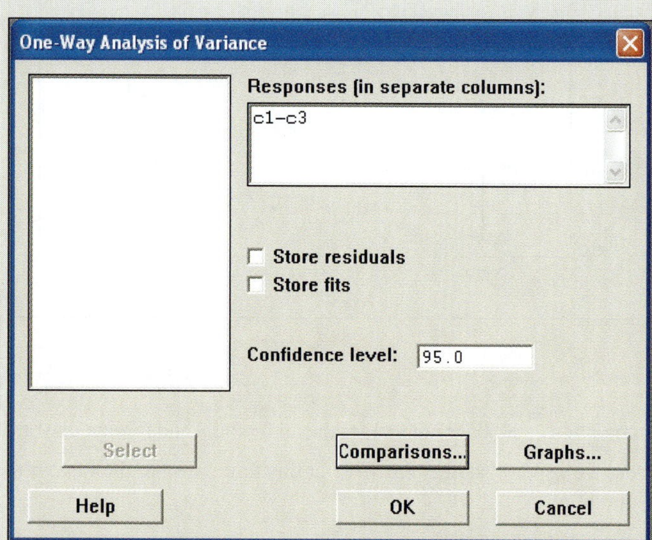

Screen 12.3

1. To perform a one-way analysis of variance on a collection of samples, enter the data for samples into columns.
2. Select **Stat>ANOVA>One-way (Unstacked)**.
3. Enter the names of the columns and select **OK**. (See **Screen 12.3**.)
4. The results include the components of the ANOVA, including the p-value, as well as the 95% confidence interval for each population mean using a pooled estimate of the variance.

Excel

1. Excel does not provide any built-in functions for performing an ANOVA test.

TECHNOLOGY ASSIGNMENTS

TA12.1 Refer to Exercise 12.18. Solve that exercise.

TA12.2 Refer to Exercise 12.26. Solve that exercise.

Chapter 13

Simple Linear Regression

13.1 Simple Linear Regression Model

13.2 Simple Linear Regression Analysis

Case Study 13–1 Regression of Heights and Weights of NBA Players

13.3 Standard Deviation of Random Errors

13.4 Coefficient of Determination

13.5 Inferences About *B*

13.6 Linear Correlation

13.7 Regression Analysis: A Complete Example

13.8 Using the Regression Model

13.9 Cautions in Using Regression

How much do you pay for your automobile insurance? A February 28, 2001, *New York Post* article reported that New York state was headed toward having the nation's highest average auto insurance premium, with an average annual rate of $1100 in 1999. The article reported that the high premium was primarily the result of fraudulent auto insurance claims. Phony auto insurance claims discovered by regulators in New York rose to 12,372 in 2000, nearly triple the 1995 figure of 4393, according to Robert Hartwig, chief economist with the Insurance Information Institute. How can you determine the impact of any one variable on insurance premiums? See Example 13–8.

This chapter considers the relationship between two variables in two ways: (1) by using regression analysis and (2) by computing the correlation coefficient. By using the regression model, we can evaluate the magnitude of change in one variable due to a certain change in another variable. For example, an economist can estimate the amount of change in food expenditure due to a certain change in the income of a household by using the regression model. A sociologist may want to estimate the increase in the crime rate due to a particular increase in the unemployment rate. Besides answering these questions, a regression model also helps predict the value of one variable for a given value of another variable. For example, by using the regression line, we can predict the (approximate) food expenditure of a household with a given income.

The correlation coefficient, on the other hand, simply tells us how strongly two variables are related. It does not provide any information about the size of the change in one variable as a result of a certain change in the other variable. For example, the correlation coefficient tells us how strongly income and food expenditure or crime rate and unemployment rate are related.

13.1 Simple Linear Regression Model

Only simple linear regression will be discussed in this chapter.[1] In the next two subsections the meaning of the words *simple* and *linear* as used in *simple linear regression* is explained.

13.1.1 Simple Regression

Let us return to the example of an economist investigating the relationship between food expenditure and income. What factors or variables does a household consider when deciding how much money it should spend on food every week or every month? Certainly, income of the household is one factor. However, many other variables also affect food expenditure. For instance, the assets owned by the household, the size of the household, the preferences and tastes of household members, and any special dietary needs of household members are some of the variables that influence a household's decision about food expenditure. These variables are called **independent** or **explanatory variables** because they all vary independently, and they explain the variation in food expenditures among different households. In other words, these variables explain why different households spend different amounts of money on food. Food expenditure is called the **dependent variable** because it depends on the independent variables. Studying the effect of two or more independent variables on a dependent variable using regression analysis is called **multiple regression**. However, if we choose only one (usually the most important) independent variable and study the effect of that single variable on a dependent variable, it is called a **simple regression**. Thus, a simple regression includes only two variables: one independent and one dependent. Note that whether it is a simple or a multiple regression analysis, it always includes one and only one dependent variable. It is the number of independent variables that changes in simple and multiple regressions.

Definition

Simple Regression A regression model is a mathematical equation that describes the relationship between two or more variables. A *simple regression* model includes only two variables: one independent and one dependent. The dependent variable is the one being explained, and the independent variable is the one used to explain the variation in the dependent variable.

13.1.2 Linear Regression

The relationship between two variables in a regression analysis is expressed by a mathematical equation called a **regression equation** or **model**. A regression equation, when plotted, may assume one of many possible shapes, including a straight line. A regression equation that gives a straight-line relationship between two variables is called a **linear regression model**; otherwise, the model is called a **nonlinear regression model**. In this chapter, only linear regression models are studied.

Definition

Linear Regression A (simple) regression model that gives a straight-line relationship between two variables is called a *linear regression model*.

[1]The term *regression* was first used by Sir Francis Galton (1822–1911), who studied the relationship between the heights of children and the heights of their parents.

The two diagrams in Figure 13.1 show a linear and a nonlinear relationship between the dependent variable food expenditure and the independent variable income. A linear relationship between income and food expenditure, shown in Figure 13.1*a*, indicates that as income increases, the food expenditure always increases at a constant rate. A nonlinear relationship between income and food expenditure, as depicted in Figure 13.1*b*, shows that as income increases, the food expenditure increases, although, after a point, the rate of increase in food expenditure is lower for every subsequent increase in income.

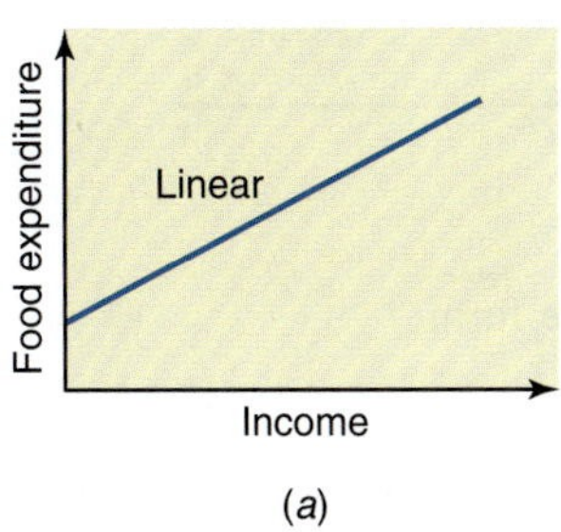

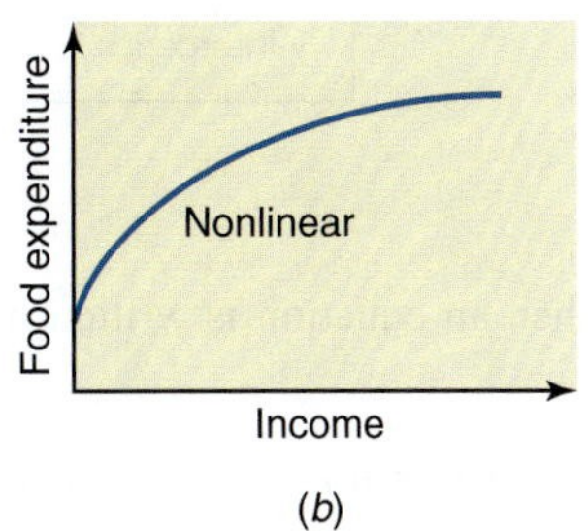

Figure 13.1 Relationship between food expenditure and income. (a) Linear relationship. (b) Nonlinear relationship.

The **equation of a linear relationship** between two variables x and y is written as

$$y = a + bx$$

Each set of values of a and b gives a different straight line. For instance, when $a = 50$ and $b = 5$, this equation becomes

$$y = 50 + 5x$$

To plot a straight line, we need to know two points that lie on that line. We can find two points on a line by assigning any two values to x and then calculating the corresponding values of y. For the equation $y = 50 + 5x$,

1. When $x = 0$, then $y = 50 + 5(0) = 50$.
2. When $x = 10$, then $y = 50 + 5(10) = 100$.

These two points are plotted in Figure 13.2. By joining these two points, we obtain the line representing the equation $y = 50 + 5x$.

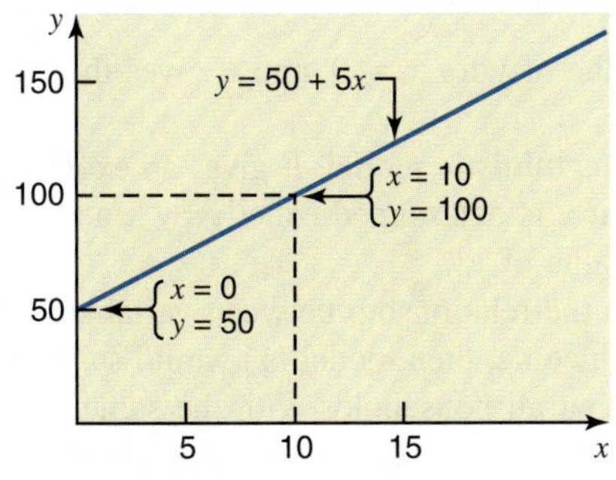

Figure 13.2 Plotting a linear equation.

Note that in Figure 13.2 the line intersects the y (vertical) axis at 50. Consequently, 50 is called the **y-intercept**. The y-intercept is given by the constant term in the equation. It is the value of y when x is zero.

In the equation $y = 50 + 5x$, 5 is called the **coefficient of *x*** or the **slope** of the line. It gives the amount of change in y due to a change of one unit in x. For example,

$$\text{If } x = 10, \text{ then } y = 50 + 5(10) = 100.$$

$$\text{If } x = 11, \text{ then } y = 50 + 5(11) = 105.$$

Hence, as x increases by 1 unit (from 10 to 11), y increases by 5 units (from 100 to 105). This is true for any value of x. Such changes in x and y are shown in Figure 13.3.

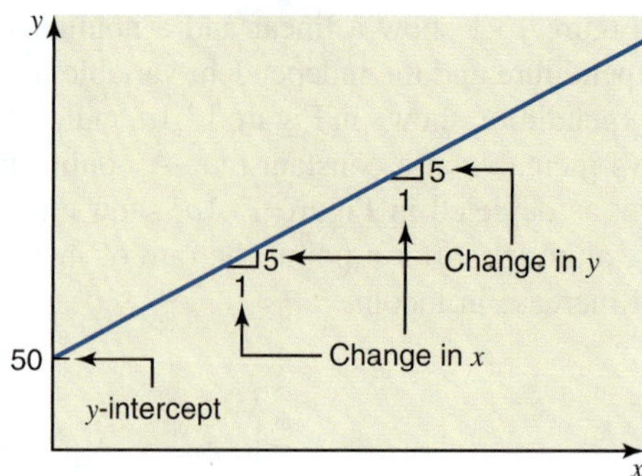

Figure 13.3 y-intercept and slope of a line.

In general, when an equation is written in the form

$$y = a + bx$$

a gives the y-intercept and b represents the slope of the line. In other words, a represents the point where the line intersects the y-axis and b gives the amount of change in y due to a change of one unit in x. Note that b is also called the coefficient of x.

13.2 Simple Linear Regression Analysis

In a regression model, the independent variable is usually denoted by x and the dependent variable is usually denoted by y. The x variable, with its coefficient, is written on the right side of the = sign, whereas the y variable is written on the left side of the = sign. The y-intercept and the slope, which we earlier denoted by a and b, respectively, can be represented by any of the many commonly used symbols. Let us denote the y-intercept (which is also called the *constant term*) by $\boldsymbol{A}$, and the slope (or the coefficient of the x variable) by $\boldsymbol{B}$. Then, our simple linear regression model is written as

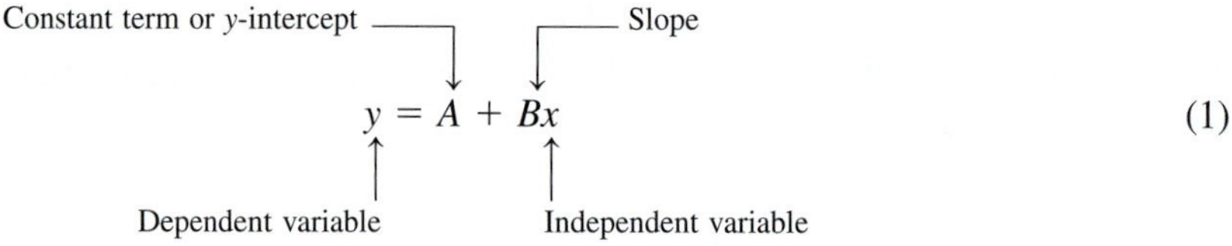

(1)

In model (1), A gives the value of y for $x = 0$, and B gives the change in y due to a change of one unit in x.

Model (1) is called a **deterministic model**. It gives an **exact relationship** between x and y. This model simply states that y is determined exactly by x and for a given value of x there is one and only one (unique) value of y.

However, in many cases the relationship between variables is not exact. For instance, if y is food expenditure and x is income, then model (1) would state that food expenditure is determined by income only and that all households with the same income spend the same amount on food. But as mentioned earlier, food expenditure is determined by many variables, only one of which is included in model (1). In reality, different households with the same income spend different amounts of money on food because of the differences in the sizes of the household, the assets they own, and their preferences and tastes. Hence, to take these variables into consideration and to make our model complete, we add another term to the right side of model (1). This term is called the **random error term**. It is denoted by ϵ (Greek letter *epsilon*). The complete regression model is written as

$$y = A + Bx + \epsilon \qquad (2)$$

↑ Random error term

The regression model (2) is called a **probabilistic model** or a **statistical relationship**.

Definition

Equation of a Regression Model In the *regression model* $y = A + Bx + \epsilon$, A is called the y-intercept or constant term, B is the slope, and ϵ is the random error term. The dependent and independent variables are y and x, respectively.

The random error term ϵ is included in the model to represent the following two phenomena.

1. *Missing or omitted variables.* As mentioned earlier, food expenditure is affected by many variables other than income. The random error term ϵ is included to capture the effect of all those missing or omitted variables that have not been included in the model.
2. *Random variation.* Human behavior is unpredictable. For example, a household may have many parties during one month and spend more than usual on food during that month. The same household may spend less than usual during another month because it spent quite a bit of money to buy furniture. The variation in food expenditure for such reasons may be called random variation.

In model (2), A and B are the **population parameters**. The regression line obtained for model (2) by using the population data is called the **population regression line**. The values of A and B in the population regression line are called the **true values of the y-intercept and slope**.

However, population data are difficult to obtain. As a result, we almost always use sample data to estimate model (2). The values of the y-intercept and slope calculated from sample data on x and y are called the **estimated values of A and B** and are denoted by a and b. Using a and b, we write the estimated regression model as

$$\hat{y} = a + bx \tag{3}$$

where $\hat{y}$ (read as *y hat*) is the **estimated** or **predicted value of y** for a given value of x. Equation (3) is called the **estimated regression model**; it gives the **regression of y on x**.

Definition

Estimates of A and B In the model $\hat{y} = a + bx$, a and b, which are calculated using sample data, are called the *estimates of A and B*.

13.2.1 Scatter Diagram

Suppose we take a sample of seven households from a low- to moderate-income neighborhood and collect information on their incomes and food expenditures for the past month. The information obtained (in hundreds of dollars) is given in Table 13.1.

Table 13.1 Incomes and Food Expenditures of Seven Households

Income	Food Expenditure
35	9
49	15
21	7
39	11
15	5
28	8
25	9

In Table 13.1, we have a pair of observations for each of the seven households. Each pair consists of one observation on income and a second on food expenditure. For example, the first household's income for the past month was $3500 and its food expenditure was $900. By plotting all seven pairs of values, we obtain a **scatter diagram** or **scatterplot**. Figure 13.4 gives the scatter diagram for the data of Table 13.1. Each dot in this diagram represents one household. A scatter diagram is helpful in detecting a relationship between two variables. For example, by looking at the scatter diagram of Figure 13.4, we can observe that there exists a strong linear relationship between food expenditure and income. If a straight line is drawn through the points, the points will be scattered closely around the line.

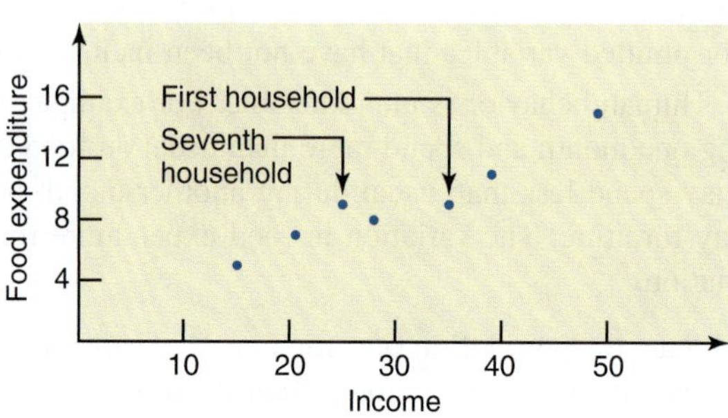

Figure 13.4 Scatter diagram.

Definition

Scatter Diagram A plot of paired observations is called a *scatter diagram.*

As shown in Figure 13.5, a large number of straight lines can be drawn through the scatter diagram of Figure 13.4. Each of these lines will give different values for a and b of model (3).

In regression analysis, we try to find a line that best fits the points in the scatter diagram. Such a line provides the best possible description of the relationship between the dependent and independent variables. The **least squares method**, discussed in the next section, gives such a line. The line obtained by using the least squares method is called the **least squares regression line**.

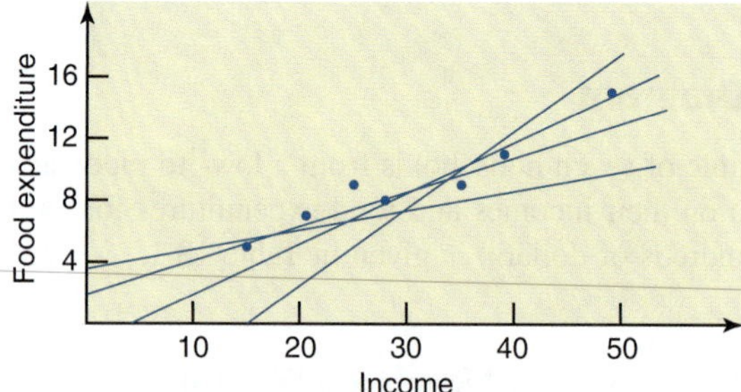

Figure 13.5 Scatter diagram and straight lines.

13.2.2 Least Squares Line

The value of y obtained for a member from the survey is called the **observed or actual value of y**. As mentioned earlier in Section 13.2, the value of y, denoted by $\hat{y}$, obtained for a given x by using the regression line is called the **predicted value of y**. The random error ϵ denotes the difference between the actual value of y and the predicted value of y for population data. For example, for a given household, ϵ is the difference between what this household actually spent on food during the past month and what is predicted using the population regression line. The ϵ is also called the *residual* because it measures the surplus (positive or negative) of actual food expenditure over what is predicted by using the regression model. If we estimate model (2) by using sample data, the difference between the actual y and the predicted y based on this estimation cannot be denoted by ϵ. *The random error for the sample regression model is denoted by* $\boldsymbol{e}$. Thus, e is an estimator of ϵ. If we estimate model (2) using sample data, then the value of e is given by

$$e = \text{Actual food expenditure} - \text{Predicted food expenditure} = y - \hat{y}$$

In Figure 13.6, e is the vertical distance between the actual position of a household and the point on the regression line. Note that in such a diagram, we always measure the dependent variable on the vertical axis and the independent variable on the horizontal axis.

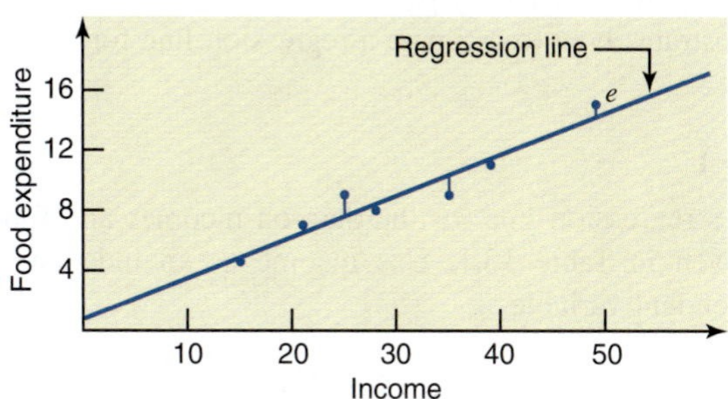

Figure 13.6 Regression line and random errors.

The value of an error is positive if the point that gives the actual food expenditure is above the regression line and negative if it is below the regression line. *The sum of these errors is always zero*. In other words, the sum of the actual food expenditures for seven households included in the sample will be the same as the sum of the food expenditures predicted from the regression model. Thus,

$$\Sigma e = \Sigma(y - \hat{y}) = 0$$

Hence, to find the line that best fits the scatter of points, we cannot minimize the sum of errors. Instead, we minimize the **error sum of squares**, denoted by **SSE**, which is obtained by adding the squares of errors. Thus,

$$\text{SSE} = \Sigma e^2 = \Sigma(y - \hat{y})^2$$

The least squares method gives the values of a and b for model (3) such that the sum of squared errors (SSE) is minimum.

Error Sum of Squares (SSE) The *error sum of squares*, denoted by SSE, is

$$\text{SSE} = \Sigma e^2 = \Sigma(y - \hat{y})^2$$

The values of a and b that give the minimum SSE are called the *least squares estimates* of A and B, and the regression line obtained with these estimates is called the *least squares line*.

The Least Squares Line For the least squares regression line $\hat{y} = a + bx$,

$$b = \frac{\text{SS}_{xy}}{\text{SS}_{xx}} \quad \text{and} \quad a = \bar{y} - b\bar{x}$$

where

$$\text{SS}_{xy} = \Sigma xy - \frac{(\Sigma x)(\Sigma y)}{n} \quad \text{and} \quad \text{SS}_{xx} = \Sigma x^2 - \frac{(\Sigma x)^2}{n}$$

and SS stands for "sum of squares." The least squares regression line $\hat{y} = a + bx$ is also called the regression of y on x.

The least squares values of a and b are computed using the formulas given on the previous page.[2] These formulas are for estimating a sample regression line. Suppose we have access to a population data set. We can find the population regression line by using the same formulas with a little adaptation. If we have access to population data, we replace a by A, b by B, and n by N in these formulas, and use the values of Σx, Σy, Σxy, and Σx^2 calculated for population data to make the required computations. The population regression line is written as

$$\mu_{y|x} = A + Bx$$

where $\mu_{y|x}$ is read as "the mean value of y for a given x." When plotted on a graph, the points on this population regression line give the average values of y for the corresponding values of x. These average values of y are denoted by $\mu_{y|x}$.

Example 13–1 illustrates how to estimate a regression line for sample data.

■ EXAMPLE 13–1

Estimating the least squares regression line.

Find the least squares regression line for the data on incomes and food expenditures on the seven households given in Table 13.1. Use income as an independent variable and food expenditure as a dependent variable.

Solution We are to find the values of a and b for the regression model $\hat{y} = a + bx$. Table 13.2 shows the calculations required for the computation of a and b. We denote the independent variable (income) by x and the dependent variable (food expenditure) by y.

Table 13.2

Income x	Food Expenditure y	xy	x^2
35	9	315	1225
49	15	735	2401
21	7	147	441
39	11	429	1521
15	5	75	225
28	8	224	784
25	9	225	625
$\Sigma x = 212$	$\Sigma y = 64$	$\Sigma xy = 2150$	$\Sigma x^2 = 7222$

The following steps are performed to compute a and b.

Step 1. *Compute Σx, Σy, $\bar{x}$, and $\bar{y}$.*

$$\Sigma x = 212 \qquad \Sigma y = 64$$

$$\bar{x} = \Sigma x/n = 212/7 = 30.2857$$

$$\bar{y} = \Sigma y/n = 64/7 = 9.1429$$

Step 2. *Compute Σxy and Σx^2.*

To calculate Σxy, we multiply the corresponding values of x and y. Then, we sum all the products. The products of x and y are recorded in the third column of Table 13.2. To compute

[2]The values of SS_{xy} and SS_{xx} can also be obtained by using the following basic formulas:

$$SS_{xy} = \Sigma(x - \bar{x})(y - \bar{y}) \quad \text{and} \quad SS_{xx} = \Sigma(x - \bar{x})^2$$

However, these formulas take longer to make calculations.

Σx^2, we square each of the x values and then add them. The squared values of x are listed in the fourth column of Table 13.2. From these calculations,

$$\Sigma xy = 2150 \quad \text{and} \quad \Sigma x^2 = 7222$$

Step 3. *Compute* SS_{xy} *and* SS_{xx}.

$$SS_{xy} = \Sigma xy - \frac{(\Sigma x)(\Sigma y)}{n} = 2150 - \frac{(212)(64)}{7} = 211.7143$$

$$SS_{xx} = \Sigma x^2 - \frac{(\Sigma x)^2}{n} = 7222 - \frac{(212)^2}{7} = 801.4286$$

Step 4. *Compute* a *and* b.

$$b = \frac{SS_{xy}}{SS_{xx}} = \frac{211.7143}{801.4286} = .2642$$

$$a = \bar{y} - b\bar{x} = 9.1429 - (.2642)(30.2857) = 1.1414$$

Thus, our estimated regression model $\hat{y} = a + bx$ is

$$\hat{y} = \mathbf{1.1414 + .2642}x$$

This regression line is called the least squares regression line. It gives the *regression of food expenditure on income.*

Note that we have rounded all calculations to four decimal places. We can round the values of a and b in the regression equation to two decimal places, but it is not done here because we will use this regression equation for prediction and estimation purposes later on. ■

Using this estimated regression model, we can find the predicted value of y for any specific value of x. For instance, suppose we randomly select a household whose monthly income is \$3500 so that $x = 35$ (recall that x denotes income in hundreds of dollars). The predicted value of food expenditure for this household is

$$\hat{y} = 1.1414 + (.2642)(35) = \$10.3884 \text{ hundred} = \$1038.84$$

In other words, based on our regression line, we predict that a household with a monthly income of \$3500 is expected to spend \$1038.84 per month on food. This value of $\hat{y}$ can also be interpreted as a point estimator of the mean value of y for $x = 35$. Thus, we can state that, on average, all households with a monthly income of \$3500 spend about \$1038.84 per month on food.

In our data on seven households, there is one household whose income is \$3500. The actual food expenditure for that household is \$900 (see Table 13.1). The difference between the actual and predicted values gives the error of prediction. Thus, the error of prediction for this household, which is shown in Figure 13.7, is

$$e = y - \hat{y} = 9.00 - 10.3884 = -\$1.3884 \text{ hundred} = -\$138.84$$

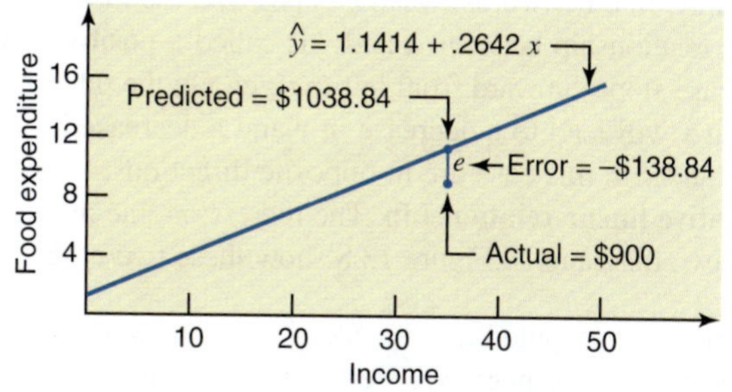

Figure 13.7 Error of prediction.

Therefore, the error of prediction is −\$138.84. The negative error indicates that the predicted value of y is greater than the actual value of y. Thus, if we use the regression model, this household's food expenditure is overestimated by \$138.84.

13.2.3 Interpretation of *a* and *b*

How do we interpret $a = 1.1414$ and $b = .2642$ obtained in Example 13–1 for the regression of food expenditure on income? A brief explanation of the *y*-intercept and the slope of a regression line was given in Section 13.1.2. Below we explain the meaning of *a* and *b* in more detail.

Interpretation of *a*

Consider a household with zero income. Using the estimated regression line obtained in Example 13–1, we get the predicted value of *y* for $x = 0$ as

$$\hat{y} = 1.1414 + .2642(0) = \$1.1414 \text{ hundred} = \$114.14$$

Thus, we can state that a household with no income is expected to spend $114.14 per month on food. Alternatively, we can also state that the point estimate of the average monthly food expenditure for all households with zero income is $114.14. Note that here we have used $\hat{y}$ as a point estimate of $\mu_{y|x}$. Thus, $a = 1.1414$ gives the predicted or mean value of *y* for $x = 0$ based on the regression model estimated for the sample data.

However, we should be very careful when making this interpretation of *a*. In our sample of seven households, the incomes vary from a minimum of $1500 to a maximum of $4900. (Note that in Table 13.1, the minimum value of *x* is 15 and the maximum value is 49.) Hence, our regression line is valid only for the values of *x* between 15 and 49. If we predict *y* for a value of *x* outside this range, the prediction usually will not hold true. Thus, since $x = 0$ is outside the range of household incomes that we have in the sample data, the prediction that a household with zero income spends $114.14 per month on food does not carry much credibility. The same is true if we try to predict *y* for an income greater than $4900, which is the maximum value of *x* in Table 13.1.

Interpretation of *b*

The value of *b* in a regression model gives the change in *y* (dependent variable) due to a change of one unit in *x* (independent variable). For example, by using the regression equation obtained in Example 13–1, we see:

$$\text{When } x = 30, \quad \hat{y} = 1.1414 + .2642(30) = 9.0674$$

$$\text{When } x = 31, \quad \hat{y} = 1.1414 + .2642(31) = 9.3316$$

Hence, when *x* increased by one unit, from 30 to 31, $\hat{y}$ increased by $9.3316 - 9.0674 = .2642$, which is the value of *b*. Because our unit of measurement is hundreds of dollars, we can state that, on average, a $100 increase in income will result in a $26.42 increase in food expenditure. We can also state that, on average, a $1 increase in income of a household will increase the food expenditure by $.2642. Note the phrase "on average" in these statements. The regression line is seen as a measure of the mean value of *y* for a given value of *x*. If one household's income is increased by $100, that household's food expenditure may or may not increase by $26.42. However, if the incomes of all households are increased by $100 each, the average increase in their food expenditures will be very close to $26.42.

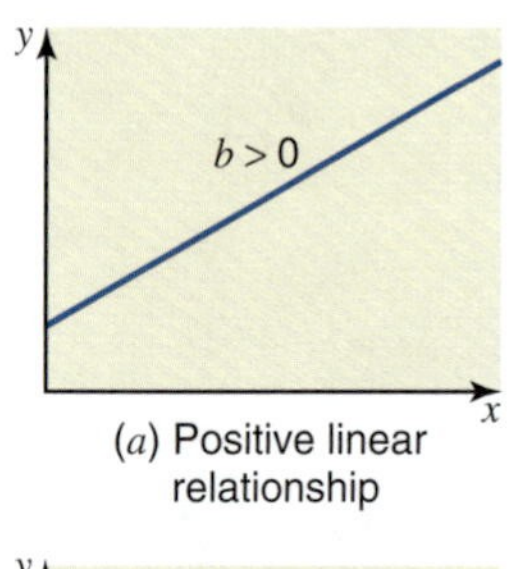

(*a*) Positive linear relationship

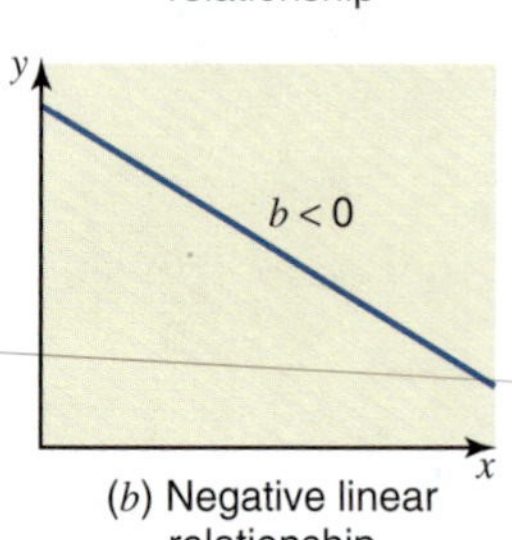

(*b*) Negative linear relationship

Figure 13.8 Positive and negative linear relationships between *x* and *y*.

Note that when *b* is positive, an increase in *x* will lead to an increase in *y* and a decrease in *x* will lead to a decrease in *y*. In other words, when *b* is positive, the movements in *x* and *y* are in the same direction. Such a relationship between *x* and *y* is called a **positive linear relationship**. The regression line in this case slopes upward from left to right. On the other hand, if the value of *b* is negative, an increase in *x* will lead to a decrease in *y* and a decrease in *x* will cause an increase in *y*. The changes in *x* and *y* in this case are in opposite directions. Such a relationship between *x* and *y* is called a **negative linear relationship**. The regression line in this case slopes downward from left to right. The two diagrams in Figure 13.8 show these two cases.

Remember ▶ For a regression model, *b* is computed as $b = SS_{xy}/SS_{xx}$. The value of SS_{xx} is always positive and that of SS_{xy} can be positive or negative. Hence, the sign of *b* depends on the sign of SS_{xy}. If SS_{xy} is positive (as in our example on the incomes and food expenditures of seven households), then *b* will be positive, and if SS_{xy} is negative, then *b* will be negative.

Case Study 13–1 illustrates the difference between the population regression line and a sample regression line.

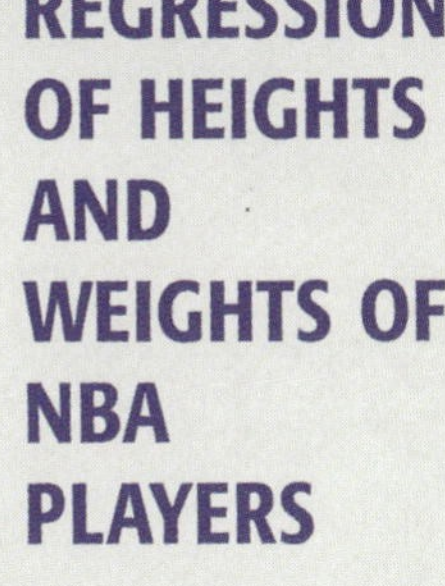

REGRESSION OF HEIGHTS AND WEIGHTS OF NBA PLAYERS

Data Set III that accompanies this text lists the heights and weights of all National Basketball Association players who were on the rosters of all NBA teams as of May 2005. These data comprise the population of NBA players for that point in time. We postulate the following simple linear regression model for these data:

$$y = A + Bx + \epsilon$$

where y is the weight (in pounds) and x is the height (in inches) of an NBA player.

Using the population data, we obtain the following regression line:

$$\mu_{y|x} = -286 + 6.44x$$

This equation gives the population regression line because it is obtained by using the population data. (Note that in the population regression line we write $\mu_{y|x}$ instead of $\hat{y}$.) Thus, the true values of A and B are

$$A = -286 \quad \text{and} \quad B = 6.44$$

The value of B indicates that for every one-inch increase in the height of an NBA player, weight increases on average by 6.44 pounds. However, $A = -286$ does not make any sense. It states that the weight of a player with zero height is -286 pounds. (Recall from Section 13.2.3 that we cannot apply the regression equation to predict y for values of x outside the range of data used to find the regression line.) Figure 13.9 gives the scatter diagram and the regression line for the heights and weights of all NBA players.

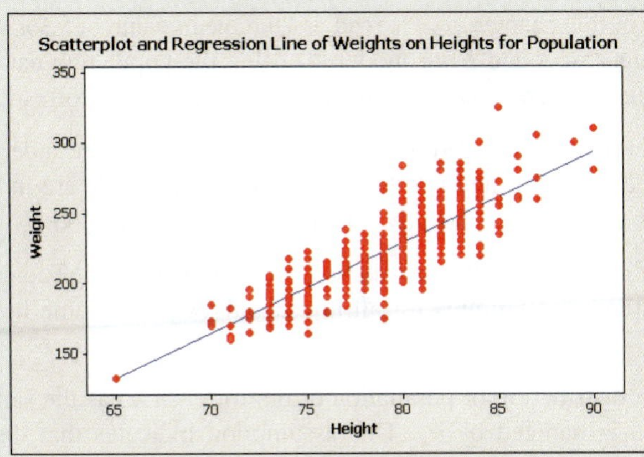

Figure 13.9 Scatter diagram for the data on heights and weights of all NBA players.

Next, we selected a random sample of 50 players and estimated the regression model for this sample. The estimated regression line for the sample is $\hat{y} = -309 + 6.71x$.

The values of a and b are $a = -309$ and $b = 6.71$. These values of a and b give the estimates of A and B based on sample data. The scatter diagram and the regression line for the sample observations on heights and weights is given in Figure 13.10.

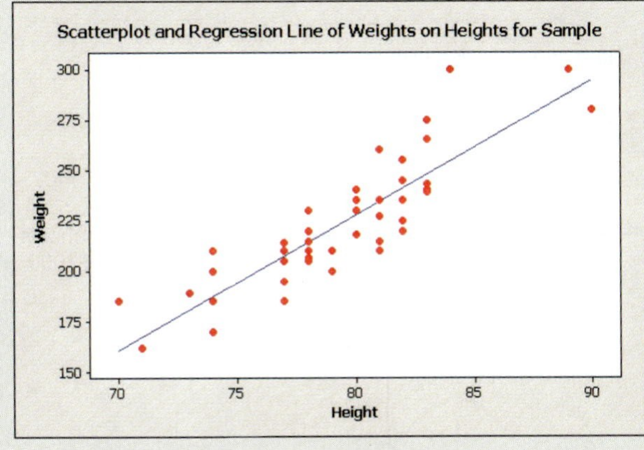

Figure 13.10 Scatter diagram for the data on heights and weights of 50 NBA players.

As we can observe from Figures 13.9 and 13.10, the scatter diagrams for population and sample data both show a (positive) linear relationship between the heights and weights of NBA players.

Source: National Basketball Association.

13.2.4 Assumptions of the Regression Model

Like any other theory, the linear regression analysis is also based on certain assumptions. Consider the population regression model

$$y = A + Bx + \epsilon \qquad (4)$$

Four assumptions are made about this model. These assumptions are explained next with reference to the example on incomes and food expenditures of households. Note that these assumptions are made about the population regression model and not about the sample regression model.

Assumption 1: The random error term ϵ has a mean equal to zero for each x. In other words, among all households with the same income, some spend more than the predicted food expenditure (and, hence, have positive errors) and others spend less than the predicted food expenditure (and, consequently, have negative errors). This assumption simply states that the sum of the positive errors is equal to the sum of the negative errors, so that the mean of errors for all households with the same income is zero. Thus, when the mean value of ϵ is zero, the mean value of y for a given x is equal to $A + Bx$ and it is written as

$$\mu_{y|x} = A + Bx$$

As mentioned earlier in this chapter, $\mu_{y|x}$ is read as "the mean value of y for a given value of x." When we find the values of A and B for model (4) using the population data, the points on the regression line give the average values of y, denoted by $\mu_{y|x}$, for the corresponding values of x.

Assumption 2: The errors associated with different observations are independent. According to this assumption, the errors for any two households in our example are independent. In other words, all households decide independently how much to spend on food.

Assumption 3: For any given x, the distribution of errors is normal. The corollary of this assumption is that the food expenditures for all households with the same income are normally distributed.

Assumption 4: The distribution of population errors for each x has the same (constant) standard deviation, which is denoted by σ_ϵ. This assumption indicates that the spread of points around the regression line is similar for all x values.

Figure 13.11 illustrates the meanings of the first, third, and fourth assumptions for households with incomes of \$2000 and \$3500 per month. The same assumptions hold true for any other income level. In the population of all households, there will be many households with a monthly income of \$2000. Using the population regression line, if we calculate the errors for all these households and prepare the distribution of these errors, it will look like the distribution given in Figure 13.11*a*. Its standard deviation will be σ_ϵ. Similarly, Figure 13.11*b* gives the

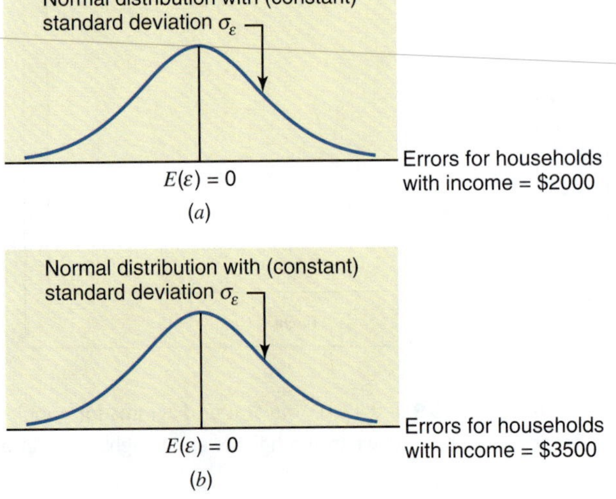

Figure 13.11 (*a*) Errors for households with an income of \$2000 per month. (*b*) Errors for households with an income of \$3500 per month.

distribution of errors for all those households in the population whose monthly income is \$3500. Its standard deviation is also σ_ϵ. Both these distributions are identical. Note that the mean of both of these distributions is $E(\epsilon) = 0$.

Figure 13.12 shows how the distributions given in Figure 13.11 look when they are plotted on the same diagram with the population regression line. The points on the vertical line through $x = 20$ give the food expenditures for various households in the population, each of which has the same monthly income of \$2000. The same is true about the vertical line through $x = 35$ or any other vertical line for some other value of x.

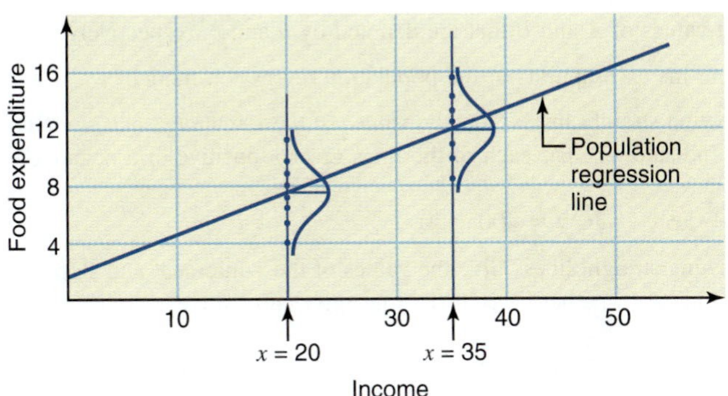

Figure 13.12 Distribution of errors around the population regression line.

13.2.5 A Note on the Use of Simple Linear Regression

We should apply linear regression with caution. When we use simple linear regression, we assume that the relationship between two variables is described by a straight line. In the real world, the relationship between variables may not be linear. Hence, before we use a simple linear regression, it is better to construct a scatter diagram and look at the plot of the data points. We should estimate a linear regression model only if the scatter diagram indicates such a relationship. The scatter diagrams of Figure 13.13 give two examples for which the relationship between x and y is not linear. Consequently, fitting linear regression in such cases would be wrong.

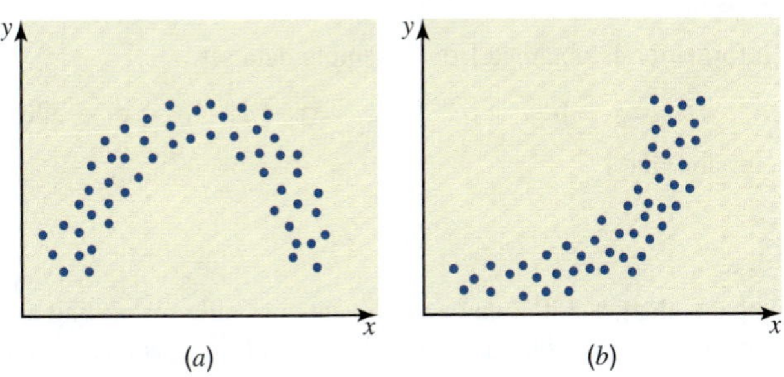

Figure 13.13 Nonlinear relationship between x and y.

EXERCISES

CONCEPTS AND PROCEDURES

13.1 Explain the meaning of the words *simple* and *linear* as used in *simple linear regression.*

13.2 Explain the meaning of independent and dependent variables for a regression model.

13.3 Explain the difference between exact and nonexact relationships between two variables.

13.4 Explain the difference between linear and nonlinear relationships between two variables.

13.5 Explain the difference between a simple and a multiple regression model.

13.6 Briefly explain the difference between a deterministic and a probabilistic regression model.

13.7 Why is the random error term included in a regression model?

13.8 Explain the least squares method and least squares regression line. Why are they called by these names?

13.9 Explain the meaning and concept of SSE. You may use a graph for illustration purposes.

13.10 Explain the difference between y and $\hat{y}$.

13.11 Two variables x and y have a positive linear relationship. Explain what happens to the value of y when x increases.

13.12 Two variables x and y have a negative linear relationship. Explain what happens to the value of y when x increases.

13.13 Explain the following.

a. Population regression line
b. Sample regression line
c. True values of A and B
d. Estimated values of A and B that are denoted by a and b, respectively

13.14 Briefly explain the assumptions of the population regression model.

13.15 Plot the following straight lines. Give the values of the y-intercept and slope for each of these lines and interpret them. Indicate whether each of the lines gives a positive or a negative relationship between x and y.

a. $y = 100 + 5x$ **b.** $y = 400 - 4x$

13.16 Plot the following straight lines. Give the values of the y-intercept and slope for each of these lines and interpret them. Indicate whether each of the lines gives a positive or a negative relationship between x and y.

a. $y = -60 + 8x$ **b.** $y = 300 - 6x$

13.17 A population data set produced the following information.

$$N = 250, \quad \Sigma x = 9880, \quad \Sigma y = 1456, \quad \Sigma xy = 85{,}080, \quad \Sigma x^2 = 485{,}870$$

Find the population regression line.

13.18 A population data set produced the following information.

$$N = 460, \quad \Sigma x = 3920, \quad \Sigma y = 2650, \quad \Sigma xy = 26{,}570, \quad \Sigma x^2 = 48{,}530$$

Find the population regression line.

13.19 The following information is obtained from a sample data set.

$$n = 10, \quad \Sigma x = 100, \quad \Sigma y = 220, \quad \Sigma xy = 3680, \quad \Sigma x^2 = 1140$$

Find the estimated regression line.

13.20 The following information is obtained from a sample data set.

$$n = 12, \quad \Sigma x = 66, \quad \Sigma y = 588, \quad \Sigma xy = 2244, \quad \Sigma x^2 = 396$$

Find the estimated regression line.

APPLICATIONS

13.21 A car rental company charges \$40 a day and 20 cents per mile for renting a car. Let y be the total rental charges (in dollars) for a car for one day and x be the miles driven. The equation for the relationship between x and y is

$$y = 40 + .20x$$

a. How much will a person pay who rents a car for one day and drives it 100 miles?
b. Suppose each of 20 persons rents a car from this agency for one day and drives it 100 miles. Will each of them pay the same amount for renting a car for a day or do you expect each person to pay a different amount? Explain.
c. Is the relationship between x and y exact or nonexact?

13.22 Bob's Pest Removal Service specializes in removing wild creatures (skunks, bats, reptiles, etc.) from private homes. He charges \$50 to go to a house plus \$20 per hour for his services. Let y be the total amount (in dollars) paid by a household using Bob's services and x the number of hours Bob spends capturing and removing the animal(s). The equation for the relationship between x and y is

$$y = 50 + 20x$$

a. Bob spent three hours removing a coyote from under Alice's house. How much will he be paid?
b. Suppose nine persons called Bob for assistance during a week. Strangely enough, each of these jobs required exactly three hours. Will each of these clients pay Bob the same amount, or do you expect each one to pay a different amount? Explain.
c. Is the relationship between x and y exact or nonexact?

13.23 A researcher took a sample of 25 electronics companies and found the following relationship between x and y where x is the amount of money (in millions of dollars) spent on advertising by a company in 2005 and y represents the total gross sales (in millions of dollars) of that company for 2005.

$$\hat{y} = 3.6 + 11.75x$$

a. An electronics company spent \$2 million on advertising in 2005. What are its expected gross sales for 2005?
b. Suppose four electronics companies spent \$2 million each on advertising in 2005. Do you expect these four companies to have the same actual gross sales for 2005? Explain.
c. Is the relationship between x and y exact or nonexact?

13.24 A researcher took a sample of 10 years and found the following relationship between x and y, where x is the number of major natural calamities (such as tornadoes, hurricanes, earthquakes, floods, etc.) that occurred during a year and y represents the average annual total profits (in millions of dollars) of all insurance companies in the United States.

$$\hat{y} = 342.6 - 2.10x$$

a. A randomly selected year had 24 major calamities. What are the expected average profits of U.S. insurance companies for that year?
b. Suppose the number of major calamities was the same for each of three years. Do you expect the average profits for all U.S. insurance companies to be the same for each of these three years? Explain.
c. Is the relationship between x and y exact or nonexact?

13.25 An auto manufacturing company wanted to investigate how the price of one of its car models depreciates with age. The research department at the company took a sample of eight cars of this model and collected the following information on the ages (in years) and prices (in hundreds of dollars) of these cars.

Age	8	3	6	9	2	5	6	3
Price	18	94	50	21	145	42	36	99

a. Construct a scatter diagram for these data. Does the scatter diagram exhibit a linear relationship between ages and prices of cars?
b. Find the regression line with price as a dependent variable and age as an independent variable.
c. Give a brief interpretation of the values of a and b calculated in part b.
d. Plot the regression line on the scatter diagram of part a and show the errors by drawing vertical lines between scatter points and the regression line.
e. Predict the price of a 7-year-old car of this model.
f. Estimate the price of an 18-year-old car of this model. Comment on this finding.

13.26 The owner of Red's Towing and Garage Service is interested in finding the relationship between the lowest temperature on a winter day and the number of emergency road service calls his shop receives. The following table gives the lowest temperature (in degrees Fahrenheit) on seven winter days and the number of emergency road service calls received on those days.

Lowest temperature	15	0	24	−10	30	9	36
Number of calls	12	22	16	31	7	24	6

a. Construct a scatter diagram for these data. Does the scatter diagram exhibit a linear relationship between lowest temperatures and number of calls?
b. Find the regression of number of calls on lowest temperatures.
c. Give a brief interpretation of the values of a and b calculated in part b.
d. Plot the regression line on the scatter diagram of part a and show the errors by drawing vertical lines between scatter points and the regression line.
e. Predict the number of calls on a day with a lowest temperature of 20 degrees.
f. Estimate the number of calls on a day with a lowest temperature of −20 degrees. Comment on this finding.

13.27 An insurance company wants to know how the amounts of life insurance depend on the incomes of persons. The research department at the company collected information on six persons. The table on the next page lists the annual incomes (in thousands of dollars) and amounts (in thousands of dollars) of life insurance policies for these six persons.

Annual income	62	78	41	53	85	34
Life insurance	250	300	100	150	500	75

a. Construct a scatter diagram for these data. Does the scatter diagram show a linear relationship between annual incomes and amounts of life insurance policies?
b. Find the regression line $\hat{y} = a + bx$ with annual income as an independent variable and amount of life insurance policy as a dependent variable.
c. Give a brief interpretation of the values of a and b calculated in part b.
d. Plot the regression line on the scatter diagram of part a and show the errors by drawing vertical lines between the scatter points and the regression line.
e. What is the estimated value of life insurance for a person with an annual income of $55,000?
f. One of the persons in our sample has an annual income of $78,000 and $300,000 of life insurance. What is the predicted value of life insurance for this person? Find the error for this observation.

13.28 A consumer welfare agency wants to investigate the relationship between the sizes of houses and the rents paid by tenants in a small city. The agency collected the following information on the sizes (in hundreds of square feet) of six houses and the monthly rents (in dollars) paid by tenants.

Size of house	21	13	19	27	34	23
Monthly rent	1000	780	1120	1500	1850	1200

a. Construct a scatter diagram for these data. Does the scatter diagram show a linear relationship between sizes of houses and monthly rents?
b. Find the regression line $\hat{y} = a + bx$ with the size of a house as an independent variable and monthly rent as a dependent variable.
c. Give a brief interpretation of the values of a and b calculated in part b.
d. Plot the regression line on the scatter diagram of part a and show the errors by drawing vertical lines between the scatter points and the regression line.
e. Predict the monthly rent for a house with 2500 square feet.
f. One of the houses in our sample is 2700 square feet and its rent is $1500. What is the predicted rent for this house? Find the error of estimation for this observation.

13.29 The following table gives the total 2004 payroll (on the opening day of the season, rounded to the nearest million dollars) and the percentage of games won during the 2004 season by each of the National League baseball teams.

Team	Total Payroll (millions of dollars)	Percentage of Games Won
Arizona Diamondbacks	70	31.5
Atlanta Braves	90	59.3
Chicago Cubs	91	54.9
Cincinnati Reds	47	46.9
Colorado Rockies	65	42.0
Florida Marlins	42	51.2
Houston Astros	75	56.8
Los Angeles Dodgers	93	57.4
Milwaukee Brewers	28	41.6
Montreal Expos*	41	41.4
New York Mets	97	43.8
Philadelphia Phillies	93	53.1
Pittsburgh Pirates	32	44.7
St. Louis Cardinals	83	64.8
San Diego Padres	55	53.7
San Francisco Giants	82	56.2

*Now Washington Nationals.

a. Find the least squares regression line with total payroll as an independent variable and percentage of games won as a dependent variable.
b. Is the regression line obtained in part a the population regression line? Why or why not? Do the values of the y-intercept and the slope of the regression line give A and B or a and b?
c. Give a brief interpretation of the values of the y-intercept and the slope.
d. Predict the percentage of games won for a team with a total payroll of $60 million.

13.30 The following table gives the total 2004 payroll (on the opening day of the season, rounded to the nearest million dollars) and the percentage of games won during the 2004 season by each of the American League baseball teams.

Team	Total Payroll (millions of dollars)	Percentage of Games Won
Anaheim Angels*	101	56.8
Baltimore Orioles	52	48.1
Boston Red Sox	127	60.5
Chicago White Sox	65	51.2
Cleveland Indians	34	49.4
Detroit Tigers	47	44.4
Kansas City Royals	48	35.8
Minnesota Twins	54	56.8
New York Yankees	184	62.3
Oakland A's	59	56.2
Seattle Mariners	82	38.9
Tampa Bay Devil Rays	30	43.5
Texas Rangers	55	54.9
Toronto Blue Jays	50	41.6

*Now Los Angeles Angels of Anaheim

a. Find the least squares regression line with total payroll as an independent variable and percentage of games won as a dependent variable.
b. Is the regression line obtained in part a the population regression line? Why or why not? Do the values of the y-intercept and the slope in the regression line give A and B or a and b?
c. Give a brief interpretation of the values of the y-intercept and the slope.
d. Predict the percentage of games won for a team with a total payroll of $70 million.

13.3 Standard Deviation of Random Errors

When we consider incomes and food expenditures, all households with the same income are expected to spend different amounts on food. Consequently, the random error ϵ will assume different values for these households. The standard deviation σ_ϵ measures the spread of these errors around the population regression line. The **standard deviation of errors** tells us how widely the errors and, hence, the values of y are spread for a given x. In Figure 13.12, which is reproduced as Figure 13.14 on the next page, the points on the vertical line through $x = 20$ give the monthly food expenditures for all households with a monthly income of $2000. The distance of each dot from the point on the regression line gives the value of the corresponding error. The standard deviation of errors σ_ϵ measures the spread of such points around the population regression line. The same is true for $x = 35$ or any other value of x.

Note that σ_ϵ denotes the standard deviation of errors for the population. However, usually σ_ϵ is unknown. In such cases, it is estimated by s_e, which is the standard deviation of errors for the sample data. The following is the basic formula to calculate s_e:

$$s_e = \sqrt{\frac{\text{SSE}}{n-2}} \quad \text{where} \quad \text{SSE} = \Sigma(y - \hat{y})^2$$

Figure 13.14 Spread of errors for $x = 20$ and $x = 35$.

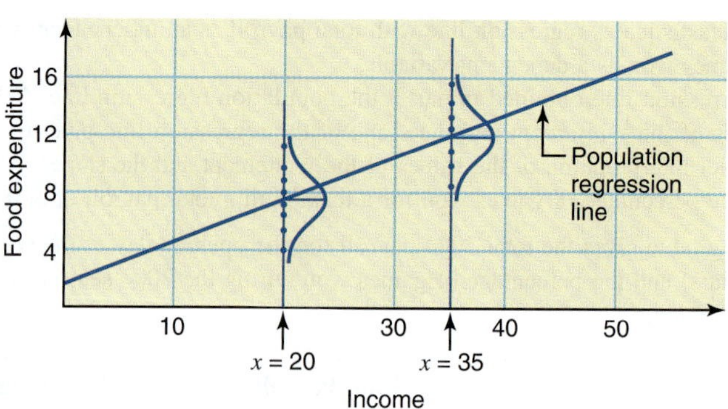

In this formula, $n - 2$ represents the **degrees of freedom** for the regression model. The reason $df = n - 2$ is that we lose one degree of freedom to calculate $\bar{x}$ and one for $\bar{y}$.

Degrees of Freedom for a Simple Linear Regression Model The *degrees of freedom* for a simple linear regression model are

$$df = n - 2$$

For computational purposes, it is more convenient to use the following formula to calculate the standard deviation of errors s_e.

Standard Deviation of Errors The *standard deviation of errors* is calculated as[3]

$$s_e = \sqrt{\frac{SS_{yy} - b\, SS_{xy}}{n - 2}}$$

where

$$SS_{yy} = \Sigma y^2 - \frac{(\Sigma y)^2}{n}$$

The calculation of SS_{xy} was discussed earlier in this chapter.[4]

Like the value of SS_{xx}, the value of SS_{yy} is always positive.

Example 13–2 illustrates the calculation of the standard deviation of errors for the data of Table 13.1.

EXAMPLE 13–2

Calculating the standard deviation of errors.

Compute the standard deviation of errors s_e for the data on monthly incomes and food expenditures of the seven households given in Table 13.1.

Solution To compute s_e, we need to know the values of SS_{yy}, SS_{xy}, and b. Earlier, in Example 13–1, we computed SS_{xy} and b. These values are

$$SS_{xy} = 211.7143 \quad \text{and} \quad b = .2642$$

[3]If we have access to population data, the value of σ_ϵ is calculated using the formula

$$\sigma_\epsilon = \sqrt{\frac{SS_{yy} - B\, SS_{xy}}{N}}$$

[4]The basic formula to calculate SS_{yy} is $\Sigma(y - \bar{y})^2$.

To compute SS_{yy}, we calculate Σy^2 as shown in Table 13.3.

Table 13.3

Income x	Food Expenditure y	y^2
35	9	81
49	15	225
21	7	49
39	11	121
15	5	25
28	8	64
25	9	81
$\Sigma x = 212$	$\Sigma y = 64$	$\Sigma y^2 = 646$

The value of SS_{yy} is

$$SS_{yy} = \Sigma y^2 - \frac{(\Sigma y)^2}{n} = 646 - \frac{(64)^2}{7} = 60.8571$$

Hence, the standard deviation of errors is

$$s_e = \sqrt{\frac{SS_{yy} - b\,SS_{xy}}{n-2}} = \sqrt{\frac{60.8571 - .2642(211.7143)}{7-2}} = \mathbf{.9922}$$ ■

13.4 Coefficient of Determination

We may ask the question: How good is the regression model? In other words: How well does the independent variable explain the dependent variable in the regression model? The *coefficient of determination* is one concept that answers this question.

For a moment, assume that we possess information only on the food expenditures of households and not on their incomes. Hence, in this case, we cannot use the regression line to predict the food expenditure for any household. As we did in earlier chapters, in the absence of a regression model, we use $\bar{y}$ to estimate or predict every household's food expenditure. Consequently, the error of prediction for each household is now given by $y - \bar{y}$, which is the difference between the actual food expenditure of a household and the mean food expenditure. If we calculate such errors for all households in the sample and then square and add them, the resulting sum is called the **total sum of squares** and is denoted by **SST**. Actually SST is the same as SS_{yy} and is defined as

$$SST = SS_{yy} = \Sigma(y - \bar{y})^2$$

However, for computational purposes, SST is calculated using the following formula.

Total Sum of Squares (SST) The *total sum of squares*, denoted by SST, is calculated as

$$SST = \Sigma y^2 - \frac{(\Sigma y)^2}{n}$$

Note that this is the same formula that we used to calculate SS_{yy}.

The value of SS_{yy}, which is 60.8571, was calculated in Example 13–2. Consequently, the value of SST is

$$SST = 60.8571$$

From Example 13–1, $\bar{y} = 9.1429$. Figure 13.15 shows the error for each of the seven households in our sample using the scatter diagram of Figure 13.4 and using $\bar{y}$.

Figure 13.15 Total errors.

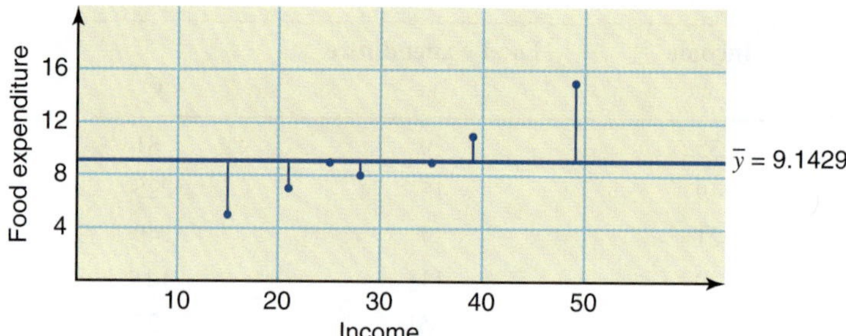

Now suppose we use the simple linear regression model to predict the food expenditure of each of the seven households in our sample. In this case, we predict each household's food expenditure by using the regression line we estimated earlier in Example 13–1, which is

$$\hat{y} = 1.1414 + .2642x$$

The predicted food expenditures, denoted by $\hat{y}$, for the seven households are listed in Table 13.4. Also given are the errors and error squares.

Table 13.4

x	y	$\hat{y} = 1.1414 + .2642x$	$e = y - \hat{y}$	$e^2 = (y - \hat{y})^2$
35	9	10.3884	−1.3884	1.9277
49	15	14.0872	.9128	.8332
21	7	6.6896	.3104	.0963
39	11	11.4452	−.4452	.1982
15	5	5.1044	−.1044	.0109
28	8	8.5390	−.5390	.2905
25	9	7.7464	1.2536	1.5715
				$\Sigma e^2 = \Sigma(y - \hat{y})^2 = 4.9283$

We calculate the values of $\hat{y}$ (given in the third column of Table 13.4) by substituting the values of x in the estimated regression model. For example, the value of x for the first household is 35. Substituting this value of x in the regression equation, we obtain

$$\hat{y} = 1.1414 + .2642(35) = 10.3884$$

Similarly we find the other values of $\hat{y}$. The error sum of squares SSE is given by the sum of the fifth column in Table 13.4. Thus,

$$\text{SSE} = \Sigma(y - \hat{y})^2 = 4.9283$$

The errors of prediction for the regression model for the seven households are shown in Figure 13.16.

Figure 13.16 Errors of prediction when regression model is used.

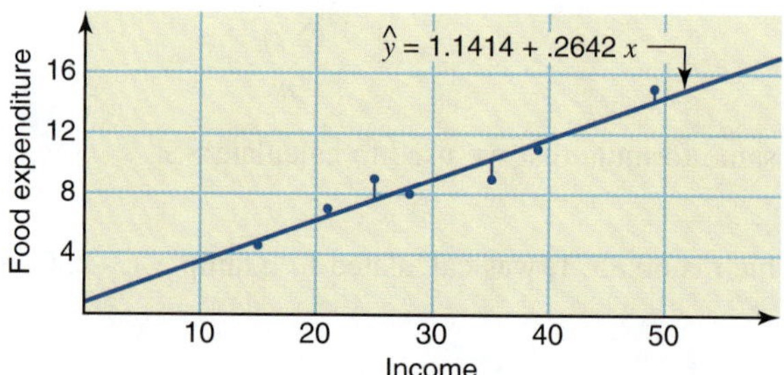

Thus, from the foregoing calculations,

$$\text{SST} = 60.8571 \quad \text{and} \quad \text{SSE} = 4.9283$$

These values indicate that the sum of squared errors decreased from 60.8571 to 4.9283 when we used $\hat{y}$ in place of $\bar{y}$ to predict food expenditures. This reduction in squared errors is called the **regression sum of squares** and is denoted by **SSR**. Thus,

$$\text{SSR} = \text{SST} - \text{SSE} = 60.8571 - 4.9283 = 55.9288$$

The value of SSR can also be computed by using the formula

$$\text{SSR} = \Sigma(\hat{y} - \bar{y})^2$$

Regression Sum of Squares (SSR) The *regression sum of squares*, denoted by SSR, is

$$\text{SSR} = \text{SST} - \text{SSE}$$

Thus, SSR is the portion of SST that is explained by the use of the regression model, and SSE is the portion of SST that is not explained by the use of the regression model. The sum of SSR and SSE is always equal to SST. Thus,

$$\text{SST} = \text{SSR} + \text{SSE}$$

The ratio of SSR to SST gives the **coefficient of determination**. The coefficient of determination calculated for population data is denoted by $\boldsymbol{\rho^2}$ (ρ is the Greek letter *rho*) and the one calculated for sample data is denoted by $\boldsymbol{r^2}$. The coefficient of determination gives the proportion of SST that is explained by the use of the regression model. The value of the coefficient of determination always lies in the range zero to one. The coefficient of determination can be calculated by using the formula

$$r^2 = \frac{\text{SSR}}{\text{SST}} \quad \text{or} \quad \frac{\text{SST} - \text{SSE}}{\text{SST}}$$

However, for computational purposes, the formula given below is more efficient to use to calculate the coefficient of determination.

Coefficient of Determination The *coefficient of determination*, denoted by r^2, represents the proportion of SST that is explained by the use of the regression model. The computational formula for r^2 is[5]

$$r^2 = \frac{b\,\text{SS}_{xy}}{\text{SS}_{yy}}$$

and

$$0 \le r^2 \le 1$$

Example 13–3 illustrates the calculation of the coefficient of determination for a sample data set.

■ EXAMPLE 13–3

Calculating the coefficient of determination.

For the data of Table 13.1 on monthly incomes and food expenditures of seven households, calculate the coefficient of determination.

[5]If we have access to population data, the value of ρ^2 is calculated using the formula

$$\rho^2 = \frac{B\,\text{SS}_{xy}}{\text{SS}_{yy}}$$

The values of SS_{xy} and SS_{yy} used here are calculated for the population data set.

Solution From earlier calculations made in Examples 13–1 and 13–2,

$$b = .2642, \quad SS_{xy} = 211.7143, \quad \text{and} \quad SS_{yy} = 60.8571$$

Hence,
$$r^2 = \frac{b\, SS_{xy}}{SS_{yy}} = \frac{(.2642)(211.7143)}{60.8571} = \mathbf{.92}$$

Thus, we can state that SST is reduced by approximately 92% (from 60.8571 to 4.9283) when we use $\hat{y}$, instead of $\bar{y}$, to predict the food expenditures of households. Note that r^2 is usually rounded to two decimal places. ■

The total sum of squares SST is a measure of the total variation in food expenditures, the regression sum of squares SSR is the portion of total variation explained by the regression model (or by income), and the error sum of squares SSE is the portion of total variation not explained by the regression model. Hence, for Example 13–3 we can state that 92% of the total variation in food expenditures of households occurs because of the variation in their incomes, and the remaining 8% is due to randomness and other variables.

Usually, the higher the value of r^2, the better the regression model. This is so because if r^2 is larger, a greater portion of the total errors is explained by the included independent variable and a smaller portion of errors is attributed to other variables and randomness.

EXERCISES

CONCEPTS AND PROCEDURES

13.31 What are the degrees of freedom for a simple linear regression model?

13.32 Explain the meaning of coefficient of determination.

13.33 Explain the meaning of SST and SSR. You may use graphs for illustration purposes.

13.34 A population data set produced the following information.

$$N = 250, \quad \Sigma x = 9880, \quad \Sigma y = 1456, \quad \Sigma xy = 85{,}080,$$
$$\Sigma x^2 = 485{,}870, \quad \text{and} \quad \Sigma y^2 = 135{,}675$$

Find the values of σ_ϵ and ρ^2.

13.35 A population data set produced the following information.

$$N = 460, \quad \Sigma x = 3920, \quad \Sigma y = 2650, \quad \Sigma xy = 26{,}570,$$
$$\Sigma x^2 = 48{,}530, \quad \text{and} \quad \Sigma y^2 = 39{,}347$$

Find the values of σ_ϵ and ρ^2.

13.36 The following information is obtained from a sample data set.

$$n = 10, \quad \Sigma x = 100, \quad \Sigma y = 220, \quad \Sigma xy = 3680,$$
$$\Sigma x^2 = 1140, \quad \text{and} \quad \Sigma y^2 = 25{,}272$$

Find the values of s_e and r^2.

13.37 The following information is obtained from a sample data set.

$$n = 12, \quad \Sigma x = 66, \quad \Sigma y = 588, \quad \Sigma xy = 2244,$$
$$\Sigma x^2 = 396, \quad \text{and} \quad \Sigma y^2 = 58{,}734$$

Find the values of s_e and r^2.

APPLICATIONS

13.38 A high school counselor wanted to know how the number of hours students spend at work during the school year affects their scholastic performance. The following table gives the number of hours worked per week at jobs by seven students and their grade point averages (GPAs).

Hours worked	10	8	20	15	18	5	10
GPA	3.5	3.7	3.0	2.8	2.1	4.0	3.6

Find the following.

a. SS_{xx}, SS_{yy}, and SS_{xy} b. Standard deviation of errors
c. SST, SSE, and SSR d. Coefficient of determination

13.39 The following table gives information on the average saturated fat (in grams) consumed per day and the cholesterol level (in milligrams per hundred milliliters) for eight men.

Fat consumption	55	68	50	34	43	58	77	36
Cholesterol level	180	215	195	165	170	204	235	150

Compute the following.

a. SS_{xx}, SS_{yy}, and SS_{xy} b. Standard deviation of errors
c. SST, SSE, and SSR d. Coefficient of determination

13.40 Refer to Exercise 13.25. The following table, which gives the ages (in years) and prices (in hundreds of dollars) of eight cars of a specific model, is reproduced from that exercise.

Age	8	3	6	9	2	5	6	3
Price	18	94	50	21	145	42	36	99

a. Calculate the standard deviation of errors.
b. Compute the coefficient of determination and give a brief interpretation of it.

13.41 The following table, reproduced from Exercise 13.26, gives the lowest temperature (in degrees Fahrenheit) on seven winter days and the number of emergency road service calls to Red's Towing and Garage Service on each of those days.

Lowest temperature	15	0	24	−10	30	9	36
Number of calls	12	22	16	31	7	24	6

a. Determine the standard deviation of errors.
b. Find the coefficient of determination and give a brief interpretation of it.

13.42 The following data set on annual incomes (in thousands of dollars) and amounts (in thousands of dollars) of life insurance policies for six persons is reproduced from Exercise 13.27.

Annual income	62	78	41	53	85	34
Life insurance	250	300	100	150	500	75

a. Find the standard deviation of errors.
b. Compute the coefficient of determination. What percentage of the variation in life insurance amounts is explained by the annual incomes? What percentage of this variation is not explained?

13.43 The following table, reproduced from Exercise 13.28, lists the sizes of six houses (in hundreds of square feet) and the monthly rents (in dollars) paid by tenants for those houses.

Size of house	21	13	19	27	34	23
Monthly rent	1000	780	1120	1500	1850	1200

a. Compute the standard deviation of errors.
b. Calculate the coefficient of determination. What percentage of the variation in monthly rents is explained by the sizes of the houses? What percentage of this variation is not explained?

13.44 Refer to data given in Exercise 13.29 on the total 2004 payroll and the percentage of games won during the 2004 season by each of the National League baseball teams.

a. Find the standard deviation of errors, σ_ϵ. (Note that this data set belongs to a population.)
b. Compute the coefficient of determination, ρ^2.

13.45 Refer to data given in Exercise 13.30 on the total 2004 payroll and the percentage of games won during the 2004 season by each of the American League baseball teams.

a. Find the standard deviation of errors, σ_ϵ. (Note that this data set belongs to a population.)
b. Compute the coefficient of determination, ρ^2.

13.5 Inferences About *B*

This section is concerned with estimation and tests of hypotheses about the population regression slope *B*. We can also make confidence intervals and test hypotheses about the *y*-intercept *A* of the population regression line. However, making inferences about *A* is beyond the scope of this text.

13.5.1 Sampling Distribution of *b*

One of the main purposes for determining a regression line is to find the true value of the slope *B* of the population regression line. However, in almost all cases, the regression line is estimated using sample data. Then, based on the sample regression line, inferences are made about the population regression line. The slope *b* of a sample regression line is a point estimator of the slope *B* of the population regression line. The different sample regression lines estimated for different samples taken from the same population will give different values of *b*. If only one sample is taken and the regression line for that sample is estimated, the value of *b* will depend on which elements are included in the sample. Thus, *b* is a random variable, and it possesses a probability distribution that is more commonly called its sampling distribution. The shape of the sampling distribution of *b*, its mean, and standard deviation are given next.

Mean, Standard Deviation, and Sampling Distribution of *b* Because of the assumption of normally distributed random errors, the sampling distribution of *b* is normal. The mean and standard deviation of *b*, denoted by μ_b and σ_b, respectively, are

$$\mu_b = B \quad \text{and} \quad \sigma_b = \frac{\sigma_\epsilon}{\sqrt{\text{SS}_{xx}}}$$

However, usually the standard deviation of population errors σ_ϵ is not known. Hence, the sample standard deviation of errors s_e is used to estimate σ_ϵ. In such a case, when σ_ϵ is unknown, the standard deviation of *b* is estimated by s_b, which is calculated as

$$s_b = \frac{s_e}{\sqrt{\text{SS}_{xx}}}$$

If σ_ϵ is known, the normal distribution can be used to make inferences about *B*. However, if σ_ϵ is not known, the normal distribution is replaced by the *t* distribution to make inferences about *B*.

13.5.2 Estimation of *B*

The value of *b* obtained from the sample regression line is a point estimate of the slope *B* of the population regression line. As mentioned in Section 13.5.1, if σ_ϵ is not known, the *t* distribution is used to make a confidence interval for *B*.

Confidence Interval for *B* The $(1 - \alpha)100\%$ *confidence interval for B* is given by

$$b \pm t s_b$$

where

$$s_b = \frac{s_e}{\sqrt{\text{SS}_{xx}}}$$

and the value of *t* is obtained from the *t* distribution table for $\alpha/2$ area in the right tail of the *t* distribution and $n - 2$ degrees of freedom.

Example 13–4 describes the procedure for making a confidence interval for *B*.

■ EXAMPLE 13–4

Constructing a confidence interval for B.

Construct a 95% confidence interval for B for the data on incomes and food expenditures of seven households given in Table 13.1.

Solution From the given information and earlier calculations in Examples 13–1 and 13–2,

$$n = 7, \quad b = .2642, \quad SS_{xx} = 801.4286, \quad \text{and} \quad s_e = .9922$$

The confidence level is 95%.

$$s_b = \frac{s_e}{\sqrt{SS_{xx}}} = \frac{.9922}{\sqrt{801.4286}} = .0350$$

$$df = n - 2 = 7 - 2 = 5$$

$$\alpha/2 = (1 - .95)/2 = .025$$

From the t distribution table, the value of t for 5 df and .025 area in the right tail of the t distribution curve is 2.571. The 95% confidence interval for B is

$$b \pm ts_b = .2642 \pm 2.571(.0350) = .2642 \pm .0900 = \mathbf{.17 \text{ to } .35}$$

Thus, we are 95% confident that the slope B of the population regression line is between .17 and .35. ■

13.5.3 Hypothesis Testing About B

Testing a hypothesis about B when the null hypothesis is $B = 0$ (that is, the slope of the regression line is zero) is equivalent to testing that x does not determine y and that the regression line is of no use in predicting y for a given x. However, we should remember that we are testing for a linear relationship between x and y. It is possible that x may determine y nonlinearly. Hence, a nonlinear relationship may exist between x and y.

To test the hypothesis that x does not determine y linearly, we will test the null hypothesis that the slope of the regression line is zero; that is, $B = 0$. The alternative hypothesis can be: (1) x determines y; that is, $B \neq 0$; (2) x determines y positively; that is, $B > 0$; or (3) x determines y negatively; that is, $B < 0$.

The procedure used to make a hypothesis test about B is similar to the one used in earlier chapters. It involves the same five steps.

Test Statistic for b The value of the *test statistic t for b* is calculated as

$$t = \frac{b - B}{s_b}$$

The value of B is substituted from the null hypothesis.

Example 13–5 illustrates the procedure for testing a hypothesis about B.

■ EXAMPLE 13–5

Conducting a test of hypothesis about B.

Test at the 1% significance level whether the slope of the regression line for the example on incomes and food expenditures of seven households is positive.

Solution From the given information and earlier calculations in Examples 13–1 and 13–4,

$$n = 7, \quad b = .2642, \quad \text{and} \quad s_b = .0350$$

Step 1. *State the null and alternative hypotheses.*

We are to test whether or not the slope B of the population regression line is positive. Hence, the two hypotheses are

$$H_0: B = 0 \quad \text{(The slope is zero)}$$

$$H_1: B > 0 \quad \text{(The slope is positive)}$$

Note that we can also write the null hypothesis as H_0: $B \leq 0$, which states that the slope is either zero or negative.

Step 2. *Select the distribution to use.*

Here, σ_ϵ is not known. Hence, we will use the t distribution to make the test about B.

Step 3. *Determine the rejection and nonrejection regions.*

The significance level is .01. The $>$ sign in the alternative hypothesis indicates that the test is right-tailed. Therefore,

$$\text{Area in the right tail of the } t \text{ distribution} = \alpha = .01$$

$$df = n - 2 = 7 - 2 = 5$$

From the t distribution table, the critical value of t for 5 df and .01 area in the right tail of the t distribution is 3.365, as shown in Figure 13.17.

Figure 13.17

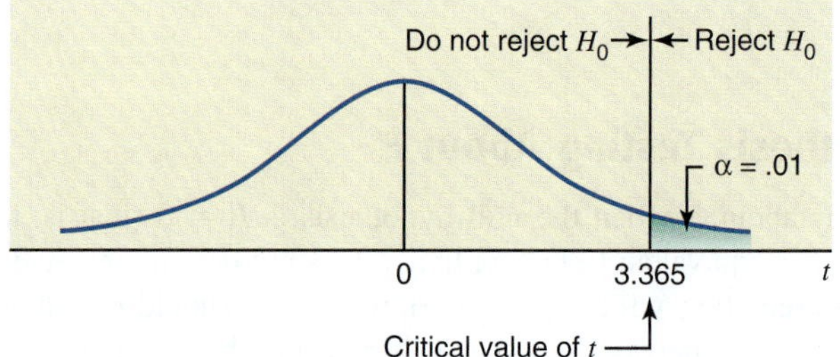

Step 4. *Calculate the value of the test statistic.*

The value of the test statistic t for b is calculated as follows:

From H_0

$$t = \frac{b - B}{s_b} = \frac{.2642 - 0}{.0350} = 7.549$$

Step 5. *Make a decision.*

The value of the test statistic $t = 7.549$ is greater than the critical value of $t = 3.365$, and it falls in the rejection region. Hence, we reject the null hypothesis and conclude that x (income) determines y (food expenditure) positively. That is, food expenditure increases with an increase in income and it decreases with a decrease in income.

Using the *p*-Value to Make a Decision

We can find the range for the p-value (as we did in Chapters 9 and 10) from the t distribution table, Table V of Appendix C, and make a decision by comparing that p-value with the significance level. For this example, $df = 5$ and the observed value of t is 7.549. From Table V (the t distribution table) in the row of $df = 5$, the largest value of t is 5.893 for which the area in the right tail of the t distribution is .001. Since our observed value of $t = 7.549$ is larger than 5.893, the p-value for $t = 7.549$ is less than .001, i.e.,

$$p\text{-value} < .001$$

Note that if we use technology to find this p-value, we will obtain a p-value of .000. Thus, we can state that for any α equal to or greater than .001 (the upper limit of the p-value range), we will reject the null hypothesis. For our example, $\alpha = .01$, which is greater than the p-value of .001. As a result, we reject the null hypothesis. ■

Note that the null hypothesis does not always have to be $B = 0$. We may test the null hypothesis that B is equal to a certain value. See Exercises 13.47 to 13.50, 13.57, and 13.58 for such cases.

EXERCISES

CONCEPTS AND PROCEDURES

13.46 Describe the mean, standard deviation, and shape of the sampling distribution of the slope b of the simple linear regression model.

13.47 The following information is obtained for a sample of 16 observations taken from a population.

$$SS_{xx} = 340.700, \quad s_e = 1.951, \quad \text{and} \quad \hat{y} = 12.45 + 6.32x$$

a. Make a 99% confidence interval for B.
b. Using a significance level of .025, can you conclude that B is positive?
c. Using a significance level of .01, can you conclude that B is different from zero?
d. Using a significance level of .02, test whether B is different from 4.50. (*Hint:* The null hypothesis here will be H_0: $B = 4.50$, and the alternative hypothesis will be H_1: $B \neq 4.50$. Notice that the value of $B = 4.50$ will be used to calculate the value of the test statistic t.)

13.48 The following information is obtained for a sample of 25 observations taken from a population.

$$SS_{xx} = 274.600, \quad s_e = .932, \quad \text{and} \quad \hat{y} = 280.56 - 3.77x$$

a. Make a 95% confidence interval for B.
b. Using a significance level of .01, test whether B is negative.
c. Testing at the 5% significance level, can you conclude that B is different from zero?
d. Test if B is different from -5.20. Use $\alpha = .01$.

13.49 The following information is obtained for a sample of 100 observations taken from a population.

$$SS_{xx} = 524.884 \quad s_e = 1.464, \quad \text{and} \quad \hat{y} = 5.48 + 2.50x$$

a. Make a 98% confidence interval for B.
b. Test at the 2.5% significance level whether B is positive.
c. Can you conclude that B is different from zero? Use $\alpha = .01$.
d. Using a significance level of .01, test whether B is greater than 1.75.

13.50 The following information is obtained for a sample of 80 observations taken from a population.

$$SS_{xx} = 380.592, \quad s_e = .961, \quad \text{and} \quad \hat{y} = 160.24 - 2.70x$$

a. Make a 97% confidence interval for B.
b. Test at the 1% significance level whether B is negative.
c. Can you conclude that B is different from zero? Use $\alpha = .01$.
d. Using a significance level of .02, test whether B is less than -1.25.

APPLICATIONS

13.51 Refer to Exercise 13.25. The data on ages (in years) and prices (in hundreds of dollars) for eight cars of a specific model are reproduced from that exercise.

Age	8	3	6	9	2	5	6	3
Price	18	94	50	21	145	42	36	99

a. Construct a 95% confidence interval for B. You can use results obtained in Exercises 13.25 and 13.40 here.
b. Test at the 5% significance level whether B is negative.

13.52 The data given in the table below are the midterm scores in a course for a sample of 10 students and the scores of student evaluations of the instructor. (In the instructor evaluation scores, 1 is the lowest and 4 is the highest score.)

Instructor score	3	2	3	1	2	4	3	4	4	2
Midterm score	90	75	97	64	47	99	75	88	93	81

a. Find the regression of instructor scores on midterm scores.
b. Construct a 99% confidence interval for B.
c. Test at the 1% significance level whether B is positive.

13.53 The following data give the experience (in years) and monthly salaries (in hundreds of dollars) of nine randomly selected secretaries.

Experience	14	3	5	6	4	9	18	5	16
Monthly salary	42	24	33	31	29	39	47	30	43

a. Find the least squares regression line with experience as an independent variable and monthly salary as a dependent variable.
b. Construct a 98% confidence interval for B.
c. Test at the 2.5% significance level whether B is greater than zero.

13.54 The following table, reproduced from Exercise 13.26, gives the lowest temperature (in degrees Fahrenheit) on seven winter days and the number of emergency road service calls to Red's Towing and Garage Service on each of those days.

Lowest temperature	15	0	24	−10	30	9	36
Number of calls	12	22	16	31	7	24	6

a. Make a 95% confidence interval for B. You can use the calculations made in Exercises 13.26 and 13.41 here.
b. Test at the 2.5% significance level whether B is negative.

13.55 The following data set on annual incomes (in thousands of dollars) and amounts (in thousands of dollars) of life insurance policies for six persons is reproduced from Exercise 13.27.

Annual income	62	78	41	53	85	34
Life insurance	250	300	100	150	500	75

a. Construct a 99% confidence interval for B. You can use the calculations made in Exercises 13.27 and 13.42 here.
b. Test at the 1% significance level whether B is different from zero.

13.56 The data on the sizes of six houses (in hundreds of square feet) and the monthly rents (in dollars) paid by tenants for those houses are reproduced from Exercise 13.28.

Size of house	21	13	19	27	34	23
Monthly rent	1000	780	1120	1500	1850	1200

a. Construct a 98% confidence interval for B. You can use the calculations made in Exercises 13.28 and 13.43 here.
b. Testing at the 5% significance level, can you conclude that B is different from zero?

13.57 Refer to Exercise 13.38. The following table gives the number of hours worked per week at jobs by seven high school students and their grade point averages (GPAs).

Hours worked	10	8	20	15	18	5	10
GPA	3.5	3.7	3.0	2.8	2.1	4.0	3.6

a. Find the least squares regression line with hours worked as an independent variable and GPA as a dependent variable.
b. Make a 95% confidence interval for B. You may use the results obtained in Exercise 13.38 here.
c. An earlier study claims that $B = -.04$ for the relationship between hours worked and GPA. Test at the 5% significance level whether B is lower than $-.04$. (*Hint:* Here the null hypothesis will be H_0: $B = -.04$ and the alternative hypothesis will be H_1: $B < -.04$. Notice that the value of $B = -.04$ will be used to calculate the value of the test statistic t.)

13.58 The following table, reproduced from Exercise 13.39, gives information on the average saturated fat (in grams) consumed per day and the cholesterol level (in milligrams per hundred milliliters) of eight men.

Fat consumption	55	68	50	34	43	58	77	36
Cholesterol level	180	215	195	165	170	204	235	150

a. Find the regression line $\hat{y} = a + bx$, where x is the fat consumption and y is the cholesterol level. You can use the results obtained in Exercise 13.39 here.
b. Construct a 90% confidence interval for B.
c. An earlier study claims that B is 1.75. Test at the 5% significance level whether B is different from 1.75.

13.6 Linear Correlation

This section describes the meaning and calculation of the linear correlation coefficient and the procedure to conduct a test of hypothesis about it.

13.6.1 Linear Correlation Coefficient

Another measure of the relationship between two variables is the correlation coefficient. This section describes the simple linear correlation, for short **linear correlation**, which measures the strength of the linear association between two variables. In other words, the linear correlation coefficient measures how closely the points in a scatter diagram are spread around the regression line. The correlation coefficient calculated for the population data is denoted by $\boldsymbol{\rho}$ (Greek letter *rho*) and the one calculated for sample data is denoted by $\boldsymbol{r}$. (Note that the square of the correlation coefficient is equal to the coefficient of determination.)

Value of the Correlation Coefficient The *value of the correlation coefficient* always lies in the range -1 to 1; that is,

$$-1 \leq \rho \leq 1 \quad \text{and} \quad -1 \leq r \leq 1$$

Although we can explain the linear correlation using the population correlation coefficient ρ, we will do so using the sample correlation coefficient r.

If $r = 1$, it is said to be a *perfect positive linear correlation.* In such a case, all points in the scatter diagram lie on a straight line that slopes upward from left to right, as shown in Figure 13.18*a*. If $r = -1$, the correlation is said to be a *perfect negative linear correlation.* In this case, all points in the scatter diagram fall on a straight line that slopes downward from left to right, as shown in Figure 13.18*b*. If the points are scattered all over the diagram, as shown in Figure 13.18*c*, then there is *no linear correlation* between the two variables and consequently r is close to 0. Note that here r is *not* equal to zero but very *close* to zero.

We do not usually encounter an example with perfect positive or perfect negative correlation. What we observe in real-world problems is either a positive linear correlation with $0 < r < 1$ (that is, the correlation coefficient is greater than zero but less than 1) or a negative linear correlation with $-1 < r < 0$ (that is, the correlation coefficient is greater than -1 but less than zero).

If the correlation between two variables is positive and close to 1, we say that the variables have a *strong positive linear correlation.* If the correlation between two variables is positive but close to zero, then the variables have a *weak positive linear correlation.* In contrast, if the correlation between two variables is negative and close to -1, then the variables are said to have a *strong negative linear correlation.* If the correlation between two variables is negative but close to zero, there exists a *weak negative linear correlation* between the variables. Graphically, a strong correlation indicates that the points in the scatter diagram are very close to the regression line, and a weak correlation indicates that the points in the scatter diagram are widely spread around the regression line. These four cases are shown in Figure 13.19*a–d*.

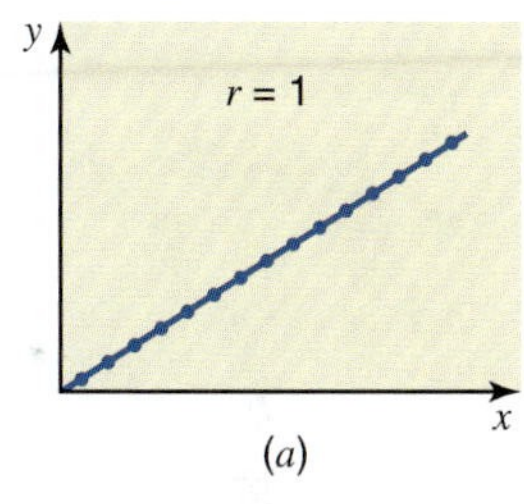

(*a*)

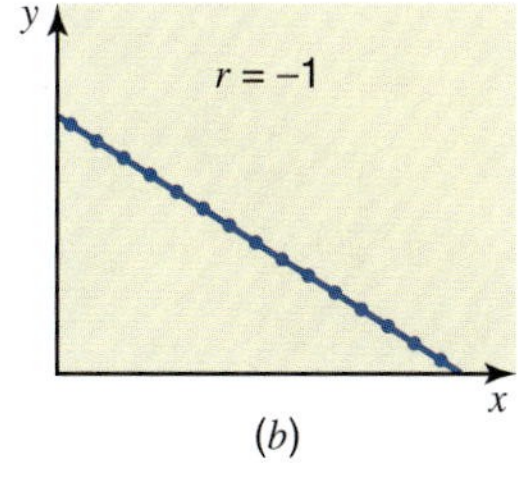

(*b*)

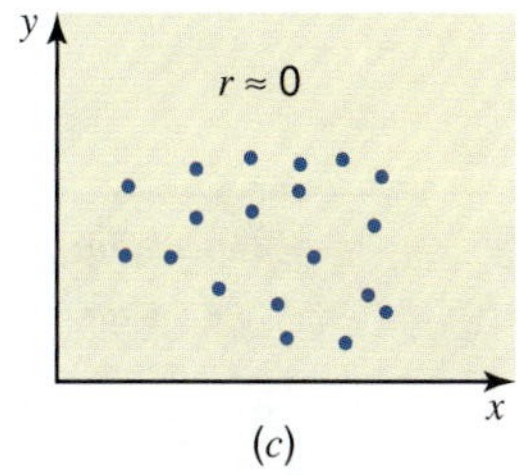

(*c*)

Figure 13.18 Linear correlation between two variables. (*a*) Perfect positive linear correlation, $r = 1$. (*b*) Perfect negative linear correlation, $r = -1$. (*c*) No linear correlation, $r \approx 0$.

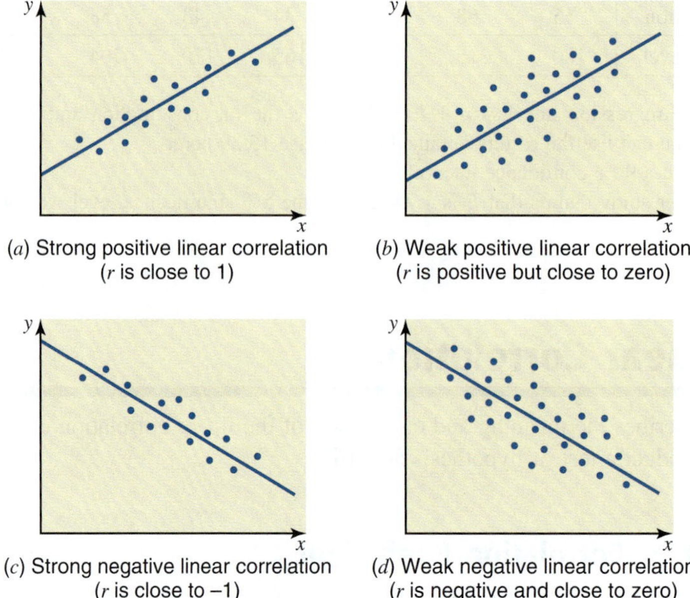

y
x
(a) Strong positive linear correlation (r is close to 1)
(b) Weak positive linear correlation (r is positive but close to zero)
(c) Strong negative linear correlation (r is close to −1)
(d) Weak negative linear correlation (r is negative and close to zero)

Figure 13.19 Linear correlation between two variables.

The linear correlation coefficient is calculated by using the following formula. (This correlation coefficient is also called the *Pearson product moment correlation coefficient.*)

Linear Correlation Coefficient The *simple linear correlation coefficient,* denoted by r, measures the strength of the linear relationship between two variables for a sample and is calculated as[6]

$$r = \frac{SS_{xy}}{\sqrt{SS_{xx}\, SS_{yy}}}$$

Because both SS_{xx} and SS_{yy} are always positive, the sign of the correlation coefficient r depends on the sign of SS_{xy}. If SS_{xy} is positive, then r will be positive, and if SS_{xy} is negative, then r will be negative. Another important observation to remember is that *r and b, calculated for the same sample, will always have the same sign. That is, both r and b are either positive or negative*. This is so because both r and b provide information about the relationship between x and y. Likewise, the corresponding population parameters ρ and B will always have the same sign.

Example 13–6 illustrates the calculation of the linear correlation coefficient r.

■ EXAMPLE 13–6

Calculating the linear correlation coefficient.

Calculate the correlation coefficient for the example on incomes and food expenditures of seven households.

Solution From earlier calculations made in Examples 13–1 and 13–2,

$$SS_{xy} = 211.7143, \quad SS_{xx} = 801.4286, \quad \text{and} \quad SS_{yy} = 60.8571$$

[6]If we have access to population data, the value of ρ is calculated using the formula

$$\rho = \frac{SS_{xy}}{\sqrt{SS_{xx}\, SS_{yy}}}$$

Here the values of SS_{xy}, SS_{xx}, and SS_{yy} are calculated using the population data.

Substituting these values in the formula for r, we obtain

$$r = \frac{SS_{xy}}{\sqrt{SS_{xx}\, SS_{yy}}} = \frac{211.7143}{\sqrt{(801.4286)(60.8571)}} = \mathbf{.96}$$

Thus, the linear correlation coefficient is .96. The correlation coefficient is usually rounded to two decimal places. ■

The linear correlation coefficient simply tells us how strongly the two variables are (linearly) related. The correlation coefficient of .96 for incomes and food expenditures of seven households indicates that income and food expenditure are very strongly and positively correlated. This correlation coefficient does not, however, provide us with any more information.

The square of the correlation coefficient gives the coefficient of determination, which was explained in Section 13.4. Thus, $(.96)^2$ is .92, which is the value of r^2 calculated in Example 13–3.

Sometimes the calculated value of r may indicate that the two variables are very strongly linearly correlated but in reality they are not. For example, if we calculate the correlation coefficient between the price of Coke and the size of families in the United States using data for the past 30 years, we will find a strong negative linear correlation. Over time, the price of Coke has increased and the size of families has decreased. This finding does not mean that family size and the price of Coke are related. As a result, before we calculate the correlation coefficient, we must seek help from a theory or from common sense to postulate whether or not the two variables have a causal relationship.

Another point to note is that in a simple regression model, one of the two variables is categorized as an independent variable and the other is classified as a dependent variable. However, no such distinction is made between the two variables when the correlation coefficient is calculated.

13.6.2 Hypothesis Testing About the Linear Correlation Coefficient

This section describes how to perform a test of hypothesis about the population correlation coefficient ρ using the sample correlation coefficient r. We can use the t distribution to make this test. However, to use the t distribution, both variables should be normally distributed.

Usually (although not always), the null hypothesis is that the linear correlation coefficient between the two variables is zero, that is $\rho = 0$. The alternative hypothesis can be one of the following: (1) the linear correlation coefficient between the two variables is less than zero, that is $\rho < 0$; (2) the linear correlation coefficient between the two variables is greater than zero, that is $\rho > 0$; or (3) the linear correlation coefficient between the two variables is not equal to zero, that is $\rho \neq 0$.

Test Statistic for *r* If both variables are normally distributed and the null hypothesis is H_0: $\rho = 0$, then the value of the test statistic t is calculated as

$$t = r\sqrt{\frac{n-2}{1-r^2}}$$

Here $n - 2$ are the degrees of freedom.

Example 13–7 describes the procedure to perform a test of hypothesis about the linear correlation coefficient.

■ EXAMPLE 13–7

Using the 1% level of significance and the data from Example 13–1, test whether the linear correlation coefficient between incomes and food expenditures is positive. Assume that the populations of both variables are normally distributed.

Performing a test of hypothesis about the correlation coefficient.

Solution From Examples 13–1 and 13–6:

$$n = 7 \quad \text{and} \quad r = .96$$

Below we use the five steps to perform this test of hypothesis.

Step 1. *State the null and alternative hypotheses.*

We are to test whether the linear correlation coefficient between incomes and food expenditures is positive. Hence, the null and alternative hypotheses are:

H_0: $\rho = 0$ (The linear correlation coefficient is zero)

H_1: $\rho > 0$ (The linear correlation coefficient is positive)

Step 2. *Select the distribution to use.*

The population distributions for both variables are normally distributed. Hence, we can use the t distribution to perform this test about the linear correlation coefficient.

Step 3. *Determine the rejection and nonrejection regions.*

The significance level is 1%. From the alternative hypothesis we know that the test is right-tailed. Hence,

$$\text{Area in the right tail of the } t \text{ distribution} = .01$$

$$df = n - 2 = 7 - 2 = 5$$

From the t distribution table, the critical value of t is 3.365. The rejection and nonrejection regions for this test are shown in Figure 13.20.

Figure 13.20

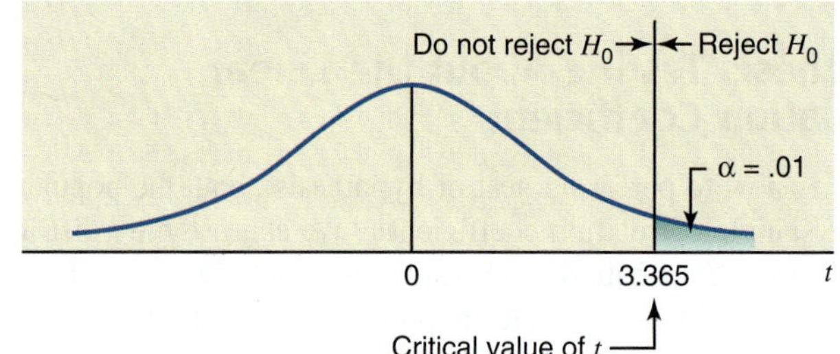

Step 4. *Calculate the value of the test statistic.*

The value of the test statistic t for r is calculated as follows:

$$t = r\sqrt{\frac{n-2}{1-r^2}} = .96\sqrt{\frac{7-2}{1-(.96)^2}} = 7.667$$

Step 5. *Make a decision.*

The value of the test statistic $t = 7.667$ is greater than the critical value of $t = 3.365$ and it falls in the rejection region. Hence, we reject the null hypothesis and conclude that there is a positive linear relationship between incomes and food expenditures.

Using the *p*-Value to Make a Decision

We can find the range for the p-value from the t distribution table (Table V of Appendix C) and make a decision by comparing that p-value with the significance level. For this example, $df = 5$ and the observed value of t is 7.667. From Table V (the t distribution table) in the row of $df = 5$, the largest value of t is 5.893 for which the area in the right tail of the t distribution is .001. Since our observed value of $t = 7.667$ is larger than 5.893, the p-value for $t = 7.667$ is less than .001, i.e.,

$$p\text{-value} < .001$$

Thus, we can state that for any α equal to or greater than .001 (the upper limit of the p-value range), we will reject the null hypothesis. For our example, $\alpha = .01$, which is greater than the p-value of .001. As a result, we reject the null hypothesis. ■

EXERCISES

CONCEPTS AND PROCEDURES

13.59 What does a linear correlation coefficient tell about the relationship between two variables? Within what range can a correlation coefficient assume a value?

13.60 What is the difference between ρ and r? Explain.

13.61 Explain each of the following concepts. You may use graphs to illustrate each concept.

- **a.** Perfect positive linear correlation
- **b.** Perfect negative linear correlation
- **c.** Strong positive linear correlation
- **d.** Strong negative linear correlation
- **e.** Weak positive linear correlation
- **f.** Weak negative linear correlation
- **g.** No linear correlation

13.62 Can the values of B and ρ calculated for the same population data have different signs? Explain.

13.63 For a sample data set, the linear correlation coefficient r has a positive value. Which of the following is true about the slope b of the regression line estimated for the same sample data?

- **a.** The value of b will be positive.
- **b.** The value of b will be negative.
- **c.** The value of b can be positive or negative.

13.64 For a sample data set, the slope b of the regression line has a negative value. Which of the following is true about the linear correlation coefficient r calculated for the same sample data?

- **a.** The value of r will be positive.
- **b.** The value of r will be negative.
- **c.** The value of r can be positive or negative.

13.65 For a sample data set on two variables, the value of the linear correlation coefficient is zero. Does this mean that these variables are not related? Explain.

13.66 Will you expect a positive, zero, or negative linear correlation between the two variables for each of the following examples?

- **a.** Grade of a student and hours spent studying
- **b.** Incomes and entertainment expenditures of households
- **c.** Ages of women and makeup expenses per month
- **d.** Price of a computer and consumption of Coke
- **e.** Price and consumption of wine

13.67 Will you expect a positive, zero, or negative linear correlation between the two variables for each of the following examples?

- **a.** SAT scores and GPAs of students
- **b.** Stress level and blood pressure of individuals
- **c.** Amount of fertilizer used and yield of corn per acre
- **d.** Ages and prices of houses
- **e.** Heights of husbands and incomes of their wives

13.68 A population data set produced the following information.

$$N = 250, \quad \Sigma x = 9880, \quad \Sigma y = 1456, \quad \Sigma xy = 85{,}080,$$
$$\Sigma x^2 = 485{,}870, \quad \text{and} \quad \Sigma y^2 = 135{,}675$$

Find the linear correlation coefficient ρ.

13.69 A population data set produced the following information.

$$N = 460, \quad \Sigma x = 3920, \quad \Sigma y = 2650, \quad \Sigma xy = 26{,}570,$$
$$\Sigma x^2 = 48{,}530, \quad \text{and} \quad \Sigma y^2 = 39{,}347$$

Find the linear correlation coefficient ρ.

13.70 A sample data set produced the following information.

$$n = 10, \quad \Sigma x = 100, \quad \Sigma y = 220, \quad \Sigma xy = 3680,$$
$$\Sigma x^2 = 1140, \quad \text{and} \quad \Sigma y^2 = 25{,}272$$

a. Calculate the linear correlation coefficient r.
b. Using the 2% significance level, can you conclude that ρ is different from zero?

13.71 A sample data set produced the following information.

$$n = 12, \quad \Sigma x = 66, \quad \Sigma y = 588, \quad \Sigma xy = 2244,$$
$$\Sigma x^2 = 396, \quad \text{and} \quad \Sigma y^2 = 58{,}734$$

a. Calculate the linear correlation coefficient r.
b. Using the 1% significance level, can you conclude that ρ is negative?

APPLICATIONS

13.72 Refer to Exercise 13.25. The data on ages (in years) and prices (in hundreds of dollars) for eight cars of a specific model are reproduced from that exercise.

Age	8	3	6	9	2	5	6	3
Price	18	94	50	21	145	42	36	99

a. Do you expect the ages and prices of cars to be positively or negatively related? Explain.
b. Calculate the linear correlation coefficient.
c. Test at the 2.5% significance level whether ρ is negative.

13.73 The following table, reproduced from Exercise 13.53, gives the experience (in years) and monthly salaries (in hundreds of dollars) of nine randomly selected secretaries.

Experience	14	3	5	6	4	9	18	5	16
Monthly salary	42	24	33	31	29	39	47	30	43

a. Do you expect the experience and monthly salaries to be positively or negatively related? Explain.
b. Compute the linear correlation coefficient.
c. Test at the 5% significance level whether ρ is positive.

13.74 The following table lists the midterm and final exam scores for seven students in a statistics class.

Midterm score	79	95	81	66	87	94	59
Final exam score	85	97	78	76	94	84	67

a. Do you expect the midterm and final exam scores to be positively or negatively related?
b. Plot a scatter diagram. By looking at the scatter diagram, do you expect the correlation coefficient between these two variables to be close to zero, 1, or -1?
c. Find the correlation coefficient. Is the value of r consistent with what you expected in parts a and b?
d. Using the 1% significance level, test whether the linear correlation coefficient is positive.

13.75 The following data give the ages of husbands and wives for six couples.

Husband's age	43	57	28	19	35	39
Wife's age	37	51	32	20	33	38

a. Do you expect the ages of husbands and wives to be positively or negatively related?
b. Plot a scatter diagram. By looking at the scatter diagram, do you expect the correlation coefficient between these two variables to be close to zero, 1, or -1?
c. Find the correlation coefficient. Is the value of r consistent with what you expected in parts a and b?
d. Using the 5% significance level, test whether the correlation coefficient is different from zero.

13.76 The following table, reproduced from Exercise 13.26, gives the lowest temperature (in degrees Fahrenheit) on seven winter days and the number of emergency road service calls made to Red's Towing and Garage Service on each of those days.

Lowest temperature	15	0	24	−10	30	9	36
Number of calls	12	22	16	31	7	24	6

a. Find the correlation coefficient. Is the sign of the correlation coefficient the same as that of b calculated in Exercise 13.26?

b. Test at the 2.5% significance level whether the linear correlation coefficient is negative. Is your decision the same as in the test of B in Exercise 13.54?

13.77 The following table, reproduced from Exercise 13.58, gives information on average saturated fat (in grams) consumed per day and cholesterol level (in milligrams per hundred milliliters) of eight men.

Fat consumption	55	68	50	34	43	58	77	36
Cholesterol level	180	215	195	165	170	204	235	150

a. Find the correlation coefficient. Is the sign of the correlation coefficient the same as that of b calculated in Exercise 13.58?

b. Test at the 1% significance level whether ρ is different from zero.

13.78 Refer to data given in Exercise 13.29 on the total 2004 payroll and the percentage of games won during the 2004 season by each of the National League baseball teams. Compute the linear correlation coefficient, ρ.

13.79 Refer to data given in Exercise 13.30 on the total 2004 payroll and the percentage of games won during the 2004 season by each of the American League baseball teams. Compute the linear correlation coefficient, ρ.

13.7 Regression Analysis: A Complete Example

This section works out an example that includes all the topics we have discussed so far in this chapter.

■ EXAMPLE 13–8

A complete example of regression analysis.

A random sample of eight drivers insured with a company and having similar auto insurance policies was selected. The following table lists their driving experiences (in years) and monthly auto insurance premiums.

Driving Experience (years)	Monthly Auto Insurance Premium
5	\$64
2	87
12	50
9	71
15	44
6	56
25	42
16	60

(a) Does the insurance premium depend on the driving experience or does the driving experience depend on the insurance premium? Do you expect a positive or a negative relationship between these two variables?

(b) Compute SS_{xx}, SS_{yy}, and SS_{xy}.

(c) Find the least squares regression line by choosing appropriate dependent and independent variables based on your answer in part a.

(d) Interpret the meaning of the values of a and b calculated in part c.

(e) Plot the scatter diagram and the regression line.

(f) Calculate r and r^2 and explain what they mean.

(g) Predict the monthly auto insurance premium for a driver with 10 years of driving experience.

(h) Compute the standard deviation of errors.

(i) Construct a 90% confidence interval for B.

(j) Test at the 5% significance level whether B is negative.

(k) Using $\alpha = .05$, test whether ρ is different from zero.

Solution

(a) Based on theory and intuition, we expect the insurance premium to depend on driving experience. Consequently, the insurance premium is a dependent variable and driving experience is an independent variable in the regression model. A new driver is considered a high risk by the insurance companies, and he or she has to pay a higher premium for auto insurance. On average, the insurance premium is expected to decrease with an increase in the years of driving experience. Therefore, we expect a negative relationship between these two variables. In other words, both the population correlation coefficient ρ and the population regression slope B are expected to be negative.

(b) Table 13.5 shows the calculation of Σx, Σy, Σxy, Σx^2, and Σy^2.

Table 13.5

Experience x	Premium y	xy	x^2	y^2
5	64	320	25	4096
2	87	174	4	7569
12	50	600	144	2500
9	71	639	81	5041
15	44	660	225	1936
6	56	336	36	3136
25	42	1050	625	1764
16	60	960	256	3600
$\Sigma x = 90$	$\Sigma y = 474$	$\Sigma xy = 4739$	$\Sigma x^2 = 1396$	$\Sigma y^2 = 29{,}642$

The values of $\bar{x}$ and $\bar{y}$ are

$$\bar{x} = \Sigma x/n = 90/8 = 11.25$$

$$\bar{y} = \Sigma y/n = 474/8 = 59.25$$

The values of SS_{xy}, SS_{xx}, and SS_{yy} are computed as follows:

$$SS_{xy} = \Sigma xy - \frac{(\Sigma x)(\Sigma y)}{n} = 4739 - \frac{(90)(474)}{8} = \mathbf{-593.5000}$$

$$SS_{xx} = \Sigma x^2 - \frac{(\Sigma x)^2}{n} = 1396 - \frac{(90)^2}{8} = \mathbf{383.5000}$$

$$SS_{yy} = \Sigma y^2 - \frac{(\Sigma y)^2}{n} = 29{,}642 - \frac{(474)^2}{8} = \mathbf{1557.5000}$$

(c) To find the regression line, we calculate a and b as follows:

$$b = \frac{SS_{xy}}{SS_{xx}} = \frac{-593.5000}{383.5000} = -1.5476$$

$$a = \bar{y} - b\bar{x} = 59.25 - (-1.5476)(11.25) = 76.6605$$

Thus, our estimated regression line $\hat{y} = a + bx$ is

$$\hat{y} = \mathbf{76.6605 - 1.5476}x$$

(d) The value of $a = 76.6605$ gives the value of $\hat{y}$ for $x = 0$; that is, it gives the monthly auto insurance premium for a driver with no driving experience. However, as mentioned earlier in this chapter, we should not attach much importance to this statement because the sample contains drivers with only two or more years of experience. The value of b gives the change in $\hat{y}$ due to a change of one unit in x. Thus, $b = -1.5476$ indicates that, on average, for every extra year of driving experience, the monthly auto insurance premium decreases by \$1.55. Note that when b is negative, y decreases as x increases.

(e) Figure 13.21 shows the scatter diagram and the regression line for the data on eight auto drivers. Note that the regression line slopes downward from left to right. This result is consistent with the negative relationship we anticipated between driving experience and insurance premium.

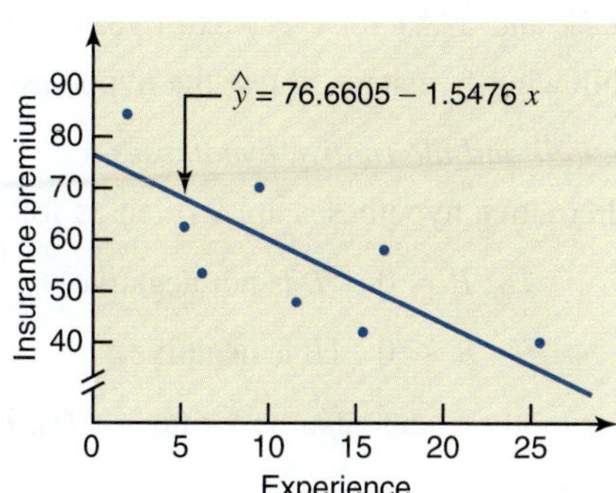

Figure 13.21 Scatter diagram and the regression line.

(f) The values of r and r^2 are computed as follows:

$$r = \frac{SS_{xy}}{\sqrt{SS_{xx}\,SS_{yy}}} = \frac{-593.5000}{\sqrt{(383.5000)(1557.5000)}} = \mathbf{-.77}$$

$$r^2 = \frac{b\,SS_{xy}}{SS_{yy}} = \frac{(-1.5476)(-593.5000)}{1557.5000} = \mathbf{.59}$$

The value of $r = -.77$ indicates that the driving experience and the monthly auto insurance premium are negatively related. The (linear) relationship is strong but not very strong. The value of $r^2 = .59$ states that 59% of the total variation in insurance premiums is explained by years of driving experience and 41% is not. The low value of r^2 indicates that there may be many other important variables that contribute to the determination of auto insurance premiums. For example, the premium is expected to depend on the driving record of a driver and the type and age of the car.

(g) Using the estimated regression line, we find the predicted value of y for $x = 10$ is

$$\hat{y} = 76.6605 - 1.5476x = 76.6605 - 1.5476(10) = \mathbf{\$61.18}$$

Thus, we expect the monthly auto insurance premium of a driver with 10 years of driving experience to be \$61.18.

(h) The standard deviation of errors is

$$s_e = \sqrt{\frac{SS_{yy} - b\,SS_{xy}}{n-2}} = \sqrt{\frac{1557.5000 - (-1.5476)(-593.5000)}{8-2}} = \mathbf{10.3199}$$

(i) To construct a 90% confidence interval for B, first we calculate the standard deviation of b:

$$s_b = \frac{s_e}{\sqrt{SS_{xx}}} = \frac{10.3199}{\sqrt{383.5000}} = .5270$$

For a 90% confidence level, the area in each tail of the t distribution is

$$\alpha/2 = (1 - .90)/2 = .05$$

The degrees of freedom are

$$df = n - 2 = 8 - 2 = 6$$

From the t distribution table, the t value for .05 area in the right tail of the t distribution and 6 df is 1.943. The 90% confidence interval for B is

$$b \pm ts_b = -1.5476 \pm 1.943(.5270)$$
$$= -1.5476 \pm 1.0240 = \mathbf{-2.57 \text{ to } -.52}$$

Thus, we can state with 90% confidence that B lies in the interval -2.57 to $-.52$. That is, on average, the monthly auto insurance premium of a driver decreases by an amount between \$.52 and \$2.57 for every extra year of driving experience.

(j) We perform the following five steps to test the hypothesis about B.

Step 1. *State the null and alternative hypotheses.*

The null and alternative hypotheses are written as follows:

$$H_0: B = 0 \quad (B \text{ is not negative})$$
$$H_1: B < 0 \quad (B \text{ is negative})$$

Note that the null hypothesis can also be written as H_0: $B \geq 0$.

Step 2. *Select the distribution to use.*

Because σ_ϵ is not known, we use the t distribution to make the hypothesis test.

Step 3. *Determine the rejection and nonrejection regions.*

The significance level is .05. The < sign in the alternative hypothesis indicates that it is a left-tailed test.

$$\text{Area in the left tail of the } t \text{ distribution} = \alpha = .05$$
$$df = n - 2 = 8 - 2 = 6$$

From the t distribution table, the critical value of t for .05 area in the left tail of the t distribution and 6 df is -1.943, as shown in Figure 13.22.

Figure 13.22

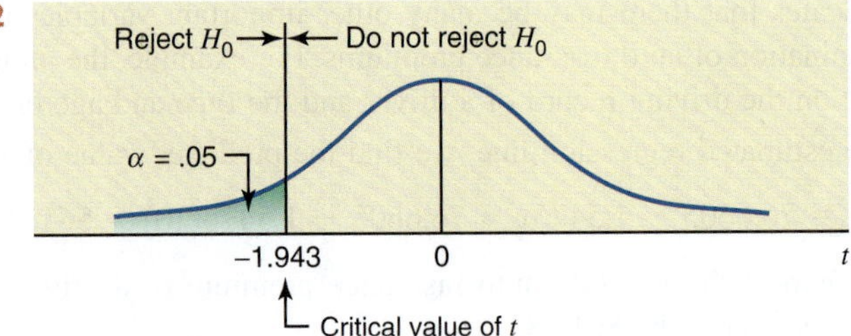

Step 4. *Calculate the value of the test statistic.*

The value of the test statistic t for b is calculated as follows:

$$t = \frac{b - B}{s_b} = \frac{-1.5476 - \overset{\substack{\text{From } H_0 \\ \downarrow}}{0}}{.5270} = -2.937$$

Step 5. *Make a decision.*

The value of the test statistic $t = -2.937$ falls in the rejection region. Hence, we reject the null hypothesis and conclude that B is negative. That is, the monthly auto insurance premium decreases with an increase in years of driving experience.

Using the *p*-Value to Make a Decision

We can find the range for the p-value from the t distribution table (Table V of Appendix C) and make a decision by comparing that p-value with the significance level. For this example, $df = 6$ and the observed value of t is -2.937. From Table V (the t distribution table) in the row of $df = 6$, 2.937 is between 2.447 and 3.143. The corresponding areas in the right tail of the t distribution are .025 and .01. But our test is left-tailed and the observed value of t is negative. Thus, $t = -2.937$ lies between -2.447 and -3.143. The corresponding areas in the left tail of the t distribution are .025 and .01. Therefore the range of the p-value is

$$.01 < p\text{-value} < .025$$

Thus, we can state that for any α equal to or greater than .025 (the upper limit of the p-value range), we will reject the null hypothesis. For our example, $\alpha = .05$, which is greater than the upper limit of the p-value of .025. As a result, we reject the null hypothesis.

Note that if we use technology to find this p-value, we will obtain a p-value of .013. Then we can reject the null hypothesis for any $\alpha > .013$.

(k) We perform the following five steps to test the hypothesis about the linear correlation coefficient ρ.

Step 1. *State the null and alternative hypotheses.*

The null and alternative hypotheses are:

H_0: $\rho = 0$ (The linear correlation coefficient is zero)

H_1: $\rho \neq 0$ (The linear correlation coefficient is different from zero)

Step 2. *Select the distribution to use.*

Assuming that variables x and y are normally distributed, we will use the t distribution to perform this test about the linear correlation coefficient.

Step 3. *Determine the rejection and nonrejection regions.*

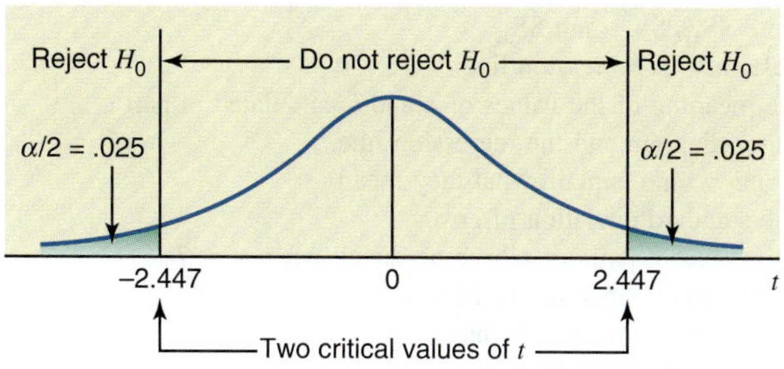

Figure 13.23

The significance level is 5%. From the alternative hypothesis we know that the test is two-tailed. Hence,

$$\text{Area in each tail of the } t \text{ distribution} = .05/2 = .025$$

$$df = n - 2 = 8 - 2 = 6$$

From the t distribution table, Table V of Appendix C, the critical values of t are -2.447 and 2.447. The rejection and nonrejection regions for this test are shown in Figure 13.23.

Step 4. *Calculate the value of the test statistic.*

The value of the test statistic t for r is calculated as follows:

$$t = r\sqrt{\frac{n-2}{1-r^2}} = (-.77)\sqrt{\frac{8-2}{1-(-.77)^2}} = -2.956$$

Step 5. *Make a decision.*

The value of the test statistic $t = -2.956$ falls in the rejection region. Hence, we reject the null hypothesis and conclude that the linear correlation coefficient between driving experience and auto insurance premium is different from zero.

Using the *p*-Value to Make a Decision

We can find the range for the p-value from the t distribution table and make a decision by comparing that p-value with the significance level. For this example, $df = 6$ and the observed value of t is -2.956. From Table V (the t distribution table) in the row of $df = 6$, $t = 2.956$ is between 2.447 and 3.143. The corresponding areas in the right tail of the t distribution curve are .025 and.01. Since the test is two tailed, the range of the p-value is

$$2(.01) < p\text{-value} < 2(.025) \quad \textbf{or} \quad .02 < p\text{-value} < .05$$

Thus, we can state that for any α equal to or greater than .05 (the upper limit of the p-value range), we will reject the null hypothesis. For our example, $\alpha = .05$, which is equal to the upper limit of the p-value. As a result, we reject the null hypothesis. ■

EXERCISES

APPLICATIONS

13.80 The owner of a small factory that produces working gloves is concerned about the high cost of air conditioning in the summer but is afraid that keeping the temperature in the factory too high will lower productivity. During the summer, he experiments with temperature settings from 68 to 81 degrees Fahrenheit and measures each day's productivity. The following table gives the temperature and the number of pairs of gloves (in hundreds) produced on each of the eight randomly selected days.

Temperature	72	71	78	75	81	77	68	76
Pairs of gloves	37	37	32	36	33	35	39	34

a. Do the pairs of gloves produced depend on temperature or does temperature depend on pairs of gloves produced? Do you expect a positive or a negative relationship between these two variables?
b. Taking temperature as an independent variable and pairs of gloves produced as a dependent variable, compute SS_{xx}, SS_{yy}, and SS_{xy}.
c. Find the least squares regression line.
d. Interpret the meaning of the values of a and b calculated in part c.
e. Plot the scatter diagram and the regression line.
f. Calculate r and r^2, and explain what they mean.
g. Compute the standard deviation of errors.
h. Predict the number of pairs of gloves produced when $x = 74$.
i. Construct a 99% confidence interval for B.
j. Test at the 5% significance level whether B is negative.
k. Using $\alpha = .01$ can you conclude that ρ is negative?

13.81 The following table gives information on ages and cholesterol levels for a random sample of 10 men.

Age	58	69	43	39	63	52	47	31	74	36
Cholesterol level	189	235	193	177	154	191	213	165	198	181

a. Taking age as an independent variable and cholesterol level as a dependent variable, compute SS_{xx}, SS_{yy}, and SS_{xy}.
b. Find the regression of cholesterol level on age.
c. Briefly explain the meaning of the values of a and b calculated in part b.
d. Calculate r and r^2 and explain what they mean.
e. Plot the scatter diagram and the regression line.
f. Predict the cholesterol level of a 60-year-old man.
g. Compute the standard deviation of errors.
h. Construct a 95% confidence interval for B.
i. Test at the 5% significance level if B is positive.
j. Using $\alpha = .025$, can you conclude that the linear correlation coefficient is positive?

13.82 A farmer wanted to find the relationship between the amount of fertilizer used and the yield of corn. He selected seven acres of his land on which he used different amounts of fertilizer to grow corn. The following table gives the amount (in pounds) of fertilizer used and the yield (in bushels) of corn for each of the seven acres.

Fertilizer Used	Yield of Corn
120	142
80	112
100	132
70	96
88	119
75	104
110	136

a. With the amount of fertilizer used as an independent variable and yield of corn as a dependent variable, compute SS_{xx}, SS_{yy}, and SS_{xy}.
b. Find the least squares regression line.
c. Interpret the meaning of the values of a and b calculated in part b.
d. Calculate r and r^2 and explain what they mean.
e. Compute the standard deviation of errors.
f. Predict the yield of corn per acre for $x = 105$.
g. Construct a 98% confidence interval for B.
h. Test at the 5% significance level whether B is different from zero.
i. Using $\alpha = .05$, can you conclude that ρ is different from zero?

13.83 The following table gives information on the incomes (in thousands of dollars) and charitable contributions (in hundreds of dollars) for the past year for a random sample of 10 households.

Income	Charitable Contributions
56	10
36	4
102	29
72	23
52	3
91	28
50	8
62	16
80	18
40	2

a. With income as an independent variable and charitable contributions as a dependent variable, compute SS_{xx}, SS_{yy}, and SS_{xy}.
b. Find the regression of charitable contributions on income.
c. Briefly explain the meaning of the values of a and b.
d. Calculate r and r^2 and briefly explain what they mean.
e. Compute the standard deviation of errors.
f. Construct a 99% confidence interval for B.
g. Test at the 1% significance level whether B is positive.
h. Using the 1% significance level, can you conclude that the linear correlation coefficient is different from zero?

13.84 The following data give information on the lowest cost ticket price (in dollars) and the average attendance (rounded to the nearest thousand) for the past year for six football teams.

Ticket price	28.50	16.50	24.00	35.50	39.50	26.00
Attendance	56	65	71	69	55	42

a. Taking ticket price as an independent variable and attendance as a dependent variable, compute SS_{xx}, SS_{yy}, and SS_{xy}.
b. Find the least squares regression line.
c. Briefly explain the meaning of the values of a and b calculated in part b.
d. Calculate r and r^2 and briefly explain what they mean.
e. Compute the standard deviation of errors.
f. Construct a 90% confidence interval for B.
g. Test at the 2.5% significance level whether B is negative.
h. Using the 2.5% significance level, test whether ρ is negative.

13.85 The following table gives information on GPAs and starting salaries (rounded to the nearest thousand dollars) of seven recent college graduates.

GPA	2.90	3.81	3.20	2.42	3.94	2.05	2.25
Starting salary	38	48	38	35	50	31	37

a. With GPA as an independent variable and starting salary as a dependent variable, compute SS_{xx}, SS_{yy}, and SS_{xy}.
b. Find the least squares regression line.
c. Interpret the meaning of the values of a and b calculated in part b.
d. Calculate r and r^2 and briefly explain what they mean.
e. Compute the standard deviation of errors.
f. Construct a 95% confidence interval for B.
g. Test at the 1% significance level whether B is different from zero.
h. Test at the 1% significance level whether ρ is positive.

13.8 Using the Regression Model

Let us return to the example on incomes and food expenditures to discuss two major uses of a regression model:

1. Estimating the mean value of y for a given value of x. For instance, we can use our food expenditure regression model to estimate the mean food expenditure of all households with a specific income (say, $3500 per month).
2. Predicting a particular value of y for a given value of x. For instance, we can determine the expected food expenditure of a randomly selected household with a particular monthly income (say, $3500) using our food expenditure regression model.

13.8.1 Using the Regression Model for Estimating the Mean Value of y

Our population regression model is

$$y = A + Bx + \epsilon$$

As mentioned earlier in this chapter, the mean value of y for a given x is denoted by $\mu_{y|x}$, read as "the mean value of y for a given value of x." Because of the assumption that the mean value of ϵ is zero, the mean value of y is given by

$$\mu_{y|x} = A + Bx$$

Our objective is to estimate this mean value. The value of $\hat{y}$, obtained from the sample regression line by substituting the value of x, is the *point estimate of* $\mu_{y|x}$ for that x.

For our example on incomes and food expenditures, the estimated sample regression line (from Example 13–1) is

$$\hat{y} = 1.1414 + .2642x$$

Suppose we want to estimate the mean food expenditure for all households with a monthly income of \$3500. We will denote this population mean by $\mu_{y|x=35}$ or $\mu_{y|35}$. Note that we have written $x = 35$ and not $x = 3500$ in $\mu_{y|35}$ because the units of measurement for the data used to estimate the above regression line in Example 13–1 were hundreds of dollars. Using the regression line, we find that the point estimate of $\mu_{y|35}$ is

$$\hat{y} = 1.1414 + .2642(35) = \$10.3884 \text{ hundred}$$

Thus, based on the sample regression line, the point estimate for the mean food expenditure $\mu_{y|35}$ for all households with a monthly income of \$3500 is \$1038.84 per month.

However, suppose we take a second sample of seven households from the same population and estimate the regression line for this sample. The point estimate of $\mu_{y|35}$ obtained from the regression line for the second sample is expected to be different. All possible samples of the same size taken from the same population will give different regression lines as shown in Figure 13.24, and, consequently, a different point estimate of $\mu_{y|x}$. Therefore, a confidence interval constructed for $\mu_{y|x}$ based on one sample will give a more reliable estimate of $\mu_{y|x}$ than will a point estimate.

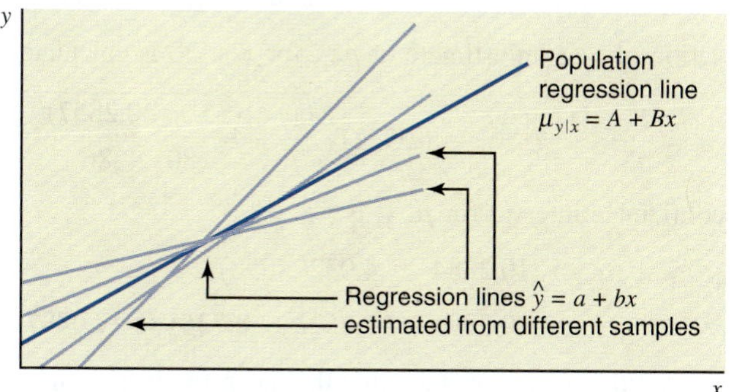

Figure 13.24 Population and sample regression lines.

To construct a confidence interval for $\mu_{y|x}$, we must know the mean, the standard deviation, and the shape of the sampling distribution of its point estimator $\hat{y}$.

The point estimator $\hat{y}$ of $\mu_{y|x}$ is normally distributed with a mean of $A + Bx$ and a standard deviation of

$$\sigma_{\hat{y}_m} = \sigma_\epsilon \sqrt{\frac{1}{n} + \frac{(x_0 - \bar{x})^2}{SS_{xx}}}$$

where $\sigma_{\hat{y}_m}$ is the standard deviation of $\hat{y}$ when it is used to estimate $\mu_{y|x}$, x_0 is the value of x for which we are estimating $\mu_{y|x}$, and σ_ϵ is the population standard deviation of ϵ.

However, usually σ_ϵ is not known. Rather, it is estimated by the standard deviation of sample errors s_e. In this case, we replace σ_ϵ by s_e and $\sigma_{\hat{y}_m}$ by $s_{\hat{y}_m}$ in the foregoing expression. To make a confidence interval for $\mu_{y|x}$, we use the t distribution because σ_ϵ is not known.

Confidence Interval for $\mu_{y|x}$ The $(1 - \alpha)100\%$ *confidence interval for* $\mu_{y|x}$ for $x = x_0$ is

$$\hat{y} \pm t s_{\hat{y}_m}$$

where the value of t is obtained from the t distribution table for $\alpha/2$ area in the right tail of the t distribution curve and $df = n - 2$. The value of $s_{\hat{y}_m}$ is calculated as follows:

$$s_{\hat{y}_m} = s_e \sqrt{\frac{1}{n} + \frac{(x_0 - \bar{x})^2}{SS_{xx}}}$$

Example 13–9 illustrates how to make a confidence interval for the mean value of y, $\mu_{y|x}$.

■ EXAMPLE 13–9

Constructing a confidence interval for the mean value of y.

Refer to Example 13–1 on incomes and food expenditures. Find a 99% confidence interval for the mean food expenditure for all households with a monthly income of $3500.

Solution Using the regression line estimated in Example 13–1, we find the point estimate of the mean food expenditure for $x = 35$ is

$$\hat{y} = 1.1414 + .2642(35) = \$10.3884 \text{ hundred}$$

The confidence level is 99%. Hence, the area in each tail of the t distribution is

$$\alpha/2 = (1 - .99)/2 = .005$$

The degrees of freedom are

$$df = n - 2 = 7 - 2 = 5$$

From the t distribution table, the t value for .005 area in the right tail of the t distribution and 5 df is 4.032. From calculations in Examples 13–1 and 13–2, we know that

$$s_e = .9922, \quad \bar{x} = 30.2857, \quad \text{and} \quad SS_{xx} = 801.4286$$

The standard deviation of $\hat{y}$ as an estimate of $\mu_{y|x}$ for $x = 35$ is calculated as follows:

$$s_{\hat{y}_m} = s_e \sqrt{\frac{1}{n} + \frac{(x_0 - \bar{x})^2}{SS_{xx}}} = (.9922)\sqrt{\frac{1}{7} + \frac{(35 - 30.2857)^2}{801.4286}} = .4098$$

Hence, the 99% confidence interval for $\mu_{y|35}$ is

$$\hat{y} \pm t s_{\hat{y}_m} = 10.3884 \pm 4.032(.4098)$$
$$= 10.3884 \pm 1.6523 = \mathbf{8.7361 \text{ to } 12.0407}$$

Thus, with 99% confidence we can state that the mean food expenditure for all households with a monthly income of $3500 is between $873.61 and $1204.07. ■

13.8.2 Using the Regression Model for Predicting a Particular Value of *y*

The second major use of a regression model is to predict a particular value of y for a given value of x—say, x_0. For example, we may want to predict the food expenditure of a randomly selected household with a monthly income of $3500. In this case, we are not interested in the mean food expenditure of all households with a monthly income of $3500 but in the food expenditure of one particular household with a monthly income of $3500. This predicted value of y is denoted by y_p. Again, to predict a single value of y for $x = x_0$ from the estimated sample regression line, we use the value of $\hat{y}$ *as a point estimate of* y_p. Using the estimated regression line, we find that $\hat{y}$ for $x = 35$ is

$$\hat{y} = 1.1414 + .2642(35) = \$10.3884 \text{ hundred}$$

Thus, based on our regression line, the point estimate for the food expenditure of a given household with a monthly income of \$3500 is \$1038.84 per month. Note that $\hat{y} = 1038.84$ is the point estimate for the mean food expenditure for all households with $x = 35$ as well as for the predicted value of food expenditure of one household with $x = 35$.

Different regression lines estimated by using different samples of seven households each taken from the same population will give different values of the point estimator for the predicted value of y for $x = 35$. Hence, a confidence interval constructed for y_p based on one sample will give a more reliable estimate of y_p than will a point estimate. The confidence interval constructed for y_p is more commonly called a **prediction interval**.

The procedure to construct a prediction interval for y_p is similar to that for constructing a confidence interval for $\mu_{y|x}$ except that the standard deviation of $\hat{y}$ is larger when we predict a single value of y than when we estimate $\mu_{y|x}$.

The point estimator $\hat{y}$ of y_p is normally distributed with a mean of $A + Bx$ and a standard deviation of

$$\sigma_{\hat{y}_p} = \sigma_\epsilon \sqrt{1 + \frac{1}{n} + \frac{(x_0 - \bar{x})^2}{\mathrm{SS}_{xx}}}$$

where $\sigma_{\hat{y}_p}$ is the standard deviation of the predicted value of y, x_0 is the value of x for which we are predicting y, and σ_ϵ is the population standard deviation of ϵ.

However, usually σ_ϵ is not known. In this case, we replace σ_ϵ by s_e and $\sigma_{\hat{y}_p}$ by $s_{\hat{y}_p}$ in the foregoing expression. To make a prediction interval for y_p, we use the t distribution when σ_ϵ is not known.

Prediction Interval for y_p The $(1 - \alpha)100\%$ *prediction interval* for the predicted value of y, denoted by y_p, for $x = x_0$ is

$$\hat{y} \pm t s_{\hat{y}_p}$$

where the value of t is obtained from the t distribution table for $\alpha/2$ area in the right tail of the t distribution curve and $df = n - 2$. The value of $s_{\hat{y}_p}$ is calculated as follows:

$$s_{\hat{y}_p} = s_e \sqrt{1 + \frac{1}{n} + \frac{(x_0 - \bar{x})^2}{\mathrm{SS}_{xx}}}$$

Example 13–10 illustrates the procedure to make a prediction interval for a particular value of y.

■ EXAMPLE 13–10

Making a prediction interval for a particular value of y.

Refer to Example 13–1 on incomes and food expenditures. Find a 99% prediction interval for the predicted food expenditure for a randomly selected household with a monthly income of \$3500.

Solution Using the regression line estimated in Example 13–1, we find the point estimate of the predicted food expenditure for $x = 35$:

$$\hat{y} = 1.1414 + .2642(35) = \$10.3884 \text{ hundred}$$

The area in each tail of the t distribution for a 99% confidence level is

$$\alpha/2 = (1 - .99)/2 = .005$$

The degrees of freedom are

$$df = n - 2 = 7 - 2 = 5$$

From the t distribution table, the t value for .005 area in the right tail of the t distribution curve and 5 df is 4.032. From calculations in Examples 13–1 and 13–2,

$$s_e = .9922, \quad \bar{x} = 30.2857, \quad \text{and} \quad \mathrm{SS}_{xx} = 801.4286$$

The standard deviation of $\hat{y}$ as an estimator of y_p for $x = 35$ is calculated as follows:

$$s_{\hat{y}_p} = s_e\sqrt{1 + \frac{1}{n} + \frac{(x_0 - \bar{x})^2}{SS_{xx}}}$$

$$= (.9922)\sqrt{1 + \frac{1}{7} + \frac{(35 - 30.2857)^2}{801.4286}} = 1.0735$$

Hence, the 99% prediction interval for y_p for $x = 35$ is

$$\hat{y} \pm ts_{\hat{y}_p} = 10.3884 \pm 4.032(1.0735)$$

$$= 10.3884 \pm 4.3284 = \mathbf{6.0600 \text{ to } 14.7168}$$

Thus, with 99% confidence we can state that the predicted food expenditure of a household with a monthly income of $3500 is between $606.00 and $1471.68. ■

As we can observe, this interval is much wider than the one for the mean value of y for $x = 35$ calculated in Example 13–9, which was $873.61 to $1204.07. This is always true. The prediction interval for predicting a single value of y is always larger than the confidence interval for estimating the mean value of y for a certain value of x.

13.9 Cautions in Using Regression

When carefully applied, regression is a very helpful technique for making predictions and estimations about one variable for a certain value of another variable. However, we need to be cautious when using the regression analysis, for it can give us misleading results and predictions. The following are the two most important points to remember when using regression.

Extrapolation

The regression line estimated for the sample data is true only for the range of x values observed in the sample. For example, the values of x in our example on incomes and food expenditures vary from a minimum of 15 to a maximum of 49. Hence, our estimated regression line is applicable only for values of x between 15 and 49; that is, we should use this regression line to estimate the mean food expenditure or to predict the food expenditure of a single household only for income levels between $1500 and $4900. If we estimate or predict y for a value of x either less than 15 or greater than 49, it is called *extrapolation*. This does not mean that we should never use the regression line for extrapolation. Instead, we should interpret such predictions cautiously and not attach much importance to them.

Similarly, if the data used for the regression estimation are time-series data (see Exercises 13.99 to 13.101), the predicted values of y for periods outside the time interval used for the estimation of the regression line should be interpreted very cautiously. When using the estimated regression line for extrapolation, we are assuming that the same linear relationship between the two variables holds true for values of x outside the given range. It is possible that the relationship between the two variables may not be linear outside that range. Nonetheless, even if it is linear, adding a few more observations at either end will probably give a new estimation of the regression line.

Causality

The regression line does not prove causality between two variables; that is, it does not predict that a change in y is *caused* by a change in x. The information about causality is based on theory or common sense. A regression line describes only whether or not a significant quantitative relationship between x and y exists. Significant relationship means that we reject the null hypothesis H_0: $B = 0$ at a given significance level. The estimated regression line gives the change in y due to a change of one unit in x. Note that it does not indicate that the reason y has changed is that x has changed. In our example on incomes and food expenditures, it is economic theory and common sense, not the regression line, that tell us that food expenditure depends on income. The regression analysis simply helps determine whether or not this dependence is significant.

EXERCISES

CONCEPTS AND PROCEDURES

13.86 Briefly explain the difference between estimating the mean value of y and predicting a particular value of y using a regression model.

13.87 Construct a 99% confidence interval for the mean value of y and a 99% prediction interval for the predicted value of y for the following.

a. $\hat{y} = 3.25 + .80x$ for $x = 15$ given $s_e = .954$, $\bar{x} = 18.52$, $SS_{xx} = 144.65$, and $n = 10$

b. $\hat{y} = -27 + 7.67x$ for $x = 12$ given $s_e = 2.46$, $\bar{x} = 13.43$, $SS_{xx} = 369.77$, and $n = 10$

13.88 Construct a 95% confidence interval for the mean value of y and a 95% prediction interval for the predicted value of y for the following.

a. $\hat{y} = 13.40 + 2.58x$ for $x = 8$ given $s_e = 1.29$, $\bar{x} = 11.30$, $SS_{xx} = 210.45$, and $n = 12$

b. $\hat{y} = -8.6 + 3.72x$ for $x = 24$ given $s_e = 1.89$, $\bar{x} = 19.70$, $SS_{xx} = 315.40$, and $n = 10$

APPLICATIONS

13.89 Refer to Exercise 13.53. Construct a 90% confidence interval for the mean monthly salary of all secretaries with 10 years of experience. Construct a 90% prediction interval for the monthly salary of a randomly selected secretary with 10 years of experience.

13.90 Refer to the data on temperature settings and pairs of gloves produced for eight days given in Exercise 13.80. Construct a 99% confidence interval for $\mu_{y|x}$ for $x = 77$ and a 99% prediction interval for y_p for $x = 77$.

13.91 Refer to Exercise 13.82. Construct a 99% confidence interval for the mean yield of corn per acre for all acres on which 90 pounds of fertilizer are used. Determine a 99% prediction interval for the yield of corn for a randomly selected acre on which 90 pounds of fertilizer are used.

13.92 Using the data on ages and cholesterol levels of 10 men given in Exercise 13.81, find a 95% confidence interval for the mean cholesterol level for all 53-year-old men. Make a 95% prediction interval for the cholesterol level for a randomly selected 53-year-old man.

13.93 Refer to Exercise 13.83. Construct a 95% confidence interval for the mean charitable contributions made by all households with an income of $64,000. Make a 95% prediction interval for the charitable contributions made by a randomly selected household with an income of $64,000.

13.94 Refer to Exercise 13.85. Construct a 98% confidence interval for the mean starting salary of recent college graduates with a GPA of 3.15. Construct a 98% prediction interval for the starting salary of a randomly selected recent college graduate with a GPA of 3.15.

USES AND MISUSES

1. PROCESSING ERRORS

Stuck on the far right side of the linear regression model is the Greek letter epsilon, ϵ. Despite its diminutive size, proper respect for the error term is critical to good linear regression modeling and analysis.

One interpretation of the error term is that it is a process. Imagine you are a chemist and you have to weigh a number of chemicals for an experiment. The balance that you use in your laboratory is very accurate—so accurate, in fact, that shuffling your feet, exhaling near it, or the rumble of trucks on the road outside can cause the reading to fluctuate. Because the value of the measurement that you take will be affected by a number of factors out of your control, you must make several measurements for each chemical, note each measurement, and then take the means and standard deviations of your samples. The distribution of measurements around a mean is the result of a random error process dependent on a number of factors out of your control; each time you use the balance, the measurement you take is the sum of the actual mass of the chemical and a "random" error. In this example, the measurements will most likely be normally distributed around the mean.

Linear regression analysis makes the same assumption about the two variables you are comparing: The value of the dependent variable is a linear function of the independent variable, plus a little bit of error that you cannot control. Unfortunately, when working with economic or survey data, you rarely can duplicate an experiment to identify the error model. But as a statistician, you can use the errors to help you refine your model of the relationship among the variables and to guide your collection of new data. For example, if the errors are skewed to the right for moderate values of the independent variable and skewed to the left for small and large values of the independent variable, you can modify your model to account for this difference. Or you can think about other relationships among the variables that might explain this particular distribution of errors. A detailed analysis of the error in your model can be just as instructive as analysis of the slope and *y*-intercept of the identified model.

2. OUTLIERS AND CORRELATION

In Chapter 3 we learned that outliers can affect the values of some of the summary measures such as the mean and range. Note that although outliers do affect many other summary measures, these two are affected substantially. Here we will see that just looking at a number that represents the correlation coefficient does not provide the entire story. A very famous data set for demonstrating this concept was created by F. J. Anscombe (Anscombe, F. J., Graphs in Statistical Analysis, *American Statistician*, 27, pp. 17–21). He created four pairs of data sets on *x* and *y* variables, each of which has a correlation of .816. To the novice, it may seem that the scatterplots for these four data sets should look virtually the same. But that may not be true. Look at the four scatterplots shown below in Figure 13.25.

No two of these scatterplots are even remotely close to being the same or even similar. The data used in the upper left plot are linearly associated, as are the data in the lower left plot. But the plot of *y*3 versus *x*3 contains an outlier. Without this outlier, the correlation between *x*3 and *y*3 would be 1. On the other hand, there is much more variability in the relationship between *x*1 and *y*1. As far as *x*4 and *y*4 are concerned, the strong correlation is defined by the single point in the upper right corner of the scatterplot. Without this point, there would be no variability among the *x*4 values and the correlation would be undefined. Lastly, the scatterplot of *y*2 versus *x*2 reveals that there is an extremely well-defined relationship between these variables, but it is not linear. Being satisfied that the correlation coefficient is close to 1.0 between variables *x*2 and *y*2 implies that there is a strong linear association between the variables when actually we are fitting a straight line to a set of data that should be represented by another type of mathematical function.

As we have mentioned before, the process of making a graph may seem trivial, but the importance of graphs in our analysis can never be overstated.

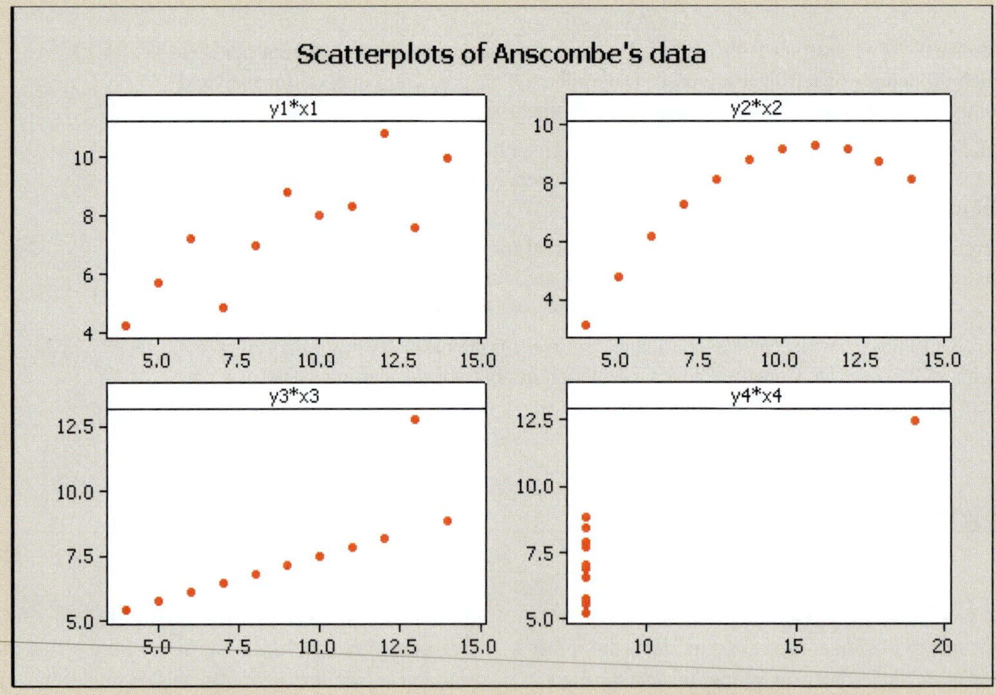

Figure 13.25 Four scatterplots with the same correlation coefficient.

Glossary

Coefficient of determination A measure that gives the proportion (or percentage) of the total variation in a dependent variable that is explained by a given independent variable.

Degrees of freedom for a simple linear regression model Sample size minus 2; that is, $n - 2$.

Dependent variable The variable to be predicted or explained.

Deterministic model A model in which the independent variable determines the dependent variable exactly. Such a model gives an exact relationship between two variables.

Estimated or **predicted value of y** The value of the dependent variable, denoted by $\hat{y}$, that is calculated for a given value of x using the estimated regression model.

Independent or **explanatory variable** The variable included in a model to explain the variation in the dependent variable.

Least squares estimates of A and B The values of a and b that are calculated by using the sample data.

Least squares method The method used to fit a regression line through a scatter diagram such that the error sum of squares is minimum.

Least squares regression line A regression line obtained by using the least squares method.

Linear correlation coefficient A measure of the strength of the linear relationship between two variables.

Linear regression model A regression model that gives a straight-line relationship between two variables.

Multiple regression model A regression model that contains two or more independent variables.

Negative relationship between two variables The value of the slope in the regression line and the correlation coefficient between two variables are both negative.

Nonlinear (simple) regression model A regression model that does not give a straight-line relationship between two variables.

Population parameters for a simple regression model The values of A and B for the regression model $y = A + Bx + \epsilon$ that are obtained by using population data.

Positive relationship between two variables The value of the slope in the regression line and the correlation coefficient between two variables are both positive.

Prediction interval The confidence interval for a particular value of y for a given value of x.

Probabilistic or **statistical model** A model in which the independent variable does not determine the dependent variable exactly.

Random error term (ϵ) The difference between the actual and predicted values of y.

Scatter diagram or **scatterplot** A plot of the paired observations of x and y.

Simple linear regression A regression model with one dependent and one independent variable that assumes a straight-line relationship.

Slope The coefficient of x in a regression model that gives the change in y for a change of one unit in x.

SSE (error sum of squares) The sum of the squared differences between the actual and predicted values of y. It is the portion of the SST that is not explained by the regression model.

SSR (regression sum of squares) The portion of the SST that is explained by the regression model.

SST (total sum of squares) The sum of the squared differences between actual y values and $\bar{y}$.

Standard deviation of errors A measure of spread for the random errors.

y-intercept The point at which the regression line intersects the vertical axis on which the dependent variable is marked. It is the value of y when x is zero.

Supplementary Exercises

13.95 The following data give information on the ages (in years) and the numbers of breakdowns during the past month for a sample of seven machines at a large company.

Age	12	7	2	8	13	9	4
Number of breakdowns	10	5	1	4	12	7	2

a. Taking age as an independent variable and number of breakdowns as a dependent variable, what is your hypothesis about the sign of B in the regression line? (In other words, do you expect B to be positive or negative?)
b. Find the least squares regression line. Is the sign of b the same as you hypothesized for B in part a?
c. Give a brief interpretation of the values of a and b calculated in part b.
d. Compute r and r^2 and explain what they mean.
e. Compute the standard deviation of errors.
f. Construct a 99% confidence interval for B.
g. Test at the 2.5% significance level whether B is positive.
h. At the 2.5% significance level, can you conclude that ρ is positive? Is your conclusion the same as in part g?

13.96 The health department of a large city has developed an air pollution index that measures the level of several air pollutants that cause respiratory distress in humans. The table on the next page gives the pollution index (on a scale of 1 to 10, here 10 being the worst) for seven randomly selected summer days and the number of patients with acute respiratory problems admitted to the emergency rooms of the city's hospitals.

Air pollution index	4.5	6.7	8.2	5.0	4.6	6.1	3.0
Emergency admissions	53	82	102	60	39	42	27

a. Taking the air pollution index as an independent variable and the number of emergency admissions as a dependent variable, do you expect B to be positive or negative in the regression model $y = A + Bx + \epsilon$?
b. Find the least squares regression line. Is the sign of b the same as you hypothesized for B in part a?
c. Compute r and r^2, and explain what they mean.
d. Compute the standard deviation of errors.
e. Construct a 90% confidence interval for B.
f. Test at the 5% significance level whether B is positive.
g. Test at the 5% significance level whether ρ is positive. Is your conclusion the same as in part f?

13.97 The management of a supermarket wants to find if there is a relationship between the number of times a specific product is promoted on the intercom system in the store and the number of units of that product sold. To experiment, the management selected a product and promoted it on the intercom system for seven days. The following table gives the number of times this product was promoted each day and the number of units sold.

Number of Promotions Per Day	Number of Units Sold Per Day (hundreds)
15	11
22	22
42	30
30	26
18	17
12	15
38	23

a. With the number of promotions as an independent variable and the number of units sold as a dependent variable, what do you expect the sign of B in the regression line $y = A + Bx + \epsilon$ will be?
b. Find the least squares regression line $\hat{y} = a + bx$. Is the sign of b the same as you hypothesized for B in part a?
c. Give a brief interpretation of the values of a and b calculated in part b.
d. Compute r and r^2 and explain what they mean.
e. Predict the number of units of this product sold on a day with 35 promotions.
f. Compute the standard deviation of errors.
g. Construct a 98% confidence interval for B.
h. Testing at the 1% significance level, can you conclude that B is positive?
i. Using $\alpha = .02$, can you conclude that the correlation coefficient is different from zero?

13.98 The following table gives information on the temperature (in degrees Fahrenheit) in a city and the volume of ice cream (in pounds) sold at an ice cream parlor for a random sample of eight days during the summer of 2005.

Temperature	93	86	77	89	98	102	87	79
Ice cream sold	208	175	123	198	232	277	158	117

a. Find the least squares regression line $\hat{y} = a + bx$. Take temperature as an independent variable and volume of ice cream sold as a dependent variable.
b. Give a brief interpretation of the values of a and b.
c. Compute r and r^2 and explain what they mean.
d. Predict the amount of ice cream sold on a day with a temperature of 95°.
e. Compute the standard deviation of errors.
f. Construct a 99% confidence interval for B.
g. Testing at the 1% significance level, can you conclude that B is different from zero?
h. Using $\alpha = .01$, can you conclude that the correlation coefficient is different from zero?

13.99 The following table gives the number of Americans (in millions) who took cruises during 1995–2004.

Year	Number of Americans Who Took Cruises (in millions)
1995	4.4
1996	4.7
1997	5.1
1998	5.4
1999	5.9
2000	6.9
2001	6.9
2002	7.6
2003	8.2
2004	9.0

Source: Cruise Lines International.
USA TODAY, February 4, 2005.

a. Assign a value of zero to 1995, 1 to 1996, 2 to 1997, and so on. Call this new variable *Time*. Make a new table with the variables *Time* and the *Number of Americans Who Took Cruises*.
b. With time as an independent variable and the number of Americans who took cruises as the dependent variable, compute SS_{xx}, SS_{yy}, and SS_{xy}.
c. Construct a scatter diagram for these data. Does the scatter diagram exhibit a linear positive relationship between time and the number of Americans who took cruises?
d. Find the least squares regression line $\hat{y} = a + bx$.
e. Give a brief interpretation of the values of a and b calculated in part d.
f. Compute the correlation coefficient r.
g. Predict the number of Americans who you expect to take cruises in 2009. Comment on this prediction.

13.100 The following table gives the total daily U.S. crude oil imports (in millions of barrels, rounded to the nearest million) for the years 1995–2004 (Energy Information Administration). Here, the number for 2004 is an estimate based on the first seven months.

Year	Daily U.S. Crude Oil Imports (millions of barrels)
1995	7.2
1996	7.5
1997	8.2
1998	8.7
1999	8.7
2000	9.1
2001	9.3
2002	9.1
2003	9.7
2004	9.9

a. Assign a value of 0 to 1995, 1 to 1996, 2 to 1997, and so on. Call this new variable *Time*. Make a new table with the variables *Time* and *Daily U.S. Crude Oil Imports*.
b. With time as an independent variable and the daily U.S. crude oil imports as the dependent variable, compute SS_{xx}, SS_{yy}, and SS_{xy}.
c. Construct a scatter diagram for these data. Does the scatter diagram exhibit a linear positive relationship between time and daily U.S. crude oil imports?
d. Find the least squares regression line $\hat{y} = a + bx$.
e. Give a brief interpretation of the values of a and b calculated in part d.
f. Compute the correlation coefficient r.
g. Predict the daily U.S. crude oil imports for $x = 15$. Comment on this prediction.

13.101 The following table gives the times for winners in the women's 100-meter freestyle swimming finals in the Summer Olympic Games from 1972 to 2004. The times are in seconds rounded to the nearest tenth of a second.

Year	Time (seconds)
1972	58.6
1976	55.7
1980	54.8
1984	55.9
1988	54.9
1992	54.6
1996	54.5
2000	53.8
2004	53.8

Source: Sports Illustrated 2005 Almanac.

a. Assign a value of 0 to 1972, 1 to 1976, 2 to 1980, and so on. Call this new variable *Year.* Make a new table with the variables *Year* and *Time.*
b. With year as an independent variable and time as the dependent variable, compute SS_{xx}, SS_{yy}, and SS_{xy}.
c. Construct a scatter diagram for these data. Does the scatter diagram exhibit a linear negative relationship between year and time?
d. Find the least squares regression line $\hat{y} = a + bx$.
e. Give a brief interpretation of the values of a and b calculated in part d.
f. Compute the correlation coefficient r.
g. Predict the time for the year 2012. Comment on this prediction.

13.102 Refer to the data on ages and numbers of breakdowns for seven machines given in Exercise 13.95. Construct a 99% confidence interval for the mean number of breakdowns per month for all machines with an age of 8 years. Find a 99% prediction interval for the number of breakdowns per month for a randomly selected machine with an age of 8 years.

13.103 Refer to the data on the air pollution index and the number of emergency hospital admissions for acute respiratory problems given in Exercise 13.96. Determine a 95% confidence interval for the mean number of such emergency admissions on all days with an air pollution index of 7.0. Make a 95% prediction interval for the number of such emergency admissions on a day when the air pollution index is 7.0.

13.104 Refer to the data given in Exercise 13.97 on the number of times a specific product is promoted on the intercom system in a supermarket and the number of units of that product sold. Make a 90% confidence interval for the mean number of units of that product sold on days with 35 promotions. Construct a 90% prediction interval for the number of units of that product sold on a randomly selected day with 35 promotions.

13.105 Refer to the data given in Exercise 13.98 on temperatures and the volumes of ice cream sold at an ice cream parlor for a sample of eight days. Construct a 98% confidence interval for the mean volume of ice cream sold at this parlor on all days with a temperature of 95°. Determine a 98% prediction interval for the volume of ice cream sold at this parlor on a randomly selected day with a temperature of 95°.

Advanced Exercises

13.106 Consider the data given in the following table.

x	10	20	30	40	50	60
y	12	15	19	21	25	30

a. Find the least squares regression line and the linear correlation coefficient r.
b. Suppose that each value of y given in the table is increased by 5 and the x values remain unchanged. Would you expect r to increase, decrease, or remain the same? How do you expect the least squares regression line to change?
c. Increase each value of y given in the table by 5 and find the new least squares regression line and the correlation coefficient r. Do these results agree with your expectation in part b?

13.107 Suppose that you work part-time at a bowling alley that is open daily from noon to midnight. Although business is usually slow from noon to 6 P.M., the owner has noticed that it is better on hotter days during the summer, perhaps because the premises are comfortably air-conditioned. The owner shows you some data that she gathered last summer. This data set includes the maximum temperature and the number of lines bowled between noon and 6 P.M. for each of 20 days. (The maximum temperatures ranged from 77° to 95° Fahrenheit during this period.) The owner would like to know if she can estimate tomorrow's business from noon to 6 P.M. by looking at tomorrow's weather forecast. She asks you to analyze the data. Let x be the maximum temperature for a day and y the number of lines bowled between noon and 6 P.M. on that day. The computer output based on the data for 20 days provided the following results:

$$\hat{y} = -432 + 7.7x, \quad s_e = 28.17, \quad SS_{xx} = 607, \quad \text{and} \quad \bar{x} = 87.5$$

Assume that the weather forecasts are reasonably accurate.

a. Does the maximum temperature seem to be a useful predictor of bowling activity between noon and 6 P.M.? Use an appropriate statistical procedure based on the information given. Use $\alpha = .05$.
b. The owner wants to know how many lines of bowling she can expect, on average, for days with a maximum temperature of 90°. Answer using a 95% confidence level.
c. The owner has seen tomorrow's weather forecast, which predicts a high of 90°. About how many lines of bowling can she expect? Answer using a 95% confidence level.
d. Give a brief commonsense explanation to the owner for the difference in the interval estimates of parts b and c.
e. The owner asks you how many lines of bowling she could expect if the high temperature were 100°. Give a point estimate, together with an appropriate warning to the owner.

13.108 An economist is studying the relationship between the incomes of fathers and their sons or daughters. Let x be the annual income of a 30-year-old person and let y be the annual income of that person's father at age 30, adjusted for inflation. A random sample of 300 thirty-year-olds and their fathers yields a linear correlation coefficient of .60 between x and y. A friend of yours, who has read about this research, asks you several questions, such as: Does the positive value of the correlation coefficient suggest that the 30-year-olds tend to earn more than their fathers? Does the correlation coefficient reveal anything at all about the difference between the incomes of 30-year-olds and their fathers? If not, what other information would we need from this study? What does the correlation coefficient tell us about the relationship between the two variables in this example? Write a short note to your friend answering these questions.

13.109 For the past 25 years Burton Hodge has been keeping track of how many times he mows his lawn and the average size of the ears of corn in his garden. Hearing about the Pearson correlation coefficient from a statistician buddy of his, Burton decides to substantiate his suspicion that the more often he mows his lawn, the bigger the ears of corn are. He does so by computing the correlation coefficient. Lo and behold, Burton finds a .93 coefficient of correlation! Elated, he calls his friend the statistician to thank him and announce that next year he will have prize-winning ears of corn because he plans to mow his lawn every day. Do you think Burton's logic is correct? If not, how would you explain to Burton the mistake he is making in his presumption (without eroding his new opinion of statistics)? Suggest what Burton could do next year to make the ears of corn large and relate this to the Pearson correlation coefficient.

13.110 It seems reasonable that the more hours per week a full-time college student works at a job, the less time he or she will have to study and, consequently, the lower his or her GPA would be.

a. Assuming a linear relationship, suggest specifically what the equation relating x and y would be, where x is the average number of hours a student works per week and y represents a student's GPA. Try several values of x and see if your equation gives reasonable values of y.
b. Using the following observations taken from 10 randomly selected students, compute the regression equation and compare it to yours of part a.

Average number of hours worked	20	28	10	35	5	14	0	40	8	23
GPA	2.8	2.5	3.1	2.1	3.4	3.3	2.8	2.5	3.6	1.8

13.111 Was Leo Durocher right when he said, "Nice guys finish last"? Cornell University and the "head hunters" Ray & Berndtson gave a personality test to 3600 U.S. and European executives (*Business Week*, July 27, 1998). One aspect of this test was a measure of how "agreeable" each executive was, measured on a scale of 1 to 60. U.S. executives' average score was 44 points. It was calculated that every five points scored above the average were associated with a loss in annual salary of $16,836 for U.S. executives. Suppose that these conclusions were based on a random sample of U.S. executives and that their mean annual salary was $200,000. Let x be an executive's "agreeableness" score on such a test and let y be the executive's annual salary.

a. Write a regression equation that is consistent with the information above.
b. Over what range of values of x would your equation be valid?

13.112 Consider the formulas for calculating a prediction interval for a new (specific) value of y. For each of the changes mentioned in parts a through c below, state the effect on the width of the confidence interval (increase, decrease, or no change) and why it happens. Note that besides the change mentioned in each part, everything else such as the values of a, b, $\bar{x}$, s_e, and SS_{xx} remains unchanged.

a. The confidence level is increased.
b. The sample size is increased.
c. The value of x_0 is moved farther away from the value of $\bar{x}$.
d. What will the value of the margin of error be if x_0 equals $\bar{x}$?

13.113 For each of the regression lines in Exercises 13.53 through 13.56, interpret the slope in terms of the application of that exercise. In addition, state whether the value of the intercept is logical, and why it is or is not logical. If it is logical, state what the value of the intercept represents in terms of the specific application of that exercise.

13.114 Consider the following data

x	−5	−4	−3	−2	−1	0	1	2	3	4	5
y	−125	−64	−27	−8	−1	0	1	8	27	64	125

a. Calculate the correlation between x and y and perform a hypothesis test to determine if the correlation is significantly greater than zero. Use a significance level of 5%.
b. Are you willing to conclude that there is a strong linear association between the two variables? Use at least one graph to support your answer, and to explain why or why not.

Self-Review Test

1. A simple regression is a regression model that contains
   a. only one independent variable
   b. only one dependent variable
   c. more than one independent variable
   d. both a and b

2. The relationship between independent and dependent variables represented by the (simple) linear regression is that of
   a. a straight line b. a curve c. both a and b

3. A deterministic regression model is a model that
   a. contains the random error term
   b. does not contain the random error term
   c. gives a nonlinear relationship

4. A probabilistic regression model is a model that
   a. contains the random error term
   b. does not contain the random error term
   c. shows an exact relationship

5. The least squares regression line minimizes the sum of
   a. errors b. squared errors c. predictions

6. The degrees of freedom for a simple regression model are
   a. $n - 1$ b. $n - 2$ c. $n - 5$

7. Indicate if the following statement is true or false.

   The coefficient of determination gives the proportion of total squared errors (SST) that is explained by the use of the regression model.

8. Indicate if the following statement is true or false.

   The linear correlation coefficient measures the strength of the linear association between two variables.

9. The value of the coefficient of determination is always in the range
   a. 0 to 1 b. −1 to 1 c. −1 to 0

10. The value of the correlation coefficient is always in the range
    a. 0 to 1   b. −1 to 1   c. −1 to 0

11. Explain why the random error term ϵ is added to the regression model.

12. Explain the difference between A and a and between B and b for a regression model.

13. Briefly explain the assumptions of a regression model.

14. Briefly explain the difference between the population regression line and a sample regression line.

15. The following table gives the temperatures (in degrees Fahrenheit) at 6 P.M. and the attendance (rounded to hundreds) at a minor league baseball team's night games on seven randomly selected evenings in May.

Temperature	61	70	50	65	48	75	53
Attendance	10	16	12	15	8	20	18

a. Do you think temperature depends on attendance or attendance depends on temperature?
b. With temperature as an independent variable and attendance as a dependent variable, what is your hypothesis about the sign of B in the regression model?
c. Construct a scatter diagram for these data. Does the scatter diagram exhibit a linear relationship between the two variables?
d. Find the least squares regression line. Is the sign of b the same as the one you hypothesized for B in part b?
e. Give a brief interpretation of the values of the y-intercept and slope calculated in part d.
f. Compute r and r^2, and explain what they mean.
g. Predict the attendance at a night game in May for a temperature of 60 degrees.
h. Compute the standard deviation of errors.
i. Construct a 99% confidence interval for B.
j. Testing at the 1% significance level, can you conclude that B is positive?
k. Construct a 95% confidence interval for the mean attendance at a night game in May when the temperature is 60 degrees.
l. Make a 95% prediction interval for the attendance at a night game in May when the temperature is 60 degrees.
m. Using the 1% significance level, can you conclude that the linear correlation coefficient is positive?

Mini-Projects

MINI-PROJECT 13–1

Using the weather sections from back issues of a local newspaper or some other source, do the following for a period of 30 or more days. For each day record the predicted maximum temperature for the next day, and then find the actual maximum temperature in the next day's newspaper. Thus, you will have the predicted and actual maximum temperatures for 30 or more days.

a. Make a scatter diagram for your data.
b. Find the regression line with actual maximum temperature as a dependent variable and predicted maximum temperature as an independent variable.
c. Using the 1% significance level, can you conclude that the slope of the regression line is different from zero?
d. If the actual maximum temperature were exactly the same as the predicted maximum temperature for each day, what would the value of the correlation coefficient be?
e. Find the correlation coefficient between the predicted and actual maximum temperatures for your data.
f. Using the 1% significance level, can you conclude that the linear correlation coefficient is positive?

MINI-PROJECT 13–2

Two friends are arguing about the relationship between the prices of soft drinks and wine in U.S. cities. Justin thinks that the prices of any two types of beverages (a soft drink and wine) should be positively related. Ivan disagrees, arguing that the prices of alcoholic beverages in a city depend primarily on state and local taxes.

a. Take a random sample of 15 U.S. cities from CITY DATA that accompany this text. Let x be the price of a 2-liter bottle of Coca-Cola and y the price of a 1.5-liter bottle of Livingston Cellars or Gallo Chablis or Chenin Blanc wine. Calculate the linear correlation coefficient between x and y.
b. Does your value of r suggest a positive linear relationship between x and y?
c. Do you think finding a regression line makes sense here?
d. Using the 1% level of significance, can you conclude that the linear correlation coefficient is positive?

MINI-PROJECT 13–3

Visit a grocery store and choose 30 different types of food items that include nutrition information on the packaging. For each food, identify the amount of fat (in grams) and the sodium content (in milligrams) per serving. Make sure that you pick a wide variety of foods in order to get a wide variety of values of these two variables. For example, selecting 30 different diet sodas would not make for an interesting analysis.

a. Calculate the linear correlation coefficient between the two variables. Do you find a positive or a negative association between sodium content and the amount of fat?
b. Create a scatterplot of these data using the amount of fat as the x variable. Does your scatterplot suggest that creating a regression line to represent these data makes sense?
c. Find a regression line for your data. If it makes sense to fit a line, interpret the values of the slope and intercept. If it does not make sense, explain why these numbers could be misleading.

DECIDE FOR YOURSELF

Does Regression Equation Always Make Sense?

Regression is a very powerful statistical tool. However, like any other tool, a failure to understand both its uses as well as its limitations can lead to ridiculous, if not disastrous, results. To demonstrate this, we took the data on two variables—the year of the Olympics from 1928 to 2004 as the independent variable and winner's time (in seconds) in men's 100 meter dash (race) as a dependent variable. Figure 13.26 shows the scatterplot and the regression line for these data.

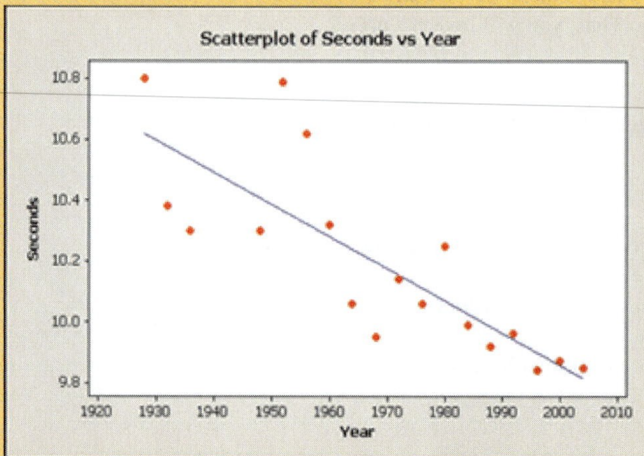

Figure 13.26 Scatterplot and Regression Line.

Looking at this scatterplot, it seems reasonable to use a regression line to explain the relationship between the year of Olympics and the winning time in the 100 meter dash. Specifically, the equation of that regression line is

$$\text{Seconds} = 31.1 - .0106 \text{ Year}$$

To calculate this regression line, we used the actual years of the Olympiad for the independent variable. Theoretically, we could use this regression equation to estimate the winning times for the years when Olympics are not held. We could also use it to predict the future winning times or to calculate what would have happened in the past. Answer the following questions to see how reasonable this process is.

1. Based on this regression equation, what is the change in the winning time per Olympic period (4 years)? Does the change represent an increase or a decrease?

2. Find the predicted winning times for the years 2200, 2600, and 3000. Using these predicted times, determine the winners' speeds (in miles per hour) for the years 2200 and 2600. Does it make sense to believe that this pattern will continue in the future? Explain.

3. A similar analysis could be done in the reverse direction. A recent (2005) scientific discovery states that fossils from 35,000-year-old *modern* humans were found in Transylvania (http://www.theglobeandmail.com/servlet/story/RTGAM.20040306.wfossil0306/BNStory/specialScienceandHealth/). Using the above regression equation, calculate the winning time for the 100 meter dash at this point in history. Does this number make sense? Why or why not?

TECHNOLOGY INSTRUCTION Simple Linear Regression

TI-84

Screen 13.1

1. To construct a simple linear regression equation, enter the independent and dependent variable values into lists. Select **STAT>CALC>LinReg(a+bx)**, press **Enter**, then enter (separated by commas) the name of your independent variable list, the name of your dependent variable list, and **Y1**. (**Y1** can be found by selecting **VARS>Y-VARS> Function>Y1**.) Then press **Enter**. (See **Screen 13.1**.) The result includes the slope and intercept of the regression equation.
2. To find the correlation coefficient, select **VARS>Statistics>EQ>r**. To find the coefficient of determination, square the correlation coefficient.
3. To find a fitted value for a given value of x, type **Y1(x)**.
4. To test that the slope of the line is non-zero, select **STATS>TESTS>LinRegTTest**. (Note that this set of commands will give you the output obtained under 1 and 2 above.) Enter the names of the lists. Leave **Freq:1**. Choose the alternative hypothesis. Select **Calculate**. The result includes a t-statistic value and a p-value.

MINITAB

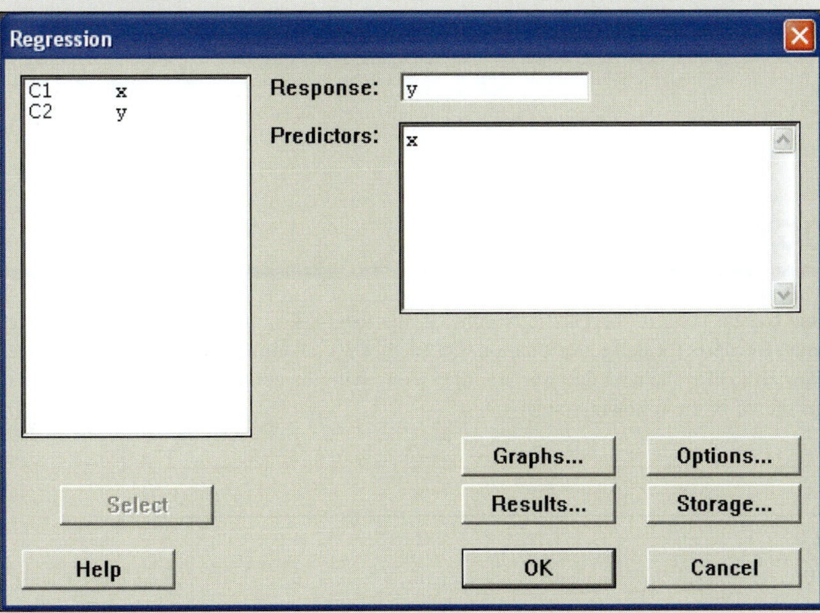

Screen 13.2

1. To construct and analyze a simple linear regression equation, enter the independent and dependent variable values into columns.
2. Select **Stat>Regression>Regression**.
3. Enter the dependent variable's column name in the **Response** box.
4. Enter the independent variable's column name in the **Predictors** box. (See **Screen 13.2**.)
5. Select **Options** if you wish to predict a value with the equation, and enter the value of the independent variable in the entry marked **Prediction intervals for new observations**. Enter the **Confidence level** and select **OK**.
6. Select **Results**, and choose **Regression equation, . . .** . Select **OK** for each dialog box.
7. The output includes the regression equation, t-statistics and p-values for tests on both the slope and intercept to find out if they are zero, the coefficient of determination (as **R-sq**), and, if requested, the fitted value as well as confidence and prediction intervals for the fitted value.

Excel

	A	B	C	D
1	x	y		
2				
3	1	2		
4	2	3		
5	3	5		
6	4	7		
7	5	11		
8				
9	slope	=SLOPE(B3:B7,A3:A7)		
10		SLOPE(known_y's, known_x's)		

Screen 13.3

1. To just calculate the coefficients of a simple linear regression equation, enter the independent and dependent variable values into ranges of cells.

2. Type **=SLOPE(dep, indep)** to find the slope of the regression equation, where **dep** is the range of cells containing the dependent variable and **indep** is the range of cells containing the independent variable. (See **Screen Shot 13.3** and **13.4**.)

3. Type **=INTERCEPT(dep, indep)** to find the value of the constant term in the regression equation.

4. To find a single predicted value, type **=FORECAST(x, dep, indep)** where x is the corresponding value of the independent variable.

5. To perform a more complete analysis, select a range of cells with two columns and five rows, type **=LINEST(dep, indep, true, true)** and press **Shift-Control-Enter**.

6. The first row of the result contains the slope and intercept; the second row of the result contains the standard deviation of the slope and the standard deviation of the intercept; the third row contains the coefficient of determination and the standard deriva-tion of the sample errors; the fourth row contains an *F*-statistic value, which is the square of the *t*-statistic for testing the coefficient, and the degrees of freedom; the fifth row contains the regression sum of squares and the residual sum of squares.

	A	B
1	x	y
2		
3	1	2
4	2	3
5	3	5
6	4	7
7	5	11
8		
9	slope	2.2
10		

Screen 13.4

TECHNOLOGY ASSIGNMENTS

TA 13.1 In a rainy coastal town in the Pacific Northwest, the local TV weatherman is often criticized for making inaccurate forecasts for daily precipitation. On each of 30 randomly selected days last winter, his precipitation forecast (x) for the next day was recorded along with the actual precipitation (y) for that day. These data are shown in the following table.

x	y	x	y	x	y
1.0	.6	0	0	.4	.2
0	.1	0	.1	.2	.5
.2	0	.1	.2	.1	.1
0	0	.2	.2	0	.2
.5	.3	.1	0	.1	0
1.0	1.4	2.0	2.1	.2	.1
.5	.3	.4	.2	1.4	1.2
.1	.1	.2	.1	.5	1.0
0	.1	0	0	0	.5
2.0	.3	.3	.2	0	0

Do the following.

a. Construct a scatter diagram for these data.

b. Find the correlation coefficient between the two variables.

c. Find the regression line with actual precipitation as a dependent variable and predicted precipitation as an independent variable.

d. Make a 95% confidence interval for *B*.

e. Test at the 1% significance level whether *B* is positive.

f. Using the 1% significance level, can you conclude that the linear correlation coefficient is positive?

TA13.2 Refer to Data Set III on NBA players. Select a random sample of 30 players from that population. Do the following for the data on heights and weights of these 30 players.

a. Construct a scatter diagram for these data.

b. Find the correlation between these two variables.

c. Find the regression line with weight as a dependent variable and height as an independent variable.

d. Make a 90% confidence interval for *B*.

e. Test at the 5% significance level whether *B* is positive.

f. Make a 95% confidence interval for the mean weight of all NBA players who are 78 inches tall. Construct a 95% prediction interval for the weight of a randomly selected NBA player with a height of 78 inches.

TA13.3 Refer to the data on the ages and the numbers of breakdowns for a sample of seven machines given in Exercise 13.95. Answer the following questions.

a. Construct a scatter diagram for these data.

b. Find the least squares regression line with age as an independent variable and the number of breakdowns as a dependent variable.

c. Compute the correlation coefficient.

d. Construct a 99% confidence interval for *B*.

e. Test at the 2.5% significance level whether *B* is positive.

Chapter

14

Multiple Regression

14.1 Multiple Regression Analysis

14.2 Assumptions of the Multiple Regression Model

14.3 Standard Deviation of Errors

14.4 Coefficient of Multiple Determination

14.5 Computer Solution of Multiple Regression

This chapter is not included in this text but is available for download on the Web site at www.wiley.com/college/mann.

Chapter 15

Nonparametric Methods

This chapter is not included in this text but is available for download on the Web site at www.wiley.com/college/mann.

15.1 The Sign Test

15.2 The Wilcoxon Signed-Rank Test for Two Dependent Samples

15.3 The Wilcoxon Rank Sum Test for Two Independent Samples

15.4 The Kruskal-Wallis Test

15.5 The Spearman Rho Rank Correlation Coefficient Test

15.6 The Runs Test for Randomness

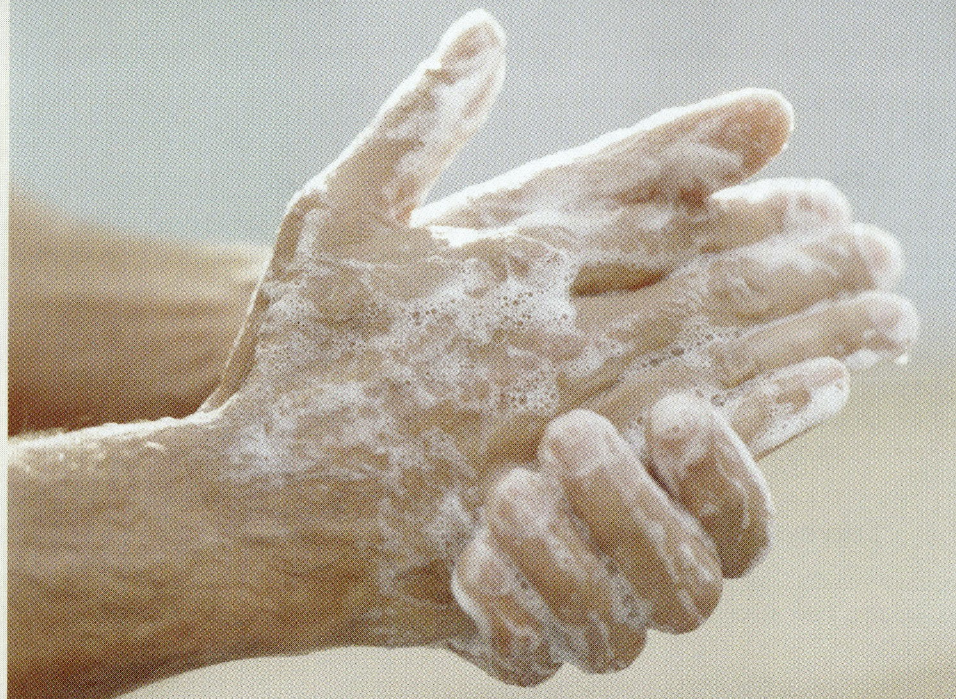

Appendix

A

Sample Surveys, Sampling Techniques, and Design of Experiments

The current American fear of germs is evident in the booming sales of antibacterial soaps. They now represent 78% of the liquid soap market, in spite of a lack of evidence that they are any better than regular soaps. Do these antibacterial soaps work, or is it just a fad? Are people using them only because it is their perception that they work to kill germs, or do they really work? (See Case Study A–2.) Proper sampling techniques employed in research studies or surveys can help accurately answer questions like these.

A.1 Sources of Data

Case Study A–1 Is It a Simple Question?

A.2 Sample Surveys and Sampling Techniques

A.3 Design of Experiments

Case Study A–2 Do Antibacterial Soaps Work?

A.1 Sources of Data

The availability of accurate data is essential for deriving reliable results and making accurate decisions. As the truism "garbage in, garbage out" (GIGO) indicates, policy decisions based on the results of poor data may prove to be disastrous.

Data sources can be divided into three categories: internal sources, external sources, and surveys and experiments.

A.1.1 Internal Sources

Many times data come from **internal sources**, such as a company's own personnel files or accounting records. A company that wants to forecast the future sales of its products might use data from its own records for previous periods. A police department might use data that exist in its own records to analyze changes in the nature of crimes over a period of time.

A.1.2 External Sources

All needed data may not be available from internal sources. Hence, to obtain data we may have to depend on sources outside the company, called **external sources**. Data obtained from external sources may be primary or secondary data. Data obtained from the organization that originally collected them are called **primary data**. If we obtain data from the Bureau of Labor Statistics that were collected by this organization, then these are primary data. Data obtained from a source that did not originally collect them are called **secondary data**. For example, data originally collected by the Bureau of Labor Statistics and published in the *Statistical Abstract of the United States* are secondary data.

A.1.3 Surveys and Experiments

Sometimes the data we need may not be available from internal or external sources. In such cases, we may have to obtain data by conducting our own survey or experiment.

Surveys

In a **survey** we do not exercise any control over the factors when we collect information.

Definition

Survey In a *survey*, data are collected from the members of a population or sample with no particular control over the factors that may affect the characteristic of interest or the results of the survey.

For example, if we want to collect data on the money various families spent last month on clothes, we will ask each of the families included in the survey how much it spent last month on clothes. Then we will record this information.

A survey may be a census or a sample survey.

(i) Census

A **census** includes every member of the population of interest, which is called the **target population**.

Definition

Census A survey that includes every member of the population is called a *census*.

In practice, a census is rarely taken because it is very expensive and time consuming. Furthermore, in many cases it is impossible to identify each member of the target population. We discuss these reasons in more detail in Section A.2.1.

(ii) Sample Survey

Usually, to conduct research, we select a portion of the target population. This portion of the population is called a **sample**. Then we collect the required information from the elements included in the sample.

Definition

Sample Survey The technique of collecting information from a portion of the population is called a *sample survey*.

A survey can be conducted by personal interviews, by telephone, or by mail. The personal interview technique has the advantages of a high response rate and a high quality of

A–1

CASE STUDY

IS IT A SIMPLE QUESTION?

Even the seemingly simplest of questions can yield complex answers. "Do you own a car?" asks Stanley Presser, a sociologist at the National Science Foundation in Washington, D.C. "That sounds like an awfully simple question. But is it really? What does 'you' mean? Suppose a wife is answering the poll, and the car is registered in her husband's name. How is she supposed to answer? What does 'own' mean? What if the car is on a long-term lease? What does 'car' mean? What if they have one of those new little vans, or a four-wheel-drive vehicle? My God, that sounds like a simple question! You can imagine how diverse the factors become in a more complicated one."

Suppose, however, that the question about car ownership had been preceded by a series of related questions: "Are you married? Does your spouse drive an automotive vehicle? Is it a car, a van or some other sort of vehicle? Is it leased, or does your spouse own it? Now about you—do you own a car?" Such a series of questions would serve to clarify the intended meaning of the one about car ownership.

Source: Rich Jaroslovsky, "What's on Your Mind, America?" *Psychology Today*, July–August 1988, 54–59. Copyright © 1988 Sussex Publishers, Inc. Reprinted with permission.

answers obtained. However, it is the most expensive and time-consuming technique. The telephone survey also gives a high response rate. It is less expensive and less time-consuming than personal interviews. Nonetheless, a problem with telephone surveys is that many people do not like to be called at home, and those who do not have a phone are left out of the survey. A survey conducted by mail is the least expensive method, but the response rate is usually very low. Many people included in such a survey do not return the questionnaires.

Conducting a survey that gives accurate and reliable results is not an easy task. To quote Warren Mitofsky, director of Elections and Surveys for CBS News, "Any damn fool with 10 phones and a typewriter thinks he can conduct a poll."[1] Preparing a questionnaire is probably the most difficult part of a survey. The way a question is phrased can affect the results of the survey. Case Study A–1, which is excerpted from an article published in *Psychology Today*, shows that writing questions for a questionnaire is a much more complex task than is usually thought.

Section A.2 discusses sample surveys and sampling techniques in detail.

Experiments

In an **experiment**, we exercise control over some factors when we collect information.

Definition

Experiment In an *experiment*, data are collected from members of a population or sample with some control over the factors that may affect the characteristic of interest or the results of the experiment.

For example, how is a new drug to be tested to find out whether or not it cures a disease? It is done by designing an experiment in which the patients under study are divided into two groups as follows:

1. The **treatment group**—the members of this group receive the actual drug.
2. The **control group**—the members of this group do not receive the actual drug but are given a substitute (called a placebo) that appears to be the actual drug.

The two groups are formed in such a way that the patients in one group are similar to the patients in the other group. This is done by making random assignments of patients to the two groups. Neither the doctors nor the patients know to which group a patient belongs. Such an experiment is called a **double-blind experiment**. Then, after a comparison of the percentage of patients cured in each of the two groups, a decision is made about the effectiveness or noneffectiveness of the new drug. For more on experiments, refer to Section A.3 on experimental design.

[1]"The Numbers Racket: How Polls and Statistics Lie," *U.S. News & World Report*, July 11, 1988.

A.2 Sample Surveys and Sampling Techniques

In this section first we discuss the reasons sample surveys are preferred over a census, and then we discuss a representative sample, random and nonrandom samples, sampling and nonsampling errors, and random sampling techniques.

A.2.1 Why Sample?

As mentioned in the previous section, most of the time surveys are conducted by using samples and not a census of the population. Three of the main reasons for conducting a sample survey instead of a census are listed next.

Time

In most cases, the size of the population is quite large. Consequently, conducting a census takes a long time, whereas a sample survey can be conducted very quickly. It is time-consuming to interview or contact hundreds of thousands or even millions of members of a population. On the other hand, a survey of a sample of a few hundred elements may be completed in little time. In fact, because of the amount of time needed to conduct a census, by the time the census is completed the results may be obsolete.

Cost

The cost of collecting information from all members of a population may easily fall outside the limited budget of most, if not all, surveys. Consequently, to stay within the available resources, conducting a sample survey may be the best approach.

Impossibility of Conducting a Census

Sometimes it is impossible to conduct a census. First, it may not be possible to identify and access each member of the population. For example, if a researcher wants to conduct a survey about homeless people, it is not possible to locate each member of the population and include him or her in the survey. Second, sometimes conducting a survey means destroying the items included in the survey. For example, to estimate the mean life of lightbulbs would necessitate burning out all the bulbs included in the survey. The same is true about finding the average life of batteries. In such cases, only a portion of the population can be selected for the survey.

A.2.2 Random and Nonrandom Samples

Depending on how a sample is drawn, it may be a **random sample** or a **nonrandom sample**.

Definition

Random and Nonrandom Samples A *random sample* is a sample drawn in such a way that each member of the population has some chance of being selected in the sample. In a *nonrandom sample*, some members of the population may not have any chance of being selected in the sample.

Suppose we have a list of 100 students and we want to select 10 of them. If we write the names of all 100 students on pieces of paper, put them in a hat, mix them, and then draw 10 names, the result will be a random sample of 10 students. However, if we arrange the names of these 100 students alphabetically and pick the first 10 names, it will be a nonrandom sample because the students who are not among the first 10 have no chance of being selected in the sample.

A random sample is usually a representative sample. Note that for a random sample, each member of the population may or may not have the same chance of being included in the sample. Four types of random samples are discussed in Section A.2.4.

Two types of nonrandom samples are a *convenience sample* and a *judgment sample.* In a **convenience sample**, the most accessible members of the population are selected to obtain the results quickly. For example, an opinion poll may be conducted in a few hours by collecting information from certain shoppers at a single shopping mall. In a **judgment sample**, the members are selected from the population based on the judgment and prior knowledge of an expert. Although such a sample may happen to be a representative sample, the chances of it being so are small. If the population is large, it is not an easy task to select a representative sample based on judgment.

The so-called *pseudo polls* are examples of nonrepresentative samples. For instance, a survey conducted by a magazine that includes only its own readers does not usually involve a representative sample. Similarly, a poll conducted by a television station giving two separate 900 telephone numbers for *yes* and *no* votes is not based on a representative sample. In these two examples, respondents will be only those people who read that magazine or watch that television station, who do not mind paying the postage and telephone charges, or who feel emotionally compelled to respond.

Another kind of sample is the **quota sample**. To draw such a sample we divide the target population into different subpopulations based on certain characteristics. Then a subsample is selected from each subpopulation in such a way that each subpopulation is represented in the sample in exactly the same proportion as in the target population. As an example of a quota sample, suppose we want to select a sample of 1000 persons from a city whose population has 48% men and 52% women. To select a quota sample, we choose 480 men from the male population and 520 women from the female population. The sample selected in this way will contain exactly 48% men and 52% women. Another way to select a quota sample is to select from the population one person at a time until we have exactly 480 men and 520 women.

Until the 1948 presidential election in the United States, quota sampling was the most commonly used sampling procedure to conduct opinion polls. The voters included in the samples were selected in such a way that they represented the population proportions of voters based on age, sex, education, income, race, and so on. However, this procedure was abandoned after the 1948 presidential election in which the underdog, Harry Truman, defeated Thomas E. Dewey, who was heavily favored based on the opinion polls. First, the quota samples failed to be representative because the interviewers were allowed to fill their quotas by choosing voters based on their own judgments. This caused the selection of more upper-income and highly educated people, who happened to be Republicans. Thus, the quota samples were unrepresentative of the population because Republicans were overrepresented in these samples. Second, the results of the opinion polls based on quota sampling happened to be false because a large number of factors differentiate voters, but the pollsters considered only a few of those factors. A quota sample based on a few factors will skew the results. A random sample (that is not based on quotas) has a much better chance of being representative of the population of all voters than a quota sample based on a few factors.

A.2.3 Sampling and Nonsampling Errors

The results obtained from a sample survey may contain two types of errors: sampling and nonsampling errors. The sampling error is also called the chance error, and nonsampling errors are also called the systematic errors.

Sampling or Chance Error

Usually, all samples taken from the same population will give different results because they contain different elements of the population. Moreover, the results obtained from any one sample will not be exactly the same as the ones obtained from a census. The difference between a sample result and the result we would have obtained by conducting a census is called the **sampling error**, assuming that the sample is random and no nonsampling error has been made.

Definition

Sampling Error The *sampling error* is the difference between the result obtained from a sample survey and the result that would have been obtained if the whole population had been included in the survey.

The sampling error occurs because of chance, and it cannot be avoided. A sampling error can occur only in a sample survey. It does not occur in a census. Sampling error is discussed in detail in Section 7.2 of Chapter 7, and an example of it is given there.

Nonsampling or Systematic Errors

Nonsampling errors can occur both in a sample survey and in a census. Such errors occur because of human mistakes and not chance.

Definition

Nonsampling Errors The errors that occur in the collection, recording, and tabulation of data are called *nonsampling errors*.

Nonsampling errors occur because of human mistakes and not chance. Nonsampling errors can be minimized if questions are prepared carefully and data are handled cautiously. Many types of systematic errors or biases can occur in a survey, including selection error, nonresponse error, response error, and voluntary response error. The following chart shows the types of errors.

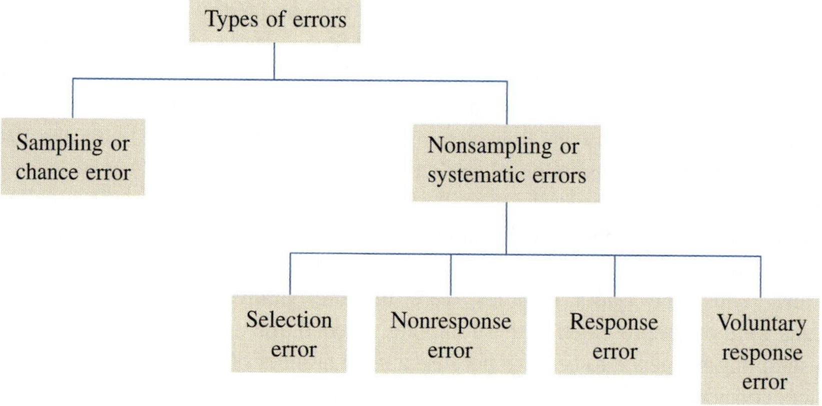

(i) Selection Error

When we need to select a sample, we use a list of elements from which we draw a sample, and this list usually does not include many members of the target population. Most of the time it is not feasible to include every member of the target population in this list. This list of members of the population that is used to select a sample is called the **sampling frame**. For example, if we use a telephone directory to select a sample, the list of names that appears in this directory makes the sampling frame. In this case we will miss the people who are not listed in the telephone directory. The people we miss, for example, will be poor people (including homeless people) who do not have telephones and people who do not want to be listed in the directory. Thus, the sampling frame that is used to select a sample may not be representative of the population. This may cause the sample results to be different from the population results. The error that occurs because the sampling frame is not representative of the population is called the **selection error**.

Definition

Selection Error The list of members of the target population that is used to select a sample is called the sampling frame. The error that occurs because the sampling frame is not representative of the population is called the *selection error*.

If a sample is nonrandom (and, hence, nonrepresentative), the sample results may be quite different from the census results.

(ii) Nonresponse Error

Even if our sampling frame and, consequently, the sample are representative of the population, **nonresponse error** may occur because many of the people included in the sample did not respond to the survey.

Definition

Nonresponse Error The error that occurs because many of the people included in the sample do not respond to a survey is called the *nonresponse error.*

This type of error occurs especially when a survey is conducted by mail. A lot of people do not return the questionnaires. It has been observed that families with low and high incomes do not respond to surveys by mail. Consequently, such surveys overrepresent middle-income families. This kind of error occurs in other types of surveys, too. For instance, in a face-to-face survey where the interviewer interviews people in their homes, many people may not be home when the interviewer visits their homes. The people who are home at the time the interviewer visits and the ones who are not home at that time may differ in many respects, causing a bias in the survey results. This kind of error may also occur in a telephone survey. Many people may not be home when the interviewer calls. This may distort the results. To avoid the nonresponse error, every effort should be made to contact all people included in the survey.

(iii) Response Error

The **response error** occurs when the answer given by a person included in the survey is not correct. This may happen for many reasons. One reason is that the respondent may not have understood the question. Thus, the wording of the question may have caused the respondent to answer incorrectly. It has been observed that when the same question is worded differently, many people do not respond the same way. Usually such an error on the part of respondents is not intentional.

Definition

Response Error The *response error* occurs when people included in the survey do not provide correct answers.

Sometimes the respondents do not want to give correct information when answering a question. For example, many respondents will not disclose their true incomes on questionnaires or in interviews. When information on income is provided, it is almost always biased in the upward direction.

Sometimes the race of the interviewer may affect the answers of respondents. This is especially true if the questions asked are about race relations. The answers given by respondents may differ depending on the race of the interviewer.

(iv) Voluntary Response Error

Another source of systematic error is a survey based on a voluntary response sample.

Definition

Voluntary Response Error *Voluntary response error* occurs when a survey is not conducted on a randomly selected sample but a questionnaire is published in a magazine or newspaper and people are invited to respond to that questionnaire.

The polls conducted based on samples of readers of magazines and newspapers suffer from **voluntary response error** or **bias**. Usually only those readers who have very strong opinions about the issues involved respond to such surveys. Surveys in which the respondents are required to call 900 telephone numbers also suffer from this type of error. Here, to participate, a respondent must pay for the call, and many people do not want to bear this cost. Consequently, the sample is usually neither random nor representative of the target population because participation is voluntary.

A.2.4 Random Sampling Techniques

There are many ways to select a random sample. Four of these techniques are discussed next.

Simple Random Sampling

Under this sampling technique, each sample of the same size selected from the same population has the same probability of being selected.

Definition

Simple Random Sampling A sampling technique under which each sample of the same size has the same probability of being selected. Such a sample is called a simple random sample.

One way to select a simple random sample is by a lottery or drawing. For example, if we need to select 5 students from a class of 50, we write each of the 50 names on a separate piece of paper. Then, we place all 50 names in a hat and mix them thoroughly. Next, we draw one name randomly from the hat. We repeat this experiment four more times. The five drawn names make up a simple random sample.

The second procedure to select a simple random sample is to use a table of random numbers, which has become an outdated procedure. In this age of technology, it is much easier to use a statistical package, such as MINITAB, to select a simple random sample.

Systematic Random Sampling

The simple random sampling procedure becomes very tedious if the size of the population is large. For example, if we need to select 150 households from a list of 45,000, it is very time-consuming either to write the 45,000 names on pieces of paper and then select 150 households or to use a table of random numbers. In such cases, it is more convenient to use **systematic random sampling**.

The procedure to select a systematic random sample is as follows. In the example just mentioned, we would arrange all 45,000 households alphabetically (or based on some other characteristic). Since the sample size should equal 150, the ratio of population to sample size is $45{,}000/150 = 300$. Using this ratio, we randomly select one household from the first 300 households in the arranged list using either method. Suppose by using either of the methods, we select the 210th household. We then select every 210th household from every 300 households in the list. In other words, our sample includes the households with numbers 210, 510, 810, 1110, 1410, 1710, and so on.

Definition

Systematic Random Sample In *systematic random sampling*, we first randomly select one member from the first k units. Then every kth member, starting with the first selected member, is included in the sample.

Stratified Random Sampling

Suppose we need to select a sample from the population of a city and we want households with different income levels to be proportionately represented in the sample. In this case, instead of selecting a simple random sample or a systematic random sample, we may prefer to apply a different technique. First, we divide the whole population into different groups based on income levels. For example, we may form three groups of low-, medium-, and high-income households. We will now have three *subpopulations*, which are usually called **strata**. We then select one sample from each subpopulation or stratum. The collection of all three samples selected from three strata gives the required sample, called the **stratified random sample**. Usually, the sizes of the samples selected from different strata are proportionate to the sizes of the subpopulations in these strata. Note that the elements of each stratum are identical with regard to the possession of a characteristic.

Definition

Stratified Random Sample In a *stratified random sample*, we first divide the population into subpopulations, which are called strata. Then, one sample is selected from each of these strata. The collection of all samples from all strata gives the stratified random sample.

Thus, whenever we observe that a population differs widely in the possession of a characteristic, we may prefer to divide it into different strata and then select one sample from each stratum. We can divide the population on the basis of any characteristic, such as income, expenditure, sex, education, race, employment, or family size.

Cluster Sampling

Sometimes the target population is scattered over a wide geographical area. Consequently, if a simple random sample is selected, it may be costly to contact each member of the sample. In such a case, we divide the population into different geographical groups or clusters and as a first step select a random sample of certain clusters from all clusters. We then take a random sample of certain elements from each selected cluster. For example, suppose we are to conduct a survey of households in the state of New York. First, we divide the whole state of New York into, say, 40 regions, which are called **clusters** or **primary units**. We make sure that all clusters are similar and, hence, representative of the population. We then select at random, say, 5 clusters from 40. Next, we randomly select certain households from each of these 5 clusters and conduct a survey of these selected households. This is called **cluster sampling**. Note that all clusters must be representative of the population.

Definition

Cluster Sampling In *cluster sampling*, the whole population is first divided into (geographical) groups called clusters. Each cluster is representative of the population. Then a random sample of clusters is selected. Finally, a random sample of elements from each of the selected clusters is selected.

A.3 Design of Experiments

As mentioned earlier, to use statistical methods to make decisions, we need access to data. Consider the following examples about decision making.

1. A government agency wants to find the average income of households in the United States.
2. A company wants to find the percentage of defective items produced on a machine.
3. A researcher wants to know if there is an association between eating unhealthy food and cholesterol level.
4. A pharmaceutical company has invented a new medicine for a disease and it wants to check if this medicine cures the disease.

All of these cases relate to decision making. We cannot reach a conclusion in these examples unless we have access to data. Data can be obtained from observational studies, experiments, or surveys. This section is devoted mainly to controlled experiments. However, it also explains observational studies and how they differ from surveys.

Suppose two diets, Diet 1 and Diet 2, are being promoted by two different companies, and each of these companies claims that its diet is successful in reducing weight. A research nutritionist wants to compare these diets with regard to their effectiveness for losing weight. Following are the two alternatives for the researcher to conduct this research.

1. The researcher contacts the persons who are using these diets and collects information on their weight loss. The researcher may contact as many persons as she has the time and financial resources for. Based on this information, the researcher makes a decision about the comparative effectiveness of these diets.
2. The researcher selects a sample of persons who want to lose weight, divides them randomly into two groups, and assigns each group to one of the two diets. Then she compares these two groups with regard to the effectiveness of these diets.

The first alternative is an example of an **observational study**, and the second is an example of a **controlled experiment**.

Definition

Treatment A condition (or a set of conditions) that is imposed on a group of elements by the experimenter is called a *treatment.*

In an observational study the investigator does not impose a treatment on subjects or elements included in the study. For instance, in the first alternative, the researcher simply collects information from the persons who are currently using these diets. In this case, the persons were not assigned to the two diets at random; instead, they chose the diets voluntarily. In this situation the researcher's conclusion about the comparative effectiveness of the two diets may not be valid because the effects of the diets will be **confounded** with many other factors or variables. When the effects of one factor cannot be separated from the effects of some other factors, the effects are said to be confounded. The persons who chose Diet 1 may be completely different with regard to age, gender, and eating and exercise habits from the persons who chose Diet 2. Thus, the weight loss may not be due entirely to the diet but to other factors or variables as well. Persons in one group may aggressively manage both diet and exercise, for example, whereas persons in the second group may depend entirely on diet. Thus, the effects of these other variables will get mixed up (confounded) with the effect of the diets.

Under the second alternative, the researcher selects a group of people, say 100, and randomly assigns them to two diets. One way to make random assignments is to write the name of each of these persons on a piece of paper, put them in a hat, and then randomly draw 50 names from this hat. These 50 persons will be assigned to one of the two diets, say Diet 1. The remaining 50 persons will be assigned to the second diet, Diet 2. This procedure is called **randomization**. Note that random assignments can also be made by using other methods such as a table of random numbers or technology.

Definition

Randomization The procedure in which elements are assigned to different groups at random is called *randomization.*

When people are assigned to one or the other of two diets at random, the other differences among people in the two groups almost disappear. In this case these groups will not differ very much with regard to such factors as age, gender, and eating and exercise habits. The two groups will be very similar to each other. By using the random process to assign people to one or the other of two diets, we have *controlled* the other factors that can affect the weights of people. Consequently, this is an example of a **designed experiment**.

As mentioned earlier, a condition (or a set of conditions) that is imposed on a group of elements by the experimenter is called a treatment. In the example on diets, each of the two diet types is called a treatment. The experimenter randomly assigns the elements to these two treatments. Again, in such cases the study is called a designed experiment.

Definition

Designed Experiment and Observational Study When the experimenter controls the (random) assignment of elements to different treatment groups, the study is said to be a *designed experiment*. In contrast, in an *observational study* the assignment of elements to different treatments is voluntary and the experimenter simply observes the results of the study.

The group of people who receive a treatment is called the **treatment group**, and the group of people who do not receive a treatment is called the **control group**. In our example on diets, both groups are treatment groups because each group is assigned to one of the two types of diet. That example does not contain a control group.

Definition

Treatment and Control Groups The group of elements that receives a treatment is called the *treatment group*, and the group of elements that does not receive a treatment is called the *control group*.

EXAMPLE A–1

An example of an observational study.

Suppose a pharmaceutical company has invented a new medicine to cure a disease. To see whether or not this medicine is effective in curing this disease, it will have to be tested on a group of humans. Suppose there are 100 persons who have this disease; 50 of them voluntarily decide to take this medicine and the remaining 50 decide not to take it. The researcher then compares the cure rates for the two groups of patients. Is this an example of a designed experiment or an observational study?

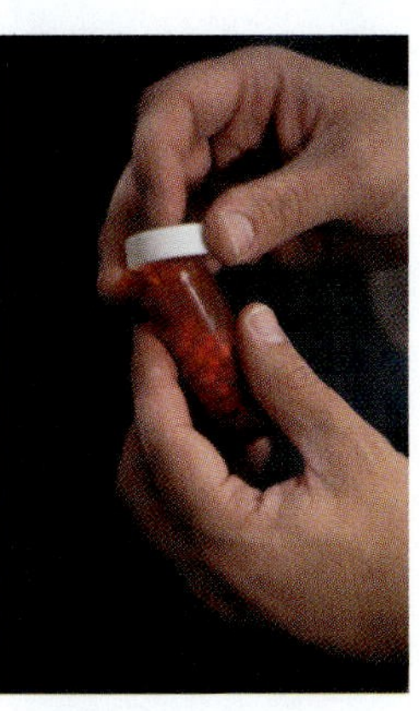

Solution This is an example of an observational study because 50 patients voluntarily joined the treatment group; they were not randomly selected. In this case, the results of the study may not be valid because the effects of the medicine will be confounded with other variables. All of the patients who decided to take the medicine may not be similar to the ones who decided not to take it. It is possible that the persons who decided to take the medicine are in the advanced stages of the disease. Consequently, they do not have much to lose by being in the treatment group. The patients in the two groups may also differ with regard to other factors such as age, gender, and so on. ■

EXAMPLE A–2

An example of a designed experiment.

Reconsider Example A–1. Now, suppose that out of the 100 people who have this disease, 50 are selected at random. These 50 people make up one group, and the remaining 50 belong to the second group. One of these groups is the treatment group, and the second is the control group. The researcher then compares the cure rates for the two groups of patients. Is this an example of a designed experiment or an observational study?

Solution In this case, the two groups will be very similar to each other. Note that we do not expect the two groups to be exactly identical. However, when randomization is used, the two groups will be very similar. After these two groups have been formed, one group will be given the actual medicine. This group is called the treatment group. The other group will be administered a placebo (a dummy medicine that looks exactly like the actual medicine). This group is called the control group. This is an example of a designed experiment because the patients are assigned to one of two groups—the treatment or the control group—randomly. ■

Usually in an experiment like the one in Example A–2, patients do not know which group they belong to. Most of the time the experimenters do not know which group a patient belongs to. This is done to avoid any bias or distortion in the results of the experiment. When neither patients nor experimenters know who is taking the real medicine and who is taking the placebo, it is called a **double-blind experiment**. For the results of the study to be unbiased and valid, an experiment must be a double-blind designed experiment. Note that if either experimenters or patients or both have access to information regarding which patients belong to treatment or control groups, it will no longer be a double-blind experiment.

The use of placebos in medical experiments is very important. A placebo is just a dummy pill that looks exactly like the real medicine. Often, patients respond to any kind of medicine. Many studies have shown that even when the patients were given sugar pills (and did not know it), many of them indicated a decrease in pain. Patients respond to placebos because they have confidence in their physicians and medicines. This is called the **placebo effect**.

Note that there can be more than two groups of elements in an experiment. For example, an investigator may need to compare three diets with regard to weight gain for chickens. Here, in a designed experiment, the chickens will be randomly assigned to one of the three diets, which are the three treatments.

In some instances we have to base our research on observational studies because it is not feasible to conduct a designed experiment. For example, suppose a researcher wants to compare the starting salaries of business and psychology majors. The researcher will have to depend on an observational study. She will select two samples, one of recent business majors and another of recent psychology majors. Based on the starting salaries of these two groups, the researcher will make a decision. Note that, here, the effects of the majors on the starting salaries of the two groups of graduates will be confounded with other variables. One of these other factors is that the business and psychology majors may be different in regard to intelligence level, which may affect their salaries. However, the researcher cannot conduct a designed experiment in this case. She cannot select a group of persons randomly and ask them to major in business and select another group and ask them to major in psychology. Instead, persons voluntarily choose their majors.

In a survey we do not exercise any control over the factors when we collect information. This characteristic of a survey makes it very close to an observational study. However, a survey may be based on a probability sample, which differentiates it from an observational study.

If an observational study or a survey indicates that two variables are related, it does not mean that there is a cause-and-effect relationship between them. For example, if an economist takes a sample of families, collects data on the incomes and rents paid by these families, and establishes an association between these two variables, it does not necessarily mean that families with higher incomes pay higher rents. Here the effects of many variables on rents are confounded. A family may pay a higher rent not because of higher income but because of various other factors, such as family size, preferences, or place of residence. We cannot make a statement about the cause-and-effect relationship between incomes and rents paid by families unless we control for these other variables. The association between incomes and rents paid by families may fit any of the following scenarios.

A–2 DO ANTIBACTERIAL SOAPS WORK?

Antibacterial soaps are no better than regular soap. Experts have said so for years. But that has not stopped millions of Americans from snapping up the supposedly superior germ killers—now 76 percent of the liquid-soap market. Part of the problem was the lack of rigorous studies to back up the experts' claims. But last week [*end of October 2002*] at the annual meeting of the Infectious Diseases Society of America, Elaine Larson, associate dean for research at Columbia University's School of Nursing, came up with the goods. In a randomized, double-blind, controlled study—the type of trial used to test pharmaceuticals—she surveyed 224 New York City homemakers. Half were given ordinary liquid soaps for a full year and the other half received antibacterial soaps. All participants' hands were cultured for germs at the beginning and the end of the study.

The results? At the outset, all participants' hands were teeming with 800,000 to 1 million bacteria. "That's normal," says Larson. "People can have up to 10 million on their hands." By the end of the year, tests revealed that they had just 300,000 or so. It didn't matter whether they used antibacterial soap or not. The difference was that they were taking more time to wash their hands thoroughly, particularly the fingers, which come in contact with the most foreign objects during the day.

Why don't antibacterial soaps do better? "The antimicrobial agent triclosan requires several minutes of contact to work," says Dr. Stuart Levy of Tufts University, author of "The Antibiotic Paradox." "Most people wash their hands for three to five seconds." Unfortunately, residues of antimicrobial soaps do linger on sinks and countertops, where Levy says they may contribute to the development of drug-resistant bacteria. A better solution for people with babies or immune-compromised patients at home is to use an alcohol-based gel, which kills germs by drying them out. Last week [*end of October 2002*] the CDC recommended these waterless germicides even in hospitals. Now, that's what the doctor ordered.

Source: Anne Underwood, "The Real Dirt on Antibacterial Soaps." *Newsweek*, November 4, 2002. Reproduced with permission.

1. These two variables have a cause-and-effect relationship. Families that have higher incomes do pay higher rents. A change in incomes of families causes a change in rents paid.
2. The incomes and rents paid by families do not have a cause-and-effect relationship. Both of these variables have a cause-and-effect relationship with a third variable. Whenever that third variable changes, these two variables change.
3. The effect of income on rent is confounded with other variables, and this indicates that income affects rent paid by families.

If our purpose in a study is to establish a cause-and-effect relationship between two variables, we must control for the effects of other variables. In other words, we must conduct a designed study.

EXERCISES

A.1 Briefly describe the various sources of data.

A.2 What is the difference between internal and external sources of data? Explain.

A.3 Explain the difference between a sample survey and a census. Why is a sample survey usually preferred over a census?

A.4 What is the difference between a survey and an experiment? Explain.

A.5 Explain the following.

- **a.** Random sample
- **b.** Nonrandom sample
- **c.** Convenience sample
- **d.** Judgment sample
- **e.** Quota sample

A.6 Explain briefly the following four sampling techniques.

- **a.** Simple random sampling
- **b.** Systematic random sampling
- **c.** Stratified random sampling
- **d.** Cluster sampling

A.7 In which sampling technique do all samples of the same size selected from a population have the same chance of being selected?

A.8 A statistics professor wanted to find out the average GPA (grade point average) for all students at her university. She used all students enrolled in her statistics class as a sample and collected information on their GPAs to find the average GPA.

- **a.** Is this sample a random or a nonrandom sample? Explain.
- **b.** What kind of sample is it? In other words, is it a simple random sample, a systematic sample, a stratified sample, a cluster sample, a convenience sample, a judgment sample, or a quota sample? Explain.
- **c.** What kind of systematic error, if any, will be made with this kind of sample? Explain.

A.9 A professor wanted to select 20 students from his class of 300 students to collect detailed information on the profiles of his students. He used his knowledge and expertise to select these 20 students.

- **a.** Is this sample a random or a nonrandom sample? Explain.
- **b.** What kind of sample is it? In other words, is it a simple random sample, a systematic sample, a stratified sample, a cluster sample, a convenience sample, a judgment sample, or a quota sample? Explain.
- **c.** What kind of systematic error, if any, will be made with this kind of sample? Explain.

A.10 Refer to Exercise A.8. Suppose the professor obtains a list of all students enrolled at the university from the registrar's office and then selects 150 students at random from this list using a statistical software package such as MINITAB.

- **a.** Is this sample a random or a nonrandom sample? Explain.
- **b.** What kind of sample is it? In other words, is it a simple random sample, a systematic sample, a stratified sample, a cluster sample, a convenience sample, a judgment sample, or a quota sample? Explain.
- **c.** Do you think any systematic error will be made in this case? Explain.

A.11 Refer to Exercise A.9. Suppose the professor enters the names of all students enrolled in his class on a computer. He then selects a sample of 20 students at random using a statistical software package such as MINITAB.

- **a.** Is this sample a random or a nonrandom sample? Explain.
- **b.** What kind of sample is it? In other words, is it a simple random sample, a systematic sample, a stratified sample, a cluster sample, a convenience sample, a judgment sample, or a quota sample? Explain.
- **c.** Do you think any systematic error will be made in this case? Explain.

A.12 A company has 1000 employees, of whom 58% are men and 42% are women. The research department at the company wanted to conduct a quick survey by selecting a sample of 50 employees and asking them about their opinions on an issue. They divided the population of employees into two groups, men and women, and then selected 29 men and 21 women from these respective groups. The interviewers were free to choose any 29 men and 21 women they wanted. What kind of sample is it? Explain.

A.13 A magazine published a questionnaire for its readers to fill out and mail to the magazine's office. In the questionnaire, cell phone owners were asked how much they would have to be paid to do without their cell phones for one month. The magazine received responses from 5439 cell phone owners.

- **a.** Based on the discussion of types of samples in Section A.2.2, what type of sample is this? Explain.
- **b.** To what kind(s) of systematic error, if any, would this survey be subject?

A.14 A researcher wanted to conduct a survey of major companies to find out what benefits are offered to their employees. She mailed questionnaires to 2500 companies and received questionnaires back from 493 companies. What kind of systematic error does this survey suffer from? Explain.

A.15 An opinion poll agency conducted a survey based on a random sample in which the interviewers called the parents included in the sample and asked them the following questions:

- **i.** Do you believe in spanking children?
- **ii.** Have you ever spanked your children?
- **iii.** If the answer to the second question is yes, how often?

What kind of systematic error, if any, does this survey suffer from? Explain.

A.16 A survey, based on a random sample taken from a borough of New York City, showed that 65% of the people living there would prefer to live somewhere other than New York City if they had the opportunity to do so. Based on this result, can the researcher say that 65% of people living in New York City would prefer to live somewhere else if they had the opportunity to do so? Explain.

A.17 In March 2005, the *New England Journal of Medicine* published the results of a 10-year clinical trial of low-dose aspirin therapy for the cardiovascular health of women (*Time*, March 21, 2005). The study was based on 40,000 healthy women, most of whom were in their 40s and 50s when the trial began. Half of these women were administered 100 mg of aspirin every other day, and the others were given a placebo. Assume that the women were assigned randomly to these two groups.

- **a.** Is this an observational study or a designed experiment? Explain.
- **b.** From the information given above, can you determine whether or not this is a double-blind study? Explain. If not, what additional information would you need?

A.18 Refer to Exercise A.17. That study also looked at the incidences of heart attacks in the two groups of women. Overall the study did not find a statistically significant difference in heart attacks between the two groups of women. However, the study noted that among women who were at least 65 years old when the study began, there was a lower incidence of heart attack for those who took aspirin than for those who took a placebo. Suppose that some medical researchers want to study this phenomenon more closely. They recruit 2000 healthy women aged 65 and older, and randomly divide them into two groups. One group takes 100 mg of aspirin every other day while the other group takes a placebo. The women did not know to which group they belonged, but the doctors who conducted the study had access to this information.

a. In this an observational study or a designed experiment? Explain.
b. Is this a double-blind study? Explain.

A.19 Refer to Exercise A.18. Now suppose that neither patients nor doctors knew what group patients belonged to.

a. Is this an observational study or a designed experiment? Explain.
b. Is this study a double-blind study? Explain.

A.20 A federal government think tank wanted to investigate whether a job training program helps the families who are on welfare to get off the welfare program. The researchers at this agency selected 5000 volunteer families who were on welfare and offered the adults in those families free job training. The researchers selected another group of 5000 volunteer families who were on welfare and did not offer them such job training. After three years the two groups were compared in regard to the percentage of families who got off welfare. Is this an observational study or a designed experiment? Explain.

A.21 Refer to Exercise A.20. Now suppose the agency selected 10,000 families at random from the list of all families that were on welfare. Of these 10,000 families, the agency randomly selected 5000 families and offered them free job training. The remaining 5000 families were not offered such job training. After three years the two groups were compared in regard to the percentage of families who got off welfare. Is this an observational study or a designed experiment? Explain.

A.22 Refer to Exercise A.20. Based on that study, the researchers concluded that the job training program causes (helps) families who are on welfare to get off the welfare program. Do you agree with this conclusion? Explain.

A.23 Refer to Exercise A.21. Based on that study, the researchers concluded that the job training program causes (helps) families who are on welfare to get off the welfare program. Do you agree with this conclusion? Explain.

A.24 A researcher advertised for volunteers to study the relationship between the amount of meat consumed and cholesterol level. In response to this advertisement, 3476 persons volunteered. The researcher collected information on the meat consumption and cholesterol level of each of these persons. Based on these data, the researcher concluded that there is a very strong positive association between these two variables.

a. Is this an observational study or a designed experiment? Explain.
b. Based on this study, can the researcher conclude that consumption of meat increases cholesterol level? Explain why or why not.

A.25 A pharmaceutical company invented a new medicine for compulsive behavior. To test this medicine on humans, the company advertised for volunteers who were suffering from this disease and wanted to participate in the study. As a result, 1820 persons responded. Using their own judgment, the group of physicians who were conducting this study assigned 910 of these patients to the treatment group and the remaining 910 to the control group. The patients in the treatment group were administered the actual medicine, and the patients in the control group were given a placebo. Six months later the conditions of the patients in the two groups were examined and compared. Based on this comparison, the physicians concluded that this medicine improves the condition of patients suffering from compulsive behavior.

a. Comment on this study and its conclusion.
b. Is this an observational study or a designed experiment? Explain.
c. Is this a double-blind study? Explain.

A.26 Refer to Exercise A.25. Suppose the physicians conducting this study obtained a list of all patients suffering from compulsive behavior who were being treated by doctors in all hospitals in the country. Further assume that this list is representative of the population of all such patients. The physicians then randomly selected 1820 patients from this list. Of these 1820, a randomly selected group of 910 patients were assigned to the treatment group, and the remaining 910 patients were assigned to the control group. The patients did not know which group they belonged to, but the doctors had access to such information. Six months later the conditions of the patients in the two groups were examined and compared. Based on this comparison, the physicians concluded that this medicine improves the condition of patients suffering from compulsive behavior.

a. Comment on this study and its conclusion.
b. Is this an observational study or a designed experiment? Explain.
c. Is this a double-blind study? Explain.

A.27 Refer to Exercise A.26. Now suppose that neither patients nor doctors knew what group the patients belonged to.

a. Is this an observational study or a designed experiment? Explain.
b. Is this a double-blind study? Explain.

A.28 The Centre for Nutrition and Food Research at Queen Margaret University College in Edinburgh studied the relationship between sugar consumption and weight gain (*Fitness*, May 2002). All the people who participated in the study were divided into two groups, and both of these groups were put on low-calorie, low-fat diets. The diet of the people in the first group was low in sugar, but the people in the second group received as much as 10% of their calories from sucrose. Both groups stayed on their respective diets for eight weeks. During these eight weeks, participants in both groups lost one-half to three-quarters of a pound per week.

a. Was this a designed experiment or an observational study?
b. Was there a control group in this study?
c. Was this a double-blind experiment?

A.29 A psychologist needs 10 pigs for a study of the intelligence of pigs. She goes to a pig farm where there are 40 young pigs in a large pen. Assume that these pigs are representative of the population of all pigs. She selects the first 10 pigs she can catch and uses them for her study.

a. Do these 10 pigs make a random sample?
b. Are these 10 pigs likely to be representative of the entire population? Why or why not?
c. If these 10 pigs do not form a random sample, what type of sample is it?
d. Can you suggest a better procedure for selecting a sample of 10 from the 40 pigs in the pen?

A.30 A newspaper wants to conduct a poll to estimate the percentage of its readers who favor a gambling casino in their city. People register their opinions by placing a phone call that costs them $1.

a. Is this method likely to produce a random sample?
b. Which, if any, of the types of biases listed in this appendix are likely to be present and why?

ADVANCED EXERCISES

A.31 A researcher sent out questionnaires to 5000 randomly chosen members of HMOs (health maintenance organizations). Only 1200 of these members completed their questionnaires and returned them. Seventy-eight percent of the respondents reported that they had experienced denial of claims by their HMOs. Of those who experienced such denials, 25% had been unable to resolve the problem to their satisfaction in at least one such instance. Write an article for a business magazine summarizing the results of the survey and cautioning the readers about possible bias in the results. Indicate which types of biases are likely to be present, how they could arise, and whether the percentages given above are likely to overestimate the true percentages of all HMO members who have experienced the denial of claims by HMOs.

A.32 A college is planning to finance an expansion of its student center through a special $20 annual fee to be levied on each student for the next four years. Because the project will take two years to complete, the students who are currently juniors or seniors will not benefit from the expansion. The campus newspaper wants to conduct a poll to seek the opinions of students on this expansion. Such opinions of students are likely to depend on their current class status, so the newspaper decides to use a stratified random sample with four class levels (freshmen, sophomores, juniors, seniors) as strata. The current student body consists of 4000 freshmen, 3200 sophomores, 2800 juniors, and 2000 seniors. The sample will contain a total of 300 students, and the size of the sample from each stratum is to be proportional to the size of the subpopulation in each stratum.

a. How many freshmen should be in the sample?
b. How many students should be chosen from each of the other three class levels?

A.33 A college mailed a questionnaire to all 5432 of its alumni who graduated in the past five years. One of the questions was about the current annual incomes of these alumni. Only 1620 of these alumni returned the completed questionnaires and 1240 of them answered that question. The current mean annual income of these 1240 respondents was $61,200.

a. Do you think $61,200 is likely to be an unbiased estimate of the current mean annual income of all 5432 alumni? If so, explain why.
b. If you think that $61,200 is probably a biased estimate of the current mean annual income of all 5432 alumni, what sources of systematic errors discussed in Section A.2.3 do you think are present here?
c. Do you expect the estimate of $61,200 to be above or below the current mean annual income of all 5432 alumni? Explain.

A.34 A group of veterinarians wants to test a new canine vaccine for Lyme disease. (Lyme disease is transmitted by the bite of an infected tick.) One hundred dogs are randomly selected (with their owners' permission), from an area that has a high incidence of Lyme disease, to receive the vaccine. These dogs are examined by veterinarians for symptoms of Lyme disease once a month for a period of 12 months. During this 12-month period, 10 of these 100 dogs are diagnosed with Lyme disease. During the same 12-month period, 18% of the unvaccinated dogs in the area are found to have contracted Lyme disease.

a. Does this experiment have a control group?
b. Is this a double-blind experiment?
c. Identify any potential sources of bias in this experiment.
d. Explain how this experiment could have been designed to reduce or eliminate the bias pointed out in part c.

Glossary

Census A survey conducted by including every element of the population.

Cluster A subgroup (usually geographical) of the population that is representative of the population.

Cluster sampling The sampling technique in which the population is divided into clusters and a sample is chosen from one or a few clusters.

Control group The group on which no condition is imposed.

Convenience sample A sample that includes the most accessible members of the population.

Designed experiment A study in which the experimenter controls the assignment of elements to different treatment groups.

Double-blind experiment Experiment in which neither the doctors (or researchers) nor the patients (or members) know to which group a patient (or member) belongs.

Experiment Method of collecting data by controlling some or all factors.

Judgment sample A sample that includes the elements of the population selected based on the judgment and prior knowledge of an expert.

Nonresponse error The error that occurs because many of the people included in the sample do not respond.

Nonsampling or systematic errors The errors that occur in the collection, recording, and tabulation of data.

Observational study A study in which the assignment of elements to different treatments is voluntary and the researcher simply observes the results of the study.

Quota sample A sample selected in such a way that each group or subpopulation is represented in the sample in exactly the same proportion as in the target population.

Random sample A sample that assigns some chance of being selected in the sample to each member of the population.

Randomization The procedure in which elements are assigned to different (treatment and control) groups at random.

Representative sample A sample that contains the characteristics of the population as closely as possible.

Response error The error that occurs because people included in the survey do not provide correct answers.

Sample A portion of the population of interest.

Sample survey A survey that includes elements of a sample.

Sampling frame The list of elements of the target population that is used to select a sample.

Sampling or chance error The difference between the result obtained from a sample survey and the result that would be obtained from the census.

Selection error The error that occurs because the sampling frame is not representative of the population.

Simple random sampling If all samples of the same size selected from a population have the same chance of being selected, it is called simple random sampling. Such a sample is called a simple random sample.

Stratified random sampling The sampling technique in which the population is divided into different strata and a sample is chosen from each stratum.

Stratum A subgroup of the population whose members are identical with regard to the possession of a characteristic.

Survey Collecting data from the elements of a population or sample.

Systematic random sampling Sampling method used to choose a sample by selecting every *k*th unit from the list.

Target population The collection of all subjects of interest.

Treatment A condition (or a set of conditions) that is imposed on a group of elements by the experimenter. This group is called the **treatment group**.

Voluntary response error The error that occurs because a survey is not conducted on a randomly selected sample but people are invited to respond voluntarily to the survey.

Explanation of Data Sets

This textbook accompanies eight large data sets that can be used for statistical analysis using technology. These data sets are:

Data Set I	City Data
Data Set II	Data on States
Data Set III	NBA Data
Data Set IV	Population Data on Manchester Road Race
Data Set V	Sample of 500 Observations Selected From Manchester Road Race Data
Data Set VI	Data on Movies
Data Set VII	Standard & Poor's 100 Index Data
Data Set VIII	McDonald's Data

These data sets are available in MINITAB, Excel, and Text format on the Web site for this text, **www.wiley.com/college/mann**. These data sets can be downloaded from this Web site. If you need more information on these data sets, you may either contact John Wiley's area representative or send an email to the author (see Preface). The above-mentioned Web site contains the following files:

1. CITYDATA (This file contains Data Set I)
2. STATEDATA (This file contains Data Set II)
3. NBA (This file contains Data Set III)
4. ROADRACE (This file contains the population data for Data Set IV)
5. RRSAMPLE (This file contains Data Set V)
6. MOVIEDATA (This file contains Data Set VI)
7. S&PDATA (This file contains Data Set VII)
8. MCDONALDDATA (This file contains Data Set VIII)

The extensions MTW, XLS, and TXT indicate that the files are in MINITAB, Excel, and text formats, respectively.

The following are the explanations of these data sets.

Data Set I: City Data

This data set contains prices (in dollars) of selected products for selected cities across the country. This data set is reproduced from ACCRA Cost of Living Index Survey for the first quarter 2005. It is reproduced with the permission of American Chamber of Commerce Researchers Association. This data set has 18 columns that contain the following variables.

C1 Name of the city
C2 Price of T-bone steak per pound

C3 Price of sausage per pound, Jimmy Dean or Owens brand, 100% pork

C4 Price of half-gallon carton of whole milk

C5 Price of parmesan cheese, grated 8 oz. canister, Kraft brand

C6 Price of potatoes, 10 pounds, white or red

C7 Price of fresh orange juice, 64 oz., Tropicana or Florida Natural brand

C8 Price of 75 oz. Cascade dishwashing powder

C9 Price of 16 oz. whole kernel frozen corn, lowest price

C10 Price of 2 liter Coca Cola, excluding any deposit

C11 Monthly rent of an unfurnished two-bedroom apartment (excluding all utilities except water) 1½ or 2 baths, 950 sq. ft.

C12 Purchase price of 2400 square feet living area new house, on 8000 square feet lot in urban area with all utilities.

C13 Monthly telephone charges for a private residential line; customer owns instruments. Price includes: basic monthly rate; additional local use charges, if any, incurred by a family of four; TouchTone fee; all other mandatory monthly charges, such as long distance access fee and 911 fee; and all taxes on the foregoing.

C14 Price of one gallon regular unleaded gas, national brand, including all taxes; cash price at self-service pump if available

C15 Price of a woman's shampoo, trim, and blow-dry

C16 Price of dry cleaning, man's two-piece suit

C17 Price of first-run movie, indoor, evening, no discount

C18 Price of 1.5-liter bottle of wine, Livingston Cellars or Gallo Chablis or Chenin Blanc.

Data Set II: Data on States

This data set contains information on different variables for all 50 states in the United States. This data set has eight columns that contain the following variables.

C1 Name of the state

C2 Per capita personal income (in current dollars), 2003 (Source: U.S. Bureau of Economic Analysis)

C3 Traffic fatalities, 2002 (Source: U.S. National Highway Safety Traffic Administration)

C4 Female labor force participation rate (in percent), 2003 (Source: U.S. Bureau of Labor Statistics)

C5 Average salaries of teachers (in dollars), 2003 (Source: Current NEA Estimates Data Base)

C6 Percent of the population (25 years and over) with a Bachelor's degree or higher, 2004 (U.S. Census Bureau)

Data Set III: NBA Data

This data set contains information on NBA players who were on the rosters of National Basketball Association teams as of May 2005. This data set has 13 columns that contain the following variables.

C1 Team name

C2 Name of player

C3 Player's annual salary
C4 Length of player's contract (in years)
C5 Total value of contract over the length specified in C4
C6 Year contract expires
C7 Number of years of experience in NBA prior to 2004–2005 season (0 implies a rookie)
C8 Primary position (C = Center, F = Forward, G = Guard)
C9 Secondary position (blank implies no secondary position)
C10 Height (in inches) of player
C11 Weight (in pounds) of player
C12 Age of player
C13 Identifies whether player came to NBA from college, a foreign country, or high school

Data Set IV: Manchester (Connecticut) Road Race Data

This data set contains time (in minutes) taken by each of the participants to complete the Sixty-eigth Road Race held on Thanksgiving Day, November 25, 2004, in Manchester (Connecticut). The total distance for this race is 4.748 miles, and it is held every year on Thanksgiving Day. A total of 8911 participants were able to finish this race. Consequently, this data set contains 8911 observations. It has one column that contains completion time in minutes.

Data Set V: Sample of 500 Observations Selected From Data Set IV

This data set contains a random sample of 500 observations selected from Data Set IV above. It has one column that contains the completion time (in minutes) for 500 runners.

Data Set VI: Data on Movies

This data set contains information on the top 150 films from 2004 in terms of gross revenue in the United States. This data set contains 8 columns that contain the following variables (Source: http://www.boxofficemojo.com).

C1 Rank
C2 Movie title
C3 Name of studio where the film was produced
C4 Gross revenue during entire theater release period
C5 Number of theaters that showed the film during release period
C6 Gross revenue during first week of theater release
C7 Number of theaters that showed the film during first week of theater release
C8 Length of release period (in days)

Data Set VII: Standard & Poor's 100 Index Data

This data set contains trading and value information on the 100 stocks in the Standard & Poor's 100 Index as of Monday, August 29, 2005. This data set has 10 columns that contain the following variables (Source: http://finance.yahoo.com).

C1 Company's stock exchange symbol
C2 Company name
C3 Company's economic sector (e.g., manufacturing)
C4 Stock price at close of business on Friday, August 26, 2005
C5 Stock price at close of business on Monday, August 29, 2005
C6 Change in stock price from close of business on 8/26/2005 to 8/29/2005
C7 Opening bid for stock price on 8/29/2005
C8 Highest stock price attained on 8/29/2005
C9 Lowest stock price attained on 8/29/2005
C10 Number of shares traded on 8/29/2005

Data Set VIII: McDonald's Data

This data set contains information on the nutritional aspects of McDonald's food. This data set is reproduced from McDonald'd Web site (http//www.mcdonalds.com/usa/eat/nutrition_info.html). The only alteration involves the approximation of the dietery fiber content of four food items listed as having less than 1 gram of dietary fiber each, which were all changed to .5 grams. Condiments (ketchup, salad dressing, dipping sauces, and so forth) and liquids (sodas, shakes, and so forth) are not included. This data set has 25 columns that contain the following variables.

C1 Menu item
C2 Serving size (in ounces)
C3 Serving size (in grams)
C4 Calories
C5 Calories from fat
C6 Total fat (in grams)
C7 Percent daily value of fat
C8 Saturated fat (in grams)
C9 Percent daily value of saturated fat
C10 Trans fat (in grams)
C11 Cholesterol (in milligrams)
C12 Percent daily value of cholesterol
C13 Sodium (in milligrams)
C14 Percent daily value of sodium
C15 Carbohydrates (in milligrams)
C16 Percent daily value of carbohydrates
C17 Dietary fiber (in grams)
C18 Percent daily value of dietary fiber
C19 Sugars (in grams)
C20 Protein (in grams)
C21 Percent daily value of Vitamin A
C22 Percent daily value of Vitamin C
C23 Percent daily value of Calcium
C24 Percent daily value of Iron
C25 Menu category (e.g., sandwich, non-sandwich chicken, breakfast, and so forth)

Appendix

Statistical Tables

Table I Table of Binomial Probabilities

Table II Values of $e^{-\lambda}$

Table III Table of Poisson Probabilities

Table IV Standard Normal Distribution Table

Table V The *t* Distribution Table

Table VI Chi-Square Distribution Table

Table VII The *F* Distribution Table

Note: The following tables are on the Web site of the text along with Chapters 14 and 15.

Table VIII Critical Values of *X* for the Sign Test

Table IX Critical Values of *T* for the Wilcoxon Signed-Rank Test

Table X Critical Values of *T* for the Wilcoxon Rank Sum Test

Table XI Critical Values for the Spearman Rho Rank Correlation Coefficient Test

Table XII Critical Values for a Two-Tailed Runs Test with $\alpha = .05$

Table I Table of Binomial Probabilities

		p										
n	x	.05	.10	.20	.30	.40	.50	.60	.70	.80	.90	.95
1	0	.9500	.9000	.8000	.7000	.6000	.5000	.4000	.3000	.2000	.1000	.0500
	1	.0500	.1000	.2000	.3000	.4000	.5000	.6000	.7000	.8000	.9000	.9500
2	0	.9025	.8100	.6400	.4900	.3600	.2500	.1600	.0900	.0400	.0100	.0025
	1	.0950	.1800	.3200	.4200	.4800	.5000	.4800	.4200	.3200	.1800	.0950
	2	.0025	.0100	.0400	.0900	.1600	.2500	.3600	.4900	.6400	.8100	.9025
3	0	.8574	.7290	.5120	.3430	.2160	.1250	.0640	.0270	.0080	.0010	.0001
	1	.1354	.2430	.3840	.4410	.4320	.3750	.2880	.1890	.0960	.0270	.0071
	2	.0071	.0270	.0960	.1890	.2880	.3750	.4320	.4410	.3840	.2430	.1354
	3	.0001	.0010	.0080	.0270	.0640	.1250	.2160	.3430	.5120	.7290	.8574
4	0	.8145	.6561	.4096	.2401	.1296	.0625	.0256	.0081	.0016	.0001	.0000
	1	.1715	.2916	.4096	.4116	.3456	.2500	.1536	.0756	.0256	.0036	.0005
	2	.0135	.0486	.1536	.2646	.3456	.3750	.3456	.2646	.1536	.0486	.0135
	3	.0005	.0036	.0256	.0756	.1536	.2500	.3456	.4116	.4096	.2916	.1715
	4	.0000	.0001	.0016	.0081	.0256	.0625	.1296	.2401	.4096	.6561	.8145
5	0	.7738	.5905	.3277	.1681	.0778	.0312	.0102	.0024	.0003	.0000	.0000
	1	.2036	.3280	.4096	.3602	.2592	.1562	.0768	.0284	.0064	.0005	.0000
	2	.0214	.0729	.2048	.3087	.3456	.3125	.2304	.1323	.0512	.0081	.0011
	3	.0011	.0081	.0512	.1323	.2304	.3125	.3456	.3087	.2048	.0729	.0214
	4	.0000	.0004	.0064	.0283	.0768	.1562	.2592	.3601	.4096	.3281	.2036
	5	.0000	.0000	.0003	.0024	.0102	.0312	.0778	.1681	.3277	.5905	.7738
6	0	.7351	.5314	.2621	.1176	.0467	.0156	.0041	.0007	.0001	.0000	.0000
	1	.2321	.3543	.3932	.3025	.1866	.0937	.0369	.0102	.0015	.0001	.0000
	2	.0305	.0984	.2458	.3241	.3110	.2344	.1382	.0595	.0154	.0012	.0001
	3	.0021	.0146	.0819	.1852	.2765	.3125	.2765	.1852	.0819	.0146	.0021
	4	.0001	.0012	.0154	.0595	.1382	.2344	.3110	.3241	.2458	.0984	.0305
	5	.0000	.0001	.0015	.0102	.0369	.0937	.1866	.3025	.3932	.3543	.2321
	6	.0000	.0000	.0001	.0007	.0041	.0156	.0467	.1176	.2621	.5314	.7351
7	0	.6983	.4783	.2097	.0824	.0280	.0078	.0016	.0002	.0000	.0000	.0000
	1	.2573	.3720	.3670	.2471	.1306	.0547	.0172	.0036	.0004	.0000	.0000
	2	.0406	.1240	.2753	.3177	.2613	.1641	.0774	.0250	.0043	.0002	.0000
	3	.0036	.0230	.1147	.2269	.2903	.2734	.1935	.0972	.0287	.0026	.0002
	4	.0002	.0026	.0287	.0972	.1935	.2734	.2903	.2269	.1147	.0230	.0036
	5	.0000	.0002	.0043	.0250	.0774	.1641	.2613	.3177	.2753	.1240	.0406
	6	.0000	.0000	.0004	.0036	.0172	.0547	.1306	.2471	.3670	.3720	.2573
	7	.0000	.0000	.0000	.0002	.0016	.0078	.0280	.0824	.2097	.4783	.6983
8	0	.6634	.4305	.1678	.0576	.0168	.0039	.0007	.0001	.0000	.0000	.0000
	1	.2793	.3826	.3355	.1977	.0896	.0312	.0079	.0012	.0001	.0000	.0000

Table I Table of Binomial Probabilities (continued)

		p										
n	*x*	.05	.10	.20	.30	.40	.50	.60	.70	.80	.90	.95
	2	.0515	.1488	.2936	.2965	.2090	.1094	.0413	.0100	.0011	.0000	.0000
	3	.0054	.0331	.1468	.2541	.2787	.2187	.1239	.0467	.0092	.0004	.0000
	4	.0004	.0046	.0459	.1361	.2322	.2734	.2322	.1361	.0459	.0046	.0004
	5	.0000	.0004	.0092	.0467	.1239	.2187	.2787	.2541	.1468	.0331	.0054
	6	.0000	.0000	.0011	.0100	.0413	.1094	.2090	.2965	.2936	.1488	.0515
	7	.0000	.0000	.0001	.0012	.0079	.0312	.0896	.1977	.3355	.3826	.2793
	8	.0000	.0000	.0000	.0001	.0007	.0039	.0168	.0576	.1678	.4305	.6634
9	0	.6302	.3874	.1342	.0404	.0101	.0020	.0003	.0000	.0000	.0000	.0000
	1	.2985	.3874	.3020	.1556	.0605	.0176	.0035	.0004	.0000	.0000	.0000
	2	.0629	.1722	.3020	.2668	.1612	.0703	.0212	.0039	.0003	.0000	.0000
	3	.0077	.0446	.1762	.2668	.2508	.1641	.0743	.0210	.0028	.0001	.0000
	4	.0006	.0074	.0661	.1715	.2508	.2461	.1672	.0735	.0165	.0008	.0000
	5	.0000	.0008	.0165	.0735	.1672	.2461	.2508	.1715	.0661	.0074	.0006
	6	.0000	.0001	.0028	.0210	.0743	.1641	.2508	.2668	.1762	.0446	.0077
	7	.0000	.0000	.0003	.0039	.0212	.0703	.1612	.2668	.3020	.1722	.0629
	8	.0000	.0000	.0000	.0004	.0035	.0176	.0605	.1556	.3020	.3874	.2985
	9	.0000	.0000	.0000	.0000	.0003	.0020	.0101	.0404	.1342	.3874	.6302
10	0	.5987	.3487	.1074	.0282	.0060	.0010	.0001	.0000	.0000	.0000	.0000
	1	.3151	.3874	.2684	.1211	.0403	.0098	.0016	.0001	.0000	.0000	.0000
	2	.0746	.1937	.3020	.2335	.1209	.0439	.0106	.0014	.0001	.0000	.0000
	3	.0105	.0574	.2013	.2668	.2150	.1172	.0425	.0090	.0008	.0000	.0000
	4	.0010	.0112	.0881	.2001	.2508	.2051	.1115	.0368	.0055	.0001	.0000
	5	.0001	.0015	.0264	.1029	.2007	.2461	.2007	.1029	.0264	.0015	.0001
	6	.0000	.0001	.0055	.0368	.1115	.2051	.2508	.2001	.0881	.0112	.0010
	7	.0000	.0000	.0008	.0090	.0425	.1172	.2150	.2668	.2013	.0574	.0105
	8	.0000	.0000	.0001	.0014	.0106	.0439	.1209	.2335	.3020	.1937	.0746
	9	.0000	.0000	.0000	.0001	.0016	.0098	.0403	.1211	.2684	.3874	.3151
	10	.0000	.0000	.0000	.0000	.0001	.0010	.0060	.0282	.1074	.3487	.5987
11	0	.5688	.3138	.0859	.0198	.0036	.0005	.0000	.0000	.0000	.0000	.0000
	1	.3293	.3835	.2362	.0932	.0266	.0054	.0007	.0000	.0000	.0000	.0000
	2	.0867	.2131	.2953	.1998	.0887	.0269	.0052	.0005	.0000	.0000	.0000
	3	.0137	.0710	.2215	.2568	.1774	.0806	.0234	.0037	.0002	.0000	.0000
	4	.0014	.0158	.1107	.2201	.2365	.1611	.0701	.0173	.0017	.0000	.0000
	5	.0001	.0025	.0388	.1321	.2207	.2256	.1471	.0566	.0097	.0003	.0000
	6	.0000	.0003	.0097	.0566	.1471	.2256	.2207	.1321	.0388	.0025	.0001
	7	.0000	.0000	.0017	.0173	.0701	.1611	.2365	.2201	.1107	.0158	.0014
	8	.0000	.0000	.0002	.0037	.0234	.0806	.1774	.2568	.2215	.0710	.0137
	9	.0000	.0000	.0000	.0005	.0052	.0269	.0887	.1998	.2953	.2131	.0867

Table I Table of Binomial Probabilities (continued)

n	x	p = .05	.10	.20	.30	.40	.50	.60	.70	.80	.90	.95
	10	.0000	.0000	.0000	.0000	.0007	.0054	.0266	.0932	.2362	.3835	.3293
	11	.0000	.0000	.0000	.0000	.0000	.0005	.0036	.0198	.0859	.3138	.5688
12	0	.5404	.2824	.0687	.0138	.0022	.0002	.0000	.0000	.0000	.0000	.0000
	1	.3413	.3766	.2062	.0712	.0174	.0029	.0003	.0000	.0000	.0000	.0000
	2	.0988	.2301	.2835	.1678	.0639	.0161	.0025	.0002	.0000	.0000	.0000
	3	.0173	.0852	.2362	.2397	.1419	.0537	.0125	.0015	.0001	.0000	.0000
	4	.0021	.0213	.1329	.2311	.2128	.1208	.0420	.0078	.0005	.0000	.0000
	5	.0002	.0038	.0532	.1585	.2270	.1934	.1009	.0291	.0033	.0000	.0000
	6	.0000	.0005	.0155	.0792	.1766	.2256	.1766	.0792	.0155	.0005	.0000
	7	.0000	.0000	.0033	.0291	.1009	.1934	.2270	.1585	.0532	.0038	.0002
	8	.0000	.0000	.0005	.0078	.0420	.1208	.2128	.2311	.1329	.0213	.0021
	9	.0000	.0000	.0001	.0015	.0125	.0537	.1419	.2397	.2362	.0852	.0173
	10	.0000	.0000	.0000	.0002	.0025	.0161	.0639	.1678	.2835	.2301	.0988
	11	.0000	.0000	.0000	.0000	.0003	.0029	.0174	.0712	.2062	.3766	.3413
	12	.0000	.0000	.0000	.0000	.0000	.0002	.0022	.0138	.0687	.2824	.5404
13	0	.5133	.2542	.0550	.0097	.0013	.0001	.0000	.0000	.0000	.0000	.0000
	1	.3512	.3672	.1787	.0540	.0113	.0016	.0001	.0000	.0000	.0000	.0000
	2	.1109	.2448	.2680	.1388	.0453	.0095	.0012	.0001	.0000	.0000	.0000
	3	.0214	.0997	.2457	.2181	.1107	.0349	.0065	.0006	.0000	.0000	.0000
	4	.0028	.0277	.1535	.2337	.1845	.0873	.0243	.0034	.0001	.0000	.0000
	5	.0003	.0055	.0691	.1803	.2214	.1571	.0656	.0142	.0011	.0000	.0000
	6	.0000	.0008	.0230	.1030	.1968	.2095	.1312	.0442	.0058	.0001	.0000
	7	.0000	.0001	.0058	.0442	.1312	.2095	.1968	.1030	.0230	.0008	.0000
	8	.0000	.0000	.0011	.0142	.0656	.1571	.2214	.1803	.0691	.0055	.0003
	9	.0000	.0000	.0001	.0034	.0243	.0873	.1845	.2337	.1535	.0277	.0028
	10	.0000	.0000	.0000	.0006	.0065	.0349	.1107	.2181	.2457	.0997	.0214
	11	.0000	.0000	.0000	.0001	.0012	.0095	.0453	.1388	.2680	.2448	.1109
	12	.0000	.0000	.0000	.0000	.0001	.0016	.0113	.0540	.1787	.3672	.3512
	13	.0000	.0000	.0000	.0000	.0000	.0001	.0013	.0097	.0550	.2542	.5133
14	0	.4877	.2288	.0440	.0068	.0008	.0001	.0000	.0000	.0000	.0000	.0000
	1	.3593	.3559	.1539	.0407	.0073	.0009	.0001	.0000	.0000	.0000	.0000
	2	.1229	.2570	.2501	.1134	.0317	.0056	.0005	.0000	.0000	.0000	.0000
	3	.0259	.1142	.2501	.1943	.0845	.0222	.0033	.0002	.0000	.0000	.0000
	4	.0037	.0349	.1720	.2290	.1549	.0611	.0136	.0014	.0000	.0000	.0000
	5	.0004	.0078	.0860	.1963	.2066	.1222	.0408	.0066	.0003	.0000	.0000
	6	.0000	.0013	.0322	.1262	.2066	.1833	.0918	.0232	.0020	.0000	.0000
	7	.0000	.0002	.0092	.0618	.1574	.2095	.1574	.0618	.0092	.0002	.0000
	8	.0000	.0000	.0020	.0232	.0918	.1833	.2066	.1262	.0322	.0013	.0000
	9	.0000	.0000	.0003	.0066	.0408	.1222	.2066	.1963	.0860	.0078	.0004
	10	.0000	.0000	.0000	.0014	.0136	.0611	.1549	.2290	.1720	.0349	.0037

Table I Table of Binomial Probabilities (continued)

		p										
n	*x*	.05	.10	.20	.30	.40	.50	.60	.70	.80	.90	.95
	11	.0000	.0000	.0000	.0002	.0033	.0222	.0845	.1943	.2501	.1142	.0259
	12	.0000	.0000	.0000	.0000	.0005	.0056	.0317	.1134	.2501	.2570	.1229
	13	.0000	.0000	.0000	.0000	.0001	.0009	.0073	.0407	.1539	.3559	.3593
	14	.0000	.0000	.0000	.0000	.0000	.0001	.0008	.0068	.0440	.2288	.4877
15	0	.4633	.2059	.0352	.0047	.0005	.0000	.0000	.0000	.0000	.0000	.0000
	1	.3658	.3432	.1319	.0305	.0047	.0005	.0000	.0000	.0000	.0000	.0000
	2	.1348	.2669	.2309	.0916	.0219	.0032	.0003	.0000	.0000	.0000	.0000
	3	.0307	.1285	.2501	.1700	.0634	.0139	.0016	.0001	.0000	.0000	.0000
	4	.0049	.0428	.1876	.2186	.1268	.0417	.0074	.0006	.0000	.0000	.0000
	5	.0006	.0105	.1032	.2061	.1859	.0916	.0245	.0030	.0001	.0000	.0000
	6	.0000	.0019	.0430	.1472	.2066	.1527	.0612	.0116	.0007	.0000	.0000
	7	.0000	.0003	.0138	.0811	.1771	.1964	.1181	.0348	.0035	.0000	.0000
	8	.0000	.0000	.0035	.0348	.1181	.1964	.1771	.0811	.0138	.0003	.0000
	9	.0000	.0000	.0007	.0116	.0612	.1527	.2066	.1472	.0430	.0019	.0000
	10	.0000	.0000	.0001	.0030	.0245	.0916	.1859	.2061	.1032	.0105	.0006
	11	.0000	.0000	.0000	.0006	.0074	.0417	.1268	.2186	.1876	.0428	.0049
	12	.0000	.0000	.0000	.0001	.0016	.0139	.0634	.1700	.2501	.1285	.0307
	13	.0000	.0000	.0000	.0000	.0003	.0032	.0219	.0916	.2309	.2669	.1348
	14	.0000	.0000	.0000	.0000	.0000	.0005	.0047	.0305	.1319	.3432	.3658
	15	.0000	.0000	.0000	.0000	.0000	.0000	.0005	.0047	.0352	.2059	.4633
16	0	.4401	.1853	.0281	.0033	.0003	.0000	.0000	.0000	.0000	.0000	.0000
	1	.3706	.3294	.1126	.0228	.0030	.0002	.0000	.0000	.0000	.0000	.0000
	2	.1463	.2745	.2111	.0732	.0150	.0018	.0001	.0000	.0000	.0000	.0000
	3	.0359	.1423	.2463	.1465	.0468	.0085	.0008	.0000	.0000	.0000	.0000
	4	.0061	.0514	.2001	.2040	.1014	.0278	.0040	.0002	.0000	.0000	.0000
	5	.0008	.0137	.1201	.2099	.1623	.0667	.0142	.0013	.0000	.0000	.0000
	6	.0001	.0028	.0550	.1649	.1983	.1222	.0392	.0056	.0002	.0000	.0000
	7	.0000	.0004	.0197	.1010	.1889	.1746	.0840	.0185	.0012	.0000	.0000
	8	.0000	.0001	.0055	.0487	.1417	.1964	.1417	.0487	.0055	.0001	.0000
	9	.0000	.0000	.0012	.0185	.0840	.1746	.1889	.1010	.0197	.0004	.0000
	10	.0000	.0000	.0002	.0056	.0392	.1222	.1983	.1649	.0550	.0028	.0001
	11	.0000	.0000	.0000	.0013	.0142	.0666	.1623	.2099	.1201	.0137	.0008
	12	.0000	.0000	.0000	.0002	.0040	.0278	.1014	.2040	.2001	.0514	.0061
	13	.0000	.0000	.0000	.0000	.0008	.0085	.0468	.1465	.2463	.1423	.0359
	14	.0000	.0000	.0000	.0000	.0001	.0018	.0150	.0732	.2111	.2745	.1463
	15	.0000	.0000	.0000	.0000	.0000	.0002	.0030	.0228	.1126	.3294	.3706
	16	.0000	.0000	.0000	.0000	.0000	.0000	.0003	.0033	.0281	.1853	.4401
17	0	.4181	.1668	.0225	.0023	.0002	.0000	.0000	.0000	.0000	.0000	.0000
	1	.3741	.3150	.0957	.0169	.0019	.0001	.0000	.0000	.0000	.0000	.0000
	2	.1575	.2800	.1914	.0581	.0102	.0010	.0001	.0000	.0000	.0000	.0000

Table I Table of Binomial Probabilities (continued)

		p										
n	*x*	.05	.10	.20	.30	.40	.50	.60	.70	.80	.90	.95
	3	.0415	.1556	.2393	.1245	.0341	.0052	.0004	.0000	.0000	.0000	.0000
	4	.0076	.0605	.2093	.1868	.0796	.0182	.0021	.0001	.0000	.0000	.0000
	5	.0010	.0175	.1361	.2081	.1379	.0472	.0081	.0006	.0000	.0000	.0000
	6	.0001	.0039	.0680	.1784	.1839	.0944	.0242	.0026	.0001	.0000	.0000
	7	.0000	.0007	.0267	.1201	.1927	.1484	.0571	.0095	.0004	.0000	.0000
	8	.0000	.0001	.0084	.0644	.1606	.1855	.1070	.0276	.0021	.0000	.0000
	9	.0000	.0000	.0021	.0276	.1070	.1855	.1606	.0644	.0084	.0001	.0000
	10	.0000	.0000	.0004	.0095	.0571	.1484	.1927	.1201	.0267	.0007	.0000
	11	.0000	.0000	.0001	.0026	.0242	.0944	.1839	.1784	.0680	.0039	.0001
	12	.0000	.0000	.0000	.0006	.0081	.0472	.1379	.2081	.1361	.0175	.0010
	13	.0000	.0000	.0000	.0001	.0021	.0182	.0796	.1868	.2093	.0605	.0076
	14	.0000	.0000	.0000	.0000	.0004	.0052	.0341	.1245	.2393	.1556	.0415
	15	.0000	.0000	.0000	.0000	.0001	.0010	.0102	.0581	.1914	.2800	.1575
	16	.0000	.0000	.0000	.0000	.0000	.0001	.0019	.0169	.0957	.3150	.3741
	17	.0000	.0000	.0000	.0000	.0000	.0000	.0002	.0023	.0225	.1668	.4181
18	0	.3972	.1501	.0180	.0016	.0001	.0000	.0000	.0000	.0000	.0000	.0000
	1	.3763	.3002	.0811	.0126	.0012	.0001	.0000	.0000	.0000	.0000	.0000
	2	.1683	.2835	.1723	.0458	.0069	.0006	.0000	.0000	.0000	.0000	.0000
	3	.0473	.1680	.2297	.1046	.0246	.0031	.0002	.0000	.0000	.0000	.0000
	4	.0093	.0700	.2153	.1681	.0614	.0117	.0011	.0000	.0000	.0000	.0000
	5	.0014	.0218	.1507	.2017	.1146	.0327	.0045	.0002	.0000	.0000	.0000
	6	.0002	.0052	.0816	.1873	.1655	.0708	.0145	.0012	.0000	.0000	.0000
	7	.0000	.0010	.0350	.1376	.1892	.1214	.0374	.0046	.0001	.0000	.0000
	8	.0000	.0002	.0120	.0811	.1734	.1669	.0771	.0149	.0008	.0000	.0000
	9	.0000	.0000	.0033	.0386	.1284	.1855	.1284	.0386	.0033	.0000	.0000
	10	.0000	.0000	.0008	.0149	.0771	.1669	.1734	.0811	.0120	.0002	.0000
	11	.0000	.0000	.0001	.0046	.0374	.1214	.1892	.1376	.0350	.0010	.0000
	12	.0000	.0000	.0000	.0012	.0145	.0708	.1655	.1873	.0816	.0052	.0002
	13	.0000	.0000	.0000	.0002	.0045	.0327	.1146	.2017	.1507	.0218	.0014
	14	.0000	.0000	.0000	.0000	.0011	.0117	.0614	.1681	.2153	.0700	.0093
	15	.0000	.0000	.0000	.0000	.0002	.0031	.0246	.1046	.2297	.1680	.0473
	16	.0000	.0000	.0000	.0000	.0000	.0006	.0069	.0458	.1723	.2835	.1683
	17	.0000	.0000	.0000	.0000	.0000	.0001	.0012	.0126	.0811	.3002	.3763
	18	.0000	.0000	.0000	.0000	.0000	.0000	.0001	.0016	.0180	.1501	.3972
19	0	.3774	.1351	.0144	.0011	.0001	.0000	.0000	.0000	.0000	.0000	.0000
	1	.3774	.2852	.0685	.0093	.0008	.0000	.0000	.0000	.0000	.0000	.0000
	2	.1787	.2852	.1540	.0358	.0046	.0003	.0000	.0000	.0000	.0000	.0000
	3	.0533	.1796	.2182	.0869	.0175	.0018	.0001	.0000	.0000	.0000	.0000
	4	.0112	.0798	.2182	.1491	.0467	.0074	.0005	.0000	.0000	.0000	.0000

Table I Table of Binomial Probabilities (continued)

		p										
n	*x*	.05	.10	.20	.30	.40	.50	.60	.70	.80	.90	.95
	5	.0018	.0266	.1636	.1916	.0933	.0222	.0024	.0001	.0000	.0000	.0000
	6	.0002	.0069	.0955	.1916	.1451	.0518	.0085	.0005	.0000	.0000	.0000
	7	.0000	.0014	.0443	.1525	.1797	.0961	.0237	.0022	.0000	.0000	.0000
	8	.0000	.0002	.0166	.0981	.1797	.1442	.0532	.0077	.0003	.0000	.0000
	9	.0000	.0000	.0051	.0514	.1464	.1762	.0976	.0220	.0013	.0000	.0000
	10	.0000	.0000	.0013	.0220	.0976	.1762	.1464	.0514	.0051	.0000	.0000
	11	.0000	.0000	.0003	.0077	.0532	.1442	.1797	.0981	.0166	.0002	.0000
	12	.0000	.0000	.0000	.0022	.0237	.0961	.1797	.1525	.0443	.0014	.0000
	13	.0000	.0000	.0000	.0005	.0085	.0518	.1451	.1916	.0955	.0069	.0002
	14	.0000	.0000	.0000	.0001	.0024	.0222	.0933	.1916	.1636	.0266	.0018
	15	.0000	.0000	.0000	.0000	.0005	.0074	.0467	.1491	.2182	.0798	.0112
	16	.0000	.0000	.0000	.0000	.0001	.0018	.0175	.0869	.2182	.1796	.0533
	17	.0000	.0000	.0000	.0000	.0000	.0003	.0046	.0358	.1540	.2852	.1787
	18	.0000	.0000	.0000	.0000	.0000	.0000	.0008	.0093	.0685	.2852	.3774
	19	.0000	.0000	.0000	.0000	.0000	.0000	.0001	.0011	.0144	.1351	.3774
20	0	.3585	.1216	.0115	.0008	.0000	.0000	.0000	.0000	.0000	.0000	.0000
	1	.3774	.2702	.0576	.0068	.0005	.0000	.0000	.0000	.0000	.0000	.0000
	2	.1887	.2852	.1369	.0278	.0031	.0002	.0000	.0000	.0000	.0000	.0000
	3	.0596	.1901	.2054	.0716	.0123	.0011	.0000	.0000	.0000	.0000	.0000
	4	.0133	.0898	.2182	.1304	.0350	.0046	.0003	.0000	.0000	.0000	.0000
	5	.0022	.0319	.1746	.1789	.0746	.0148	.0013	.0000	.0000	.0000	.0000
	6	.0003	.0089	.1091	.1916	.1244	.0370	.0049	.0002	.0000	.0000	.0000
	7	.0000	.0020	.0545	.1643	.1659	.0739	.0146	.0010	.0000	.0000	.0000
	8	.0000	.0004	.0222	.1144	.1797	.1201	.0355	.0039	.0001	.0000	.0000
	9	.0000	.0001	.0074	.0654	.1597	.1602	.0710	.0120	.0005	.0000	.0000
	10	.0000	.0000	.0020	.0308	.1171	.1762	.1171	.0308	.0020	.0000	.0000
	11	.0000	.0000	.0005	.0120	.0710	.1602	.1597	.0654	.0074	.0001	.0000
	12	.0000	.0000	.0001	.0039	.0355	.1201	.1797	.1144	.0222	.0004	.0000
	13	.0000	.0000	.0000	.0010	.0146	.0739	.1659	.1643	.0545	.0020	.0000
	14	.0000	.0000	.0000	.0002	.0049	.0370	.1244	.1916	.1091	.0089	.0003
	15	.0000	.0000	.0000	.0000	.0013	.0148	.0746	.1789	.1746	.0319	.0022
	16	.0000	.0000	.0000	.0000	.0003	.0046	.0350	.1304	.2182	.0898	.0133
	17	.0000	.0000	.0000	.0000	.0000	.0011	.0123	.0716	.2054	.1901	.0596
	18	.0000	.0000	.0000	.0000	.0000	.0002	.0031	.0278	.1369	.2852	.1887
	19	.0000	.0000	.0000	.0000	.0000	.0000	.0005	.0068	.0576	.2702	.3774
	20	.0000	.0000	.0000	.0000	.0000	.0000	.0000	.0008	.0115	.1216	.3585
21	0	.3406	.1094	.0092	.0006	.0000	.0000	.0000	.0000	.0000	.0000	.0000
	1	.3764	.2553	.0484	.0050	.0003	.0000	.0000	.0000	.0000	.0000	.0000
	2	.1981	.2837	.1211	.0215	.0020	.0001	.0000	.0000	.0000	.0000	.0000

Table I Table of Binomial Probabilities (continued)

		p										
n	*x*	.05	.10	.20	.30	.40	.50	.60	.70	.80	.90	.95
	3	.0660	.1996	.1917	.0585	.0086	.0006	.0000	.0000	.0000	.0000	.0000
	4	.0156	.0998	.2156	.1128	.0259	.0029	.0001	.0000	.0000	.0000	.0000
	5	.0028	.0377	.1833	.1643	.0588	.0097	.0007	.0000	.0000	.0000	.0000
	6	.0004	.0112	.1222	.1878	.1045	.0259	.0027	.0001	.0000	.0000	.0000
	7	.0000	.0027	.0655	.1725	.1493	.0554	.0087	.0005	.0000	.0000	.0000
	8	.0000	.0005	.0286	.1294	.1742	.0970	.0229	.0019	.0000	.0000	.0000
	9	.0000	.0001	.0103	.0801	.1677	.1402	.0497	.0063	.0002	.0000	.0000
	10	.0000	.0000	.0031	.0412	.1342	.1682	.0895	.0176	.0008	.0000	.0000
	11	.0000	.0000	.0008	.0176	.0895	.1682	.1342	.0412	.0031	.0000	.0000
	12	.0000	.0000	.0002	.0063	.0497	.1402	.1677	.0801	.0103	.0001	.0000
	13	.0000	.0000	.0000	.0019	.0229	.0970	.1742	.1294	.0286	.0005	.0000
	14	.0000	.0000	.0000	.0005	.0087	.0554	.1493	.1725	.0655	.0027	.0000
	15	.0000	.0000	.0000	.0001	.0027	.0259	.1045	.1878	.1222	.0112	.0004
	16	.0000	.0000	.0000	.0000	.0007	.0097	.0588	.1643	.1833	.0377	.0028
	17	.0000	.0000	.0000	.0000	.0001	.0029	.0259	.1128	.2156	.0998	.0156
	18	.0000	.0000	.0000	.0000	.0000	.0006	.0086	.0585	.1917	.1996	.0660
	19	.0000	.0000	.0000	.0000	.0000	.0001	.0020	.0215	.1211	.2837	.1981
	20	.0000	.0000	.0000	.0000	.0000	.0000	.0003	.0050	.0484	.2553	.3764
	21	.0000	.0000	.0000	.0000	.0000	.0000	.0000	.0006	.0092	.1094	.3406
22	0	.3235	.0985	.0074	.0004	.0000	.0000	.0000	.0000	.0000	.0000	.0000
	1	.3746	.2407	.0406	.0037	.0002	.0000	.0000	.0000	.0000	.0000	.0000
	2	.2070	.2808	.1065	.0166	.0014	.0001	.0000	.0000	.0000	.0000	.0000
	3	.0726	.2080	.1775	.0474	.0060	.0004	.0000	.0000	.0000	.0000	.0000
	4	.0182	.1098	.2108	.0965	.0190	.0017	.0001	.0000	.0000	.0000	.0000
	5	.0034	.0439	.1898	.1489	.0456	.0063	.0004	.0000	.0000	.0000	.0000
	6	.0005	.0138	.1344	.1808	.0862	.0178	.0015	.0000	.0000	.0000	.0000
	7	.0001	.0035	.0768	.1771	.1314	.0407	.0051	.0002	.0000	.0000	.0000
	8	.0000	.0007	.0360	.1423	.1642	.0762	.0144	.0009	.0000	.0000	.0000
	9	.0000	.0001	.0140	.0949	.1703	.1186	.0336	.0032	.0001	.0000	.0000
	10	.0000	.0000	.0046	.0529	.1476	.1542	.0656	.0097	.0003	.0000	.0000
	11	.0000	.0000	.0012	.0247	.1073	.1682	.1073	.0247	.0012	.0000	.0000
	12	.0000	.0000	.0003	.0097	.0656	.1542	.1476	.0529	.0046	.0000	.0000
	13	.0000	.0000	.0001	.0032	.0336	.1186	.1703	.0949	.0140	.0001	.0000
	14	.0000	.0000	.0000	.0009	.0144	.0762	.1642	.1423	.0360	.0007	.0000
	15	.0000	.0000	.0000	.0002	.0051	.0407	.1314	.1771	.0768	.0035	.0001
	16	.0000	.0000	.0000	.0000	.0015	.0178	.0862	.1808	.1344	.0138	.0005
	17	.0000	.0000	.0000	.0000	.0004	.0063	.0456	.1489	.1898	.0439	.0034
	18	.0000	.0000	.0000	.0000	.0001	.0017	.0190	.0965	.2108	.1098	.0182
	19	.0000	.0000	.0000	.0000	.0000	.0004	.0060	.0474	.1775	.2080	.0726

Table I Table of Binomial Probabilities (continued)

		p										
n	*x*	.05	.10	.20	.30	.40	.50	.60	.70	.80	.90	.95
	20	.0000	.0000	.0000	.0000	.0000	.0001	.0014	.0166	.1065	.2808	.2070
	21	.0000	.0000	.0000	.0000	.0000	.0000	.0002	.0037	.0406	.2407	.3746
	22	.0000	.0000	.0000	.0000	.0000	.0000	.0000	.0004	.0074	.0985	.3235
23	0	.3074	.0886	.0059	.0003	.0000	.0000	.0000	.0000	.0000	.0000	.0000
	1	.3721	.2265	.0339	.0027	.0001	.0000	.0000	.0000	.0000	.0000	.0000
	2	.2154	.2768	.0933	.0127	.0009	.0000	.0000	.0000	.0000	.0000	.0000
	3	.0794	.2153	.1633	.0382	.0041	.0002	.0000	.0000	.0000	.0000	.0000
	4	.0209	.1196	.2042	.0818	.0138	.0011	.0000	.0000	.0000	.0000	.0000
	5	.0042	.0505	.1940	.1332	.0350	.0040	.0002	.0000	.0000	.0000	.0000
	6	.0007	.0168	.1455	.1712	.0700	.0120	.0008	.0000	.0000	.0000	.0000
	7	.0001	.0045	.0883	.1782	.1133	.0292	.0029	.0001	.0000	.0000	.0000
	8	.0000	.0010	.0442	.1527	.1511	.0584	.0088	.0004	.0000	.0000	.0000
	9	.0000	.0002	.0184	.1091	.1679	.0974	.0221	.0016	.0000	.0000	.0000
	10	.0000	.0000	.0064	.0655	.1567	.1364	.0464	.0052	.0001	.0000	.0000
	11	.0000	.0000	.0019	.0332	.1234	.1612	.0823	.0142	.0005	.0000	.0000
	12	.0000	.0000	.0005	.0142	.0823	.1612	.1234	.0332	.0019	.0000	.0000
	13	.0000	.0000	.0001	.0052	.0464	.1364	.1567	.0655	.0064	.0000	.0000
	14	.0000	.0000	.0000	.0016	.0221	.0974	.1679	.1091	.0184	.0002	.0000
	15	.0000	.0000	.0000	.0004	.0088	.0584	.1511	.1527	.0442	.0010	.0000
	16	.0000	.0000	.0000	.0001	.0029	.0292	.1133	.1782	.0883	.0045	.0001
	17	.0000	.0000	.0000	.0000	.0008	.0120	.0700	.1712	.1455	.0168	.0007
	18	.0000	.0000	.0000	.0000	.0002	.0040	.0350	.1332	.1940	.0505	.0042
	19	.0000	.0000	.0000	.0000	.0000	.0011	.0138	.0818	.2042	.1196	.0209
	20	.0000	.0000	.0000	.0000	.0000	.0002	.0041	.0382	.1633	.2153	.0794
	21	.0000	.0000	.0000	.0000	.0000	.0000	.0009	.0127	.0933	.2768	.2154
	22	.0000	.0000	.0000	.0000	.0000	.0000	.0001	.0027	.0339	.2265	.3721
	23	.0000	.0000	.0000	.0000	.0000	.0000	.0000	.0003	.0059	.0886	.3074
24	0	.2920	.0798	.0047	.0002	.0000	.0000	.0000	.0000	.0000	.0000	.0000
	1	.3688	.2127	.0283	.0020	.0001	.0000	.0000	.0000	.0000	.0000	.0000
	2	.2232	.2718	.0815	.0097	.0006	.0000	.0000	.0000	.0000	.0000	.0000
	3	.0862	.2215	.1493	.0305	.0028	.0001	.0000	.0000	.0000	.0000	.0000
	4	.0238	.1292	.1960	.0687	.0099	.0006	.0000	.0000	.0000	.0000	.0000
	5	.0050	.0574	.1960	.1177	.0265	.0025	.0001	.0000	.0000	.0000	.0000
	6	.0008	.0202	.1552	.1598	.0560	.0080	.0004	.0000	.0000	.0000	.0000
	7	.0001	.0058	.0998	.1761	.0960	.0206	.0017	.0000	.0000	.0000	.0000
	8	.0000	.0014	.0530	.1604	.1360	.0438	.0053	.0002	.0000	.0000	.0000
	9	.0000	.0003	.0236	.1222	.1612	.0779	.0141	.0008	.0000	.0000	.0000
	10	.0000	.0000	.0088	.0785	.1612	.1169	.0318	.0026	.0000	.0000	.0000
	11	.0000	.0000	.0028	.0428	.1367	.1488	.0608	.0079	.0002	.0000	.0000

Table I Table of Binomial Probabilities (continued)

		p										
n	*x*	.05	.10	.20	.30	.40	.50	.60	.70	.80	.90	.95
	12	.0000	.0000	.0008	.0199	.0988	.1612	.0988	.0199	.0008	.0000	.0000
	13	.0000	.0000	.0002	.0079	.0608	.1488	.1367	.0428	.0028	.0000	.0000
	14	.0000	.0000	.0000	.0026	.0318	.1169	.1612	.0785	.0088	.0000	.0000
	15	.0000	.0000	.0000	.0008	.0141	.0779	.1612	.1222	.0236	.0003	.0000
	16	.0000	.0000	.0000	.0002	.0053	.0438	.1360	.1604	.0530	.0014	.0000
	17	.0000	.0000	.0000	.0000	.0017	.0206	.0960	.1761	.0998	.0058	.0001
	18	.0000	.0000	.0000	.0000	.0004	.0080	.0560	.1598	.1552	.0202	.0008
	19	.0000	.0000	.0000	.0000	.0001	.0025	.0265	.1177	.1960	.0574	.0050
	20	.0000	.0000	.0000	.0000	.0000	.0006	.0099	.0687	.1960	.1292	.0238
	21	.0000	.0000	.0000	.0000	.0000	.0001	.0028	.0305	.1493	.2215	.0862
	22	.0000	.0000	.0000	.0000	.0000	.0000	.0006	.0097	.0815	.2718	.2232
	23	.0000	.0000	.0000	.0000	.0000	.0000	.0001	.0020	.0283	.2127	.3688
	24	.0000	.0000	.0000	.0000	.0000	.0000	.0000	.0002	.0047	.0798	.2920
25	0	.2774	.0718	.0038	.0001	.0000	.0000	.0000	.0000	.0000	.0000	.0000
	1	.3650	.1994	.0236	.0014	.0000	.0000	.0000	.0000	.0000	.0000	.0000
	2	.2305	.2659	.0708	.0074	.0004	.0000	.0000	.0000	.0000	.0000	.0000
	3	.0930	.2265	.1358	.0243	.0019	.0001	.0000	.0000	.0000	.0000	.0000
	4	.0269	.1384	.1867	.0572	.0071	.0004	.0000	.0000	.0000	.0000	.0000
	5	.0060	.0646	.1960	.1030	.0199	.0016	.0000	.0000	.0000	.0000	.0000
	6	.0010	.0239	.1633	.1472	.0442	.0053	.0002	.0000	.0000	.0000	.0000
	7	.0001	.0072	.1108	.1712	.0800	.0143	.0009	.0000	.0000	.0000	.0000
	8	.0000	.0018	.0623	.1651	.1200	.0322	.0031	.0001	.0000	.0000	.0000
	9	.0000	.0004	.0294	.1336	.1511	.0609	.0088	.0004	.0000	.0000	.0000
	10	.0000	.0001	.0118	.0916	.1612	.0974	.0212	.0013	.0000	.0000	.0000
	11	.0000	.0000	.0040	.0536	.1465	.1328	.0434	.0042	.0001	.0000	.0000
	12	.0000	.0000	.0012	.0268	.1140	.1550	.0760	.0115	.0003	.0000	.0000
	13	.0000	.0000	.0003	.0115	.0760	.1550	.1140	.0268	.0012	.0000	.0000
	14	.0000	.0000	.0001	.0042	.0434	.1328	.1465	.0536	.0040	.0000	.0000
	15	.0000	.0000	.0000	.0013	.0212	.0974	.1612	.0916	.0118	.0001	.0000
	16	.0000	.0000	.0000	.0004	.0088	.0609	.1511	.1336	.0294	.0004	.0000
	17	.0000	.0000	.0000	.0001	.0031	.0322	.1200	.1651	.0623	.0018	.0000
	18	.0000	.0000	.0000	.0000	.0009	.0143	.0800	.1712	.1108	.0072	.0001
	19	.0000	.0000	.0000	.0000	.0002	.0053	.0442	.1472	.1633	.0239	.0010
	20	.0000	.0000	.0000	.0000	.0000	.0016	.0199	.1030	.1960	.0646	.0060
	21	.0000	.0000	.0000	.0000	.0000	.0004	.0071	.0572	.1867	.1384	.0269
	22	.0000	.0000	.0000	.0000	.0000	.0001	.0019	.0243	.1358	.2265	.0930
	23	.0000	.0000	.0000	.0000	.0000	.0000	.0004	.0074	.0708	.2659	.2305
	24	.0000	.0000	.0000	.0000	.0000	.0000	.0000	.0014	.0236	.1994	.3650
	25	.0000	.0000	.0000	.0000	.0000	.0000	.0000	.0001	.0038	.0718	.2774

Table II Values of $e^{-\lambda}$

λ	$e^{-\lambda}$	λ	$e^{-\lambda}$
0.0	1.00000000	3.9	.02024191
0.1	.90483742	4.0	.01831564
0.2	.81873075	4.1	.01657268
0.3	.74081822	4.2	.01499558
0.4	.67032005	4.3	.01356856
0.5	.60653066	4.4	.01227734
0.6	.54881164	4.5	.01110900
0.7	.49658530	4.6	.01005184
0.8	.44932896	4.7	.00909528
0.9	.40656966	4.8	.00822975
1.0	.36787944	4.9	.00744658
1.1	.33287108	5.0	.00673795
1.2	.30119421	5.1	.00609675
1.3	.27253179	5.2	.00551656
1.4	.24659696	5.3	.00499159
1.5	.22313016	5.4	.00451658
1.6	.20189652	5.5	.00408677
1.7	.18268352	5.6	.00369786
1.8	.16529889	5.7	.00334597
1.9	.14956862	5.8	.00302755
2.0	.13533528	5.9	.00273944
2.1	.12245643	6.0	.00247875
2.2	.11080316	6.1	.00224287
2.3	.10025884	6.2	.00202943
2.4	.09071795	6.3	.00183630
2.5	.08208500	6.4	.00166156
2.6	.07427358	6.5	.00150344
2.7	.06720551	6.6	.00136037
2.8	.06081006	6.7	.00123091
2.9	.05502322	6.8	.00111378
3.0	.04978707	6.9	.00100779
3.1	.04504920	7.0	.00091188
3.2	.04076220	7.1	.00082510
3.3	.03688317	7.2	.00074659
3.4	.03337327	7.3	.00067554
3.5	.03019738	7.4	.00061125
3.6	.02732372	7.5	.00055308
3.7	.02472353	7.6	.00050045
3.8	.02237077	7.7	.00045283

Table II Values of $e^{-\lambda}$ (continued)

λ	$e^{-\lambda}$	λ	$e^{-\lambda}$
7.8	.00040973	9.5	.00007485
7.9	.00037074	9.6	.00006773
8.0	.00033546	9.7	.00006128
8.1	.00030354	9.8	.00005545
8.2	.00027465	9.9	.00005017
8.3	.00024852	10.0	.00004540
8.4	.00022487	11.0	.00001670
8.5	.00020347	12.0	.00000614
8.6	.00018411	13.0	.00000226
8.7	.00016659	14.0	.00000083
8.8	.00015073	15.0	.00000031
8.9	.00013639	16.0	.00000011
9.0	.00012341	17.0	.00000004
9.1	.00011167	18.0	.000000015
9.2	.00010104	19.0	.000000006
9.3	.00009142	20.0	.000000002
9.4	.00008272		

Table III Table of Poisson Probabilities

x	λ 0.1	0.2	0.3	0.4	0.5	0.6	0.7	0.8	0.9	1.0
0	.9048	.8187	.7408	.6703	.6065	.5488	.4966	.4493	.4066	.3679
1	.0905	.1637	.2222	.2681	.3033	.3293	.3476	.3595	.3659	.3679
2	.0045	.0164	.0333	.0536	.0758	.0988	.1217	.1438	.1647	.1839
3	.0002	.0011	.0033	.0072	.0126	.0198	.0284	.0383	.0494	.0613
4	.0000	.0001	.0003	.0007	.0016	.0030	.0050	.0077	.0111	.0153
5	.0000	.0000	.0000	.0001	.0002	.0004	.0007	.0012	.0020	.0031
6	.0000	.0000	.0000	.0000	.0000	.0000	.0001	.0002	.0003	.0005
7	.0000	.0000	.0000	.0000	.0000	.0000	.0000	.0000	.0000	.0001

x	λ 1.1	1.2	1.3	1.4	1.5	1.6	1.7	1.8	1.9	2.0
0	.3329	.3012	.2725	.2466	.2231	.2019	.1827	.1653	.1496	.1353
1	.3662	.3614	.3543	.3452	.3347	.3230	.3106	.2975	.2842	.2707
2	.2014	.2169	.2303	.2417	.2510	.2584	.2640	.2678	.2700	.2707
3	.0738	.0867	.0998	.1128	.1255	.1378	.1496	.1607	.1710	.1804
4	.0203	.0260	.0324	.0395	.0471	.0551	.0636	.0723	.0812	.0902
5	.0045	.0062	.0084	.0111	.0141	.0176	.0216	.0260	.0309	.0361
6	.0008	.0012	.0018	.0026	.0035	.0047	.0061	.0078	.0098	.0120
7	.0001	.0002	.0003	.0005	.0008	.0011	.0015	.0020	.0027	.0034
8	.0000	.0000	.0001	.0001	.0001	.0002	.0003	.0005	.0006	.0009
9	.0000	.0000	.0000	.0000	.0000	.0000	.0001	.0001	.0001	.0002

x	λ 2.1	2.2	2.3	2.4	2.5	2.6	2.7	2.8	2.9	3.0
0	.1225	.1108	.1003	.0907	.0821	.0743	.0672	.0608	.0550	.0498
1	.2572	.2438	.2306	.2177	.2052	.1931	.1815	.1703	.1596	.1494
2	.2700	.2681	.2652	.2613	.2565	.2510	.2450	.2384	.2314	.2240
3	.1890	.1966	.2033	.2090	.2138	.2176	.2205	.2225	.2237	.2240
4	.0992	.1082	.1169	.1254	.1336	.1414	.1488	.1557	.1622	.1680
5	.0417	.0476	.0538	.0602	.0668	.0735	.0804	.0872	.0940	.1008
6	.0146	.0174	.0206	.0241	.0278	.0319	.0362	.0407	.0455	.0504
7	.0044	.0055	.0068	.0083	.0099	.0118	.0139	.0163	.0188	.0216
8	.0011	.0015	.0019	.0025	.0031	.0038	.0047	.0057	.0068	.0081
9	.0003	.0004	.0005	.0007	.0009	.0011	.0014	.0018	.0022	.0027
10	.0001	.0001	.0001	.0002	.0002	.0003	.0004	.0005	.0006	.0008
11	.0000	.0000	.0000	.0000	.0000	.0001	.0001	.0001	.0002	.0002
12	.0000	.0000	.0000	.0000	.0000	.0000	.0000	.0000	.0000	.0001

Table III Table of Poisson Probabilities (continued)

	λ									
x	3.1	3.2	3.3	3.4	3.5	3.6	3.7	3.8	3.9	4.0
0	.0450	.0408	.0369	.0334	.0302	.0273	.0247	.0224	.0202	.0183
1	.1397	.1304	.1217	.1135	.1057	.0984	.0915	.0850	.0789	.0733
2	.2165	.2087	.2008	.1929	.1850	.1771	.1692	.1615	.1539	.1465
3	.2237	.2226	.2209	.2186	.2158	.2125	.2087	.2046	.2001	.1954
4	.1733	.1781	.1823	.1858	.1888	.1912	.1931	.1944	.1951	.1954
5	.1075	.1140	.1203	.1264	.1322	.1377	.1429	.1477	.1522	.1563
6	.0555	.0608	.0662	.0716	.0771	.0826	.0881	.0936	.0989	.1042
7	.0246	.0278	.0312	.0348	.0385	.0425	.0466	.0508	.0551	.0595
8	.0095	.0111	.0129	.0148	.0169	.0191	.0215	.0241	.0269	.0298
9	.0033	.0040	.0047	.0056	.0066	.0076	.0089	.0102	.0116	.0132
10	.0010	.0013	.0016	.0019	.0023	.0028	.0033	.0039	.0045	.0053
11	.0003	.0004	.0005	.0006	.0007	.0009	.0011	.0013	.0016	.0019
12	.0001	.0001	.0001	.0002	.0002	.0003	.0003	.0004	.0005	.0006
13	.0000	.0000	.0000	.0000	.0001	.0001	.0001	.0001	.0002	.0002
14	.0000	.0000	.0000	.0000	.0000	.0000	.0000	.0000	.0000	.0001

	λ									
x	4.1	4.2	4.3	4.4	4.5	4.6	4.7	4.8	4.9	5.0
0	.0166	.0150	.0136	.0123	.0111	.0101	.0091	.0082	.0074	.0067
1	.0679	.0630	.0583	.0540	.0500	.0462	.0427	.0395	.0365	.0337
2	.1393	.1323	.1254	.1188	.1125	.1063	.1005	.0948	.0894	.0842
3	.1904	.1852	.1798	.1743	.1687	.1631	.1574	.1517	.1460	.1404
4	.1951	.1944	.1933	.1917	.1898	.1875	.1849	.1820	.1789	.1755
5	.1600	.1633	.1662	.1687	.1708	.1725	.1738	.1747	.1753	.1755
6	.1093	.1143	.1191	.1237	.1281	.1323	.1362	.1398	.1432	.1462
7	.0640	.0686	.0732	.0778	.0824	.0869	.0914	.0959	.1002	.1044
8	.0328	.0360	.0393	.0428	.0463	.0500	.0537	.0575	.0614	.0653
9	.0150	.0168	.0188	.0209	.0232	.0255	.0281	.0307	.0334	.0363
10	.0061	.0071	.0081	.0092	.0104	.0118	.0132	.0147	.0164	.0181
11	.0023	.0027	.0032	.0037	.0043	.0049	.0056	.0064	.0073	.0082
12	.0008	.0009	.0011	.0014	.0016	.0019	.0022	.0026	.0030	.0034
13	.0002	.0003	.0004	.0005	.0006	.0007	.0008	.0009	.0011	.0013
14	.0001	.0001	.0001	.0001	.0002	.0002	.0003	.0003	.0004	.0005
15	.0000	.0000	.0000	.0000	.0001	.0001	.0001	.0001	.0001	.0002

	λ									
x	5.1	5.2	5.3	5.4	5.5	5.6	5.7	5.8	5.9	6.0
0	.0061	.0055	.0050	.0045	.0041	.0037	.0033	.0030	.0027	.0025
1	.0311	.0287	.0265	.0244	.0225	.0207	.0191	.0176	.0162	.0149

Table III Table of Poisson Probabilities (continued)

	λ									
x	5.1	5.2	5.3	5.4	5.5	5.6	5.7	5.8	5.9	6.0
2	.0793	.0746	.0701	.0659	.0618	.0580	.0544	.0509	.0477	.0446
3	.1348	.1293	.1239	.1185	.1133	.1082	.1033	.0985	.0938	.0892
4	.1719	.1681	.1641	.1600	.1558	.1515	.1472	.1428	.1383	.1339
5	.1753	.1748	.1740	.1728	.1714	.1697	.1678	.1656	.1632	.1606
6	.1490	.1515	.1537	.1555	.1571	.1584	.1594	.1601	.1605	.1606
7	.1086	.1125	.1163	.1200	.1234	.1267	.1298	.1326	.1353	.1377
8	.0692	.0731	.0771	.0810	.0849	.0887	.0925	.0962	.0998	.1033
9	.0392	.0423	.0454	.0486	.0519	.0552	.0586	.0620	.0654	.0688
10	.0200	.0220	.0241	.0262	.0285	.0309	.0334	.0359	.0386	.0413
11	.0093	.0104	.0116	.0129	.0143	.0157	.0173	.0190	.0207	.0225
12	.0039	.0045	.0051	.0058	.0065	.0073	.0082	.0092	.0102	.0113
13	.0015	.0018	.0021	.0024	.0028	.0032	.0036	.0041	.0046	.0052
14	.0006	.0007	.0008	.0009	.0011	.0013	.0015	.0017	.0019	.0022
15	.0002	.0002	.0003	.0003	.0004	.0005	.0006	.0007	.0008	.0009
16	.0001	.0001	.0001	.0001	.0001	.0002	.0002	.0002	.0003	.0003
17	.0000	.0000	.0000	.0000	.0000	.0001	.0001	.0001	.0001	.0001

	λ									
x	6.1	6.2	6.3	6.4	6.5	6.6	6.7	6.8	6.9	7.0
0	.0022	.0020	.0018	.0017	.0015	.0014	.0012	.0011	.0010	.0009
1	.0137	.0126	.0116	.0106	.0098	.0090	.0082	.0076	.0070	.0064
2	.0417	.0390	.0364	.0340	.0318	.0296	.0276	.0258	.0240	.0223
3	.0848	.0806	.0765	.0726	.0688	.0652	.0617	.0584	.0552	.0521
4	.1294	.1249	.1205	.1162	.1118	.1076	.1034	.0992	.0952	.0912
5	.1579	.1549	.1519	.1487	.1454	.1420	.1385	.1349	.1314	.1277
6	.1605	.1601	.1595	.1586	.1575	.1562	.1546	.1529	.1511	.1490
7	.1399	.1418	.1435	.1450	.1462	.1472	.1480	.1486	.1489	.1490
8	.1066	.1099	.1130	.1160	.1188	.1215	.1240	.1263	.1284	.1304
9	.0723	.0757	.0791	.0825	.0858	.0891	.0923	.0954	.0985	.1014
10	.0441	.0469	.0498	.0528	.0558	.0588	.0618	.0649	.0679	.0710
11	.0244	.0265	.0285	.0307	.0330	.0353	.0377	.0401	.0426	.0452
12	.0124	.0137	.0150	.0164	.0179	.0194	.0210	.0227	.0245	.0263
13	.0058	.0065	.0073	.0081	.0089	.0099	.0108	.0119	.0130	.0142
14	.0025	.0029	.0033	.0037	.0041	.0046	.0052	.0058	.0064	.0071
15	.0010	.0012	.0014	.0016	.0018	.0020	.0023	.0026	.0029	.0033
16	.0004	.0005	.0005	.0006	.0007	.0008	.0010	.0011	.0013	.0014
17	.0001	.0002	.0002	.0002	.0003	.0003	.0004	.0004	.0005	.0006
18	.0000	.0001	.0001	.0001	.0001	.0001	.0001	.0002	.0002	.0002
19	.0000	.0000	.0000	.0000	.0000	.0000	.0001	.0001	.0001	.0001

Table III Table of Poisson Probabilities (continued)

x	λ 7.1	7.2	7.3	7.4	7.5	7.6	7.7	7.8	7.9	8.0
0	.0008	.0007	.0007	.0006	.0006	.0005	.0005	.0004	.0004	.0003
1	.0059	.0054	.0049	.0045	.0041	.0038	.0035	.0032	.0029	.0027
2	.0208	.0194	.0180	.0167	.0156	.0145	.0134	.0125	.0116	.0107
3	.0492	.0464	.0438	.0413	.0389	.0366	.0345	.0324	.0305	.0286
4	.0874	.0836	.0799	.0764	.0729	.0696	.0663	.0632	.0602	.0573
5	.1241	.1204	.1167	.1130	.1094	.1057	.1021	.0986	.0951	.0916
6	.1468	.1445	.1420	.1394	.1367	.1339	.1311	.1282	.1252	.1221
7	.1489	.1486	.1481	.1474	.1465	.1454	.1442	.1428	.1413	.1396
8	.1321	.1337	.1351	.1363	.1373	.1381	.1388	.1392	.1395	.1396
9	.1042	.1070	.1096	.1121	.1144	.1167	.1187	.1207	.1224	.1241
10	.0740	.0770	.0800	.0829	.0858	.0887	.0914	.0941	.0967	.0993
11	.0478	.0504	.0531	.0558	.0585	.0613	.0640	.0667	.0695	.0722
12	.0283	.0303	.0323	.0344	.0366	.0388	.0411	.0434	.0457	.0481
13	.0154	.0168	.0181	.0196	.0211	.0227	.0243	.0260	.0278	.0296
14	.0078	.0086	.0095	.0104	.0113	.0123	.0134	.0145	.0157	.0169
15	.0037	.0041	.0046	.0051	.0057	.0062	.0069	.0075	.0083	.0090
16	.0016	.0019	.0021	.0024	.0026	.0030	.0033	.0037	.0041	.0045
17	.0007	.0008	.0009	.0010	.0012	.0013	.0015	.0017	.0019	.0021
18	.0003	.0003	.0004	.0004	.0005	.0006	.0006	.0007	.0008	.0009
19	.0001	.0001	.0001	.0002	.0002	.0002	.0003	.0003	.0003	.0004
20	.0000	.0000	.0001	.0001	.0001	.0001	.0001	.0001	.0001	.0002
21	.0000	.0000	.0000	.0000	.0000	.0000	.0000	.0000	.0001	.0001

x	λ 8.1	8.2	8.3	8.4	8.5	8.6	8.7	8.8	8.9	9.0
0	.0003	.0003	.0002	.0002	.0002	.0002	.0002	.0002	.0001	.0001
1	.0025	.0023	.0021	.0019	.0017	.0016	.0014	.0013	.0012	.0011
2	.0100	.0092	.0086	.0079	.0074	.0068	.0063	.0058	.0054	.0050
3	.0269	.0252	.0237	.0222	.0208	.0195	.0183	.0171	.0160	.0150
4	.0544	.0517	.0491	.0466	.0443	.0420	.0398	.0377	.0357	.0337
5	.0882	.0849	.0816	.0784	.0752	.0722	.0692	.0663	.0635	.0607
6	.1191	.1160	.1128	.1097	.1066	.1034	.1003	.0972	.0941	.0911
7	.1378	.1358	.1338	.1317	.1294	.1271	.1247	.1222	.1197	.1171
8	.1395	.1392	.1388	.1382	.1375	.1366	.1356	.1344	.1332	.1318
9	.1255	.1269	.1280	.1290	.1299	.1306	.1311	.1315	.1317	.1318
10	.1017	.1040	.1063	.1084	.1104	.1123	.1140	.1157	.1172	.1186
11	.0749	.0775	.0802	.0828	.0853	.0878	.0902	.0925	.0948	.0970
12	.0505	.0530	.0555	.0579	.0604	.0629	.0654	.0679	.0703	.0728

Table III Table of Poisson Probabilities (continued)

x	8.1	8.2	8.3	8.4	λ 8.5	8.6	8.7	8.8	8.9	9.0
13	.0315	.0334	.0354	.0374	.0395	.0416	.0438	.0459	.0481	.0504
14	.0182	.0196	.0210	.0225	.0240	.0256	.0272	.0289	.0306	.0324
15	.0098	.0107	.0116	.0126	.0136	.0147	.0158	.0169	.0182	.0194
16	.0050	.0055	.0060	.0066	.0072	.0079	.0086	.0093	.0101	.0109
17	.0024	.0026	.0029	.0033	.0036	.0040	.0044	.0048	.0053	.0058
18	.0011	.0012	.0014	.0015	.0017	.0019	.0021	.0024	.0026	.0029
19	.0005	.0005	.0006	.0007	.0008	.0009	.0010	.0011	.0012	.0014
20	.0002	.0002	.0002	.0003	.0003	.0004	.0004	.0005	.0005	.0006
21	.0001	.0001	.0001	.0001	.0001	.0002	.0002	.0002	.0002	.0003
22	.0000	.0000	.0000	.0000	.0001	.0001	.0001	.0001	.0001	.0001

x	9.1	9.2	9.3	9.4	λ 9.5	9.6	9.7	9.8	9.9	10
0	.0001	.0001	.0001	.0001	.0001	.0001	.0001	.0001	.0001	.0000
1	.0010	.0009	.0009	.0008	.0007	.0007	.0006	.0005	.0005	.0005
2	.0046	.0043	.0040	.0037	.0034	.0031	.0029	.0027	.0025	.0023
3	.0140	.0131	.0123	.0115	.0107	.0100	.0093	.0087	.0081	.0076
4	.0319	.0302	.0285	.0269	.0254	.0240	.0226	.0213	.0201	.0189
5	.0581	.0555	.0530	.0506	.0483	.0460	.0439	.0418	.0398	.0378
6	.0881	.0851	.0822	.0793	.0764	.0736	.0709	.0682	.0656	.0631
7	.1145	.1118	.1091	.1064	.1037	.1010	.0982	.0955	.0928	.0901
8	.1302	.1286	.1269	.1251	.1232	.1212	.1191	.1170	.1148	.1126
9	.1317	.1315	.1311	.1306	.1300	.1293	.1284	.1274	.1263	.1251
10	.1198	.1209	.1219	.1228	.1235	.1241	.1245	.1249	.1250	.1251
11	.0991	.1012	.1031	.1049	.1067	.1083	.1098	.1112	.1125	.1137
12	.0752	.0776	.0799	.0822	.0844	.0866	.0888	.0908	.0928	.0948
13	.0526	.0549	.0572	.0594	.0617	.0640	.0662	.0685	.0707	.0729
14	.0342	.0361	.0380	.0399	.0419	.0439	.0459	.0479	.0500	.0521
15	.0208	.0221	.0235	.0250	.0265	.0281	.0297	.0313	.0330	.0347
16	.0118	.0127	.0137	.0147	.0157	.0168	.0180	.0192	.0204	.0217
17	.0063	.0069	.0075	.0081	.0088	.0095	.0103	.0111	.0119	.0128
18	.0032	.0035	.0039	.0042	.0046	.0051	.0055	.0060	.0065	.0071
19	.0015	.0017	.0019	.0021	.0023	.0026	.0028	.0031	.0034	.0037
20	.0007	.0008	.0009	.0010	.0011	.0012	.0014	.0015	.0017	.0019
21	.0003	.0003	.0004	.0004	.0005	.0006	.0006	.0007	.0008	.0009
22	.0001	.0001	.0002	.0002	.0002	.0002	.0003	.0003	.0004	.0004
23	.0000	.0001	.0001	.0001	.0001	.0001	.0001	.0001	.0002	.0002
24	.0000	.0000	.0000	.0000	.0000	.0000	.0000	.0001	.0001	.0001

Table III Table of Poisson Probabilities (continued)

					λ					
x	11	12	13	14	15	16	17	18	19	20
0	.0000	.0000	.0000	.0000	.0000	.0000	.0000	.0000	.0000	.0000
1	.0002	.0001	.0000	.0000	.0000	.0000	.0000	.0000	.0000	.0000
2	.0010	.0004	.0002	.0001	.0000	.0000	.0000	.0000	.0000	.0000
3	.0037	.0018	.0008	.0004	.0002	.0001	.0000	.0000	.0000	.0000
4	.0102	.0053	.0027	.0013	.0006	.0003	.0001	.0001	.0000	.0000
5	.0224	.0127	.0070	.0037	.0019	.0010	.0005	.0002	.0001	.0001
6	.0411	.0255	.0152	.0087	.0048	.0026	.0014	.0007	.0004	.0002
7	.0646	.0437	.0281	.0174	.0104	.0060	.0034	.0019	.0010	.0005
8	.0888	.0655	.0457	.0304	.0194	.0120	.0072	.0042	.0024	.0013
9	.1085	.0874	.0661	.0473	.0324	.0213	.0135	.0083	.0050	.0029
10	.1194	.1048	.0859	.0663	.0486	.0341	.0230	.0150	.0095	.0058
11	.1194	.1144	.1015	.0844	.0663	.0496	.0355	.0245	.0164	.0106
12	.1094	.1144	.1099	.0984	.0829	.0661	.0504	.0368	.0259	.0176
13	.0926	.1056	.1099	.1060	.0956	.0814	.0658	.0509	.0378	.0271
14	.0728	.0905	.1021	.1060	.1024	.0930	.0800	.0655	.0514	.0387
15	.0534	.0724	.0885	.0989	.1024	.0992	.0906	.0786	.0650	.0516
16	.0367	.0543	.0719	.0866	.0960	.0992	.0963	.0884	.0772	.0646
17	.0237	.0383	.0550	.0713	.0847	.0934	.0963	.0936	.0863	.0760
18	.0145	.0255	.0397	.0554	.0706	.0830	.0909	.0936	.0911	.0844
19	.0084	.0161	.0272	.0409	.0557	.0699	.0814	.0887	.0911	.0888
20	.0046	.0097	.0177	.0286	.0418	.0559	.0692	.0798	.0866	.0888
21	.0024	.0055	.0109	.0191	.0299	.0426	.0560	.0684	.0783	.0846
22	.0012	.0030	.0065	.0121	.0204	.0310	.0433	.0560	.0676	.0769
23	.0006	.0016	.0037	.0074	.0133	.0216	.0320	.0438	.0559	.0669
24	.0003	.0008	.0020	.0043	.0083	.0144	.0226	.0328	.0442	.0557
25	.0001	.0004	.0010	.0024	.0050	.0092	.0154	.0237	.0336	.0446
26	.0000	.0002	.0005	.0013	.0029	.0057	.0101	.0164	.0246	.0343
27	.0000	.0001	.0002	.0007	.0016	.0034	.0063	.0109	.0173	.0254
28	.0000	.0000	.0001	.0003	.0009	.0019	.0038	.0070	.0117	.0181
29	.0000	.0000	.0001	.0002	.0004	.0011	.0023	.0044	.0077	.0125
30	.0000	.0000	.0000	.0001	.0002	.0006	.0013	.0026	.0049	.0083
31	.0000	.0000	.0000	.0000	.0001	.0003	.0007	.0015	.0030	.0054
32	.0000	.0000	.0000	.0000	.0001	.0001	.0004	.0009	.0018	.0034
33	.0000	.0000	.0000	.0000	.0000	.0001	.0002	.0005	.0010	.0020
34	.0000	.0000	.0000	.0000	.0000	.0000	.0001	.0002	.0006	.0012
35	.0000	.0000	.0000	.0000	.0000	.0000	.0000	.0001	.0003	.0007
36	.0000	.0000	.0000	.0000	.0000	.0000	.0000	.0001	.0002	.0004
37	.0000	.0000	.0000	.0000	.0000	.0000	.0000	.0000	.0001	.0002
38	.0000	.0000	.0000	.0000	.0000	.0000	.0000	.0000	.0000	.0001
39	.0000	.0000	.0000	.0000	.0000	.0000	.0000	.0000	.0000	.0001

Table IV Standard Normal Distribution Table

The entries in this table give the cumulative area under the standard normal curve to the left of *z* with the values of *z* equal to 0 or negative.

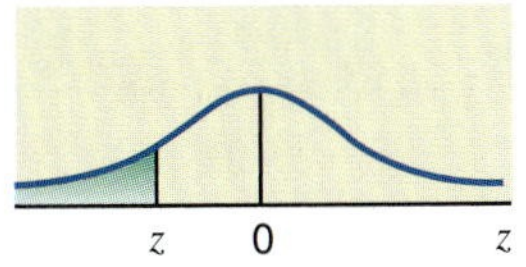

z	.00	.01	.02	.03	.04	.05	.06	.07	.08	.09
−3.4	.0003	.0003	.0003	.0003	.0003	.0003	.0003	.0003	.0003	.0002
−3.3	.0005	.0005	.0005	.0004	.0004	.0004	.0004	.0004	.0004	.0003
−3.2	.0007	.0007	.0006	.0006	.0006	.0006	.0006	.0005	.0005	.0005
−3.1	.0010	.0009	.0009	.0009	.0008	.0008	.0008	.0008	.0007	.0007
−3.0	.0013	.0013	.0013	.0012	.0012	.0011	.0011	.0011	.0010	.0010
−2.9	.0019	.0018	.0018	.0017	.0016	.0016	.0015	.0015	.0014	.0014
−2.8	.0026	.0025	.0024	.0023	.0023	.0022	.0021	.0021	.0020	.0019
−2.7	.0035	.0034	.0033	.0032	.0031	.0030	.0029	.0028	.0027	.0026
−2.6	.0047	.0045	.0044	.0043	.0041	.0040	.0039	.0038	.0037	.0036
−2.5	.0062	.0060	.0059	.0057	.0055	.0054	.0052	.0051	.0049	.0048
−2.4	.0082	.0080	.0078	.0075	.0073	.0071	.0069	.0068	.0066	.0064
−2.3	.0107	.0104	.0102	.0099	.0096	.0094	.0091	.0089	.0087	.0084
−2.2	.0139	.0136	.0132	.0129	.0125	.0122	.0119	.0116	.0113	.0110
−2.1	.0179	.0174	.0170	.0166	.0162	.0158	.0154	.0150	.0146	.0143
−2.0	.0228	.0222	.0217	.0212	.0207	.0202	.0197	.0192	.0188	.0183
−1.9	.0287	.0281	.0274	.0268	.0262	.0256	.0250	.0244	.0239	.0233
−1.8	.0359	.0351	.0344	.0336	.0329	.0322	.0314	.0307	.0301	.0294
−1.7	.0446	.0436	.0427	.0418	.0409	.0401	.0392	.0384	.0375	.0367
−1.6	.0548	.0537	.0526	.0516	.0505	.0495	.0485	.0475	.0465	.0455
−1.5	.0668	.0655	.0643	.0630	.0618	.0606	.0594	.0582	.0571	.0559
−1.4	.0808	.0793	.0778	.0764	.0749	.0735	.0721	.0708	.0694	.0681
−1.3	.0968	.0951	.0934	.0918	.0901	.0885	.0869	.0853	.0838	.0823
−1.2	.1151	.1131	.1112	.1093	.1075	.1056	.1038	.1020	.1003	.0985
−1.1	.1357	.1335	.1314	.1292	.1271	.1251	.1230	.1210	.1190	.1170
−1.0	.1587	.1562	.1539	.1515	.1492	.1469	.1446	.1423	.1401	.1379
−0.9	.1841	.1814	.1788	.1762	.1736	.1711	.1685	.1660	.1635	.1611
−0.8	.2119	.2090	.2061	.2033	.2005	.1977	.1949	.1922	.1894	.1867
−0.7	.2420	.2389	.2358	.2327	.2296	.2266	.2236	.2206	.2177	.2148
−0.6	.2743	.2709	.2676	.2643	.2611	.2578	.2546	.2514	.2483	.2451
−0.5	.3085	.3050	.3015	.2981	.2946	.2912	.2877	.2843	.2810	.2776
−0.4	.3446	.3409	.3372	.3336	.3300	.3264	.3228	.3192	.3156	.3121
−0.3	.3821	.3783	.3745	.3707	.3669	.3632	.3594	.3557	.3520	.3483
−0.2	.4207	.4168	.4129	.4090	.4052	.4013	.3974	.3936	.3897	.3859
−0.1	.4602	.4562	.4522	.4483	.4443	.4404	.4364	.4325	.4286	.4247
0.0	.5000	.4960	.4920	.4880	.4840	.4801	.4761	.4721	.4681	.4641

Table IV Standard Normal Distribution Table (continued)

The entries in this table give the cumulative area under the standard normal curve to the left of *z* with the values of *z* equal to 0 or positive.

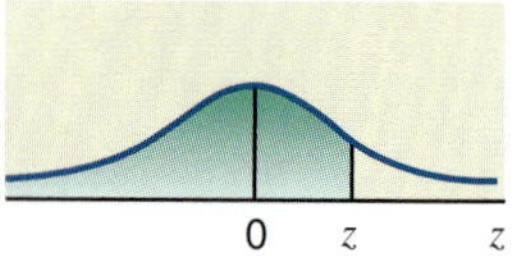

z	.00	.01	.02	.03	.04	.05	.06	.07	.08	.09
0.0	.5000	.5040	.5080	.5120	.5160	.5199	.5239	.5279	.5319	.5359
0.1	.5398	.5438	.5478	.5517	.5557	.5596	.5636	.5675	.5714	.5753
0.2	.5793	.5832	.5871	.5910	.5948	.5987	.6026	.6064	.6103	.6141
0.3	.6179	.6217	.6255	.6293	.6331	.6368	.6406	.6443	.6480	.6517
0.4	.6554	.6591	.6628	.6664	.6700	.6736	.6772	.6808	.6844	.6879
0.5	.6915	.6950	.6985	.7019	.7054	.7088	.7123	.7157	.7190	.7224
0.6	.7257	.7291	.7324	.7357	.7389	.7422	.7454	.7486	.7517	.7549
0.7	.7580	.7611	.7642	.7673	.7704	.7734	.7764	.7794	.7823	.7852
0.8	.7881	.7910	.7939	.7967	.7995	.8023	.8051	.8078	.8106	.8133
0.9	.8159	.8186	.8212	.8238	.8264	.8289	.8315	.8340	.8365	.8389
1.0	.8413	.8438	.8461	.8485	.8508	.8531	.8554	.8577	.8599	.8621
1.1	.8643	.8665	.8686	.8708	.8729	.8749	.8770	.8790	.8810	.8830
1.2	.8849	.8869	.8888	.8907	.8925	.8944	.8962	.8980	.8997	.9015
1.3	.9032	.9049	.9066	.9082	.9099	.9115	.9131	.9147	.9162	.9177
1.4	.9192	.9207	.9222	.9236	.9251	.9265	.9279	.9292	.9306	.9319
1.5	.9332	.9345	.9357	.9370	.9382	.9394	.9406	.9418	.9429	.9441
1.6	.9452	.9463	.9474	.9484	.9495	.9505	.9515	.9525	.9535	.9545
1.7	.9554	.9564	.9573	.9582	.9591	.9599	.9608	.9616	.9625	.9633
1.8	.9641	.9649	.9656	.9664	.9671	.9678	.9686	.9693	.9699	.9706
1.9	.9713	.9719	.9726	.9732	.9738	.9744	.9750	.9756	.9761	.9767
2.0	.9772	.9778	.9783	.9788	.9793	.9798	.9803	.9808	.9812	.9817
2.1	.9821	.9826	.9830	.9834	.9838	.9842	.9846	.9850	.9854	.9857
2.2	.9861	.9864	.9868	.9871	.9875	.9878	.9881	.9884	.9887	.9890
2.3	.9893	.9896	.9898	.9901	.9904	.9906	.9909	.9911	.9913	.9916
2.4	.9918	.9920	.9922	.9925	.9927	.9929	.9931	.9932	.9934	.9936
2.5	.9938	.9940	.9941	.9943	.9945	.9946	.9948	.9949	.9951	.9952
2.6	.9953	.9955	.9956	.9957	.9959	.9960	.9961	.9962	.9963	.9964
2.7	.9965	.9966	.9967	.9968	.9969	.9970	.9971	.9972	.9973	.9974
2.8	.9974	.9975	.9976	.9977	.9977	.9978	.9979	.9979	.9980	.9981
2.9	.9981	.9982	.9982	.9983	.9984	.9984	.9985	.9985	.9986	.9986
3.0	.9987	.9987	.9987	.9988	.9988	.9989	.9989	.9989	.9990	.9990
3.1	.9990	.9991	.9991	.9991	.9992	.9992	.9992	.9992	.9993	.9993
3.2	.9993	.9993	.9994	.9994	.9994	.9994	.9994	.9995	.9995	.9995
3.3	.9995	.9995	.9995	.9996	.9996	.9996	.9996	.9996	.9996	.9997
3.4	.9997	.9997	.9997	.9997	.9997	.9997	.9997	.9997	.9997	.9998

Table V The *t* Distribution Table

The entries in this table give the critical values of *t* for the specified number of degrees of freedom and areas in the right tail.

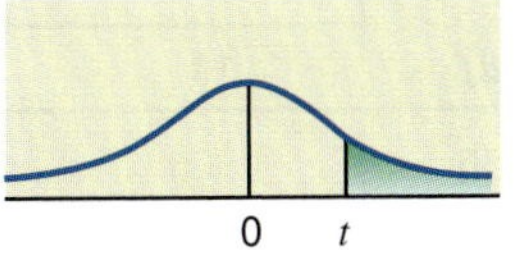

df	Area in the Right Tail under the *t* Distribution Curve					
	.10	**.05**	**.025**	**.01**	**.005**	**.001**
1	3.078	6.314	12.706	31.821	63.657	318.309
2	1.886	2.920	4.303	6.965	9.925	22.327
3	1.638	2.353	3.182	4.541	5.841	10.215
4	1.533	2.132	2.776	3.747	4.604	7.173
5	1.476	2.015	2.571	3.365	4.032	5.893
6	1.440	1.943	2.447	3.143	3.707	5.208
7	1.415	1.895	2.365	2.998	3.499	4.785
8	1.397	1.860	2.306	2.896	3.355	4.501
9	1.383	1.833	2.262	2.821	3.250	4.297
10	1.372	1.812	2.228	2.764	3.169	4.144
11	1.363	1.796	2.201	2.718	3.106	4.025
12	1.356	1.782	2.179	2.681	3.055	3.930
13	1.350	1.771	2.160	2.650	3.012	3.852
14	1.345	1.761	2.145	2.624	2.977	3.787
15	1.341	1.753	2.131	2.602	2.947	3.733
16	1.337	1.746	2.120	2.583	2.921	3.686
17	1.333	1.740	2.110	2.567	2.898	3.646
18	1.330	1.734	2.101	2.552	2.878	3.610
19	1.328	1.729	2.093	2.539	2.861	3.579
20	1.325	1.725	2.086	2.528	2.845	3.552
21	1.323	1.721	2.080	2.518	2.831	3.527
22	1.321	1.717	2.074	2.508	2.819	3.505
23	1.319	1.714	2.069	2.500	2.807	3.485
24	1.318	1.711	2.064	2.492	2.797	3.467
25	1.316	1.708	2.060	2.485	2.787	3.450
26	1.315	1.706	2.056	2.479	2.779	3.435
27	1.314	1.703	2.052	2.473	2.771	3.421
28	1.313	1.701	2.048	2.467	2.763	3.408
29	1.311	1.699	2.045	2.462	2.756	3.396
30	1.310	1.697	2.042	2.457	2.750	3.385
31	1.309	1.696	2.040	2.453	2.744	3.375
32	1.309	1.694	2.037	2.449	2.738	3.365
33	1.308	1.692	2.035	2.445	2.733	3.356
34	1.307	1.691	2.032	2.441	2.728	3.348
35	1.306	1.690	2.030	2.438	2.724	3.340

Table V The *t* Distribution Table (continued)

	Area in the Right Tail under the *t* Distribution Curve					
df	.10	.05	.025	.01	.005	.001
36	1.306	1.688	2.028	2.434	2.719	3.333
37	1.305	1.687	2.026	2.431	2.715	3.326
38	1.304	1.686	2.024	2.429	2.712	3.319
39	1.304	1.685	2.023	2.426	2.708	3.313
40	1.303	1.684	2.021	2.423	2.704	3.307
41	1.303	1.683	2.020	2.421	2.701	3.301
42	1.302	1.682	2.018	2.418	2.698	3.296
43	1.302	1.681	2.017	2.416	2.695	3.291
44	1.301	1.680	2.015	2.414	2.692	3.286
45	1.301	1.679	2.014	2.412	2.690	3.281
46	1.300	1.679	2.013	2.410	2.687	3.277
47	1.300	1.678	2.012	2.408	2.685	3.273
48	1.299	1.677	2.011	2.407	2.682	3.269
49	1.299	1.677	2.010	2.405	2.680	3.265
50	1.299	1.676	2.009	2.403	2.678	3.261
51	1.298	1.675	2.008	2.402	2.676	3.258
52	1.298	1.675	2.007	2.400	2.674	3.255
53	1.298	1.674	2.006	2.399	2.672	3.251
54	1.297	1.674	2.005	2.397	2.670	3.248
55	1.297	1.673	2.004	2.396	2.668	3.245
56	1.297	1.673	2.003	2.395	2.667	3.242
57	1.297	1.672	2.002	2.394	2.665	3.239
58	1.296	1.672	2.002	2.392	2.663	3.237
59	1.296	1.671	2.001	2.391	2.662	3.234
60	1.296	1.671	2.000	2.390	2.660	3.232
61	1.296	1.670	2.000	2.389	2.659	3.229
62	1.295	1.670	1.999	2.388	2.657	3.227
63	1.295	1.669	1.998	2.387	2.656	3.225
64	1.295	1.669	1.998	2.386	2.655	3.223
65	1.295	1.669	1.997	2.385	2.654	3.220
66	1.295	1.668	1.997	2.384	2.652	3.218
67	1.294	1.668	1.996	2.383	2.651	3.216
68	1.294	1.668	1.995	2.382	2.650	3.214
69	1.294	1.667	1.995	2.382	2.649	3.213
70	1.294	1.667	1.994	2.381	2.648	3.211
71	1.294	1.667	1.994	2.380	2.647	3.209
72	1.293	1.666	1.993	2.379	2.646	3.207
73	1.293	1.666	1.993	2.379	2.645	3.206
74	1.293	1.666	1.993	2.378	2.644	3.204
75	1.293	1.665	1.992	2.377	2.643	3.202
∞	1.282	1.645	1.960	2.326	2.576	3.090

Table VI Chi-Square Distribution Table

The entries in this table give the critical values of χ^2 for the specified number of degrees of freedom and areas in the right tail.

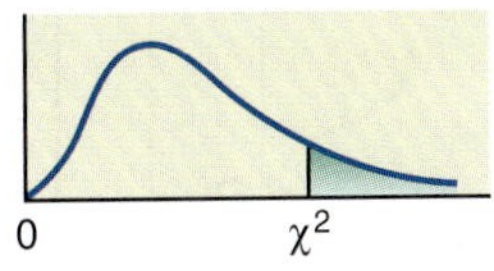

	Area in the Right Tail under the Chi-square Distribution Curve									
df	.995	.990	.975	.950	.900	.100	.050	.025	.010	.005
1	0.000	0.000	0.001	0.004	0.016	2.706	3.841	5.024	6.635	7.879
2	0.010	0.020	0.051	0.103	0.211	4.605	5.991	7.378	9.210	10.597
3	0.072	0.115	0.216	0.352	0.584	6.251	7.815	9.348	11.345	12.838
4	0.207	0.297	0.484	0.711	1.064	7.779	9.488	11.143	13.277	14.860
5	0.412	0.554	0.831	1.145	1.610	9.236	11.070	12.833	15.086	16.750
6	0.676	0.872	1.237	1.635	2.204	10.645	12.592	14.449	16.812	18.548
7	0.989	1.239	1.690	2.167	2.833	12.017	14.067	16.013	18.475	20.278
8	1.344	1.646	2.180	2.733	3.490	13.362	15.507	17.535	20.090	21.955
9	1.735	2.088	2.700	3.325	4.168	14.684	16.919	19.023	21.666	23.589
10	2.156	2.558	3.247	3.940	4.865	15.987	18.307	20.483	23.209	25.188
11	2.603	3.053	3.816	4.575	5.578	17.275	19.675	21.920	24.725	26.757
12	3.074	3.571	4.404	5.226	6.304	18.549	21.026	23.337	26.217	28.300
13	3.565	4.107	5.009	5.892	7.042	19.812	22.362	24.736	27.688	29.819
14	4.075	4.660	5.629	6.571	7.790	21.064	23.685	26.119	29.141	31.319
15	4.601	5.229	6.262	7.261	8.547	22.307	24.996	27.488	30.578	32.801
16	5.142	5.812	6.908	7.962	9.312	23.542	26.296	28.845	32.000	34.267
17	5.697	6.408	7.564	8.672	10.085	24.769	27.587	30.191	33.409	35.718
18	6.265	7.015	8.231	9.390	10.865	25.989	28.869	31.526	34.805	37.156
19	6.844	7.633	8.907	10.117	11.651	27.204	30.144	32.852	36.191	38.582
20	7.434	8.260	9.591	10.851	12.443	28.412	31.410	34.170	37.566	39.997
21	8.034	8.897	10.283	11.591	13.240	29.615	32.671	35.479	38.932	41.401
22	8.643	9.542	10.982	12.338	14.041	30.813	33.924	36.781	40.289	42.796
23	9.260	10.196	11.689	13.091	14.848	32.007	35.172	38.076	41.638	44.181
24	9.886	10.856	12.401	13.848	15.659	33.196	36.415	39.364	42.980	45.559
25	10.520	11.524	13.120	14.611	16.473	34.382	37.652	40.646	44.314	46.928
26	11.160	12.198	13.844	15.379	17.292	35.563	38.885	41.923	45.642	48.290
27	11.808	12.879	14.573	16.151	18.114	36.741	40.113	43.195	46.963	49.645
28	12.461	13.565	15.308	16.928	18.939	37.916	41.337	44.461	48.278	50.993
29	13.121	14.256	16.047	17.708	19.768	39.087	42.557	45.722	49.588	52.336
30	13.787	14.953	16.791	18.493	20.599	40.256	43.773	46.979	50.892	53.672
40	20.707	22.164	24.433	26.509	29.051	51.805	55.758	59.342	63.691	66.766
50	27.991	29.707	32.357	34.764	37.689	63.167	67.505	71.420	76.154	79.490
60	35.534	37.485	40.482	43.188	46.459	74.397	79.082	83.298	88.379	91.952
70	43.275	45.442	48.758	51.739	55.329	85.527	90.531	95.023	100.425	104.215
80	51.172	53.540	57.153	60.391	64.278	96.578	101.879	106.629	112.329	116.321
90	59.196	61.754	65.647	69.126	73.291	107.565	113.145	118.136	124.116	128.299
100	67.328	70.065	74.222	77.929	82.358	118.498	124.342	129.561	135.807	140.169

Table VII The *F* Distribution Table

The entries in this table give the critical values of *F* for .01 area in the right tail under the *F* distribution curve and specified degrees of freedom for the numerator and denominator.

Degrees of Freedom for the Numerator (columns); Degrees of Freedom for the Denominator (rows)

	1	2	3	4	5	6	7	8	9	10	11	12	15	20	25	30	40	50	100
1	4052	5000	5403	5625	5764	5859	5928	5981	6022	6056	6083	6106	6157	6209	6240	6261	6287	6303	6334
2	98.50	99.00	99.17	99.25	99.30	99.33	99.36	99.37	99.39	99.40	99.41	99.42	99.43	99.45	99.46	99.47	99.47	99.48	99.49
3	34.12	30.82	29.46	28.71	28.24	27.91	27.67	27.49	27.35	27.23	27.13	27.05	26.87	26.69	26.58	26.50	26.41	26.35	26.24
4	21.20	18.00	16.69	15.98	15.52	15.21	14.98	14.80	14.66	14.55	14.45	14.37	14.20	14.02	13.91	13.84	13.75	13.69	13.58
5	16.26	13.27	12.06	11.39	10.97	10.67	10.46	10.29	10.16	10.05	9.96	9.89	9.72	9.55	9.45	9.38	9.29	9.24	9.13
6	13.75	10.92	9.78	9.15	8.75	8.47	8.26	8.10	7.98	7.87	7.79	7.72	7.56	7.40	7.30	7.23	7.14	7.09	6.99
7	12.25	9.55	8.45	7.85	7.46	7.19	6.99	6.84	6.72	6.62	6.54	6.47	6.31	6.16	6.06	5.99	5.91	5.86	5.75
8	11.26	8.65	7.59	7.01	6.63	6.37	6.18	6.03	5.91	5.81	5.73	5.67	5.52	5.36	5.26	5.20	5.12	5.07	4.96
9	10.56	8.02	6.99	6.42	6.06	5.80	5.61	5.47	5.35	5.26	5.18	5.11	4.96	4.81	4.71	4.65	4.57	4.52	4.41
10	10.04	7.56	6.55	5.99	5.64	5.39	5.20	5.06	4.94	4.85	4.77	4.71	4.56	4.41	4.31	4.25	4.17	4.12	4.01
11	9.65	7.21	6.22	5.67	5.32	5.07	4.89	4.74	4.63	4.54	4.46	4.40	4.25	4.10	4.01	3.94	3.86	3.81	3.71
12	9.33	6.93	5.95	5.41	5.06	4.82	4.64	4.50	4.39	4.30	4.22	4.16	4.01	3.86	3.76	3.70	3.62	3.57	3.47
13	9.07	6.70	5.74	5.21	4.86	4.62	4.44	4.30	4.19	4.10	4.02	3.96	3.82	3.66	3.57	3.51	3.43	3.38	3.27
14	8.86	6.51	5.56	5.04	4.69	4.46	4.28	4.14	4.03	3.94	3.86	3.80	3.66	3.51	3.41	3.35	3.27	3.22	3.11
15	8.68	6.36	5.42	4.89	4.56	4.32	4.14	4.00	3.89	3.80	3.73	3.67	3.52	3.37	3.28	3.21	3.13	3.08	2.98
16	8.53	6.23	5.29	4.77	4.44	4.20	4.03	3.89	3.78	3.69	3.62	3.55	3.41	3.26	3.16	3.10	3.02	2.97	2.86
17	8.40	6.11	5.18	4.67	4.34	4.10	3.93	3.79	3.68	3.59	3.52	3.46	3.31	3.16	3.07	3.00	2.92	2.87	2.76
18	8.29	6.01	5.09	4.58	4.25	4.01	3.84	3.71	3.60	3.51	3.43	3.37	3.23	3.08	2.98	2.92	2.84	2.78	2.68
19	8.18	5.93	5.01	4.50	4.17	3.94	3.77	3.63	3.52	3.43	3.36	3.30	3.15	3.00	2.91	2.84	2.76	2.71	2.60
20	8.10	5.85	4.94	4.43	4.10	3.87	3.70	3.56	3.46	3.37	3.29	3.23	3.09	2.94	2.84	2.78	2.69	2.64	2.54
21	8.02	5.78	4.87	4.37	4.04	3.81	3.64	3.51	3.40	3.31	3.24	3.17	3.03	2.88	2.79	2.72	2.64	2.58	2.48
22	7.95	5.72	4.82	4.31	3.99	3.76	3.59	3.45	3.35	3.26	3.18	3.12	2.98	2.83	2.73	2.67	2.58	2.53	2.42
23	7.88	5.66	4.76	4.26	3.94	3.71	3.54	3.41	3.30	3.21	3.14	3.07	2.93	2.78	2.69	2.62	2.54	2.48	2.37
24	7.82	5.61	4.72	4.22	3.90	3.67	3.50	3.36	3.26	3.17	3.09	3.03	2.89	2.74	2.64	2.58	2.49	2.44	2.33
25	7.77	5.57	4.68	4.18	3.85	3.63	3.46	3.32	3.22	3.13	3.06	2.99	2.85	2.70	2.60	2.54	2.45	2.40	2.29
30	7.56	5.39	4.51	4.02	3.70	3.47	3.30	3.17	3.07	2.98	2.91	2.84	2.70	2.55	2.45	2.39	2.30	2.25	2.13
40	7.31	5.18	4.31	3.83	3.51	3.29	3.12	2.99	2.89	2.80	2.73	2.66	2.52	2.37	2.27	2.20	2.11	2.06	1.94
50	7.17	5.06	4.20	3.72	3.41	3.19	3.02	2.89	2.78	2.70	2.63	2.56	2.42	2.27	2.17	2.10	2.01	1.95	1.82
100	6.90	4.82	3.98	3.51	3.21	2.99	2.82	2.69	2.59	2.50	2.43	2.37	2.22	2.07	1.97	1.89	1.80	1.74	1.60

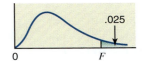

The entries in this table give the critical values of *F* for .025 area in the right tail under the *F* distribution curve and specified degrees of freedom for the numerator and denominator.

Degrees of Freedom for the Denominator	Degrees of Freedom for the Numerator																		
	1	2	3	4	5	6	7	8	9	10	11	12	15	20	25	30	40	50	100
1	647.8	799.5	864.2	899.6	921.8	937.1	948.2	956.7	963.3	968.6	973.0	976.7	984.9	993.1	998.1	1001	1006	1008	1013
2	38.51	39.00	39.17	39.25	39.30	39.33	39.36	39.37	39.39	39.40	39.41	39.41	39.43	39.45	39.46	39.46	39.47	39.48	39.49
3	17.44	16.04	15.44	15.10	14.88	14.73	14.62	14.54	14.47	14.42	14.37	14.34	14.25	14.17	14.12	14.08	14.04	14.01	13.96
4	12.22	10.65	9.98	9.61	6.36	9.20	9.07	8.98	8.90	8.84	8.79	8.75	8.66	8.56	8.50	8.46	8.41	8.38	8.32
5	10.01	8.43	7.76	7.39	7.15	6.98	6.85	6.76	6.68	6.62	6.57	6.52	6.43	6.33	6.27	6.23	6.18	6.14	6.08
6	8.81	7.26	6.60	6.23	5.99	5.82	5.70	5.60	5.52	5.46	5.41	5.37	5.27	5.17	5.11	5.07	5.01	4.98	4.92
7	8.07	6.54	5.89	5.52	5.29	5.12	4.99	4.90	4.82	4.76	4.71	4.67	4.57	4.47	4.40	4.36	4.31	4.28	4.21
8	7.57	6.06	5.42	5.05	4.82	4.65	4.53	4.43	4.36	4.30	4.24	4.20	4.10	4.00	3.94	3.89	3.84	3.81	3.74
9	7.21	5.72	5.08	4.72	4.48	4.32	4.20	4.10	4.03	3.96	3.91	3.87	3.77	3.67	3.60	3.56	3.51	3.47	3.40
10	6.94	5.46	4.83	4.47	4.24	4.07	3.95	3.85	3.78	3.72	3.66	3.62	3.52	3.42	3.35	3.31	3.26	3.22	3.15
11	6.72	5.26	4.63	4.28	4.04	3.88	3.76	3.66	3.59	3.53	3.47	3.43	3.33	3.23	3.16	3.12	3.06	3.03	2.96
12	6.55	5.10	4.47	4.12	3.89	3.73	3.61	3.51	3.44	3.37	3.32	3.28	3.18	3.07	3.01	2.96	2.91	2.87	2.80
13	6.41	4.97	4.35	4.00	3.77	3.60	3.48	3.39	3.31	3.25	3.20	3.15	3.05	2.95	2.88	2.84	2.78	2.74	2.67
14	6.30	4.86	4.24	3.89	3.66	3.50	3.38	3.29	3.21	3.15	3.09	3.05	2.95	2.84	2.78	2.73	2.67	2.64	2.56
15	6.20	4.77	4.15	3.80	3.58	3.41	3.29	3.20	3.12	3.06	3.01	2.96	2.86	2.76	2.69	2.64	2.59	2.55	2.47
16	6.12	4.69	4.08	3.73	3.50	3.34	3.22	3.12	3.05	2.99	2.93	2.89	2.79	2.68	2.61	2.57	2.51	2.47	2.40
17	6.04	4.62	4.01	3.66	3.44	3.28	3.16	3.06	2.98	2.92	2.87	2.82	2.72	2.62	2.55	2.50	2.44	2.41	2.33
18	5.98	4.56	3.95	3.61	3.38	3.22	3.10	3.01	2.93	2.87	2.81	2.77	2.67	2.56	2.49	2.44	2.38	2.35	2.27
19	5.92	4.51	3.90	3.56	3.33	3.17	3.05	2.96	2.88	2.82	2.76	2.72	2.62	2.51	2.44	2.39	2.33	2.30	2.22
20	5.87	4.46	3.86	3.51	3.29	3.13	3.01	2.91	2.84	2.77	2.72	2.68	2.57	2.46	2.40	2.35	2.29	2.25	2.17
21	5.83	4.42	3.82	3.48	3.25	3.09	2.97	2.87	2.80	2.73	2.68	2.64	2.53	2.42	2.36	2.31	2.25	2.21	2.13
22	5.79	4.38	3.78	3.44	3.22	3.05	2.93	2.84	2.76	2.70	2.65	2.60	2.50	2.39	2.32	2.27	2.21	2.17	2.09
23	5.75	4.35	3.75	3.41	3.18	3.02	2.90	2.81	2.73	2.67	2.62	2.57	2.47	2.36	2.29	2.24	2.18	2.14	2.06
24	5.72	4.32	3.72	3.38	3.15	2.99	2.87	2.78	2.70	2.64	2.59	2.54	2.44	2.33	2.26	2.21	2.15	2.11	2.02
25	5.69	4.29	3.69	3.35	3.13	2.97	2.85	2.75	2.68	2.61	2.56	2.51	2.41	2.30	2.23	2.18	2.12	2.08	2.00
30	5.57	4.18	3.59	3.25	3.03	2.87	2.75	2.65	2.57	2.51	2.46	2.41	2.31	2.20	2.12	2.07	2.01	1.97	1.88
40	5.42	4.05	3.46	3.13	2.90	2.74	2.62	2.53	2.45	2.39	2.33	2.29	2.18	2.07	1.99	1.94	1.88	1.83	1.74
50	5.34	3.97	3.39	3.05	2.83	2.67	2.55	2.46	2.38	2.32	2.26	2.22	2.11	1.99	1.92	1.87	1.80	1.75	1.66
100	5.18	3.83	3.25	2.92	2.70	2.54	2.42	2.32	2.24	2.18	2.12	2.08	1.97	1.85	1.77	1.71	1.64	1.59	1.48

Table VII The F Distribution Table (continued)

The entries in this table give the critical values of F for .05 area in the right tail under the F distribution curve and specified degrees of freedom for the numerator and denominator.

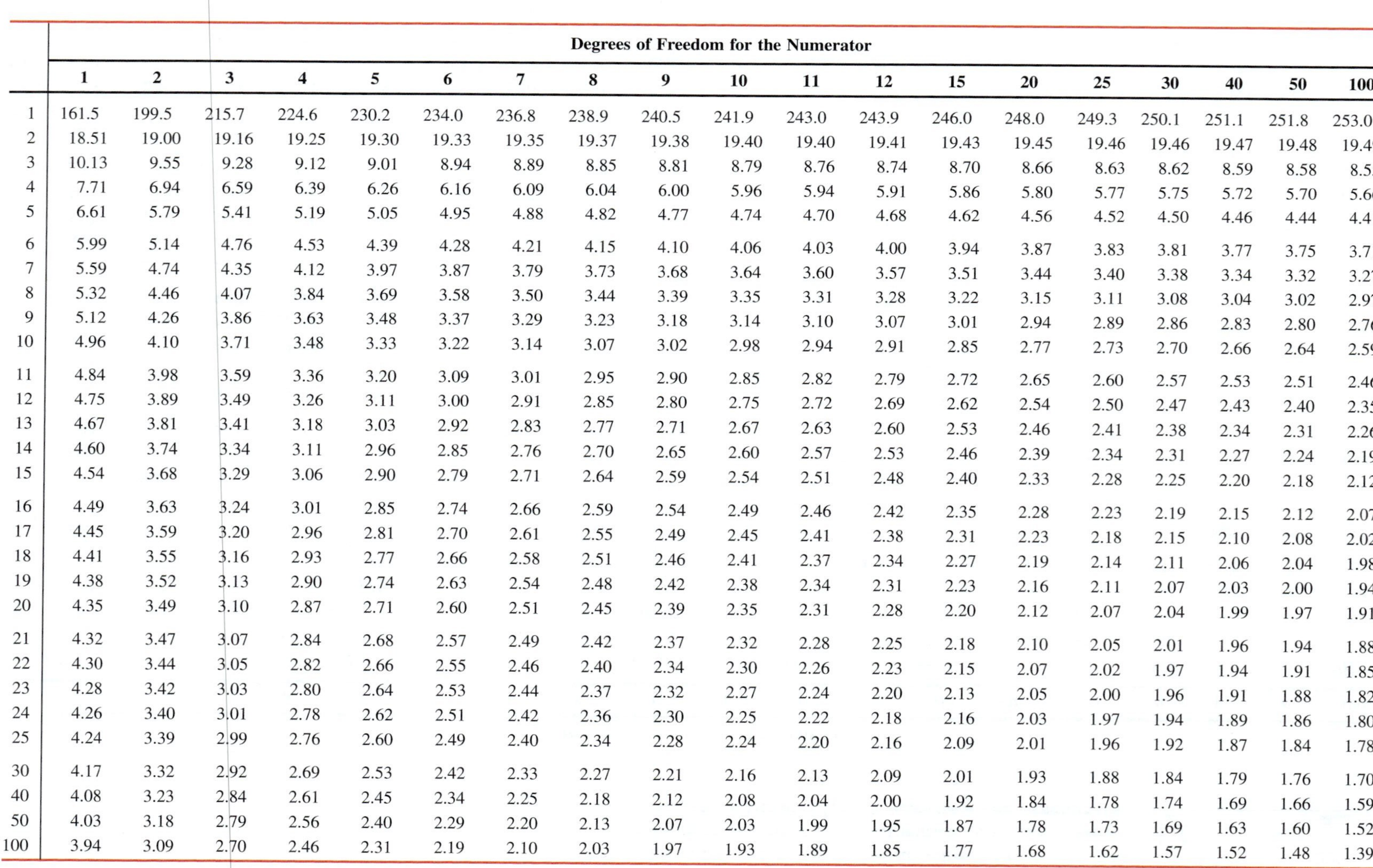

Degrees of Freedom for the Numerator (columns); Degrees of Freedom for the Denominator (rows)

	1	2	3	4	5	6	7	8	9	10	11	12	15	20	25	30	40	50	100
1	161.5	199.5	215.7	224.6	230.2	234.0	236.8	238.9	240.5	241.9	243.0	243.9	246.0	248.0	249.3	250.1	251.1	251.8	253.0
2	18.51	19.00	19.16	19.25	19.30	19.33	19.35	19.37	19.38	19.40	19.40	19.41	19.43	19.45	19.46	19.46	19.47	19.48	19.49
3	10.13	9.55	9.28	9.12	9.01	8.94	8.89	8.85	8.81	8.79	8.76	8.74	8.70	8.66	8.63	8.62	8.59	8.58	8.55
4	7.71	6.94	6.59	6.39	6.26	6.16	6.09	6.04	6.00	5.96	5.94	5.91	5.86	5.80	5.77	5.75	5.72	5.70	5.66
5	6.61	5.79	5.41	5.19	5.05	4.95	4.88	4.82	4.77	4.74	4.70	4.68	4.62	4.56	4.52	4.50	4.46	4.44	4.41
6	5.99	5.14	4.76	4.53	4.39	4.28	4.21	4.15	4.10	4.06	4.03	4.00	3.94	3.87	3.83	3.81	3.77	3.75	3.71
7	5.59	4.74	4.35	4.12	3.97	3.87	3.79	3.73	3.68	3.64	3.60	3.57	3.51	3.44	3.40	3.38	3.34	3.32	3.27
8	5.32	4.46	4.07	3.84	3.69	3.58	3.50	3.44	3.39	3.35	3.31	3.28	3.22	3.15	3.11	3.08	3.04	3.02	2.97
9	5.12	4.26	3.86	3.63	3.48	3.37	3.29	3.23	3.18	3.14	3.10	3.07	3.01	2.94	2.89	2.86	2.83	2.80	2.76
10	4.96	4.10	3.71	3.48	3.33	3.22	3.14	3.07	3.02	2.98	2.94	2.91	2.85	2.77	2.73	2.70	2.66	2.64	2.59
11	4.84	3.98	3.59	3.36	3.20	3.09	3.01	2.95	2.90	2.85	2.82	2.79	2.72	2.65	2.60	2.57	2.53	2.51	2.46
12	4.75	3.89	3.49	3.26	3.11	3.00	2.91	2.85	2.80	2.75	2.72	2.69	2.62	2.54	2.50	2.47	2.43	2.40	2.35
13	4.67	3.81	3.41	3.18	3.03	2.92	2.83	2.77	2.71	2.67	2.63	2.60	2.53	2.46	2.41	2.38	2.34	2.31	2.26
14	4.60	3.74	3.34	3.11	2.96	2.85	2.76	2.70	2.65	2.60	2.57	2.53	2.46	2.39	2.34	2.31	2.27	2.24	2.19
15	4.54	3.68	3.29	3.06	2.90	2.79	2.71	2.64	2.59	2.54	2.51	2.48	2.40	2.33	2.28	2.25	2.20	2.18	2.12
16	4.49	3.63	3.24	3.01	2.85	2.74	2.66	2.59	2.54	2.49	2.46	2.42	2.35	2.28	2.23	2.19	2.15	2.12	2.07
17	4.45	3.59	3.20	2.96	2.81	2.70	2.61	2.55	2.49	2.45	2.41	2.38	2.31	2.23	2.18	2.15	2.10	2.08	2.02
18	4.41	3.55	3.16	2.93	2.77	2.66	2.58	2.51	2.46	2.41	2.37	2.34	2.27	2.19	2.14	2.11	2.06	2.04	1.98
19	4.38	3.52	3.13	2.90	2.74	2.63	2.54	2.48	2.42	2.38	2.34	2.31	2.23	2.16	2.11	2.07	2.03	2.00	1.94
20	4.35	3.49	3.10	2.87	2.71	2.60	2.51	2.45	2.39	2.35	2.31	2.28	2.20	2.12	2.07	2.04	1.99	1.97	1.91
21	4.32	3.47	3.07	2.84	2.68	2.57	2.49	2.42	2.37	2.32	2.28	2.25	2.18	2.10	2.05	2.01	1.96	1.94	1.88
22	4.30	3.44	3.05	2.82	2.66	2.55	2.46	2.40	2.34	2.30	2.26	2.23	2.15	2.07	2.02	1.97	1.94	1.91	1.85
23	4.28	3.42	3.03	2.80	2.64	2.53	2.44	2.37	2.32	2.27	2.24	2.20	2.13	2.05	2.00	1.96	1.91	1.88	1.82
24	4.26	3.40	3.01	2.78	2.62	2.51	2.42	2.36	2.30	2.25	2.22	2.18	2.16	2.03	1.97	1.94	1.89	1.86	1.80
25	4.24	3.39	2.99	2.76	2.60	2.49	2.40	2.34	2.28	2.24	2.20	2.16	2.09	2.01	1.96	1.92	1.87	1.84	1.78
30	4.17	3.32	2.92	2.69	2.53	2.42	2.33	2.27	2.21	2.16	2.13	2.09	2.01	1.93	1.88	1.84	1.79	1.76	1.70
40	4.08	3.23	2.84	2.61	2.45	2.34	2.25	2.18	2.12	2.08	2.04	2.00	1.92	1.84	1.78	1.74	1.69	1.66	1.59
50	4.03	3.18	2.79	2.56	2.40	2.29	2.20	2.13	2.07	2.03	1.99	1.95	1.87	1.78	1.73	1.69	1.63	1.60	1.52
100	3.94	3.09	2.70	2.46	2.31	2.19	2.10	2.03	1.97	1.93	1.89	1.85	1.77	1.68	1.62	1.57	1.52	1.48	1.39

Table VII The *F* Distribution Table (continued)

The entries in this table give the critical values of *F* for .10 area in the right tail under the *F* distribution curve and specified degrees of freedom for the numerator and denominator.

Degrees of Freedom for the Denominator	Degrees of Freedom for the Numerator																		
	1	2	3	4	5	6	7	8	9	10	11	12	15	20	25	30	40	50	100
1	39.86	49.50	53.59	55.83	57.24	58.20	58.91	59.44	59.86	60.19	60.47	60.71	61.22	61.74	62.05	62.26	62.53	62.69	63.01
2	8.53	9.00	9.16	9.24	9.29	9.33	9.35	9.37	9.38	9.39	9.40	9.41	9.42	9.44	9.45	9.46	9.47	9.47	9.48
3	5.54	5.46	5.39	5.34	5.31	5.28	5.27	5.25	5.24	5.23	5.22	5.22	5.20	5.18	5.17	5.17	5.16	5.15	5.14
4	4.54	4.32	4.19	4.11	4.05	4.01	3.98	3.95	3.94	3.92	3.91	3.90	3.87	3.84	3.83	3.82	3.80	3.80	3.78
5	4.06	3.78	3.62	3.52	3.45	3.40	3.37	3.34	3.32	3.30	3.28	3.27	3.24	3.21	3.19	3.17	3.16	3.15	3.13
6	3.78	3.46	3.29	3.18	3.11	3.05	3.01	2.98	2.96	2.94	2.92	2.90	2.87	2.84	2.81	2.80	2.78	2.77	2.75
7	3.59	3.26	3.07	2.96	2.88	2.83	2.78	2.75	2.72	2.70	2.68	2.67	2.63	2.59	2.57	2.56	2.54	2.52	2.50
8	3.46	3.11	2.92	2.81	2.73	2.67	2.62	2.59	2.56	2.54	2.52	2.50	2.46	2.42	2.40	2.38	2.36	2.35	2.32
9	3.36	3.01	2.81	2.69	2.61	2.55	2.51	2.47	2.44	2.42	2.40	2.38	2.34	2.30	2.27	2.25	2.23	2.22	2.19
10	3.29	2.92	2.73	2.61	2.52	2.46	2.41	2.38	2.35	2.32	2.30	2.28	2.24	2.20	2.17	2.16	2.13	2.12	2.09
11	3.23	2.86	2.66	2.54	2.45	2.39	2.34	2.30	2.27	2.25	2.23	2.21	2.17	2.12	2.10	2.08	2.05	2.04	2.01
12	3.18	2.81	2.61	2.48	2.39	2.33	2.28	2.24	2.21	2.19	2.17	2.15	2.10	2.06	2.03	2.01	1.99	1.97	1.94
13	3.14	2.76	2.56	2.43	2.35	2.28	2.23	2.20	2.16	2.14	2.12	2.10	2.05	2.01	1.98	1.96	1.93	1.92	1.88
14	3.10	2.73	2.52	2.39	2.31	2.24	2.19	2.15	2.12	2.10	2.07	2.05	2.01	1.96	1.93	1.91	1.89	1.87	1.83
15	3.07	2.70	2.49	2.36	2.27	2.21	2.16	2.12	2.09	2.06	2.04	2.02	1.97	1.92	1.89	1.87	1.85	1.83	1.79
16	3.05	2.67	2.46	2.33	2.24	2.18	2.13	2.09	2.06	2.03	2.01	1.99	1.94	1.89	1.86	1.84	1.81	1.79	1.76
17	3.03	2.64	2.44	2.31	2.22	2.15	2.10	2.06	2.03	2.00	1.98	1.96	1.91	1.86	1.83	1.81	1.78	1.76	1.73
18	3.01	2.62	2.42	2.29	2.20	2.13	2.08	2.04	2.00	1.98	1.95	1.93	1.89	1.84	1.80	1.78	1.75	1.74	1.70
19	2.99	2.61	2.40	2.27	2.18	2.11	2.06	2.02	1.98	1.96	1.93	1.91	1.86	1.81	1.78	1.76	1.73	1.71	1.67
20	2.97	2.59	2.38	2.25	2.16	2.09	2.04	2.00	1.96	1.94	1.91	1.89	1.84	1.79	1.76	1.74	1.71	1.69	1.65
21	2.96	2.57	2.36	2.23	2.14	2.08	2.02	1.98	1.95	1.92	1.90	1.87	1.83	1.78	1.74	1.72	1.69	1.67	1.63
22	2.95	2.56	2.35	2.22	2.13	2.06	2.01	1.97	1.93	1.90	1.88	1.86	1.81	1.76	1.73	1.70	1.67	1.65	1.61
23	2.94	2.55	2.34	2.21	2.11	2.05	1.99	1.95	1.92	1.89	1.87	1.84	1.80	1.74	1.71	1.69	1.66	1.64	1.59
24	2.93	2.54	2.33	2.19	2.10	2.04	1.98	1.94	1.91	1.88	1.85	1.83	1.78	1.73	1.70	1.67	1.64	1.62	1.58
25	2.92	2.53	2.32	2.18	2.09	2.02	1.97	1.93	1.89	1.87	1.84	1.82	1.77	1.72	1.68	1.66	1.63	1.61	1.56
30	2.88	2.49	2.28	2.14	2.05	1.98	1.93	1.88	1.85	1.82	1.79	1.77	1.72	1.67	1.63	1.61	1.57	1.55	1.51
40	2.84	2.44	2.23	2.09	2.00	1.93	1.87	1.83	1.79	1.76	1.74	1.71	1.66	1.61	1.57	1.54	1.51	1.48	1.43
50	2.81	2.41	2.20	2.06	1.97	1.90	1.84	1.80	1.76	1.73	1.70	1.68	1.63	1.57	1.53	1.50	1.46	1.44	1.39
100	2.76	2.36	2.14	2.00	1.91	1.83	1.78	1.73	1.69	1.66	1.64	1.61	1.56	1.49	1.45	1.42	1.38	1.35	1.29

Appendix C

Statistical Tables on the Web Site

Note: The following tables are on the Web Site of the text along with Chapters 14 and 15.

Table VIII Critical Values of *X* for the Sign Test

Table IX Critical Values of *T* for the Wilcoxon Signed-Rank Test

Table X Critical Values of *T* for the Wilcoxon Rank Sum Test

Table XI Critical Values for the Spearman Rho Rank Correlation Coefficient Test

Table XII Critical Values for a Two-Tailed Runs Test with $\alpha = .05$

ANSWERS TO SELECTED ODD-NUMBERED EXERCISES AND SELF-REVIEW TESTS

(Note: Due to differences in rounding, the answers obtained by readers may differ slightly from the ones given in this Appendix.)

Chapter 1

1.7 **a.** population **b.** sample **c.** population **d.** sample **e.** population
1.11 **a.** number of dog bites **b.** six observations **c.** six elements
1.15 **a.** quantitative **b.** quantitative **c.** qualitative **d.** quantitative **e.** quantitative
1.17 **a.** discrete **b.** continuous **d.** discrete **e.** continuous
1.21 **a.** cross-section data **b.** cross-section data **c.** time-series data **d.** time-series data
1.23 **a.** $\Sigma f = 69$ **b.** $\Sigma m^2 = 1363$ **c.** $\Sigma mf = 922$ **d.** $\Sigma m^2 f = 17{,}128$
1.25 **a.** $\Sigma x = 88$ **b.** $\Sigma y = 58$ **c.** $\Sigma xy = 855$ **d.** $\Sigma x^2 = 1590$ **e.** $\Sigma y^2 = 622$
1.27 **a.** $\Sigma y = 529$ **b.** $(\Sigma y)^2 = 279{,}841$ **c.** $\Sigma y^2 = 80{,}199$
1.29 **a.** $\Sigma x = 148$ **b.** $(\Sigma x)^2 = 21{,}904$ **c.** $\Sigma x^2 = 4486$
1.33 **a.** sample **b.** population for the year **c.** sample **d.** population
1.35 **a.** sampling without replacement **b.** sampling with replacement
1.37 **a.** $\Sigma x = 47$ **b.** $(\Sigma x)^2 = 2209$ **c.** $\Sigma x^2 = 443$
1.39 **a.** $\Sigma m = 59$ **b.** $\Sigma f^2 = 2662$ **c.** $\Sigma mf = 1508$ **d.** $\Sigma m^2 f = 24{,}884$ **e.** $\Sigma m^2 = 867$

Self-Review Test

1. b **2.** c **3.** **a.** sampling without replacement **b.** sampling with replacement
4. **a.** qualitative **b.** quantitative (continuous) **c.** quantitative (discrete) **d.** qualitative
6. **a.** $\Sigma x = 29$ **b.** $(\Sigma x)^2 = 841$ **c.** $\Sigma x^2 = 231$
7. **a.** $\Sigma m = 45$ **b.** $\Sigma f = 112$ **c.** $\Sigma m^2 = 495$ **d.** $\Sigma mf = 975$ **e.** $\Sigma m^2 f = 9855$ **f.** $\Sigma f^2 = 2994$

Chapter 2

2.3 **c.** 26.7% **d.** 73.4%
2.5 **c.** 52% **2.7** **c.** 50% **2.15** **d.** 62%
2.17 **a.** class limits: \$1–\$25, \$26–\$50, \$51–\$75, \$76–\$100, \$101–\$125, \$126–\$150 **b.** class boundaries: \$.5, \$25.5, \$50.5, \$75.5, \$100.5, \$125.5, \$150.5; width = \$25 **c.** class midpoints: \$13, \$38, \$63, \$88, \$113, \$138
2.19 **d.** 30% **2.29** **c.** 11
2.35 **c.** 38% **e.** about 52% **2.43** 7 teams
2.47 218, 245, 256, 329, 367, 383, 397, 404, 427, 433, 471, 523, 537, 551, 563, 581, 592, 622, 636, 647, 655, 678, 689, 810, 841
2.67 **d.** 23.3% **2.69** **c.** 16.7% **2.71** **c.** 56.7%
2.73 **d.** Boundaries of the fourth class are \$4200.5 and \$5600.5; width = \$1400.
2.87 No. The older group may drive more miles per week than the younger group.

Self-Review Test

2. **a.** 5 **b.** 7 **c.** 17 **d.** 6.5 **e.** 13 **f.** 90 **g.** .30
4. **c.** 35% **5.** **c.** 70.8%
8. 30, 33, 37, 42, 44, 46, 47, 49, 51, 53, 53, 56, 60, 67, 67, 71, 79

Chapter 3

3.5 mode **3.9** mean = 3.00; median = 3.50; no mode
3.11 mean = \$16,434.8; median = \$16,134.5
3.13 **a.** mean = 1128.92 thousand workers; median = 749 thousand workers; **b.** mode = 1200 thousand workers
3.15 mean = \$109,417.85; median = \$98,500
3.17 mean = \$327.83 million; median = \$175 million; no mode
3.19 mean = 2.92 power outages; median = 2.5 power outages; mode = 2 power outages
3.21 mean = 29.4; median = 28.5; mode = 23
3.23 **a.** mean = 40.27; median = 13 **b.** outlier = 256; when the outlier is dropped: mean = 18.7; median = 12.5; mean changes by a larger amount **c.** median
3.25 combined mean = \$99 **3.27** total = \$1055
3.29 age of the sixth person = 48 years
3.31 mean for data set I = 24.60; mean for data set II = 31.60 The mean of the second data set is equal to the mean of the first data set plus 7.
3.33 10% trimmed mean = 38.25 years **3.35** weighted mean = 77.5
3.41 range = 25; $\sigma^2 = 61.5$; $\sigma = 7.84$
3.43 **a.** $\bar{x} = 9$; deviations from the mean: −2, 1, −1, −6, 6, 3, −3, 2. The sum of these deviations is zero. **b.** range = 12; $s^2 = 14.2857$; $s = 3.78$

3.45 range = 13; $s^2 = 13.8409$; $s = 3.72$
3.47 range = 27 pieces; $s^2 = 78.1$; $s = 8.84$ pieces
3.49 range = 7 stings; $s^2 = 4.5769$; $s = 2.14$ stings
3.51 range = 30; $s^2 = 107.4286$; $s = 10.36$
3.53 range = \$28.2 billion; $s^2 = 106.9107$; $s = \$10.34$ billion
3.55 $s = 0$
3.57 CV for salaries = 10.94%; CV for years of schooling = 13.33%; The relative variation in salaries is lower.
3.59 $s = 14.64$ for both data sets
3.63 $\bar{x} = 9.40$; $s^2 = 37.7114$; $s = 6.14$
3.65 $\mu = 14$ hours; $\sigma^2 = 51.9167$; $\sigma = 7.21$ hours
3.67 $\bar{x} = 19.67$; $s^2 = 67.6979$; $s = 8.23$
3.69 $\bar{x} = 36.80$ minutes; $s^2 = 597.7143$; $s = 24.45$ minutes
3.71 **a.** $\bar{x} = \$51.76$ **c.** $\bar{x} = \$51.73$
3.75 at least 75%; at least 84%; at least 89%
3.77 68%; 95%; 99.7%
3.79 **a.** at least 75% **b.** at least 84% **c.** at least 89%
3.81 **a. i.** at least 75% **ii.** at least 89% **b.** \$1515 to \$3215
3.83 **a.** 95% **b.** 68% **c.** 99.7%
3.85 **a. i.** 99.7% **ii.** 68% **b.** 66 to 78 mph
3.91 **a.** $Q_1 = 69$; $Q_2 = 73$; $Q_3 = 76.5$; $IQR = 7.5$ **b.** $P_{35} = 71$ **c.** 30.77%
3.93 **a.** $Q_1 = 5$; $Q_2 = 6.5$; $Q_3 = 8$; $IQR = 3$ **b.** $P_{63} = 8$ **c.** 83.33%
3.95 **a.** $Q_1 = 25$; $Q_2 = 28.5$; $Q_3 = 33$; $IQR = 8$ **b.** $P_{65} = 31$ **c.** 33.33%
3.97 **a.** $Q_1 = 543.5$; $Q_2 = 686.5$; $Q_3 = 798.5$; $IQR = 255$ **b.** $P_{77} = 789$ **c.** 55%
3.99 no outlier
3.109 **a.** mean = \$58.7 thousand; median = \$56.5 thousand **b.** outlier = 104; when the outlier is dropped: mean = \$53.67 thousand ; median = \$56 thousand; mean changes by a larger amount **c.** median
3.111 **a.** mean = 184.9 yards; median = 175 yards; mode = 167 and 170 **b.** range = 66 yards; $s^2 = 487.6556$; $s = 22.08$ yards
3.113 $\bar{x} = 5.08$ inches; $s^2 = 6.8506$; $s = 2.62$ inches
3.115 **a. i.** at least 75% **ii.** at least 89% **b.** 160 to 240 minutes
3.117 **a. i.** 68% **ii.** 95% **b.** 140 to 260 minutes
3.119 **a.** $Q_1 = 44$; $Q_2 = 56.5$; $Q_3 = 68$; $IQR = 24$ **b.** $P_{70} = 59$ **c.** 30%
3.121 The data set is skewed slightly to the right; 135 is an outlier.
3.123 The minimum score is 169.
3.125 **a.** new mean = 76.4 inches; new median = 78 inches; new range = 13 inches **b.** new mean = 75.2 inches
3.127 mean = \$54.46 per barrel
3.129 **a.** trimmed mean = 9.5 **b.** 14.3%
3.131 **a.** age 30 and under: rate for A = 25; rate for B = 20 **b.** age 31 and over: rate for A = 100; rate for B = 85.7 **c.** overall: rate for A = 50; rate for B = 58.3 **d.** Country A has the lower overall average because 66.67% of its population is under 30.
3.133 **a.** $k = 1.41$ **b.** $k = 2.24$ **3.135** **b.** median
3.137 **b.** For men: mean = 82, median = 79, modes = 75, 79, and 92, $s = 12.08$, $Q_1 = 73.5$, $Q_3 = 89.5$, and $IQR = 16$. For women: mean = 97.53, median = 98, modes = 94 and 100, $s = 8.44$, $Q_1 = 94$, $Q_3 = 101$, and $IQR = 7$
3.139 **a.** mean = 30 **b.** mean = 50
3.141 **a.** at least 55.56% **b.** 1 to 11 inches **c.** 2.66 to 9.34 inches
3.143 **a.** For men: mean = 174.91 lbs = 76,189.05 grams = 12.49 stone, median = 179 lbs = 77,970.61 grams = 12.79 stone, and st. dev. = 19.12 lbs = 8328.48 grams = 1.37 stone. For women: mean = 124.95 lbs = 54,426.97 grams = 8.93 stone, median = 123 lbs = 53,577.57 grams = 8.79 stone, st. dev. = 17.48 lbs = 7614.11 grams = 1.25 stone. **b.** see answer to a, as answers are identical. **c.** yes **d** & **e.** Smaller unit has more volatility.
3.145 108 to 111

Self-Review Test

1. b **2.** a and d **3.** c **4.** c **5.** b **6.** b **7.** a **8.** a **9.** b **10.** a **11.** b **12.** c **13.** a **14.** a
15. mean = 10.9; median = 8; mode = 6; range = 26; $s^2 = 65.2111$; $s = 8.08$
19. **b.** $\bar{x} = 19.46$; $s^2 = 44.0400$; $s = 6.64$
20. **a. i.** at least 84% **ii.** at least 89% **b.** 2.9 to 11.7 years
21. **a. i.** 68% **ii.** 99.7% **b.** 2.9 to 11.7 years
22. **a.** $Q_1 = 3$; $Q_2 = 8$; $Q_3 = 13$; $IQR = 10$ **b.** $P_{60} = 10$ **c.** 66.67%
23. Data are skewed slightly to the right.
24. combined mean = \$466.43
25. GPA of fifth student = 3.17
26. 10% trimmed mean = 197.75; trimmed mean is a better measure
27. **a.** mean for data set I = 19.75; mean for data set II = 16.75. The mean of the second data set is equal to the mean of the first data set minus 3. **b.** $s = 11.32$ for both data sets.

Chapter 4

4.3 $S = \{AB, AC, BA, BC, CA, CB\}$
4.5 four possible outcomes; $S = \{LL, LI, IL, II\}$
4.7 four possible outcomes; $S = \{DD, DG, GD, GG\}$
4.9 $S = \{HHH, HHT, HTH, HTT, THH, THT, TTH, TTT\}$
4.11 **a.** {LI and IL}; a compound event **b.** {LL, LI, and IL}; a compound event **c.** {II, IL, and LI}; a compound event **d.** {LI}; a simple event
4.13 **a.** {DG, GD, and GG}; a compound event **b.** {DG and GD}; a compound event **c.** {GD}; a simple event **d.** {DD, DG, and GD}; a compound event
4.19 −.55, 1.56, 5/3, −2/7
4.21 not equally likely events; use relative frequency approach
4.23 subjective probability
4.25 **a.** .450 **b.** .550 **4.27** .560 **4.29** .580
4.31 **a.** .200 **b.** .800 **4.33** .6667; .3333
4.35 .325; .675 **4.37** **a.** .0939 **b.** .5
4.39 use relative frequency approach **4.45** 1296
4.47 **a.** no **b.** no **c.** $\bar{A} = \{1, 3, 4, 6, 8\}$; $\bar{B} = \{1, 3, 5, 6, 7\}$; $P(\bar{A}) = .625$; $P(\bar{B}) = .625$
4.49 50 **4.51** 960 **4.53** **a. i.** .600 **ii.** .600 **iii.** .375 **iv.** .583 **b.** Events "male" and "female" are mutually exclusive. Events "have shopped"

and "male" are not mutually exclusive. **c.** Events "female" and "have shopped" are dependent.

4.55 **a.** **i.** .3475 **ii.** .5425 **iii.** .2727 **iv.** .4545 **b.** Events "male" and "in favor" are not mutually exclusive. Events "in favor" and "against" are mutually exclusive. **c.** Events "female" and "no opinion" are dependent.

4.57 **a.** **i.** .1012 **ii.** .4835 **iii.** .5524 **iv.** .1014 **b.** Events "Airline A" and "more than 1 hour late" are not mutually exclusive. Events "less than 30 minutes late" and "more than one hour late" are mutually exclusive. **c.** Events "Airline B" and "30 minutes to 1 hour late" are dependent.

4.59 Events "female" and "pediatrician" are dependent but not mutually exclusive.

4.61 Events "female" and "business major" are dependent but not mutually exclusive.

4.63 $P(A) = .3333$; $P(\bar{A}) = .6667$ **4.65** .88

4.71 **a.** .4543 **b.** .0980

4.73 **a.** .1520 **b.** .1824

4.75 **a.** .2462 **b.** .1086

4.77 .6923 **4.79** .725

4.81 **a.** **i.** .3844 **ii.** .1590 **b.** .0000

4.83 **a.** **i.** .350 **ii.** .150

4.85 **a.** **i.** .225 **ii.** .035 **b.** .0000

4.87 .3529 **4.89** .2667 **4.91** .1600

4.93 **a.** .0025 **b.** .9025 **4.95** .5120

4.97 .5278 **4.99** .40

4.105 **a.** .56 **b.** .76 **4.107** **a.** .52 **b.** .67

4.109 **a.** .6358 **b.** .9075

4.111 **a.** .750 **b.** .750 **c.** 1.0

4.113 **a.** .780 **b.** .550 **c.** .790

4.115 .910 **4.117** .77

4.119 .830 **4.121** .80 **4.123** .9744

4.125 **a.** .2571 **b.** .1429

4.127 **a.** **i.** .4360 **ii.** .4800 **iii.** .3462 **iv.** .6809 **v.** .3400 **vi.** .6600 **b.** Events "female" and "prefers watching sports" are dependent but not mutually exclusive.

4.129 **a.** **i.** .750 **ii.** .700 **iii.** .225 **iv.** .775 **b.** Events "student athlete" and "should be paid" are dependent but not mutually exclusive.

4.131 **a.** .2601 **b.** .7399 **4.133** .0605

4.135 .0048 **4.137** **a.** 17,576,000 **b.** 5200

4.139 **a.** 1/120,526,770 = .0000000083 **b.** 1/2,939,677 = .00000034

4.141 **a.** .5000 **b.** .3333 **c.** No; the sixth toss is independent of the first five tosses. Equivalent to part a.

4.143 **a.** .030 **b.** .150

4.145 **a.** .50 **b.** .50 **4.147** **a.** .8333 **b.** .1667

4.149 **a.** .01% **b.** **i.** .0048 **ii.** .0028 **iii.** .0278 **iv.** .0111 **4.151** **a.** .8851 **b.** .0035

Self-Review Test

1. a **2.** b **3.** c **4.** a **5.** a **6.** b
7. c **8.** b **9.** b **10.** c **11.** b
12. 120 **13.** **a.** .3333 **b.** .6667
14. **a.** Events "female" and "out of state" are dependent but not mutually exclusive. **b.** **i.** .4500 **ii.** .6364
15. .825 **16.** .3894 **17.** .4225 **18.** .40; .60
19. **a.** .279 **b.** .829
20. **a.** **i.** .358 **ii.** .405 **iii.** .235 **iv.** .5593 **b.** Events "woman" and "yes" are dependent but not mutually exclusive.

Chapter 5

5.3 **a.** discrete random variable **b.** continuous random variable **c.** continuous random variable **d.** discrete random variable **e.** discrete random variable **f.** continuous random variable

5.5 discrete random variable

5.9 **a.** not a valid probability distribution **b.** a valid probability distribution **c.** not a valid probability distribution

5.11 **a.** .17 **b.** .20 **c.** .58 **d.** .42 **e.** .42 **f.** .27 **g.** .68

5.13 **b.** **i.** .25 **ii.** .24 **iii.** .51 **iv.** .69

5.15 **a.**

x	1	2	3	4	5
$P(x)$	.10	.25	.30	.20	.15

b. approximate **c.** **i.** .30 **ii.** .65 **iii.** .75 **iv.** .65

5.17

x	0	1	2
$P(x)$	.3025	.4950	.2025

5.19

x	0	1	2
$P(x)$	.0841	.4118	.5041

5.21

x	0	1	2
$P(x)$	.4789	.4422	.0789

5.23 **a.** $\mu = 1.590$; $\sigma = .960$ **b.** $\mu = 7.070$; $\sigma = 1.061$

5.25 $\mu = .440$ error; $\sigma = .852$ error

5.27 $\mu = 2.94$ camcorders; $\sigma = 1.441$ camcorders

5.29 $\mu = 1.00$ head; $\sigma = .707$ head

5.31 $\mu = 1.896$ sets; $\sigma = 1.079$ sets

5.33 $\mu = .100$ car; $\sigma = .308$ car

5.35 $\mu = \$3.9$ million; $\sigma = \$3.015$ million

5.37 $\mu = .500$ person; $\sigma = .584$ person

5.39 $3! = 6$; $(9 - 3)! = 720$; $9! = 362{,}880$; $(14 - 12)! = 2$; ${}_5C_3 = 10$; ${}_7C_4 = 35$; ${}_9C_3 = 84$; ${}_4C_0 = 1$; ${}_3C_3 = 1$; ${}_6P_2 = 30$; ${}_8P_4 = 1680$

5.41 ${}_9C_2 = 36$; ${}_9P_2 = 72$ **5.43** ${}_{12}C_3 = 220$; ${}_{12}P_3 = 1320$ **5.45** ${}_{20}C_6 = 38{,}760$; ${}_{20}P_6 = 27{,}907{,}200$

5.47 167,960

5.51 **a.** not a binomial experiment **b.** a binomial experiment **c.** a binomial experiment

5.53 **a.** .2541 **b.** .1536 **c.** .3241

5.55 **b.** $\mu = 2.100$; $\sigma = 1.212$

5.59 **a.** 0, 1, 2, 3, 4, 5, 6, 7, 8, 9, 10 **b.** .2394

5.61 **a.** .1423 **b.** .5271 **c.** .6912

5.63 **a.** .0981 **b.** .0000 **c.** .2241

5.65 **a.** .2725 **b.** .0839

5.67 **a.** $\mu = 5.6$ customers; $\sigma = 1.058$ customers **b.** .1147

5.69 **a.** $\mu = 5.600$ customers; $\sigma = 1.296$ customers **b.** .0467

5.71 **a.** .4286 **b.** .0714 **c.** .5

5.73 **a.** .3818 **b.** .0030 **c.** .5303

5.75 **a.** .4747; **b.** .0440 **c.** .3407

5.77 **a.** .2545 **b.** .2787 **c.** .2121

5.81 a. .0404 b. .2565

5.83 a. $\mu = 1.3$; $\sigma^2 = 1.3$; $\sigma = 1.140$ b. $\mu = 2.1$; $\sigma^2 = 2.1$; $\sigma = 1.449$

5.85 .1496 5.87 .1185

5.89 a. .1162 b. i. .6625 ii. .1699 iii. .4941

5.91 a. .3033 b. i. .0900 ii. .0018 iii. .9098

5.93 a. .0031 b. i. .0039 ii. .4911

5.95 a. .2466 c. $\mu = 1.4$ $\sigma^2 = 1.4$ $\sigma = 1.183$

5.97 a. .0446 b. i. .0390 ii. .2580 iii. .0218

5.99 $\mu = 4.11$; $\sigma = 1.019$; This mechanic repairs, on average, 4.11 cars per day

5.101 b. $\mu = \$557{,}000$; $\sigma = \$1{,}288{,}274$; μ gives the company's expected profit.

5.103 a. .0000 b. .0351 c. .7214

5.105 a. .9246 b. .0754

5.107 a. .3692 b. .1429 c. .0923

5.109 a. .8643 b. .1357

5.111 a. .0912 b. i. .5502 ii. .0817 iii. .2933

5.113 a. .2466

5.115 $\Sigma x\, P(x) = -2.22$. This game is not fair to you and you should not play as you expect to lose $2.22.

5.117 a. .0625 b. .125 c. .3125

5.119 c. .7149 d. 3 nights

5.121 8 cheesecakes

5.123 a. 35 b. 10 c. .2857 5.127 $6

5.129 a. .0211 b. .0475 c. .4226

Self-Review Test

2. probability distribution table
3. a 4. b 5. a 7. b 8. a
9. b 10. a 11. c 13. a
15. $\mu = 2.040$ sales; $\sigma = 1.449$ sales
16. a. i. .2128 ii. .8418 iii. .0153
    b. $\mu = 7.2$ adults; $\sigma = 1.697$ adults
17. a. .4525 b. .0646 c. .0666
18. a. i. .0521 ii. .2203 iii. .2013

Chapter 6

6.11 .8664 6.13 .9876

6.15 a. .4744 b. .4678 c. .1162 d. .0610 e. .9452

6.17 a. .0594 b. .0244 c. .9798 d. .9686

6.19 a. .5 approximately b. .5 approximately c. .00 approximately d. .00 approximately

6.21 a. .9613 b. .4783 c. .4767 d. .0694

6.23 a. .0162 b. .2450 c. .1570 d. .9625

6.25 a. .8365 b. .8762 c. approximately .5 d. approximately .5 e. approximately .00 f. approximately .00

6.27 a. 1.80 b. −2.60 c. −1.60 d. 2.40

6.29 a. .4599 b. .1210 c. .2223

6.31 a. .3336 b. .9564 c. .9564 d. approximately .00

6.33 a. .2178 b. .5997

6.35 a. .8212 b. .3085 c. .0401 d. .7486

6.37 a. .0287 b. .2345

6.39 a. .1093 b. .6902

6.41 a. 93.32% b. 14.65%

6.43 a. .0344 b. .9379

6.45 a. .7357 b. 15.64%

6.47 a. .1190 or about 12% b. .0475 or about 5%

6.49 a. 9.51% b. 3.59% c. 84.69% d. 18.81%

6.51 2.64%

6.53 a. 2.00 b. −2.02 approximately c. −.37 approximately d. 1.02 approximately

6.55 a. approximately 1.65 b. −1.96 c. −2.33 approximately d. 2.58 approximately

6.57 a. 208.50 b. 241.25 c. 178.50 d. 145.75 e. 158.25 f. 251.25

6.59 19 minutes approximately

6.61 2060 kilowatt-hours

6.63 $82.02 approximately

6.65 $np > 5$ and $nq > 5$

6.67 a. .7688 b. .7697; difference is .0009

6.69 a. $\mu = 72$; $\sigma = 5.36656315$ b. .5359 c. .4564

6.71 a. .0764 b. .6793 c. .8413 d. .8238

6.73 .1692 6.75 a. .0381 b. .1230 c. .7013

6.77 a. .0454 b. .0516 c. .8646

6.79 a. .7549 b. .2451

6.81 a. .1093 b. 9.31% c. 57.33% d. It is possible, but its probability is close to zero.

6.83 .0124 or 1.24%

6.85 a. 848 hours b. 792 hours approximately

6.87 a. .0454 b. .0838 c. .8861 d. .2477

6.89 $2136 6.91 a. 85.08% b. $4000

6.93 .0091

6.95 a. at most .0062 b. 65 mph

6.97 8.16 ounces

6.99 company A: $.0490 company B: $.0508

6.101 a. .7967 b. 62

6.105 .1064

Self-Review Test

1. a 2. a 3. d 4. b 5. a 6. c
7. b 8. b
9. a. .1878 b. .9304 c. .0985 d. .7704
10. a. −1.28 approximately b. .61 c. 1.65 approximately d. −1.07 approximately
11. a. .8466 b. .0571 c. .1539 d. .0320
12. a. 14.5 minutes approximately b. 29.15 minutes approximately
13. a. i. .0354 ii. .8660 iii. .2327 iv. .9345 v. .1735 b. .9345 c. .3236

Chapter 7

7.5 a. 16.60 b. sampling error = −.27
c. sampling error = −.27; nonsampling error = 1.11
d. $\bar{x}_1 = 16.22$; $\bar{x}_2 = 15.67$; $\bar{x}_3 = 17.00$; $\bar{x}_4 = 16.33$; $\bar{x}_5 = 17.44$; $\bar{x}_6 = 16.78$; $\bar{x}_7 = 17.22$; $\bar{x}_8 = 17.67$; $\bar{x}_9 = 16.56$; $\bar{x}_{10} = 15.11$

7.7 **b.** $\bar{x}_1 = 28.4$; $\bar{x}_2 = 28.8$; $\bar{x}_3 = 33.8$; $\bar{x}_4 = 34.4$; $\bar{x}_5 = 35.2$; $\bar{x}_6 = 36.4$; **c.** $\mu = 32.83$

7.13 **a.** $\mu_{\bar{x}} = 60$; $\sigma_{\bar{x}} = 2.357$ **b.** $\mu_{\bar{x}} = 60$; $\sigma_{\bar{x}} = 1.054$

7.15 **a.** $\sigma_{\bar{x}} = 1.400$ **b.** $\sigma_{\bar{x}} = 2.500$

7.17 **a.** $n = 100$ **b.** $n = 256$

7.19 $\mu_{\bar{x}} = 3$ hours; $\sigma_{\bar{x}} = .092$ hour

7.21 $\mu_{\bar{x}} = \$320$; $\sigma_{\bar{x}} = \$14.40$ **7.23** $n = 121$

7.25 **a.** $\mu_{\bar{x}} = 80.60$ **b.** $\sigma_{\bar{x}} = 3.302$ **d.** $\sigma_{\bar{x}} = 3.302$

7.33 $\mu_{\bar{x}} = 46$ miles per hour; $\sigma_{\bar{x}} = .671$ mile per hour; the normal distribution

7.35 $\mu_{\bar{x}} = 3.020$; $\sigma_{\bar{x}} = .042$; approximately normal distribution

7.37 $\mu_{\bar{x}} = \$96$; $\sigma_{\bar{x}} = \$2.846$; approximately normal distribution

7.39 $\mu_{\bar{x}} = 62$ minutes; $\sigma_{\bar{x}} = .7$ minute; approximately normal distribution

7.41 86.64%

7.43 **a.** $z = 2.44$ **b.** $z = -7.25$ **c.** $z = -3.65$ **d.** $z = 5.82$

7.45 **a.** .1940 **b.** .8749

7.47 **a.** .00 approximately **b.** .9292

7.49 **a.** .1093 **b.** .0322 **c.** .7776

7.51 **a.** .8254 **b.** .7888 **c.** .2033

7.53 **a.** .8109 **b.** .0322

7.55 **a.** .1464 **b.** .9624 **c.** .0418

7.57 **a.** .3085 **b.** .8543 **c.** .8664 **d.** .1056

7.59 .0124 **7.61** $p = .12$; $\hat{p} = .15$

7.63 7125 subjects in the population; 312 subjects in the sample

7.65 sampling error $= -.05$

7.71 **a.** $\mu_{\hat{p}} = .21$; $\sigma_{\hat{p}} = .020$ **b.** $\mu_{\hat{p}} = .21$; $\sigma_{\hat{p}} = .015$

7.73 **a.** $\sigma_{\hat{p}} = .051$ **b.** $\sigma_{\hat{p}} = .071$

7.77 **a.** $p = .667$ **b.** 6 **d.** $-.067, -.067, .133, .133, -.067, -.067$

7.79 $\mu_{\hat{p}} = .19$; $\sigma_{\hat{p}} = .018$; approximately normal distribution

7.81 $\mu_{\hat{p}} = .17$; $\sigma_{\hat{p}} = .048$; approximately normal distribution

7.83 95.44%

7.85 **a.** $z = -.61$ **b.** $z = 1.83$ **c.** $z = -1.22$ **d.** $z = 1.22$

7.87 **a.** .6575 **b.** .3409

7.89 **a.** .6126 **b.** .0078

7.91 .2005

7.93 $\mu_{\bar{x}} = 750$ hours; $\sigma_{\bar{x}} = 11$ hours; the normal distribution

7.95 **a.** .9131 **b.** .1698 **c.** .8262 **d.** .0344

7.97 **a.** .9484 **b.** .2800 **c.** .9426 **d.** .0516

7.99 $\mu_{\hat{p}} = .88$; $\sigma_{\hat{p}} = .036$; approximately normal distribution

7.101 **a.** **i.** .0146 **ii.** .0907 **b.** .9912 **c.** .0146

7.103 .4108

7.105 10 approximately

7.107 **a.** .8023 **b.** 754 approximately

7.109 .0035

Self-Review Test

1. b **2.** b **3.** a **4.** a **5.** b
6. b **7.** c **8.** a **9.** a
10. a **11.** c **12.** a

14. **a.** $\mu_{\bar{x}} = 145$ pounds; $\sigma_{\bar{x}} = 3.600$ pounds; approximately normal distribution **b.** $\mu_{\bar{x}} = 145$ pounds; $\sigma_{\bar{x}} = 1.800$ pounds; approximately normal distribution

15. **a.** $\mu_{\bar{x}} = 11$ minutes; $\sigma_{\bar{x}} = .54$ minute; unknown distribution **b.** $\mu_{\bar{x}} = 11$ minutes; $\sigma_{\bar{x}} = .312$ minute; approximately normal distribution

16. **a.** .2366 **b.** .8414 **c.** .2389 **d.** .9037 **e.** .7611

17. **a.** **i.** .1203 **ii.** .1335 **iii.** .7486 **b.** .9736 **c.** .0013

18. **a.** $\mu_{\hat{p}} = .49$; $\sigma_{\hat{p}} = .079$; approximately normal distribution **b.** $\mu_{\hat{p}} = .49$; $\sigma_{\hat{p}} = .050$; approximately normal distribution **c.** $\mu_{\hat{p}} = .49$; $\sigma_{\hat{p}} = .022$; approximately normal distribution

19. **a.** **i.** .1401 **ii.** .7835 **iii.** .7642 **iv.** .2200 **b.** .7924 **c.** .1841 **d.** .1401

Chapter 8

8.11 **a.** 24.5 **b.** 22.71 to 26.29 **c.** ±1.79

8.13 **a.** 70.59 to 79.01 **b.** 69.80 to 79.80 **c.** 68.22 to 81.38 **d.** yes

8.15 **a.** 77.84 to 85.96 **b.** 78.27 to 85.53 **c.** 78.65 to 85.15 **d.** yes

8.17 **a.** 38.34 **b.** 37.30 to 39.38 **c.** ±1.04

8.19 **a.** $n = 167$ **b.** $n = 65$

8.21 **a.** $n = 299$ **b.** $n = 126$ **c.** $n = 61$

8.23 $265,146.86 to $274,293.14

8.25 **a.** 63.73 to 76.27 hours

8.27 31.87 to 32.01 ounces

8.29 **a.** $1532.41 to $1617.59

8.31 $n = 167$

8.33 $n = 61$

8.41 **a.** $t = -1.325$ **b.** $t = 2.160$ **c.** $t = 3.281$ **d.** $t = -2.715$

8.43 **a.** $\alpha \approx .10$, left tail **b.** $\alpha = .005$, right tail **c.** $\alpha = .10$, right tail **d.** $\alpha \approx .01$ left tail

8.45 **a.** $t = 2.080$ **b.** $t = 1.671$ **c.** $t = 2.807$

8.47 **a.** 44.10 **b.** 38.62 to 49.58 **c.** ±5.48

8.49 **a.** 24.06 to 26.94 **b.** 23.58 to 27.42 **c.** 23.73 to 27.27

8.51 **a.** 91.03 to 93.87 **b.** 90.06 to 93.44 **c.** 88.07 to 91.19 **d.** confidence intervals of parts b and c cover μ, that of part a does not

8.53 40.04 to 42.36 bushels

8.55 31.89 to 32.07 ounces

8.57 18.64 to 25.36 minutes

8.59 **a.** 21.56 to 24.44 hours

8.61 4.88 to 11.12 hours

8.63 $1194.22 to $1305.78

8.65 **a.** $21,213 **b.** $20,855.33 to 21,570.67; margin of error: ±$357.67

8.71 **a.** no, sample size is not large **b.** no, sample size is not large **c.** yes, sample size is large **d.** yes, sample size is large

8.73 **a.** .294 to .346 **b.** .334 to .386 **c.** .275 to .325 **d.** confidence intervals of parts a and b cover p, but that of part c does not
8.75 **a.** .189 to .351 **b.** .202 to .338 **c.** .218 to .322 **d.** yes
8.77 **a.** .284 to .336 **b.** .269 to .351 **c.** .209 to .411 **d.** yes
8.79 **a.** $n = 668$ **b.** $n = 671$
8.81 **a.** $n = 1610$ **b.** $n = 413$ **c.** $n = 1569$
8.83 **a.** .29 to .45
8.85 **a.** 40% **b.** 33.1% to 46.9%; margin of error = ±6.9%
8.87 **a.** 20.3% to 55.7% **8.89** **a.** .169 to .191
8.91 **a.** .333 **b.** 8.5% to 58.1%
8.93 $n = 1084$
8.95 $n = 1849$
8.99 **a.** \$2640 **b.** \$2514.57 to \$2765.43
8.101 3.969 to 4.011 inches; the machine needs to be adjusted
8.103 12.82 to 17.56 hours
8.105 21.76 to 26.24 minutes
8.107 4.4 to 4.6 hours
8.109 144.33 to 158.47 calories
8.111 **a.** 44% **b.** 40.3% to 47.7%
8.113 6.1% to 56.5%
8.115 $n = 221$ **8.117** $n = 359$
8.121 $n = 65$
8.123 **a.** $n = 20$ days **b.** 90% **c.** ±75 cars

Self-Review Test

1. **a.** population parameter; sample statistic **b.** sample statistic; population parameter **c.** sample statistic; population parameter
2. b **3.** a **4.** a **5.** c **6.** b
7. **a.** \$159,000 **b.** \$147,390 to \$170,610; margin of error = ± \$11,610
8. \$379,539.30 to \$441,310.70 **9.** **a.** .508 **b.** .483 to .533; margin of error = ± .025
10. $n = 45$ **11.** $n = 273$ **12.** $n = 229$

Chapter 9

9.5 **a.** a left-tailed test **b.** a right-tailed test **c.** a two-tailed test
9.7 **a.** Type II error **b.** Type I error
9.9 **a.** H_0: $\mu = 20$ hours; H_1: $\mu \neq 20$ hours; a two-tailed test **b.** H_0: $\mu = 10$ hours; H_1: $\mu > 10$ hours; a right-tailed test **c.** H_0: $\mu = 3$ years; H_1: $\mu \neq 3$ years; a two-tailed test **d.** H_0: $\mu = \$1000$; H_1: $\mu < \$1000$; a left-tailed test **e.** H_0: $\mu = 12$ minutes; H_1: $\mu > 12$ minutes; a right-tailed test
9.17 **a.** p-value = .0188 **b.** p-value = .0116 **c.** p-value = .0087
9.19 **a.** p-value = .0166 **b.** no, do not reject H_0 **c.** yes, reject H_0
9.21 **a.** rejection region is to the left of −2.58 and to the right of 2.58; nonrejection region is between −2.58 and 2.58 **b.** rejection region is to the left of −2.58; nonrejection region is to the right of −2.58 **c.** rejection region is to the right of 1.96; nonrejection region is to the left of 1.96
9.23 Statistically not significant
9.25 **a.** .10 **b.** .02 **c.** .005
9.27 **a.** observed value of z is −6.61; critical value of z is −2.33 **b.** observed value of z is −6.61; critical values of z are −2.58 and 2.58
9.29 **a.** reject H_0 if $z < -2.33$ **b.** reject H_0 if $z < -2.58$ or $z > 2.58$ **c.** reject H_0 if $z > 2.33$
9.31 **a.** critical value: $z = -1.96$; test statistic: $z = -2.67$; reject H_0 **b.** critical value: $z = -1.96$; test statistic: $z = -1.00$; do not reject H_0
9.33 **a.** critical values: $z = -1.65$ and 1.65; test statistic: $z = -1.34$; do not reject H_0 **b.** critical value: $z = -2.33$; test statistic: $z = -6.44$; reject H_0 **c.** critical value: $z = 1.65$; test statistic: $z = 8.70$; reject H_0
9.35 **a.** $H_0 : \mu = 45$; $H_1 : \mu < 45$ months; p value = .0170; if $\alpha = .025$ reject H_0 **b.** test statisitc: $z = -2.12$; Critical value: $z = -1.96$; reject H_0
9.37 **a.** $H_0 : \mu = 2320$ square feet; $H_1 : \mu > 2320$ square feet; p value = .0020; if $\alpha = .02$ reject H_0 **b.** Critical value: $z = 2.05$; observed value: $z = 2.88$; Reject H_0
9.39 **a.** H_0: $\mu = 10$ minutes; H_1: $\mu \neq 10$ minutes; test statisitc: $z = -2.11$; p value = .0348 If $\alpha = .02$, do not reject H_0. If $\alpha = .05$ Reject H_0. **b.** Observed value $z = -2.11$; If $\alpha = .02$, critical values: $z = -2.33$ and 2.33; do not reject H_0. If $\alpha = .05$, critical values: $z = -1.96$ and 1.96 reject H_0.
9.41 **a.** test statistic: $z = -2.33$; p value = .0198; If $\alpha = .01$; do not reject H_0: If $\alpha = .05$, reject H_0. **b.** Observed value $z = -2.33$; If $\alpha = .01$, critical values: $z = -2.58$ and 2.58, do not reject H_0: If $\alpha = .05$, critical values: $z = -1.96$ and 1.96; reject H_0.
9.43 **a.** H_0: $\mu \geq \$35{,}000$; H_1: $\mu < \$35{,}000$; critical value: $z = -2.33$; test statistic: $z = -3.63$; reject H_0 **b.** do not reject H_0.
9.45 **a.** H_0: $\mu \geq 8$ hours; H_1: $\mu < 8$ hours; critical value: $z = -2.33$; $\alpha = .01$; test statistic: $z = -.68$; p value = .2483 do not reject H_0. **b.** critical value: $z = -1.96$; test statistic: $z = -.68$; do not reject H_0.
9.49 **a.** reject H_0 if $t < -2.977$ or $t > 2.977$ **b.** reject H_0 if $t < -2.797$ **c.** reject H_0 if $t > 2.080$
9.51 **a.** critical value: $t = -1.753$; observed value: $t = -1.800$; $.025 < p\text{-value} < .05$ **b.** critical values: $t = -2.131$ and 2.131; observed value: $t = -1.800$; $.05 < p\text{-value} < .10$
9.53 **a.** reject H_0 if $t < -1.675$ **b.** reject H_0 if $t < -2.008$ or $t > 2.008$ **c.** reject H_0 if $t > 1.675$
9.55 **a.** critical value: $t = 1.998$; test statistic: $t = 4.800$; reject H_0 **b.** critical value: $t = 1.998$; test statistic: $t = 1.143$; do not reject H_0
9.57 **a.** critical values: $t = -2.160$ and 2.160; test statistic: $t = -1.247$; do not reject H_0 **b.** critical value: $t = -2.807$; test statistic: $t = -6.351$; reject H_0 **c.** critical value: $t = 3.646$; test statistic: $t = 2.121$; do not reject H_0
9.59 H_0: $\mu \leqslant \$220{,}680$; H_1: $\mu > \$220{,}680$; critical value: $t = 2.602$; test statistic: $t = 4.891$; reject H_0; p value < .001; reject H_0

9.61 H_0: $\mu = \$850$; H_1: $\mu < \$850$; critical value: $t = -2.397$; test statistic: $t = -2.257$; do not reject H_0; if $\alpha = .025$, critical value $= -2.005$; reject H_0

9.63 H_0: $\mu = \$17{,}989$; H_1: $\mu \neq \$17{,}989$; test statistic: $t = -2.866$; $.002 < p\text{-value} < .010$; reject H_0; for $\alpha = .02$, critical values: $t = -2.378$ and $t = 2.378$; test statistic: $t = -2.866$; reject H_0

9.65 **a.** H_0: $\mu \geqslant \$150$; H_1: $\mu < \$150$; test statistic: $t = -1.964$; $.025 < p\text{-value} < .050$; do not reject H_0; for $\alpha = .01$, critical value: $t = -2.492$; test statistic: $t = -1.964$; do not reject H_0 **b.** $\alpha = .01$

9.67 **a.** H_0: $\mu = 58$ years; H_1: $\mu \neq 58$ years; if $\alpha = 0$, do not reject H_0 **b.** test statistic: $t = -4.183$; $p\text{-value} < .002$; for $\alpha = .01$, reject H_0; critical values: $t = -2.649$ and 2.649; test statistic: $t = -4.183$; reject H_0

9.69 H_0: $\mu = \$65$; H_1: $\mu > \$65$; critical value: $t = 2.718$; test statistic: $t = .889$; do not reject H_0

9.71 H_0: $\mu = 231$ minutes; H_1: $\mu \neq 231$ minutes; test statistic: $t = 9.771$; $p\text{-value} < .002$; for $\alpha = .02$, reject H_0; critical values: $t = -2.326$ and 2.326; reject H_0.

9.75 **a.** not large enough **b.** large enough **c.** not large enough **d.** large enough

9.77 **a.** reject H_0 if $z < -1.65$ or $z > 1.65$ **b.** reject H_0 if $z < -2.33$ **c.** reject H_0 if $z > 1.65$

9.79 **a.** critical value: $z = 1.65$; observed value: $z = 3.90$ **b.** critical values: $z = -1.96$ and 1.96; observed value: $z = 3.90$

9.81 **a.** reject H_0 if $z < -1.65$ **b.** reject H_0 if $z < -1.96$ or $z > 1.96$ **c.** reject H_0 if $z > 1.65$

9.83 **a.** critical values: $z = -2.58$ and 2.58; test statistic: $z = -1.07$; do not reject H_0 **b.** critical values: $z = -2.58$ and 2.58; test statistic: $z = 3.21$; reject H_0

9.85 **a.** critical values: $z = -1.65$ and 1.65; test statistic: $z = .80$; do not reject H_0 **b.** critical value: $z = -1.65$; test statistic: $z = -4.71$; reject H_0 **c.** critical value: $z = 2.33$; test statistic: $z = .93$; do not reject H_0

9.87 H_0: $p = .22$; H_1: $p \neq .22$; test statistic: $z = -1.62$; p-value $= .1052$; for $\alpha = .05$, do not reject H_0; critical values for $\alpha = .05$: $z = -1.96$ and 1.96; test statistic: $z = -1.62$; do not reject H_0

9.89 H_0: $p = .41$; H_1: $p < .41$; critical value: $z = -1.96$; test statistic: $z = -3.05$; reject H_0; p-value $= .0011$; for $\alpha = .025$, reject H_0

9.91 **a.** H_0: $p = .60$; H_1: $p > .60$; critical value: $z = 2.05$; test statistic: $z = .71$; do not reject H_0 **b.** P(Type I error) $= .02$ **c.** p value $= .2389$; do not reject H_0

9.93 **a.** H_0: $p \geq .35$; H_1: $p < .35$; critical value: $z = -1.96$; test statistic: $z = -2.94$; reject H_0 **b.** do not reject H_0 **c.** $\alpha = .025$; p value $= .0016$; reject H_0

9.95 **a.** critical value: $z = 1.96$; test statistic: $z = 2.27$; reject H_0; adjust machine **b.** critical value: $z = 2.33$; test statistic: $z = 2.27$; do not reject H_0; do not adjust the machine

9.99 **a.** critical value: $z = 1.96$; test statistic: $z = 2.10$; reject H_0 **b.** P(Type I error) $= .025$ **c.** p-value $= .0179$; do not reject H_0 if $\alpha = .01$; reject H_0 if $\alpha = .05$

9.101 **a.** critical values: $z = -2.33$ and 2.33; test statistic: $z = 2.55$; reject H_0 **b.** P(Type I error) $= .02$ **c.** p-value $= .0108$; reject H_0 if $\alpha = .025$; do not reject H_0 if $\alpha = .005$

9.103 **a.** H_0: $\mu = 377$ minutes; H_1: $\mu \neq 377$ minutes; test statistic: $z = 5.29$; p-value $= .0000$; if $\alpha = .05$, reject H_0 **b.** critical values =: z -1.96 and 1.96; test statistic: $z = 5.29$; reject H_0

9.105 **a.** H_0: $\mu = 50$; H_1: $\mu < 50$; critical value of $z = -1.96$; test statistic: $z = -3.00$; reject H_0 **b.** P(Type I error) $= .025$ **c.** do not reject H_0 **d.** p-value $= .0013$; for $\alpha = .025$, reject H_0

9.107 **a.** H_0: $\mu \leqslant 2400$ square feet; H_1: $\mu > 2400$ square feet; critical value: $t = 1.677$; test statistic: $t = 2.097$; reject H_0 **b.** for $\alpha = .01$ critical value: $t = 2.405$; test statistic: $t = 2.097$; do not reject H_0

9.109 H_0: $\mu \leqslant 15$ minutes; H_1: $\mu > 15$ minutes; critical value: $t = 2.438$; test statistic: $t = 1.875$; do not reject H_0

9.111 H_0: $\mu = 25$ minutes; H_1: $\mu \neq 25$ minutes; critical values: $t = -2.947$ and 2.947; test statistic: $t = 2.083$; do not reject H_0

9.113 **a.** H_0: $\mu \leqslant 2$ hours; H_1: $\mu > 2$ hours; critical value: $t = 2.718$; test statistic: $t = 1.679$; do not reject H_0

9.115 **a.** H_0: $p = .156$; H_1: $p > .156$; critical value: $z = 2.05$; test statistic: $z = 2.29$; reject H_0 **b.** P(Type I error) $= .02$ **c.** $\alpha = .02$; p value $= .0110$; reject H_0

9.117 H_0: $p = .40$; H_1: $p \neq .40$; critical values: $z = -2.58$ and 2.58; test statistic: $z = -1.62$; do not reject H_0; p-value $= .1052$; do not reject H_0

9.119 **a.** H_0: $p = .80$; H_1: $p < .80$; critical value: $z = -2.33$; test statistic: $z = -.79$; do not reject H_0 **b.** do not reject H_0

9.121 **a.** .0238 **b.** $\alpha = .0238$

9.123 $\alpha = .3446$

9.125 H_0: $\mu = 750$ hours; H_1: $\mu < 750$ hours; reject H_0 if $\bar{x} < 735$: $\alpha = .0082$; reject H_0 if $\bar{x} < 700$: $\alpha = .0000$

9.129 **a.** 29 or more, or 11 or less **b.** 226 or more, or 174 or less **c.** 2081 or more, or 1919 or less

Self-Review Test

1. a **2.** b **3.** a **4.** b **5.** a **6.** a **7.** a **8.** b **9.** c **10.** a **11.** c **12.** b **13.** c **14.** a **15.** b

16. **a.** H_0: $\mu = 52$ minutes; H_1: $\mu \neq 52$ minutes; critical values: $z = -2.58$ and 2.58; test statistic: $z = 18.71$; reject H_0 **b.** H_0: $\mu = 52$ minutes; H_1: $\mu > 52$ minutes; critical value: $z = 1.96$; test statistic: $z = 18.71$; reject H_0 **c.** in part a, $\alpha = .01$; in part b, $\alpha = .025$ **d.** p-value $= .0000$, reject H_0 **e.** p-value $= .0000$, reject H_0

17. **a.** H_0: $\mu = 185$; H_1: $\mu < 185$; critical value: $t = -2.438$ test statistic: $t = -3.000$; reject H_0 **b.** P(Type I error) $= .01$ **c.** do not reject H_0 **d.** $.001 < p\text{-value} < .005$; for $\alpha = .01$, reject H_0

18. **a.** H_0: $\mu \geq 31$ months; H_1: $\mu < 31$ months; critical value: $t = -2.131$; test statistic: $t = -3.333$; reject H_0 **b.** P(Type I error) $= .025$ **c.** critical value: $t = -3.733$; do not reject H_0

19. **a.** H_0: $p = .5$; H_1: $p < .5$; critical value: $z = -1.65$; test statistic: $z = -3.16$; reject H_0 **b.** P(Type I error) $= .05$ **c.** do not reject H_0 **d.** p-value $= .0008$; reject H_0 if $\alpha = .05$; reject H_0 if $\alpha = .01$

Chapter 10

10.3 **a.** .76; **b.** −.41 to 1.93; margin of error = ±1.17

10.5 H_0: $\mu_1 - \mu_2 = 0$; H_1: $\mu_1 - \mu_2 \neq 0$; critical values: $z = -1.96$ and 1.96; test statistic: $z = 1.67$; do not reject H_0

10.7 H_0: $\mu_1 - \mu_2 = 0$; H_1: $\mu_1 - \mu_2 < 0$; critical value: $z = -1.65$; test statistic: $z = -13.96$; reject H_0

10.9 **a.** 9 hours **b.** 1.65 to 16.35 hours; **c.** H_0 : $\mu_1 - \mu_2 = 0$; H_1: $\mu_1 - \mu_2 \neq 0$; critical values: $z = -2.33$ and 2.33; test statistic: $z = 2.66$; reject H_0; p-value = .0078; for $\alpha = .02$, reject H_0

10.11 **a.** −3.64 to −2.16 days **b.** H_0: $\mu_1 - \mu_2 = 0$; H_1: $\mu_1 - \mu_2 < 0$; critical value: $z = -1.96$; test statistic: $z = -9.15$; reject H_0 **c.** P(Type I error) = .025

10.13 **a.** −$1024.54 to −$75.46 **b.** H_0: $\mu_1 - \mu_2 = 0$; H_1: $\mu_1 - \mu_2 \neq 0$; critical values: $z = -2.58$ and 2.58; test statistic: $z = -2.99$; reject H_0 **c.** do not reject H_0

10.15 **a.** −6.87 to .87 calories **b.** H_0: $\mu_1 - \mu_2 = 0$; H_1: $\mu_1 - \mu_2 < 0$; critical value: $z = -2.58$; test statistic: $z = -1.81$; do not reject H_0 **c.** p value = .0351; do not reject H_0 for $\alpha = .005$; do not reject H_0 for $\alpha = .025$

10.17 **a.** −2.10 **b.** −4.20 to 0

10.19 H_0: $\mu_1 - \mu_2 = 0$; H_1: $\mu_1 - \mu_2 \neq 0$; critical values: $t = -2.017$ and 2.017; test statistic $t = -2.013$; do not reject H_0

10.21 H_0: $\mu_1 - \mu_2 = 0$; H_1: $\mu_1 - \mu_2 < 0$; critical value: $t = -2.416$; test statistic: $t = -2.013$; do not reject H_0

10.23 **a.** 4.12 **b.** −.13 to 8.37 **c.** H_0: $\mu_1 - \mu_2 = 0$; H_1: $\mu_1 - \mu_2 > 0$; critical value: $t = 2.500$; test statistic: $t = 2.426$; do not reject H_0

10.25 **a.** −$25.75 to − $6.25; **b.** H_0: $\mu_1 - \mu_2 = 0$; H_1: $\mu_1 - \mu_2 < 0$; critical value: $t = -1.993$; test statistic: $t = -4.342$; reject H_0

10.27 **a.** 2.29 to 5.71 mph **b.** H_0: $\mu_1 - \mu_2 = 0$; H_1: $\mu_1 - \mu_2 > 0$; critical value: $t = 2.416$; test statistic: $t = 5.658$; reject H_0

10.29 **a.** −12.95 to 2.95 minutes **b.** H_0: $\mu_1 - \mu_2 = 0$; H_1: $\mu_1 - \mu_2 < 0$; critical value: $t = -2.412$; test statistic: $t = -1.691$; do not reject H_0

10.31 **a.** −.61 to −.39 **b.** H_0: $\mu_1 - \mu_2 = 0$; H_1: $\mu_1 - \mu_2 \neq 0$; critical values: $t = -2.576$ and 2.576; test statistic: $t = -10.130$; reject H_0

10.33 −5.21 to 1.01

10.35 H_0: $\mu_1 - \mu_2 = 0$; H_1: $\mu_1 - \mu_2 \neq 0$; critical values: $t = -2.056$ and 2.056; test statistic: $t = -1.387$; do not reject H_0

10.37 H_0: $\mu_1 - \mu_2 = 0$; H_1: $\mu_1 - \mu_2 < 0$; critical value: $t = -2.479$; test statistic: $t = -1.387$; do not reject H_0

10.39 **a.** −$25.90 to −$6.10 **b.** H_0: $\mu_1 - \mu_2 = 0$; H_1: $\mu_1 - \mu_2 < 0$; critical value: $t = -1.997$; test statistic: $t = -4.286$; reject H_0

10.41 **a.** 2.23 to 5.77 mph **b.** H_0: $\mu_1 - \mu_2 = 0$; H_1: $\mu_1 - \mu_2 > 0$; critical value: $t = 2.445$; test statistic: $t = 5.513$; reject H_0

10.43 **a.** −12.86 to 2.86 minutes **b.** H_0: $\mu_1 - \mu_2 = 0$; H_1: $\mu_1 - \mu_2 < 0$; critical value: $t = -2.414$; test statistic: $t = -1.713$; do not reject H_0

10.45 **a.** −.61 to −.39 **b.** H_0: $\mu_1 - \mu_2 = 0$; H_1: $\mu_1 - \mu_2 \neq 0$; critical values: $t = -2.576$ and 2.576; test statistic: $t = -10.162$; reject H_0

10.49 **a.** 11.85 to 23.15 **b.** 50.08 to 61.72; **c.** 25.66 to 32.94

10.51 **a.** critical values: $t = -2.060$ and 2.060; test statistic: $t = 12.551$; reject H_0 **b.** critical value: $t = 2.624$; test statistic: $t = 7.252$; reject H_0 **c.** critical value: $t = -1.328$; test statistic: $t = -14.389$; reject H_0

10.53 **a.** −2.98 to 9.84 minutes **b.** H_0: $\mu_d = 0$; H_1: $\mu_d > 0$; critical value: $t = 2.447$; test statistic: $t = 1.983$; do not reject H_0

10.55 **a.** 2.28 to 17.38 pounds **b.** H_0: $\mu_d = 0$; H_1: $\mu_d > 0$; critical value: $t = 3.365$; test statistic: $t = 3.347$; do not reject H_0

10.57 **a.** −5.43 to 2.17 minutes **b.** H_0: $\mu_d = 0$; H_1: $\mu_d \neq 0$; critical values: $t = -2.365$ and 2.365; test statistic: $t = -1.287$; do not reject H_0

10.61 −.062 to .142

10.63 H_0: $p_1 - p_2 = 0$; H_1: $p_1 - p_2 \neq 0$; critical values: $z = -1.96$ and 1.96; test statistic: $z = .76$; do not reject H_0

10.65 H_0: $p_1 - p_2 = 0$; H_1: $p_1 - p_2 > 0$; critical value: $z = 2.05$; test statistic: $z = .76$; do not reject H_0

10.67 **a.** −.04 **b.** −.086 to .006 **c.** rejection region to the left of $z = -2.33$; non-rejection region to the right of $z = -2.33$ **d.** test statistic: $z = -2.02$ **e.** do not reject H_0

10.69 **a.** −.082 to .012 **b.** H_0: $p_1 - p_2 = 0$; H_1: $p_1 - p_2 < 0$; critical value: $z = -2.33$; test statistic: $z = -1.73$; do not reject H_0 **c.** p-value = .0418; for $\alpha = .01$, do not reject H_0

10.71 **a.** .024 **b.** −.020 to .068 **c.** H_0: $p_1 - p_2 = 0$; H_1: $p_1 - p_2 \neq 0$; critical values: $z = -1.96$ and 1.96; test statistic: $z = 1.09$; do not reject H_0; p-value = .2758; for $\alpha = .05$, do not reject H_0

10.73 **a.** .10 **b.** .018 to .182 **c.** H_0: $p_1 - p_2 = 0$; H_1: $p_1 - p_2 \neq 0$; critical values: $z = -2.58$ and 2.58; test statistic: $z = 3.04$; reject H_0

10.75 **a.** −.013 to .093 **b.** H_0: $p_1 - p_2 = 0$; H_1: $p_1 - p_2 > 0$; critical value: $z = 2.33$ test statistic: $z = 1.75$; do not reject H_0

10.77 **a.** −$119.16 to −$114.84 **b.** H_0: $\mu_1 - \mu_2 = 0$; H_1: $\mu_1 - \mu_2 < 0$; critical value: $z = -1.96$; test statistic: $z = -106.25$; reject H_0

10.79 **a.** −$2.63 to $18.63 **b.** H_0: $\mu_1 - \mu_2 = 0$; H_1: $\mu_1 - \mu_2 > 0$; critical value: $t = 2.416$; test statistic: $t = 2.027$; do not reject H_0

10.81 **a.** −$119.04 to −$44.96 **b.** H_0: $\mu_1 - \mu_2 = 0$; H_1: $\mu_1 - \mu_2 < 0$; critical value: $t = -2.326$; test statistic: $t = -5.702$; reject H_0

10.83 **a.** −$2.46 to −$18.46 **b.** H_0: $\mu_1 - \mu_2 = 0$; H_1: $\mu_1 - \mu_2 > 0$; critical value: $t = 2.418$; test statistic: $t = 2.063$; do not reject H_0

10.85 **a.** −$118.95 to −$45.05 **b.** H_0: $\mu_1 - \mu_2 = 0$; H_1: $\mu_1 - \mu_2 < 0$; critical value: $t = -2.326$; test statistic: $t = -5.717$; reject H_0

10.87 **a.** −9.54 to −.24 **b.** H_0: $\mu_d = 0$; H_1: $\mu_d < 0$; critical value: $t = -2.896$; test statistic: $t = -2.425$; do not reject H_0

10.89 **a.** .025 to .095 **b.** H_0: $p_1 - p_2 = 0$; H_1: $p_1 - p_2 > 0$; critical value: $z = 2.33$; test statistic: $z = 4.37$; reject H_0 **c.** p-value = .0000; reject H_0

10.91 **a.** .053 to .147 **b.** H_0: $p_1 - p_2 = 0$; H_1: $p_1 - p_2 > 0$; critical value: $z = 1.96$; test statistic: $z = 4.14$; reject H_0

10.93 **a.** .2611
10.95 $n = 9$
10.97 **a.** $n = 545$ **b.** .8708
10.101 **a.** .3564 **b.** .0793 **c.** .0013

Self-Review Test

1. a
3. **a.** 1.62 to 2.78 **b.** H_0: $\mu_1 - \mu_2 = 0$; H_1: $\mu_1 - \mu_2 > 0$; critical value: $z = 1.96$; test statistic: $z = 9.86$; reject H_0
4. **a.** -2.72 to -1.88 hours **b.** H_0: $\mu_1 - \mu_2 = 0$; H_1: $\mu_1 - \mu_2 < 0$; critical value: $t = -2.416$; test statistic: $t = -10.997$; reject H_0
5. **a.** -2.70 to -1.90 hours **b.** H_0: $\mu_1 - \mu_2 = 0$; H_1: $\mu_1 - \mu_2 < 0$; critical value: $t = -2.421$; test statistic: $t = -11.474$; reject H_0
6. **a.** $-\$53.60$ to $\$186.18$ **b.** H_0: $\mu_d = 0$; H_1: $\mu_d \neq 0$; critical values: $t = -2.447$ and 2.447; test statistic: $t = 2.050$; do not reject H_0
7. **a.** $-.052$ to .092 **b.** H_0: $p_1 - p_2 = 0$; H_1: $p_1 - p_2 \neq 0$; critical values: $z = -2.58$ and 2.58; test statistic: $z = .60$; do not reject H_0

Chapter 11

11.3 $\chi^2 = 41.337$ **11.5** $\chi^2 = 41.638$
11.7 **a.** $\chi^2 = 5.009$ **b.** $\chi^2 = 3.565$
11.13 critical value: $\chi^2 = 11.070$; test statistic: $\chi^2 = 5.200$; do not reject H_0
11.15 critical value: $\chi^2 = 11.070$; test statistic: $\chi^2 = 5.663$; do not reject H_0
11.17 critical value: $\chi^2 = 9.488$; test statistic: $\chi^2 = 6.752$; do not reject H_0
11.19 critical value: $\chi^2 = 17.275$; test statistic: $\chi^2 = 25.209$; reject H_0
11.21 critical value: $\chi^2 = 9.348$; test statistic: $\chi^2 = 65.087$; reject H_0
11.27 **a.** H_0: the proportion in each row is the same for all four populations;
H_1: the proportion in each row is not the same for all four populations
c. critical value: $\chi^2 = 14.449$ **d.** test statistic: $\chi^2 = 52.451$ **e.** reject H_0
11.29 critical value: $\chi^2 = 5.024$; test statistic: $\chi^2 = .953$; do not reject H_0
11.31 critical value: $\chi^2 = 3.841$; test statistic: $\chi^2 = 3.205$; do not reject H_0
11.33 critical value: $\chi^2 = 7.815$; test statistic: $\chi^2 = 2.587$; do not reject H_0
11.35 critical value: $\chi^2 = 6.635$; test statistic: $\chi^2 = 8.178$; reject H_0
11.37 critical value: $\chi^2 = 9.210$; test statistic: $\chi^2 = 28.942$; reject H_0
11.39 critical value: $\chi^2 = 7.378$; test statistic: $\chi^2 = 2.404$; do not reject H_0
11.41 **a.** 18.4376 to 84.9686 **b.** 21.3393 to 67.7365
c. 23.0674 to 60.6586
11.43 **a.** H_0: $\sigma^2 = .80$; H_1: $\sigma^2 > .80$
b. reject H_0 if $\chi^2 > 30.578$
c. test statistic: $\chi^2 = 20.625$
d. do not reject H_0
11.45 **a.** H_0: $\sigma^2 = 2.2$; H_1: $\sigma^2 \neq 2.2$ **b.** reject H_0 if $\chi^2 < 7.564$ or $\chi^2 > 30.191$ **c.** test statistic: $\chi^2 = 35.545$ **d.** reject H_0
11.47 **a.** 1.4743 to 6.2251; 1.214 to 2.495 **b.** H_0: $\sigma^2 \leq 2$; H_1: $\sigma^2 > 2$; critical value: $\chi^2 = 40.289$; test statistic: $\chi^2 = 29.700$; do not reject H_0
11.49 **a.** 2739.3051 to 12,623.9126; 52.338 to 112.356
b. H_0: $\sigma^2 = 4200$; H_1: $\sigma^2 \neq 4200$; critical values: $\chi^2 = 12.401$ and 39.364; test statistic: $\chi^2 = 29.714$; do not reject H_0
11.51 critical value: $\chi^2 = 7.815$; test statistic: $\chi^2 = 10.464$; reject H_0
11.53 critical value: $\chi^2 = 12.592$; test statistic: $\chi^2 = 13.585$; reject H_0
11.55 critical value: $\chi^2 = 11.345$; test statistic: $\chi^2 = 15.920$; reject H_0
11.57 critical value: $\chi^2 = 9.348$; test statistic: $\chi^2 = 22.675$; reject H_0
11.59 critical value: $\chi^2 = 4.605$; test statistic: $\chi^2 = 13.593$; reject H_0
11.61 critical value: $\chi^2 = 16.812$; test statistic: $\chi^2 = 10.181$; do not reject H_0
11.63 **a.** 3.4064 to 24.0000; 1.846 to 4.899 **b.** 8.3336 to 33.2628; 2.887 to 5.767
11.65 H_0: $\sigma^2 = 1.1$; H_1: $\sigma^2 > 1.1$; critical value: $\chi^2 = 28.845$; test statistic: $\chi^2 = 24.727$; do not reject H_0
11.67 H_0: $\sigma^2 = 10.4$; H_1: $\sigma^2 \neq 10.4$; critical values: $\chi^2 = 7.564$ and 30.191; test statistic: $\chi^2 = 24.192$; do not reject H_0
11.69 **a.** H_0: $\sigma^2 = 5000$; H_1: $\sigma^2 < 5000$; critical value: $\chi^2 = 8.907$; test statistic: $\chi^2 = 12.065$; do not reject H_0
b. 1666.8509 to 7903.1835; 40.827 to 88.900
11.71 **a.** .1001 to .4613; .316 to .679 **b.** H_0: $\sigma^2 = .13$; H_1: $\sigma^2 \neq .13$; critical values: $\chi^2 = 9.886$ and 45.559; test statistic: $\chi^2 = 35.077$; do not reject H_0
11.73 **a.** $s^2 = 1107.4107$ **b.** 484.0989 to 4586.9082; 22.002 to 67.727 **c.** H_0: $\sigma^2 = 500$; H_1: $\sigma^2 \neq 500$; critical values: $\chi^2 = 1.690$ and 16.013; test statistic: $\chi^2 = 15.504$; do not reject H_0
11.75 critical value: $\chi^2 = 5.991$; test statistic: $\chi^2 = 12.931$; reject H_0
11.77 critical value: $\chi^2 = 9.488$; test statistic: $\chi^2 = 11.822$; reject H_0
11.79 critical value: $\chi^2 = 16.919$; test statistic: $\chi^2 = 215.568$; reject H_0
11.81 **a.** test statistic: $\chi^2 = 2.480$; p-value $> .10$ **b.** no

Self-Review Test

1. b 2. a 3. c 4. a 5. b 6. b
7. c 8. b 9. a
10. critical value: $\chi^2 = 9.488$; test statistic: $\chi^2 = 10.458$; do not reject H_0
11. critical value: $\chi^2 = 11.345$; test statistic: $\chi^2 = 31.188$; reject H_0
12. critical value: $\chi^2 = 9.488$; test statistic: $\chi^2 = 82.450$; reject H_0
13. **a.** .2364 to 1.3326; .486 to 1.154 **b.** H_0: $\sigma^2 = .25$; H_1: $\sigma^2 > .25$; critical value: $\chi^2 = 36.191$; test statistic: $\chi^2 = 36.480$; reject H_0

Chapter 12

12.3 **a.** 3.73 **b.** 3.61 **c.** 5.37
12.5 **a.** 2.39 **b.** 2.27 **c.** 3.28
12.7 **a.** 9.96 **b.** 6.57 **12.9** **a.** 4.85 **b.** 3.22
12.13 **a.** $\bar{x}_1 = 15$; $\bar{x}_2 = 11$; $s_1 = 4.50924975$; $s_2 = 4.39696865$ **b.** H_0: $\mu_1 = \mu_2$; H_1: $\mu_1 \neq \mu_2$; critical values: $t = -2.179$ and 2.179; test statistic: $t = 1.680$; do not reject H_0 **c.** critical value: $F = 4.75$; test statistic: $F = 2.82$; do not reject H_0 **d.** conclusions are the same
12.15 **b.** critical value: $F = 3.29$; test statistic: $F = 4.07$; reject H_0
12.17 **a.** H_0: $\mu_1 = \mu_2 = \mu_3$; H_1: all three population means are not equal **b.** numerator: $df = 2$; denominator: $df = 22$ **c.** $SSB = 243.1844$; $SSW = 561.0556$; $SST = 804.2400$ **d.** reject H_0 if $F > 5.72$ **e.** $MSB = 121.5922$; $MSW = 25.5025$ **f.** critical value: $F = 5.72$ **g.** test statistic: $F = 4.77$ **i.** do not reject H_0
12.19 critical value: $F = 3.55$; test statistic: $F = 2.09$; do not reject H_0
12.21 critical value: $F = 3.72$; test statistic: $F = 5.44$; reject H_0
12.23 **a.** critical value: $F = 3.34$; test statistic: $F = 6.08$; reject H_0 **b.** .05
12.25 **a.** critical value: $F = 6.93$; test statistic: $F = 1.24$; do not reject H_0
12.27 **a.** critical value: $F = 3.89$; test statistic: $F = 4.89$; reject H_0 **b.** do not reject H_0
12.29 critical value: F is 5.29; test statistic: $F = .57$; do not reject H_0
12.33 **a.** 5 groups with 10 members each. **b.** 36 members each.

Self-Review Test

1. a **2.** b **3.** c **4.** a **5.** a
6. a **7.** b **8.** a
10. **a.** critical value: $F = 3.10$; test statistic: $F = 4.46$; reject H_0 **b.** Type I error

Chapter 13

13.15 **a.** y-intercept = 100; slope = 5; positive relationship **b.** y-intercept = 400; slope = −4; negative relationship
13.17 $\mu_{y|x} = -5.5815 + .2886x$
13.19 $\hat{y} = -83.7140 + 10.5714x$
13.21 **a.** \$60.00 **b.** the same amount **c.** exact relationship
13.23 **a.** \$27.10 million **b.** different amounts **c.** nonexact relationship
13.25 **b.** $\hat{y} = 150.4136 - 16.6264x$ **e.** \$3402.88 **f.** −\$14,886.16
13.27 **b.** $\hat{y} = -208.4001 + 7.4374x$ **e.** \$200,656.90 **f.** error = −\$71,717.10
13.29 **a.** $\mu_{y|x} = 38.0933 + .1751x$ **b.** population regression line because data set includes all 16 National League teams; values of A and B **d.** 48.60%
13.35 $\sigma_\epsilon = 7.0756$; $\rho^2 = .04$
13.37 $s_e = 4.7117$; $r^2 = .99$
13.39 **a.** $SS_{xx} = 1587.8750$; $SS_{yy} = 5591.5000$; $SS_{xy} = 2842.7500$ **b.** $s_e = 9.1481$ **c.** SST = 5591.5000; SSE = 502.1652; SSR = 5089.3348 **d.** $r^2 = .91$
13.41 **a.** $s_e = 3.6590$ **b.** $r^2 = .87$
13.43 **a.** $s_e = 93.4608$ **b.** $r^2 = .95$
13.45 **a.** $\sigma_\epsilon = 6.3086$ **b.** $\rho^2 = .37$
13.47 **a.** 6.01 to 6.63 **b.** H_0: $B = 0$; H_1: $B > 0$; critical value: $t = 2.145$; test statistic: $t = 59.792$; reject H_0 **c.** H_0: $B = 0$; H_1: $B \neq 0$; critical values: $t = -2.977$ and 2.977; test statistic: $t = 59.792$; reject H_0 **d.** H_0: $B = 4.50$; H_1: $B \neq 4.50$; critical values: $t = -2.624$ and 2.624; test statistic: $t = 17.219$; reject H_0
13.49 **a.** 2.35 to 2.65 **b.** H_0: $B = 0$; H_1: $B > 0$; critical value: $t = 1.960$; test statistic: $t = 39.124$; reject H_0; **c.** H_0: $B = 0$; H_1: $B \neq 0$; critical values: $t = -2.576$ and 2.576; test statistic: $t = 39.124$; reject H_0; **d.** H_0: $B \leq 1.75$; H_1: $B > 1.75$; critical value: $t = 2.326$; test statistic: $t = 11.737$; reject H_0
13.51 **a.** −23.5406 to −9.7122 **b.** H_0: $B = 0$; H_1: $B < 0$; critical value: $t = -1.943$; test statistic: $t = -5.884$; reject H_0
13.53 **a.** $\hat{y} = 23.7297 + 1.3054x$ **b.** .88 to 1.73 **c.** H_0: $B = 0$; H_1: $B > 0$; critical value: $t = 2.365$; test statistic: $t = 9.193$; reject H_0
13.55 **a.** 1.57 to 13.31 **b.** H_0: $B = 0$; H_1: $B \neq 0$; critical values: $t = -4.604$ and 4.604; test statistic: $t = 5.838$; reject H_0
13.57 **a.** $\hat{y} = 4.4948 - .1019x$ **b.** −.17 to −.03 **c.** H_0: $B = -.04$; H_1: $B < -.04$; critical value: $t = -2.015$; test statistic: $t = -2.310$; reject H_0
13.63 a **13.67** **a.** positive **b.** positive **c.** positive **d.** negative **e.** zero
13.69 $\rho = .21$
13.71 **a.** $r = -.996$ **b.** H_0: $\rho = 0$; H_1: $\rho < 0$; critical value: $t = -2.764$; test statistic: $t = -35.249$; reject H_0
13.73 **a.** positively **b.** $r = .96$ **c.** H_0: $\rho = 0$; H_1: $\rho > 0$; critical value: $t = 1.895$; test statistic: $t = 9.071$; reject H_0
13.75 **a.** positively **b.** close to 1 **c.** $r = .97$ **d.** H_0: $\rho = 0$; H_1: $\rho \neq 0$; critical values: $t = -2.776$ and 2.776; test statistic: $t = 7.980$; reject H_0
13.77 **a.** $r = .95$ **b.** H_0: $\rho = 0$; H_1: $\rho \neq 0$; critical values: $t = -3.707$ and 3.707; test statistic: $t = 7.452$; reject H_0
13.79 $\rho = .61$
13.81 **a.** $SS_{xx} = 1895.6000$; $SS_{yy} = 4798.4000$; $SS_{xy} = 1231.8000$ **b.** $\hat{y} = 156.3302 + .6498x$ **d.** $r = .41$; $r^2 = .17$ **f.** 195.3182 **g.** $s_e = 22.3550$ **h.** −.53 to 1.83 **i.** H_0: $B = 0$; H_1: $B > 0$; critical value: $t = 1.860$; test statistic: $t = 1.265$; do not reject H_0 **j.** H_0: $\rho = 0$; H_1: $\rho > 0$; critical value: $t = 2.306$; test statistic: $t = 1.271$; do not reject H_0
13.83 **a.** $SS_{xx} = 4260.9000$; $SS_{yy} = 938.9000$; $SS_{xy} = 1895.9000$ **b.** $\hat{y} = -14.4245 + .4450x$ **d.** $r = .95$; $r^2 = .90$ **e.** $s_e = 3.4501$ **f.** .27 to .62 **g.** H_0: $B = 0$; H_1: $B > 0$; critical value: $t = 2.896$; test statistic: $t = 8.412$; reject H_0 **h.** H_0: $\rho = 0$; H_1: $\rho \neq 0$; critical values: $t = -3.355$ and 3.355; test statistic: $t = 8.605$; reject H_0
13.85 **a.** $SS_{xx} = 3.3647$; $SS_{yy} = 285.7143$; $SS_{xy} = 29.1957$ **b.** $\hat{y} = 14.0729 + 8.6771x$ **d.** $r = .94$;

$r^2 = .89$ **e.** $s_e = 2.5448$ **f.** 5.11 to 12.24 **g.** H_0: $B = 0$; H_1: $B \neq 0$; critical values: $t = -4.032$ and 4.032; test statistic: $t = 6.255$; reject H_0 **h.** H_0: $\rho = 0$; H_1: $\rho > 0$; critical value: $t = 3.365$; test statistic: $t = 6.161$; reject H_0

13.87 **a.** 13.8708 to 16.6292; 11.7648 to 18.7352 **b.** 62.3590 to 67.7210; 56.3623 to 73.7177

13.89 35.3155 to 38.2519; 32.2279 to 41.3395

13.91 112.9148 to 124.0182; 102.8425 to 134.0905

13.93 \$1153.97 to \$1657.13; \$571.12 to \$2239.98

13.95 **a.** positive relationship **b.** $\hat{y} = -1.9175 + .9895x$ **d.** $r = .97$; $r^2 = .94$ **e.** $s_e = 1.0941$ **f.** .54 to 1.44 **g.** H_0: $B = 0$; H_1: $B > 0$; critical value: $t = 2.571$; test statistic: $t = 8.811$; reject H_0 **h.** H_0: $\rho = 0$; H_1: $\rho > 0$; critical value: $t = 2.571$; test statistic: $t = 8.922$; reject H_0; same conclusion

13.97 **a.** positive **b.** $\hat{y} = 7.8299 + .5039x$ **d.** $r = .89$; $r^2 = .79$ **e.** 2547 **f.** $s_e = 3.3525$ **g.** .11 to .90 **h.** H_0: $B = 0$; H_1: $B > 0$; critical value: $t = 3.365$; test statistic: $t = 4.278$; reject H_0 **i.** H_0: $\rho = 0$; H_1: $\rho \neq 0$; critical values: $t = -3.365$ and 3.365; test statistic: $t = 4.365$; reject H_0

13.99 **b.** $SS_{xx} = 82.5000$; $SS_{yy} = 21.7690$; $SS_{xy} = 41.9500$ **c.** yes **d.** $\hat{y} = 4.1218 + .5085x$ **f.** $r = .99$ **g.** 11.24 million

13.101 **b.** $SS_{xx} = 60.0000$; $SS_{yy} = 17.3156$; $SS_{xy} = -26.8000$ **c.** yes **d.** $\hat{y} = 56.9646 - .4467x$ **f.** $r = -.83$ **g.** 49.82 seconds

13.103 60.7337 to 97.3719; 40.0140 to 118.0916

13.105 207.3197 to 240.1195; 183.2730 to 264.1662

13.107 **a.** yes **b.** 246.4670 to 275.5330 lines **c.** 200.0567 to 321.9433 lines **e.** 338 lines

13.111 **a.** $\hat{y} = 348{,}156.8 - 3367.20x$

Self-Review Test

1. d **2.** a **3.** b **4.** a **5.** b **6.** b **7.** true **8.** true **9.** a **10.** b

15. **a.** The attendence depends on temperature. **b.** positive **d.** $\hat{y} = -2.2247 + .2715x$ **f.** $r = .65$; $r^2 = .42$ **g.** 1407 people **h.** $s_e = 3.6172$ **i.** $-.30$ to .84 **j.** H_0: $B = 0$; H_1: $B > 0$; critical value: $t = 3.365$; test statistic: $t = 1.904$; do not reject H_0 **k.** 1055 to 1758 **l.** 412 to 2401 **m.** H_0: $\rho = 0$; H_1: $\rho > 0$; critical value: $t = 3.365$; test statistic: $t = 1.913$; do not reject H_0

Appendix A

A.7 simple random sample **A.9** **a.** nonrandom sample **b.** judgment sample **c.** selection error

A.11 **a.** random sample **b.** simple random sample **c.** no

A.13 **a.** nonrandom sample **b.** voluntary response error and selection error

A.15 response error

A.17 **a.** designed experiment **b.** no; would need to know if the women or the doctors who evaluated their health knew which women took aspirin and which were in the control group

A.19 **a.** designed experiment **b.** double-blind study

A.21 designed experiment **A.23** yes

A.25 **b.** observational study **c.** not a double-blind study

A.27 **a.** designed experiment **b.** double-blind study

A.29 **a.** no **b.** no **c.** convenience sample

A.33 **a.** no **b.** nonresponse error and response error **c.** above

Photo Credits

Chapter 1
Page 1: Jaume Gual/Age Fotostock America, Inc. Page 16: PhotoDisc, Inc./Getty Images.

Chapter 2
Page 26: Blend Images, Inc./Getty Images. Page 28: PhotoDisc, Inc./Getty Images. Page 43: Photo Disc, Inc./Getty Images. Page 57: Mark Harmel/Stone/ Getty Images.

Chapter 3
Page 74: Al Bello/Getty Images Sport Services. Page 79: Ariel Skelley/Corbis Stock Market. Page 89: Corbis Digital Stock. Page 102: PhotoDisc, Inc./Getty Images.

Chapter 4
Page 132: Tom Mihalek/AFP/Getty Images. Page 140: PhotoDisc, Inc./Getty Images. Page 146: PhotoDisc, Inc./Getty Images. Page 168: PhotoDisc, Inc./Getty Images.

Chapter 5
Page 188: Richard Drury/The Image Bank/Getty Images. Page 201: Corbis Digital Stock. Page 215: PhotoDisc, Inc./Getty Images.

Chapter 6
Page 247: V.C.L./Taxi/Getty Images. Page 271: PhotoDisc, Inc./Getty Images. Page 273: Courtesy Texas Instruments, Inc.

Chapter 7
Page 296: Rob Crandall/The Image Works. Page 304: PhotoDisc, Inc./Getty Images. Page 313: Corbis Digital Stock. Page 325: Image State.

Chapter 8
Page 337: © David Young-Wolff/Alamy. Page 343: © Corbis Digital Stock. Page 348: PhotoDisc, Inc./Getty Images. Page 355: PhotoDisc, Inc./Getty Images.

Chapter 9
Page 378: Stuart O'Sullivan/The Image Bank/Getty Images. Page 390: Nova Stock/Photo Researchers. Page 405: Cohen/Ostrow/Digital Vision. Page 406: PhotoDisc, Inc./Getty Images.

Chapter 10
Page 433: Bryce Duffy/Stone+/Getty Images. Page 443: PhotoDisc, Inc./Getty Images. Page 446: Corbis Digital Stock. Page 468: PhotoDisc, Inc./Getty Images.

Chapter 11
Page 489: Cosmo Condina/Stone/Getty Images. Page 495: PhotoDisc, Inc./Getty Images. Page 507: PhotoDisc, Inc./Getty Images. Page 518: Corbis Digital Stock.

Chapter 12
Page 532: Steven W. Jones/Taxi/Getty Images. Page 537: PhotoDisc, Inc./Getty Images. Page 541: PhotoDisc, Inc./Getty Images.

Chapter 13
Page 555: Image State. Page 580: PhotoDisc, Inc./Getty Images. Page 589: PhotoDisc, Inc./Getty Images.

Chapter 14
Page 614: PhotoDisc, Inc./Getty Images.

Chapter 15
Page 615: Stephen Marks/The Image Bank/Getty Images.

Appendix A
Page A1: Laura Lane/Taxi/Getty Images. Page A11: PhotoDisc, Inc./Getty Images.

INDEX

actual value of *y*, 560–561
addition rule of probability, 169
  for mutually exclusive events, 171–173
  for probability of union of events, 169–170
alpha (α, significance level), 340
  confidence interval for mean, 342
  confidence level and interval, 340–341
  probability of making a Type 1 error, 381–382
alternative hypothesis, 379–380
  critical value, 380–381
  tails of test, 382–383
analysis of variance (ANOVA), one-way, 535–537, 547
  alternative hypothesis, 535, 538–539
  assumptions of, 536–537
  between-samples sum of squares (SSB), 537–538
  null hypothesis, 535–536, 539–540
  table, 539
  test statistic *F*, 537–358
  total sum of squares (SST), 538–539
  treatments, 537
  variance (or mean square)
    between samples (MSB), 536, 537–539
    within samples (MSW), 536, 537–539
analysis of variance (*continued*)
  within-samples sum of squares (SSW), 537–538
ANOVA. *See* analysis of variance
Anscombe's data, 602
applied statistics, 2
arithmetic mean, 75–78
arrangements, 209–210
average. *See* mean
average versus typical, 18

bar graphs, 30,193
bell-shaped distribution. *See also* normal distribution; *t* distribution
  empirical rule for, 103–104, 261
  grading exams on a curve, 114
Bernoulli trial, 211
beta (β, Type II error), 381–382
bias, 328, A7–8
bimodal distribution, 81
binomial distribution, 211
  mean and standard deviation, 220–221
  normal distribution as approximation to, 280–285
  Poisson distribution as approximation to, 229–230
  probability of failure, 211

binomial distribution (*continued*)
probability of success, 211, 219–220, 317
shape of, 219–220
table, C2–10, 217–219
binomial experiment, 211–213, 493
conditions of, 211–213
binomial formula, 213–216
binomial parameters, 213
binomial random variable, 213, 221
boundaries, class, 34–35, 38
in cumulative frequency distributions, 52
sensitivity and robustness of, 62
box-and-whisker plots, 111–113

categorical data. *See* qualitative data
categorical variable, 12
causality, regression contrasted to, 600
cell, in contingency tables, 146
census and census data, 6, A2, A4
estimation versus, 338
nonsampling errors in, 300, A6
sampling errors and, A5
sources of, 14
central limit theorem
for sample mean, 309
for sample proportion, 321–323, 325–327
and sample size, 368
central tendency measures. *See* measures of central tendency
chance experiment, 189
chance (random) variables, 189–190, 221, 298
Chebyshev's theorem, 101–103, 203
chi-square distribution, 490–492
table, C23, 490–492
chi-square tests
hypothesis tests, 489, 521
about population variance, 516–519
goodness-of-fit, 493–500, 521
test of homogeneity, 508–510
test of independence, 503–508, 521
population variance, 514–516
classes, 34–35
boundaries, 34–35, 38
in cumulative frequency distributions, 52–53
sensitivity and robustness of, 62
for histograms, 39–40
less than method for writing, 42–43
midpoint or mark, 34, 35
open-ended, 40
for polygons, 41
single-valued, 43–44
width/size, 35, 36–37
classical probability rule, 139–140
cluster sampling technique, A9
coefficient of determination, 573–576
coefficient of variation, 93
population, 93
sample, 93
coefficient of *x*, linear regression line, 557–558
column of frequencies, 28–29
combinations, 206–209, 235–237
combined mean, 85
complementary events, 153–154
composite (compound) event, 135–137, 140
compound event, 135–1, 140
conditional probability, 147–149
calculating, 161
confidence coefficient, 340
confidence interval, 340
for difference between two population means, 436
standard deviations known, large samples, 436–437
standard deviations unknown and unequal, 452–453
standard deviations unknown but equal, 442–443, 448–449
for difference between two population proportions, large and independent samples, 468–469
lower limit, 340
and margin of error, 368
confidence interval (*continued*)
for mean of the paired differences, 458
for mean value of *y* for a given *x*, 598
in one-way ANOVA, 553
for population mean
known standard deviation, 341–348
unknown standard deviation, 351–356
using *t* distribution, 354–356, 357
for population variance and standard deviation, 515–516
for proportion for large sample, 344–345, 346
for true value of slope in population regression line, 578–579
upper limit, 340
width of, 345, 347
confidence level, 340, 483
confounding, of variables in experiments, A10
consistent estimators
sample mean, 304
sample proportion, 321
constant, 9, 296
constant term, 558
contingency tables, 146, 502
test of homogeneity, 508–509
test of independence, 503
continuity correction factor, 282–285
continuous probability distribution, 248–251
continuous random variables, 190, 247. *See also* normal distribution
continuous variables, 11–12
control group, A3, All–12
controlled experiment, A10–12
convenience samples, A5
correction factor
continuity, 282–285
finite population, 303, 321
correlation, 583–587
correlation coefficient, 583–585
hypothesis testing about, 585–587
and outliers, 602

counting rule to find total outcomes, 145–146
critical value of *F*, 534
critical value (point), 380–381
cross-section data, 13
cross tabulation tables, 502
cumulative frequency distributions, 51–53
cumulative percentage distributions, 52–53
cumulative probability, 256–257
cumulative relative frequency distributions, 52–53

data, 9–10. *See also* qualitative or categorical data; quantitative data; sample data; ungrouped data, grouped data
  cross-section, 13
  primary, A2
  ranked, 78–79, 106, 108–109
  raw, 27
  secondary, A2
  sources of, 14–15, A1–3
  time-series, 13–14
data sets, 3, 10. *See also* census and census data
  sample and population, 75
  uni-, bi-, and multimodal, 81
  for use with text, Bl–4
degrees of freedom, 352–356
  chi-square distribution, 490
  F distribution, 533–534
  goodness-of-fit test, 494
  linear regression model, 572
  paired samples, 442
  pooled sample standard deviation, 442
  t distribution, 352–353, 449
  test of homogeneity, 509
  test of independence, 503
dependent events, 151–153
dependent samples. *See* paired samples
dependent variables, 556, 602–603
descriptive statistics, 3
designed experiments, Al1–12
determining sample size. *See* sample size
deterministic regression model, 558–559
deviation of *x* value from the mean, 88
difference between two population means, independent samples
  standard deviations known, 434–439
  standard deviations unknown and unequal, 451–455
  standard deviations unknown but equal, 441–447, 448–449
difference between two population means, paired samples, 457–465
difference between two population proportions, large and independent samples, 467–472, 474–475
differences. *See* hypothesis tests for differences
discrete random variables, 188, 189–190
discrete variables, 11
dispersion, measures of, 87–91
distribution. *See also* binomial distribution; hypergeometric distribution; normal distribution; relative frequency distribution; sampling distributions; *t* distribution
  bell-shaped, 103–104, 114
  chi-square, 490–492
  cumulative frequency, 51–53
  cumulative percentage, 52–53
  determining shape of, 293, 306–310, 321–323 *F*, 533–534
  frequency, 28, 34
  percentage, 29, 38, 44, 52–53
  Poisson probability, 227–230
  standard normal, 256–262, 264–265, 269
dotplots, 58–60
double-blind experiment, A3, A12

element
  of data set, 3
  of sample, 8–9
  of sample space, 133
elementary events, 135
empirical rule for standard deviation, 103–104, 261
equally likely outcomes or events, 139–140
equation of a regression model, 559
equation of linear relationships, 557
equation of the normal distribution, 255
errors
  error of prediction, 563, 573–574
  nonresponse, A7
  nonsampling, 299–301, A5–A8
  random error term, 558–559, 560–561, 566
  response, A7–8
  sampling, 299–301, A5–6, A5–A8
    statistical significance and, 397
  selection, A6–7
  standard error of sample mean, 302
  sum of squares (SSE), 561–562
  Type I, 381–382
  Type II, 381–382
  voluntary response, A7–8
estimate, point and interval, 339–341. *See also* point estimate
estimated mean value of *y* for a given *x*, 596–598
estimated regression model, 559
estimated value of *y*, 559
estimation of regression coefficients, 561
estimation, 337–339
estimators, 303–304
  consistent, 304, 321
  unbiased, 303, 320
events. *See also* probability distribution
  complementary, 153–154
  compound, 135–137, 140
  dependent, 151–153
  elementary, 135
  equally likely, 139–140
  impossible, 138
  independent, 151–153, 162–163

events (*continued*)
intersection of, 157–158
independent events, 162–163
joint probability, 158–161
mutually exclusive, 150–151
addition rule for, 171–173
joint probability of, 163–164
mutually nonexclusive, 150
simple, 135, 136–137, 139–140
sure, 138
union of, 167–173
exact relationship, deterministic model, 558–159
Excel
chi-square tests, 530
combinations, binomial distributions, and Poisson distributions, 245
confidence intervals and hypothesis testing for two populations, 486
confidence intervals for population means and proportions, 376
entering and saving data, 23–24
hypothesis testing, 431
normal and inverse normal probabilities, 294
numerical descriptive measures, 130
organizing data, 73
random number generation, 186–187
sampling distribution of means, 336
simple linear regression, 612
expected frequencies
goodness-of-fit test, 493–495, 498–499
test of homogeneity, 510
test of independence, 503, 504
expected value of a discrete random variable, 198
experiments, 133–137, A3. *See also* binomial experiment
binomial, 211–213
control group, A3
controlled, A10–12
designed, All–12
double-blind, A3, A12
experiments (*continued*)
multinomial, 493
placebo effect, A12
random, 189
treatment group, A3
explanatory variables (independent), 556, 601–602
exponential distribution, 287
external data sources, A2
extrapolation, 600
extreme values. *See* outliers

factorials, 205–206
failure. *See* probability of failure
F distribution, 533–534
degrees of freedom, 533–534
table, C24–27, 533–534
fence
lower inner, 111–112
lower outer, 113
upper inner, 111–112
upper outer, 113
final outcomes. *See* outcomes
finite population correction factor, 303, 321
first quartile, 106
and box-and-whisker plot, 111, 113
Fisher, Sir Ronald, 533
frequency, 34
of a category, 28, 29
of a class, 30, 53, 97, 252
expected
goodness-of-fit test, 493–495, 498–499
test of homogeneity, 510
test of independence, 503, 504
histogram, 39–40
joint, 502
observed, 493–494, 503, 505
polygon, 40–41 -
relative, 29, 140–142
frequency density, relative, 249, 252
frequency distribution curve, 41
relationship of mean, median, and mode, 82
symmetric and skewed, 46
frequency distributions, 27–28, 34
cumulative, 51–53
frequency distributions (*continued*)
cumulative relative, 52–53
less than method, 42–43
for qualitative data, 28–29
for quantitative data, 34–38
relative, 29
for continuous random variables, 248–249
single-valued classes, 43–44
frequency distribution table
column of frequencies, 28–29
constructing, 28–29, 35–38
for qualitative data, 28–29
for quantitative data, 34, 35–38
frequency histograms, 39–40
frequency polygons, 40–41

Galton, Sir Francis, 556
gambling, 176–77, 200–201, 209, 235–237
testing for fairness of equipment, 529
geometric mean, 86
goodness-of-fit test, 493–495, 521
degrees of freedom, 494
equal proportions for all categories, 495–497
expected frequencies, 493–494
observed frequencies, 493–494
test statistic (χ^2), 494
Gosset, W. S., 352
grading on a curve, 114
grouped data. *See* quantitative data

histograms, 39–40
decision-making with, 127
frequency, 40
percentage, 40
relationship of mean, median, and mode in, 82
relative frequency, 39–40, 248–249
shapes of, 45–46
skewed to the right or left, 45–46
symmetric, 45–46
truncation, 40, 47
uniform or rectangular, 45
homogeneity, chi-square test of, 508–510
hypergeometric distribution, 223–226

hypotheses. *See also* tests of hypotheses
null and alternative, 379–380
rejection and nonrejection regions, 380–382
impossible events, 138
independence, chi-square test of. *See* test of independence
independence of occurrences, 227
independent events, 151–153
multiplication rule for, 162–163
independent samples, 434
for ANOVA test, 536
independent variables in a regression model, 556, 601–602
inferential (inductive) statistics, 3–4
inner fence, 111–112
internal data sources, Al
interpretation *of a* and *b,* 564, 591
interquartile range, 106–108
intersection of events, 157–158
independent events, 162–163
joint probability, 158–161
interval estimate, point estimate and, 339–341
investments, 104, 244

joint frequencies, 502
joint probability, 158
of independent events, 151
multiplication rule, 158–161
mutually exclusive events, 163–164
tree diagram of, 162–163
judgment samples, A5

Law of Large Numbers, 142
least squares estimates of *A* and *B*, 559
calculating, 591
interpretation of, 564–565
SSE and least squares values, 561, 562–563
least squares method and scatter diagram, 560
least squares regression line, 560–562, 591
error sum of squares (SSE), 561
estimating, 562–563, 578–579
left-tailed hypothesis tests, 384, 385
line graph, 193
linear correlation, 583–587
linear correlation coefficient, 583–585
hypothesis testing about, 585–587
and outliers, 602
linear regression. *See* regression
linear regression line. *See* regression line
linear regression model. *See* regression model
linear relationship, 557, 564
lottery, 200–201, 209
lower boundary of a class, 34–35
lower inner fence, 111–112
lower limit of class, 34–35, 36
lower limit of a confidence interval, 340
lower outer fence, 113

mail, surveys by, A3
marginal probability, 146–147
margin of error, 5, 340
matched samples. *See* paired samples
mean, 75–78, 94–96. *See also* population mean; sample mean; sampling distributions
arithmetic, 75–78
binomial probability distribution, 220–221
combined, 85
confidence interval for
known standard deviation, 341–348
unknown standard deviation, 351–356
using *t* distribution, 354–356, 357
deviation of *x* value from the, 88
of discrete random variables, 198–199, 203
for grouped data, 94–96
median, mode, and, 82
normal distribution and, 254–255
of paired difference for two samples, 458–460
mean (*continued*)
of Poisson distribution, 232–233
of sample mean for a nonnormal population, 310
of sample mean for a normal population, 307–308, 312–315
of sampling distribution
of sample mean, 302–304, 434–436
of sample proportion, 320, 321–323
of *t* distribution, 353
for ungrouped data, 75–78
x values in relation to, 268–269, 271–273
z values in relation to, 266–267
mean square between samples (MSB), 536, 537–539
mean square within samples (MSW), 536, 537–539
measurement, 9
measures of central tendency. *See also* mean
median, 78–80, 82
mode, 80–82
trimmed mean, 86
for ungrouped data, 75
mean, 75, 76–77
median, 78–80
mode, 80–82
outliers, effect on mean of, 77–78
relationships among mean, median, and mode, 82
sample mean, 76–77
weighted mean, 86
measures of dispersion. See *also* standard deviation; variance
coefficient of variation, 93
range, 87–88
for ungrouped data, 87–91
measures of position
percentile rank, 109–110
percentiles, 108–109
quartiles, 106–108
measures of spread, 87
median
and box-and-whisker plot, 111
mean, mode, and, 82
for ungrouped data, 78–80

member of a sample, 8–9
memoryless distribution, 287
method of least squares. *See* least squares
mild outliers, 113
MINITAB
  analysis of variance, 554
  chi-square tests, 529–530
  combinations, binomial distributions, and Poisson distributions, 245
  confidence intervals and hypothesis testing for two populations, 484–486
  confidence intervals for population means and proportions, 375–376
  entering and saving data, 23–24
  hypothesis testing, 430
  normal and inverse normal probabilities, 293–294
  numerical descriptive measures, 128–130
  organizing data, 72
  random number generation, 186
  sampling distribution of means, 335
  simple linear regression, 612
mode for ungrouped data, 80–82
MSB (mean square between samples), 536, 537–539
MSW (mean square within samples), 536, 537–539
multimodal distribution, 81
multinomial experiments, 493
multiple regression model, 556
multiplication rule, 158–163
  for independent events, 162–163
  for joint probability, 158–161
multistage sampling procedure, 375
mutual funds, 104, 244
mutually exclusive events, 150–151
  addition rule for, 171–173
  joint probability of, 163–164
mutually nonexclusive events, 150

negative linear correlation, 583
negative linear relationship, 564
n factorial, 207
no linear correlation, 583
nonlinear regression model, 556–557
nonlinear relationship, 557, 567
nonparametric method, 351
nonrandom sample, A6–7
nonrejection region, 380–382
nonresponse errors, A7
nonsampling errors, 299–301, A5–8
normal distribution, 251–255. *See also* standard normal distribution
  applications of, 270–273
  as approximation to binomial, 230, 280–285
  as approximation to *t* distribution, 356
  as bell-shaped, 254, 281
  central limit theorem, 312–315
  characteristics of, 254–255
  and continuous random variables, 251–255, 268–269, 270–273
  empirical rule for standard deviation, 103–104
  equation of the, 255
  formula for, 255
  and hypothesis tests, 421, 461–465
    p-value approach, 389–391, 396
  mean of the, 254–255
  parameters of the, 255
  sampling distribution of sample mean for, 306–308
  for standard deviation known, large sample, 396
  standardizing, 264
  table of the standard, C19–20
  using table of standard, 256–262
  x and *z* values for known area, 275–279
normal distribution curve, 254–255
normal random variables, 254, 264, 267–268
not significantly different, 397
null hypothesis, 379–382. *See also* analysis of variance
number of combinations, 207
numerical descriptive measures, 74. *See also* measures of central tendency; measures of dispersion; measures of position
numerical precision, hypothesis testing and, 421

observation, 3, 9, 368
observational study, A10–12
observed frequencies
  goodness-of-fit test, 493–494
  test of homogeneity, 510
  test of independence, 503, 505
observed value of *t*, 402
observed value of the sample mean, 388, 393
observed value of *y*, 560–561
observed value of *z*, 389, 394, 411–413
occurrence, 227
odds-makers (gambling), 176–177, 236–237
ogives, 52–53
one-tailed hypothesis tests, 383
one-way analysis of variance, 535–537
  assumptions of, 536
outcomes (basic or final), 133–137. *See also* probability of failure
  counting rule to find total, 145–146
  equally likely, 139–140
  probability of success in binomial experiments, 211, 219–220, 317
outer fence, 113
outliers (extreme values), 58–59
  box-and-whisker plots and, 111–113
  and correlation in regression analysis, 602
  mean and, 77–78, 82
  median and, 80, 82
  mild, 113
  range and, 88

paired differences
  confidence interval for the mean of, 458, 459–461

paired differences (*continued*)
hypothesis tests, 461–463, 463–465
mean of the sample, 458–460
sampling distributions, 459
standard deviation, 458–459, 459
test statistic *t* for mean of, 461
between two population means, 457–465
paired observations, 560
paired samples, 434, 457–458
degrees of freedom, 442
hypothesis tests, 457–465
pairwise comparison procedures, 553
parallel systems, 185
parameters of the binomial probability distribution, 213
parameters of the normal distribution, 255
parameter of the Poisson probability distribution, 228
percentage, 29
histogram, 39–40
polygon, 41
percentage distributions
cumulative, 52–53
with open-ended multi-valued class, 44
qualitative data, 29
quantitative data, 38
percentile rank, 109–110
percentiles, 108–109
perfect negative linear correlation, 583
perfect positive linear correlation, 583
permutations, 209–210
personal interview surveys, A2–3
pie charts, 31–32
placebo effect, A12
point estimate, 339, 339–341, 360–361
confidence interval and, 340
sample mean as, 346
for small, normal population, 343
of standard deviation of sample proportion, 360–361
Poisson, Simeon D., 227
Poisson distribution, 227–228
binomial distribution compared to, 229–230
conditions for applying, 227–228
formula, 228–230
limits to, 287
mean and standard deviation, 232–233
parameter of the, 228
table, C13–19, 230–232
poll results
and election projections, 334
viability of, 375
polygons, 40–41
frequency, 40–41
percentage, 41
relative frequency, 41, 248–249
pooled sample proportion, 469, 475
pooled sample standard deviation, 442
population, 3
distribution, 297
sample versus, 5–8
target population versus, 5–6, A2, A5–6, A8
population correlation coefficient, 583
population parameters, 91
estimate and estimators of, 338–339
mean, 76–77
proportion, 317–318
for regression model, 558–559
variance, 88–91, 96–98, 122–123, 514–519
population regression line, 559, 578–581
estimating the mean value of *y*, 597
sample regression line compared to, 565
sampling distributions, 578
population regression model, 566–567
position, measure of, 106
positive linear relationship, 564
power of the test, 382
predicted value of *y* for a given *x*, 559, 560–561
finding, 563, 591, 598–600
prediction interval, predicting a particular value of *y* for a given *x*, 599–600
primary data, A2
primary unit sampling technique, A9
probabilistic regression model, 558–559
probability, 4, 132, 188
approximating by relative frequency, sample data, 141–143
as area under the curve, 249–250
classical approach, 139–140
complementary events, 153–154
conditional, 147–149, 161
confidence coefficient, 340
dependent events, 151–153
gambling and, 200–201, 209, 235–237
independent events, 151–153
intersection of events, 157–158
joint, 158–161
Law of Large Numbers, 142
marginal, 146–147
mutually exclusive events, 150–151
properties of, 138–139
relative frequency concept of, 140–142
of sample mean in an interval, 313–315
of sample proportion in an interval, 325–326
statistics versus, 176–177
subjective approach, 142–143
of Type I and Type II errors, 381–382
union of events, 167–169
probability density function, 249
probability distribution. *See also* binomial distribution; hypergeometric distribution; normal distribution; Poisson distribution; sampling distributions
of continuous random variables, 248–251

probability (*continued*)
- of discrete random variables, 191–195, 203
- exponential, 287
- mean of, for discrete random variables, 197–199
- population distribution and, 297
- of sample statistics, 298
- standard deviation of, for discrete random variables, 199–203
- probability distribution curve area under the, 249–250
- of continuous random variables, 248–251, 252–253

probability of failure
- binomial experiments, 211–213
- binomial formula, 213, 215–216
- estimating proportion, 362–365
- mean and standard deviation of a binomial distribution, 220–221
- normal approximation to the binomial, 281–282, 284–285
- and standard deviation of the sample proportion, 321–323

probability of success, 211, 219–220, 317
production processes, 185
proportions. *See* population proportion; sample proportions
p-value, 483
- for one-tailed test, 388
- for two-tailed test, 388
- range of, 402–403

p-value approach to hypothesis testing, 386
- population mean, standard deviation known, 388–391
- population mean, standard deviation unknown, 402–405
- population proportion, large sample, 411–414, 417

qualitative or categorical data, 12
- finding mode tor, 81
- frequency distribution for, 28–29
- graphing, 30–32
- mode for, 81
- organizing, 27–29

qualitative (*continued*)
- percentage distribution for, 29
- relative frequency distribution for, 29

qualitative or categorical variables, 12
quantitative data, 11, 34
- cumulative frequency distribution, 51–53
- cumulative percentage distribution, 52–53
- cumulative relative frequency distribution, 52–53
- dotplots, 58–60
- frequency distribution, 34–38, 42–43
- graphing, 39–41, 55–57, 58–60
- less than method frequency distribution, 42–43
- measures of central tendency, 94–96
- measures of dispersion, 96–98
- organizing, 34–38
- percentage distribution, 38
- relative frequency distribution, 52–53
- sensitivity of, 62
- standard deviation, 96–98, 123–124
- stem-and-leaf displays, 55–57

quantitative variables, 11
quartiles, 106–108
questionnaires, survey, A3
quota samples, A5

random error, standard deviation of, 571–573
random error term, 558–559, 560–561, 566
random experiment, 189
random number table, A8
random occurrences, 227, 558–559
random sample, 6, A6–7
- for ANOVA test, 536

random sampling techniques
- cluster sampling, A9
- simple random sampling, A8
- stratified random sampling, A9
- systematic random sampling, A8

random variables, 189–190, 221, 247, 298. *See also* sample statistic
- continuous, 190
- discrete, 189

random variation, 559
range for ungrouped data, 87–88
rank, percentile, 109–110
ranked data, 78–79, 106, 108–109
raw data, 27
real class limits, 34
rectangular histograms, 45
regression, 555, 556–559, 589–594
- cautions about, 567, 600, 610–611
- coefficient of *x*, 557–558
- coefficient of determination, 575
- confidence interval for *B*, 578–579, 592
- dependent variable, 556
- errors, processing, 601–602
- hypothesis test about true value of slope in population regression line, 579–581
- independent variable, 556
- interpretation of *a* and *b*, 564, 591
- least squares estimates of *A* and *B*, 559
  - calculating, 591
  - interpretation of, 564–565
  - SSE and least squares values, 561, 562–563
- least squares method and scatter diagram, 560
- overview of entire process, 589–594
- predicted value of *y*, 559, 560–561
  - finding, 563, 591, 598–600
- random error, 558–559, 560–561
- random error term, 558–559, 560–561, 566
- scatter diagram, 559–560, 565, 591
- significance level, 579–580, 585–586, 592–593
- standard deviation of errors, 572–573, 592

regression, simple, 556

regression line
  coefficient of x, 557–558
  least squares, 560–562, 591
    error sum of squares (SSE), 561
    estimating, 562–563, 578–579
  sample regression line, 578
    estimating the mean value of y and, 597
    least squares line, 560–562
    population regression line compared to, 565
    predicting a value of y and, 598–599
regression line slope, 559, 564–565
  confidence interval for the, 578–579
  hypothesis tests about the, 579–581
  mean, standard deviation, and sampling distribution of b, 557
regression model, multiple, 556
regression model, simple linear, 556–558
  assumptions of, 566–567
  coefficient of x (slope of the line), 557–558
  degrees of freedom, 572
  estimating the mean value of y, 596–598
  population parameters for, 558–559
  predicting a particular value of y, 598–600
regression of y on x, 559
regression sum of squares (SSR), 575–576
rejection region, 380–383
relative frequency, 140–141
  approximating probability with, 141–142
  proportion equated to, 317
relative frequency densities, 249, 252
relative frequency distribution, 29, 38
  for continuous random variables, 248–249
  cumulative, 52–53
relative frequency (*continued*)
  histograms, 39–40, 248–249
  and percentage distributions, 29
  polygons, 41, 248–249
  for qualitative data, 29
  for quantitative data, 52–53
right-tailed hypothesis tests, 384–385
right tail of F distribution curve, 533–534
  and one-way ANOVA test, 537
robustness versus sensitivity, 62

sample, 3, A2. *See also* sample size; sample statistic
  dependent versus independent, 434
  mean, 75
  population versus, 5–8
  random, 6
  representative, 6
  simple random, 6
  standard deviation, 89, 96, 354, 354, 360–361, 360–361
  variance, 88–90, 96, 98, 514–519
sample correlation coefficient, 583–584
sample data, 38, 75
  coefficient of determination from, 575
  empirical rule for standard deviation, 103–104
  estimates of a and b, 559
  interpretation of a and b, 564, 591
  least squares line, 560–561
  standard deviation of errors, 571–573
  variance and standard deviation for, 88
sample mean, 75. *See also* sampling distributions
  as consistent estimator, 304
  as estimator, 303–304, 328, 338–339
  for grouped data, 94, 95–96
  interval, probability of sample mean in, 313–315
  standard deviations of two, 442
  standard error of, 302
  for ungrouped data, 76–77
sample points, 133
sample proportion, 317–18. *See also* sampling distributions
  central limit theorem for, 321–323, 325–327
  as consistent estimator, 321
  as estimator, 320–321, 338–339
  mean, 320
  probability that sample proportion is in an interval, 325–326
  probability that sample proportion is less than a certain value, 326–327
  sampling distribution, 319–323, 325–327 shape of, 321
  standard deviation, 321
sample regression line, 578
  estimating the mean value of y, 597
  least squares line, 560–562
  population regression line compared to, 565
  predicting a value of y, 598–599
sample regression model, 561
sample size, 458
  ANOVA test, one-way, 540–543
  for estimation of mean, 347–348
  for estimation of proportion, 362–365
  large, 341, 344–45, 346, 387
  in paired samples, 458
  t distribution table and large samples, 356, 407–408, 447
  and width of confidence interval, 345, 347
sample space, 133–137
sample standard deviation, 89, 96
  as estimator, 354, 360–361
sample standard deviation of errors, 578
sample statistic, 91. *See also* sample mean; sample proportion
  probability distribution of, 298
  sample variance, 88–90, 96, 98, 514–519

sample surveys, 6, A2. *See also* sampling techniques
convenience samples, A5
judgment samples, A5
nonsampling errors, A6–8
quota samples, A5
random and nonrandom, A4–5
sampling errors, 299–301, A5–6
statistical significance and, 397
sample variance, 88–90, 96, 98, 514–519
samples
dependent, 434
independent, 434
paired or matched, 434, 457–465
sampling distribution, 296
applications of, 312–315
difference between mean, standard deviations, and
population proportions, 467–468
sample means, 434–436
mean, and standard deviation,
of estimated values of *B,* 578
of paired differences, 459
of sample mean, 302–304, 307–308, 310
of sample proportion, 319–323
population and, 297–299
of sample proportion, 319–323, 325–327
of sample variance, 514
shape of
sample mean, 306–210
sample proportion, 321–223
sample regression line, 578
sampling errors, 299–301, A5–6
statistical significance and, 397
sampling frame, A6
sampling techniques, A4
multistage procedure, 375
with replacement, 6, 8, 303
without replacement, 8, 303
scatter diagrams, 559–560, 565, 591
scatterplots, 602, 610
secondary data, A2
second quartile, 106
selection errors, A6–7
sensitivity versus robustness, 62
sequential processes, 185
series systems, 185
significance level, 340, 381
significantly different, 397
simple events, 135, 136–137,139–140
simple linear correlation, 583–587. *See also* linear correlation coefficient
simple linear regression, *See also* regression
simple probability, 146–147
simple random sample, 6, 339, 386
multistage sampling versus, 375
simple random sampling technique, A8
simple regression, 556. *See also* regression
single-valued classes, 43–44
skewed-to-the-teft histograms, 45, 82
skewed-to-the-right histograms, 45, 82
slope. *See* regression line slope
sources of data, 14–15
space interval, 227
specification, 420–421
stacked dotplots, 60
standard deviation, 88–91, 96–98, 122–124. *See also* sampling distributions
and area under normal distribution curve, 261
basic formulas, 122–124
binomial probability distribution, 220–121
Chebyshev's theorem, 101–103
confidence level and, 340
of difference between two sample means, 435, 442
of discrete random variables, 199–203
estimation of population mean with known, 341–348
estimation of population mean with unknown, 351–356
for grouped data, 96–98, 123–124
standard deviation (*continued*)
inferences about population variance, 514–519
of mutual funds' total returns, 104
normal distribution and, 254–255
of paired differences for two samples, 458–459
pooled sample standard deviation, 442
of random errors, 571–573
of sample mean, 302–304, 347, 435, 442
of sample mean for a nonnormal population, 310
of sample mean for a normal population, 307–308
of sample proportion, 321–323, 360–361, 362
of sampling distributions of sample mean, 302–304
of *t* distribution, 352
for ungrouped data, 88–91, 122–123
standard deviation of errors, 571–573, 592
standard error of sample mean, 302
standardizing a normal distribution, 264–269
standard normal distribution, 256–262, 264–265, 269. *See also* normal distribution; *z* values
curve, 256
empirical rule, 261
table, C19–20, 342–343
confidence levels and, 361
standard units. *See z* values
statistical analysis, apples-to-apples, 476–477
statistical model, 558–559
statistical relationship, 558–559
statistical significance, 397, 429
statistics, 1–8
descriptive, 3
inferential, 3–4
probability versus, 176–177
terminology, 8–10,18
stem-and-leaf displays, 55–57
grouped, 57
ranked, 56

stratified random sampling technique, A9
strong positive linear correlation, 583
Student's *t* distribution. *See t* distribution
Sturge's formula, 36
subjective probability, 142–143
success, probability of, 211, 219–220, 317
summation notation, 15–17
sum of squares
  between-samples (SSB), 537–538
  error (SSE), 561–562
  regression (SSR), 575–576
  total (SST), 538–539, 573–576 (*See also* SS_{yy})
  within-samples (SSW), 537–538
sure events, 138
surveys, 6, A2–3, A4. *See also* census and census data
  sample surveys, , 6, A2
    nonsampling errors, A6–8
symmetric distributions. *See also* normal distribution; *t* distribution
  bell-shaped, 103–104, 114, 254, 281
  binomial, 219
  chi-square for large *df*, 490
  frequency curves of, 46, 82
  histograms., 45, 82
systematic random sampling technique, A8

tables
  analysis of variance, 539
  binomial probabilities, C2–10, 217–219
  chi-square distribution, C23, 490–492
  contingency, 146, 502
    chi-square tests, 502, 503, 508–509
  cross tabulation, 502
  F distribution, C24–27, 533–534
  Poisson distribution, C13–19, 230–232
  standard normal distribution, C19–20, 342–343, 361
tables (*continued*)
  t distribution, C21–22
tails of hypothesis tests, 382–383
target population, 5, 5–6, A2, A5–6, A8
t distribution, 352–353
  conditions for using, 351
  degrees of freedom, 352–353, 449
t distribution table, C21–22
  using the table, 353–354, 356
telephone surveys, A3
test of homogeneity, 508–510
test of hypotheses, 378, 379–380, 433, 453–454. *See also* test statistic, tests of hypotheses
  about linear correlation coefficient, 585–587
  about population mean
    standard deviation known, 387–397
    standard deviation unknown, 401–408
  about population proportion, 411–417
  about true value of slope in population regression line, 579–581
  chi-square, 489, 521
  critical-value approach, 391–396, 405–408
  errors, Type I and Type II, 381–382
  goodness-of-fit, 493–500, 521
  homogeneity, 508–510
  independence, 503–508, 521
  left-tailed, 384, 385
  one-tailed, 383
  power of the test, 382
  p-value approach, 388–391,396, 402–405
  rejection and nonrejection regions, 380–381
  right-tailed, 384–385
  significantly different and not significantly different, 397
  tails of, 382–383
  two-tailed, 383–384, 385
  Type I and Type II errors, 381–382
  uses and misuses, 420–421, 429
test of hypothesis for differences
  two population means, independent samples
    standard deviations known, 434–439
    standard deviations unknown and unequal, 451–455
    standard deviations unknown but equal, 441–447, 448–449
  two population means, paired samples, 457–465
  two population proportions, large and independent samples, 467–472, 474–475
test of independence, 503–508, 521
test statistic, tests of hypothesis
  ANOVA, 537
  difference between two means
    standard deviations known, 438
    standard deviations unknown and unequal, 455
    standard deviations unknown but equal, 444
  difference between two population proportions
    large and independent samples, 469
  goodness-of-fit, 494
  independence or homogeneity, 503
  linear correlation coefficient, 585–587
  mean, large samples, 392, 394, 402
  mean of paired differences, 461
  proportion, large samples, 411
  slope of regression line, 517
  variance, 517
third quartile, 106
  box-and-whisker plot, 111
TI-84
  analysis of variance, 554
  chi-square tests, 529
  combinations, binomial distributions, and Poisson distributions, 245
  confidence intervals and hypothesis testing for two populations, 484

TI (*continued*)
confidence intervals for population means and proportions, 375
entering and saving data, 22–23
hypothesis testing, 429
normal and inverse normal probabilities, 293
numerical descriptive measures, 128
organizing data, 71
random number generation, 186
sampling distribution of means, 334–335
simple linear regression, 611
time-series data, 13–14
total sum of squares (SST), 563, 590
coefficient of determination, 573–576
one-way ANOVA, 538–539
treatment group, A3
tree diagrams, 133–137
trials, 211
trimmed mean, 86
true population mean, 338, 340
true population proportion, 338, 340
true values of the *y*-intercept and slope, 559
truncation, 39, 40, 46, 47
t value, finding the, 353–354, 356
two-tailed hypothesis tests, 383–384, 385
Type I error, 381–382
Type II error, 381–382
typical values, 75
unbiased estimator, 303, 320
ungrouped data, 27
mean, 75–78
measures of central tendency, 75
measures of dispersion, 87–91
median, 78–80
mode, 80–82
range, 87–88
standard deviation, 88–91, 122–123
variance, 88–91, 122–123
uniform histograms, 45
unimodal distribution, 81
union of events, 167–169
upper boundary of a class, 34–35
upper inner fence, 111–112
upper limit of a class, 34–35
upper limit of a confidence interval, 340
upper outer fence, 113

variables, 9, 10–12
binomial random, 213
categorical, 12
continuous, 11–12
dependent, 556, 602–603
discrete, 11
independent, 556, 601–602
linear correlation between two, 583–584
missing or omitted, 559
normal random, 254, 264, 267–268
positive/negative linear relationship, 564
variables (*continued*)
qualitative, 12
quantitative, 11
random, 189
variance, 88–89. *See also* analysis of variance
basic formulas, 122–124
between samples, 537, 539
confidence interval for, 515–516
of discrete random variable, 200
for grouped data, 96–98
hypothesis tests about, 514–519, 537–543
population, 88–91, 96–98, 122–123, 514–519
sample, 96–98,123–124, 514–519
within samples, 536, 537–539
for ungrouped data, 88–91, 122–123
Venn diagrams, 133–137
voluntary response errors, A7–8

weak negative linear correlation, 583
weak positive linear correlation, 583
weighted mean, 86
whiskers, 112–113
width of a confidence interval, 345, 347

y-intercept in regression model, 557–558, 564

z values or *z* scores, 256–257
and standard normal distribution table, C19–20, 342–343, 361